W9-AUX-267

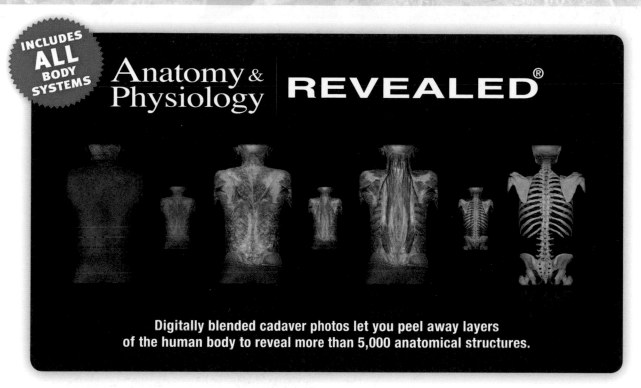

INCLUDES ALL BODY SYSTEMS

Anatomy & Physiology | **REVEALED**®

Digitally blended cadaver photos let you peel away layers of the human body to reveal more than 5,000 anatomical structures.

This amazing multimedia tool is designed to help students learn and review human anatomy using cadaver specimens. Detailed cadaver photographs blended together with a state-of-the-art layering technique provide a uniquely interactive dissection experience.

Available at **www.aprevealed.com**

Electronic Books

GO GREEN!

Go Green. Save Green. McGraw-Hill eBooks.

Green. . . it's on everybody's minds these days. It's not only about saving trees; it's also about saving money. At 55% of the bookstore price, **McGraw-Hill eBooks** are an eco-friendly and cost-saving alternative to the traditional print textbook. So, you do some good for the environment. . . and you do some good for your wallet.

Visit **www.mhhe.com/ebooks** for details.

Kenneth S. Saladin

Georgia College & State University

HUMAN ANATOMY

THIRD EDITION

The McGraw-Hill Companies

Mc Graw Hill

Connect
Learn
Succeed™

HUMAN ANATOMY, THIRD EDITION

Published by McGraw-Hill, a business unit of The McGraw-Hill Companies, Inc., 1221 Avenue of the Americas, New York, NY 10020. Copyright © 2011 by The McGraw-Hill Companies, Inc. All rights reserved. Previous editions © 2008 and 2005. No part of this publication may be reproduced or distributed in any form or by any means, or stored in a database or retrieval system, without the prior written consent of The McGraw-Hill Companies, Inc., including, but not limited to, in any network or other electronic storage or transmission, or broadcast for distance learning.

Some ancillaries, including electronic and print components, may not be available to customers outside the United States.

This book is printed on acid-free paper.

1 2 3 4 5 6 7 8 9 0 WDQ/WDQ 1 0 9 8 7 6 5 4 3 2 1 0

ISBN 978–0–07–352560–0
MHID 0–07–352560–X

Vice President & Editor-in-Chief: *Marty Lange*
Vice President, EDP: *Kimberly Meriwether-David*
Director of Development: *Kristine Tibbetts*
Executive Editor: *Colin H. Wheatley*
Senior Developmental Editor: *Kristine A. Queck*
Marketing Manager: *Denise M. Massar*
Senior Project Manager: *Vicki Krug*
Senior Production Supervisor: *Laura Fuller*
Lead Media Project Manager: *Stacy A. Patch*
Senior Designer: *David W. Hash*
Cover Designer: *Greg Nettles/Squarecrow Design*
(USE) Cover Image: *colored angiogram (x-ray) of a healthy kidney, ©James Cavallini/Photo Researchers, Inc.; colored scanning micrograph (SEM) of a kidney glomerulus, ©Susumu Nishinaga/Photo Researchers, Inc.; colored scanning electron micrograph (SEM) of kidney blood vessels, ©Susumu Nishinaga/Photo Researchers, Inc.*
Senior Photo Research Coordinator: *John C. Leland*
Photo Research: *Mary Reeg*
Supplement Producer: *Mary Jane Lampe*
Compositor: *Electronic Publishing Services Inc., NYC*
Typeface: *10/12 Melior*
Printer: *Worldcolor*

All credits appearing on page or at the end of the book are considered to be an extension of the copyright page.

Library of Congress Cataloging-in-Publication Data

Saladin, Kenneth S.
   Human anatomy / Kenneth S. Saladin. -- 3rd ed.
     p. cm.
   Includes index.
   ISBN 978-0-07-352560-0 — ISBN 0-07-352560-X (hard copy : alk. paper)
   1. Human anatomy--Textbooks. I. Title.
QM23.2.S25 2011
611--dc22   2009014672

Brief Contents

About the Author

This book is dedicated to my students worldwide, including those who rarely pass through my office door but who are always on my mind, inspiring me to give them my best.

KEN SALADIN has taught since 1977 at Georgia College & State University in Milledgeville, Georgia. He earned a B.S. in zoology at Michigan State University and Ph.D. in parasitology at Florida State University, with interests especially in the sensory ecology of freshwater invertebrates. In addition to human anatomy and physiology, his teaching experience includes histology, parasitology, animal behavior, sociobiology, introductory biology, general zoology, biological etymology, and study abroad in the Galápagos Islands. Nine times over the years, outstanding students inducted into Phi Kappa Phi have tapped Ken for recognition as their most significant undergraduate mentor. He <u>has</u> received the university's Excellence in Research and Publication Award, and was named Distinguished Professor in 2001.

Ken is a member of the Human Anatomy and Physiology Society, the Society for Integrative and Comparative Biology, the American Association of Anatomists, and the American Association for the Advancement of Science. He served as a developmental reviewer and wrote supplements for several other McGraw-Hill anatomy and physiology textbooks for a number of years before becoming a textbook writer.

Ken is married to Diane Saladin, a registered nurse. They have two adult children.

Reviewers

Ben F. Brammell
Morehead State University

Jennifer K. Brueckner
University of Kentucky Lexington

Ty W. Bryan
Bossier Parish Community College

Sarah C. Cotton
Chaffey College

D. Dodenhoff
College of Sequoias

Cammie K. Emory
Bossier Parish Community College

Gibril O. Fadika
Hampton University

Ray Fagenbaum
University of Iowa

Michael E. Fultz
Morehead State University

Becky L. Green-Marroquin
Los Angeles Valley College

Eric Hall
Rhode Island College

Wanda G. Hargroder
Louisiana State University

Candi K. Heimgartner
University of Idaho

Roishene Johnson
Bossier Parish Community College

Marie Kelly-Worden
Ball State University

Ronald L. Koller
Los Angeles Pierce College

Suzanne Koch-Krueger
College of Sequoias

Dan Miska
Wright State University

Rae Osborn
Northwestern State University

John Pellegrini
College of Saint Katherine

Mark Schlueter
College of Saint Mary

Rachel D. Smetanka
Southern Utah University

Leeann S. Sticker
Northwestern State University of Louisiana

MaryJo A. Witz
Monroe Community College

Michele Zimmerman
Indiana University Southeast

EVOLUTION OF A
Storyteller

Ken Saladin's penchant for writing began early. For his 10th-grade biology class, he wrote a 318-page monograph on hydras with 53 original India ink drawings and 10 original photomicrographs. We at McGraw-Hill think of this as Ken's "first book." At a young age, Ken already was developing his technical writing style, research habits, and illustration skills.

Ken Saladin's "first book,"
Hydra Ecology (1965)

Some of Ken's first pen-and-ink artwork (1965)

Ken in 1964

Ken served as an A&P textbook reviewer and testbank writer for several years and then embarked on his first book for McGraw-Hill in 1993. He published the first edition of *Anatomy and Physiology: The Unity of Form and Function* in 1997 and his first edition of *Human Anatomy* in 2004. The story continues with *Human Anatomy,* third edition.

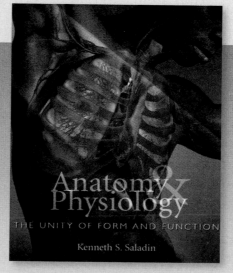

Ken's first textbook published in 1997

The story continues in 2009

GUIDE TO
Human Anatomy, Third Edition

Saladin's *Human Anatomy* is more than a description of body structure. It is a story that weaves together basic science, clinical applications, the history of medicine, and the evolutionary basis of human structure. Saladin combines this humanistic perspective with vibrant photos and art to convey the beauty and excitement of the subject to beginning students.

> *"Saladin's* Human Anatomy *is the easiest textbook to read and comprehend on the market today. I have used this textbook for 8 semesters with all of my students. Students consistently tell me that Saladin's* Human Anatomy *textbook is their favorite science textbook to read. They find it interesting and easy to understand."*
>
> —Mark Schlueter
> *College of Saint Mary*

The Third Edition—What's New?

New Organization

The introductory chapter of the previous edition has now been substantially condensed and merged with the general orientation to the human body that formerly constituted Atlas A. Students now get a broad introduction to regional terminology in chapter 1. Cadaver photos from the former Atlas A are now combined with the former Atlas B into a single "Atlas of Regional and Surface Anatomy" (p. 329). Other chapters have also been condensed, with the result that the third edition is 32 pages shorter than the second edition.

New Science

The third edition embraces both new findings in anatomical science and new incorporation of established knowledge, in such topics as:

- radiologic anatomy
- cardiac tamponade
- mitochondrial DNA
- basement membrane function
- tissue engineering
- albinism
- intercostal muscle function
- microglia and satellite cell functions
- vaccine for shingles
- meningitis
- empty sella syndrome
- galactorrhea
- brain natriuretic peptide
- osteocalcin
- cord blood transplants
- platelet production
- regulatory T cells
- brainstem respiratory centers
- ghrelin and appetite regulation

New Writing

In response to feedback from anatomy instructors, reviewers, and students, Ken has paid particular attention to rewriting and clarifying many topics in this edition, including

- adipose tissue, bone, and blood (chapter 3)
- keratinocytes, skin color, and skin cancer (chapter 5)
- the pelvic girdle and pelvis (chapter 7)
- anatomical disorders of the endocrine system (chapter 18)
- RBC morphology and transfusion compatibility (chapter 19)
- the myocardial vortex and cardiac conduction system (chapter 20)
- upper respiratory anatomy (chapter 23)
- regional anatomy and histology of the GI tract (chapter 24)
- renal innervation and the nephron loop (chapter 25)
- oogenesis, folliculogenesis, and andropause (chapter 26)

"Dr. Saladin seems very open to suggestions from instructors and users of his text for further improvement. This open attitude is very important to me as an instructor. For the students, Saladin's text presents a significant advantage over other authors in his ease of readability and presentation. Students actually understand and comprehend passages read from his text and find it interesting!"

—Candi K. Heimgartner
University of Idaho

New Balance

With the help of reviewers, Saladin has struck a new balance between anatomy and physiology—just enough physiology to lend meaning to the anatomy, but not so much as to be excessive for a textbook dedicated to anatomy.

New Photographs

The following photographs are new to this edition.

- 3-dimensional fetal sonogram (figure 1.4)
- Wilms tumor of the kidney (figure 2.20)
- transitional epithelium (figure 3.11)
- reticular tissue (figure 3.15)
- hyaline cartilage (figure 3.19)
- elastic cartilage (figure 3.20)
- fibrocartilage (figure 3.21)
- fracture X-rays (figure 6.14)
- male and female pelvic girdles (figure 8.9)
- external anatomy of the eye (figure 17.19)
- the myocardial vortex (figure 20.6)
- endoscopic view of aortic valve (figure 20.8)
- corrosion cast of coronary blood vessels (figure 20.11)
- pulmonary histology (figure 23.10a)
- stages of folliculogenesis (figure 26.14)

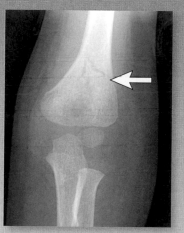
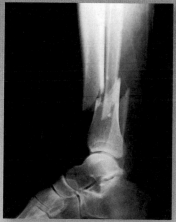
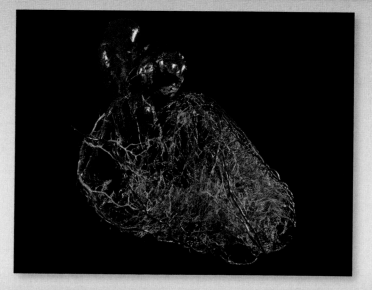

"The illustrations and photographs in . . . Saladin are a huge improvement from our previous text. His use of dissection photos and histological photos is the best I have seen in an undergraduate text."

—Rachel D. Smetanka, *Southern Utah University*

New Art

Many of the existing illustrations have been improved, and the following drawings are entirely new to the third edition.

- cell junctions (figure 2.14)
- the cytoskeleton (figure 2.15)
- development of exocrine and endocrine glands (figure 3.28)
- serous membrane histology (figure 3.31b)
- epidermal histology (figure 5.3)
- medial view of elbow joint (figure 8.4c)
- medial view of foot skeleton (figure 8.14c)

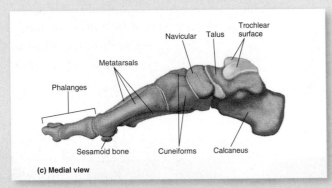

(c) Medial view

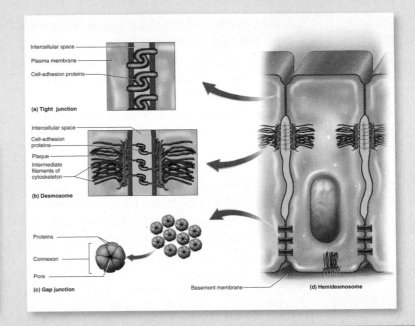

(a) Tight junction

(b) Desmosome

(c) Gap junction

(d) Hemidesmosome

New Pedagogy

The third edition features new devices, in addition to pedagogical aids used in the second edition, for more effective teaching and for student self-assessment.

Study Guide A renamed chapter review. When students ask instructors if they will provide a study guide for the next exam, the answer now is easy: There's already one at the end of every chapter!

Assess Your Learning Outcomes A chapter study outline that doesn't paraphrase the book, but prompts the student to extract, organize, and paraphrase the information for him- or herself. Discourages passivity and shortcutting and makes the student an active learner. (Chapter summaries in paraphrased form are available for those who want them at www.mhhe.com/saladinha3).

Building Your Medical Vocabulary A new exercise at the end of each chapter serves to strengthen the student's familiarity with and application of widely used word roots in medical terminology.

Study Guide

Assess Your Learning Outcomes

You should have a good understanding of this chapter if you can accurately address the following issues.

9.1 Joints and Their Classification (p. 205)
1. The definition of *joint (articulation)*
2. Names of the sciences concerned with joint structure and movement
3. The general rule for how joints are commonly named
4. The criteria used to classify joints into anatomical and functional categories
5. The distinguishing characteristics of bony joints, fibrous joints, and cartilaginous joints; synonyms for these terms; and examples of joints in each category
6. The three subclasses of fibrous joints and three types of sutures, and examples of each
7. The two subclasses of cartilaginous joints, and examples of each

9.2 Synovial Joints (p. 209)
1. The definition of *synovial joint*
2. The anatomical features of a generalized synovial joint
3. The functions of articular discs and menisci at certain synovial joints, where they can be found, and their appearance
4. The defining characteristics of tendons, ligaments, and bursae, and the roles they play at the joints; how tendon sheaths differ from other bursae
5. The distinction between monaxial, biaxial, and multiaxial joints

6. The six types of synovial joints and where they can be found
7. The definitions of joint *flexion, extension,* and *hyperextension;* some everyday scenarios in which these movements occur; and the ability to demonstrate them with your own body
8. The same for *abduction, adduction, hyperabduction,* and *hyperadduction*
9. The same for *elevation* and *depression*
10. The same for *protraction* and *retraction*
11. The same for *circumduction*
12. The same for *medial (internal)* and *lateral (external) rotation*
13. The same for *supination* and *pronation* of the forearm
14. The same for *flexion, extension, hyperextension,* and *lateral flexion* of the vertebral column
15. The same for *rotation* of the head or torso
16. The same for *lateral* and *medial excursion* of the mandible
17. The same for wrist *flexion* and *extension* anteriorly and posteriorly, and *ulnar* and *radial flexion* from side to side
18. The same for the thumb movements of *radial abduction, palmar abduction, opposition,* and *reposition*
19. The same for the ankle or foot movements of *dorsiflexion, plantar flexion, inversion,* and *eversion,* and how several of these movements are combined in foot pronation and supination
20. How a joint's range of motion (ROM) is measured and what anatomical features govern the ROM

9.3 Anatomy of Selected Synovial Joints (p. 219)
1. Special functional qualities of the temporomandibular joint (TMJ); its major anatomical features; and two common disorders of the TMJ
2. Special functional qualities of the glenohumeral joint; its major anatomical features; and two of its common injuries
3. The names of the three joints that occur at the elbow; how they enable the varied movements of the forearm; and the major anatomical features of the elbow joints
4. Special functional qualities of the coxal joint; its major anatomical features; and the actions of the ligaments at this joint when a person stands
5. Special functional qualities of the tibiofemoral joint; its major anatomical features (especially its menisci and cruciate ligaments); and the common injuries of this joint
6. Special functional qualities of the talocrural joint; its major anatomical features; and the nature of sprains at this joint

9.4 Clinical Perspectives (p. 228)
1. The range of disorders included in the concept of rheumatism, and the related term for physicians who specialize in joint disorders
2. The general meaning of *arthritis,* and the pathology and distinctions between osteoarthritis and rheumatoid arthritis
3. Joint prostheses and arthroplasty

A Storytelling Writing Style

Students and instructors routinely cite Saladin's prose style as the number one attraction of this book. Students doing blind comparisons of Ken Saladin's chapters and those of other anatomy books routinely choose Saladin hands down, finding Saladin

- more clearly written; easier to understand
- fun and interesting to read
- stimulating; not just a dry recitation of fact after fact

Innovative Perspectives

Instructors often say even they learn new facts and novel ways of looking at and teaching things from Saladin's creative perspectives.

filaments, and *microtubules.* **Microfilaments (thin filaments)** are about 6 nm thick and are made of the protein actin. They fo rm a fibrous **terminal web (membrane skeleton)** on the cytoplasmic side of the plasma membrane. The lipids of the plasma membrane are spread out over the terminal web like butter on a slice of bread. The web, like the bread, provides physical support, whereas the lipids, like butter, provide a permeability barrier. It is thought that, without this support by the terminal web, the lipids would break up into little droplets and the plasma membrane would not

round cells), and **simple columnar** (tall narrow cells). In the fourth type, **pseudostratified columnar epithelium,** not all cells reach the free surface; the taller cells cover the shorter ones. This e pithelium looks stratified in most tissue sections, but careful examination, especially with the electron microscope, shows that every cell reaches the basement membrane—like trees in a forest, where some grow taller than others but all are anchored in the soil below.

Simple columnar and pseudostratified columnar epithelia often have wineglass-shaped **goblet cells** that produce protective mucus

2. Eventually, the perichondrium stops producing chondrocytes and begins producing osteoblasts. These deposit a thin collar of bone around the middle of the cartilage model, encircling it like a napkin ring and providing physical reinforcement. The former perichondrium is now considered to be a periosteum.

chapter 17. The **hyoid**[28] **bone** is a slender U-shaped bone between the chin and larynx (fig. 7.16). It is one of the few bones that does not articulate with any other. The hyoid is suspended from the styloid processes of the skull, somewhat like a hammock, by the small *stylohyoid muscles* and *stylohyoid ligaments.* The medial **body** of the hyoid is flanked on either side by projections called the **greater** and

Ken Saladin explains his approach to writing ... "I remember how difficult it was for me to understand some complicated concepts as an undergraduate. When I proofread my own writing, I try to put myself back in that student frame of mind and write as if I were tutoring my 18-year-old self. I choose my words and paragraph structure carefully, aiming for the clarity that I would have appreciated as a student back then. I write both to reach the cognitive level of the average beginning student, but also to elevate that cognitive level by the time the course is over."

"While reading through the chapters, I can easily visualize the structures in my head and I am confident that an undergraduate anatomy student could do the same."

—Michele Zimmerman
Indiana University Southeast

Fresh Analogies

Saladin's analogy-rich writing enables students to easily visualize abstract concepts in terms of everyday experience.

Neurosomas range from 5 to 135 μm in diameter, whereas axons range from 1 to 20 μm in diameter and from a few millimeters to more than a meter long. Such dimensions are more impressive when we scale them up to the size of familiar objects. If the soma of a spinal motor neuron were the size of a tennis ball, its dendrites would form a huge bushy mass that could fill a 30-seat classroom from floor to ceiling. Its axon would be up to a mile long but a little narrower than a garden hose. This is quite a point to ponder. The neuron must assemble molecules and organelles in its "tennis ball" soma and deliver them through its "mile-long garden hose" to the end of the axon. In a process called *axonal transport,* neurons employ *motor proteins* that can carry organelles and macromolecules as they crawl along the cytoskeleton of the nerve fiber to distant destinations in the cell.

Artwork That Piques Interest and Clarifies Ideas

Saladin's portfolio of stunning illustrations and photos draws in students who regard themselves as "visual learners."

Vivid Illustrations with rich textures and shading and bold, bright colors bring anatomy to life.

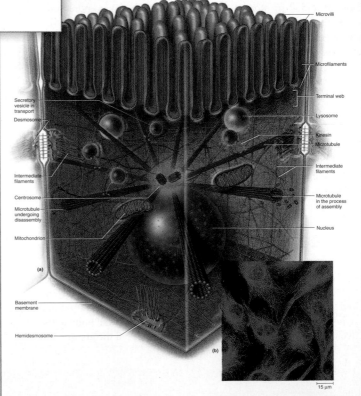

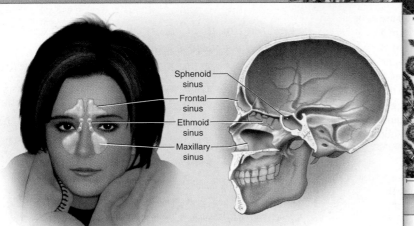

Ken Saladin explains his approach to helping visual learners . . . "The visual appeal of nature is immensely important in motivating one to study it. We certainly see this at work in human anatomy—in the countless students who describe themselves as visual learners; in the millions of laypeople who flock to museums and popular exhibitions such as Body Worlds; and in all those who find anatomy atlases so intriguing. I have illustrated Human Anatomy not only to visually explain concepts, but also to appeal to this sense of the aesthetics of the human body."

Process Figures relate numbered steps in the art with correspondingly numbered text descriptions.

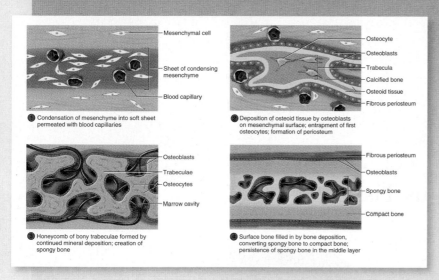

Intramembranous Ossification

Intramembranous[23] (IN-tra-MEM-bruh-nus) **ossification** produces the flat bones of the skull and most of the clavicle (collarbone). Such bones develop within a fibrous sheet similar to the dermis of the skin, so they are sometimes called *dermal bones*. Figure 6.8 shows the stages of the process.

1. An area of the embryonic connective tissue (mesenchyme) condenses into a layer of soft tissue with a dense supply of blood capillaries. The mesenchymal cells enlarge and differentiate into osteogenic cells, and regions of mesenchyme become a network of soft sheets called trabeculae.

2. Osteogenic cells gather on these trabeculae and differentiate into osteoblasts. These cells deposit an organic matrix called **osteoid**[24] **tissue**—soft collagenous tissue similar to bone except for a lack of minerals (fig. 6.9). As the trabeculae grow thicker, calcium phosphate is deposited in the matrix. Some osteoblasts become trapped in the matrix and are now osteocytes. Mesenchyme close to the surface of a trabecula remains uncalcified, but becomes denser and more fibrous, forming a periosteum.

3. Osteoblasts continue to deposit minerals, producing a honeycomb of bony trabeculae. Some trabeculae persist as permanent spongy bone, while osteoclasts resorb and remodel others to form a marrow cavity in the middle of the bone.

4. Trabeculae at the surface continue to calcify until the spaces between them are filled in, converting the spongy bone to compact bone. This process gives rise to the sandwichlike arrangement typical of mature flat bones.

Orientation Tools clarify the perspective from which a structure is viewed.

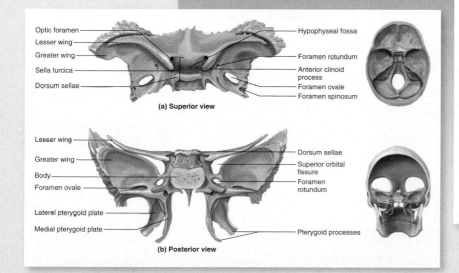

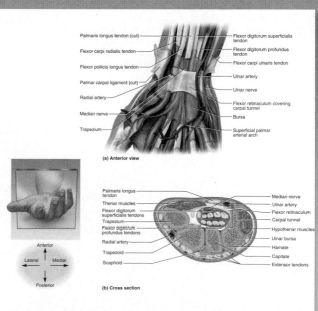

The Art of Teaching—Pedagogical Features

Having taught human anatomy for 32 years, Saladin knows what works in the classroom and brings those approaches to *Human Anatomy*.

Chapters Laid Out for Preview and Review

- Chapters begin with a preview of topics to be covered, as well as a reminder to review previously read material that is relevant to the current chapter.
- Chapters are divided into manageable sections conducive to limited blocks of study time.

Chapter Outline provides a content preview and facilitates review and study.

Brushing Up reminds students of the relevance of earlier chapters to the one on which they are newly embarking.

Deeper Insights pique the interest of health-science students by showing the clinical relevance of the core science.

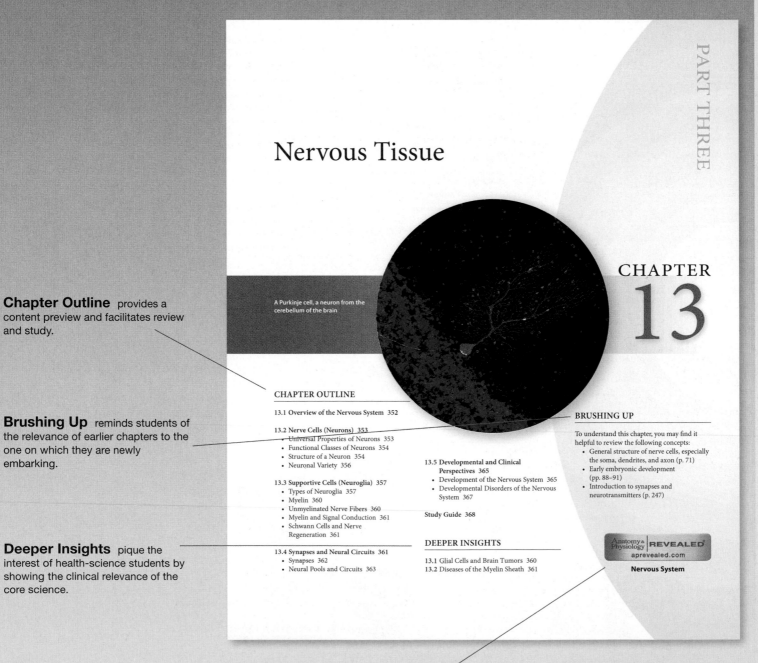

PART THREE

Nervous Tissue

A Purkinje cell, a neuron from the cerebellum of the brain

CHAPTER
13

BRUSHING UP

To understand this chapter, you may find it helpful to review the following concepts:
- General structure of nerve cells, especially the soma, dendrites, and axon (p. 71)
- Early embryonic development (pp. 88–91)
- Introduction to synapses and neurotransmitters (p. 247)

Anatomy & Physiology REVEALED
aprevealed.com

Nervous System

Anatomy & Physiology REVEALED icons indicate which area of this interactive cadaver dissection program corresponds to the chapter topic.

The Bookends of Knowledge

- Each section is a conceptually unified topic framed between a pair of learning "bookends"—a set of learning objectives at the beginning and a set of review and self-testing questions at the end.
- Each section is numbered for easy reference in lecture, assignments, and ancillary materials.

Expected Learning Outcomes give the student a preview of key points to be learned within the next few pages.

(b)

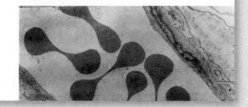

19.2 Erythrocytes

▶ Expected Learning Outcomes

When you have completed this section, you should be able to

- describe the morphology and functions of erythrocytes (RBCs);
- explain some clinical measurements of RBC and hemoglobin quantities;
- describe the structure and function of hemoglobin;
- discuss the formation, life span, death, and disposal of RBCs; and
- explain the chemical and immunologic basis and the clinical significance of blood types.

Erythrocytes, or **red blood cells (RBCs),** have two principal functions: (1) to pick up oxygen from the lungs and deliver it to tissues elsewhere, and (2) to pick up carbon dioxide from the tissues and unload it in the lungs. RBCs are the most abundant formed elements of the blood and therefore the most obvious things one sees upon its

---

Before You Go On prompts the student to pause and spot-check his or her mastery of the previous few pages before going on to new material.

wide range of other blood diseases. Efforts are being made to further improve the procedure by stimulating placental stem cells to multiply before the transplant, and by removing placental T cells that may react against the recipient.

AB, and O) and Rh group (with blood types Rh-positive and Rh-negative). These types differ with respect to the chemical composition of glycolipids on the RBC surface; figure 19.6 shows how the ABO types differ in this regard. The glycolipids act as *antigens,* substances capable of evoking an immune reaction. The blood plasma contains *antibodies* that react against incompatible antigens on foreign RBCs.

RBC antigens and plasma antibodies determine the compatibility of donor and recipient blood in transfusions. For example, a person with blood type A has anti-B antibodies in the blood plasma. If this person is mistakenly given a transfusion of type B blood, those antibodies attack the donor RBCs. The RBCs agglutinate—they form clumps that obstruct the circulation in small blood vessels, with devastating consequences for such crucial organs as the brain, heart, lungs, and kidneys. The agglutinated RBCs also rupture *(hemolyze)* and release their hemoglobin. This free hemoglobin is filtered out by the kidneys, clogs the microscopic kidney tubules, and can cause death from renal failure within a week or so.

The same can happen if a type B person receives type A blood, or if a type O person receives either type A or type B. Incompatibility of Rh types between the mother and fetus sometimes causes severe anemia in the newborn infant *(hemolytic disease of the newborn).*

Apply What You Know

Why might a court of law be interested even in human blood types that have no connection to disease?

Before You Go On

Answer the following questions to test your understanding of the preceding section:

5. What are the two main functions of RBCs?

6. Define *hematocrit* and *RBC count,* and state some normal clinical values for each.

7. Describe the structure of a hemoglobin molecule. Explain where O_2 and CO_2 are carried on a hemoglobin molecule.

8. Name the stages in the production of an RBC, and state the differences between them.

9. Explain what plasma and RBC components are responsible for blood types, and why blood types are clinically important.

Vocabulary Building

Several features help build a student's level of comfort with medical vocabulary.

Pronunciation Guides Students can't very well remember or spell terms if they can't pronounce them in the first place. Saladin gives simple, intuitive "pro-NUN-see-AY-shun" guides to help students over this hurdle and widen the student's comfort zone for medical vocabulary.

Word Origins Accurate spelling and insight into medical terms are greatly enhanced by a familiarity with commonly used word roots, prefixes, and suffixes.. Footnotes throughout the chapters will help build the student's working lexicon of word elements.

Building Your Medical Vocabulary An exercise at the end of each chapter helps students creatively use their knowledge of new medical word elements.

The **foramen spinosum,** about the diameter of a pencil lead, provides passage for an artery of the meninges. An irregular gash called the **foramen lacerum**[18] (LASS-eh-rum) occurs at the junction of the sphenoid, temporal, and occipital bones. It is filled with cartilage in life and transmits no major vessels or nerves.

In an inferior view of the skull, the sphenoid can be seen just anterior to the basilar part of the occipital bone (see fig. 7.5a). The internal openings of the nasal cavity seen here are called the **posterior nasal apertures,** or **choanae**[19] (co-AH-nee). Lateral to each aperture, the sphenoid bone exhibits a pair of parallel plates—the **medial** and **lateral pterygoid**[20] (TERR-ih-goyd) **plates.** Each plate has a narrow inferior extension called the *pterygoid process.* The plates provide attachment for some of the jaw muscles. The sphenoid sinus occurs within the body of the sphenoid bone.

Ethmoid Bone

The **ethmoid**[21] (ETH-moyd) bone is an anterior cranial bone located between the eyes (fig. 7.12). It contributes to the medial wall of the orbit, the roof and walls of the nasal cavity, and the nasal septum. It is a very porous and delicate bone, with three major portions:

1. The vertical **perpendicular plate,** a thin median plate of bone that forms the superior two-thirds of the nasal septum (see fig. 7.4b). (The lower part is formed by the vomer, discussed

---

[18]*lacerum* = torn, lacerated
[19]*choana* = funnel
[20]*pterygo* = wing
[21]*ethmo* = sieve, strainer + *oid* = resembling

Building Your Medical Vocabulary

State a medical meaning of each of the following word elements, and give a term in which it is used.

1. cyto-
2. squam-
3. -form
4. poly-
5. -philic
6. phago-
7. endo-
8. glyco-
9. chromo-
10. meta-

Answers in the Appendix

Desktop Experiments

Many chapters offer simple experiments and palpations a student can do at his or her desk, with no equipment, to help visualize chapter concepts.

- **Tactile (Meissner[4]) corpuscles.** These are receptors for light touch and texture. They are tall, ovoid to pear-shaped, and consist of two or three nerve fibers meandering upward through a mass of flattened Schwann cells. They occur in the dermal papillae of the skin and are limited to sensitive hairless areas such as the fingertips, palms, eyelids, lips, nipples, and parts of the genitals. Drag your fingernails lightly across your forearm and then across your palm. The difference in sensation you feel is due to the high density of tactile corpuscles in your palmar skin. Tactile corpuscles enable you to tell the difference between silk and sandpaper, for example, by light strokes of your fingertips.

Self-Assessment Tools

Saladin provides students with abundant opportunities to evaluate their comprehension of concepts. A wide variety of questions from simple recall to analytical evaluation cover all six cognitive levels of Bloom's Taxonomy of Educational Objectives.

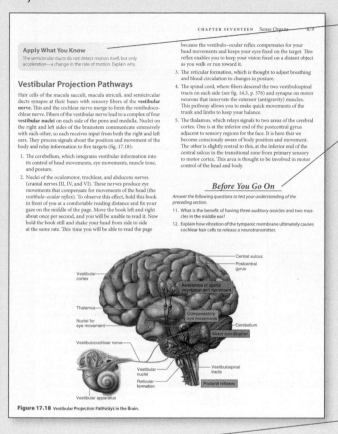

Apply What You Know

The semicircular ducts do not detect motion itself, but only acceleration—a change in the rate of motion. Explain why.

Vestibular Projection Pathways

Hair cells of the macula sacculi, macula utriculi, and semicircular ducts synapse at their bases with sensory fibers of the **vestibular nerve.** This and the cochlear nerve merge to form the vestibulocochlear nerve. Fibers of the vestibular nerve lead to a complex of four **vestibular nuclei** on each side of the pons and medulla. Nuclei on the right and left sides of the brainstem communicate extensively with each other, so each receives input from both the right and left ears. They process signals about the position and movement of the body and relay information to five targets (fig. 17.18):

1. The cerebellum, which integrates vestibular information into its control of head movements, eye movements, muscle tone, and posture.
2. Nuclei of the oculomotor, trochlear, and abducens nerves (cranial nerves III, IV, and VI). These nerves produce eye movements that compensate for movements of the head (the *vestibulo–ocular reflex*). To observe this reflex, hold this book in front of you at a comfortable reading distance and fix your gaze on the middle of the page. Move the book left and right about once per second, and you will be unable to read it. Now hold the book still and shake your head from side to side at the same rate. This time you will be able to read the page

because the vestibulo–ocular reflex compensates for your head movements and keeps your eyes fixed on the target. This reflex enables you to keep your vision fixed on a distant object as you walk or run toward it.
3. The reticular formation, which is thought to adjust breathing and blood circulation to changes in posture.
4. The spinal cord, where fibers descend the two vestibulospinal tracts on each side (see fig. 14.3, p. 376) and synapse on motor neurons that innervate the extensor (antigravity) muscles. This pathway allows you to make quick movements of the trunk and limbs to keep your balance.
5. The thalamus, which relays signals to two areas of the cerebral cortex. One is at the inferior end of the postcentral gyrus adjacent to sensory regions for the face. It is here that we become consciously aware of body position and movement. The other is slightly rostral to this, at the inferior end of the central sulcus in the transitional zone from primary sensory to motor cortex. This area is thought to be involved in motor control of the head and body.

Before You Go On

Answer the following questions to test your understanding of the preceding section.

11. What is the benefit of having three auditory ossicles and two muscles in the middle ear?
12. Explain how vibration of the tympanic membrane ultimately causes cochlear hair cells to release a neurotransmitter.

Figure 17.18 Vestibular Projection Pathways in the Brain.

Apply What You Know Questions interjected in the chapter spot-test a student's ability to think of the deeper implications or clinical applications of a point he or she just read.

Before You Go On Tests simple recall and lower-level interpretation of information read within the last few pages

Figure Legend Questions Questions posed in many of the figure legends prompt the student to interpret the art and apply it to the reading.

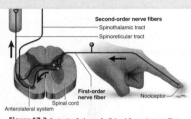

Figure 17.3 Projection Pathways for Pain. A first-order nerve fiber conducts a pain signal to the posterior horn of the spinal cord, a second-order fiber conducts it to the thalamus, and a third-order fiber conducts it to the cerebral cortex. Signals from the spinothalamic tract pass through the thalamus. Signals from the spinoreticular tract bypass the thalamus on the way to the sensory cortex.
● *What is the name of the gyrus that contains the primary somatosensory cortex? To which of the five lobes of the cerebrum does this gyrus belong?*

Testing Your Recall

1. When a conceptus arrives in the uterus, it is at what stage of development?
   a. zygote
   b. morula
   c. blastomere
   d. blastocyst
   e. embryo

2. The entry of a sperm nucleus into an egg must be preceded by
   a. the cortical reaction.
   b. the acrosomal reaction.
   c. the fast block.
   d. implantation.
   e. cleavage.

3. The primitive gut develops as a result of
   a. gastrulation.
   b. cleavage.
   c. embryogenesis.
   d. embryonic folding.
   e. aneuploidy.

4. Chorionic villi develop from
   a. the zona pellucida.
   b. the endometrium.
   c. the syncytiotrophoblast.
   d. the embryoblast.
   e. the epiblast.

5. Which of these results from aneuploidy?
   a. Turner syndrome
   b. fetal alcohol syndrome
   c. nondisjunction
   d. mutation
   e. rubella

6. Fetal urine accumulates in the _____ and contributes to the fluid there.
   a. placental sinus d. chorion
   b. yolk sac   e. amnion
   c. allantois

7. A preembryo has
   a. a neural tube.
   b. a heart bulge.
   c. a cytotrophoblast.
   d. a coelom.
   e. decidual cells.

8. The feature that distinguishes a fetus from an embryo is that the fetus has
   a. all of the organ systems.
   b. three germ layers.
   c. a placenta.
   d. an amnion.
   e. arm and leg buds.

9. The first blood and future egg and sperm cells come from
   a. the mesoderm.
   b. the hypoblast.
   c. the syncytiotrophoblast.
   d. the placenta.
   e. the yolk sac.

10. For the first 8 weeks of gestation, a conceptus is nourished mainly by
    a. the placenta.
    b. amniotic fluid.
    c. colostrum.
    d. decidual cells.
    e. yolk cytoplasm.

11. Viruses and chemicals that cause congenital anatomical deformities are called _____.

12. Aneuploidy is caused by _____, the failure of a pair of chromosomes to separate in meiosis.

13. The brain and spinal cord develop from a longitudinal ectodermal channel called the _____.

14. Attachment of the conceptus to the uterine wall is called _____.

15. Fetal _____

16. The _____ pene orga

17. Ferti repr

18. Bon segm

19. The sperm

20. A de a/an laye

Testing Your Recall Twenty simple recall questions at the end of each chapter test retention of terminology and basic ideas.

Building Your Medical Vocabulary

State a medical meaning of each of the following word elements, and give a term in which it is used.

1. haplo-
2. gameto-
3. zygo-
4. tropho-
5. cephalo-
6. gyneco-
7. -genesis
8. syn-
9. meso-
10. terato-

Answers in the Appendix

True or False

Determine which five of the following statements are false, and briefly explain why.

1. Freshly ejaculated sperm are more capable of fertilizing an egg than are sperm several hours old.
2. Fertilization normally occurs in the uterus.
3. An egg is usually fertilized by the first sperm that contacts it.
4. By the time a conceptus reaches the uterus, it has already undergone several cell divisions and consists of 16 cells or more.
5. The individual is first considered a fetus when all of the organ systems are present.
6. The placenta becomes increasingly permeable as it develops.
7. During cleavage, the preembryo acquires a greater number of cells but does not increase in size.
8. In oogenesis, a germ cell divides into four equal-sized egg cells.
9. The stage of the conceptus that implants in the uterine wall is the blastocyst.
10. The energy for sperm motility comes from its acrosome.

Answers in the Appendix

Testing Your Comprehension

1. Only one sperm is needed to fertilize an egg, yet a man who ejaculates fewer than 10 million sperm is usually infertile. Explain this apparent contradiction. Supposing 10 million sperm were ejaculated, predict how many would come within close range of the egg. How likely is it that any one of these sperm would fertilize it?

2. What is the difference between embryology and teratology?

3. At what point in the timeline of table 4.4 do you think thalidomide exerts its teratogenic effect? Explain your reasoning.

4. A teratologist is studying the cytology of a fetus that aborted spontaneously at 12 weeks. She concludes that the fetus was triploid. What do you think this term means? How many chromosomes do you think she found in each of the fetus's cells? To produce this state, what normal process of human development apparently failed?

5. A young woman finds out she is about 4 weeks pregnant. She tells her doctor that she drank heavily at a party 3 weeks earlier, and she is worried about the possible effects of this on her baby. If you were the doctor, would you tell her that there is serious cause for concern? Why or why not?

Answers at www.mhhe.com/saladinha3

Improve Your Grade at www.mhhe.com/saladinha3

Practice quizzes, labeling activities, and games provide fun ways to master concepts. You can also download image PowerPoint files for each chapter to create a study guide or for taking notes during lecture.

Testing Your Comprehension Clinical application and other interpretive essay questions require the student to apply the chapter's basic science to new clinical or other scenarios.

True or False These statements require students not only to evaluate their truth but also to concisely explain why the false statements are untrue or rephrase them in a way that makes them true.

Making It Relevant

Students understandably want to know how the basic science of anatomy relates to their career goals.

Apply What You Know prompts the student to think of the deeper implications or clinical applications of what he or she has just read.

Apply What You Know

Spinal cord injuries commonly result from fractures of vertebrae C5 to C6, but never from fractures of L3 to L5. Explain both observations.

Apply What You Know

Why does it make more functional sense for the collecting ducts to connect to the subclavian veins than it would for them to connect to the subclavian arteries?

Apply What You Know

What effect would you expect from a small brain tumor that blocked the left interventricular foramen?

Deeper Insights are brief side essays on the clinical application of the basic science.

Other Deeper Insight boxes highlight medical history and evolutionary interpretations of human structure and function.

DEEPER INSIGHT 4.2

Morning Sickness

A woman's earliest sign of pregnancy is often *morning sickness*, a nausea that sometimes progresses to vomiting. Severe and prolonged vomiting, called *hyperemesis gravidarum*,[17] can necessitate hospitalization for fluid therapy to restore electrolyte and acid–base balance. The physiological cause of morning sickness is unknown[...] the steroids of pregnancy inhibiting intestinal motili[...] tain whether it is merely an undesirable effect of preg[...] it has a biological purpose. An evolutionary hypothe[...] sickness is an adaptation to protect the embryo fro[...] bryo is most vulnerable to toxins at the same time t[...] ness peaks, and women with morning sickness ten[...] foods and to avoid spicy and pungent foods, which a[...] compounds. Pregnant women tend also to be espe[...] flavors and odors that suggest spoiled food. Women[...] rience morning sickness are more likely to miscarr[...] with birth defects.

which gives rise to the dermis of the skin and[...] subcutaneous tissue.

At 5 weeks, the embryo exhibits a prominent[...] cephalic end and a pair of optic vesicles destine[...] eyes. A large **heart bulge** contains a heart, which[...] since day 22. The **arm buds** and **leg buds**, the[...]

[17]*hyper* = excessive + *emesis* = vomiting + *gravida* = pregnant w[...]

DEEPER INSIGHT 6.1

Radioactivity and Bone Cancer

Radioactivity captured the public imagination when Marie and Pierre Curie and Henri Becquerel shared the 1903 Nobel Prize for its discovery. Not for several decades, however, did anyone realize its dangers. Factories employed women to paint luminous numbers on watch and clock dials with radium paint. As they moistened the paint brushes with their tongues to keep them finely pointed, the women ingested radium. Their bones readily absorbed it and many of the women developed *osteosarcoma*, the most common and deadly form of bone cancer.

Even more horrific, in the wisdom of hindsight, was a deadly health fad in which people drank "tonics" made of radium-enriched water. One famous enthusiast was the champion golfer and millionaire playboy Eben Byers, who drank several bottles of radium tonic each day and praised its virtues as a wonder drug and aphrodisiac. Like the factory women, Byers contracted osteosarcoma. By the time of his death, holes had formed in his skull and doctors had removed his entire up[...] to halt the spreading[...] ive they could expose[...] amage left him unable[...] e bitter end. His tragic[...] nd put an end to the

DEEPER INSIGHT 11.3

Hernias

A hernia is any condition in which the viscera protrude through a weak point in the muscular wall of the abdominopelvic cavity. The most common type to require treatment is an *inguinal hernia* (fig. 11.18). In the male fetus, each testis descends from the pelvic cavity into the scrotum by way of a passage called the *inguinal canal* through the muscles of the groin. This canal remains a weak point in the pelvic floor, especially in infants and children. When pressure rises in the abdominal cavity, it can force part of the intestine or bladder into this canal or even into the scrotum. This also sometimes occurs in men who hold their breath while lifting heavy weights. When the diaphragm and abdominal muscles contract, pressure in the abdominal cavity can soar to 1,500 pounds per square inch—more than 100 times the normal pressure and quite sufficient to produce an inguinal hernia, or "rupture." Inguinal hernias rarely occur in women.

Two other sites of hernia are the diaphragm and navel. A *hiatal hernia* is a condition in which part of the stomach protrudes through the diaphragm into the thoracic cavity. This is most common in overweight people over age 40. It may cause heartburn due to the regurgitation of stomach acid into the esophagus, but most cases go undetected. In an *umbilical hernia*, abdominal viscera protrude through the navel.

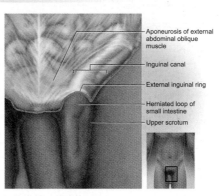

- Aponeurosis of external abdominal oblique muscle
- Inguinal canal
- External inguinal ring
- Herniated loop of small intestine
- Upper scrotum

Figure 11.18 Inguinal Hernia. A loop of small intestine has protruded through the inguinal canal into a space beneath the skin.

DEEPER INSIGHT 1.2

Cardiac Tamponade

Being confined by the pericardium can cause a problem for the heart under some circumstances. If a heart wall weakened by disease should rupture, blood spurts from the heart chamber into the pericardial cavity, filling the cavity more and more with each heartbeat. Diseased hearts also sometimes seep serous fluid into the pericardial sac. Either way, the effect is the same: the pericardial sac has little room to expand, so the accumulating fluid puts pressure on the heart, squeezing it and preventing it from refilling between beats. This condition is called *cardiac tamponade*. If the heart chambers cannot refill, then cardiac output declines and a person may die of catastrophic circulatory failure. A similar situation occurs if serous fluid or air accumulates in the pleural cavity, causing collapse of a lung.

"The clinical applications and evolutionary medicine sections are...nice feature(s) that provide interesting supplemental information."

—Ben F. Brammell
Morehead State University

Engaging Presentation Materials

Create customized lectures, visually enhanced tests and quizzes, compelling course websites, or attractive printed support material using McGraw-Hill's Presentation Assets

Assets available to instructors at www.mhhe.com/saladinha3 include illustrations with and without labels, photos, and tables, as well as PowerPoint lecture outlines and animations.

NEW! A complete set of animations embedded in PowerPoint slides is now available!

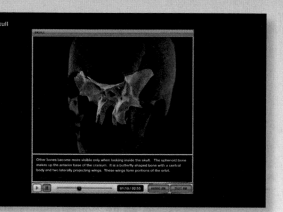

Other Resources

Instructor Materials on Textbook Website

- Test Bank—A comprehensive bank of test questions is provided as simple Word files, and also via McGraw-Hill's EZ Test. EZ Test allows instructors to search for questions by topic, format, or difficulty level; edit existing questions, or add new ones; and create multiple versions of a test. Any test can be exported for use with course management systems such as BlackBoard.
- Instructor's Manual—This handy guide includes chapter overviews, discussion topics, learning strategies, and other helpful materials specific to the textbook.

Laboratory Manual

The *Human Anatomy Laboratory Manual,* by Eric Wise of Santa Barbara City College, is expressly written to coincide with the chapters of Saladin's *Human Anatomy.* It contains a variety of laboratory exercises designed to provide a comprehensive overview of the human body, and is appropriate for laboratory courses using cats or cadavers as dissection specimens. An instructor's manual containing answers to laboratory reports is available.

Measure
Your Students'
Progress

McGRAW-HILL CONNECT ANATOMY & PHYSIOLOGY

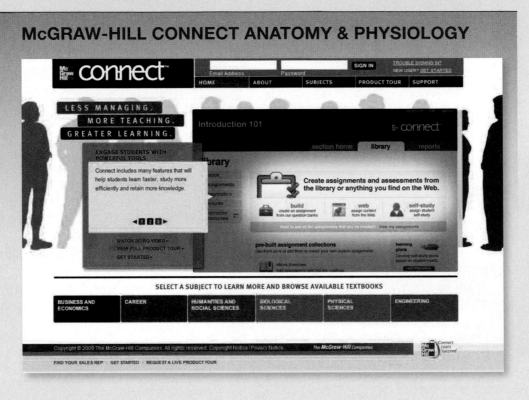

McGraw-Hill *Connect*™ *Anatomy* & Physiology is a Web-based assignment and assessment platform that gives students the means to better connect with their coursework, with their instructors, and with the important concepts that they will need to know for success now and in the future.

With *Connect,* instructors can deliver assignments, quizzes, and tests online. Questions are presented in an auto-gradable format and tied to the organization of the textbook. Instructors can edit existing questions and author entirely new problems. Track individual student performance—by question, assignment, or in relation to the class overall—with detailed grade reports. Integrate grade reports easily with learning management systems (LMS) such as WebCT and Blackboard. And much more.

By choosing *Connect,* instructors are providing their students with a powerful tool for improving academic performance and truly mastering course material. *Connect* allows students to practice important skills at their own pace and on their own schedule. Importantly, students' assessment results and instructors' feedback are all saved online—so students can continually review their progress and plot their course to success.

Some instructors may also choose *ConnectPlus*™ for their students. Like *Connect, ConnectPlus* provides students with online assignments and assessments, plus 24/7 online access to an eBook—an online edition of the text—to aid them in successfully completing their work, wherever and whenever they choose.

Performance-Enhancing Study Tools and Low-Cost Textbook Alternatives

Anatomy & Physiology REVEALED

This amazing multimedia tool is designed to help students learn and review human anatomy using cadaver specimens. Detailed cadaver photographs blended with a state-of-the-art layering technique provide a uniquely interactive dissection experience for all body systems. Anatomy & Physiology REVEALED features the following sections:

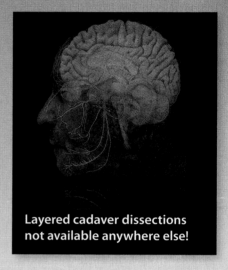

Layered cadaver dissections not available anywhere else!

- Dissection—Peel away layers of the human body to reveal structures beneath the surface. Structures can be pinned and labeled, just like in a real dissection lab. Each labeled structure is accompanied by detailed information and an audio pronunciation. Dissection images can be captured and saved.
- Animation—Compelling animations demonstrate muscle actions, clarify anatomical relationships, and explain difficult concepts.
- Histology—Labeled micrographs presented with each body system allow students to study tissues at their own pace.
- Imaging—Labeled X-ray, MRI, and CT images familiarize students with the appearance of key anatomical structures as seen through different medical imaging techniques.
- Self-Test—Challenging exercises let students test their ability to identify anatomical structures in a timed practical exam format or traditional multiple choice. A results page provides analysis of test scores plus links back to all incorrectly identified structures for review.
- Anatomy Terms—This visual glossary includes directional and regional terms, as well as planes and terms of movement.

Visit www.aprevealed.com to learn more.

Electronic Books

Go Green. Save Green.

McGraw-Hill eBooks.

Go Green!

Green ... it's on everybody's minds these days. It's not only about saving trees; it's also about saving money. At 50% of the bookstore price, McGraw-Hill eBooks are an eco-friendly and cost-saving alternative to the traditional print textbook. So, you do some good for the environment . . . and you do some good for your wallet.

Visit www.mhhe.com/ebooks for details.

Table of Contents

PART FIVE

Reproduction

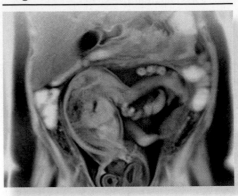

Letter to Students

When I was a young boy, I became interested in what I then called "nature study" for two reasons. One was the sheer beauty of nature. I reveled in children's books with abundant, colorful drawings and photographs of animals, plants, minerals, and gems. It was this aesthetic appreciation of nature that made me want to learn more about it and made me happily surprised to discover I could make a career of it. At a slightly later age, another thing that drew me still deeper into biology was to discover writers who had a way with words—who could captivate my imagination and curiosity with their elegant prose. Once I was old enough to hold part-time jobs, I began buying zoology and anatomy books that mesmerized me with their gracefulness of writing and fascinating art and photography. I wanted to write and draw like that myself, and I began teaching myself by learning from "the masters." I spent many late nights in my room peering into my microscope and jars of pond water, typing page after page of manuscript, and trying pen and ink as a medium. In short, I was the ultimate nerd. My "first book" was a 318-page paper on some little pond animals called hydras, with 53 India ink illustrations, that I wrote for my tenth-grade biology class when I was 16.

Fast forward almost 40 years, and I found myself bringing that same enjoyment of writing and illustrating to the first edition of the book you are now holding. Why? Not only for its intrinsic creative satisfaction, but because I'm guessing that you're like I was—you can appreciate a book that does more than simply give you the information you need. You appreciate, I trust, a writer who makes it enjoyable for you through his scientific, storytelling prose and his concept of the way things should be illustrated to spark interest and facilitate understanding. Some of you probably think of yourselves as "visual learners" and others as "verbal learners." Either way, I hope this book will serve your learning style.

I know from my own students, however, that you need more than captivating illustrations and enjoyable reading. Let's face it—human anatomy is a complex subject and it may seem a formidable task to acquire even a basic knowledge of the human body. It was difficult even for me to learn (and the learning never ends). So in addition to simply writing this book, I've given a lot of thought to what we call pedagogy—the art of teaching. I've designed my chapters to make them easier for you to study and to give you abundant opportunity to check whether you've understood what you read—to test yourself (as I advise my own students) before the instructor tests you.

Each chapter is broken down into short, digestible bits with a set of learning goals (Expected Learning Outcomes) at the beginning of each section, and self-testing questions (Before You Go On) just a few pages later. Even if you have just 30 minutes to read during a lunch break or a bus ride, you can easily read or review one of these brief sections. At the end of each chapter, you will find a Study Guide with a set of learning goals (Assess Your Learning Outcomes) parallel to the chapter organization, and a variety of self-testing questions. You will also find additional self-testing questions in the body of each chapter in the figure legends and Apply What You Know feature. The questions cover a broad range of cognitive skills, from simple recall of a term to your ability to evaluate, analyze, and apply what you've learned to new clinical situations or other problems.

The pages that precede this "letter" take you through the learning aids we've created for you within the book itself and at our website, www.mhhe.com/saladinha3. I hope you will take a little time to look through these pages and the website to see what we have to offer you. It will give you some idea of my thinking behind the book's design and content and will also help you get more out of your experience.

I hope you enjoy your study of this book, but I know there are always ways to make it even better. Indeed, what quality you may find in this edition owes a great deal to feedback I've received from students all over the world. If you find any typos or other errors, if you have any suggestions for improvement, if I can clarify a concept for you, or even if you just want to comment on something you really like about the book, I hope you'll feel free to write to me. I correspond quite a lot with students and would enjoy hearing from you.

Ken Saladin
Georgia College & State University
Milledgeville, GA 31061 (USA)
ken.saladin@gcsu.edu

PART ONE

The Study of Human Anatomy

A new life begins—a human embryo
on the point of a pin (scanning
electron micrograph)

CHAPTER OUTLINE

DEEPER INSIGHTS

This book is an introduction to the structure of the human body. It is meant primarily to provide a foundation for advanced study in fields related to health and fitness. Beyond that purpose, however, the study of anatomy can also provide a satisfying sense of self-understanding. Even as children, we are curious about what's inside the body. Dried skeletons, museum exhibits, and beautifully illustrated atlases of the body have long elicited widespread public fascination.

This chapter lays a foundation for our study of anatomy by considering some broad themes. We will consider what this science encompasses and what methods are used for the study of anatomy. We will lay out a general "roadmap" of the human body to provide a context for the chapters that follow. We will also get some insights into how a beginning anatomy student can become comfortable with medical terminology.

1.1 The Scope of Human Anatomy

▶ Expected Learning Outcomes

When you have completed this section, you should be able to

- define *anatomy* and some of its subdisciplines;
- name and describe some approaches to studying anatomy;
- describe some methods of medical imaging; and
- discuss the variability of human anatomy.

Human anatomy is the study of the structure of the human body. It provides an essential foundation for understanding **physiology,** the study of function; anatomy and physiology together are the bedrock of the health sciences. You can study human anatomy from an atlas; yet as beautiful, fascinating, and valuable as many anatomy atlases are, they teach almost nothing but the locations, shapes, and names of things. This book is different; it deals with what biologists call **functional morphology**[1]—not just the structure of organs, but the functional reasons behind that structure..

Anatomy and physiology complement each other; each makes sense of the other, and each molds the other in the course of human development and evolution. We cannot delve into the details of physiology in this book, but enough will be said of function to help you make sense of human structure and to more deeply appreciate the beauty of human form.

The Anatomical Sciences

Anatomy is an ancient human interest, undoubtedly older than any written language we know. We can only guess when people began deliberately cutting into human bodies out of curiosity, simply to know what was inside. Some of the earliest and most influential books of anatomy were written by the Greek philosopher Aristotle

(384–322 BCE), the Greek physician Galen (129–c. 199 CE), and Persian physician Avicenna (Ibn Sina, 980–1037 CE). For nearly 1,500 years, medical professors in Europe practically idolized these "ancient masters" and considered their works above reproach. Modern human anatomy, however, dates to the sixteenth century, when Flemish physician and professor Andreas Vesalius (1514–64) questioned the accuracy of the earlier authorities and commissioned the first accurate anatomical illustrations for his book, *De Humani Corporis Fabrica* (*On the Structure of the Human Body,* 1543) (fig. 1.1). The tradition begun by Vesalius has been handed down to us through such famous contemporary works as *Gray's Anatomy,* Frank Netter's *Atlas of Human Anatomy,* and many others, to the richly illustrated textbooks used by college students today.

For all its attention to the deceased body, or **cadaver,**[2] human anatomy is hardly a "dead science." New techniques of study continually produce exciting new insights into human structure, and anatomists have discovered far more about the human body in the last century than in the 2,500 years before. Anatomy now embraces several subdisciplines that study human structure from different perspectives. **Gross anatomy** is the study of structure visible to the naked eye, using methods such as surface observation, dissection, X-rays, and MRI scans. **Surface anatomy** is the external structure of the body, and is especially important in conducting a physical examination of a patient. **Radiologic anatomy** is the study of internal structure, using X-rays and other medical imaging techniques described in the next section. In many cases, such as MRI scans, this entails examination of a two-dimensional image of a thin "slice" through the body.

Systemic anatomy is the study of one organ system at a time and is the approach taken by most introductory textbooks such as this one. **Regional anatomy** is the study of multiple organ systems at once in a given region of the body, such as the head or chest. (See the Atlas of Regional and Surface Anatomy on p. 329.) Medical schools typically teach anatomy from a regional perspective, because it is more practical to dissect all structures of the head and neck, the chest, or a limb, than it would be to try to dissect the entire digestive system, then the cardiovascular system, and so forth. Dissecting one system almost invariably destroys organs of another system that stand in the way. Furthermore, as surgeons operate on a particular area of the body, they must think from a regional perspective and attend to the interrelationships of all structures in that area.

Ultimately, the structure and function of the body result from its individual cells. To see those, we usually take tissue specimens, thinly slice and stain them, and observe them under the microscope. This approach is called **microscopic anatomy (histology).** **Histopathology**[3] is the microscopic examination of tissues for signs of disease. **Cytology**[4] is the study of the structure and function of individual cells. **Ultrastructure** refers to fine detail, down to the molecular level, revealed by the electron microscope.

Anatomy, of course, is not limited to the study of humans, but extends to all living organisms. Even students of human structure benefit from **comparative anatomy**—the study of more than one

---

[1] *morpho* = form, structure + *logy* = study of

[2] from *cadere* = to fall down or die
[3] *histo* = tissue + *patho* = disease + *logy* = study of
[4] *cyto* = cell + *logy* = study of

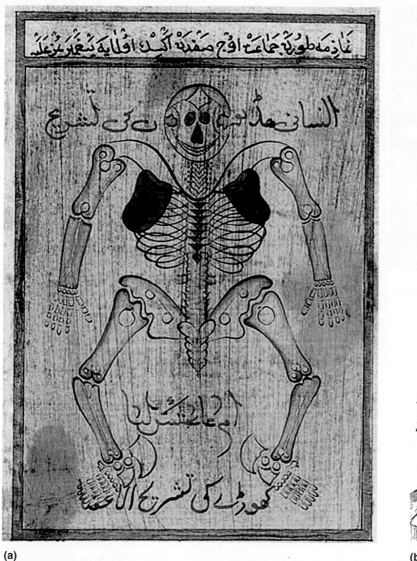

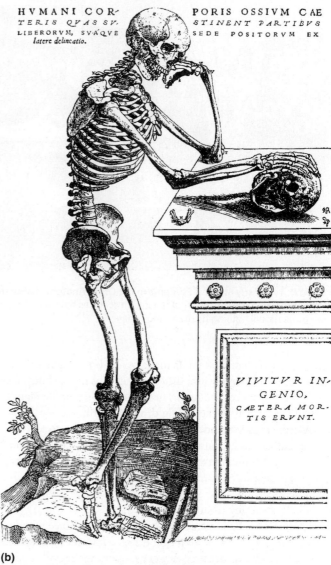

(a)

(b)

Figure 1.1 **The Skeletal System.** Illustrations from an eleventh-century work attributed to Persian physician Avicenna and from Vesalius's *De Humani Corporis Fabrica* (1543).

species in order to examine structural similarities and differences and analyze evolutionary trends. Anatomy students often begin by dissecting other animals with which we share a common ancestry and many structural similarities. Indeed, many of the reasons for human structure become apparent only when we look at the structure of other animals. In chapter 25, for example, you will see that physiologists had little idea of the purpose of certain tubular loops in the kidney (*nephron loops*) until they compared human kidneys with those of desert and aquatic animals, which have greater and lesser needs to conserve water. The greater an animal's need to conserve water (the drier its habitat), the longer these loops are. Thus, comparative anatomy hinted at the function of the nephron loop, which could then be confirmed through experimental physiology. Such are the insights that can be gained by comparing different species with each other.

Methods of Study

There are several ways to examine the structure of the human body. The simplest is **inspection**—simply looking at the body's appearance in careful detail, as in performing a physical examination or making a clinical diagnosis from surface appearance. Observations of the skin and nails, for example, can provide clues to such underlying problems as vitamin deficiencies, anemia, heart disease, and liver disease. Physical examinations involve not only looking at the body for signs of normalcy or disease, but also touching and listening to it. **Palpation**[5] means feeling a structure with the hands, such as palpating a swollen lymph node or taking a pulse. **Auscultation**[6] (AWS-cul-TAY-shun) is

---

[5]*palp* = touch, feel + *ation* = process
[6]*auscult* = listen + *ation* = process

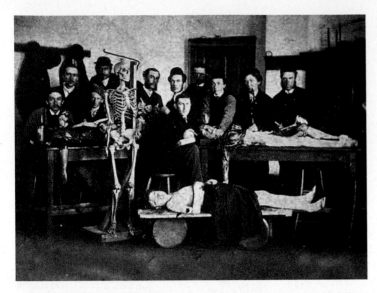

Figure 1.2 Early Medical Students in the Gross Anatomy Laboratory with Three Cadavers. Students of the health sciences have long begun their professional training by dissecting cadavers.

listening to the natural sounds made by the body, such as heart and lung sounds. In **percussion,** the examiner taps on the body, feels for abnormal resistance, and listens to the emitted sound for signs of abnormalities such as pockets of fluid, air, or scar tissue.

A deeper understanding of the body depends on **dissection**—the careful cutting and separation of tissues to reveal their relationships. The very words *anatomy*[7] and *dissection*[8] both mean "cutting apart"; until the nineteenth century, dissection was called "anatomizing." In many schools of health science, cadaver dissection is one of the first steps in the training of students (fig. 1.2).

Dissection, of course, is not the method of choice when studying a living person! Not long ago, it was common to diagnose disorders through **exploratory surgery**—opening the body and taking a look inside to see what was wrong and what could be done about it. Any breach of the body cavities is risky, however, and most exploratory surgery has now been replaced by **medical imaging** techniques—methods of viewing the inside of the body without surgery (fig. 1.3). The branch of medicine concerned with imaging is called **radiology.** Anatomy learned in this way is called **radiologic anatomy,** and those who use radiologic methods for clinical purposes include **radiologists** and **radiologic technicians.**

Some radiologic methods involve high-energy **ionizing radiation** such as X-rays or particles called positrons. These penetrate the tissues and can be used to produce images on X-ray film or through electronic detectors. The benefits of ionizing radiation must always be weighed against its risks. It is called *ionizing* because it ejects electrons from the atoms and molecules it strikes. This effect can cause mutation and trigger cancer. Thus, ionizing radiation cannot be used indiscriminately. Used judiciously, however, the benefits of a mammogram or dental X-ray substantially outweigh the small risk.

Some of the imaging methods to follow are considered *noninvasive* because they do not involve any penetration of the skin or body orifices. *Invasive* imaging techniques may entail inserting ultrasound probes into the esophagus, vagina, or rectum to get closer to the organ to be imaged, or injecting substances into the bloodstream or body passages to enhance image formation.

Any anatomy student today must be acquainted with the basic techniques of radiology and their respective advantages and limitations. Many of the images printed in this book have been produced by the following techniques.

Radiography

Radiography, first performed in 1895, is the process of photographing internal structures with X-rays. Until the 1960s, this was the only widely available imaging method; even today, it accounts for more than 50% of all clinical imaging. X-rays penetrate soft tissues of the body and darken photographic film on the other side. They are absorbed, however, by dense tissues such as bones, teeth, tumors, and tuberculosis nodules, which leave the film lighter in these areas (fig. 1.3a). The term *X-ray* also applies to a photograph (*radiograph*) made by this method. Radiography is commonly used in dentistry, mammography, diagnosis of fractures, and examination of the chest. Hollow organs can be visualized by filling them with a *radiopaque* substance that absorbs X-rays. Barium sulfate, for example, is given orally for examination of the esophagus, stomach, and small intestine, or by enema for examination of the large intestine. Other substances are given by injection for *angiography,* the examination of blood vessels (fig. 1.3b). Some disadvantages of radiography are that images of overlapping organs can be confusing and slight differences in tissue density are not easily detected. In addition, X-rays present the aforementioned risks of ionizing radiation.

Sonography

Sonography[9] is the second oldest and second most widely used method of imaging. It is an outgrowth of sonar technology developed in World War II. A handheld device held firmly against the skin emits high-frequency ultrasound waves and receives the signals reflected back from internal organs. Sonography avoids the harmful effects of X-rays, and the equipment is relatively inexpensive and portable. Its primary disadvantage is that it does not produce a very sharp image. Although sonography was first used medically in the 1950s, images of significant clinical value had to wait until computer technology had developed enough to analyze differences in the way tissues reflect ultrasound. Sonography is not very useful for examining bones or lungs, but it is the method of choice in obstetrics, where the image (*sonogram*) can be used to locate the placenta and evaluate fetal age, position, and development (fig. 1.4). *Echocardiography* is the sonographic examination of the beating heart.

[7]*ana* = apart + *tom* = cut
[8]*dis* = apart + *sect* = cut
[9]*sono* = sound + *graphy* = recording process

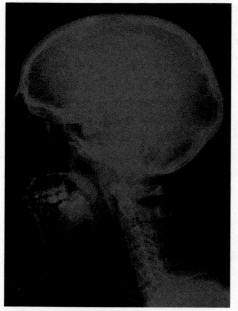

(a) X-ray (radiograph)

(b) Cerebral angiogram

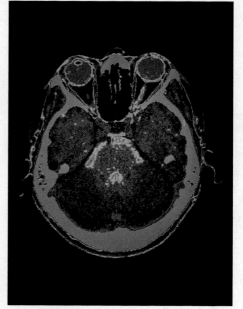

(c) Computed tomographic (CT) scan

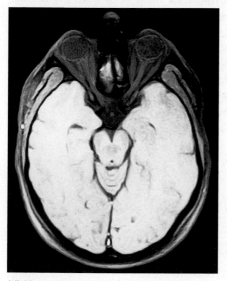

(d) Magnetic resonance image (MRI)

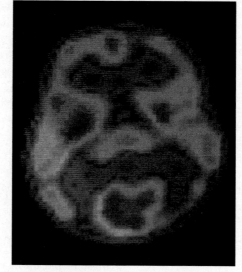

(e) Positron emission tomographic (PET) scan

Figure 1.3 Radiologic Images of the **Head.** (a) X-ray (radiograph) showing the bones and teeth. (b) An angiogram of the cerebral blood vessels. The arteries are enhanced with false color. (c) A CT scan. The eyes and skin are shown in blue, bone in pink, and the brain in green. (d) An MRI scan at the level of the eyes. The optic nerves appear in red, and the muscles that move the eyes appear in green. (e) A PET scan of the brain of an unmedicated schizophrenic patient. Red areas indicate regions of high metabolic rate. In this patient, the visual center of the brain at the rear of the head (bottom of photo) was especially active during the scan.
• *What structures are seen better by MRI than by X-ray? What structures are seen better by X-ray than by PET?*

Computed Tomography

Computed tomography (a **CT scan**), formerly called a *computerized axial tomographic*[10] *(CAT) scan,* is a more sophisticated application of X-rays developed in 1972. The patient is moved through a ring-shaped machine that emits low-intensity X-rays on one side and receives them with a detector on the opposite side. A computer analyzes signals from the detector and produces an image of a "slice" of the body about as thin as a coin (fig. 1.3c). The computer can "stack" a series of these images to construct a three-dimensional image of the body. CT scanning has the advantage of imaging thin sections of the body, so there is little organ overlap and the image is much sharper than a conventional X-ray. It requires extensive knowledge of cross-sectional anatomy to interpret the images. CT scanning is useful for identifying tumors, aneurysms, cerebral hemorrhages, kidney stones, and other abnormalities.

Magnetic Resonance Imaging

Magnetic resonance imaging (MRI) was conceived as a technique superior to CT for visualizing soft tissues (fig. 1.3d). The patient lies in a chamber surrounded by a large electromagnet that creates a magnetic field 3,000 to 60,000 times as strong as the earth's. Hydrogen atoms in the tissues align themselves with the magnetic field. The radiologic technologist then turns on a field of radio waves, causing the hydrogen atoms to absorb additional energy and align in a different direction.

---

[10]*tomo* = section, cut, slice + *graphic* = pertaining to a recording

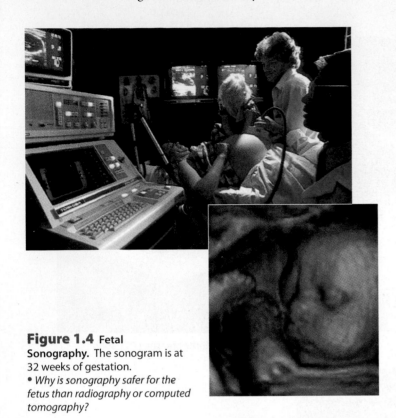

Figure 1.4 Fetal **Sonography.** The sonogram is at 32 weeks of gestation.
• *Why is sonography safer for the fetus than radiography or computed tomography?*

When the radio waves are turned off, the hydrogen atoms abruptly realign to the magnetic field, giving off their excess energy at rates that depend on the type of tissue. A computer analyzes the emitted energy to produce an image of the body. MRI can "see" clearly through the skull and vertebral column to produce images of the nervous tissue. Moreover, it is better than CT for distinguishing between soft tissues such as the white and gray matter of the brain. It also avoids the harmful effects of X-rays. *Functional MRI (fMRI)* is a form of MRI that visualizes moment-to-moment changes in tissue function; fMRI scans of the brain, for example, show shifting patterns of activity as the brain applies itself to a specific task.

Apply What You Know

The concept of MRI was conceived in 1948 but could not be put into clinical practice until the 1970s. Speculate on a possible reason for this delay.

Positron Emission Tomography

Positron emission tomography (the **PET scan**), developed in the 1970s, is used to assess the metabolic state of a tissue and to distinguish which tissues are most active at a given moment (fig. 1.3e). The procedure begins with an injection of radioactively labeled glucose, which emits positrons (electron-like particles with a positive charge). When a positron and electron meet, they annihilate each other and give off gamma rays that can be detected by sensors and processed by computer. The result is a color image that shows which tissues were using the most glucose at the moment. In cardiology, PET scans can show the extent of damaged heart tissue.

Since damaged tissue consumes little or no glucose, it appears dark. In neuroscience, PET scans are used, like fMRI, to show which regions of the brain are most active when a person performs a specific task. The PET scan is an example of **nuclear medicine**—the use of radioisotopes to treat disease or to form diagnostic images of the body.

Variation in Human Structure

A quick look around any classroom is enough to show that no two humans look exactly alike; on close inspection, even identical twins exhibit differences. Anatomy atlases and textbooks can easily give you the impression that everyone's internal anatomy is the same, but this simply is not true. Books such as this one can teach you only the most common structure—the anatomy seen in approximately 70% or more of people. Someone who thinks that all human bodies are the same internally would make a very confused medical student or an incompetent surgeon.

Some people completely lack certain organs. For example, most of us have a *palmaris longus* muscle in the forearm and a *plantaris* muscle in the leg, but not everyone. Most of us have five lumbar vertebrae (bones of the lower spine), but some have four and some have six. Most of us have one spleen, but some people have two. Most have two kidneys, but some have only one. Most kidneys are supplied by a single *renal artery* and drained by one *ureter,* but in some people, a single kidney has two renal arteries or ureters. Figure 1.5 shows some common variations in human anatomy, and Deeper Insight 1.1 describes a particularly dramatic variation.

DEEPER INSIGHT   1.1

Situs Inversus and Other Unusual Anatomy

In most people, the heart tilts toward the left, the spleen and sigmoid colon are on the left, the liver lies mainly on the right, the gallbladder and appendix are on the right, and so forth. This normal arrangement of the viscera is called *situs* (SITE-us) *solitus.* About 1 in 8,000 people are born, however, with a striking developmental abnormality called *situs inversus*—the organs of the thoracic and abdominal cavities are reversed between right and left. A selective left–right reversal of the heart is called *dextrocardia.* In *situs perversus,* a single organ occupies an atypical position, not necessarily a left–right reversal—for example, a kidney located low in the pelvic cavity instead of high in the abdominal cavity.

Some conditions, such as *dextrocardia* in the absence of complete situs inversus, can cause serious medical problems. Complete situs inversus, however, usually causes no functional problems because all of the viscera, though reversed, maintain their normal relationships to each other. Situs inversus is often diagnosed prenatally by sonography, but many people remain unaware of their condition for several decades until it is discovered by medical imaging, on physical examination, or in surgery. However, you can easily imagine the importance of such conditions in diagnosing appendicitis, performing gallbladder surgery, interpreting an X-ray, auscultating the heart valves, or recording an electrocardiogram.

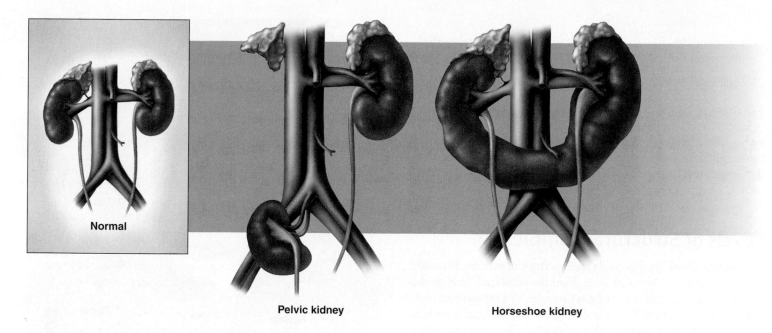

Normal

Pelvic kidney

Horseshoe kidney

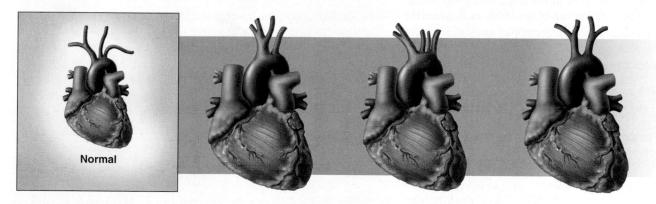

Normal

Variations in branches of the aorta

Figure 1.5 Variations in Anatomy of the Kidneys and Major Arteries near the Heart.

Apply What You Know

People who are allergic to penicillin or aspirin often wear Medic Alert bracelets or necklaces that note this fact in case they need emergency medical treatment and are unable to communicate. Why would it be important for a person with situs inversus to have this noted on a Medic Alert bracelet?

Before You Go On

Answer the following questions to test your understanding of the preceding section:

1. How does functional morphology differ from the sort of anatomy taught by a photographic atlas of the body?

2. Why would regional anatomy be a better learning approach than systemic anatomy for a cadaver dissection course?

3. What is the difference between radiology and radiography?

4. What are some reasons that sonography would be unsuitable for examining the size and location of a brain tumor?

1.2 General Plan of the Human Body

Expected Learning Outcomes

When you have completed this section, you should be able to

- list in proper order the levels of structural complexity of the body, from atom to organism;

- name the 11 human organ systems and state the basic functions and major components of each one;

- describe the anatomical position and explain why it is important in medical language;

- identify the three fundamental anatomical planes of the body;

- define several terms that describe the locations of structures relative to each other;

- identify the major body regions and their subdivisions;

- name and describe the body cavities and the membranes that line them; and
- explain what a potential space is, and give some examples.

The chapters that follow assume a certain core, common language of human structure. You will need to know what we mean by the names for the major body cavities and regions, know the difference between a tissue and an organ, and know where to look if you read that structure X is distal or medial to structure Y, for example. This section introduces this core terminology.

Levels of Structural Complexity

Although this book is concerned mainly with gross anatomy, the study of human structure spans all levels from the subatomic level to the organism, the body taken as a whole. Consider for a moment an analogy to human structure: The English language, like the human body, is very complex, yet an endless array of ideas can be conveyed with a limited number of words. All words in the English language are, in turn, composed of various combinations of just 26 letters. Between the alphabet and a book are successively more complex levels of organization: syllables, words, sentences, paragraphs, and chapters. Humans have an analogous hierarchy of complexity (fig. 1.6), as follows:

Subatomic particles compose the atoms,

atoms compose molecules,

molecules compose organelles,

organelles and other matter compose cells,

cells compose our tissues,

tissues compose organs,

organs compose organ systems, and

11 organ systems compose the organism.

All **atoms** consist of the same three fundamental **subatomic particles**—protons, neutrons, and electrons—assembled in various proportions. Whenever two or more atoms bond with each other, they form a **molecule.** Much of the study of human anatomy is an examination of the body's molecular structure. The largest molecules, such as proteins, fats, and DNA, are called **macromolecules.**

Molecules, in turn, are the building blocks of microscopic structures within our cells called **organelles**[11]—the nucleus, mitochondria, lysosomes, centrioles, and so forth, which carry out individual functions of a cell much like organs such as the heart, liver, and kidneys carry out individual functions of the whole body.

Cells are the smallest entities considered to be alive. A cell is enclosed in a *plasma membrane* composed of lipids and protein, and it usually has one nucleus, an organelle that contains most of its DNA. *Cytology,* the study of cells and organelles, is the subject of chapter 2.

A **tissue** is a mass of similar cells and cell products that forms a discrete region of an organ and performs a specific function. The

[11]*elle* = little

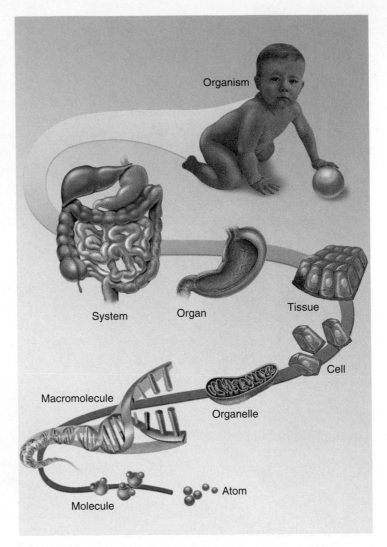

Figure 1.6 **The Body's Structural Hierarchy.** Each level depends on the structure and function of the level below it.

body is composed of only four primary classes of tissue—epithelial, connective, nervous, and muscular tissues. *Histology,* the study of tissues, is the subject of chapter 3.

An **organ** is any structure that has definite anatomical boundaries, is visually distinguishable from adjacent organs, and is composed of two or more tissue types working together to carry out a particular function. Most organs and higher levels of structure are within the domain of gross anatomy. However, there are organs within organs—the large organs visible to the naked eye often contain smaller organs visible only with the microscope. The skin, for example, is the body's largest organ. Included within it are thousands of smaller organs: each hair follicle, nail, sweat gland, nerve, and blood vessel of the skin is an organ in itself.

An **organ system** is a group of organs that carry out a basic function such as circulation, respiration, or digestion. The human body has 11 organ systems (fig. 1.7), defined and illustrated in the next section. Usually, the organs of a system are physically interconnected, such as the kidneys, ureters, urinary bladder, and urethra that compose the urinary system. The endocrine system, however,

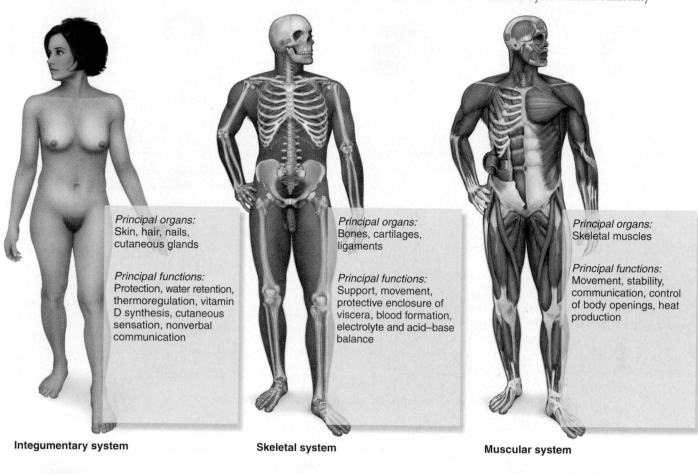

Principal organs:
Skin, hair, nails, cutaneous glands

Principal functions:
Protection, water retention, thermoregulation, vitamin D synthesis, cutaneous sensation, nonverbal communication

Integumentary system

Principal organs:
Bones, cartilages, ligaments

Principal functions:
Support, movement, protective enclosure of viscera, blood formation, electrolyte and acid–base balance

Skeletal system

Principal organs:
Skeletal muscles

Principal functions:
Movement, stability, communication, control of body openings, heat production

Muscular system

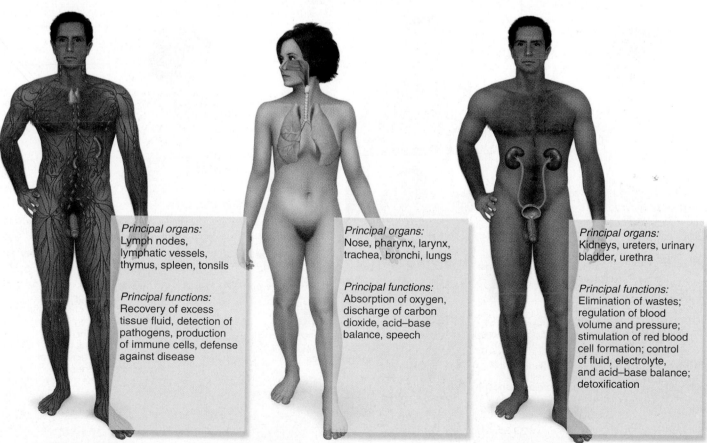

Principal organs:
Lymph nodes, lymphatic vessels, thymus, spleen, tonsils

Principal functions:
Recovery of excess tissue fluid, detection of pathogens, production of immune cells, defense against disease

Lymphatic system

Principal organs:
Nose, pharynx, larynx, trachea, bronchi, lungs

Principal functions:
Absorption of oxygen, discharge of carbon dioxide, acid–base balance, speech

Respiratory system

Principal organs:
Kidneys, ureters, urinary bladder, urethra

Principal functions:
Elimination of wastes; regulation of blood volume and pressure; stimulation of red blood cell formation; control of fluid, electrolyte, and acid–base balance; detoxification

Urinary system

Figure 1.7 The Human Organ Systems.

(continued)

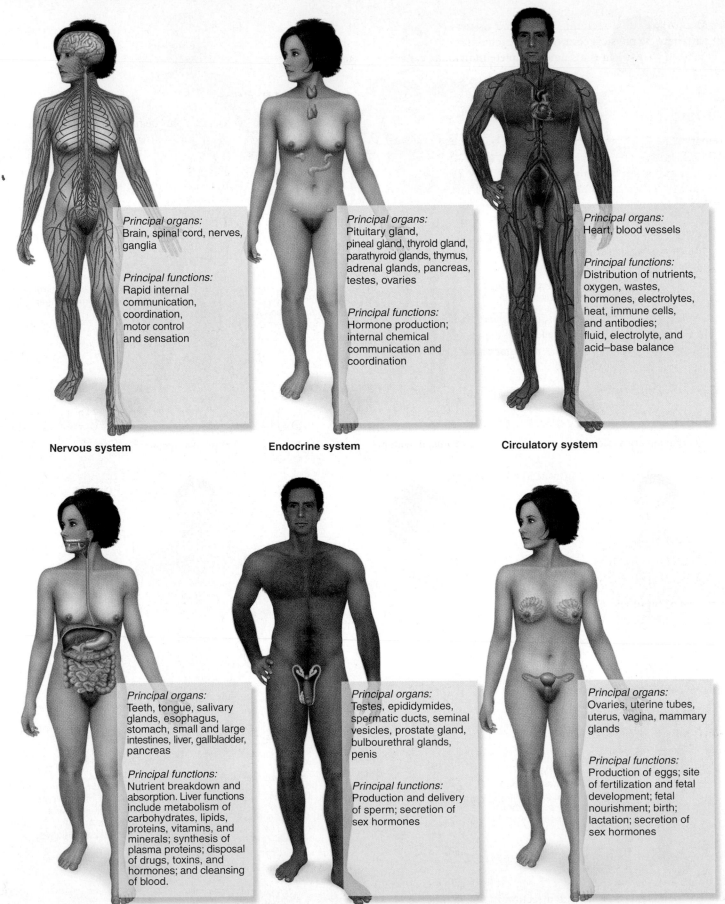

Principal organs:
Brain, spinal cord, nerves, ganglia

Principal functions:
Rapid internal communication, coordination, motor control and sensation

Nervous system

Principal organs:
Pituitary gland, pineal gland, thyroid gland, parathyroid glands, thymus, adrenal glands, pancreas, testes, ovaries

Principal functions:
Hormone production; internal chemical communication and coordination

Endocrine system

Principal organs:
Heart, blood vessels

Principal functions:
Distribution of nutrients, oxygen, wastes, hormones, electrolytes, heat, immune cells, and antibodies; fluid, electrolyte, and acid–base balance

Circulatory system

Principal organs:
Teeth, tongue, salivary glands, esophagus, stomach, small and large intestines, liver, gallbladder, pancreas

Principal functions:
Nutrient breakdown and absorption. Liver functions include metabolism of carbohydrates, lipids, proteins, vitamins, and minerals; synthesis of plasma proteins; disposal of drugs, toxins, and hormones; and cleansing of blood.

Digestive system

Principal organs:
Testes, epididymides, spermatic ducts, seminal vesicles, prostate gland, bulbourethral glands, penis

Principal functions:
Production and delivery of sperm; secretion of sex hormones

Male reproductive system

Principal organs:
Ovaries, uterine tubes, uterus, vagina, mammary glands

Principal functions:
Production of eggs; site of fertilization and fetal development; fetal nourishment; birth; lactation; secretion of sex hormones

Female reproductive system

Figure 1.7 The Human Organ Systems. (continued)

is a group of hormone-secreting glands and tissues that, for the most part, have no physical connection to each other.

Finally, the **organism** is a single, complete individual, capable of acting separately from other individuals.

The Human Organ Systems

As remarked earlier, human structure can be learned from the perspective of regional anatomy or systemic anatomy. This book takes the systemic approach, in which we will fully examine one organ system at a time. There are 11 organ systems in the human body, as well as an *immune system,* which is better described as a population of cells that inhabit multiple organs rather than as an organ system. The organ systems are illustrated and summarized in figure 1.7 in the order that they are covered by this book. They are classified in the following list by their principal functions, although this is an unavoidably flawed classification. Some organs belong to two or more systems—for example, the male urethra is part of both the urinary and reproductive systems; the pharynx is part of the digestive and respiratory systems; and the mammary glands belong to both the integumentary and female reproductive systems.

Systems of Protection, Support, and Movement

Integumentary system

Skeletal system

Muscular system

Systems of Internal Communication and Integration

Nervous system

Endocrine system

Systems of Fluid Transport

Circulatory system

Lymphatic system

Systems of Input and Output

Respiratory system

Digestive system

Urinary system

Systems of Reproduction

Reproductive system

Some medical terms combine the names of two functionally related systems—for example, the *musculoskeletal system, cardiopulmonary system,* and *urogenital (genitourinary) system.* Such terms serve to call attention to the close anatomical or physiological relationship between two systems, but these are not literally individual organ systems.

The Terminology of Body Orientation

When anatomists describe the body, they must indicate where one structure is relative to another, the direction in which a nerve or blood vessel travels, the directions in which body parts move, and so forth. Clear communication on such points requires a universal terminology and frame of reference.

Anatomical Position

The directional language of anatomy begins with a certain assumption about the position of the body under consideration. **Anatomical position** is a stance in which a person stands erect with the feet flat on the floor and close together, arms at the sides, and the palms and face directed forward (fig. 1.8). Without such a frame of reference, to say that a structure such as the sternum, thymus, or aorta is "above the heart" would be vague, since it would depend on whether the subject was standing, lying face down, or lying face up.

Figure 1.8 Anatomical Position.

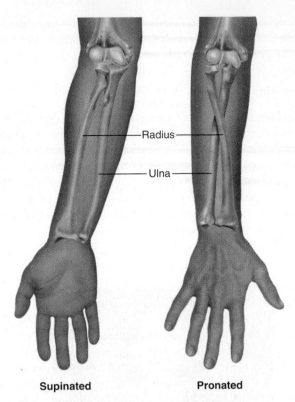

Supinated   **Pronated**

Figure 1.9 Supination and Pronation of the Forearm. When the forearm is supinated, the palm faces anteriorly; when pronated, it faces posteriorly. Note the differences in the relationship of the radius to the ulna.

From the perspective of anatomical position, however, we can describe the thyroid gland as *superior* to the heart, the sternum as *anterior* to it, and the aorta as *posterior* to it. These descriptions remain valid regardless of the subject's position. Even if the body is lying down, such as a cadaver on the medical student's dissection table, to say that the sternum is anterior to the heart invites the viewer to imagine the body standing in anatomical position and not to call it "above the heart" simply because that is the way the body happens to be lying.

Unless stated otherwise, assume that all anatomical descriptions refer to anatomical position. Bear in mind that if a subject is facing you in anatomical position, the subject's left will be on your right, and vice versa. In most anatomical illustrations, for example, the left atrium of the heart appears toward the right side of the page, and although the appendix is located in the right lower quadrant of the abdomen, it appears on the left side of most illustrations.

The forearm is said to be **supinated** when the palms face up or anteriorly and **pronated** when they face down or posteriorly (fig. 1.9); in anatomical position, the forearm is supinated. The difference is particularly important to descriptions of anatomy of this region. In the supinated position, the two forearm bones (radius and ulna) are parallel and the radius is lateral to the ulna. In the pronated position, the radius and ulna cross; the radius is lateral to the ulna at the elbow but medial to it at the wrist. Descriptions of nerves, muscles, blood vessels, and other structures of the arm assume that the arm is supinated.

The words *prone* and *supine* seem similar to these but have an entirely different meaning. A person is **prone** if lying face down, and **supine** if lying face up.

Anatomical Planes

Many views of the body are based on real or imaginary "slices" called sections or planes. *Section* implies an actual cut or slice to reveal internal anatomy, whereas *plane* implies an imaginary flat surface passing through the body. The three major anatomical planes are *sagittal, frontal,* and *transverse* (fig. 1.10).

A **sagittal**[12] (SADJ-ih-tul) **plane** extends vertically and divides the body or an organ into right and left portions. The **median (midsagittal) plane** passes through the midline of the body and

---

[12]*sagitta* = arrow

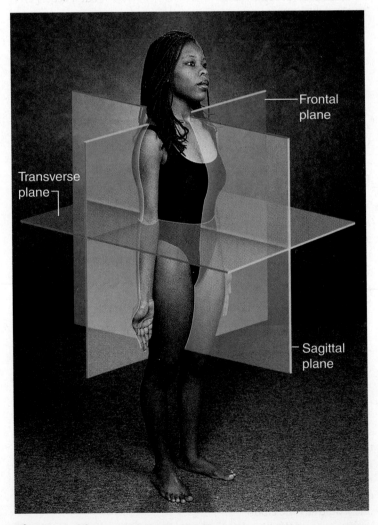

Transverse plane — Frontal plane — Sagittal plane

Figure 1.10 Anatomical Planes of Reference.
• *What is another name for the specific sagittal plane shown here?*

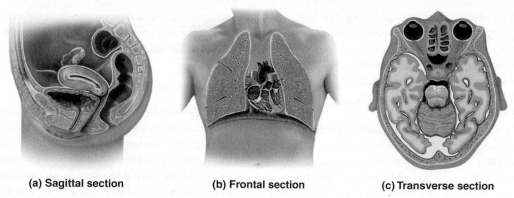

|(a) Sagittal section|(b) Frontal section|(c) Transverse section|

Figure 1.11 **Views of the Body in the Three Primary Anatomical Planes.** (a) Sagittal section of the pelvic region. (b) Frontal section of the thoracic region. (c) Transverse section of the head at the level of the eyes.

divides it into *equal* right and left halves. Other sagittal planes parallel to this (off center) divide the body into unequal right and left portions. The head and pelvic organs are commonly illustrated in median views (fig. 1.11a).

A **frontal (coronal) plane** also extends vertically, but it is perpendicular to the sagittal plane and divides the body into anterior (front) and posterior (back) portions. A frontal section of the head, for example, would divide it into one portion bearing the face and another bearing the back of the head. Contents of the thoracic and abdominal cavities are commonly shown in frontal section (fig. 1.11b).

A **transverse (horizontal) plane** passes across the body or an organ perpendicular to its long axis (fig. 1.11c); it divides the body into superior (upper) and inferior (lower) portions. CT scans are typically transverse sections (see fig. 1.3c), but not always.

Directional Terms

Words that describe the location of one structure relative to another are generally called the **directional terms** of anatomy. Table 1.1 summarizes those most frequently used. Most of these terms exist in pairs with opposite meanings: *anterior* versus *posterior*, *superior* versus *inferior*, *medial* versus *lateral*, *proximal* versus *distal*, and *superficial* versus *deep*. Intermediate directions are often indicated by combinations of these terms. For example, one structure may be described as *anterolateral* to another (toward the front and side).

The terms *proximal* and *distal* are used especially in the anatomy of the limbs, with *proximal* used to denote something relatively close to the limb's point of attachment (the shoulder or hip joint) and *distal* to denote something farther away. These terms do have some applications to anatomy of the trunk of the body, however—for example, in referring to certain aspects of the intestines and the microscopic structure of the kidneys. But when describing the trunk and referring to a structure that lies above or below another in anatomical position, *superior* and *inferior* are the preferred terms. These terms are not usually used for the limbs. Although it may be technically correct, one would not generally say the elbow is superior to the wrist; rather, one would say the elbow is proximal to the wrist.

Because of the bipedal, upright stance of humans, some directional terms have different meanings for humans than they do for other animals. *Anterior*, for example, denotes the region of the body that leads the way in normal locomotion. For a four-legged animal such as a cat, this is the head end of the body; for a human, however, it is the front of the chest and abdomen. What we call *anterior* in a human would be called *ventral* in a cat. *Posterior* denotes the region that comes last in normal locomotion—the tail end of a cat but the back of a human. In the anatomy of most other animals, *ventral* denotes the surface of the body closest to the ground and *dorsal* denotes the surface farthest away from the ground. These two words are too entrenched in human anatomy to completely ignore them, but we will minimize their use in this book to avoid confusion. You must keep such differences in mind, however, when dissecting other animals for comparison to human anatomy.

One vestige of the term *dorsal* is **dorsum,** used to denote the upper surface of the foot and the back of the hand. If you consider how a quadrupedal animal stands, the corresponding surfaces of its paws are both uppermost, facing the same direction as the dorsal side of its trunk. Although these surfaces of the human hand and foot face entirely different directions in anatomical position, the term *dorsum* is still used.

Major Body Regions

Knowledge of the external anatomy and landmarks of the body is important in performing a physical examination and many other clinical procedures. For purposes of study, the body is divided into two major regions called the *axial* and *appendicular regions*. Smaller areas within the major regions are described in the following paragraphs and illustrated in figure 1.12.

Axial Region

The **axial region** consists of the **head, neck (cervical**[13] **region), and trunk.** The trunk is further divided into the **thoracic region** above the diaphragm and the **abdominal region** below it.

[13]*cervic* = neck

One way of referring to the locations of abdominal structures is to divide the region into quadrants. Two perpendicular lines intersecting at the umbilicus (navel) divide the abdomen into a **right upper quadrant (RUQ)**, **right lower quadrant (RLQ)**, **left upper quadrant (LUQ)**, and **left lower quadrant (LLQ)** (fig. 1.13a, b). The quadrant scheme is often used to describe the site of an abdominal pain or abnormality.

The abdomen also can be divided into nine regions defined by four lines that intersect like a tic-tac-toe grid (fig. 1.13c, d). Each vertical line is called a *midclavicular line* because it passes through the midpoint of the clavicle (collarbone). The superior horizontal line is called the *subcostal*[14] *line* because it connects the inferior borders of the lowest costal cartilages (cartilage connecting the tenth rib on each side to the inferior end of the sternum). The inferior horizontal line is called the *intertubercular*[15] *line* because it passes from left to right between the tubercles (*anterior superior spines*) of the pelvis—two points of bone located about where the front pockets open on most pants. The lateral regions of this grid, from upper to lower, are the left and right **hypochondriac,**[16] **lumbar,** and **inguinal**[17] **regions.** The three medial regions from upper to lower are the **epigastric,**[18] **umbilical,** and **hypogastric (pubic) regions.**

Appendicular Region

The **appendicular** (AP-en-DIC-you-lur) **region** of the body consists of the **upper limbs** and **lower limbs** (also called *appendages* or *extremities*). The upper limb includes the **arm** (**brachial region,** BRAY-kee-ul), **forearm** (**antebrachial region,**[19] AN-teh-BRAY-kee-ul), **wrist** (**carpal region**), **hand** (**manual region**), and **fingers (digits)**. The lower limb includes the **thigh (femoral region)**, **leg (crural region,** CROO-rul), **ankle (tarsal region), foot (pedal region,** PEE-dul**)**, and **toes (digits)**. In strict anatomical terms, *arm* refers only to that part of the upper limb between the shoulder and elbow. *Leg* refers only to that part of the lower limb between the knee and ankle.

---

[14]*sub* = below + *cost* = rib
[15]*inter* = between + *tubercul* = little swelling

[16]*hypo* = below + *chondr* = cartilage
[17]*inguin* = groin
[18]*epi* = above, over + *gastr* = stomach
[19]*ante* = fore, before + *brachi* = arm

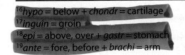

Describe how a part is related to another

TABLE 1.1	Directional Terms in Human Anatomy	
Term	**Meaning**	**Examples of Usage**
Anterior	Toward the front of the body	The sternum is *anterior* to the heart.
Posterior	Toward the back of the body	The esophagus is *posterior* to the trachea.
Ventral	Toward the anterior side*	The abdomen is the *ventral* side of the body.
Dorsal	Toward the posterior side*	The scapulae are *dorsal* to the rib cage.
Superior	Above	The heart is *superior* to the diaphragm.
Inferior	Below	The liver is *inferior* to the diaphragm.
Cephalic	Toward the head or superior end	The *cephalic* end of the embryonic neural tube develops into the brain.
Rostral	Toward the forehead or nose	The forebrain is *rostral* to the brainstem.
Caudal	Toward the tail or inferior end	The spinal cord is *caudal* to the brain.
Medial	Toward the midline of the body	The heart is *medial* to the lungs.
Lateral	Away from the midline of the body	The eyes are *lateral* to the nose.
Proximal	Closer to the point of attachment or origin	The elbow is *proximal* to the wrist.
Distal	Farther from the point of attachment or origin	The fingernails are at the *distal* ends of the fingers.
Ipsilateral	On the same side of the body	The liver is *ipsilateral* to the appendix.
Contralateral	On opposite sides of the body	The spleen is *contralateral* to the liver.
Superficial	Closer to the body surface	The skin is *superficial* to the muscles.
Deep	Farther from the body surface	The bones are *deep* to the muscles.

*In humans only; definition differs for other animals. In human anatomy, *anterior* and *posterior* are usually used in place of *ventral* and *dorsal*.

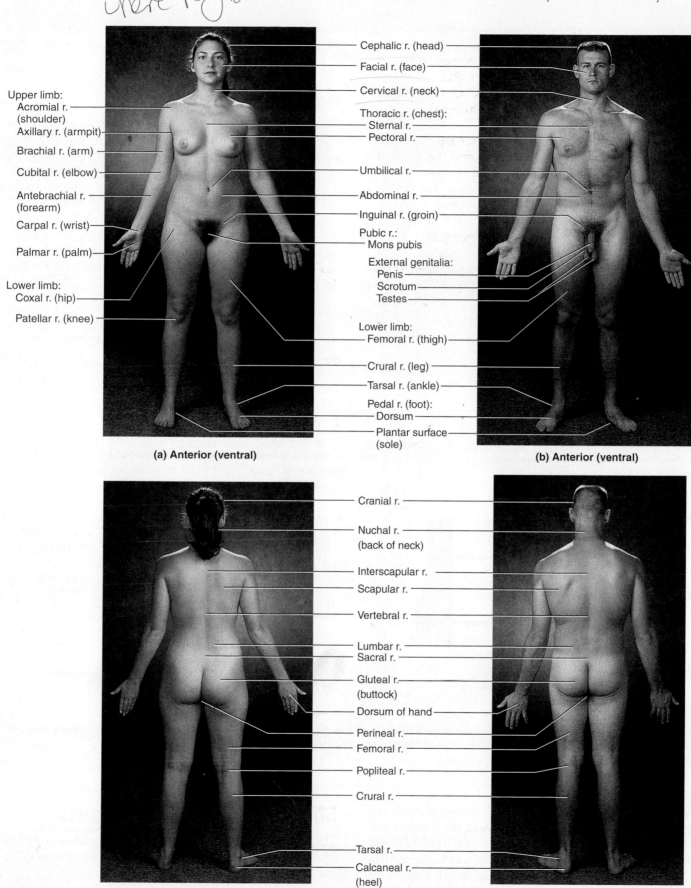

Cephalic r. (head)
Facial r. (face)
Cervical r. (neck)
Thoracic r. (chest):
Sternal r.
Pectoral r.
Umbilical r.
Abdominal r.
Inguinal r. (groin)
Pubic r.:
Mons pubis
External genitalia:
Penis
Scrotum
Testes
Lower limb:
Femoral r. (thigh)
Crural r. (leg)
Tarsal r. (ankle)
Pedal r. (foot):
Dorsum
Plantar surface
(sole)

Upper limb:
Acromial r.
(shoulder)
Axillary r. (armpit)
Brachial r. (arm)
Cubital r. (elbow)
Antebrachial r.
(forearm)
Carpal r. (wrist)
Palmar r. (palm)
Lower limb:
Coxal r. (hip)
Patellar r. (knee)

(a) Anterior (ventral)

(b) Anterior (ventral)

Cranial r.
Nuchal r.
(back of neck)
Interscapular r.
Scapular r.
Vertebral r.
Lumbar r.
Sacral r.
Gluteal r.
(buttock)
Dorsum of hand
Perineal r.
Femoral r.
Popliteal r.
Crural r.
Tarsal r.
Calcaneal r.
(heel)

(c) Posterior (dorsal)

(d) Posterior (dorsal)

Figure 1.12 The Adult Female and Male Bodies. (r. = region)

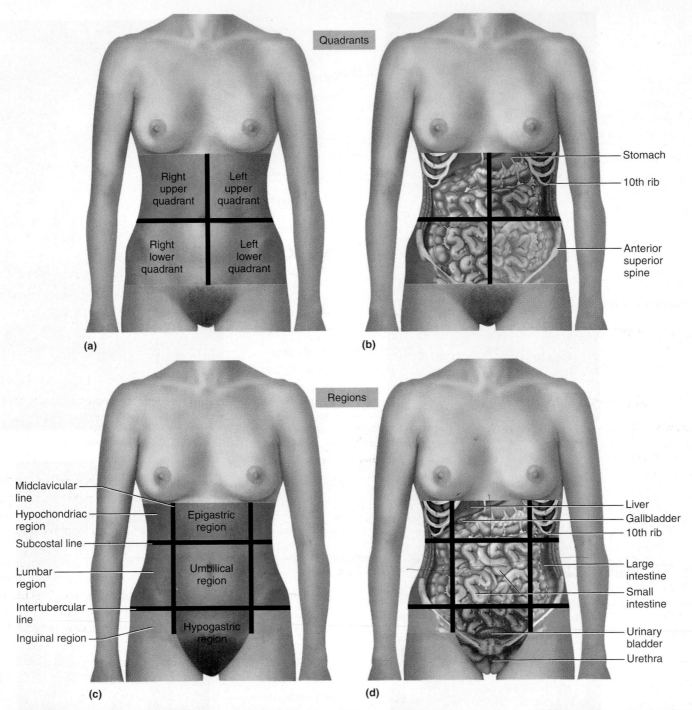

Figure 1.13 **The Four Quadrants and Nine Regions of the Abdomen.** (a) External division into four quadrants. (b) Internal anatomy correlated with the quadrants. (c) External division into nine regions. (d) Internal anatomy correlated with the nine regions.

A **segment** of a limb is a region between one joint and the next. The arm, for example, is the segment between the shoulder and elbow joints, and the forearm is the segment between the elbow and wrist joints. Slightly flexing your fingers, you can easily see that your thumb has two segments (proximal and distal), whereas the other four digits have three segments (proximal, middle, and distal). The segment concept is especially useful in describing the locations of bones and muscles and the movements of the joints.

Body Cavities and Membranes

The body wall encloses several **body cavities,** each lined by a membrane and containing internal organs called the **viscera** (VISS-er-uh) (singular, *viscus*[20]) (fig. 1.14, table 1.2).

---

[20]*viscus* = body organ

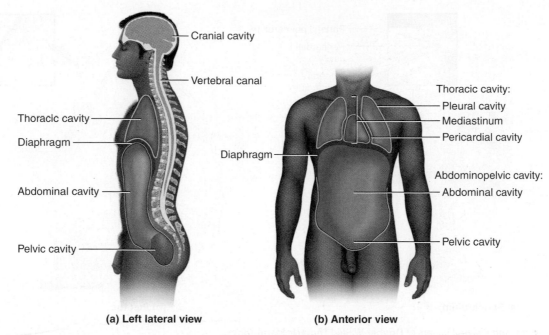

(a) **Left lateral view**

- Cranial cavity
- Vertebral canal
- Thoracic cavity
- Diaphragm
- Abdominal cavity
- Pelvic cavity

(b) **Anterior view**

- Thoracic cavity:
  - Pleural cavity
  - Mediastinum
  - Pericardial cavity
- Diaphragm
- Abdominopelvic cavity:
  - Abdominal cavity
- Pelvic cavity

Figure 1.14 The Major Body Cavities.

The Cranial Cavity and Vertebral Canal

The **cranial** (CRAY-nee-ul) **cavity** is enclosed by the cranial bones (braincase) of the skull and contains the brain. The **vertebral canal** is enclosed by the vertebral column (spine, backbone) and contains the spinal cord. The two are continuous with each other and lined by three membrane layers called the **meninges** (meh-NIN-jeez). Among other functions, the meninges protect the delicate nervous tissue from the hard protective bone that encloses it, and anchor the spinal cord to the vertebral column and limit its movement.

The Thoracic Cavity

During embryonic development, a space called the **coelom** (SEE-loam) forms within the trunk (see fig. 4.5c, p. 89). It subsequently becomes partitioned by a muscular sheet, the **diaphragm**, into a superior **thoracic cavity** and an inferior **abdominopelvic cavity**. Both cavities are lined with thin **serous membranes**, which secrete a lubricating film of moisture similar to blood serum (hence their name).

The thoracic cavity is divided by a thick partition called the **mediastinum**[21] (ME-dee-ah-STY-num) (fig. 1.14b). This is the region between the lungs, extending from the base of the neck to the diaphragm, occupied by the heart, the major blood vessels connected to it, the esophagus, the trachea and bronchi, and a gland called the *thymus*.

A two-layered serous membrane called the **pericardium**[22] wraps around the heart. The inner layer of the pericardium forms the surface of the heart itself and is called the **visceral** (VISS-er-ul) **pericardium (epicardium)**. The outer layer is called the **parietal**[23] (pa-RY-eh-tul) **pericardium (pericardial sac)**. It is separated from the visceral pericardium by a space called the **pericardial cavity** (fig. 1.15a) (see Deeper Insight 1.2). This space is lubricated by a thin film of **pericardial fluid.**

The right and left sides of the thoracic cavity contain the lungs. A serous membrane called the **pleura**[24] (PLOOR-uh) wraps around

TABLE 1.2	Body Cavities and Membranes	
Name of Cavity	**Associated Viscera**	**Membranous Lining**
Cranial cavity	Brain	Meninges
Vertebral canal	Spinal cord	Meninges
Thoracic cavity		
Pleural cavities (2)	Lungs	Pleura
Pericardial cavity	Heart	Pericardium
Abdominopelvic cavity		
Abdominal cavity	Digestive organs, spleen, kidneys, ureters	Peritoneum
Pelvic cavity	Bladder, rectum, reproductive organs	Peritoneum

---

[21]*mediastinum* = in the middle
[22]*peri* = around + *cardi* = heart
[23]*pariet* = wall
[24]*pleur* = rib, side

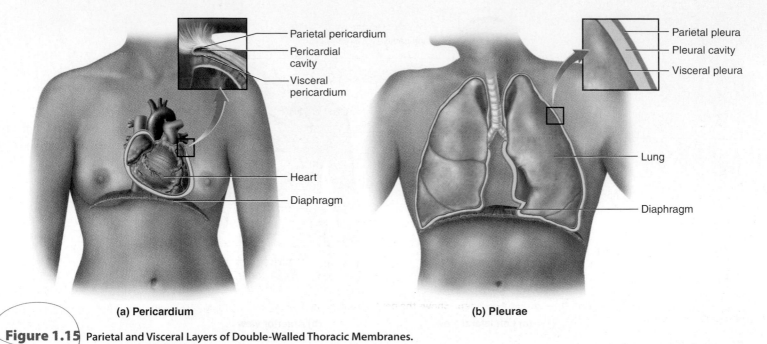

(a) Pericardium

- Parietal pericardium
- Pericardial cavity
- Visceral pericardium
- Heart
- Diaphragm

(b) Pleurae

- Parietal pleura
- Pleural cavity
- Visceral pleura
- Lung
- Diaphragm

Figure 1.15   Parietal and Visceral Layers of Double-Walled Thoracic Membranes.

each lung (fig. 1.15b). Like the pericardium, the pleura has visceral (inner) and parietal (outer) layers. The **visceral pleura** forms the external surface of the lung, and the **parietal pleura** lines the inside of the rib cage. The narrow space between them is called the **pleural cavity** (see fig. A.11, p. 340). It is lubricated by slippery **pleural fluid.**

Note that in both the pericardium and pleura, the visceral layer of the membrane *covers* the surface of an organ and the parietal layer *lines* the inside of a body cavity. We will see this pattern repeated elsewhere, including the abdominopelvic cavity.

DEEPER INSIGHT 1.2

Cardiac Tamponade

Being confined by the pericardium can cause a problem for the heart under some circumstances. If a heart wall weakened by disease should rupture, blood spurts from the heart chamber into the pericardial cavity, filling the cavity more and more with each heartbeat. Diseased hearts also sometimes seep serous fluid into the pericardial sac. Either way, the effect is the same: the pericardial sac has little room to expand, so the accumulating fluid puts pressure on the heart, squeezing it and preventing it from refilling between beats. This condition is called *cardiac tamponade*. If the heart chambers cannot refill, then cardiac output declines and a person may die of catastrophic circulatory failure. A similar situation occurs if serous fluid or air accumulates in the pleural cavity, causing collapse of a lung.

The Abdominopelvic Cavity

The abdominopelvic cavity consists of the **abdominal cavity** superiorly and the **pelvic cavity** inferiorly. The abdominal cavity contains most of the digestive organs as well as the spleen, kidneys, and ureters. It extends inferiorly to the level of a bony landmark called the *brim* of the pelvis (see fig. 8.6, p. 189, and fig. A.7, p. 337). The pelvic cavity, below the brim, is continuous with the abdominal cavity, but is markedly narrower and tilts posteriorly (see fig. 1.14a). It contains the rectum, urinary bladder, urethra, and reproductive organs.

The abdominopelvic cavity contains a two-layered serous membrane called the **peritoneum**[25] (PERR-ih-toe-NEE-um). The **parietal peritoneum** lines the cavity wall. The **visceral peritoneum** turns inward from the body wall, wraps around the abdominal viscera, binds them to the body wall or suspends them from it, and holds them in their proper place. The **peritoneal cavity** is the space between the parietal and visceral layers. It is lubricated by **peritoneal fluid.**

Some organs of the abdominal cavity lie against the posterior body wall and are covered by peritoneum only on the side facing the peritoneal cavity. They are said to have a **retroperitoneal**[26] position (fig. 1.16). These include the kidneys; ureters; adrenal glands; most of the pancreas; and abdominal portions of two major blood vessels, the aorta and inferior vena cava (see fig. A.6, p. 336). Organs that are encircled by peritoneum and connected to the posterior body wall by peritoneal sheets are described as **intraperitoneal.**[27]

The intestines are suspended from the posterior abdominal wall by a translucent membrane called the **posterior mesentery**[28]

---

[25]*peri* = around + *tone* = stretched
[26]*retro* = behind
[27]*intra* = within
[28]*mes* = in the middle + *enter* = intestine

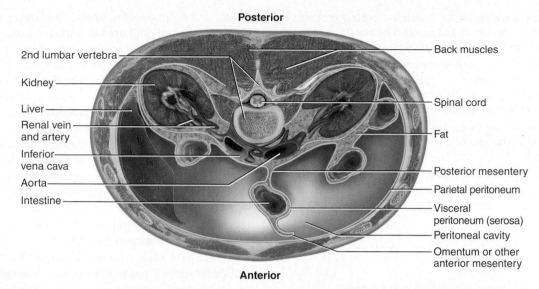

Posterior

2nd lumbar vertebra

Kidney

Liver

Renal vein
and artery

Inferior
vena cava

Aorta

Intestine

Back muscles

Spinal cord

Fat

Posterior mesentery

Parietal peritoneum

Visceral
peritoneum (serosa)

Peritoneal cavity

Omentum or other
anterior mesentery

Anterior

Figure 1.16 **Transverse Section Through the Abdomen.** Shows the peritoneum, peritoneal cavity (with most viscera omitted), and some retroperitoneal organs.

(MESS-en-tare-ee), an infolding of the peritoneum. The posterior mesentery of the large intestine is called the **mesocolon.** In some places, after wrapping around the intestines or other viscera, the mesentery continues toward the anterior body wall as the **anterior mesentery.** The most prominent example of this is a fatty membrane called the **greater omentum,**[29] which hangs like an apron from the inferolateral margin of the stomach and overlies the intestines (fig. 1.17; see also fig. A.4, p. 334). It is unattached at its inferior border and can be lifted to reveal the intestines. A smaller **lesser omentum** extends from the superomedial margin of the stomach to the liver.

Where the visceral peritoneum meets an organ such as the stomach or small intestine, it divides and enfolds it, forming an outer layer of the organ called the **serosa** (seer-OH-sa) (fig. 1.16). The visceral peritoneum thus consists of the mesenteries and serosae.

Potential Spaces

Some of the spaces between body membranes are considered to be **potential spaces,** so named because under normal conditions, the membranes are pressed firmly together and there is no actual space between them. The membranes are not physically attached, however, and under unusual conditions, they may separate and create a space filled with fluid or other matter. Thus, there is only a potential for the membranes to separate and create a space.

The pleural cavity is one example. Normally, the parietal and visceral pleurae are pressed together without a gap between them, but under pathological conditions, air or serous fluid can accumulate between the membranes and open up a space. Another example is the internal cavity *(lumen)* of the uterus. In a nonpregnant uterus, the mucous membranes of the opposite walls are pressed together, so there is little or no open space in the organ. In pregnancy, of course, a growing fetus occupies this space and pushes the mucous membranes apart.

Before You Go On

Answer the following questions to test your understanding of the preceding section:

5. Put the following list in order from the largest and most complex to the smallest and least complex components of the human body: cells, molecules, organelles, organs, organ systems, tissues.

6. Name the organ system responsible for each of the following functions: (a) movement and distribution of blood; (b) water

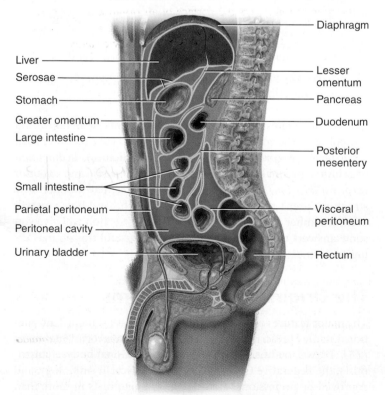

Diaphragm

Liver

Serosae

Stomach

Greater omentum

Large intestine

Small intestine

Parietal peritoneum

Peritoneal cavity

Urinary bladder

Lesser
omentum

Pancreas

Duodenum

Posterior
mesentery

Visceral
peritoneum

Rectum

Figure 1.17 **Serous Membranes of the Abdominal Cavity.** Sagittal section, left lateral view.
• *Is the urinary bladder in the peritoneal cavity?*

[29]omentum = covering

retention, sensation, and protection from infection; (c) hormone secretion; (d) nutrient breakdown and absorption; and (e) recovery of excess tissue fluid and detection of pathogens in the tissues.

7. State the directional term that describes the position of: (a) the spinal cord relative to the heart; (b) the eyes relative to the nose; (c) the urinary bladder relative to the intestines; (d) the diaphragm relative to the liver; and (e) skin relative to the muscles.

8. State the alternative anatomical terms for the regions commonly known as the neck, the sole of the foot, the lower back, the buttocks, and the calf.

9. Name the membranes that enclose the brain, the heart, the lungs, and the abdominal cavity.

1.3 The Language of Anatomy

▶ Expected Learning Outcomes

When you have completed this section, you should be able to

- explain why modern anatomical terminology is so heavily based on Greek and Latin;

- recognize eponyms when you see them;

- describe the efforts to achieve an internationally uniform anatomical terminology;

- discuss the Greek, Latin, or other derivations of medical terms;

- state some reasons why the literal meaning of a word may not lend insight into its definition;

- relate singular noun forms to their plural forms; and

- discuss why precise spelling is important in medical communication.

One of the greatest challenges faced by students of anatomy and physiology is the vocabulary. In this book, you will encounter such Latin terms as *corpus callosum* (a brain structure), *ligamentum arteriosum* (a small fibrous band near the heart), and *extensor carpi radialis longus* (a forearm muscle). You may wonder why structures aren't named in "just plain English," and how you will ever remember such formidable names. This section will give you some answers to these questions and some useful tips on mastering anatomical terminology.

The Origins of Medical Terms

The major features of human gross anatomy have standard international names prescribed by a book titled *Terminologia Anatomica (TA). TA* was codified in 1998 by an international body of anatomists, the Federative Committee on Anatomical Terminology, and approved by professional associations of anatomists in more than 50 countries.

About 90% of today's medical terms are formed from about 1,200 Greek and Latin roots. Scientific investigation began in ancient Greece and soon spread to Rome. The Greeks and Romans coined many of the words still used in human anatomy today: *duodenum, uterus, prostate, cerebellum, diaphragm, sacrum, amnion,* and others. In the Renaissance, the fast pace of anatomical discovery required a profusion of new terms to describe things. Anatomists in different countries began giving different names to the same structures. Adding to the confusion, they often named new structures and diseases in honor of their esteemed teachers and predecessors, giving us such nondescriptive terms as the *fallopian tube* and *duct of Santorini.* Terms coined from the names of people, called **eponyms,**[30] afford little clue as to what a structure or condition is.

In hopes of resolving this growing confusion, anatomists began meeting as early as 1895 to try to devise a uniform international terminology. After several false starts, they agreed on a list of terms titled *Nomina Anatomica (NA). NA* rejected all eponyms as unofficial and gave each structure a unique Latin name to be used worldwide. Even if you were to look at an anatomy atlas in Japanese or Arabic, the illustrations might be labeled with the same Latin terms as in an English-language atlas. *NA* served for many decades until replaced by *TA,* which prescribes both Latin names and accepted English equivalents. The terminology in this book conforms to *TA* except where undue confusion would result from abandoning widely used, yet unofficial terms.

Analyzing Medical Terms

The task of learning anatomical terminology seems overwhelming at first, but as you study this book, there is a simple habit that can quickly make you more comfortable with the technical language of medicine—read the footnotes, which explain the roots and origins of the words. People who find scientific terms confusing and difficult to pronounce, spell, and remember usually feel more confident once they realize the logic of how terms are composed. A term such as *hyponatremia* is less forbidding once we recognize that it is composed of three common word elements: *hypo-* (below normal), *natr-* (sodium), and *-emia* (blood condition). Thus, hyponatremia is a deficiency of sodium in the blood. Those three word elements appear over and over in many other medical terms: *hypothermia, natriuretic, anemia,* and so on. Once you learn the meanings of *hypo-, natri-,* and *-emia,* you already have the tools to at least partially understand hundreds of other biomedical terms. Inside the back cover, you will find a lexicon of the 400 word elements most commonly footnoted in this book.

Scientific terms are typically composed of one or more of the following elements:

- At least one *root (stem)* that bears the core meaning of the word. In *cardiology,* for example, the root is *cardi* (heart). Many words have two or more roots. In *adipocyte,* the roots are *adip* (fat) and *cyte* (cell).

- *Combining vowels,* which are often inserted to join roots and make the word easier to pronounce. The letter *o* is the most common combining vowel (as in *adipocyte*), but all vowels are used in this way, such as *a* in *ligament, e* in *vitreous, i* in *fusiform, u* in *ovulation,* and *y* in *tachycardia.* Some words have no combining vowels. A combination of a root and combining

[30]*epo* = after, related to + *nym* = name

vowel is called a *combining form:* for example, *odont* (tooth) + *o* (the combining vowel) make the combining form *odonto,* as in *odontoblast* (a cell that produces the dentin of a tooth).

- A *prefix* may be present to modify the core meaning of the word. For example, *gastric* (pertaining to the stomach or to the belly of a muscle) takes on a wide variety of new meanings when prefixes are added to it: *epigastric* (above the stomach), *hypogastric* (below the stomach), *endogastric* (within the stomach), and *digastric* (a muscle with two bellies).

- A *suffix* may be added to the end of a word to modify its core meaning. For example, *microscope, microscopy, microscopic,* and *microscopist* have different meanings because of their suffixes alone. Often two or more suffixes, or a root and suffix, occur together so often that they are treated jointly as a *compound suffix;* for example, *log* (study) + *y* (process) form the compound suffix *-logy* (the study of).

To summarize these basic principles, consider the word *gastro-enterology,* a branch of medicine dealing with the stomach and small intestine. It breaks down into

gastro/entero/logy

gastro = a combining form meaning "stomach"
entero = a combining form meaning "small intestine"
logy = a compound suffix meaning "the study of"

"Dissecting" words in this way and paying attention to the word-origin footnotes throughout this book will help make you more comfortable with the language of anatomy. Knowing how a word breaks down and knowing the meaning of its elements make it far easier to pronounce a word, spell it, and remember its definition.

There are a few unfortunate exceptions, however. The path from original meaning to current usage has often become obscured by history (see Deeper Insight 1.3). The foregoing approach also is no help with eponyms or with **acronyms**—words composed of the first letter, or first few letters, of a series of words. For example, *calmodulin,* a calcium-binding protein found in many cells, is cobbled together from a few letters of the three words, *cal*cium *modul*ating prote*in.*

Variant Forms of Medical Terms

A point of confusion for many beginning students is how to recognize the plural forms of medical terms. Few people would fail to recognize that *ovaries* is the plural of *ovary,* but the connection is harder to make in other cases: For example, the plural of *cortex* is *cortices* (COR-ti-sees), the plural of *corpus* is *corpora,* and the plural of *epididymis* is *epididymides* (EP-ih-DID-ih-MID-eez). Table 1.3 will help you make the connection between common singular and plural noun terminals.

In some cases, what appears to the beginner to be two completely different words may be only the noun and adjective forms of the same word. For example, *brachium* denotes the arm, and *brachii* (as in the muscle name *biceps brachii*) means "of the arm." *Carpus* denotes the wrist, and *carpi,* a word used in several muscle names, means "of the wrist." Adjectives can also take different forms for the singular and plural and for different degrees of comparison. The *digits* are the fingers and toes. The word *digiti* in a muscle name

Obscure Word Origins

The literal translation of a word doesn't always provide great insight into its modern meaning. The history of language is full of twists and turns that are fascinating in their own right and say much about the history of the whole of human culture, but they can create confusion for students.

For example, the *amnion* is a transparent sac that forms around the developing fetus. The word is derived from *amnos,* from the Greek for "lamb." From this origin, *amnos* came to mean a bowl for catching the blood of sacrificial lambs, and from there the word found its way into biomedical usage for the membrane that emerges (quite bloody) as part of the afterbirth. The *acetabulum,* the socket of the hip joint, literally means "vinegar cup." Apparently the hip socket reminded an anatomist of the little cups used to serve vinegar as a condiment on dining tables in ancient Rome. The word *testicles* literally means "little witnesses." The history of medical language has some amusing conjectures as to why this word was chosen to name the male gonads.

means "of a single finger (or toe)," whereas *digitorum* is the plural, meaning "of multiple fingers (or toes)." Thus the *extensor digiti minimi* muscle extends only the little finger, whereas the *extensor digitorum* muscle extends all fingers except the thumb.

The English words *large, larger,* and *largest* are examples of the positive, comparative, and superlative degrees of comparison. In Latin, these are *magnus, major* (from *maior*), and *maximus.* We find these in the muscle names *adductor magnus* (a *large* muscle of the thigh), the *pectoralis major* (the *larger* of two *pectoralis* muscles of the chest), and *gluteus maximus* (the *largest* of the three gluteal muscles of the buttock).

Some noun variations indicate the possessive, such as the *rectus abdominis,* a straight (*rectus*) muscle of the abdomen (*abdominis,* "of the abdomen"), and the *erector spinae,* a muscle that straightens (*erector*) the spinal column (*spinae,* "of the spine").

Anatomical terminology also follows the Greek and Latin practice of placing the adjective after the noun. Thus, we have such names as the *stratum lucidum* for a clear (*lucidum*) layer (*stratum*) of the epidermis, the *foramen magnum* for a large (*magnum*) hole (*foramen*) in the skull, and the aforementioned *pectoralis major* muscle of the chest.

This is not to say that you must be conversant in Latin or Greek grammar to proceed with your study of anatomy. These few examples, however, may alert you to some patterns to watch for in the terminology you study, and ideally, will make your encounters with anatomical terminology less confusing.

The Importance of Precision

A final word of advice for your study of anatomy: Be precise in your use of anatomical terms. It may seem trivial if you misspell *trapezius* as *trapezium,* but in doing so, you would be changing the name of a back muscle to the name of a wrist bone. Similarly, changing *occipitalis* to *occipital* or *zygomaticus* to *zygomatic* changes other muscle names to bone names. A "little" error such as misspelling *ileum* as *ilium* changes the name of part of the small

TABLE 1.3	Singular and Plural Forms of Some Noun Terminals	
Singular Ending	Plural Ending	Examples
-a	-ae	axilla, axillae
-ax	-aces	thorax, thoraces
-en	-ina	lumen, lumina
-ex	-ices	cortex, cortices
-is	-es	diagnosis, diagnoses
-is	-ides	epididymis, epididymides
-ix	-ices	appendix, appendices
-ma	-mata	carcinoma, carcinomata
-on	-a	ganglion, ganglia
-um	-a	septum, septa
-us	-era	viscus, viscera
-us	-i	villus, villi
-us	-ora	corpus, corpora
-x	-ges	phalanx, phalanges
-y	-ies	ovary, ovaries
-yx	-yces	calyx, calyces

intestine to the name of a hip bone. Changing *malleus* to *malleolus* changes the name of a middle-ear bone to the name of a bony protuberance of the ankle. *Elephantiasis is* a disease that produces an elephant-like thickening of the limbs and skin. Many people misspell this *elephantitis,* if such a word existed, it would mean inflammation of an elephant.

The health professions demand the utmost attention to detail and precision—people's lives may one day be in your hands. The habit of carefulness must extend to your use of language as well. Many patients die because of miscommunication in the hospital.

Before You Go On

Answer the following questions to test your understanding of the preceding section:

10. Explain why modern anatomical terminology is so heavily based on Greek and Latin.

11. Distinguish between an eponym and an acronym, and explain why both of these present difficulties for interpreting anatomical terms.

12. Break the following words down into their roots, prefixes, and suffixes and state their meanings, following the example of *gastroenterology* analyzed earlier: *pericardium, appendectomy, subcutaneous, arteriosclerosis, hypercalcemia.* Consult the list of word elements inside the back cover of the book for help.

13. Write the singular form of each of the following words: *pleurae, gyri, nomina, ganglia, fissures.* Write the plural form of each of the following: *villus, tibia, encephalitis, cervix, stoma.*

Study Guide

Assess Your Learning Outcomes

You should have a good understanding of this chapter if you can accurately address the following issues.

1.1   The Scope of Human Anatomy (p. 2)

1. The distinction between the sciences of anatomy and physiology, and the way functional morphology unites the two

2. The distinctions between gross, microscopic, surface, radiologic, systemic, regional, and comparative anatomy

3. Examples of what a physician might be looking for when he or she employs simple inspection, palpation, auscultation, and percussion with a patient

4. Ways in which dissection differs from exploratory surgery, and why exploratory surgery is far less common now than it was in the 1950s

5. The principles behind radiography, computed tomography (CT), magnetic resonance imaging (MRI), positron emission tomography (PET), and sonography

6. Differences between invasive and noninvasive methods of medical imaging

7. Reasons why the anatomy presented in this book may not apply to every human being

1.2   General Plan of the Human Body (p. 7)

1. The successive levels of human structural complexity from atom to organism

2. Correlation between the levels of human structure and the sciences of gross anatomy, histology, cytology, and ultrastructure

3. The 11 human organ systems, including the basic functions and major organs of each

4. Anatomical position and why it is important in anatomical communication

5. What it means to say the forearm is pronated or supinated, and how this differs from the meanings of *prone* and *supine*

6. The three primary anatomical planes, and what a given region of the body (such as midthoracic) would look like in each of these planes

7. The distinctions between *anterior* and *posterior; cephalic, rostral,* and *caudal; superior* and *inferior; medial* and *lateral; proximal* and *distal; ipsilateral* and *contralateral;* and *superficial* and *deep;* and the ability to use these terms correctly in descriptive anatomical sentences

8. Why the words *anterior* and *posterior* are preferable to *ventral* and *dorsal* for most purposes in human anatomy, and why *ventral* and *dorsal* would be more relevant to dissection of a cat than to dissection of the human cadaver

9. The principal body parts of the axial region and the appendicular region

10. The landmarks used to divide the abdomen into four quadrants, and the name of each quadrant

11. The landmarks used to divide the abdomen into a 3 × 3 grid, and the names of each of the 9 resulting regions

12. Names of the cavities that house the brain and spinal cord, and of the membranes that line these cavities

13. Landmarks that divide the thoracic, abdominal, and pelvic cavities from each other

14. The names of the cavities that enfold the heart and lungs; names of the membranes that line these cavities; names of the relatively superficial and deep layers of each of these two-layered membranes; and names of the fluids that lubricate these membranes and allow for painless heart and lung movements

15. The name of the membrane that lines the abdominal cavity; the name of its lubricating fluid; and the term for organs that lie between this membrane and the body wall

16. The name of the serous membranes that suspend and bind the abdominal organs, and the name for the outer surface of an organ formed by this membrane passing around it

17. The meaning of *potential spaces,* and some examples

1.3   The Language of Anatomy (p. 20)

1. The reason so many medical terms are based on Latin and Greek

2. The role of *Terminologia Anatomica (TA)* in modern medical terminology, and the problems that it is meant to solve

3. How to divide medical terms such as *histology, cardiovascular, anatomy, endometrium, pseudostratified, subcutaneous, corticospinal,* and *hypodermic* into their prefixes, roots, combining forms, and suffixes, and how to recognize combining vowels where they exist

4. The differences between an eponym and an acronym, and between an acronym and an abbreviation, with medical examples of each

5. Recognition of the singular and plural forms of the same term, as in *extensor digiti* and *extensor digitorum*

6. Recognition of the positive, comparative, and superlative forms of the same term, as in the second word of *adductor magnus, pectoralis major,* and *gluteus maximus*

7. The importance of accurate spelling; why even one-letter or other trivial seeming errors may be very significant in clinical practice; and examples of where this may apply

Testing Your Recall

1. Structure that can be observed with the naked eye is called
   a. gross anatomy.
   b. ultrastructure.
   c. microscopic anatomy.
   d. macroscopic anatomy.
   e. cytology.

2. Which of the following techniques requires an injection of radioisotopes into a patient's bloodstream?
   a. sonography   d. a CT scan
   b. a PET scan   e. an MRI scan
   c. radiography

3. The simplest structures considered to be alive are
   a. organs.   d. organelles.
   b. tissues.   e. proteins.
   c. cells.

4. The tarsal region is _____ to the popliteal region.
   a. medial   d. dorsal
   b. superficial   e. distal
   c. superior

5. The _____ region is immediately medial to the coxal region.
   a. inguinal
   b. hypochondriac
   c. umbilical
   d. popliteal
   e. cubital

6. Which of these regions is *not* part of the upper limb?
   a. plantar   d. brachial
   b. carpal   e. palmar
   c. cubital

7. Which of these organs is intraperitoneal?
   a. urinary bladder d. small intestine
   b. kidney   e. brain
   c. heart

8. Which of these is *not* an organ system?
   a. muscular system
   b. integumentary system
   c. endocrine system
   d. lymphatic system
   e. immune system

9. The term *histology* is most nearly equivalent to
   a. histopathology.
   b. microscopic anatomy.
   c. cytology.
   d. ultrastructure.
   e. systemic anatomy.

10. An imaging technique that exposes the patient to no harmful radiation is
    a. radiography.
    b. positron emission tomography (PET).
    c. computed tomography (CT).
    d. magnetic resonance imaging (MRI).
    e. angiography.

11. Cutting and separating tissues to reveal their structural relationships is called _____.

12. The forearm is said to be _____ when the palms are facing forward.

13. The relatively superficial layer of the pleura is called the _____ pleura.

14. Abdominal organs that lie against the posterior abdominal wall and are covered with peritoneum only on the anterior side are said to have a/an _____ position.

15. _____ is a science that doesn't merely describe bodily structure but interprets structure in terms of its function.

16. When a doctor presses on the upper abdomen to feel the size and texture of the liver, he or she is using a technique of physical examination called _____.

17. _____ is a method of medical imaging that uses X-rays and a computer to generate images of thin slices of the body.

18. A/An _____ is the simplest body structure to be composed of two or more types of tissue.

19. The left hand and left foot are _____ to each other, whereas the left hand and right hand are _____ to each other.

20. The anterior pit of the elbow is called the _____ region, whereas the corresponding (but posterior) pit of the knee is called the _____ region.

Answers in the Appendix

Building Your Medical Vocabulary

State a medical meaning of each of the following word elements, and give a term in which it is used.

1. ana-
2. -graphy
3. morpho-
4. hypo-
5. -ation
6. -elle
7. palp-
8. ante-
9. intra-
10. auscult-

Answers in the Appendix

True or False

Determine which five of the following statements are false, and briefly explain why.

1. Regional anatomy is a variation of gross anatomy.
2. A single sagittal section through the body could show one lung but not both.
3. Abnormal skin color or dryness could be one piece of diagnostic information gained by auscultation.
4. Radiology refers only to those medical imaging methods that use radioisotopes.
5. It is more harmful to have only the heart reversed from left to right than to have all of the thoracic and abdominal organs reversed.
6. There are more cells than organelles in the body.
7. The diaphragm is ventral to the lungs.
8. It would be possible to see both eyes in a single frontal section of the head.
9. Each lung is enclosed in a space between the parietal and visceral pleura.
10. The word *scuba*, derived from the words *self-contained underwater breathing apparatus*, is an acronym.

Answers in the Appendix

Testing Your Comprehension

1. Classify each of the following radiologic techniques as invasive or noninvasive and explain your reasoning for each: angiography, sonography, CT, MRI, and PET.
2. Beginning medical students are always told to examine multiple cadavers and not confine their study to just one. Other than the obvious purpose of studying both male and female anatomy, why is this instruction so important in medical education?
3. Identify which anatomical plane—sagittal, frontal, or transverse—is the only one that could *not* show (a) both the brain and tongue; (b) both eyes; (c) both the heart and uterus; (d) both the hypogastric and gluteal regions; (e) both kidneys; and (f) both the sternum and vertebral column.
4. Lay people often misunderstand medical terminology. What do you think people really mean when they say they have "planter's warts"?
5. Why do you think the writers of *Terminologia Anatomica* decided to reject eponyms? Do you agree with that decision? Why do you think they decided to name structures in Latin? Do you agree with that decision? Explain your reasons for agreeing or disagreeing with each.

Answers at www.mhhe.com/saladinha3

Improve Your Grade at **www.mhhe.com/saladinha3**

Practice quizzes, labeling activities, and games provide fun ways to master concepts. You can also download image PowerPoint files for each chapter to create a study guide or for taking notes during lecture.

Cytology—The Study of Cells

A mitochondrion photographed with a transmission electron microscope (TEM)

CHAPTER

2

BRUSHING UP

To understand this chapter, you may find it helpful to review the following concept:
- Levels of human structure (p. 8)

The most important revolution in the history of medicine was the realization that all bodily functions result from cellular activity. By extension, nearly every dysfunction of the body is now recognized as stemming from a dysfunction at the cellular level. The majority of new medical research articles published every week are on cellular function, and all drug development is based on an intimate knowledge of how cells work. The cellular perspective has thus become indispensable to any true understanding of the structure and function of the human body, the mechanisms of disease, and the rationale of therapy.

This chapter therefore begins our study of anatomy at the cellular level. We will see how continued developments in microscopy have deepened our insight into cell structure, examine the structural components of cells, and briefly survey two aspects of cellular function—transport through the plasma membrane and the cell life cycle. It is the derangement of that life cycle that gives rise to one of the most dreaded of human diseases, cancer.

2.1 The Study of Cells

▶ Expected Learning Outcomes

When you have completed this section, you should be able to

- state some tenets of the cell theory;
- discuss the way that developments in microscopy have changed our view of cell structure;
- outline the major structural components of a cell;
- identify cell shapes from their descriptive terms; and
- state the size range of human cells and explain why cell size is limited.

The scientific study of cellular structure and function is called **cytology.**[1] Some historians date the birth of this science to April 15, 1663, when English inventor Robert Hooke employed his newly created microscope to observe the little boxes formed by the cell walls of cork. He named them *cellulae.* Cytology was greatly advanced by refinements in microscope technology in the nineteenth century. By 1900, it was established beyond reasonable doubt that every living organism is made of cells; that cells now arise only through the division of preexisting cells rather than springing spontaneously from nonliving matter; and that all cells have the same basic chemical components, such as carbohydrates, lipids, proteins, and nucleic acids. Cells are the simplest entities considered to be alive; no one molecule such as DNA or an enzyme is alive in itself. These and other principles have been codified as the **cell theory.**

---

[1]*cyto* = cell + *logy* = study of

Microscopy

Cytology would not exist without the microscope. Throughout this book, you will find many **photomicrographs**—photos of tissues and cells taken through the microscope. The microscopes used to produce them fall into three basic categories: the light microscope, transmission electron microscope, and scanning electron microscope.

The **light microscope (LM)** uses visible light to produce its images. It is the least expensive type of microscope, the easiest to use, and the most often used, but it is also the most limited in the amount of useful magnification it can produce. Light microscopes today magnify up to 1,200 times. There are several varieties of light microscopes, including the fluorescence microscope used to produce figure 2.15b.

Most of the structure we study in this chapter is invisible to the LM, not because the LM cannot magnify enough but because it cannot reveal enough detail. The most important thing about a good microscope is not magnification but **resolution**—the ability to reveal detail. Any image can be photographed and enlarged as much as we wish, but if enlargement fails to reveal greater detail, it is *empty magnification.* A large blurry image is not nearly as informative as one that is small and sharp. For reasons of physics beyond the scope of this chapter, it is the wavelength of light that places a limit on resolution. Visible light has wavelengths ranging from about 400 to 700 nanometers (nm). At these wavelengths, the LM cannot distinguish between two objects any closer together than 200 nm (0.2 micrometers, or μm).

Resolution improves when objects are viewed with radiation of shorter wavelengths. *Electron microscopes* achieve higher resolution by using not visible light but beams of electrons with very short wavelength (0.005 nm). The **transmission electron microscope (TEM),** invented in the mid-twentieth century, is usually used to study specimens that have been sliced ultrathin with diamond knives and stained with heavy metals such as osmium, which absorbs electrons. The TEM resolves details as small as 0.5 nm and attains useful magnifications of biological material up to 600,000 times. This is good enough to see proteins, nucleic acids, and other large molecules. Such fine detail is called cell *ultrastructure.* Even at the same magnifications as the LM, the TEM reveals far more detail (fig. 2.1). It usually produces two-dimensional black-and-white images, but electron photomicrographs are often colorized for instructional purposes.

The **scanning electron microscope (SEM)** uses a specimen coated with vaporized metal (usually gold). The electron beam strikes the specimen and discharges secondary electrons from the metal coating. These electrons then strike a fluorescent screen and produce an image. The SEM yields less resolution than the TEM and is used at lower magnification, but it produces dramatic three-dimensional images that are sometimes more informative than TEM images, and it does not require that the specimen be cut into thin slices. The SEM can view only the surfaces of specimens; it does not see through an object as the LM or TEM does. Cell interiors can be viewed, however, by a *freeze-fracture* method in which a cell is frozen, cracked open, coated with gold vapor, and then viewed by either TEM or SEM. Figure 2.2 compares red blood cells photographed with the LM, TEM, and SEM.

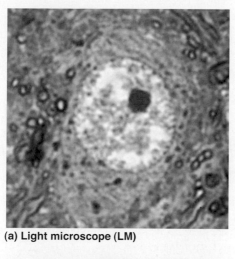

(a) Light microscope (LM)

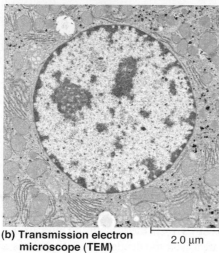

(b) Transmission electron microscope (TEM)   2.0 μm

Figure 2.1 **Magnification Versus Resolution.** These cell nuclei were photographed at the same magnification (about × 750) through (a) a light microscope (LM) and (b) a transmission electron microscope (TEM). Note the finer detail visible with the TEM.

Apply What You Know

Beyond figure 2.2, list all of the photomicrographs in this chapter that you believe were made with the LM, with the TEM, and with the SEM.

Cell Shapes and Sizes

We will shortly examine the structure of a generic cell, but the generalizations we draw should not blind you to the diversity of cellular form and function in humans. There are about 200 kinds of cells in the human body, with a variety of shapes, sizes, and functions. Descriptions of organ and tissue structure often refer to the shapes of cells by the following terms (fig. 2.3):

- *Squamous*[2] (SQUAY-mus)—a thin, flat, scaly shape, often with a bulge where the nucleus is—much like the shape of a fried egg "sunny side up." Squamous cells line the esophagus and form the surface layer (epidermis) of the skin.

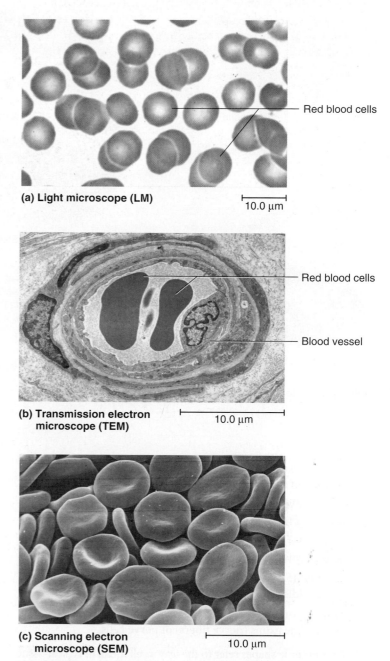

(a) Light microscope (LM)   10.0 μm — Red blood cells

(b) Transmission electron microscope (TEM)   10.0 μm — Red blood cells — Blood vessel

(c) Scanning electron microscope (SEM)   10.0 μm

Figure 2.2 **Images of Red Blood Cells Produced by Three Kinds of Microscopes.**
• *Based on the SEM image (c), can you explain why the cells in part (a) have such pale centers?*

- *Cuboidal*[3] (cue-BOY-dul)—squarish-looking in frontal tissue sections and about equal in height and width; liver cells are a good example.
- *Columnar*—distinctly taller than wide, such as the inner lining cells of the stomach and intestines.
- *Polygonal*[4]—having irregularly angular shapes with four, five, or more sides.

---

[2]*squam* = scale + *ous* = characterized by

[3]*cub* = cube-shaped + *oidal* = like, resembling
[4]*poly* = many + *gon* = angles

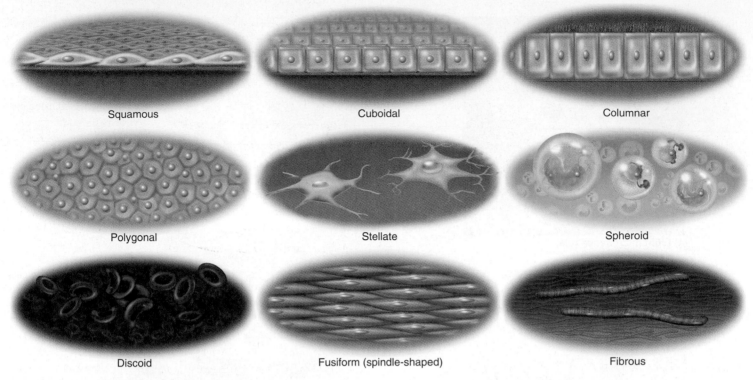

Figure 2.3 Common Cell Shapes.

Draw

- *Stellate*[5]—having multiple pointed processes projecting from the body of a cell, giving it a somewhat starlike shape. The cell bodies of many nerve cells are stellate.
- *Spheroidal* to *ovoid*—round to oval, as in egg cells and white blood cells.
- *Discoid*—disc-shaped, as in red blood cells.
- *Fusiform*[6] (FEW-zih-form)—spindle-shaped; elongated, with a thick middle and tapered ends, as in smooth muscle cells.
- *Fibrous*—long, slender, and threadlike, as in skeletal muscle cells and the axons (nerve fibers) of nerve cells.

Some of these shapes refer to the way a cell looks in typical tissue sections, not to the complete three-dimensional shape of the cell. A cell that looks squamous, cuboidal, or columnar in a tissue section, for example, usually looks polygonal if viewed from its upper surface.

In some cells, it is important to distinguish one surface from another, because cell surfaces may differ in function and membrane composition. This is especially true in *epithelia,* cell layers that cover organ surfaces. An epithelial cell rests on a lower **basal surface** often attached to an extracellular *basement membrane* (see chapter 3). The upper surface of the cell is called the **apical surface.** Its sides are **lateral surfaces.**

The most useful unit of measurement for designating cell sizes is the **micrometer (μm),** formerly called the *micron*—one-millionth (10^{-6}) of a meter, one-thousandth (10^{-3}) of a millimeter. The smallest objects most people can see with the naked eye are about 100 μm, which is about one-quarter the size of the period at the end of this sentence. A few human cells fall within this range, such as the egg cell and some fat cells, but most human cells are about 10 to 15 μm wide. The longest human cells are nerve cells (sometimes over a meter long) and muscle cells (up to 30 cm long), but both are usually too slender to be seen with the naked eye.

There are several factors that limit the size of cells. If a cell swells to excessive size, it ruptures like an overfilled water balloon. Also, if a cell were too large, molecules could not diffuse from place to place fast enough to support its metabolism. The time required for diffusion is proportional to the square of distance, so if cell diameter doubled, the travel time for molecules within the cell would increase fourfold. A nucleus can therefore effectively control only a limited volume of cytoplasm.

In addition, cell size is limited by the relationship between its volume and surface area. The surface area of a cell is proportional to the square of its diameter, while volume is proportional to the cube of diameter. Thus, for a given increase in diameter, cell volume increases much faster than surface area. Picture a cuboidal cell 10 μm on each side (fig. 2.4). It would have a surface area of 600 μm² (10 μm × 10 μm × 6 sides) and a volume of 1,000 μm³ (10 × 10 × 10 μm). Now, suppose it grew by another 10 μm on each side. Its new surface area would be 20 μm × 20 μm × 6 = 2,400 μm², and its volume would be 20 × 20 × 20 μm = 8,000 μm³. The 20 μm cell has eight times as much cytoplasm needing nourishment and waste removal, but only four times as much membrane surface through which wastes and nutrients can be exchanged. In short, a cell that is too big cannot support itself.

Apply What You Know

Can you conceive of some other reasons for an organ to consist of many small cells rather than fewer large ones?

---

[5]*stell* = star + *ate* = characterized by
[6]*fusi* = spindle + *form* = shape

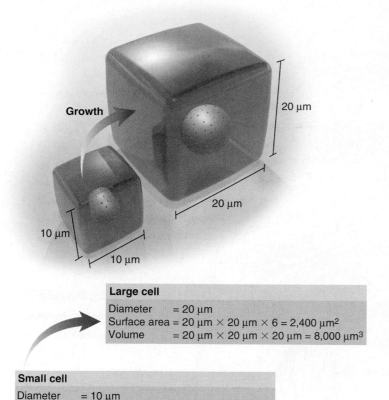

Large cell

Diameter = 20 μm
Surface area = 20 μm × 20 μm × 6 = 2,400 μm²
Volume = 20 μm × 20 μm × 20 μm = 8,000 μm³

Small cell

Diameter = 10 μm
Surface area = 10 μm × 10 μm × 6 = 600 μm²
Volume = 10 μm × 10 μm × 10 μm = 1,000 μm³

Effect of cell growth:

Diameter (D) increased by a factor of 2
Surface area increased by a factor of 4 (= D^2)
Volume increased by a factor of 8 (= D^3)

Figure 2.4 **The Relationship Between Cell Surface Area and Volume.** As a cell doubles in width, its volume increases eightfold, but its surface area increases only fourfold. A cell that is too large may have too little plasma membrane to support the metabolic needs of its volume of cytoplasm.

Basic Components of a Cell

Before electron microscopy, little was known about structural cytology except that cells were enclosed in a membrane and contained a nucleus. The material between the nucleus and surface membrane was thought to be little more than a gelatinous mixture of chemicals and vaguely defined particles. But the electron microscope revealed that the cytoplasm is crowded with a maze of passages, compartments, and filaments (fig. 2.5). Earlier microscopists were little aware of this detail simply because most of these structures are too small to be resolved by the light microscope (table 2.1).

We now regard cells as having the following major components:

Plasma membrane
Cytoplasm
  Cytoskeleton
  Organelles
  Inclusions
  Cytosol
Nucleoplasm

TABLE 2.1	Sizes of Biological Structures in Relation to the Resolution of the Eye, Light Microscope, and Transmission Electron Microscope
Object	**Size**
Visible with the Eye (Resolution 70–100 μm)	
Human egg, diameter	100 μm
Visible with the Light Microscope (Resolution 200 μm)	
Most human cells, diameter	10–15 μm
Cilia, length	7–10 μm
Mitochondria, width × length	0.2 × 4 μm
Bacteria (*Escherichia coli*), length	1–3 μm
Microvilli, length	1–2 μm
Visible with the Transmission Electron Microscope (Resolution 0.5 μm)	
Nuclear pores, diameter	30–100 μm
Ribosomes, diameter	15 μm
Globular proteins, diameter	5–10 μm
Plasma membrane, thickness	7.5 μm
DNA molecule, diameter	2.0 μm
Plasma membrane channels, diameter	0.8 μm

The **plasma membrane (cell membrane)** forms the surface boundary of the cell. The material between the plasma membrane and the nucleus is the **cytoplasm,**[7] and the material within the nucleus is the **nucleoplasm.** The cytoplasm contains the *cytoskeleton*, a supportive framework of protein filaments and tubules; an abundance of *organelles*, diverse structures that perform various metabolic tasks for the cell; and *inclusions*, which are foreign matter or stored cell products. The cytoskeleton, organelles, and inclusions are embedded in a clear gel called the **cytosol.**

The cytosol is also called the **intracellular fluid (ICF).** All body fluids not contained in the cells are collectively called the **extracellular fluid (ECF).** The extracellular fluid located amid the cells is also called **tissue (interstitial) fluid.** Some other extracellular fluids include blood plasma, lymph, and cerebrospinal fluid.

Before You Go On

Answer the following questions to test your understanding of the preceding section:

1. State some tenets of the cell theory.

2. What is the main advantage of an electron microscope over a light microscope?

3. Explain why cells cannot grow to unlimited size.

4. Define *cytoplasm, cytosol,* and *organelle.*

---

[7]*cyto* = cell + *plasm* = formed, molded

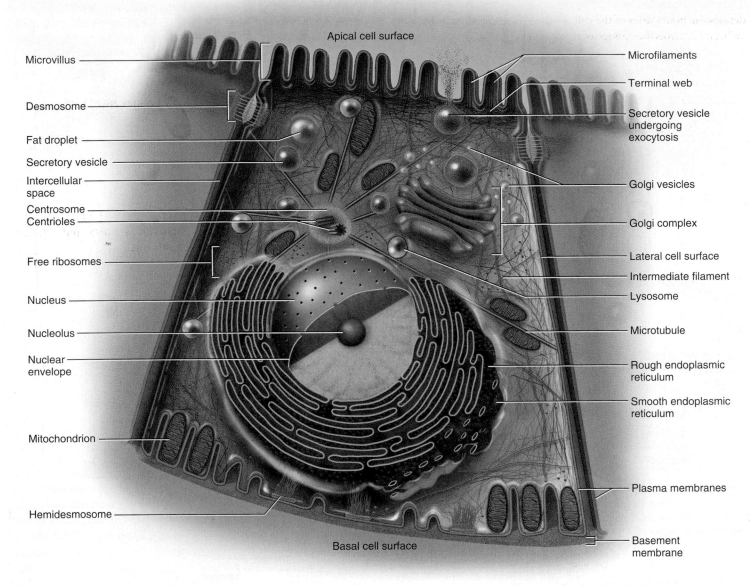

Figure 2.5 **Structure of a Generalized Cell.** The cytoplasm is usually more crowded with organelles than shown here. The organelles are not all drawn to the same scale.

2.2 The Cell Surface

▶ Expected Learning Outcomes

When you have completed this section, you should be able to

- describe the structure of the plasma membrane;
- explain the functions of the lipid, protein, and carbohydrate components of the plasma membrane;
- describe the processes for moving material into and out of a cell; and
- describe the structure and function of microvilli, cilia, flagella, and cell junctions.

A great deal of human physiology takes place at the cell surface—for example, the binding of signaling molecules such as hormones, the stimulation of cellular activity, the attachment of cells to each other, and the transport of materials into and out of cells. This, then, is where we begin our study of cellular structure and function. In this section, we examine the plasma membrane that defines the outer boundary of a cell; a carbohydrate coating called the *glycocalyx* on the membrane surface; hairlike extensions of the cell surface; the attachment of cells to each other and to extracellular materials; and the processes of membrane transport. Like explorers of a new continent, we will explore the interior only after we have explored the borders.

The Plasma Membrane

The electron microscope reveals that the cell and many of the organelles within it are bordered by a **unit membrane,** which appears as a pair of dark parallel lines with a total thickness of about 7.5 nm (fig. 2.6a). The plasma membrane is the unit membrane at the cell surface. It defines the boundaries of the cell, governs its interactions with other cells, and controls the passage of materials into and out of the cell. The side that faces the cytoplasm is the **intracellular face** of the membane, and the side that faces outward is the **extracellular face.**

The plasma membrane is an oily, two-layered lipid film with proteins embedded in it (fig. 2.6b). By weight, it is about half lipid and half protein. Since the lipid molecules are smaller and lighter, however, they constitute about 90% to 99% of the molecules in the membrane.

Membrane Lipids

About 75% of the membrane lipid molecules are phospholipids. A **phospholipid** (fig. 2.7) consists of a three-carbon backbone called glycerol, with fatty acid tails attached to two of the carbons and a phosphate-containing head attached to the third. The two fatty acid tails are *hydrophobic*[8] (water-repellent) and the head is *hydrophilic*[9] (attracted to water). The heads of the phospholipids face the ECF and ICF, whereas the tails form the middle of the "sandwich," as far away from the surrounding water as possible. The phospholipids drift laterally from place to place, spin on their axes, and flex their tails. These movements keep the membrane fluid.

[8]*hydro* = water + *phobic* = fearing, repelled by
[9]*hydro* = water + *philic* = loving, attracted to

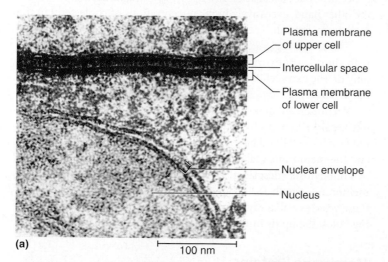

Plasma membrane of upper cell
Intercellular space
Plasma membrane of lower cell
Nuclear envelope
Nucleus

(a)

100 nm

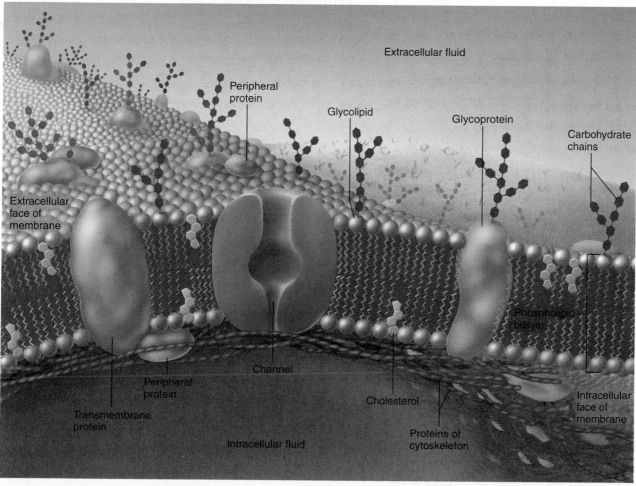

Extracellular fluid

Peripheral protein

Glycolipid

Glycoprotein

Carbohydrate chains

Extracellular face of membrane

Phospholipid bilayer

Peripheral protein

Channel

Cholesterol

Intracellular face of membrane

Transmembrane protein

Proteins of cytoskeleton

Intracellular fluid

(b)

Figure 2.6 **The Plasma Membrane.** (a) Plasma membranes of two adjacent cells (TEM). Note also that the nuclear envelope is composed of two *unit membranes*, each of which is similar to a plasma membrane. (b) Molecular organization of the plasma membrane.

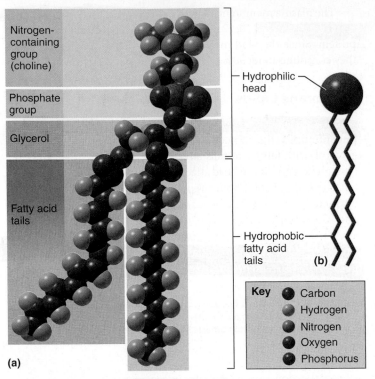

Nitrogen-containing group (choline)

Phosphate group

Glycerol

Fatty acid tails

(a)

Hydrophilic head

Hydrophobic fatty acid tails

(b)

Key
Carbon
Hydrogen
Nitrogen
Oxygen
Phosphorus

Figure 2.7 **Phospholipid Structure and Symbol.** (a) Molecular model of a phospholipid. (b) Common symbol used to represent phospholipids in diagrams of cell membranes.

Fat-soluble substances such as steroid hormones, oxygen, and carbon dioxide easily pass into and out of the cell through the phospholipid bilayer. However, the phospholipids severely restrict the movement of water-soluble substances such as glucose, salts, and

water itself, which must pass primarily through the membrane proteins discussed shortly.

About 20% of the lipid molecules are **cholesterol.** Cholesterol has an important impact on the fluidity of the membrane. If there is too little cholesterol, plasma membranes become excessively fragile. People with abnormally low cholesterol levels suffer an increased incidence of strokes because of the rupturing of fragile blood vessels. On the other hand, excessively high concentrations of cholesterol in the membrane can inhibit the action of its enzymes and other proteins.

The remaining 5% of the lipids are **glycolipids**—phospholipids with short carbohydrate chains bound to them. Glycolipids occur only on the extracellular face of the membrane. They contribute to the **glycocalyx,** a sugary cell coating discussed later.

An important quality of the plasma membrane is its capacity for self-repair. When a physiologist inserts a probe into a cell, it does not pop the cell like a balloon. The probe slips through the oily film and the membrane seals itself around it. When cells take in matter by endocytosis (described later), they pinch off bits of their own membrane, which form bubblelike vesicles in the cytoplasm. As these vesicles pull away from the membrane, they do not leave gaping holes; the lipids immediately flow together to seal the break.

Membrane Proteins

Proteins constitute from 1% to 10% of the membrane molecules. Some of them, called **transmembrane proteins,** pass all the way through the membrane. They have hydrophilic regions in contact with the cytoplasm and extracellular fluid, and hydrophobic regions that pass back and forth through the membrane lipid (fig. 2.8). Most of the transmembrane proteins are **glycoproteins,** which, like glycolipids, have carbohydrate chains linked to them and help form the glycocalyx. **Peripheral proteins** are those that do not protrude

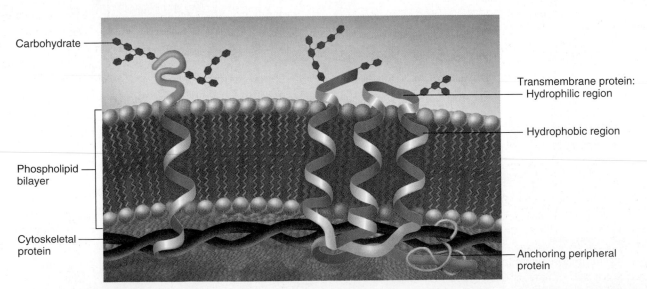

Carbohydrate

Phospholipid bilayer

Cytoskeletal protein

Transmembrane protein: Hydrophilic region

Hydrophobic region

Anchoring peripheral protein

Figure 2.8 **Transmembrane Proteins.** A transmembrane protein has hydrophobic regions embedded in the phospholipid bilayer and hydrophilic regions projecting into the extracellular and intracellular fluids. The protein may cross the membrane once (left) or multiple times (right). The intracellular "domain" of the protein is often anchored to the cytoskeleton by peripheral proteins.
• *What other regions of the protein on the right would be hydrophilic in addition to the one labeled?*

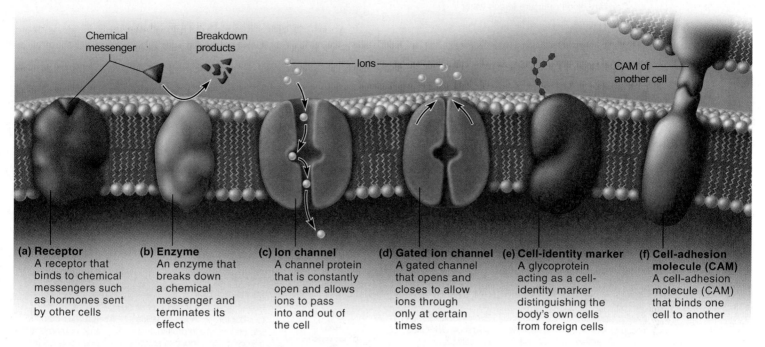

Figure 2.9 Some Functions of Plasma Membrane Proteins.

(a) Receptor
A receptor that binds to chemical messengers such as hormones sent by other cells

(b) Enzyme
An enzyme that breaks down a chemical messenger and terminates its effect

(c) Ion channel
A channel protein that is constantly open and allows ions to pass into and out of the cell

(d) Gated ion channel
A gated channel that opens and closes to allow ions through only at certain times

(e) Cell-identity marker
A glycoprotein acting as a cell-identity marker distinguishing the body's own cells from foreign cells

(f) Cell-adhesion molecule (CAM)
A cell-adhesion molecule (CAM) that binds one cell to another

into the phospholipid layer but adhere to either face of the membrane, usually the intracellular face. Some transmembrane proteins drift about freely in the plasma membrane, while others are anchored to the cytoskeleton and thus held in one place. Most peripheral proteins are anchored to the cytoskeleton and associated with transmembrane proteins.

The functions of membrane proteins are very diverse and are among the most interesting aspects of cell physiology. These proteins serve in the following roles:

- **Receptors** (fig. 2.9a). Cells communicate with each other by chemical signals such as hormones and neurotransmitters. Some of these messengers (epinephrine, for example) cannot enter their target cells but can only "knock on the door" with their message. They bind to a membrane protein called a **receptor,** and the receptor triggers physiological changes inside the cell.

- **Enzymes** (fig. 2.9b). Some membrane proteins are enzymes that carry out chemical reactions at the cell surface. Some of these break down chemical messengers after the message has been received. Enzymes in the plasma membranes of intestinal cells carry out the final stages of starch and protein digestion.

- **Channel proteins** (fig. 2.9c). Some membrane proteins have tunnels through them that allow water and hydrophilic solutes to enter or leave a cell. These are called **channel proteins.** Some channels are always open, whereas others, called **gates** (fig. 2.9d), open or close when they are stimulated and thus allow things to enter or leave the cell only at appropriate times. Membrane gates are responsible for firing of the heart's pacemaker, muscle contraction, and most of our sensory processes, among other functions.

- **Transport proteins** (see fig. 2.10c, d). Some membrane proteins, called **transport proteins (carriers),** don't merely open to allow substances through—they actively bind to a substance on one side of the membrane and release it on the other side. Carriers are responsible for transporting glucose, amino acids, sodium, potassium, calcium, and many other substances into and out of cells.

- **Cell-identity markers** (fig. 2.9e). The glycoproteins and glycolipids of the membrane are like genetic identification tags, unique to an individual (or to identical twins). They enable the body to distinguish what belongs to it from what does not—especially from foreign invaders such as bacteria and parasites.

- **Cell-adhesion molecules** (fig. 2.9f). Cells adhere to each other and to extracellular material through membrane proteins called **cell-adhesion molecules (CAMs).** With few exceptions (such as blood cells and metastasizing cancer cells), cells do not grow or survive normally unless they are mechanically linked to the extracellular material. Special events such as sperm–egg binding and the binding of an immune cell to a cancer cell also require CAMs.

Membrane Transport

One of the most important functions of the plasma membrane is to control the passage of materials into and out of the cell. Figure 2.10 illustrates three methods of movement through plasma membranes, as well as filtration, an important mode of transport across the walls of certain epithelia.

Filtration

Filtration (fig. 2.10a) is a process in which a physical pressure forces fluid through a membrane, like the weight of water forcing it through the paper filter in a drip coffeemaker. In the body, the prime example of filtration is blood pressure forcing fluid to seep through the walls of the blood capillaries into the tissue fluid. This is how water, salts,

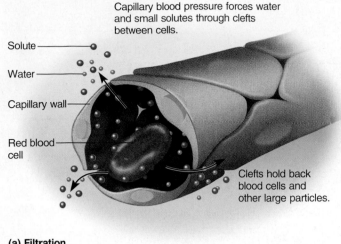

Capillary blood pressure forces water
and small solutes through clefts
between cells.

Solute

Water

Capillary wall

Red blood
cell

Clefts hold back
blood cells and
other large particles.

(a) Filtration

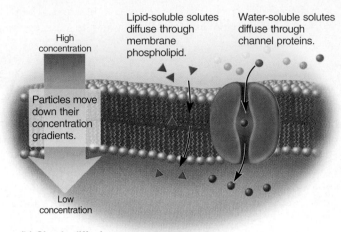

High
concentration

Lipid-soluble solutes
diffuse through
membrane
phospholipid.

Water-soluble solutes
diffuse through
channel proteins.

Particles move
down their
concentration
gradients.

Low
concentration

(b) Simple diffusion

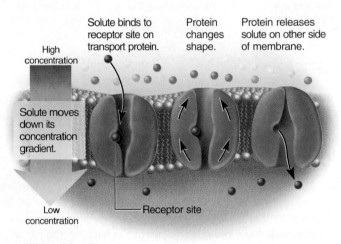

Solute binds to
receptor site on
transport protein.

Protein
changes
shape.

Protein releases
solute on other side
of membrane.

High
concentration

Solute moves
down its
concentration
gradient.

Low
concentration

Receptor site

(c) Facilitated diffusion

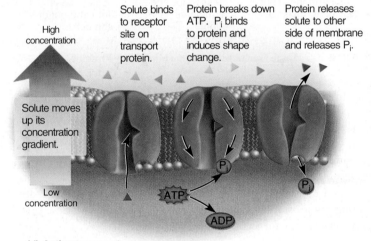

Solute binds
to receptor
site on
transport
protein.

Protein breaks down
ATP. P_i binds
to protein and
induces shape
change.

Protein releases
solute to other
side of membrane
and releases P_i.

High
concentration

Solute moves
up its
concentration
gradient.

Low
concentration

P_i

P_i

ATP

ADP

(d) Active transport

Figure 2.10   Modes of Membrane Transport.

organic nutrients, and other solutes pass from the bloodstream to the tissue fluid, where they can get to the cells surrounding a blood vessel. This is also how the kidneys filter wastes from the blood. Capillaries hold back large particles such as blood cells and proteins. Filtration generally involves substances passing through an epithelium between cells or by way of large **filtration pores** through the cells rather than through the plasma membrane.

Simple Diffusion

Simple diffusion (fig. 2.10b) is the net movement of particles from a place of high concentration to a place of low concentration—in other words, *down a concentration gradient*. Diffusion is how oxygen and steroid hormones enter cells and potassium ions leave, for example. The cell does not have to expend any energy to achieve this; all molecules are in spontaneous random motion, and this alone provides the energy for their diffusion through space. Molecules diffuse through air, liquids, and solids. They can penetrate both living membranes (the plasma membrane) and nonliving ones (such as dialysis tubing and cellophane) if the membrane has large

enough gaps or pores. We say that the plasma membrane is *selectively permeable* because it lets some particles through but holds back larger ones.

Osmosis

Osmosis[10] (oz-MO-sis) is the net flow of water through a selectively permeable membrane from the side with less dissolved matter (more water) to the side with more dissolved matter (less water). Water molecules tend to associate with particles of dissolved matter and thus resist going back through the membrane in the opposite direction—hence the net accumulation of water on the side with more solute. An exception to this is **reverse osmosis,** in which a physical force drives water back to the more dilute side of the membrane. This principle is used to desalinate seawater, converting it to drinkable freshwater, in arid regions and ships at sea. In the human body, the force generated by the heartbeat causes reverse osmosis at the blood capillaries, driving water out of the blood and

[10]*osm* = push, thrust + *osis* = process

into the tissue fluid; but where the blood pressure is lower, capillaries absorb tissue fluid by osmosis. Many cells have membrane channel proteins called **aquaporins** that allow water to pass easily through the membrane.

Facilitated Diffusion

The next two processes, facilitated diffusion and active transport, are called *carrier-mediated transport* because they employ transport proteins in the plasma membrane. **Facilitated**[11] **diffusion** (fig. 2.10c) can be defined as the movement of a solute through a unit membrane, down its concentration gradient, with the aid of a carrier. The carrier transports solutes such as glucose that cannot pass through the membrane unaided. It binds to a particle on one side of a membrane, where the solute is more concentrated, and releases it on the other side, where it is less concentrated. The process requires no expenditure of metabolic energy by the cell. One use of facilitated diffusion is to absorb the sugars and amino acids from digested food.

Active Transport

Active transport (fig. 2.10d) is the carrier-mediated transport of a solute through a unit membrane *up its concentration gradient*, with the expenditure of energy provided by adenosine triphosphate (ATP). ATP is essential to this process because moving particles up a gradient requires an energy input, like getting a wagon to roll uphill. If a cell dies and stops producing ATP, active transport ceases immediately. One use of active transport is to pump calcium out of cells. Calcium is already more concentrated in the ECF than in the ICF, so pumping even more calcium into the ECF is an "uphill" movement.

An especially well-known active transport process is the **sodium–potassium (Na⁺−K⁺) pump,** which binds three sodium ions from the ICF and ejects them from the cell, then binds two potassium ions from the ECF and releases these into the cell. The Na⁺−K⁺ pump plays roles in controlling cell volume; generating body heat; maintaining the electrical excitability of your nerves, muscles, and heart; and providing energy for other transport pumps to draw upon in moving such solutes as glucose through the plasma membrane. About half of the calories that you "burn" every day are used just to operate your Na⁺−K⁺ pumps.

Vesicular Transport

All of the processes discussed up to this point move molecules or ions individually through the plasma membrane. In **vesicular transport,** however, cells move much larger particles or droplets of fluid through the membrane in bubblelike *vesicles*. Vesicular processes that bring matter into a cell are called **endocytosis**[12] (EN-doe-sy-TOE-sis), and those that release material from a cell are called **exocytosis**[13] (EC-so-sy-TOE-sis). Like active transport, all forms of vesicular transport require ATP. There are three forms of endocytosis: *phagocytosis, pinocytosis,* and *receptor-mediated endocytosis* (fig 2.11).

---

[11]*facil* = easy
[12]*endo* = into + *cyt* = cell + *osis* = process
[13]*exo* = out of + *cyt* = cell + *osis* = process

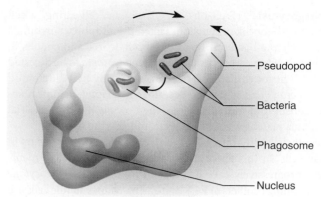

(a) Phagocytosis

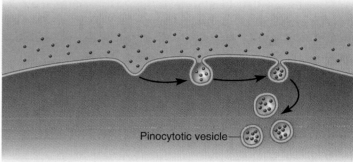

(b) Pinocytosis

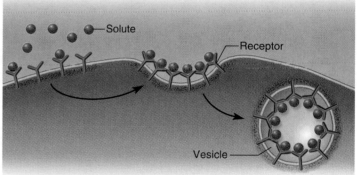

(c) Receptor-mediated endocytosis

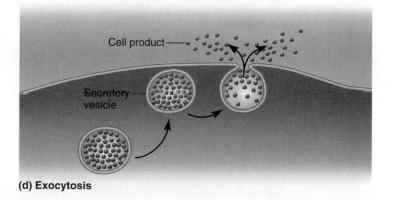

(d) Exocytosis

Figure 2.11 **Modes of Vesicular Transport.** (a) Phagocytosis. A white blood cell engulfing bacteria with its pseudopods. (b) Pinocytosis. A cell imbibing droplets of extracellular fluid. (c) Receptor-mediated endocytosis. The Ys in the plasma membrane represent membrane receptors. The receptors bind a solute in the extracellular fluid, then cluster together. The membrane caves in at that point until a vesicle pinches off into the cytoplasm bearing the receptors and bound solute. (d) Exocytosis. A cell releasing a secretion or waste product.

In **phagocytosis**[14] (FAG-oh-sy-TOE-sis), or "cell eating," a cell reaches out with footlike extensions called **pseudopods,** surrounds a particle such as a bacterium or a bit of cell debris, and engulfs it, taking it into a cytoplasmic vesicle called a *phagosome* to be digested (fig. 2.11a). Phagocytosis is carried out especially by white blood cells and *macrophages,* which are described in chapter 3. Some macrophages consume as much as 25% of their own volume in material per hour, thus playing a vital role in cleaning up the tissues.

Pinocytosis[15] (PIN-oh-sy-TOE-sis), or "cell drinking," occurs in all human cells. In this process, dimples form in the plasma membrane and progressively cave in until they pinch off as *pinocytotic vesicles* containing droplets of ECF (fig. 2.11b). Kidney tubule cells use this method to reclaim the small amount of protein that filters out of the blood, thus preventing the protein from being lost in the urine.

Receptor-mediated endocytosis (fig. 2.11c) is more selective. It enables a cell to take in specific molecules from the ECF with a minimum of unnecessary fluid. Molecules in the ECF bind to specific receptor proteins on the plasma membrane. The receptors then cluster together and the membrane sinks in at this point, creating a pit. The pit soon pinches off to form a vesicle in the cytoplasm. Cells use receptor-mediated endocytosis to absorb cholesterol and insulin from the blood. Hepatitis, polio, and AIDS viruses trick our cells into admitting them by receptor-mediated endocytosis.

Exocytosis (fig. 2.11d) is the process of discharging material from a cell. It is used, for example, by digestive glands to secrete enzymes, by breast cells to secrete milk, and by sperm cells to release enzymes for penetrating an egg. It resembles endocytosis in reverse. A *secretory vesicle* in the cell migrates to the surface and fuses with the plasma membrane. A pore opens up that releases the products from the cell, and the empty vesicle usually becomes part of the plasma membrane. In addition to releasing cell products, exocytosis is the cell's way of replacing the bits of membrane removed by endocytosis.

The Glycocalyx

The carbohydrate components of the glycoproteins and glycolipids of the plasma membrane form a fuzzy, sugary coating called the **glycocalyx**[16] (GLY-co-CAY-licks) on every cell surface (fig. 2.12). The glycocalyx has multiple functions. It cushions the plasma membrane and protects it from physical and chemical injury. It functions in cell identity and thus in the body's ability to distinguish its own healthy cells from diseased cells, invading organisms, and transplanted tissues. Human blood types and transfusion compatibility are determined by the glycocalyx. The glycocalyx also includes the cell-adhesion molecules described earlier and, thus, helps to bind tissues together and enables a sperm to bind to an egg and fertilize it.

---

[14]*phago* = eating + *cyt* = cell + *osis* = process
[15]*pino* = drinking + *cyt* = cell + *osis* = process

[16]*glyco* = sugar + *calyx* = cup, vessel

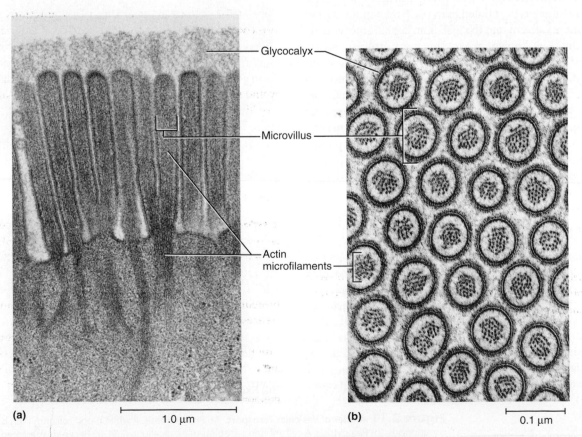

(a)   1.0 μm

(b)   0.1 μm

Figure 2.12 **Microvilli and the Glycocalyx.** The microvilli are anchored by bundles of actin microfilaments, which occupy the core of each microvillus and project into the cytoplasm. (a) Longitudinal sections, perpendicular to cell surface. (b) Cross sections.
• *Identify three functions served by the actin bundles.*

Microvilli, Cilia, and Flagella

Many cells have surface extensions called *microvilli, cilia,* and *flagella*. These aid in absorption, movement, and sensory processes.

Microvilli

Microvilli[17] (MY-cro-VIL-eye; singular, *microvillus*) are extensions of the plasma membrane that serve primarily to increase its surface area (fig. 2.12). They are best developed in cells specialized for absorption, such as the epithelial cells of the intestines and kidney tubules. An absorptive cell of the small intestine has about 3,000 microvilli on its surface; microvilli of the intestine number about 200 million/mm^2. They give such cells far more absorptive surface area than they would have if their apical surfaces were flat. On many cells, microvilli are little more than tiny bumps on the plasma membrane. On cells of the taste buds and inner ear, they are well developed but serve sensory rather than absorptive functions.

Individual microvilli cannot be distinguished very well with the light microscope because they are only 1 to 2 μm long. On some cells, they are very dense and appear as a fringe called the **brush border** at the apical cell surface. With the scanning electron microscope, they resemble a deep-pile carpet. With the transmission electron microscope, microvilli typically look like finger-shaped projections of the cell surface. They show little internal structure, but often have a bundle of stiff supportive filaments of a protein called *actin*. Actin filaments attach to the inside of the plasma membrane at the tip of the microvillus and serve to stiffen and support it. At its base, they extend a little way into the cell and anchor the microvillus to a protein mesh called the *terminal web*. When tugged by another protein in the cytoplasm, actin can shorten a microvillus to milk its absorbed contents downward into the cell.

Cilia

Cilia (SIL-ee-uh; singular, *cilium*[18]) are hairlike processes about 7 to 10 μm long. Nearly every cell has a solitary, nonmotile *primary cilium* a few micrometers long. Its function in some cases is still a mystery, but apparently many of them are sensory, serving as the cell's "antenna" for monitoring nearby conditions. The light-absorbing parts of the retinal cells in the eye are modified primary cilia; in the inner ear, they play a role in the senses of motion and balance; and in kidney tubules, they are thought to monitor fluid flow. Odor molecules bind to nonmotile cilia on the sensory cells of the nose.

Motile cilia are less widespread, occurring mainly in the respiratory tract, uterine (fallopian) tubes, internal cavities of the brain and spinal cord, and some male reproductive ducts. Cells here typically have 50 to 200 cilia each (fig. 2.13a). These cilia beat in waves that sweep across the surface of an epithelium, always in the same direction, moving substances such as fluid, mucus, and egg cells.

Cilia possess a central core called the **axoneme**[19] (ACK-so-neem), an orderly array of thin protein cylinders called *microtu-*

bules. There are two central microtubules surrounded by a pinwheel-like ring of nine microtubule pairs (fig. 2.13b–d). The central microtubules stop at the cell surface, but the peripheral microtubules continue a short distance into the cell as part of a **basal body** that anchors the cilium. In each pair of peripheral microtubules, one tubule has paired *dynein* (DINE-een) *arms* along its length. **Dynein,**[20] a *motor protein,* uses energy from ATP to "crawl" up the adjacent pair of microtubules. When microtubules on the front of the cilium crawl up the microtubules behind them, the cilium bends toward the front.

Flagella

There is only one functional **flagellum**[21] (fla-JEL-um) in humans— the whiplike tail of a sperm. It is much longer than a cilium and has an identical axoneme, but between the axoneme and plasma membrane it also has a complex sheath of coarse cytoskeletal filaments that stiffen the tail and give it more propulsive power.

> **Apply What You Know**
>
> Kartagener syndrome is a hereditary disease in which dynein is lacking from cilia and flagella. How do you think Kartagener syndrome will affect a man's ability to father a child? How might it affect his respiratory health? Explain your answers.

Cell Junctions *Questions about these ter are based on function*

Also at the cell surface are certain **cell junctions** that link cells together and attach them to the extracellular material. Such attachments enable cells to grow and divide normally, resist stress, communicate with each other, and control the movement of substances through the gaps between cells. Without them, cardiac muscle cells would pull apart when they contracted, and every swallow of food would scrape away the lining of the esophagus. We will examine three types of junctions—*tight junctions, desmosomes,* and *gap junctions* (fig. 2.14). Each type serves a different purpose, and two or more types often occur in a single cell.

Tight Junctions

A **tight junction** completely encircles an epithelial cell near its apical surface and joins it tightly to the neighboring cells like the plastic harness on a six-pack of soda cans. At a tight junction, the plasma membranes of two adjacent cells come very close together and are linked by transmembrane cell adhesion proteins. These proteins seal off the intercellular space and make it difficult for substances to pass between the cells. In the stomach and intestines, for example, tight junctions prevent digestive juices from seeping between epithelial cells and digesting the underlying connective tissue. They also help to prevent intestinal bacteria from invading the tissues, and they ensure that most digested nutrients pass *through* the epithelial cells and not *between* them.

---

[17]*micro* = small + *villi* = hairs
[18]*cilium* = eyelash
[19]*axo* = axis + *neme* = thread

[20]*dyn* = power, energy + *in* = protein
[21]*flagellum* = whip

Figure 2.13 **The Structure of Cilia.** (a) Inner surface of the trachea (SEM). Several nonciliated, mucus-secreting goblet cells are visible among the ciliated cells. The goblet cells have short microvilli on their surface. (b) Three-dimensional structure of a cilium and its basal body. (c) Cross section of a few cilia and microvilli. (d) Cross-sectional structure of a cilium.
• *Describe as many structural differences between cilia and microvilli as you can.*

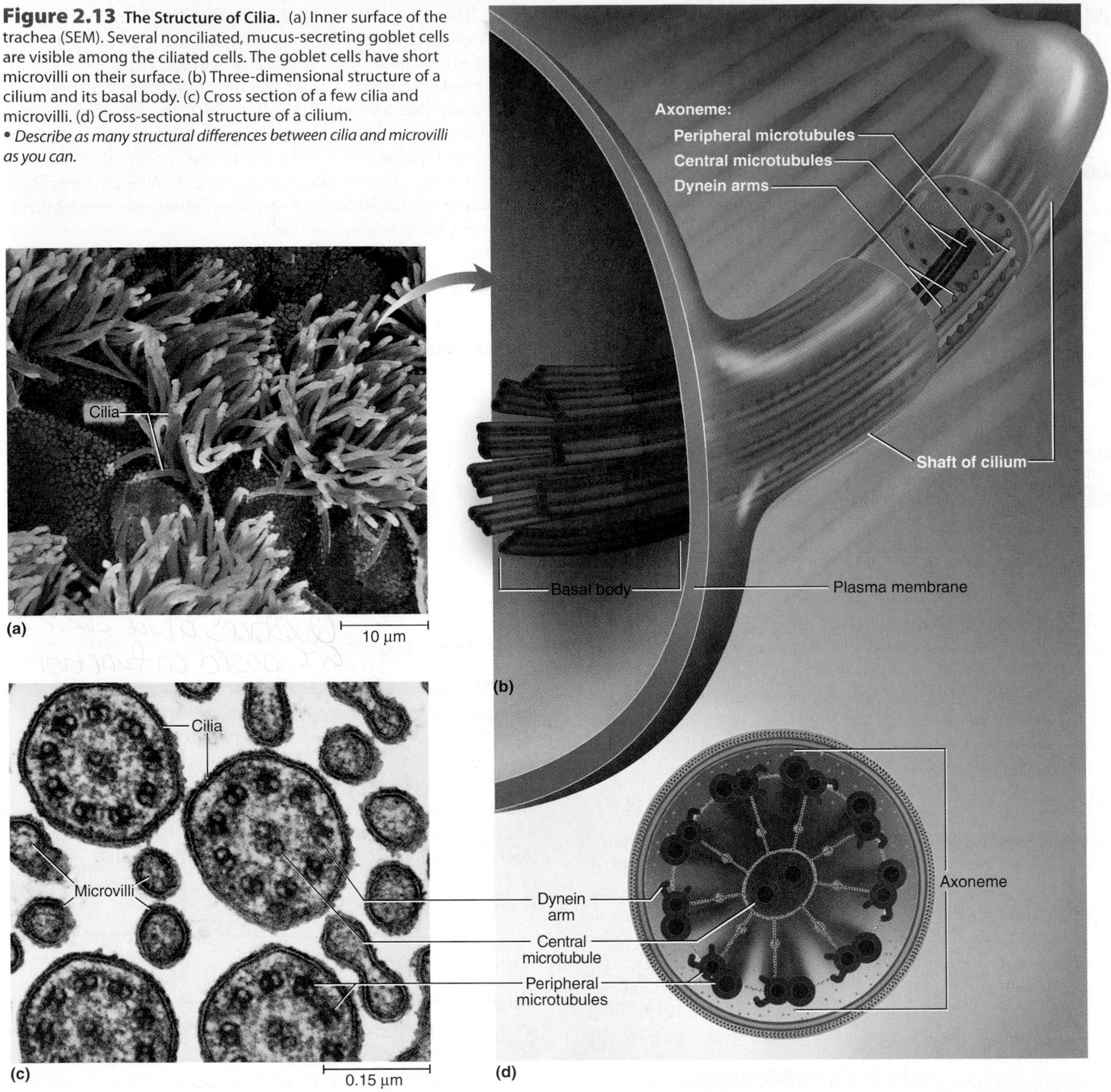

(a)

10 µm

Axoneme:
Peripheral microtubules
Central microtubules
Dynein arms

Shaft of cilium

Basal body   Plasma membrane

(b)

(c)

Cilia

Microvilli

0.15 µm

Dynein arm

Central microtubule

Peripheral microtubules

Axoneme

(d)

Desmosomes

A **desmosome**[22] (DEZ-mo-some) is a patch that holds cells tightly together somewhat like a snap on a pair of pants. Desmosomes are not continuous and therefore cannot prevent substances from passing around them and going between cells. They serve to keep cells from pulling apart and thus enable a tissue to resist mechanical stress. Desmosomes are common in the epidermis, the epithe-

---

[22]*desmo* = band, bond, ligament + *som* = body

lium of the uterine cervix, other epithelia, and cardiac muscle. Hooklike J-shaped proteins approach the cell surface from within and penetrate into a thick protein plaque on the inner face of the plasma membrane, and then the short arm of the J turns back into the cell—thus anchoring the plaque to the cytoskeleton. Proteins of the plaque are linked to transmembrane proteins which, in turn, are linked to transmembrane proteins of the next cell, forming a zone of strong cell adhesion. Each cell mirrors the other and contributes half of the desmosome. Such connections among neighboring cells create a strong structural network that binds

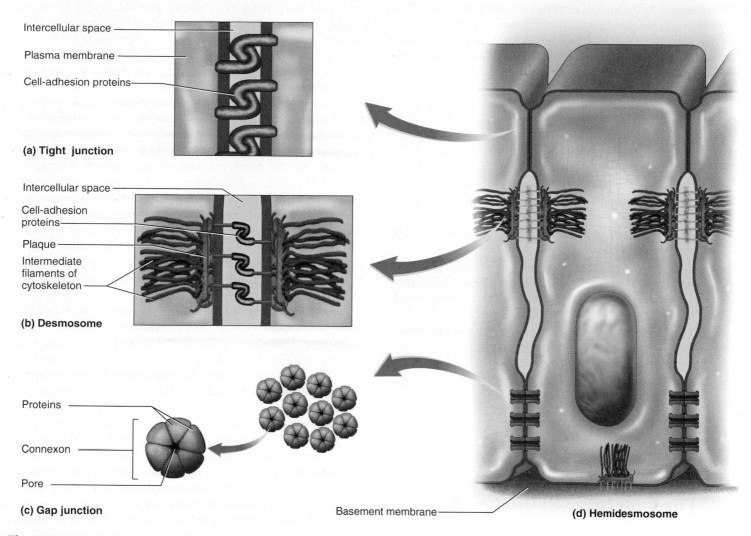

Intercellular space

Plasma membrane

Cell-adhesion proteins

(a) Tight junction

Intercellular space

Cell-adhesion proteins

Plaque

Intermediate filaments of cytoskeleton

(b) Desmosome

Proteins

Connexon

Pore

(c) Gap junction

Basement membrane

(d) Hemidesmosome

Figure 2.14 Types of Cell Junctions.
• *Which of these junctions allows material to pass from one cell directly into the next?*

cells together throughout the tissue (see Deeper Insight 2.1). The basal cells of an epithelium are similarly linked to the underlying basement membrane by half desmosomes called *hemidesmosomes,* so an epithelium cannot easily peel away from the underlying tissue.

> ### Apply What You Know
>
> Why would desmosomes not be suitable as the sole type of cell junction between epithelial cells of the stomach?

Gap Junctions

A **gap (communicating) junction** is formed by a *connexon,* which consists of six transmembrane proteins arranged in a ring somewhat like the segments of an orange, surrounding a central water-filled channel. Ions, glucose, amino acids, and other small solutes can diffuse through the channel directly from the cytoplasm of one

DEEPER INSIGHT   2.1

When Desmosomes Fail

We often get our best insights into the importance of a structure from the dysfunctions that occur when it breaks down. Desmosomes are destroyed in a disease called *pemphigus vulgaris*[23] (PEM-fih-gus vul-GAIR-iss), in which misguided antibodies (defensive proteins) called *autoantibodies* attack the desmosome proteins, especially in the skin. The resulting breakdown of desmosomes between the epidermal cells leads to widespread blistering of the skin and oral mucosa, loss of tissue fluid, and sometimes death. The condition can be controlled with drugs that suppress the immune system, but such drugs compromise the body's ability to fight off infections.

[23]*pemphigus* = blistering + *vulgaris* = common

cell into the next. In the human embryo, nutrients pass from cell to cell through gap junctions until the circulatory system forms and takes over the role of nutrient distribution. In cardiac muscle, gap junctions allow electrical excitation to pass directly from cell to cell so that the cells contract in near unison.

Before You Go On

Answer the following questions to test your understanding of the preceding section:

5. Generally speaking, what sort of substances can enter a cell by diffusing through its phospholipid membrane? What sort of substances must enter primarily through the channel proteins?

6. Compare the structure and function of transmembrane proteins with peripheral proteins.

7. What membrane transport processes get all the necessary energy from the spontaneous movement of molecules? What ones require ATP as a source of energy? What membrane transport processes are carrier mediated? What ones are not?

8. Identify several reasons why the glycocalyx is important to human survival.

9. How do microvilli and cilia differ in structure and function?

10. What are the functional differences between tight junctions, gap junctions, and desmosomes?

2.3 The Cell Interior

▶ Expected Learning Outcomes

When you have completed this section, you should be able to

- describe the cytoskeleton and its functions;
- list the main organelles of a cell and explain their functions; and
- give some examples of cell inclusions and explain how inclusions differ from organelles.

We now probe more deeply into the cell to study its internal structures. These are classified into three groups—cytoskeleton, organelles, and inclusions—all embedded in the clear, gelatinous cytosol.

The Cytoskeleton

The **cytoskeleton** is a network of protein filaments and tubules that structurally support a cell, determine its shape, organize its contents, direct the movement of substances within the cell, and contribute to movements of the cell as a whole. It can form a very dense supportive web in the cytoplasm (fig. 2.15). It is connected to transmembrane proteins of the plasma membrane. They, in turn, are connected to proteins external to the cell, so there is a strong struc-

tural continuity from extracellular material to the cytoplasm. Cytoskeletal elements may even connect to chromosomes in the nucleus, enabling physical tension on a cell to move nuclear contents and mechanically stimulate genetic function.

The cytoskeleton is composed of *microfilaments, intermediate filaments,* and *microtubules.* **Microfilaments (thin filaments)** are about 6 nm thick and are made of the protein actin. They form a fibrous **terminal web (membrane skeleton)** on the cytoplasmic side of the plasma membrane. The lipids of the plasma membrane are spread out over the terminal web like butter on a slice of bread. The web, like the bread, provides physical support, whereas the lipids, like butter, provide a permeability barrier. It is thought that, without this support by the terminal web, the lipids would break up into little droplets and the plasma membrane would not be able to hold together. As described earlier, actin microfilaments also form the supportive cores of the microvilli and play a role in cell movement. Muscle cells especially are packed with actin, which is pulled upon by the motor protein myosin to make muscle contract.

Intermediate filaments (8–10 nm in diameter) are thicker and stiffer than microfilaments. They give a cell its shape, resist stress, and participate in the junctions that attach cells to their neighbors. In epidermal cells, they are made of the tough protein *keratin* and occupy most of the cytoplasm. They are responsible for the strength of hair and fingernails. The J-shaped filaments of desmosomes are also in this category.

A **microtubule** (25 nm in diameter) is a hollow cylinder made of 13 parallel strands called *protofilaments.* Each protofilament is a long chain of globular proteins called *tubulin* (fig. 2.16). Microtubules radiate from the centrosome (see p. 45) and hold organelles in place, form bundles that maintain cell shape and rigidity, and act somewhat like railroad tracks to guide organelles and molecules to specific destinations in a cell. They form the ciliary and flagellar basal bodies and axonemes described earlier, and as discussed later in relation to organelles and mitosis, they form the centrioles and mitotic spindle involved in cell division. Microtubules are not permanent structures. They appear and disappear moment by moment as tubulin molecules assemble into a tubule and then suddenly break apart again to be used somewhere else in the cell (fig. 2.15a). The double and triple sets of microtubules in cilia, flagella, basal bodies, and centrioles, however, are more stable.

Organelles

The minute, metabolically active structures within a cell are called **organelles** (literally, "little organs") because they are to the cell what organs are to the body—structures that play individual physiological roles in the survival of the whole (see fig. 2.5). A cell may have 10 billion protein molecules, some of which are powerful enzymes with the potential to destroy the cell if they are not contained and isolated from other cellular components. You can imagine the enormous problem of keeping track of all this material, directing molecules to the correct destinations, and maintaining order against the incessant tug of disorder. Cells maintain order partly by compartmentalizing their contents in organelles. Figure 2.17 shows some major organelles.

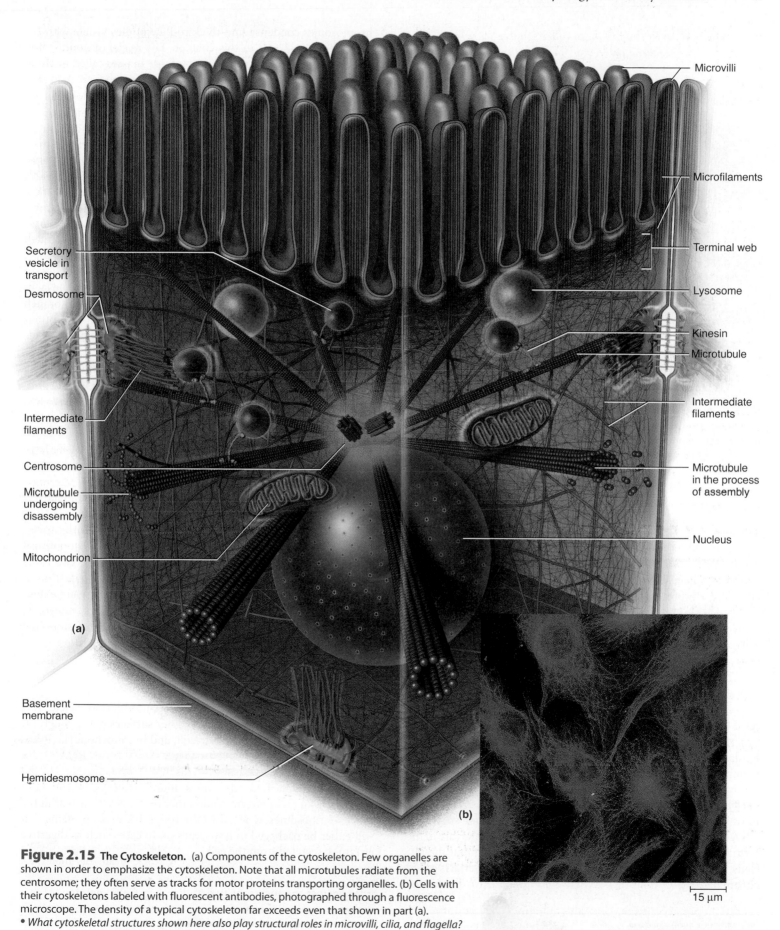

Figure 2.15 **The Cytoskeleton.** (a) Components of the cytoskeleton. Few organelles are shown in order to emphasize the cytoskeleton. Note that all microtubules radiate from the centrosome; they often serve as tracks for motor proteins transporting organelles. (b) Cells with their cytoskeletons labeled with fluorescent antibodies, photographed through a fluorescence microscope. The density of a typical cytoskeleton far exceeds even that shown in part (a).

• *What cytoskeletal structures shown here also play structural roles in microvilli, cilia, and flagella?*

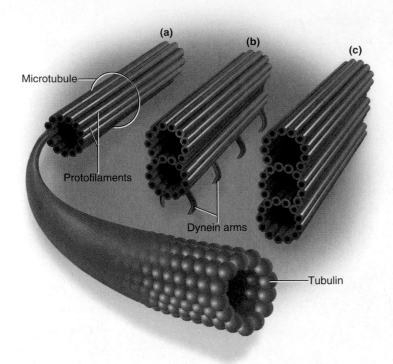

Figure 2.16 Microtubules. (a) A single microtubule is composed of 13 protofilaments. Each protofilament is a helical chain of globular proteins called tubulin. (b) One of the nine microtubule pairs of a cilium. (c) One of the nine microtubule triplets of a centriole.

The Nucleus

The nucleus (fig. 2.17a) is the largest organelle and usually the only one visible with the light microscope. It contains the cell's chromosomes and is therefore the genetic control center of cellular activity. Granular organelles called ribosomes are produced here, and the early steps in protein synthesis occur here under the direction of the genes. Most cells have only one nucleus, but there are exceptions. Mature red blood cells have none; they are *anuclear*.[24] A few cell types are *multinucleate*—having 2 to 50 nuclei—including some liver cells, skeletal muscle cells, and certain bone-dissolving cells.

The nucleus is usually spheroidal to elliptical in shape and averages about 5 μm in diameter. It is surrounded by a **nuclear envelope** consisting of two parallel unit membranes. The envelope is perforated with **nuclear pores,** about 30 to 100 nm in diameter, formed by a ring-shaped complex of proteins. These proteins regulate molecular traffic into and out of the nucleus and bind the two unit membranes together. The inside of the nuclear envelope is lined by a web of intermediate filaments, like a cage enclosing the DNA.

The material within the nucleus is called the *nucleoplasm*. In most human cells, it contains 46 **chromosomes**[25]—long strands of DNA and protein. In nondividing cells, the chromosomes are in the form of very fine filaments broadly dispersed throughout the nucleus, visible only with the TEM. Collectively, this material is called **chromatin** (CRO-muh-tin). When cells prepare to divide, the

chromosomes condense into thick rodlike bodies visible with the LM, as described later in this chapter. The nuclei of nondividing cells also usually exhibit one or more dense masses called **nucleoli** (singular, *nucleolus*), where subunits of the ribosomes are made before they are transported out to the cytoplasm.

Endoplasmic Reticulum

The term *endoplasmic reticulum (ER)* literally means "little network within the cytoplasm." It is a system of interconnected channels called **cisternae**[26] (sis-TUR-nee) enclosed by a unit membrane (fig. 2.17b). In areas called **rough endoplasmic reticulum,** the network consists of parallel, flattened cisternae covered with ribosomes, which give it its rough or granular appearance. The rough ER is continuous with the outer membrane of the nuclear envelope, and adjacent cisternae are connected by perpendicular bridges. In areas called **smooth endoplasmic reticulum,** the membrane lacks ribosomes, the cisternae are more tubular in shape, and they branch more extensively. The cisternae of the smooth ER are continuous with those of the rough ER, so the two are different parts of the same cytoplasmic network.

The endoplasmic reticulum synthesizes steroids and other lipids, detoxifies alcohol and other drugs, and manufactures all of the membranes of the cell. Rough ER produces the phospholipids and proteins of the plasma membrane. It also synthesizes proteins that are either secreted from the cell or packaged in organelles called *lysosomes*. Rough ER is most abundant in cells that synthesize large amounts of protein, such as antibody-producing cells and cells of the digestive glands.

Most cells have only a scanty smooth ER, but it is relatively abundant in cells that engage extensively in detoxification, such as liver and kidney cells. Long-term abuse of alcohol, barbiturates, and other drugs leads to tolerance partly because the smooth ER proliferates and detoxifies the drugs more quickly. Smooth ER is also abundant in cells of the testes and ovaries that synthesize steroid hormones. Skeletal muscle and cardiac muscle contain extensive networks of modified smooth ER called *sarcoplasmic reticulum*, which releases calcium to trigger muscle contraction and stores the calcium between contractions.

Ribosomes

Ribosomes are small granules of protein and ribonucleic acid (RNA) found in the cytosol, on the outer surfaces of the rough ER and nuclear envelope, in the nucleoli, and in mitochondria. Ribosomes "read" coded genetic messages (messenger RNA) from the nucleus and assemble amino acids into proteins specified by the code. The unattached ribosomes found scattered throughout the cytoplasm make enzymes and other proteins for use within the cell. The ribosomes attached to the rough ER make proteins that will either be packaged in lysosomes or, in cases such as digestive enzymes, secreted from the cell.

---

[24]*a* = without + *nucle* = nucleus

[25]*chromo* = color + *some* = body
[26]*cistern* = reservoir

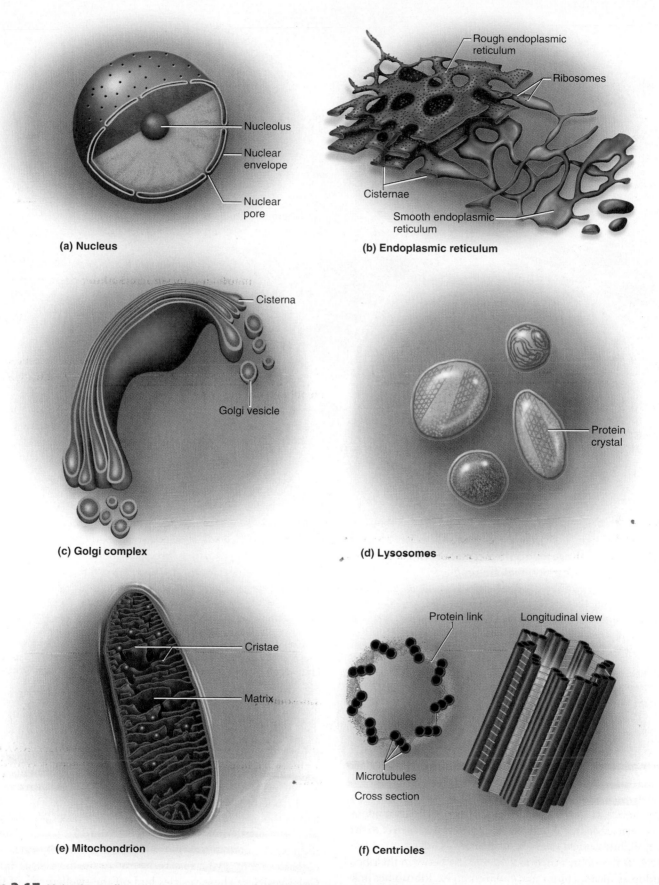

(a) **Nucleus**
- Nucleolus
- Nuclear envelope
- Nuclear pore

(b) **Endoplasmic reticulum**
- Rough endoplasmic reticulum
- Ribosomes
- Cisternae
- Smooth endoplasmic reticulum

(c) **Golgi complex**
- Cisterna
- Golgi vesicle

(d) **Lysosomes**
- Protein crystal

(e) **Mitochondrion**
- Cristae
- Matrix

(f) **Centrioles**
- Protein link
- Longitudinal view
- Microtubules
- Cross section

Figure 2.17 Major Organelles. (a) Nucleus. (b) Endoplasmic reticulum, showing rough and smooth regions. (c) Golgi complex and Golgi vesicles. (d) Lysosomes. (e) Mitochondrion. (f) A pair of centrioles. Centrioles are typically found in pairs, perpendicular to each other so that an electron micrograph shows one in cross section and one in longitudinal section.

• *With rare exceptions, only one of these organelles can normally be seen with a typical student light microscope. Which one?*

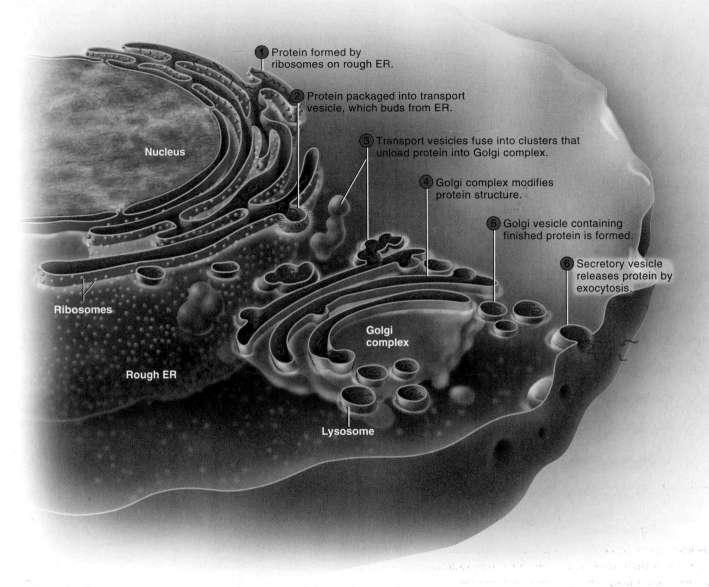

① Protein formed by ribosomes on rough ER.

② Protein packaged into transport vesicle, which buds from ER.

③ Transport vesicles fuse into clusters that unload protein into Golgi complex.

④ Golgi complex modifies protein structure.

⑤ Golgi vesicle containing finished protein is formed.

⑥ Secretory vesicle releases protein by exocytosis.

Nucleus

Ribosomes

Rough ER

Golgi complex

Lysosome

Figure 2.18 Organelle Collaboration in Protein Production. The steps in protein synthesis and secretion are numbered 1 through 6. (1) Acting on instructions from the nucleus, each ribosome assembles amino acids in the correct sequence to make a particular protein. The protein is threaded into the cisterna of the rough ER as it is synthesized. The rough ER cuts and splices proteins and may make other modifications. (2) The rough ER packages the modified protein into transport vesicles that carry it to the Golgi complex. (3) The transport vesicle fuses with a cisterna of the Golgi complex and unloads its protein. (4) The Golgi complex may further modify the protein. (5) The Golgi complex buds off Golgi vesicles containing the finished protein. (6) Some Golgi vesicles become secretory vesicles, which travel to the plasma membrane and release the product by exocytosis.

Golgi Complex

The **Golgi**[27] (GOAL-jee) **complex** is a small system of cisternae that synthesizes carbohydrates and certain lipids and puts the finishing touches on protein and glycoprotein synthesis (fig. 2.17c). The complex resembles a stack of pita bread. Typically, it consists of about six cisternae, slightly separated from each other, each of them a flattened, slightly curved sac with swollen edges.

Figure 2.18 shows the functional interaction between the ribosomes, endoplasmic reticulum, and Golgi complex. Ribosomes link amino acids together in a genetically specified order to make a particular protein. This new protein threads its way into the cisterna of the rough ER, where enzymes trim and modify it. The altered protein is then shuffled into a little **transport vesicle,** a small, spheroidal organelle that buds off the ER and carries the protein to the nearest cisterna of the Golgi complex. The Golgi complex sorts these proteins, passes them along from one cisterna to the next, cuts and splices some of them, adds carbohydrates to some of them, and finally packages the proteins in membrane-bounded **Golgi vesicles.** These vesicles bud off the swollen rim of the cisterna farthest from the ER. They are seen in abundance in the neighborhood of the Golgi complex.

---

[27]Camillo Golgi (1843–1926), Italian histologist

Some Golgi vesicles become lysosomes, the organelles discussed next; some migrate to the plasma membrane and fuse with it, contributing fresh protein and phospholipid to the membrane; and some become **secretory vesicles** that store a cell product, such as breast milk, mucus, or digestive enzymes, for later release by exocytosis.

Lysosomes

A **lysosome**[28] (LY-so-some) (fig. 2.17d) is a package of enzymes contained in a single unit membrane. Although often round or oval, lysosomes are extremely variable in shape. When viewed with the TEM, they often exhibit dark gray contents devoid of structure, but sometimes show crystals or parallel layers of protein. At least 50 lysosomal enzymes have been identified. They break down proteins, nucleic acids, carbohydrates, phospholipids, and other substances. White blood cells called *neutrophils* phagocytize bacteria and digest them with the enzymes of their lysosomes. Lysosomes also digest and dispose of worn-out mitochondria and other organelles; this process is called *autophagy*[29] (aw-TOFF-uh-jee). They also aid in a process of "cell suicide" called *apoptosis* (AP-op-TOE-sis), or *programmed cell death,* in which cells that are no longer needed undergo a prearranged death. The uterus, for example, weighs about 900 g at full-term pregnancy and, through apoptosis, shrinks to 60 g within 5 or 6 weeks after birth.

Peroxisomes

Peroxisomes resemble lysosomes (and are not illustrated) but contain different enzymes and are not produced by the Golgi complex. Their general function is to use molecular oxygen (O_2) to oxidize organic molecules. These reactions produce hydrogen peroxide (H_2O_2), hence the name of the organelle. H_2O_2 is then used to oxidize other molecules, and the excess is broken down to water and oxygen by an enzyme called *catalase.*

Peroxisomes occur in nearly all cells but are especially abundant in liver and kidney cells. They neutralize free radicals and detoxify alcohol, other drugs, and a variety of blood-borne toxins. Peroxisomes also decompose fatty acids into two-carbon fragments that the mitochondria use as an energy source for ATP synthesis.

Mitochondria

Mitochondria[30] (MY-toe-CON-dree-uh) (singular, *mitochondrion*) are organelles specialized for a process called *aerobic respiration,* which synthesizes most of the body's ATP. They have a variety of shapes: spheroidal, rod-shaped, bean-shaped, or threadlike (fig. 2.17e). Like the nucleus, a mitochondrion is surrounded by a double unit membrane. The inner membrane usually has folds called **cristae**[31] (CRIS-tee), which project like shelves across the organelle. Cristae bear the enzymes that produce most of the ATP. The space between the cristae is called the **mitochondrial matrix.** It contains enzymes,

ribosomes, and a small, circular DNA molecule called *mitochondrial DNA (mtDNA),* which is genetically different from the DNA in the cell's nucleus (see Deeper Insight 2.2). Mutations in mtDNA are responsible for some muscle, heart, and eye diseases.

Centrioles

A **centriole** (SEN-tree-ole) is a short cylindrical assembly of microtubules arranged in nine groups of three microtubules each (fig. 2.17f). Near the nucleus, most cells have a small, clear patch of cytoplasm called the **centrosome**[32] containing a pair of mutually perpendicular centrioles (see fig. 2.5). These centrioles play a role in cell division described later. In ciliated cells, each cilium also has a **basal body** composed of a single centriole oriented perpendicular to the plasma membrane. Two microtubules of each triplet form the peripheral microtubules of the axoneme of the cilium.

Inclusions

Inclusions are of two kinds: accumulated cell products such as pigments, fat droplets, and granules of glycogen (a starchlike carbohydrate); and internalized foreign matter such as dust, viruses,

[31]*crista* = crest
[32]*centro* = central + *some* = body

DEEPER INSIGHT 2.2

The Origin of Mitochondria

There is compelling evidence that mitochondria evolved from bacteria that either invaded or were engulfed by other ancient cells, then evolved a mutually beneficial metabolic relationship with them. In size and physiology, mitochondria resemble certain bacteria that live within other cells in a state of symbiosis. Mitochondrial DNA (mtDNA) resembles bacterial DNA more than it does nuclear DNA, and it is replicated independently of nuclear DNA.

While nuclear DNA is reshuffled in each generation by the process of sexual reproduction, mtDNA remains unchanged from generation to generation except by the slow pace of random mutation. Mitochondrial DNA is inherited primarily from the individual's mother, because it is passed to the developing embryo mainly from the cytoplasm of the egg cell, with little contribution from sperm mitochondria. Biologists and anthropologists have used mtDNA as a "molecular clock" to trace evolutionary lineages in humans and other species. The differences in mtDNA between one individual and another, or between one species and another, can be combined with the rate of mutation to estimate how long ago two individuals or species arose from the same ancestor. Geneticists have gained evidence, although still controversial, that of all females who lived in Africa about 200,000 years ago, only one left any descendants still living today—everyone now on earth is descended from this "mitochondrial Eve," a woman of unparalleled genetic success.

[28]*lyso* = loosen, dissolve + *some* = body
[29]*auto* = self + *phagy* = eating
[30]*mito* = thread + *chondr* = grain

and bacteria. Inclusions are never enclosed in a unit membrane, and unlike the organelles and cytoskeleton, they are not essential to cell survival.

Before You Go On

Answer the following questions to test your understanding of the preceding section:

11. Describe at least three functions of the cytoskeleton.

12. Briefly state how each of the following cell components can be recognized in electron micrographs: the nucleus, a mitochondrion, a lysosome, and a centriole. What is the primary function of each?

13. Distinguish between organelles and inclusions. State two examples of each.

14. What three organelles are involved in protein synthesis?

15. Define *centriole, microtubule, cytoskeleton,* and *axoneme.* How are these structures related to each other?

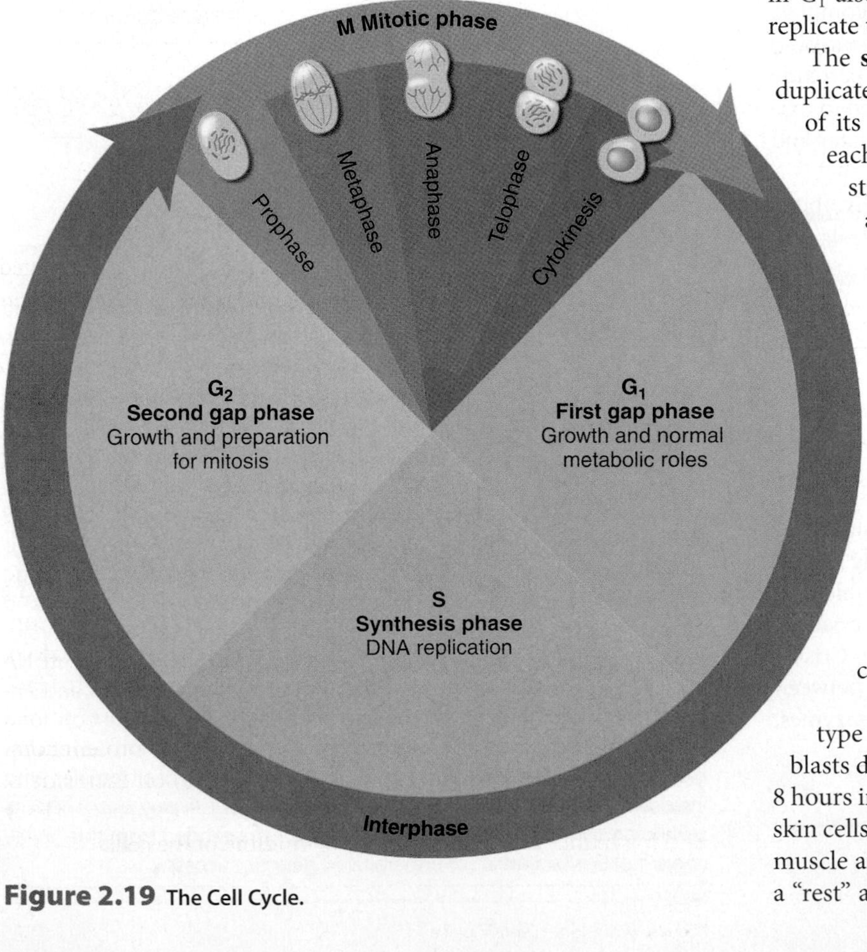

Figure 2.19 The Cell Cycle.

2.4 The Cell Life Cycle

▌ Expected Learning Outcomes

When you have completed this section, you should be able to

- describe the life cycle of a cell;
- name the stages of mitosis and describe the events that occur in each one; and
- discuss the types and clinical uses of stem cells.

This chapter concludes with an examination of the typical life cycle of human cells, including the process of cell division. Finally, we examine an issue of current controversy, the therapeutic use of embryonic stem cells.

The Cell Cycle

Most cells periodically divide into two daughter cells, so a cell has a life cycle extending from one division to the next. This **cell cycle** is divided into four main phases: G_1, S, G_2, and M *(fig. 2.19)*.

The **first gap (G_1) phase** is an interval between cell division and DNA replication. During this time, a cell synthesizes proteins, grows, and carries out its predestined tasks for the body. Most human physiology pertains to what cells do in the G_1 phase. Cells in G_1 also accumulate the materials needed in the next phase to replicate their DNA.

The **synthesis (S) phase** is the period in which a cell makes duplicate copies of its centrioles and all of its nuclear DNA. Each of its DNA molecules uncoils into two separate strands, and each strand acts as a template for the synthesis of the missing strand. A cell begins the S phase with 46 molecules of DNA and ends it with 92. The cell then has two identical sets of DNA molecules, which are available to be divided up between daughter cells at the next cell division.

The **second gap (G_2) phase** is a relatively brief interval between DNA replication and cell division. In G_2, a cell finishes replicating its centrioles and synthesizes enzymes that control cell division. It also checks the accuracy of its DNA replication and usually repairs any errors that are detected.

The **mitotic (M) phase** is the period in which a cell replicates its nucleus, divides its DNA into two identical sets (one per nucleus), and pinches in two to form two genetically identical daughter cells. The details of this phase are considered in the next section. Phases G_1, S, and G_2 are collectively called **interphase**—the time between M phases.

The length of the cell cycle varies greatly from one cell type to another. Cultured connective tissue cells called fibroblasts divide about once a day and spend 8 to 10 hours in G_1, 6 to 8 hours in S, 4 to 6 hours in G_2, and 1 to 2 hours in M. Stomach and skin cells divide rapidly, bone and cartilage cells slowly, and skeletal muscle and nerve cells not at all. Some cells leave the cell cycle for a "rest" and cease to divide for days, years, or the rest of one's life.

DEEPER INSIGHT   2.3

Cancer

A *tumor (neoplasm*[33]*)* is a mass of tissue produced when the rate of cell division exceeds the rate of cell death. A *malignant*[34] tumor, or *cancer* (fig. 2.20), is especially fast-growing, lacks a confining fibrous capsule, and has cells that are capable of breaking free and spreading to other organs *(metastasizing*[35]*)*. Cancer was named by Hippocrates, who compared the distended veins in some breast tumors to the outstretched legs of a crab.[36]

All cancer is caused by mutations (changes in DNA or chromosome structure), which can be induced by chemicals, viruses, or radiation, or simply occur through errors in DNA replication in the cell cycle. Agents that cause mutation are called *mutagens,*[37] and those that induce cancer are also called *carcinogens.*[38] Many forms of cancer stem from mutations in two gene families, the oncogenes and tumor-suppressor genes. *Oncogenes*[39] are mutated genes that promote the synthesis of excessive amounts of growth factors (chemicals that stimulate cell division) or excessive sensitivity of target cells to growth factors. *Tumor-suppressor (TS) genes* inhibit the development of cancer by opposing oncogenes, promoting DNA repair, and other means. Cancer occurs when TS genes are unable to perform this function. Oncogenes are like an accelerator to the cell cycle, and TS genes are like a brake.

Untreated cancer is almost always fatal. Tumors destroy healthy tissue; they can grow to block major blood vessels or respiratory airways; they can damage blood vessels and cause hemorrhaging; they can compress and kill brain tissue; and they tend to drain the body of nutrients and energy as they "hungrily" consume a disproportionate share of the body's oxygen and nutrients.

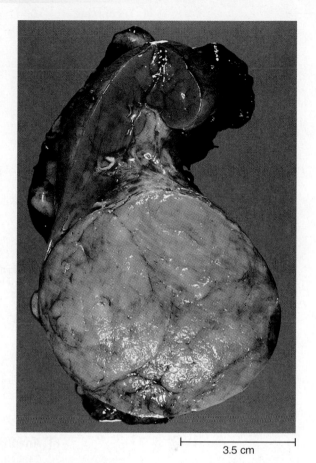

3.5 cm

Figure 2.20 Wilms Tumor. This is a malignant tumor of the kidney occurring especially in children.

Such cells are said to be in the **G₀ (G-zero) phase.** The balance between cells that are actively cycling and those standing by in G_0 is an important factor in determining the number of cells in the body. An inability to stop cycling and enter G_0 is characteristic of cancer cells (see Deeper Insight 2.3).

Cell Division

Cells divide by two mechanisms called mitosis and meiosis. Meiosis, however, is restricted to one purpose, the production of eggs and sperm, and is therefore treated in chapter 26 on reproduction. **Mitosis** serves all the other functions of cell division: the development of an individual, composed of some 40 trillion cells, from a one-celled fertilized egg; continued growth of all the organs after birth; the replacement of cells that die; and the repair of damaged tissues. Four phases of mitosis are recognizable—*prophase, metaphase, anaphase,* and *telophase* (fig. 2.21).

In **prophase,**[40] at the outset of mitosis, the chromosomes coil into short, compact rods that are easier to distribute to daughter cells than the long, delicate chromatin of interphase. At this stage, a chromosome consists of two genetically identical bodies called **chromatids,** joined together at a pinched spot called the **centromere** (fig. 2.22). There are 46 chromosomes, two chromatids per chromosome, and one molecule of DNA in each chromatid. The nuclear envelope disintegrates during prophase and releases the chromosomes into the cytosol. The centrioles begin to sprout elongated microtubules called **spindle fibers,** which push the centrioles apart as they grow. Eventually, a pair of centrioles comes to lie at each pole of the cell. Spindle fibers grow toward the chromosomes, where some of them become attached to a platelike protein complex called the **kinetochore**[41] (kih-NEE-toe-core) on each side of the centromere. The spindle fibers then tug the chromosomes back and forth until they line up along the midline of the cell.

---

[33]*neo* = new + *plasm* = growth, formation
[34]*mal* = bad, evil
[35]*meta* = beyond + *stas* = being stationary
[36]*cancer* = crab
[37]*muta* = change + *gen* = to produce
[38]*carcino* = cancer + *gen* = to produce
[39]*onco* = tumor

[40]*pro* = first
[41]*kineto* = motion + *chore* = place

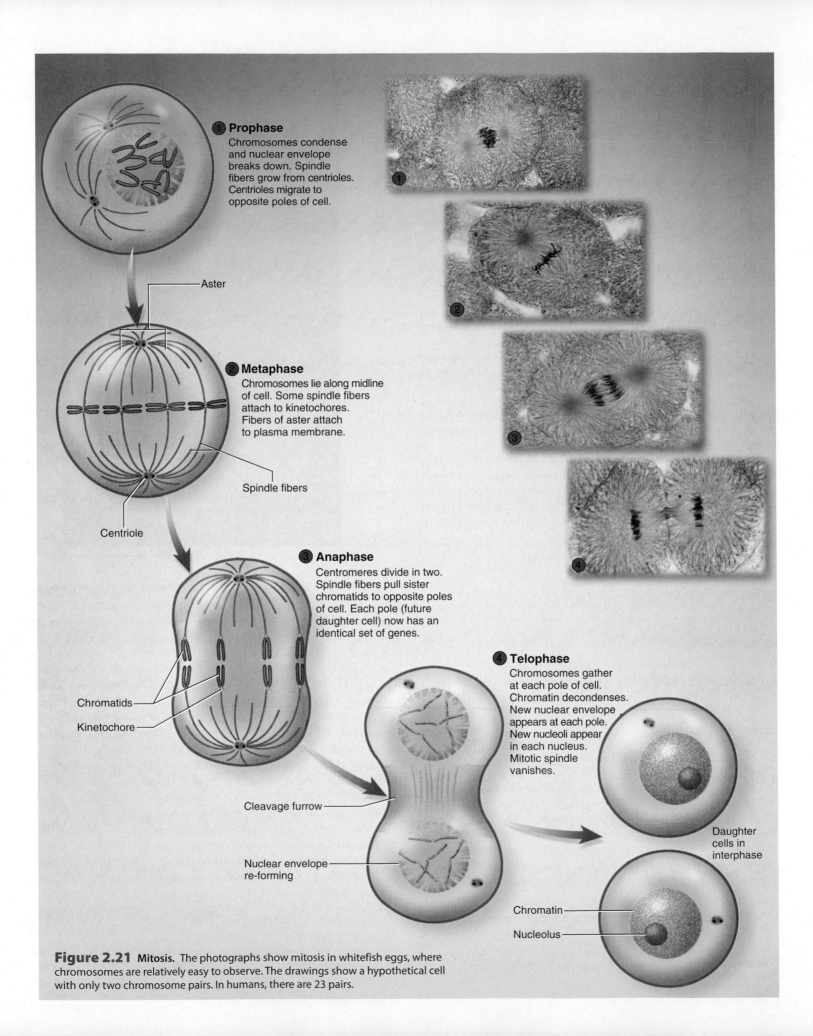

① Prophase
Chromosomes condense and nuclear envelope breaks down. Spindle fibers grow from centrioles. Centrioles migrate to opposite poles of cell.

Aster

② Metaphase
Chromosomes lie along midline of cell. Some spindle fibers attach to kinetochores. Fibers of aster attach to plasma membrane.

Spindle fibers

Centriole

③ Anaphase
Centromeres divide in two. Spindle fibers pull sister chromatids to opposite poles of cell. Each pole (future daughter cell) now has an identical set of genes.

Chromatids

Kinetochore

④ Telophase
Chromosomes gather at each pole of cell. Chromatin decondenses. New nuclear envelope appears at each pole. New nucleoli appear in each nucleus. Mitotic spindle vanishes.

Cleavage furrow

Nuclear envelope re-forming

Daughter cells in interphase

Chromatin

Nucleolus

Figure 2.21 Mitosis. The photographs show mitosis in whitefish eggs, where chromosomes are relatively easy to observe. The drawings show a hypothetical cell with only two chromosome pairs. In humans, there are 23 pairs.

In **metaphase,**[42] the chromosomes are aligned on the cell equator, oscillating slightly and awaiting a signal that stimulates each chromosome to split in two at the centromere. The spindle fibers now form a lemon-shaped array called the **mitotic spindle.** Long microtubules reach out from each centriole to the chromosomes, and shorter microtubules form a starlike *aster*[43] that anchors the assembly to the inside of the plasma membrane at each end of the cell.

Anaphase[44] begins with activation of an enzyme that cleaves the two sister chromatids from each other at the centromere. Each chromatid is now regarded as a separate, single-stranded *daughter chromosome.* One daughter chromosome migrates to each pole of the cell, with its centromere leading the way and the arms trailing behind. Migration is achieved by means of motor proteins in the kinetochore crawling along the spindle fiber as the fiber itself is "chewed up" and disassembled at the chromosomal end. Since sister chromatids are genetically identical and since each daughter cell receives one chromatid from each chromosome, the daughter cells of mitosis are genetically identical.

In **telophase,**[45] the chromatids cluster on each side of the cell. The rough ER produces a new nuclear envelope around each cluster, and the chromatids begin to uncoil and return to the thinly dispersed chromatin form. The mitotic spindle breaks up and vanishes. Each new nucleus forms nucleoli, indicating it has already begun making RNA and preparing for protein synthesis.

Telophase is the end of nuclear division but overlaps with **cytokinesis**[46] (SY-toe-kih-NEE-sis), division of the cytoplasm into two cells. Early traces of cytokinesis appear even in anaphase. It is achieved by the motor protein myosin pulling on microfilaments of actin in the terminal web. This creates a crease called the *cleavage furrow* around the equator of the cell, and the cell eventually pinches in two. Interphase has now begun for these new cells.

Stem Cells

One of the most controversial scientific issues in the last several years has been stem-cell research. **Stem cells** are immature cells with the ability to develop into one or more types of mature, specialized cells. Their ability to give rise to a diversity of mature cell types is called **developmental plasticity. Adult stem (AS) cells** exist in most of the body's organs. Despite the name, they are not limited to adults, but are found also in fetuses, infants, and children. They multiply and replace older cells that are lost to damage or normal cellular turnover. Some AS cells are **unipotent,** able to develop into only one mature cell type, such as the cells that develop into sperm or epidermal squamous cells. Some are **multipotent,** able to differentiate into multiple mature cell types, such as certain bone marrow cells that give rise to multiple types of white blood cells.

Embryonic stem (ES) cells comprise human embryos (technically, preembryos; see chapter 4) of up to 150 cells. They are **pluripotent,** able to develop into any type of embryonic or adult cell. ES cells are easily obtained from the excess embryos created in fertility clinics when a couple attempts to conceive a child by *in vitro fertilization* (IVF). In IVF, eggs are fertilized in glassware and

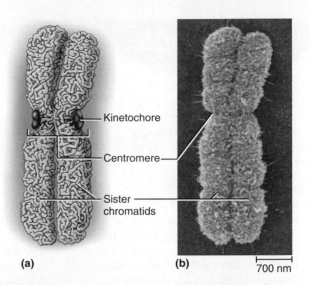

(a)   **(b)**   700 nm

Figure 2.22 Chromosomes. (a) Diagram of a chromosome in metaphase. From the end of the S phase of the cell cycle to the beginning of anaphase in mitosis, a chromosome consists of two genetically identical chromatids. (b) SEM of a metaphase chromosome.

allowed to develop to about 8 to 16 cells. Some of these are then transplanted into the mother's uterus. Excess embryos are created to compensate for the low probability of success. Those that are not transplanted to the uterus are usually destroyed, but they present a potential source of stem cells for research and therapy.

Skin and bone marrow adult stem cells have been used in therapy for many years. There is hope that stem cells can be manipulated to replace a broader range of tissues, such as cardiac muscle damaged by a heart attack; injured spinal cords; the brain cells lost in Parkinson and Alzheimer diseases; or the insulin-secreting cells needed by people with diabetes mellitus. Adult stem cells, however, seem to have limited developmental potential and to be unable to produce all cell types needed to treat a broad range of diseases. In addition, they are present in very small numbers and are difficult to isolate and culture in the quantities needed for therapy. Embryonic stem cells are easier to obtain and culture and have more developmental flexibility, but their use remains embroiled in political, religious, and ethical debate. Some would argue that if the excess embryos from IVF are destined to be destroyed, it would seem sensible to use them for beneficial purposes. Others argue, however, that potential medical benefits cannot justify the destruction of a human embryo or even a preembryo of scarcely more than 100 cells.

Before You Go On

Answer the following questions to test your understanding of the preceding section:

16. State what occurs in each of the four phases of the cell cycle.

17. State what occurs in each of the four phases of mitosis.

18. Explain how a cell ensures that each of its daughter cells gets an identical set of genes.

19. Define *unipotent, multipotent,* and *pluripotent stem cells.* Give an example of each.

20. Discuss the advantages and disadvantages of adult and embryonic stem cells for therapy.

---

[42]*meta* = next in a series
[43]*aster* = star
[44]*ana* = apart
[45]*telo* = end, final
[46]*cyto* = cell + *kinesis* = action, motion

Study Guide

Assess Your Learning Outcomes

You should have a good understanding of this chapter if you can accurately address the following issues.

2.1 The Study of Cells (p. 26)

1. The meaning of *cytology* and why it is important in medical anatomy and physiology
2. How the light microscope, transmission electron microscope, and scanning electron microscope differ, and the relative importance of magnification and resolution
3. The cell shapes denoted by the terms *squamous, cuboidal, columnar, polygonal, stellate, spheroidal, ovoid, discoid, fusiform,* and *fibrous*
4. The distinction between the basal, apical, and lateral surfaces of an epithelial cell, and why it is important to differentiate them
5. The size of a micrometer and some common and extreme human cell sizes in terms of micrometers
6. Reasons why cells cannot grow indefinitely large; limitations on cell size
7. The meaning of *plasma membrane, cytoplasm, nucleoplasm, cytoskeleton, organelles, inclusions,* and *cytosol*
8. Terminology of the fluids on the inside and outside of a cell

2.2 The Cell Surface (p. 30)

1. Why the cell surface is so important for an understanding of human function

2. The molecular composition and organization of the plasma membrane
3. The general functions of membrane phospholipids, cholesterol, proteins, glycoproteins, and glycolipids
4. How transmembrane proteins differ from the peripheral proteins of a plasma membrane
5. The diverse physiological roles of membrane proteins, and terms for the proteins that play each role
6. The importance of filtration through biological membranes, and the principal place in the body where this is relevant
7. The processes of simple diffusion, osmosis, reverse osmosis, facilitated diffusion, and active transport in relation to the plasma membrane; the relationship of each process to concentration gradients, the involvement of membrane proteins, and the necessity for ATP
8. The methods of vesicular transport through the plasma membrane: endocytosis (by phagocytosis, pinocytosis, and receptor-mediated endocytosis) and exocytosis; how each process is carried out; and the purposes for which each process may be used
9. The composition, location, and functions of a cell's glycocalyx
10. The structural and functional differences between microvilli, cilia, and flagella
11. The cell junctions that bind cells to each

other and to extracellular material—tight junctions, desmosomes, hemidesmosomes, and gap junctions—and the structural and functional differences between them

2.3 The Cell Interior (p. 40)

1. The three components of the cytoskeleton, and how they differ in composition and function
2. Structure and function of a cell's nucleus, rough and smooth endoplasmic reticulum, ribosomes, the Golgi complex, lysosomes, peroxisomes, mitochondria, and centrioles
3. The two types of cellular inclusions, where they come from, and how they differ from organelles

2.4 The Cell Life Cycle (p. 46)

1. The four stages of the cell cycle and what events occur in each one
2. The four stages of mitosis, what events occur in each one, and how the chromosomes behave in such a way as to produce two genetically identical daughter cells
3. The process of cytokinesis and how it overlaps with mitosis
4. The structure of a chromosome as seen from prophase to metaphase
5. The meaning of *stem cells;* their usefulness in medicine; the distinction between adult and embryonic stem cells; and the various degrees of developmental plasticity seen in stem cells

Testing Your Recall

1. The clear, structureless gel in a cell is its
   a. nucleoplasm.   d. neoplasm.
   b. endoplasm.   e. cytosol.
   c. cytoplasm.

2. New nuclei form and a cell pinches in two during
   a. prophase.   d. telophase.
   b. metaphase.   e. anaphase.
   c. interphase.

3. The amount of _____ in a plasma membrane affects its stiffness versus fluidity.
   a. phospholipid   d. glycoprotein
   b. cholesterol   e. transmembrane
   c. glycolipid   protein

4. Cells specialized for absorption of matter from the ECF are likely to show an abundance of
   a. lysosomes.   d. secretory vesicles.
   b. microvilli.   e. ribosomes.
   c. mitochondria.

5. Osmosis is a special case of
   a. pinocytosis.
   b. carrier-mediated transport.
   c. active transport.
   d. facilitated diffusion.
   e. simple diffusion.

6. Embryonic stem cells are best described as
   a. pluripotent.
   b. multipotent.
   c. unipotent.
   d. more developmentally limited than adult stem cells.
   e. more difficult to culture and harvest than adult stem cells.

7. The amount of DNA in a cell doubles during
   a. prophase.   d. the S phase.
   b. metaphase.   e. the G_2 phase.
   c. anaphase.

8. Fusion of a secretory vesicle with the plasma membrane followed by release of the vesicle's contents is
   a. exocytosis.
   b. receptor-mediated endocytosis.
   c. active transport.
   d. pinocytosis.
   e. phagocytosis.
9. Most cellular membranes are made by
   a. the nucleus.
   b. the cytoskeleton.
   c. enzymes in the peroxisomes.
   d. the endoplasmic reticulum.
   e. replication of existing membranes.
10. Matter can leave a cell by any of the following means *except*
    a. active transport. d. simple diffusion.
    b. pinocytosis.   e. exocytosis.
    c. facilitated diffusion.
11. Most human cells are 10 to 15 _____ wide.
12. When a hormone cannot enter a cell, it binds to a _____ at the cell surface.
13. _____ are channels in the plasma membrane that open or close in response to various stimuli.
14. Most ATP is produced by organelles called _____.
15. Leakage between cells is restricted by cell junctions called _____.
16. Thin scaly cells are described by the term _____.
17. Two human organelles that are surrounded by a double unit membrane are the _____ and _____.
18. Liver cells can detoxify alcohol with two organelles, the _____ and _____.
19. Cells adhere to each other and to extracellular material by means of membrane proteins called _____.
20. A macrophage would use the process of _____ to engulf a dying tissue cell.

Answers in the Appendix

Building Your Medical Vocabulary

State a medical meaning of each of the following word elements, and give a term in which it is used.

1. cyto-
2. squam-
3. -form
4. poly-
5. -philic
6. phago-
7. endo-
8. glyco-
9. chromo-
10. meta-

Answers in the Appendix

True or False

Determine which five of the following statements are false, and briefly explain why.

1. The shape of a cell is determined mainly by its cytoskeleton.
2. The plasma membrane is composed mostly of protein.
3. A plasma membrane is too thin to be seen with the light microscope.
4. The hydrophilic heads of membrane phospholipids are in contact with both the ECF and ICF.
5. Water-soluble substances usually must pass through channel proteins to enter a cell.
6. Cells must use ATP to move substances down a concentration gradient.
7. Osmosis is a type of active transport involving water.
8. Cilia and flagella have an axoneme, but microvilli do not.
9. Desmosomes enable substances to pass from cell to cell.
10. A nucleolus is an organelle within the nucleoplasm.

Answers in the Appendix

Testing Your Comprehension

1. What would probably happen to the plasma membrane of a cell if it were composed of hydrophilic molecules such as carbohydrates?
2. Since electron microscopes are capable of much more resolution than light microscopes, why do you think biologists go on using light microscopes? Why are students in introductory biology courses not provided with electron microscopes?
3. This chapter mentions that the polio virus enters cells by means of receptor-mediated endocytosis. Why do you think the viruses don't simply enter through the channels in the plasma membrane? Cite some specific facts from this chapter to support your conjecture.
4. A major tenet of the cell theory is that all bodily structure and function result from the function of cells. Yet the structural properties of bone are due more to its extracellular material than to its cells. Is this an exception to the cell theory? Why or why not?
5. If a cell were poisoned so its mitochondria ceased to function, what membrane transport processes would immediately stop? What ones could continue?

Answers at www.mhhe.com/saladinha3

Improve Your Grade at www.mhhe.com/saladinha3

Practice quizzes, labeling activities, and games provide fun ways to master concepts. You can also download image PowerPoint files for each chapter to create a study guide or for taking notes during lecture.

Histology—The Study of Tissues

CHAPTER

3

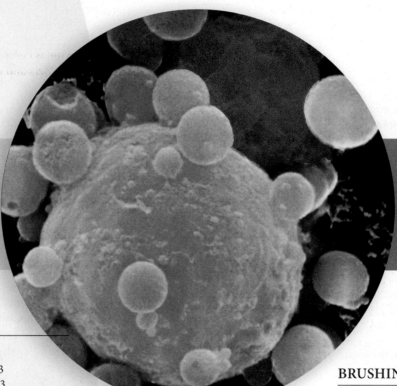

A cancer cell (mauve) undergoing apoptosis (cell suicide) under attack by an immune cell (orange) (SEM)

CHAPTER OUTLINE

DEEPER INSIGHTS

BRUSHING UP

To understand this chapter, you may find it helpful to review the following concepts:

With its 50 trillion cells and thousands of organs, the human body may seem to be a structure of forbidding complexity. Fortunately for our health, longevity, and self-understanding, the biologists of past generations were not discouraged by this complexity, but discovered patterns that made it more understandable. One pattern is the fact that these trillions of cells belong to only 200 different types or so, and these cells are organized into tissues that fall into just 4 broad categories—*epithelial, connective, nervous,* and *muscular tissue.*

Here we study the four tissue classes; variations within each class; how to recognize tissue types microscopically and relate their microscopic anatomy to their function; how tissues are arranged to form an organ; and how tissues change as they grow, shrink, or change from one type to another over the life of the individual. This chapter describes only mature tissue types. Embryonic tissues are discussed in chapter 4.

3.1 The Study of Tissues

▶ Expected Learning Outcomes

When you have completed this section, you should be able to

- name the four primary classes into which all adult tissues are classified; and
- visualize the three-dimensional shape of a structure from a two-dimensional tissue section.

Histology[1] **(microscopic anatomy)** is the study of tissues and how they are arranged into organs. Histology bridges the gap between the *cytology* of the preceding chapter and the *organ system* approach of the chapters that follow.

---

[1]*histo* = tissue + *logy* = study of

The Primary Tissue Classes

A **tissue** is a mass of similar cells and cell products that forms a discrete region of an organ and performs a specific function. The four **primary tissues** are epithelial, connective, nervous, and muscular (table 3.1). These tissues differ from each other in the types and functions of their cells, the characteristics of the **matrix (extracellular material)** that surrounds the cells, and the relative amount of space occupied by cells versus matrix.

In epithelial and muscular tissue, the cells are so close together that the matrix is barely visible, whereas in connective tissue, the matrix usually occupies more space than the cells do. The matrix is composed of fibrous proteins and **ground substance.** The latter is also variously known as the **extracellular fluid (ECF), interstitial**[2] **fluid,** or **tissue fluid,** although in cartilage and bone, the matrix is rubbery or stony in consistency.

In summary, a tissue is composed of cells and matrix, and the matrix is composed of fibers and ground substance.

Interpreting Tissue Sections

In your study of histology, you may be presented with various tissue preparations mounted on microscope slides. Most such preparations are thin slices called **histological sections,** and are artificially colored to bring out detail. The best anatomical insight depends on an ability to deduce the three-dimensional structure of an organ from these two-dimensional sections. This ability, in turn, depends on an awareness of how tissues are prepared for study.

Histologists use a variety of techniques to preserve, section (slice), and stain tissues to show their structural details as clearly as possible. Tissue specimens are preserved in a **fixative**—a chemical such as formalin that prevents decay and makes the tissue more firm. After fixation, most tissues are sectioned by a machine called a *microtome,* which makes slices that are typically only one or two cells thick. This is necessary so the light of a

---

[2]*inter* = between + *stit* = to stand

TABLE 3.1	The Four Primary Tissue Classes	
Type	**Definition**	**Representative Locations**
Epithelial	Tissue composed of layers of closely spaced cells that cover organ surfaces or form glands; serves for protection, secretion, and absorption	Epidermis Inner lining of digestive tract Liver and other glands
Connective	Tissue with usually more matrix than cell volume; often specialized to support, bind, and protect organs	Tendons and ligaments Cartilage and bone Blood and lymph
Nervous	Tissue containing excitable cells specialized for rapid transmission of coded information to other cells	Brain Spinal cord Nerves
Muscular	Tissue composed of elongated, excitable cells specialized for contraction	Skeletal muscles Heart (cardiac muscle) Walls of viscera (smooth muscle)

microscope can pass through and so the image is not confused by too many superimposed layers of cells. The sections are then mounted on slides and colored with histological **stains** to enhance detail. If they were not stained, most tissues would appear pale gray. With stains that bind to different components of a tissue, however, you may see pink cytoplasm; violet nuclei; and pink, blue, green, or golden-brown protein fibers, depending on the stain used.

When viewing such sections, you must try to translate the microscopic image into a mental image of the whole structure. Like the boiled egg and elbow macaroni in figure 3.1, an object may look quite different when it is cut at various levels, or *planes of section*. A coiled tube, such as a gland of the uterus (fig. 3.1c), is often broken up into multiple portions since it meanders in and out of the plane of section. An experienced viewer, however, would recognize that the separated pieces are parts of a single tube winding its way to the organ surface. Note that a grazing slice through a boiled egg might miss the yolk (fig. 3.1a). Similarly, a grazing slice through a cell may miss the nucleus and give the false impression that the cell does not have one. In some tissue sections, you are likely to see many cells with nuclei and many others in which the nucleus did not fall in the plane of section and is therefore absent.

Many anatomical structures are significantly longer in one direction than another—the humerus and esophagus, for example. A tissue cut in the long direction is called a **longitudinal section (l.s.),** and one cut perpendicular to this is a **cross section (c.s.** or **x.s.),** or **transverse section (t.s.).** A section cut on a slant between a longitudinal and cross section is an **oblique section.** Figure 3.2 shows how certain organs look when sectioned on each of these planes.

Not all histological preparations are sections. Liquid tissues such as blood and soft tissues such as spinal cord may be prepared as **smears,** in which the tissue is rubbed or spread across the slide rather than sliced. Some membranes and cobwebby tissues like the *areolar tissue* described later in this chapter are sometimes mounted as **spreads,** in which the tissue is laid out on the slide, like placing a small square of tissue paper or a tuft of lint on a sheet of glass.

Before You Go On

Answer the following questions to test your understanding of the preceding section:

1. Define *tissue* and distinguish a tissue from a cell and an organ.

2. Classify each of the following into one of the four primary tissue classes: the skin surface, fat, the spinal cord, most heart tissue, bones, tendons, blood, and the inner lining of the stomach.

3. What are tissues composed of in addition to cells?

4. What is the term for a thin, stained slice of tissue mounted on a microscope slide?

5. Sketch what a pencil would look like in a longitudinal section, cross section, and oblique section.

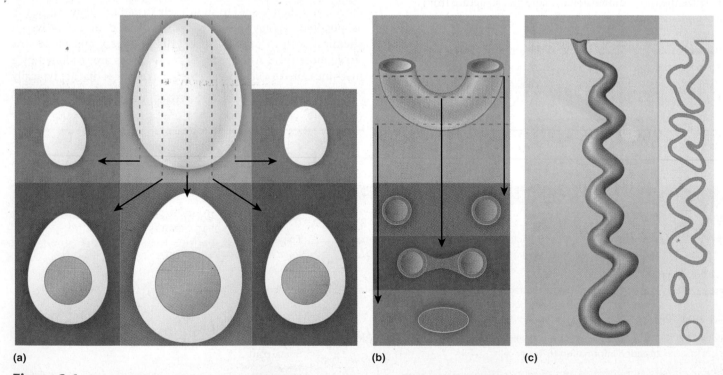

(a)   (b)   (c)

Figure 3.1 **Three-Dimensional Interpretation of Two-Dimensional Images.**   (a) A boiled egg. Note that the grazing sections (upper left and right) miss the yolk, just as a tissue section may miss a nucleus or other structure. (b) Elbow macaroni, which resembles many curved ducts and tubules. A section far from the bend would give the impression of two separate tubules; a section near the bend would show two interconnected lumina (cavities); and a section still farther down could miss the lumen completely. (c) A coiled gland in three dimensions and as it would look in a vertical tissue section.
● *Consider the microtubule in figure 2.16a (p. 42). Sketch what you think it would look like in a median longitudinal section.*

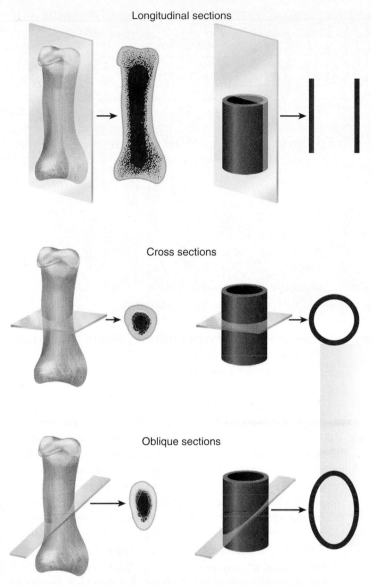

Longitudinal sections

Cross sections

Oblique sections

Figure 3.2 **Longitudinal, Cross, and Oblique Sections.** Note the effect of the plane of section on the two-dimensional appearance of elongated structures such as bones and blood vessels.

3.2 Epithelial Tissue

▶ Expected Learning Outcomes

When you have completed this section, you should be able to

- describe the properties that distinguish epithelium from other tissue classes;
- list and classify eight types of epithelium, distinguish them from each other, and state where each type can be found in the body;

- explain how the structural differences between epithelia relate to their functional differences; and
- visually recognize each epithelial type from specimens or photographs.

Epithelium[3] is a sheet of tissue composed of one or more layers of closely adhering cells, usually serving as the internal lining of a hollow organ or body cavity or the external surface of an organ. Thus, we find epithelia lining the pleural, pericardial, and peritoneal cavities; lining the inner passages of the respiratory, digestive, urinary, and reproductive tracts; and covering the outer surfaces of the stomach and intestines. Epithelia also constitute the secretory tissue and ducts of the glands; most tissue of the liver, pancreas, and kidneys is epithelium. The body's most extensive epithelium is the epidermis of the skin. Epithelia typically serve for protection, secretion of products such as mucus and digestive enzymes, excretion of wastes, and absorption of materials such as oxygen and nutrients.

The cells and extracellular material of an epithelium can be loosely compared to the bricks and mortar of a wall. The extracellular material ("mortar") of an epithelium is so thin, however, that it is barely visible with a light microscope, and the cells appear pressed very closely together. Epithelia are *avascular*[4]—there is no room between the cells for blood vessels. Other avascular tissues, such as cartilages and tendons, have low metabolic rates and are slow to heal when injured. Epithelia, however, almost always lie on a vessel-rich layer of loose connective tissue, which furnishes them with nutrients and waste removal. Epithelial cells closest to the connective tissue layer typically exhibit a high rate of mitosis. This allows epithelia to repair themselves very quickly—an ability of special importance in protective epithelia that are highly vulnerable to such injuries as skin abrasions and erosion by digestive enzymes and acid.

Between an epithelium and the underlying connective tissue is a layer called the **basement membrane.** It contains collagen, glycoproteins, and other protein–carbohydrate complexes, and gradually blends with other proteins in the underlying connective tissue. The basement membrane serves to anchor an epithelium to the connective tissue below it; it regulates the exchange of materials between the epithelium and the tissues below; and it binds growth factors from below that regulate epithelial development. The surface of an epithelial cell that faces the basement membrane is its *basal surface,* and the one that faces away from the basement membrane toward the internal cavity (lumen) of an organ is the *apical surface.*

Epithelia are classified into two broad categories, simple and stratified, with four types in each category:

Simple epithelia
- Simple squamous epithelium
- Simple cuboidal epithelium
- Simple columnar epithelium
- Pseudostratified columnar epithelium

---

[3]*epi* = upon + *thel* = nipple, female
[4]*a* = without + *vas* = vessels

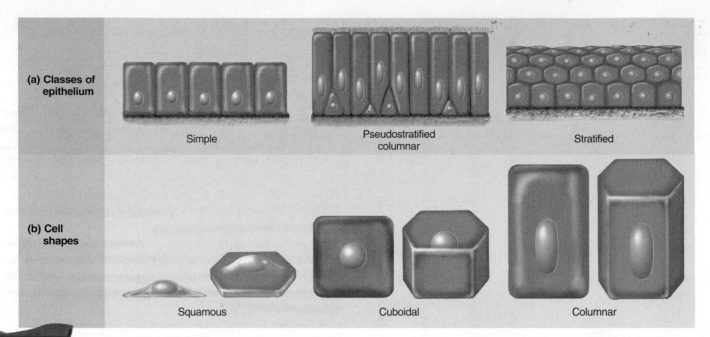

Figure 3.3 Cell Shapes and Epithelial Types. Pseudostratified columnar epithelium is a special type of simple epithelium that gives a false appearance of multiple cell layers.

Stratified epithelia

- Stratified squamous epithelium
- Stratified cuboidal epithelium
- Stratified columnar epithelium
- Transitional epithelium

In a *simple epithelium,* every cell rests on the basement membrane, whereas in a *stratified epithelium,* some cells rest on top of others and do not touch it (fig. 3.3a). The *pseudostratified columnar* type is included among simple epithelia, although it presents a false appearance of stratification, for reasons explained shortly. Individual characteristics of all eight epithelia follow (tables 3.2 and 3.3).

Simple Epithelia

Generally, a simple epithelium has only one layer of cells, although this is a somewhat debatable point in the pseudostratified columnar type. Three types of simple epithelia are named for their cell shapes: simple squamous (thin scaly cells), simple cuboidal (squarish or round cells), and simple columnar (tall narrow cells). In the fourth type, pseudostratified columnar epithelium, not all cells reach the free surface; the taller cells cover the shorter ones. This epithelium looks stratified in most tissue sections, but careful examination, especially with the electron microscope, shows that every cell reaches the basement membrane—like trees in a forest, where some grow taller than others but all are anchored in the soil below.

Simple columnar and pseudostratified columnar epithelia often have wineglass-shaped goblet cells that produce protective mucus over the mucous membranes. These cells have an expanded apical end filled with secretory vesicles; their product becomes mucus when it is

secreted and absorbs water. The basal part of the cell is a narrow stem, like that of a wine glass, that reaches to the basement membrane.

Table 3.2 illustrates and summarizes the structural and functional differences among the four simple epithelia. In this and the subsequent tables, each tissue is represented by a photograph and a corresponding line drawing with labels. The drawings help to clarify cell boundaries and other relevant features that may otherwise be difficult to see or identify in photographs or through the microscope. Each figure indicates the approximate magnification at which the original photograph was made. Each is enlarged much more than this when printed in the book, but selecting the closest magnifications on a microscope should enable you to see a comparable level of detail (resolution).

Stratified Epithelia

Stratified epithelia range from 2 to 20 or more layers of cells, with some cells resting directly on others and only the deepest layer resting on the basement membrane. Three of the stratified epithelia are named for the shapes of their surface cells: stratified squamous, stratified cuboidal, and stratified columnar. The deeper cells, however, may be of a different shape than the surface cells. The fourth type, transitional epithelium, was named when it was thought to represent a transitional stage between stratified squamous and stratified columnar epithelium. This is now known to be untrue, but the name has persisted.

Stratified columnar epithelium is rare and of relatively minor importance—seen only in short stretches where two other epithelial types meet, as in limited regions of the pharynx, larynx, anal canal, and male urethra. We will not consider this type any further. The other three types are illustrated and summarized in table 3.3.

TABLE 3.2   Simple Epithelia

Simple Squamous Epithelium

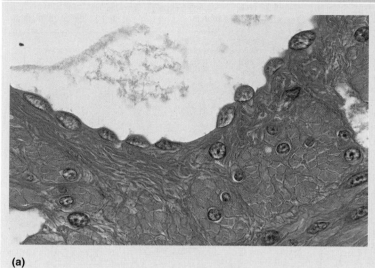

(a)

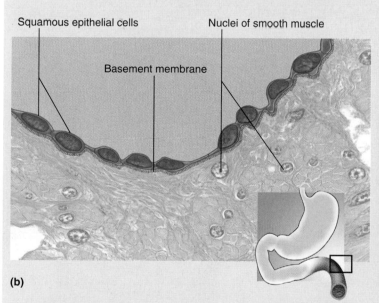

Squamous epithelial cells   Nuclei of smooth muscle

Basement membrane

(b)

Figure 3.4 Simple Squamous Epithelium in the Serosa of the Small Intestine (×400).

Microscopic appearance: Single layer of thin cells, shaped like fried eggs with bulge where nucleus is located; nucleus flattened in the plane of the cell, like an egg yolk; cytoplasm may be so thin it is hard to see in tissue sections; in surface view, cells have angular contours and nuclei appear round

Representative locations: Air sacs (alveoli) of lungs; glomerular capsules of kidneys; some kidney tubules; inner lining (endothelium) of heart and blood vessels; serous membranes of stomach, intestines, and some other viscera; surface mesothelium of pleurae, pericardium, peritoneum, and mesenteries

Functions: Allows rapid diffusion or transport of substances through membrane; secretes lubricating serous fluid

Simple Cuboidal Epithelium

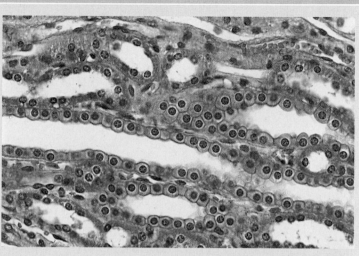

(a)

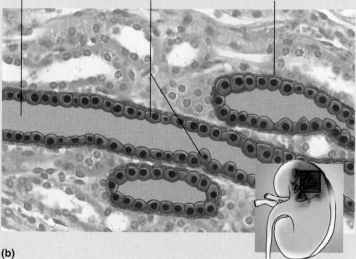

Lumen of kidney tubule   Cuboidal epithelial cells   Basement membrane

(b)

Figure 3.5 Simple Cuboidal Epithelium in Kidney Tubules (×400).

Microscopic appearance: Single layer of square or round cells; in glands, cells often pyramidal and arranged like segments of an orange around a central space; spherical, centrally placed nuclei; often with a brush border of microvilli in some kidney tubules; ciliated in bronchioles of lung

Representative locations: Liver; thyroid, mammary, salivary, and other glands; most kidney tubules; bronchioles

Functions: Absorption and secretion; production and movement of respiratory mucus

TABLE 3.2	Simple Epithelia (continued)

Simple Columnar Epithelium

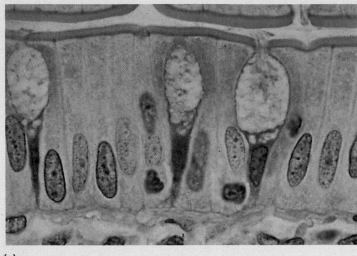

(a)

Brush border (microvilli) Connective tissue Basement membrane Nuclei Goblet cell Columnar cells

(b)

Figure 3.6 Simple Columnar Epithelium in the Mucosa of the Small Intestine (×400).

Microscopic appearance: Single layer of tall, narrow cells; oval or sausage-shaped nuclei, vertically oriented, usually in basal half of cell; apical portion of cell often shows secretory vesicles visible with the transmission electron microscope (TEM); often shows a brush border of microvilli; ciliated in some organs; may possess goblet cells
Representative locations: Inner lining of stomach, intestines, gallbladder, uterus, and uterine tubes; some kidney tubules
Functions: Absorption; secretion of mucus and other products; movement of egg and embryo in uterine tube

Pseudostratified Columnar Epithelium

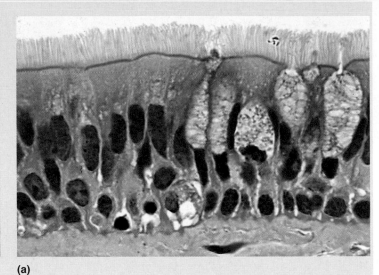

(a)

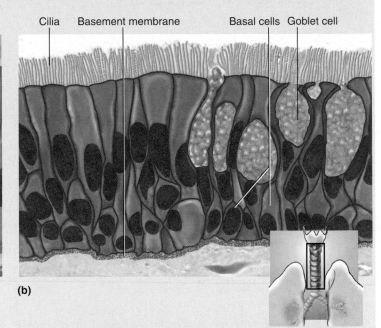

Cilia   Basement membrane   Basal cells   Goblet cell

(b)

Figure 3.7 Ciliated Pseudostratified Columnar Epithelium in the Mucosa of the Trachea (×400).

Microscopic appearance: Looks multilayered; some cells do not reach free surface but all cells reach basement membrane; nuclei at several levels in deeper half of epithelium; often with goblet cells; often ciliated
Representative locations: Respiratory tract from nasal cavity to bronchi; portions of male urethra
Functions: Secretes and propels mucus

TABLE 3.3   Stratified Epithelia

Stratified Squamous Epithelium—Keratinized	Stratified Squamous Epithelium—Nonkeratinized

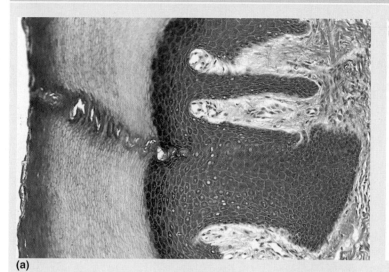

(a)

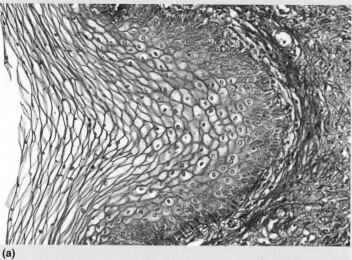

(a)

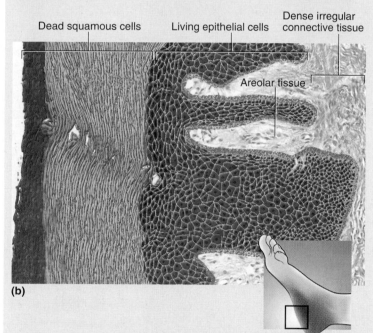

Dead squamous cells   Living epithelial cells   Dense irregular connective tissue

Areolar tissue

(b)

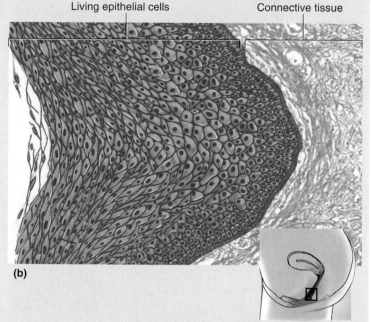

Living epithelial cells   Connective tissue

(b)

Figure 3.8 Keratinized Stratified Squamous Epithelium in the Sole of the Foot (×400).

Microscopic appearance: Multiple cell layers with cells becoming increasingly flat and scaly toward surface; surface covered with a layer of compact dead cells without nuclei; basal cells may be cuboidal to columnar

Representative locations: Epidermis; palms and soles are especially heavily keratinized

Functions: Resists abrasion; retards water loss through skin; resists penetration by pathogenic organisms

Figure 3.9 Nonkeratinized Stratified Squamous Epithelium in the Mucosa of the Vagina (×400).

Microscopic appearance: Same as keratinized epithelium but without the surface layer of dead cells

Representative locations: Tongue, esophagus, anal canal, vagina

Functions: Resists abrasion and penetration by pathogenic organisms

TABLE 3.3	Stratified Epithelia (continued)
Stratified Cuboidal Epithelium	**Transitional Epithelium**

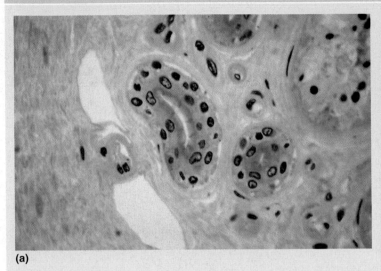

(a)

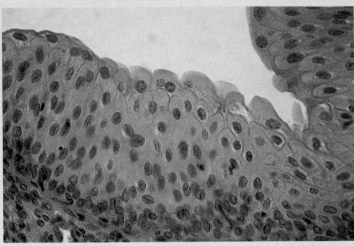

(a)

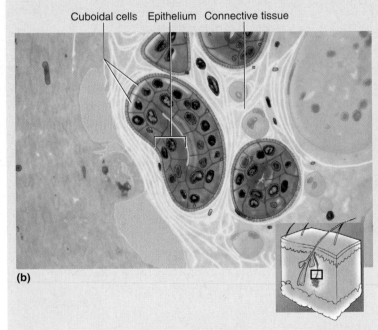

Cuboidal cells   Epithelium   Connective tissue

(b)

Basement membrane   Connective tissue   Binucleate epithelial cell

(b)

Figure 3.10 Stratified Cuboidal Epithelium in the Duct of a Sweat Gland (×400).

Microscopic appearance: Two or more layers of cells; surface cells roughly square or round

Representative locations: Sweat gland ducts; egg-producing vesicles (follicles) of ovaries; sperm-producing ducts (seminiferous tubules) of testis

Functions: Contributes to sweat secretion; secretes ovarian hormones; produces sperm

Figure 3.11 Transitional Epithelium in the Kidney (×400).

Microscopic appearance: Somewhat resembles stratified squamous epithelium, but surface cells are rounded, not flattened, and often bulge above surface; typically five or six cells thick when relaxed, two or three cells thick when stretched; cells may be flatter and thinner when epithelium is stretched (as in a distended bladder); some cells have two nuclei

Representative locations: Urinary tract—part of kidney, ureter, bladder, part of urethra; allantoic duct and external surface of umbilical cord

Function: Stretches to allow filling of urinary tract

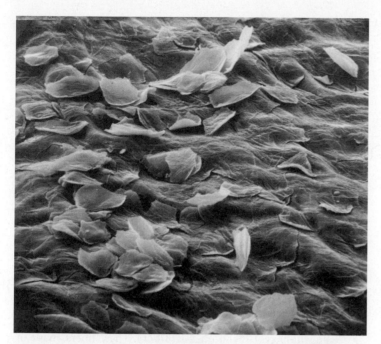

Figure 3.12 **Exfoliation of Squamous Cells from the Vaginal Mucosa (SEM).** A Pap smear is prepared from similar loose cells scraped from the epithelial surface of the cervix. [From R. G. Kessel and R. H. Kardon, *Tissues and Organs: A Text-Atlas of Scanning Electron Microscopy* (W. H. Freeman, 1979)]

The most widespread epithelium in the body is stratified squamous, which warrants further discussion. In the deepest layer, the cells are cuboidal to columnar and undergo continual mitosis. Their daughter cells push toward the surface and become flatter (more squamous) as they migrate farther upward, until they die and flake off. Their separation from the surface is called **exfoliation (desquamation)** (fig. 3.12); the study of exfoliated cells is called *exfoliate cytology.* You can easily study exfoliated cells by scraping your gums with a toothpick, smearing this material on a slide, and staining it. A similar procedure is used in the *Pap smear,* an examination of exfoliated cells from the cervix for signs of uterine cancer (see fig. 26.26, p. 735).

Stratified squamous epithelia are of two kinds—keratinized and nonkeratinized. A **keratinized (cornified)** epithelium, found on the skin surface (epidermis), is covered with a layer of compact, dead squamous cells (see fig. 3.8). These cells are packed with the durable protein **keratin** and coated with a water-repellent glycolipid. The skin surface is therefore relatively dry; it retards water loss from the body; and it resists penetration by disease organisms. (Keratin is also the protein of which animal horns are made, hence its name.[5]) The tongue, esophagus, vagina, and a few other internal surfaces are covered with the nonkeratinized type, which lacks the layer of dead cells (see fig. 3.9). This type provides a surface that is, again, abrasion-resistant, but also moist and slippery. These characteristics are well suited to resist stress produced by the chewing and swallowing of food and by sexual intercourse and childbirth.

[5]*kerat* = horn

Before You Go On

Answer the following questions to test your understanding of the preceding section:

6. Distinguish between simple and stratified epithelia, and explain why pseudostratified columnar epithelium belongs in the former category.

7. Explain how to distinguish a stratified squamous epithelium from a transitional epithelium.

8. What function do keratinized and nonkeratinized stratified squamous epithelia have in common? What is the structural difference between these two? How is this structural difference related to a functional difference between them?

9. How do the epithelia of the esophagus and stomach differ? How does this relate to their respective functions?

10. Explain why epithelial tissue is able to repair itself quickly even though it is avascular.

3.3 Connective Tissue

▌ **Expected Learning Outcomes**

When you have completed this section, you should be able to

- describe the properties that most connective tissues have in common;

- discuss the types of cells found in connective tissue;

- explain what the matrix of a connective tissue is and describe its components;

- name 10 types of connective tissue, describe their cellular components and matrix, and explain what distinguishes them from each other; and

- visually recognize each connective tissue type from specimens or photographs.

Overview

Connective tissue is a type of tissue in which cells usually occupy less space than the extracellular material, and that serves in most cases to support and protect organs or to bind organs to each other (for example, the way a tendon connects muscle to bone) or to support and protect organs. Most cells of a connective tissue are not in direct contact with each other, but are well separated by extracellular material. Most kinds of connective tissue are highly vascular—richly supplied with blood vessels. Connective tissue is the most abundant, widely distributed, and histologically variable of the primary tissues. Mature connective tissues fall into four broad categories: *fibrous connective tissue, adipose tissue, supportive connective tissue* (cartilage and bone), and *fluid connective tissue* (blood).

The functions of connective tissue include the following:

- **Binding of organs.** Tendons bind muscle to bone, ligaments bind one bone to another, fat holds the kidneys and eyes in place, and fibrous tissue binds the skin to underlying muscle.
- **Support.** Bones support the body, and cartilage supports the ears, nose, trachea, and bronchi.
- **Physical protection.** The cranium, ribs, and sternum protect delicate organs such as the brain, lungs, and heart; fatty cushions around the kidneys and eyes protect these organs.
- **Immune protection.** Connective tissue cells attack foreign invaders, and connective tissue fiber forms a "battlefield" under the skin and mucous membranes where immune cells can be quickly mobilized against disease agents.
- **Movement.** Bones provide the lever system for body movement, cartilages are involved in movement of the vocal cords, and cartilages on bone surfaces ease joint movements.
- **Storage.** Fat is the body's major energy reserve; bone is a reservoir of calcium and phosphorus that can be drawn upon when needed.
- **Heat production.** Metabolism of brown fat generates heat in infants and children.
- **Transport.** Blood transports gases, nutrients, wastes, hormones, and blood cells.

Fibrous Connective Tissue

Fibrous connective tissue is the most diverse type of connective tissue. It is also called *fibroconnective tissue* or *connective tissue proper*. Nearly all connective tissues contain fibers, but the tissues considered here are classified together because the fibers are so conspicuous. The tissue, of course, also includes cells and ground substance. Before examining specific types of fibrous connective tissue, we will examine these components.

Components of Fibrous Connective Tissue

Cells   The cells of fibrous connective tissue include the following types:

- **Fibroblasts.**[6] These are large, fusiform cells that often show slender, wispy branches. They produce the fibers and ground substance that form the matrix of the tissue.
- **Macrophages.**[7] These are large phagocytic cells that wander through the connective tissues. They phagocytize and destroy bacteria, other foreign matter, and dead or dying cells of our own body. They also activate the immune system when they sense foreign matter called *antigens.* They arise from certain white blood cells called *monocytes* or from the stem cells that produce monocytes.
- **Leukocytes,**[8] or **white blood cells (WBCs).** WBCs travel briefly in the bloodstream, then crawl out through the capillary

walls and spend most of their time in the connective tissues. The two most common types are *neutrophils,* which wander about in search of bacteria, and *lymphocytes,* which react against bacteria, toxins, and other foreign agents. Lymphocytes often form dense patches in the mucous membranes.

- **Plasma cells.** Certain lymphocytes turn into plasma cells when they detect foreign agents. The plasma cells then synthesize disease-fighting proteins called *antibodies.* Plasma cells are rarely seen except in inflamed tissue and the walls of the intestines.
- **Mast cells.** These cells, found especially alongside blood vessels, secrete a chemical called *heparin* that inhibits blood clotting, and one called *histamine* that increases blood flow by dilating blood vessels.
- **Adipocytes** (AD-ih-po-sites), or **fat cells.** These appear in small clusters in some fibrous connective tissues. When they dominate an area, the tissue is called *adipose tissue.*

Fibers   Three types of protein fibers are found in fibrous connective tissues:

- **Collagenous** (col-LADJ-eh-nus) **fibers.** These fibers, made of collagen, are tough and flexible and resist stretching. Collagen is the body's most abundant protein, constituting about 25% of the total. It is the base of such animal products as gelatin, leather, and glue.[9] In fresh tissue, collagenous fibers have a glistening white appearance, as seen in tendons and some cuts of meat (fig. 3.13); thus, they are often called *white fibers.* In tissue sections, collagen forms coarse, wavy bundles, often dyed pink, blue, or green by the most common histological stains. Tendons, ligaments, and the deep layer of the skin (the dermis) are made mainly of collagen. Less visibly, collagen pervades the matrix of cartilage and bone.
- **Reticular**[10] **fibers.** These are thin collagen fibers coated with glycoprotein. They form a spongelike framework for such organs as the spleen and lymph nodes.
- **Elastic fibers.** These are thinner than collagenous fibers, and they branch and rejoin each other along their course. They are made of a protein called *elastin,* whose coiled structure allows it to stretch and recoil like a rubber band. Elastic fibers account for the ability of the skin, lungs, and arteries to spring back after they are stretched. (Elasticity is not the ability to stretch, but the tendency to recoil when tension is released.) Fresh elastic fibers are yellowish and therefore often called *yellow fibers.*

Ground Substance   Amid the cells and fibers in some connective tissue sections, there appears to be a lot of empty space. In life, this space is occupied by the featureless ground substance. Ground substance usually has a gelatinous consistency, resulting from three classes of large molecules composed of protein and carbohydrate: *glycosaminoglycans (GAGs), proteoglycans,* and *adhesive glycoproteins.* Some of these molecules are up to 20 μm long—larger than some cells. The ground substance absorbs compressive forces and, like the

---

[6]*fibro* = fiber + *blast* = producing
[7]*macro* = big + *phage* = eater
[8]*leuko* = white + *cyte* = cell

[9]*colla* = glue + *gen* = producing
[10]*ret* = network + *icul* = little

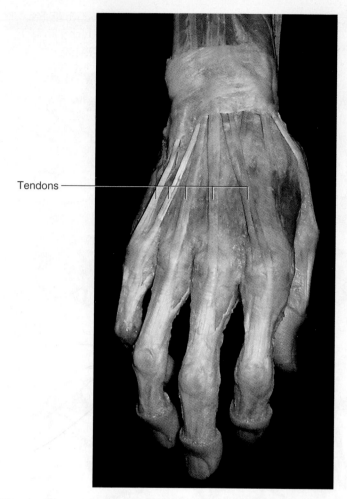

Tendons

Figure 3.13 Tendons of the Hand. The white glistening appearance results from the collagen of which tendons are composed. The braceletlike band across the wrist is also composed of collagen.
● *Which category of connective tissue composes these tendons?*

styrofoam packing in a shipping carton, it protects the more delicate cells from mechanical injury. GAGs also form a very slippery lubricant in the joints and constitute much of the jellylike *vitreous body* of the eyeball. In connective tissue, such molecules form a gel that slows down the spread of bacteria and other pathogens (disease-causing agents). Adhesive glycoproteins bind plasma membrane proteins to collagen and proteoglycans outside the cell. They bind all the components of a tissue together and mark pathways that guide migrating embryonic cells to their destinations in a tissue.

Types of Fibrous Connective Tissue

Fibrous connective tissue is divided into two broad categories according to the relative abundance of fiber: *loose* and *dense connective tissue.* In loose connective tissue, much of the space is occupied by ground substance, which is dissolved out of the tissue during histological fixation and leaves empty space in prepared tissue sections. The loose connective tissues we will discuss are *areolar tissue* and *reticular tissue* (table 3.4). In dense connective tissue,

DEEPER INSIGHT   3.1

Collagen Diseases

The gene for collagen is especially subject to mutation, and there are consequently several diseases that stem from hereditary defects in collagen synthesis. Since collagen is such a widespread protein in the body, the effects are very diverse. People with *Ehlers–Danlos*[11] syndrome have abnormally long, loose collagen fibers, which show their effects in unusually stretchy skin; loose joints; slow wound healing; and abnormalities in the blood vessels, intestines, and urinary bladder. Infants with this syndrome are often born with dislocated hips. *Osteogenesis imperfecta* is a hereditary collagen disease that affects bone development (see Deeper Insight 3.3).

Not all collagen diseases are hereditary, however. *Scurvy,* for example, results from a dietary deficiency of vitamin C (ascorbic acid). Ascorbic acid is a cofactor needed for the metabolism of proline and lysine, two amino acids that are especially abundant in collagen. The signs of scurvy include bleeding gums, loose teeth, subcutaneous and intramuscular hemorrhages, and poor wound healing.

fiber occupies more space than the cells and ground substance, and appears closely packed in tissue sections. The two dense connective tissues we will discuss are *dense regular* and *dense irregular connective tissue* (table 3.5)

Areolar[12] (AIR-ee-OH-lur) **tissue** exhibits loosely organized fibers, abundant blood vessels, and a lot of seemingly empty space. It possesses all six of the aforementioned cell types. Its fibers run in random directions and are mostly collagenous, but elastic and reticular fibers are also present. Areolar tissue is highly variable in appearance. In many serous membranes, it looks like figure 3.14, but in the skin and mucous membranes, it is more compact (see fig. 3.8b) and is sometimes difficult to distinguish from dense irregular connective tissue. Some advice on how to tell them apart is given after the discussion of dense irregular connective tissue (p. 66).

Areolar tissue is found in histological sections from almost every part of the body. It surrounds blood vessels and nerves and penetrates with them even into the small spaces of muscles, tendons, and other organs. Nearly every epithelium rests on a layer of areolar tissue, whose blood vessels provide the epithelium with nutrition, waste removal, and a ready supply of infection-fighting leukocytes in times of need. Because of the abundance of open, fluid-filled space, leukocytes can move about freely in areolar tissue and can easily find and destroy pathogens.

Reticular tissue (fig. 3.15) is a mesh of reticular fibers and fibroblasts. It forms the structural framework (stroma) of such organs as the lymph nodes, spleen, thymus, and bone marrow. The space amid the fibers is filled with blood cells. Imagine a kitchen sponge soaked with blood; the sponge fibers are analogous to the reticular tissue stroma.

[11]Edward L. Ehlers (1863–1937), Danish dermatologist; Henri A. Danlos (1844–1912), French dermatologist
[12]*areola* = little space

TABLE 3.4	Loose Connective Tissues

Areolar Tissue	Reticular Tissue Adipose Tissue

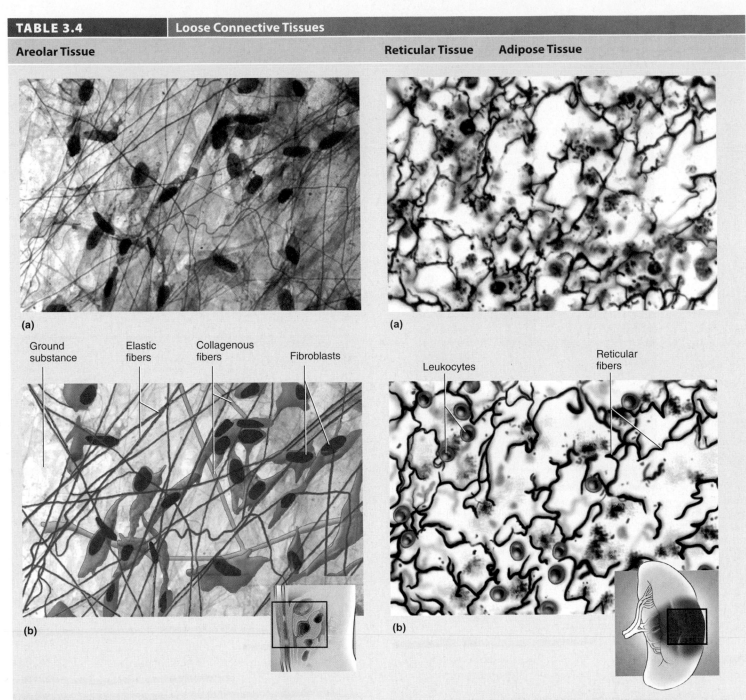

Figure 3.14 Areolar Tissue in a Spread of the Mesentery (×400).

Figure 3.15 Reticular Tissue in the Spleen (×400).

Microscopic appearance: Loose arrangement of collagenous and elastic fibers; scattered cells of various types; abundant ground substance; numerous blood vessels

Representative locations: Underlying nearly all epithelia; surrounding blood vessels, nerves, esophagus, and trachea; fasciae between muscles; mesenteries; visceral layers of pericardium and pleura

Functions: Loosely binds epithelia to deeper tissues; allows passage of nerves and blood vessels through other tissues; provides an arena for immune defense; provides nutrients and waste removal for overlying epithelia

Microscopic appearance: Loose network of reticular fibers and cells, infiltrated with numerous lymphocytes and other blood cells

Representative locations: Lymph nodes, spleen, thymus, bone marrow

Functions: Supportive stroma (framework) for lymphatic organs

TABLE 3.5	Dense Connective Tissues
Dense Regular Connective Tissue	**Dense Irregular Connective Tissue**

(a)

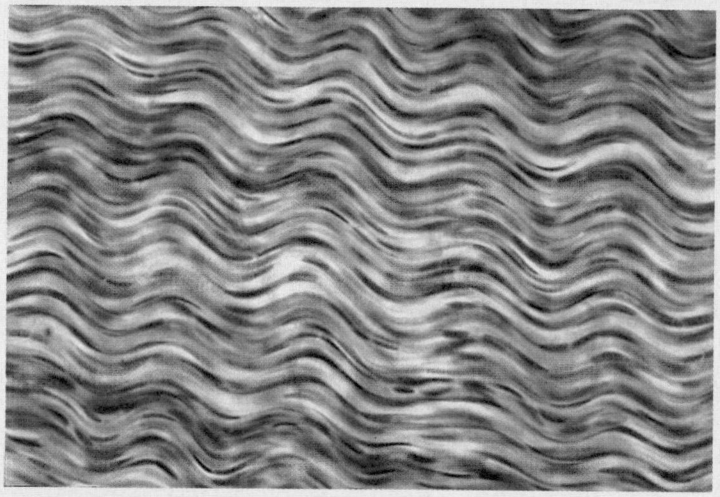

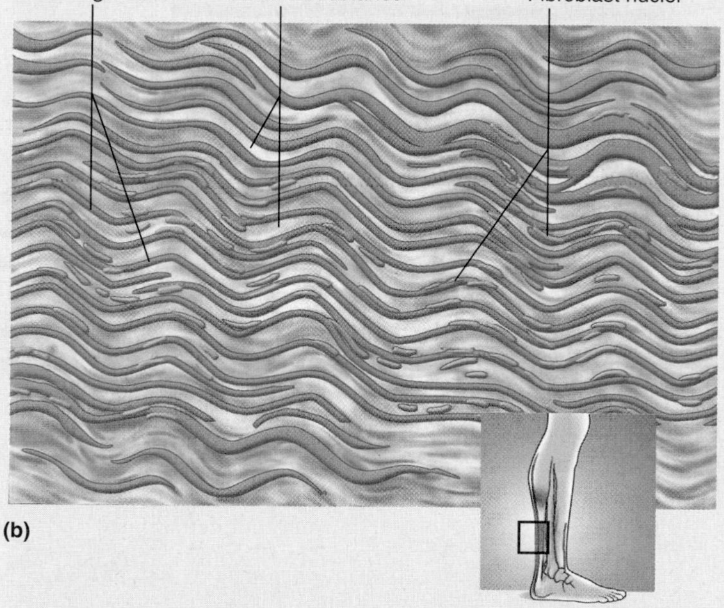

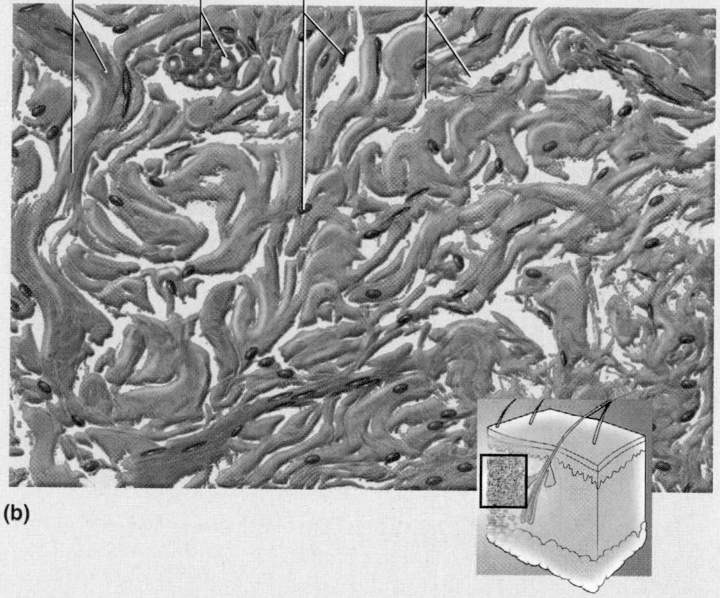

(a)

Collagen fibers   Ground substance   Fibroblast nuclei

(b)

Bundles of collagen fibers   Gland ducts   Fibroblast nuclei   Ground substance

(b)

Figure 3.16 Dense Regular Connective Tissue in a Tendon (×400).

Microscopic appearance: Densely packed, parallel, often wavy collagen fibers; slender fibroblast nuclei compressed between collagen bundles; scanty open space (ground substance); scarcity of blood vessels

Representative locations: Tendons and ligaments

Functions: Ligaments tightly bind bones together and resist stress; tendons attach muscle to bone and move the bones when the muscles contract

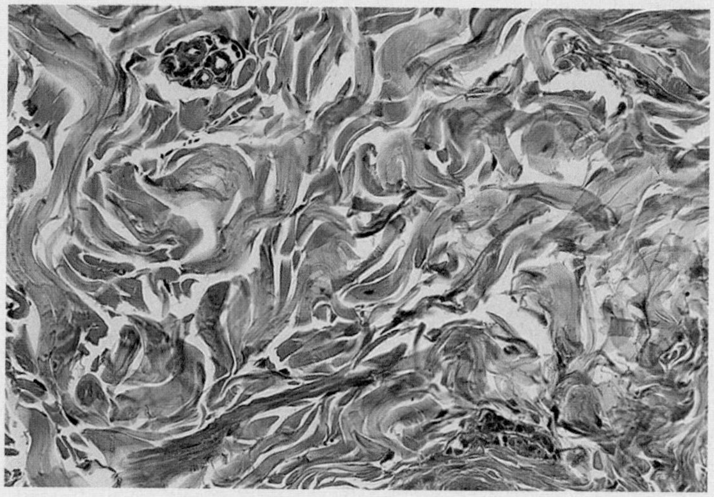

Figure 3.17 Dense Irregular Connective Tissue in the Dermis of the Skin (×400).

Microscopic appearance: Densely packed collagen fibers running in random directions; scanty open space (ground substance); few visible cells; scarcity of blood vessels

Representative locations: Deeper portion of dermis of skin; capsules around viscera such as liver, kidney, spleen; fibrous sheaths around muscles, nerves, cartilages, and bones

Functions: Durable, hard to tear; withstands stresses applied in unpredictable directions

Dense regular connective tissue is named for two properties: (1) the collagen fibers are closely packed and leave relatively little open space, and (2) the fibers are parallel to each other. It is found especially in tendons and ligaments. The parallel arrangement of fibers is an adaptation to the fact that tendons and ligaments are pulled in predictable directions. With some minor exceptions such as blood vessels and sensory nerve fibers, the only cells in this tissue are fibroblasts, visible by their slender, violet-staining nuclei squeezed between bundles of collagen. This type of tissue has few blood vessels and receives a meager supply of oxygen and nutrients, so injured tendons and ligaments are slow to heal.

The vocal cords, suspensory ligament of the penis, and some ligaments of the vertebral column are made of a type of dense regular connective tissue called **elastic tissue**. In addition to the densely packed collagen fibers, it exhibits branching elastic fibers and more fibroblasts. The fibroblasts have larger, more conspicuous nuclei than seen in most dense regular connective tissue.

Elastic tissue takes the form of wavy sheets in the walls of the large and medium arteries. When the heart pumps blood into the arteries, these sheets enable them to expand and relieve some of the pressure on smaller vessels downstream. When the heart relaxes, the arterial wall springs back and keeps the blood pressure from dropping too low between heartbeats. The importance of this elastic tissue becomes especially clear in diseases such as atherosclerosis, in which the tissue is stiffened by lipid and calcium deposits, and Marfan syndrome, a genetic defect in elastin synthesis (see Deeper Insight 3.2).

Dense irregular connective tissue also has thick bundles of collagen and relatively little room for cells and ground substance, but the collagen bundles run in seemingly random directions. This arrangement enables the tissue to resist unpredictable stresses. Dense irregular connective tissue constitutes most of the dermis,

DEEPER INSIGHT   3.2

The Consequences of Defective Elastin

Marfan[13] syndrome is a hereditary defect in elastin fibers, usually resulting from a mutation in the gene for fibrillin, a glycoprotein that forms the structural scaffold for elastin. Clinical signs of Marfan syndrome include hyperextensible joints, hernias of the groin, and vision problems resulting from abnormally elongated eyes and deformed lenses. People with Marfan syndrome typically show unusually tall stature, long limbs, spidery fingers, abnormal spinal curvature, and a protruding "pigeon breast." More serious problems are weakened heart valves and arterial walls. The aorta, where blood pressure is highest, is sometimes enormously dilated close to the heart and may rupture. Marfan syndrome is present in about 1 out of 20,000 live births, and most victims die by their mid-30s. Some authorities speculate that Abraham Lincoln's tall, gangly physique and spindly fingers were signs of Marfan syndrome, which may have ended his life prematurely had he not been assassinated. A number of star athletes have died at a young age of Marfan syndrome, including Olympic volleyball champion Flo Hyman, who died of a ruptured aorta during a game in Japan in 1986, at the age of 31.

[13]Antoine Bernard-Jean Marfan (1858–1942), French physician

where it binds the skin to the underlying muscle and connective tissue. It forms a protective capsule around organs such as the kidneys, testes, and spleen and a tough fibrous sheet around the bones, nerves, and most cartilages.

It is sometimes difficult to judge whether a tissue is areolar or dense irregular. In the dermis, for example, these tissues occur side by side, and the transition from one to the other is not at all obvious (see fig. 3.8b). A relatively large amount of clear space suggests areolar tissue, and thicker bundles of collagen with relatively little clear space suggest dense irregular tissue.

Adipose Tissue

Adipose tissue, or **fat,** is tissue in which adipocytes are the dominant cell type (table 3.6). Adipocytes may also occur singly or in small clusters in areolar tissue. The space between adipocytes is occupied by areolar tissue, reticular tissue, and blood capillaries. Adipocytes usually range from 70 to 120 μm in diameter, but they may be five times as large in obese people. Increases in the amount of body fat result from the enlargement of existing fat cells, not from increases in the number of cells. Mature adipocytes cannot undergo mitosis.

Fat is the body's primary energy reservoir. The quantity of stored fat and the number of adipocytes are quite stable in a person, but this does not mean stored fat is stagnant. New fat molecules are constantly synthesized and stored as others are broken down and released into circulation. Thus, there is a constant turnover of stored fat, with an equilibrium between synthesis and breakdown, energy storage and energy use. Adipose tissue also provides thermal insulation, anchors and cushions such organs as the eyeballs and kidneys, and contributes to sex-specific body contours such as the female breasts and hips. On average, women have more fat relative to body weight than men do. Female body fat helps to meet the caloric needs of pregnancy and nursing an infant, and having too little fat can reduce female fertility.

Most adipose tissue is of a type called *white (or yellow) fat,* but fetuses, infants, and children also have a heat-generating tissue called *brown fat,* which accounts for up to 6% of an infant's weight. Brown fat gets its color from an unusual abundance of blood vessels and certain enzymes in its mitochondria. It stores fat in the form of multiple droplets rather than one large one. Brown fat has numerous mitochondria, but their metabolism is not linked to ATP synthesis. Therefore, when these cells oxidize fats, they release all of the energy as heat. Hibernating animals accumulate brown fat in preparation for winter.

Apply What You Know

Why would infants and children have more need for brown fat than adults do? (*Hint:* Smaller bodies have a higher ratio of surface area to volume than larger bodies do.)

Cartilage

Cartilage (table 3.7) is a relatively stiff connective tissue with a flexible rubbery matrix; you can feel its texture by folding and releasing the external ear or palpating the tip of the nose or the "Adam's apple"

TABLE 3.6	**Adipose Tissue**

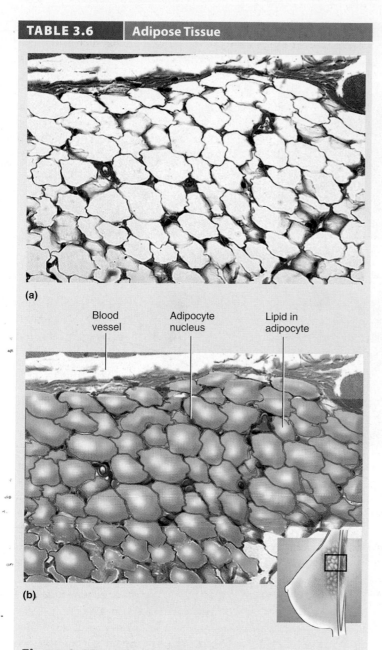

(a)

Blood vessel   Adipocyte nucleus   Lipid in adipocyte

(b)

Figure 3.18 Adipose Tissue in the Breast (×100).

Microscopic appearance: Dominated by adipocytes—large, empty-looking cells with thin margins and nucleus pressed against inside of plasma membrane; cells usually shriveled by histological fixative; tissue sections often pale because of scarcity of stained cytoplasm; blood vessels often visible

Representative locations: Subcutaneous fat beneath skin; breast; heart surface; mesenteries; surrounding such organs as kidneys and eyes

Functions: Energy storage; thermal insulation; heat production by brown fat; protective cushion for some organs; filling space, shaping body

Cartilage is produced by stem cells called **chondroblasts**[14] (CON-dro-blasts), which secrete the matrix and surround themselves with it until they become trapped in little cavities called **lacunae**[15] (la-CUE-nee). Once enclosed in lacunae, they are called **chondrocytes** (CON-dro-sites). Cartilage only rarely exhibits blood vessels, and even when it does, these vessels are just passing through without giving off capillaries to nourish the tissue. Therefore, nutrition and waste removal depend on solute diffusion through the stiff matrix. Because this is a slow process, chondrocytes have low rates of metabolism and cell division, and injured cartilage heals slowly.

The matrix is rich in glycosaminoglycans and contains collagen fibers that range from invisibly fine to conspicuously coarse. Differences in the fibers provide a basis for classifying cartilage into three types: *hyaline cartilage, elastic cartilage,* and *fibrocartilage.*

> ### Apply What You Know
>
> When the following tissues are injured, which do you think is the fastest to heal, and which do you think is the slowest—cartilages, adipose tissue, or tendons? Explain your reasoning.

Hyaline[16] (HY-uh-lin) **cartilage** is named for its clear, glassy, microscopic appearance, which stems from the usually invisible fineness of its collagen fibers. **Elastic cartilage** is named for its conspicuous elastic fibers, and **fibrocartilage** for its coarse, readily visible bundles of collagen. Elastic cartilage and most hyaline cartilage are surrounded by a sheath of dense irregular connective tissue called the **perichondrium**[17] (PAIR-ih-CON-dree-um). A reserve population of chondroblasts between the perichondrium and cartilage contributes to cartilage growth throughout life. Perichondrium is lacking from fibrocartilage and from some hyaline cartilage, such as the cartilage caps at the ends of the long bones.

Bone

Bone, or **osseous tissue,** is a hard, calcified connective tissue that composes the skeleton. The term *bone* has two meanings in anatomy—an entire organ such as the femur or mandible, or just the osseous tissue. The bones of the skeleton are composed of not only osseous tissue but also cartilage, bone marrow, dense irregular connective tissue, and other tissue types. Bones support the body as a whole and give protective enclosure for the brain, spinal cord, heart, lungs, and pelvic organs.

There are two forms of osseous tissue: (1) **Spongy bone** fills the heads of the long bones (see fig. 6.5a, p. 135) and forms the middle layer of flat bones such as the sternum and cranial bones. Although it is calcified and hard, its delicate slivers and plates give it a spongy appearance. (2) **Compact (dense) bone** is a denser calcified tissue with no spaces visible to the naked eye. It forms the external surfaces of all bones, so spongy bone, when present, is always covered by a shell of compact bone.

(*thyroid cartilage* of the larynx). It is also easily seen in many grocery items—it is the milky-colored gristle at the end of pork ribs and on chicken leg and breast bones, for example. Among other functions, cartilages shape and support the nose and ears and partially enclose the larynx (voice box), trachea (windpipe), and thoracic cavity.

---

[14]*chondro* = cartilage, gristle + *blast* = forming
[15]*lacuna* = lake, cavity
[16]*hyal* = glass
[17]*peri* = around + *chondri* = cartilage

TABLE 3.7 | Types of Cartilage

Hyaline Cartilage	Elastic Cartilage	Fibrocartilage

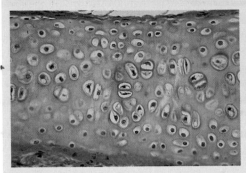

(a)

Matrix Cell nest Perichondrium Lacunae Chondrocytes

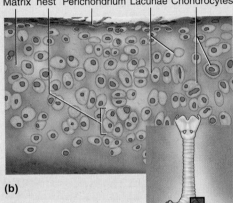

(b)

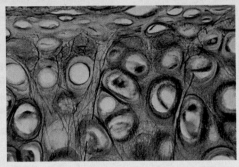

(a)

Perichondrium Elastic fibers Lacunae Chondrocytes

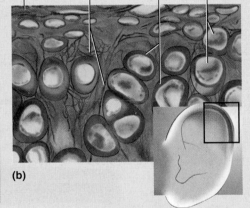

(b)

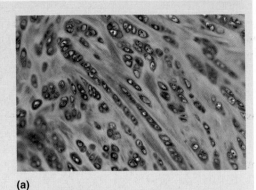

(a)

Collagen fibers Chondrocytes

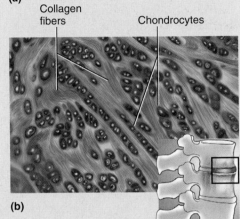

(b)

Figure 3.19 Hyaline Cartilage in the Bronchus (×400).

Microscopic appearance: Clear, glassy matrix, often stained light blue or pink in tissue sections; fine, dispersed collagen fibers, not usually visible; chondrocytes often in small clusters of three or four cells (*cell nests*), enclosed in lacunae; usually covered by perichondrium

Representative locations: Forms a thin *articular cartilage,* lacking perichondrium, over the ends of bones at movable joints; forms supportive rings and plates around trachea and bronchi; forms a boxlike enclosure around the larynx; a *costal cartilage* attaches the end of a rib to the breastbone; forms much of the fetal skeleton

Functions: Eases joint movements; holds airway open during respiration; moves vocal cords during speech; a precursor of bone in the fetal skeleton and forms the growth zones of long bones of children

Figure 3.20 Elastic Cartilage in the External Ear (×1,000).

Microscopic appearance: Elastic fibers form weblike mesh amid lacunae; always covered by perichondrium

Representative locations: External ear; epiglottis

Functions: Provides flexible, elastic support

Figure 3.21 Fibrocartilage in an Intervertebral Disc (×400).

Microscopic appearance: Parallel collagen fibers similar to those of tendon; rows of chondrocytes in lacunae between collagen fibers; never has a perichondrium

Representative locations: Pubic symphysis (anterior joint between two halves of pelvic girdle); intervertebral discs that separate bones of spinal column; menisci, or pads of shock-absorbing cartilage, in knee joint; at points where tendons insert on bones near articular hyaline cartilage

Functions: Resists compression and absorbs shock in some joints; often a transitional tissue between dense connective tissue and hyaline cartilage (for example, at some tendon-bone junctions)

Further differences between compact and spongy bone are described in chapter 6. Here, we examine only compact bone (table 3.8). Most specimens you study will probably be chips of dried compact bone ground to microscopic thinness. Most compact bone is arranged as cylinders of tissue that surround **central (haversian**[18]**) canals,** which run longitudinally through the shafts of long bones such as the femur. The bone matrix is deposited in concentric **lamellae**[19]—onionlike layers around each central canal. A central canal and its surrounding lamellae are called an **osteon.** Tiny lacunae between the lamellae are occupied by mature bone cells, or **osteocytes.**[20] Delicate canals called **canaliculi**[21] radiate from each lacuna to its neighbors and allow the osteocytes to contact each other. The bone as a whole is covered with a tough fibrous **periosteum**[22] (PERR-ee-OSS-tee-um) similar to the perichondrium of cartilage.

In the living state, blood vessels and nerves run through the central canals and osteocytes occupy the lacunae and canaliculi. The osseous tissue most commonly presented for introductory laboratory study, however, is dead and dried—devoid of cells, blood vessels, and nerves. For contrast, these bone chips are often saturated with India ink to blacken the central canals, lacunae, and canaliculi (as in fig. 3.22).

About one-third of the dry weight of bone is composed of collagen fibers and glycosaminoglycans, which enable bone to bend slightly under stress (see Deeper Insight 3.3). Two-thirds consists of minerals (mainly calcium and phosphate salts) that enable bones to withstand compression by the weight of the body.

Blood

Blood (table 3.9) is a fluid connective tissue that travels through tubular blood vessels. Its primary function is to transport cells and dissolved matter from place to place. It may seem odd that a tissue as fluid as blood and another as rock hard as bone are both considered connective tissues, but they have more in common than first meets the eye. Like other connective tissues, blood is composed of more ground substance than cells. Its ground substance is the **blood plasma,** and its cellular components are collectively called the **formed elements.** Unlike other connective tissues, blood does not usually exhibit fibers, yet protein fibers do appear when the blood

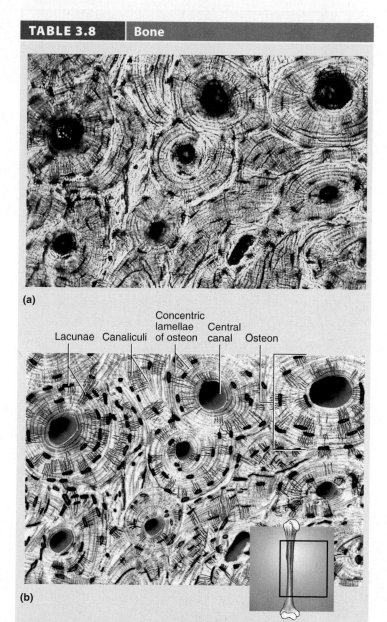

TABLE 3.8	Bone

(a)

Figure 3.22 Compact Bone (×100).

Concentric lamellae of osteon · Central canal · Osteon · Lacunae · Canaliculi

(b)

Microscopic appearance (compact bone): Calcified matrix arranged in concentric lamellae around central canals; osteocytes occupy lacunae between adjacent lamellae; lacunae interconnected by delicate canaliculi
Representative location: Skeleton
Functions: Physical support of body; leverage for muscle action; protective enclosure of viscera; reservoir of calcium and phosphorus

DEEPER INSIGHT   3.3

Brittle Bone Disease

Osteogenesis[23] *imperfecta* (OI) is a hereditary defect of collagen deposition in the bones. Collagen-deficient bones are exceptionally brittle, and so this disorder is also called *brittle bone disease.* Children with OI often exhibit bone fractures even at birth, and they suffer frequent spontaneous fractures throughout childhood. They may have deformed teeth as well as hearing impairments resulting from deformity of the middle-ear bones. Children with OI are sometimes mistaken for battered children, and their parents falsely accused, before the disease is diagnosed. In severe cases, the child is stillborn or dies soon after birth, whereas other persons live with OI into adulthood. Little can be done for children with this disease except for very careful handling, prompt treatment of fractures, and orthopedic braces to minimize skeletal deformity.

[18]Clopton Havers (1650–1702), English anatomist
[19]*lam* = plate + *ella* = little
[20]*osteo* = bone + *cyte* = cell
[21]*canal* = canal, channel + *icul* = little
[22]*peri* = around + *oste* = bone
[23]*osteo* = bone + *genesis* = formation

clots. Yet another factor placing blood in the connective tissue category is that it is produced by the connective tissues of the bone marrow and lymphatic organs.

The formed elements are of three kinds—erythrocytes, leukocytes, and platelets. **Erythrocytes**[24] (eh-RITH-ro-sites), or **red blood cells (RBCs),** are the most abundant. In stained blood smears, they

---

[24]*erythro* = red + *cyte* = cell

---

look like pink discs with thin, pale centers and no nuclei. Erythrocytes transport oxygen and carbon dioxide. **Leukocytes,** or **white blood cells (WBCs),** serve various roles in defense against infection and other diseases. They travel from one organ to another in the bloodstream and lymph but spend most of their lives in the connective tissues. Leukocytes are somewhat larger than erythrocytes and have conspicuous nuclei that usually appear violet in stained preparations. There are five kinds, distinguished partly by variations in nuclear shape: *neutrophils, eosinophils, basophils, lymphocytes,* and *monocytes.* Their individual characteristics are considered in detail in chapter 19. **Platelets** are small cell fragments scattered amid the blood cells. They are involved in clotting and other mechanisms for minimizing blood loss, and they secrete growth factors that promote blood vessel growth and maintenance.

TABLE 3.9	Blood

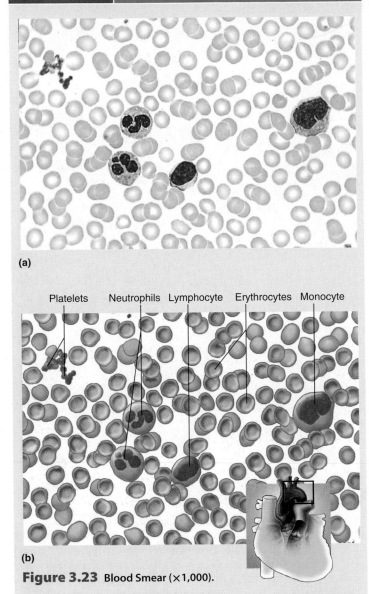

(a)

Platelets   Neutrophils   Lymphocyte   Erythrocytes   Monocyte

(b)

Figure 3.23 Blood Smear (×1,000).

Microscopic appearance: RBCs appear as pale pink discs with light centers and no nuclei; WBCs are slightly larger, are much fewer, and have variously shaped nuclei, which usually stain violet; platelets are cell fragments with no nuclei, about one-quarter the diameter of erythrocytes

Representative locations: Contained in heart and blood vessels

Functions: Transports gases, nutrients, wastes, chemical signals, and heat throughout body; provides defensive WBCs; contains clotting agents to minimize bleeding; platelets secrete growth factors that promote tissue maintenance and repair

Before You Go On

Answer the following questions to test your understanding of the preceding section:

11. What features do most or all connective tissues have in common to set this class apart from nervous, muscular, and epithelial tissue?

12. List the cell and fiber types found in fibrous connective tissues and state their functional differences.

13. What substances account for the gelatinous consistency of the ground substance of fibrous connective tissue?

14. What is areolar tissue? How can it be distinguished from any other kind of connective tissue?

15. Discuss the difference between dense regular and dense irregular connective tissue as an example of the relationship between form and function.

16. Describe some similarities, differences, and functional relationships between hyaline cartilage and bone.

17. What are the three basic kinds of formed elements in blood, and what are their respective functions?

3.4 Nervous and Muscular Tissue—Excitable Tissues

▌ Expected Learning Outcomes

When you have completed this section, you should be able to

- explain what distinguishes excitable tissues from other tissues;

- name the cell types that compose nervous tissue;

- identify the major parts of a nerve cell;

- name the three kinds of muscular tissue and describe the differences between them; and

- visually identify nervous and muscular tissues from specimens or photographs.

Excitability is a characteristic of all living cells, but it is developed to its highest degree in nervous and muscular tissues, which are therefore described as **excitable tissues.** The basis for their excitation is an electrical charge difference (voltage) called the *membrane potential,* which occurs across the plasma membranes of all cells. Nervous and muscular tissues respond quickly to outside stimuli by means of changes in membrane potential. In nerve cells, these changes result in the rapid transmission of signals to other cells. In muscle cells, they result in contraction, or shortening of the cell.

Nervous Tissue

Nervous tissue is specialized for communication by means of electrical and chemical signals. It consists of **neurons** (NOOR-ons), or nerve cells, and a much greater number of supportive **neuroglia** (noo-ROG-lee-uh), or **glial** (GLEE-ul) **cells,** which protect and assist the neurons (table 3.10). Neurons detect stimuli, respond quickly, and transmit information to other cells. Each neuron has a prominent **neurosoma,** or cell body, that houses the nucleus and most other organelles. This is the cell's center of genetic control and protein synthesis. Neurosomas are usually round, ovoid, or stellate in shape. Extending from the neurosoma, there are usually multiple short, branched processes called **dendrites,**[25] which receive signals from other cells and conduct messages to the neurosoma, and a single, much longer **axon (nerve fiber),** which sends outgoing signals to other cells. Some axons are more than a meter long and extend from the brainstem to the foot.

Glial cells constitute most of the volume of the nervous tissue. They are usually much smaller than neurons. There are six types of glial cells, described in chapter 13, which provide a variety of supportive, protective, and "housekeeping" functions for the nervous system. Although they communicate with neurons and each other, they do not transmit long-distance signals.

Nervous tissue is found in the brain, spinal cord, nerves, and ganglia (aggregations of neurosomas forming knotlike swellings in nerves). Local variations in the structure of nervous tissue are described in chapters 13 to 16.

Muscular Tissue

Muscular tissue is specialized to contract when it is stimulated, and thus to exert a physical force on other tissues, organs, or fluids; for example, a skeletal muscle pulls on a bone, the heart contracts and expels blood, and the bladder contracts and expels urine. Not only do movements of the body and its limbs depend on muscular tissue, but so do such processes as digestion, waste elimination, breathing, speech, and blood circulation. The muscles are also an important source of body heat.

There are three types of muscular tissue—*skeletal, cardiac,* and *smooth*—which differ in appearance, physiology, and function (table 3.11). **Skeletal muscle** consists of long, threadlike cells called **muscle fibers.** Most of it is attached to bones, but there are exceptions in the tongue, upper esophagus, some facial muscles, and some **sphincter**[26] (SFINK-tur) muscles (ringlike or cufflike muscles that open and close body passages). Each cell contains multiple

---

[25]*dendr* = tree + *ite* = little
[26]*sphinc* = squeeze, bind tightly

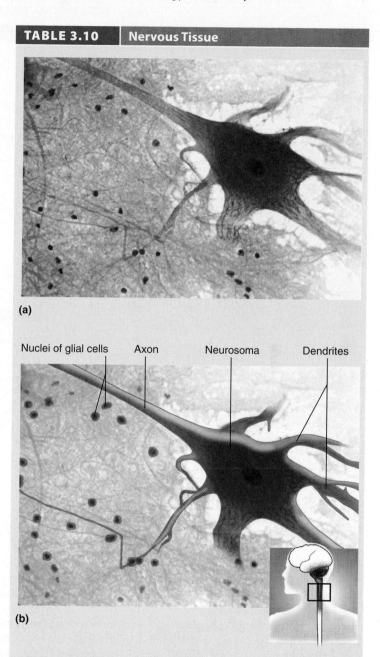

| TABLE 3.10 | Nervous Tissue |

(a)

Nuclei of glial cells   Axon   Neurosoma   Dendrites

(b)

Figure 3.24 A Neuron and Glial Cells of a Spinal Cord Smear (×400).

Microscopic appearance: Most sections show a few large neurons, usually with rounded or stellate cell bodies (neurosomas) and fibrous processes (axon and dendrites) extending from the neurosomas; neurons are surrounded by a greater number of much smaller glial cells, which lack dendrites and axons
Representative locations: Brain, spinal cord, nerves, ganglia
Function: Internal communication

nuclei adjacent to the plasma membrane. Skeletal muscle is described as *striated* and *voluntary.* The first term refers to alternating light and dark bands, or **striations** (stry-AY-shuns), created by the overlapping pattern of cytoplasmic protein filaments that cause muscle contraction. The second term, **voluntary,** refers to the fact that we usually have conscious control over skeletal muscles.

TABLE 3.11   Muscular Tissue

Skeletal Muscle	Cardiac Muscle	Smooth Muscle

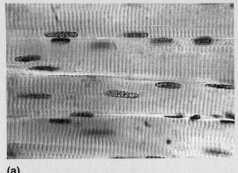

(a)

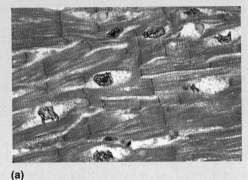

(a)

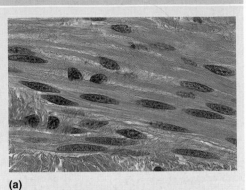

(a)

Nuclei   Striations   Muscle fiber

Intercalated discs   Striations   Glycogen

Nuclei   Muscle cells

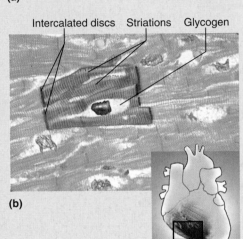

(b)

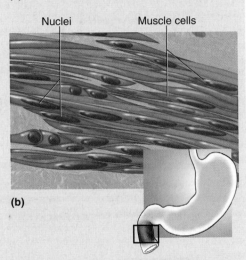

(b)

(b)

Figure 3.25 Skeletal Muscle (×400).

Figure 3.26 Cardiac Muscle (×400).

Figure 3.27 Smooth Muscle in the Intestinal Wall (×1,000).

Microscopic appearance: Long, threadlike, unbranched cells (fibers), relatively parallel in longitudinal tissue sections; striations; multiple nuclei per cell, near plasma membrane
Representative locations: Skeletal muscles, mostly attached to bones but also including voluntary sphincters of the eyelids, urethra, and anus; diaphragm; tongue; some muscles of esophagus
Functions: Body movements, facial expression, posture, breathing, speech, swallowing, control of urination and defecation, and childbirth; under voluntary control

Microscopic appearance: Short branched cells (myocytes); less parallel appearance than other muscle types in tissue sections; striations; intercalated discs; one nucleus per cell, centrally located and often surrounded by a light zone
Representative location: Heart
Functions: Pumping of blood; under involuntary control

Microscopic appearance: Short fusiform cells overlapping each other; nonstriated; one nucleus per cell, centrally located
Representative locations: Usually found as sheets of tissue in walls of viscera and blood vessels; also in iris and associated with hair follicles; involuntary sphincters of urethra and anus
Functions: Swallowing; contractions of stomach and intestines; expulsion of feces and urine; labor contractions; control of blood pressure and flow; control of respiratory airflow; control of pupillary diameter; erection of hairs; under involuntary control

Apply What You Know

How does the meaning of the word *fiber* differ in the following uses: *muscle fiber*, *nerve fiber*, and *connective tissue fiber*?

Cardiac muscle is limited to the heart. It, too, is striated, but it differs from skeletal muscle in its other features. Cardiac muscle is considered **involuntary** because it is not usually under conscious control; it contracts even if all nerve connections to it are severed. Its cells are much shorter, so they are commonly called **myocytes**[27] or **cardiocytes** rather than fibers. The myocytes are branched or notched at the ends. They contain only one nucleus, which is located

[27]*myo* = muscle + *cyte* = cell

near the center and is often surrounded by a light-staining region of glycogen, a starchlike energy source. Cardiac myocytes are joined end to end by junctions called **intercalated**[28] (in-TUR-ku-LAY-ted) **discs**, which appear as dark transverse lines separating each myocyte from the next. They may be only faintly visible, however, unless the tissue has been specially stained for them. Gap junctions in the discs enable a wave of excitation to travel rapidly from cell to cell so that all the myocytes of a heart chamber are stimulated and contract almost simultaneously.

Smooth muscle lacks striations and is involuntary. Smooth muscle cells, also called **myocytes**, are tapered at the ends (fusiform) and relatively **short**. They have a **single centrally placed nucleus**. Small amounts of smooth muscle are found in the iris of the eye and in the skin, but most of it, called **visceral muscle**, forms layers in the walls of the digestive, respiratory, and urinary tracts; the uterus; blood vessels; and other organs. In locations such as the esophagus and small intestine, smooth muscle forms adjacent layers, with the cells of one layer encircling the organ and the cells of the other layer running longitudinally. When the circular smooth muscle contracts, it may propel contents such as food through the organ. When the longitudinal layer contracts, it makes the organ shorter and thicker. By regulating the diameter of blood vessels, smooth muscle is very important in controlling blood pressure and flow. Both smooth and skeletal muscle form sphincters that control the emptying of the bladder and rectum.

Before You Go On

Answer the following questions to test your understanding of the preceding section:

18. What do nervous and muscular tissue have in common? What is the primary function of each?

19. What are the two basic types of cells in nervous tissue, and how can they be distinguished from each other?

20. Name the three kinds of muscular tissue, describe how to distinguish them from each other in microscopic appearance, and state a location and function for each one.

3.5 Glands and Membranes

▶ Expected Learning Outcomes

When you have completed this section, you should be able to

- describe or define different types of glands;
- describe the typical anatomy of a gland;
- name and compare different modes of glandular secretion;
- describe the way tissues are organized to form the body's membranes; and

- name and describe the major types of membranes in the body.

We have surveyed all of the fundamental categories of human tissue, and we will now look at the way in which multiple tissue types are assembled to form the body's glands and membranes.

Glands

A **gland** is a cell or organ that secretes substances for use elsewhere in the body or for elimination as waste. The gland product may be something synthesized by the gland cells (such as digestive enzymes) or something removed from the tissues and modified by the gland (such as urine). Its product is called a **secretion** if it is useful to the body (such as a digestive enzyme or hormone) and an **excretion** if it is a waste product (such as urine). Glands are composed predominantly of epithelial tissue, but usually have a supporting connective tissue framework and capsule.

Endocrine and Exocrine Glands

Glands are broadly classified as endocrine or exocrine. Both types originate as invaginations of a surface epithelium (fig. 3.28). Multicellular **exocrine**[29] (EC-so-crin) **glands** maintain their contact with the surface by way of a **duct**, an epithelial tube that conveys their secretion to the surface. The secretion may be released to the body surface, as in the case of sweat, mammary, and tear glands, but more often it is released into the lumen (cavity) of another organ such as the mouth or intestine. **Endocrine**[30] (EN-do-crin) **glands** lose their contact with the surface and have no ducts. They do, however, have a high density of blood capillaries and secrete their products directly into the blood. The secretions of endocrine glands, called **hormones,** function as chemical messengers to stimulate cells elsewhere in the body. Endocrine glands are the subject of chapter 18.

The exocrine–endocrine distinction is not always clear. The liver is an exocrine gland that secretes one of its products, bile, through a system of ducts, but secretes hormones, albumin, and other products directly into the bloodstream. Several glands, such as the pancreas, testis, ovary, and kidney, have both exocrine and endocrine components. Nearly all of the viscera have at least some cells that secrete hormones, even though most of these organs are not usually thought of as glands (for example, the brain and heart).

Unicellular glands are secretory cells found in an epithelium that is predominantly nonsecretory. They can be endocrine or exocrine. For example, the respiratory tract, which is lined mainly by ciliated cells, also has a liberal scattering of nonciliated, mucus-secreting **goblet cells**, which are exocrine (see fig. 3.7). The digestive tract has many scattered endocrine cells, which secrete hormones that coordinate digestive processes.

Exocrine Gland Structure

Figure 3.29 shows a generalized multicellular exocrine gland—a structural arrangement found in such organs as the mammary

---

[28]*inter* = between + *calated* = inserted

[29]*exo* = out + *crin* = to separate, secrete
[30]*endo* = in, into + *crin* = to separate, secrete

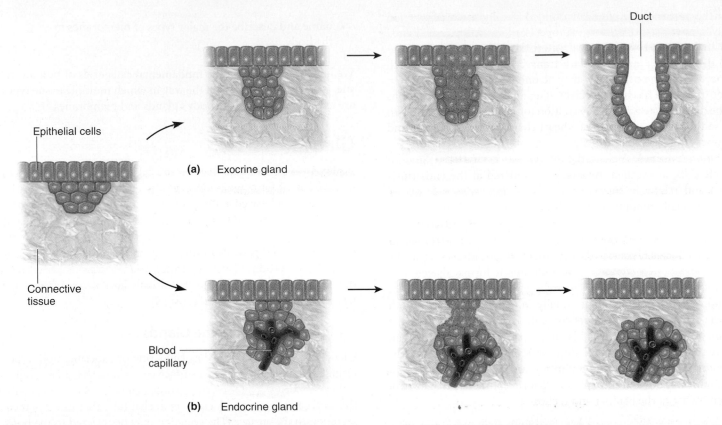

(a) Exocrine gland

Epithelial cells

Connective tissue

Blood capillary

Duct

(b) Endocrine gland

Figure 3.28 **Development of Exocrine and Endocrine Glands.** (a) An exocrine gland begins with epithelial cells proliferating into the connective tissue below. Apoptosis of the cells in the core hollows out a duct. The gland remains connected to the surface for life by way of this duct and releases its secretions onto the epithelial surface. (b) An endocrine gland begins similarly, but the cells connecting it to the surface degenerate while the secretory tissue becomes infiltrated with blood capillaries. The secretory cells will secrete their products (hormones) into the blood.

gland, pancreas, and salivary glands. Most glands are enclosed in a fibrous **capsule.** The capsule often gives off extensions called **septa,**[31] or **trabeculae**[32] (trah-BEC-you-lee), that divide the interior of the gland into compartments called **lobes,** which are visible to the naked eye. Finer connective tissue septa may further subdivide each lobe into microscopic **lobules** (LOB-yools). Blood vessels, nerves, and the gland's ducts generally travel through these septa. The connective tissue framework of the gland, called its **stroma,** supports and organizes the glandular tissue. The cells that perform the tasks of synthesis and secretion are collectively called the **parenchyma**[33] (pa-REN-kih-muh). This is typically simple cuboidal or simple columnar epithelium.

Exocrine glands are classified as **simple** if they have a single unbranched duct and **compound** if they have a branched duct (fig. 3.30). If the duct and secretory portion are of uniform diameter, the gland is called **tubular.** If the secretory cells form a dilated sac, the gland is called **acinar** and the sac is an **acinus**[34] (ASS-ih-nus), or **alveolus**[35] (AL-vee-OH-lus). A gland in which both the acini and tubules secrete a product is called a **tubuloacinar gland.**

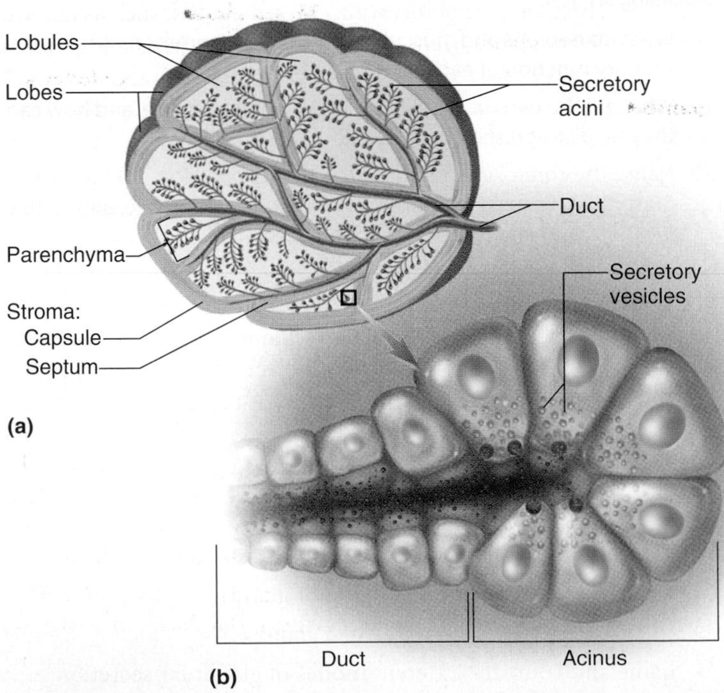

Lobules

Lobes

Parenchyma

Stroma:

Capsule

Septum

Secretory acini

Duct

Secretory vesicles

(a)

Duct

Acinus

(b)

Figure 3.29 **General Structure of an Exocrine Gland.** (a) The gland duct branches repeatedly, following the connective tissue septa, until its finest divisions end in saccular acini of secretory cells. (b) Detail of an acinus and the beginning of a duct.
• *What membrane transport process (review pp. 33–36) are the cells of this acinus carrying out?*

---

[31]*septum* = wall
[32]*trab* = plate + *cula* = little
[33]*par* = beside + *enchym* = pour in
[34]*acinus* = berry
[35]*alveol* = cavity, pit

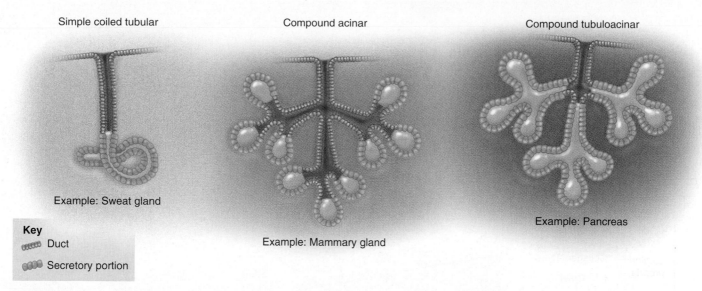

Simple coiled tubular

Example: Sweat gland

Compound acinar

Example: Mammary gland

Compound tubuloacinar

Example: Pancreas

Key

Duct

Secretory portion

Figure 3.30 Some Types of Exocrine Glands. Glands are classified according to the branching of their ducts and the appearance and extent of the secretory portions.

Types of Exocrine Secretions

Exocrine glands are classified not only by their structure but also by the nature of their secretions. **Serous (SEER-us) glands** produce relatively thin, watery fluids such as perspiration, milk, tears, and digestive juices. **Mucous glands,** found in the tongue and roof of the mouth among other places, secrete a glycoprotein called *mucin* (MEW-sin). After it is secreted, mucin absorbs water and forms the sticky product *mucus*. (Note that *mucus,* the secretion, is spelled differently from *mucous,* the adjective form of the word.) **Mixed glands,** such as the two pairs of salivary glands in the chin, contain both serous and mucous cells and produce a mixture of the two types of secretions. **Cytogenic[36] glands** release whole cells. The only examples of these are the testes and ovaries, which produce sperm and egg cells.

Methods of Exocrine Secretion

Glands are classified as merocrine or holocrine depending on how they produce their secretions. **Merocrine[37]** (MERR-oh-crin) **glands,** also called **eccrine[38]** (EC-rin) **glands,** release their secretion by exocytosis, as described in chapter 2. These include the tear glands, pancreas, gastric glands, and most others. In **holocrine[39] glands,** cells accumulate a product and then the entire cell disintegrates, so the secretion is a mixture of cell fragments and the substance the cell had synthesized prior to its disintegration. Only a few glands use this mode of secretion, such as the oil-producing glands of the scalp and certain glands of the eyelid.

Some glands, such as the axillary (armpit) sweat glands and mammary glands, are named **apocrine[40] glands** from a former belief that the secretion was composed of blobs of apical cytoplasm

that broke away from the cell surface. Closer study showed this to be untrue; these glands are primarily merocrine in their mode of secretion. They are nevertheless different from other merocrine glands in function and histological appearance, so they are still referred to as apocrine glands.

Membranes

Chapter 1 describes the major body cavities and the membranes that line them and cover their viscera. We now consider some histological aspects of these membranes.

The largest membrane of the body is the **cutaneous** (cue-TAY-nee-us) **membrane**—or more simply, the skin (detailed in chapter 5). It consists of a stratified squamous epithelium (epidermis) resting on a layer of connective tissue (dermis).

The two principal kinds of internal membranes are mucous and serous membranes. A **mucous membrane (mucosa)** (mew-CO-sa) lines passageways that open to the exterior: the digestive, respiratory, urinary, and reproductive tracts (fig. 3.31a). A mucosa consists of two to three layers: (1) an epithelium; (2) an areolar connective tissue layer called the **lamina propria[41]** (LAM-ih-nuh PRO-pree-uh); and sometimes (3) a layer of smooth muscle called the **muscularis mucosae** (MUSK-you-LAIR-iss mew-CO-see). Mucous membranes have absorptive, secretory, and protective functions. They are often covered with mucus secreted by **goblet cells,** multicellular mucous glands, or both. The mucus traps bacteria and foreign particles, which keeps them from invading the tissues and aids in their removal from the body. The epithelium of a mucous membrane may also include absorptive, ciliated, and other types of cells.

A **serous membrane (serosa)** is composed of a **simple squamous epithelium** resting on a thin layer of areolar connective

---

[36]*cyto* = cell + *genic* = producing
[37]*mero* = part + *crin* = to separate, secrete
[38]*ec* = *ex* = out + *crin* = to separate, secrete
[39]*holo* = whole, entire + *crin* = to separate, secrete
[40]*apo* = from, off, away + *crin* = to separate, secrete

[41]*lamina* = layer + *propria* = of one's own

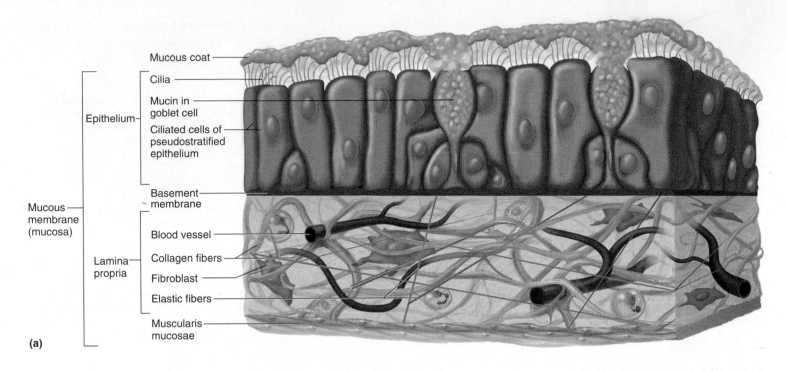

Epithelium
- Mucous coat
- Cilia
- Mucin in goblet cell
- Ciliated cells of pseudostratified epithelium

Mucous membrane (mucosa)

- Basement membrane

Lamina propria
- Blood vessel
- Collagen fibers
- Fibroblast
- Elastic fibers

- Muscularis mucosae

(a)

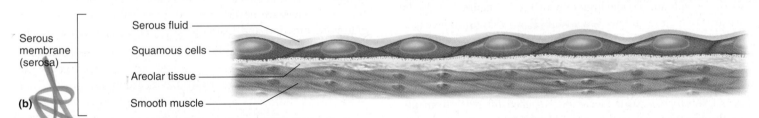

Serous membrane (serosa)
- Serous fluid
- Squamous cells
- Areolar tissue
- Smooth muscle

(b)

Figure 3.31 **Mucous and Serous Membranes.** (a) Histology of a mucous membrane such as the inner lining of the trachea. (b) Histology of a serous membrane such as the outer surface of the small intestine.
● *Identify another specific place in the body where you could find a membrane of each type.*

tissue (fig. 3.31b). In the linings of the pleural, pericardial, and peritoneal cavities, the epithelial component is called **mesothelium.** Serous membranes produce watery **serous** (SEER-us) **fluid,** which arises from the blood and derives its name from the fact that it is similar to blood serum in composition. Serous membranes line the insides of some body cavities and form a smooth surface on the outer surfaces of some of the viscera, such as the digestive tract. The pleurae, pericardium, and peritoneum described in chapter 1 are serous membranes.

The circulatory system is lined with a simple squamous epithelium called **endothelium.** The endothelium rests on a thin layer of areolar tissue, which often rests in turn on an elastic sheet. Collectively, these tissues make up a membrane called the *tunica interna* of the blood vessels and *endocardium* of the heart.

Some joints of the skeletal system are enclosed by fibrous **synovial** (sih-NO-vee-ul) **membranes,** made only of connective tissue. These membranes span the gap from one bone to the next and secrete slippery *synovial fluid* into the joint.

Before You Go On

Answer the following questions to test your understanding of the preceding section:

21. Distinguish between a simple gland and a compound gland, and give an example of each. Distinguish between a tubular gland and an acinar gland, and give an example of each.

22. Contrast the merocrine and holocrine methods of secretion, and name a gland product produced by each method.

23. Describe the differences between a mucous and a serous membrane.

24. Name the layers of a mucous membrane, and state which of the four primary tissue classes composes each layer.

3.6 Tissue Growth, Development, Repair, and Death

▶ **Expected Learning Outcomes**

When you have completed this section, you should be able to

- name and describe the modes of tissue growth;
- name and describe the ways that a tissue can change from one type to another;
- name and describe the ways the body repairs damaged tissues; and
- name and describe the modes and causes of tissue shrinkage and death.

Tissue Growth

Tissues grow because their cells increase in number or size. Most embryonic and childhood growth occurs by **hyperplasia**[42] (HY-pur-PLAY-zhuh), tissue growth through cell multiplication. Skeletal muscle and adipose tissue grow, however, through **hypertrophy**[43] (hy-PUR-truh-fee), the enlargement of preexisting cells. Even a very muscular or obese adult has essentially the same number of muscle fibers or adipocytes as he or she had in late childhood, but the cells may be substantially larger. **Neoplasia**[44] (NEE-oh-PLAY-zhuh) is the development of a tumor (neoplasm)—whether benign or malignant—composed of abnormal, nonfunctional tissue.

Changes in Tissue Type

You have studied the form and function of more than two dozen discrete types of human tissue in this chapter. You should not leave this subject, however, with the impression that once these tissue types are established, they never change. Tissues are, in fact, capable of changing from one type to another within certain limits. Most obviously, unspecialized tissues of the embryo develop into more diverse and specialized types of mature tissue—embryonic *mesenchyme* to muscle, for example. This development of a more specialized form and function is called **differentiation.**

Epithelia sometimes exhibit **metaplasia,**[45] a change from one type of mature tissue to another. For example, the vagina of a young girl is lined with a simple cuboidal epithelium. At puberty, it changes to a stratified squamous epithelium, better adapted to the future demands of intercourse and childbirth. The nasal cavity is lined with ciliated pseudostratified columnar epithelium. However, if we block one nostril and breathe through the other one for several days, the epithelium in the unblocked passage changes to stratified squamous.

In smokers, the ciliated pseudostratified columnar epithelium of the bronchi may transform into a stratified squamous epithelium.

Apply What You Know

What functions of a ciliated pseudostratified columnar epithelium could not be served by a stratified squamous epithelium? In light of this, what might be some consequences of bronchial metaplasia in heavy smokers?

Tissue Repair

Damaged tissues can be repaired in two ways: regeneration or fibrosis. **Regeneration** is the replacement of dead or damaged cells by the same type of cells as before. Regeneration restores normal function to the organ. Most skin injuries (cuts, scrapes, and minor burns) heal by regeneration. The liver also regenerates remarkably well. **Fibrosis** is the replacement of damaged tissue with scar tissue, composed mainly of collagen produced by fibroblasts. Scar tissue helps to hold an organ together, but it does not restore normal function. Examples include the healing of severe cuts and burns, the healing of muscle injuries, and scarring of the lungs in tuberculosis.

Tissue Shrinkage and Death

Atrophy[46] (AT-ruh-fee) is the shrinkage of a tissue through a loss in cell size or number. It results from both normal aging *(senile atrophy)* and lack of use of an organ *(disuse atrophy).* Muscles that are not exercised exhibit disuse atrophy as their cells become smaller. This was a serious problem for the first astronauts who participated in prolonged microgravity space flights. Upon return to normal gravity, they were sometimes too weak from muscular atrophy to walk. Space stations and shuttles now include exercise equipment to maintain the crew's muscular condition. Disuse atrophy also occurs when a limb is immobilized in a cast or by paralysis.

Necrosis[47] (neh-CRO-sis) is the premature, pathological death of tissue due to trauma, toxins, infection, and so forth. **Infarction** is the sudden death of tissue, such as cardiac muscle *(myocardial infarction)* or brain tissue *(cerebral infarction),* that occurs when its blood supply is cut off. **Gangrene** is any tissue necrosis resulting from an insufficient blood supply, usually involving infection. *Dry gangrene* often occurs in diabetics, especially in the feet. A lack of sensation due to diabetic nerve damage can make a person oblivious to injury and infection, and poor blood circulation due to diabetic arterial damage results in slow healing and rapid spread of infection. This often necessitates the amputation of toes, feet, or legs. A **decubitus ulcer** (bed sore) is a form of dry gangrene that occurs when immobilized persons, such as those confined to a hospital bed or wheelchair, are unable to move, and continual pressure on the skin cuts off blood flow to an area. *Gas gangrene* is necrosis resulting from infection with

---

[42]*hyper* = excessive + *plas* = growth
[43]*hyper* = excessive + *trophy* = nourishment
[44]*neo* = new + *plas* = form, growth
[45]*meta* = change + *plas* = form, growth

[46]*a* = without + *trophy* = nourishment
[47]*necr* = death + *osis* = process

certain bacteria of the genus *Clostridium,* usually introduced when a wound is contaminated with soil. The disorder is named for bubbles of gas (mainly hydrogen) that accumulate in the tissues. This is a deadly condition that requires immediate intervention, often including amputation.

Cells dying by necrosis usually swell, exhibit *blebbing* (bubbling) of their plasma membranes, and then rupture. The cell contents released into the tissues trigger an inflammatory response in which macrophages phagocytize the cellular debris.

Apoptosis[48] (AP-op-TOE-sis), or **programmed cell death,** is the normal death of cells that have completed their function and best serve the body by dying and getting out of the way. Cells undergoing apoptosis shrink and are quickly phagocytized by macrophages and other cells. The cell contents never escape, so there is no inflammatory response. Although billions of cells die every hour by apoptosis, they are engulfed so quickly that they are almost never seen except within macrophages.

One example of apoptosis is that in embryonic development, we produce about twice as many neurons as we need. Those that make connections with target cells survive, while the excess 50% die. Apoptosis also dissolves the webbing between the fingers and toes during embryonic development, it frees the earlobe from the side of the head in people with detached (pendulous) earlobes, and it causes shrinkage of the uterus after pregnancy ends. Immune cells can stimulate cancer cells to "commit suicide" by apoptosis (see photo on p. 52).

Before You Go On

Answer the following questions to test your understanding of the preceding section:

25. Tissues can grow through an increase in cell size or cell number. What are the respective terms for these two kinds of growth?

26. Distinguish between *differentiation* and *metaplasia.* Give an example of a developmental process involving each of these.

27. Distinguish between *regeneration* and *fibrosis.* Which process restores normal cellular function? What good is the other process if it does not restore function?

28. Distinguish between *atrophy, necrosis,* and *apoptosis,* and describe a circumstance under which each of these forms of tissue loss may occur.

---

[48]*apo* = away + *ptosis* = falling

DEEPER INSIGHT 3.4

Tissue Engineering

Tissue repair is not only a natural process but also a lively area of research in biotechnology. *Tissue engineering* is the artificial production of tissues and organs in the laboratory for implantation in the human body. The process commonly begins with a synthetic scaffold of collagen or polyester, sometimes in the shape of a desired organ such as a blood vessel or ear. The scaffold is then seeded with human cells and allowed to grow until it reaches a point suitable for implantation. Tissue-engineered skin grafts have been on the market for several years. Scientists have also grown liver, bone, tendon, breast, and other tissues in the laboratory. Researchers have grown "human" ears by seeding polymer scaffolds with human cartilage cells and growing them on the backs of immunodeficient mice unable to reject the human tissue (fig. 3.32). It is hoped that ears and noses grown in this way may be used for the cosmetic treatment of children with birth defects or people who have suffered injuries from accidents or animal bites. In recent years, tissue engineering has been used to construct a new bronchus and new urinary bladders in several patients, seeding a nonliving protein scaffold with cells taken from elsewhere in the patients' bodies.

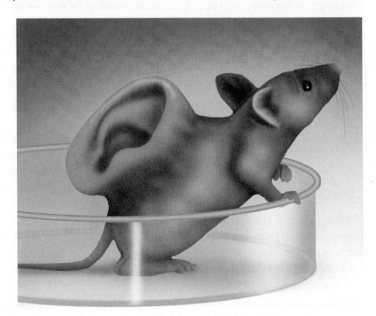

Figure 3.32 Tissue Engineering. Scientists have grown an external ear from human tissue on the back of an immunodeficient mouse. The ear can be removed without killing the mouse and used for implantation into a human patient. In the future, such artificial organs might be used to improve the appearance of patients with facial deformities or injuries.

Study Guide

Assess Your Learning Outcomes

You should have a good understanding of this chapter if you can accurately address the following issues.

3.1 The Study of Tissues (p. 53)

1. The scope of histology and how it relates to cytology and gross anatomy
2. The definition of *tissue*
3. The four primary tissue types, the ways in which they differ from each other, and some places where each type can be found
4. The definitions of *matrix* and *ground substance* and their place in tissue structure
5. The preparation of tissues as stained histological sections
6. The common planes of section in histology and their relevance to one's three-dimensional interpretation of two-dimensional histological images
7. Methods of histological tissue preparation other than sectioning, and the kinds of tissue for which these other methods are commonly applied

3.2 Epithelial Tissue (p. 55)

1. The defining characteristics of epithelial tissue as a class, and general locations in which epithelia are found
2. The location, composition, and functions of a basement membrane
3. The defining characteristics of the two major classes of epithelium and of the four types in each class
4. The appearance and function of goblet cells
5. The appearance, representative locations, and functions of the four types of simple epithelium: simple squamous, simple cuboidal, simple columnar, and pseudostratified columnar
6. The appearance, representative locations, and functions of the four types of stratified epithelium: stratified squamous, stratified cuboidal, stratified columnar, and transitional
7. Differences in structure, location, and function between the nonkeratinized and keratinized forms of stratified squamous epithelium; and the functions of keratin
8. The process of exfoliation and a clinical application of exfoliate cytology

3.3 Connective Tissue (p. 61)

1. The defining characteristics of connective tissue as a class, and general locations in which connective tissues are found
2. The diverse functions of connective tissues and the types of connective tissue that perform such roles
3. The types of connective tissue classified as fibrous connective tissues; the types of cells that are common in fibrous connective tissue, and their functions; and the three types of protein fibers common in fibrous connective tissue and their functional differences
4. The meaning of *ground substance* and its chemical makeup in connective tissue
5. The distinction between loose and dense fibrous connective tissue
6. The appearance, representative locations, and functions of two types of loose fibrous connective tissue: areolar and reticular tissue
7. The appearance, representative locations, and functions of two types of dense fibrous connective tissue: dense regular and dense irregular connective tissue
8. The defining characteristics, microscopic appearance, representative locations, and functions of adipose tissue, and the distinctions between white fat and brown fat
9. The defining characteristics of cartilage as a class, and the composition and organization of its cells and matrix
10. The relationship of the perichondrium to cartilage, and places where the perichondrium is absent
11. The appearance, representative locations, and functions of the three types of cartilage: hyaline cartilage, elastic cartilage, and fibrocartilage
12. The defining characteristics of bone as a class, and the composition and organization of its cells and matrix
13. The relationship of the periosteum to bone
14. The spongy and compact types of bone and how they differ in appearance and location
15. The microscopic appearance, location, and functions of compact bone
16. Why blood is considered a connective tissue
17. The two principal constituents of blood, and its three types of formed elements and how they can be microscopically distinguished from each other
18. The functions of blood

3.4 Nervous and Muscular Tissue— Excitable Tissues (p. 70)

1. The meaning of cell excitability, and why nervous and muscular tissue are called excitable tissues in view of the fact that all living cells are excitable
2. The general function of nervous tissue, its two basic cell types, and the functional distinction between the two cell types
3. The parts of a neuron and the function of each part
4. The microscopic appearance and locations of nervous tissue
5. The defining characteristics of muscular tissue as a class, and the diverse functions of muscular tissue
6. The difference between striated and non-striated muscle, and between voluntary and involuntary muscle
7. The meanings of *sphincter, cardiocyte, myocyte, intercalated disc,* and *visceral muscle*
8. The microscopic appearance, representative locations, and functions of skeletal, cardiac, and smooth muscle

3.5 Glands and Membranes (p. 73)

1. The definition and general composition of a gland
2. The distinction between exocrine and endocrine glands and why it is not always easy to make a clear distinction between them
3. Examples and locations of unicellular glands
4. The generalized structure of a multicellular exocrine gland, including structural and functional interpretations of its lobes and lobules, capsule and septa, stroma and parenchyma
5. The system of classifying exocrine glands according to whether their ducts are branched or unbranched and whether or not their secretory cells are aggregated into dilated sacs at the ends of the ducts
6. The distinctions between serous, mucous, mixed, and cytogenic glands, and examples of each

7. The distinctions between merocrine and holocrine glands, and examples of each; the basis for referring to some glands as apocrine
8. The distinctions between cutaneous, mucous, and serous membranes, and locations and functions of each
9. The tissue layers of a mucous membrane and of a serous membrane
10. The nature and locations of endothelium, mesothelium, and synovial membranes

3.6 Tissue Growth, Development, Repair, and Death (p. 77)
1. Three types of tissue growth—hyperplasia, hypertrophy, and neoplasia—including examples of each, how they differ from each other, and whether each is normal or pathological
2. The difference between differentiation and metaplasia, and examples of each
3. Two ways in which the body repairs damaged tissues, and how they differ

4. The meaning of tissue atrophy, its causes, and examples
5. Some forms of tissue necrosis, and examples
6. The process of apoptosis, how it differs from necrosis, and examples of situations in which apoptosis occurs

Testing Your Recall

1. Transitional epithelium is found in
   a. the urinary system.
   b. the respiratory system.
   c. the digestive system.
   d. the reproductive system.
   e. all of the above.

2. The external surface of the stomach is covered by
   a. a mucosa.
   b. a serosa.
   c. the parietal peritoneum.
   d. a lamina propria.
   e. a basement membrane.

3. The interior of the respiratory tract is lined with
   a. a serosa.   d. endothelium.
   b. mesothelium.   e. peritoneum.
   c. a mucosa.

4. A seminiferous tubule of the testis is lined with _____ epithelium.
   a. simple cuboidal
   b. pseudostratified columnar ciliated
   c. stratified squamous
   d. transitional
   e. stratified cuboidal

5. When the blood supply to a tissue is cut off, the tissue is most likely to undergo
   a. metaplasia.   d. necrosis.
   b. hyperplasia.   e. hypertrophy.
   c. apoptosis.

6. A fixative serves to
   a. stop tissue decay.
   b. improve contrast.
   c. repair a damaged tissue.
   d. bind epithelial cells together.
   e. bind cardiac myocytes together.

7. The collagen of areolar tissue is produced by
   a. macrophages.   d. leukocytes.
   b. fibroblasts.   e. chondrocytes.
   c. mast cells.

8. Tendons are composed of _____ connective tissue.
   a. skeletal
   b. areolar
   c. dense irregular
   d. yellow elastic
   e. dense regular

9. The shape of the external ear is due to
   a. skeletal muscle.
   b. elastic cartilage.
   c. fibrocartilage.
   d. articular cartilage.
   e. hyaline cartilage.

10. The most abundant formed element(s) of blood is/are
    a. plasma.   d. leukocytes.
    b. erythrocytes.   e. proteins.
    c. platelets.

11. The prearranged death of a cell that has completed its task is called _____.

12. The simple squamous epithelium that lines the peritoneal cavity is called _____.

13. Osteocytes and chondrocytes occupy little cavities called _____.

14. Muscle cells and axons are often called _____ because of their shape.

15. Tendons and ligaments are made mainly of the protein _____.

16. _____ is a type of tissue necrosis associated with poor circulation, often combined with infection.

17. An epithelium rests on a layer called the _____ between its deepest cells and the underlying connective tissue.

18. Fibers and ground substance make up the _____ of a connective tissue.

19. In _____ glands, the secretion is formed by the complete disintegration of the gland cells.

20. Any epithelium in which every cell touches the basement membrane is called a _____ epithelium.

Answers in the Appendix

Building Your Medical Vocabulary

State a medical meaning of each of the following word elements, and give a term in which it is used.

1. histo-
2. inter-
3. lam-
4. -blast
5. reticul-
6. chondro-
7. peri-
8. thel-
9. exo-
10. necro-

Answers in the Appendix

True or False

Determine which five of the following statements are false, and briefly explain why.

1. If we assume that the aorta is cylindrical, an oblique section of it would have an oval shape.
2. Everything in a tissue that is not a cell is classified as ground substance.
3. The colors seen in prepared histology slides are not the natural colors of those tissues.
4. The parenchyma of the liver is a simple cuboidal epithelium.
5. The tongue is covered with keratinized stratified squamous epithelium.
6. Macrophages are large phagocytic cells that develop from lymphocytes.
7. Most of the body's protein is collagen.
8. Brown fat produces more ATP than white fat.
9. After tissue differentiation is complete, a tissue cannot change type.
10. Erythrocytes have no nuclei.

Answers in the Appendix

Testing Your Comprehension

1. A woman in labor is often told to push. In doing so, is she consciously contracting her uterus to expel the baby? Justify your answer based on the muscular composition of the uterus.
2. The Deeper Insights in this chapter describe some hereditary defects in collagen and elastin. Predict some pathological consequences that might result from a hereditary defect in keratin.
3. When cartilage is compressed, water is squeezed out of it, and when pressure is taken off, water flows back into the matrix. This being the case, why do you think cartilage at weight-bearing joints such as the knees can degenerate from lack of exercise?
4. The epithelium of the respiratory tract is mostly of the pseudostratified ciliated type, but in the alveoli—the tiny air sacs where oxygen and carbon dioxide are exchanged between the blood and inhaled air—the epithelium is simple squamous. Explain the functional significance of this histological difference. That is, why don't the alveoli have the same kind of epithelium as the rest of the respiratory tract?
5. Suppose you cut your finger on a broken bottle. How might mast cells contribute to the healing?

Answers at www.mhhe.com/saladinha3

Improve Your Grade at www.mhhe.com/saladinha3

Practice quizzes, labeling activities, and games provide fun ways to master concepts. You can also download image PowerPoint files for each chapter to create a study guide or for taking notes during lecture.

Human Development

CHAPTER
4

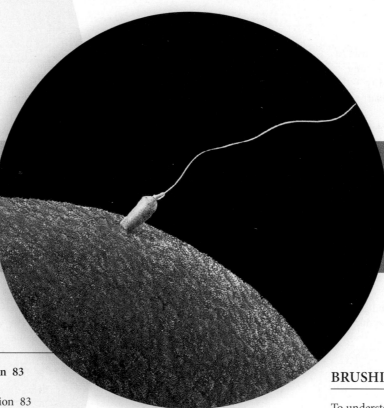

Boy meets girl: the union of sperm and egg (SEM)

BRUSHING UP

To understand this chapter, you may find it helpful to review the following concepts:
- Fetal sonography (p. 4)
- Cell division (p. 47)

Perhaps the most dramatic, seemingly miraculous aspect of human life is the transformation of a one-celled fertilized egg into an independent, fully developed individual. From the beginning of recorded thought, people have pondered how a baby forms in the mother's body and how two parents can produce another human being who, although unique, possesses characteristics of each. Aristotle dissected the embryos of various birds, established their sequence of organ development, and speculated that the hereditary traits of a child resulted from the mixing of the male's semen with the female's menstrual blood. Misconceptions about human development persisted for many centuries. Scientists of the seventeenth century thought that all the features of the infant existed in a preformed state in the egg or the sperm and that they simply unfolded and expanded as the embryo developed. Some thought that the head of the sperm had a miniature human curled up in it, while others thought the miniature person existed in the egg, and the sperm were merely parasites in the semen. Modern developmental biology was not born until the nineteenth century, largely because darwinism at last gave biologists a systematic framework for asking the right questions and discovering unifying themes in the development of diverse species of animals, including humans. Recent decades have brought dramatic leaps in understanding as geneticists have discovered how genes direct the intricate patterns and processes of human development.

This chapter describes only a few of the earliest and most general developments of the human embryo and fetus. Later chapters in the book describe some major features in the specialized development of each organ system. Knowledge of each system's prenatal development provides a deeper understanding of its mature anatomy.

4.1 Gametogenesis and Fertilization

▶ Expected Learning Outcomes

When you have completed this section, you should be able to

- describe the major features of sperm and egg production;
- explain how sperm migrate to the egg and acquire the capacity to fertilize it; and
- describe the fertilization process and how an egg prevents fertilization by more than one sperm.

Embryology, the study of prenatal development, embraces processes extending from the production of sperm and eggs (**gametogenesis**[1]) to fertilization, embryonic and fetal development, and birth. In this section, we will deal with gametogenesis and fertilization.

Gametogenesis

Sexual reproduction has a great advantage over asexual reproduction in that it produces genetically varied offspring, a crucial key to species survival in a challenging and ever-changing environment. Yet sexual reproduction also presents a problem. If offspring are to be produced by the union of cells from two parents, and if human cells normally have 46 chromosomes each, it might seem that the union of sperm and egg would produce a fertilized egg (**zygote**) with 92 chromosomes. All cells descended from the zygote by mitosis would also have 92. Then in that generation, a sperm with 92 chromosomes would fertilize an egg with 92, and the next generation would have 184 chromosomes in each cell, and so on. Obviously, if sexual reproduction is to combine cells from two parents in each generation, there must be a mechanism for maintaining the normal chromosome number. The solution is to reduce the chromosome number by half as the sex cells (**gametes**) are formed; this function is achieved by a special form of cell division called **meiosis**[2] **(reduction division).**

Gametogenesis and meiosis are detailed in chapter 26 because they are best understood in relation to the structure of the testes and ovaries. However, a few basic facts are needed here in order to best understand fertilization and the beginning of human development. One of the most important properties of the gametes is that they have only 23 chromosomes—half as many as other cells of the body. They are called **haploid** for this reason, whereas the other cells of the body are **diploid.** Diploid cells have two complete sets of 23 chromosomes (46 in all), one set from the mother and one from the father.

Gametogenesis begins with diploid stem cells that sustain their numbers through mitosis. Some of these cells then set off on a path that leads to egg or sperm cells through meiosis. Meiosis consists of two cell divisions (meiosis I and II) that have the effects of creating new genetic variety in the chromosomes and halving the chromosome number (see p. 711 for details).

In sperm production (*spermatogenesis*), meiosis I and II result in four equal-sized cells called spermatids, which then grow tails and develop into sperm without further division. In egg production (*oogenesis*), however, each meiotic division produces one large cell and one much smaller cell. The small cell, called a *polar body,* is merely a way of disposing of the excess chromosomes, and it soon dies. The large cell goes on to become the egg. This uneven division produces an egg with as much cytoplasm as possible—the raw material for early development. In oogenesis, an unfertilized egg dies after meiosis I. Meiosis II never occurs unless the egg is fertilized.

Sperm Migration and Capacitation

The human ovary usually releases one egg (oocyte) per month, around day 14 of a typical 28-day ovarian cycle. This egg is swept into the *uterine (fallopian) tube* by the beating of cilia on the tube's epithelial cells and begins a 3-day trip down the tube toward the uterus. If the egg is not fertilized, it dies within 24 hours and gets no more than one-third of the way to the uterus. Therefore, if a sperm is to fertilize an egg, it must migrate up the tube to meet it. The vast majority of sperm never make it. Although a typical ejaculation may contain 300 million sperm, many of these are destroyed by vaginal acid or drain out of the vagina; others fail to get through

---

[1]*gameto* = marriage, union + *genesis* = production

[2]*meio* = less, fewer

the cervical canal into the uterus; still more are destroyed by leukocytes in the uterus; and half of the survivors of all these ordeals are likely to go up the wrong uterine tube. Finally, 2,000 to 3,000 spermatozoa (0.001%) reach the general vicinity of the egg.

Freshly ejaculated sperm cannot immediately fertilize an egg. They must undergo a process called **capacitation,** which takes about 10 hours and occurs during their migration in the female reproductive tract. In fresh sperm, the plasma membrane is toughened by cholesterol. During capacitation, cholesterol is leached from the membrane by fluids of the female tract. As a result, the membrane becomes more fragile so it can break open more easily upon contact with the egg. It also becomes more permeable to calcium ions, which diffuse into the sperm and stimulate more powerful lashing of the tail.

The anterior tip of the sperm contains a specialized lysosome called the **acrosome,** a packet of enzymes used to penetrate the egg and certain barriers around it (see fig. 26.8, p. 713). When the sperm contacts an egg, the acrosome undergoes exocytosis—the *acrosomal reaction*—releasing these enzymes (fig. 4.1). But the first sperm to reach an egg is not the one to fertilize it. The egg is surrounded by a gelatinous membrane called the *zona pellucida* and,

outside this, a layer of small *granulosa cells.* It may require hundreds of sperm to clear a path through these barriers before one of them can penetrate into the egg itself.

Fertilization

When a sperm contacts the egg's plasma membrane, it digests a hole into the membrane, the sperm and egg membranes fuse, and the sperm nucleus and midpiece enter the egg (fig. 4.1). The *midpiece,* a short segment of the tail behind the head, contains sperm mitochondria, the "powerhouses" that synthesize ATP for sperm motility. The egg, however, usually destroys all sperm mitochondria, so only the mother's mitochondria (and mitochondrial DNA) pass to the offspring.

It is important that only one sperm be permitted to fertilize an egg. If two or more sperm did so—an event called **polyspermy**—the fertilized egg would have 69 or more chromosomes and would fail to develop. The egg has two mechanisms for preventing such a wasteful fate: (1) In the *fast block to polyspermy,* binding of the sperm to the egg opens pores called sodium channels in the egg

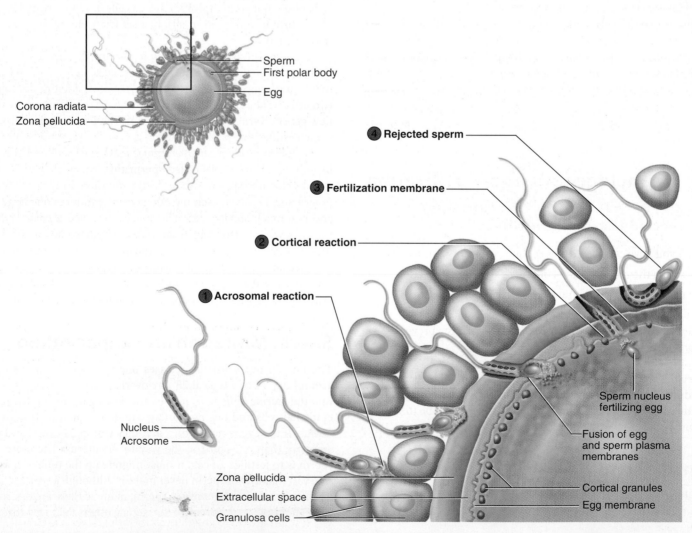

Figure 4.1 Fertilization and the Slow Block to Polyspermy.

membrane. Sodium ions flow rapidly into the egg and cause a change in the electrical charge on the membrane, which inhibits the binding of any more sperm. (2) This is followed by a *slow block to polyspermy*, in which sperm penetration triggers an inflow of calcium ions. Calcium stimulates a *cortical reaction*—the exocytosis of secretory vesicles called **cortical granules** just beneath the egg membrane. The secretion from these granules swells with water, pushes all remaining sperm away from the egg, and creates an impenetrable **fertilization membrane** between the egg and zona pellucida (fig. 4.1, steps 2–4).

Upon fertilization, the egg completes meiosis II and discards a second polar body. The sperm and egg nuclei swell and become **pronuclei.** A mitotic spindle forms between them, each pronucleus ruptures, and the chromosomes of the two gametes mix into a single diploid set. This mingling of the maternal and paternal chromosomes is called **amphimixis.**[3] The cell, now called a **zygote,**[4] is ready for its first mitotic division.

Before You Go On

Answer the following questions to test your understanding of the preceding section:

1. Why is it necessary for gametogenesis to reduce the chromosome number of the sex cells by one-half?
2. Explain why sperm cannot fertilize an egg immediately after ejaculation.
3. Explain why nearly all of an individual's mitochondria originate from the mother.
4. Describe two ways a fertilized egg prevents the entry of excess sperm.

4.2 Stages of Prenatal Development

▶ Expected Learning Outcomes

When you have completed this section, you should be able to

- name and define the three basic stages of prenatal development;
- describe the implantation of a conceptus in the uterine wall;
- describe the major events that transform a fertilized egg into an embryo;
- define and describe the membranes associated with the embryo;
- describe three ways in which the conceptus is nourished during its development;

- describe the formation and functions of the placenta; and
- describe some major developments in the fetal stage.

Human **gestation** (pregnancy) lasts an average of 266 days (38 weeks) from **conception** (fertilization) to **parturition** (childbirth). Since the date of conception is seldom known with certainty, the gestational calendar is usually measured from the day a woman's last menstrual period (LMP) began, and birth is predicted to occur about 280 days (40 weeks) thereafter. Time periods in this chapter, however, are measured from the date of conception.

All the products of conception are collectively called the **conceptus.** This includes all developmental stages from zygote through fetus, and the associated structures such as the umbilical cord, placenta, and amniotic sac.

Clinically, the course of a pregnancy is divided into 3-month intervals called **trimesters:**

1. The **first trimester** (first 12 weeks) extends from fertilization through the first month of fetal life. This is the most precarious stage of development; more than half of all embryos die in the first trimester. Stress, drugs, and nutritional deficiencies are most threatening to the conceptus during this time.
2. The **second trimester** (weeks 13 through 24) is a period in which the organs complete most of their development. It becomes possible with sonography to see good anatomical detail in the fetus. By the end of this trimester, the fetus looks distinctly human, and with intensive clinical care, infants born at the end of the second trimester have a chance of survival.
3. In the **third trimester** (week 25 to birth), the fetus grows rapidly and the organs achieve enough cellular differentiation to support life outside the womb. Some organs, such as the brain, liver, and kidneys, however, require further differentiation after birth to become fully functional. At 35 weeks from fertilization, the fetus typically weighs about 2.5 kg (5.5 lb). It is considered mature at this weight, and usually survives if born early. Most twins are born at about 35 weeks' gestation and solitary infants at 40 weeks.

From a more biological than clinical standpoint, human development is divided into three stages called the preembryonic, embryonic, and fetal stages (table 4.1).

1. The **preembryonic stage** begins with the zygote and lasts about 16 days. It involves three main processes: (1) *cleavage,* or cell division; (2) *implantation,* in which the conceptus becomes embedded in the mucosal lining (*endometrium*) of the uterus; and (3) *embryogenesis,* in which the embryonic cells migrate and differentiate into three tissue layers called the *ectoderm, mesoderm,* and *endoderm*—collectively known as the **primary germ layers.** Once these layers exist, the individual is called an *embryo.*
2. The **embryonic stage** extends from day 17 until the end of week 8. It is a stage in which the primary germ layers develop into the rudiments of all the organ systems. When all organ systems are represented (even though not yet functional), the individual is considered a *fetus.*

---

[3]*amphi* = both + *mixis* = mingling
[4]*zygo* = union

TABLE 4.1	The Stages of Prenatal Development	
Stage	**Age***	**Major Developments and Defining Characteristics**
Preembryonic stage		
Zygote	0–30 hours	A single diploid cell formed by the union of egg and sperm
Cleavage	30–72 hours	Mitotic division of the zygote into smaller, identical blastomeres
Morula	3–4 days	A spherical stage consisting of 16 or more blastomeres
Blastocyst	4–16 days	A fluid-filled, spherical stage with an outer mass of trophoblast cells and inner mass of embryoblast cells; becomes implanted in the endometrium; inner mass forms an embryonic disc and differentiates into the three primary germ layers
Embryonic stage	16 days–8 weeks	A stage in which the primary germ layers differentiate into organs and organ systems; ends when all organ systems are present
Fetal stage	8–38 weeks	A stage in which organs grow and mature at a cellular level to the point of being capable of supporting life independently of the mother

*From the time of fertilization

3. The **fetal stage** of development extends from the beginning of week 9 until birth. This is a stage in which the organs grow, differentiate, and become capable of functioning outside the mother's body.

Preembryonic Stage

The three major events of the preembryonic stage are cleavage, implantation, and embryogenesis.

Cleavage

Cleavage consists of mitotic divisions that occur in the first 3 days after fertilization, dividing the zygote into smaller and smaller cells called **blastomeres.**[5] It begins as the conceptus migrates down the uterine tube (fig. 4.2). The first cleavage occurs in about 30 hours. Blastomeres divide again at shorter and shorter time intervals, doubling the number of cells each time. In the early divisions, the blastomeres divide simultaneously, but as cleavage progresses, they become less synchronized.

By the time the conceptus arrives in the uterus about 72 hours after ovulation, it consists of 16 or more cells and has a bumpy surface similar to a mulberry—hence it is called a **morula.**[6] The morula is no larger than the zygote; cleavage merely produces smaller and smaller blastomeres. This increases the ratio of cell surface area to volume, which favors efficient nutrient uptake and waste removal, and it produces a larger number of cells from which to form different embryonic tissues.

The morula lies free in the uterine cavity for 4 or 5 days and divides into 100 cells or so. It becomes a hollow sphere called the **blastocyst,** with an internal cavity called the **blastocoel** (BLAST-oh-seal). The wall of the blastocyst is a layer of squamous cells

called the **trophoblast,**[7] which is destined to form part of the placenta and plays an important role in nourishing the embryo. On one side of the blastocoel, adhering to the inside of the trophoblast, is an inner cell mass called the **embryoblast,** which is destined to become the embryo itself.

Implantation

About a week after ovulation, the blastocyst attaches to the endometrium, usually on the "ceiling" or on the posterior wall of the uterus. The process of attachment, called **implantation,** begins when the blastocyst adheres to the endometrium (fig. 4.3). The trophoblast cells on this side separate into two layers. In the superficial layer, in contact with the endometrium, the plasma membranes break down and the trophoblast cells fuse into a multinucleate mass called the **syncytiotrophoblast**[8] (sin-SISH-ee-oh-TRO-foe-blast). (A *syncytium* is any body of protoplasm containing multiple nuclei.) The deep layer, close to the embryoblast, is called the **cytotrophoblast**[9] because it retains individual cells divided by membranes.

The syncytiotrophoblast grows into the uterus like little roots, digesting endometrial cells along the way. The endometrium reacts to this injury by growing over the trophoblast and eventually covering it, so the conceptus becomes completely buried in endometrial tissue. Implantation takes about a week and is completed about the time the next menstrual period would have occurred if the woman had not become pregnant.

Embryogenesis

During implantation, the embryoblast undergoes **embryogenesis,** culminating with arrangement of the blastomeres into the three primary germ layers. At the beginning of this phase, the embryoblast

[5]*blast* = bud, precursor + *mer* = segment, part
[6]*mor* = mulberry + *ula* = little

[7]*troph* = food, nourishment + *blast* = to produce
[8]*syn* = together + *cyt* = cell
[9]*cyto* = cell

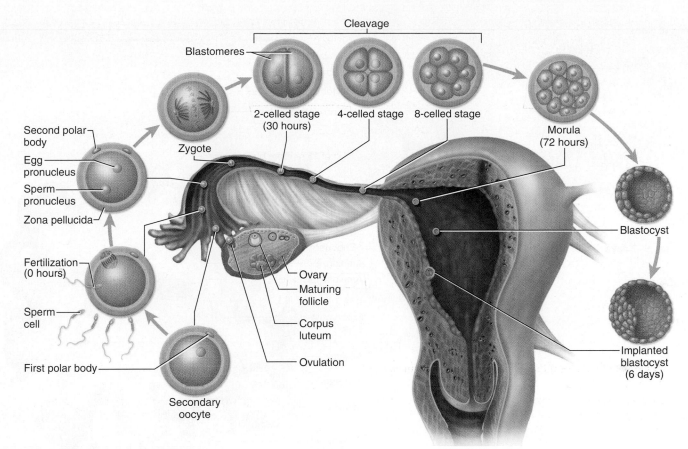

Figure 4.2 **Migration of the Conceptus.** The egg is fertilized in the distal end of the uterine tube, and the preembryo begins cleavage as it migrates to the uterus.

● *Why can't the egg be fertilized in the uterus?* → The egg doesn't last long enough to make the trip.

separates slightly from the trophoblast, creating a narrow space between them called the **amniotic cavity**. The embryoblast flattens into an **embryonic disc (blastodisc)** composed of two cell layers: the **epiblast** facing the amniotic cavity and the **hypoblast** facing away. Some hypoblast cells multiply and form a membrane called the *yolk sac* enclosing the blastocoel. Now the embryonic disc is flanked by two spaces: the amniotic cavity on one side and the yolk sac on the other (fig. 4.3c).

Meanwhile, the disc elongates and, around day 15, a groove called the **primitive streak** forms along the midline of the epiblast. These events make the embryo bilaterally symmetric and define its future right and left sides, dorsal and ventral surfaces, and **cephalic**[10] and **caudal**[11] ends.

The next step is **gastrulation**—multiplying epiblast cells migrate medially toward the primitive streak and down into it (fig. 4.4). These cells replace the original hypoblast with a layer now called the **endoderm**, which will become the inner lining of the digestive tract, among other things. A day later, migrating epiblast cells form a third layer between the first two, called the **mesoderm**. Once this is formed, the epiblast is called **ectoderm**. Thus, all three primary germ layers arise from the original epiblast. Some mesoderm overflows the embryonic disc and becomes an extensive *extraembryonic mesoderm*, which contributes to formation of the placenta (fig. 4.3c).

DEEPER INSIGHT 4.1

Ectopic Pregnancy

In about 1 out of 300 pregnancies, the blastocyst implants somewhere other than the uterus, producing an *ectopic*[12] *pregnancy*. Most cases are *tubal pregnancies,* implantation in the uterine tube. This usually occurs because the conceptus encounters an obstruction such as a constriction resulting from earlier pelvic inflammatory disease, tubal surgery, previous ectopic pregnancies, or repeated miscarriages. The uterine tube cannot expand enough to accommodate the growing conceptus for long; if the situation is not detected and treated early, the tube usually ruptures within 12 weeks, potentially killing the mother. Occasionally, a conceptus implants in the abdominopelvic cavity, producing an *abdominal pregnancy*. It can grow anywhere it finds an adequate blood supply, such as on the outside of the uterus, colon, or bladder. About 1 pregnancy in 7,000 is abdominal. Abdominal pregnancy is a serious threat to the mother's life and usually requires abortion, but about 9% of abdominal pregnancies end in live birth by cesarean section.

---

[10]*cephal* = head
[11]*caud* = tail

---

[12]*ec* = outside + *top* = place

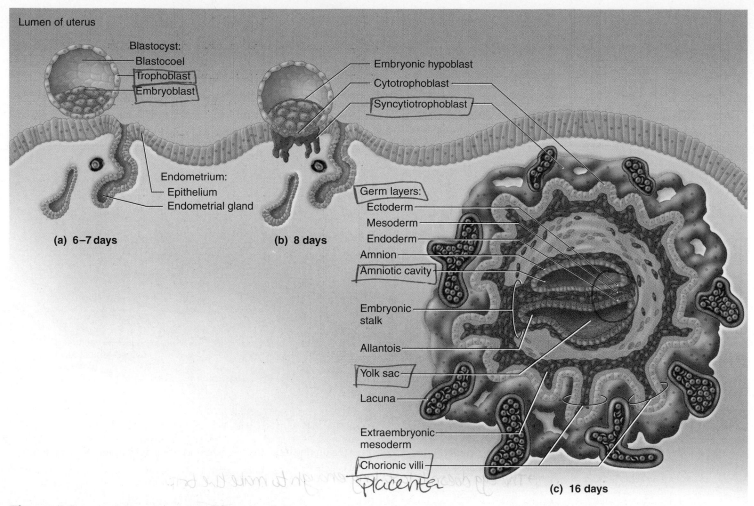

Figure 4.3 Implantation. (a) Structure of the blastocyst 6 to 7 days after ovulation, when it first adheres to the uterine wall. (b) The progress of implantation about 1 day later. The syncytiotrophoblast has begun growing rootlets, which penetrate the endometrium. (c) By 16 days, the conceptus is completely covered by endometrial tissue. The embryo is now flanked by a yolk sac and amnion and is composed of three primary germ layers.

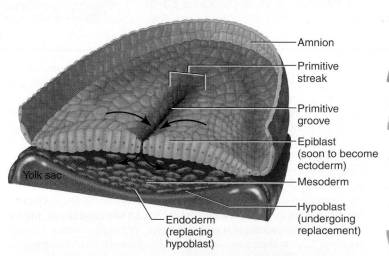

Figure 4.4 Formation of the Primary Germ Layers (Gastrulation). Composite view of the embryonic disc at 15 to 16 days. Epiblast cells migrate over the surface and down into the primitive groove, first replacing the hypoblast cells with endoderm, then filling the space with mesoderm. Upon completion of this process, the uppermost layer is considered ectoderm.

The ectoderm and endoderm are epithelia composed of tightly joined cells, but the mesoderm is a more loosely organized tissue. It later differentiates into a loose fetal connective tissue called **mesenchyme,**[13] which gives rise to such tissues as muscle, bone, and blood. Mesenchyme is composed of a loose network of wispy *mesenchymal cells* embedded in a gelatinous ground substance.

Once the three primary germ layers are formed, embryogenesis is complete and the individual is considered an **embryo.** It is about 2 mm long and 16 days old at this point.

Embryonic Stage

The embryonic stage of development begins around day 16 and extends to the end of week 8. During this time, the placenta and other accessory structures develop, the embryo begins receiving nutrition primarily from the placenta, and the germ layers differentiate into organs and organ systems. Although these organs are still far from functional, it is their presence at 8 weeks that marks the transition from the embryonic stage to the fetal stage.

---

[13]*mes* = middle + *enchym* = poured into

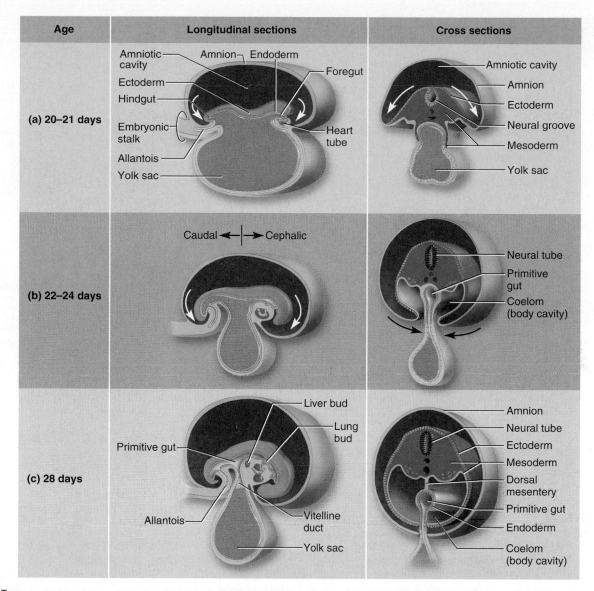

Age	Longitudinal sections	Cross sections
(a) 20–21 days	Amniotic cavity, Amnion, Endoderm, Ectoderm, Foregut, Hindgut, Embryonic stalk, Heart tube, Allantois, Yolk sac	Amniotic cavity, Amnion, Ectoderm, Neural groove, Mesoderm, Yolk sac
(b) 22–24 days	Caudal ←	→ Cephalic	Neural tube, Primitive gut, Coelom (body cavity)
(c) 28 days	Liver bud, Lung bud, Primitive gut, Allantois, Vitelline duct, Yolk sac	Amnion, Neural tube, Ectoderm, Mesoderm, Dorsal mesentery, Primitive gut, Endoderm, Coelom (body cavity)

Figure 4.5 **Embryonic Folding.** The left-hand figures are longitudinal sections with the cephalic (head) end facing right. The right-hand figures are cross sections cut about midway along the figures on the left. Part (a) corresponds to figure 4.3c at a slightly later stage of development. Note the general trend for the cephalic and caudal ends of the embryo to curl toward each other (left-hand figures) until the embryo assumes a C shape, and for the flanks of the embryo to fold laterally (right-hand figures), converting the flat embryonic disc into a more cylindrical body and eventually enclosing a body cavity (c).

Embryonic Folding and Organogenesis

In weeks 3 to 4, the embryo grows rapidly and folds around the yolk sac, converting the flat embryonic disc into a somewhat cylindrical form. As the cephalic and caudal ends curve around the ends of the yolk sac, the embryo becomes C-shaped, with the head and tail almost touching (fig. 4.5). The lateral margins of the disc fold around the sides of the yolk sac to form the ventral surface of the embryo. This lateral folding encloses a longitudinal channel, the *primitive gut,* which later becomes the digestive tract.

As a result of embryonic folding, the entire surface is covered with ectoderm, which later produces the epidermis of the skin. In the meantime, the mesoderm splits into two layers. One adheres to the ectoderm and the other to the endoderm, thus opening a space

called the **coelom** (SEE-loam) between them (fig. 4.5b, c). The coelom becomes divided into the thoracic cavity and peritoneal cavity by a wall, the diaphragm. By the end of week 5, the thoracic cavity further subdivides into pleural and pericardial cavities.

The formation of organs and organ systems during this time is called **organogenesis.** Table 4.2 lists the major tissues and organs that arise from each primary germ layer.

Apply What You Know

List the four primary tissue types of the adult body (see chapter 3), and identify which of the three primary germ layers of the embryo predominantly gives rise to each.

TABLE 4.2	Derivatives of the Three Primary Germ Layers
Layer	**Major Derivatives**
Ectoderm	Epidermis; hair follicles and piloerector muscles; cutaneous glands; nervous system; adrenal medulla; pineal and pituitary glands; lens, cornea, and intrinsic muscles of the eye; internal and external ear; salivary glands; epithelia of the nasal cavity, oral cavity, and anal canal
Mesoderm	Dermis; skeleton; skeletal, cardiac, and most smooth muscle; cartilage; adrenal cortex; middle ear; blood and lymphatic vessels; blood; bone marrow; lymphoid tissue; epithelium of kidneys, ureters, gonads, and genital ducts; mesothelium of abdominal and thoracic cavities
Endoderm	Most of the mucosal epithelium of the digestive and respiratory tracts; mucosal epithelium of urinary bladder and parts of urethra; epithelial components of accessory reproductive and digestive glands (except salivary glands); thyroid and parathyroid glands; thymus

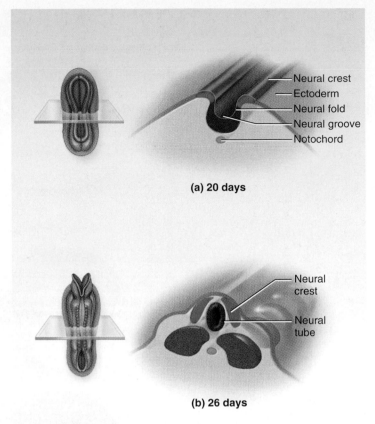

(a) 20 days

(b) 26 days

Figure 4.6 **Neurulation.** (a) The neural groove at 20 days. (b) The neural tube at 26 days.

Three major events of organogenesis are especially important for understanding organ development in later chapters: development of the *neural tube,* outpocketing of the throat region to form *pharyngeal pouches,* and the appearance of body segments called *somites.*

The formation of the **neural tube** is called **neurulation.** This process is detailed in chapter 13, but a few essential points are needed here. By week 3, a thick ridge of ectoderm called the *neural plate* appears along the midline of the embryonic disc. This is the source of the entire nervous system. As development progresses, the neural plate sinks and becomes a *neural groove,* with a raised edge called the *neural fold* on each side (fig. 4.6a). Cells along the lateral margins of the fold become specialized *neural crest* tissue. Next, the edges of the fold meet and close, somewhat like a zipper, beginning in the middle of the embryo and progressing toward both ends. By 4 weeks, this process creates an enclosed channel, the neural tube (fig. 4.6b).

Neural crest cells dissociate from the overlying ectoderm and sink a little more deeply into the embryo to flank the neural tube. As described in chapter 13, neural crest cells later migrate to various positions in the embryo and give rise to several other components of the nervous system and to other tissues. By week 4, the cephalic end of the neural tube develops bulges or *vesicles* that develop into different regions of the brain, and the more caudal part becomes the spinal cord. Neurulation is one of the most sensitive periods of prenatal development. Abnormal developments called *neural tube defects* are among the most common and devastating birth defects (see p. 367).

Pharyngeal (branchial) pouches are five pairs of pockets that form in the walls of the future throat of the embryo at around 4 to 5 weeks' gestation (fig. 4.7). They are separated by **pharyngeal arches,** which appear as external bulges in the neck region (fig. 4.8b). Pharyngeal pouches are among the basic defining characteristics of all vertebrates from fish to mammals. In humans, they give rise to the middle-ear cavity, palatine tonsil, thymus, parathyroid glands, and part of the thyroid gland.

Somites are bilaterally paired blocks of mesoderm that give the embryo a segmented appearance (fig. 4.8a, b). They represent a primitive vertebrate segmentation that is more distinctly visible in fish, snakes, and other lower vertebrates than in mammals. Humans, however, show traces of this segmentation in the linear series of vertebrae, ribs, spinal nerves, and trunk muscles. Somites begin to appear by day 20 and number 42 to 44 pairs by day 35. Beginning in week 4, each somite subdivides into three tissue masses: a **sclerotome,**[14] which surrounds the neural tube and gives rise to bone tissue of the vertebral column; a **myotome,**[15] which gives rise to muscles of the trunk; and a **dermatome,**[16]

[14]*sclero* = hard + *tom* = segment
[15]*myo* = muscle + *tom* = segment
[16]*derma* = skin + *tom* = segment

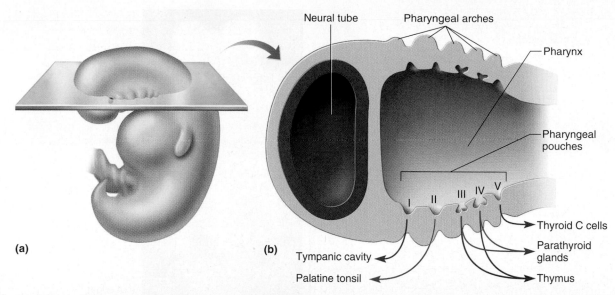

Figure 4.7 The Pharyngeal Pouches. (a) Level at which the section in part (b) is taken. (b) Superior view of the pharyngeal region showing the five pairs of pharyngeal pouches (I–V) and their developmental fates. Compare figure 4.8b.

DEEPER INSIGHT 4.2

Morning Sickness

A woman's earliest sign of pregnancy is often *morning sickness,* a nausea that sometimes progresses to vomiting. Severe and prolonged vomiting, called *hyperemesis gravidarum,*[17] can necessitate hospitalization for fluid therapy to restore electrolyte and acid–base balance. The physiological cause of morning sickness is unknown; it may result from the steroids of pregnancy inhibiting intestinal motility. It is also uncertain whether it is merely an undesirable effect of pregnancy or whether it has a biological purpose. An evolutionary hypothesis is that morning sickness is an adaptation to protect the embryo from toxins. The embryo is most vulnerable to toxins at the same time that morning sickness peaks, and women with morning sickness tend to prefer bland foods and to avoid spicy and pungent foods, which are highest in toxic compounds. Pregnant women tend also to be especially sensitive to flavors and odors that suggest spoiled food. Women who do not experience morning sickness are more likely to miscarry or bear children with birth defects.

which gives rise to the dermis of the skin and to its associated subcutaneous tissue.

At 5 weeks, the embryo exhibits a prominent **head bulge** at the cephalic end and a pair of optic vesicles destined to become the eyes. A large **heart bulge** contains a heart, which has been beating since day 22. The **arm buds** and **leg buds,** the future limbs, are

[17]*hyper* = excessive + *emesis* = vomiting + *gravida* = pregnant woman + *arum* = of

present at 24 and 28 days, respectively. Figure 4.8 shows the external appearance of embryos from 3 to 7 weeks.

Extraembryonic Membranes

The conceptus develops a number of accessory organs external to the embryo itself. These include the placenta, umbilical cord, and four **extraembryonic membranes**—the *amnion, yolk sac, allantois,* and *chorion* (fig. 4.9). To understand these membranes, it helps to realize that all mammals evolved from egg-laying reptiles. Within the shelled, self-contained egg of a reptile, the embryo rests atop a yolk, which is enclosed in the yolk sac; it is suspended in a pool of liquid contained in the amnion; it stores toxic wastes in another sac, the allantois; and to breathe, it has a chorion permeable to gases. All of these membranes persist in mammals, including humans, but are modified in their functions.

The **amnion** is a transparent sac that develops from epiblast cells of the embryonic disc. It grows to completely enclose the embryo and is penetrated only by the umbilical cord (figs. 4.9a, b; 4.12a, c). The amnion fills with **amniotic fluid,** which enables the embryo to develop symmetrically; keeps its surface tissues from adhering to each other; protects it from trauma, infection, and temperature fluctuations; allows the freedom of movement important to muscle development; and plays a role in lung development as the fetus "breathes" the fluid. At first, the amniotic fluid forms by filtration of the mother's blood plasma, but beginning at 8 to 9 weeks, the fetus urinates into the amniotic cavity about once an hour and contributes substantially to the fluid volume. The volume remains stable, however, because the fetus swallows amniotic fluid at a comparable rate. By the end of gestation, the amnion contains 700 to 1,000 mL of fluid.

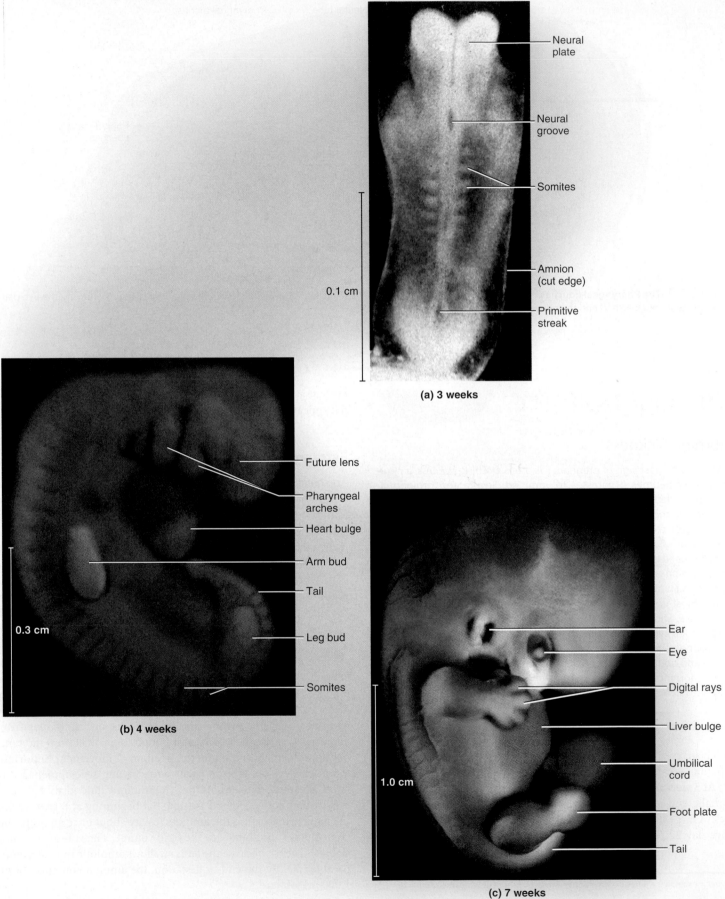

Neural plate

Neural groove

Somites

Amnion (cut edge)

Primitive streak

0.1 cm

(a) 3 weeks

Future lens

Pharyngeal arches

Heart bulge

Arm bud

Tail

Leg bud

Somites

0.3 cm

(b) 4 weeks

Ear

Eye

Digital rays

Liver bulge

Umbilical cord

Foot plate

Tail

1.0 cm

(c) 7 weeks

Figure 4.8 Human Embryos from 3 to 7 Weeks.

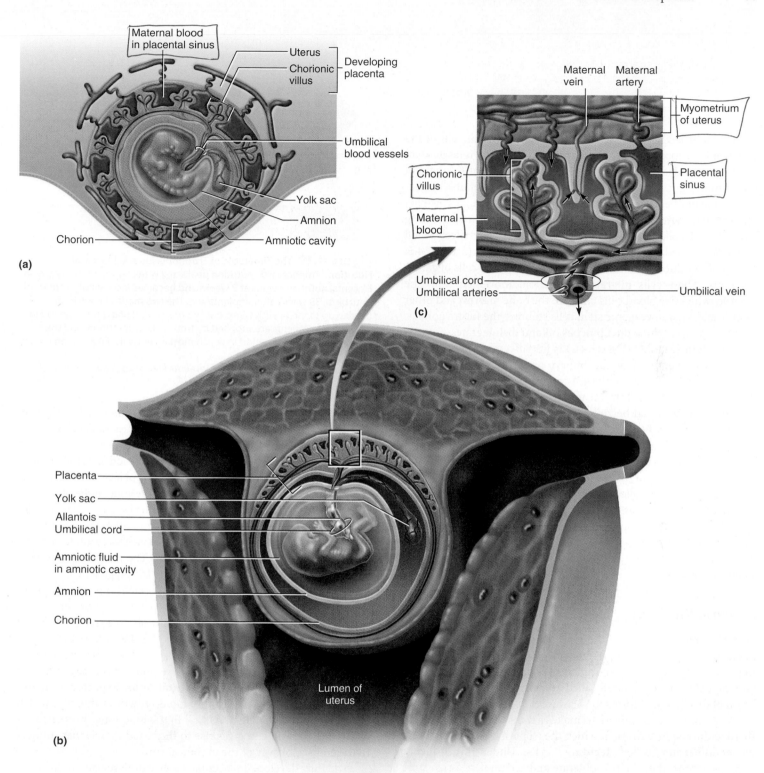

Figure 4.9 **The Placenta and Extraembryonic Membranes.** (a) Embryo at 4 weeks, enclosed in the amnion and chorion and surrounded by a developing placenta. (b) Fetus at 12 weeks. The placenta is now complete and lies on only one side of the fetus. (c) A portion of the mature placenta and umbilical cord, showing the relationship between fetal and maternal circulation.

Apply What You Know

Oligohydramnios[18] is an abnormally low volume of amniotic fluid. *Renal agenesis*[19] is a failure of the fetal kidneys to develop. Which of these do you think is most likely to cause the other one? Explain why. What could be some consequences of oligohydramnios to fetal development?

The **yolk sac,** as we have already seen, arises from cells of the embryonic hypoblast opposite the amnion. Initially, it is larger than the embryo and is broadly connected to almost the entire length of the primitive gut. During embryonic folding, however, its connection to the gut becomes constricted and reduced to a narrow passage called the **vitelline**[20] **duct.** Since the embryo continues growing long after the yolk sac stops, the yolk sac becomes a relatively small pouch suspended from the ventral side of the embryo (see figs. 4.5; 4.9a, b). It produces the first blood cells and the stem cells of gametogenesis. These cells migrate by ameboid movement into the embryo, where the blood cells colonize the bone marrow and other tissues, and the gametogenic stem cells colonize the future gonads. Eventually, the vitelline duct pinches off and disintegrates.

The **allantois** (ah-LON-toe-iss) is initially an outpocketing of the yolk sac; eventually, as the embryo grows, it becomes an outgrowth of the caudal end of the gut connected to it by the **allantoic duct** (see figs. 4.5 and 4.9b). It forms a foundation for growth of the umbilical cord and becomes part of the urinary bladder. The allantoic duct can be seen in histological cross sections of the umbilical cord if they are cut close enough to the fetal end.

The **chorion** (CORE-ee-on) is the outermost membrane, enclosing all the rest of the membranes and the embryo. Initially, it has shaggy processes called **chorionic villi** around its entire surface (fig. 4.9a). As the pregnancy advances, the villi on the placental side grow and branch, and this surface is then called the *villous chorion.* The villous chorion forms the fetal portion of the placenta. The villi degenerate over the rest of the surface, which is then called the *smooth chorion.*

Prenatal Nutrition

Over the course of gestation, the conceptus is nourished in three different, overlapping ways. As it travels down the uterine tube and lies free in the uterine cavity before implantation, it absorbs a glycogen-rich secretion of the uterine glands called **uterine milk.** It is the accumulation of this fluid that forms the blastocoel in figure 4.3a.

As it implants, the conceptus makes a transition to **trophoblastic (deciduous) nutrition,** in which the trophoblast digests cells of the endometrium called **decidual**[21] **cells.** Under the influence of progesterone, these cells proliferate and accumulate a rich store of glycogen, proteins, and lipids. As the conceptus burrows into the endometrium, the syncytiotrophoblast digests them and supplies the nutrients to the embryoblast. Trophoblastic nutrition is the only mode of nutrition for the first week after implantation. It remains the dominant source of nutrients through the end of week 8; the period from implantation through week 8 is therefore called the

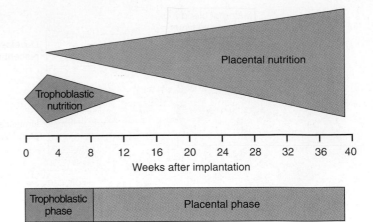

Figure 4.10 **The Timetable of Trophoblastic and Placental Nutrition.** Trophoblastic nutrition peaks at 2 weeks and ends by 12 weeks. Placental nutrition begins at 2 weeks and becomes increasingly important until birth, 39 weeks after implantation. The two modes of nutrition overlap up to the twelfth week, but the *trophoblast phase* is the period in which most nutrients are supplied by trophoblastic nutrition, and the *placental phase* is the period in which most (eventually all) nutrition comes from the placenta.
● *At what point do the two modes contribute equally to prenatal nutrition?*

trophoblastic phase of the pregnancy. Trophoblastic nutrition wanes as placental nutrition takes over, and ceases entirely by the end of week 12 (fig. 4.10).

In **placental nutrition,** nutrients from the mother's blood diffuse through the placenta into the fetal blood. The **placenta**[22] is a vascular organ attached to the uterine wall on one side and, on the other side, connected to the fetus by way of the **umbilical cord.** It begins to develop about 11 days after conception, becomes the dominant mode of nutrition at the beginning of week 9, and is the sole mode of nutrition from the end of week 12 until birth. The period from week 9 until birth is called the **placental phase** of the pregnancy.

Placental development begins when extensions of the syncytiotrophoblast, the first chorionic villi, penetrate more and more deeply into the endometrium, like the roots of a tree penetrating into the nourishing "soil" of the uterus (fig. 4.9a, c). As they digest their way through uterine blood vessels, the villi become surrounded by pools of free blood. These pools eventually merge to form a blood-filled cavity, the **placental sinus.** Exposure to maternal blood stimulates increasingly rapid growth of the villi, which become branched and treelike. Extraembryonic mesenchyme grows into the villi and gives rise to the blood vessels that connect to the embryo by way of the umbilical cord.

The fully developed placenta is a disc of tissue about 20 cm in diameter and 3 cm thick (fig. 4.11). At birth, it weighs about one-sixth as much as the baby. The surface facing the fetus is smooth and gives rise to the umbilical cord. The surface facing the uterine wall is rougher. It consists of the chorionic villi, which are contributed by the fetus, and a region of the mother's endometrium called the *decidua basalis.*

The umbilical cord contains two **umbilical arteries** and one **umbilical vein.** Pumped by the fetal heart, blood flows into the placenta by way of the umbilical arteries and then returns to the

---

[18]*oligo* = few, little + *hydr* = water, fluid + *amnios* = amniotic
[19]*a* = without + *genesis* = formation, development
[20]*vitell* = yolk
[21]*decid* = falling off

[22]*placenta* = flat cake

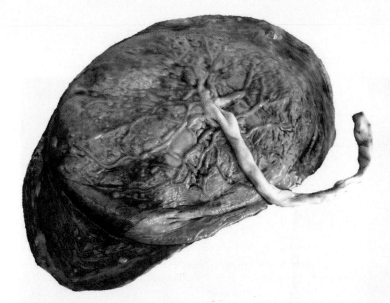

(a) Fetal side

(b) Maternal (uterine) side

Figure 4.11 **The Placenta and Umbilical Cord.** (a) The fetal side, showing blood vessels, the umbilical cord, and some of the amniotic sac attached to the lower left margin. (b) The maternal (uterine) side, where chorionic villi give the placenta a rougher texture.
● *How many arteries and how many veins are found in the umbilical cord? Which vessels carry blood with the higher level of oxygen?*

fetus by way of the umbilical vein. The chorionic villi are *filled with* fetal blood and *surrounded by* maternal blood; the two blood-streams do not mix unless there is damage to the placental barrier. The barrier, however, is only 3.5 μm thick—half the diameter of a red blood cell. Early in development, the chorionic villi have thick membranes that are not very permeable to nutrients and wastes, and their total surface area is relatively small. As the villi grow and branch, their surface area increases and the membranes become thinner and more permeable. Thus there is a dramatic increase in *placental conductivity*, the rate at which substances diffuse through the membrane. Oxygen and nutrients pass from the maternal blood to the fetal blood, while fetal wastes pass the other way to be eliminated by the mother. Unfortunately, the placenta is also permeable to nicotine, alcohol, and most other drugs in the maternal blood-stream. Nutrition, excretion, and other functions of the placenta are summarized in table 4.3.

TABLE 4.3	Functions of the Placenta
Nutritional roles	Transports nutrients such as glucose, amino acids, fatty acids, minerals, and vitamins from the maternal blood to the fetal blood; stores nutrients such as carbohydrates, protein, iron, and calcium in early pregnancy and releases them to the fetus later, when fetal demand is greater than the mother can absorb from the diet
Excretory roles	Transports nitrogenous wastes such as ammonia, urea, uric acid, and creatinine from the fetal blood to the maternal blood
Respiratory roles	Transports O_2 from mother to fetus, and CO_2 from fetus to mother
Endocrine roles	Secretes hormones (estrogen, progesterone, relaxin, human chorionic gonadotropin, and human chorionic somatomammotropin); allows other hormones synthesized by the conceptus to pass into the mother's blood and maternal hormones to pass into the fetal blood
Immune role	Transports maternal antibodies into fetal blood to confer immunity on fetus

DEEPER INSIGHT   4.3

Placental Disorders

The two primary causes of third-trimester bleeding are placental disorders called *placenta previa* and *abruptio placentae*. These are similar and easily mistaken for each other. A suspicion of either condition calls for a sonogram to differentiate the two and decide on a course of action.

The conceptus usually implants high on the body of the uterus or on its ceiling. In about 0.5% of births, however, the placenta is so low on the uterine wall that it partially or completely blocks the cervical canal. This condition, called *placenta previa*, makes it impossible for the infant to be born without the placenta separating from the uterine wall first. Thus, there is a possibility of life-threatening hemorrhaging during pregnancy or birth. If placenta previa is detected by sonography, the infant is delivered by cesarean (C) section.

Abruptio placentae (ah-BRUP-she-oh pla-SEN-tee) is the premature partial or total separation of the placenta from the uterine wall. It occurs in 0.4% to 3.5% of pregnancies. Slight separations may require no more than bed rest and observation, but more severe cases can threaten the life of the mother, fetus, or both. Such cases require early delivery, usually by C section.

Fetal Stage

At the end of 8 weeks, all of the organ systems are present, the individual is about 3 cm long, and it is now considered a **fetus** (fig. 4.12). Its bones have just begun to calcify and the skeletal muscles exhibit spontaneous contractions, although these are too weak to be felt by

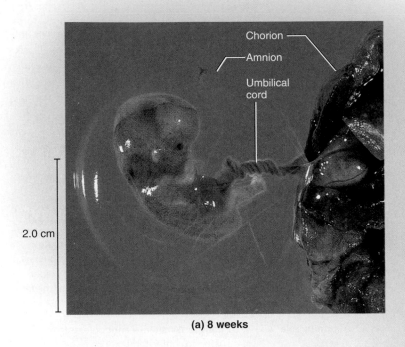

Chorion

Amnion

Umbilical cord

2.0 cm

(a) 8 weeks

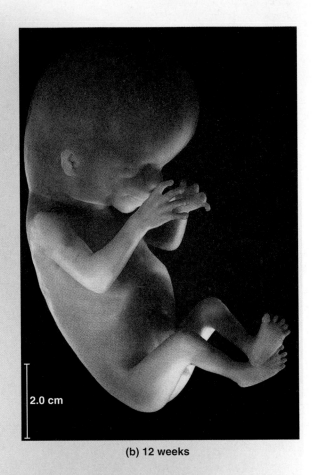

2.0 cm

(b) 12 weeks

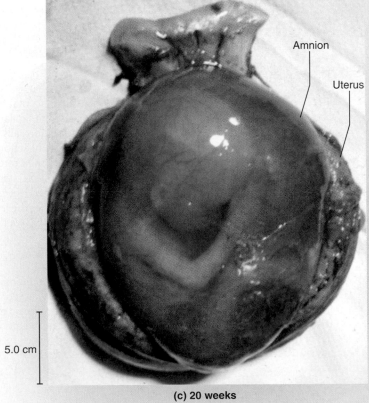

Amnion

Uterus

5.0 cm

(c) 20 weeks

Figure 4.12 Human Fetuses from 8 to 20 Weeks.

TABLE 4.4	Major Events of Prenatal Development, with Emphasis on the Fetal Stage	
End of Week	**Crown-to-Rump Length; Weight**	**Developmental Events**
4	0.6 cm; <1 g	Vertebral column and central nervous system begin to form; limbs represented by small limb buds; heart begins beating around day 22; no visible eyes, nose, or ears
8	3 cm; 1 g	Eyes form, eyelids fused shut; nose flat, nostrils evident but plugged with mucus; head nearly as large as the rest of the body; brain waves detectable; bone calcification begins; limb buds form paddlelike hands and feet with ridges called *digital rays,* which then separate into distinct fingers and toes; blood cells and major blood vessels form; genitals present but sexes not yet distinguishable
12	9 cm; 45 g	Eyes well developed, facing laterally; eyelids still fused; nose develops bridge; external ears present; limbs well formed, digits exhibit nails; fetus swallows amniotic fluid and produces urine; fetus moves, but too weakly for mother to feel it; liver is prominent and produces bile; palate is fusing; sexes can be distinguished
16	14 cm; 200 g	Eyes face anteriorly, external ears stand out from head, face looks more distinctly human; body larger in proportion to head; skin is bright pink, scalp has hair; joints forming; lips exhibit sucking movements; kidneys well formed; digestive glands forming and *meconium*[23] (fetal feces) accumulating in intestine; heartbeat can be heard with a stethoscope
20	19 cm; 460 g	Body covered with fine hair called *lanugo*[24] and cheeselike sebaceous secretion called *vernix caseosa,*[25] which protects it from amniotic fluid; skin bright pink; brown fat forms and will be used for postpartum heat production; fetus is now bent forward into "fetal position" because of crowding; *quickening* occurs—mother can feel fetal movements
24	23 cm; 820 g	Eyes partially open; skin wrinkled, pink, and translucent; lungs begin producing *surfactant,* a fluid that aids postpartum respiration; rapid weight gain
28	27 cm; 1,300 g	Eyes fully open; skin wrinkled and red; full head of hair present; eyelashes formed; fetus turns to upside-down *vertex position;* testes begin to descend into scrotum; may survive if born at 28 weeks
32	30 cm; 2,100 g	Subcutaneous fat deposition gives fetus a more plump, babyish appearance, with lighter, less wrinkled skin; testes descending
36	34 cm; 2,900 g	More subcutaneous fat deposited, body plump; lanugo is shed; nails extend to fingertips; limbs flexed; firm hand grip
38	36 cm; 3,400 g	Prominent chest, protruding breasts; testes in inguinal canal or scrotum; fingernails extend beyond fingertips

the mother. The heart, which has been beating since the fourth week, now circulates blood. The heart and liver are very large and form the prominent ventral bulges seen in figure 4.8b and c. The head is nearly half the body length.

The primary changes in the fetal period are that the organ systems become functional and the fetus rapidly gains weight and becomes more human looking. Full-term fetuses average about 36 cm from the crown of the head to the curve of the buttock in a sitting position *(crown-to-rump length, CRL).* Most neonates (newborn infants) weigh between 3.0 and 3.4 kg (6.6–7.5 lb). About 50% of this weight is gained in the last 10 weeks. Most neonates weighing 1.5 to 2.5 kg are viable, but with difficulty. Neonates weighing under 500 g rarely survive.

The face acquires a more distinctly human appearance during the last trimester. The head grows more slowly than the rest of the body, so its relative length drops from one-half of the CRL at 8 weeks to one-fourth at birth. The skull has the largest circumference of any body region at term (about 10 cm), and passage of the head is therefore the most difficult part of labor. The limbs grow more rapidly than the trunk during the fetal stage, and achieve their final relative proportions to the trunk by 20 weeks (fig. 4.12c). Other highlights of fetal development are described in table 4.4, and the development of individual organ systems is detailed in the chapters that follow (table 4.5).

TABLE 4.5	Information on Further Development of the Organ Systems
Integumentary system	p. 121
Bone tissue	p. 137
Skeletal system, axial	p. 175
Skeletal system, appendicular	p. 198
Muscular system	p. 256
Central nervous system	p. 365
Autonomic nervous system	p. 456
Sense organs	p. 490
Endocrine system	p. 512
Circulatory system, the heart	p. 557
Circulatory system, blood vessels	p. 601
Lymphatic system	p. 626
Respiratory system	p. 648
Digestive system	p. 679
Urinary system	p. 698
Reproductive system	p. 729

[23]*mecon* = poppy juice, opium
[24]*lan* = down, wool
[25]*vernix* = varnish + *caseo* = cheese

DEEPER INSIGHT 4.4

Kindred Genes for Embryonic Development

Ever since scientists first observed sperm, eggs, and zygotes, they have been intrigued with how a single cell could develop into something as complex as the human body. As noted at the beginning of this chapter, seventeenth-century microscopists imagined that they saw tiny, fully formed human bodies curled up in the sperm or egg. Toward the end of the twentieth century, however, geneticists made a major breakthrough. They discovered a set of genes, now called *homeobox (Hox) genes,* that control anatomical development. By regulating other genes, homeobox genes indirectly determine what proteins the body's cells produce and when they produce them. They act as master switches that determine such fundamental aspects of anatomy as limb position, shapes of the bones, segmentation of the mesoderm into somites, the body's dorsal–ventral and cephalic–caudal axes, and its basic left–right symmetry. Mutations in homeobox genes are now known to underlie some developmental defects in the brain, head, and limbs.

Homeobox genes have become a key concept in the understanding of both the embryonic and evolutionary histories of an organism, thus launching a new branch of science called *evolutionary-developmental (evo-devo) biology.* Homeobox genes were discovered in fruit flies, but the DNA sequences of these genes are almost identical even in such distantly related species as mice. Fruit flies, frogs, mice, and humans have the same homeobox gene for making an eye, for example. Despite vast differences between the *compound eye* of an insect and the *camera eye* of a mammal, evolution has conserved the same gene for eye development for hundreds of millions of years. This is a spectacular example of how genes not only regulate basic embryonic development, but also reconfirm the kinship of living species.

Before You Go On

Answer the following questions to test your understanding of the preceding section:

5. What is the criterion for classifying a developing individual as an embryo? What is the criterion for classifying it as a fetus? At what gestational ages are these stages reached?

6. In the blastocyst, what are the cells called that eventually give rise to the embryo? What are the cells that carry out implantation?

7. Name and define the three principal processes that occur in the preembryonic stage.

8. Name the three primary germ layers and explain how they develop in the embryonic disc.

9. Distinguish between trophoblastic and placental nutrition.

10. State the functions of the placenta, amnion, chorion, yolk sac, and allantois.

11. Define and describe the neural tube, primitive gut, somites, and pharyngeal pouches.

4.3 Clinical Perspectives

▶ Expected Learning Outcomes

When you have completed this section, you should be able to

- discuss the frequency and some causes of early spontaneous abortion;

- discuss some types of birth defects and major categories of their causes;

- describe some syndromes that result from chromosomal nondisjunction; and

- explain what teratogens are and describe some of their effects.

Expectant parents worry a great deal about the possibilities of miscarriage or birth defects. It is estimated that, indeed, more than half of all pregnancies end in miscarriage, often without the parents realizing that a pregnancy had even begun, and 2% to 3% of infants born in the United States have clinically significant birth defects.

Spontaneous Abortion

Most miscarriages are *early spontaneous abortions,* occurring within 3 weeks of fertilization. Such abortions are easily mistaken for a late and unusually heavy menstrual period. One investigator estimated that 25% to 30% of blastocysts fail to implant; 42% of implanted blastocysts die by the end of the second week; and 16% of those that make it through 2 weeks are seriously abnormal and abort within the next week. Another study found that 61% of early spontaneous abortions were due to chromosomal abnormalities.

Even later in development, spontaneously aborted fetuses show a significantly higher incidence of neural tube defects, cleft lip, and cleft palate than do newborns or fetuses of induced abortions. Spontaneous abortion may in fact be a natural mechanism for preventing the development of nonviable fetuses or the birth of severely deformed infants.

Birth Defects

A birth defect, or **congenital[26] anomaly,** is the abnormal structure or position of an organ at birth, resulting from a defect in prenatal development. The study of birth defects is called **teratology.**[27] Birth defects are the single most common cause of infant mortality in North America. Not all congenital anomalies are noticeable at birth; some are detected months to years later. Thus, by the age of 2 years, 6% of children are diagnosed with congenital anomalies, and by age 5, the incidence is 8%. The following sections discuss some known causes of congenital anomalies, but in 50% to 60% of cases, the cause is unknown.

---

[26]*con* = with + *gen* = born
[27]*terato* = monster + *logy* = study of

Mutagens and Genetic Anomalies

Genetic anomalies are the most common known cause of birth defects, accounting for an estimated one-third of all cases and 85% of those with an identifiable cause. One cause of genetic defects is **mutations,** or changes in DNA structure. Among other disorders, mutations cause achondroplastic dwarfism (see Deeper Insight 6.3, p. 142), microcephaly (abnormal smallness of the head), stillbirth, and childhood cancer. Mutations can occur through errors in DNA replication during the cell cycle or under the influence of environmental agents called **mutagens,** including some chemicals, viruses, and radiation.

Some of the most common genetic disorders result not from mutagens, however, but from **aneuploidy**[28] (AN-you-ploy-dee), an abnormal number of chromosomes in the zygote. Aneuploidy results from **nondisjunction,** a failure of one of the 23 pairs of chromosomes to separate during meiosis, so that both members of the pair go to the same daughter cell. For example, suppose nondisjunction resulted in an egg with 24 chromosomes instead of the normal 23. If this egg were fertilized by a normal sperm, the zygote would have 47 chromosomes instead of the usual 46.

Figure 4.13 compares normal disjunction of the X chromosomes with some effects of nondisjunction. In nondisjunction, an egg cell may receive both X chromosomes. If it is fertilized by an X-bearing sperm, the result is an XXX zygote and a suite of defects called the **triplo-X syndrome.** Triplo-X females are sometimes infertile and may have mild intellectual impairments. If an XX egg is fertilized by a Y-bearing sperm, the result is an XXY combination, causing **Klinefelter**[29] **syndrome.** People with Klinefelter syndrome are sterile males, usually of average intelligence, but with undeveloped testes, sparse body hair, unusually long arms and legs, and enlarged breasts *(gynecomastia*[30]*).* This syndrome often goes undetected until puberty, when failure to develop secondary sex characteristics may prompt genetic testing.

The other possible outcome of X chromosome nondisjunction is that an egg cell may receive no X chromosome (both X chromosomes are discarded in the first polar body). If fertilized by a Y-bearing sperm, such an egg dies for lack of the indispensable genes on the X chromosome. If it is fertilized by an X-bearing sperm, however, the result is a female

with **Turner**[31] **syndrome,** with an XO combination (*O* represents the absence of one sex chromosome). Only 3% of fetuses with Turner syndrome survive to birth. Girls who survive show no serious impairments as children, but tend to have a webbed neck and widely spaced nipples. At puberty, the secondary sex characteristics fail to develop. The ovaries are nearly absent, the girl remains sterile, and she usually has a short stature.

---

[31]Henry H. Turner (1892–1970), American endocrinologist

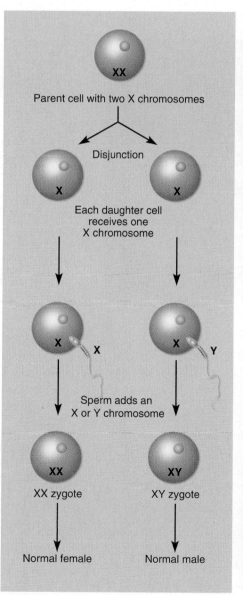

(a) Normal disjunction of X chromosomes

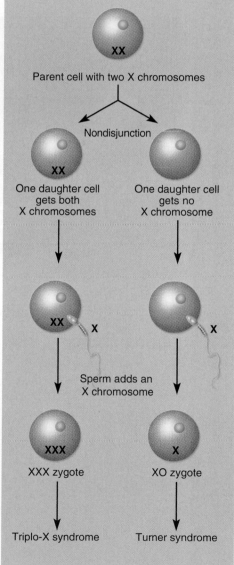

(b) Nondisjunction of X chromosomes

Figure 4.13 Disjunction and Nondisjunction. (a) The outcome of normal disjunction and fertilization by X- or Y-bearing sperm. (b) Two of the possible outcomes of nondisjunction followed by fertilization with an X-bearing sperm.
● *In the right half of the figure, what would be the two outcomes if the sperm carried a Y chromosome instead of an X?*

---

[28]*an* = not + *eu* = good, normal + *ploid* = form
[29]Harry F. Klinefelter, Jr. (1912–), American physician
[30]*gyneco* = female + *mast* = breast + *ia* = condition

The other 22 pairs of chromosomes (the *autosomes*) are also subject to nondisjunction. Nondisjunction of chromosomes 13 and 18 results in **Patau syndrome (trisomy-13)** and **Edward syndrome (trisomy-18)**, respectively. Affected individuals have three copies of the respective chromosome. Nearly all fetuses with these trisomies die before birth. Live-born infants with these syndromes are severely deformed, and fewer than 5% survive for one year. The most common autosomal anomaly is **Down**[32] **syndrome**

(**trisomy-21**). Its signs include retarded physical development; short stature; a relatively flat face with a flat nasal bridge; low-set ears; *epicanthal folds* at the median corners of the eyes; an enlarged, protruding tongue; stubby fingers; and a short broad hand with only one palmar crease (fig. 4.14). People with Down syndrome tend to have outgoing, affectionate personalities. Mental retardation is common and sometimes severe, but is not inevitable. Down syndrome occurs in about 1 out of 700 to 800 live births in the United States and increases in proportion to the age of the mother. The chance of having a child with Down syndrome is about 1 in 3,000 for a woman under 30, 1 in 365 by age 35, and 1 in 9 by age 48.

[32]John Langdon H. Down (1828–96), British physician

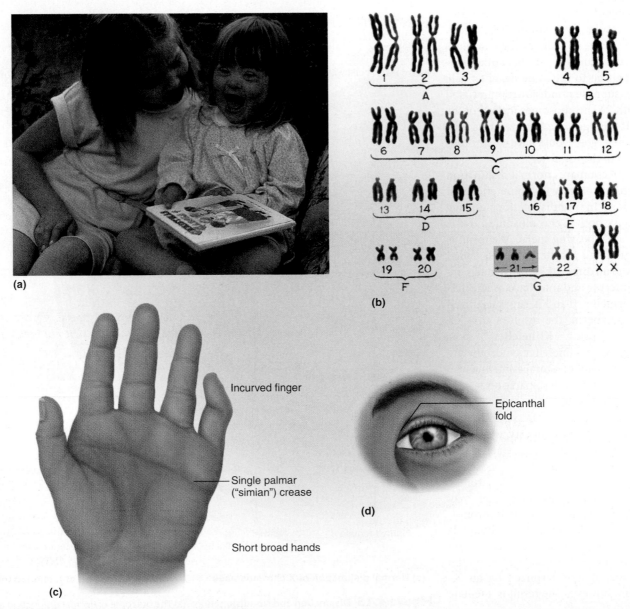

(a)

(b)

(c)

Incurved finger

Single palmar ("simian") crease

Short broad hands

Epicanthal fold

(d)

Figure 4.14 **Down Syndrome.** (a) A child with Down syndrome (right) and her sister. (b) The karyotype (chart of the chromosomes arranged in pairs) seen in Down syndrome, showing the trisomy of chromosome 21. (c) Characteristics of the hand seen in Down syndrome. (d) The epicanthal fold over the medial commissure (canthus) of the left eye.
• *What was the sex of the person from whom the karyotype in part (b) was obtained?*

About 75% of the victims of trisomy-21 die before birth, and about 20% die before the age of 10 years. Typical causes of death include immune deficiency and abnormalities of the heart or kidneys. For those who survive beyond 10 years, modern medical care has extended life expectancy to about 60. After the age of 40, however, many of these people develop early-onset Alzheimer disease, linked to a gene on chromosome 21.

Teratogens

Teratogens[33] are agents that cause anatomical deformities in the fetus. They fall into three major classes: drugs and other chemicals, radiation, and infectious diseases. The effect of a teratogen depends on the genetic susceptibility of the embryo, the dosage of the teratogen, and the time of exposure. Teratogen exposure during the first 2 weeks usually does not cause birth defects, but may cause spontaneous abortion. Teratogens can exert destructive effects at any stage of development, but the period of greatest vulnerability is weeks 3 through 8. Different organs have different critical periods. For example, limb abnormalities are most likely to result from teratogen exposure at 24 to 36 days, and brain abnormalities from exposure at 3 to 16 weeks.

Perhaps the most notorious teratogenic drug is thalidomide, a sedative first marketed in 1957. Thalidomide was taken by women in early pregnancy, often before they knew they were pregnant; it caused over 5,000 babies to be born with unformed arms or legs (fig. 4.15) and often with defects of the ears, heart, and intestines. It was taken off the American market in 1961, but has recently been reintroduced for more limited purposes. People still take thalidomide for leprosy and AIDS in some Third World countries, where it has resulted in an upswing in severe birth defects. A general lesson to be learned from the thalidomide tragedy and other cases is that pregnant women should avoid all sedatives, barbiturates, and opiates. Even the acne medicine isotretinoin (Accutane) has caused severe birth defects. Many teratogens produce less obvious or delayed effects, including physical or mental retardation, hyperirritability, inattention, strokes, seizures, respiratory arrest, crib death, and cancer.

Alcohol causes more birth defects than any other teratogen. Even one drink a day can have adverse effects on fetal and childhood development, some of which are not noticed until a child begins school. Alcohol abuse during pregnancy can cause **fetal alcohol syndrome (FAS),** characterized by a small head, malformed facial features, cardiac and central nervous system defects, stunted growth, and behavioral symptoms such as hyperactivity, nervousness, and a poor attention span. Cigarette smoking also contributes to fetal and infant mortality, ectopic pregnancy, anencephaly (failure of the cerebrum to develop), cleft lip and palate, and cardiac abnormalities.

The effects of ionizing radiation on an embryo were tragically demonstrated by the many grotesque birth defects caused by the 1986 accident at the nuclear power plant at Chernobyl in the

Figure 4.15 **The Teratogenic Effects of Thalidomide.** The infant in this photo was born in 2004 in Kenya to a woman believed to have used thalidomide during her pregnancy. The infant has no arms or legs and only rudimentary hands and feet. His birth father wanted to kill him, a common fate for deformed infants in Kenya, but he was adopted and taken to England. On the right is Mr. Freddie Astbury, president of Thalidomide UK in Liverpool. Mr. Astbury also was born with rudimentary limbs because of thalidomide.

Ukraine. Even diagnostic medical X-rays, however, can be teratogenic and should be minimized or avoided during pregnancy.

Apply What You Know

Martha is showing a sonogram of her unborn baby to her coworkers. Her friend Betty tells her she shouldn't have sonograms made because X-rays can cause birth defects. Is Betty's concern well founded? Explain.

Infectious diseases are largely beyond the scope of this book, but it must be noted at least briefly that several microorganisms can cross the placenta and cause serious congenital anomalies, stillbirth, or neonatal death. Common viral infections of the fetus and newborn include herpes simplex, rubella, cytomegalovirus, and human immunodeficiency virus (HIV). Congenital bacterial infections include gonorrhea and syphilis. *Toxoplasma*, a protozoan contracted from meat, unpasteurized milk, and house cats, is another common cause of fetal deformity. Some of these pathogens have relatively mild effects on adults, but because of its immature immune system, the fetus is vulnerable to devastating effects such as blindness, hydrocephalus, cerebral

---

[33]*terato* = monster + *gen* = producing

TABLE 4.6	Some Disorders of Human Development
Anencephaly	Lack of a forebrain due to failure of the cranial roof to form, leaving the forebrain exposed. The exposed tissue dies, and the fetus is born (or stillborn) with only a brainstem. Live-born anencephalic infants are very short-lived.
Cleft lip and palate	Failure of the right and left sides of the lip or palate to fuse medially, resulting in abnormal facial appearance, defective speech, and difficulty suckling.
Clubfoot (talipes)	Deformity of the foot involving an ankle bone, the talus. The sole of the foot is commonly turned medially, and as a child grows, he or she may walk on the ankles rather than on the soles.
Cri du chat[34]	A congenital anomaly due to deletion of a portion of chromosome 5. Infants with cri du chat have microcephaly, congenital heart disease, profound mental retardation, and a weak catlike cry.
Hydrocephalus	Abnormal accumulation of cerebrospinal fluid in the brain. When it occurs in the fetus, the cranial bones separate, the head becomes abnormally large, and the face looks disproportionately small. May reduce the cerebrum to a thin shell of nervous tissue. Fatal for about half of patients but can be treated by inserting a shunt that drains fluid from the brain to a vein in the neck.
Meromelia	Partial absence of limbs (as in fig. 4.15), such as the lack of some digits, a hand, or a forearm. Complete absence of a limb is *amelia*.

Disorders Described Elsewhere

Abruptio placentae 95
Achondroplastic dwarfism 142
Birthmarks 114
Congenital defects of the kidney 699
Cryptorchidism 717, 734
Dextrocardia 6
Down syndrome 100
Ectopic pregnancy 87

Edward syndrome 100
Fetal alcohol syndrome 101
Hypospadias 717, 734
Klinefelter syndrome 99
Osteogenesis imperfecta 69
Patau syndrome 100
Patent ductus arteriosus 559
Placenta previa 95

Premature birth 649
Respiratory distress syndrome 649
Situs inversus 6
Situs perversus 6
Spina bifida 367
Spontaneous abortion 98, 736
Triplo-X syndrome 99
Turner syndrome 99

palsy, seizures, and profound physical and mental retardation. Infections of the fetus and newborn are treated in greater detail in microbiology textbooks.

Some congenital anomalies and other developmental disorders are described in table 4.6.

[34]*cri du chat* = cry of the cat (French)

Before You Go On

Answer the following questions to test your understanding of the preceding section:

12. In what sense can spontaneous abortion be considered a protective mechanism?

13. What is the difference between mutation and nondisjunction?

14. Name and describe two birth defects resulting from nondisjunction of autosomes and two from nondisjunction of sex chromosomes.

15. Name three distinctly different classes of teratogens and give one example from each class.

Study Guide

Assess Your Learning Outcomes

You should have a good understanding of this chapter if you can accurately address the following issues.

4.1   Gametogenesis and Fertilization (p. 83)

1. The meaning of *gametes* and *gametogenesis*
2. How gametes differ from the body's other cells in chromosome number, why this is a necessity of sexual reproduction, and what type of cell division produces this change in chromosome number
3. How the products of spermatogenesis differ from the products of oogenesis
4. The time required for an egg to travel to the uterus, the time limit on fertilization, and what this implies for sperm migration
5. The purpose and process of sperm capacitation
6. The fertilization process, and the reason that many sperm must collaborate so that one of them can fertilize the egg
7. The egg's mechanisms for preventing polyspermy, and why this is important
8. Events involving the sperm and egg nuclei and chromosomes immediately after sperm penetration

4.2   Stages of Prenatal Development (p. 85)

1. The duration of pregnancy and how the date of childbirth is predicted
2. The components of the conceptus
3. The three trimesters of pregnancy and the developmental milestones of each
4. The preembryonic, embryonic, and fetal stages of development; the developmental milestones that define them; and the timetable of these transitions
5. The three major events of the preembryonic stage; the developmental age (in days) at which this stage is concluded; and the product of this stage
6. The process of cleavage; the appearance of the morula and blastocyst; and the two cell masses of the blastocyst and their respective fates
7. The process of implantation, and the origins and respective roles of the syncytiotrophoblast and cytotrophoblast
8. The process of embryogenesis, and the names of the cell layers and extraembryonic membranes that result from it
9. The age at which the developing individual is considered an embryo; its size at that time; and what anatomical feature defines it as an embryo
10. The major changes that occur during the embryonic stage of development
11. How embryonic folding converts the flat embryonic disc into a cylindrical and C-shaped form, and how it gives rise to the primitive gut
12. How the coelom forms, how it divides into the peritoneal and thoracic cavities, and what cavities arise by further division of the thoracic cavity
13. The meaning of *organogenesis,* and some tissues and organs that arise from each of the three primary germ layers of the embryo
14. The process of neurulation, including the progression from neural plate, to neural groove and folds, to neural tube
15. The origin, location, and fate of the neural crests
16. The location and fate of the pharyngeal pouches
17. The location, three subdivisions, and fate of the myotomes
18. The early appearances of the head, heart, and limbs
19. The appearances, relative locations, and functions of the four extraembryonic membranes—amnion, yolk sac, allantois, and chorion—and functions of the amniotic fluid
20. The three modes in which the conceptus is nourished from the time of its migration down the uterine tube to the time of birth; the timetable and overlap between trophoblastic and placental nutrition
21. The development and mature structure of the placenta; the fetal and maternal components of the placenta; and how nutrients, wastes, hormones, and other materials are exchanged through the placenta while preventing the mixing of maternal and fetal blood
22. The relationship of the placenta to the umbilical cord and its three blood vessels
23. The various functions of the placenta, including but not limited to fetal nutrition and waste removal
24. The time at which the individual is considered to be a fetus, and what distinguishes a fetus from an embryo
25. Major events that occur from the beginning of the fetal stage to the time of birth

4.3   Clinical Perspectives (p. 98)

1. The incidence and common causes of early spontaneous abortion
2. The definitions of *teratology* and *congenital anomaly,* and examples of congenital anomalies that are not evident until months to years after birth
3. The distinction between mutation and aneuploidy as genetic causes of birth defects, and examples of birth defects resulting from each of these
4. The meaning of *nondisjunction,* and some types of aneuploidy that result from it
5. The cause and characteristics of Down syndrome (trisomy-21)
6. The meaning of *teratogens;* the three common categories of teratogens; and examples of teratogenic effects in each category
7. The cause and characteristics of fetal alcohol syndrome

Testing Your Recall

1. When a conceptus arrives in the uterus, it is at what stage of development?
   a. zygote
   b. morula
   c. blastomere
   d. blastocyst
   e. embryo

2. The entry of a sperm nucleus into an egg must be preceded by
   a. the cortical reaction.
   b. the acrosomal reaction.
   c. the fast block.
   d. implantation.
   e. cleavage.

3. The primitive gut develops as a result of
   a. gastrulation.
   b. cleavage.
   c. embryogenesis.
   d. embryonic folding.
   e. aneuploidy.

4. Chorionic villi develop from
   a. the zona pellucida.
   b. the endometrium.
   c. the syncytiotrophoblast.
   d. the embryoblast.
   e. the epiblast.

5. Which of these results from aneuploidy?
   a. Turner syndrome
   b. fetal alcohol syndrome
   c. nondisjunction
   d. mutation
   e. rubella

6. Fetal urine accumulates in the _____ and contributes to the fluid there.
   a. placental sinus   d. chorion
   b. yolk sac   e. amnion
   c. allantois

7. A preembryo has
   a. a neural tube.
   b. a heart bulge.
   c. a cytotrophoblast.
   d. a coelom.
   e. decidual cells.

8. The feature that distinguishes a fetus from an embryo is that the fetus has
   a. all of the organ systems.
   b. three germ layers.
   c. a placenta.
   d. an amnion.
   e. arm and leg buds.

9. The first blood and future egg and sperm cells come from
   a. the mesoderm.
   b. the hypoblast.
   c. the syncytiotrophoblast.
   d. the placenta.
   e. the yolk sac.

10. For the first 8 weeks of gestation, a conceptus is nourished mainly by
    a. the placenta.
    b. amniotic fluid.
    c. colostrum.
    d. decidual cells.
    e. yolk cytoplasm.

11. Viruses and chemicals that cause congenital anatomical deformities are called _____.

12. Aneuploidy is caused by _____, the failure of a pair of chromosomes to separate in meiosis.

13. The brain and spinal cord develop from a longitudinal ectodermal channel called the _____.

14. Attachment of the conceptus to the uterine wall is called _____.

15. Fetal blood flows through growths called _____, which project into the placental sinus.

16. The enzymes with which a sperm penetrates an egg are contained in an organelle called the _____.

17. Fertilization occurs in a part of the female reproductive tract called the _____.

18. Bone, muscle, and dermis arise from segments of mesoderm called _____.

19. The egg cell has fast and slow blocks to _____, or fertilization by more than one sperm.

20. A developing individual is first classified as a/an _____ when the three primary germ layers have formed.

Answers in the Appendix

Building Your Medical Vocabulary

State a medical meaning of each of the following word elements, and give a term in which it is used.

1. haplo-
2. gameto-
3. zygo-
4. tropho-
5. cephalo-
6. gyneco-
7. -genesis
8. syn-
9. meso-
10. terato-

Answers in the Appendix

True or False

Determine which five of the following statements are false, and briefly explain why.

1. Freshly ejaculated sperm are more capable of fertilizing an egg than are sperm several hours old.
2. Fertilization normally occurs in the uterus.
3. An egg is usually fertilized by the first sperm that contacts it.
4. By the time a conceptus reaches the uterus, it has already undergone several cell divisions and consists of 16 cells or more.
5. The individual is first considered a fetus when all of the organ systems are present.
6. The placenta becomes increasingly permeable as it develops.
7. During cleavage, the preembryo acquires a greater number of cells but does not increase in size.
8. In oogenesis, a germ cell divides into four equal-sized egg cells.
9. The stage of the conceptus that implants in the uterine wall is the blastocyst.
10. The energy for sperm motility comes from its acrosome.

Answers in the Appendix

Testing Your Comprehension

1. Only one sperm is needed to fertilize an egg, yet a man who ejaculates fewer than 10 million sperm is usually infertile. Explain this apparent contradiction. Supposing 10 million sperm were ejaculated, predict how many would come within close range of the egg. How likely is it that any one of these sperm would fertilize it?

2. What is the difference between embryology and teratology?

3. At what point in the timeline of table 4.4 do you think thalidomide exerts its teratogenic effect? Explain your reasoning.

4. A teratologist is studying the cytology of a fetus that aborted spontaneously at 12 weeks. She concludes that the fetus was triploid. What do you think this term means? How many chromosomes do you think she found in each of the fetus's cells? To produce this state, what normal process of human development apparently failed?

5. A young woman finds out she is about 4 weeks pregnant. She tells her doctor that she drank heavily at a party 3 weeks earlier, and she is worried about the possible effects of this on her baby. If you were the doctor, would you tell her that there is serious cause for concern? Why or why not?

Answers at www.mhhe.com/saladinha3

Improve Your Grade at www.mhhe.com/saladinha3

Practice quizzes, labeling activities, and games provide fun ways to master concepts. You can also download image PowerPoint files for each chapter to create a study guide or for taking notes during lecture.

PART TWO

The Integumentary System

CHAPTER

5

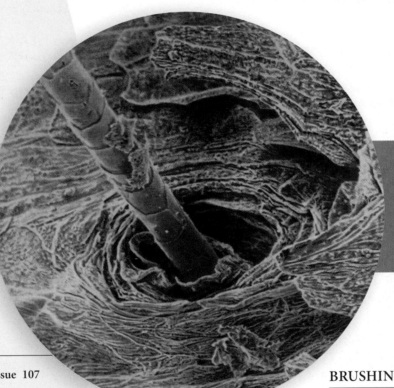

A human hair emerging from its follicle (SEM)

CHAPTER OUTLINE

DEEPER INSIGHTS

BRUSHING UP

To understand this chapter, you may find it helpful to review the following concepts:

Anatomy & Physiology REVEALED®
aprevealed.com

Integumentary System

The first organ system we deal with in this book is also the most visible one—the *integumentary system,* composed of the skin and its glands, hair, and nails. People pay more attention to this organ system than to any other. Being so visible, its appearance strongly affects our social interactions. Few people venture out of the house without first looking in a mirror to see if their skin and hair are presentable. In the United States alone, we spend billions of dollars annually on skin- and hair-care products and cosmetics. A health-care practitioner must not dismiss this as mere vanity, for a positive self-image is important to the attitudes that promote overall health. Care of the integumentary system must be considered as a particularly important part of the total plan of patient care.

The appearance of the skin, hair, and nails is a matter of more than aesthetics—their inspection is a significant part of a physical examination. They can provide clues not only to their own health, but also to deeper disorders such as liver cancer, anemia, and heart failure. The skin also is the most vulnerable of our organs, exposed to radiation, trauma, infection, and injurious chemicals. Consequently, it needs and receives more medical attention than any other organ system.

5.1 The Skin and Subcutaneous Tissue

▶ **Expected Learning Outcomes**

When you have completed this section, you should be able to

- list the functions of the skin and relate them to its structure;

- describe the histological structure of the epidermis, dermis, and subcutaneous tissue;

- describe the normal and pathological colors that the skin can have and explain their causes; and

- describe the common markings of the skin.

The skin, hair, nails, and cutaneous glands (sweat glands and others) constitute the **integumentary**[1] **system;** the skin alone is called the **integument.** The treatment of this system is a branch of medicine called **dermatology.**[2]

The skin is the body's largest and heaviest organ. In adults, it covers an area of 1.5 to 2.0 m^2 and accounts for about 15% of the body weight. It consists of two layers: a stratified squamous epithelium called the *epidermis* and a deeper connective tissue layer called the *dermis* (fig. 5.1). Below the dermis is another connective tissue layer, the *hypodermis,* which is not part of the skin but is customarily studied in conjunction with it.

Most of the skin is 1 to 2 mm thick, but it ranges from less than 0.5 mm on the eyelids to 6 mm between the shoulder blades. The

difference is due mainly to variation in the thickness of the dermis, although skin is classified as thick or thin based on the relative thickness of the epidermis alone. **Thick skin** covers the palms, soles, and corresponding surfaces of the fingers and toes. Its epidermis alone is about 0.5 mm thick, due to a very thick surface layer of dead cells called the *stratum corneum.* This layer serves to resist the pressure and friction to which the palms and soles are especially subjected. Thick skin has sweat glands but no hair follicles or sebaceous (oil) glands. The rest of the body is covered with **thin skin,** which has an epidermis about 0.1 mm thick, with a thin stratum corneum. It possesses hair, sebaceous glands, and sweat glands.

Functions of the Skin

The skin is much more than a container for the body. It has a variety of important functions that go well beyond appearance, as we will see here.

1. **Resistance to trauma and infection.** The skin bears the brunt of most physical injuries to the body, but it resists and recovers from trauma better than other organs do. The epidermal cells are packed with the tough protein **keratin** and linked by strong desmosomes that give the epithelium its durability. Few infectious organisms can penetrate the intact skin. Bacteria and fungi colonize the surface, but their numbers are kept in check by its relative dryness and slight acidity (pH 4–6). Its protective acidic film is called the *acid mantle.* Immune cells called *dendritic cells* in the epidermis stand guard against pathogens that do breach the surface.

2. **Water retention.** The skin is a barrier to water. It prevents the body from absorbing excess water when you are swimming or bathing, but even more importantly, it prevents the body from losing excess water.

3. **Vitamin D synthesis.** The skin carries out the first step in the synthesis of vitamin D, which is needed for bone development and maintenance. The liver and kidneys complete the process.

4. **Sensation.** The skin is our most extensive sense organ. It is equipped with a variety of nerve endings that react to heat, cold, touch, texture, pressure, vibration, and tissue injury (see chapter 17). These sensory receptors are especially abundant on the face, palms, fingers, soles, nipples, and genitals. There are relatively few on the back and in skin overlying joints such as the knees and elbows.

5. **Thermoregulation.** Cutaneous nerve endings called **thermoreceptors** monitor the body surface temperature. In response to chilling, we can better retain body heat by constricting blood vessels of the dermis (*cutaneous vasoconstriction*), keeping warm blood deeper in the body, not so close to the skin surface. In response to overheating, we can lose excess heat by dilating the dermal blood vessels (*cutaneous vasodilation*), allowing more blood to flow close to the surface and lose heat through the skin. If this is insufficient to restore normal temperature, the sweat glands secrete perspiration. The evaporation of sweat can have a powerful cooling effect. Thus, the skin plays key roles in both warming and cooling the body.

---

[1]*integument* = covering
[2]*dermat* = skin + *logy* = study of

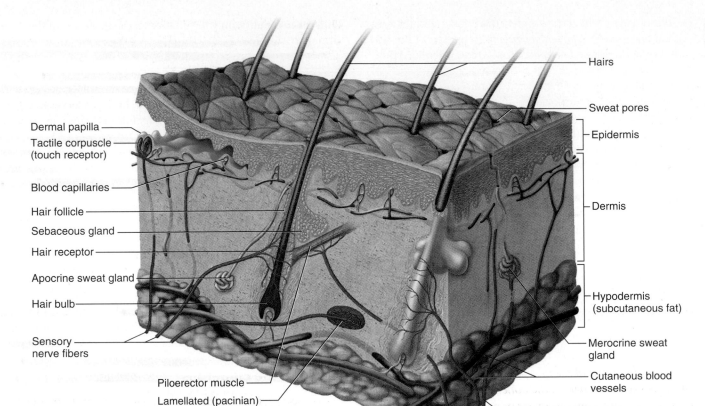

Figure 5.1 Structure of the Skin and Subcutaneous Tissue.

6. **Nonverbal communication.** The skin is an important means of communication. Humans, like other primates, have much more expressive faces than most mammals. Complex skeletal muscles insert on dermal collagen fibers and move the skin to create subtle and varied facial expressions (fig. 5.2).

The Epidermis

The **epidermis**[3] is a keratinized stratified squamous epithelium, as discussed in chapter 3. That is, its surface consists of dead cells packed with keratin. Like other epithelia, it lacks blood vessels and depends on the diffusion of nutrients from the underlying connective tissue. It has sparse nerve endings for touch and pain, but most sensations of the skin are due to nerve endings in the dermis.

Cells of the Epidermis

The epidermis is composed of five types of cells (fig. 5.3):

1. **Stem cells** are undifferentiated cells that divide and produce the keratinocytes described next. They are found only in the deepest layer of the epidermis, called the *stratum basale* (described later).

2. **Keratinocytes** (keh-RAT-ih-no-sites) are the great majority of epidermal cells. They are named for their role in synthesizing keratin. In ordinary histological specimens, nearly all visible epidermal cells are keratinocytes.

3. **Melanocytes** also occur only in the stratum basale, amid the stem cells and deepest keratinocytes. They synthesize the brown to black pigment *melanin*. They have long branching processes that spread among the keratinocytes and continually shed melanin-containing fragments from their tips. The keratinocytes phagocytize these fragments and accumulate melanin granules on the "sunny side" of the nucleus. Like a parasol, the pigment shields the DNA from ultraviolet radiation. Melanocytes are discussed later in relation to ethnic differences in skin color.

4. **Tactile (Merkel[4]) cells,** relatively few in number, are receptors for touch. They, too, are found in the basal layer of the epidermis and are associated with an underlying dermal nerve fiber. The tactile cell and its nerve fiber are collectively called a *tactile (Merkel) disc*.

5. **Dendritic[5] (Langerhans[6]) cells** are found in two layers of the epidermis called the *stratum spinosum* and *stratum*

---

[3]*epi* = above, upon + *derm* = skin

[4]F. S. Merkel (1845–1919), German anatomist
[5]*dendr* = tree, branch
[6]Paul Langerhans (1847–88), German anatomist

(a)

(b)

Figure 5.2 **Importance of the Skin in Nonverbal Expression.** Primates differ from other mammals in having very expressive faces due to facial muscles that insert on collagen fibers of the dermis and move the skin.

granulosum (described in the next section). They are macrophages that originate in the bone marrow and migrate to the epidermis and epithelia of the oral cavity, esophagus, and vagina. The epidermis has as many as 800 dendritic cells per square millimeter. They stand guard against toxins, microbes, and other pathogens that penetrate into the skin. When they detect such invaders, they carry fragments of the foreign matter to the lymph nodes and alert the immune system so the body can defend itself.

Layers of the Epidermis

Cells of the epidermis are usually arranged in four to five zones, or strata (five in thick skin), shown in figure 5.3. The following description progresses from deep to superficial, and from the youngest to the oldest keratinocytes.

1. The **stratum basale** (bah-SAY-lee) consists mainly of a single layer of cuboidal to low columnar stem cells and keratinocytes resting on the basement membrane. Scattered among these are

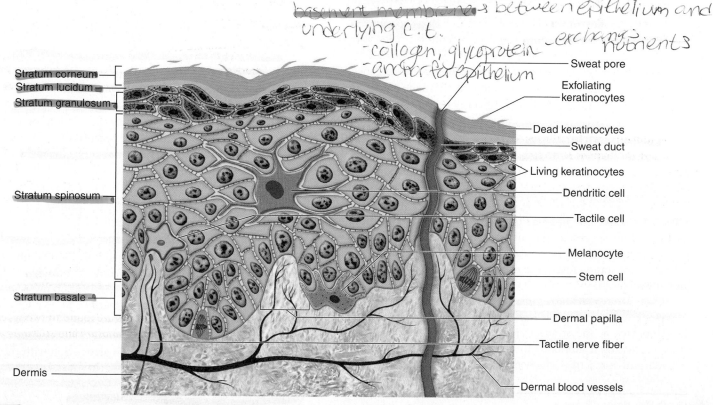

Figure 5.3 **Strata and Cell Types of the Epidermis.**

the melanocytes, tactile cells, and stem cells. As the stem cells divide, they give rise to keratinocytes that migrate toward the skin surface and replace lost epidermal cells. The life history of these cells is described in the next section.

2. The **stratum spinosum** (spy-NO-sum) consists of several layers of keratinocytes; in most places, this is the thickest stratum, but on the palms and soles it is usually exceeded by the stratum corneum. The deepest cells of the stratum corneum remain capable of mitosis, but as they are pushed farther upward, they cease dividing. Instead, they produce more and more keratin filaments, which cause the cells to flatten. Therefore, the higher up you look in the stratum spinosum, the flatter the cells appear. Dendritic cells are also found throughout the stratum spinosum, but are not usually visible in tissue sections.

The stratum spinosum is named for an artificial appearance (*artifact*) created by the histological fixation of tissue specimens. Keratinocytes are firmly attached to each other by numerous desmosomes, which partly account for the toughness of the epidermis. Histological fixatives shrink the keratinocytes, so they pull away from each other; however, they remain attached by the desmosomes—like two people holding hands while they step farther apart. The desmosomes thus create bridges from cell to cell, giving each cell a spiny appearance from which we derive the word *spinosum*.

3. The **stratum granulosum** consists of three to five layers of flat keratinocytes—more in thick skin than in thin skin—and some dendritic cells. The keratinocytes of this layer contain coarse, dark-staining *keratohyalin granules* that give the layer its name. The functional significance of these granules will be explained shortly.

4. The **stratum lucidum**[7] (LOO-sih-dum) is a thin translucent zone seen only in thick skin. Here, the keratinocytes are densely packed with a clear protein named *eleidin* (ee-LEE-ih-din). The cells have no nuclei or other organelles. Because organelles are absent and eleidin does not stain well, this zone has a pale, featureless appearance with indistinct cell boundaries.

5. The **stratum corneum** consists of up to 30 layers of dead, scaly, keratinized cells that form a durable surface layer. It is especially resistant to abrasion, penetration, and water loss.

The Life History of a Keratinocyte

Dead cells constantly flake off the skin surface. They float around as tiny white specks in the air, settling on household surfaces and forming much of the house dust that accumulates there. Because we constantly lose these epidermal cells, they must be continually replaced.

Keratinocytes are produced deep in the epidermis by the mitosis of stem cells in the stratum basale. Some of the deepest keratinocytes in the stratum spinosum also remain mitotic and thus increase their number. Mitosis requires an abundant supply of oxygen and nutrients, which these deep cells acquire from the

blood vessels in the nearby dermis. Once the epidermal cells migrate more than two or three cells away from the dermis, their mitosis ceases. Mitosis is seldom seen in prepared slides of the skin, because it occurs mainly at night whereas most histological specimens are taken during the day.

As new keratinocytes are formed, they push the older ones upward. In 30 to 40 days, a keratinocyte makes its way to the skin surface and flakes off. This migration is slower in old age and faster in skin that has been injured or stressed. Injured epidermis regenerates more rapidly than any other tissue in the body. Mechanical stress from manual labor or tight shoes accelerates keratinocyte multiplication and results in *calluses* or *corns,* thick accumulations of dead keratinocytes on the hands or feet.

As keratinocytes are shoved upward by the dividing cells below, they flatten and produce more keratin filaments and lipid-filled **membrane-coating vesicles.** In the stratum granulosum, three important developments occur: (1) The keratinocyte nuclei and other organelles degenerate and the cells die. (2) The keratohyalin granules release a protein that binds the keratin filaments together into coarse, tough bundles. (3) The membrane-coating vesicles release a lipid mixture that spreads out over the cell surface and waterproofs it.

An **epidermal water barrier** forms between the stratum granulosum and the stratum spinosum. It consists of the lipids secreted by the keratinocytes, tight junctions between the keratinocytes, and a thick layer of insoluble protein on the inner surfaces of their plasma membranes. This barrier is crucial to retaining water in the body and preventing dehydration. Cells above the barrier quickly die because it cuts them off from the supply of nutrients below. Thus, the stratum corneum consists of compact layers of dead keratinocytes and keratinocyte fragments. Dead keratinocytes exfoliate (flake off) from the epidermal surface as tiny specks called **dander.** *Dandruff* is composed of clumps of dander stuck together by sebum (oil).

A curious effect of the epidermal water barrier is the way our skin wrinkles when we linger in the bath or a lake. The keratin of the stratum corneum absorbs water and swells, but the deeper layers of the skin do not. The thickening of the stratum corneum forces it to wrinkle. This is especially conspicuous on the tips of the fingers and toes ("prune fingers") because they have such a thick stratum corneum and they lack the sebaceous glands that produce water-resistant oil elsewhere on the body.

The Dermis

Beneath the epidermis is a connective tissue layer, the **dermis.** It ranges from 0.2 mm thick in the eyelids to about 4 mm thick in the palms and soles. It is composed mainly of collagen, but also contains elastic and reticular fibers, fibroblasts, and the other cells typical of fibrous connective tissue (described in chapter 3). It is well supplied with blood vessels, cutaneous glands, and nerve endings. The hair follicles and nail roots are embedded in the dermis. The dermis contains smooth muscles in association with the hair follicles, as described later. In the face, skeletal muscles attach to dermal collagen fibers and produce such expressions as a smile, a wrinkle of the forehead, or the lifting of an eyebrow.

---

[7]*lucid* = light, clear

The boundary between the epidermis and dermis is histologically conspicuous and usually wavy (see fig. 5.1). The upward waves are fingerlike extensions of the dermis called **dermal papillae**,[8] and the downward waves are extensions of the epidermis called **epidermal ridges**. The dermal and epidermal boundaries thus interlock like corrugated cardboard, an arrangement that resists slippage of the epidermis across the dermis. If you look closely at your hand and wrist, you will see delicate furrows that divide the skin into tiny rectangular to rhomboidal areas. The dermal papillae produce the raised areas between the furrows. On the fingertips, this wavy boundary forms the *friction ridges* that form one's fingerprints. In highly sensitive areas such as the lips and genitals, exceptionally tall dermal papillae allow nerve fibers and blood capillaries to reach close to the skin surface.

Apply What You Know

Dermal papillae are relatively high and numerous in palmar and plantar skin but low and few in number in the face and abdomen. What do you think is the functional significance of this difference?

There are two zones of dermis called the papillary and reticular layers (fig. 5.4). The **papillary** (PAP-ih-lerr-ee) **layer** is a thin zone

---

[8]*pap* = nipple + *illa* = little

DEEPER INSIGHT 5.1

Tension Lines and Surgery

The collagen bundles in the dermis are arranged mostly in parallel rows that run longitudinally to obliquely in the limbs, but encircle the neck, trunk, wrists, and a few other areas. They keep the skin under constant tension and are thus called *tension lines (Langer[9] lines)*. If an incision is made in the skin, especially if it is perpendicular to the tension lines, the wound gapes because the collagen bundles pull the edges of the incision apart. Even if the skin is punctured with a circular object such as an ice pick, the wound gapes with a lemon-shaped opening, the direction of the wound axis being perpendicular to the tension lines. Such gaping wounds are relatively difficult to close and tend to heal with excessive scarring. Surgeons make incisions parallel to the tension lines—for example, making a transverse incision when delivering a baby by cesarean section—so that the incisions will gape less and heal with less scarring.

---

of areolar tissue in and near the dermal papillae. It is especially rich in small blood vessels. The loosely organized tissue of the papillary layer allows for mobility of leukocytes and other defenses against organisms introduced through breaks in the epidermis.

---

[9]Karl Langer (1819–87), Austrian physician

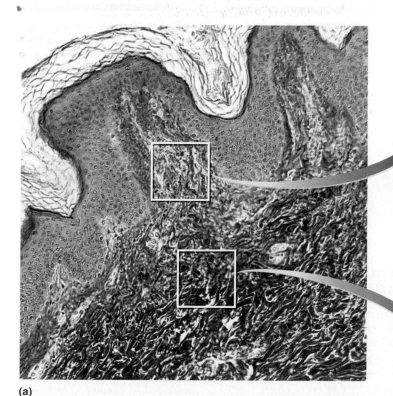

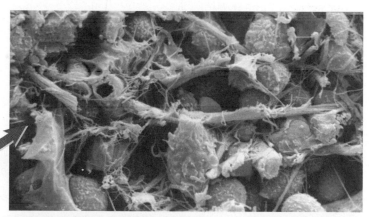

(b) Papillary layer of dermis

(c) Reticular layer of dermis

(a)

Figure 5.4 The Dermis. (a) Light micrograph of axillary skin, with the collagen stained blue. (b) The papillary layer, made of loose (areolar) tissue, forms the dermal papillae. (c) The reticular layer, made of dense irregular connective tissue, forms the deeper four-fifths of the dermis.
[Parts (b) and (c) from R. G. Kessel and R. H. Kardon, *Tissues and Organs: A Text-Atlas of Scanning Electron Microscopy* (W. H. Freeman, 1979).]

The **reticular**[10] **layer** of the dermis is deeper and much thicker. It consists of dense irregular connective tissue. Leather is composed of the reticular layer of animal skin. The boundary between the papillary and reticular layers is often vague. In the reticular layer, the collagen forms thicker bundles with less room for ground substance, and there are often small clusters of adipocytes. Stretching of the skin in obesity and pregnancy can tear the collagen fibers and produce *striae* (STRY-ee), or stretch marks. These occur especially in areas most stretched by weight gain: the thighs, buttocks, abdomen, and breasts.

The Hypodermis

Beneath the skin is a layer called the **hypodermis**[11] (**subcutaneous tissue**). The boundary between the dermis and hypodermis is indistinct, but the hypodermis generally has more areolar and adipose tissue. It pads the body and binds the skin to the underlying tissues. Drugs are introduced here by hypodermic injection because the subcutaneous tissue is highly vascular and absorbs them quickly.

Subcutaneous fat is hypodermis composed predominantly of adipose tissue. It serves as an energy reservoir and thermal insulation. It is not uniformly distributed; for example, it is virtually absent from the scalp but relatively abundant in the breasts, abdomen, hips, and thighs. The subcutaneous fat averages about 8% thicker in women than in men, and varies with age. Infants and elderly people have less subcutaneous fat than other people and are therefore more sensitive to cold.

Table 5.1 summarizes the layers of the skin and hypodermis.

Skin Color

The most significant factor in skin color is melanin, which is produced by melanocytes but accumulates in the keratinocytes of the stratum basale and the stratum spinosum (fig. 5.5). There are two forms of melanin—a brownish black **eumelanin**[12] and a reddish yellow sulfur-containing pigment, **pheomelanin.**[13] People of different skin colors have essentially the same number of melanocytes, but in dark-skinned people, the melanocytes produce greater quantities of melanin; the melanin granules are more spread out than tightly clumped; and the melanin breaks down more slowly. Thus, melanized cells may be seen throughout the epidermis, from stratum basale to stratum corneum. In light-skinned people, the melanin is clumped near the keratinocyte nucleus, so it imparts less color to the cells. It also breaks down more rapidly, so little of it is seen beyond the stratum basale, if even there.

Skin color also varies with exposure to the ultraviolet (UV) rays of sunlight, which stimulate melanin synthesis and darken the skin. A suntan fades as melanin is degraded in older keratinocytes and the keratinocytes migrate to the surface and exfoliate. The amount of melanin also varies substantially from place to place on the body. It is relatively concentrated in freckles and moles; on the dorsal surfaces of the hands and feet as compared with the palms and soles; in the nipple and areola of the breast; around the anus; on the scrotum and penis; and on the lateral surfaces of the female genital folds *(labia majora)*. The contrast between heavily melanized and lightly melanized regions of the skin is more pronounced in some people than in others, but it exists to some extent in nearly everyone. Variation in ancestral exposure to UV radiation is the primary reason for the geographic and ethnic variation in skin color seen today (see Deeper Insight 5.2).

Other factors in skin color are hemoglobin and carotene. **Hemoglobin,** the red pigment of blood, imparts reddish to pinkish hues to the skin. Its color is lightened by the white of the dermal collagen. The skin is redder in places such as the lips, where blood capillaries come closer to the surface and the hemoglobin shows through more vividly. **Carotene**[14] is a yellow pigment acquired from egg yolks

---

[10]*reti* = network + *cul* = little
[11]*hypo* = below + *derm* = skin
[12]*eu* = true + *melan* = black
[13]*pheo* = dusky + *melan* = black
[14]*carot* = carrot

TABLE 5.1	Stratification of the Skin and Hypodermis
Layer	**Description**
Epidermis	Keratinized stratified squamous epithelium
Stratum corneum	Dead, keratinized cells of the skin surface
Stratum lucidum	Clear, featureless, narrow zone seen only in thick skin
Stratum granulosum	Two to five layers of cells with dark-staining keratohyalin granules; scanty in thin skin
Stratum spinosum	Many layers of keratinocytes, typically shrunken in fixed tissues but attached to each other by desmosomes, which give them a spiny look; progressively flattened the farther they are from the dermis. Dendritic cells are abundant here but are not distinguishable in routinely stained preparations.
Stratum basale	Single layer of cuboidal to columnar cells resting on basement membrane; site of most mitosis; consists of stem cells, keratinocytes, melanocytes, and tactile cells, but these are not all distinguishable with routine stains. Melanin is conspicuous in keratinocytes of this layer in black to brown skin.
Dermis	Fibrous connective tissue, richly endowed with blood vessels and nerve endings. Sweat glands and hair follicles originate here and in hypodermis.
Papillary layer	Superficial one-fifth of dermis; composed of areolar tissue; often extends upward as dermal papillae
Reticular layer	Deeper four-fifths of dermis; dense irregular connective tissue
Hypodermis	Areolar or adipose tissue between skin and muscle

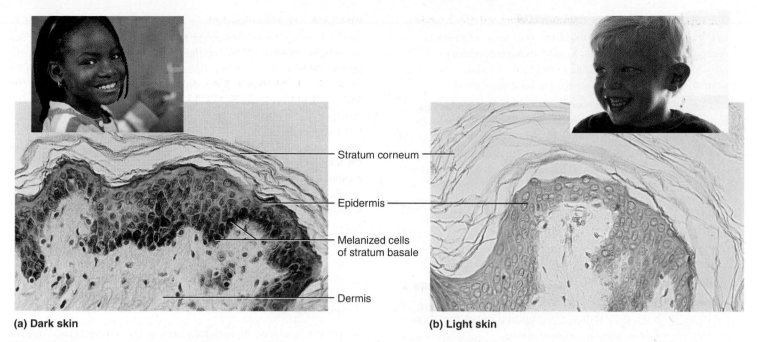

(a) Dark skin

(b) Light skin

Figure 5.5 Variation in Skin Pigmentation. (a) The stratum basale shows heavy deposits of melanin in dark skin. (b) Light skin shows little or no visible melanin.
● *Which of the five types of epidermal cells are the melanized cells in part (a)?*

and yellow and orange vegetables. Depending on the diet, it concentrates to various degrees in the stratum corneum and subcutaneous fat. It is often most conspicuous in skin of the heel and in calluses of the feet because this is where the stratum corneum is thickest.

The skin may also exhibit abnormal colors of diagnostic value:

- **Cyanosis**[15] is blueness of the skin resulting from a deficiency of oxygen in the circulating blood. Oxygen deficiency turns the hemoglobin a reddish violet color, which is lightened to blue-violet as it shows through the white dermal collagen. Oxygen deficiency can result from conditions that prevent the blood from picking up a normal load of oxygen in the lungs, such as airway obstructions in drowning and choking, lung diseases such as emphysema, and respiratory arrest. Cyanosis also occurs in situations such as cold weather and cardiac arrest, when blood flows so slowly through the skin that the tissues consume most of its oxygen faster than freshly oxygenated blood arrives.

- **Erythema**[16] (ERR-ih-THEE-muh) is abnormal redness of the skin. It occurs in such situations as exercise, hot weather, sunburns, anger, and embarrassment. Erythema is caused by increased blood flow in dilated cutaneous blood vessels or by dermal pooling of red blood cells that have escaped from abnormally permeable capillaries, as in sunburn.

- **Pallor** is a pale or ashen color that occurs when there is so little blood flow through the skin that the white color of the dermal collagen shows through. It can result from emotional stress, low blood pressure, circulatory shock, cold temperatures, or severe anemia.

[15]*cyan* = blue + *osis* = condition
[16]*eryth* = red + *em* = blood

DEEPER INSIGHT 5.2

The Evolution of Skin Color

One of the most conspicuous signs of human variation is skin color, which can range from the color of espresso or milk chocolate to cafe au lait or light peach. Such variation results from a combination of evolutionary selection pressures, especially differences in exposure to ultraviolet radiation (UVR).

UVR can have two adverse effects: It causes skin cancer and it breaks down folic acid, a B vitamin needed for normal cell division, fertility, and fetal development. It also has a desirable effect: It stimulates keratinocytes to synthesize vitamin D, which is needed for the absorption of dietary calcium and thus for healthy bone development. Too much UVR and one is at risk of infertility and fetal deformities such as spina bifida; too little and one is at risk of bone deformities such as rickets. Consequently, populations native to the tropics and people descended from them tend to have well-melanized skin to screen out excessive UVR. Populations native to far northern and southern latitudes, where the sunlight is weak, tend to have light skin to allow for adequate UVR penetration. Ancestral skin color is thus partly a compromise between vitamin D and folic acid requirements. Worldwide, women have skin averaging about 4% lighter than men do, perhaps because of their greater need for vitamin D and calcium to support pregnancy and lactation.

But for multiple reasons, there are exceptions to this trend. UVR exposure is determined by more than latitude. It increases at higher elevations and in dry air, because the thinner, drier atmosphere filters out less UVR. This helps to explain the dark skin of people in such localities as the Andes and the high plateaus of Tibet and Ethiopia. UVR levels account for up to 77% of the variation in human skin color. Some other exceptions may be the result of human migrations from one latitude to another occurring too recently for their skin color to have adapted to the new level of UVR exposure. Variation may also result from cultural differences in clothing and shelter, intermarriage among people of different geographic ancestries, and darwinian sexual selection—a preference in mate choice for partners of light or dark complexion.

- **Albinism**[17] is a genetic lack of melanin that usually results in milky white hair and skin, and blue-gray eyes. Melanin is synthesized from the amino acid tyrosine by the enzyme tyrosinase. People with albinism have inherited a recessive, nonfunctional tyrosinase gene from both parents.
- **Jaundice**[18] is a yellowing of the skin and whites of the eyes resulting from high levels of bilirubin in the blood. Bilirubin is a hemoglobin breakdown product. When erythrocytes get old, they disintegrate and release their hemoglobin. The liver and spleen convert hemoglobin to bilirubin and other pigments, which the liver excretes in the bile. Bilirubin can accumulate enough to discolor the skin, however, in such situations as a rapid rate of erythrocyte destruction; when diseases such as cancer, hepatitis, and cirrhosis compromise liver function; and in premature infants, whose liver is not well enough developed to dispose of bilirubin efficiently.
- A **hematoma**,[19] or bruise, is a mass of clotted blood showing through the skin. It is usually due to accidental trauma (blows to the skin), but it may indicate hemophilia, other metabolic or nutritional disorders, or physical abuse.

Apply What You Know

An infant brought to a clinic shows abnormally yellow skin. What sign could you look for to help decide whether this was due to jaundice or to a large amount of carotene from strained vegetables in the diet?

Skin Markings

The skin is marked by many lines, creases, ridges, and patches of accentuated pigmentation. **Friction ridges** are the markings on the fingertips that leave distinctive oily fingerprints on surfaces we touch. They are characteristic of most primates. They enable us to manipulate small objects more easily, and recent evidence indicates that they enhance the texture sensitivity of cutaneous nerve endings. Friction ridges form during fetal development and remain essentially unchanged for life. Everyone has a unique pattern of friction ridges; not even identical twins have identical fingerprints.

Flexion lines (flexion creases) are lines on the flexor surfaces of the digits, palms, wrists, elbows, and other places (see fig. A.19, p. 345). They mark sites where the skin folds during flexion of the joints. The skin is tightly bound to the deeper connective tissue along these lines.

Freckles and moles are tan to black aggregations of melanocytes. **Freckles** are flat, melanized patches that vary with heredity and exposure to the sun. A **mole (nevus)** is an elevated patch of melanized skin, often with hair. Moles are harmless and sometimes even regarded as "beauty marks," but they should be watched for changes in color, diameter, or contour that may suggest malignancy (skin cancer).

Birthmarks, or **hemangiomas**,[20] are patches of discolored skin caused by benign tumors of the blood capillaries. *Capillary hemangiomas* (strawberry birthmarks) usually develop about a month after birth. They become bright red to deep purple and develop small capillary-dense elevations that give them a strawberry-like appearance. About 90% of capillary hemangiomas disappear by the age of 5 or 6 years. *Cavernous hemangiomas* are flatter and duller in color. They are present at birth, enlarge up to 1 year of age, and then regress. About 90% disappear by the age of 9 years. A *port-wine stain* is flat and pinkish to dark purple in color. It can be quite large and remains for life.

Before You Go On

Answer the following questions to test your understanding of the preceding section:

1. What is the major histological difference between thick and thin skin? Where on the body is each type of skin found?
2. How does the skin help to adjust body temperature?
3. List the five cell types of the epidermis. Describe their locations and functions.
4. List the five layers of epidermis from deep to superficial. What are the distinctive features of each layer?
5. What are the two layers of the dermis? What type of tissue composes each layer?
6. Name the pigments responsible for normal skin colors, and explain how certain conditions can produce discolorations of the skin.

5.2 Hair and Nails

Expected Learning Outcomes

When you have completed this section, you should be able to

- distinguish between three types of hair;
- describe the histology of a hair and its follicle;
- discuss some theories of the purposes served by various kinds of hair;
- describe the life cycle of a hair; and
- describe the structure and function of nails.

The hair, nails, and cutaneous glands are the **accessory organs (appendages)** of the skin. Hair and nails consist mainly of dead, keratinized cells. The stratum corneum of the skin is made of pliable *soft keratin,* but the hair and nails are composed mostly of *hard keratin.* Hard keratin is more compact and is toughened by numerous cross-linkages between the keratin molecules.

Hair

A hair is also known as a **pilus** (PY-lus); in the plural, *pili* (PY-lye). It is a slender filament of keratinized cells that grows from an oblique tube in the skin called a **hair follicle** (fig. 5.6).

---

[17]*alb* = white + *ism* = state, condition
[18]*jaun* = yellow
[19]*hemat* = blood + *oma* = mass
[20]*hem* = blood + *angi* = vessel + *oma* = tumor

Distribution and Types

Hair occurs almost everywhere on the body except the lips, nipples, parts of the genitals, palms and soles, ventral and lateral surfaces of the fingers and toes, and distal segment of the fingers. The trunk and limbs have about 55 to 70 hairs per square centimeter, and the face has about 10 times as many. There are about 30,000 hairs in a man's beard and about 100,000 in the average person's scalp. The density of hair does not differ much from one person to another or even between the sexes; indeed, it is virtually the same in humans, chimpanzees, and gorillas. Differences in apparent hairiness are due mainly to differences in texture and pigmentation.

Not all hair is alike, even on one person. Over the course of our lives, we grow three kinds of hair: lanugo, vellus, and terminal hair. **Lanugo**[21] is fine, downy, unpigmented hair of the fetus. By the time of birth, most of it is replaced by **vellus,**[22] a similarly fine, pale hair. Vellus constitutes about two-thirds of the hair of women, one-tenth of the hair of men, and all of the hair of children except for the eyebrows, eyelashes, and hair of the scalp. **Terminal hair** is longer, coarser, and usually more heavily pigmented. It forms the eyebrows and eyelashes; covers the scalp; and after puberty, it forms the axillary and pubic hair, the male facial hair, and some of the hair on the trunk and limbs.

[21]*lan* = down, wool
[22]*vellus* = fleece

TABLE 5.2 — Functions of Hair

Hair of the torso and limbs	Vestigial, but serves a sensory purpose as in detection of small insects crawling on the skin
Scalp hair	Heat retention, protection from sun
Beard, pubic, and axillary (armpit) hair	Advertises sexual maturity; associated with apocrine scent glands in these areas and modulates the dispersal of sexual scents (pheromones) from these glands
Guard hairs (vibrissae)	Help keep foreign objects out of nostrils and auditory canal; eyelashes help keep debris from eyes
Eyebrows	Enhance facial expression, may reduce glare of sun and help keep forehead perspiration from eyes

Functions of Hair

In most mammals, hair serves to retain body heat. Humans have too little hair to serve this purpose except on the scalp, where there is no insulating fat. Hair elsewhere on the body plays a variety of roles that are somewhat speculative, but probably best inferred by comparison to the specialized types and patches of hair in other mammals (table 5.2).

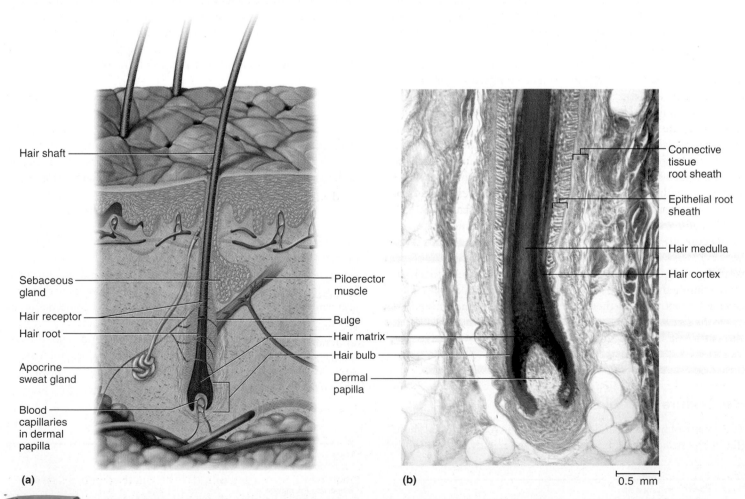

Figure 5.6 Structure of a Hair and Follicle. (a) Anatomy of the follicle and associated structures. (b) Light micrograph of the base of a hair follicle.

Structure of the Hair and Follicle

A hair is divisible into three zones along its length: (1) the **bulb,** a swelling at the base where the hair originates in the dermis or hypodermis; (2) the **root,** which is the remainder of the hair within the follicle; and (3) the **shaft,** which is the portion above the skin surface. The only living cells of a hair are in and near the bulb. The bulb grows around a bud of vascular connective tissue called the **dermal papilla,** which provides the hair with its sole source of nutrition. Immediately above the papilla is a region of mitotically active cells, the **hair matrix,** which is the hair's growth center. All cells higher up are dead.

In cross section, a hair reveals up to three layers. From the inside out, these are the medulla, cortex, and cuticle. The **medulla** is a core of loosely arranged cells and air spaces. It is most prominent in thick hairs such as those of the eyebrows, but narrower in hairs of medium thickness and absent from the thinnest hairs of the scalp and elsewhere. The **cortex** constitutes most of the bulk of a hair. It consists of several layers of elongated keratinized cells that appear cuboidal to flattened in cross sections. The **cuticle** is composed of multiple layers of very thin, scaly, surface cells that overlap each other like roof shingles, with their free edges directed upward (see photo on p. 106). Cells lining the follicle are like shingles facing in the opposite direction. They interlock with the scales of the hair cuticle and resist pulling on the hair. When a hair is pulled out, this layer of follicle cells comes with it.

The follicle is a diagonal tube that dips deeply into the dermis and sometimes extends as far as the hypodermis. It has two principal layers: an **epithelial root sheath** and a **connective tissue root sheath** (fig. 5.6a, b). The epithelial root sheath, which is an extension of the epidermis, lies immediately adjacent to the hair root. Toward the deep end of the follicle, it widens to form a **bulge,** a source of stem cells for follicle growth. The connective tissue root sheath, derived from the dermis, surrounds the epithelial sheath and is somewhat denser than the adjacent dermal connective tissue.

Associated with the follicle are nerve and muscle fibers. Nerve fibers called **hair receptors** entwine each follicle and respond to hair movements. You can feel their effect by carefully moving a single hair with a pin or by lightly running your finger over the hairs of your arm without touching the skin. Associated with each hair is a **piloerector muscle**—also known as a **pilomotor muscle** or **arrector pili**[23]—a bundle of smooth muscle cells extending from dermal collagen fibers to the connective tissue root sheath of the follicle (see figs. 5.1 and 5.6). In response to cold, fear, touch, or other stimuli, the sympathetic nervous system stimulates the piloerector to contract and thereby makes the hair stand on end. In other mammals, this traps an insulating layer of warm air next to the skin or makes the animal appear larger and less vulnerable to a potential enemy. In humans, it pulls the follicles into a vertical position and causes "goose bumps," but serves no useful purpose.

Hair Texture and Color

The texture of hair is related to differences in cross-sectional shape (fig. 5.7)—straight hair is round, wavy hair is oval, and tightly curly hair is relatively flat. Hair color is due to pigment granules in the cells of the cortex. Brown and black hair are rich in eumelanin. Red hair has less eumelanin but a high concentration of pheomelanin. Blond hair has an intermediate amount of pheomelanin but very little eumelanin. Gray and white hair result from a scarcity or absence of melanins in the cortex and the presence of air in the medulla.

Hair Growth and Loss

A given hair goes through a **hair cycle** consisting of three developmental stages: anagen, catagen, and telogen (fig. 5.8). At any given time, about 90% of scalp follicles are in the **anagen**[24] phase. In this phase, stem cells from the bulge in the follicle multiply and travel downward, pushing the dermal papilla deeper into the skin and forming the epithelial root sheath. Root sheath cells directly above the papilla form the hair matrix. Here, sheath cells transform into hair cells, which synthesize keratin and then die as they are pushed upward away from the papilla. The new hair grows up the follicle, often alongside an old *club hair* left from the previous cycle.

In the **catagen**[25] phase, mitosis in the hair matrix ceases and sheath cells below the bulge die. The follicle shrinks and the dermal papilla draws up toward the bulge. The base of the hair keratinizes into a hard club and the hair, now known as a **club hair,** loses its anchorage. Club hairs are easily pulled out by brushing the hair, and the hard club can be felt at the hair's end. When the papilla reaches the bulge, the hair goes into a resting period called the **telogen**[26] **phase.** Eventually, anagen begins anew and the cycle repeats itself. A club hair may fall out during catagen, telogen, or as it is pushed out by the new hair in the next anagen phase. We lose about 50 to 100 scalp hairs daily.

In a young adult, scalp follicles typically spend 6 to 8 years in anagen, 2 to 3 weeks in catagen, and 1 to 3 months in telogen. Scalp hairs grow at a rate of about 1 mm per 3 days (10–18 cm/yr) in the anagen phase. Hair grows fastest from adolescence until the 40s. After that, an increasing percentage of follicles are in catagen and telogen rather than the growing anagen phase. Follicles also shrink and begin producing wispy vellus hairs instead of thicker terminal hairs.

Thinning of the hair, or baldness, is called **alopecia**[27] (AL-oh-PEE-she-uh). It occurs to some degree in both sexes and may be worsened by disease, poor nutrition, fever, emotional stress, radiation, or chemotherapy. In the great majority of cases, however, it is simply a matter of aging. **Pattern baldness** is the condition in which hair is lost from select regions of the scalp rather than thinning uniformly across the entire scalp. It results from a combination of genetic and hormonal influences.

Contrary to popular misconceptions, hair and nails do not continue to grow after a person dies, cutting hair does not make it grow faster or thicker, and emotional stress cannot turn the hair white overnight.

23*arrect* = erect + *pili* = of a hair

24*ana* = up + *gen* = build, produce
25*cata* = down
26*telo* = end
27*alopecia* = fox mange

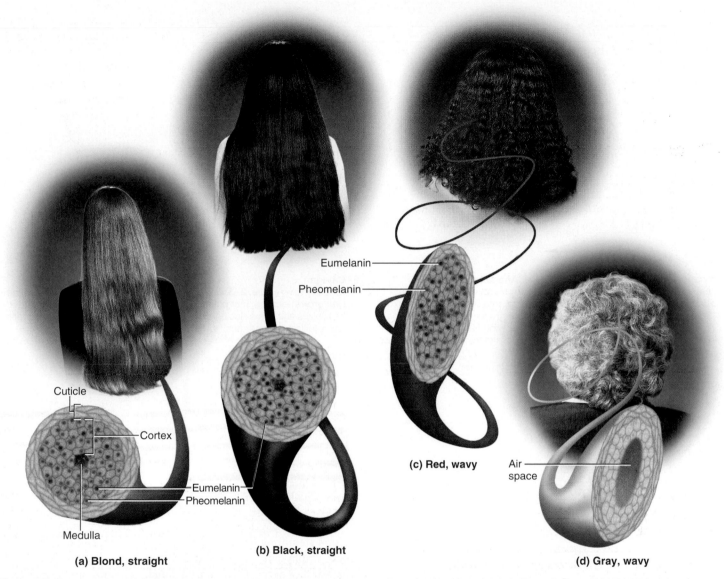

Cuticle

Cortex

Eumelanin
Pheomelanin

Medulla

(a) Blond, straight

Eumelanin

Pheomelanin

Eumelanin
Pheomelanin

(b) Black, straight

(c) Red, wavy

Air
space

(d) Gray, wavy

Figure 5.7 **The Basis of Hair Color and Texture.** Straight hair (a and b) is round in cross section whereas curly hair (c and d) is flatter. Blonde hair (a) has scanty eumelanin and a moderate amount of pheomelanin. Eumelanin predominates in black and brown hair (b). Red hair (c) derives its color predominantly from pheomelanin. Gray and white hair (d) lack pigment and have air in the medulla.
● *Which of the hair layers illustrated here corresponds to the scales seen on the hair shafts in the photo on page 106?*

Nails

Fingernails and toenails are clear, hard derivatives of the stratum corneum. They are composed of very thin, dead, scaly cells, densely packed together and filled with parallel fibers of hard keratin. Most mammals have claws, whereas flat nails are one of the distinguishing characteristics of humans and other primates. Flat nails allow for more fleshy and sensitive fingertips, while they also serve as strong keratinized "tools" that can be used for grooming, picking apart food, and other manipulations.

The hard part of the nail is the **nail plate,** which includes the **free edge** overhanging the tip of the finger or toe; the **nail body,** which is the visible attached part of the nail; and the **nail root,** which extends proximally under the overlying skin (fig. 5.9). The surrounding skin rises a bit above the nail as a **nail fold,** separated from the margin of the nail plate by a **nail groove.** The groove and the space under the free edge accumulate dirt and bacteria and require special attention when scrubbing for duty in an operating room or nursery. The skin underlying the nail plate is the **nail bed;** its epidermis is called the **hyponychium**[28] (HIPE-o-NICK-ee-um). At the proximal end of the nail, the stratum basale thickens into a growth zone called the **nail matrix.** Mitosis here in the matrix accounts for the growth of the nail—about 1 mm per week in the fingernails and slightly slower in the toenails. The thickness of the

---

[28]*hypo* = below + *onych* = nail

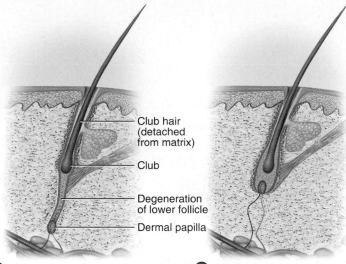

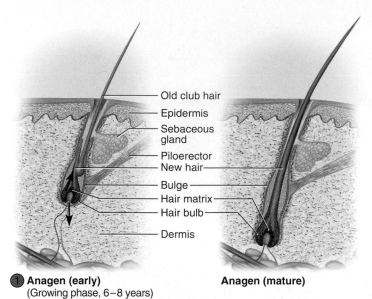

(1) Anagen (early)
(Growing phase, 6–8 years)
Stem cells multiply and follicle grows deeper into dermis; hair matrix cells multiply and keratinize, causing hair to grow upward; old club hair may persist temporarily alongside newly growing hair.

Anagen (mature)

(2) Catagen
(Degenerative phase, 2–3 weeks)
Hair growth ceases; hair bulb keratinizes and forms club hair; lower follicle degenerates.

(3) Telogen
(Resting phase, 1–3 months)
Dermal papilla has ascended to level of bulge; club hair falls out, usually in telogen or next anagen.

Figure 5.8 The Hair Cycle.

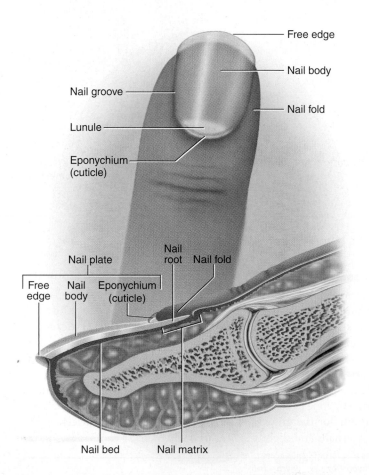

Figure 5.9 Anatomy of a Fingernail.

matrix obscures the underlying dermal blood vessels and is the reason why an opaque white crescent, the **lunule**[29] (LOON-yule), often appears at the proximal end of a nail. A narrow zone of dead skin, the **cuticle** or **eponychium**[30] (EP-o-NICK-ee-um), commonly overhangs this end of the nail.

The appearance of the fingertips and nails can be valuable in medical diagnosis. The fingertips become swollen or *clubbed* in response to long-term hypoxemia—a deficiency of oxygen in the blood stemming from conditions such as congenital heart defects and emphysema. Dietary deficiencies may be reflected in the appearance of the nails. An iron deficiency, for example, may cause them to become flat or concave (spoonlike) rather than convex. Contrary to popular belief, adding gelatin to the diet has no effect on the growth or hardness of the nails.

Before You Go On

Answer the following questions to test your understanding of the preceding section:

7. What is the difference between vellus and terminal hair?

8. Describe the three regions of a hair from its base to its tip, and the three layers of a hair seen in cross section.

9. State the function of the dermal papilla, hair receptor, and piloerector muscle associated with a hair follicle.

10. State a reasonable theory for the different functions of hair of the eyebrows, eyelashes, scalp, nostrils, and axilla.

11. Describe some similarities between a nail and a hair.

[29]*lun* = moon + *ule* = little
[30]*ep* = above + *onych* = nail

5.3 Cutaneous Glands

▶ Expected Learning Outcomes

When you have completed this section, you should be able to

- name two types of sweat glands and describe the structure and function of each;

- describe the location, structure, and function of sebaceous and ceruminous glands; and

- discuss the distinction between breasts and mammary glands, and explain their respective functions.

The skin has five types of glands: *apocrine sweat glands, merocrine sweat glands, sebaceous glands, ceruminous glands,* and *mammary glands.*

Sweat Glands

Sweat glands, or **sudoriferous**[31] (soo-dor-IF-er-us) **glands,** are of two kinds, apocrine and merocrine. **Apocrine glands** (fig. 5.10a) occur in the groin, anal region, axilla, and areola, and in mature males, in the beard area. They are absent from the axillary region of Koreans and sparse in the Japanese. Their ducts lead into nearby hair follicles, rather than directly to the skin surface. Both apocrine and merocrine glands produce their secretion by exocytosis. The secretory part of an apocrine gland, however, has a much larger lumen than that of a merocrine gland, so these glands continue to be called apocrine glands to distinguish them functionally and histologically from the merocrine type. Apocrine sweat is thicker and more milky than merocrine sweat because it has more fatty acids in it.

Apocrine sweat glands are scent glands that respond especially to stress and sexual stimulation. They are not activated until puberty,

---

[31]*sudor* = sweat + *fer* = carry, bear

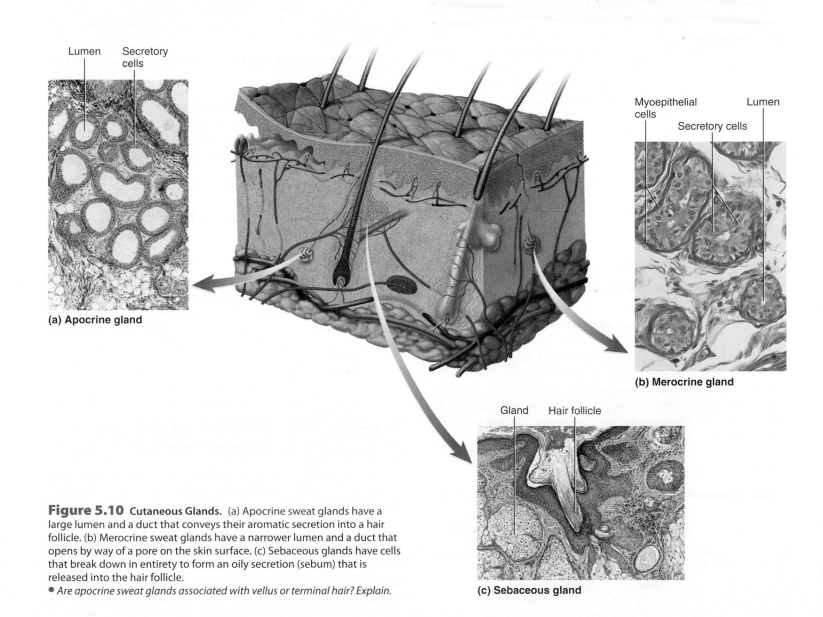

Figure 5.10 **Cutaneous Glands.** (a) Apocrine sweat glands have a large lumen and a duct that conveys their aromatic secretion into a hair follicle. (b) Merocrine sweat glands have a narrower lumen and a duct that opens by way of a pore on the skin surface. (c) Sebaceous glands have cells that break down in entirety to form an oily secretion (sebum) that is released into the hair follicle.

● *Are apocrine sweat glands associated with vellus or terminal hair? Explain.*

Lumen   **Secretory cells**

(a) Apocrine gland

Myoepithelial cells   **Lumen**   **Secretory cells**

(b) Merocrine gland

Gland   **Hair follicle**

(c) Sebaceous gland

and in women, they enlarge and shrink in phase with the menstrual cycle. These facts, as well as experimental evidence, suggest that their function is to secrete *sex pheromones*—chemicals that exert subtle effects on the sexual behavior and physiology of other people. They apparently correspond to the scent glands that develop in other mammals on attainment of sexual maturity. Fresh apocrine sweat does not have a disagreeable odor and, indeed, it is considered attractive or arousing in some cultures. Stale apocrine sweat acquires a rancid odor from the action of bacteria on the lipids in the perspiration. Disagreeable body odor is called *bromhidrosis*.[32] It occasionally indicates a metabolic disorder, but more often reflects inadequate hygiene.

Many mammals have apocrine scent glands associated with specialized tufts of hair. In humans, they occur almost exclusively in regions covered by the pubic hair, axillary hair, and beard, suggesting that they are similar to other mammalian scent glands in function. The hair serves to retain the aromatic secretion and regulate its rate of evaporation from the skin. Thus, it seems no mere coincidence that women's faces lack both apocrine scent glands and a beard.

Merocrine[33] (**eccrine**[34]) **glands** (fig. 5.10b) are widely distributed over the entire body, but are especially abundant on the palms, soles, and forehead. Each is a simple tubular gland with a twisted coil in the dermis or hypodermis and an undulating or coiled duct leading to a sweat pore on the skin surface. This duct is lined by a stratified cuboidal epithelium in the dermis and by keratinocytes in the epidermis. Amid the secretory cells at the deep end of the gland, there are specialized **myoepithelial**[35] **cells** with properties similar to smooth muscle. They contract in response to the sympathetic nervous system and squeeze perspiration up the duct. These cells exist and function similarly in apocrine sweat glands as well. Merocrine perspiration functions to cool the body. It contains sodium chloride, ammonia, urea, and uric acid—wastes that are present also in the urine. There are 3 to 4 million merocrine sweat glands in the skin, with a total mass about equal to that of one kidney.

Sebaceous Glands

Sebaceous[36] (seh-BAY-shus) **glands** (fig. 5.10c) produce an oily secretion called **sebum** (SEE-bum). They occur everywhere except in thick skin, but are most abundant on the scalp and face. They are flask-shaped and usually clustered around a hair follicle, with short ducts that open into the follicle. Some, however, open directly onto the skin surface. They are holocrine glands with little visible lumen. Their secretion consists of broken-down cells that are replaced by mitosis around the perimeter of the gland. Sebum keeps the skin and hair from becoming dry, brittle, and cracked. The sheen of well-brushed hair is due to sebum distributed by the hairbrush.

Ceruminous Glands

Ceruminous (seh-ROO-mih-nus) **glands** are found only in the auditory (external ear) canal, where their secretion combines with sebum and dead epidermal cells to form earwax, or **cerumen**.[37] They are simple, coiled, tubular glands with ducts leading to the skin surface. Cerumen keeps the eardrum pliable, waterproofs the canal, kills bacteria, and coats the guard hairs of the ear, making them sticky and more effective in blocking foreign particles from entering the canal.

Mammary Glands

Mammary glands are milk-producing glands that develop within the breasts (*mammae*) under conditions of pregnancy and lactation. They are not synonymous with the breasts, which are present in both sexes and which, even in females, usually contain only small traces of mammary gland. Mammary glands are modified apocrine sweat glands that produce a richer secretion than other apocrine glands and channel it through ducts to a nipple for more efficient conveyance to the offspring. The mammary glands are discussed in more detail in chapter 26. Table 5.3 summarizes the cutaneous glands.

Before You Go On

Answer the following questions to test your understanding of the preceding section:

12. How do merocrine and apocrine sweat glands differ in structure and function?
13. What types of hair are associated with apocrine glands? Why?
14. What other type of gland is associated with hair follicles? How does its mode of secretion differ from that of sweat glands?
15. What is the difference between a breast and mammary gland? What other type of cutaneous gland is most closely related to mammary glands?

DEEPER INSIGHT   5.3

Extra Nipples—An Evolutionary Throwback

Humans normally have only one pair of nipples. This is a trait that we share with other primates, related to the fact that most primates have only one infant at a time rather than having to nurse a whole litter. Some women and men, however, develop one or more additional nipples, either on the same breast or slightly superior or inferior to it. This condition is called *polythelia*.[38] The extra nipple often is so little developed that it is mistaken for a mole. In a few cases, fully formed additional breasts develop inferior to the primary ones—a condition called *polymastia*.[39] In medieval Europe and colonial America, polythelia was sometimes used to incriminate women as supposed witches and used as a pretext to put some women to death.

Why would such a thing as polythelia occur? It is an example of what evolutionary biologists call an *atavism*, a "throwback" to an ancestral condition. Most mammals develop two rows of mammary glands along lines called the *milk lines* or *mammary ridges,* which extend from the axillary to the inguinal region. This enables a female to nurse a litter of young. Primates have dispensed with all but the most anterior pair, but still posses the genes for development of more. These genes are normally inactive in humans, but when activated, additional nipples or breasts appear along the milk line. Polythelia is a further testimony to our ancestry and kinship with other mammals.

---

[32]*brom* = stench + *hidros* = sweat
[33]*mero* = part + *crin* = to separate, secrete
[34]*ec* = out + *crin* = to separate, secrete
[35]*myo* = muscle
[36]*seb* = fat, tallow + *aceous* = possessing
[37]*cer* = wax

[38]*poly* = many, multiple + *thel* = nipples + *ia* = condition
[39]*poly* = many, multiple + *mast* = breasts + *ia* = condition

TABLE 5.3	Cutaneous Glands
Gland Type	**Definition**
Sudoriferous glands	Sweat glands
Apocrine	Sweat glands that function as scent glands; found in the regions covered by the pubic, axillary, and male facial hair; open by ducts into hair follicles
Merocrine	Sweat glands that function in evaporative cooling; widely distributed over the body surface; open by ducts onto the skin surface
Sebaceous glands	Oil-producing glands associated with hair follicles
Ceruminous glands	Glands of the ear canal that produce cerumen (earwax)
Mammary glands	Milk-producing glands located in the breasts

5.4 Developmental and Clinical Perspectives

▶ Expected Learning Outcomes

When you have completed this section, you should be able to

- describe the prenatal development of the skin, hair, and nails;
- describe the three most common forms of skin cancer; and
- discuss the three classes of burns and the priorities in burn treatment.

Prenatal Development of the Integumentary System

Skin

The epidermis develops from the embryonic ectoderm, and the dermis from the mesoderm. In week 4 of embryonic development, ectodermal cells multiply and organize into two layers—a superficial *periderm* of squamous cells and a deeper *basal layer* (fig. 5.11). In week 11, the basal layer gives rise to a new *intermediate layer* of cells between these two. From then until birth, the basal layer is known as the *germinative layer*. Its cells remain for life as the stem cells of the stratum basale. Cells of the intermediate layer synthesize keratin and become the first keratinocytes. These cells organize into three layers—the stratum spinosum, granulosum, and corneum—as the periderm is sloughed off into the amniotic fluid. By week 21, the periderm is gone and the stratum corneum is the outermost layer of the fetal integument.

Beneath the developing epidermis, the mesoderm differentiates into a gelatinous connective tissue called mesenchyme. Mesenchymal cells begin producing collagenous and elastic fibers by week 11, and the mesenchyme takes on the characteristics of typical fibrous connective tissue. Dermal papillae appear along the dermal–

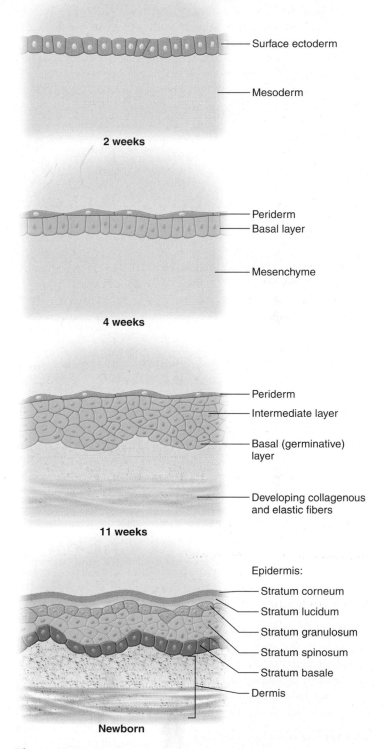

2 weeks — Surface ectoderm / Mesoderm

4 weeks — Periderm / Basal layer / Mesenchyme

11 weeks — Periderm / Intermediate layer / Basal (germinative) layer / Developing collagenous and elastic fibers

Newborn — Epidermis: Stratum corneum / Stratum lucidum / Stratum granulosum / Stratum spinosum / Stratum basale / Dermis

Figure 5.11   Prenatal Development of the Epidermis and Dermis.

epidermal boundary in the third month. Blood vessels appear in the dermis late in week 6. At birth, the skin has 20 times as many blood vessels as it needs to support its metabolism. The excess may help to regulate the body temperature of the newborn.

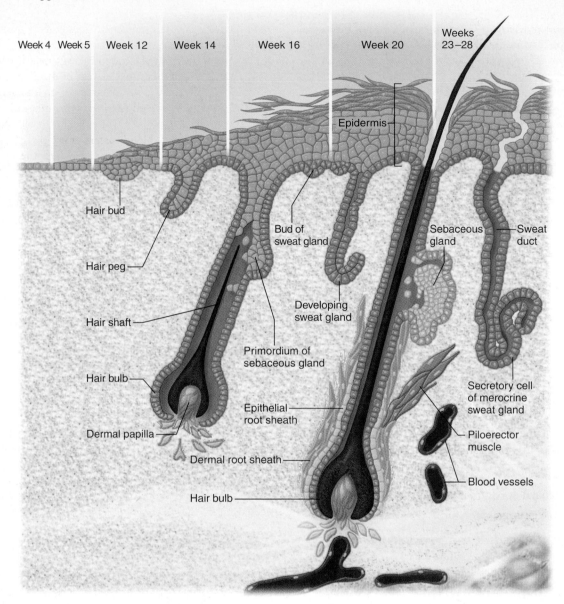

Week 4 Week 5   Week 12   Week 14   Week 16   Week 20   Weeks 23–28

Epidermis

Hair bud

Bud of sweat gland

Sebaceous gland   Sweat duct

Hair peg

Hair shaft

Developing sweat gland

Primordium of sebaceous gland

Hair bulb

Secretory cell of merocrine sweat gland

Epithelial root sheath

Dermal papilla

Piloerector muscle

Dermal root sheath

Blood vessels

Hair bulb

Figure 5.12 Prenatal Development of a Hair Follicle and Cutaneous Glands.

Hair and Nails

The first hair follicles appear around the end of the second month on the eyebrows, eyelids, upper lip, and chin; follicles do not appear elsewhere until the fourth month. At birth, there are about 5 million hair follicles in both sexes; no additional follicles form after birth.

A hair follicle begins as a cluster of ectodermal cells called a *hair bud,* which pushes down into the dermis and elongates into a rodlike *hair peg* (fig. 5.12). The lower end of the peg expands into a *hair bulb.* The dermal papilla first appears as a small mound of tissue just below the bulb, then expands into the bulb itself. Ectodermal cells overlying the papilla form the *germinal matrix,* a mass of mitotically active cells that produce the hair root. The first hair to develop in the fetus is lanugo, which appears in week 12 and is abundant by week 20. By the time of birth, most lanugo is replaced by vellus.

The first indications of nail development are epidermal thickenings that appear on the ventral surfaces of the fingers around 10 weeks and on the toes around 14 weeks. They soon migrate to the dorsal surfaces of the digits, where they form a shallow depression called the

primary nail field. The margins of the nail field are the nail folds. In the proximal nail fold of each digit, the germinal layer of epidermis develops into the nail root. Mitosis in the root produces the keratinocytes that become compressed into the hard nail plate. The nail plate reaches the fingertips by 8 months and the toe tips by birth.

Glands

Sebaceous glands begin to bud from the sides of a hair follicle about 4 weeks after the hair bulb begins to elongate (fig. 5.12). Mature sebaceous glands are present on the face by 6 months and secrete very actively before birth. Their sebum mixes with epidermal and peridermal cells to form a white, greasy skin coating called the *vernix caseosa.*[40] The vernix protects the skin from abrasions and from the amniotic fluid, which can otherwise cause the fetal skin to chap and harden. Its slipperiness also aids in the birth passage through the vagina. The vernix is anchored to the skin by the lanugo and later by the vellus.

---

[40]*vernix* = varnish + *case* = cheese + *osa* = having the qualities of

Sebaceous glands become largely dormant by the time of birth, and are reactivated at puberty under the influence of the sex hormones.

Apocrine sweat glands also develop as outgrowths from the hair follicles. They appear over most of the body at first, but then degenerate except in the limited areas described earlier—especially in the axillary and genital regions. Like the sebaceous glands, they become active at puberty.

Merocrine sweat glands develop as buds of the embryonic germinative layer that grow and push their way down into the dermis (fig. 5.12). These buds develop at first into solid cords of epithelial tissue, but cells in the center of the cord later degenerate to form the lumen of the sweat duct, while cells at the lower end differentiate into secretory and myoepithelial cells.

The Aging Integumentary System

Senescence (age-related degeneration) of the integumentary system often becomes noticeable by the late 40s. The hair turns grayer and thinner as melanocyte stem cells die out, mitosis slows down, and dead hairs are not replaced. Atrophy of the sebaceous glands leaves the skin and hair drier. As epidermal mitosis declines and collagen is lost from the dermis, the skin becomes almost paper-thin and translucent. It becomes looser because of a loss of elastic fibers and flattening of the dermal papillae. If you pinch a fold of skin on the back of a child's hand, it quickly springs back when you let go; do the same on an elderly person, and the skin fold remains longer. Because of its loss of elasticity, aged skin sags to various degrees and may hang loosely from the arms and other places.

Cutaneous blood vessels become fewer and more fragile in old age. The skin may redden as broken vessels leak into the connective tissue, and aged skin bruises more easily. Many older people exhibit *rosacea*—patchy networks of tiny, dilated blood vessels visible especially on the nose and cheeks. Injured skin heals slowly in old age because of diminished circulation and a relative scarcity of immune cells and fibroblasts. Dendritic cells decline by as much as 40% in the aged epidermis, leaving the skin more susceptible to recurring infections.

Thermoregulation can be a problem in old age because of the atrophy of cutaneous blood vessels, sweat glands, and subcutaneous fat. Older people are more vulnerable to hypothermia in cold weather and heatstroke in hot weather. Heat waves and cold spells take an especially heavy toll among elderly poor people who suffer from a combination of reduced homeostasis and inadequate housing.

Degeneration of the skin is accelerated by excessive exposure to the ultraviolet radiation of sunlight. This *photoaging* accounts for more than 90% of the changes that people find medically troubling or cosmetically disagreeable: skin cancer; yellowing and mottling of the skin; age spots, which resemble enlarged freckles on the back of the hand and other sun-exposed areas; and wrinkling, which especially affects the most exposed areas of skin (face, hands, and arms). Sun-damaged skin shows many malignant and premalignant cells; extensive damage to the dermal blood vessels; and dense masses of coarse, frayed elastic fibers underlying the surface wrinkles and creases.

Skin Disorders

Because it is the most exposed of all our organs, skin is not only the most vulnerable to injury and disease, but is also the one place where we are most likely to notice anything out of the ordinary. We focus here on two particularly common and serious disorders, skin cancer and burns. Other skin diseases are briefly summarized in table 5.4.

TABLE 5.4	Some Disorders of the Integumentary System
Acne	Inflammation of the sebaceous glands, especially beginning at puberty; follicle becomes blocked with keratinocytes and sebum and develops into a blackhead *(comedo)* composed of these and bacteria; continued inflammation of follicle results in pus production and pimples
Dermatitis	Any inflammation of the skin, typically marked by itching and redness; often *contact dermatitis,* caused by exposure to toxins such as poison ivy
Eczema (ECK-zeh-mah)	Itchy, red, "weeping" skin lesions caused by an allergy, usually beginning before age 5; may progress to thickened, leathery, darkly pigmented patches of skin
Psoriasis (so-RY-ah-sis)	Recurring, reddened plaques covered with silvery scale; sometimes disfiguring; possibly caused by an autoimmune response; runs in families
Ringworm	A fungal infection of the skin (not a worm) that sometimes grows in a circular pattern; common in moist areas such as the axilla, groin, and foot *(athlete's foot)*
Rosacea (ro-ZAY-she-ah)	A red rashlike area, often in the area of the nose and cheeks, marked by fine networks of dilated blood vessels; worsened by hot drinks, alcohol, and spicy food
Warts	Benign, elevated, rough lesions caused by human papillomaviruses (HPV). *Common warts* appear most frequently in late childhood on the fingers, elbows, and other areas of skin subject to stress. *Plantar warts* occur on the soles and *venereal warts* on the genitals. Warts can be treated by freezing with liquid nitrogen, electric cauterization (burning), laser vaporization, surgical excision, and some medicines such as salicylic acid.

Disorders Described Elsewhere

Skin Cancer

Skin cancer is induced by the ultraviolet rays of the sun. It occurs most often on the head and neck, where exposure is greatest. It is most common in fair-skinned people and the elderly, who have had the longest lifetime UV exposure and have less melanin to shield the keratinocyte DNA from radiation. The popularity of sun tanning, however, has caused an alarming increase in skin cancer among younger people. While sunscreens protect against sunburn, there is no evidence that they afford protection from skin cancer. Skin cancer is one of the most common cancers, but it is also one of the easiest to treat and has one of the highest survival rates when it is detected and treated early.

There are three types of skin cancer named for the epidermal cells in which they originate: basal cell carcinoma, squamous cell carcinoma, and malignant melanoma. The three types are also distinguished from each other by the appearance of their **lesions**[41] (zones of tissue injury).

Basal cell carcinoma[42] is the most common type. It is the least deadly because it seldom metastasizes, but if neglected, it can cause severe facial disfigurement. It arises from cells of the stratum basale and eventually invades the dermis. On the surface, the lesion first appears as a small, shiny bump. As the bump enlarges, it often develops a central depression and a beaded "pearly" edge (fig. 5.13a).

Squamous cell carcinoma arises from keratinocytes of the stratum spinosum. The lesion has a raised, reddened, scaly appearance and later forms a concave ulcer with raised edges (fig. 5.13b). The chance of recovery is good with early detection and surgical removal, but if it goes unnoticed or is neglected, this cancer tends to metastasize to the lymph nodes and can be lethal.

Malignant melanoma is a skin cancer that arises from the melanocytes, often in a preexisting mole. It accounts for no more than 5% of skin cancers, but it is the most deadly form. It can be treated surgically if it is caught early, but if it metastasizes—which it does quickly—it is unresponsive to chemotherapy and usually fatal. The average person with metastatic melanoma lives only 6 months from diagnosis, and only 5% to 14% of patients survive with it for 5 years. The greatest risk factor for malignant melanoma next to UV exposure is a family history of the disease. It has a relatively high incidence in men, in redheads, and in people who experienced severe sunburns in childhood.

It is important to distinguish a mole from malignant melanoma. A mole usually has a uniform color and even contour, and it is no larger in diameter than the end of a pencil eraser (about 6 mm). If it becomes malignant, however, it forms a large, flat, spreading lesion with a scalloped border (fig. 5.13c). The American Cancer Society suggests an "ABCD rule" for recognizing malignant melanoma: *A* for asymmetry (one side of the lesion looks different from the other); *B* for border irregularity (the contour is not uniform but wavy or scalloped); *C* for color (often a mixture of brown, black, tan, and sometimes red and blue); and *D* for diameter (greater than 6 mm).

Skin cancer is treated by surgical excision, radiation therapy, or destruction of the lesion by heat (electrodesiccation) or cold (cryosurgery).

---

[41]*lesio* = injure
[42]*carcin* = cancer + *oma* = tumor

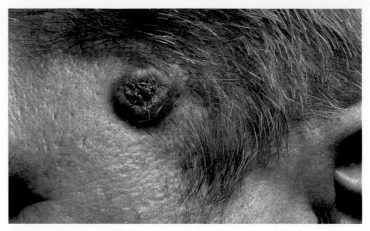

(a) Basal cell carcinoma

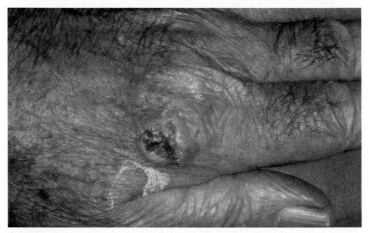

(b) Squamous cell carcinoma

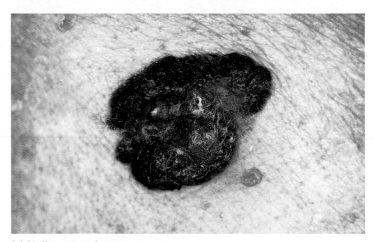

(c) Malignant melanoma

Figure 5.13 Typical Lesions of the Three Forms of Skin Cancer.
● *Which of the ABCD rules can you identify in part (c)?*

Burns

Burns are the leading cause of accidental death. They are usually caused by UV radiation, fires, kitchen spills, or excessively hot bath water, but they also can be caused by other forms of radiation, strong acids and bases, or electrical shock. Burn deaths result pri-

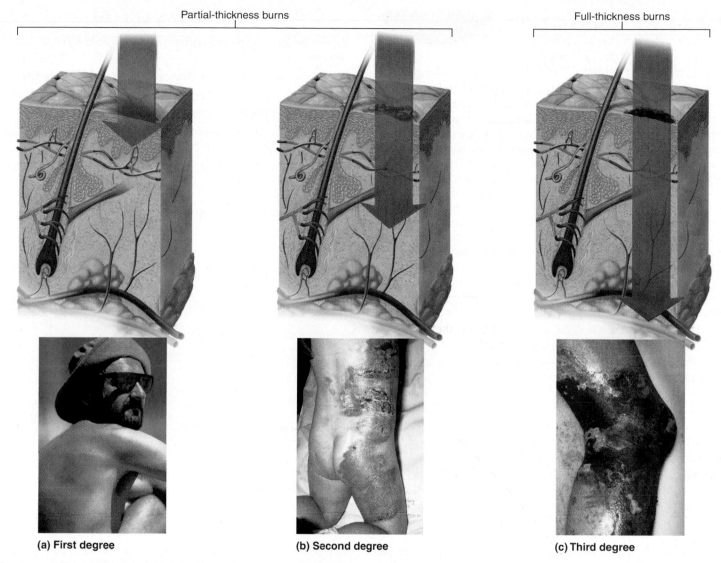

Partial-thickness burns

Full-thickness burns

(a) First degree　　　　**(b) Second degree**　　　　**(c) Third degree**

Figure 5.14 **Three Degrees of Burns.** (a) First-degree burn, involving only the epidermis. (b) Second-degree burn, involving the epidermis and part of the dermis. (c) Third-degree burn, extending through the entire dermis and often involving even deeper tissue.

marily from fluid loss, infection, and the toxic effects of **eschar**[43] (ESS-car)—the burned, dead tissue.

Burns are classified according to the depth of tissue involvement (fig. 5.14). **First-degree burns** involve only the epidermis and are marked by redness, slight edema, and pain. They heal in a few days and seldom leave scars. Most sunburns are first-degree burns.

Second-degree burns involve the epidermis and part of the dermis but leave at least some of the dermis intact. First- and second-degree burns are therefore also known as **partial-thickness burns.** A second-degree burn may be red, tan, or white and is blistered and very painful. It may take from 2 weeks to several months to heal and may leave scars. The epidermis regenerates by division of epithelial cells in the hair follicles and sweat glands and those around the edges of the lesion. Severe sunburns and many scalds are second-degree burns.

Third-degree burns are also called **full-thickness burns** because the epidermis and dermis are completely destroyed. Sometimes even deeper tissue is damaged (hypodermis, muscle, and bone). Since no dermis remains, the skin can regenerate only from the edges of the wound. Third-degree burns often require skin grafts (see Deeper Insight 5.4). If a third-degree burn is left to itself to heal, contracture (abnormal connective tissue fibrosis) and severe disfigurement may result.

Apply What You Know

A third-degree burn may be surrounded by painful areas of first- and second-degree burns, but the region of the third-degree burn is painless. Explain the reason for this lack of pain.

---

[43]*eschar* = scab

DEEPER INSIGHT   5.4

Skin Grafts and Artificial Skin

Third-degree burns leave no dermis to regenerate what was lost and, therefore, require skin grafts. Ideally, these should come from elsewhere on the same patient's body (autografts[44]) so there is no problem with immune rejection, but this may not be feasible in patients with extensive burns. A skin graft from another person (called an *allograft*[45] or *homograft*[46]) or even skin from another species (called a *heterograft*[47] or *xenograft*[48]), such as pig skin, may be used, but they present problems with immune rejection. At least two bioengineering companies produce artificial skin as a temporary burn covering. One such product is made by culturing fibroblasts on a collagen gel to produce a dermis, then culturing keratinocytes on this substrate to produce an epidermis. This is used to treat not only burn patients but also patients with leg and foot ulcers caused by diabetes mellitus.

[44]*auto* = self
[45]*allo* = different, other
[46]*homo* = same
[47]*hetero* = different
[48]*xeno* = strange, alien

Before You Go On

Answer the following questions to test your understanding of the preceding section:

16. What adult skin layer arises from the germinative layer of the fetus?
17. What is the vernix caseosa of the fetus? What purpose does it serve?
18. What types of cells are involved in each type of skin cancer?
19. Which type of skin cancer is most dangerous? What are its early warning signs?
20. What is the difference between a first-, second-, and third-degree burn?
21. What are the two most urgent priorities in treating a burn victim? How are these needs dealt with?

Study Guide

Assess Your Learning Outcomes

You should have a good understanding of this chapter if you can accurately address the following issues.

5.1 The Skin and Subcutaneous Tissue (p. 107)

1. The difference between the integumentary system and integument, and the branch of medicine that deals with this system
2. The two principal layers of the skin and alternative terms for the connective tissue beneath the skin
3. The range of thicknesses of the skin, the basis for distinguishing thick skin from thin skin, and where those two types of skin are located
4. The multiple functions of the skin and what aspects of skin structure contribute to these functions
5. The five epidermal cell types and their respective functions
6. The four to five strata seen in thin and thick skin, their order of occurrence, and the distinguishing histological features of each stratum
7. The histological distinctions between thin and thick skin
8. The life history of a keratinocyte from the time of its mitotic birth to the time it dies and flakes off the skin surface, and how the stages of its development correlate with the histological appearance of the epidermal strata
9. The significance of the epidermal water barrier, what it is composed of, and how keratinocytes produce it
10. The composition of the dermis, including its fiber and cell types and the diverse small organs that it contains
11. The structure of the dermal–epidermal boundary, the names of its interlocking troughs and ridges, and how and why the appearance of this boundary differs from one region of the body to another
12. The difference between the papillary and reticular layers of the dermis—where they are seen, their tissue composition, and their functional difference
13. The histological composition of the hypodermis and how it differs from the dermis
14. The pigments responsible for normal skin colors, the two types of melanin, and reasons for the differences between light and dark skin
15. A variety of pathological skin colors and what causes such variations
16. The various kinds of lines, creases, and other markings of the skin

5.2 Hair and Nails (p. 114)

1. The protein composition of hair and nails and how this compares the dominant epidermal protein
2. The distinction between a hair and its follicle, and their general structural relationship
3. The three types of human hair and their differences in appearance, bodily location, and occurrence over the human life span
4. The functions of human hair of various types and bodily locations
5. The three regions of a hair from its base to the portion above the skin; where it gets its nourishment; and which region serves as the hair's growth zone

6. The three zones of a hair seen in cross section, from its core to its surface, including the differences between these zones in cell morphology
7. The layers of a hair follicle and the functional significance of its bulge
8. Nerves and muscles associated with a hair, and their functions
9. Factors that account for differences in hair color and texture (straight, wavy, or curly)
10. Stages in the life cycle of a hair, the principal processes that occur in each stage, and approximately how long each stage lasts
11. Types of hair thinning and factors that contribute to it
12. The morphology of the nails, and how variations in appearance can be of diagnostic value in certain diseases and disorders

5.3 Cutaneous Glands (p. 119)
1. The two types of sweat glands and how they differ in histological appearance, bodily distribution, function, and development over the human life span
2. The function and locations of sebaceous glands, and how they differ from sweat glands in their method of secretion
3. The location of the ceruminous glands, how cerumen differs from the product secreted by these glands, and what functions are served by cerumen
4. Why the terms *mammary gland* and *breast* are not synonymous, and how mammary glands compare and contrast with apocrine sweat glands

5.4 Developmental and Clinical Perspectives (p. 121)
1. Stages in the embryonic development of the epidermis from the ectoderm, and how the dermis differs from the epidermis in its origin and mode of development
2. How the hair and nails develop from the embryonic epidermis
3. Where and how sebaceous glands arise in the embryo; the nature of the vernix caseosa; and how the vernix caseosa, along with the fetal lanugo and vellus, protect the fetus
4. How the two types of sweat glands differ in their prenatal development and how they differ in childhood versus in and beyond puberty
5. Ways in which the adult skin changes with age, especially in the elderly, and how this is influenced by one's lifetime history of UV exposure
6. The three types of skin cancer and how they differ in the cells of origin, relative frequency in the population, and relative risk of metastasis and mortality
7. Distinguishing characteristics of the three degrees of burns

Testing Your Recall

1. Cells of the _____ are keratinized and dead.
   a. papillary layer
   b. stratum spinosum
   c. stratum basale
   d. stratum corneum
   e. stratum granulosum

2. The epidermal water barrier forms at the point where epidermal cells
   a. enter the telogen stage.
   b. pass from stratum basale to stratum spinosum.
   c. pass from stratum spinosum to stratum granulosum.
   d. form the epidermal ridges.
   e. exfoliate.

3. Which of the following skin conditions or appearances would most likely result from liver failure?
   a. pallor
   b. erythema
   c. pemphigus vulgaris
   d. jaundice
   e. melanization

4. All of the following interfere with microbial invasion of the skin *except*
   a. the acid mantle.
   b. melanin.
   c. cerumen.
   d. keratin.
   e. sebum.

5. The hair on a 6-year-old's arms is
   a. vellus.
   b. lanugo.
   c. pilorum.
   d. terminal hair.
   e. rosacea.

6. Which of the following terms is *least* related to the rest?
   a. lunule
   b. nail plate
   c. hyponychium
   d. free edge
   e. cortex

7. Which of the following is a scent gland?
   a. an eccrine gland
   b. a sebaceous gland
   c. an apocrine gland
   d. a ceruminous gland
   e. a merocrine gland

8. _____ are skin cells with a sensory role.
   a. Tactile cells
   b. Dendritic cells
   c. Langerhans cells
   d. Melanocytes
   e. Keratinocytes

9. The embryonic periderm becomes part of
   a. the vernix caseosa.
   b. the lanugo.
   c. the stratum corneum.
   d. the stratum basale.
   e. the dermis.

10. Which of the following skin cells alert the immune system to pathogens?
    a. fibroblasts
    b. melanocytes
    c. keratinocytes
    d. dendritic cells
    e. tactile cells

11. Two common word roots that refer to the skin in medical terminology are _____ and _____.

12. A muscle that causes a hair to stand on end is called a/an _____.

13. The most abundant protein of the epidermis is _____, while the most abundant protein of the dermis is _____.

14. Blueness of the skin due to low oxygen concentration in the blood is called _____.

15. Projections of the dermis toward the epidermis are called _____.

16. Cerumen is more commonly known as _____.

17. The holocrine glands that secrete into a hair follicle are called _____.

18. The scaly outermost layer of a hair is called the _____.

19. A hair is nourished by blood vessels in a connective tissue projection called the _____.

20. A _____ burn destroys part of the dermis, but not all of it.

Answers in the Appendix

Building Your Medical Vocabulary

State a medical meaning of each of the following word elements, and give a term in which it is used.

1. dermato-
2. epi-
3. sub-
4. pap-
5. melano-
6. cyano-
7. lucid-
8. -illa
9. pilo-
10. carcino-

Answers in the Appendix

True or False

Determine which five of the following statements are false, and briefly explain why.

1. Dander consists of dead keratinocytes.
2. The term *integument* means only the skin, but *integumentary system* refers also to the hair, nails, and cutaneous glands.
3. The dermis is composed mainly of keratin.
4. Vitamin D is synthesized by certain cutaneous glands.
5. Cells of the stratum granulosum cannot undergo mitosis.
6. Dermal papillae are better developed in skin that is subject to a lot of mechanical stress than in skin that is subject to less stress.
7. The three layers of the skin are the epidermis, dermis, and hypodermis.
8. People of African descent have a much higher density of epidermal melanocytes than do people of northern European descent.
9. Malignant melanoma is the most common and deadly form of skin cancer.
10. Apocrine scent glands are activated at the same time in life as the pubic and axillary hair begin to grow.

Answers in the Appendix

Testing Your Comprehension

1. Many organs of the body contain numerous smaller organs, perhaps even thousands. Describe an example of this in the integumentary system.
2. Certain aspects of human form and function are easier to understand when viewed from the perspective of comparative anatomy and evolution. Discuss examples of this in the integumentary system.
3. Explain how the complementarity of form and function is reflected in the fact that the dermis has two histological layers and not just one.
4. Cold weather does not normally interfere with oxygen uptake by the blood, but it can cause cyanosis anyway. Why?
5. Why is it important for the epidermis to be effective, but not *too* effective, in screening out UV radiation?

Answers at www.mhhe.com/saladinha3

Improve Your Grade at www.mhhe.com/saladinha3

Practice quizzes, labeling activities, and games provide fun ways to master concepts. You can also download image PowerPoint files for each chapter to create a study guide or for taking notes during lecture.

Bone Tissue

Spongy bone of the human femur

CHAPTER OUTLINE

DEEPER INSIGHTS

BRUSHING UP

To understand this chapter, you may find it helpful to review the following concepts:
- Stem cells (p. 49)
- General properties of connective tissues (p. 61)
- Hyaline cartilage (p. 67)
- Introduction to bone histology (p. 67)

Anatomy & Physiology | REVEALED®
aprevealed.com

Skeletal System

In art and history, nothing has symbolized death so much as a skull or skeleton.[1] Bones and teeth are the most durable remains of a once-living body and the most vivid reminder of the impermanence of life.

The dry bones presented for laboratory study may wrongly suggest that the skeleton is a nonliving scaffold for the body, like the steel girders of a building. Seeing it in such a sanitized form makes it easy to forget that the living skeleton is made of dynamic tissues, full of cells—that it continually remodels itself and interacts physiologically with all of the other organ systems. The skeleton is permeated with nerves and blood vessels, evidence of its sensitivity and metabolic activity.

Bone is the subject of chapters 6 through 8. In this chapter, we study bone as a tissue—its composition, development, and growth. This will provide a basis for understanding the skeleton, joints, and muscles in the chapters that follow.

6.1 Tissues and Organs of the Skeletal System

▶ Expected Learning Outcomes

When you have completed this section, you should be able to

- name the tissues and organs that compose the skeletal system;
- state several functions of the skeletal system;
- distinguish between bone as a tissue and as an organ;
- describe how bones are classified by shape; and
- describe the general features of a long bone.

The **skeletal system** consists of bones, cartilages, and ligaments tightly joined to form a strong, flexible framework for the body. Cartilage, the embryonic forerunner of most bones, covers many joint surfaces in the mature skeleton. Ligaments hold bones together at the joints and are discussed in chapter 9. Tendons are structurally similar to ligaments but attach muscles to bones; they are discussed with the muscular system in chapters 11 and 12.

Functions of the Skeleton

The skeleton obviously provides the body with physical support, but it plays many other roles that go unnoticed by most people. Its functions include the following.

- **Support.** Bones of the legs, pelvis, and vertebral column hold up the body; the jaw bones support the teeth; and nearly all bones provide support for muscles.
- **Movement.** Skeletal muscles would serve little purpose if not for their attachment to the bones and ability to move them.

- **Protection.** Bones enclose and protect such delicate organs and tissues as the brain, spinal cord, lungs, heart, pelvic viscera, and bone marrow.
- **Blood formation.** Red bone marrow is the major producer of blood cells, including most cells of the immune system.
- **Electrolyte balance.** The skeleton is the body's main reservoir of calcium and phosphate. It stores these minerals and releases them when needed for other purposes.
- **Acid–base balance.** Bone buffers the blood against excessive pH changes by absorbing or releasing alkaline salts such as calcium phosphate.
- **Detoxification.** Bone tissue absorbs heavy metals and other foreign elements from the blood and thus mitigates their toxic effects on other tissues. It can later release these contaminants more slowly for excretion. The tendency of bone to absorb foreign elements can, however, have terrible consequences (see Deeper Insight 6.1).

Bones and Osseous Tissue

The study of **bone,** or **osseous**[2] **tissue,** is called **osteology.**[3] Bone is a connective tissue in which the matrix is hardened by the deposition of calcium phosphate and other minerals. The hardening process is called **mineralization** or **calcification.** Osseous tissue, however, is only one of the components of a bone. Also present are blood, bone

---

DEEPER INSIGHT   6.1

Radioactivity and Bone Cancer

Radioactivity captured the public imagination when Marie and Pierre Curie and Henri Becquerel shared the 1903 Nobel Prize for its discovery. Not for several decades, however, did anyone realize its dangers. Factories employed women to paint luminous numbers on watch and clock dials with radium paint. As they moistened the paint brushes with their tongues to keep them finely pointed, the women ingested radium. Their bones readily absorbed it and many of the women developed *osteosarcoma,* the most common and deadly form of bone cancer.

Even more horrific, in the wisdom of hindsight, was a deadly health fad in which people drank "tonics" made of radium-enriched water. One famous enthusiast was the champion golfer and millionaire playboy Eben Byers, who drank several bottles of radium tonic each day and praised its virtues as a wonder drug and aphrodisiac. Like the factory women, Byers contracted osteosarcoma. By the time of his death, holes had formed in his skull and doctors had removed his entire upper jaw and most of his mandible in an effort to halt the spreading cancer. Byers's bones and teeth were so radioactive they could expose photographic film in complete darkness. Brain damage left him unable to speak, but he remained mentally alert to the bitter end. His tragic decline and death in 1932 shocked the world and put an end to the radium tonic fad.

---

[1]*skelet* = dried up

[2]*os, osse, oste* = bone
[3]*osteo* = bone + *logy* = study of

marrow, cartilage, adipose tissue, nervous tissue, and fibrous connective tissue. The word *bone* can denote an organ composed of all these components, or it can denote just the osseous tissue.

The Shapes of Bones

Bones are classified into four groups according to their shapes and corresponding functions (fig. 6.1):

1. **Long bones** are roughly cylindrical in shape and significantly longer than wide. Like crowbars, they serve as rigid levers that are acted upon by the skeletal muscles to produce body movements. Long bones include the humerus of the arm, the radius and ulna of the forearm, the metacarpals and phalanges of the hand, the femur of the thigh, the tibia and fibula of the leg, and the metatarsals and phalanges of the feet.

2. **Short bones** are more nearly equal in length and width. They include the carpal (wrist) and tarsal (ankle) bones. They have limited motion and merely glide across one another, enabling the ankles and wrists to flex in multiple directions.

3. **Flat bones** enclose and protect soft organs and provide broad surfaces for muscle attachment. They include most cranial bones, the ribs, the sternum (breastbone), the scapula (shoulder blade), and the hip bones. Most of them are not truly flat, but conspicuously curved.

4. **Irregular bones** have elaborate shapes that do not fit into any of the preceding categories. They include the vertebrae and some skull bones, such as the sphenoid and ethmoid bones.

General Features of Bones

Bones have an outer shell of dense white osseous tissue called **compact (dense) bone,** usually enclosing a more loosely organized form of osseous tissue called **spongy (cancellous) bone** (figs. 6.2 and 6.5). The skeleton is about three-quarters compact bone and one-quarter spongy bone by weight. Compact and spongy bone are described later in more detail.

Figure 6.2 shows longitudinal sections of a long bone. The principal features of a long bone are its shaft, called the **diaphysis**[4] (dy-AF-ih-sis), and an expanded head at each end, called the **epiphysis**[5] (eh-PIF-ih-sis). The diaphysis consists largely of a cylinder of compact bone enclosing a space called the **marrow cavity** or **medullary**[6] (MED-you-lerr-ee) **cavity.** The epiphysis is filled with spongy bone. Bone marrow occupies the medullary cavity and the spaces amid the spongy bone of the epiphysis. The diaphysis of a long bone provides leverage, whereas the epiphysis is enlarged to strengthen the joint and provide added surface area for the attachment of tendons and ligaments.

In children and adolescents, an **epiphyseal** (EP-ih-FIZZ-ee-ul) **plate** of hyaline cartilage separates the marrow cavities of the epi-

---

[4]*dia* = across + *physis* = growth; originally named for a ridge on the shaft of the tibia
[5]*epi* = upon, above + *physis* = growth
[6]*medulla* = marrow

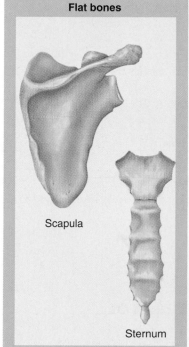
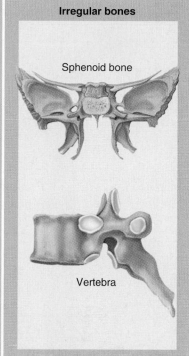
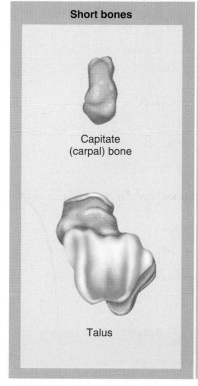
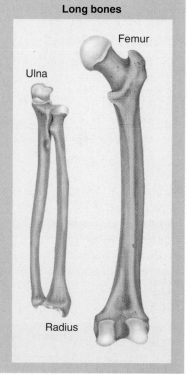

Figure 6.1 Classification of Bones by Shape. The light-blue areas are articular cartilages.

physis and diaphysis. On X-rays, it appears as a transparent line at the end of a long bone (see fig. 6.11). The epiphyseal plate is a zone where the bones grow in length. In adults, the plate is depleted and the bones can grow no longer, but an *epiphyseal line* on the bone surface marks its former location.

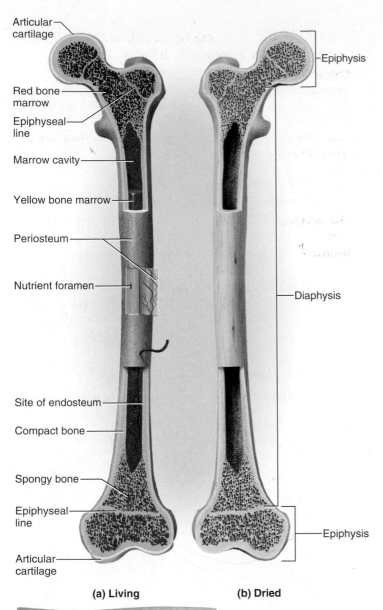

(a) Living   **(b) Dried**

Figure 6.2 **Anatomy of a Long Bone.** (a) The femur, with its soft tissues including bone marrow, articular cartilage, blood vessels, and periosteum. (b) A dried femur in longitudinal section.
● *What is the functional significance of a long bone being wider at the epiphyses than at the diaphysis?*

Externally, most of the bone is covered with a sheath called the **periosteum.**[7] This has a tough, outer *fibrous layer* of collagen and an inner *osteogenic layer* of bone-forming cells. Some collagen fibers of the outer layer are continuous with the tendons that bind muscle to bone, and some penetrate into the bone matrix as **perforating (Sharpey**[8]**) fibers.** The periosteum thus provides strong attachment and continuity from muscle to tendon to bone. The osteogenic layer is important to the growth of bone and healing of fractures. Blood vessels of the periosteum penetrate into the bone through minute holes called **nutrient foramina** (for-AM-ih-nuh); we will trace where they go when we consider bone histology. The internal surface of a bone is

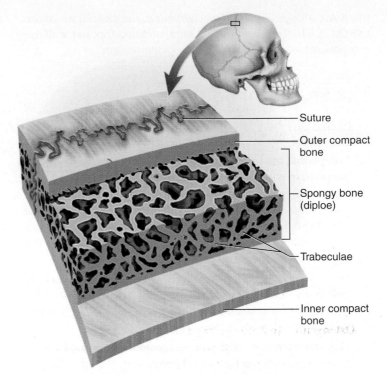

Figure 6.3 **Anatomy of a Flat Bone.**

lined with **endosteum,**[9] a thin layer of reticular connective tissue with cells that deposit osseous tissue and others that dissolve it.

At most joints, the ends of the adjoining bones have no periosteum but rather a thin layer of hyaline cartilage, the **articular**[10] **cartilage.** Together with a lubricating fluid secreted between the bones, this cartilage enables a joint to move far more easily than it would if one bone rubbed directly against the other.

Flat bones have a sandwichlike construction, with two layers of compact bone enclosing a middle layer of spongy bone (fig. 6.3). In the skull, the spongy layer is called the **diploe**[11] (DIP-lo-ee). A moderate blow to the skull can fracture the outer layer of compact bone, but the diploe can sometimes absorb the impact and leave the inner layer of compact bone unharmed.

Before You Go On

Answer the following questions to test your understanding of the preceding section:

1. Name five tissues found in a bone.

2. List three or more functions of the skeletal system other than supporting the body and protecting some of the internal organs.

3. Name the four bone shapes and give an example of each.

4. Explain the difference between compact and spongy bone, and describe their spatial relationship to each other.

5. State the anatomical terms for the shaft, head, growth zone, and fibrous covering of a long bone.

---

[7]*peri* = around + *oste* = bone
[8]William Sharpey (1802–80), Scottish histologist

[9]*endo* = within + *oste* = bone
[10]*artic* = joint
[11]*diplo* = double

6.2 Histology of Osseous Tissue

▶ Expected Learning Outcomes

When you have completed this section, you should be able to

- list and describe the cells, fibers, and ground substance of bone tissue;
- state the functional importance of each constituent of bone tissue;
- compare the histology of the two types of bone tissue; and
- distinguish between two types of bone marrow.

Bone Cells

Like any other connective tissue, bone consists of cells, fibers, and ground substance. There are four types of bone cells (fig. 6.4):

1. **Osteogenic[12] (osteoprogenitor) cells** are stem cells found in the endosteum, the inner layer of the periosteum, and within

the *central canals,* to be described shortly. They arise from embryonic mesenchyme. Osteogenic cells multiply continually and give rise to the *osteoblasts* described next.

2. **Osteoblasts[13]** are bone-forming cells that synthesize the organic matter of the matrix and help to mineralize the bone. They line up in rows in the endosteum and inner layer of periosteum and resemble a cuboidal epithelium on the bone surface (see fig. 6.9). Osteoblasts are nonmitotic, so the only source of new osteoblasts is the osteogenic cells. Stress and fractures stimulate accelerated mitosis of those cells and therefore a rapid rise in the number of osteoblasts, which then reinforce or rebuild the bone.

3. **Osteocytes** are former osteoblasts that have become trapped in the matrix they deposited. They live in tiny cavities called **lacunae,**[14] which are connected to each other by slender channels called **canaliculi**[15] (CAN-uh-LIC-you-lye). Each osteocyte has delicate cytoplasmic processes that reach into the canaliculi to meet the processes of neighboring osteocytes. Adjacent osteocytes are joined by gap junctions at the tip

---

[12]*osteo* = bone + *genic* = producing

[13]*osteo* = bone + *blast* = form, produce
[14]*lac* = lake, hollow + *una* = little
[15]*canal* = canal, channel + *icul* = little

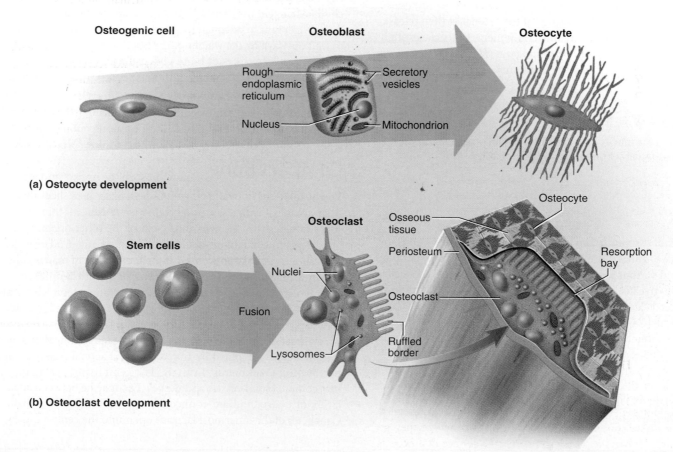

(a) Osteocyte development

(b) Osteoclast development

Figure 6.4 Bone Cells and Their Development. (a) Osteogenic cells give rise to osteoblasts, which deposit matrix around themselves and transform into osteocytes. (b) Bone marrow stem cells fuse to form osteoclasts.

of these processes. These junctions allow osteocytes to pass nutrients and chemical signals to each other and to transfer wastes to the nearest blood vessels for disposal. Osteocytes also communicate by gap junctions with the osteoblasts on the bone surface.

Osteocytes have multiple functions. Some resorb bone matrix and others deposit it, so they contribute to the homeostatic maintenance of both bone density and blood concentrations of calcium and phosphate ions. Perhaps even more importantly, they are strain sensors. When a load is applied to a bone, it produces a flow in the extracellular fluid of the lacunae and canaliculi. This stimulates the osteocytes to secrete biochemical signals that may regulate bone remodeling—adjustments in bone shape and density to adapt to stress.

4. **Osteoclasts**[16] are bone-dissolving macrophages found on bone surfaces. They develop from the same bone marrow stem cells that give rise to blood cells. Several stem cells fuse with each other to form an osteoclast; thus, osteoclasts are unusually large (up to 150 μm in diameter) and typically have 3 or 4 nuclei, but sometimes up to 50. The side of the osteoclast facing the bone has a *ruffled border* with many deep infoldings of the plasma membrane, increasing its surface area. Hydrogen pumps in the ruffled border secrete hydrogen ions (H^+) into the extracellular fluid, and chloride ions (Cl^-) follow by electrical attraction; thus, the space between the osteoclast and the bone becomes filled with hydrochloric acid (HCl). The HCl, with a pH of about 4, dissolves the minerals of the adjacent bone. Lysosomes of the osteoclast then release enzymes that digest the organic component. Osteoclasts often reside in little pits called *resorption bays (Howship*[17] *lacunae)* that they have etched into the bone surface.

Apply What You Know

Considering the function of osteoblasts, what organelles do you think are especially abundant in their cytoplasm?

Matrix

The matrix of osseous tissue is, by dry weight, about one-third organic and two-thirds inorganic matter. The organic matter includes collagen and various large protein–carbohydrate complexes called glycosaminoglycans, proteoglycans, and glycoproteins. The inorganic matter is about 85% **hydroxyapatite,** a crystallized calcium phosphate salt [$Ca_{10}(PO_4)_6(OH)_2$], 10% calcium carbonate ($CaCO_3$), and lesser amounts of magnesium, sodium, potassium, fluoride, sulfate, carbonate, and hydroxide ions.

The minerals and collagen form a composite that gives bones a combination of flexibility and strength similar to fiberglass (see Deeper Insight 6.2). The minerals resist compression (crumbling or sagging when weight is applied). When bones are deficient in cal-

---
[16]*osteo* = bone + *clast* = destroy, break down
[17]J. Howship (1781–1841), English surgeon

Polymers, Ceramics, and Bones

The physical properties of bone can be understood by analogy to some principles of engineering. Engineers use four kinds of construction materials: metals, ceramics (stone, glass, cement), polymers (rubber, plastic, cellulose), and composites (mixtures of two or more of the other classes). Bone is a composite of polymer (protein) and ceramic (mineral). The protein gives it flexibility and resistance to tension, while the mineral gives it resistance to compression. Owing to the mineral component, a bone can support the weight of the body without sagging, and owing to the protein component, it can bend a little when subjected to stress.

Bone is somewhat like a fiberglass fishing rod, which is made of a ceramic (glass fibers) embedded in a polymer (resin). The fibers alone would be too flexible and limp to serve the purpose of a fishing rod, while the resin alone would be too brittle and would easily break. The combination of the two, however, gives the rod strength and flexibility. Unlike fiberglass, however, the ratio of ceramic to polymer in a bone varies from one location to another, adapting osseous tissue to different amounts of tension and compression exerted on different parts of the skeleton.

---

cium salts, they become soft and bend easily. Soft bones are characteristic of a childhood disease called **rickets,** which occurs when a child is deficient in vitamin D and therefore cannot absorb enough dietary calcium to adequately harden the bones. The legs become bowed outward by the weight of the body.

The collagen fibers of bone give it the ability to resist tension, so the bone can bend slightly without snapping. Without collagen, the bones become very brittle, as in brittle bone disease (see Deeper Insight 3.3, p. 69). Without collagen, a jogger's bones would shatter under the impact of running.

Compact Bone

The histological study of compact bone usually uses slices that have been dried, cut with a saw, and ground to translucent thinness. This procedure destroys the cells and much of the other organic content but reveals fine details of the inorganic matrix (fig. 6.5d). Such sections show onionlike **concentric lamellae**—layers of matrix concentrically arranged around a **central (haversian**[18] **or osteonic) canal.** A central canal and its lamellae constitute an **osteon (haversian system)**—the basic structural unit of compact bone. In longitudinal views and three-dimensional reconstructions, we find that an osteon is a cylinder of tissue surrounding a central canal. Along their length, central canals are joined by transverse or diagonal passages called **perforating (Volkmann**[19]**) canals.** The central and perforating canals contain blood vessels and nerves. Lacunae lie between adjacent layers of matrix and are connected with each other by canaliculi. Canaliculi of the innermost lacunae open into the central canal.

---
[18]Clopton Havers (1650–1702), English anatomist
[19]Alfred Volkmann (1800–77), German physiologist

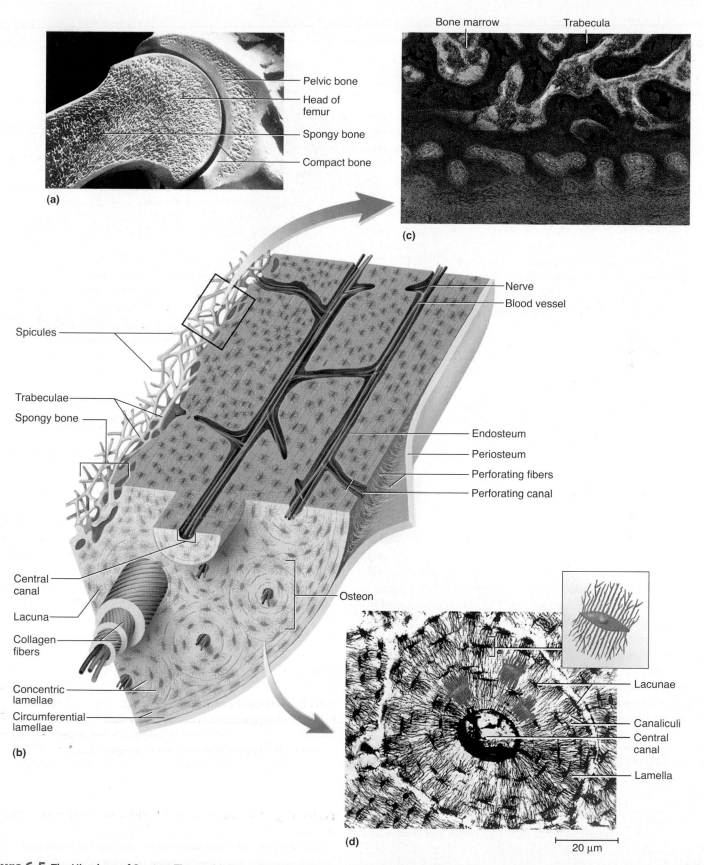

Figure 6.5 **The Histology of Osseous Tissue.** (a) Compact and spongy bone in a frontal section of the hip joint. (b) The three-dimensional structure of compact bone. Lamellae of the uppermost osteon are telescoped to show their alternating arrangement of collagen fibers. (c) Microscopic appearance of a cross section of compact bone. (d) Microscopic appearance of spongy bone.

• *Which type of osseous tissue has more surface area exposed to osteoclast action?*

In each lamella, the collagen fibers "corkscrew" down the matrix in a helical pattern like the threads of a screw. The helices coil in one direction in one lamella and in the opposite direction in the next lamella. Like alternating layers of a sheet of plywood, this makes the bone stronger and enables it to resist tension in multiple directions. In areas where the bone must resist tension (bending), the helix is loosely coiled like the threads on a wood screw and the fibers are more stretched out on the longitudinal axis of the bone. In weight-bearing areas, where the bone must resist compression, the helix is more tightly coiled like the closely spaced threads on a bolt, and the fibers are more nearly transverse.

The skeleton receives about half a liter of blood per minute. Blood vessels, along with nerves, enter the bone tissue through nutrient foramina on the surface. These open into perforating canals that cross the matrix and lead to the central canals. The wall of a central canal looks as if it were pierced with innumerable pinholes. These are the openings from the canaliculi of the innermost lacunae. The osteocytes closest to the central canal receive nutrients from the blood and pass them along through their gap junctions to neighboring osteocytes. They also receive wastes from their neighbors and convey them to the central canal for removal by the bloodstream. Thus, the cytoplasmic processes of the osteocytes maintain a two-way flow of nutrients and wastes between the central canal and the outermost cells of the osteon.

Not all of the matrix is organized into osteons. The inner and outer boundaries of dense bone are arranged in *circumferential lamellae* that run parallel to the bone surface. Between osteons, we can find irregular patches of *interstitial lamellae,* the remains of old osteons that broke down as the bone grew and remodeled itself.

Spongy Bone

Spongy bone consists of a lattice of delicate-looking slivers of bone called **spicules**[20] (rods or spines, as in the photo on p. 129) and **trabeculae**[21] (thin plates as in fig. 6.5c). Although calcified and hard, spongy bone is named for its porous appearance; it is permeated by spaces filled with bone marrow. The matrix is arranged in lamellae like those of compact bone, but they are not arranged in concentric layers, and there are few osteons. Central canals are not needed here because no osteocyte is very far from the blood supply in the marrow. Spongy bone is well designed to impart strength to a bone with a minimum of weight. Its trabeculae are not randomly arranged, as they might seem at a glance, but develop along the bone's lines of stress (fig. 6.6).

Bone Marrow

Bone marrow is a general term for the soft material that occupies the marrow cavity of a long bone, the spaces amid the trabeculae of spongy bone, and the largest of the central canals. In a child, the medullary cavity of nearly every bone is filled with **red bone marrow,** which gets its color from an abundance of red blood cells. Although it is called *myeloid tissue,* red bone marrow is probably best

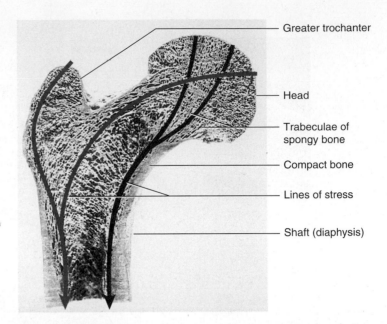

Figure 6.6 **Spongy Bone Structure in Relation to Mechanical Stress.** In this frontal section of the femur, the trabeculae of spongy bone can be seen oriented along lines of mechanical stress applied by the weight of the body.

Labels on figure:
- Greater trochanter
- Head
- Trabeculae of spongy bone
- Compact bone
- Lines of stress
- Shaft (diaphysis)

regarded as an organ. Chapter 22 (p. 617) describes the microscopic structure that justifies that assessment. Red bone marrow is also described as (or more accurately, it includes) *hemopoietic*[22] (HE-mo-poy-ET-ic), or blood-forming, tissue. All types of blood cells are produced here, although some types are also produced in hemopoietic tissues elsewhere, such as the lymph nodes and thymus.

With age, the red bone marrow is gradually replaced by fatty **yellow bone marrow,** like the fat seen at the center of a ham bone. By early adulthood, red bone marrow is limited to the skull, vertebrae, sternum, ribs, part of the pelvic (hip) girdle, and the proximal heads of the humerus and femur; the rest of the skeleton contains yellow marrow (fig. 6.7). Yellow bone marrow no longer produces blood, although in the event of severe or chronic anemia, it can transform back into red marrow and resume that role.

Before You Go On

Answer the following questions to test your understanding of the preceding section:

6. Suppose you had unlabeled electron micrographs of the four kinds of bone cells and their neighboring tissues. Name each of the four cells and explain how you could visually distinguish each one from the other three.

7. Name three organic components of the bone matrix.

8. What are the mineral crystals of bone called, and what are they made of?

9. Sketch a cross section of an osteon and label its major parts.

10. What are the two kinds of bone marrow? What does *hemopoietic tissue* mean? Which type of bone marrow fits this description?

[20]*spicul* = dart, little point
[21]*trabe* = plate + *cul* = little

[22]*hemo* = blood + *poietic* = forming

Intramembranous Ossification

Intramembranous[23] (IN-tra-MEM-bruh-nus) **ossification** produces the flat bones of the skull and most of the clavicle (collarbone). Such bones develop within a fibrous sheet similar to the dermis of the skin, so they are sometimes called *dermal bones*. Figure 6.8 shows the stages of the process.

1 An area of the embryonic connective tissue (mesenchyme) condenses into a layer of soft tissue with a dense supply of blood capillaries. The mesenchymal cells enlarge and differentiate into osteogenic cells, and regions of mesenchyme become a network of soft sheets called trabeculae.

2 Osteogenic cells gather on these trabeculae and differentiate into osteoblasts. These cells deposit an organic matrix called **osteoid**[24] **tissue**—soft collagenous tissue similar to bone except for a lack of minerals (fig. 6.9). As the trabeculae grow thicker, calcium phosphate is deposited in the matrix. Some osteoblasts become trapped in the matrix and are now osteocytes. Mesenchyme close to the surface of a trabecula remains uncalcified, but becomes denser and more fibrous, forming a periosteum.

3 Osteoblasts continue to deposit minerals, producing a honeycomb of bony trabeculae. Some trabeculae persist as permanent spongy bone, while osteoclasts resorb and remodel others to form a marrow cavity in the middle of the bone.

4 Trabeculae at the surface continue to calcify until the spaces between them are filled in, converting the spongy bone to compact bone. This process gives rise to the sandwichlike arrangement typical of mature flat bones.

Endochondral Ossification

Endochondral[25] (EN-doe-CON-drul) **ossification** is a process in which a bone develops from a preexisting model composed of hyaline cartilage. It begins around the sixth week of fetal development and continues into a person's 20s. Most bones of the body, including the vertebrae, ribs, sternum, scapula, pelvis, and bones of the limbs, develop in this way. Figure 6.10 shows the following steps in endochondral ossification, using the relatively simple example of a *metacarpal bone* in the palmar region of the hand. In the metacarpal bones, this occurs in only one epiphysis. In longer bones of the arms, forearms, legs, and thighs, it occurs at both ends and is relatively complex. The epiphyses of those bones are formed from several pieces of childhood bone with multiple ossification centers. Bones such as the metacarpals, metatarsals, and phalanges afford simpler illustrations of the ossification process.

1. Mesenchyme develops into a body of hyaline cartilage, covered with a fibrous perichondrium, in the location of a future bone. For a time, the perichondrium produces chondrocytes and the cartilage model grows in thickness.

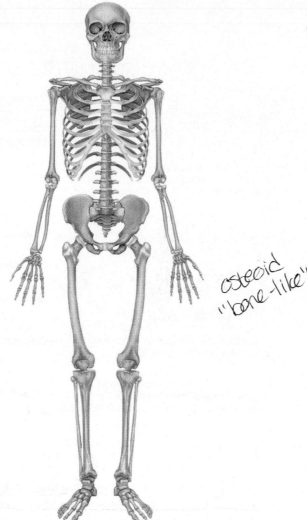

Figure 6.7 **Distribution of Red and Yellow Bone Marrow.** In an adult, red bone marrow occupies the medullary cavities in the regions colored red. Yellow bone marrow occurs in the long bones of the limbs.
● *Suppose red bone marrow was needed for transfusion to a patient. On the basis of this illustration, suggest one or more optimal anatomical sites for drawing red bone marrow from a donor.*

osteoid
"bone-like"

6.3 Bone Development

▶ Expected Learning Outcomes

When you have completed this section, you should be able to

- describe two mechanisms of bone formation;
- explain how a child grows in height; and
- explain how mature bone continues to grow and remodel itself.

The formation of bone is called **ossification** (OSS-ih-fih-CAY-shun), or **osteogenesis**. There are two methods of ossification—*intramembranous* and *endochondral*.

[23]*intra* = within + *membran* = membrane
[24]*oste* = bone + *oid* = like, resembling
[25]*endo* = within + *chondr* = cartilage

[handwritten annotations: "Soft tissue Sheet", "Flat Bones", "Matrix Starting to mineralize"]

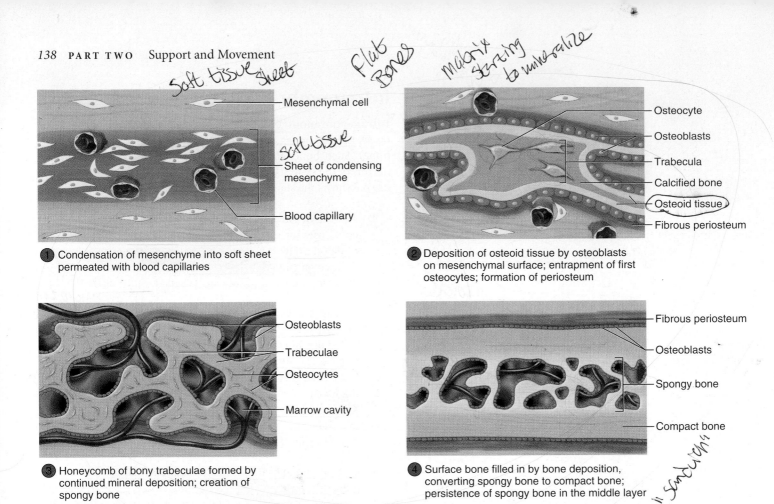

Mesenchymal cell

[handwritten: Soft tissue]

Sheet of condensing mesenchyme

Blood capillary

① Condensation of mesenchyme into soft sheet permeated with blood capillaries

Osteocyte
Osteoblasts
Trabecula
Calcified bone
Osteoid tissue
Fibrous periosteum

② Deposition of osteoid tissue by osteoblasts on mesenchymal surface; entrapment of first osteocytes; formation of periosteum

Osteoblasts
Trabeculae
Osteocytes
Marrow cavity

③ Honeycomb of bony trabeculae formed by continued mineral deposition; creation of spongy bone

Fibrous periosteum
Osteoblasts
Spongy bone
Compact bone

④ Surface bone filled in by bone deposition, converting spongy bone to compact bone; persistence of spongy bone in the middle layer

[handwritten: "Sandwich"]

Figure 6.8 Stages of Intramembranous Ossification.
● *With the aid of chapter 7, name at least two specific bones that would form by this process.*

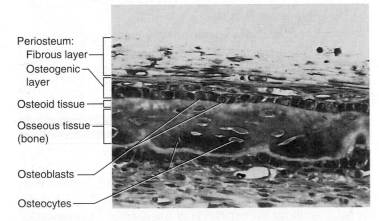

Periosteum:
Fibrous layer
Osteogenic layer
Osteoid tissue
Osseous tissue (bone)
Osteoblasts
Osteocytes

Figure 6.9 Intramembranous Ossification of a Cranial Bone of the Human Fetus. Note the layers of osteoid tissue, osteoblasts, and fibrous periosteum on both sides of the bone.

2. Eventually, the perichondrium stops producing chondrocytes and begins producing osteoblasts. These deposit a thin collar of bone around the middle of the cartilage model, encircling it like a napkin ring and providing physical reinforcement. The former perichondrium is now considered to be a periosteum. Meanwhile, chondrocytes in the middle of the model enlarge and the matrix between their lacunae is reduced to thin walls. This region of chondrocyte enlargement is called the **primary ossification center.** The walls of matrix between the lacunae

calcify and block nutrients from reaching the chondrocytes. The cells die and their lacunae merge into a single cavity in the middle of the model.

3. Blood vessels penetrate the bony collar and invade the primary ossification center. As the center of the model is hollowed out and filled with blood and stem cells, it becomes the **primary marrow cavity.** Various stem cells introduced with the blood give rise to osteoblasts and osteoclasts. Osteoblasts line the cavity, begin depositing osteoid tissue, and calcify it to form a temporary network of bony trabeculae. As the bony collar under the periosteum thickens and elongates, a wave of cartilage death progresses toward the ends of the bone. Osteoclasts in the marrow cavity follow this wave, dissolving calcified cartilage remnants and enlarging the marrow cavity of the diaphysis. The region of transition from cartilage to bone at each end of the primary marrow cavity is called a **metaphysis** (meh-TAF-ih-sis). Soon, chondrocyte enlargement and death occur in the epiphysis of the model as well, creating a **secondary ossification center.**

4. The secondary ossification center becomes hollowed out by the same process as the diaphysis, generating a **secondary marrow cavity** in the epiphysis. This cavity expands outward from the center, in all directions. At the time of birth, the bone typically looks like step 4 in figure 6.10. In bones with two secondary ossification centers, one center lags behind the other in development, so at birth there is a secondary marrow cavity at one end, but chondrocyte growth has just begun at the other. The joints of the limbs are still

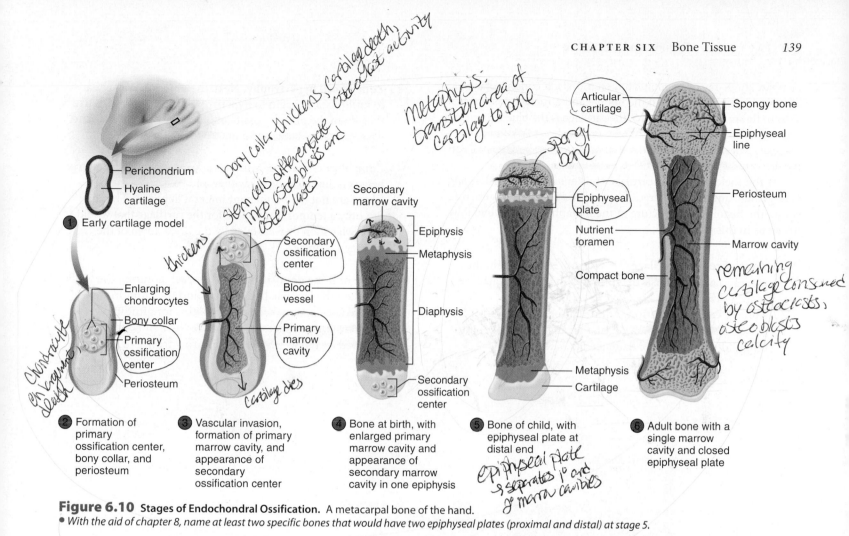

Handwritten notes:
bony collar thickens, cartilage death, osteoclast activity
stem cells differentiate into osteoblasts and osteoclasts
Metaphysis: transition area of cartilage to bone
thickens
spongy bone
Chondrocyte enlargement, death
Cartilage dies
remaining cartilage consumed by osteoclasts, osteoblasts calcify
epiphyseal plate separates 1° and 2° marrow cavities

Figure 6.10 Stages of Endochondral Ossification. A metacarpal bone of the hand.
● *With the aid of chapter 8, name at least two specific bones that would have two epiphyseal plates (proximal and distal) at stage 5.*

cartilaginous at birth, much as they are in the 12-week fetus in figure 7.30, p. 175.

5. During infancy and childhood, the epiphyses fill with spongy bone. Cartilage is then limited to the articular cartilage covering each joint surface and to the epiphyseal plate of cartilage separating the primary and secondary marrow cavities at one or both ends of the bone. The plate persists through childhood and adolescence and serves as a growth zone for bone elongation. This growth process is described in the next section.

6. By the late teens to early twenties, all remaining cartilage in the epiphyseal plate is generally consumed, and the gap between the epiphysis and diaphysis closes. The primary and secondary marrow cavities then unite into a single cavity, and the bone can no longer grow in length.

Bone Growth

Ossification does not end at birth, but continues throughout life with the growth and remodeling of bones. Bones grow in two directions: length and width.

Bone Elongation

To understand growth in length, we must return to the epiphyseal plates mentioned earlier. On an X-ray, the plate appears as a translucent line across the end of a bone, since it is not yet ossified (fig. 6.11;

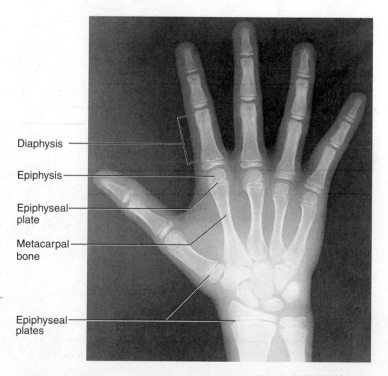

Figure 6.11 X-ray of a Child's Hand. The cartilaginous epiphyseal plates are evident at the ends of the long bones. These will disappear and the epiphyses will fuse with the diaphyses by adulthood. Long bones of the hand and fingers develop only one epiphyseal plate.

compare the X-ray of an adult hand on p. 183). It consists of a band of typical hyaline cartilage in the middle and a metaphysis on each side. In figure 6.10, steps 4 and 5, the cartilage is the blue region and each metaphysis is violet. Even if one end of a bone lacks an epiphyseal plate, it has a metaphysis—the transitional zone between the epiphyseal cartilage and diaphyseal osseous tissue.

At the metaphysis, the cartilage thickens by cell division and enlargement and then undergoes replacement by bone. Figure 6.12 shows the histological structure of the metaphysis and the following steps in this process.

1 Zone of reserve cartilage. This region, farthest from the marrow cavity, consists of typical resting hyaline cartilage.

2 Zone of cell proliferation. A little closer to the marrow cavity, chondrocytes multiply and arrange themselves into longitudinal columns of flattened lacunae.

3 Zone of cell hypertrophy. Next, the chondrocytes cease to multiply and begin to hypertrophy (enlarge), much like they do in the primary ossification center of the fetus. The walls of matrix between lacunae become very thin.

4 Zone of calcification. Minerals are deposited in the matrix between the columns of lacunae and calcify the cartilage. These are not the permanent mineral deposits of bone, but only a temporary support for the cartilage that would otherwise soon be weakened by the breakdown of the enlarged lacunae.

5 Zone of bone deposition. Within each column, the walls between the lacunae break down and the chondrocytes die. This converts each column into a longitudinal channel (white spaces in the figure), which is immediately invaded by blood vessels and marrow from the marrow

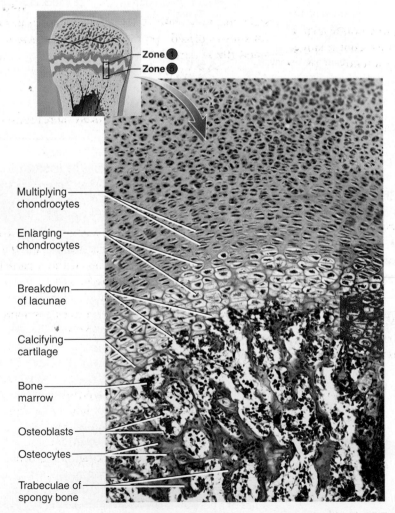

Zone 4
Zone 5

1 Zone of reserve cartilage
Typical histology of resting hyaline cartilage

2 Zone of cell proliferation
Chondrocytes multiplying and lining up in rows of small flattened lacunae

3 Zone of cell hypertrophy
Cessation of mitosis; enlargement of chondrocytes and thinning of lacuna walls

4 Zone of calcification
Temporary calcification of cartilage matrix between columns of lacunae

5 Zone of bone deposition
Breakdown of lacuna walls, leaving open channels; death of chondrocytes; bone deposition by osteoblasts, forming trabeculae of spongy bone

Multiplying chondrocytes

Enlarging chondrocytes

Breakdown of lacunae

Calcifying cartilage

Bone marrow

Osteoblasts

Osteocytes

Trabeculae of spongy bone

Figure 6.12 Zones of the Metaphysis. This micrograph shows the replacement of cartilage with bone in the growth zone of a long bone.
• *Which two zones in this figure account for the growth of a child in height?*

cavity. Osteoblasts line up along the walls of these channels and begin depositing concentric lamellae of matrix, while osteoclasts dissolve the temporarily calcified cartilage.

The process of bone deposition in zone 5 creates a region of spongy bone at the end of the marrow cavity facing the metaphysis. This spongy bone remains for life, although with extensive lifelong remodeling. But around the perimeter of the marrow cavity, continuing ossification converts this spongy bone to compact bone. Osteoblasts lining the aforementioned channels deposit layer after layer of bone matrix, so the channel grows narrower and narrower. These layers become the concentric lamellae of an osteon. Finally only a slender channel persists, the central canal of a new osteon. As usual, osteoblasts trapped in the matrix become osteocytes.

Apply What You Know

In a given osteon, which lamellae are the oldest—those immediately adjacent to the central canal or those around the perimeter of the osteon? Explain your answer.

How does a child or adolescent grow in height? Chondrocyte multiplication in zone 2 and hypertrophy in zone 3 continually push the zone of reserve cartilage (1) toward the ends of the bone, so the bone elongates. In the lower limbs, this process causes a person to grow in height, while bones of the upper limbs grow proportionately.

Thus, bone elongation is really a result of cartilage growth. Cartilage growth from within, by the multiplication of chondrocytes and deposition of new matrix in the interior, is called **interstitial growth.**[26] The most common form of dwarfism results from a failure of cartilage growth in the long bones (see Deeper Insight 6.3).

In the late teens to early twenties, all the cartilage of the epiphyseal plate is depleted. The primary and secondary marrow cavities now unite into one cavity. The junctional region where they meet is filled with spongy bone, and the site of the original epiphyseal plate is marked with a line of slightly denser spongy bone called the **epiphyseal line** (see figs. 6.2 and 6.10, step 6). Often a delicate ridge on the bone surface marks the location of this line. When the epiphyseal plate is depleted, we say that the epiphyses have "closed" because no gap between the epiphysis and diaphysis is visible on an X-ray. Once the epiphyses have all closed in the lower limbs, a person can grow no taller. The epiphyseal plates close at different ages in different bones and in different regions of the same bone. The state of closure in various bones is often used in forensic science to estimate the age at death of a subadult skeleton.

Bone Widening and Thickening

Bones also grow throughout life in diameter and thickness. This involves a process called **appositional growth,**[27] the deposition of new

tissue at the surface. Cartilages can enlarge by both interstitial and appositional growth, but osseous tissue is limited to the appositional method. Embedded in a calcified matrix, the osteocytes have little room to spare for the deposition of more matrix internally.

Appositional growth is similar to intramembranous ossification. Osteoblasts in the inner layer of periosteum deposit osteoid tissue on the bone surface, calcify it, and become trapped in it as osteocytes—much like the process in figure 6.9. They lay down matrix in layers parallel to the surface, not in cylindrical osteons like those deeper in the bone. This process produces the surface layers of bone called *circumferential lamellae,* described earlier. As a bone increases in diameter, its marrow cavity also widens. This is achieved by osteoclasts of the endosteum dissolving tissue on the inner bone surface.

Bone Remodeling

In addition to their growth, bones are continually remodeled throughout life by the absorption of old bone and deposition of new. This process replaces about 10% of the skeletal tissue per year. It repairs microfractures, releases minerals into the blood, and reshapes bones in response to use and disuse. **Wolff's**[28] **law of bone** states that the architecture of a bone is determined by the mechanical stresses placed upon it, and the bone thereby adapts to withstand those stresses. Wolff's law is a fine example of the complementarity of form and function, showing that the form of a bone is shaped by its functional experience. The effect of stress on bone development is quite evident in tennis players, for example, in whom the bones of the racket arm are more robust than those of the other arm. Wolff's law is admirably demonstrated by figure 6.6, in which we see that the trabeculae of spongy bone have developed along the lines of stress placed on the femur. Wolff observed that these stress lines were very similar to the ones that engineers knew of in mechanical cranes.

Bone remodeling comes about through the collaborative action of osteoblasts and osteoclasts. If a bone is little used, osteoclasts remove matrix and get rid of unnecessary mass. If a bone is heavily used or a stress is consistently applied to a particular region of a bone, osteoblasts deposit new osseous tissue and thicken the bone. Consequently, the comparatively smooth bones of an infant or toddler develop a variety of surface bumps, ridges, and spines (described in chapter 7) as the child begins to walk. The greater trochanter of the femur, for example (see fig. 6.6, p. 136; and fig. 8.10, p. 193), is a massive outgrowth of bone stimulated by the pull of tendons from several powerful hip muscles employed in walking.

On average, bones have greater density and mass in athletes and people engaged in heavy manual labor than they do in sedentary people. Anthropologists who study ancient skeletal remains use evidence of this sort to help distinguish between members of different social classes, such as distinguishing royalty from laborers. Even in studying modern skeletal remains, as in investigating a suspicious death, Wolff's law comes into play as the bones give evidence of a person's sex, race, height, weight, work or exercise habits, nutritional status, and medical history.

---

[26]*inter* = between + *stit* = to place, stand
[27]*ap* = *ad* = to, near + *posit* = to place

[28]Julius Wolff (1836–1902), German anatomist and surgeon

Achondroplastic Dwarfism

Achondroplastic[29] (ah-con-dro-PLAS-tic) *dwarfism* is a condition in which the long bones of the limbs stop growing in childhood, while the growth of other bones is unaffected. As a result, a person has a short stature but a normal-sized head and trunk (fig. 6.13). As its name implies, achondroplastic dwarfism results from a failure of cartilage growth—specifically, failure of the chondrocytes in zones 2 and 3 of the metaphysis to multiply and enlarge. This is different from *pituitary dwarfism,* in which a deficiency of growth hormone stunts the growth of all of the bones and a person has short stature but normal proportions throughout the skeletal system.

Achondroplastic dwarfism results from a spontaneous mutation that can arise any time DNA is replicated. Two people of normal height with no family history of dwarfism can therefore have a child with achondroplastic dwarfism. The mutant allele is dominant, so the children of a heterozygous achondroplastic dwarf have at least a 50% chance of exhibiting dwarfism, depending on the genotype of the other parent.

Figure 6.13 **Achondroplastic Dwarfism.** The student on the right, pictured with her roommate of normal height, is an achondroplastic dwarf with a height of about 122 cm (48 in.). Her parents were of normal height. Note the normal proportion of head to trunk but shortening of the limbs.

[29]*a* = without + *chondro* = cartilage + *plast* = growth

The orderly remodeling of bone depends on a precise balance between deposition and resorption, between osteoblasts and osteoclasts. If one process outpaces the other, or both processes occur too rapidly, various bone deformities, developmental abnormalities, and other disorders occur, such as *osteitis deformans* (Paget disease), *osteogenesis imperfecta* (brittle bone disease), and *osteoporosis* (see table 6.2, p. 146; and Deeper Insight 3.3, p. 69).

Nutritional and Hormonal Factors

The balance between bone deposition and resorption is influenced by nearly two dozen nutrients, hormones, and growth factors. The most important factors that promote bone deposition are as follows.

- **Calcium** and **phosphate** are needed as raw materials for the calcified ground substance of bone.

- **Vitamin A** promotes synthesis of the glycosaminoglycans (GAGs) of the bone matrix.

- **Vitamin C (ascorbic acid)** promotes the cross-linking of collagen molecules in bone and other connective tissues.

- **Vitamin D (calcitriol)** is necessary for calcium absorption by the small intestine, and it reduces the urinary loss of calcium and phosphate. Vitamin D is synthesized by one's own body. The process begins when the ultraviolet radiation in sunlight acts on a cholesterol derivative (7-dehydrocholesterol) in the keratinocytes of the epidermis. The product is picked up by the bloodstream, and the liver and kidneys complete its conversion to vitamin D.

- **Calcitonin,** a hormone secreted by the thyroid gland, stimulates osteoblast activity. It functions chiefly in children and pregnant women; it seems to be of little significance in nonpregnant adults.

- **Growth hormone** promotes intestinal absorption of calcium, the proliferation of cartilage at the epiphyseal plates, and the elongation of bones.

- **Sex steroids** (estrogen and testosterone) stimulate osteoblasts and promote the growth of long bones, especially in adolescence.

Bone deposition is also promoted by thyroid hormone, insulin, and local *growth factors* produced within the bone itself. Bone resorption is stimulated mainly by one hormone:

- **Parathyroid hormone (PTH)** is produced by four small *parathyroid glands,* which adhere to the back of the thyroid gland in the neck. The parathyroid glands secrete PTH in response to a drop in blood calcium level. PTH stimulates osteoblasts, which then secrete an *osteoclast-stimulating factor* that promotes bone resorption by the osteoclasts. The principal purpose of this response is not to maintain bone composition but to maintain an appropriate level of blood calcium, without which a person can suffer fatal muscle spasms. PTH also reduces urinary calcium losses and promotes calcitriol synthesis.

The Aging Skeletal System

The predominant effect of aging on the skeleton is a loss of bone mass and strength. After age 35 or 40, osteoblasts become less active than osteoclasts. The imbalance between deposition and resorption leads to **osteopenia**,[30] the loss of bone; when the loss is severe enough to compromise physical activity and health, it is called *osteoporosis* (discussed in the next section). After age 40, women lose about 8% of their bone mass per decade, and men lose about 3%. Bone loss from the jaws is a contributing factor in tooth loss. Not only does bone density decline with age, but the bones become more brittle as the osteoblasts synthesize less protein. Fractures occur more easily and heal more slowly. Arthritis, a family of joint disorders associated with aging, is discussed in chapter 9.

Before You Go On

Answer the following questions to test your understanding of the preceding section:

11. Describe the stages of intramembranous ossification. Name a bone that is formed in this way.

12. Describe the five zones of a metaphysis and the major distinctions between them.

13. How does Wolff's law explain some of the structural differences between the bones of a young child and the bones of a young adult?

14. Identify the nutrients most important to bone growth.

15. Identify the principal hormones that stimulate bone growth.

[30]*osteo* = bone + *penia* = lack, deficiency

6.4 Structural Disorders of Bone

Expected Learning Outcomes

When you have completed this section, you should be able to

- name and describe the types of fractures;
- explain how a fracture is repaired;
- discuss the causes and effects of osteoporosis; and
- briefly describe a few other structural defects of the skeleton.

Most people probably give little thought to their skeletal systems unless they break a bone, although many women know to be more concerned about a far more common bone disorder, osteoporosis. We end this chapter with a consideration of these two pathologies.

Fractures

There are multiple ways of classifying bone fractures. A **stress fracture** is a break caused by abnormal trauma to a bone, such as fractures incurred in falls, athletics, and military combat. A **pathological fracture** is a break in a bone weakened by some other disease, such as bone cancer or osteoporosis, usually due to a stress that would not normally fracture a bone. Fractures are also classified according to the direction of the fracture line, whether or not the skin is broken, and whether a bone is merely cracked or is broken into separate pieces (table 6.1; fig. 6.14).

TABLE 6.1	Classification of Fractures
Type	**Description**
Closed	Skin is not broken (formerly called a *simple* fracture)
Open	Skin is broken; bone protrudes through skin or wound extends to fractured bone (formerly called a *compound* fracture)
Complete	Bone is broken into two or more pieces
Incomplete	Partial fracture that extends only partway across bone; pieces remain joined
Nondisplaced	The portions of bone are still in correct anatomical alignment (fig. 6.14a)
Displaced	The portions of bone are out of anatomical alignment (fig. 6.14b)
Comminuted	Bone is broken into three or more pieces (fig. 6.14c)
Greenstick	Bone is bent toward one side and has incomplete fracture on opposite side (fig. 6.14d)
Hairline	Fine crack in which sections of bone remain aligned; common in skull
Impacted	One bone fragment is driven into the marrow cavity or spongy bone of the other
Depressed	Broken portion of bone forms a concavity, as in skull fractures
Linear	Fracture parallel to long axis of bone
Transverse	Fracture perpendicular to long axis of bone
Oblique	Diagonal fracture, between linear and transverse
Spiral	Fracture spirals around axis of long bone as the result of a twisting stress such as a skiing accident

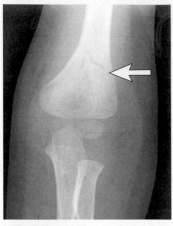

(a) Nondisplaced

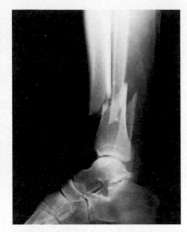

(b) Displaced

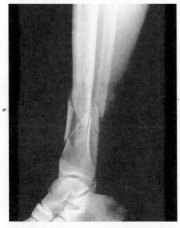

(c) Comminuted

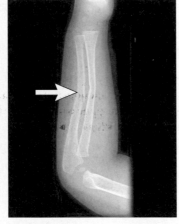

(d) Greenstick

Figure 6.14 X-rays of Representative Fracture Types. (a) Nondisplaced fracture of the distal humerus in a 3-year-old. (b) Displaced fracture of the tibia and fibula. (c) Comminuted fracture of the tibia and fibula. (d) Greenstick fracture of the radius and ulna.

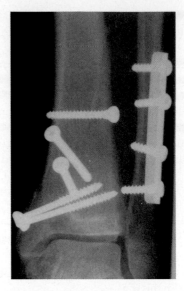

Figure 6.15 Open Reduction of Ankle Fractures. Fractures of the tibia and fibula have been set by surgically exposing the bone and realigning the fragments with plates and screws.

DEEPER INSIGHT   **6.4**

When *Not* to Eat Your Spinach

Many a child has been exhorted, "Eat your spinach! It's good for you." Popeye the Sailor assured cartoon fans that it made him strong to the finish. There is one time, however, when eating spinach isn't such a good idea. People with healing bone fractures are advised to avoid it. Why? Spinach is rich in oxalate, an organic compound that binds calcium and magnesium in the digestive tract and interferes with their absorption. Consequently, oxalate can deprive a fractured bone of the free calcium that it needs in order to heal. There are about 571 milligrams of oxalate per 100 grams of spinach. Some other foods high in oxalate are cocoa (623 mg), rhubarb (447 mg), and beets (109 mg).

Osteoporosis

Osteoporosis[31] (OSS-tee-oh-pore-OH-sis) is a disease in which bone density declines to the extent that the bones become brittle and subject to pathological fractures. It involves the loss of proportionate amounts of organic matrix and minerals, especially from the spongy bone, since this is the most metabolically active osseous tissue and has the greatest surface area exposed to osteoclast action. The bone that remains is histologically normal, but insufficient in quantity to support the body's weight (fig. 6.17a).

People with osteoporosis are especially subject to fractures of the hip, wrist, and vertebral column. Their bones may break under stresses as slight as sitting down too quickly. Among the elderly, slowly healing hip fractures can impose prolonged immobility which, in turn, may lead to fatal complications such as pneumonia.

Most fractures are set by *closed reduction,* a procedure in which the bone fragments are manipulated into their normal positions without surgery. *Open reduction* involves the surgical exposure of the bone and the use of plates, screws, or pins to realign the fragments (fig. 6.15). To stabilize the bone during healing, fractures are often set in fiberglass casts. *Traction* is used to treat fractures of the femur in children. It aids in the alignment of the bone fragments by overriding the force of the strong thigh muscles. Traction is rarely used for elderly patients, however, because the risks from long-term confinement to bed outweigh the benefits. Hip fractures are usually pinned, and early ambulation (walking) is encouraged because it promotes blood circulation and healing.

An uncomplicated fracture heals in 8 to 12 weeks, but complex fractures take longer, and all fractures heal more slowly in older people. Figure 6.16 shows the healing process. Usually, a healed fracture leaves a slight thickening of the bone visible by X-ray, but in some cases, healing is so complete that no trace of the fracture can be found.

---

[31]*osteo* = bone + *por* = porous + *osis* = condition

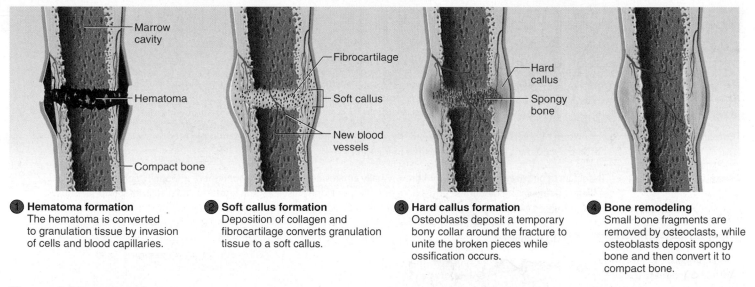

1 Hematoma formation
The hematoma is converted to granulation tissue by invasion of cells and blood capillaries.

2 Soft callus formation
Deposition of collagen and fibrocartilage converts granulation tissue to a soft callus.

3 Hard callus formation
Osteoblasts deposit a temporary bony collar around the fracture to unite the broken pieces while ossification occurs.

4 Bone remodeling
Small bone fragments are removed by osteoclasts, while osteoblasts deposit spongy bone and then convert it to compact bone.

Figure 6.16 The Healing of a Bone Fracture.

Those who survive often face a long, costly recovery. Spinal deformity is also a common consequence of osteoporosis. As the bodies of the vertebrae lose spongy bone, they become compressed by body weight (fig. 6.17b). People commonly lose height after middle age because of this, but those with osteoporosis often develop a more noticeable spinal deformity called **kyphosis,**[32] an exaggerated thoracic curvature ("widow's hump" or "dowager's hump") (fig. 6.17c).

Those at greatest risk for osteoporosis are postmenopausal white women, especially those of light body build. Compared to men, they have less initial bone mass and begin to lose it at an earlier age. Bone loss accelerates after menopause, when the estrogen level declines sharply. Estrogen supports bone mass by inhibiting osteoclast activity (bone resorption), but the ovaries no longer secrete it after menopause and the rate of bone resorption rises. By age 70, the average white woman has lost 30% of her bone mass, and some as much as 50%. Osteoporosis is less common among black women. Young black women have denser bone on average, and even though they too lose bone density after menopause, the loss usually doesn't reach the threshold for osteoporosis and pathological fractures. About 20% of osteoporosis sufferers are men. Men produce estrogen in both the adrenal glands and testes. By age 70, 50% of men have estrogen levels below the threshold needed to maintain bone density. Osteoporosis also occurs among young female runners, dancers, and gymnasts in spite of their vigorous exercise. Their percentage of body fat is so low that they may stop ovulating, and their ovaries secrete unusually low levels of estrogen.

[32]*kypho* = bent, humpbacked + *osis* = condition

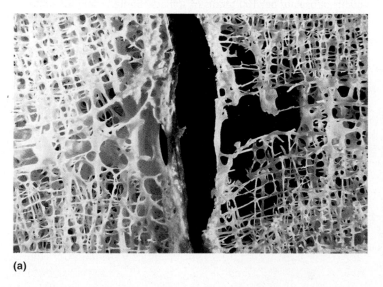

(a)

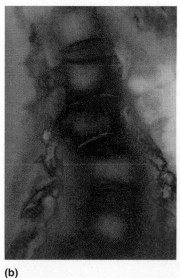

(b)

(c)

Figure 6.17 Spinal Osteoporosis. (a) Spongy bone in the body of a vertebra in good health (left) and with osteoporosis (right). (b) Colorized X-ray of lumbar vertebrae severely damaged by osteoporosis. (c) Abnormal thoracic spinal curvature (kyphosis) due to compression of thoracic vertebrae with osteoporosis.

TABLE 6.2	Structural Disorders of Bone
Acromegaly[33]	A result of adult growth hormone hypersecretion, resulting in thickening of the bones and soft tissues, especially noticeable in the face, hands, and feet.
Osteitis deformans (Paget[34] disease)	Excessive osteoclast proliferation and bone resorption, with osteoblasts attempting to compensate by depositing extra bone. This results in rapid, disorderly bone remodeling and weak, deformed bones. Osteitis deformans usually passes unnoticed, but in some cases it causes pain, disfiguration, and fractures. It is most common in males over the age of 50.
Osteomalacia	Adult form of rickets, most common in poorly nourished women who have had multiple pregnancies. Bones become softened, deformed, and more susceptible to fractures.
Rickets	Defective mineralization of bone in children, usually as a result of insufficient sunlight or vitamin D, sometimes due to a dietary deficiency of calcium or phosphate or to liver or kidney diseases that interfere with calcitriol synthesis. Causes bone softening and deformity, especially in the weight-bearing bones of the lower limbs.

Disorders Described Elsewhere

Achondroplastic dwarfism 142	Osteopenia 143
Brittle bone disease 69	Osteoporosis 144
Fractures 143	

Other risk factors for osteoporosis include dietary deficiencies of calcium, vitamin D, and protein; inadequate exercise; smoking; and diabetes mellitus.

Osteoporosis is now diagnosed with *dual-energy X-ray absorptiometry (DEXA),* which uses low-dose X-rays to measure bone density. DEXA has allowed for earlier diagnosis and more effective drug treatment. However, the severity of osteoporosis depends not on bone density alone, but also on the degree of connectivity between the spongy bone trabeculae, which is lost as trabeculae deteriorate. Neither DEXA nor any other diagnostic method yet available can detect this.

Treatment of osteoporosis is aimed at slowing the rate of bone resorption. Estrogen-replacement therapy has fallen out of favor because it raises the risk of breast cancer, stroke, and coronary artery disease. One of the current preferred treatments is a family of osteoclast-destroying drugs called *bis-phosphonates* (trade names Fosamax, Actonel). They can increase bone mass and cut the risk of fractures in half, but their safety is still being evaluated. Pulsed doses of parathyroid hormone or derivatives such as teriparatide (trade name Forteo) can also increase bone mass by stimulating osteoblasts, but cannot be used for longer than 2 years because of a risk of bone cancer.

The risk of osteoporosis is best minimized by weight-bearing exercise and ample calcium and protein intake early in life, especially in the 20s and 30s when the skeleton is building to its maximum mass.

Other Structural Disorders

Several additional bone disorders are summarized in table 6.2. **Orthopedics**[35] is the branch of medicine that deals with the prevention and correction of injuries and disorders of the bones, joints, and muscles. As the word suggests, this field originated as the treatment of skeletal deformities in children, but it is now much more extensive. It includes the design of artificial joints and limbs and the treatment of athletic injuries.

Before You Go On

Answer the following questions to test your understanding of the preceding section:

16. Name and describe any five types of bone fractures.

17. What is a callus? How does it contribute to fracture repair?

18. List the major risk factors for osteoporosis and describe some ways of preventing it.

---

[33]*acro* = extremity + *megaly* = abnormal enlargement
[34]Sir James Paget (1814–99), English surgeon
[35]*ortho* = straight + *ped* = child, foot

Study Guide

Assess Your Learning Outcomes

You should have a good understanding of this chapter if you can accurately address the following issues.

6.1   Tissues and Organs of the Skeletal System (p. 130)

1. Components of the skeletal system, including but not limited to the bones
2. Seven functions of the skeletal system
3. The constituents of a bone, including but not limited to osseous tissue
4. How bones are classified by shape
5. The spatial relationship between compact and spongy bone
6. The parts of a typical long bone
7. The structure of a typical flat bone

6.2   Histology of Osseous Tissue (p. 133)

1. The four kinds of cells found in a bone, and their respective origins and functions
2. The specialized structure of an osteocyte and how this relates to the microscopic appearance of the bone matrix
3. The specialized structure of an osteoclast and how this relates to its function
4. The composition of the bone matrix and the complementary functional importance of its organic and inorganic components

5. The structural organization of the matrix of compact bone, including the names of the features seen with the microscope
6. The organization of spongy bone and how it combines strength with lightness
7. The two kinds of bone marrow, their locations in the adult skeleton, and their functional differences

6.3   Bone Development (p. 137)

1. The two modes of bone development (ossification) and some bones that develop by each method
2. The stages of intramembranous ossification, converting a soft mesenchymal sheet into a mature (usually flat) bone
3. The stages of endochondral ossification, replacing a hyaline cartilage model with a mature bone
4. Structural differences between the endochondral bones of adults and children, especially with regard to the epiphyseal plates
5. The zones of tissue found in the metaphysis of a child or adolescent; how they relate to the replacement of cartilage by bone; how they relate to the person's growth in height; and why a person cannot grow any taller after the end of adolescence

6. How the endochondral development of compact bone results in osteons with concentric lamellae arranged around the central canal
7. How bones grow in thickness and change shape even after they can no longer grow in length
8. What Wolff's law implies about the ability of bone to adapt to changes in stress
9. The nutrients needed for bone growth and maintenance
10. Hormones that regulate bone growth and remodeling
11. The effects of aging on the skeleton

6.4   Structural Disorders of Bone (p. 143)

1. The difference between stress fractures and pathological fractures
2. Terms for various types of fractures
3. The two basic approaches to clinical treatment of a fracture
4. The definition of *osteoporosis* and why this disease can be such a serious concern for the elderly
5. Osteoporosis risk factors, prevention, diagnosis, and treatment

Testing Your Recall

1. Which cells have a ruffled border and secrete hydrochloric acid?
   a. chondrocytes   d. osteoblasts
   b. osteocytes   e. osteoclasts
   c. osteogenic cells

2. The medullary cavity of a child's bone may contain
   a. red bone marrow.
   b. hyaline cartilage.
   c. periosteum.
   d. osteocytes.
   e. articular cartilages.

3. The long bones of the limbs grow in length by cell proliferation and hypertrophy in
   a. the epiphysis.
   b. the epiphyseal line.
   c. the dense bone.
   d. the metaphysis.
   e. the spongy bone.

4. Osteoclasts are most closely related by common descent to
   a. osteocytes.   d. fibroblasts.
   b. osteogenic cells. e. osteoblasts.
   c. monocytes.

5. The walls between cartilage lacunae break down in the zone of
   a. cell proliferation.
   b. calcification.
   c. reserve cartilage.
   d. bone deposition.
   e. cell hypertrophy.

6. Which of these does *not* promote bone deposition?
   a. dietary calcium
   b. vitamin D
   c. parathyroid hormone
   d. calcitonin
   e. testosterone

7. A child jumps to the ground from the top of a playground "jungle gym." His leg bones do not shatter mainly because they contain
   a. an abundance of glycosaminoglycans.
   b. young, resilient osteocytes.
   c. an abundance of calcium phosphate.
   d. collagen fibers.
   e. hydroxyapatite crystals.

8. One long bone meets another at its
   a. diaphysis.   d. metaphysis.
   b. epiphyseal plate. e. epiphysis.
   c. periosteum.

9. Calcitriol is made from
   a. calcitonin.
   b. 7-dehydrocholesterol.
   c. hydroxyapatite.
   d. estrogen.
   e. PTH.

10. One sign of osteoporosis is
    a. osteitis deformans.
    b. osteomalacia.
    c. a stress fracture.
    d. kyphosis.
    e. a calcium deficiency.

11. Calcium phosphate crystallizes in bone as a mineral called _____.

12. Osteocytes contact each other through channels called _____ in the bone matrix.

13. A bone increases in diameter only by _____ growth, the addition of new surface lamellae.

14. Most compact bone is organized in cylindrical units called _____, composed of lamellae encircling a central canal.

15. The _____ glands secrete a hormone that stimulates cells to resorb bone and return its minerals to the blood.

16. The ends of a bone are covered with a layer of hyaline cartilage called the _____.

17. The cells that deposit new bone matrix are called _____.

18. The most common bone disease is _____.

19. The transitional region between epiphyseal cartilage and the primary marrow cavity of a young bone is called the _____.

20. The cranial bones develop from a flat sheet of condensed mesenchyme in a process called _____.

Answers in the Appendix

Building Your Medical Vocabulary

State a medical meaning of each of the following word elements, and give an example of a term in which it is used.

1. osteo-
2. diplo-
3. lac-
4. -clast
5. -osis
6. dia-
7. -logy
8. artic-
9. -icul
10. -oid

Answers in the Appendix

True or False

Determine which five of the following statements are false, and briefly explain why.

1. Spongy bone is normally covered by compact bone.
2. Most bones develop from hyaline cartilage.
3. Fractures are the most common bone disorder.
4. The growth zone of the long bones of adolescents is the articular cartilage.
5. Osteoclasts develop from osteoblasts.
6. Osteocytes develop from osteoblasts.
7. The protein of the bone matrix is called hydroxyapatite.
8. Blood vessels travel through the central canals of compact bone.
9. Yellow bone marrow has a hemopoietic function.
10. Parathyroid hormone promotes bone resorption and raises blood calcium concentration.

Answers in the Appendix

Testing Your Comprehension

1. Most osteocytes of an osteon are far removed from blood vessels, but still receive blood-borne oxygen and nutrients. Explain how this is possible.
2. Predict what symptoms a person might experience if he or she suffered a degenerative disease in which the articular cartilages were worn away and the fluid between the bones dried up.
3. One of the more common fractures in children and adolescents is an *epiphyseal fracture,* in which the epiphysis of a long bone separates from the diaphysis. Explain why this would be more common in children than in adults.
4. Describe how the arrangement of trabeculae in spongy bone demonstrates the complementarity of form and function.
5. Identify two bone diseases you would expect to see if the epidermis were a completely effective barrier to UV radiation and a person took no dietary supplements to compensate for this. Explain your answer.

Answers at www.mhhe.com/saladinha3

Improve Your Grade at **www.mhhe.com/saladinha3**

Practice quizzes, labeling activities, and games provide fun ways to master concepts. You can also download image PowerPoint files for each chapter to create a study guide or for taking notes during lecture.

The Axial Skeleton

Anterosuperior view of the
thoracic cage and pectoral girdle
(colorized CT scan)

CHAPTER OUTLINE

BRUSHING UP

To understand this chapter, you may find it
helpful to review the following concepts:
- Directional terms (p. 13)
- The axial and appendicular body regions (p. 13)
- The neural tube, somites, and pharyngeal arches of the embryo (p. 90)
- General features of bones (p. 131)
- Intramembranous and endochondral ossification (p. 137)

Anatomy &
Physiology | **REVEALED**®
aprevealed.com

Skeletal System

Aknowledge of skeletal anatomy will be useful as you study later chapters. It provides a point of reference for studying the gross anatomy of other organ systems because many organs are named for their relationships to nearby bones. The subclavian artery and vein, for example, are adjacent to the clavicle; the temporalis muscle is attached to the temporal bone; the ulnar nerve and radial artery travel beside the ulna and radius of the forearm; and the frontal, parietal, temporal, and occipital lobes of the brain are named for adjacent bones of the cranium. An understanding of how the muscles produce body movements also depends on knowledge of skeletal anatomy. In addition, the positions, shapes, and processes of bones can serve as landmarks for a clinician in determining where to give an injection or record a pulse, what to look for in an X-ray, or how to perform physical therapy and other medical procedures.

7.1 Overview of the Skeleton

▶ Expected Learning Outcomes

When you have completed this section, you should be able to

- define the two subdivisions of the skeleton;
- state the approximate number of bones in the adult body;
- explain why this number varies with age and from one person to another; and
- define several terms that denote surface features of bones.

The skeleton (fig. 7.1) is divided into two regions: axial and appendicular. The **axial skeleton,** studied in this chapter, forms the central supporting axis of the body and includes the skull, vertebral column, and thoracic cage (ribs and sternum). The **appendicular skeleton,** studied in chapter 8, includes the bones of the upper limb and pectoral girdle, and bones of the lower limb and pelvic girdle.

Bones of the Skeletal System

It is often stated that there are 206 bones in the skeleton, but this is only a typical adult count, not an invariable number. At birth, there are about 270, and even more bones form during childhood. With age, however, the number decreases as separate bones fuse. For example, each side of a child's pelvic girdle has three bones—the *ilium,*

ischium, and *pubis*—but in adults, these are fused into a single *hip (coxal) bone* on each side. The fusion of several bones, completed by late adolescence to the mid-20s, brings about the average adult number of 206. These bones are listed in table 7.1.

The number varies even among adults. One reason is the development of **sesamoid**[1] **bones**—bones that form within certain tendons in response to stress. The patella (kneecap) is the largest of these; most of the others are small, rounded bones in such locations as the hands and feet (see fig. 8.14c, p. 196). Another reason for adult variation is that some people have extra bones in the skull called **sutural** (SOO-chure-ul), or **wormian,**[2] **bones** (see fig. 7.6).

Anatomical Features of Bones

A bone may exhibit a variety of ridges, spines, bumps, depressions, canals, pores, slits, cavities, and articular surfaces, often called *bone markings.* It is important to know the names of these features because later descriptions of joints, muscle attachments, and the routes traveled by nerves and blood vessels are based on this terminology. The terms for the most common of these features are listed in table 7.2, and several of them are illustrated in figure 7.2.

As you study the skeleton, use yourself as a model. You can easily palpate (feel) many of the bones and some of their details through the skin. Rotate your forearm, cross your legs, palpate your skull and wrist, and think about what is happening beneath the surface or what you can feel through the skin. You will gain the most from this chapter (and indeed, the entire book) if you are conscious of your own body in relation to what you are studying.

Before You Go On

Answer the following questions to test your understanding of the preceding section:

1. Name the major components of the axial skeleton. Name those of the appendicular skeleton.

2. Explain why an adult does not have as many bones as a child does. Explain why one adult may have more bones than another adult of the same age.

3. Briefly describe each of the following bone features: condyle, epicondyle, process, tubercle, fossa, sulcus, and foramen.

---

[1]*sesam* = sesame seed + *oid* = resembling
[2]Ole Worm (1588–1654), Danish physician

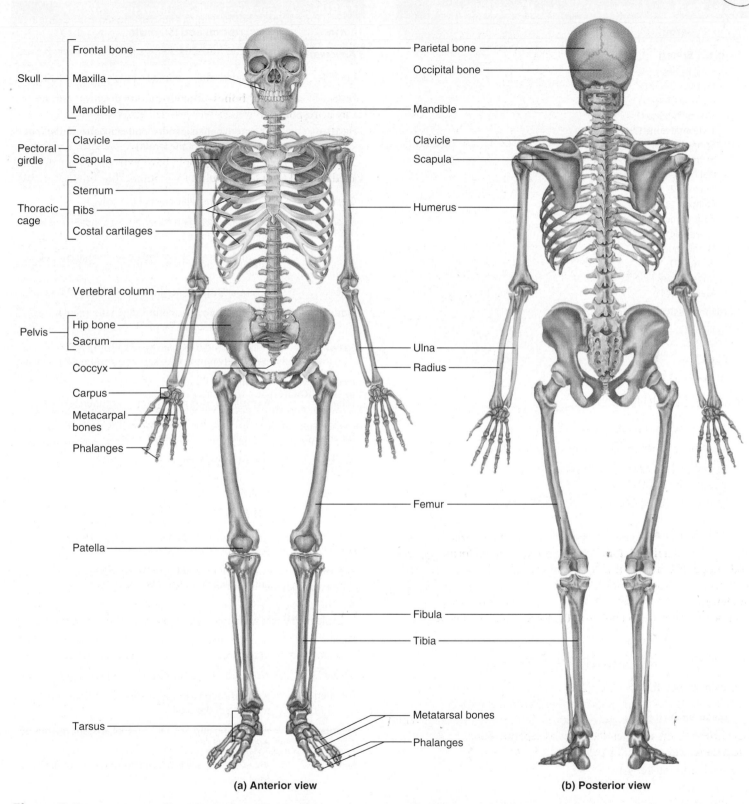

(a) Anterior view

(b) Posterior view

Figure 7.1 **The Adult Skeleton.** The appendicular skeleton is colored green, and the rest is axial skeleton.

TABLE 7.1	Bones of the Adult Skeletal System

Axial Skeleton

Skull (22 Bones)

Cranial bones
    Frontal bone (1)
    Parietal bone (2)
    Occipital bone (1)
    Temporal bone (2)
    Sphenoid bone (1)
    Ethmoid bone (1)
Facial bones
    Maxilla (2)
    Palatine bone (2)
    Zygomatic bone (2)
    Lacrimal bone (2)
    Nasal bone (2)
    Vomer (1)
    Inferior nasal concha (2)
    Mandible (1)

Auditory Ossicles (6 Bones)

Malleus (2)
Incus (2)
Stapes (2)

Hyoid Bone (1 Bone)

Vertebral Column (26 Bones)

Cervical vertebrae (7)
Thoracic vertebrae (12)
Lumbar vertebrae (5)
Sacrum (1)
Coccyx (1)

Thoracic Cage (25 Bones Plus Thoracic Vertebrae)

Ribs (24)
Sternum (1)

Appendicular Skeleton

Pectoral Girdle (4 Bones)

Scapula (2)
Clavicle (2)

Upper Limbs (60 Bones)

Humerus (2)
Radius (2)
Ulna (2)
Carpals (16)
Metacarpals (10)
Phalanges (28)

Hip Bone (2 Bones)

Lower Limbs (60 Bones)

Femur (2)
Patella (2)
Tibia (2)
Fibula (2)
Tarsals (14)
Metatarsals (10)
Phalanges (28)

Grand Total: 206 Bones

TABLE 7.2	Surface Features (Markings) of Bones

Term	Description and Example
Articulations	
Condyle	A rounded knob (occipital condyles of the skull)
Facet	A smooth, flat, slightly concave or convex articular surface (articular facets of the vertebrae)
Head	The prominent expanded end of a bone, sometimes rounded (head of the femur)
Extensions and Projections	
Crest	A narrow ridge (iliac crest of the pelvis)
Epicondyle	An expanded region superior to a condyle (medial epicondyle of the femur)
Line	A slightly raised, elongated ridge (nuchal lines of the skull)
Process	Any bony prominence (mastoid process of the skull)
Protuberance	A bony outgrowth or protruding part (mental protuberance of the chin)
Spine	A sharp, slender, or narrow process (mental spines of the mandible)
Trochanter	Two massive processes unique to the femur
Tubercle	A small, rounded process (greater tubercle of the humerus)
Tuberosity	A rough surface (tibial tuberosity)
Depressions	
Alveolus	A pit or socket (tooth socket)
Fossa	A shallow, broad, or elongated basin (mandibular fossa)
Fovea	A small pit (fovea capitis of the femur)
Sulcus	A groove for a tendon, nerve, or blood vessel (intertubercular sulcus of the humerus)
Passages and Cavities	
Canal	A tubular passage or tunnel in a bone (auditory canal of the skull)
Fissure	A slit through a bone (orbital fissures behind the eye)
Foramen	A hole through a bone, usually round (foramen magnum of the skull)
Meatus	An opening into a canal (external acoustic meatus of the ear)
Sinus	An air-filled space in a bone (frontal sinus of the forehead)

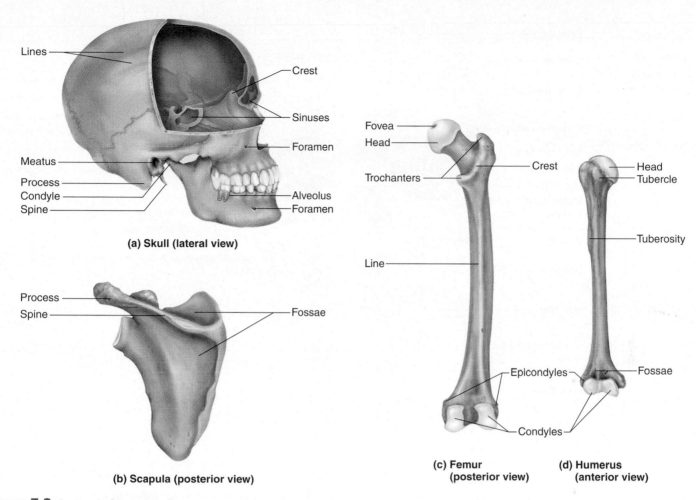

Figure 7.2 **Anatomical Features of Bones.** Most of these also occur on many other bones of the body.

7.2 The Skull

▶ Expected Learning Outcomes

When you have completed this section, you should be able to

- distinguish between cranial bones and facial bones;
- name the bones of the skull and identify the anatomical features of each;
- identify the cavities within the skull and in some of its individual bones;
- name the principal sutures that join bones of the skull;
- describe some bones that are closely associated with the skull; and
- describe some adaptations of the skull for upright locomotion.

The skull is the most complex part of the skeleton. Figures 7.3 through 7.6 present an overview of its general anatomy. Although the skull may seem to consist of only the mandible (lower jaw) and "the rest," it is composed of 22 bones and sometimes more. Most of them are rigidly joined by **sutures** (SOO-chures), joints that appear as seams on the cranial surface (fig. 7.4). These are important landmarks in the descriptions that follow.

The skull contains several prominent cavities (fig. 7.7). The largest, with an adult volume of about 1,300 mL, is the **cranial cavity**, which encloses the brain. Other cavities include the **orbits** (eye sockets), **nasal cavity, paranasal sinuses, oral cavity** (mouth or buccal cavity), and **middle-** and **inner-ear cavities.** The paranasal sinuses are named for the bones in which they occur (fig. 7.8)—the **frontal, ethmoid, sphenoid,** and **maxillary sinuses.** These sinuses are connected with the nasal cavity, lined by a mucous membrane, and filled with air. They lighten the anterior portion of the skull and act as chambers that add resonance to the voice.

Bones of the skull have especially conspicuous **foramina**—singular, *foramen* (fo-RAY-men)—holes that allow passage for nerves and blood vessels. The major foramina are summarized in table 7.3. The details of this table will mean more to you when you study cranial nerves and blood vessels in later chapters.

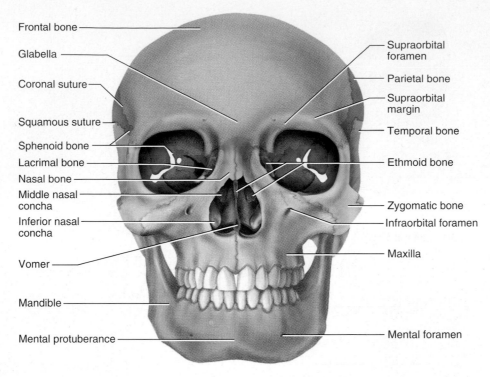

Figure 7.3 The Skull, Anterior View.

Labels (left side, top to bottom): Frontal bone, Glabella, Coronal suture, Squamous suture, Sphenoid bone, Lacrimal bone, Nasal bone, Middle nasal concha, Inferior nasal concha, Vomer, Mandible, Mental protuberance

Labels (right side, top to bottom): Supraorbital foramen, Parietal bone, Supraorbital margin, Temporal bone, Ethmoid bone, Zygomatic bone, Infraorbital foramen, Maxilla, Mental foramen

Cranial Bones

Cranial bones are those that enclose the brain; collectively, they compose the **cranium**[3] (braincase). The delicate brain tissue does not come directly into contact with the cranial bones, but is separated from them by three membranes called the *meninges* (meh-NIN-jeez) (see chapter 15). The thickest and toughest of these, the *dura mater*[4] (DUE-rah MAH-tur), lies loosely against the inside of the cranium in most places but is firmly attached to it at a few points.

The cranium is a rigid structure with an opening, the *foramen magnum* (literally "big hole"), where the spinal cord meets the brain. The cranium consists of two major parts—the calvaria and the base. The **calvaria**[5] (skullcap) is not a single bone but simply the dome of the top of the head, composed of parts of multiple bones (see fig. 7.6). In study skulls, it is often sawed so that the top can be lifted off for examination of the interior. This reveals the **base** (floor) of the cranial cavity (see fig. 7.5b), which exhibits three depressions called **cranial fossae**. These correspond to the contour of the inferior surface of the brain (fig. 7.9). The relatively shallow **anterior cranial fossa** is crescent-shaped and accommodates the frontal lobes of the brain. The **middle cranial fossa**, which drops abruptly deeper, is shaped like a pair of outstretched bird wings and accommodates the temporal lobes. The **posterior cranial fossa** is deepest and houses a large posterior division of the brain called the cerebellum.

There are eight cranial bones:

1 frontal bone	1 occipital bone
2 parietal bones	1 sphenoid bone
2 temporal bones	1 ethmoid bone

Frontal Bone

The **frontal bone** extends from the forehead back to a prominent *coronal suture*, which crosses the crown of the head from right to left and joins the frontal bone to the parietal bones (see figs. 7.3 and 7.4). The frontal bone forms the anterior wall and about one-third of the roof of the cranial cavity, and it turns inward to form nearly all of the anterior cranial fossa and the roof of the orbit. Deep to the eyebrows it has a ridge called the **supraorbital margin**. The center of each margin is perforated by a single **supraorbital foramen** (see figs. 7.3 and 7.14), which provides passage for a nerve, artery, and vein. In some people, the edge of this foramen breaks through the margin of the orbit and forms a *supraorbital notch*. The smooth area of the frontal bone just above the root of the nose is called the

[3]*crani* = helmet
[4]*dura* = tough, strong + *mater* = mother
[5]*calvar* = bald, skull

(*Text continues on p. 159*)

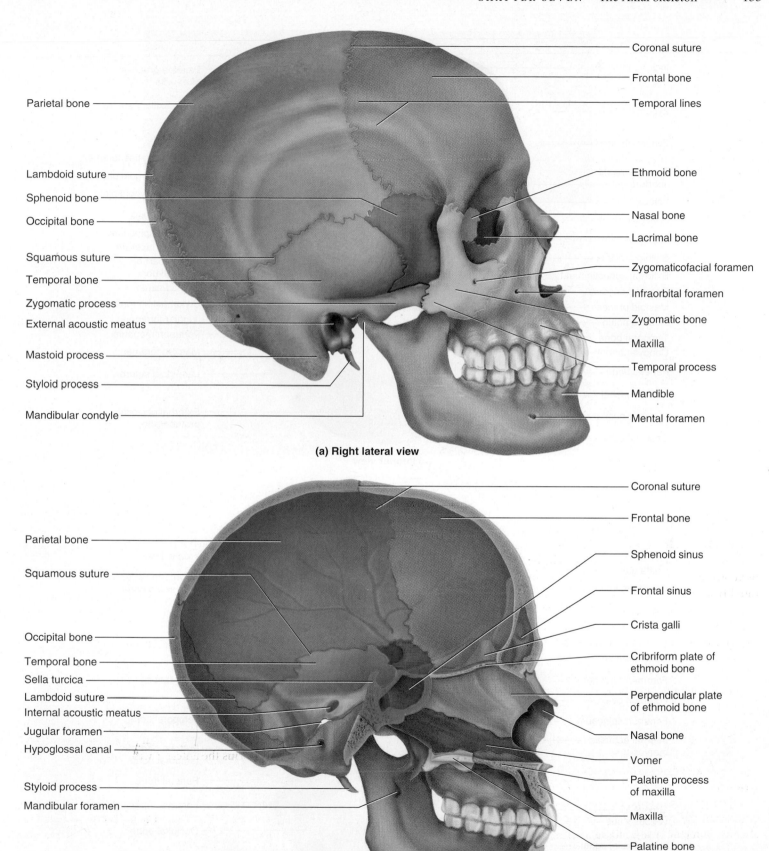

(a) Right lateral view

Figure 7.4 The Skull, Lateral Views
(External and Internal).

(b) Median section

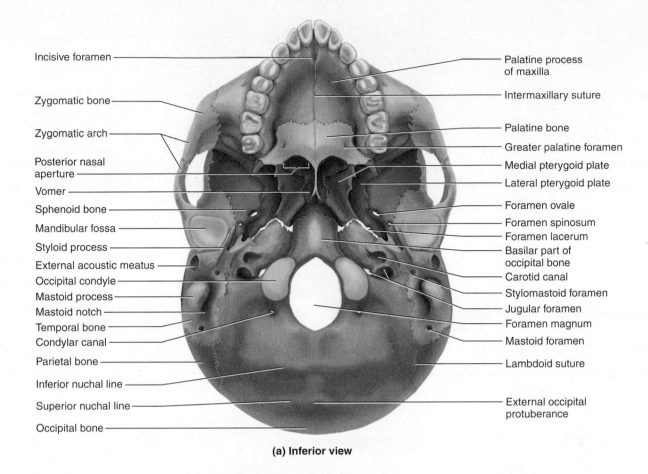

Incisive foramen

Zygomatic bone

Zygomatic arch

Posterior nasal aperture

Vomer

Sphenoid bone

Mandibular fossa

Styloid process

External acoustic meatus

Occipital condyle

Mastoid process

Mastoid notch

Temporal bone

Condylar canal

Parietal bone

Inferior nuchal line

Superior nuchal line

Occipital bone

Palatine process of maxilla

Intermaxillary suture

Palatine bone

Greater palatine foramen

Medial pterygoid plate

Lateral pterygoid plate

Foramen ovale

Foramen spinosum

Foramen lacerum

Basilar part of occipital bone

Carotid canal

Stylomastoid foramen

Jugular foramen

Foramen magnum

Mastoid foramen

Lambdoid suture

External occipital protuberance

(a) Inferior view

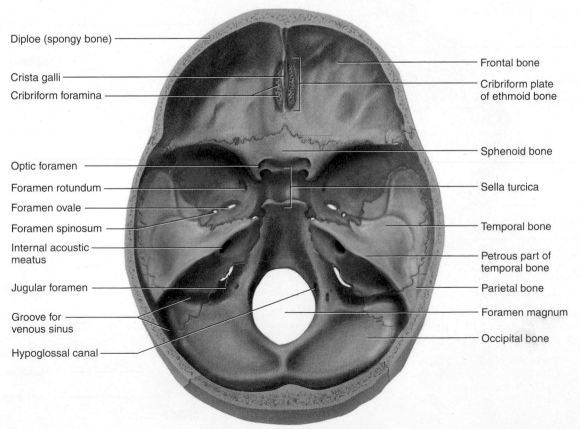

Diploe (spongy bone)

Crista galli

Cribriform foramina

Optic foramen

Foramen rotundum

Foramen ovale

Foramen spinosum

Internal acoustic meatus

Jugular foramen

Groove for venous sinus

Hypoglossal canal

Frontal bone

Cribriform plate of ethmoid bone

Sphenoid bone

Sella turcica

Temporal bone

Petrous part of temporal bone

Parietal bone

Foramen magnum

Occipital bone

Figure 7.5 The Base of the Skull (External and Internal).

(b) Superior view of cranial floor

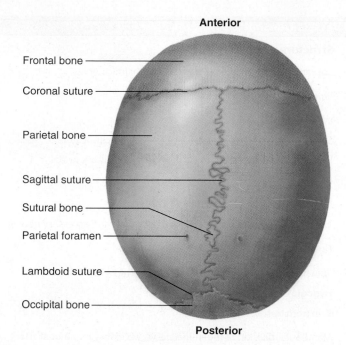

Anterior

Frontal bone

Coronal suture

Parietal bone

Sagittal suture

Sutural bone

Parietal foramen

Lambdoid suture

Occipital bone

Posterior

Figure 7.6 The Calvaria, Superior View.

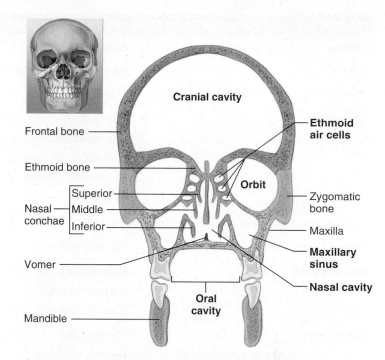

Cranial cavity

Frontal bone

Ethmoid bone

Ethmoid air cells

Orbit

Superior
Nasal — Middle
conchae
Inferior

Zygomatic bone

Vomer

Maxilla

Maxillary sinus

Nasal cavity

Oral cavity

Mandible

Figure 7.7 Major Cavities of the Skull, Frontal Section.
● *What is the function of the nasal conchae?*

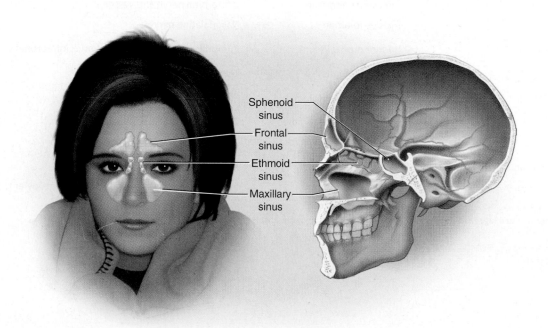

Sphenoid sinus

Frontal sinus

Ethmoid sinus

Maxillary sinus

Figure 7.8 The Paranasal Sinuses.
● *If these sinuses did not exist, it would require significantly more effort to hold the head erect. Explain.*

TABLE 7.3	Foramina of the Skull and the Nerves and Blood Vessels Transmitted Through Them	
Bones	**Foramina***	**Structures Transmitted**
Frontal bone	Supraorbital foramen or notch	Supraorbital nerve, artery, and vein; ophthalmic nerve
Parietal bone	Parietal foramen	Emissary vein of superior sagittal sinus
Temporal bone	Carotid canal	Internal carotid artery
	External acoustic meatus	Sound waves to eardrum
	Internal acoustic meatus	Vestibulocochlear and facial nerves; internal auditory vessels
	Stylomastoid foramen	Facial nerve
	Mastoid foramen	Meningeal artery; vein from sigmoid sinus
Temporal–occipital region	Jugular foramen	Internal jugular vein; glossopharyngeal, vagus, and accessory nerves
Temporal–occipital–sphenoid region	Foramen lacerum	No major nerves or vessels; closed by cartilage
Occipital bone	Foramen magnum	Spinal cord; accessory nerve; vertebral arteries
	Hypoglossal canal	Hypoglossal nerve to muscles of tongue
	Condylar canal	Vein from transverse sinus
Sphenoid bone	Foramen ovale	Mandibular division of trigeminal nerve; accessory meningeal artery
	Foramen rotundum	Maxillary division of trigeminal nerve
	Foramen spinosum	Middle meningeal artery; spinosal nerve; part of trigeminal nerve
	Optic foramen	Optic nerve; ophthalmic artery
	Superior orbital fissure	Oculomotor, trochlear, and abducens nerves; ophthalmic division of trigeminal nerve; ophthalmic veins
Ethmoid bone	Olfactory foramina	Olfactory nerves
Maxilla	Infraorbital foramen	Infraorbital nerve and vessels
	Incisive foramen	Nasopalatine nerves
Maxilla–sphenoid region	Inferior orbital fissure	Infraorbital nerve; zygomatic nerve; infraorbital vessels
Lacrimal bone	Lacrimal foramen	Tear duct leading to nasal cavity
Palatine bone	Greater palatine foramen	Palatine nerves
Zygomatic bone	Zygomaticofacial foramen	Zygomaticofacial nerve
	Zygomaticotemporal foramen	Zygomaticotemporal nerve
Mandible	Mental foramen	Mental nerve and vessels
	Mandibular foramen	Inferior alveolar nerves and vessels to the lower teeth

*When two or more bones are listed together (for example, temporal–occipital), it indicates that the foramen passes between them.

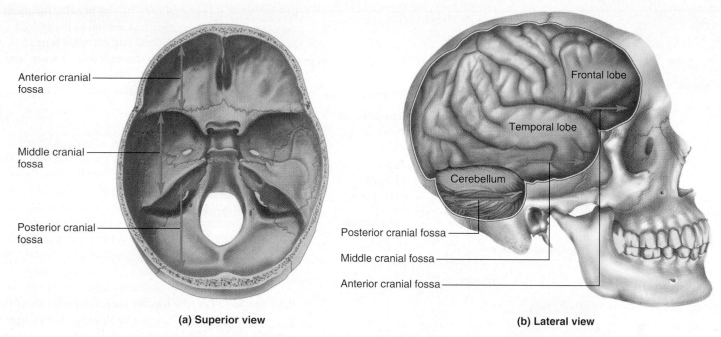

(a) Superior view

(b) Lateral view

Figure 7.9 **Cranial Fossae.** The three fossae conform to the contours of the base of the brain.

glabella.[6] The frontal bone also contains the frontal sinus. You may not see this sinus on all study skulls. On some skulls, the calvaria is cut too high to show the sinus, and some people simply do not have one. Along the cut edge of the calvaria, you can also see the *diploe* (DIP-lo-ee)—the layer of spongy bone in the middle of the cranial bones (see fig. 7.5b).

Parietal Bones

The right and left **parietal** (pa-RYE-eh-tul) **bones** form most of the cranial roof and part of its walls (see figs. 7.4 and 7.6). Each is bordered by four sutures that join it to the neighboring bones: (1) the **sagittal suture** between the parietal bones; (2) the **coronal**[7] **suture** at the anterior margin; the **lambdoid**[8] (LAM-doyd) **suture** at the posterior margin; and (4) the **squamous suture** laterally. Small sutural (wormian) bones are often seen along the sagittal and lambdoid sutures, like little islands of bone with the suture lines passing around them. Internally, the parietal and frontal bones have markings that look a bit like aerial photographs of river tributaries (see fig. 7.4b). These represent places where the bone has been molded around blood vessels of the meninges.

Externally, the parietal bones have few features. A **parietal foramen** sometimes occurs near the corner of the lambdoid and sagittal sutures (see fig. 7.6). A pair of slight thickenings, the superior and inferior **temporal lines,** forms an arc across the parietal and frontal bones (see fig. 7.4a). The temporal lines mark the attachment of the large, fan-shaped *temporalis muscle,* a chewing muscle that inserts on the mandible.

Temporal Bones

If you palpate your skull just above and anterior to the ear—that is, the temporal region—you can feel the **temporal bone,** which forms the lower wall and part of the floor of the cranial cavity (fig. 7.10). The temporal bone derives its name from the fact that people often develop their first gray hairs on the temples with the passage of time.[9] The relatively complex shape of the temporal bone is best understood by dividing it into four parts:

1. The **squamous**[10] **part** (which you just palpated) is relatively flat and vertical. It is encircled by the squamous suture. It bears two prominent features: (a) the **zygomatic process,** which extends anteriorly to form part of the **zygomatic arch** (cheekbone); and (b) the **mandibular fossa,** a depression where the mandible articulates with the cranium.

2. The **tympanic**[11] **part** is a small plate of bone that borders the **external acoustic meatus** (me-AY-tus), the opening into the ear canal. It has a pointed spine on its inferior surface, the **styloid process,** named for its resemblance to the stylus used by ancient Greeks and Romans to write on wax tablets. The styloid process provides attachment for muscles of the tongue, pharynx, and hyoid bone.

3. The **mastoid**[12] **part** lies posterior to the tympanic part. It bears a heavy **mastoid process,** which you can palpate as a prominent lump behind the earlobe. It is filled with small air sinuses that communicate with the middle-ear cavity.

---

[6]*glab* = smooth
[7]*corona* = crown
[8]*lambd* = the Greek letter lambda (λ) + *oid* = resembling

[9]*tempor* = time
[10]*squam* = flat + *ous* = characterized by
[11]*tympan* = drum (eardrum) + *ic* = pertaining to
[12]*mast* = breast + *oid* = resembling

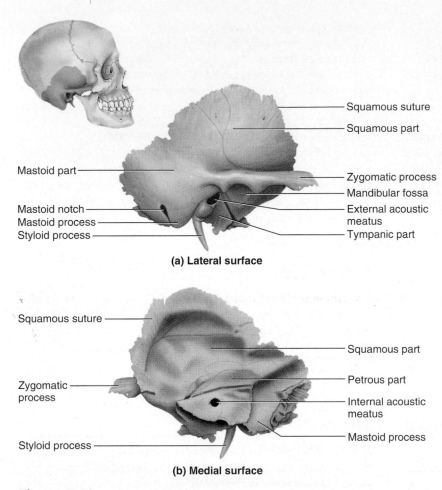

(a) Lateral surface

- Squamous suture
- Squamous part
- Mastoid part
- Zygomatic process
- Mandibular fossa
- Mastoid notch
- External acoustic meatus
- Mastoid process
- Styloid process
- Tympanic part

(b) Medial surface

- Squamous suture
- Squamous part
- Zygomatic process
- Petrous part
- Internal acoustic meatus
- Styloid process
- Mastoid process

Figure 7.10 **The Right Temporal Bone.** The lateral surface faces the scalp and external ear; the medial surface faces the brain.

These sinuses are subject to infection and inflammation (mastoiditis), which can erode the bone and spread to the brain. Inferiorly, there is a groove called the **mastoid notch** medial to the mastoid process (see fig. 7.5a). It is the origin of the *digastric muscle,* a muscle that opens the mouth. The notch is perforated by the **stylomastoid foramen** at its anterior end and the **mastoid foramen** at its posterior end.

4. The **petrous**[13] **part** can be seen in the cranial floor, where it resembles a little mountain range separating the middle cranial fossa from the posterior fossa (see figs. 7.5b and 7.10b). It houses the middle- and inner-ear cavities. The **internal acoustic meatus,** an opening on its posteromedial surface, allows passage of the *vestibulocochlear* (vess-TIB-you-lo-COC-lee-ur) *nerve,* which carries sensations of hearing and balance from the inner ear to the brain. On the inferior surface of the petrous part are two prominent foramina named for the major blood vessels that pass through them (see fig. 7.5a): (a) The **carotid canal** is a passage for the internal carotid artery, a major blood supply to the brain. This artery is so close to the inner ear that you may be able to hear the pulsing of its blood when your ear is resting on a pillow

or your heart is beating hard. (b) The **jugular foramen** is a large, irregular opening just medial to the styloid process, between the temporal and occipital bones. Blood from the brain drains through this foramen into the internal jugular vein of the neck. Three cranial nerves also pass through this foramen.

Occipital Bone

The **occipital** (oc-SIP-ih-tul) **bone** forms the rear of the skull (*occiput*) and much of its base (see fig. 7.5). Its most conspicuous feature is the **foramen magnum.** An important consideration in head injuries is swelling of the brain. Since the cranium cannot enlarge, swelling puts pressure on the brain and results in even more tissue damage. Severe swelling may force the brainstem out through the foramen magnum, usually with fatal consequences.

The occipital bone continues anterior to this as a thick median plate, the **basilar part.** On each side of the foramen magnum is a smooth knob called the **occipital condyle** (CON-dile), where the skull rests on the vertebral column. At the anterolateral edge of each condyle is a **hypoglossal**[14] **canal,** named for the *hypoglossal nerve* that passes through it to innervate the muscles of the tongue. In some people, a **condylar** (CON-dih-lur) **canal** occurs posterior to each occipital condyle.

Internally, the occipital bone displays impressions left by large venous sinuses that drain blood from the brain (see fig. 7.5b). One of these grooves travels along the midsagittal line. Just before reaching the foramen magnum, it branches into right and left grooves that wrap around the occipital bone like outstretched arms before terminating at the jugular foramina. The sinuses that occupy these grooves are described in table 21.3 (p. 580).

Other features of the occipital bone can be palpated on the back of your head. One is a prominent median bump called the **external occipital protuberance**—the attachment for the **nuchal**[15] (NEW-kul) **ligament,** which binds the skull to the vertebral column. A ridge, the **superior nuchal line,** can be traced horizontally from this protuberance toward the mastoid process (see fig. 7.5a). It defines the superior limit of the neck and provides attachment for several neck and back muscles to the skull. It forms the boundary where, in palpating the upper neck, you feel the transition from muscle to bone. By pulling down on the occipital bone, some of these muscles help to keep the head erect. The deeper **inferior nuchal line** provides attachment for some of the deep neck muscles. This inconspicuous ridge cannot be palpated on the living body but is visible on an isolated skull.

Sphenoid Bone

The **sphenoid**[16] (SFEE-noyd) **bone** has a complex shape with a thick median **body** and outstretched **greater** and **lesser wings,** which give the bone as a whole a ragged mothlike shape. Most of it is best seen from the superior perspective (fig. 7.11a). In this view,

---

[13]*petr* = stone, rock + *ous* = like

[14]*hypo* = below + *gloss* = tongue
[15]*nucha* = back of the neck
[16]*sphen* = wedge + *oid* = resembling

the lesser wings form the posterior margin of the anterior cranial fossa and end at a sharp bony crest, where the sphenoid drops abruptly to the greater wings. The greater wings form about half of the middle cranial fossa (the temporal bone forming the rest) and are perforated by several foramina, to be discussed shortly.

The greater wing forms part of the lateral surface of the cranium just anterior to the temporal bone (see fig. 7.4a). The lesser wing forms the posterior wall of the orbit and contains the **optic foramen,** which permits passage of the optic nerve and ophthalmic artery (see fig. 7.14). Superiorly, a pair of bony spines of the lesser wing called the **anterior clinoid processes** appears to guard the optic foramina. A gash in the posterior wall of the orbit, the **superior orbital fissure,** angles upward lateral to the optic foramen. It serves as a passage for three nerves that supply the muscles of eye movement.

The body of the sphenoid bone contains the sphenoid sinus and has a saddlelike surface feature named the **sella turcica**[17] (SEL-la TUR-sih-ca). The sella turcica consists of a deep pit called the *hypophyseal fossa,* which houses the pituitary gland (hypophysis); a raised anterior margin called the *tuberculum sellae* (too-BUR-cu-lum SEL-lee); and a posterior margin called the *dorsum sellae.* In life, the dura mater stretches over the sella turcica and attaches to the anterior clinoid processes. A stalk penetrates the dura to connect the pituitary gland to the floor of the brain.

Lateral to the sella turcica, the sphenoid is perforated by several foramina (see fig. 7.5). The **foramen rotundum** and **foramen ovale** (oh-VAY-lee) are passages for two branches of the trigeminal nerve.

The **foramen spinosum,** about the diameter of a pencil lead, provides passage for an artery of the meninges. An irregular gash called the **foramen lacerum**[18] (LASS-eh-rum) occurs at the junction of the sphenoid, temporal, and occipital bones. It is filled with cartilage in life and transmits no major vessels or nerves.

In an inferior view of the skull, the sphenoid can be seen just anterior to the basilar part of the occipital bone (see fig. 7.5a). The internal openings of the nasal cavity seen here are called the **posterior nasal apertures,** or **choanae**[19] (co-AH-nee). Lateral to each aperture, the sphenoid bone exhibits a pair of parallel plates—the **medial** and **lateral pterygoid**[20] (TERR-ih-goyd) **plates.** Each plate has a narrow inferior extension called the *pterygoid process.* The plates provide attachment for some of the jaw muscles. The sphenoid sinus occurs within the body of the sphenoid bone.

Ethmoid Bone

The **ethmoid**[21] (ETH-moyd) bone is an anterior cranial bone located between the eyes (fig. 7.12). It contributes to the medial wall of the orbit, the roof and walls of the nasal cavity, and the nasal septum. It is a very porous and delicate bone, with three major portions:

1. The vertical **perpendicular plate,** a thin median plate of bone that forms the superior two-thirds of the nasal septum (see fig. 7.4b). (The lower part is formed by the vomer, discussed

---

[17]*sella* = saddle + *turcica* = Turkish

[18]*lacerum* = torn, lacerated
[19]*choana* = funnel
[20]*pterygo* = wing
[21]*ethmo* = sieve, strainer + *oid* = resembling

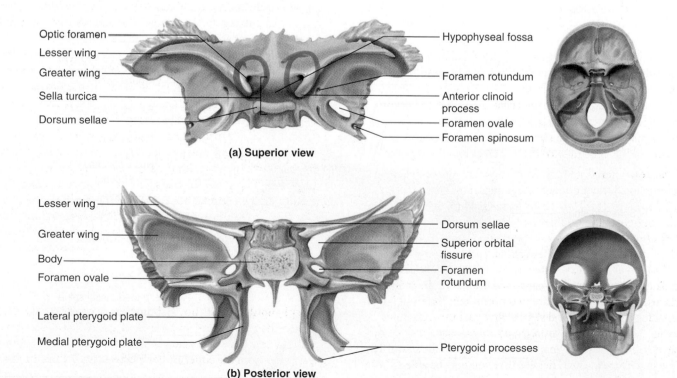

(a) Superior view

Optic foramen — Lesser wing — Greater wing — Sella turcica — Dorsum sellae

Hypophyseal fossa — Foramen rotundum — Anterior clinoid process — Foramen ovale — Foramen spinosum

(b) Posterior view

Lesser wing — Greater wing — Body — Foramen ovale — Lateral pterygoid plate — Medial pterygoid plate

Dorsum sellae — Superior orbital fissure — Foramen rotundum — Pterygoid processes

Figure 7.11 The Sphenoid Bone.

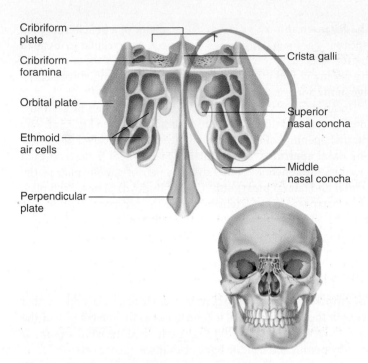

Cribriform plate

Cribriform foramina

Orbital plate

Ethmoid air cells

Perpendicular plate

Crista galli

Superior nasal concha

Middle nasal concha

Figure 7.12 The Ethmoid Bone, Anterior View.

later.) The septum divides the nasal cavity into right and left air spaces called the **nasal fossae** (FOSS-ee). The septum is often curved, or deviated, toward one nasal fossa or the other.

2. A horizontal **cribriform**[22] (CRIB-rih-form) **plate,** which forms the roof of the nasal cavity. This plate has a median crest called

---

[22]*cribri* = sieve + *form* = in the shape of

the **crista galli**[23] (GAL-eye), an attachment point for the dura mater. On each side of the crista is an elongated depressed area perforated with numerous holes, the **cribriform (olfactory) foramina**. A pair of *olfactory bulbs* of the brain, concerned with the sense of smell, rests in these depressions, and the foramina allow passage for olfactory nerves from the nasal cavity to the bulbs (see Deeper Insight 7.1).

3. The **labyrinth,** a large mass on each side of the perpendicular plate. The labyrinth is named for the fact that internally, it has a maze of air spaces called the **ethmoidal cells.** Collectively, these constitute the *ethmoid sinus* discussed earlier. The lateral surface of the labyrinth is a smooth, slightly concave **orbital plate** seen on the medial wall of the orbit (see fig. 7.14). The medial surface of the labyrinth gives rise to two curled, scroll-like plates of bone called the **superior** and **middle nasal conchae**[24] (CON-kee), or **turbinates.** These project into the nasal fossa from its lateral wall toward the septum (see figs. 7.7 and 7.13). There is also a separate *inferior nasal concha* discussed later. The three conchae occupy most of the space in the nasal cavity. By filling space and creating turbulence in the flow of inhaled air, they ensure that the air contacts the mucous membranes that cover these bones, which cleanses, humidifies, and warms the inhaled air before it reaches the lungs. The superior concha and adjacent part of the nasal septum also bear the sensory cells of smell.

Usually, all that can be seen of the ethmoid is the perpendicular plate, by looking into the nasal cavity (see fig. 7.3); the orbital plate, by looking at the medial wall of the orbit (fig. 7.14); and the crista galli and cribriform plate, viewed from within the cranial cavity (see fig. 7.5b).

---

[23]*crista* = crest + *galli* = of a rooster
[24]*conchae* = conchs (large marine snails)

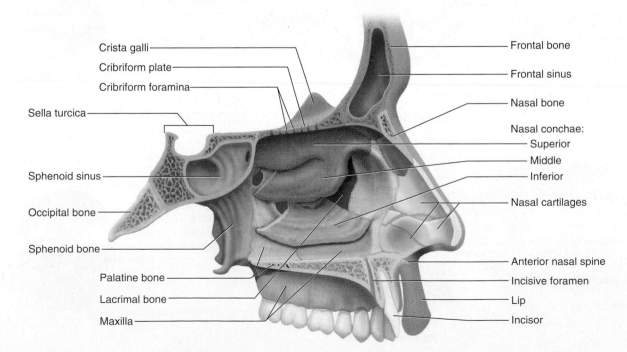

Crista galli

Cribriform plate

Cribriform foramina

Sella turcica

Sphenoid sinus

Occipital bone

Sphenoid bone

Palatine bone

Lacrimal bone

Maxilla

Frontal bone

Frontal sinus

Nasal bone

Nasal conchae:
Superior
Middle
Inferior

Nasal cartilages

Anterior nasal spine

Incisive foramen

Lip

Incisor

Figure 7.13 The Left Nasal Fossa, Sagittal Section.
● *What bone(s), or parts of bone(s), in figure 7.4b would have to be removed to see the structures in this figure?*

DEEPER INSIGHT 7.1

Injury to the Ethmoid Bone

The ethmoid bone is very delicate and easily injured by a sharp upward blow to the nose, such as a person might suffer by striking an automobile dashboard in a collision. The force of a blow can drive bone fragments through the cribriform plate into the meninges or brain tissue. Such injuries are often evidenced by leakage of cerebrospinal fluid into the nasal cavity, and may be followed by the spread of infection from the nasal cavity to the brain. Blows to the head can also shear off the olfactory nerves that pass through the ethmoid bone and cause *anosmia*, an irreversible loss of the sense of smell and a great reduction in the sense of taste (most of which depends on smell). This not only deprives life of some of its pleasures, but can also be dangerous, as when a person fails to smell smoke, gas, or spoiled food.

Facial Bones

Facial bones do not enclose the brain but lie anterior to the cranial cavity. They support the orbital, nasal, and oral cavities; give shape to the face; and provide attachment for the muscles of facial expression and mastication. There are 14 facial bones:

2 maxillae	2 nasal bones
2 palatine bones	2 inferior nasal conchae
2 zygomatic bones	1 vomer
2 lacrimal bones	1 mandible

Maxillae

The **maxillae** (mac-SILL-ee) are the largest facial bones. They form the upper jaw and meet each other at a median **intermaxillary suture** (see figs. 7.3, 7.4a, and 7.5a). Small points of maxillary bone called **alveolar processes** grow into the spaces between the bases of the teeth. The root of each tooth is inserted into a deep socket, or **alveolus.** Even though the teeth are preserved with the skull, they are not bones. They are discussed in detail in chapter 24.

Each maxilla extends from the teeth upward to the inferomedial wall of the orbit. Just below the orbit, it exhibits an **infraorbital foramen,** which provides passage for a blood vessel to the face and a nerve that receives sensations from the nasal region and cheek. This nerve emerges through the foramen rotundum into the cranial cavity. The maxilla forms part of the floor of the orbit, where it exhibits a gash called the **inferior orbital fissure** that angles downward and medially (fig. 7.14). The inferior and superior orbital fissures form a sideways V whose apex lies near the optic foramen. The inferior orbital fissure is a passage for blood vessels and sensory nerves from the face.

The **palate** forms the roof of the mouth and floor of the nasal cavity. It consists of a bony **hard palate** in front and a fleshy **soft palate** in the rear. Most of the hard palate is formed by horizontal extensions of the maxillae called **palatine** (PAL-uh-tine) **processes** (see fig. 7.5a). Just behind the incisors (front teeth) is a pair of **incisive foramina.** The palatine processes normally meet at the intermaxillary suture at about 12 weeks of fetal development. Failure to join causes cleft palate (see table 7.5, p. 178).

Palatine Bones

The **palatine bones** are located in the posterior nasal cavity (fig. 7.13). Each one has an L-shape formed by a *horizontal plate* and a *perpendicular plate.* The horizontal plates form the posterior one-third of

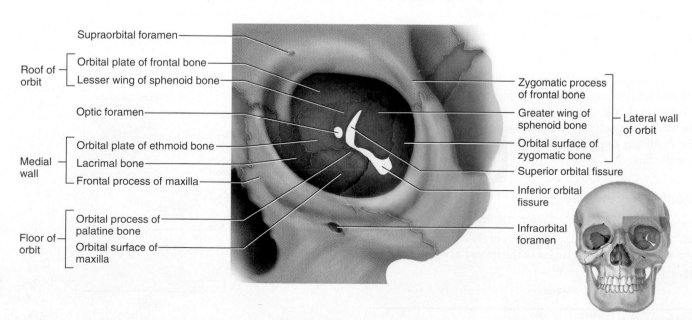

Figure 7.14 The Left Orbit, Anterior View.

DEEPER INSIGHT 7.2

Evolutionary Significance of the Palate

In most vertebrates, the nasal passages open into the oral cavity. Mammals, by contrast, have a palate that separates the nasal and oral cavities. In order to maintain our high metabolic rate, we must digest our food rapidly; in order to do this, we chew it thoroughly to break it up into small, easily digested particles before swallowing it. We would be unable to breathe freely during this prolonged chewing if we lacked a palate to separate the airflow from the oral cavity.

the bony palate (see fig. 7.5a). Each is marked by a large greater palatine foramen, a nerve passage to the palate. The perpendicular plate is a thin, delicate, irregularly shaped plate that forms part of the wall between the nasal cavity and the orbit (fig. 7.12).

Zygomatic Bones

The **zygomatic**[25] **bones** form the angles of the cheeks inferolateral to the eyes and form part of the lateral wall of each orbit; they extend about halfway to the ear (see figs. 7.4a and 7.5a). Each zygomatic bone has an inverted T shape and usually a small **zygomaticofacial** (ZY-go-MAT-ih-co-FAY-shul) **foramen** near the intersection of the stem and crossbar of the T. The prominent zygomatic arch that flares from each side of the skull is formed by the union of the zygomatic bone, temporal bone, and maxilla.

Lacrimal Bones

The **lacrimal**[26] (LACK-rih-mul) **bones** form part of the medial wall of each orbit (fig. 7.14). These are the smallest bones of the skull, about the size of a small fingernail. A depression called the **lacrimal fossa** houses a membranous *lacrimal sac* in life. Tears from the eye collect in this sac and drain into the nasal cavity.

Nasal Bones

Two small rectangular **nasal bones** form the bridge of the nose (see fig. 7.3) and support cartilages that shape the lower portion of the nose. They are only slightly larger than the lacrimal bones. If you palpate the bridge, you can easily feel where the nasal bones end and the cartilages begin. The nasal bones are often fractured by blows to the nose.

Inferior Nasal Conchae

There are three conchae in the nasal cavity. The superior and middle conchae, as discussed earlier, are parts of the ethmoid bone. The **inferior nasal concha**—the largest of the three—is a separate bone (see fig. 7.13).

[25]*zygo* = to join, unite
[26]*lacrim* = tear, to cry

Vomer

The **vomer** forms the inferior portion of the nasal septum (see figs. 7.3 and 7.4b). Its name literally means "plowshare," which refers to its resemblance to the blade of a plow. The superior half of the nasal septum is formed by the perpendicular plate of the ethmoid bone, as mentioned earlier. The vomer and perpendicular plate support a wall of *septal cartilage* that forms most of the anterior part of the septum.

Mandible

The **mandible** (fig. 7.15) is the strongest bone of the skull and the only one that can move significantly. It supports the lower teeth and provides attachment for muscles of mastication and facial expression. The horizontal portion is called the **body;** the vertical to oblique posterior portion is the **ramus** (RAY-mus)—plural, *rami* (RAY-my); and these two portions meet at a corner called the **angle.** The point of the chin is the **mental protuberance.** The inner (posterior) surface of the mandible in this region has a pair of small points, the **mental spines,** which serve for attachment of certain chin muscles (see fig. 7.4b). On the anterolateral surface of the body, the **mental foramen** permits the passage of nerves and blood vessels of the chin. The inner surface of the body has a number of shallow depressions and ridges to accommodate muscles and salivary glands. The angle of the mandible has a rough lateral surface for insertion of the *masseter,* a muscle of mastication. Like the maxilla, the mandible has pointed alveolar processes between the teeth.

The ramus is somewhat Y-shaped. Its posterior branch, called the **condylar** (CON-dih-lur) **process,** bears the **mandibular condyle**—an oval knob that articulates with the mandibular fossa of the temporal bone. The hinge of the mandible is the **temporomandibular joint (TMJ).** The anterior branch of the ramus, called the **coronoid process,** is the point of insertion for the temporalis muscle, which pulls the mandible upward when you bite. The U-shaped arch between the two processes is called the **mandibular**

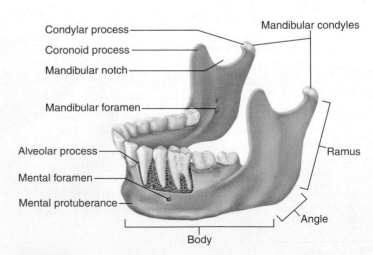

Condylar process
Coronoid process
Mandibular notch
Mandibular foramen
Alveolar process
Mental foramen
Mental protuberance
Mandibular condyles
Ramus
Angle
Body

Figure 7.15 The Mandible.

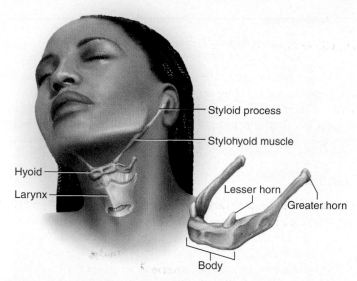

Figure 7.16 The Hyoid Bone.
• *Why would a fractured hyoid bone be a common finding in strangulation cases?*

notch. Just below the notch, on the medial surface of the ramus, is the **mandibular foramen,** a passage for a nerve and blood vessels that enter the bone here to reach the lower teeth (see fig. 7.4b).

Bones Associated with the Skull

Seven bones are closely associated with the skull but not considered part of it. These are the three auditory ossicles in each middle-ear cavity and the hyoid bone beneath the chin. The **auditory ossicles**[27]—named the **malleus** (hammer), **incus** (anvil), and **stapes** (STAY-peez) (stirrup)—are discussed in connection with hearing in chapter 17. The **hyoid**[28] **bone** is a slender U-shaped bone between the chin and larynx (fig. 7.16). It is one of the few bones that does not articulate with any other. The hyoid is suspended from the styloid processes of the skull, somewhat like a hammock, by the small *stylohyoid muscles* and *stylohyoid ligaments.* The medial **body** of the hyoid is flanked on either side by projections called the **greater** and **lesser horns.** The hyoid bone serves for attachment of several muscles that control the mandible, tongue, and larynx. Forensic pathologists look for a fractured hyoid as evidence of strangulation.

Adaptations of the Skull for Bipedalism

Some mammals can stand, hop, or walk briefly on their hind legs, but humans are the only mammals that are habitually bipedal. Efficient bipedal locomotion is possible only because of several adaptations of the feet, legs, vertebral column, and skull. The human head is balanced on the vertebral column with the gaze directed forward. This was made possible in part by an evolutionary remodeling of the skull. The foramen magnum moved to a more inferior location in

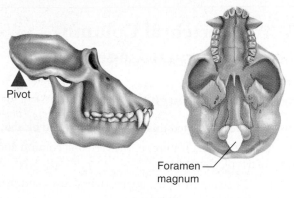

Chimpanzee

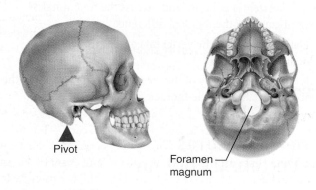

Human

Figure 7.17 Adaptations of the Skull for Bipedalism. Comparison of chimpanzee and human skulls. The foramen magnum is shifted rostrally and the face is flatter in humans. Thus, the skull is balanced on the vertebral column and the gaze is directed forward when a person is standing.

the course of human evolution, and the face is much flatter than an ape's face, so there is less weight anterior to the occipital condyles and the head has less tendency to tip forward (fig. 7.17).

Before You Go On

Answer the following questions to test your understanding of the preceding section:

4. Name the paranasal sinuses and state their locations. Name any four other cavities in the skull.

5. Explain the difference between a cranial bone and a facial bone. Give four examples of each.

6. Draw an oval representing a superior view of the calvaria. Draw lines representing the coronal, lambdoid, and sagittal sutures. Label the four bones separated by these sutures.

7. State which bone has each of these features: a squamous part, hypoglossal foramen, greater horn, greater wing, condylar process, and cribriform plate.

8. Palpate as many of the following structures as possible, and identify which ones cannot normally be palpated on a living person: the mastoid process, crista galli, superior orbital fissure, palatine processes, zygomatic bone, mental protuberance, and stapes.

[27]*os* = bone + *icle* = little
[28]*hy* = the letter *U* + *oid* = resembling

7.3 The Vertebral Column and Thoracic Cage

▶ Expected Learning Outcomes

When you have completed this section, you should be able to

- describe the general features of the vertebral column and those of a typical vertebra;
- describe the structure of the intervertebral discs and their relationship to the vertebrae;
- describe the special features of vertebrae in different regions of the vertebral column, and discuss the functional significance of the regional differences;
- relate the shape of the vertebral column to upright locomotion; and
- describe the anatomy of the sternum and ribs and how the ribs articulate with the thoracic vertebrae.

General Features of the Vertebral Column

The **vertebral column (spine)** physically supports the skull and trunk, allows for their movement, protects the spinal cord, and absorbs stresses produced by walking, running, and lifting. It also provides attachment for the limbs, thoracic cage, and postural muscles. Although commonly called the backbone, it consists not of a single bone but a chain of 33 **vertebrae** with **intervertebral discs** of fibrocartilage between most of them. The adult vertebral column averages about 71 cm (28 in.) long, with the 23 intervertebral discs accounting for about one-quarter of the length. Most people are about 1% shorter when they go to bed at night than when they first rise in the morning. This is because during the day, the weight of the body compresses the intervertebral discs and squeezes water out of them. When one is sleeping, with the weight off the spine, the discs reabsorb water and swell.

As shown in figure 7.18, the vertebrae are divided into five groups: 7 *cervical* (SUR-vih-cul) *vertebrae* in the neck, 12 *thoracic vertebrae* in the chest, 5 *lumbar vertebrae* in the lower back, 5 *sacral vertebrae* at the base of the spine, and 4 tiny *coccygeal* (coc-SIDJ-ee-ul) *vertebrae*. To help remember the numbers of cervical, thoracic, and lumbar vertebrae—7, 12, and 5—you might think of a typical work day: go to work at 7, have lunch at 12, and go home at 5.

Variations in this arrangement occur in about 1 person in 20. For example, the last lumbar vertebra is sometimes incorporated into the sacrum, producing 4 lumbar and 6 sacral vertebrae. In other cases, the first sacral vertebra fails to fuse with the second, producing 6 lumbar and 4 sacral vertebrae. The cervical and thoracic vertebrae are more constant in number.

Beyond the age of 3 years, the vertebral column is slightly S-shaped, with four bends called the **cervical, thoracic, lumbar,** and **pelvic curvatures** (fig. 7.19). The thoracic and pelvic curvatures are

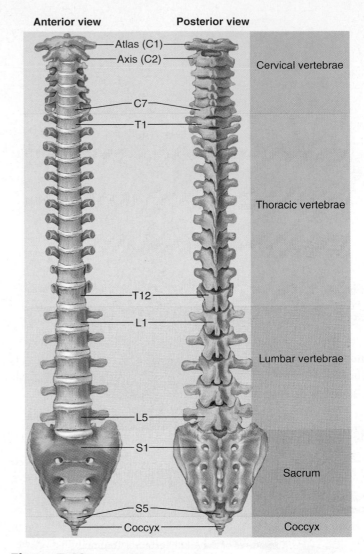

Figure 7.18 The Vertebral Column.

called *primary curvatures* because they are present at birth, when the spine has a single C-shaped curvature. The cervical and lumbar curvatures are called *secondary curvatures* because they develop later, in the child's first few years of crawling and walking, as described later in this chapter. The resulting S shape makes sustained bipedal walking possible because the trunk of the body does not lean forward as it does in primates, such as a chimpanzee; the head is balanced over the body's center of gravity; and the eyes are directed straight forward (fig. 7.20). Abnormal lateral or anterior–posterior spinal curvatures are among the most common back problems (see Deeper Insight 7.3).

General Structure of a Vertebra

A representative vertebra and intervertebral disc are shown in figure 7.22. The most obvious feature of a vertebra is the **body,** or **centrum**—a mass of spongy bone and red bone marrow covered

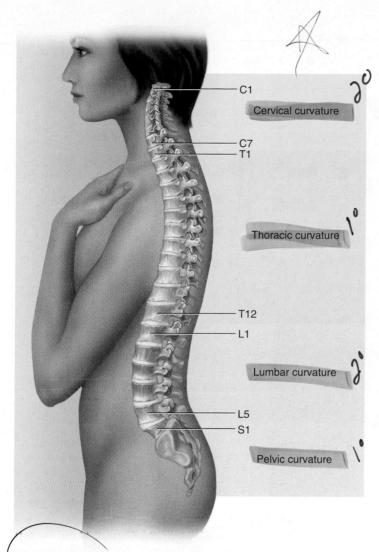

Figure 7.19 Curvatures of the Adult Vertebral Column.

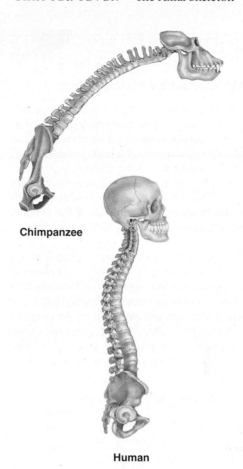

Figure 7.20 Comparison of Chimpanzee and Human Vertebral Columns. The S-shaped human vertebral column Is an adaptation for bipedal locomotion.

with a thin shell of compact bone. This is the weight-bearing portion of the vertebra. Its rough superior and inferior surfaces provide firm attachment for the intervertebral discs.

Apply What You Know

The lower we look on the vertebral column, the larger the vertebral bodies and intervertebral discs are. What is the functional significance of this trend?

Posterior to the body of each vertebra is an ovoid to triangular space called the **vertebral foramen**. Collectively, these foramina form the **vertebral canal,** a passage for the spinal cord. The foramen is bordered by a bony **vertebral arch** composed of two parts on each side, a pillarlike **pedicle**[29] and platelike **lamina.**[30] Extending from the apex of the arch, a projection called the **spinous pro-**

---

[29]*ped* = foot + *icle* = little
[30]*lamina* = layer, plate

cess is directed toward the rear and downward. You can see and feel the spinous processes as a row of bumps along the spine. A **transverse process** extends laterally from the point where the pedicle and lamina meet. The spinous and transverse processes provide points of attachment for spinal muscles and ligaments.

A pair of **superior articular processes** project upward from one vertebra and meet a similar pair of **inferior articular processes** that project downward from the vertebra above (fig. 7.23a). Each process has a flat articular surface (facet) facing that of the adjacent vertebra. These processes restrict twisting of the vertebral column, which could otherwise severely damage the spinal cord.

Where two vertebrae are joined, they exhibit an opening between their pedicles called the **intervertebral foramen.** This allows passage for spinal nerves that connect with the spinal cord at regular intervals. Each foramen is formed by an **inferior vertebral notch** in the pedicle of the superior vertebra and a **superior vertebral notch** in the pedicle of the one below it (fig. 7.23b).

Intervertebral Discs

An **intervertebral disc** is a cartilaginous pad located between the bodies of two adjacent vertebrae. It consists of an inner gelatinous **nucleus pulposus** surrounded by a ring of fibrocartilage called the

DEEPER INSIGHT 7.3

Abnormal Spinal Curvatures

Abnormal spinal curvatures (fig. 7.21) can result from disease, weakness or paralysis of the trunk muscles, poor posture, or congenital defects in vertebral anatomy. The most common deformity is an abnormal lateral curvature called *scoliosis*. It occurs most often in the thoracic region, particularly among adolescent girls. It sometimes results from a developmental abnormality in which the body and arch of a vertebra fail to develop on one side. If the person's skeletal growth is not yet complete, scoliosis can be corrected with a back brace.

An exaggerated thoracic curvature is called *kyphosis* (hunchback, in lay language). It is usually a result of osteoporosis, but it also occurs in people with osteomalacia or spinal tuberculosis and in adolescents who engage heavily in such sports as wrestling and weightlifting. An exaggerated lumbar curvature is called *lordosis* (swayback). It can have the same causes as kyphosis, or it can result from added abdominal weight in pregnancy or obesity.

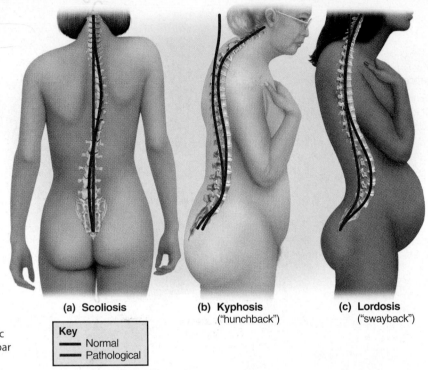

(a) **Scoliosis**

(b) **Kyphosis** ("hunchback")

(c) **Lordosis** ("swayback")

Key
— Normal
— Pathological

Figure 7.21 **Abnormal Spinal Curvatures.** (a) Scoliosis, an abnormal lateral deviation. (b) Kyphosis, an exaggerated thoracic curvature common in old age. (c) Lordosis, an exaggerated lumbar curvature common in pregnancy and obesity.

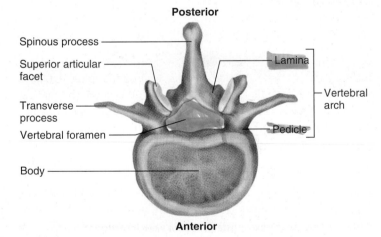

Posterior

Spinous process

Superior articular facet

Lamina

Transverse process

Vertebral arch

Vertebral foramen

Pedicle

Body

Anterior

(a) 2nd lumbar vertebra (L2)

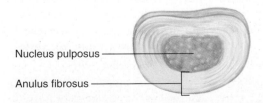

Nucleus pulposus

Anulus fibrosus

(b) Intervertebral disc

Figure 7.22 **A Representative Vertebra and Intervertebral Disc, Superior Views.** (a) A typical vertebra. (b) An intervertebral disc oriented the same way as the vertebral body in part (a) for comparison. (See figs. 7.23 and 7.25b for lateral views.)

anulus fibrosus (fig. 7.22). There are 23 discs—the first one between cervical vertebrae 2 and 3 and the last one between the last lumbar vertebra and the sacrum. The discs help to bind adjacent vertebrae together, enhance spinal flexibility, support the weight of the body, and absorb shock. Under stress—for example, when you lift a heavy weight—the discs bulge laterally. Excessive stress can cause a *herniated disc* (see p. 179).

Regional Characteristics of Vertebrae

We are now prepared to consider how vertebrae differ from one region of the vertebral column to another and from the generalized anatomy just described. Knowing these variations will enable you to identify the region of the spine from which an isolated vertebra was taken. More importantly, these modifications in form reflect functional differences among the vertebrae.

Cervical Vertebrae

The **cervical vertebrae (C1–C7)** are relatively small. Their function is to support the head and allow for its movements. The first two (C1 and C2) have unique structures for this purpose (fig. 7.24). Vertebra C1 is called the **atlas** because it supports the head in a manner reminiscent of the Titan of Greek mythology who was condemned by Zeus to carry the world on his shoulders. It scarcely resembles the typical vertebra; it has no body and is little more than a delicate ring surrounding a large vertebral foramen. On each side is a **lateral mass** with a deeply concave

Transverse process

Body (centrum)

Intervertebral disc

Inferior articular process of L2

Superior articular process of L3

Lamina

L2

L3

(a) Posterior (dorsal) view

Superior notch

Superior articular process of L1

Inferior vertebral notch of L1

Superior vertebral notch of L2

Intervertebral foramen

Spinous process

Intervertebral disc

Inferior articular process of L3

Pedicle

L1

L2

L3

(b) Left lateral view

Figure 7.23 Articulated Vertebrae.

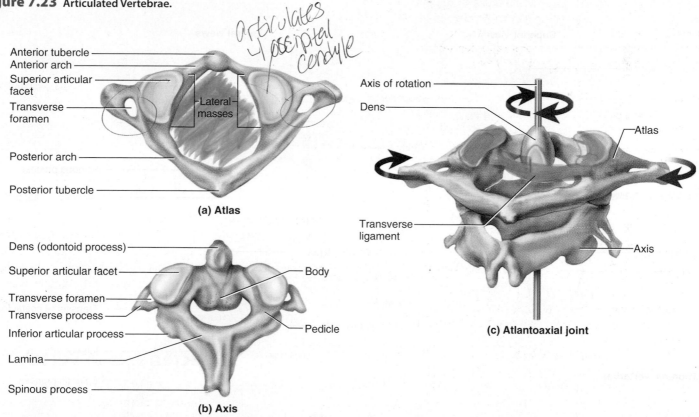

articulates w/ occipital condyle

Anterior tubercle

Anterior arch

Superior articular facet

Transverse foramen

Lateral masses

Posterior arch

Posterior tubercle

(a) Atlas

Dens (odontoid process)

Superior articular facet

Transverse foramen

Transverse process

Inferior articular process

Lamina

Spinous process

Body

Pedicle

(b) Axis

Axis of rotation

Dens

Atlas

Transverse ligament

Axis

(c) Atlantoaxial joint

Figure 7.24 **The Atlas and Axis, Cervical Vertebrae C1 and C2.** (a) The atlas, superior view. (b) The axis, posterosuperior view. (c) Articulation of the atlas and axis and rotation of the atlas. This movement turns the head from side to side, as in gesturing "no." Note the transverse ligament holding the dens of the axis in place.

● *What serious consequence could result if the transverse ligament were ruptured, allowing the dens to slip anteriorly?*

superior articular facet, which articulates with the occipital condyle of the skull. In nodding motions of the skull, as in gesturing "yes," the occipital condyles rock back and forth on these facets. The **inferior articular facets,** which are comparatively flat or only slightly concave, articulate with C2. The lateral masses are connected by an **anterior arch** and a **posterior arch,** which bear slight protuberances called the **anterior** and **posterior tubercle,** respectively.

Vertebra C2, the **axis,** allows rotation of the head as in gesturing "no." Its most distinctive feature is a prominent anterior knob

called the **dens** (pronounced "denz"), or **odontoid**[31] **process.** No other vertebra has a dens. It begins to form as an independent ossification center during the first year of life and fuses with the axis by the age of 3 to 6 years. It projects into the vertebral foramen of the atlas, where it is nestled in a facet and held in place by a **transverse ligament** (fig. 7.24c). A heavy blow to the top of the head can cause a fatal injury in which the dens is driven through

[31]*dens* = odont = tooth + *oid* = resembling

the foramen magnum into the brainstem. The articulation between the atlas and the cranium is called the **atlanto–occipital joint;** the one between the atlas and axis is called the **atlantoaxial joint.**

The axis is the first vertebra that exhibits a spinous process. In vertebrae C2 through C6, the process is forked, or *bifid*,[32] at its tip (fig. 7.25a). This fork provides attachment for the *nuchal ligament* of the back of the neck. All seven cervical vertebrae have a prominent round **transverse foramen** in each transverse process. These foramina provide passage and protection for the *vertebral arteries,* which supply blood to the brain, and *vertebral veins,* which drain blood from various neck structures. Transverse foramina occur in no other vertebrae and thus provide an easy means of recognizing a cervical vertebra.

---

[32]*bifid* = cleft into two parts

Apply What You Know

How would head movements be affected if vertebrae C1 and C2 had the same structure as C3? What is the functional advantage of the lack of a spinous process in C1?

Cervical vertebrae C3 through C6 are similar to the typical vertebra described earlier, with the addition of the transverse foramina and bifid spinous processes. Vertebra C7 is a little different—its spinous process is not bifid, but it is especially long and forms a prominent bump on the lower back of the neck. C7 is sometimes called the *vertebra prominens* because of this especially conspicuous spinous process. This feature is a convenient landmark for counting vertebrae. One can easily identify the largest bump on the neck as C7, then count up or down from there to identify others.

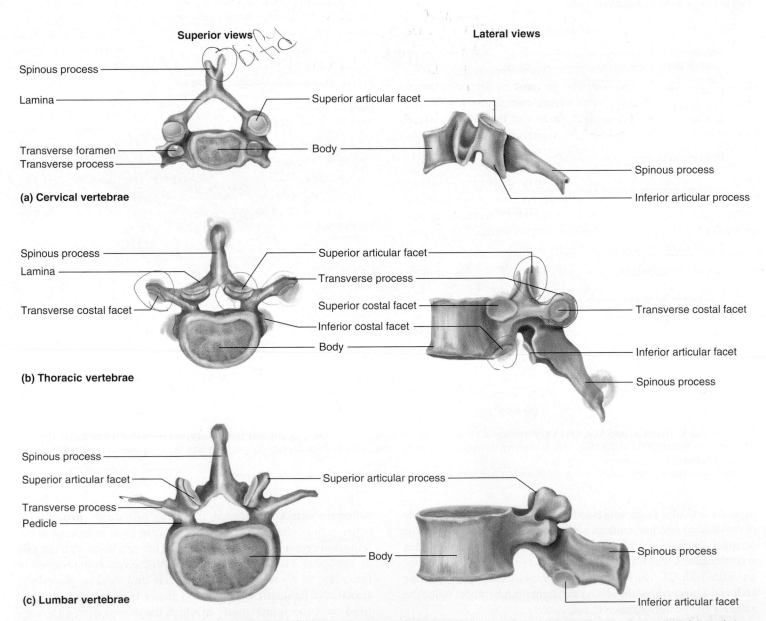

Figure 7.25 Typical Cervical, Thoracic, and Lumbar Vertebrae. The left-hand figures are superior views, and the right-hand figures are left lateral views.

Thoracic Vertebrae

There are 12 **thoracic vertebrae (T1–T12),** corresponding to the 12 pairs of ribs attached to them; no other vertebrae have ribs. One function of the thoracics is to support the thoracic cage enclosing the heart and lungs. They lack the transverse foramina and bifid processes that distinguish the cervicals, but possess the following distinctive features of their own (fig. 7.25b):

- The spinous processes are relatively pointed and angle sharply downward.
- The body is somewhat heart-shaped and more massive than in the cervical vertebrae, but less than in the lumbar vertebrae.
- The body has small, smooth, slightly concave spots called *costal facets* for attachment of the ribs.
- Vertebrae T1 through T10 have a shallow, cuplike **transverse costal**[33] **facet** at the end of each transverse process. These provide a second point of articulation for ribs 1 through 10. There are no transverse costal facets on T11 and T12, and ribs 11 and 12 attach only to the bodies of the vertebrae.

Thoracic vertebrae vary among themselves mainly in the mode of articulation with the ribs. In most cases, a rib inserts between two vertebrae, so each vertebra contributes one-half of the articular surface—the rib articulates with the **inferior costal facet** of the upper vertebra and the **superior costal facet** of the vertebra below that. This terminology may be a little confusing, but note that the facets are named for their position on the vertebral body, not for which part of the rib's articulation they provide. Vertebrae T1, T10, T11, and T12, however, have complete costal facets on the bodies for ribs 1 and 10 through 12, which articulate on the vertebral bodies instead of between vertebrae. These varia-

tions will be more functionally understandable after you have studied the anatomy of the ribs, so we will return then to the details of these articular surfaces.

Vertebra T12 differs in its articular processes from those above it. Its superior articular processes face posteriorly to meet the anteriorly facing inferior processes of T11, but the inferior articular processes of T12 face laterally like those of the lumbar vertebrae, described next. T12 thus represents a transition between the thoracic and lumbar pattern.

Lumbar Vertebrae

There are five **lumbar vertebrae (L1–L5).** Their most distinctive features are a thick, stout body and a blunt, squarish spinous process (fig. 7.25c). In addition, their articular processes are oriented differently than those of other vertebrae. In thoracic vertebrae, the superior processes face posteriorly and the inferior processes face anteriorly. In lumbar vertebrae, the superior processes face medially (like the palms of your hands about to clap), and the inferior processes face laterally, toward the superior processes of the next vertebra. This arrangement resists twisting of the lower spine. These differences are best observed on an articulated (assembled) skeleton.

Sacral Vertebrae

The five **sacral vertebrae (S1–S5)** of a child begin to fuse around age 16, and by age 26 they are usually fused into a single bony plate, the **sacrum** (SACK-rum or SAY-krum) (fig. 7.26). The sacrum forms the posterior wall of the pelvic cavity and protects the organs within. It is named for the fact that it was once considered the seat of the soul.[34]

---

[33]*costa* = rib + *al* = pertaining to

[34]*sacr* = sacred

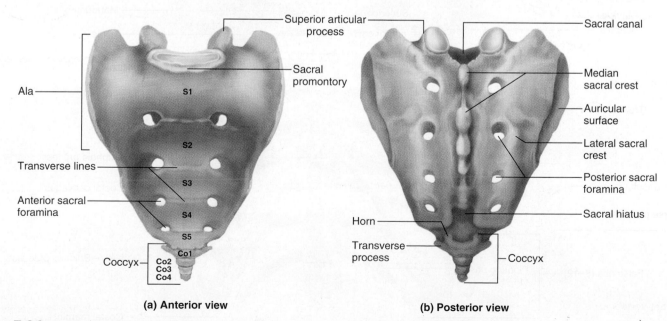

(a) Anterior view

(b) Posterior view

Figure 7.26 **The Sacrum and Coccyx.** (a) The anterior surface, which faces the viscera of the pelvic cavity. (b) The posterior surface, whose surface features can be palpated in the sacral region.

The anterior surface of the sacrum is relatively smooth and concave and has four transverse lines that indicate where the five vertebrae have fused. This surface exhibits four pairs of large **anterior sacral (pelvic) foramina,** which allow for the passage of nerves and arteries to the pelvic organs. The posterior surface of the sacrum is very rough. The spinous processes of the vertebrae fuse into a posterior ridge called the **median sacral crest.** The transverse processes fuse into a less prominent **lateral sacral crest** on each side of the median crest. Again on the posterior side of the sacrum, there are four pairs of openings for spinal nerves, the **posterior sacral foramina.** The nerves that emerge here supply the gluteal region and lower limbs.

A **sacral canal** runs through the sacrum and ends in an inferior opening called the **sacral hiatus** (hy-AY-tus). This canal contains spinal nerve roots in life. On each side of the sacrum is an ear-shaped region called the **auricular**[35] (aw-RIC-you-lur) **surface.** This articulates with a similarly shaped surface on the hip bone and forms the strong, nearly immovable **sacroiliac** (SAY-cro-ILL-ee-ac) **(SI) joint.** The body of vertebra S1 juts anteriorly to form a **sacral promontory,** which supports the body of vertebra L5. Lateral to the median sacral crest, S1 also has a pair of **superior articular processes** that articulate with L5. Lateral to these are a pair of large, rough, winglike extensions called the **alae**[36] (AIL-ee).

---

[35]*auri* = ear + *cul* = little + *ar* = pertaining to
[36]*alae* = wings

Coccygeal Vertebrae

Four (sometimes five) tiny **coccygeal vertebrae** (Co1 to Co4 or Co5) fuse by the age of 20 to form the **coccyx**[37] (fig. 7.26), colloquially called the tailbone. Although it is indeed the vestige of a tail, it is not entirely useless; it provides attachment for muscles of the pelvic floor. Vertebra Co1 has a pair of **horns (cornua),** which serve as attachment points for ligaments that bind the coccyx to the sacrum. The coccyx can be fractured by a difficult childbirth or a hard fall on the buttocks.

The Thoracic Cage

The **thoracic cage** (fig. 7.27) consists of the thoracic vertebrae, sternum, and ribs. It forms a roughly conical enclosure for the lungs and heart and provides attachment for the pectoral girdle and upper limb. It has a broad base and a narrower superior apex. Its inferior border is the arc of the lower ribs, called the **costal margin.** The cage protects not only the thoracic organs but also the spleen, most of the liver, and to some extent the kidneys. Most important is its role in breathing; it is rhythmically expanded by the respiratory muscles to create a vacuum that draws air into the lungs, and then compressed to expel air.

---

[37]*coccyx* = cuckoo (named for resemblance to a cuckoo's beak)

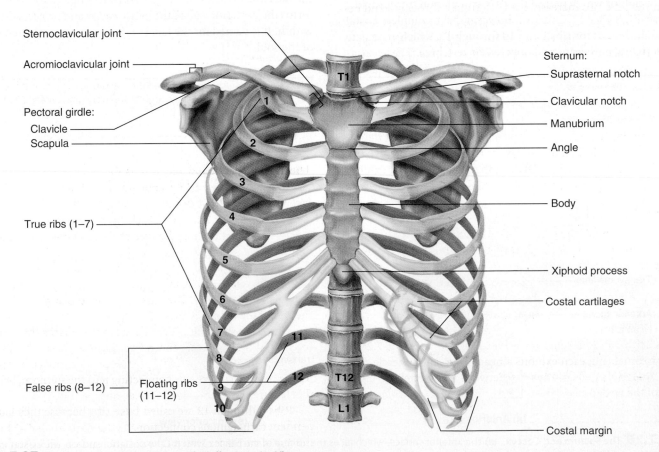

Figure 7.27 The Thoracic Cage and Pectoral Girdle, Anterior View.

The Sternum

The **sternum** (breastbone) is a bony plate anterior to the heart (fig. 7.27). It is subdivided into three regions: the *manubrium, body,* and *xiphoid process.* The **manubrium**[38] (ma-NOO-bree-um) is the broad superior portion. It has a median **suprasternal notch (jugular notch),** which you can easily palpate between your clavicles (collarbones), and right and left **clavicular notches** where it articulates with the clavicles. The **body,** or **gladiolus,**[39] is the longest part of the sternum. It joins the manubrium at the **sternal angle,** which can be palpated as a transverse ridge at the point where the sternum projects farthest forward. In some people, however, the angle is rounded or concave. The second rib attaches here, making the sternal angle a useful landmark for counting ribs in a physical examination. The manubrium and body have scalloped lateral margins where cartilages of the ribs are attached. At the inferior end is a small, pointed **xiphoid**[40] (ZIF-oyd) **process** that provides attachment for some of the abdominal muscles. In cardiopulmonary resuscitation, improperly performed chest compression can drive the xiphoid process into the liver and cause a fatal hemorrhage.

The Ribs

There are 12 pairs of **ribs,** with no difference between the sexes. Each is attached at its posterior (proximal) end to the vertebral column, and most of them are also attached at the anterior (distal) end to the sternum. The anterior attachment is by way of a long strip of hyaline cartilage called the **costal cartilage.**

As a rule, the ribs increase in length from 1 through 7 and become progressively smaller again through rib 12. They are increasingly oblique (slanted) in orientation from 1 through 9, then less so from 10 through 12. They also differ in their individual structure and attachments at different levels of the thoracic cage, so we will examine them in order as we descend the chest, taking note of their universal characteristics as well as their individual variations.

Rib 1 is peculiar. On an articulated skeleton, you must look for its vertebral attachment just below the base of the neck; much of this rib lies above the level of the clavicle (fig. 7.27). It is a short, flat, C-shaped plate of bone (fig. 7.28a). At the vertebral end, it exhibits a knobby **head** that articulates with the body of vertebra T1. On an isolated vertebra, you can find a smooth costal facet for the attachment on the middle of the body. Immediately distal to the head, the rib narrows to a **neck** and then widens again to form a rough area called the **tubercle.** This is its point of attachment to the transverse costal facet of the same vertebra. Beyond the tubercle, the rib flattens and widens into a gently sloping bladelike **shaft.** The shaft ends distally in a squared-off, rough area. In the living individual, the costal cartilage begins here and spans the rest of the distance to the upper sternum.

Ribs 2 through 7 present a more typical appearance (fig. 7.28b). At the proximal end, each exhibits a head, neck, and tubercle. The head is wedge-shaped and inserts between two vertebrae. Each margin of the wedge has a smooth surface called an *articular facet.*

---

[38]*manubrium* = handle
[39]*gladiolus* = sword
[40]*xipho* = sword + *oid* = resembling

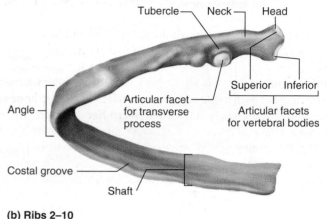

(a) Rib 1

(b) Ribs 2–10

(c) Ribs 11–12

Figure 7.28 Anatomy of the Ribs. (a) Rib 1 is an atypical flat plate. (b) Typical features of ribs 2 through 10. (c) Appearance of the floating ribs, 11 and 12.

The **superior articular facet** joins the inferior costal facet of the vertebra above; the **inferior articular facet** joins the superior costal facet of the vertebra below. The tubercle of the rib articulates with the transverse costal facet of each same-numbered vertebra. Figure 7.29 details the three rib–vertebra attachments typical of this region of the rib cage.

Beyond the tubercle, each rib makes a sharp curve around the side of the chest and then progresses anteriorly to approach the sternum (see fig. 7.27). The curve is called the **angle** of the rib and the rest of the bony blade distal to it is called the shaft. The inferior margin of the shaft has a **costal groove** that marks the path of the intercostal blood vessels and nerve. Each of these ribs, like rib 1, ends in a blunt, rough area where the costal cartilage begins. Each has its own costal cartilage connecting it to the sternum; because of this feature, ribs 1 through 7 are called **true ribs.**

Ribs 8 through 12 are called **false ribs** because they lack independent cartilaginous connections to the sternum. In 8 through 10, the costal cartilages sweep upward and end on the costal cartilage of rib 7 (see fig. 7.27). Rib 10 also differs from 2 through 9 in that it

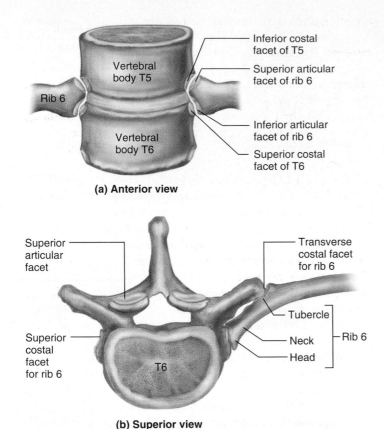

(a) Anterior view

Inferior costal facet of T5
Superior articular facet of rib 6
Vertebral body T5
Rib 6
Inferior articular facet of rib 6
Vertebral body T6
Superior costal facet of T6

Superior articular facet
Transverse costal facet for rib 6
Superior costal facet for rib 6
Tubercle
Neck
Head
T6
Rib 6

(b) Superior view

Figure 7.29 Articulation of Rib 6 with Vertebrae T5 and T6.
(a) Anterior view. Note the relationship of the articular facets of the rib with the costal facets of the two vertebrae. (b) Superior view. Note that the rib articulates with a vertebra at two points: the costal facet of the vertebral body and the transverse costal facet on the transverse process.

attaches to the body of a single vertebra (T10) rather than between vertebrae. Thus, vertebra T10 has a complete costal facet on its body for rib 10.

Ribs 11 and 12 are again unusual (fig. 7.28c). Posteriorly, they articulate with the bodies of vertebrae T11 and T12, but they do not have tubercles and do not attach to the transverse processes of the vertebrae. Those two vertebrae therefore have no transverse costal facets. At the distal end, these two relatively small, delicate ribs taper to a point and are capped by a small cartilaginous tip, but there is no cartilaginous connection to the sternum or to any of the higher costal cartilages. The ribs are merely embedded in lumbar muscle at this end. Consequently, 11 and 12 are also called **floating ribs.** Among the Japanese and some other people, rib 10 is also usually floating.

Table 7.4 summarizes these variations in rib anatomy and their vertebral and sternal attachments.

Before You Go On

Answer the following questions to test your understanding of the preceding section:

9. Discuss the contributions of the intervertebral discs to the length and flexibility of the spine.

10. Make a table with three columns headed "cervical," "thoracic," and "lumbar." In each column, list the distinctive characteristics of each type of vertebra.

11. Name the three parts of the sternum. How many ribs attach (directly or indirectly) to each part?

12. Describe how rib 5 articulates with the spine. How do ribs 1 and 12 differ from this and from each other in their modes of articulation?

TABLE 7.4	Articulations of the Ribs				
Rib	Type	Costal Cartilage	Articulating Vertebral Bodies	Articulating with a Transverse Costal Facet?	Rib Tubercle
1	True	Individual	T1	Yes	Present
2	True	Individual	T1 and T2	Yes	Present
3	True	Individual	T2 and T3	Yes	Present
4	True	Individual	T3 and T4	Yes	Present
5	True	Individual	T4 and T5	Yes	Present
6	True	Individual	T5 and T6	Yes	Present
7	True	Individual	T6 and T7	Yes	Present
8	False	Shared with rib 7	T7 and T8	Yes	Present
9	False	Shared with rib 7	T8 and T9	Yes	Present
10	False	Shared with rib 7	T10	Yes	Present
11	False, floating	None	T11	No	Absent
12	False, floating	None	T12	No	Absent

13. Distinguish between true, false, and floating ribs. Which ribs fall into each category?

14. Palpate as many of the following structures as possible, and identify which ones cannot normally be palpated on a living person: the dens; the spinous process of vertebra C7; the transverse process of vertebra T12; the median sacral crest; the coccyx; the manubrium; the xiphoid process; and the costal cartilage of rib 5.

7.4 Developmental and Clinical Perspectives

▶ Expected Learning Outcomes

When you have completed this section, you should be able to

- describe the prenatal development of the axial skeleton; and
- describe some common disorders of the axial skeleton.

Development of the Axial Skeleton

The axial skeleton develops primarily by endochondral ossification. This is a two-step process: (1) **chondrification,** in which embryonic mesenchyme condenses and differentiates into hyaline cartilage; and (2) **ossification,** in which the cartilage is replaced by bone, as described in chapter 6. Parts of the skull develop by intramembranous ossification, with no cartilage precursor.

The Skull

Development of the skull is very complex, and we will take only a broad overview of the process here. We can view the skull as developing in three major parts: the base, the calvaria, and the facial bones. The base and calvaria are collectively called the **neurocranium** because they enclose the brain; the facial skeleton is called the **viscerocranium** because it develops from the pharyngeal (visceral) arches described on page 90. Both the neurocranium and viscerocranium have regions of cartilaginous and membranous origin. The cartilaginous neurocranium is also called the **chondrocranium.**

The base of the cranium develops from several pairs of cartilaginous plates. They give rise to most parts of the sphenoid, ethmoid, temporal, and occipital bones. The flat bones of the calvaria form, in contrast, by the intramembranous method. They begin to ossify in week 9, slightly later than the cranial base. As a membranous bone ossifies, trabeculae and spicules of osseous tissue first appear in the center and then spread toward the edges of the bone (fig. 7.30).

Facial bones develop mainly from the first two pharyngeal arches. Although these arches are initially supported by cartilage, that cartilage does not transform into bone. It becomes surrounded by developing membranous bone, and while some of the cartilages

become middle-ear bones and part of the hyoid bone, others simply degenerate and disappear. Thus, the facial bones are built *around* cartilages but develop by the intramembranous process.

The skull therefore develops from a multitude of separate pieces. These undergo considerable fusion by the time of birth, but their fusion is by no means complete then. At birth, the frontal bone is still paired. The right and left halves usually fuse by the age of 5 or 6 years, but in some people a *metopic*[41] *suture* persists in the forehead between them. Traces of this suture are evident in some adult skulls.

The cranial bones are separated at birth by gaps called **fontanels,**[42] bridged by fibrous membranes (fig. 7.31). The term refers to the fact that pulsation of the infant's blood can be felt there. Fontanels permit the bones to shift as the infant squeezes through the birth canal. This shifting may deform the infant's head, but it usually assumes a normal shape within a few days after birth. Four of the fontanels are especially prominent and regular in location: the **anterior, posterior, sphenoid,** and **mastoid fontanels.** The fontanels close by intramembranous ossification. Most are fully ossified by 12 months of age, but the largest one, the anterior fontanel, does not close for 18 to 24 months. Up to that age, a "soft spot" can be palpated in the corner between the four frontal and parietal bones.

The mandible changes markedly with age. At birth, it consists of right and left bones joined medially by the *mental symphysis*, a zone of cartilage and fibrous connective tissue. The halves begin to fuse in

---

[41]*met* = beyond + *op* = the eyes
[42]*fontan* = fountain + *el* = little

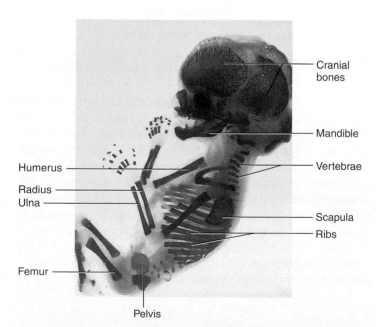

Figure 7.30 The Fetal Skeleton at 12 Weeks. The red-stained regions are ossified at this age, whereas the elbow, wrist, knee, and ankle joints appear translucent because they are still cartilaginous. The cranial bones are still widely separated.

● *Why are the joints of an infant weaker than those of an older child?*

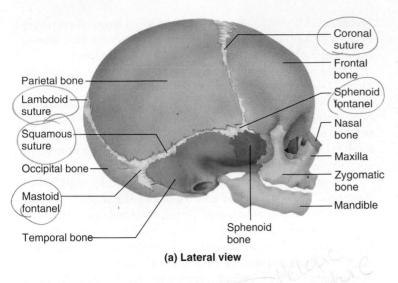

(a) Lateral view

Coronal suture
Frontal bone
Sphenoid fontanel
Nasal bone
Maxilla
Zygomatic bone
Mandible
Sphenoid bone

Parietal bone
Lambdoid suture
Squamous suture
Occipital bone
Mastoid fontanel
Temporal bone

Metopic suture

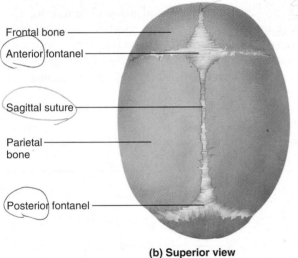

(b) Superior view

Frontal bone
Anterior fontanel
Sagittal suture
Parietal bone
Posterior fontanel

Figure 7.31 The Fetal Skull Near the Time of Birth.

the first year and become fully united into a single bone by age 3. The body of the mandible is very slender at birth, and the ramus is not strongly developed (fig. 7.31a). In early childhood, the mandible grows in a generally downward and forward direction, making the ramus longer and the chin more pronounced. The deciduous ("baby") teeth begin to erupt at about 7 months and continue through the second year, while the body of the mandible widens to accommodate their roots. The deciduous teeth are replaced with permanent teeth mostly between the ages of 6 and 13 years, although the third molar, if it emerges at all, may come as late as age 25 (see fig. 24.6, p. 661). If teeth are lost in old age, the alveoli are resorbed and the body of the mandible becomes narrower, much as it was in infancy.

Apply What You Know

Suppose you were studying a skull with some teeth missing. How could you tell whether the teeth had been lost after the person's death or years before it?

The face of a newborn is flat and small compared to the large cranium. It enlarges as the mandible, teeth, and paranasal sinuses develop. To accommodate the growing brain, a child's skull grows more rapidly than the rest of the skeleton. It reaches about half its adult size by age 9 months, three-quarters by age 2, and nearly final size by 8 or 9 years. The heads of babies and young children are therefore much larger in proportion to the trunk than are the heads of adults—an attribute thoroughly exploited by cartoonists and advertisers who draw big-headed characters to give them a more endearing or immature appearance. In humans and other animals, the large, rounded heads of the young are thought to promote survival by stimulating parental caregiving instincts.

The Vertebral Column

One of the universal characteristics of all chordate animals, including humans, is the **notochord,** a flexible, middorsal rod of mesodermal tissue. In humans, the notochord is evident inferior to the neural tube in the third week of development. Segments of embryonic mesoderm called *somites* lie on each side of the notochord and neural tube (see p. 90 and fig. 4.8, p. 92). In week 4, part of each somite becomes a sclerotome. This gives rise to vertebral cartilage, which is replaced by bone through endochondral ossification. The sclerotomes are temporarily separated by zones of looser mesenchyme (fig. 7.32a).

As shown in figure 7.32b, each vertebral body arises from portions of two adjacent sclerotomes and the loose mesenchyme between them. The midportion of each sclerotome gives rise to the anulus fibrosus of the intervertebral disc. The notochord degenerates and disappears in the regions of the developing vertebral bodies, but persists and expands between the vertebrae to form the nucleus pulposus of the intervertebral discs.

Meanwhile, mesenchyme surrounding the neural tube condenses and forms the vertebral arches of the vertebrae. Approaching the end of the embryonic stage, the mesenchyme of the sclerotomes forms the cartilaginous forerunners of the vertebral bodies (fig. 7.33a). The two halves of the arch fuse with each other and with the body, and the spinous and transverse processes grow outward from the arch. Thus, a complete cartilaginous vertebral column is established.

Ossification of the vertebrae begins during the embryonic period but is not completed until age 25. Each vertebra develops three primary ossification centers: one in the body and one in each half of the vertebral arch. At birth, these three bony parts are still connected by hyaline cartilage (fig. 7.33b). The two halves of the arch finish ossifying and fuse around 3 to 5 years of age, beginning in the lumbar region and progressing rostrally. The attachments of the arch to the body remain cartilaginous for a time in order to allow for growth of the spinal cord. These attachments ossify at 3 to 6 years. Secondary ossification centers form in puberty at the tips of the spinous and transverse processes and in a ring encircling the body. They unite with the rest of the vertebra by age 25.

At birth, the vertebral column exhibits one continuous C-shaped curve (fig. 7.34), as it does in monkeys, apes, and most other four-legged animals. As an infant begins to crawl and lift its head, the

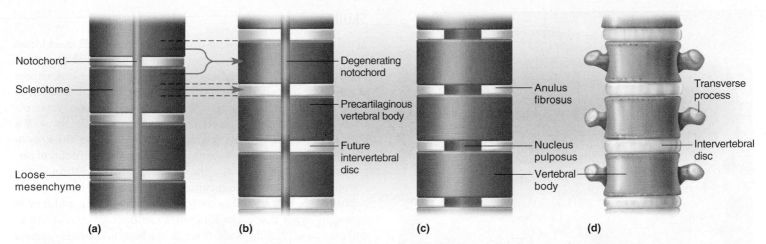

Figure 7.32 Development of the Vertebrae and Intervertebral Discs. (a) The notochord is flanked by sclerotomes, which are separated by zones of loose mesenchyme. (b) Each vertebral body forms by the condensation of parts of two sclerotomes and the loose mesenchyme between them. The midregion of each sclerotome remains less condensed and forms the anulus fibrosus of the intervertebral disc. The notochord degenerates in the regions of condensing mesenchyme but persists between vertebral bodies as the nucleus pulposus. Dashed lines indicate which regions of the sclerotomes in part (a) give rise to the vertebral body and intervertebral disc in part (b). (c) Further condensation of the vertebral bodies. The notochord has now disappeared except at the nucleus pulposus of each disc. (d) Chondrification and ossification give rise to the fully developed vertebral bodies.

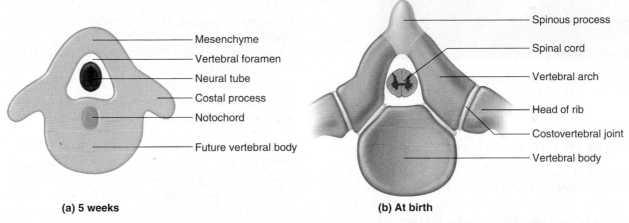

Figure 7.33 Development of a Thoracic Vertebra. (a) At 5 weeks. (b) At birth. In part (a), the vertebra is composed of mesenchyme surrounding the neural tube. The notochord is still present. The costal process is the forerunner of the rib. In part (b), the vertebra shows three centers of ossification at birth—the body and the two vertebral arches. Hyaline cartilage (blue) still composes the spinous process, the joints between the vertebral arches and the body, and the joints between the ribs and the vertebra.

cervical curvature forms, enabling an infant on its belly to look forward. The lumbar curvature—completing the S shape of the spine—starts to appear when a toddler begins walking.

The Ribs and Sternum

At 5 weeks, a developing thoracic vertebra consists of a body of mesenchyme with a vertebral body, vertebral foramen, and a pair of winglike lateral extensions called the **costal processes** (see fig. 7.33a), which soon give rise to the ribs. At 6 weeks, a chondrification center develops at the base of each process. At 7 weeks, these centers begin to undergo endochondral ossification. A *costovertebral joint* now appears at the base of the process, separating it from the vertebral body (see fig. 7.33b). By this time, the first seven ribs (true ribs) connect to the sternum by way of costal cartilages.

An ossification center soon appears at the angle of the rib, and endochondral ossification proceeds from there to the distal end of the shaft. Secondary ossification centers appear in the tubercle and head of the rib during adolescence.

The sternum begins as a pair of longitudinal strips of condensed mesenchyme called the **sternal bars.** These form initially in the anterolateral body wall and migrate medially during chondrification. The right and left sternal bars begin to fuse in week 7 as the most superior ribs contact them. Fusion of the sternal bars progresses downward, ending with the formation of the xiphoid process in week 9. The sternal bones form by endochondral ossification beginning superiorly and progressing toward the xiphoid. Ossification begins in month 5 and is completed shortly after birth. In some cases, the sternal bars fail to fuse completely at the inferior end, so the infant xiphoid process is forked or perforated.

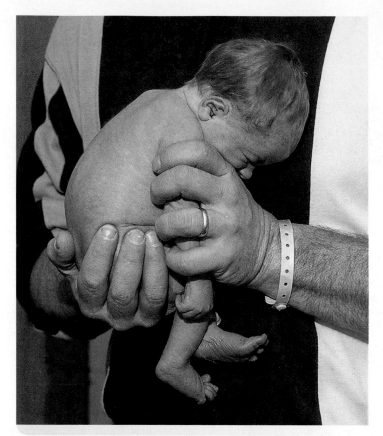

Figure 7.34 **Spinal Curvature of the Newborn Infant.** At this age, the spine forms a single C-shaped curve.

Pathology of the Axial Skeleton

Disorders that affect all parts of the skeleton are discussed in chapter 6, especially fractures and osteoporosis. Table 7.5 lists some disorders that affect especially the axial skeleton. We will consider in slightly more depth skull fractures, vertebral fractures and dislocations, and herniated intervertebral discs.

Skull Fractures

The domed shape of the skull distributes the strain of most blows to the head and tends to minimize their effects. Hard blows can nevertheless fracture the calvaria (fig. 7.35a). Most cranial fractures are *linear fractures* (elongated cracks), which can radiate away from the point of impact. In a *depressed fracture,* the cranium caves inward and may compress and damage underlying brain tissue. If a blow occurs in an area where the calvaria is especially thick, as in the occipital region, the bone may bend inward at the point of impact without breaking, but as the strain is distributed through the cranium, it can generate a fracture some distance away, even on the opposite side of the skull (a *contrafissura fracture*). In addition to damaging brain tissue, skull fractures can damage cranial nerves and meningeal blood vessels. A break in a blood vessel may cause a hematoma (mass of clotted blood) that compresses the brain tissue, potentially leading to death within a few hours.

Facial trauma can produce linear *Le Fort*[43] *fractures,* which predictably follow lines of weakness in the facial bones. The three typical Le Fort fractures are shown in figure 7.35b. The type II Le Fort fracture separates the entire central region of the face from the rest of the skull.

> ### Apply What You Know
>
> Describe two accidents or other incidents in a person's life that could result in a depressed fracture of the calvaria, and two that could result in a Le Fort facial fracture.

Vertebral Fractures and Dislocations

Injury to the cervical vertebrae (a "broken neck") often results from violent blows to the head, as in diving, motorcycle, and equestrian accidents, and sudden flexion or extension of the neck, as in automobile accidents. Such injuries often crush the body or arches of a

---

[43]Léon C. Le Fort (1829–93), French surgeon and gynecologist

TABLE 7.5	Disorders of the Axial Skeleton
Cleft palate	Failure of the palatine processes of the maxilla to fuse during fetal development, resulting in a fissure connecting the oral and nasal cavities; often accompanied by cleft lip. Causes difficulty for an infant in nursing. Can be surgically corrected with good cosmetic results.
Craniosynostosis	Premature closure of the cranial sutures within the first 2 years after birth, resulting in skull asymmetry, deformity, and sometimes mental retardation. Cause is unknown. Surgery can limit brain damage and improve appearance.
Spinal stenosis	Abnormal narrowing of the vertebral canal or intervertebral foramina caused by hypertrophy of the vertebral bone. Most common in middle-aged and older people. May compress spinal nerves and cause low back pain or muscle weakness.
Spondylosis	A defect of the laminae of the lumbar vertebrae. Defective vertebrae may shift anteriorly, especially at the L5 through S1 level. Stress on the bone may cause microfractures in the laminae and eventual dissolution of the laminae. May be treated by nonsurgical manipulation or by surgery, depending on severity.

Disorders Described Elsewhere

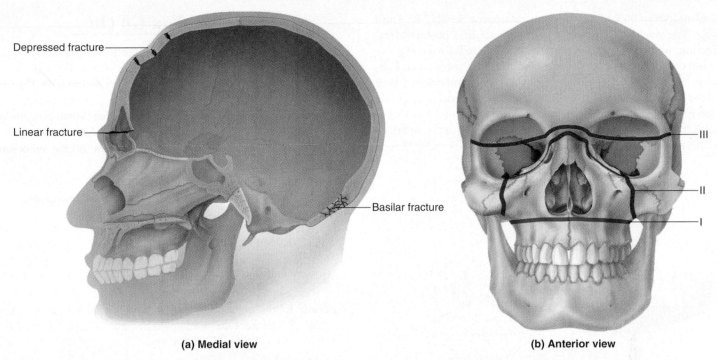

(a) Medial view

(b) Anterior view

Figure 7.35 Skull Fractures. (a) Medial view showing linear and depressed fractures of the frontal bone and a basilar fracture of the occipital bone. (b) The three types of Le Fort fractures of the facial bones.

vertebra or cause one vertebra to slip forward relative to the one below it. The dislocation of one vertebra relative to the next can cause irreparable damage to the spinal cord. "Whiplash" often results from rear-end automobile collisions causing violent hyperextension of the neck (backward jerking of the head). This stretches or tears the *anterior longitudinal ligament* that courses along the front of the vertebral bodies, and it may fracture the vertebral body (fig. 7.36a). Dislocations are relatively rare in the thoracic and lumbar regions because of the way the vertebrae are tightly interlocked. When fractures occur in these regions ("broken back"), they most often involve vertebra T11 or T12, at the transition from the thoracic to lumbar spine.

Herniated Discs

A **herniated** ("slipped" or "ruptured") **disc** is cracking of the anulus fibrosis of an intervertebral disc under strain, often caused by violent flexion of the vertebral column or by lifting heavy weights. Cracking of the anulus allows the gelatinous nucleus pulposus to ooze out, sometimes putting pressure on a spinal nerve root or the spinal cord (fig. 7.36b). Back pain results from both pressure on the nervous tissue and inflammation stimulated by the nucleus

Figure 7.36 Injuries to the Vertebral Column. (a) Whiplash injury. Violent hyperextension of the neck has torn the anterior longitudinal ligament and fractured the vertebral body. (b) Herniated intervertebral disc. The nucleus pulposus is oozing into the vertebral canal and compressing a bundle of spinal nerve roots that passes through the lumbar vertebrae.
● *Herniated discs are more common in the lumbar region than in the cervical region. Explain.*

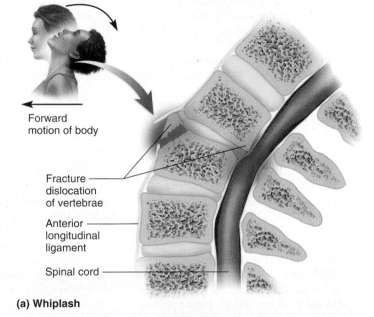

(a) Whiplash

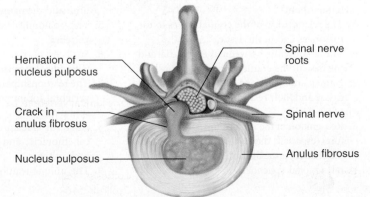

(b) Herniated disc

pulposus. About 95% of disc herniations occur at levels L4/L5 and L5/S1. The nucleus pulposus usually escapes in a posterolateral direction, where the anulus is thinnest. Herniated discs rarely occur in young people because their discs are well hydrated and absorb pressure well. As people get older, the discs become dehydrated and they degenerate and grow thinner, becoming more susceptible to herniation. After middle age, however, the anulus fibrosus becomes thicker and tougher, and the nucleus pulposus is smaller, so disc herniations again become less common.

Before You Go On

Answer the following questions to test your understanding of the preceding section:

15. Define *chondrocranium* and *viscerocranium,* and explain why each of them is named that.

16. What is the functional significance of fontanels? When does the last fontanel close?

17. What structure in the adult is a remnant of the embryonic notochord?

18. What is a Le Fort fracture? What is whiplash?

19. Explain why a herniated disc can cause nerve pain (neuralgia).

Study Guide

Assess Your Learning Outcomes

You should have a good understanding of this chapter if you can accurately address the following issues.

7.1 Overview of the Skeleton (p. 150)

1. The distinction between the axial and appendicular skeletons

2. Changes in the number of bones with age; the typical adult number; and reasons why the adult number varies from one person to another

3. Terminology of the surface features of bones

7.2 The Skull (p. 153)

1. Names and locations of the cavities in the skull

2. Definitions of *suture* and *foramen* in skull anatomy

3. The distinction between cranial bones and facial bones

4. Names and locations of the cranial bones, and names of the sutures that form their boundaries (to the extent that this book names them)

5. The relationship of the cranial bones to the meninges and brain tissue

6. The distinction between the calvaria and the base of the cranium

7. Names and locations of the three cranial fossae and their relationship to the anatomy of the brain

8. Recognition of the important anatomical features (especially those boldfaced in the text) of the cranial bones: the frontal, parietal, temporal, occipital, sphenoid, and ethmoid bones

9. Recognition of the important anatomical features (especially those boldfaced in the text) of the facial bones: the maxilla; inferior nasal concha; vomer; mandible; and the palatine, zygomatic, lacrimal, and nasal bones

10. The location, anatomy, and function of the hyoid bone

11. The names, locations, and anatomy of the three auditory ossicles, and their collective function

12. How the location of the foramen magnum and relative flatness of the human face are related to the bipedal stance of humans

7.3 The Vertebral Column and Thoracic Cage (p. 166)

1. Functions of the vertebral column

2. The five classes of vertebrae; and the number of vertebrae in each class

3. The number of intervertebral discs; which vertebrae have discs between them and which ones do not; and the structure of a disc

4. The overall shape of the adult vertebral column and the names of its four curvatures

5. The anatomical features of a generalized vertebra

6. Surfaces of the vertebrae that articulate with adjacent vertebrae and with the ribs

7. The relationships of the vertebral and intervertebral foramina to the spinal cord and spinal nerves

8. Anatomical features that distinguish cervical, thoracic, and lumbar vertebrae from each other

9. The unique features and names of each of the first two cervical vertebrae (C1–C2); names of the joints between the skull and C1 and between C1 and C2; and how the features of these two vertebrae relate to head movements

10. Anatomy of the articulations between the ribs and thoracic vertebrae

11. Anatomical features of the sacrum and coccyx and of the articulations between the sacrum and hip bones

12. Components of the thoracic cage

13. The three parts of the sternum; its other features; and its articulations with the clavicles and ribs

14. The categories of ribs; how each category is defined; the total number of ribs; and which ribs belong to each category

15. The generalized anatomy of a rib; how rib 1 and ribs 11 and 12 anatomically differ from the others; and the relationships of the ribs to the costal cartilages, sternum, and vertebral bodies and transverse processes

7.4 Developmental and Clinical Perspectives (p. 175)

1. The complementary roles of intramembranous and endochondral ossification in the prenatal development of the skull

2. The portions of the skull that constitute the neurocranium, viscerocranium, and chondrocranium

3. The role of pharyngeal arches in development of the skull

4. Developmental changes in the skull after birth

5. The names and locations of the fontanels in the fetal skull, why they exist, and what becomes of them
6. The role of the embryonic notochord and sclerotomes in development of the vertebral column
7. The ossification process of a vertebra
8. Development of the ribs from the embryonic vertebrae
9. Embryonic development of the sternum
10. The common types of skull fractures: linear, depressed, contrafissura, and Le Fort fractures
11. Causes and effects of vertebral fractures and dislocations
12. The causes, anatomical aspects, and effects of herniated intervertebral discs

Testing Your Recall

1. Which of these is *not* a paranasal sinus?
   a. frontal
   b. temporal
   c. sphenoid
   d. ethmoid
   e. maxillary

2. Which of these is a facial bone?
   a. frontal
   b. ethmoid
   c. occipital
   d. temporal
   e. lacrimal

3. What occupies the transverse foramina seen in certain vertebrae?
   a. vertebral arteries
   b. nucleus pulposus
   c. spinal nerves
   d. carotid arteries
   e. internal jugular veins

4. All of the following are groups of vertebrae *except* for _____, which is a spinal curvature.
   a. thoracic
   b. cervical
   c. lumbar
   d. pelvic
   e. sacral

5. Thoracic vertebrae do *not* have
   a. transverse foramina.
   b. costal facets.
   c. transverse costal facets.
   d. transverse processes.
   e. pedicles.

6. Which of these bones forms by intramembranous ossification?
   a. a vertebra
   b. a parietal bone
   c. the occipital bone
   d. the sternum
   e. a rib

7. The viscerocranium includes
   a. the maxilla.
   b. the parietal bones.
   c. the occipital bone.
   d. the temporal bone.
   e. the atlas.

8. Which of these is *not* a suture?
   a. parietal
   b. coronal
   c. lambdoid
   d. sagittal
   e. squamous

9. The sella turcica contains
   a. the pituitary gland.
   b. the auditory ossicles.
   c. air cells.
   d. the foramen lacerum.
   e. the lacrimal sac.

10. The nasal septum is composed partly of the same bone as
    a. the zygomatic arch.
    b. the hard palate.
    c. the cribriform plate.
    d. the nasal concha.
    e. the centrum.

11. Gaps between the cranial bones of an infant are called _____.

12. The external acoustic meatus is an opening in the _____ bone.

13. Bones of the skull are joined along lines called _____.

14. The _____ bone has greater and lesser wings and protects the pituitary gland.

15. A herniated disc occurs when a ring called the _____ cracks.

16. The transverse ligament of the atlas holds the _____ of the axis in place.

17. The sacroiliac joint is formed where the _____ surface of the sacrum articulates with that of the ilium.

18. We have five pairs of _____ ribs and two pairs of _____ ribs.

19. Ribs 1 through 10 are joined to the sternum by way of strips of connective tissue called _____.

20. The point at the inferior end of the sternum is the _____.

Answers in the Appendix

Building Your Medical Vocabulary

State a medical meaning of each of the following word elements, and give a term in which it is used.

1. crani-
2. tempor-
3. masto-
4. petr-
5. lamina
6. pterygo-
7. crista
8. lacrimo-
9. costo-
10. ped-

Answers in the Appendix

True or False

Determine which five of the following statements are false, and briefly explain why.

1. The bodies of the vertebrae are derived from the notochord of the embryo.
2. Adults have more bones than children do.
3. A smooth round knob on a bone is called a condyle.
4. The zygomatic arch consists entirely of the zygomatic bone.
5. The dura mater adheres tightly to the entire inner surface of the cranial cavity.
6. The sphenoid bone forms part of the orbit.
7. The nasal septum is not entirely bony.
8. Not everyone has a frontal sinus.
9. The anterior surface of the sacrum is smoother than the posterior surface.
10. The lumbar vertebrae do not articulate with any ribs and therefore do not have transverse processes.

Answers in the Appendix

Testing Your Comprehension

1. A child was involved in an automobile collision. She was not wearing a safety restraint, and her chin struck the dashboard hard. When the physician looked into her auditory canal, he could see into her throat. What do you infer from this about the nature of her injury?
2. Chapter 1 noted that there are significant variations in the internal anatomy of different people (p. 6). Give some examples from this chapter other than pathological cases (such as cleft palate) and normal age-related differences.
3. Vertebrae T12 and L1 look superficially similar and are easily confused. Explain how to tell the two apart.
4. What effect would you predict if an ossification disorder completely closed off the superior orbital fissure?
5. For each of the following bones, name all the other bones with which it articulates: parietal, zygomatic, temporal, and ethmoid bones.

Answers at www.mhhe.com/saladinha3

Improve Your Grade at www.mhhe.com/saladinha3

Practice quizzes, labeling activities, and games provide fun ways to master concepts. You can also download image PowerPoint files for each chapter to create a study guide or for taking notes during lecture.

The Appendicular Skeleton

X-ray of an adult hand

CHAPTER
8

BRUSHING UP

To understand this chapter, you may find it helpful to review the following concepts:
- Terminology of the appendicular region (p. 14)
- General features of bones (p. 131)
- Intramembranous and endochondral ossification (p. 137)

Skeletal System

In this chapter, we turn our attention to the *appendicular skeleton*—the bones of the upper and lower limbs and the pectoral and pelvic girdles that attach them to the axial skeleton. We depend so heavily on the limbs for mobility and the ability to manipulate objects that deformities and injuries to the appendicular skeleton are more disabling than most disorders of the axial skeleton. Hand injuries, especially, can disable a person far more than a comparable amount of tissue injury elsewhere on the body. Injuries to the appendicular skeleton are especially common in athletics, recreation, and the workplace. A knowledge of the anatomy of the appendicular skeleton is therefore especially important to such professionals as athletic trainers, physical therapists, and other health-care providers.

8.1 The Pectoral Girdle and Upper Limb

▶ Expected Learning Outcomes

When you have completed this section, you should be able to

- identify and describe the features of the clavicle, scapula, humerus, radius, ulna, and bones of the wrist and hand; and
- describe the evolutionary innovations of the human forelimb.

Pectoral Girdle

The **pectoral**[1] **girdle** (shoulder girdle) supports the arm and links it to the axial skeleton. It consists of two bones on each side of the body: the *clavicle* (collarbone) and *scapula* (shoulder blade). The medial end of the clavicle articulates with the sternum at the **sternoclavicular joint,** and its lateral end articulates with the scapula at the **acromioclavicular**[2] **joint** (see fig. 7.27, p. 172). The scapula also articulates with the humerus at the **glenohumeral**[3] (shoulder) **joint.** These are loose attachments that result in a shoulder far more flexible than that of most other mammals, but they also make the shoulder joint easy to dislocate.

Clavicle

The **clavicle**[4] (fig. 8.1) is a slightly S-shaped bone, somewhat flattened from the upper to lower surface, and easily seen and palpated on the upper thorax (see fig. A.1b, p. 331). The superior surface is relatively smooth and rounded, whereas the inferior surface is flatter and marked by grooves and ridges for muscle attachment. The medial **sternal end** has a rounded, hammerlike head, and the lateral **acromial end** is markedly flattened. Near the acromial end is a rough tuberosity called the **conoid**[5] **tubercle**—a ligament attachment that faces posteriorly and slightly downward.

---

[1]*pect* = chest + *oral* = pertaining to
[2]*acr* = extremity, peak + *omo* = shoulder
[3]*gleno* = socket
[4]*clav* = hammer, club + *icle* = little
[5]*con* = cone + *oid* = shaped

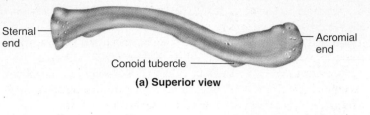

(a) Superior view

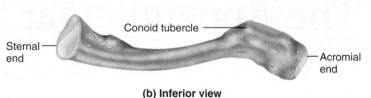

(b) Inferior view

Figure 8.1 The Right Clavicle (Collarbone).
● *The clavicle is broken more often than any other bone in the body. State some reasons why.*

The clavicle braces the shoulder, keeping the upper limb away from the midline of the body. It is thickened in people who do heavy manual labor, and because most people are right-handed, the right clavicle is usually stronger and shorter than the left. Without the clavicles, the pectoralis major muscles would pull the shoulders forward and medially—which indeed happens when a clavicle is fractured (see Deeper Insight 8.1).

Scapula

The **scapula**[6] (fig. 8.2) is a triangular plate that overlies ribs 2 through 7. The three sides of the triangle are called the **superior, medial (vertebral),** and **lateral (axillary) borders,** and its three angles are

DEEPER INSIGHT   8.1

Clavicular Fractures

The clavicle is the most frequently broken bone of the body. It can break when one falls directly on the shoulder or thrusts out an arm to break a fall and the force of the fall is transmitted through the limb to the pectoral girdle. Fractures most often occur at a weak point about one-third of the way from the lateral end. When the clavicle is broken, the shoulder tends to drop, the sternocleidomastoid muscle of the neck elevates the medial fragment, and the pectoralis major muscle of the chest may pull the lateral fragment toward the sternum. In wide-shouldered infants, the clavicle sometimes fractures during birth, but these neonatal fractures heal quickly. In children, clavicular fractures are often of the greenstick type.

---

[6]*scap* = spade, shovel + *ula* = little

the **superior, inferior,** and **lateral angles.** A conspicuous **suprascapular notch** in the superior border provides passage for a nerve. The broad anterior surface of the scapula, called the **subscapular fossa,** is slightly concave and relatively featureless. The posterior surface has a transverse shelflike ridge called the **spine,** a deep indentation superior to the spine called the **supraspinous fossa,** and a broad surface inferior to it called the **infraspinous fossa.**[7]

The most complex region of the scapula is the lateral angle, which has three main features:

1. The **acromion** (ah-CRO-me-on) is a platelike extension of the scapular spine that forms the apex of the shoulder. It articulates with the clavicle, which forms the sole bridge from the appendicular to the axial skeleton.

2. The **coracoid**[8] (COR-uh-coyd) **process** is shaped like a bent finger but named for a vague resemblance to a crow's beak; it provides attachment for tendons of the *biceps brachii* and other muscles of the arm.

3. The **glenoid** (GLEN-oyd) **cavity** is a shallow socket that articulates with the head of the humerus, forming the glenohumeral joint.

Apply What You Know

What part of the scapula do you think is most commonly fractured? Why?

---

[7]*supra* = above; *infra* = below
[8]*corac* = crow + *oid* = shaped, resembling

Upper Limb

The upper limb is divided into 4 segments containing a total of 30 bones per limb:

1. The **brachium**[9] (BRAY-kee-um), or arm proper, extends from shoulder to elbow. It contains only one bone, the *humerus.*

2. The **antebrachium,**[10] or forearm, extends from elbow to wrist and contains two bones—the *radius* and *ulna.* In anatomical position, these bones are parallel and the radius is lateral to the ulna.

3. The **carpus,**[11] or wrist, contains eight small *carpal bones* arranged in two rows.

4. The **manus,**[12] or hand, contains 19 bones in two groups— 5 *metacarpals* in the palm and 14 *phalanges* in the digits.

Humerus

The **humerus** has a hemispherical **head** that articulates with the glenoid cavity of the scapula (fig. 8.3). The smooth surface of the head (covered with articular cartilage in life) is bordered by a groove called the **anatomical neck.** Other prominent features of the proximal end are muscle attachments called the **greater** and **lesser tubercles** and an **intertubercular sulcus (groove)** between them that accommodates a tendon of the biceps muscle. The **surgical neck,** a common fracture site, is a narrowing of the bone just distal

---

[9]*brachi* = arm
[10]*ante* = before
[11]*carp* = wrist
[12]*man* = hand

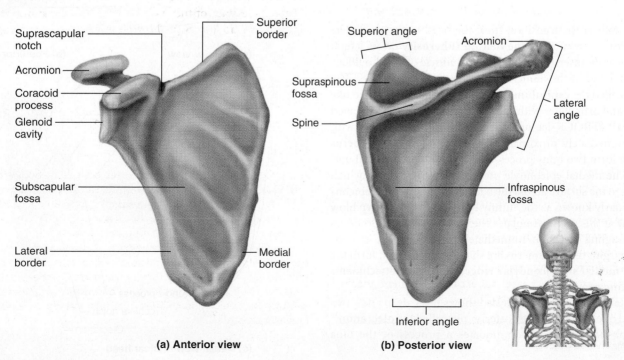

(a) Anterior view   **(b) Posterior view**

Figure 8.2 **The Right Scapula.**
• *Identify the two points where this bone is attached to the rest of the skeleton.*

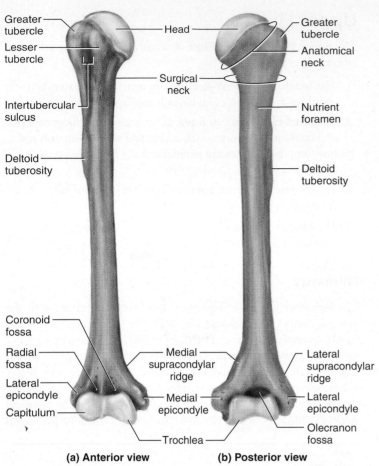

(a) Anterior view   **(b) Posterior view**

Figure 8.3 The Right Humerus.

to the tubercles, at the transition from the head to the shaft. The shaft has a rough area called the **deltoid tuberosity** on its lateral surface. This is an insertion for the *deltoid muscle* of the shoulder.

The distal end of the humerus has two smooth condyles. The lateral one, called the **capitulum**[13] (ca-PIT-you-lum), is shaped like a fat wheel and articulates with the radius. The medial one, called the **trochlea**[14] (TROCK-lee-uh), is pulleylike and articulates with the ulna. Immediately proximal to these condyles, the humerus flares out to form two bony processes, the **lateral** and **medial epicondyles.** The medial epicondyle protects the *ulnar nerve,* which passes close to the surface across the back of the elbow. This epicondyle is popularly known as the "funny bone" because a sharp blow to the elbow at this point stimulates the ulnar nerve and produces an intense tingling sensation. Immediately proximal to the epicondyles, the margins of the humerus are sharply angular and form the **lateral** and **medial supracondylar ridges.** These are attachments for certain forearm muscles.

The distal end of the humerus also shows three deep pits—two anterior and one posterior. The posterior pit, called the **olecranon**[15] (oh-LEC-ruh-non) **fossa,** accommodates a process of the ulna

---

[13]*capit* = head + *ulum* = little
[14]*troch* = wheel, pulley
[15]*olecranon* = elbow

---

called the olecranon when the elbow is extended. On the anterior surface, a medial pit called the **coronoid**[16] **fossa** accommodates the coronoid process of the ulna when the elbow is flexed. The lateral pit is the **radial fossa,** named for the nearby head of the radius.

---

[16]*coron* = something curved + *oid* = shaped

(a) Anterior view   **(b) Posterior view**

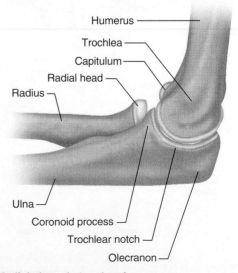

(c) Medial view of ulnar head

Figure 8.4 The Right Radius and Ulna. Part (c) shows the relationship of the trochlear notch of the ulna to the trochlea of the humerus, with the elbow flexed 90°.

Radius

The **radius** has a distinctive discoidal **head** at its proximal end (fig. 8.4). When the forearm is rotated so the palm turns forward and back, the circular superior surface of this disc spins on the capitulum of the humerus, and the edge of the disc spins on the radial notch of the ulna. Immediately distal to the head, the radius has a narrower **neck** and then widens to a rough prominence, the **radial tuberosity,** on its medial surface. The distal tendon of the biceps muscle terminates on this tuberosity.

The distal end of the radius has the following features, from lateral to medial:

1. A bony point, the **styloid[17] process,** can be palpated proximal to the thumb.

2. Two shallow depressions (articular facets) articulate with the scaphoid and lunate bones of the wrist.

3. The **ulnar notch** articulates with the end of the ulna.

Ulna

At the proximal end of the **ulna[18]** (fig. 8.4) is a deep, C-shaped **trochlear notch** that wraps around the trochlea of the humerus. The posterior side of this notch is formed by a prominent **olecranon**—the

---

[17]*styl* = pillar + *oid* = shaped
[18]*ulna* = elbow

bony point where you rest your elbow on a table. The anterior side is formed by a less prominent **coronoid process.** Laterally, the head of the ulna has a less conspicuous **radial notch,** which accommodates the edge of the head of the radius. At the distal end of the ulna is a medial **styloid process.** The bony lumps you can palpate on each side of your wrist are the styloid processes of the radius and ulna.

The radius and ulna are attached along their shafts by a ligament called the **interosseous[19]** (IN-tur-OSS-ee-us) **membrane (IM),** which is attached to an angular ridge called the **interosseous border** on the facing margins of each bone. Most fibers of the IM are oriented obliquely, slanting upward from the ulna to the radius. If you lean forward on a table supporting your weight on your hands, about 80% of the force is borne by the radius. This tenses the IM, which pulls the ulna upward and transfers some of this force through the ulna to the humerus. The IM therefore enables the two elbow joints (between humerus and radius, and humerus and ulna) to share the load and reduces the wear and tear that one joint would otherwise have to bear alone. The IM also serves as an attachment for several forearm muscles.

Carpal Bones

The **carpal bones,** which form the wrist, are arranged in two rows of four bones each (fig. 8.5). These short bones allow movements of the wrist from side to side and forward and back (in anatomical

---

[19]*inter* = between + *osse* = bones

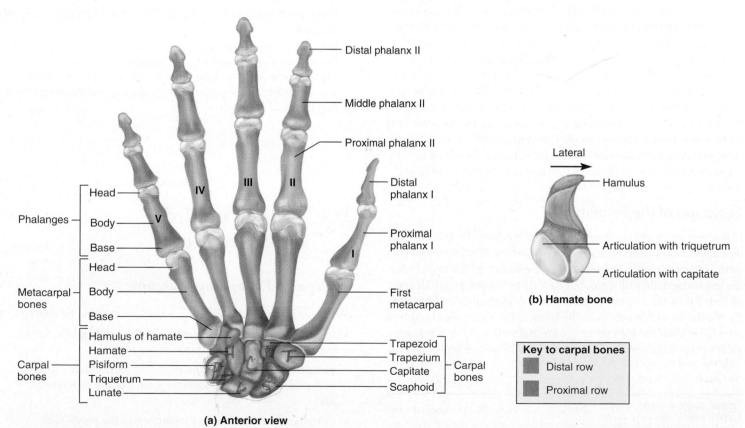

(a) Anterior view

(b) Hamate bone

Figure 8.5 **The Right Wrist and Hand, Anterior (Palmar) View.** (a) Some people remember the names of the carpal bones with the mnemonic, "Sally left the party to take Charlie home." The first letters of these words correspond to the first letters of the carpal bones, from lateral to medial, proximal row first. (b) The right hamate bone, viewed from the proximal side of the wrist to show its distinctive hook. This unique bone is a useful landmark for locating the others when studying the skeleton.

position). The carpal bones of the proximal row, starting at the lateral (thumb) side, are the **scaphoid, lunate, triquetrum** (tri-QUEE-trum), and **pisiform** (PY-sih-form)—Latin for boat-, moon-, triangle-, and pea-shaped, respectively. Unlike the other carpal bones, the pisiform is a sesamoid bone; it is not present at birth but develops around the age of 9 to 12 years within the tendon of the *flexor carpi ulnaris muscle.*

The bones of the distal row, again starting on the lateral side, are the **trapezium,**[20] **trapezoid, capitate,**[21] and **hamate.**[22] The hamate can be recognized by a prominent hook, or **hamulus,** on the palmar side. The hamulus is an attachment for the *flexor retinaculum,* a fibrous sheet in the wrist that covers the carpal tunnel (see fig. 12.9, p. 307).

Metacarpal Bones

Bones of the palm are called **metacarpals.**[23] Metacarpal I is located proximal to the thumb and metacarpal V is proximal to the little finger (fig. 8.5). On a skeleton, the metacarpals look like extensions of the fingers, so the fingers seem much longer than they really are. The proximal end of a metacarpal bone is called the **base,** the shaft is called the **body,** and the distal end is called the **head.** The heads of the metacarpals form the knuckles when you clench your fist.

Phalanges

The bones of the fingers are called **phalanges** (fah-LAN-jeez); in the singular, *phalanx*[24] (FAY-lanks). There are two phalanges in the **pollex** (thumb) and three in each of the other digits (fig. 8.5). Phalanges are identified by roman numerals preceded by *proximal, middle,* and *distal.* For example, proximal phalanx I is in the basal segment of the thumb (the first segment beyond the web between the thumb and palm); the left proximal phalanx IV is where people usually wear wedding rings; and distal phalanx V forms the tip of the little finger. The three parts of a phalanx are the same as in a metacarpal: *base, body,* and *head.* The ventral surface of a phalanx is slightly concave from end to end and flattened from side to side; the dorsal surface is rounder and slightly convex from end to end.

Evolution of the Forelimb

Elsewhere in chapters 7 and 8, we examine how the evolution of bipedal locomotion in humans has affected the skull, vertebral column, and lower limb. The effects of bipedalism on the upper limb are less immediately obvious, but nevertheless substantial. In apes, all four limbs are adapted primarily for walking and climbing, and the forelimbs are longer than the hindlimbs. Thus, the shoulders are higher than the hips when the animal walks. When some apes such as orangutans and gibbons walk bipedally, they typically hold their long forelimbs over their heads to prevent them from dragging on the ground. By contrast, the human forelimbs are adapted primarily for reaching out, exploring the environment, and manipulating objects. They are shorter than the hindlimbs and far less muscular than the forelimbs of apes. No longer needed for locomotion, our forelimbs, especially the hands, have become better adapted for carrying objects, holding things closer to the eyes, and manipulating them more precisely. Although the forelimbs have the same basic bone and muscle pattern as the hindlimbs, the joints of the shoulders and hands, especially, give the forelimbs far greater mobility.

Before You Go On

Answer the following questions to test your understanding of the preceding section:

1. Describe how to distinguish the medial and lateral ends of the clavicle from each other and how to distinguish its superior and inferior surfaces.

2. Name the three fossae of the scapula and describe the location of each.

3. What three bones meet at the elbow? Identify the fossae, articular surfaces, and processes of this joint and state to which bone each of these features belongs.

4. Name the carpal bones of the proximal row from lateral to medial, and then the bones of the distal row in the same order.

5. Name the four bones from the tip of the little finger to the base of the hand.

6. Palpate as many of the following structures as possible, and identify which ones cannot normally be palpated on a living person: the inferior angle of the scapula, the subscapular fossa, the acromion, the epicondyles of the humerus, the olecranon, and the interosseous membrane of the forearm.

8.2 The Pelvic Girdle and Lower Limb

▶ Expected Learning Outcomes

When you have completed this section, you should be able to

- identify and describe the features of the pelvic girdle, femur, patella, tibia, fibula, and bones of the foot;

- compare the anatomy of the male and female pelvic girdles and explain the functional significance of the differences; and

- describe the evolutionary adaptations of the pelvis and hindlimb for bipedal locomotion.

---

[20]*trapez* = table, grinding surface
[21]*capit* = head + *ate* = possessing
[22]*ham* = hook + *ate* = possessing
[23]*meta* = beyond + *carp* = wrist
[24]*phalanx* = line of soldiers, closely knit row

Pelvic Girdle

The terms *pelvis* and *pelvic girdle* are used in contradictory ways by various authorities. Here we will follow the practice of *Gray's Anatomy* in considering the **pelvic girdle** to consist of a complete ring composed of three bones (fig. 8.6)—two **hip (coxal) bones** and the **sacrum** (which of course is also part of the vertebral column). The hip bones are also frequently called the *ossa coxae*[25] (OS-sa COC-see) (singular, *os coxae*) and sometimes *innominate*[26] bones—arguably the most self-contradictory term in human anatomy ("the bones with no name"). The **pelvis**[27] is a bowl-shaped structure composed of these bones as well as their ligaments and the muscles that line the pelvic cavity and form its floor. The pelvic girdle supports the trunk on the lower limbs and encloses and protects the viscera of the pelvic cavity—mainly the lower colon, urinary bladder, and internal reproductive organs.

[25]*os, ossa* = bone, bones + *coxae* = of the hip(s)
[26]*in* = without + *nomin* = name + *ate* = possessing
[27]*pelvis* = basin, bowl

Each hip bone is joined to the vertebral column at one point, the **sacroiliac (SI) joint,** where its **auricular**[28] **surface** matches the auricular surface of the sacrum. The two hip bones articulate with each other on the anterior side of the pelvis, where they are joined by a pad of fibrocartilage called the **interpubic disc.** The interpubic disc and the adjacent region of pubic bone on each side constitute the **pubic symphysis,**[29] which can be palpated as a hard prominence immediately above the genitalia.

The pelvis has a bowl-like shape with the broad **greater (false) pelvis** between the flare of the hips and the narrow **lesser (true) pelvis** below (fig. 8.6b). The two are separated by a somewhat round margin called the **pelvic brim.** The opening circumscribed by the brim is called the **pelvic inlet**—an entry into the lesser pelvis through which an infant's head passes during birth. The lower margin of the lesser pelvis is called the **pelvic outlet.**

[28]*aur* = ear + *icul* = little + *ar* = like
[29]*sym* = together + *physis* = growth

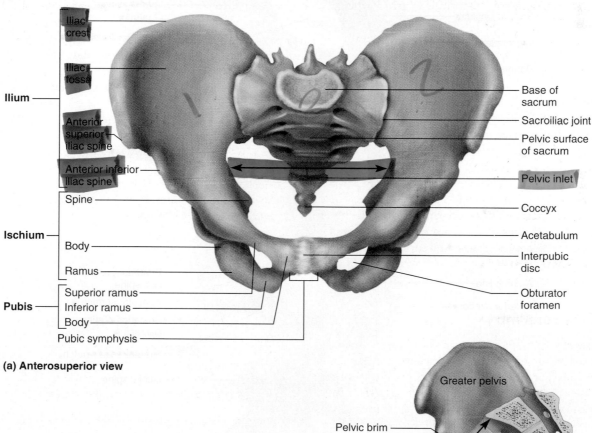

Ilium
- Iliac crest
- Iliac fossa
- Anterior superior iliac spine
- Anterior inferior iliac spine
- Spine

Ischium
- Body
- Ramus

Pubis
- Superior ramus
- Inferior ramus
- Body
- Pubic symphysis

- Base of sacrum
- Sacroiliac joint
- Pelvic surface of sacrum
- Pelvic inlet
- Coccyx
- Acetabulum
- Interpubic disc
- Obturator foramen

(a) Anterosuperior view

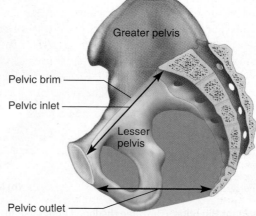

- Greater pelvis
- Pelvic brim
- Pelvic inlet
- Lesser pelvis
- Pelvic outlet

(b) Median section

Figure 8.6 **The Pelvic Girdle.** (a) Anterosuperior view, tilted slightly toward the viewer to show the base of the sacrum and the pelvic inlet. (b) Median section to show the greater and lesser pelvis and the pelvic inlet and outlet.
● *How is this anatomy related to the existence of fontanels in the skull of the newborn infant?*

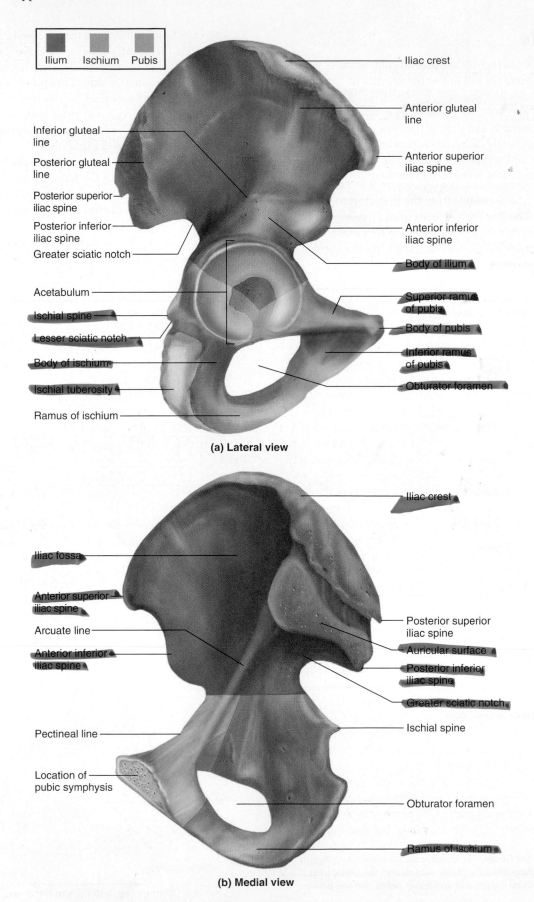

Ilium   Ischium   Pubis

Iliac crest

Inferior gluteal line

Posterior gluteal line

Posterior superior iliac spine

Posterior inferior iliac spine

Greater sciatic notch

Acetabulum

Ischial spine

Lesser sciatic notch

Body of ischium

Ischial tuberosity

Ramus of ischium

Anterior gluteal line

Anterior superior iliac spine

Anterior inferior iliac spine

Body of ilium

Superior ramus of pubis

Body of pubis

Inferior ramus of pubis

Obturator foramen

(a) Lateral view

Iliac fossa

Anterior superior iliac spine

Arcuate line

Anterior inferior iliac spine

Pectineal line

Location of pubic symphysis

Iliac crest

Posterior superior iliac spine

Auricular surface

Posterior inferior iliac spine

Greater sciatic notch

Ischial spine

Obturator foramen

Ramus of ischium

(b) Medial view

Figure 8.7 The Right Hip Bone. The three childhood bones that fuse to form the adult hip bone are identified by color.

The hip bones have three distinctive features that will serve as landmarks for further description. These are the **iliac**[30] **crest** (superior crest of the hip); **acetabulum**[31] (ASS-eh-TAB-you-lum) (the hip socket—named for its resemblance to vinegar cups used on ancient Roman dining tables); and **obturator**[32] **foramen** (a large round-to-triangular hole below the acetabulum, closed in life by a ligament called the *obturator membrane*).

The adult hip bone forms by the fusion of three childhood bones called the *ilium* (ILL-ee-um), *ischium* (ISS-kee-um), and *pubis* (PEW-biss), identified by color in figure 8.7. The largest of these is the **ilium,** which extends from the iliac crest to the center of the acetabulum. The iliac crest extends from an anterior point or angle called the **anterior superior iliac spine** to a sharp posterior angle called the **posterior superior iliac spine.** In a lean person, the anterior spines form visible anterior protrusions at a point where the pockets usually open on a pair of pants, and the posterior spines are sometimes marked by dimples above the buttocks where connective tissue attached to the spines pulls inward on the skin (see fig. A.15b , p. 343).

Below the superior spines are the **anterior** and **posterior inferior iliac spines.** Below the latter is a deep **greater sciatic**[33] (sy-AT-ic) **notch,** named for the sciatic nerve that passes through it and continues down the posterior side of the thigh.

The posterolateral surface of the ilium is relatively rough-textured because it serves for attachment of several muscles of the buttocks and thighs. The anteromedial surface, by contrast, is the smooth, slightly concave **iliac fossa,** covered in life by the broad *iliacus muscle.* Medially, the ilium exhibits an auricular surface that matches the one on the sacrum, so the two bones form the sacroiliac joint.

The **ischium**[34] forms the inferoposterior portion of the hip bone. Its heavy **body** is marked with a prominent **spine.** Inferior to the spine is a slight indentation, the **lesser sciatic notch,** and then the thick, rough-surfaced **ischial tuberosity,** which supports your body when you are sitting. The tuberosity can be palpated by sitting on your fingers. The **ramus** of the ischium joins the inferior ramus of the pubis anteriorly.

The **pubis (pubic bone)** is the most anterior portion of the hip bone. It has a **superior** and **inferior ramus** and a triangular **body.** The body of one pubis meets the body of the other at the pubic symphysis. The pubis and ischium encircle the obturator foramen. The pubis is sometimes fractured when the pelvis is subjected to violent anteroposterior compression, as in seat-belt injuries.

Human pelvic anatomy is a compromise between two requirements—bipedalism and childbirth. In apes and other quadrupedal (four-legged) mammals, the abdominal viscera are supported by the muscular wall of the abdomen. In humans, however, they bear down on the floor of the pelvic cavity, and a bowl-shaped pelvis helps support their weight. This has resulted in a narrower pelvic outlet—a condition that creates pain and difficulty in giving birth to the large-brained infants of our species. Humans must be born before the cranial bones fuse so that the head can squeeze through the outlet. This is thought to be the reason why our infants are born in such an immature state compared to those of other primates.

The largest muscle of the buttock, the *gluteus maximus,* serves in chimpanzees and other apes primarily as an abductor of the thigh—that is, it moves the lower limb laterally. In humans, however, the ilium has expanded posteriorly, so the gluteus maximus originates behind the hip joint. This changes the function of the muscle—instead of abducting the thigh, it pulls it back in the second half of a stride (pulling back on your right thigh, for example, when your left foot is off the ground and swinging forward). When one is standing, the gluteus maximus also pulls the pelvis posteriorly so that the body's weight is better balanced on the lower limbs. This helps us to stand upright without expending a great deal of energy to keep from falling forward. The posterior growth of the ilium is the reason the greater sciatic notch is so deeply concave (fig. 8.8). The ilium also curves farther anteriorly in humans than in apes. This positions other gluteal muscles (the *gluteus medius* and *minimus*) in front of the hip joint, where they rotate and balance the trunk so we do not sway from side to side like chimpanzees do when they walk bipedally. These evolutionary changes in the pelvis and associated muscles account for the smooth, efficient stride of a human as compared to the awkward, shuffling gait of a chimpanzee or gorilla when it walks upright.

The pelvis is the most *sexually dimorphic* part of the skeleton—that is, the one whose anatomy most differs between the sexes. In identifying the sex of skeletal remains, forensic scientists focus especially on the pelvis. The average male pelvis is more robust (heavier and thicker) than the female's owing to the forces exerted on the bone by stronger muscles. The female pelvis is adapted to the needs of pregnancy and childbirth. It is wider and shallower and has a larger pelvic inlet and outlet for passage of the infant's head. Table 8.1 and figure 8.9 summarize the most useful features of the pelvis in sex identification.

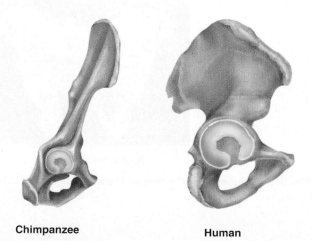

Chimpanzee   **Human**

Figure 8.8 Chimpanzee and Human Hip Bones. Lateral views of the right hip bone. The human ilium forms a more bowl-like greater pelvis and is expanded posteriorly (toward page left) so that the gluteus maximus muscle produces an effective backswing of the thigh during the stride.

---

[30]*ili* = flank, loin + *ac* = pertaining to
[31]*acetabulum* = vinegar cup
[32]*obtur* = to close, stop up + *ator* = that which
[33]*sciat* = hip or ischium + *ic* = pertaining to
[34]*ischium* = hip

TABLE 8.1	Comparison of the Male and Female Pelvis	
Feature	**Male**	**Female**
General appearance	More massive; rougher; heavier processes	Less massive; smoother; more delicate processes
Tilt	Upper end of pelvis relatively vertical	Upper end of pelvis tilted forward
Ilium	Deeper; projects farther above sacroiliac joint	Shallower; does not project as far above sacroiliac joint
Sacrum	Narrower and deeper	Wider and shallower
Coccyx	Less movable; more vertical	More movable; tilted posteriorly
Width of greater pelvis	Anterior superior spines closer together; hips less flared	Anterior superior spines farther apart; hips more flared
Pelvic inlet	Heart-shaped	Round or oval
Pelvic outlet	Smaller	Larger
Pubic symphysis	Taller	Shorter
Greater sciatic notch	Narrower	Wider
Obturator foramen	Round	Triangular to oval
Acetabulum	Larger, faces more laterally	Smaller, faces slightly anteriorly
Pubic arch	Usually 90° or less	Usually greater than 100°

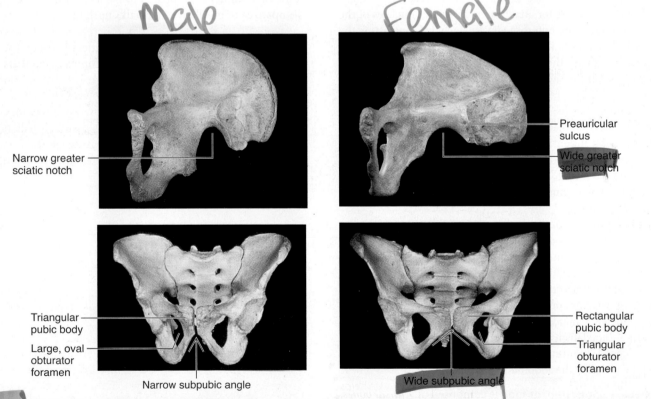

Figure 8.9 **Comparison of the Male and Female Pelvic Girdles.** Top: Medial views. Bottom: Anterior views. Left: Male. Right: Female. Compare with table 8.1.

Lower Limb

The number and arrangement of bones in the lower limb are similar to those of the upper limb. In the lower limb, however, they are adapted for weight-bearing and locomotion and are therefore shaped and articulated differently. The lower limb is divided into four regions containing a total of 30 bones per limb:

1. The **femoral region,** or thigh, extends from hip to knee and contains the *femur.* The *patella* (kneecap) is a sesamoid bone at the junction of the femoral and crural regions.

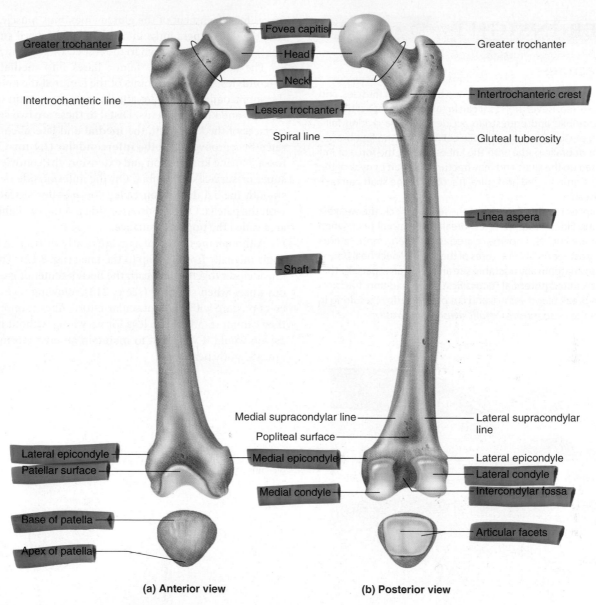

Greater trochanter

Fovea capitis

Head

Neck

Intertrochanteric line

Lesser trochanter

Spiral line

Greater trochanter

Intertrochanteric crest

Gluteal tuberosity

Linea aspera

Shaft

Medial supracondylar line

Popliteal surface

Lateral supracondylar line

Lateral epicondyle

Patellar surface

Medial epicondyle

Lateral epicondyle

Lateral condyle

Medial condyle

Intercondylar fossa

Base of patella

Apex of patella

Articular facets

(a) Anterior view   **(b) Posterior view**

Figure 8.10 **The Right Femur and Patella.**
● *How many articular surfaces can you identify on the femur? Where are they?*

2. The **crural** (CROO-rul) **region,** or leg proper, extends from knee to ankle and contains two bones, the medial *tibia* and lateral *fibula.*

3. The **tarsal region (tarsus),** or ankle, is the union of the crural region with the foot. The tarsal bones are treated as part of the foot.

4. The **pedal region (pes),** or foot, is composed of 7 *tarsal bones,* 5 *metatarsals,* and 14 *phalanges* in the toes.

Femur

The **femur** (FEE-mur) is the longest and strongest bone of the body (fig. 8.10). It has a hemispherical head that articulates with the ac-etabulum of the pelvis, forming a quintessential *ball-and-socket joint.*

A ligament extends from the acetabulum to a pit, the **fovea capitis**[35] (FOE-vee-uh CAP-ih-tiss), in the head of the femur. Distal to the head is a constricted **neck,** a common site of femoral fractures (see Deeper Insight 8.2 and fig. 8.11). Just beyond the neck are two massive, rough processes called the **greater** and **lesser trochanters**[36] (tro-CAN-turs), which are insertions for the powerful muscles of the hip. The tro-chanters are connected on the posterior side by a thick oblique ridge of bone, the **intertrochanteric crest,** and on the anterior side by a more delicate **intertrochanteric line.**

The primary feature of the shaft is a posterior ridge called the **linea aspera**[37] (LIN-ee-uh ASS-peh-ruh). At its upper end, the linea aspera forks into a medial **spiral line** and a lateral **gluteal tuberosity.** The gluteal tuberosity is a rough ridge (sometimes a depression) that

---

[35]*fovea* = pit + *capitis* = of the head

[36]*trochanter* = to run

[37]*linea* = line + *asper* = rough

DEEPER INSIGHT   8.2

Femoral Fractures

The femur is a very strong bone, well guarded by the thigh muscles, and it is not often fractured. Nevertheless, it can break in high-impact trauma suffered in automobile and equestrian accidents, figure skating falls, and so forth. If a person in an automobile collision has the feet braced against the floor or brake pedal with the knees locked, the force of impact is transmitted up the shaft and may fracture the shaft or neck of the femur (fig. 8.11). Comminuted and spiral fractures of the shaft can take up to a year to heal.

A "broken hip" is usually a fracture of the femoral neck, the weakest part of the femur. Elderly people often break the femoral neck when they stumble or are knocked down—especially women whose femurs are weakened by osteoporosis. Fractures of the femoral neck heal poorly because this is an anatomically unstable site and it has an especially thin periosteum with limited potential for ossification. In addition, fractures in this site often break blood vessels and cut off blood flow, resulting in degeneration of the head *(posttraumatic avascular necrosis).*

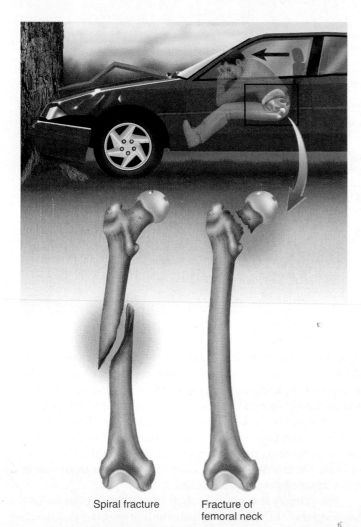

Spiral fracture   Fracture of
                       femoral neck

Figure 8.11 **Fractures of the Femur.** Violent trauma, as in automobile accidents, may cause spiral fractures of the femoral shaft. The femoral neck often fractures in elderly people as a result (or cause) of falls.

serves for attachment of the gluteus maximus muscle. At its lower end, the linea aspera forks into **medial** and **lateral supracondylar lines,** which continue down to the respective epicondyles.

The distal end of the femur flares into **medial** and **lateral epicondyles,** the widest points of the femur at the knee. These and the supracondylar lines are attachments for certain thigh and leg muscles and knee ligaments. Distal to these are two smooth round surfaces of the knee joint, the **medial** and **lateral condyles,** separated by a groove called the **intercondylar** (IN-tur-CON-dih-lur) **fossa.** During knee flexion and extension, the condyles rock on the superior surface of the tibia. On the anterior side of the femur, a smooth medial depression called the **patellar surface** articulates with the patella. On the posterior side is a flat or slightly depressed area called the **popliteal surface.**

Although the femurs of apes are nearly vertical, in humans they angle medially from the hip to the knee (fig. 8.12). This places our knees closer together, beneath the body's center of gravity. We lock our knees when standing (see p. 213), allowing us to maintain an erect posture with little muscular effort. Apes cannot do this, and they cannot stand on two legs for very long without tiring—much as you would if you tried to maintain an erect posture with your knees slightly bent.

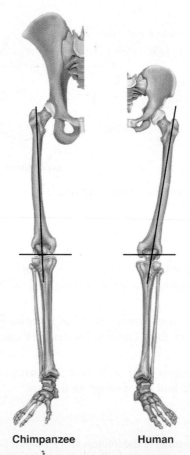

Chimpanzee   **Human**

Figure 8.12 **Adaptation of the Lower Limb for Bipedalism.** In contrast to chimpanzees, which are quadrupedal, humans have the femurs angled medially so the knees are more nearly below the body's center of gravity.

Patella

The **patella,**[38] or kneecap (see fig. 8.10), is a roughly triangular sesamoid bone embeded in the tendon of the knee. It is cartilaginous at birth and ossifies at 3 to 6 years of age. It has a broad superior **base,** a pointed inferior **apex,** and a pair of shallow **articular facets** on its posterior surface where it articulates with the femur. The lateral facet is usually larger than the medial. The *quadriceps femoris tendon* extends from the anterior *quadriceps femoris muscle* of the thigh to the patella, then continues as the *patellar ligament* from the patella to the tibia. This is a change in terminology more than a change in structure or function, as a tendon connects muscle to bone and a ligament connects bone to bone.

Tibia

The leg has two bones—a thick strong tibia (TIB-ee-uh) on the medial side and a slender fibula (FIB-you-luh) on the lateral side (fig. 8.13). The **tibia**[39] is the only weight-bearing bone of the crural region. Its broad superior head has two fairly flat articular surfaces, the **medial** and **lateral condyles,** separated by a ridge called the **intercondylar eminence.** The condyles of the tibia articulate with those of the femur. The rough anterior surface of the tibia, the **tibial tuberosity,** can be palpated just below the patella. This is where the patellar ligament inserts and the quadriceps muscle exerts its pull when it extends the leg. Distal to this, the shaft has a sharply angular **anterior crest,** which can be palpated in the shin region. At the ankle, just above the rim of a standard dress shoe, you can palpate a prominent bony knob on each side. These are the **medial** and **lateral malleoli**[40] (MAL-ee-OH-lie). The medial malleolus is part of the tibia, and the lateral malleolus is part of the fibula.

Fibula

The **fibula**[41] (fig. 8.13) is a slender lateral strut that helps to stabilize the ankle. It does not bear any of the body's weight; indeed, orthopedic surgeons sometimes remove part of the fibula and use it to replace damaged or missing bone elsewhere in the body. The fibula is somewhat thicker and broader at its proximal end, the **head,** than at the distal end. The point of the head is called the **apex.** The distal expansion is the lateral malleolus. The fibula is joined to the tibia by a ligament called the *interosseous membrane,* similar to the one that joins the ulna and radius, and by shorter ligaments at the superior and inferior ends where the head and apex contact the tibia.

[38]*pat* = pan + *ella* = little
[39]*tibia* = shinbone

[40]*malle* = hammer + *olus* = little
[41]*fib* = pin + *ula* = little

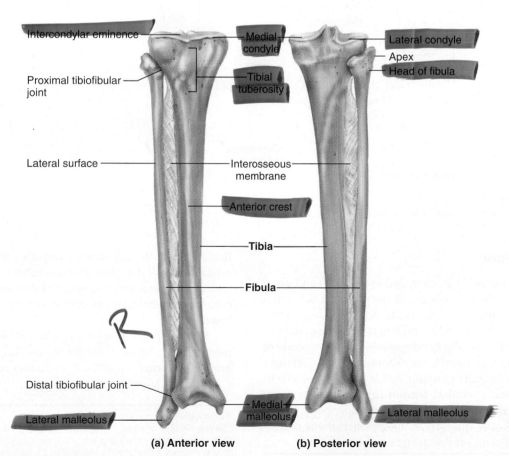

(a) Anterior view   **(b) Posterior view**

Figure 8.13 The Right Tibia and Fibula.

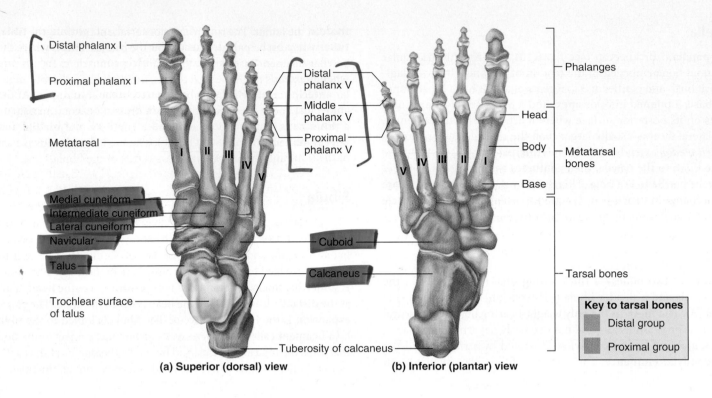

(a) Superior (dorsal) view

(b) Inferior (plantar) view

Key to tarsal bones
Distal group
Proximal group

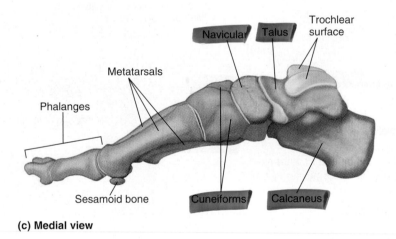

(c) Medial view

Figure 8.14 The Right Foot.
● *In part (c), what one bone is least likely to be present in a year-old infant? Explain.*

The Ankle and Foot

The **tarsal bones** of the ankle are arranged in proximal and distal groups somewhat like the carpal bones of the wrist (fig. 8.14). Because of the load-bearing role of the ankle, however, their shapes and arrangement are conspicuously different from those of the carpal bones, and they are thoroughly integrated into the structure of the foot. The largest tarsal bone is the **calcaneus**[42] (cal-CAY-nee-us), which forms the heel. Its posterior end is the point of attachment for the **calcaneal (Achilles) tendon** from the calf muscles. The second-largest tarsal bone, and the most superior, is the **talus.**[43] It has three articular surfaces: an inferoposterior one that ar-

ticulates with the calcaneus, a superior **trochlear**[44] **surface** that articulates with the tibia, and an anterior surface that articulates with a short, wide tarsal bone called the **navicular.**[45] The talus, calcaneus, and navicular are considered the proximal row of tarsal bones.

The distal group forms a row of four bones. Proceeding from medial side to lateral, these are the **medial, intermediate,** and **lateral cuneiforms**[46] (cue-NEE-ih-forms) and the **cuboid.**[47] The cuboid is the largest.

---

[42]*calc* = stone, chalk

[43]*talus* = ankle
[44]*trochle* = pully + *ar* = like
[45]*nav* = boat + *icul* = little + *ar* = like
[46]*cunei* = wedge + *form* = in the shape of
[47]*cub* = cube + *oid* = shaped

Apply What You Know

The upper and lower limbs each contain 30 bones, yet we have 8 carpal bones in the upper limb and only 7 tarsal bones in the lower limb. What makes up the difference in the lower limb?

The remaining bones of the foot are similar in arrangement and name to those of the hand. The proximal **metatarsals**[48] are similar to the metacarpals. They are **metatarsals I** through **V** from medial to lateral, metatarsal I being proximal to the great toe. Note that roman numeral I represents the *medial* group of bones in the foot but the *lateral* group in the hand. In both cases, however, it refers to the largest digit of the limb (see Deeper Insight 8.3). Metatarsals I through III articulate with the three cuneiforms; metatarsals IV and V articulate with the cuboid.

Bones of the toes, like those of the fingers, are called phalanges. The great toe is the **hallux** and contains only two bones, the proximal and distal phalanx I. The other toes each contain a proximal,

[48]*meta* = beyond + *tars* = ankle

middle, and distal phalanx. The metatarsal and phalangeal bones each have a base, body, and head, like the bones of the hand. All of them, especially the phalanges, are slightly concave on the inferior (plantar) side.

As important as the hand has been to human evolution, the foot may be an even more significant adaptation. Unlike other mammals, humans support their entire body weight on two feet. The tarsal bones are tightly articulated with each other, and the calcaneus is strongly developed. The hallux is not opposable as it is in most Old World monkeys and apes (fig. 8.15)—that is, humans cannot effectively grasp objects with the great toe—but it is highly developed so that it provides the "toe-off" that pushes the body forward in the last phase of the stride. For this reason, loss of the hallux has a more crippling effect than the loss of any other toe.

Although apes are flat-footed, humans have strong, springy foot arches that absorb shock as the body jostles up and down during walking and running (fig. 8.16). The **medial longitudinal arch,** which essentially extends from heel to hallux, is formed from the calcaneus, talus, navicular, cuneiforms, and metatarsals I through III. The **lateral longitudinal arch** extends from heel to little toe and includes the calcaneus, cuboid, and metatarsals IV and V. The

DEEPER INSIGHT  8.3

Anatomical Position—Clinical and Biological Perspectives

It may seem puzzling that we count metacarpal bones I through V from lateral to medial, but count metatarsal bones I through V from medial to lateral. This minor point of confusion is the legacy of a committee of anatomists who met in the early 1900s to define anatomical position. A controversy arose as to whether the arms should be presented with the palms forward or facing the rear in anatomical position. Veterinary anatomists argued that palms to the rear (forearms pronated) would be a more natural position comparable to forelimb orientation in other animals. It is more comfortable to stand with the forearms pronated, and when a child crawls on all fours, he or she does so with the palms on the floor, in the pronated position. In this animal-like stance, the largest digits (thumbs and great toes) are medial on all four limbs. Human clinical anatomists, however, argued that if you ask patients to "show me your arms" or "show me your hands," most present the palms forward or upward—that is, supinated. The clinical anatomists won the debate, forcing us, the heirs to this terminology, to number the hand and foot bones in a biologically less rational order.

A few other anatomical terms also reflect less than perfect logic. The *dorsum* of the foot is its superior surface—it does not face dorsally—and the *dorsal artery* and *dorsal nerve* of the penis lie along the surface that faces anteriorly (ventrally) (see fig. 26.10, p. 716). In a cat, dog, or other quadrupedal mammal, however, the dorsum of the foot and the dorsal artery and nerve of the penis do face dorsally (upward). However illogical some of these terms may seem, we inherit them from comparative anatomy and the habit of naming human structures after the corresponding structures in other species.

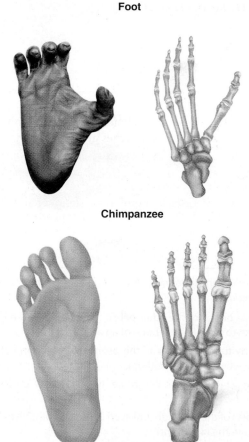

Figure 8.15 **Some Adaptations of the Foot for Bipedalism.** The great toe (hallux) of the chimpanzee is prehensile—able to encircle and grasp objects. The human great toe is nonprehensile but is more robust and is adapted for the toe-off part of the long bipedal stride.

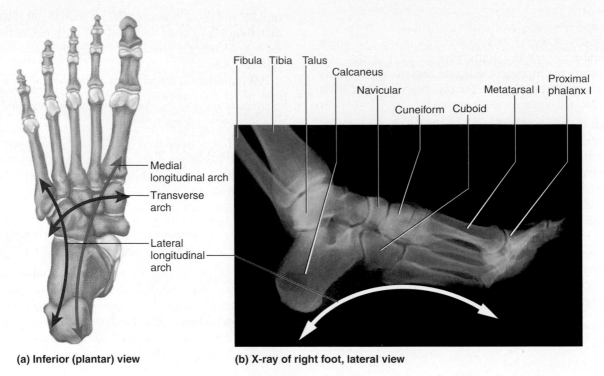

(a) Inferior (plantar) view

Medial longitudinal arch

Transverse arch

Lateral longitudinal arch

(b) X-ray of right foot, lateral view

Fibula   Tibia   Talus

Calcaneus

Navicular

Cuneiform   Cuboid

Metatarsal I

Proximal phalanx I

Figure 8.16 Arches of the Foot.

transverse arch includes the cuboid, cuneiforms, and proximal heads of the metatarsals. These arches are held together by short, strong ligaments. Excessive weight, repetitive stress, or congenital weakness of these ligaments can stretch them, resulting in *pes planus* (commonly called flat feet or fallen arches). This condition makes a person less tolerant of prolonged standing and walking.

Before You Go On

Answer the following questions to test your understanding of the preceding section:

7. Name the bones of the adult pelvic girdle. What three bones of a child fuse to form the hip bone of an adult?

8. Name any four structures of the pelvis that you can palpate, and describe where to palpate them.

9. Describe several ways in which the male and female pelvic girdles differ.

10. What parts of the femur are involved in the hip joint? What parts are involved in the knee joint?

11. Name the prominent knobs on each side of your ankle. What bones contribute to these structures?

12. Name all the bones that articulate with the talus and describe the location of each.

13. Describe several ways in which the human and ape pelvis and hindlimb differ, and the functional reason for the differences.

8.3 Developmental and Clinical Perspectives

▶ Expected Learning Outcomes

When you have completed this section, you should be able to

• describe the pre- and postnatal development of the appendicular skeleton; and

• describe some common disorders of the appendicular skeleton.

Development of the Appendicular Skeleton

With few exceptions, the bones of the limbs and their girdles form by endochondral ossification. Exceptions include the clavicle, which forms primarily by intramembranous ossification, and sesamoid bones (the pisiform and patella). Although limb and limb girdle ossification is well under way at the time of birth, it is not completed until a person is in his or her 20s.

The first sign of limb development is the appearance of upper **limb buds** around day 26 to 27 and lower limb buds 1 or 2 days later. A limb bud consists of a core of mesenchyme covered with ectoderm. The limb buds elongate as the mesenchyme proliferates. The distal ends of the limb buds flatten into paddlelike **hand** and **foot plates.** By day 38 in the hand and day 44 in the foot, these plates show parallel ridges called **digital rays,** the future fingers and toes. The mesenchyme between the digital rays breaks down by apoptosis, forming notches between the rays that deepen until, at the end of week 8, the digits are well separated.

Condensed mesenchymal models of the future limb bones first appear during week 5. Chondrification is apparent by the end of that week, and by the end of week 6, a complete cartilaginous limb skeleton is present. The long bones begin to ossify in the following week, in the manner described in chapter 6 (see fig. 6.10, p. 139). The humerus, radius, ulna, femur, and tibia develop primary ossification centers in weeks 7 to 8; the scapula and ilium in week 9; the metacarpals, metatarsals, and phalanges over the next 3 weeks; and the ischium and pubis in weeks 15 and 20, respectively. The clavicle ossifies intramembranously beginning early in week 7.

The carpal bones are still cartilaginous at birth. Some of them ossify as early as 2 months of age (the capitate) and some as late as 9 years (the pisiform). Among the tarsal bones, the calcaneus and talus begin to ossify prenatally at 3 and 6 months, respectively; the cuboid begins to ossify just before or after birth; and the cuneiforms do not ossify until 1 to 3 years of age. The patella ossifies at 3 to 6 years. The epiphyses of the long bones are cartilaginous at birth, with their secondary ossification centers just beginning to form. The epiphyseal plates persist until about age 20, at which time the epiphysis and diaphysis fuse and bone elongation ceases. The ilium, ischium, and pubis are not fully fused into a single hip bone until the age of 25.

It may seem curious that the largest digits (the thumb and great toe) are lateral in the hand but medial in the foot, and that the elbows and knees flex in opposite directions. This results from a rotation of the limbs that occurs in week 7. Early that week, the limbs extend anteriorly from the body, the hand plate shows the first traces of separation of the finger buds, and the foot is still a relatively undifferentiated foot plate (fig. 8.17a). The future thumb and great toe are both directed superiorly, and the future palms and soles face each other medially. But then the limbs rotate about 90° in opposite directions. The upper limb rotates laterally. To visualize this, hold your hands straight out in front of you with the palms facing each other as if you were about to clap. Then rotate your forearms so the thumbs face away from each other (laterally) and the palms face upward. The lower limbs rotate in the opposite direction, medially, so that the soles face downward and the great toes become medial. Thus, the thumb and great toe end up on opposite sides of the hand and foot (fig. 8.17b). This rotation also explains why the elbow and knee flex in opposite directions, and why (as you will see in chapter 12) the muscles that flex the elbow are on the anterior side of the arm, but those that flex the knee are on the posterior side of the thigh.

Pathology of the Appendicular Skeleton

The appendicular skeleton is subject to several developmental abnormalities, occurring in as many as 2 out of 1,000 live births. The most striking is **amelia,**[49] a complete absence of one or more limbs.

---

[49]*a* = without + *melia* = limb

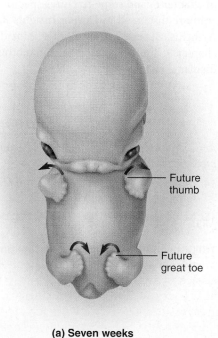

(a) Seven weeks

— Future thumb

— Future great toe

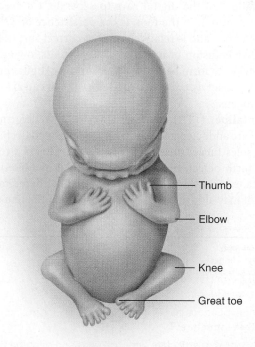

(b) Eight weeks

— Thumb

— Elbow

— Knee

— Great toe

Figure 8.17 **Embryonic Limb Rotation.** In the seventh week of development, the forelimbs and hindlimbs of the embryo rotate about 90° in opposite directions. This explains why the largest digits (digit I) are on opposite sides of the hand and foot, and why the elbow and knee flex in opposite directions.

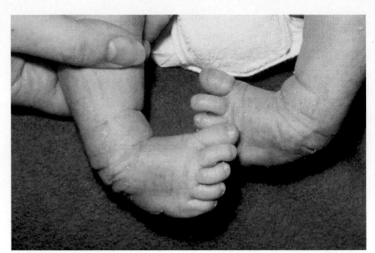

(a) Polydactyly

(b) Talipes (clubfoot)

Figure 8.18 Congenital Deformities of the Hands and Feet.

The partial absence of a limb is called **meromelia.**[50] Meromelia typically entails an absence of the long bones, with rudimentary hands or feet attached directly to the trunk (see fig. 4.15, p. 101). Such defects are often accompanied by deformities of the heart, urogenital system, or craniofacial skeleton. These abnormalities are usually hereditary, but they can be induced by teratogenic chemicals such as thalidomide. The limbs are most vulnerable to teratogens in the fourth and fifth weeks of development.

Another class of developmental limb disorders includes **polydactyly,**[51] the presence of extra fingers or toes (fig. 8.18a), and **syndactyly,**[52] the fusion of two or more digits. The latter results from a failure of the digital rays to separate. *Cutaneous syndactyly,* most common in the foot, is the persistence of a skin web between the digits, and is relatively easy to correct surgically. *Osseous syndactyly* is fusion of the bones of the digits, owing to failure of the notches to form between the embryonic digital rays. Polydactyly and syndactyly are usually hereditary, but can also be induced by teratogens.

Clubfoot, or **talipes**[53] (TAL-ih-peez) is a congenital deformity in which the feet are adducted and plantar flexed (defined in chapter 9), with the soles turned medially (fig. 8.18b). This is a relatively common birth defect, present in about 1 out of 1,000 live births, but the cause remains obscure. It is sometimes hereditary, and some think it may also result from malposition of the fetus in

the uterus, but the latter hypothesis remains unproven. Children with talipes cannot support their weight on their feet and tend to walk on their ankles. In some cases, talipes requires that the foot be manipulated and set in a new cast every week beginning in the neonatal nursery and lasting 4 to 6 months. Some cases require surgery at 6 to 9 months to release tight ligaments and tendons and realign the foot.

The most common noncongenital disorders of the appendicular skeleton are osteoporosis, fractures, dislocations, and arthritis. Even though these disorders can affect the axial skeleton as well, they are more common and more often disabling in the appendicular skeleton. The general classification of bone fractures is discussed in chapter 6, and some fractures specific to the appendicular skeleton are discussed in Deeper Insights 8.1 and 8.2 of this chapter. Arthritis, dislocation, and other joint disorders are described in chapter 9. Table 8.2 describes some other disorders of the appendicular skeleton.

Before You Go On

Answer the following questions to test your understanding of the preceding section:

14. Describe the progression from a limb bud to a hand with fully formed and separated fingers.

15. Name some appendicular bones that do not ossify until a person is at least a few years old.

16. Distinguish between amelia and meromelia, and between polydactyly and syndactyly.

---

[50]*mero* = part + *melia* = limb
[51]*poly* = many + *dactyl* = finger
[52]*syn* = together + *dactyl* = finger
[53]*tali* = heel + *pes* = foot

TABLE 8.2	Disorders of the Appendicular Skeleton
Avulsion	A fracture in which a body part, such as a finger, is completely torn from the body, as in many accidents with farm and factory machinery. The term can also refer to nonosseous structures such as the avulsion of an ear.
Calcaneal (heel) spurs	Abnormal outgrowths of the calcaneus. Often results from high-impact exercise such as aerobics and running, especially if done with inappropriate footwear. Stress on the plantar aponeurosis (a connective tissue sheet in the sole of the foot) stimulates exostosis, or growth of a bony spur, and can cause severe foot pain.
Colles[54] fracture	Pathological fracture at the distal end of the radius and ulna, often occurring when stress is placed on the wrist (as in pushing oneself up from an armchair) and the bones have been weakened by osteoporosis.
Epiphyseal fracture	Separation of the epiphysis from the diaphysis of a long bone. Common in children and adolescents because of their cartilaginous epiphyseal plates. May present a threat to normal completion of bone growth.
Pes planus[55]	"Flat feet" or "fallen arches" (absence of visible arches) in adolescents and adults. Caused by stretching of plantar ligaments due to prolonged standing or excess weight.
Pott[56] fracture	Fracture of the distal end of the tibia, fibula, or both; a sports injury common in football, soccer, and snow skiing.

Disorders Described Elsewhere

Fracture of the clavicle   184   Fracture of the femur   194

---

[54]Abraham Colles (1773–1843), Irish surgeon
[55]*pes* = foot + *planus* = flat
[56]Sir Percivall Pott (1713–88), British surgeon

Study Guide

Assess Your Learning Outcomes

You should have a good understanding of this chapter if you can accurately address the following issues.

8.1 The Pectoral Girdle and Upper Limb (p. 184)

1. The function of the pectoral girdle; the bones that compose it; and all points at which these bones articulate with each other, with the upper limb, and with the axial skeleton
2. The anatomical features of the clavicle and scapula
3. The four segments (regions) of the upper limb
4. The names and locations of all 30 bones of the upper limb, and all points at which they articulate with each other
5. The anatomical features of the humerus, radius, ulna, carpal bones (especially the hamate), metacarpal bones, and phalanges
6. How the upper limb is anatomically adapted to the bipedalism of humans

8.2 The Pelvic Girdle and Lower Limb (p. 188)

1. The distinction between the pelvic girdle and pelvis

2. The function of the pelvic girdle; the bones that compose it; and all points at which these bones articulate with each other, with the lower limb, and with the axial skeleton
3. The anatomical features of the hip bone, and the names and boundaries of the three childhood bones from which it arises
4. The meaning of *greater pelvis, lesser pelvis, pelvic brim, pelvic inlet,* and *pelvic outlet,* and the relationship of these structures to pregnancy and childbirth
5. How the pelvic girdle and gluteal muscles are anatomically adapted to the bipedalism of humans
6. Differences between the male and female pelvic girdles
7. The four segments (regions) of the lower limb
8. The names and locations of all 30 bones of the lower limb, and all points at which they articulate with each other
9. The anatomical features of the femur, patella, tibia, fibula, talus, calcaneus, navicular, cuboid, cuneiforms, metatarsals, and phalanges
10. The names and landmarks of the three foot arches

11. How the lower limb is adapted to the bipedalism of humans, including the angle of the femurs, locking of the knees, the great toe, and the foot arches

8.3 Developmental and Clinical Perspectives (p. 198)

1. What portions of the appendicular skeleton are formed by intramembranous and endochondral ossification
2. Development of the limb buds and the manner in which they differentiate into the limbs, especially the processes of hand and foot development
3. The developmental processes that result in opposite directions of knee and elbow flexion and opposite orientations of the pollex and hallux
4. Developmental abnormalities of the appendicular skeleton including amelia, meromelia, polydactyly, syndactyly, and talipes
5. The most common noncongenital disorders of the appendicular skeleton

Testing Your Recall

1. The hip bone is attached to the axial skeleton through its
   a. auricular surface.
   b. articular cartilage.
   c. pubic symphysis.
   d. conoid tubercle.
   e. coronoid process.

2. Which of these bones supports the most body weight?
   a. ilium
   b. pubis
   c. femur
   d. tibia
   e. talus

3. Which of these structures can be most easily palpated on a living person?
   a. the deltoid tuberosity
   b. the greater sciatic notch
   c. the medial malleolus
   d. the coracoid process of the scapula
   e. the glenoid cavity

4. Compared to the male pelvis, the pelvis of a female
   a. has a less movable coccyx.
   b. has a rounder pelvic inlet.
   c. is narrower between the iliac crests.
   d. has a narrower pubic arch.
   e. has a narrower sacrum.

5. The lateral and medial malleoli are most similar to
   a. the radial and ulnar styloid processes.
   b. the humeral capitulum and trochlea.
   c. the acromion and coracoid process.
   d. the base and head of a metacarpal bone.
   e. the anterior and posterior superior iliac spines.

6. When you rest your hands on your hips, you are resting them on
   a. the pelvic inlet.
   b. the pelvic outlet.
   c. the pelvic brim.
   d. the iliac crests.
   e. the auricular surfaces.

7. The disc-shaped head of the radius articulates with the _____ of the humerus.
   a. radial tuberosity
   b. trochlea
   c. capitulum
   d. olecranon process
   e. glenoid cavity

8. All of the following are carpal bones *except* the _____, which is a tarsal bone.
   a. trapezium
   b. cuboid
   c. trapezoid
   d. triquetrum
   e. pisiform

9. The bone that supports your body weight when you are sitting down is
   a. the acetabulum.
   b. the pubis.
   c. the ilium.
   d. the coccyx.
   e. the ischium.

10. Which of these is the bone of the heel?
    a. cuboid
    b. calcaneus
    c. navicular
    d. trochlear
    e. talus

11. The Latin anatomical name for the thumb is _____, and the name for the great toe is _____.

12. The acromion and coracoid process are parts of what bone?

13. How many phalanges, total, does the human body have?

14. The bony prominences on each side of your elbow are the lateral and medial _____ of the humerus.

15. One of the wrist bones, the _____, is characterized by a prominent hook.

16. The fibrocartilage pad that holds the pelvic girdle together anteriorly is called the _____.

17. The leg proper, between the knee and ankle, is called the _____ region.

18. The _____ processes of the radius and ulna form bony protuberances on each side of the wrist.

19. Two massive protuberances unique to the proximal end of the femur are the greater and lesser _____.

20. The _____ arch of the foot extends from the heel to the great toe.

Answers in the Appendix

Building Your Medical Vocabulary

State a medical meaning for each of the following word elements, and give a term in which it is used.

1. pect-
2. acro-
3. -icle
4. supra-
5. carpo-
6. capit-
7. -ulum
8. meta-
9. auro-
10. tarso-

Answers in the Appendix

True or False

Determine which five of the following statements are false, and briefly explain why.

1. There are more carpal bones than tarsal bones.
2. The hands have more phalanges than the feet.
3. The upper limb is attached to the axial skeleton at only one point, the acromioclavicular joint.
4. On a living person, it would be possible to palpate the muscles in the infraspinous fossa but not those of the subscapular fossa.
5. In strict anatomical terminology, the words *arm* and *leg* both refer to regions with only one bone.
6. If you rest your chin on your hands with your elbows on a table, the olecranon of the ulna rests on the table.
7. The most frequently broken bone in humans is the humerus.
8. The proximal end of the radius articulates with both the humerus and ulna.
9. The pisiform bone and patella are both sesamoid bones.
10. The pelvic outlet is the opening in the floor of the greater pelvis leading into the lesser pelvis.

Answers in the Appendix

Testing Your Comprehension

1. In adolescents, trauma sometimes separates the head of the femur from the neck. Why do you think this is more common in adolescents than in adults?
2. By palpating the hind leg of a cat or dog or examining a laboratory skeleton, you can see that cats and dogs stand on the heads of their metatarsal bones; the calcaneus does not touch the ground. How is this similar to the stance of a woman wearing high-heeled shoes? How is it different?
3. A deer hunter discovers a human skeleton in the woods and notifies authorities. A news report on the finding describes it as the body of an unidentified male between 17 and 20 years of age. What skeletal features would have been most useful for determining the sex and approximate age of the individual?
4. A surgeon has removed 8 cm of Joan's radius because of osteosarcoma, a bone cancer, and replaced it with a graft taken from one of the bones of Joan's lower limb. What bone do you think would most likely be used as the source of the graft? Explain your answer.
5. Andy, a 55-year-old, 75 kg (165-pound) roofer, is shingling the steeply pitched roof of a new house when he loses his footing and slides down the roof and over the edge, feet first. He braces himself for the fall, and when he hits the ground, he cries out and doubles up in excruciating pain. Emergency medical technicians called to the scene tell him he has broken his hips. Describe, more specifically, where his fractures most likely occurred. On the way to the hospital, Andy says, "You know it's funny, when I was a kid, I used to jump off roofs that high, and I never got hurt." Why do you think Andy was more at risk of a fracture as an adult than he was as a boy?

Answers at www.mhhe.com/saladinha3

Improve Your Grade at www.mhhe.com/saladinha3

Practice quizzes, labeling activities, and games provide fun ways to master concepts. You can also download image PowerPoint files for each chapter to create a study guide or for taking notes during lecture.

Joints

CHAPTER

9

X-ray of hands with severe rheumatoid arthritis

CHAPTER OUTLINE

BRUSHING UP

To understand this chapter, you may find it helpful to review the following concepts:
- Anatomical planes (p. 12)
- The distinction between hyaline cartilage and fibrocartilage (p. 67)
- Anatomy of the skeletal system (chapter 8)

Skeletal System

Joints, or articulations, link the bones of the skeletal system into a functional whole—a system that supports the body, permits effective movement, and protects the softer organs. Joints such as the shoulder, elbow, and knee are remarkable specimens of biological design—self-lubricating, almost frictionless, and able to bear heavy loads and withstand compression while executing smooth and precise movements. Yet, it is equally important that other joints be less movable or even immobile. Such joints are better able to support the body and protect delicate organs. The vertebral column, for example, is only moderately movable, for it must allow for flexibility of the torso and yet protect the delicate spinal cord and support much of the body's weight. Bones of the cranium must protect the brain and sense organs, but need not allow for movement (except during birth); thus, they are locked together by immobile joints, the sutures studied in chapter 7.

In everyday life, we take the greatest notice of the most freely movable joints of the limbs, and it is here that people feel most severely compromised by disabling diseases such as arthritis. Much of the work of physical therapists focuses on limb mobility. In this chapter, we will survey all types of joints, from the utterly immobile to the most movable, but with an emphasis on the latter. This survey of joint anatomy and movements will provide a foundation for the study of muscle actions in chapters 11 and 12.

9.1 Joints and Their Classification

▶ Expected Learning Outcomes

When you have completed this section, you should be able to

- explain what joints are, how they are named, and what functions they serve;
- name and describe the four major classes of joints;
- name some joints that become solidly fused by bone as they age;
- describe the three types of fibrous joints and give an example of each;
- distinguish between the three types of sutures; and
- describe the two types of cartilaginous joints and give an example of each.

Any point where two bones meet is called a **joint (articulation),** whether or not the bones are movable at that interface. The science of joint structure, function, and dysfunction is called **arthrology.**[1] The study of musculoskeletal movement is **kinesiology**[2] (kih-NEE-see-OL-oh-jee). This is a branch of **biomechanics,** which deals with a broad variety of movements and mechanical processes in the body, including the physics of blood circulation, respiration, and hearing.

The name of a joint is typically derived from the names of the bones involved. For example, the *atlanto–occipital joint* is where the atlas meets the occipital condyles, the *glenohumeral joint* is where the glenoid cavity of the scapula meets the humerus, and the *radioulnar joint* is where the radius meets the ulna.

Joints are classified according to the manner in which the adjacent bones are bound to each other, with corresponding differences in how freely the bones can move. Authorities differ in their classification schemes, but one common view places the joints in four major categories: *bony, fibrous, cartilaginous,* and *synovial joints.* This section will describe the first three of these and the subclasses of each. The remainder of the chapter will then be concerned primarily with synovial joints.

Bony Joints

A **bony joint,** or **synostosis**[3] (SIN-oss-TOE-sis), is an immobile joint formed when the gap between two bones ossifies and they become, in effect, a single bone. Bony joints can form by ossification of either fibrous or cartilaginous joints. An infant is born with right and left frontal and mandibular bones, for example, but these soon fuse seamlessly to form single bones. In old age, some cranial sutures become obliterated by ossification, and the adjacent cranial bones, such as the parietal bones, fuse. The epiphyses and diaphyses of the long bones are joined by cartilaginous joints in childhood and adolescence, and these become synostoses in early adulthood. The attachment of the first rib to the sternum also becomes a synostosis with age.

Fibrous Joints

A **fibrous joint** is also called a **synarthrosis**[4] (SIN-ar-THRO-sis) or **synarthrodial joint.** It is a point at which adjacent bones are bound by collagen fibers that emerge from the matrix of one bone, cross the space between them, and penetrate into the matrix of the other (fig. 9.1). There are three kinds of fibrous joints: *sutures, gomphoses,* and *syndesmoses.* In sutures and gomphoses, the fibers are very short and allow for little or no movement. In syndesmoses, the fibers are longer and the attached bones are more movable.

Sutures

Sutures are immobile or only slightly movable fibrous joints that closely bind the bones of the skull to each other; they occur nowhere else. In chapter 7, we did not take much notice of the differences between one suture and another, but some differences may have caught your attention as you studied the diagrams in that chapter or examined laboratory specimens. Sutures can be classified as *serrate, lap,* and *plane sutures.* Readers with some knowledge of woodworking may recognize that the structures and functional

---

[1]*arthro* = joint + *logy* = study of
[2]*kinesio* = movement + *logy* = study of

[3]*syn* = together + *ost* = bone + *osis* = condition
[4]*syn* = together + *arthr* = joined + *osis* = condition

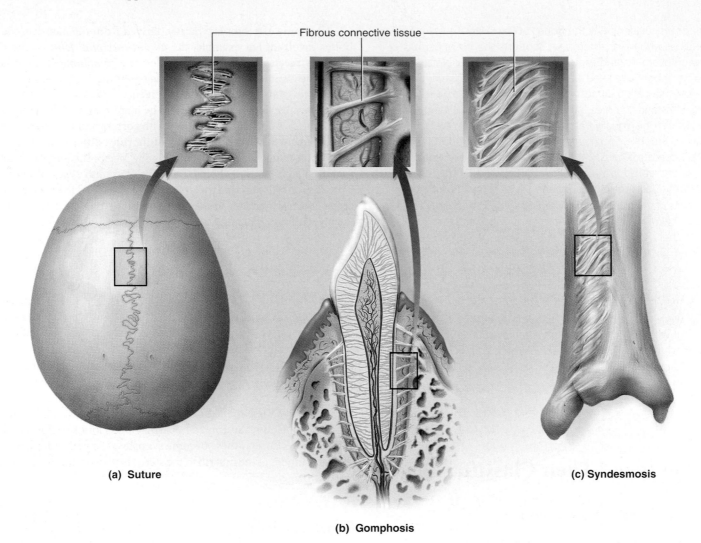

Fibrous connective tissue

(a) Suture

(b) Gomphosis

(c) Syndesmosis

Figure 9.1 **Fibrous Joints.** (a) A suture between the parietal bones. (b) A gomphosis between a tooth and the jaw. (c) A syndesmosis between the tibia and fibula.

● *Which of these is not a joint between two bones? Why?*

properties of these sutures have something in common with basic types of carpentry joints (fig. 9.2).

A **serrate suture** is one in which the adjoining bones firmly interlock by their serrated margins, like pieces of a jigsaw puzzle. It is analogous to a dovetail wood joint. On the surface, it appears as a wavy line between the two bones, as we see in the coronal, sagittal, and lambdoid sutures that border the parietal bones.

A **lap (squamous) suture** is one in which the adjacent bones have overlapping beveled edges, like a miter joint in carpentry. An example is the squamous suture that encircles most of the temporal bone. Its beveled edge can be seen in figure 7.10b (p. 160). On the surface, a lap suture appears as a relatively smooth (nonserrated) line.

A **plane (butt) suture** is one in which the adjacent bones have straight nonoverlapping edges. The two bones merely border each other, like two boards glued together in a butt joint. An example is seen between the palatine processes of the maxillae in the roof of the mouth.

Gomphoses

Even though the teeth are not bones, the attachment of a tooth to its socket is classified as a **gomphosis** (gom-FOE-sis). The term refers to its similarity to a nail hammered into wood.[5] The tooth is held firmly in place by a fibrous **periodontal ligament,** which consists of collagen fibers that extend from the bone matrix of the jaw into the dental tissue (see fig. 9.1b). The periodontal ligament allows the tooth to move or give a little under the stress of chewing. This allows us to sense how hard we are biting and to sense a particle of food stuck between the teeth.

Syndesmoses

A **syndesmosis**[6] (SIN-dez-MO-sis) is a fibrous joint at which two bones are bound by relatively long collagenous fibers. The separa-

[5]*gomph* = nail, bolt + *osis* = condition
[6]*syn* = together + *desm* = band + *osis* = condition

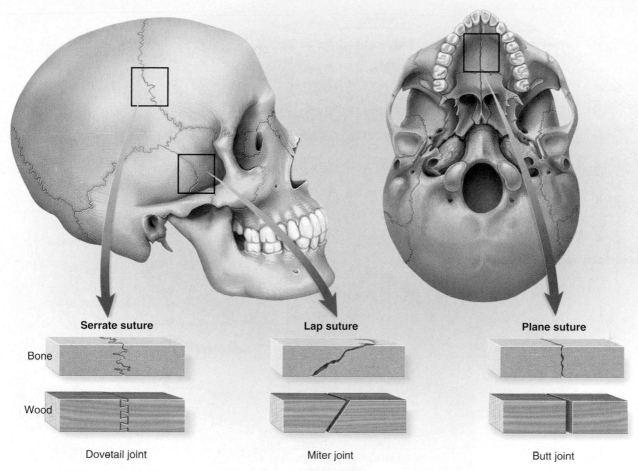

Serrate suture

Bone

Wood

Dovetail joint

Lap suture

Miter joint

Plane suture

Butt joint

Figure 9.2 Sutures. Serrate, lap, and plane sutures compared to some common wood joints.

tion between the bones and length of the fibers gives these joints more mobility than a suture or gomphosis. An especially movable syndesmosis exists between the shafts of the radius and ulna, which are joined by a broad fibrous *interosseous membrane*. This syndesmosis permits such movements as pronation and supination of the forearm. A less mobile syndesmosis is the one that binds the distal ends of the tibia and fibula together, side by side (fig. 9.1c).

Cartilaginous Joints

A **cartilaginous joint** is also called an **amphiarthrosis**[7] (AM-fee-ar-THRO-sis) or **amphiarthrodial joint**. In these joints, two bones are linked by cartilage (fig. 9.3). The two types of cartilaginous joints are *synchondroses* and *symphyses*.

Synchondroses

A **synchondrosis**[8] (SIN-con-DRO-sis) is a joint in which the bones are bound by hyaline cartilage. An example is the temporary joint between the epiphysis and diaphysis of a long bone in a child, formed by the cartilage of the epiphyseal plate. Another is the attachment of the first rib to the sternum by a hyaline costal cartilage (fig. 9.3a). (The other costal cartilages are joined to the sternum by synovial joints.)

Symphyses

In a **symphysis**[9] (SIM-fih-sis), two bones are joined by fibrocartilage (fig. 9.3b, c). One example is the pubic symphysis, in which the right and left pubic bones are joined by the cartilaginous interpubic disc. Another is the joint between the bodies of two vertebrae,

[7]*amphi* = on all sides + *arthr* = joined + *osis* = condition

[8]*syn* = together + *chondr* = cartilage + *osis* = condition
[9]*sym* = together + *physis* = growth

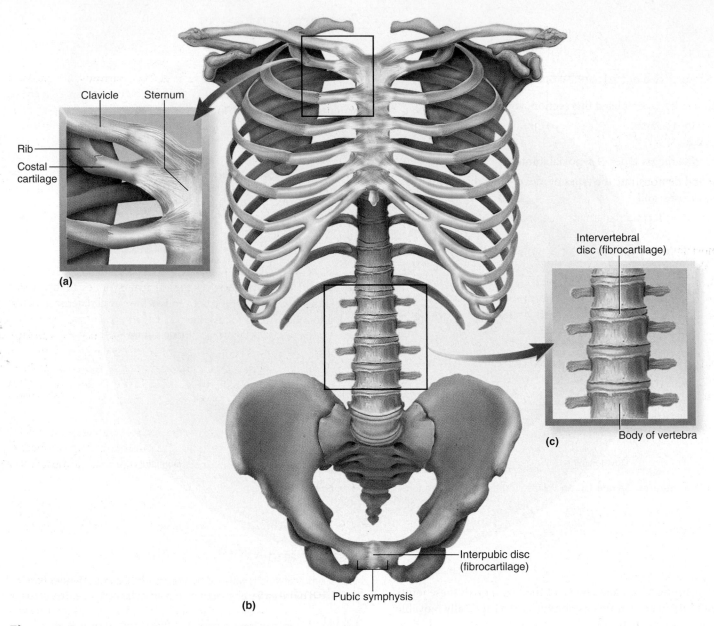

Clavicle   Sternum

Rib

Costal
cartilage

(a)

Intervertebral
disc (fibrocartilage)

Body of vertebra

(c)

Interpubic disc
(fibrocartilage)

Pubic symphysis

(b)

Figure 9.3 Cartilaginous Joints. (a) A synchondrosis, represented by the costal cartilage joining rib 1 to the sternum. (b) The pubic symphysis. (c) Intervertebral discs, which join adjacent vertebrae to each other by symphyses.
● *What is the difference between the pubic symphysis and the interpubic disc?*

united by an intervertebral disc. The surface of each vertebral body is covered with hyaline cartilage. Between the vertebrae, this cartilage becomes infiltrated with collagen bundles to form fibrocartilage. Each intervertebral disc permits only slight movement between adjacent vertebrae, but the collective effect of all 23 discs gives the spine considerable flexibility.

Apply What You Know

The intervertebral joints are symphyses only in the cervical through the lumbar region. How would you classify the intervertebral joints of the sacrum and coccyx in a middle-aged adult?

Before You Go On

Answer the following questions to test your understanding of the preceding section:

1. What is the difference between arthrology and kinesiology?

2. Explain the distinction between a synostosis, amphiarthrosis, and synarthrosis.

3. Give some examples of joints that become synostoses with age.

4. Define *suture, gomphosis,* and *syndesmosis,* and explain what these three joints have in common.

5. Name the three types of sutures and describe how they differ.

6. Name two synchondroses and two symphyses.

9.2 Synovial Joints

▶ Expected Learning Outcomes

When you have completed this section, you should be able to

- describe the anatomy of a synovial joint and its associated structures;
- describe the six types of synovial joints;
- list and demonstrate the types of movements that occur at diarthroses; and
- discuss the factors that affect the range of motion of a joint.

The most familiar type of joint is the **synovial** (sih-NO-vee-ul) **joint,** also called a **diarthrosis**[10] (DY-ar-THRO-sis) or **diarthrodial joint.** Ask most people to point out any joint in the body, and they are likely to point to a synovial joint such as the elbow, knee, or knuckles. Many synovial joints, like these examples, are freely movable. Others, such as the joints between the wrist and ankle bones and between the articular processes of the vertebrae, have more limited mobility. Synovial joints are the most structurally complex type of joint and are the most likely to develop uncomfortable and crippling dysfunctions.

---

[10]*dia* = separate, apart + *arthr* = joint + *osis* = condition

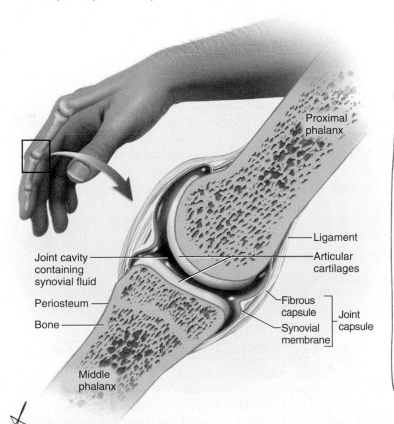

Figure 9.4 **Structure of a Simple Synovial Joint.** Most synovial joints are more complex than the interphalangeal joints.
• *Why is a meniscus unnecessary in an interphalangeal joint?*

Labels on figure:
Proximal phalanx
Ligament
Articular cartilages
Joint cavity containing synovial fluid
Periosteum
Bone
Fibrous capsule
Synovial membrane
Joint capsule
Middle phalanx

General Anatomy

In synovial joints, the facing surfaces of the two bones are covered with **articular cartilage,** a layer of hyaline cartilage usually about 2 or 3 mm thick. These surfaces are separated by a narrow space, the **joint (articular) cavity,** containing a slippery lubricant called **synovial fluid** (fig. 9.4). This fluid, for which the joint is named, is rich in albumin and hyaluronic acid, which give it a viscous, slippery texture similar to raw egg white.[11] It nourishes the articular cartilages, removes their wastes, and makes movements at synovial joints almost frictionfree. A connective tissue **joint (articular) capsule** encloses the cavity and retains the fluid. It has an outer **fibrous capsule** continuous with the periosteum of the adjoining bones, and an inner, cellular **synovial membrane.** The synovial membrane consists mainly of fibroblastlike cells that secrete the fluid, and is populated by macrophages that remove debris from the joint cavity.

In a few synovial joints, fibrocartilage grows inward from the joint capsule and forms a pad between the articulating bones. In the jaw and distal radioulnar joints, and at both ends of the clavicle (sternoclavicular and acromioclavicular joints), the pad crosses the entire joint capsule and is called an **articular disc** (see fig. 9.18c). In the knee, two cartilages extend inward from the left and right but do not entirely cross the joint (see fig. 9.23d). Each is called a **meniscus**[12] because of its crescent shape. These cartilages absorb shock and pressure, guide the bones across each other, improve the fit between the bones, and stabilize the joint, reducing the chance of dislocation.

Accessory structures associated with a synovial joint include tendons, ligaments, and bursae. A **tendon** is a strip or sheet of tough collagenous connective tissue that attaches a muscle to a

DEEPER INSIGHT 9.1

Exercise and Articular Cartilage

When synovial fluid is warmed by exercise, it becomes thinner (less viscous) and more easily absorbed by the articular cartilage. The cartilage then swells and provides a more effective cushion against compression. For this reason, a warm-up period before vigorous exercise helps protect the articular cartilage from undue wear and tear.

Because cartilage is nonvascular, its repetitive compression during exercise is important to its nutrition and waste removal. Each time a cartilage is compressed, fluid and metabolic wastes are squeezed out of it. When weight is taken off the joint, the cartilage absorbs synovial fluid like a sponge, and the fluid carries oxygen and nutrients to the chondrocytes. Lack of exercise causes the articular cartilages to deteriorate more rapidly from lack of nutrition, oxygenation, and waste removal.

Weight-bearing exercise builds bone mass and strengthens the muscles that stabilize many of the joints, thus reducing the risk of joint dislocations. Excessive joint stress, however, can hasten the progression of osteoarthritis (p. 228) by damaging the articular cartilage. Swimming is a good way of exercising the joints with minimal damage.

---

[11]*ovi* = egg
[12]*men* = moon, crescent + *iscus* = little

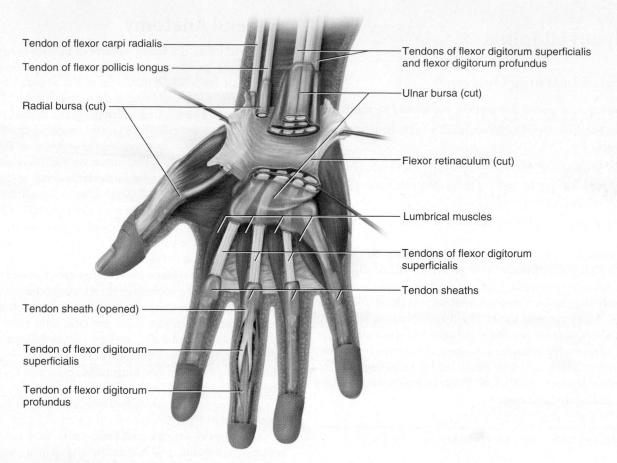

Tendon of flexor carpi radialis

Tendon of flexor pollicis longus

Radial bursa (cut)

Tendons of flexor digitorum superficialis and flexor digitorum profundus

Ulnar bursa (cut)

Flexor retinaculum (cut)

Lumbrical muscles

Tendons of flexor digitorum superficialis

Tendon sheaths

Tendon sheath (opened)

Tendon of flexor digitorum superficialis

Tendon of flexor digitorum profundus

Figure 9.5 Tendon Sheaths and Other Bursae in the Hand and Wrist.

bone. Tendons are often the most important structures in stabilizing a joint. A **ligament** is a similar tissue that attaches one bone to another. Several ligaments are named and illustrated in our later discussion of individual joints, and tendons are more fully considered in chapters 10 through 12 along with the gross anatomy of muscles.

A **bursa**[13] is a fibrous sac filled with synovial fluid, located between adjacent muscles, between bone and skin, or where a tendon passes over a bone (see fig. 9.19). Bursae cushion muscles, help tendons slide more easily over the joints, and sometimes enhance the mechanical effect of a muscle by modifying the direction in which its tendon pulls. Bursae called **tendon sheaths** are elongated cylinders wrapped around a tendon. These are especially numerous in the hand and foot (fig. 9.5). **Bursitis** is inflammation of a bursa, usually due to overexertion of a joint. **Tendinitis** is a form of bursitis in which a tendon sheath is inflamed.

Classes of Synovial Joints

There are six fundamental classes of synovial joints, distinguished by patterns of motion determined by the shapes of the articular surfaces of the bones (table 9.1). We will examine them here in descending order of mobility, from multiaxial to monaxial joints. A **multiaxial** joint is one that can move in any of the three fundamental mutually perpendicular planes (x, y, and z); a **biaxial** joint is able to move in only two planes; and a **monaxial** joint moves in only one plane. Ball-and-socket joints are the only multiaxial type; condylar, saddle, and plane joints are biaxial; and hinge and pivot joints are monaxial. As we see in figure 9.6, all six types of synovial joints have representatives in the upper limb.

1. **Ball-and-socket joints.** These are the shoulder and hip joints. In both cases, one bone (the humerus or femur) has a smooth hemispherical head that fits into a cuplike socket on the other (the glenoid cavity of the scapula or the acetabulum of the hip bone).

2. **Condylar (ellipsoid) joints.** These joints exhibit an oval convex surface on one bone that fits into a similarly shaped depression on the other. The radiocarpal joint of the wrist and metacarpophalangeal (MET-uh-CAR-po-fah-LAN-jee-ul) joints at the bases of the fingers are examples. To demonstrate their biaxial motion, hold your hand with the palm facing you. Make a fist, and these joints flex in the sagittal plane. Fan your fingers apart, and they move in the frontal plane.

[13]*burs* = purse

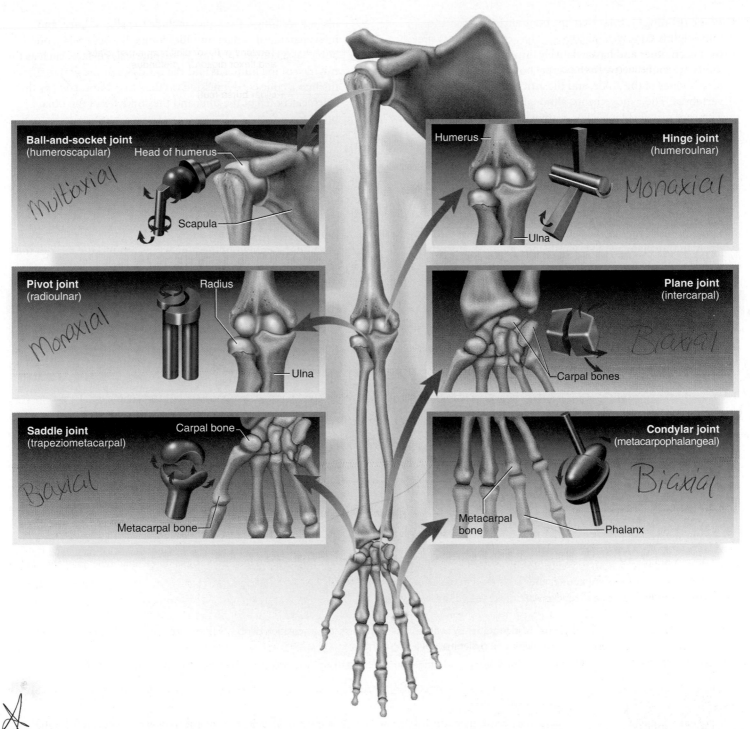

Figure 9.6 **The Six Types of Synovial Joints.** All six have representatives in the forelimb. Mechanical models show the types of motion possible at each joint.

3. **Saddle joints.** Here, both bones have a saddle-shaped surface—concave in one direction (like the front to rear curvature of a horse's saddle) and convex in the other (like the left to right curvature of a saddle). The clearest example of this is the trapeziometacarpal joint between the trapezium of the wrist and metacarpal I at the base of the thumb. Saddle joints are also biaxial. The thumb, for example, moves in a frontal plane when you spread the fingers apart, and in a sagittal plane when you move it as if to grasp a tool such as a hammer. This range of motion gives us and other primates that anatomical hallmark, the opposable thumb. Another saddle joint is the sternoclavicular joint, where the clavicle articulates with the sternum. The clavicle moves vertically in the frontal plane at this joint when you lift a suitcase, and moves horizontally in the transverse plane when you reach forward to push open a door.

4. **Plane (gliding) joints.** Here the bone surfaces are flat or only slightly concave and convex. The adjacent bones slide over each other and have relatively limited movement. Plane joints are found between the carpal bones of the wrist, the tarsal bones of the ankle, and the articular processes of the vertebrae. Their movements, although slight, are complex. They are usually biaxial. For example, when the head is tilted forward and back, the articular facets of the vertebrae slide anteriorly and posteriorly; when the head is tilted from side to side, the facets slide laterally. Although any one joint moves only slightly, the combined action of the many joints in the wrist, ankle, and vertebral column allows for a significant amount of overall movement.

5. **Hinge joints.** These are essentially monaxial joints, moving freely in one plane with very little movement in any other, like a door hinge. Examples include the elbow, knee, and interphalangeal (finger and toe) joints. In these cases, one bone has a convex (but not hemispherical) surface, such as the trochlea of the humerus and the condyles of the femur. This fits into a concave depression on the other bone, such as the trochlear notch of the ulna and the condyles of the tibia.

6. **Pivot joints.** These are monaxial joints in which a bone spins on its longitudinal axis. There are two principal examples: the radioulnar joint at the elbow and the atlantoaxial joint between the first two vertebrae. At the atlantoaxial joint, the dens of the axis projects into the vertebral foramen of the atlas and is held against its anterior arch by the transverse ligament (see fig. 7.24, p. 169). As the head rotates left and right, the skull and atlas pivot around the dens. At the radioulnar joint, the anular ligament of the ulna wraps around the neck of the

TABLE 9.1	Anatomical Classification of the Joints
Joint	**Characteristics and Examples**
Bony joint (synostosis)	Former fibrous or cartilaginous joint in which adjacent bones have become fused by ossification. Examples: median line of frontal bone; fusion of epiphysis and diaphysis of an adult long bone; and fusion of ilium, ischium, and pubis to form hip bone
Fibrous joint (synarthrosis)	Adjacent bones bound by collagen fibers extending from the matrix of one into the matrix of the other
Suture (figs. 9.1a, 9.2)	Immobile fibrous joint between cranial or facial bones
Serrate suture	Bones joined by a wavy line formed by interlocking teeth along the margins. Examples: coronal, sagittal, and lambdoid sutures
Lap suture	Bones beveled to overlap each other; superficial appearance is a smooth line. Example: squamous suture around temporal bone
Plane suture	Bones butted against each other without overlapping or interlocking. Example: palatine suture
Gomphosis (fig. 9.1b)	Insertion of a tooth into a socket, held in place by collagen fibers of periodontal ligament
Syndesmosis (fig. 9.1c)	Slightly movable joint held together by ligaments or interosseous membranes. Examples: tibiofibular joint and radioulnar joint
Cartilaginous joint (amphiarthrosis)	Adjacent bones bound by cartilage
Synchondrosis (fig. 9.3a)	Bones held together by hyaline cartilage. Examples: articulation of rib 1 with sternum, and epiphyseal plate uniting the epiphysis and diaphysis of a long bone of a child
Symphysis (fig. 9.3b, c)	Slightly movable joint held together by fibrocartilage. Examples: intervertebral discs and pubic symphysis
Synovial joint (diarthrosis) (Figs. 9.4 and 9.6)	Adjacent bones covered with hyaline articular cartilage, separated by lubricating synovial fluid and enclosed in a fibrous joint capsule
Ball-and-socket joint	Multiaxial diarthrosis in which a smooth hemispherical head of one bone fits into a cuplike depression of another. Examples: shoulder and hip joints
Condylar (ellipsoid) joint	Biaxial diarthrosis in which an oval convex surface of one bone articulates with an elliptical depression of another. Examples: radiocarpal and metacarpophalangeal joints
Saddle joint	Joint in which each bone surface is saddle-shaped (concave on one axis and convex on the perpendicular axis). Examples: trapeziometacarpal and sternoclavicular joints
Plane (gliding) joint	Usually biaxial diarthroses with slightly concave or convex bone surfaces that slide across each other. Examples: intercarpal and intertarsal joints; joints between the articular processes of the vertebrae
Hinge joint	Monaxial diarthrosis, able to flex and extend in only one plane. Examples: elbow, knee, and interphalangeal joints
Pivot joint	Joint in which a projection of one bone fits into a ringlike ligament of another, allowing one bone to rotate on its longitudinal axis. Examples: atlantoaxial joint and proximal radioulnar joint

radius. During pronation and supination of the forearm, the disclike radial head pivots like a wheel turning on its axle. The edge of the wheel spins against the radial notch of the ulna like a car tire spinning in snow.

Some joints cannot be easily classified into any one of these six categories. The jaw joint, for example, has some aspects of condylar, hinge, and plane joints. It clearly has an elongated condyle where it meets the temporal bone of the cranium, but it moves in a hingelike fashion when the mandible moves up and down in speaking, biting, and chewing; it glides slightly forward when the jaw juts (protracts) to take a bite; and it glides from side to side to grind food between the molars. The knee is a classic hinge joint, but has an element of the pivot type; when we lock our knees to stand more effortlessly,

the femur pivots slightly on the tibia. The humeroradial joint (between humerus and radius) acts as a hinge joint when the elbow flexes and a pivot joint when the forearm pronates.

Movements of Synovial Joints

Kinesiology, physical therapy, and other medical and scientific fields have a specific vocabulary for the movements of synovial joints. The following terms form a basis for describing the muscle actions in chapters 11 and 12 and may also be indispensable to your advanced coursework or intended career. This section introduces the terms for joint movements, many of which are presented in pairs or groups with opposite or contrasting meanings. This section relies on familiarity with the three cardinal anatomical planes and the directional terms in chapter 1 (p. 13). All directional terms used here refer to a person in standard anatomical position. When one is in anatomical position, each joint is said to be in its **zero position.** Joint movements can be described as deviation from the zero position or returning to it.

Flexion and Extension

Flexion (fig. 9.7) is a movement that decreases a joint angle, usually in the sagittal plane. This is particularly common at hinge joints— for example, bending of the elbow or knee—but it occurs in other types of joints as well. For example, if you hold out your hands with the palms up, flexion of the wrist tips your palms toward you. The meaning of *flexion* is perhaps least obvious in the ball-and-socket joints of the shoulder and hip. At the shoulder, it means to raise your arm as if pointing at something directly in front of you or to continue in that arc and point toward the sky. At the hip, it means to raise the thigh, for example to place your foot on the next higher step when ascending a flight of stairs.

Extension is a movement that straightens a joint and generally returns a body part to the zero position—for example, straightening the elbow, wrist, or knee, or returning the arm or thigh back to zero position. In stair climbing, both the hip and knee extend when lifting the body to the next higher step.

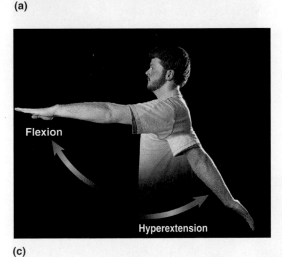

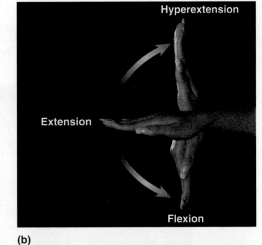

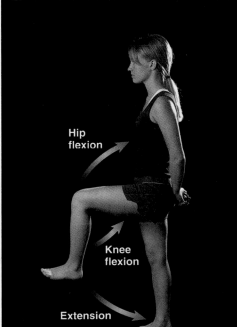

Figure 9.7 Flexion and Extension. (a) Flexion and extension of the elbow. (b) Flexion, extension, and hyperextension of the wrist. (c) Flexion and hyperextension of the shoulder. (d) Flexion and extension of the hip and knee.

Extreme extension of a joint, beyond the zero position, is called **hyperextension**.[14] For example, if you hold your hand in front of you with the palm down, then raise the back of your hand as if you were admiring a new ring, you hyperextend the wrist. Hyperextension of the upper or lower limb means to move the limb to a position behind the frontal plane of the trunk, as in reaching around with your arm to scratch your back. Each backswing of the lower limb when you walk hyperextends the hip.

Flexion and extension occur at nearly all diarthroses, but hyperextension is limited to only a few. At most diarthroses, ligaments or bone structure prevents hyperextension.

> ### Apply What You Know
>
> Try hyperextending some of your synovial joints and list a few for which this is impossible.

---

[14]*hyper* = excessive, beyond normal

Abduction and Adduction

Abduction[15] (ab-DUC-shun) (fig. 9.8a) is the movement of a body part in the frontal plane away from the midline of the body—for example, moving the feet apart to stand spread-legged, or raising an arm to one side of the body. **Adduction**[16] (fig. 9.8b) is movement in the frontal plane back toward the midline. Some joints can be **hyperadducted**, as when you stand with your ankles crossed, cross your fingers, or hyperadduct the shoulder to stand with your elbows straight and your hands clasped below your waist. You **hyperabduct** the arm if you raise it high enough to cross slightly over the front or back of your head.

Elevation and Depression

Elevation (fig. 9.9a) is a movement that raises a body part vertically in the frontal plane. **Depression** (fig. 9.9b) lowers a body part in the

---

[15]*ab* = away + *duc* = to lead or carry
[16]*ad* = toward + *duc* = to lead or carry

Figure 9.8 Abduction and Adduction.

(a) Abduction

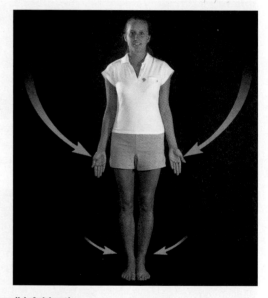

(b) Adduction

Figure 9.9 Elevation and Depression.

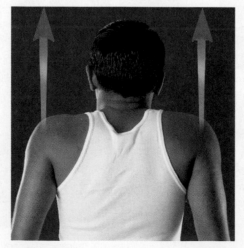

(a) Elevation

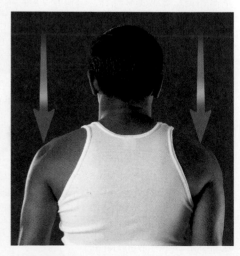

(b) Depression

(a) Protraction   (b) Retraction

Figure 9.10 Protraction and Retraction.

Figure 9.11 Circumduction.

same plane. For example, to lift a heavy suitcase from the floor, you elevate your scapula; in setting it down again, you depress the scapula. These are also important jaw movements in biting.

Protraction and Retraction

Protraction[17] (fig. 9.10a) is the anterior movement of a body part in the transverse (horizontal) plane, and **retraction**[18] (fig. 9.10b) is posterior movement. Your shoulder protracts, for example, when you reach in front of you to push a door open. It retracts when you return it to the resting (zero) position or pull the shoulders back to stand at military attention. Such exercises as rowing a boat, bench presses, and push-ups involve repeated protraction and retraction of the shoulders.

Circumduction

In **circumduction**[19] (fig. 9.11), one end of an appendage remains fairly stationary while the other end makes a circular motion. If an artist standing at an easel reaches forward and draws a circle on a canvas, she circumducts the upper limb; the shoulder remains stationary while the hand moves in a circle. A baseball player winding up for the pitch circumducts the upper limb in a more extreme "windmill" fashion. One can also circumduct an individual finger, the hand, the thigh, the foot, the trunk, and the head.

Apply What You Know

Choose any example of circumduction and explain why this motion is actually a sequence of flexion, abduction, extension, and adduction.

Rotation

In one sense, the term *rotation* applies to any bone turning around a fixed axis, as described earlier. But in the terminology of specific

joint movements, **rotation** (fig. 9.12) is a movement in which a bone spins on its longitudinal axis. For example, if you stand with bent elbow and move your forearm to place your palm against your abdomen, your humerus spins in a motion called **medial (internal) rotation.** If you make the opposite action, so the forearm points away from the trunk, your humerus undergoes **lateral (external) rotation.** If you turn your right foot so your toes are pointing away from your left foot, and then turn it so your toes are pointing toward your left foot, your femur undergoes lateral and medial rotation, respectively. Other examples are given in the ensuing discussions of forearm and head movements.

Supination and Pronation

Supination and pronation (fig. 9.13) are known primarily as forearm movements, but see also the later discussion of foot movements. **Supination**[20] (SOO-pih-NAY-shun) of the forearm is a movement that turns the palm to face anteriorly or upward; in anatomical position, the forearm is supinated and the radius is parallel to the ulna. **Pronation**[21] is the opposite movement, causing the palm to face posteriorly or downward and the radius to cross the ulna like an X. During these movements, the concave end of the disc-shaped head of the radius spins on the capitulum of the humerus, and the edge of the disc spins in the radial notch of the ulna. The ulna remains relatively stationary.

As an aid to remembering these terms, think of it this way: You are *prone* to stand in the most comfortable position, which is with the forearm *pronated*. But if you were holding a cup of *soup* in your palm, you would need to *supinate* the forearm to keep from spilling it.

Chapter 12 describes the muscles that perform these actions. Of these, the *supinator* is the most powerful. Supination is the type of movement you would usually make with your right hand to turn a doorknob clockwise or to drive a screw into a piece of wood. The threads of screws and bolts are designed with the relative strength of the supinator in mind, so the greatest power can be applied when driving them with a screwdriver in the right hand.

---

[17]*pro* = forward + *trac* = to pull or draw
[18]*re* = back + *trac* = to pull or draw
[19]*circum* = around + *duc* = to carry, lead

[20]*supin* = to lay back
[21]*pron* = to bend forward

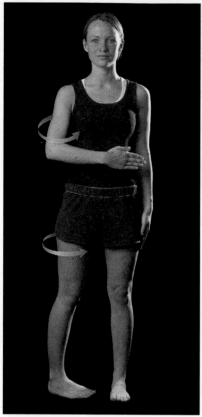

(a) Medial (internal) rotation   **(b) Lateral (external) rotation**

Figure 9.12 **Medial (Internal) and Lateral (External) Rotation.** Shows rotation of both the humerus and the femur.

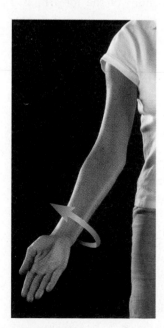

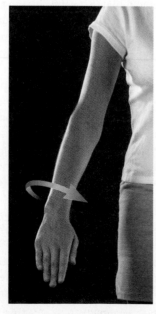

(a) Supination   **(b) Pronation**

Figure 9.13 Supination and Pronation.

We will now consider a few body regions that combine the foregoing motions or have unique movements and terminology.

Special Movements of the Head and Trunk

Flexion of the spine produces forward-bending movements, as in tilting the head forward or bending at the waist in a toe-touching exercise (fig. 9.14a). *Extension* of the spine straightens the trunk or neck, as in standing up or returning the head to a forward-looking zero position. *Hyperextension* is employed in looking toward the sky or bending over backward (fig. 9.14b).

Lateral flexion is tilting the head or the trunk to the right or left of the midline (fig. 9.14c). Twisting at the waist or turning the head is called **right rotation** or **left rotation** when the chest or the face turns to the right or left of the forward-facing zero position (fig. 9.14d, e). Powerful right and left rotation at the waist is important in baseball pitching, golf, discus throwing, and other sports.

Special Movements of the Mandible

Movements of the mandible are concerned especially with biting and chewing (fig. 9.15). Imagine taking a bite of raw carrot. Most people have some degree of overbite; at rest, the upper incisors (front teeth) overhang the lower ones. For effective biting, however, the chisel-like edges of the incisors must meet. In preparation to bite, we therefore *protract* the mandible to bring the lower incisors forward. After the bite is taken, we *retract* it. To actually take the bite, we must *depress* the mandible to open the mouth, then *elevate* it so the incisors can cut off the piece of food.

Next, to chew the food, we do not simply raise and lower the mandible as if hammering away at the food between the teeth; rather, we exercise a grinding action that shreds the food between the broad, bumpy surfaces of the premolars and molars. This entails a side-to-side movement of the mandible called **lateral excursion** (movement to the left or right of the zero position) and **medial excursion** (movement back to the median, zero position).

Special Movements of the Hand and Fingers

The hand moves anteriorly and posteriorly by flexion and extension of the wrist. It can also move in the frontal plane. **Radial flexion** tilts the hand toward the thumb, and **ulnar flexion** tilts it toward the little finger (fig. 9.16a, b). We often use such motions when waving hello to someone with a side-to-side wave of the hand, or when washing windows, polishing furniture, or keyboarding.

Movements of the fingers are more varied, especially those of the thumb (fig. 9.16c–e). *Flexion* of the fingers is curling them; *extension* is straightening them. Most people cannot hyperextend their fingers. Spreading the fingers apart is *abduction,* and bringing them together again so they touch along their surfaces is *adduction.*

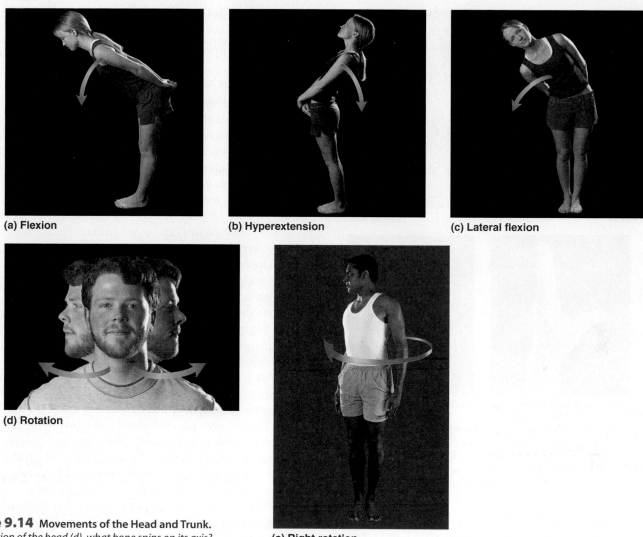

(a) Flexion

(b) Hyperextension

(c) Lateral flexion

(d) Rotation

(e) Right rotation

Figure 9.14 Movements of the Head and Trunk.
● *In rotation of the head (d), what bone spins on its axis?*

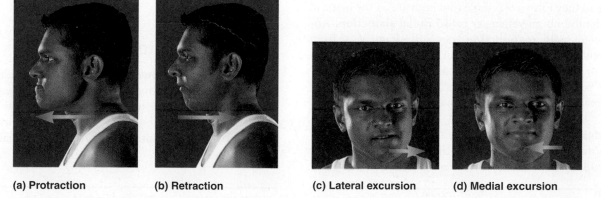

(a) Protraction

(b) Retraction

(c) Lateral excursion

(d) Medial excursion

Figure 9.15 Movements of the Mandible.

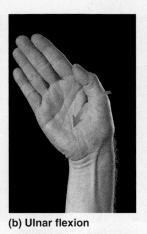

(a) Radial flexion

(b) Ulnar flexion

(c) Abduction of fingers

(d) Palmar abduction of thumb

(e) Opposition of thumb

Figure 9.16 Movements of the Hand and Fingers. (a) Radial flexion of the wrist. (b) Ulnar flexion of the wrist. (c) Abduction of the fingers. The thumb position in this figure is called *radial abduction*. Parts (a) and (b) show adduction of the fingers. (d) Palmar abduction of the thumb. (e) Opposition of the thumb (reposition is shown in parts [a] and [b]).

The thumb is different, however, because in embryonic development it rotates nearly 90° from the rest of the hand. If you hold your hand in a completely relaxed position, you will probably see that the plane that contains your thumb and index finger is about 90° to the plane that contains the index through little fingers. Much of the terminology of thumb movement therefore differs from that of the other four fingers. *Flexion* of the thumb is bending the joints so the tip of the thumb is directed toward the palm, and *extension* is straightening it. If you place the palm of your hand on a table top with all five fingers parallel and touching, the thumb is extended. Keeping your hand there, if you move your thumb away from the index finger so they form a 90° angle (but both are on the plane of the table), the thumb movement is called **radial abduction.** Another movement, **palmar abduction,** moves the thumb away from the plane of the hand so it points anteriorly, as you would do if you were about to wrap your hand around a tool handle (fig. 9.16d). From either position—radial or palmar abduction—*adduction* of the thumb means to bring it back to touch the base of the index finger.

Two additional terms are unique to the thumb: **Opposition**[22] means to move the thumb to touch the tip of any of the other four fingers (fig. 9.16e). **Reposition**[23] is the return to zero position.

Special Movements of the Foot

A few additional movement terms are unique to the foot (fig. 9.17). **Dorsiflexion** (DOR-sih-FLEC-shun) is a movement in which the toes are elevated, as one might do in applying toenail polish. In each step you take, the foot dorsiflexes as it comes forward. This prevents you from scraping your toes on the ground and results in the characteristic *heel strike* of human locomotion when the foot touches down in front of you. **Plantar flexion** is movement of the foot so the toes point downward, as in pressing the gas pedal of a car or standing on tiptoes. This motion also produces the *toe-off* in each step you take, as the heel of the foot behind you lifts off the ground. Plantar flexion can be a very powerful motion, epitomized by high jumpers and the jump shots of basketball players.

Inversion[24] is a foot movement that tips the soles medially, somewhat facing each other, and **eversion**[25] is a movement that tips them laterally, away from each other. These movements are common in fast sports such as tennis and football, and sometimes cause ankle sprains. These terms also refer to congenital deformities of the feet, which are often corrected by orthopedic shoes or braces.

Pronation and *supination,* while used mainly for forearm movements, also apply to the feet but refer here to a more complex combination of movements. Pronation of the foot is a combination of dorsiflexion, eversion, and abduction—that is, the toes are elevated and turned away from the other foot and the sole is tilted away. Supination of the foot is a combination of plantar flexion, inversion, and adduction—the toes are lowered and turned toward the other foot and the sole is tilted toward it. These may seem a little difficult to visualize and perform, but they are ordinary motions in walking, running, ballet, and crossing uneven surfaces such as stepping stones.

You can perhaps understand why these terms apply to the feet if you place the palms of your hands on a table and pretend they are your soles. Tilt your hands so the inner edge (thumb side) of each is raised from the table. This is like raising the medial edge of your foot from the ground, and as you can see, it involves a slight supination of your forearms. Resting your hands palms down on a table, your forearms are already pronated; but if you raise the outer edges of your hands (the little finger side), like pronating the feet, you will see that it involves a continuation of the pronation movement of the forearm.

Range of Motion

A joint's **range of motion (ROM)** is the number of degrees through which one bone can move relative to another at that joint. For example, the ankle has an ROM of about 74°, the first knuckle about 90°, and the knee about 130° to 140°. Range of motion obviously affects a person's functional independence and quality of life. It is

---

[22]*op* = against + *posit* = to place
[23]*re* = back + *posit* = to place
[24]*in* = inward + *version* = turning
[25]*e* = outward + *version* = turning

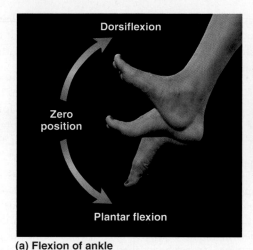

(a) Flexion of ankle

(b) Inversion

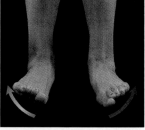

(c) Eversion

Figure 9.17 Movements of the Foot.
● *Identify some common activities in which inversion and eversion of the foot would be important.*

also an important consideration in training for athletics or dance, in clinical diagnosis, and in monitoring the progress of rehabilitation. Several factors affect the ROM and stability of a joint:

- **Structure of the articular surfaces of the bones.** In many cases, joint movement is limited by the shapes of the bone surfaces. For example, you cannot straighten your elbow beyond 180° or so because, as it straightens, the olecranon of the ulna swings into the olecranon fossa of the humerus and cannot move farther.

- **Strength and tautness of ligaments and joint capsules.** Some bone surfaces impose little if any limitation on joint movement. The articulations of the phalanges are an example; as one can see by examining a dry skeleton, an interphalangeal joint can bend through a broad arc. In life, however, these bones are joined by ligaments that limit their movement. As you flex one of your knuckles, ligaments on the anterior (palmar) side of the joint go slack, but ligaments on the posterior (dorsal) side tighten and prevent the joint from flexing beyond 90° or so. The knee is another case in point. In kicking a football, the knee rapidly extends to about 180°, but it can go no farther. Its motion is limited in part by a *cruciate ligament* and other knee ligaments described later. Gymnasts, dancers, and acrobats increase the ROM of their synovial joints by gradually stretching their ligaments during training. "Double-jointed" people have unusually large ROMs at some joints, not because the joint is actually double or fundamentally different from normal in its anatomy, but because the ligaments are unusually long or slack.

- **Action of the muscles and tendons.** Extension of the knee is also limited by the *hamstring muscles* on the posterior side of the thigh. In many other joints, too, pairs of muscles oppose each other and moderate the speed and range of joint motion. Even a resting muscle maintains a state of tension called *muscle tone,* which serves in many cases to stabilize a joint. One of the major factors preventing dislocation of the shoulder joint, for example, is tension in the *biceps brachii* muscle, whose tendons cross the joint, insert on the scapula,

and hold the head of the humerus against the glenoid cavity. The nervous system continually monitors and adjusts joint angles and muscle tone to maintain joint stability and limit unwanted movements.

Before You Go On

Answer the following questions to test your understanding of the preceding section:

7. What are the two components of a joint capsule? What is the function of each?

8. Give at least one example each of a monaxial, biaxial, and multiaxial joint, and explain the reason for its classification.

9. Name the joints that would be involved if you reached directly overhead and screwed a lightbulb into a ceiling fixture. Describe the joint actions that would occur.

9.3 Anatomy of Selected Synovial Joints

▶ **Expected Learning Outcomes**

When you have completed this section, you should be able to

- identify the major anatomical features of the jaw, shoulder, elbow, hip, knee, and ankle joints; and

- explain how the anatomical differences between these joints are related to differences in function.

We now examine the gross anatomy of certain diarthroses. It is beyond the scope of this book to discuss all of them, but the ones selected here most often require medical attention and many of them have a strong bearing on athletic performance and everyday function.

The Jaw Joint

The **temporomandibular joint (TMJ)** is the articulation of the condyle of the mandible with the mandibular fossa of the temporal bone (fig. 9.18). You can feel its action by pressing your fingertips against the jaw immediately anterior to the ear while opening and closing your mouth. This joint combines elements of condylar, hinge, and plane joints. It functions in a hingelike fashion when the mandible is elevated and depressed, it glides slightly forward when the jaw is protracted to take a bite, and it glides from side to side to grind food between the molars.

The synovial cavity of the TMJ is divided into superior and inferior chambers by an articular disc, which permits lateral and medial excursion of the mandible. Two ligaments support the joint. The **lateral ligament** prevents posterior displacement of the mandible. If the jaw receives a hard blow, this ligament normally prevents the condylar process from being driven upward and fracturing the base of the skull. The **sphenomandibular ligament** on the medial side of the joint extends from the sphenoid bone to the ramus of the mandible. A *stylomandibular ligament* extends from the styloid process to the angle of the mandible but is not part of the TMJ proper.

A deep yawn or other strenuous depression of the mandible can dislocate the TMJ by making the condyle pop out of the fossa and slip forward. The joint can be relocated by pressing down on the molars while pushing the jaw backward.

DEEPER INSIGHT   9.2

TMJ Syndrome

Temporomandibular joint (TMJ) syndrome has received medical recognition only recently, although it may affect as many as 75 million Americans. It can cause moderate intermittent facial pain, clicking sounds in the jaw, limitation of jaw movement, and in some people, more serious symptoms—severe headaches, vertigo (dizziness), tinnitus (ringing in the ears), and pain radiating from the jaw down the neck, shoulders, and back. It seems to be caused by a combination of psychological tension and malocclusion (misalignment of the teeth). Treatment may involve psychological management, physical therapy, analgesic and anti-inflammatory drugs, and sometimes corrective dental appliances to align the teeth properly.

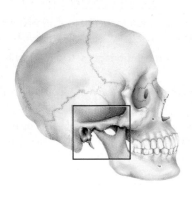

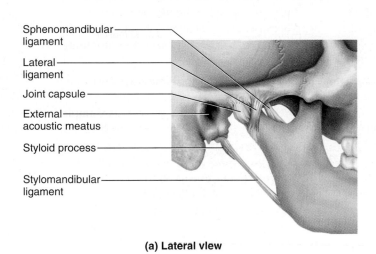

(a) Lateral view

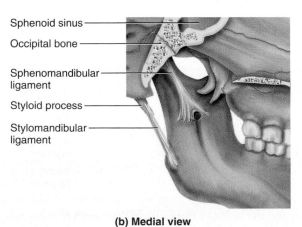

(b) Medial view

(c) Sagittal section

Figure 9.18 The Jaw (Temporomandibular) Joint.

The Shoulder Joint

The **glenohumeral (humeroscapular) joint,** or shoulder joint, is where the hemispherical head of the humerus articulates with the glenoid cavity of the scapula (fig. 9.19). Together, the shoulder and elbow joints serve to position the hand for the performance of a task; without a hand, shoulder and elbow movements are far less useful. The relatively loose shoulder joint capsule and shallow glenoid cavity sacrifice joint stability for freedom of movement. The

cavity, however, has a ring of fibrocartilage called the **glenoid labrum**[26] around its margin, making it somewhat deeper than it looks on a dried skeleton.

Five principal ligaments support this joint. The **coracohumeral ligament** extends from the coracoid process of the scapula to the greater tubercle of the humerus, and the **transverse humeral ligament** extends from the greater to the lesser tubercle of the humerus,

[26]*labrum* = lip

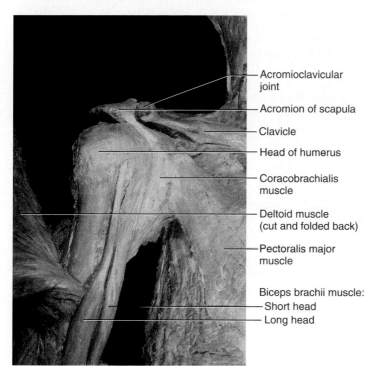

(a) Anterior dissection

(b) Anterior view

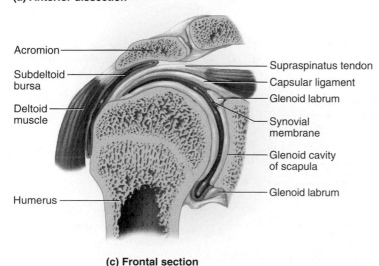

(c) Frontal section

(d) Lateral view, humerus removed

Figure 9.19 The Shoulder (Glenohumeral) Joint.
● *The socket of the shoulder joint is a little deeper in life than it appears on a dried skeleton. What structure makes it so?*

creating a tunnel, the intertubercular sulcus, through which a tendon of the biceps brachii passes. The other three ligaments, called **glenohumeral ligaments,** are relatively weak and sometimes absent.

A tendon of the biceps brachii muscle is the most important stabilizer of the shoulder. It originates on the margin of the glenoid cavity, passes through the joint capsule, and emerges into the intertubercular sulcus, where it is held by the transverse humeral ligament. Inferior to the sulcus, it merges into the biceps brachii. Thus, the tendon functions as a taut, adjustable strap that holds the humerus against the glenoid cavity.

In addition to the biceps brachii, four muscles important in stabilizing the glenohumeral joint are the *subscapularis, supraspinatus, infraspinatus,* and *teres minor.* The tendons of these four muscles form the *rotator cuff,* which is fused to the joint capsule on all sides except inferiorly. The rotator cuff is discussed more fully on page 300.

Four bursae are associated with the shoulder joint. Their names describe their locations—the **subdeltoid, subacromial, subcoracoid,** and **subscapular bursae.**

Shoulder dislocations are very painful and sometimes result in permanent damage. The most common dislocation is downward displacement of the humerus, because (1) the rotator cuff protects the joint in all directions except inferiorly, and (2) the joint is protected from above by the coracoid process, acromion, and clavicle. Dislocations most often occur when the arm is abducted and then receives a blow from above—for example, when the outstretched arm is struck by heavy objects falling off a shelf. They also occur in children who are jerked off the ground by one arm or forced to follow by a hard tug on the arm. Children are especially prone to such injury not only because of the inherent stress caused by such abuse, but also because a child's shoulder is

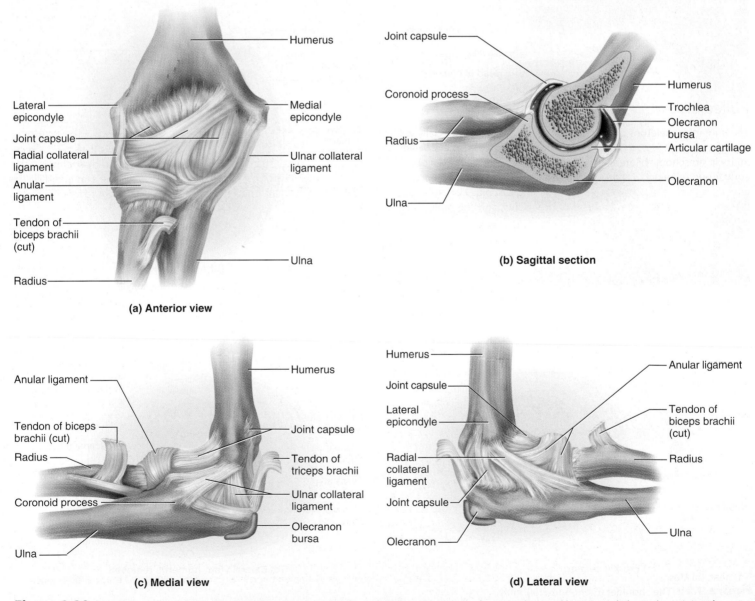

(a) Anterior view

(b) Sagittal section

(c) Medial view

(d) Lateral view

Figure 9.20 **The Elbow Joint.** This region includes two joints that form the elbow hinge—the humeroulnar and humeroradial—and one joint, the radioulnar, not involved in the hinge but involved in forearm rotation.

not fully ossified and the rotator cuff is not strong enough to withstand such stress. Because this joint is so easily dislocated, you should never attempt to move an immobilized person by pulling on his or her arm.

The Elbow Joint

The elbow is a hinge joint composed of two articulations—the **humeroulnar joint,** where the trochlea of the humerus joins the trochlear notch of the ulna, and the **humeroradial joint,** where the capitulum of the humerus meets the head of the radius (fig. 9.20). Both are enclosed in a single joint capsule. On the posterior side of the elbow, there is a prominent **olecranon bursa** to ease the movement of tendons over the elbow. Side-to-side motions of the elbow joint are restricted by a pair of ligaments, the **radial (lateral) collateral ligament** and **ulnar (medial) collateral ligament.**

Another joint occurs in the elbow region, the **proximal radioulnar joint,** but it is not involved in the hinge. At this joint, the disclike head of the radius fits into the radial notch of the ulna and is held in place by the **anular ligament**, which encircles the head of the radius and attaches at each end to the ulna.

The Hip Joint

The **coxal (hip) joint** is the point where the head of the femur inserts into the acetabulum of the hip bone (fig. 9.22). Because the coxal joints bear much of the body's weight, they have deep sockets and are much more stable than the shoulder joints. The depth of the socket is somewhat greater than you see on dried bones because a horseshoe-shaped ring of fibrocartilage, the **acetabular labrum,** is attached to its rim. A **transverse acetabular ligament** bridges a gap in the inferior margin of the acetabular labrum. Dislocations of the hip are rare, but some infants suffer congenital dislocations because the acetabulum is

DEEPER INSIGHT   9.3

Pulled Elbow

The immature skeletons of children and adolescents are especially vulnerable to injury. Pulled elbow (dislocation of the radius) is a common injury in preschool children (especially girls). It typically results from an adult lifting or jerking a child up by one arm when the arm is pronated, as in lifting a child into a high chair or shopping cart (fig. 9.21). This tears the anular ligament from the head of the radius, and the radius pulls

partially or entirely out of the ligament. The proximal part of the torn ligament is then painfully pinched between the radial head and the capitulum of the humerus. Radial dislocation is treated by supinating the forearm with the elbow flexed and then putting the arm in a sling for about 2 weeks—time enough for the anular ligament to heal.

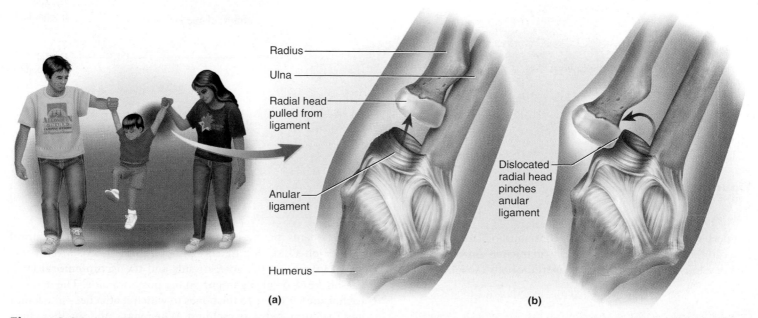

Radius

Ulna

Radial head pulled from ligament

Anular ligament

Humerus

Dislocated radial head pinches anular ligament

(a)   (b)

Figure 9.21 Pulled Elbow. Lifting a child by the arm can dislocate the radius. (a) The anular ligament tears, and the radial head is pulled from the ligament. (b) Muscle contraction pulls the radius upward. The head of the radius produces a lump on the lateral side of the elbow and may painfully pinch the anular ligament.

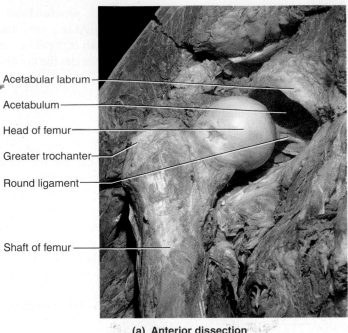

Acetabular labrum

Acetabulum

Head of femur

Greater trochanter

Round ligament

Shaft of femur

(a) Anterior dissection

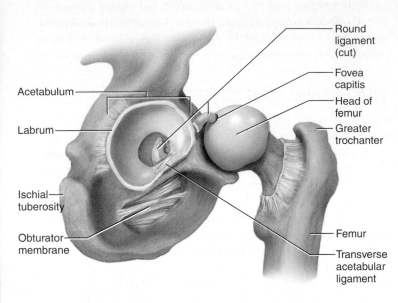

Acetabulum

Labrum

Ischial tuberosity

Obturator membrane

Round ligament (cut)

Fovea capitis

Head of femur

Greater trochanter

Femur

Transverse acetabular ligament

(b) Lateral view, femur retracted

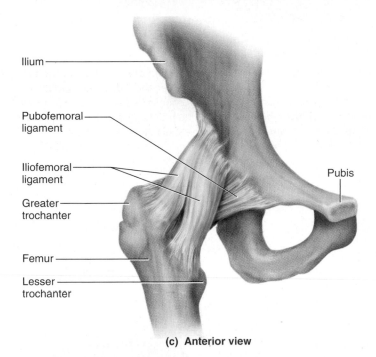

Ilium

Pubofemoral ligament

Iliofemoral ligament

Greater trochanter

Femur

Lesser trochanter

Pubis

(c) Anterior view

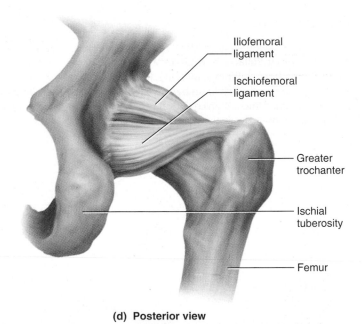

Iliofemoral ligament

Ischiofemoral ligament

Greater trochanter

Ischial tuberosity

Femur

(d) Posterior view

Figure 9.22 The Hip (Coxal) Joint.

not deep enough to hold the head of the femur in place. This condition can be treated by placing the infant in traction until the acetabulum develops enough strength to support the body's weight.

Apply What You Know

Where else in the body is there a structure similar to the acetabular labrum? What do those two locations have in common?

Ligaments that support the coxal joint include the **iliofemoral** (ILL-ee-oh-FEM-oh-rul) and **pubofemoral** (PYU-bo-FEM-or-ul) **ligaments** on the anterior side and the **ischiofemoral** (ISS-kee-oh-FEM-or-ul) **ligament** on the posterior side. The name of each ligament refers to the bones to which it attaches—the femur and the ilium, pubis, or ischium. When you stand up, these ligaments become twisted and pull the head of the femur tightly into the acetabulum. The head of the femur has a conspicuous pit called the *fovea capitis*. The **round ligament,** or **ligamentum** teres[27] (TERR-eez), arises here and attaches to the lower margin of the acetabulum. This is a relatively slack ligament, so it is doubtful that it plays a significant role in holding the femur in its socket. It does, however, contain an artery that supplies blood to the head of the femur.

---

[27]teres = round

The Knee Joint

The **tibiofemoral** (knee) **joint** is the largest and most complex diarthrosis of the body (figs. 9.23 and 9.24). It is primarily a hinge joint, but when the knee is flexed it is also capable of slight rotation and lateral gliding. The patella and its ligament also form a plane **patellofemoral joint** with the femur.

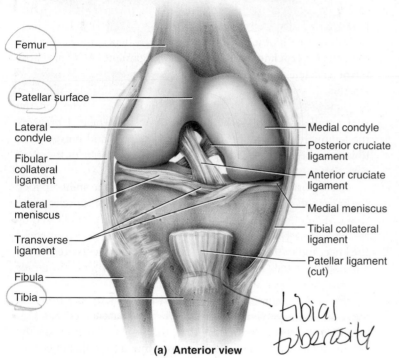

(a) Anterior view

tibial tuberosity

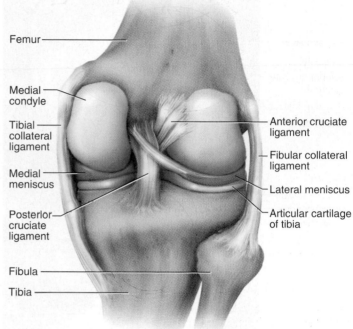

(b) Posterior view

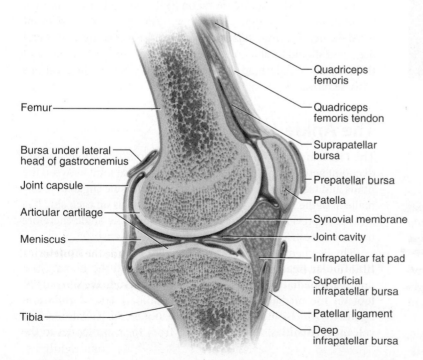

(c) Sagittal section

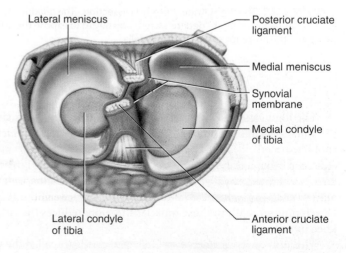

(d) Superior view of tibia and menisci

Figure 9.23 The Knee (Tibiofemoral) Joint.

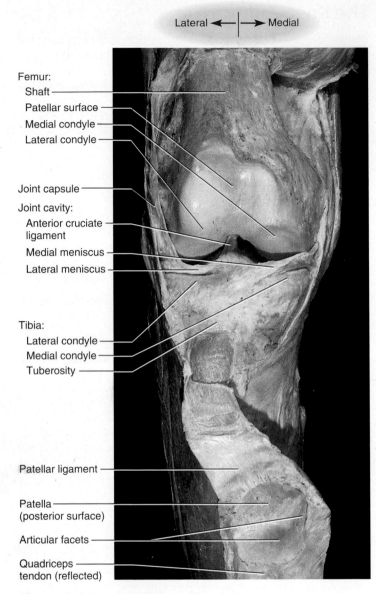

Lateral ←——|——→ Medial

Femur:
Shaft
Patellar surface
Medial condyle
Lateral condyle

Joint capsule
Joint cavity:
Anterior cruciate
ligament
Medial meniscus
Lateral meniscus

Tibia:
Lateral condyle
Medial condyle
Tuberosity

Patellar ligament

Patella
(posterior surface)

Articular facets

Quadriceps
tendon (reflected)

Figure 9.24 **The Right Knee, Anterior Dissection.** The quadriceps tendon has been cut and folded (reflected) downward to expose the joint cavity and the posterior surface of the patella.

The joint capsule encloses only the lateral and posterior sides the patellar ligament and the *lateral* and *medial patellar retinacula* (not illustrated). These are extensions of the tendon of the *quadriceps femoris* muscle, the large anterior muscle of the thigh. The knee is stabilized mainly by the quadriceps tendon in front and the tendon of the *semimembranosus* muscle on the rear of the thigh. Developing strength in these muscles therefore reduces the risk of knee injury.

The joint cavity contains two C-shaped cartilages called the **lateral** and **medial menisci** (singular, **meniscus**) joined by a **transverse ligament.** They absorb the shock of the body weight jostling up and down on the knee and prevent the femur from rocking from side to side on the tibia. The posterior side of the knee, the **popliteal**

(pop-LIT-ee-ul) **region,** is supported by a complex array of **intracapsular ligaments** within the joint capsule and **extracapsular ligaments** external to it. The extracapsular ligaments are the **oblique popliteal ligament** (an extension of the tendon of the *semimembranosus* hamstring muscle), **arcuate** (AR-cue-et) **popliteal ligament, fibular (lateral) collateral ligament,** and **tibial (medial) collateral ligament.** Only the two collateral ligaments are illustrated; they prevent the knee from rotating when the joint is extended.

There are two intracapsular ligaments deep within the joint cavity. The synovial membrane folds around them, however, so that they are excluded from the fluid-filled synovial cavity. These ligaments cross each other in the form of an X; hence, they are called the **anterior cruciate**[28] (CROO-she-ate) **ligament (ACL)** and **posterior cruciate ligament (PCL).** These are named according to whether they attach to the anterior or posterior side of the tibia, not for their attachments to the femur. When the knee is extended, the ACL is pulled tight and prevents hyperextension. The PCL prevents the femur from sliding off the front of the tibia and prevents the tibia from being displaced backward. The ACL is one of the most common sites of knee injury (see Deeper Insight 9.4).

An important aspect of human bipedalism is the ability to lock the knees and stand erect without tiring the extensor muscles of the thigh. When the knee is extended to the fullest degree allowed by the ACL, the femur rotates medially on the tibia. This action locks the knee, and in this state, all the major knee ligaments are twisted and taut. To unlock the knee, the *popliteus* muscle rotates the femur laterally, causing the ligaments to untwist.

The knee joint has at least 13 bursae. Four of these are anterior—the **superficial infrapatellar, suprapatellar, prepatellar,** and **deep infrapatellar.** Located in the popliteal region are the **popliteal bursa** and **semimembranosus bursa** (not illustrated). At least seven more bursae are found on the lateral and medial sides of the knee joint. From figure 9.23a, your knowledge of the relevant word elements (*infra-, supra-, pre-*), and the terms *superficial* and *deep,* you should be able to work out the reasoning behind most of these names and develop a system for remembering the locations of these bursae.

The Ankle Joint

The **talocrural**[29] **(ankle) joint** includes two articulations—a medial joint between the tibia and talus and a lateral joint between the fibula and talus, both enclosed in one joint capsule (fig. 9.26). The malleoli of the tibia and fibula overhang the talus on each side like a cap and prevent most side-to-side motion. The ankle therefore has a more restricted range of motion than the wrist.

The ligaments of the ankle include (1) **anterior** and **posterior tibiofibular ligaments,** which bind the tibia to the fibula; (2) a multipart **medial (deltoid) ligament,** which binds the tibia to the foot on the medial side; and (3) a multipart **lateral collateral ligament,** which binds the fibula to the foot on the lateral side. The **calcaneal (Achilles) tendon** extends from the calf muscles to the

---

[28]*cruci* = cross + *ate* = characterized by
[29]*talo* = ankle + *crural* = pertaining to the leg

DEEPER INSIGHT   9.4

Knee Injuries and Arthroscopic Surgery

Although the knee can bear a lot of weight, it is highly vulnerable to rotational and horizontal stress, especially when the knee is flexed (as in skiing or running) and receives a blow from behind or from the side (fig. 9.25). The most common injuries are to a meniscus or the anterior cruciate ligament (ACL). Knee injuries heal slowly because ligaments and tendons have a scanty blood supply and cartilage usually has no blood vessels at all.

The diagnosis and surgical treatment of knee injuries have been greatly improved by *arthroscopy,* a procedure in which the interior of a joint is viewed with a pencil-thin instrument, the *arthroscope,* inserted through a small incision. The arthroscope has a light, a lens, and fiber optics that allow a viewer to see into the cavity and take photographs or video recordings. A surgeon can also withdraw samples of synovial fluid by arthroscopy or inject saline into the joint cavity to expand it and provide a clearer view. If surgery is required, additional small incisions can be made for the surgical instruments and the procedures can be observed through the arthroscope or on a monitor. Arthroscopic surgery produces much less tissue damage than conventional surgery and enables patients to recover more quickly.

Orthopedic surgeons often replace a damaged ACL with a graft from the patellar ligament or a hamstring tendon. The surgeon "harvests" a strip from the middle of the patient's ligament or tendon, drills a hole into the femur and tibia within the joint cavity, threads the ligament through the holes, and fastens it with biodegradable screws. The grafted ligament is more taut and "competent" than the damaged ACL. It becomes ingrown with blood vessels and serves as a substrate for the deposition of more collagen, which further strengthens it in time. Following arthroscopic ACL reconstruction, a patient typically must use crutches for 7 to 10 days and undergo supervised physical therapy for 6 to 10 weeks, followed by self-directed exercise therapy. Healing is completed in about 9 months.

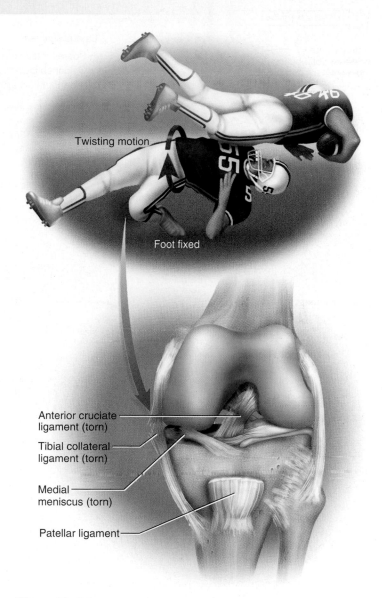

Figure 9.25 Knee Injuries.

calcaneus. It plantarflexes the foot and limits dorsiflexion. Plantar flexion is limited by extensor tendons on the anterior side of the ankle and by the anterior part of the joint capsule.

Sprains (torn ligaments and tendons) are common at the ankle, especially when the foot is suddenly inverted or everted to excess. They are painful and usually accompanied by immediate swelling. They are best treated by immobilizing the joint and reducing swelling with an ice pack, but in extreme cases may require a cast or surgery.

The synovial joints described in this section are summarized in table 9.2.

Before You Go On

Answer the following questions to test your understanding of the preceding section:

10. What keeps the mandibular condyle from slipping out of its fossa in a posterior direction?

11. List at least three ways that the shoulder joint is stabilized.

12. What keeps the femur from slipping backward off the tibia?

13. What keeps the tibia from slipping sideways off the talus?

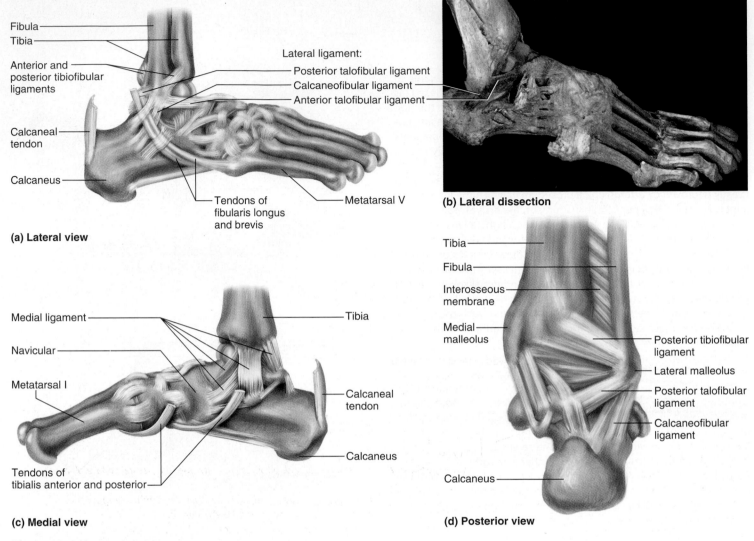

Figure 9.26 The Ankle (Talocrural) Joint and Ligaments of the Right Foot.

9.4 Clinical Perspectives

▶ Expected Learning Outcomes

When you have completed this section, you should be able to

- define *rheumatism* and describe the profession of rheumatology;
- define *arthritis* and describe its forms and causes;
- discuss the design and application of artificial joints; and
- identify several joint diseases other than arthritis.

Our quality of life depends greatly on mobility, and mobility depends on proper functioning of the diarthroses. Not surprisingly, therefore, joint dysfunctions are among the most common medical complaints. **Rheumatism** is a broad term for any pain in the supportive and lo-comotor organs of the body, including bones, ligaments, tendons, and muscles. Physicians who deal with the study, diagnosis, and treatment of joint disorders are called **rheumatologists.**

Arthritis

The most widespread crippling disorder in the United States is **arthritis,**[30] a broad term that embraces more than a hundred diseases of largely obscure or unknown causes. In general, arthritis means inflammation of a joint. Nearly everyone develops arthritis to some degree after middle age and sometimes earlier.

The most common form of arthritis is **osteoarthritis (OA),** also called "wear-and-tear arthritis" because it is apparently a normal consequence of years of wear on the joints. As joints age, the articular cartilage softens and degenerates. As the cartilage be-

---

[30]*arthr* = joint + *itis* = inflammation

TABLE 9.2	Review of the Principal Diarthroses
Joint	**Major Anatomical Features and Actions**
Jaw joint (fig. 9.18)	*Type:* condylar, hinge, and plane *Movements:* elevation, depression, protraction, retraction, lateral and medial excursion *Articulation:* condyle of mandible, mandibular fossa of temporal bone *Ligaments:* lateral, sphenomandibular *Cartilage:* articular disc
Shoulder joint (fig. 9.19)	*Type:* ball-and-socket *Movements:* adduction, abduction, flexion, extension, circumduction, medial and lateral rotation *Articulation:* head of humerus, glenoid fossa of scapula *Ligaments:* coracohumeral, transverse humeral, three glenohumerals *Tendons:* rotator cuff (tendons of subscapularis, supraspinatus, infraspinatus, teres minor), tendon of biceps brachii *Bursae:* subdeltoid, subacromial, subcoracoid, subscapular *Cartilage:* glenoid labrum
Elbow joint (fig. 9.20)	*Type:* hinge and pivot *Movements:* flexion, extension, pronation, supination, rotation *Articulation:* humeroulnar—trochlea of humerus, trochlear notch of ulna; humeroradial—capitulum of humerus, head of radius; radioulnar—head of radius, radial notch of ulna *Ligaments:* radial collateral, ulnar collateral, anular *Bursa:* olecranon
Hip joint (fig. 9.22)	*Type:* ball-and-socket *Movements:* adduction, abduction, flexion, extension, circumduction, medial and lateral rotation *Articulation:* head of femur, acetabulum of hip bone *Ligaments:* iliofemoral, pubofemoral, ischiofemoral, ligamentum teres, transverse acetabular *Cartilage:* acetabular labrum
Knee joint (fig. 9.23)	*Type:* primarily hinge *Movements:* flexion, extension, slight rotation *Articulation:* tibiofemoral, patellofemoral *Ligaments:* anterior—lateral patellar retinaculum, medial patellar retinaculum; popliteal intracapsular—anterior cruciate, posterior cruciate; popliteal extracapsular—oblique popliteal, arcuate popliteal, lateral collateral, medial collateral *Bursae:* anterior—superficial infrapatellar, suprapatellar, prepatellar, deep infrapatellar; popliteal—popliteal, semimembranosus; medial and lateral—seven other bursae not named in this chapter *Cartilages:* lateral meniscus, medial meniscus (connected by transverse ligament)
Ankle joint (fig. 9.26)	*Type:* hinge *Movements:* dorsiflexion, plantar flexion, extension *Articulation:* tibia–talus, fibula–talus, tibia–fibula *Ligaments:* anterior and posterior tibiofibular, deltoid, lateral collateral *Tendon:* calcaneal (Achilles)

comes roughened by wear, joint movement may be accompanied by crunching or crackling sounds called *crepitus.* Osteoarthritis affects especially the fingers, intervertebral joints, hips, and knees. As the articular cartilage wears away, exposed bone tissue often develops spurs that grow into the joint cavity, restrict movement, and cause pain. Though OA rarely occurs before age 40, it affects about 85% of people older than 70. It usually does not cripple, but in severe cases it can immobilize the hip.

Rheumatoid arthritis (RA), which is far more severe, results from an autoimmune attack against the joint tissues. Like other autoimmune diseases, RA is caused by an *autoantibody*—a misguided antibody that attacks the body's own tissues instead of limiting its attack to foreign matter. In RA, an autoantibody called *rheumatoid factor* attacks the synovial membranes. Inflammatory cells accu-

mulate in the synovial fluid and produce enzymes that degrade the articular cartilage. The synovial membrane thickens and adheres to the articular cartilage, fluid accumulates in the joint capsule, and the capsule is invaded by fibrous connective tissue. As articular cartilage degenerates, the joint begins to ossify, and sometimes the bones become solidly fused and immobilized, a condition called **ankylosis**[31] (fig. 9.27). RA tends to develop symmetrically—if the right wrist or hip develops RA, so does the left.

Rheumatoid arthritis tends to flare up and subside (go into remission) periodically.[32] It affects women far more than men and typically begins between the ages of 30 and 40. There is no cure, but

---

[31]*ankyl* = bent, crooked + *osis* = condition
[32]*rheumat* = tending to change

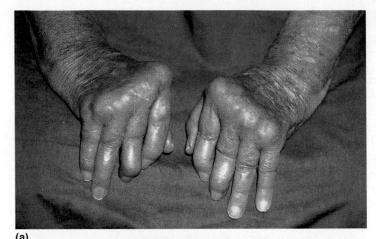

(a)

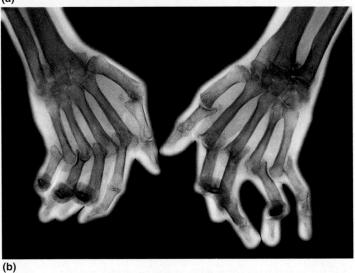

(b)

Figure 9.27 Rheumatoid Arthritis (RA). (a) A severe case with ankylosis of the joints. (b) Colorized X-ray of hands with RA.

joint damage can be slowed with hydrocortisone or other steroids. Because long-term use of steroids weakens the bone, however, aspirin is the treatment of first choice to control the inflammation. Physical therapy is also used to preserve the joint's range of motion and the patient's functional ability.

Several common pathologies of the joints are briefly described in table 9.3.

Joint Prostheses

Arthroplasty,[33] a treatment of last resort, is the replacement of a diseased joint with an artificial device called a **joint prosthesis.**[34] Joint prostheses were first developed to treat war injuries in World War II and the Korean War. Total hip replacement (THR), first performed in 1963 by English orthopedic surgeon Sir John Charnley, is now the most common orthopedic procedure for the elderly. The first knee replacements were performed in the 1970s. Joint prostheses are now available for finger, shoulder, elbow, hip, and knee joints. Arthroplasty is performed on over 250,000 patients per year in the United States, primarily to relieve pain and restore function in elderly people with OA or RA.

Arthroplasty presents ongoing challenges for biomedical engineering. An effective prosthesis must be strong, nontoxic, and corrosion-resistant. In addition, it must bond firmly to the patient's bones and enable a normal range of motion with a minimum of friction. The heads of long bones are usually replaced with prostheses made of a metal alloy such as cobalt-chrome, titanium alloy, or stainless steel. Joint sockets are made of polyethylene (fig. 9.28). Prostheses are bonded to the patient's bone with screws or bone cement.

[33]*arthro* = joint + *plasty* = surgical repair
[34]*prosthe* = something added

TABLE 9.3	Disorders of the Joints
Dislocation (luxation)	Displacement of a bone from its normal position at a joint, usually accompanied by a sprain of the adjoining connective tissues. Most common at the fingers, thumb, shoulder, and knee.
Gout	A hereditary disease, most common in men, in which uric acid crystals accumulate in the joints and irritate the articular cartilage and synovial membrane. Causes *gouty arthritis,* with swelling, pain, tissue degeneration, and sometimes fusion of the joint. Most commonly affects the great toe.
Strain	Painful overstretching of a tendon or muscle without serious tissue damage. Often results from inadequate warm-up before exercise.
Subluxation	Partial dislocation in which two bones maintain contact between their articular surfaces.
Synovitis	Inflammation of a joint capsule, often as a complication of a sprain.

Disorders Described Elsewhere

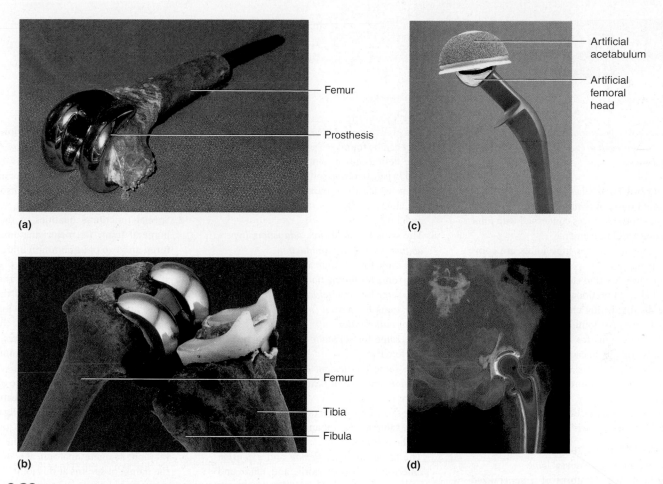

(a)

(c)

— Femur

— Prosthesis

— Artificial acetabulum

— Artificial femoral head

(b)

(d)

— Femur

— Tibia

— Fibula

Figure 9.28 **Joint Prostheses.** (a) Artificial femoral condyles affixed to the distal end of the femur. (b) An artificial knee joint bonded to a natural femur and tibia. (c) A porous-coated hip prosthesis. The caplike portion replaces the acetabulum of the hip bone, and the ball and shaft below it are bonded to the proximal end of the femur. (d) X-ray of a patient with a total hip replacement.

Over 90% of artificial knees last 10 years, 85% for 15 years, and 75% for 20 years. The most common form of failure is detachment of the prosthesis from the bone. This problem has been reduced by using *porous-coated prostheses,* which become infiltrated by the patient's own bone and create a firmer bond. A prosthesis is not as strong as a natural joint, however, and is not an option for many young, active patients.

Before You Go On

Answer the following questions to test your understanding of the preceding section:

14. Define *arthritis.* How do the causes of osteoarthritis and rheumatoid arthritis differ? Which type is more common?

15. What are the major engineering problems in the design of joint prostheses? What is the most common cause of failure of a prosthesis?

Study Guide

Assess Your Learning Outcomes

You should have a good understanding of this chapter if you can accurately address the following issues.

9.1   Joints and Their Classification (p. 205)

1. The definition of *joint (articulation)*
2. Names of the sciences concerned with joint structure and movement
3. The general rule for how joints are commonly named
4. The criteria used to classify joints into anatomical and functional categories
5. The distinguishing characteristics of bony joints, fibrous joints, and cartilaginous joints; synonyms for these terms; and examples of joints in each category
6. The three subclasses of fibrous joints and three types of sutures, and examples of each
7. The two subclasses of cartilaginous joints, and examples of each

9.2   Synovial Joints (p. 209)

1. The definition of *synovial joint*
2. The anatomical features of a generalized synovial joint
3. The functions of articular discs and menisci at certain synovial joints, where they can be found, and their appearance
4. The defining characteristics of tendons, ligaments, and bursae, and the roles they play at the joints; how tendon sheaths differ from other bursae
5. The distinction between monaxial, biaxial, and multiaxial joints

6. The six types of synovial joints and where they can be found
7. The definitions of joint *flexion, extension,* and *hyperextension;* some everyday scenarios in which these movements occur; and the ability to demonstrate them with your own body
8. The same for *abduction, adduction, hyperabduction,* and *hyperadduction*
9. The same for *elevation* and *depression*
10. The same for *protraction* and *retraction*
11. The same for *circumduction*
12. The same for *medial (internal)* and *lateral (external) rotation*
13. The same for *supination* and *pronation* of the forearm
14. The same for *flexion, extension, hyperextension,* and *lateral flexion* of the vertebral column
15. The same for *rotation* of the head or torso
16. The same for *lateral* and *medial excursion* of the mandible
17. The same for wrist *flexion* and *extension* anteriorly and posteriorly, and *ulnar* and *radial flexion* from side to side
18. The same for the thumb movements of *radial abduction, palmar abduction, opposition,* and *reposition*
19. The same for the ankle or foot movements of *dorsiflexion, plantar flexion, inversion,* and *eversion,* and how several of these movements are combined in foot pronation and supination
20. How a joint's range of motion (ROM) is measured and what anatomical features govern the ROM

9.3   Anatomy of Selected Synovial Joints (p. 219)

1. Special functional qualities of the temporomandibular joint (TMJ); its major anatomical features; and two common disorders of the TMJ
2. Special functional qualities of the glenohumeral joint; its major anatomical features; and two of its common injuries
3. The names of the three joints that occur at the elbow; how they enable the varied movements of the forearm; and the major anatomical features of the elbow joints
4. Special functional qualities of the coxal joint; its major anatomical features; and the actions of the ligaments at this joint when a person stands
5. Special functional qualities of the tibiofemoral joint; its major anatomical features (especially its menisci and cruciate ligaments); and the common injuries of this joint
6. Special functional qualities of the talocrural joint; its major anatomical features; and the nature of sprains at this joint

9.4   Clinical Perspectives (p. 228)

1. The range of disorders included in the concept of rheumatism, and the related term for physicians who specialize in joint disorders
2. The general meaning of *arthritis,* and the pathology and distinctions between osteoarthritis and rheumatoid arthritis
3. Joint prostheses and arthroplasty

Testing Your Recall

1. Lateral and medial excursion are movements unique to
   a. the ankle.
   b. the thumb.
   c. the mandible.
   d. the knee.
   e. the clavicle.

2. Which of the following is the least movable?
   a. a diarthrosis
   b. a synostosis
   c. a symphysis
   d. a syndesmosis
   e. a condylar joint

3. Which of the following movements are unique to the foot?
   a. dorsiflexion and inversion
   b. elevation and depression
   c. circumduction and rotation
   d. abduction and adduction
   e. opposition and reposition

4. Which of the following joints cannot be circumducted?
   a. trapeziometacarpal
   b. metacarpophalangeal
   c. glenohumeral
   d. coxal
   e. interphalangeal

5. Which of the following terms denotes a general condition that includes the other four?
   a. gout
   b. arthritis
   c. rheumatism
   d. osteoarthritis
   e. rheumatoid arthritis

6. In the adult, the ischium and pubis are united by
   a. a synchondrosis.
   b. a diarthrosis.
   c. a synostosis.
   d. an amphiarthrosis.
   e. a symphysis.

7. Articular discs are found only in certain
   a. synostoses.
   b. symphyses.
   c. diarthroses.
   d. synchondroses.
   e. amphiarthroses.

8. Which of the following joints has anterior and posterior cruciate ligaments?
   a. the shoulder
   b. the elbow
   c. the hip
   d. the knee
   e. the ankle

9. To bend backward at the waist involves _____ of the vertebral column.
   a. rotation
   b. hyperextension
   c. dorsiflexion
   d. abduction
   e. flexion

10. If you sit on a sofa and then raise your left arm to rest it on the back of the sofa, your left shoulder joint undergoes primarily
    a. lateral excursion.
    b. abduction.
    c. elevation.
    d. adduction.
    e. extension.

11. The lubricant of a diarthrosis is _____.

12. A fluid-filled sac that eases the movement of a tendon over a bone is called a/an _____.

13. A _____ joint allows one bone to swivel on another.

14. _____ is the science of movement.

15. The joint between a tooth and the mandible is called a/an _____.

16. In a/an _____ suture, the articulating bones have interlocking wavy margins, somewhat like a dovetail joint in carpentry.

17. In kicking a football, what type of action does the knee joint exhibit?

18. The angle through which a joint can move is called its _____.

19. A person with a degenerative joint disorder would most likely be treated by a physician called a/an _____.

20. The femur is prevented from slipping sideways off the tibia in part by a pair of cartilages called the lateral and medial _____.

Answers in the Appendix

Building Your Medical Vocabulary

State a medical meaning of each of the following word elements, and give a term in which it is used.

1. arthro-
2. re-
3. sym-
4. amphi-
5. -physis
6. circum-
7. ab-
8. ad-
9. duc-
10. kinesio-

Answers in the Appendix

True or False

Determine which five of the following statements are false, and briefly explain why.

1. More people get rheumatoid arthritis than osteoarthritis.
2. A doctor who treats arthritis is called a kinesiologist.
3. Synovial joints are also known as synarthroses.
4. Most ligaments, but not all, connect one bone to another.
5. Reaching behind you to take something out of your hip pocket involves hyperextension of the elbow.
6. The anterior cruciate ligament normally prevents hyperextension of the knee.
7. There is no meniscus in the elbow joint.
8. The knuckles are diarthroses.
9. Synovial fluid is secreted by the bursae.
10. Most condylar joints can move in more planes than a hinge joint.

Answers in the Appendix

Testing Your Comprehension

1. Why are there menisci in the knee joint but not in the elbow, the corresponding joint of the upper limb? Why is there an articular disc in the temporomandibular joint?

2. What ligaments would most likely be torn if you slipped and your foot was suddenly forced into an excessively inverted position: (a) the posterior talofibular and calcaneofibular ligaments, or (b) the medial ligament? Explain. What would the resulting condition of the ankle be called?

3. In order of occurrence, list the joint actions (flexion, pronation, etc.) and the joints where they would occur as you (a) sit down at a table, (b) reach out and pick up an apple, (c) take a bite, and (d) chew it. Assume that you start in anatomical position.

4. What structure in the elbow joint serves the same purpose as the anterior cruciate ligament (ACL) of the knee?

5. List the six types of synovial joints and for each one, if possible, identify a joint in the upper limb and a joint in the lower limb that fall into each category. Which of these six joints have no examples in the lower limb?

Answers at www.mhhe.com/saladinha3

Improve Your Grade at www.mhhe.com/saladinha3

Practice quizzes, labeling activities, games, and flashcards provide fun ways to master concepts. You can also download image PowerPoint files for each chapter to create a study guide or for taking notes during lecture.

The Muscular System—Introduction

Neuromuscular junctions (SEM). Muscle fibers are shown in blue and nerve fibers in yellow.

CHAPTER 10

CHAPTER OUTLINE

DEEPER INSIGHTS

BRUSHING UP

To understand this chapter, you may find it helpful to review the following concepts:
- Proteins of the plasma membrane (p. 32)
- Components of a neuron (p. 71)
- The three types of muscle (p. 71)
- Embryonic mesoderm, somites, and myotomes (p. 90)
- Skeletal anatomy (chapters 7 and 8)

Anatomy & Physiology | REVEALED®
aprevealed.com

Muscular System

Muscles constitute nearly half of the body's weight and occupy a place of central interest in several fields of health care and fitness. Physical and occupational therapists must be well acquainted with the muscular system to plan and carry out rehabilitation programs. Athletes and trainers, dancers and acrobats, and amateur fitness enthusiasts follow programs of resistance training to strengthen individual muscle groups through movement regimens based on knowledge of muscle, bone, and joint anatomy. Nurses employ their knowledge of the muscular system to give intramuscular injections correctly and to safely and effectively move patients who are physically incapacitated. Gerontological nurses are keenly aware of how deeply a person's muscular condition affects the quality of life in old age. The muscular system is highly important to biomedical disciplines even beyond the scope of the movement sciences. It is the primary source of body heat in the moving individual, and through its absorption, storage, and use of glucose, it is a significant factor in blood glucose level.

The next three chapters focus on the muscular system—the functional anatomy of muscular tissue in this chapter, muscles of the axial region of the body (head and trunk) in chapter 11, and muscles of the appendicular region (limbs and limb girdles) in chapter 12. These chapters draw on what we have covered in the preceding chapters—bone and joint structure—to flesh out our comprehension of body posture and movement. The current chapter also considers cardiac and smooth muscles and how they compare with skeletal muscle.

10.1 Muscle Types and Functions

▶ Expected Learning Outcomes

When you have completed this section, you should be able to
- describe the distinctions between the three types of muscular tissue; and
- list the functions of muscular tissue and the properties it must have to carry out these functions.

As we saw in chapter 3, there are three kinds of muscular tissue in the human body—skeletal, cardiac, and smooth. All types, however, are specialized for one fundamental purpose: to convert the chemical energy of ATP into the mechanical energy of motion. Muscle cells exert a useful force on other cells or tissues—either to produce desirable movements or to prevent undesirable ones.

Although we examine all three muscle types in this chapter, most of our attention will focus on the **muscular**[1] **system,** composed of the skeletal muscles only. The word *muscle* means "little mouse," apparently referring to the appearance of muscles rippling under the skin. The study of the skeletal muscles is called **myology.**[2]

---

[1]*mus* = mouse + *cul* = little + *ar* = resembling
[2]*myo* = muscle + *logy* = study of

Types of Muscle

Skeletal muscle may be defined as voluntary striated muscle that is usually attached to one or more bones. It is called *voluntary* because it is usually subject to conscious control; we can decide when to contract a skeletal muscle. It is called *striated* because it exhibits a microscopic pattern of alternating light and dark bands, or **striations,** which result from the overlapping arrangement of the contractile proteins within each cell (fig. 10.1). A typical skeletal muscle cell is about 100 μm in diameter and 3 cm (30,000 μm) long; some are as thick as 500 μm and as long as 30 cm. Because of their extraordinary length, skeletal muscle cells are usually called **muscle fibers** or **myofibers.**

Cardiac muscle is also striated, but it is **involuntary**—not normally under conscious control. Its cells are not fibrous in shape, but relatively short and stumpy, somewhat like logs with notched ends. Thus, they are called **cardiocytes** or **myocytes** rather than fibers. Cardiocytes are commonly about 80 μm long by 15 μm wide.

Smooth muscle is also involuntary, and unlike skeletal and cardiac muscle, it lacks striations—hence the description *smooth*. It contains the same contractile proteins as the other muscle types, but they are not arranged in a regularly overlapping way, so there are no striations. Its cells, also called myocytes, are fusiform in shape—thick in the middle and tapered at the ends. They average about 200 μm long by 5 μm wide at the thickest part, but they can be as short as 20 μm in small blood vessels or as long as 500 μm in the pregnant uterus.

Functions of Muscle

The functions of muscular tissue are as follows:

- **Movement.** Muscles enable us to move from place to place and to move individual body parts; they move body contents in the course of breathing, blood circulation, feeding and digestion, defecation, urination, and childbirth; and they serve various roles in communication—speech, writing, facial expressions, and other body language.

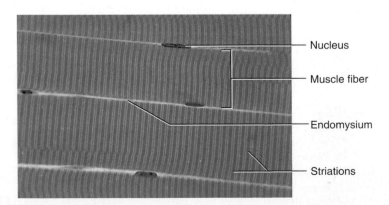

Figure 10.1 Skeletal Muscle Fibers.
● *What tissue characteristics evident in this photo distinguish this from cardiac and smooth muscle?*

Labels: Nucleus, Muscle fiber, Endomysium, Striations

- **Stability.** Muscles maintain posture by preventing unwanted movements. Some are called *antigravity muscles* because, at least part of the time, they resist the pull of gravity and prevent us from falling or slumping over. Many muscles also stabilize the joints by maintaining tension on tendons and bones.

- **Control of body openings and passages.** Muscles encircling the mouth serve not only for speech but also for food intake and retention of food while chewing. In the eyelid and pupil, they regulate the admission of light to the eye. Internal muscular rings control the movement of food, bile, blood, and other materials within the body. Muscles encircling the urethra and anus control the elimination of waste.

- **Heat production.** The skeletal muscles produce as much as 85% of one's body heat, which is vital to the functioning of enzymes and therefore to all metabolism.

- **Glycemic control.** This means the regulation of blood glucose within normal limits. The skeletal muscles absorb, store, and use a large share of one's glucose and play a significant role in stabilizing its blood concentration. In old age, in obesity, and when muscles become deconditioned and weakened, people suffer an increased risk of type 2 diabetes mellitus because of the decline in this glucose-buffering function.

Universal Properties of Muscle

To carry out the foregoing functions, muscle cells must have the following properties:

- **Excitability (responsiveness).** This is a property of all living cells, but it is developed to the highest degree in muscle and nerve cells. When stimulated by chemical signals, stretch, and other stimuli, muscle cells exhibit electrical and mechanical responses.

- **Conductivity.** The local electrical excitation produced at the point of muscle stimulation is conducted throughout the entire plasma membrane, stimulating all regions of the muscle cell and initiating the events that lead to contraction.

- **Contractility.** Muscle fibers are unique in their ability to shorten substantially when stimulated. This enables them to pull on bones and other organs to create movement.

- **Extensibility.** Most cells rupture if they are stretched even a little, but skeletal muscle fibers are unusually extensible; they can stretch to as much as three times their contracted length without harm. If it were not for this property, the muscle on one side of a joint would resist the action of a muscle on the other side. An elbow flexor such as the *biceps brachii*, for example, would resist elbow extension by the *triceps brachii* (see fig. 10.4).

- **Elasticity.** When a muscle cell is stretched and then the tension is released, it recoils to a shorter original length. If it were not for this elastic recoil, resting muscles would be flabby.

From this point on, this chapter concerns skeletal muscles unless otherwise stated.

Before You Go On

Answer the following questions to test your understanding of the preceding section:

1. What general function of muscular tissue distinguishes it from other tissue types?
2. What are the basic structural differences between skeletal, cardiac, and smooth muscle?
3. State four functions of the muscular system.
4. State five special properties of muscular tissue that enable it to perform its functions.

10.2 General Anatomy of Muscles

▶ Expected Learning Outcomes

When you have completed this section, you should be able to

- describe the connective tissues and associated structural organization of a muscle;
- explain how muscles are classified by the arrangement of their fiber bundles (fascicles);
- describe the types of muscle–bone attachments;
- describe the parts of a typical muscle;
- describe the way that muscles are arranged in groups with complementary actions at a joint;
- explain what intrinsic and extrinsic muscles are; and
- describe three types of musculoskeletal levers and their respective advantages.

Connective Tissues and Fascicles

A skeletal muscle is more than muscular tissue. It also contains connective tissue, nerves, and blood vessels. The connective tissue components, from the smallest to largest and from deep to superficial, are as follows (fig. 10.2):

- **Endomysium**[3] (EN-doe-MIZ-ee-um). This is a thin sleeve of loose connective tissue that surrounds each muscle fiber. It creates room for blood capillaries and nerve fibers to reach every muscle fiber. The endomysium also provides the extracellular chemical environment for the muscle fiber and its associated nerve ending. Excitation of a muscle fiber is based on the exchange of calcium, sodium, and potassium ions between the endomysial tissue fluid and the nerve and muscle fibers.

- **Perimysium.**[4] This is a thicker connective tissue sheath that wraps muscle fibers together in bundles called **fascicles**[5] (FASS-ih-culs). Fascicles are visible to the naked eye as

---

[3] *endo* = within + *mys* = muscle
[4] *peri* = around
[5] *fasc* = bundle + *icle* = little

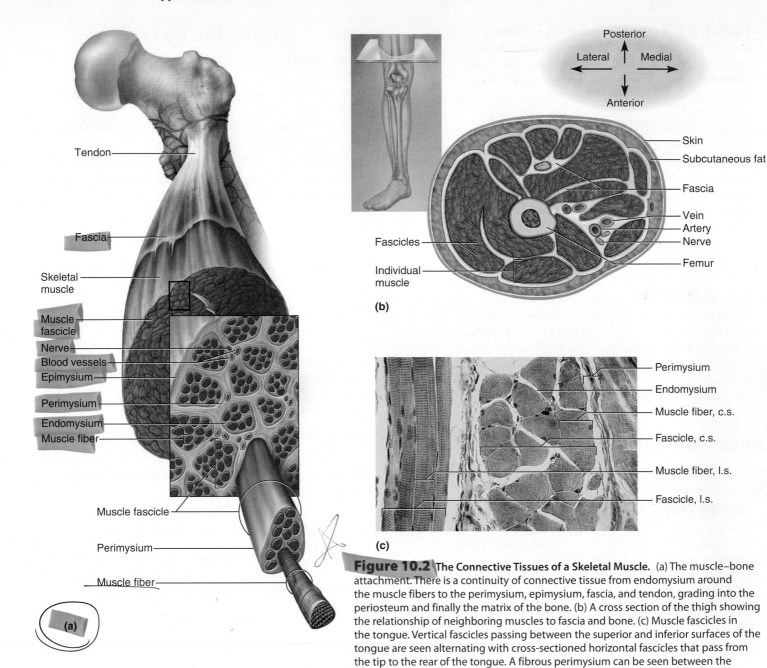

Figure 10.2 **The Connective Tissues of a Skeletal Muscle.** (a) The muscle–bone attachment. There is a continuity of connective tissue from endomysium around the muscle fibers to the perimysium, epimysium, fascia, and tendon, grading into the periosteum and finally the matrix of the bone. (b) A cross section of the thigh showing the relationship of neighboring muscles to fascia and bone. (c) Muscle fascicles in the tongue. Vertical fascicles passing between the superior and inferior surfaces of the tongue are seen alternating with cross-sectioned horizontal fascicles that pass from the tip to the rear of the tongue. A fibrous perimysium can be seen between the fascicles, and endomysium can be seen between individual muscle fibers within each fascicle. (c.s. = cross section; l.s. = longitudinal section)

parallel strands—the "grain" in a cut of meat; tender roast beef is easily pulled apart along its fascicles. The perimysium carries the larger nerves and blood vessels as well as stretch receptors called muscle spindles (see p. 463).

- **Epimysium.**[6] This is a fibrous sheath that surrounds the entire muscle. On its outer surface, the epimysium grades into the fascia, and its inner surface issues projections between the fascicles to form the perimysium.

- **Fascia** (FASH-ee-uh). This is a sheet of connective tissue that separates neighboring muscles or muscle groups from each other and from the subcutaneous tissue. Muscles are grouped in *compartments* separated from each other by fascia.

Fascicles and Muscle Shapes

The strength of a muscle and the direction of its pull are determined partly by the orientation of its fascicles. Muscles can be classified according to fascicle orientation as follows (fig. 10.3):

---

[6]*epi* = upon, above

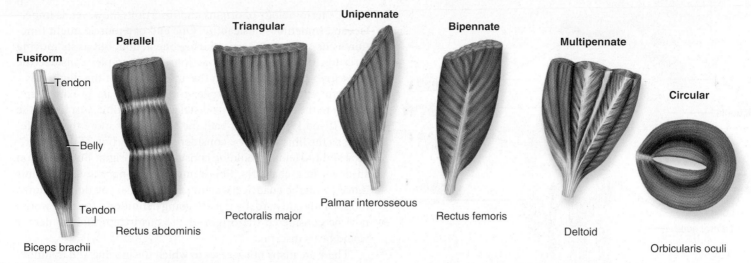

Figure 10.3 **Classification of Muscles According to Fascicle Orientation.** The fascicles are the "grain" visible in each illustration.
• *Why could some parallel muscles be stronger than some pennate muscles?*

- **Fusiform[7] muscles** are thick in the middle and tapered at each end. The *biceps brachii* of the arm and *gastrocnemius* of the calf are examples of this type. Strength is proportional to a muscle's diameter at its thickest point, and fusiform muscles are relatively strong.

- **Parallel muscles** have a fairly uniform width and parallel fascicles. Some of these are elongated straps, such as the *rectus abdominis* of the abdomen, *sartorius* of the thigh, and *zygomaticus major* of the face. Others are more squarish and are called *quadrilateral* (four-sided) muscles, such as the *masseter* of the jaw. Parallel muscles can span long distances, such as from hip to knee, and they shorten more than other muscle types. However, having fewer muscle fibers than fusiform muscles of comparable mass, they produce less force.

- **Triangular (convergent) muscles** are fan-shaped—broad at one end and narrower at the other. Examples include the *pectoralis major* in the chest and the *temporalis* on the side of the head. Despite their small localized insertions on a bone, these muscles are relatively strong because they contain a large number of fibers in the wider part of the muscle.

- **Pennate[8] muscles** are feather-shaped. Their fascicles insert obliquely on a tendon that runs the length of the muscle, like the shaft of a feather. There are three types of pennate muscles: *unipennate*, in which all fascicles approach the tendon from one side (for example, the *palmar interosseous muscles* of the hand and *semimembranosus* of the thigh); *bipennate*, in which fascicles approach the tendon from both sides (for example, the *rectus femoris* of the thigh); and *multipennate*, shaped like a bunch of feathers with their quills converging on a single point (for example, the *deltoid* of the shoulder). These muscles generate more force than the preceding types because they fit more muscle fibers into a given length of muscle.

- **Circular muscles (sphincters)** form rings around certain body openings. When they contract, they constrict the opening and tend to prevent the passage of material through it. Examples include the *orbicularis oculi* of the eyelids and the *external urethral* and *anal* sphincters. Smooth muscle can also form sphincters—for example, the *pyloric valve* at the passage from the stomach to the small intestine and sphincters of the urinary tract and anal canal.

Muscle Attachments

Skeletal muscles attach to bones through extensions of their connective tissue components. There are two forms of attachment—*indirect* and *direct*.

In an **indirect attachment,** the muscle ends conspicuously short of its bony destination, and the gap is bridged by a fibrous band or sheet called a **tendon.** See, for example, the two ends of the biceps brachii in figure 10.4 and the photographs in figures 12.10b (p. 308) and 12.16 (p. 317). You can easily palpate tendons and feel their texture just above the heel (your *calcaneal* or *Achilles tendon*) and on the anterior side of the wrist (tendons of the *palmaris longus* and *flexor carpi radialis* muscles). Collagen fibers of the muscle (the endo-, peri-, and epimysium) continue into the tendon and from there into the periosteum and matrix of the bone, creating very strong structural continuity from muscle to bone.

In some cases, the tendon is a broad sheet called an **aponeurosis[9]** (AP-oh-new-RO-sis). This term originally referred to the tendon located beneath the scalp, but now it also refers to similar tendons associated with certain abdominal, lumbar, hand, and foot muscles. For example, the palmaris longus tendon passes through the wrist and then expands into a fanlike *palmar aponeurosis* beneath the skin of the palm (see fig. 12.8a, p. 305).

---

[7]*fusi* = spindle + *form* = shape
[8]*penna* = feather + *ate* = characterized by

[9]*apo* = upon, above + *neuro* = nerve, nervous tissue

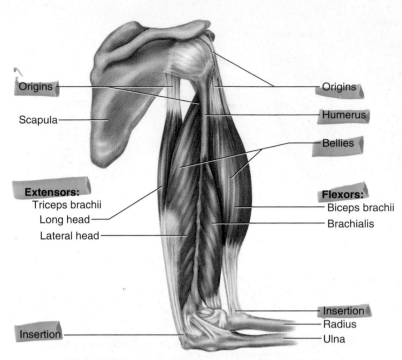

Origins

Scapula

Extensors:
Triceps brachii
Long head
Lateral head

Insertion

Origins

Humerus

Bellies

Flexors:
Biceps brachii
Brachialis

Insertion
Radius
Ulna

Figure 10.4 Muscle Groups Acting on the Elbow. The biceps brachii and brachialis are synergists in elbow flexion. The brachialis is the prime mover in flexion. The triceps brachii is an antagonist of these two muscles and is the prime mover in elbow extension.
● *Which of these muscles have direct attachments to the bones, and which have indirect attachments?*

In a **direct (fleshy) attachment,** there is so little separation between muscle and bone that to the naked eye, the red muscular tissue seems to emerge directly from the bone—for example, along the margins of the *brachialis* and lateral head of the *triceps brachii* in figure 10.4. At a microscopic level, however, the muscle fibers stop slightly short of the bone, and the gap between muscle and bone is spanned by collagen fibers.

Some authorities contend that the tendons and other collagenous tissues stretch and recoil significantly during muscle action and contribute to the power output and efficiency of a muscle. When you are running, for example, recoil of the calcaneal tendon may help to lift the heel and produce some of the thrust as your toes push off from the ground. (Such recoil contributes significantly to the long, energy-efficient leaping of kangaroos.) Others feel the elasticity of these components is negligible in humans and that the recoil is produced entirely by certain intracellular proteins of the muscle fibers themselves.

Muscle Origins and Insertions

Most skeletal muscles are attached to a different bone at each end, so either the muscle or its tendon spans at least one joint. When the muscle contracts, it moves one bone relative to the other. The bony site of attachment at the relatively stationary end is called its **origin.** The attachment site at the more mobile end is called the **insertion.** For the biceps brachii, for example, the origin is the scapula and the insertion is the radius (fig. 10.4). The *middle,* usually thicker region is called the **belly.**

The terminology of origins and insertions, however, is imperfect and sometimes misleading. One end of a muscle might function as its stationary origin during one action, but as its moving insertion during a different action. For example, consider the *quadriceps femoris* muscle on the anterior side of the thigh. It is a powerful extensor of the knee, connected at its proximal end mainly to the femur and at its distal end to the tibia, just below the knee. If you kick a soccer ball, the tibia moves more than the femur, so the tibia would be considered the insertion of the quadriceps and the femur would be considered its origin. But when you sit down in a chair, the tibia remains stationary and the femur moves, with the quadriceps acting as a brake so you don't sit down too abruptly and hard. By the foregoing definitions, the tibia would now be considered the origin of the quadriceps and the femur would be its insertion.

There are many other cases in which the moving and nonmoving ends of the muscle are reversed when different actions are performed. Consider the difference, for example, in the relative movements of the humerus and ulna when flexing the elbow to lift dumbbells as compared with flexing the elbow to perform chin-ups or scale a climbing wall. For such reasons, some anatomists are abandoning origin and insertion terminology and speaking instead of a muscle's proximal and distal or superior and inferior attachments, especially in the limbs. Nevertheless, this book uses the traditional, admittedly imperfect descriptions.

Some muscles insert not on bone but on the fascia or tendon of another muscle or on collagen fibers of the dermis of the skin. The distal tendon of the biceps brachii, for example, inserts partly on the fascia of the forearm. Many facial muscles insert in the skin, enabling them to produce expressions such as a smile.

Functional Groups of Muscles

The effect produced by a muscle, whether it is to produce or prevent a movement, is called its **action.** Skeletal muscles seldom act independently; instead, they function in groups whose combined actions produce the coordinated control of a joint. Muscles can be classified into four categories according to their actions, but it must be stressed that a particular muscle can act in a certain way during one joint action and in a different way during other actions of the same joint. The following examples are illustrated in figure 10.4:

1. The **prime mover (agonist)** is the muscle that produces most of the force during a particular joint action. In flexing the elbow, for example, the prime mover is the *brachialis*.

2. A **synergist**[10] (SIN-ur-jist) is a muscle that aids the prime mover. Two or more synergists acting on a joint can produce more power than a single larger muscle. The *biceps brachii,* for example, overlies the brachialis and works with it as a synergist to flex the elbow. The actions of a prime mover and its synergist are not necessarily identical and redundant. If the prime mover worked alone at a joint, it might cause rotation or other undesirable movements of a bone. A synergist may

[10]*syn* = together + *erg* = work

stabilize a joint and restrict these movements, or modify the direction of a movement so that the action of the prime mover is more coordinated and specific.

3. An **antagonist**[11] is a muscle that opposes the prime mover. In some cases, it relaxes to give the prime mover almost complete control over an action. More often, however, the antagonist maintains some tension on a joint and thus limits the speed or range of the prime mover, preventing excessive movement, joint injury, or inappropriate actions. If you extend your arm to reach out and pick up a cup of tea, for example, your *triceps brachii* serves as the prime mover of elbow extension, and your brachialis acts as an antagonist to slow the extension and stop it at the appropriate point. If you extend your arm rapidly to throw a dart, however, the brachialis must be quite relaxed. The brachialis and triceps represent an **antagonistic pair** of muscles that act on opposite sides of a joint. We need antagonistic pairs at a joint because a muscle can only pull, not push—for example, a single muscle cannot flex *and* extend the elbow. Which member of the pair acts as the prime mover depends on the motion under consideration. In flexion of the elbow, the brachialis is the prime mover and the triceps is the antagonist; when the elbow is extended, their roles are reversed.

4. A **fixator** is a muscle that prevents a bone from moving. To *fix* a bone means to hold it steady, allowing another muscle attached to it to pull on something else. For example, consider again the flexion of the elbow by the biceps brachii. The biceps originates on the scapula, and inserts on the radius. The scapula is loosely attached to the axial skeleton, so when the biceps contracts, it seems that it would pull the scapula laterally. However, there are fixator muscles (the *rhomboids*) that attach the scapula to the vertebral column. They contract at the same time as the biceps, holding the scapula firmly in place and ensuring that the force generated by the biceps moves the radius rather than the scapula.

Intrinsic and Extrinsic Muscles

In places such as the tongue, larynx, back, hand, and foot, anatomists distinguish between intrinsic and extrinsic muscles. An **intrinsic muscle** is entirely contained within a particular region, having both its origin and insertion there. An **extrinsic muscle** acts upon a designated region but has its origin elsewhere. For example, some movements of the fingers are produced by extrinsic muscles in the forearm, whose long tendons reach to the phalanges; other finger movements are produced by intrinsic muscles between the metacarpal bones.

Muscles, Bones, and Levers

Many bones, especially the long bones, act as levers on which the muscles exert their force. A **lever** is any elongated, rigid object that rotates around a fixed point called the **fulcrum** (fig. 10.5). Rotation

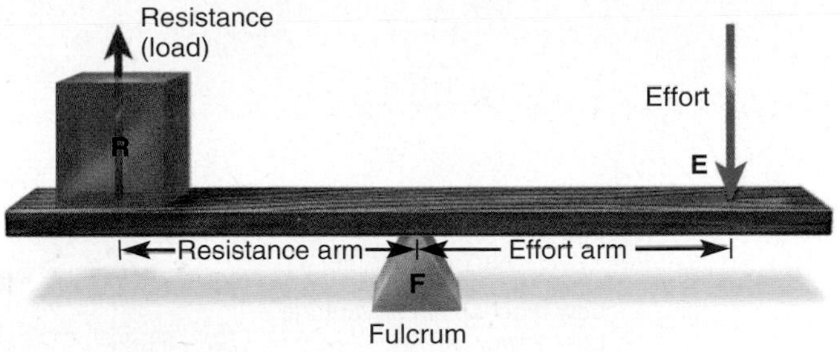

Figure 10.5 Basic Components of a Lever.
● *Is this a first-, second-, or third-class lever? How do you know?*

occurs when an **effort** applied to one point on the lever overcomes a **resistance (load)** located at some other point. The part of a lever from the fulcrum to the point of effort is called the **effort arm,** and the part from the fulcrum to the point of resistance is the **resistance arm.** In the body, a bone acts as a lever, a joint serves as the fulcrum, and the effort is generated by a muscle.

The function of a lever is to produce a gain in the speed, distance, or force of a motion—either to exert more force against a resisting object than the force applied to the lever (for example, in moving a heavy object with a crowbar) or to move the resisting object farther or faster than the effort arm is moved (as in rowing a boat). A single lever cannot confer both advantages. There is a trade-off between force on one hand and speed or distance on the other—as one increases, the other decreases.

The **mechanical advantage (MA)** of a lever is the ratio of its output force to its input force. It is equal to the length of the effort arm, L_E, divided by the length of the resistance arm, L_R; that is, MA $= L_E/L_R$. If MA is greater than 1.0, the lever produces more force, but less speed or distance, than the force exerted on it. If MA is less than 1.0, the lever produces more speed or distance, but less force, than the input. Consider the forearm, for example (fig. 10.6a). The resistance arm of the ulna is longer than the effort arm, so we know

DEEPER INSIGHT   10.1

Muscle-Bound

Any well-planned program of resistance (strength) training or bodybuilding must include exercises aimed at proportional development of the different members of a muscle group, such as flexors and extensors of the arm. Otherwise, the muscles on one side of a joint may develop out of proportion to their antagonists and restrict the joint's range of motion (ROM). If the biceps brachii is heavily developed without proportionate attention to the triceps brachii, for example, the stronger biceps will cause the elbow to be somewhat flexed constantly, and the ROM of the elbow will be restricted. The joint is then said to be "muscle-bound." People with muscle-bound joints move awkwardly and are poor at activities that require agility, such as dance and ball games.

---

[11]*ant* = against + *agonist* = actor, competitor

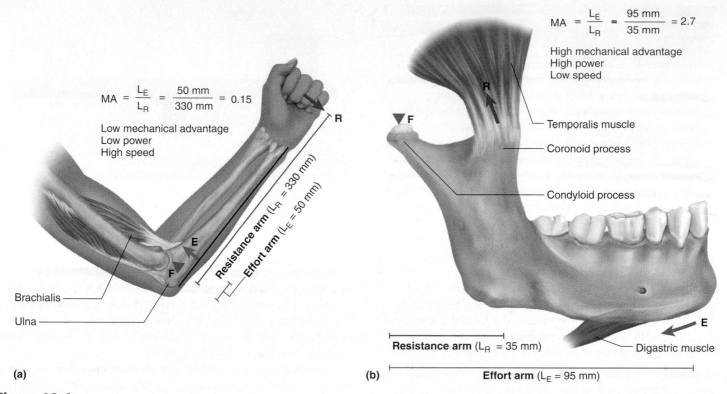

$$MA = \frac{L_E}{L_R} = \frac{50 \text{ mm}}{330 \text{ mm}} = 0.15$$

Low mechanical advantage
Low power
High speed

Resistance arm (L_R = 330 mm)

Effort arm (L_E = 50 mm)

E

F

R

Brachialis

Ulna

(a)

$$MA = \frac{L_E}{L_R} = \frac{95 \text{ mm}}{35 \text{ mm}} = 2.7$$

High mechanical advantage
High power
Low speed

R

F

Temporalis muscle

Coronoid process

Condyloid process

E

Digastric muscle

Resistance arm (L_R = 35 mm)

Effort arm (L_E = 95 mm)

(b)

Figure 10.6 **Mechanical Advantage (MA).** MA is calculated as the length of the effort arm divided by the length of the resistance arm. (a) The forearm acts as a third-class lever during flexion of the elbow. (b) The mandible acts as a second-class lever when the jaw is forcibly opened. The digastric muscle and others provide the effort, while tension in the temporalis muscle and others provides the resistance.

from the preceding formula that the mechanical advantage is less than 1.0. The figure shows some representative values for L_E and L_R that yield MA = 0.15. The brachialis puts more power into the ulna than we get out of it, but the hand moves farther and faster than the insertion of the brachialis tendon. By contrast, when the digastric muscle depresses the mandible, the MA is about 2.7. The coronoid process of the mandible moves with greater force, but a shorter distance, than the insertion of the digastric (fig. 10.6b).

As we have already seen, some joints have two or more muscles acting on them that seemingly produce the same effect, such as elbow flexion. At first, you might consider this arrangement redundant, but it makes sense if the tendinous insertions of the muscles are at slightly different places and produce different mechanical advantages. A runner taking off from the starting line, for example, uses "low-gear" (high-MA) muscles that do not generate much speed but have the power to overcome the inertia of the body. A runner then "shifts into high gear" by using muscles with different insertions that have a lower mechanical advantage but produce more speed at the feet. This is analogous to the way an automobile transmission works to get a car moving and then shifts gears to cruise at high speed.

There are three classes of levers that differ with respect to which component is in the middle—the fulcrum (F), effort (E), or resistance (R). A **first-class lever** (fig. 10.7a) is one with the fulcrum in

the middle (EFR), such as a seesaw. An anatomical example is the atlanto–occipital joint of the neck, where the muscles of the back of the neck pull down on the occipital bone of the skull and oppose the tendency of the head to tip forward. Loss of muscle tone here can be embarrassing if you nod off in class.

A **second-class lever** (fig. 10.7b) is one in which the resistance is in the middle (FRE). Lifting the handles of a wheelbarrow, for example, makes it pivot on its wheel at the opposite end and lift a load in the middle. The mandible acts as a second-class lever when the digastric muscle pulls down on the chin to open the mouth. The fulcrum is the temporomandibular joint, the effort is applied to the chin by the digastric muscle, and the resistance is the tension of muscles such as the temporalis muscle, which is used to bite and to hold the mouth closed. (This arrangement is upside down relative to a wheelbarrow, but the mechanics remain the same.)

In a **third-class lever** (fig. 10.7c), the effort is applied between the fulcrum and resistance (FER). For example, in paddling a canoe, the relatively stationary grip at the upper end of the paddle is the fulcrum, the effort is applied to the middle of the shaft, and the resistance is produced by the water against the blade. Most musculoskeletal levers are third-class. The forearm acts as a third-class lever when you flex your elbow. The fulcrum is the joint between the ulna and humerus; the effort is applied by the brachialis and biceps brachii muscles; and the resistance can be provided by

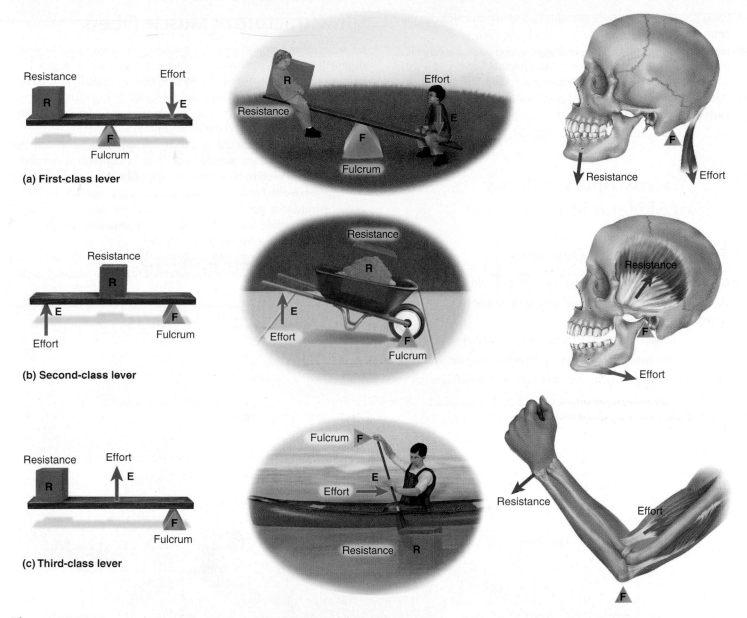

Figure 10.7 **The Three Classes of Levers.** Left: The lever classes defined by the relative positions of the resistance (load), fulcrum, and effort. Center: Mechanical examples. Right: Anatomical examples. (a) Muscles at the back of the neck pull down on the occipital bone to oppose the tendency of the head to tip forward. The fulcrum is the occipital condyles. (b) To open the mouth, the digastric muscle pulls down on the chin. It is resisted by the temporalis muscle on the side of the head. The fulcrum is the temporomandibular joint. (c) In flexing the elbow, the biceps brachii exerts an effort on the radius. Resistance is provided by the weight of the forearm or anything held in the hand. The fulcrum is the elbow joint.

any weight in the hand or the weight of the forearm itself. The mandible acts as a third-class lever when you close your mouth to bite off a piece of food. Again, the temporomandibular joint is the fulcrum, but now the temporalis muscle exerts the effort, while the resistance is supplied by the item of food being bitten.

Apply What You Know

Sit on the edge of a desk with your feet off the floor. Plantarflex your foot. Where is the effort? Where is the fulcrum? (Name the specific joint, based on chapter 9.) Where is the resistance? Which class of lever does the foot represent in plantar flexion?

Before You Go On

Answer the following questions to test your understanding of the preceding section:

5. Name the connective tissue layers of a muscle beginning with the individual muscle fiber and ending with the tissue that separates the muscles from the skin.

6. Sketch the fascicle arrangements that define a fusiform, parallel, convergent, pennate, and circular muscle.

7. Define the *origin, insertion,* and *action* of a muscle.

8. Distinguish between a prime mover, synergist, antagonist, and fixator.

9. What is the difference between intrinsic and extrinsic muscles that control the fingers?

10. What is the principal benefit of a joint action with a mechanical advantage less than 1.0? What is the principal benefit of a joint action with a mechanical advantage greater than 1.0?

11. Define a first-, second-, and third-class lever and give an example of each in the musculoskeletal system.

10.3 Microscopic Anatomy of Skeletal Muscle

▶ Expected Learning Outcomes

When you have completed this section, you should be able to

- describe the ultrastructure of a muscle fiber and its myofilaments;
- explain what accounts for the striations of skeletal muscle;
- describe the relationship of a nerve fiber to a muscle fiber;
- define a *motor unit* and discuss its functional significance; and
- describe the blood supply to a skeletal muscle.

Ultrastructure of Muscle Fibers

In order to understand muscle function, you must know how the organelles and macromolecules of a muscle fiber are arranged. Perhaps more than any other cell, a muscle fiber exemplifies the adage, form follows function. It has a complex, tightly organized internal structure in which even the spatial arrangement of protein molecules is closely tied to its contractile function (fig. 10.8).

The plasma membrane is called the **sarcolemma**[12] and the cytoplasm is called the **sarcoplasm.** The sarcoplasm is occupied mainly by long protein bundles called **myofibrils,** each about 1 μm in diameter. It also contains an abundance of **glycogen,** a starchlike carbohydrate that provides energy for the cell during periods of heightened exercise, and the red pigment **myoglobin,** which stores oxygen until needed for muscular activity.

Muscle fibers have numerous flattened or sausage-shaped nuclei pressed against the inside of the sarcolemma. This unusual multi-

---

[12]*sarco* = flesh, muscle + *lemma* = husk

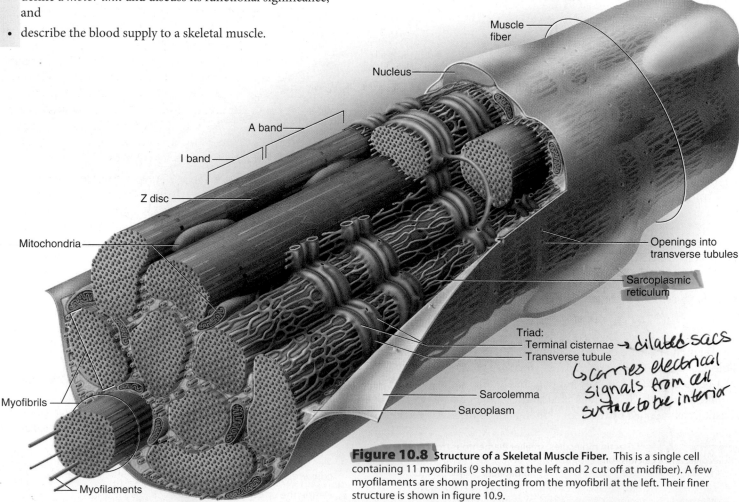

Figure 10.8 **Structure of a Skeletal Muscle Fiber.** This is a single cell containing 11 myofibrils (9 shown at the left and 2 cut off at midfiber). A few myofilaments are shown projecting from the myofibril at the left. Their finer structure is shown in figure 10.9.

● *Why is it important for the transverse tubule to be so closely associated with the terminal cisternae?*

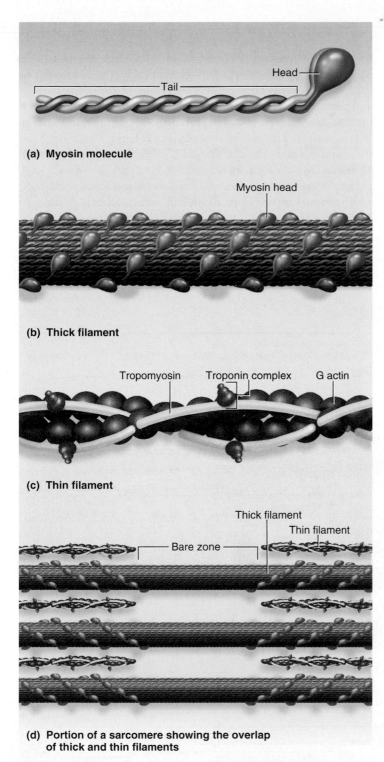

(a) Myosin molecule

(b) Thick filament

(c) Thin filament

(d) Portion of a sarcomere showing the overlap of thick and thin filaments

Figure 10.9 **Molecular Structure of Thick and Thin Filaments.** (a) A single myosin molecule consists of two intertwined proteins forming a filamentous tail and a double globular head. (b) A thick filament consists of 200 to 500 myosin molecules bundled together with the heads projecting outward In a helical array. (c) A thin filament consists of two intertwined chains of G actin molecules, smaller filamentous tropomyosin molecules, and a calcium-binding protein called troponin associated with the tropomyosin. (d) A region of overlap between the thick and thin myofilaments.

nuclear condition results from the embryonic development of the muscle fiber, as we will see near the end of this chapter. Most other organelles of the cell, such as mitochondria, are packed into the spaces between the myofibrils. The smooth endoplasmic reticulum, here called the **sarcoplasmic reticulum (SR),** forms a network around each myofibril. It periodically exhibits dilated end-sacs called **terminal cisternae,** which cross the muscle fiber from one side to the other. The sarcolemma turns inward at many points to form tunnels called **transverse (T) tubules,** which penetrate through the cell and emerge on the other side. Each T tubule is intimately associated with two terminal cisternae, which run alongside it. The T tubule carries electrical signals from the cell surface into the interior and induces gates in the SR membrane to open. The SR is a reservoir of calcium ions (Ca^{2+}). On command, it opens the gates and releases a flood of Ca^{2+} into the cytosol, and the Ca^{2+} activates muscle contraction.

Myofilaments

Most of the muscle fiber is filled with myofibrils. Each myofibril is composed of parallel protein microfilaments called **myofilaments.** The key to muscle contraction lies in the arrangement and action of these myofilaments, so we must examine these at a molecular level. There are three kinds of myofilaments:

1. **Thick filaments** (fig. 10.9a, b) are about 15 nm in diameter. Each is made of several hundred molecules of a protein called **myosin.** A myosin molecule is shaped like a golf club, with two chains that intertwine to form a shaftlike *tail* and a double globular *head* projecting from it at an angle. A thick filament may be likened to a bundle of 200 to 500 such "golf clubs," with their heads directed outward in a helical array around the bundle. The heads on one half of the thick filament angle to the left, and the heads on the other half angle to the right; in the middle is a *bare zone* with no heads.

2. **Thin filaments** (fig. 10.9c, d), 7 nm in diameter, are composed primarily of two intertwined strands of a protein called **fibrous (F) actin.** Each F actin is like a bead necklace, consisting of a string of subunits called **globular (G) actin.** Each G actin has an **active site** that can bind to the head of a myosin molecule (see fig. 10.14).

   A thin filament also has 40 to 60 molecules of yet another protein called **tropomyosin.** When a muscle fiber is relaxed, tropomyosin blocks the active sites of actin and prevents myosin from binding to them. Each tropomyosin molecule, in turn, has a smaller calcium-binding protein called **troponin** bound to it.

3. **Elastic filaments** (fig. 10.10b), 1 nm in diameter, are made of a huge springy protein called **titin**[13] **(connectin).** They flank each thick filament and anchor it to a structure called the *Z disc*. This helps to stabilize the thick filament, center it between the thin filaments, and prevent overstretching.

   Myosin and actin are called **contractile proteins** because they do the work of shortening the muscle fiber. Tropomyosin and troponin are called the **regulatory proteins** because they act like a switch to determine when the fiber can contract and when it cannot.

---

[13]*tit* = giant + *in* = protein

At least seven other accessory proteins occur in the thick and thin filaments or are associated with them. They anchor the myofilaments, regulate their length, and keep them aligned with each other for optimal contractile effectiveness. The most clinically important of these is **dystrophin,** an enormous protein located just under the sarcolemma in the vicinity of each *I band* of the striations described in the next section. It links the actin filaments to transmembrane proteins in the sarcolemma. These, in turn, are linked to proteins immediately external to the muscle fiber and ultimately to the endomysium. Thus, it is dystrophin that links the shortening of components within the fiber to a mechanical pull on the connective tissue external to the muscle fiber. Genetic defects in dystrophin are responsible for the disabling disease, muscular dystrophy (see p. 258).

Striations and Sarcomeres

Myosin and actin are not unique to muscle; these proteins occur in all cells, where they function in cellular motility, mitosis, and transport of intracellular materials. In skeletal and cardiac muscle they are especially abundant, however, and are organized into a precise array that accounts for the striations of these two muscle types (fig. 10.10).

Striated muscle has dark **A bands** alternating with lighter **I bands.** (*A* stands for *anisotropic* and *I* for *isotropic,* which refer to the way these bands affect polarized light. To help remember which band is which, think "d**A**rk" and "l**I**ght.") Each A band consists of thick filaments lying side by side. Part of the A band, where thick and thin filaments overlap, is especially dark. In this region, each thick filament is surrounded by thin filaments. In the middle of the A band, there is a lighter region called the **H band,**[14] into which the thin filaments do not reach. The thick filaments originate at a dark **M line** in the middle of the H band.

Each light I band is bisected by a dark narrow **Z disc**[15] **(Z line),** which provides anchorage for the thin filaments and elastic filaments. Each segment of a myofibril from one Z disc to the next is called a **sarcomere**[16] (SAR-co-meer), the functional contractile unit of the muscle fiber. A muscle shortens because its individual sarcomeres shorten and pull the Z discs closer to each other, and dystrophin and the linking proteins pull on the extracellular proteins of the muscle.

The terminology of muscle fiber structure is reviewed in table 10.1.

---

[14]H = *helle* = bright (German)
[15]Z = *Zwischenscheibe* = between disc (German)
[16]*sarco* = muscle + *mere* = part, segment

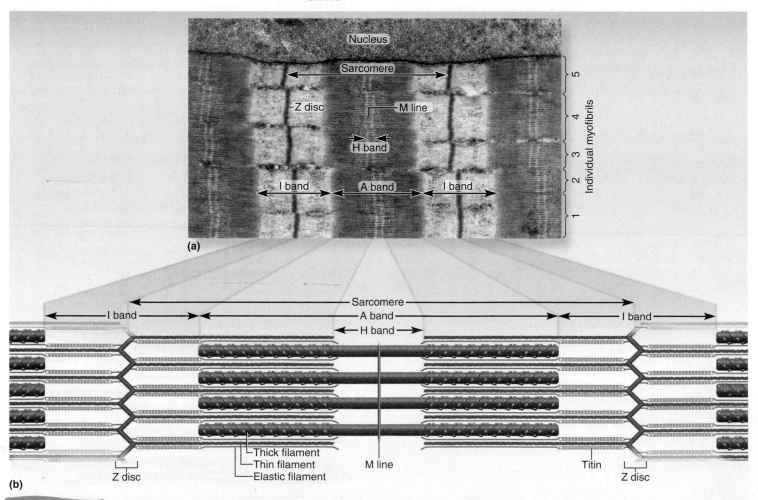

(a)

(b)

Figure 10.10 **Muscle Striations and Their Molecular Basis.** (a) Five myofibrils of a single muscle fiber, showing the striations in the relaxed state (TEM). (b) The overlapping pattern of thick and thin myofilaments that accounts for the striations seen in part (a).

The Nerve–Muscle Relationship

Skeletal muscle never contracts unless it is stimulated by a nerve (or artificially with electrodes). If its nerve connections are severed or poisoned, a muscle is paralyzed. Thus, muscle contraction cannot be understood without first understanding the relationship between nerve and muscle cells. Any point where a nerve fiber meets and stimulates another cell is called a **synapse** (SIN-aps). The other

TABLE 10.1	Structural Components of a Muscle Fiber
Term	**Definition**
General Structure and Contents of the Muscle Fiber	
Sarcolemma	The plasma membrane of a muscle fiber
Sarcoplasm	The cytoplasm of a muscle fiber
Glycogen	An energy-storage polysaccharide abundant in muscle
Myoglobin	An oxygen-storing red pigment of muscle
T tubule	A tunnel-like extension of the sarcolemma extending from one side of the muscle fiber to the other; conveys electrical signals from the cell surface to its interior
Sarcoplasmic reticulum	The smooth ER of a muscle fiber; a Ca^{2+} reservoir
Terminal cisternae	The dilated ends of sarcoplasmic reticulum adjacent to a T tubule
Myofibrils	
Myofibril	A bundle of protein microfilaments (myofilaments)
Myofilament	A threadlike complex of several hundred contractile protein molecules
Thick filament	A myofilament about 11 nm in diameter composed of bundled myosin molecules
Elastic filament	A myofilament about 1 nm in diameter composed of a giant protein, titin, that flanks a thick filament and anchors it to a Z disc; stabilizes and centers the thick filament and prevents overstretching
Thin filament	A myofilament about 5 to 6 nm in diameter composed of actin, troponin, and tropomyosin
Myosin	A protein with a long shaftlike tail and a globular head; constitutes the thick myofilament
F actin	A fibrous protein made of a long chain of G actin molecules twisted into a helix; main protein of the thin myofilament
G actin	A globular subunit of F actin with an active site for binding a myosin head
Regulatory proteins	Troponin and tropomyosin, proteins that do not directly engage in the sliding filament process of muscle contraction, but regulate myosin–actin binding
Tropomyosin	A regulatory protein that lies in the groove of F actin and, in relaxed muscle, blocks the myosin-binding active sites
Troponin	A regulatory protein associated with tropomyosin that acts as a calcium receptor
Titin	A springy protein that forms the elastic filaments and anchors the thick filaments to the Z discs
Dystrophin	A large protein that links thin filaments to transmembrane proteins, which, in turn, are linked to extracellular proteins; transfers the force of sarcomere contraction to the fibrous connective tissues of a muscle
Striations and Sarcomeres	
Striations	Alternating light and dark transverse bands across a myofibril
A band	Dark band formed by parallel thick filaments that partly overlap the thin filaments
H band	A lighter region in the middle of an A band that contains thick filaments only; thin filaments do not reach this far into the A band in relaxed muscle
M line	A dark line in the middle of an H band marking the origin of the thick filaments
I band	A light band composed of thin filaments only
Z disc	A disc of protein to which thin and elastic filaments are anchored at each end of a sarcomere; appears as a narrow dark line in the middle of the I band, and is often called the Z line
Sarcomere	A segment from one Z disc to the next; the contractile unit of a muscle fiber

cell can be another neuron, a gland cell, a muscle cell, or other type of cell. The heart of the nerve–muscle relationship is the synapse formed by a nerve cell and a skeletal muscle fiber.

Motor Neurons

Skeletal muscles are innervated by nerve cells called *somatic motor neurons* (see fig. 3.24, p. 71). The cell bodies of these neurons lie in the brainstem and spinal cord; their axons, called **somatic motor fibers,** lead to the skeletal muscles. At its distal end, each somatic motor fiber branches out to innervate a number of muscle fibers, but any given muscle fiber is supplied by only one motor neuron.

The Neuromuscular Junction

As a nerve fiber approaches an individual muscle fiber, it branches again to establish several points of contact within an ovoid region called the **neuromuscular junction (NMJ),** or **motor end plate** (fig. 10.11). Each terminal branch of the nerve fiber within the NMJ forms a synapse with the muscle fiber. The sarcolemma of the NMJ is irregularly indented, a little like a handprint pressed into soft clay. If you imagine the nerve fiber to be like your forearm and your hand to be spread out in this handprint, the individual synapses would be like the points where your fingertips contact the clay. Thus, one nerve fiber stimulates the muscle fiber at several points within each NMJ.

At each synapse, the nerve fiber ends with a bulbous swelling called a **synaptic** (sih-NAP-tic) **knob.** The knob doesn't directly touch the muscle fiber, but is separated from it by a narrow space called the **synaptic cleft,** about 60 to 100 nm wide (scarcely any wider than the thickness of one plasma membrane). A third cell, called a *Schwann cell,* envelops the entire junction and isolates it from the surrounding tissue fluid.

The synaptic knob contains spheroid organelles called **synaptic vesicles,** which are filled with a chemical called **acetylcholine (ACh)** (ASS-eh-till-CO-leen). When a nerve signal arrives at the synaptic knob, some of these vesicles release their ACh by exocytosis. ACh diffuses across the synaptic cleft and binds to membrane proteins called **ACh receptors** on the sarcolemma. These receptors respond to ACh by initiating electrical events that lead to muscle contraction. The sarcolemma has infoldings called **junctional folds** that increase the membrane surface area and allow for more ACh receptors, and thus more sensitivity of the muscle fiber to nervous stimulation.

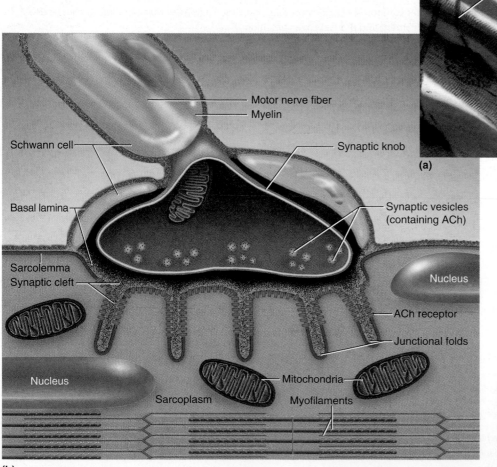

Motor nerve fibers

Neuromuscular junction

Muscle fibers

(a)

100 µm

Motor nerve fiber
Myelin

Schwann cell

Synaptic knob

Basal lamina

Synaptic vesicles (containing ACh)

Sarcolemma
Synaptic cleft

Nucleus

ACh receptor

Junctional folds

Nucleus

Mitochondria

Sarcoplasm　Myofilaments

(b)

Figure 10.11　Innervation of Skeletal Muscle. (a) Neuromuscular junctions, with muscle fibers slightly teased apart (LM). Compare the SEM photo on page 235. (b) Structure of a single neuromuscular synapse.
● *Why is a neuromuscular junction not synonymous with a neuromuscular synapse?*

DEEPER INSIGHT   10.2

Neuromuscular Toxins and Paralysis

Toxins that interfere with synaptic function can paralyze the muscles. Some pesticides, for example, contain *cholinesterase inhibitors* that bind to AChE and prevent it from degrading ACh. This causes *spastic paralysis,* a state of continual contraction of the muscle, which poses the danger of suffocation if it affects the laryngeal and respiratory muscles. *Tetanus* (lockjaw) is a form of spastic paralysis caused by the toxin of a soil bacterium, *Clostridium tetani.* In the spinal cord, a chemical called glycine normally stops motor neurons from producing unwanted muscle contractions. The tetanus toxin blocks glycine release and thus causes overstimulation and spastic paralysis of the muscles.

Flaccid paralysis is a state in which the muscles are limp and cannot contract. This too can cause respiratory arrest if it affects the thoracic muscles. Flaccid paralysis can be caused by poisons such as curare (cue-RAH-ree) that compete with ACh for receptor sites but do not stimulate the muscle. Curare is extracted from certain plants and used by some South American natives to poison blowgun darts. It has been used to treat muscle spasms in some neurological disorders and to relax abdominal muscles for surgery, but other muscle relaxants have now replaced curare for most purposes.

Botulism is a type of food poisoning caused by a neuromuscular toxin secreted by the bacterium *Clostridium botulinum.* Botulinum toxin blocks the release of ACh and causes flaccid muscle paralysis. Purified botulinum toxin was approved by the U.S. Food and Drug Administration (FDA) in 2002 for cosmetically treating "frown lines" caused by muscle tautness between the eyebrows. Marketed as Botox Cosmetic (a prescription drug despite the name), it is injected in small doses into specific facial muscles. The wrinkles gradually disappear as muscle paralysis sets in over the next few hours. The effect lasts about 4 months until the muscles retighten and the wrinkles return. Botox treatment has become the fastest-growing cosmetic medical procedure in the United States, with many people going for treatment every few months in their quest for a youthful appearance. It has begun to have some undesirable consequences, however, as it is sometimes administered by unqualified practitioners. Even some qualified physicians use it for treatments not yet approved by the FDA, and some host festive "Botox parties" for treatment of patients in assembly-line fashion.

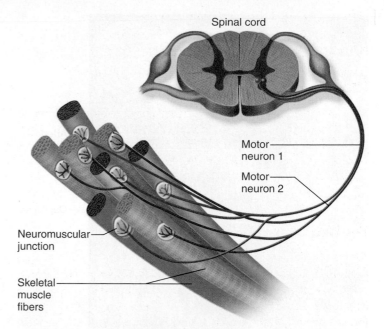

Figure 10.12 **Motor Units.** A motor unit consists of one motor neuron and all skeletal muscle fibers that it innervates. Two motor units here are represented by the red and blue nerve and muscle fibers. Note that the muscle fibers of a motor unit are not clustered together, but distributed throughout the muscle and commingled with the fibers of other motor units.

The entire muscle fiber is surrounded by a *basal lamina,* a thin layer composed partly of collagen and glycoproteins, which separates it from the surrounding connective tissue. The basal lamina passes through the synaptic cleft and virtually fills it, and covers the Schwann cell of the neuromuscular junction. An enzyme called **acetylcholinesterase (AChE)** (ASS-eh-till-CO-lin-ESS-ter-ase) is found in both the sarcolemma and basal lamina. It breaks down ACh after the ACh has stimulated the muscle cell; thus, it is important in turning off muscle contraction (see Deeper Insight 10.2).

The Motor Unit

When a nerve signal approaches the end of an axon, it spreads out over all of its terminal branches and stimulates all muscle fibers supplied by them. These muscle fibers therefore contract in unison. Since they behave as a single functional unit, one nerve fiber and all the muscle fibers innervated by it are called a **motor unit.** The muscle fibers of a single motor unit are not clustered together but are dispersed throughout a muscle (fig. 10.12). Thus, when they are stimulated, they cause a weak contraction over a wide area—not just a localized twitch in one small region. Effective muscle contraction usually requires the activation of several motor units at once.

On average, there are about 200 muscle fibers per motor neuron. But where fine control is needed, we have *small motor units.* In the muscles of eye movement, for example, there are about 3 to 6 muscle fibers per nerve fiber. Small motor units are not very strong, but they provide the fine degree of control needed for subtle movements. They also have small neurons that are easily stimulated. Where strength is more important than fine control, we have *large motor units.* The gastrocnemius muscle of the calf, for example, has about 1,000 muscle fibers per nerve fiber. Large motor units are much stronger, but have larger neurons that are harder to stimulate, and they do not produce such fine control.

One advantage of having multiple motor units in a muscle is that they are able to "work in shifts." Muscle fibers fatigue when subjected to continual stimulation. If all of the fibers in one of your postural muscles fatigued at once, for example, you might collapse. To prevent this, other motor units take over while the fatigued ones rest, and the muscle as a whole can sustain long-term contraction. Another advantage is that the strength of muscle contraction can be varied by activating more or fewer motor units.

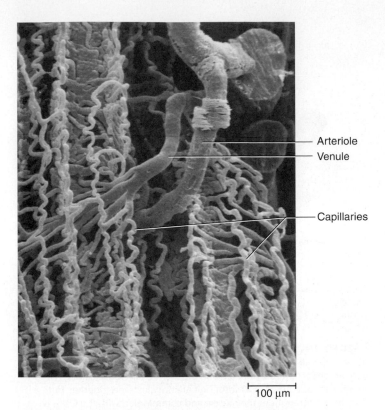

100 μm

Figure 10.13 A Vascular Cast of the Blood Vessels in a Contracted Skeletal Muscle. This cast was prepared by injecting the blood vessels with a polymer, digesting away the tissue to leave a replica of the vessels, and photographing the cast through the SEM. [From R. G. Kessell and R. H. Kardon, *Tissues and Organs: A Text-Atlas of Scanning Electron Microscopy* (W. H. Freeman & Co., 1979)]

The image is labeled: Arteriole, Venule, Capillaries

The Blood Supply

The muscular system as a whole receives about 1.25 L of blood per minute at rest—which is about one-quarter of the blood pumped by the heart. During heavy exercise, total cardiac output rises, and the muscular system's share of it is more than three-quarters, or 11.6 L/min. Working muscle has a great demand for glucose and oxygen. Blood capillaries ramify through the endomysium to reach every muscle fiber, sometimes so intimately associated with the muscle fibers that the fibers have surface indentations to accommodate them. The capillaries of skeletal muscle undulate or coil when the muscle is contracted (fig. 10.13), allowing them enough slack to stretch out straight, without breaking, when the muscle lengthens.

Before You Go On

Answer the following questions to test your understanding of the preceding section:

12. What special terms are given to the plasma membrane, cytoplasm, and smooth ER of a muscle cell?

13. Name the proteins that compose the thick and thin filaments of a muscle fiber and describe their structural arrangement.

14. Define *sarcomere.* Describe the striations of a sarcomere and sketch the arrangement of thick and thin filaments that accounts for the striations.

15. Describe the role of a synaptic knob, synaptic vesicles, synaptic cleft, and acetylcholine in neuromuscular function.

16. What is a motor unit? How do large and small motor units differ functionally?

17. Why is it important that the blood capillaries of a contracted muscle have an undulating or coiled arrangement?

10.4 Relating Structure to Function

▶ **Expected Learning Outcomes**

When you have completed this section, you should be able to

- explain how a muscle fiber contracts and relaxes, and relate this to its ultrastructure;

- describe how muscle grows and shrinks with use and disuse; and

- discuss two physiological categories of muscle fibers and their respective advantages.

Contraction and Relaxation

We noted earlier that a muscle fiber epitomizes the essential unity of form and function. Having examined the structure of a muscle fiber down to its molecular details, our task would be incomplete if we didn't ask ourselves what all of this is for. *Why* are all the molecules and organelles of the muscle fiber so closely packed and in such a precisely organized way? To answer that, we'll look briefly at the essential aspects of muscle contraction and relaxation. These form a cycle of events with four main phases:

1. **Excitation,** in which a nerve fiber releases ACh, ACh excites the muscle fiber, and a wave of electrical excitation spreads into the sarcoplasm by way of the T tubules.

2. **Excitation–contraction coupling,** consisting of events that link this electrical stimulus to the onset of muscle tension. Excitation of the T tubules induces the sarcoplasmic reticulum (SR) to release calcium ions, and calcium clears the way for the myosin of the thick filaments to bind to the actin of the thin filaments.

3. **Contraction,** in which the myosin heads repeatedly attach to the thin filaments and pull on them, causing the thin filaments to slide across the thick filaments and shorten the cell.

4. **Relaxation,** in which nerve signaling ceases, myosin is once again blocked from binding to actin, and muscle tension subsides.

Figure 10.14 shows a few of the finer steps of this process, numbered to correspond to the following description.

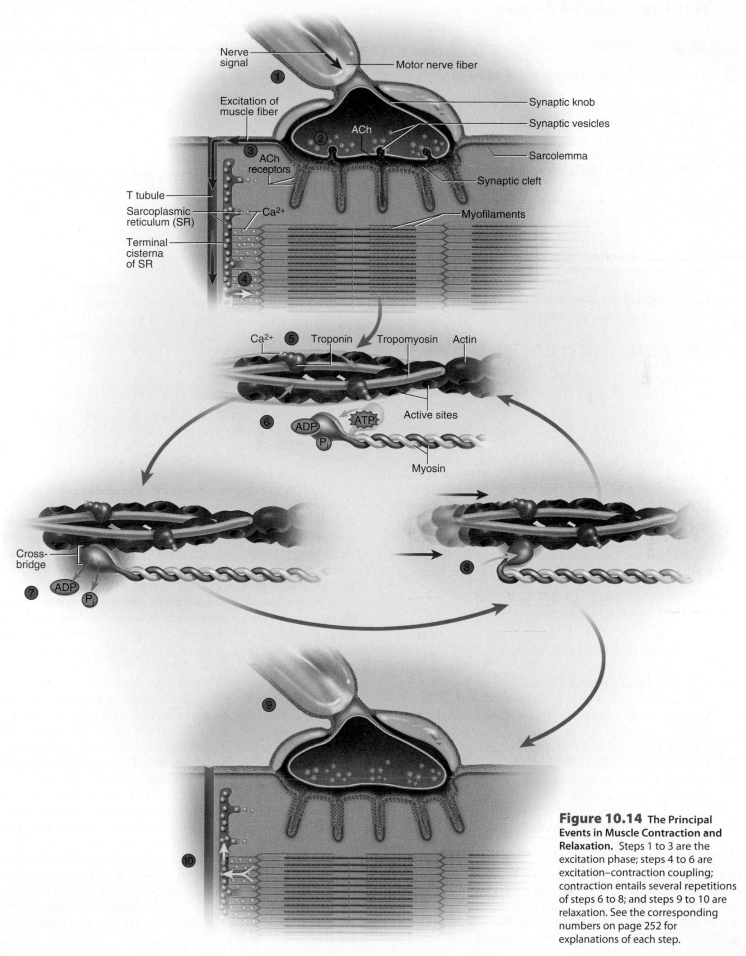

Nerve signal

① Motor nerve fiber

Synaptic knob

Synaptic vesicles

ACh

② Sarcolemma

Excitation of muscle fiber

③ ACh receptors

Synaptic cleft

T tubule

Sarcoplasmic reticulum (SR)

Ca²⁺

Myofilaments

Terminal cisterna of SR

④

Ca²⁺ ⑤ Troponin Tropomyosin Actin

Active sites

⑥ ADP ATP

Pᵢ

Myosin

Cross-bridge

⑦ ADP Pᵢ

⑧

⑨

⑩

Figure 10.14 **The Principal Events in Muscle Contraction and Relaxation.** Steps 1 to 3 are the excitation phase; steps 4 to 6 are excitation–contraction coupling; contraction entails several repetitions of steps 6 to 8; and steps 9 to 10 are relaxation. See the corresponding numbers on page 252 for explanations of each step.

Excitation

1 A nerve signal arrives at the synaptic knob.

2 The synaptic vesicles release ACh, which diffuses across the synaptic cleft and binds to receptors on the sarcolemma of the muscle fiber. This opens ion gates in the sarcolemma, resulting in sodium and potassium movements through the membrane that electrically excite the muscle fiber.

3 A self-propagating wave of excitation spreads down the length of the fiber and into the T tubules.

Excitation–Contraction Coupling

4 Electrical events in the T tubule lead to the opening of calcium gates in the sarcoplasmic reticulum. The SR releases a flood of calcium ions into the cytosol.

5 Calcium ions bind to troponin molecules in the thin myofilaments. Troponin induces the long tropomyosin molecule to shift position, sinking into the groove between the two F actin filaments and exposing the active sites on the G actin.

6 Meanwhile, myosin has been "waiting" in a flexed position, like a bent elbow, with a molecule of ATP bound to its head. The head contains an enzyme called *myosin ATPase,* which breaks down ATP into adenosine diphosphate (ADP) and an inorganic phosphate group (P_i). This is a *hydrolysis* reaction—one in which water is used to split a chemical bond in an organic molecule (the ATP). The energy liberated by this reaction straightens the myosin head into an extended, high-energy "cocked" position.

Contraction

7 Myosin forms a link, or *cross-bridge,* with one of the active sites of actin, and releases the ADP and P_i.

8 Myosin flexes and tugs on the thin filament, like flexing your elbow to pull in the rope of a boat anchor, and the thin filament slides a short distance along the thick filament. This movement is called the *power stroke.* Myosin binds a new ATP, lets go of the thin filament, breaks down the ATP, and recocks; this is the *recovery stroke.* Whenever one myosin head lets go of the thin filament, other myosin heads hold on, so the thin filament is never entirely released as long as the muscle is contracting. The myosin heads take turns pulling on the actin filament and letting go.

Steps 6 through 8 occur repeatedly. The overall effect is that the thin filament slides smoothly alongside the thick filament; this model of muscle contraction is therefore called the *sliding filament theory.* No myofilaments get shorter during contraction—they merely slide across each other. Since the thin filaments are connected through dystrophin to the sarcolemma and ultimately to the extracellular connective tissue, their sliding shortens the entire cell and its endomysium. If all the myosin heads in a muscle fiber executed a single cycle of power and recovery strokes, the muscle fiber would shorten about 1%. Through repetition of the process, however, a fiber can shorten by as much as 40%.

Relaxation

9 The first step needed to make a muscle fiber relax is to stop stimulating it. The motor neuron stops firing and releasing ACh, so the muscle fiber is no longer electrically excited.

10 The sarcoplasmic reticulum reabsorbs the calcium ions and stores them until the next time the muscle is stimulated. Without calcium, troponin moves back into the position that blocks the active sites and prevents myosin–actin cross-bridges from forming. Thus, the muscle can no longer maintain tension. It relaxes, and the action of gravity or an antagonistic muscle at the same joint stretches it back to its resting length.

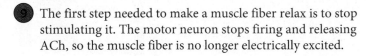

Apply What You Know

During muscle contraction, which band(s) of the muscle striations would you expect to become narrower or disappear? Which would remain the same width as in relaxed muscle? Explain.

Muscle Growth and Atrophy

It is common knowledge that muscles grow larger when exercised and shrink when they are not used. This is the basis for *resistance exercises* such as weight lifting. And yet, skeletal muscle fibers are incapable of mitosis. We have about the same number of muscle fibers in adulthood as we do in late childhood. How, then, does a muscle grow?

Exercise stimulates the muscle fiber to produce more protein myofilaments. As a result, the myofibrils grow thicker. At a certain point, a large myofibril splits longitudinally, so a well-conditioned muscle has more myofibrils per muscle fiber than does a weakly conditioned one. The entire muscle grows in bulk (thickness), not by the mitosis of existing cells (hyperplasia), but by the enlargement of cells that have existed since childhood (hypertrophy). Some authorities, however, think that entire muscle fibers (not just their myofibrils) may split longitudinally when they reach a certain size, thus giving rise to an increase in the number of fibers—not by mitosis but by a process more akin to tearing. Well-exercised muscles also develop more mitochondria, more myoglobin and glycogen, and a greater density of blood capillaries.

When a muscle is not used, it shrinks (atrophies). This can result from spinal cord injuries or other injuries that sever the nerve connections to a muscle *(denervation atrophy),* from lack of exercise *(disuse atrophy),* or from aging *(senescence atrophy).* The shrinkage of a limb that has been in a cast for several weeks is a good example of disuse atrophy. Muscle quickly regrows when exercise resumes, but if the atrophy becomes too advanced, muscle fibers die and are not replaced. Physical therapy is therefore important for maintaining muscle mass in people who are unable to use the muscles voluntarily.

Physiological Classes of Muscle Fibers

Not all muscle fibers are metabolically alike or adapted to perform the same task. Some respond slowly but are relatively resistant to fatigue, while others respond more quickly but also fatigue quickly. These fiber types also appear distinctly different when histologically stained for certain mitochondrial enzymes and other cellular components (fig. 10.15). Each primary type of fiber goes by several names. Table 10.2 summarizes their differences.

- **Slow oxidative (SO), slow-twitch, red,** or **type I fibers.** These fibers have relatively abundant mitochondria, myoglobin, and blood capillaries, and therefore a relatively deep red color. They are well adapted to aerobic respiration, a means for making ATP that requires oxygen but does not generate lactic acid, a major contributor to muscle fatigue. Thus, these fibers do not fatigue easily. However, in response to a single stimulus, they exhibit a relatively long *twitch,* or contraction, lasting about 100 milliseconds (msec). The soleus muscle of the calf and the postural muscles of the back are composed mainly of these slow-twitch, fatigue-resistant fibers.

- **Fast glycolytic (FG), fast-twitch, white,** or **type II fibers.** These fibers are rich in enzymes for anaerobic fermentation, a process that is independent of oxygen but produces lactic acid. They respond quickly, with twitches as short as 7.5 msec, but because of the lactic acid, they fatigue more easily than SO fibers. They are poorer in mitochondria, myoglobin, and blood capillaries than SO fibers, so they are relatively pale (hence the expression *white* fibers). They are well adapted for quick responses but not for endurance. Thus, they are especially important in sports such as basketball that require stop-and-go activity and frequent changes of pace. The gastrocnemius muscle of the calf, biceps brachii of the arm, and the muscles of eye movement consist mainly of FG fibers.

Some authorities recognize two subtypes of FG fibers called types IIA and IIB. Type IIB is the common type just described,

TABLE 10.2	Classification of Skeletal Muscle Fibers	
Properties	**Slow Oxidative Fibers**	**Fast Glycolytic Fibers**
Relative diameter	Smaller	Larger
ATP synthesis	Aerobic	Anaerobic
Fatigue resistance	Good	Poor
ATP hydrolysis	Slow	Fast
Glycolysis	Moderate	Fast
Myoglobin content	Abundant	Low
Glycogen content	Low	Abundant
Mitochondria	Abundant and large	Fewer and smaller
Capillaries	Abundant	Fewer
Color	Red	White, pale

Representative Muscles in Which Fiber Type Is Predominant

	Soleus	Gastrocnemius
	Erector spinae	Biceps brachii
	Quadratus lumborum	Muscles of eye movement

TABLE 10.3	Proportion of Slow Oxidative and Fast Glycolytic Fibers in the Quadriceps Femoris of Male Athletes	
Sample Population	**Slow Oxidative (SO)**	**Fast Glycolytic (FG)**
Marathon runners	82%	18%
Swimmers	74	26
Average males	45	55
Sprinters and jumpers	37	63

whereas IIA, or **intermediate fibers,** combine fast-twitch responses with aerobic fatigue-resistant metabolism. Type IIA fibers, however, are relatively rare except in some endurance-trained athletes.

Nearly all muscles are composed of both SO and FG fibers, but the proportions of these fiber types differ from one muscle to another. Muscles composed mainly of SO fibers are called *red muscles* and those composed mainly of FG fibers are called *white muscles.* People with different types and levels of physical activity differ in the proportion of one fiber type to another even in the same muscle, such as the *quadriceps femoris* of the anterior thigh (table 10.3). It is thought that people are born with a genetic predisposition for a certain ratio of fiber types. Those who go into competitive sports discover the sports at which they can excel and gravitate toward those for which heredity has best equipped them. One person might be a "born sprinter" and another a "born marathoner."

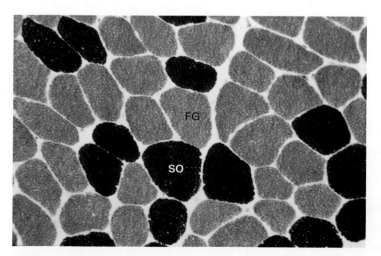

Figure 10.15 Muscle Stained to Distinguish Fast Glycolytic (FG) from Slow Oxidative (SO) Fibers. Cross section.

We noted earlier that sometimes two or more muscles act across the same joint and superficially seem to have the same function. We have already seen why such muscles are not as redundant as they seem, such as differences in mechanical advantage. Another reason is that they may differ in the proportion of SO to FG fibers. For example, the gastrocnemius and soleus muscles of the calf both insert on the calcaneus through the calcaneal tendon, so they exert the same pull on the heel. The gastrocnemius, however, is a predominantly fast glycolytic muscle adapted for quick, powerful movements such as jumping, whereas the soleus is a predominantly slow oxidative muscle that does most of the work in endurance exercises such as jogging and skiing.

Before You Go On

Answer the following questions to test your understanding of the preceding section:

18. What role does the sarcoplasmic reticulum play in muscle contraction? What role does it play in muscle relaxation?

19. Why does tropomyosin have to move before a muscle fiber can contract? What makes it move?

20. What role does ATP play in muscle contraction?

21. What is the mechanism of muscle growth? Describe the growth process in muscle and distinguish it from hyperplasia.

22. What are the basic functional differences between slow oxidative and fast glycolytic muscle fibers?

10.5 Cardiac and Smooth Muscle

▶ Expected Learning Outcomes

When you have completed this section, you should be able to

- describe cardiac muscle tissue and compare its structure and physiology to the other types; and
- describe smooth muscle tissue and compare its structure and physiology to the other types.

Cardiac and smooth muscle have special structural and physiological properties related to their distinctive functions.

Cardiac Muscle

Cardiac muscle constitutes most of the heart. Its form and function are discussed extensively in chapter 20 so that you will be able to relate these to the actions of the heart. Here, we only briefly compare it to skeletal and smooth muscle (table 10.4).

Cardiac muscle is striated like skeletal muscle, but otherwise exhibits several differences. Its *cardiocytes (myocytes)* are not long multinuclear fibers but short, stumpy, slightly branched cells. Microscopically, cardiac muscle exhibits characteristic dark lines called **intercalated** (in-TUR-kuh-LAY-ted) **discs** where the cells meet. These are steplike regions containing electrical gap junctions

TABLE 10.4	Comparison of Skeletal, Cardiac, and Smooth Muscle		
Feature	**Skeletal Muscle**	**Cardiac Muscle**	**Smooth Muscle**
Location	Associated with skeletal system	Heart	Walls of viscera and blood vessels, iris of eye, piloerectors of hair follicles
Cell shape	Long cylindrical fibers	Short branched cells	Fusiform cells
Cell length	100 μm–30 cm	50–100 μm	30–200 μm
Cell width	10–500 μm	10–20 μm	5–10 μm
Striations	Present	Present	Absent
Nuclei	Multiple nuclei, adjacent to sarcolemma	Usually one nucleus, near middle of cell	One nucleus, near middle of cell
Connective tissues	Endomysium, perimysium, epimysium	Endomysium only	Endomysium only
Sarcoplasmic reticulum	Abundant	Present	Scanty
T tubules	Present, narrow	Present, wide	Absent
Gap junctions	Absent	Present in intercalated discs	Present in single-unit smooth muscle
Ca^{2+} source	Sarcoplasmic reticulum	Sarcoplasmic reticulum and extracellular fluid	Mainly extracellular fluid
Innervation and control	Somatic motor fibers (voluntary)	Autonomic fibers (involuntary)	Autonomic fibers (involuntary)
Nervous stimulation required?	Yes	No	No
Mode of tissue repair	Limited regeneration, mostly fibrosis	Limited regeneration, mostly fibrosis	Relatively good capacity for regeneration

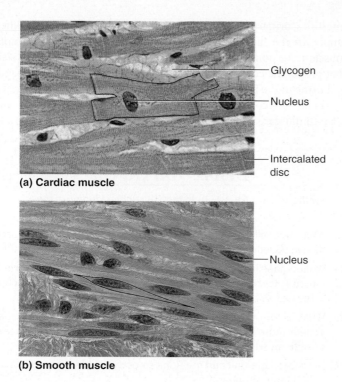

(a) Cardiac muscle

Glycogen

Nucleus

Intercalated disc

(b) Smooth muscle

Nucleus

Figure 10.16 Cardiac and Smooth Muscle.

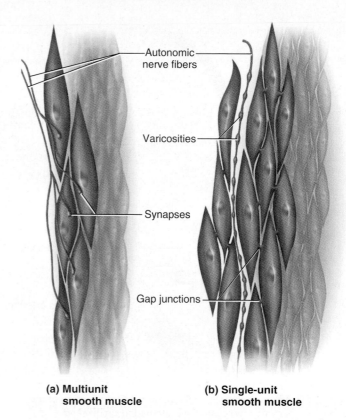

Autonomic nerve fibers

Varicosities

Synapses

Gap junctions

(a) Multiunit smooth muscle

(b) Single-unit smooth muscle

Figure 10.17 Multiunit and Single-Unit Smooth Muscle. (a) Multiunit smooth muscle, in which each muscle cell receives its own nerve supply and contracts independently. (b) Single-unit smooth muscle, in which a nerve fiber passes through the tissue without synapsing with any specific muscle cell, and muscle cells are coupled by electrical gap junctions.

that allow the cells to communicate with each other, and various mechanical junctions that prevent the cells from pulling apart when they contract (see details in chapter 20). Each myocyte can join several others at its intercalated discs.

Cardiocytes usually have only one, centrally placed nucleus (occasionally two) (fig. 10.16a), often surrounded by glycogen. Cardiac muscle is very rich in glycogen and myoglobin, and it has especially large mitochondria that fill about 25% of the cell, compared to smaller mitochondria occupying about 2% of a skeletal muscle fiber. Cardiac muscle is therefore well adapted to aerobic respiration and very resistant to fatigue, although it is highly vulnerable to interruptions in its oxygen supply. The sarcoplasmic reticulum is less developed than in skeletal muscle, but the T tubules are larger and admit supplemental calcium from the extracellular fluid. Cardiac myocytes have little capacity for mitosis; the repair of damaged cardiac muscle is primarily by fibrosis.

Cardiac muscle is innervated by the *autonomic nervous system (ANS)* rather than by somatic motor neurons. The ANS is a division of the nervous system that usually operates without one's conscious awareness or control. It does not generate the heartbeat, but it modulates the heart rate and contraction strength. Cardiocytes pulsate rhythmically even without nervous stimulation; this property is called *autorhythmicity.* In an intact heart, their beating is triggered by the heart's pacemaker.

Smooth Muscle

Smooth muscle is composed of fusiform myocytes (fig. 10.16b). There is only one nucleus, located near the middle of the cell. Thick and thin filaments are both present, but they are not aligned with

each other and produce no visible striations or sarcomeres; this is the reason for the name *smooth* muscle. Z discs are absent; instead, the thin filaments are attached by way of the cytoskeleton to **dense bodies,** little masses of protein scattered throughout the sarcoplasm and on the inner face of the sarcolemma.

The sarcoplasmic reticulum is scanty, and T tubules are absent. The calcium needed to activate smooth muscle contraction comes mainly from the extracellular fluid by way of calcium channels in the sarcolemma. During relaxation, calcium is pumped back out of the cell. Some smooth muscle has no nerve supply, but when nerve fibers are present, they come from the autonomic nervous system, like those of the heart.

Unlike skeletal and cardiac muscle, smooth muscle is capable of mitosis and hyperplasia. Thus, an organ such as the pregnant uterus can grow by adding more myocytes, and injured smooth muscle regenerates well.

There are two functional categories of smooth muscle called *multiunit* and *single-unit* types (fig. 10.17). **Multiunit smooth muscle** occurs in some of the largest arteries and pulmonary air passages, in the piloerector muscles of the hair follicles, and in the iris of the eye. Its innervation, although autonomic, is otherwise similar to that of skeletal muscle—the terminal branches of a nerve fiber synapse with individual myocytes and form a motor unit. Each motor unit contracts independently of the others, hence the name of this muscle type.

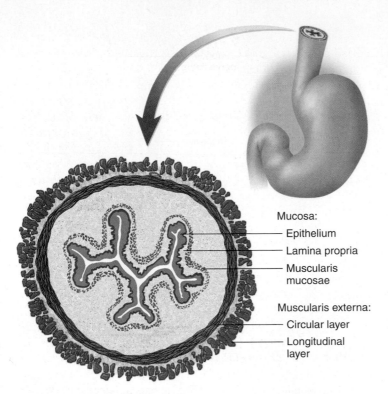

Figure 10.18 **Layers of Visceral Muscle in a Cross Section of the Esophagus.** Many hollow organs have alternating circular and longitudinal layers of smooth muscle.

Mucosa:
- Epithelium
- Lamina propria
- Muscularis mucosae

Muscularis externa:
- Circular layer
- Longitudinal layer

Single-unit smooth muscle is more widespread. It occurs in most blood vessels and in the digestive, respiratory, urinary, and reproductive tracts—thus it is also called **visceral muscle.** The nerve fibers in this type of muscle do not synapse with individual muscle cells, but pass through the tissue and exhibit swellings called **varicosities** at which they release neurotransmitters. Neurotransmitters nonselectively stimulate multiple muscle cells in the vicinity of a varicosity. The muscle cells themselves are electrically coupled to each other by gap junctions. Thus, they directly stimulate each other and a large number of cells contract as a unit, almost as if they were a single cell. This is the reason that this muscle type is called *single-unit* smooth muscle.

In many of the hollow internal organs, visceral muscle forms two or more layers—typically an inner *circular layer,* in which the fibers encircle the organ, and an outer *longitudinal layer,* in which the fibers run lengthwise along the organ (fig. 10.18). When the circular layer of muscle contracts, it narrows the organ and may make it longer (like a roll of dough squeezed in your hands); when the longitudinal muscle contracts, it makes the organ shorter and thicker.

Smooth muscle contracts and relaxes slowly, responding not only to nervous stimulation but also to chemicals and stretch. Its metabolism is mostly aerobic, but it has a very low energy requirement compared to skeletal and cardiac muscle, so it is highly fatigue-resistant. This enables smooth muscle to maintain a state of continual, partial contraction called **smooth muscle tone.** Smooth muscle tone maintains blood pressure by keeping the blood vessels partially constricted, and it prevents such organs as the stomach,

intestines, urinary bladder, and uterus from becoming flaccid when empty. In the digestive tract and some other locations, smooth muscle is responsible for waves of contraction called **peristalsis** that propel the contents through an organ (food in the esophagus and urine in the ureters, for example).

Table 10.4 compares some properties of skeletal, cardiac, and smooth muscle.

Before You Go On

Answer the following questions to test your understanding of the preceding section:

23. What organelles are more abundant and larger in cardiac muscle than in skeletal muscle? What is the functional significance of this?

24. What organelle is less developed in cardiac muscle than in skeletal muscle? How does this affect the activation of muscle contraction in the heart?

25. What factors make cardiac muscle more resistant to fatigue than skeletal muscle? What accounts for the relative fatigue resistance of smooth muscle?

26. How are single-unit and multiunit smooth muscle different? Which type is more abundant?

10.6 Developmental and Clinical Perspectives

▶ Expected Learning Outcomes

When you have completed this section, you should be able to

- describe how the three types of muscle develop in the embryo;
- describe the changes that occur in the muscular system in old age;
- discuss two muscle diseases, muscular dystrophy and myasthenia gravis; and
- briefly define and discuss several other disorders of the muscular system.

Embryonic Development of Muscle

Muscular tissue arises from embryonic mesoderm, with the exception of the piloerector muscles and the muscles within the eye. As described in chapter 4, the mesoderm of the trunk forms segmentally arranged blocks of tissue called somites, which then divide into regions called the dermatomes, sclerotomes, and myotomes. Beginning in week 4, some mesodermal cells migrate to the center of a somite and form the myotome, which will give rise to major axial muscles such as the erector spinae of the back. Others migrate away from the somite to the limb buds and body wall, where they will give rise to limb, abdominal, thoracic, and other muscles.

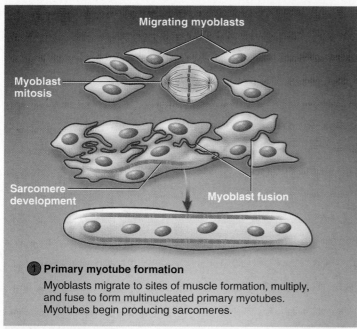

1 Primary myotube formation

Myoblasts migrate to sites of muscle formation, multiply, and fuse to form multinucleated primary myotubes. Myotubes begin producing sarcomeres.

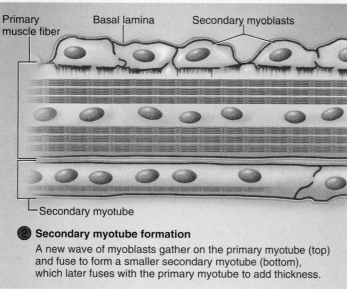

2 Secondary myotube formation

A new wave of myoblasts gather on the primary myotube (top) and fuse to form a smaller secondary myotube (bottom), which later fuses with the primary myotube to add thickness.

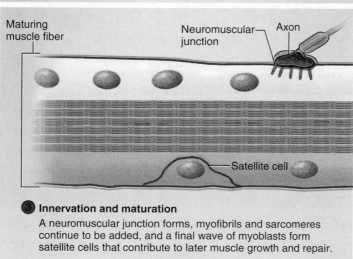

3 Innervation and maturation

A neuromuscular junction forms, myofibrils and sarcomeres continue to be added, and a final wave of myoblasts form satellite cells that contribute to later muscle growth and repair.

Figure 10.19 Embryonic Development of Skeletal Muscle Fibers.

In these locations, the mesodermal cells elongate into spindle-shaped **myoblasts**[17] and rapidly multiply (fig. 10.19). Myoblasts fuse into long multinuclear masses called the **primary myotubes,** with the nuclei in a chain down the core. These myotubes, the future muscle fibers, attach at each end to the developing tendons and skeleton. Internally, they begin assembling muscle proteins into sarcomeres, starting at the periphery of the myotube and progressing inward. Nuclei migrate to the periphery as the center becomes filled with myofibrils. Additional waves of myoblasts aggregate along the primary myotubes and form smaller *secondary* and *tertiary myotubes* (sometimes more in larger muscles) that fuse with the primary myotube to thicken the fiber.

By 9 weeks, most muscle groups are present, and nerve fibers have synapsed with the muscle fibers. The muscle fibers begin contracting in response to stimulation by week 10. By week 17, the contractions are strong enough to be felt by the mother. It was once thought that the fetus first comes alive at this time, so this stage of development is called *quickening.*[18]

Late in fetal development, a final wave of myoblasts associate with the muscle fiber and become **satellite cells.** These are stem cells that persist throughout life. Some of them fuse with growing muscle fibers and contribute nuclei to it through childhood, and satellite cells may regenerate a limited amount of damaged skeletal muscle even in adults. The development of new muscle fibers by mitosis ends, according to various estimates, from week 24 of gestation to 1 year after birth. After that, all muscle growth is by hypertrophy (enlargement of existing fibers) or longitudinal, nonmitotic splitting of large fibers. After age 25, the number of fibers in each muscle begins to decline.

Cardiac muscle develops in association with an embryonic *heart tube* described in chapter 20. Mesenchymal cells near the heart tube differentiate into myoblasts, and these proliferate mitotically as they do in the development of skeletal muscle. But in contrast to skeletal muscle development, the myoblasts do not fuse. They remain joined to each other and develop intercalated discs at their points of adhesion. The heart begins beating in week 3. Mitosis in cardiac myocytes continues after birth and is active until about age 9, although there is now evidence of limited mitotic capability even in adults. There is understandable interest in being able to stimulate this process in hopes of promoting regeneration of cardiac muscle damaged by heart attacks.

Smooth muscle develops similarly from myoblasts associated with the embryonic gut, blood vessels, and other organs. As in cardiac muscle, these myoblasts never fuse with each other, but in single-unit smooth muscle, they do become interconnected through gap junctions.

The Aging Muscular System

One of the most noticeable changes we experience with age is the loss of lean body mass (muscle) and accumulation of fat. The change is dramatically exemplified by CT scans of the thigh. In a young well-conditioned male, muscle accounts for 90% of the cross-sectional area of the midthigh; whereas in a frail 90-year-old

---

[17]*myo* = muscle + *blast* = precursor
[18]*quick* = alive

woman, it is only 30%. Muscular strength and mass peak in the 20s; by the age of 80, most people have only half as much strength and endurance. Many people over age 75 cannot lift a 4.5 kg (10 lb) weight with their arms; such simple tasks as carrying a sack of groceries into the house may become impossible. The loss of strength is a major contributor to falls, fractures, and dependence on others for the routine activities of daily living. Fast glycolytic (fast-twitch) fibers exhibit the earliest and most severe atrophy, thus increasing reaction time and reducing coordination.

There are multiple reasons for the loss of strength. Aged muscle fibers have fewer myofibrils, so they are smaller and weaker. The sarcomeres are increasingly disorganized, and muscle mitochondria are smaller and have reduced quantities of oxidative enzymes. Aged muscle has less ATP, glycogen, and myoglobin; consequently, it fatigues quickly. Muscles also exhibit more fat and fibrous tissue with age, which limits their movement and blood circulation. With reduced circulation, muscle injuries heal more slowly and with more scar tissue.

But the weakness and easy fatigue of aged muscle also stem from the aging of other organ systems. There are fewer motor neurons in the spinal cord, and some muscle shrinkage may represent denervation atrophy. The remaining neurons produce less acetylcholine and show less efficient synaptic transmission, which makes the muscles slower to respond to stimulation. As muscle atrophies, motor units have fewer muscle fibers per motor neuron, and more motor units must be recruited to perform a given task. Tasks that used to be easy, such as buttoning the clothes or eating a meal, take more time and effort. The sympathetic nervous system is also less efficient in old age, and less effective in increasing the blood flow to the muscles during exercise. This contributes to reduced endurance.

The muscles can be significantly reconditioned, however, even by exercise begun late in life. A person in his or her 90s can increase muscle strength two- or threefold in 6 months with as little as 40 minutes of exercise per week.

Diseases of the Muscular System

Diseases of muscular tissue are called **myopathies.** The muscular system suffers fewer diseases than any other organ system, but two of particular importance are muscular dystrophy and myasthenia gravis.

Muscular dystrophy[19] is a collective term for several hereditary diseases in which the skeletal muscles degenerate, lose strength, and are gradually replaced by fat and scar tissue. This new connective tissue impedes blood circulation, which in turn accelerates muscle degeneration, creating a fatal spiral of positive feedback. The most common form of the disease is *Duchenne*[20] *muscular dystrophy (DMD),* a sex-linked disease that occurs especially in males (about 1 in 3,500 live-born males). It results from a defective gene for dystrophin. In the absence of dystrophin, the sarcolemma tears and the muscle fiber dies. DMD is not evident at birth, but difficulties appear

as a child begins to walk. The child falls frequently and has difficulty standing up again. The disease affects the hips first, then the lower limbs, and progresses to the abdominal and spinal muscles. The muscles shorten as they atrophy, causing postural abnormalities such as scoliosis. DMD is incurable, but is treated with exercise to slow the atrophy and with braces to reinforce the weakened hips and correct the posture. Patients are usually confined to a wheelchair by early adolescence and rarely live beyond the age of 20.

Myasthenia gravis[21] **(MG)** (MY-ass-THEE-nee-uh GRAV-iss) is most prevalent in women from 20 to 40 years old. It is an autoimmune disease in which antibodies attack the neuromuscular junctions and trigger the destruction of ACh receptors. As a result,

---

[21]*my* = muscle + *asthen* = weakness + *grav* = severe

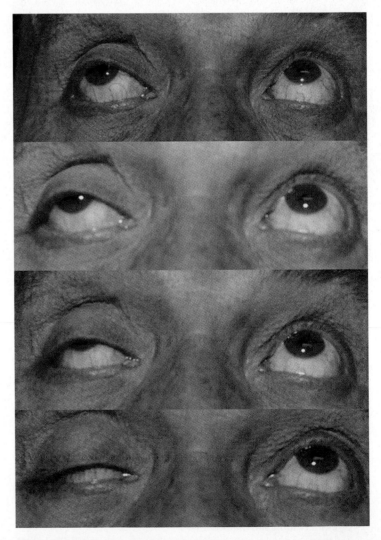

Figure 10.20 **Myasthenia Gravis (MG).** These photographs were taken when the patient was first told to gaze upward (top photo) and then at 30, 60, and 90 seconds. Note the inability to keep the right eyelid open. This drooping, or *ptosis,* is diagnostic of MG.

---

[19]*dys* = bad, abnormal + *trophy* = growth
[20]Guillaume B. A. Duchenne (1806–75), French physician

TABLE 10.5	Disorders of the Muscular System
Charley horse	Slang for any painful tear, stiffness, and blood clotting in a muscle caused by contusion (a blow to the muscle causing hemorrhaging).
Contracture	Abnormal muscle shortening not caused by nervous stimulation. Can result from a persistence of calcium in the sarcoplasm after stimulation or from contraction of scar tissue.
Crush syndrome	A shocklike state following the massive crushing of muscles, associated with a high and potentially fatal fever; cardiac irregularities caused by K^+ released from the injured muscles; and kidney failure caused by blockage of the renal tubules with myoglobin released by the traumatized muscle. Myoglobin in the urine *(myoglobinuria)* is a common sign.
Delayed onset muscle soreness	Pain and stiffness felt from several hours to a day after strenuous exercise. Associated with microtrauma to the muscles, with disrupted Z discs, myofibrils, and plasma membranes, and with elevated blood levels of myoglobin and enzymes released by damaged muscle fibers.
Rhabdomyoma	A rare, benign muscle tumor, usually occurring in the tongue, neck, larynx, nasal cavity, throat, heart, or vulva. Treated by surgical removal.
Rhabdomyosarcoma	A malignant muscle tumor; the most common form of pediatric soft-tissue sarcoma, although accounting for <3% of childhood cancers and rarely seen in adults. Results from abnormal proliferation of myoblasts. Begins as a painless mass in a muscle but metastasizes rapidly. Diagnosed by biopsy and treated with surgery, chemotherapy, or radiation therapy.

Disorders Described Elsewhere

Atrophy 252	Compartment syndrome 294	Muscular dystrophy 258	Paralysis 249, 394
Back injuries 286	Hamstring injuries 316	Myasthenia gravis 258	Sports injuries 324

the muscle fibers become less and less sensitive to ACh. The effects often appear first in the facial muscles and include drooping eyelids (fig. 10.20) and double vision (due to weakness of the eye muscles). These signs are often followed by difficulty swallowing, weakness of the limbs, and poor physical endurance. Some people with MG die quickly as a result of respiratory failure, while others have normal life spans. The symptoms can be controlled with cholinesterase inhibitors, which retard the breakdown of ACh and prolong its action on the muscles, and with drugs that suppress the immune system and thus slow the attack on ACh receptors.

Some other disorders of the muscular system in general are briefly described in table 10.5, whereas disorders more specific to the axial or appendicular musculature are described in chapters 11 and 12.

Before You Go On

Answer the following questions to test your understanding of the preceding section:

27. What cells come between mesodermal cells and the muscle fiber in the stages of skeletal muscle development? Describe how several uninuclear cells transform into a multinuclear muscle fiber.

28. What is the principal difference between the way cardiac and smooth muscle form and the way skeletal muscle forms?

29. Describe the major changes seen in the muscular system in old age.

30. What is the root cause of Duchenne muscular dystrophy? What is the normal function of dystrophin?

31. How is synaptic function altered in myasthenia gravis? How does this synaptic dysfunction affect a person with MG?

Study Guide

Assess Your Learning Outcomes

You should have a good understanding of this chapter if you can accurately address the following issues.

10.1 Muscle Types and Functions (p. 236)

1. The scope of myology and of the term *muscular system,* and what muscular tissues are not included in the muscular system
2. Differences between skeletal, cardiac, and smooth muscle with respect to the presence or absence of striations; voluntary or involuntary control; the shapes of their cells; and terms for the three types of muscle cells
3. The multiple functions of muscular tissue
4. Five physiological properties that muscle cells must have to carry out their functions

10.2 General Anatomy of Muscles (p. 237)

1. The tissues that constitute a skeletal muscle
2. The arrangement of a muscle's endomysium, perimysium, and epimysium; how these relate to the muscle fibers and fascicles; and the relationship of the connective tissue of a muscle with the periosteum of an associated bone
3. The separation of muscles by fasciae and the grouping of muscles into compartments
4. Orientation of the fascicles in fusiform, parallel, triangular, pennate, and circular muscles; and how this relates to muscle strength
5. The difference between an indirect and direct muscle attachment
6. How an aponeurosis differs from other tendons; what it has in common with them; and the locations of some aponeuroses
7. The meaning of a muscle's origin, insertion, and belly; and why the origin–insertion distinction is imperfect, with examples to support this argument
8. Tissues other than bone on which some muscles insert and act
9. The meaning of a muscle's action
10. The functional interactions of prime movers, synergists, antagonists, and fixators; and examples of each
11. The distinction between intrinsic and extrinsic muscles, and examples of each
12. The fundamental parts of a musculoskeletal lever system
13. How to calculate the mechanical advantage (MA) of a musculoskeletal lever if given the necessary measurements; the benefits and limitations of musculoskeletal levers in which the MA is greater than 1 versus those in which it is less than 1

14. The difference between a first-, second-, and third-class lever; and both a nonliving and a musculoskeletal example of each

10.3 Microscopic Anatomy of Skeletal Muscle (p. 244)

1. The internal ultrastructure of a skeletal muscle fiber; the special names given to some of its organelles and cytoskeletal components; and the functions of each component
2. The relationship between myofilaments, myofibrils, and muscle fibers
3. The purposes of glycogen and myoglobin in a muscle fiber
4. The structural network of the sarcoplasmic reticulum and its physiological function
5. The three types of myofilaments that compose a myofibril; what they are made of; and what roles they play
6. The location and role of dystrophin in a muscle fiber
7. The names of the striations in skeletal and cardiac muscle and how the overlapping myofilaments account for the striations
8. The definition of a *sarcomere;* its relationship to the Z discs; and the basic action of the sarcomeres in muscle contraction
9. The relationship of a somatic motor neuron to a group of skeletal muscle fibers
10. The structure of a neuromuscular junction and the functions of all of its components
11. The roles of acetylcholine (ACh), acetylcholine receptors, and acetylcholinesterase at a neuromuscular junction; where ACh comes from when a muscle is stimulated; and what becomes it afterward
12. The components of a motor unit; what is meant by the size of a motor unit; the respective advantages of large and small motor units; and some places where large and small motor units can be found
13. The blood supply to the skeletal muscles, and the share of the circulating blood that the muscular system receives in rest and exercise

10.4 Relating Structure to Function (p. 250)

1. The four stages of muscle contraction and relaxation
2. The events of excitation—how a signal in a motor nerve fiber results in a chain reaction of electrical excitation in the muscle fiber; the roles of acetylcholine and gated ion channels in producing these events
3. The events of excitation–contraction coupling; how electrical excitation of the muscle fiber

leads to changes in the myofilaments that make it possible for a sarcomere to contract; and the roles of the sarcoplasmic reticulum, calcium, troponin, and tropomyosin in this process
4. The repetitive cycle of myosin–actin interactions that result in muscle contraction; the role of ATP in this process
5. The events of relaxation; how a muscle fiber is allowed to stop contracting
6. The effects of regular resistance exercise on a muscle, and the mode of muscle growth
7. The definition of muscle *atrophy,* and its causes
8. Differences between slow oxidative (SO) and fast glycolytic (FG) muscle fibers; examples of some muscles in which each of these fiber types is dominant; and the properties of intermediate fibers

10.5 Cardiac and Smooth Muscle (p. 254)

1. The structure of cardiac muscle cells and how they are joined to each other
2. The spontaneous activity of cardiac muscle and how it is influenced by the nervous system
3. The special cytological properties of cardiac muscle cells that make them highly resistant to fatigue
4. Similarities and differences between smooth muscle cells and skeletal muscle fibers with respect to their organelle systems and the reason for the absence of striations in smooth muscle
5. The functional differences between multiunit and single-unit smooth muscle; how they differ in their relationships with motor nerve fibers; and where each type occurs
6. Differences in the behavior of smooth muscle versus skeletal muscle, and why smooth muscle is advantageous for certain purposes

10.6 Developmental and Clinical Perspectives (p. 256)

1. The development of embryonic mesenchymal cells into skeletal muscle fibers
2. How cardiac and smooth muscle differ from skeletal muscle in the embryonic development of their cells
3. Changes seen in the skeletal muscles in old age; and multiple reasons for the reduced strength, endurance, and efficiency of the muscles with age
4. The mode of inheritance and pathology of muscular dystrophy
5. The cause and effects of myasthenia gravis

Testing Your Recall

1. A fascicle is bounded and defined by
   a. the endomysium.
   b. the fascia.
   c. the myofilaments.
   d. the epimysium.
   e. the perimysium.

2. Muscle cells must have all of the following properties except _____ to carry out their function.
   a. extensibility
   d. contractility
   b. elasticity
   e. conductivity
   c. autorhythmicity

3. If a tendon runs longitudinally throughout a muscle and fascicles insert obliquely on it along both sides, the muscle is classified as
   a. parallel.
   d. convergent.
   b. oblique.
   e. multipennate.
   c. bipennate.

4. A muscle that holds a bone still during a particular action is called
   a. a fixator.
   b. an antagonist.
   c. an agonist.
   d. a synergist.
   e. an intrinsic muscle.

5. Which of the following muscle proteins is not intracellular?
   a. actin
   d. troponin
   b. myosin
   e. dystrophin
   c. collagen

6. Smooth muscle cells have _____, whereas skeletal muscle fibers do not.
   a. T tubules
   b. ACh receptors
   c. thick myofilaments
   d. thin myofilaments
   e. dense bodies

7. ACh receptors are found in
   a. synaptic vesicles.
   b. terminal cisternae.
   c. thick filaments.
   d. thin filaments.
   e. junctional folds.

8. Single-unit smooth muscle cells can stimulate each other because they have
   a. pacemakers.
   b. diffuse junctions.
   c. gap junctions.
   d. tight junctions.
   e. calcium pumps.

9. A second-class lever always has
   a. the fulcrum in the middle.
   b. the effort applied between the fulcrum and resistance.
   c. a mechanical advantage less than 1.
   d. a mechanical advantage greater than 1.
   e. the resistance at one end.

10. Slow oxidative muscle fibers have all of the following except
    a. an abundance of myoglobin.
    b. an abundance of glycogen.
    c. high fatigue resistance.
    d. a red color.
    e. a high capacity to synthesize ATP aerobically.

11. Acetylcholine is released from organelles called _____.

12. The region where a motor nerve fiber meets a skeletal muscle fiber is called a/an _____.

13. Parts of the sarcoplasmic reticulum called _____ lie on each side of a T tubule.

14. Thick myofilaments consist mainly of the protein _____.

15. Sheets of fibrous connective tissue called _____ separate a muscle or muscle group from neighboring muscles.

16. Muscle contains an oxygen-storage pigment called _____.

17. The _____ of skeletal muscle play the same role as dense bodies in smooth muscle.

18. A circular muscle that controls a body opening or passage is called a/an _____.

19. Skeletal muscle fibers develop by the fusion of embryonic cells called _____.

20. A wave of contraction passing along the esophagus or small intestine is called _____.

Answers in the Appendix

Building Your Medical Vocabulary

State a medical meaning of each of the following word elements, and give a term in which it is used.

1. myo-
2. fasc-
3. fusi-
4. sarco-
5. dys-
6. mys-
7. anti-
8. erg-
9. -mer
10. penna-

Answers in the Appendix

True or False

Determine which five of the following statements are false, and briefly explain why.

1. Pennate muscles are stronger than parallel muscles of comparable size.
2. A given muscle may be an agonist in one joint movement and an antagonist in a different movement of that joint.
3. Extrinsic muscles are not located entirely within the body region that they control.
4. Cardiac myocytes are of the fatigue-resistant, slow oxidative type.
5. One motor neuron can supply only one muscle fiber.
6. To initiate muscle contraction, calcium ions must bind to the myosin heads.
7. A first-class lever can have a mechanical advantage either greater than or less than 1.
8. Slow oxidative fibers are more fatigue-resistant than fast glycolytic fibers.
9. The blood vessels of a skeletal muscle are more wavy or coiled when a muscle is relaxed than when it contracts.
10. Well-exercised muscles generally gain in thickness by the addition of new muscle fibers.

Answers in the Appendix

Testing Your Comprehension

1. In a baseball game, the pitcher hits the batter in the thigh with a fastball. Which of the following conditions would this most likely cause: atrophy, charley horse, contracture, crush syndrome, or rhabdomyoma? Explain your choice.

2. What would be the consequences for muscular system function if muscle fibers were not elastic? What if they were not extensible?

3. For each of the following muscle pairs, state which muscle you think would have the higher percentage of fast glycolytic fibers: (a) Muscles that move the eyes or muscles of the upper throat that initiate swallowing. (b) The abdominal muscles employed in doing sit-ups or the muscles employed in handwriting. (c) Muscles of the tongue or the skeletal muscle sphincter of the anus. Explain each answer.

4. The forearm contains five flexor muscles that flex the wrist and fingers. Explain why it is better for them to have an indirect attachment rather than a direct attachment to the bones on which they act.

5. The brachialis muscle is the most powerful flexor of the elbow. Examine its bone attachments in figure 12.3 (p. 298) and table 12.3 (p. 301). From a functional standpoint, identify what bone attachment would be its origin and what would be its insertion in a person lifting barbells. Then identify which would be the origin and insertion in a person climbing the face of a cliff. In light of your conclusions, concisely explain the imperfection in such conventional interpretations as presented by that table.

Answers at www.mhhe.com/saladinha3

Improve Your Grade at www.mhhe.com/saladinha3

Practice quizzes, labeling activities, games, and flashcards provide fun ways to master concepts. You can also download image PowerPoint files for each chapter to create a study guide or for taking notes during lecture.

The Axial Musculature

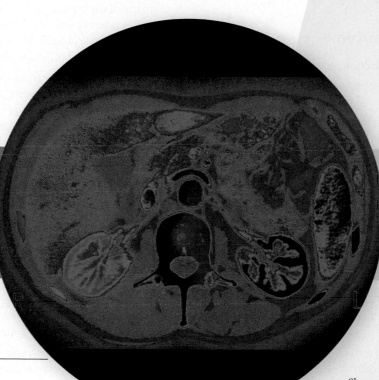

CT scan showing muscles of the body wall at the level of vertebra L1 (cross section)

CHAPTER
11

CHAPTER OUTLINE

DEEPER INSIGHTS

BRUSHING UP

To understand this chapter, you may find it helpful to review the following concepts:

Muscular System

There are about 600 skeletal muscles In the human body. Chapters 11 and 12 describe fewer than one-third of these. This chapter deals with the muscles that act on the axial division of the body—that is, on the head and trunk. Muscles that act on the limbs and limb girdles (the appendicular division) are described in chapter 12. This chapter opens with some tips to help you study the muscles more insightfully.

11.1 Learning Approaches

▶ Expected Learning Outcomes

When you have completed this section, you should be able to

- translate several Latin words commonly used in the naming of muscles;
- define the *origin, insertion, action,* and *innervation* of a muscle;
- describe the sources of the nerves to the head–neck and trunk muscles and explain the numbering system for the cranial and spinal nerves; and
- describe and practice some methods that will help in the learning of the skeletal muscles.

How Muscles Are Named

Figure 11.1 shows an overview of the major superficial muscles. Learning the names of these and other muscles may seem a forbidding task at first, especially when some of them have such long Latin names as *depressor labii inferioris* and *flexor digiti minimi brevis.* Such names, however, typically describe some distinctive aspects of the structure, location, or action of a muscle, and become very helpful once we grow familiar with a few common Latin words. For example, the depressor labii inferioris is a muscle that lowers (depresses) the bottom (inferior) lip (labium), and the flexor digiti minimi brevis is a short (brevis) muscle that flexes the smallest (minimi) finger (digit). Several of the most common words in muscle names are interpreted in table 11.1, and others are explained in footnotes throughout the chapter. Familiarity with these terms and attention to the footnotes will help you translate muscle names and remember the location, appearance, and action of the muscles. You can listen to pronunciations of these muscle names on the CD-ROM or online versions of *Anatomy & Physiology | Revealed.*

Muscle Innervation

The **innervation** of a muscle refers to the identity of the nerve that stimulates it. Knowing the innervation to each muscle enables clinicians to diagnose nerve, spinal cord, and brainstem injuries from their effects on muscle function, and to set realistic goals for reha-

bilitation. The innervations described in this chapter will be more meaningful after you have studied the peripheral nervous system (chapters 14 and 15), but a brief orientation will be helpful here. The muscles are innervated by two groups of nerves:

- **Spinal nerves** arise from the spinal cord, emerge through the intervertebral foramina, and innervate muscles below the neck. Spinal nerves are identified by letters and numbers that refer to the adjacent vertebrae—for example, T6 for the sixth thoracic nerve and S2 for the second sacral nerve. Immediately after emerging from an intervertebral foramen, each spinal nerve branches into a *posterior* and *anterior ramus.*[1] You will note references to nerve numbers and rami in many of the muscle tables. The term *plexus* in some of the tables refers to weblike networks of spinal nerves adjacent to the vertebral column. All of the spinal nerves named here are illustrated—and most are also discussed—in chapter 14.

- **Cranial nerves** arise from the base of the brain, emerge through the skull foramina, and innervate muscles of the head and neck. Cranial nerves are identified by roman numerals (CN I–XII) and by names given in chapter 15, although not all 12 of them innervate skeletal muscles.

A Learning Strategy

The following suggestions can help you develop a rational strategy for learning the skeletal muscles as you first encounter them in the textbook and laboratory:

- Examine models, cadavers, dissected animals, or an anatomical atlas as you read about the muscles. Visual images are often easier to remember than words, and direct observation of a muscle may stick in your memory better than descriptive text or two-dimensional drawings.

- When studying a particular muscle, palpate it on yourself if possible. Contract the muscle to feel it bulge and sense its action. Doing so will make muscle locations and actions less abstract. The atlas following chapter 12 shows where you can see and palpate several muscles on the living body.

- Locate the origins and insertions of muscles on an articulated skeleton. Some study skeletons are painted and labeled to show these. This will help you visualize the locations of muscles and understand how they produce particular joint actions.

- Study the derivations of the muscle names; look for descriptive meaning in their names.

- Say the names *aloud* to yourself or a study partner. It is harder to remember and spell terms you cannot pronounce, and silent pronunciation is not nearly as effective as speaking and hearing the names. Pronunciation guides are provided in the muscle tables for all but the most obvious cases.

---

[1] *ramus* = branch

TABLE 11.1	Words Commonly Used to Name Muscles	
Criterion	**Term and Meaning**	**Examples of Usage**
Size	Brevis (short)	Extensor pollicis brevis
	Longus (long)	Abductor pollicis longus
	Major (large)	Pectoralis major
	Maximus (largest)	Gluteus maximus
	Minimus (smallest)	Gluteus minimus
	Minor (small)	Pectoralis minor
Shape	Deltoid (triangular)	Deltoid
	Quadratus (four-sided)	Pronator quadratus
	Rhomboideus (rhomboidal)	Rhomboideus major
	Teres (round, cylindrical)	Pronator teres
	Trapezius (trapezoidal)	Trapezius
Location	Abdominis (of the abdomen)	Rectus abdominis
	Brachii (of the arm)	Biceps brachii
	Capitis (of the head)	Splenius capitis
	Carpi (of the wrist)	Flexor carpi ulnaris
	Cervicis (of the neck)	Semispinalis cervicis
	Digiti (of a finger or toe, singular)	Extensor digiti minimi
	Digitorum (of the fingers or toes, plural)	Flexor digitorum profundus
	Femoris (of the femur, or thigh)	Quadriceps femoris
	Fibularis (of the fibula)	Fibularis longus
	Hallucis (of the great toe)	Abductor hallucis
	Indicis (of the index finger)	Extensor indicis
	Intercostal (between the ribs)	External intercostals
	Lumborum (of the lower back)	Quadratus lumborum
	Pectoralis (of the chest)	Pectoralis major
	Pollicis (of the thumb)	Opponens pollicis
	Profundus (deep)	Flexor digitorum profundus
	Superficialis (superficial)	Flexor digitorum superficialis
	Thoracis (of the thorax)	Spinalis thoracis
Number of heads	Biceps (two heads)	Biceps femoris
	Triceps (three heads)	Triceps brachii
	Quadriceps (four heads)	Quadriceps femoris
Orientation	Oblique (slanted)	External abdominal oblique
	Rectus (straight)	Rectus abdominis
	Transversus (transverse)	Transversus abdominis
Action	Abductor	Abductor digiti minimi
	Adductor	Adductor pollicis
	Depressor	Depressor anguli oris
	Extensor	Extensor carpi radialis
	Flexor	Flexor carpi radialis
	Levator	Levator scapulae
	Pronator	Pronator teres
	Supinator	Supinator

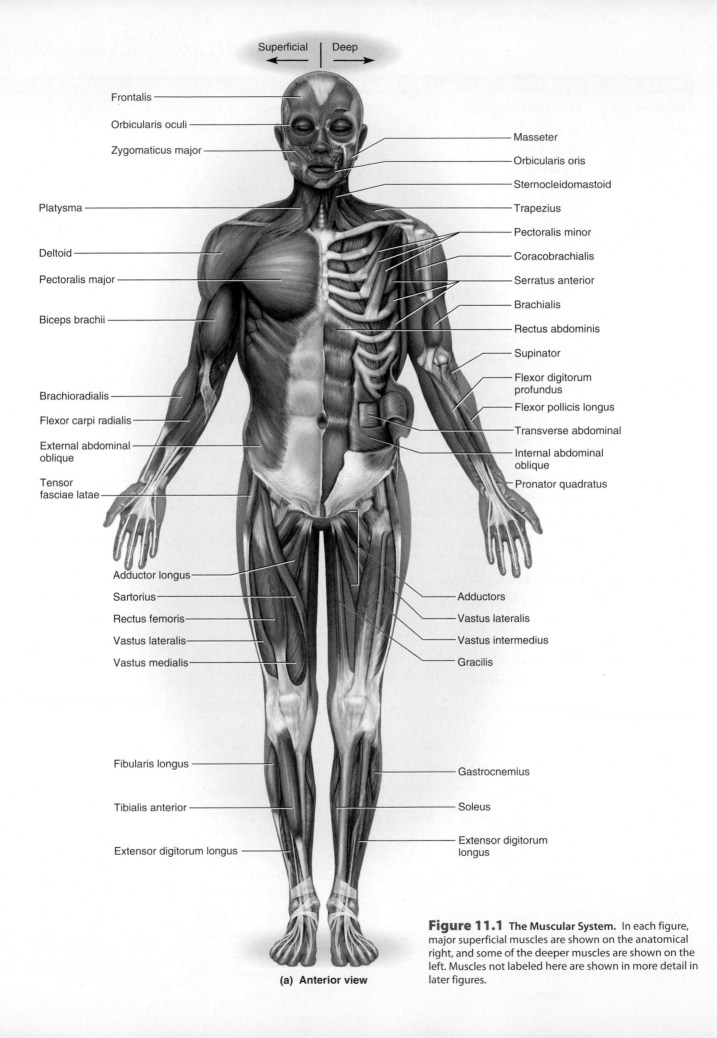

Superficial | Deep

Frontalis

Orbicularis oculi

Zygomaticus major

Masseter

Orbicularis oris

Sternocleidomastoid

Platysma

Trapezius

Pectoralis minor

Deltoid

Coracobrachialis

Pectoralis major

Serratus anterior

Biceps brachii

Brachialis

Rectus abdominis

Supinator

Flexor digitorum
profundus

Brachioradialis

Flexor carpi radialis

Flexor pollicis longus

External abdominal
oblique

Transverse abdominal

Internal abdominal
oblique

Tensor
fasciae latae

Pronator quadratus

Adductor longus

Sartorius

Adductors

Rectus femoris

Vastus lateralis

Vastus lateralis

Vastus intermedius

Vastus medialis

Gracilis

Fibularis longus

Gastrocnemius

Tibialis anterior

Soleus

Extensor digitorum
longus

Extensor digitorum longus

(a) Anterior view

Figure 11.1 The Muscular System. In each figure, major superficial muscles are shown on the anatomical right, and some of the deeper muscles are shown on the left. Muscles not labeled here are shown in more detail in later figures.

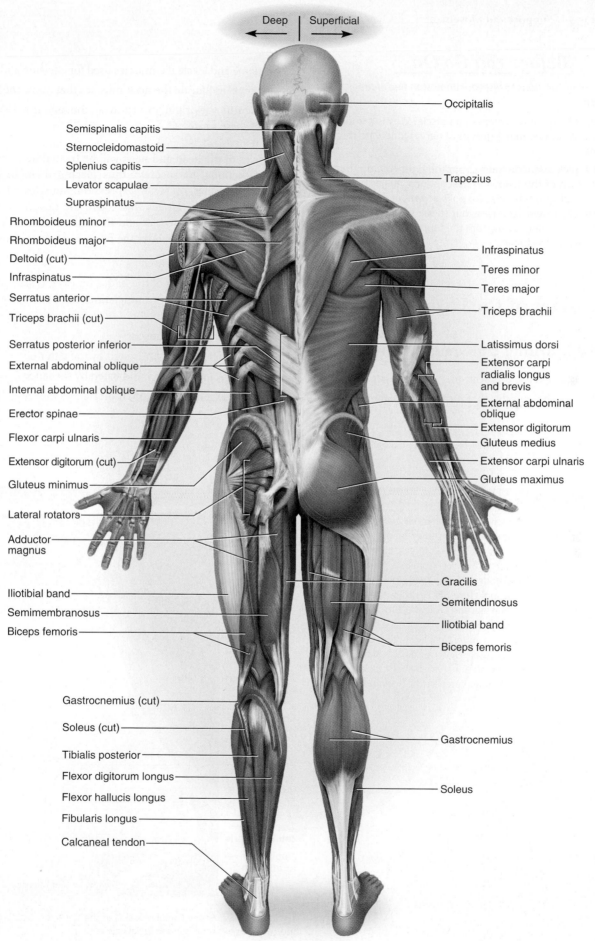

Deep | Superficial

Occipitalis

Semispinalis capitis

Sternocleidomastoid

Splenius capitis

Levator scapulae

Supraspinatus

Rhomboideus minor

Rhomboideus major

Deltoid (cut)

Infraspinatus

Serratus anterior

Triceps brachii (cut)

Serratus posterior inferior

External abdominal oblique

Internal abdominal oblique

Erector spinae

Flexor carpi ulnaris

Extensor digitorum (cut)

Gluteus minimus

Lateral rotators

Adductor magnus

Iliotibial band

Semimembranosus

Biceps femoris

Gastrocnemius (cut)

Soleus (cut)

Tibialis posterior

Flexor digitorum longus

Flexor hallucis longus

Fibularis longus

Calcaneal tendon

Trapezius

Infraspinatus

Teres minor

Teres major

Triceps brachii

Latissimus dorsi

Extensor carpi radialis longus and brevis

External abdominal oblique

Extensor digitorum

Gluteus medius

Extensor carpi ulnaris

Gluteus maximus

Gracilis

Semitendinosus

Iliotibial band

Biceps femoris

Gastrocnemius

Soleus

(b) Posterior view

Before You Go On

Answer the following questions to test your understanding of the preceding section:

1. What is meant by the innervation of a muscle? Why is it important to know this? What two major groups of nerves innervate the skeletal muscles?

2. In table 11.1, pick a muscle name from the right column that you think meets each of the following descriptions: (a) lies beside the radius and straightens the wrist; (b) pulls down the corners of your mouth when you frown; (c) raises your shoulder blades; (d) moves your little finger laterally, away from the fourth digit; (e) is the largest muscle deep to the breast.

11.2 Muscles of the Head and Neck

▶ Expected Learning Outcomes

When you have completed this section, you should be able to

- name and locate the muscles that produce facial expressions;
- name and locate the muscles used for chewing and swallowing;
- name and locate the neck muscles that move the head; and
- identify the origin, insertion, action, and innervation of any of these muscles.

Muscles of the head and neck will be treated here from a regional and functional perspective, thus placing them in the following groups: muscles of facial expression, muscles of chewing and swallowing, and muscles that move the head as a whole (tables 11.2–11.4). In these tables and throughout the rest of the chapter, each muscle entry provides the following information:

- the name of the muscle;
- the pronunciation of the name, unless it is self-evident or uses words whose pronunciations have been provided in a recent entry;
- the actions of the muscle;
- the muscle's origin, indicated by the letter *O*;
- the muscle's insertion, indicated by the letter *I*; and
- the muscle's innervation, indicated by the letter *N*.

TABLE 11.2	Muscles of Facial Expression

Humans have much more expressive faces than other mammals because of a complex array of muscles that insert in the dermis and subcutaneous tissues (figs. 11.2 and 11.3). These muscles tense the skin and produce such expressions as a pleasant smile, a threatening scowl, a puzzled frown, or a flirtatious wink (fig. 11.4). They add subtle shades of meaning to our spoken words. Facial muscles also contribute directly to speech, chewing, and other oral functions. All but one of these muscles are innervated by the facial nerve (CN VII). This nerve is especially vulnerable to injury from lacerations and skull fractures, which can paralyze the muscles and cause parts of the face to sag. The only muscle in this table not innervated by the facial nerve is the levator palpebrae superioris, innervated by the oculomotor nerve (CN III).

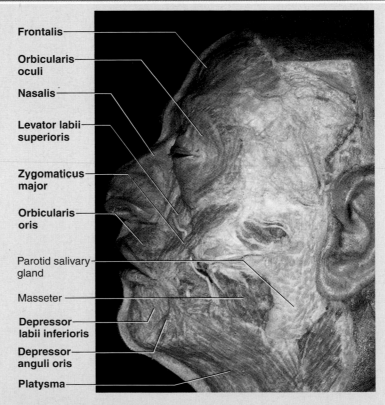

Frontalis
Orbicularis oculi
Nasalis
Levator labii superioris
Zygomaticus major
Orbicularis oris
Parotid salivary gland
Masseter
Depressor labii inferioris
Depressor anguli oris
Platysma

Figure 11.2 Some Facial Muscles of the Cadaver. Boldface labels indicate muscles employed in facial expression.

TABLE 11.2   **Muscles of Facial Expression (continued)**

Figure 11.3 **Muscles of Facial Expression.** Boldface labels indicate muscles employed in facial expression.

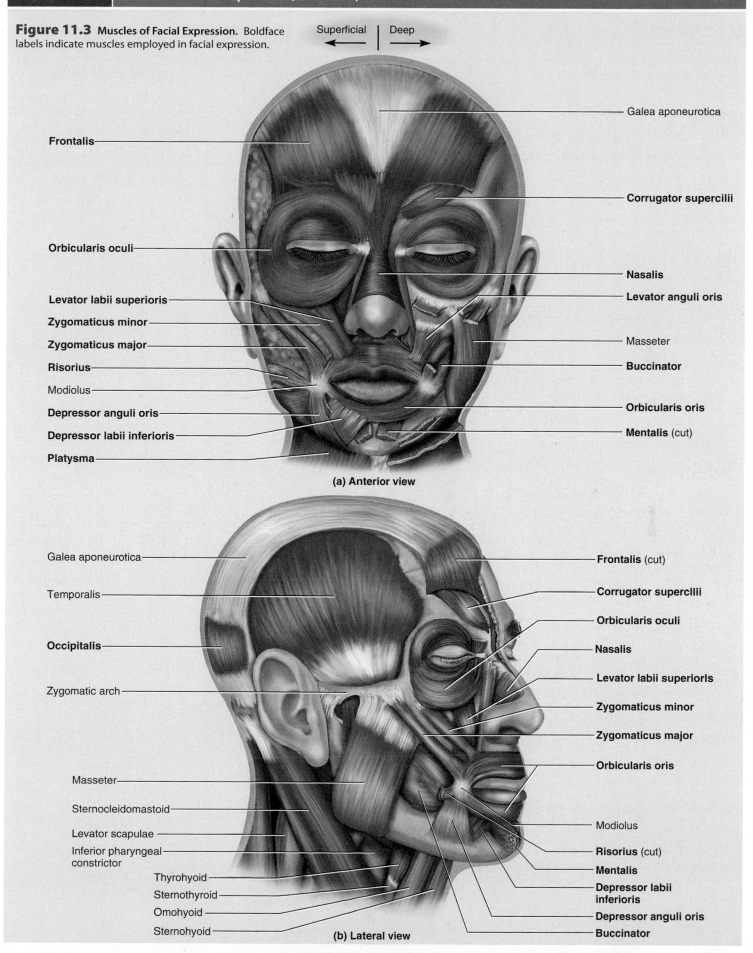

Superficial ← | → Deep

Galea aponeurotica

Frontalis

Corrugator supercilii

Orbicularis oculi

Nasalis

Levator anguli oris

Levator labii superioris

Zygomaticus minor

Zygomaticus major

Masseter

Risorius

Buccinator

Modiolus

Depressor anguli oris

Orbicularis oris

Depressor labii inferioris

Mentalis (cut)

Platysma

(a) Anterior view

Galea aponeurotica

Frontalis (cut)

Temporalis

Corrugator supercilii

Occipitalis

Orbicularis oculi

Nasalis

Zygomatic arch

Levator labii superioris

Zygomaticus minor

Zygomaticus major

Orbicularis oris

Masseter

Sternocleidomastoid

Levator scapulae

Modiolus

Inferior pharyngeal constrictor

Risorius (cut)

Thyrohyoid

Mentalis

Sternothyroid

Depressor labii inferioris

Omohyoid

Depressor anguli oris

Sternohyoid

(b) Lateral view

Buccinator

TABLE 11.2	**Muscles of Facial Expression (continued)**

The Scalp.   The *occipitofrontalis* overlies the dome of the cranium. It is divided into the *frontalis* of the forehead and *occipitalis* at the rear of the head, named for the frontal and occipital bones underlying them. They are connected to each other by a broad aponeurosis, the **galea aponeurotica**[2] (GAY-lee-uh AP-oh-new-ROT-ih-cuh).

Frontalis (frun-TAY-lis)   Elevates eyebrows in glancing upward and expressions of surprise or fright; draws scalp forward and wrinkles skin of forehead

> O: Galea aponeurotica   I: Subcutaneous tissue of eyebrows   **N:** Facial n.

Occipitalis (oc-SIP-ih-TAY-lis)   Retracts scalp; fixes galea aponeurotica so frontalis can act on eyebrows

> O: Superior nuchal line and temporal bone   I: Galea aponeurotica   **N:** Facial n.

The Orbital and Nasal Regions.   The *orbicularis oculi* is a sphincter of the eyelid that encircles and closes the eye. The *levator palpebrae superioris* lies deep to the orbicularis oculi, in the eyelid and orbital cavity (see fig. 17.20, p. 481), and opens the eye. Other muscles in this group move the eyelids and skin of the forehead and dilate the nostrils. Muscles within the orbit that move the eyeball itself are discussed in chapter 17.

Orbicularis Oculi[3] (or-BIC-you-LERR-is OC-you-lye)   Sphincter of the eyelids; closes eye in blinking, squinting, and sleep; aids in flow of tears across eye

> O: Lacrimal bone, adjacent regions of frontal bone and   I: Upper and lower eyelids, skin around margin of orbit   **N:** Facial n.
> maxilla, medial angle of eyelids

Levator Palpebrae Superioris[4] (leh-VAY-tur pal-PEE-bree soo-PEER-ee-OR-is)   Elevates upper eyelid, opens eye

> O: Lesser wing of sphenoid in posterior wall of orbit   I: Upper eyelid   **N:** Oculomotor n.

Corrugator Supercilii[5] (COR-oo-GAY-tur SOO-per-SIL-ee-eye)   Draws eyebrows medially and downward in frowning and concentration; reduces glare of bright sunlight

> O: Medial end of supraorbital margin   I: Skin of eyebrow   **N:** Facial n.

Nasalis[6] (nay-ZAIL-is)   Widens nostrils; narrows internal air passage between vestibule and nasal cavity

> O: Maxilla just lateral to nose   I: Bridge and alar cartilages of nose   **N:** Facial n.

The Oral Region.   The mouth is the most expressive part of the face, and lip movements are necessary for intelligible speech; thus, it is not surprising that the muscles here are especially diverse. The *orbicularis oris* is a complex of muscles in the lips that encircle the mouth; until recently it was misinterpreted as a sphincter, or circular muscle, but it is actually composed of four independent quadrants that interlace and give only an appearance of circularity. Other muscles in this region approach the lips from all directions and thus draw the lips or angles (corners) of the mouth upward, laterally, and downward. Some of these have origins or insertions in a complex cord called the **modiolus**[7] just lateral to each angle of the lips (see fig. 11.3). Named for the hub of a cartwheel, the modiolus is a point of convergence of several muscles of the lower face. You can palpate it by inserting one finger just inside the corner of your lips and pinching the corner between the finger and thumb, feeling for a thick knot of tissue.

Orbicularis Oris[8] (or-BIC-you-LERR-is OR-is)   Encircles mouth, closes lips, protrudes lips as in kissing; uniquely developed in humans for speech

> O: Modiolus of mouth   I: Submucosa and dermis of lips   **N:** Facial n.

Levator Labii Superioris[9] (leh-VAY-tur LAY-bee-eye soo-PEER-ee-OR-is)   Elevates and everts upper lip in sad, sneering, or serious expressions

> O: Zygomatic bone and maxilla near inferior margin of orbit   I: Muscles of upper lip   **N:** Facial n.

Levator Anguli Oris[10] (leh-VAY-tur ANG-you-lye OR-is)   Elevates angle of mouth as in smiling

> O: Maxilla just below infraorbital foramen   I: Muscles at angle of mouth   **N:** Facial n.

Zygomaticus[11] Major (ZY-go-MAT-ih-cus)   Draws angle of mouth upward and laterally as in laughing

> O: Zygomatic bone   I: Superolateral angle of mouth   **N:** Facial n.

Zygomaticus Minor   Elevates upper lip, exposes upper teeth in smiling or sneering

> O: Zygomatic bone   I: Muscles of upper lip   **N:** Facial n.

Risorius[12] (rih-SOR-ee-us)   Draws angle of mouth laterally in expressions of laughing, horror, or disdain

> O: Zygomatic arch, fascia near ear   I: Modiolus of mouth   **N:** Facial n.

---

[2]*galea* = helmet + *apo* = above + *neuro* = nerves, brain
[3]*orb* = circle + *ocul* = eye
[4]*levat* = to raise + *palpebr* = eyelid + *superior* = upper
[5]*corrug* = wrinkle + *supercil* = eyebrow
[6]*nas* = nose
[7]*modiolus* = hub

[8]*orb* = circle + *or* = mouth
[9]*levat* = to raise + *labi* = lip + *superior* = upper
[10]*angul* = angle, corner + *or* = mouth
[11]*zygo* = join, unite (refers to zygomatic bone)
[12]*risor* = laughter

TABLE 11.2   Muscles of Facial Expression (continued)

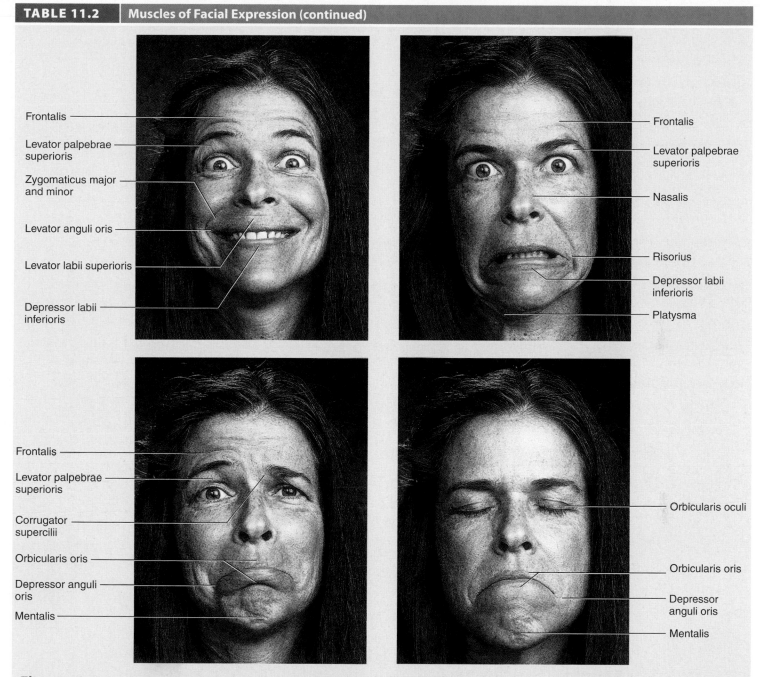

Figure 11.4 Expressions Produced by Several of the Facial Muscles. The ordinary actions of these muscles are usually more subtle than these demonstrations.

• *Name an antagonist of each of these muscles: the depressor anguli oris, orbicularis oculi, and levator labii superioris.*

TABLE 11.2 | **Muscles of Facial Expression (continued)**

Depressor Anguli Oris[13]   Draws angle of mouth laterally and downward in opening mouth or sad expressions

   O: Inferior margin of mandibular body   **I:** Modiolus of mouth   **N:** Facial n.

Depressor Labii Inferioris[14]   Draws lower lip downward and laterally in chewing and expressions of melancholy or doubt

   O: Near mental protuberance   **I:** Skin and mucosa of lower lip   **N:** Facial n.

The Mental and Buccal Regions.   Adjacent to the oral orifice are the mental region (chin) and buccal region (cheeks). In addition to muscles already discussed that act directly on the lower lip, the mental region has a pair of small *mentalis* muscles extending from the upper margin of the mandible to the skin of the chin. In some people, these muscles are especially thick and have a visible dimple between them called the *mental cleft*. The *buccinator* is the muscle in the cheek. It has multiple functions in chewing, sucking, and blowing. If the cheek is inflated with air, compression of the buccinator blows it out. Sucking is achieved by contracting the buccinators to draw the cheeks inward, and then relaxing them. This action is especially important to nursing infants. To feel this action, hold your fingertips lightly on your cheeks as you make a kissing noise. You will notice the relaxation of the buccinators at the moment air is sharply drawn in through the pursed lips.

Mentalis (men-TAY-lis)   Elevates and protrudes lower lip in drinking, pouting, and expressions of doubt or disdain; elevates and wrinkles skin of chin

   O: Mandible near inferior incisors   **I:** Skin of chin at mental protuberance   **N:** Facial n.

Buccinator[15] **(BUC-sin-AY-tur)**   Compresses cheek against teeth and gums; directs food between molars; retracts cheek from teeth when mouth is closing to prevent biting cheek; expels air and liquid

   O: Alveolar processes on lateral surfaces of maxilla and mandible   **I:** Orbicularis oris; submucosa of cheek and lips   **N:** Facial n.

The Cervical and Mental Region.   The *platysma* is a thin superficial muscle of the upper chest and lower face. It is relatively unimportant, but when men shave, they tend to tense the platysma to make the concavity between the jaw and neck shallower and the skin tauter.

Platysma[16] **(plah-TIZ-muh)**   Draws lower lip and angle of mouth downward in expressions of horror or surprise; may aid in opening mouth widely

   O: Fascia of deltoid and pectoralis major   **I:** Mandible; skin and subcutaneous tissue of lower face   **N:** Facial n.

---

[13]*depress* = to lower + *angul* = angle, corner + *or* = mouth
[14]*labi* = lip + *inferior* = lower

[15]*buccinator* = trumpeter
[16]*platy* = flat

TABLE 11.3	**Muscles of Chewing and Swallowing**

The following muscles contribute to facial expression and speech, but are primarily concerned with food manipulation.

Extrinsic Muscles of the Tongue.   The tongue is a very agile organ. It pushes food between the molars for chewing *(mastication)* and later forces the food into the pharynx for swallowing *(deglutition);* it is also, of course, of crucial importance to speech. Both intrinsic and extrinsic muscles are responsible for its complex movements. The intrinsic muscles consist of a variable number of vertical fascicles that extend from the superior to the inferior sides of the tongue, transverse fascicles that extend from right to left, and longitudinal fascicles that extend from root to tip (see figs. 10.2c, p. 238, and 24.5b, p. 660). The extrinsic muscles listed here connect the tongue to other structures in the head (fig. 11.5). Three of these are innervated by the hypoglossal nerve (CN XII), whereas the fourth is innervated by both the vagus (CN X) and accessory (CN XI) nerves.

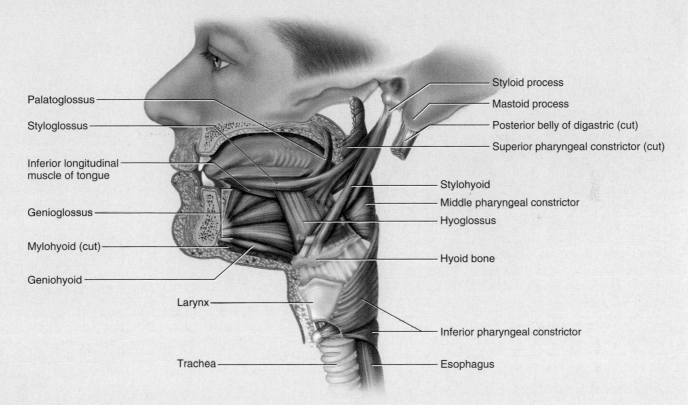

Figure 11.5 Muscles of the Tongue and Pharynx.

Genioglossus[17] **(JEE-nee-oh-GLOSS-us)**   Unilateral action draws tongue to one side; bilateral action depresses midline of tongue or protrudes tongue

  O: Superior mental spine on posterior surface of mental protuberance   **I:** Inferior surface of tongue from root to apex   **N:** Hypoglossal n.

Hyoglossus[18] **(HI-oh-GLOSS-us)**   Depresses tongue

  O: Body and greater horn of hyoid bone   **I:** Lateral and inferior surfaces of tongue   **N:** Hypoglossal n.

Styloglossus[19] **(STY-lo-GLOSS-us)**   Draws tongue upward and posteriorly

  O: Styloid process of temporal bone and ligament from styloid process to mandible   **I:** Superolateral surface of tongue   **N:** Hypoglossal n.

Palatoglossus[20] **(PAL-a-toe-GLOSS-us)**   Elevates root of tongue and closes oral cavity off from pharynx; forms palatoglossal arch at rear of oral cavity

  O: Soft palate   **I:** Lateral surface of tongue   **N:** Vagus and accessory nn.

---

[17]*genio* = chin + *gloss* = tongue
[18]*hyo* = hyoid bone + *gloss* = tongue

[19]*stylo* = styloid process + *gloss* = tongue
[20]*palato* = palate + *gloss* = tongue

TABLE 11.3	**Muscles of Chewing and Swallowing (continued)**

Muscles of Chewing. Four pairs of muscles produce the biting and chewing movements of the mandible: the *temporalis, masseter,* and two pairs of *pterygoid muscles* (fig. 11.6). Their actions include *depression* to open the mouth for receiving food; *elevation* for biting off a piece of food or crushing it between the teeth; *protraction* so that the incisors meet in cutting off a piece of food, and *retraction* to draw the lower incisors behind the upper incisors and make the rear teeth meet; and *lateral* and *medial excursion,* the side-to-side movements that grind food between the rear teeth. The last four of these movements are shown in figure 9.15 (p. 217). All of these muscles are innervated by the mandibular nerve, which is a branch of the trigeminal (CN V).

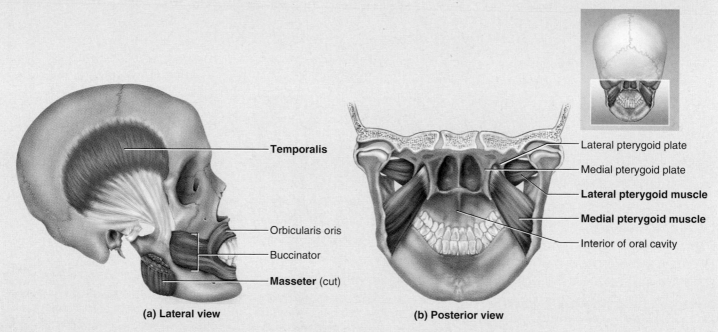

(a) Lateral view   **(b) Posterior view**

Figure 11.6 Muscles of Chewing. Boldface labels indicate muscles that act on the mandible in its chewing movements. (a) Right lateral view. In order to expose the insertion of the temporalis muscle on the mandible, part of the zygomatic arch and masseter muscle are removed. (b) View of the pterygoid muscles looking into the oral cavity from behind the skull.
● *Explain how the medial pterygoid muscles can produce both lateral and medial excursion of the mandible.*

Temporalis[21] (TEM-po-RAY-liss)   Elevation, retraction, and lateral and medial excursion of the mandible

O: Temporal lines and temporal fossa of cranium   I: Coronoid process and anterior border of mandibular ramus   N: Mandibular n.

Masseter[22] (ma-SEE-tur)   Elevation of the mandible, with smaller roles in protraction, retraction, and lateral and medial excursion

O: Zygomatic arch   I: Lateral surface of mandibular ramus and angle   N: Mandibular n.

Medial Pterygoid[23] (TERR-ih-goyd)   Elevation, protraction, and lateral and medial excursion of the mandible

O: Medial surface of lateral pterygoid plate, palatine bone, lateral surface of maxilla near molar teeth   I: Medial surface of mandibular ramus and angle   N: Mandibular n.

Lateral Pterygoid   Depression (in wide opening of the mouth), protraction, and lateral and medial excursion of the mandible

O: Lateral surfaces of lateral pterygoid plate and greater wing of sphenoid   I: Neck of mandible (just below condyle); articular disc and capsule of temporomandibular joint   N: Mandibular n.

---

[21]*temporalis* = of the temporal region of the head
[22]*masset* = chew

[23]*pteryg* = wing + *oid* = resembling (refers to pterygoid plate of sphenoid bone)

TABLE 11.3   **Muscles of Chewing and Swallowing (continued)**

Hyoid Muscles—Suprahyoid Group.   Several aspects of chewing, swallowing, and vocalizing are aided by eight pairs of *hyoid muscles* associated with the hyoid bone (fig. 11.7). The *suprahyoid group* is composed of the four pairs superior to the hyoid—the *digastric, geniohyoid, mylohyoid,* and *stylohyoid.* The digastric is an unusual muscle, named for its two bellies. Its *posterior belly* arises from the mastoid notch of the cranium and slopes downward and forward. The *anterior belly* arises from a trench called the *digastric fossa* on the inner surface of the mandibular body. It slopes downward and backward. The two bellies meet at a constriction, the *intermediate tendon.* This tendon passes through a connective tissue loop, the *fascial sling,* attached to the hyoid bone. Thus, when the two bellies of the digastric contract, they pull upward on the hyoid; but if the hyoid is fixed from below, the digastric aids in wide opening of the mouth. The lateral pterygoids are more important in wide mouth opening, with the digastrics coming into play only in extreme opening, as in yawning or taking a large bite of an apple. Cranial nerves V (trigeminal), VII (facial), and XII (hypoglossal) innervate these muscles; the trigeminal nerve gives rise to the mylohyoid nerve of the digastric and mylohyoid muscles.

Digastric[24]   Depresses mandible when hyoid is fixed; opens mouth widely, as when ingesting food or yawning; elevates hyoid when mandible is fixed

O: Mastoid notch of temporal bone, digastric fossa of mandible	**I:** Hyoid bone, via fascial sling	**N:** Posterior belly: facial n. Anterior belly: mylohyoid n. (a branch of the trigeminal n.)

Geniohyoid[25] **(JEE-nee-oh-HY-oyd)**   Depresses mandible when hyoid is fixed; elevates and protracts hyoid when mandible is fixed

O: Inferior mental spine of mandible	**I:** Hyoid bone	**N:** Spinal nerve C1 via hypoglossal n.

Mylohyoid[26]   Spans mandible from side to side and forms floor of mouth; elevates floor of mouth in initial stage of swallowing

O: Mylohyoid line near inferior margin of mandible	**I:** Hyoid bone	**N:** Mylohyoid n.

Stylohyoid   Elevates and retracts hyoid, elongating floor of mouth; roles in speech, chewing, and swallowing are not yet clearly understood

O: Styloid process of temporal bone	**I:** Hyoid bone	**N:** Facial n.

Hyoid Muscles—Infrahyoid Group.   The infrahyoid muscles are inferior to the hyoid bone. By fixing the hyoid from below, they enable the suprahyoid muscles to open the mouth. The *omohyoid* is unusual in that it arises from the shoulder, passes under the sternocleidomastoid, and then ascends to the hyoid bone. Like the digastric, it has two bellies. The *thyrohyoid,* named for the hyoid bone and the large shield-shaped *thyroid cartilage* of the larynx, helps prevent choking. It elevates the larynx during swallowing so that its superior opening is sealed by a flap of tissue, the *epiglottis.* You can feel this effect by placing your finger on the "Adam's apple" (the anterior prominence of the thyroid cartilage) and feeling it bob up as you swallow. The *sternothyroid* muscle then pulls the larynx down again so you can resume breathing; it is the only infrahyoid muscle with no connection to the hyoid bone. The sternohyoid lowers the hyoid bone after it has been elevated.

The infrahyoid muscles that act on the larynx are regarded as the extrinsic muscles of the larynx. The *intrinsic muscles,* considered in chapter 23, are concerned with control of the vocal cords and laryngeal opening. The *ansa cervicalis,*[27] which innervates three of these muscles, is a loop of nerve on the side of the neck formed by certain fibers from cervical nerves 1 to 3 (see fig. 14.13, p. 386). Cranial nerves IX (glossopharyngeal), X (vagus), and XII (hypoglossal) also innervate these muscles.

Omohyoid[28]   Depresses hyoid after it has been elevated

O: Superior border of scapula	**I:** Hyoid bone	**N:** Ansa cervicalis

Sternohyoid[29]   Depresses hyoid after it has been elevated

O: Manubrium of sternum, medial end of clavicle	**I:** Hyoid bone	**N:** Ansa cervicalis

Thyrohyoid[30]   Depresses hyoid; with hyoid fixed, elevates larynx as in singing high notes

O: Thyroid cartilage of larynx	**I:** Hyoid bone	**N:** Spinal nerve C1 via hypoglossal n.

Sternothyroid   Depresses larynx after it has been elevated in swallowing and vocalization; aids in singing low notes

O: Manubrium of sternum, costal cartilage 1	**I:** Thyroid cartilage of larynx	**N:** Ansa cervicalis

Muscles of the Pharynx.   The three pairs of *pharyngeal constrictors* encircle the pharynx on its posterior and lateral sides, forming a muscular funnel that aids in swallowing (see fig. 11.5).

Pharyngeal Constrictors (three muscles)   During swallowing, contract in order from *superior* to *middle* to *inferior constrictor* to drive food into esophagus

O: Medial pterygoid plate, mandible, hyoid bone, stylohyoid ligament, cricoid and thyroid cartilages of larynx	**I:** Median pharyngeal raphe (seam on posterior side of pharynx); basilar part of occipital bone	**N:** Glossopharyngeal and vagus nn.

---

[24]*di* = two + *gastr* = bellies
[25]*genio* = chin
[26]*mylo* = mill, molar tooth
[27]*ansa* = handle + *cervic* = neck + *alis* = of, belonging to

[28]*omo* = shoulder
[29]*sterno* = chest, sternum
[30]*thyro* = shield (refers to thyroid cartilage)

TABLE 11.4   **Muscles Acting on the Head (continued)**

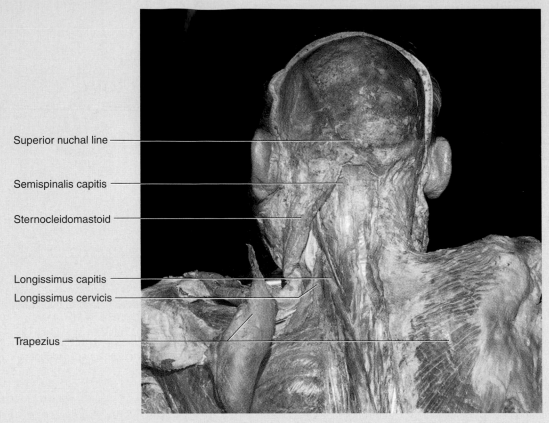

Superior nuchal line

Semispinalis capitis

Sternocleidomastoid

Longissimus capitis
Longissimus cervicis

Trapezius

Figure 11.9 Muscles of the Shoulder and Nuchal Regions.

Splenius Capitis[36] (SPLEE-nee-us CAP-ih-tiss) and Splenius Cervicis[37] (SIR-vih-sis)   Acting unilaterally, produce ipsilateral flexion and slight rotation of head; extend head when acting bilaterally

O: Inferior half of nuchal ligament, spinous processes of vertebrae C7–T6

I: Mastoid process and occipital bone just inferior to superior nuchal line; cervical vertebrae C1–C2 or C3

N: Posterior rami of middle cervical nn.

Semispinalis Capitis (SEM-ee-spy-NAY-lis) and Semispinalis Cervicis   Extend and contralaterally rotate head

O: Articular processes of vertebrae C4–C7, transverse processes of T1–T6

I: Occipital bone between nuchal lines, spinous processes of vertebrae C2–C5

N: Posterior rami of cervical and thoracic nn.

Apply What You Know

Of the muscles you have studied so far, name three that you would consider intrinsic muscles of the head and three that you would classify as extrinsic. Explain your reason for each.

Before You Go On

Answer the following questions to test your understanding of the preceding section:

3. Name two muscles that elevate the upper lip and two that depress the lower lip.

4. Name the four paired muscles of mastication and state where they insert on the mandible.

5. Distinguish between the functions of the suprahyoid and infrahyoid muscles.

6. List the prime movers of neck extension and flexion.

---

[36]splenius = bandage + capitis = of the head
[37]cervicis = of the neck

11.3 Muscles of the Trunk

▶ Expected Learning Outcomes

When you have completed this section, you should be able to
- name and locate the muscles of respiration and explain how they affect airflow and abdominal pressure;
- name and locate the muscles of the abdominal wall, back, and pelvic floor; and

- identify the origin, insertion, action, and innervation of any of these muscles.

In this section, we will examine muscles of the trunk of the body in three functional groups concerned with respiration, support of the abdominal wall and pelvic floor, and movement of the vertebral column (tables 11.5–11.8). In the illustrations, you will note some major muscles that are not discussed in the associated tables—for example, the pectoralis major and serratus anterior. Although they are *located in* the trunk, they *act upon* the limbs and limb girdles and are therefore discussed in chapter 12.

TABLE 11.5	Muscles of Respiration

We breathe primarily by means of muscles that enclose the thoracic cavity—the diaphragm and the external intercostal, internal intercostal, and innermost intercostal muscles (fig. 11.10).

The *diaphragm* is a muscular dome between the thoracic and abdominal cavities, bulging upward against the base of the lungs. It has openings for passage of the esophagus, major blood and lymphatic vessels, and nerves between the two cavities. Its fibers converge from the margins toward a fibrous **central tendon.** When the diaphragm contracts, it flattens slightly and enlarges the thoracic cavity, causing air intake *(inspiration);* when it relaxes, it rises and shrinks the thoracic cavity, expelling air *(expiration).*

Three layers of muscle lie between the ribs: the external, internal, and innermost intercostal muscles. The 11 pairs of *external intercostal muscles* constitute the most superficial layer. They extend from the rib tubercle posteriorly almost to the beginning of the costal cartilage anteriorly. Each one slopes downward and anteriorly from one rib to the next inferior one. The 11 pairs of *internal intercostal muscles* lie deep to the external intercostals and extend from the margin of the sternum to the angles of the ribs. They are thickest in the region between the costal cartilages and grow thinner in the region where they overlap the external intercostals. Their fibers slope downward and posteriorly from each rib to the one below, at nearly right angles to the external intercostals. Each is divided into an *intercartilaginous part* between the costal cartilages and an *interosseous part* between the ribs proper. The two parts differ in their respiratory roles. The *innermost intercostals* vary in number, as they are sometimes absent from the upper thoracic cage. Their fibers run in the same direction as the internal intercostals, and they are presumed to serve the same function. Intercostal nerves and blood vessels travel through the fascia between the internal and innermost intercostals (see fig. 14.12, p. 385).

The primary function of the intercostal muscles is to stiffen the thoracic cage during respiration so that it does not cave inward when the diaphragm descends. However, they also contribute to enlargement and contraction of the thoracic cage and thus add to the air volume that ventilates the lungs.

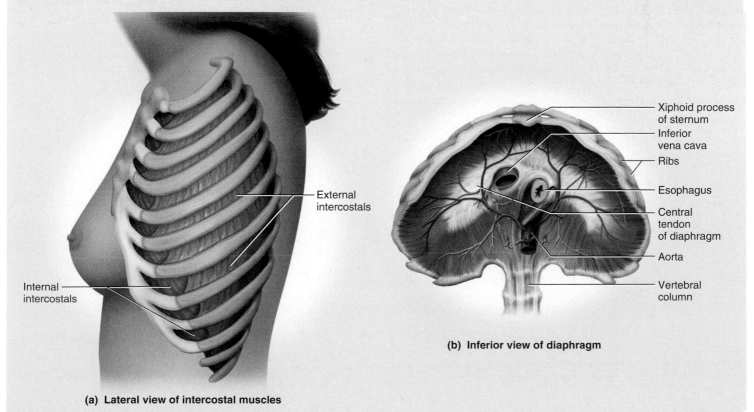

(a) Lateral view of intercostal muscles

External intercostals

Internal intercostals

(b) Inferior view of diaphragm

Xiphoid process of sternum

Inferior vena cava

Ribs

Esophagus

Central tendon of diaphragm

Aorta

Vertebral column

Figure 11.10 Muscles of Respiration.

TABLE 11.5	Muscles of Respiration (continued)

Diaphragm[38] **(DY-ah-fram)**   Prime mover of inspiration (responsible for about two-thirds of air intake); contracts in preparation for sneezing, coughing, crying, laughing, and weight lifting; contraction compresses abdominal viscera and aids in childbirth and expulsion of urine and feces

O: Xiphoid process of sternum; ribs and costal cartilages 7–12; lumbar vertebrae	**I:** Central tendon of diaphragm	**N:** Phrenic nn.

External Intercostals[39] **(IN-tur-COSS-tul)**   When scalenes fix rib 1, external intercostals elevate and protract ribs 2–12; this expands the thoracic cavity and creates a partial vacuum, causing inflow of air. Exercise a braking action during expiration so that expiration is not overly abrupt.

O: Inferior margins of ribs 1–11	**I:** Superior margin of next lower rib	**N:** Intercostal nn.

Internal Intercostals   In inspiration, the intercartilaginous part aids in elevating the ribs and expanding the thoracic cavity. In expiration, the interosseous part depresses and retracts the ribs, compressing the thoracic cavity and expelling air. The latter occurs only in forceful expiration, not in relaxed breathing.

O: Superior margins and costal cartilages of ribs 2–12; margin of sternum	**I:** Inferior margin of next higher rib	**N:** Intercostal nn.

Innermost Intercostals   Presumed to have the same action as the internal intercostals

O: Superomedial surface of ribs 2–12; may be absent from upper ribs	**I:** Medial edge of costal groove of next higher rib	**N:** Intercostal nn.

Many other muscles of the chest and abdomen contribute significantly to breathing (see Deeper Insight 11.1): the sternocleidomastoid and scalenes of the neck; pectoralis major and serratus anterior of the chest; latissimus dorsi of the lower back; internal and external abdominal obliques and transverse abdominal muscle; and even some of the anal muscles. The respiratory actions of all these muscles are described in chapter 23.

DEEPER INSIGHT    11.1

Difficulty Breathing

Asthma, emphysema, heart failure, and other conditions can cause *dyspnea,* difficulty catching one's breath. People with dyspnea make increased use of accessory muscles to aid the diaphragm and intercostals in breathing, and often lean on a table or chair back to breathe more deeply. This action fixes the clavicles and scapulae so that accessory muscles such as the *pectoralis major* and *serratus anterior* (see chapter 12) move the ribs instead of the bones of the pectoral girdle.

Apply What You Know

What muscles are eaten as "spare ribs"? What is the tough fibrous membrane between the meat and the bone?

---

[38]*dia* = across + *phragm* = partition
[39]*inter* = between + *costa* = rib

TABLE 11.6	Muscles of the Anterior Abdominal Wall

Unlike the thoracic cavity, the abdominal cavity has little skeletal support. It is enclosed, however, in layers of broad flat muscles whose fibers run in different directions, strengthening the abdominal wall on the same principle as the alternating layers of plywood. Three layers of muscle enclose the lumbar region and extend about halfway across the anterior abdomen (fig. 11.11). The most superficial layer is the *external abdominal oblique.* Its fibers pass downward and anteriorly. The next deeper layer is the *internal abdominal oblique,* whose fibers pass upward and anteriorly, roughly perpendicular to those of the external oblique. The deepest layer is the *transverse abdominal (transversus abdominis),* with horizontal fibers. Anteriorly, a pair of vertical *rectus abdominis muscles* extend from sternum to pubis. These are divided into segments by three transverse **tendinous intersections,** giving them an appearance that bodybuilders nickname the "six pack."

The tendons of the oblique and transverse muscles are *aponeuroses*—broad fibrous sheets that continue medially and inferiorly (figs. 11.12 and 11.13). At the rectus abdominis, they diverge and pass around its anterior and posterior sides, enclosing the muscle in a vertical sleeve called the **rectus sheath.** They meet again at a median line called the **linea alba** between the rectus muscles. Another line, the **linea semilunaris,** marks the lateral boundary where the rectus sheath meets the aponeurosis. The aponeurosis of the external oblique also forms a cordlike **inguinal ligament** at its inferior margin. This extends obliquely from the anterior superior spine of the ilium to the pubis. The linea alba, linea semilunaris, and inguinal ligament are externally visible on a person with good muscle definition (see fig. A.8, p. 338).

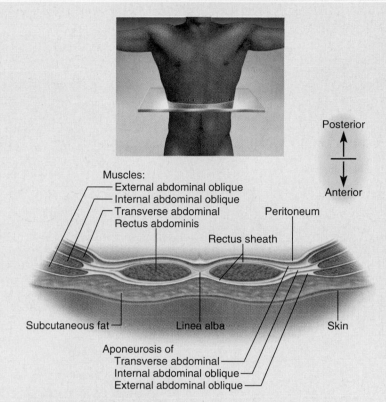

Figure 11.11 **Cross Section of the Anterior Abdominal Wall.**

External Abdominal Oblique   Supports abdominal viscera against pull of gravity; stabilizes vertebral column during heavy lifting; maintains posture; compresses abdominal organs, thus aiding in forceful expiration; aids in childbirth, urination, defecation, and vomiting. Unilateral contraction causes contralateral rotation of waist.

O: Ribs 5–12   **I:** Anterior half of iliac crest, pubic symphysis, and superior margin of pubis   **N:** Anterior rami of spinal nerves T7–T12

Internal Abdominal Oblique   Same as external oblique except that unilateral contraction causes ipsilateral rotation of waist

O: Inguinal ligament, iliac crest, and thoracolumbar fascia   **I:** Ribs 10–12, costal cartilages 7–10, pubis   **N:** Anterior rami of spinal nerves T7–L1

Transverse Abdominal   Compresses abdominal contents, with same effects as external oblique, but does not contribute to movements of vertebral column

O: Inguinal ligament, iliac crest, thoracolumbar fascia, costal cartilages 7–12   **I:** Linea alba, pubis, aponeurosis of internal oblique   **N:** Anterior rami of spinal nerves T7–L1

Rectus[40] Abdominis (REC-tus ab-DOM-ih-nis)   Flexes lumbar region of vertebral column, producing forward bending at the waist

O: Pubic symphysis and superior margin of pubis   **I:** Xiphoid process, costal cartilages 5–7   **N:** Anterior rami of spinal nerves T6–T12

[40]*rectus* = straight

TABLE 11.6 | **Muscles of the Anterior Abdominal Wall (continued)**

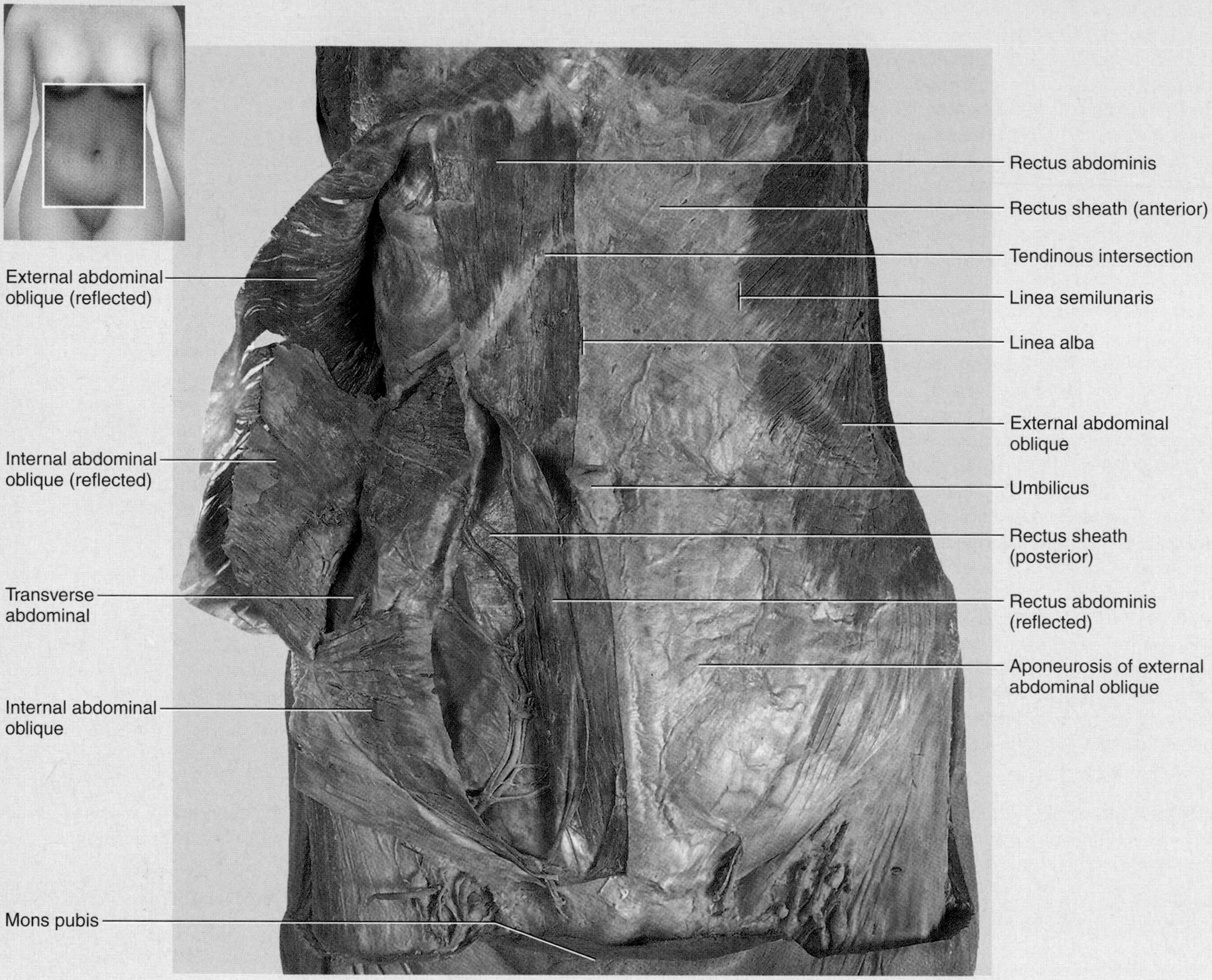

External abdominal oblique (reflected)

Internal abdominal oblique (reflected)

Transverse abdominal

Internal abdominal oblique

Mons pubis

Rectus abdominis

Rectus sheath (anterior)

Tendinous intersection

Linea semilunaris

Linea alba

External abdominal oblique

Umbilicus

Rectus sheath (posterior)

Rectus abdominis (reflected)

Aponeurosis of external abdominal oblique

Figure 11.12 **Thoracic and Abdominal Muscles of the Cadaver.** The rectus sheath is removed on the anatomical right to expose the right rectus abdominis muscle. Inset shows area of dissection.

TABLE 11.6 | **Muscles of the Anterior Abdominal Wall (continued)**

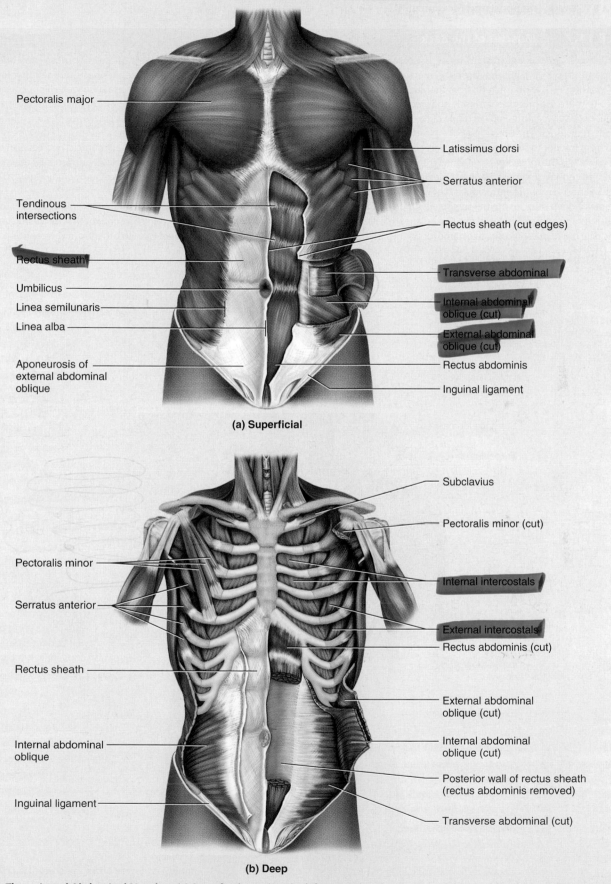

Pectoralis major

Latissimus dorsi

Serratus anterior

Tendinous intersections

Rectus sheath (cut edges)

Rectus sheath

Transverse abdominal

Umbilicus

Internal abdominal oblique (cut)

Linea semilunaris

External abdominal oblique (cut)

Linea alba

Aponeurosis of external abdominal oblique

Rectus abdominis

Inguinal ligament

(a) Superficial

Subclavius

Pectoralis minor (cut)

Pectoralis minor

Internal intercostals

Serratus anterior

External intercostals

Rectus abdominis (cut)

Rectus sheath

External abdominal oblique (cut)

Internal abdominal oblique

Internal abdominal oblique (cut)

Posterior wall of rectus sheath (rectus abdominis removed)

Inguinal ligament

Transverse abdominal (cut)

(b) Deep

Figure 11.13 **Thoracic and Abdominal Muscles.** (a) Superficial muscles. The left rectus sheath is cut away to expose the rectus abdominis muscle. (b) Deep muscles. On the anatomical right, the external abdominal oblique has been removed to expose the internal abdominal oblique, and the pectoralis major has been removed to expose the pectoralis minor. On the anatomical left, the internal abdominal oblique has been cut to expose the transverse abdominal, and the middle of the rectus abdominis has been cut out to expose the posterior rectus sheath.
● *Name at least three muscles that lie deep to the pectoralis major.*

TABLE 11.8   Muscles of the Pelvic Floor (continued)

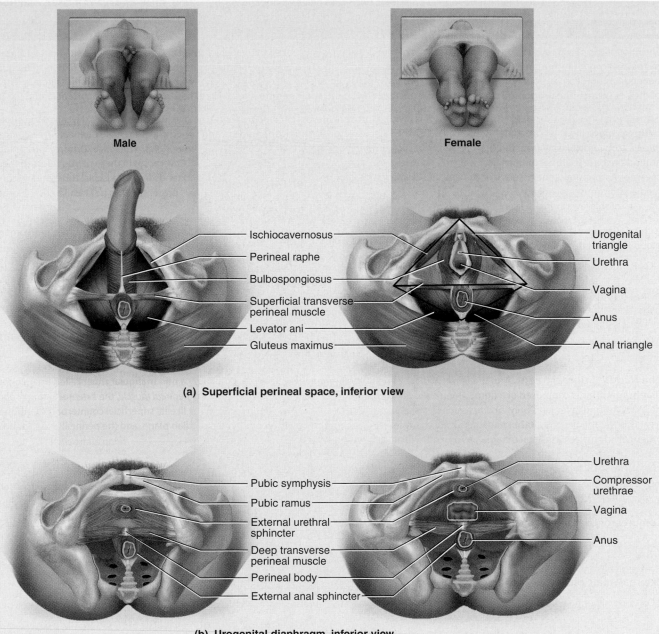

Male   **Female**

Ischiocavernosus
Perineal raphe
Bulbospongiosus
Superficial transverse perineal muscle
Levator ani
Gluteus maximus

Urogenital triangle
Urethra
Vagina
Anus
Anal triangle

(a) Superficial perineal space, inferior view

Pubic symphysis
Pubic ramus
External urethral sphincter
Deep transverse perineal muscle
Perineal body
External anal sphincter

Urethra
Compressor urethrae
Vagina
Anus

(b) Urogenital diaphragm, inferior view

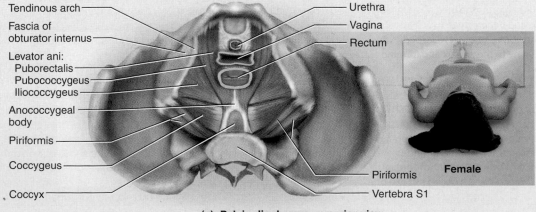

Tendinous arch
Fascia of obturator internus
Levator ani:
  Puborectalis
  Pubococcygeus
  Iliococcygeus
Anococcygeal body
Piriformis
Coccygeus
Coccyx

Urethra
Vagina
Rectum

Piriformis
Vertebra S1

Female

(c) Pelvic diaphragm, superior view

Figure 11.17 **Muscles of the Pelvic Floor.** (a) The superficial perineal space, with triangles of the perineum marked on the right. (b) The urogenital diaphragm; this is the next deeper layer after the muscles in part (a). (c) Superior view of the female pelvic diaphragm, the deepest layer, seen from within the pelvic cavity.

DEEPER INSIGHT   11.3

Hernias

A hernia is any condition in which the viscera protrude through a weak point in the muscular wall of the abdominopelvic cavity. The most common type to require treatment is an *inguinal hernia* (fig. 11.18). In the male fetus, each testis descends from the pelvic cavity into the scrotum by way of a passage called the *inguinal canal* through the muscles of the groin. This canal remains a weak point in the pelvic floor, especially in infants and children. When pressure rises in the abdominal cavity, it can force part of the intestine or bladder into this canal or even into the scrotum. This also sometimes occurs in men who hold their breath while lifting heavy weights. When the diaphragm and abdominal muscles contract, pressure in the abdominal cavity can soar to 1,500 pounds per square inch—more than 100 times the normal pressure and quite sufficient to produce an inguinal hernia, or "rupture." Inguinal hernias rarely occur in women.

Two other sites of hernia are the diaphragm and navel. A *hiatal hernia* is a condition in which part of the stomach protrudes through the diaphragm into the thoracic cavity. This is most common in overweight people over age 40. It may cause heartburn due to the regurgitation of stomach acid into the esophagus, but most cases go undetected. In an *umbilical hernia,* abdominal viscera protrude through the navel.

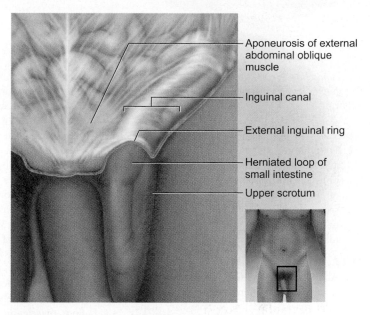

Aponeurosis of external abdominal oblique muscle

Inguinal canal

External inguinal ring

Herniated loop of small intestine

Upper scrotum

Figure 11.18 **Inguinal Hernia.** A loop of small intestine has protruded through the inguinal canal into a space beneath the skin.

Before You Go On

Answer the following questions to test your understanding of the preceding section:

7. Which muscles are used more often, the external intercostals or internal intercostals? Explain.

8. Explain how pulmonary ventilation affects abdominal pressure, and vice versa.

9. Name a major superficial muscle and two major deep muscles of the back.

10. Define *perineum, urogenital triangle,* and *anal triangle.*

11. Name one muscle in the superficial perineal space, one in the urogenital diaphragm, and one in the pelvic diaphragm. State the function of each.

Study Guide

Assess Your Learning Outcomes

You should have a good understanding of this chapter if you can accurately address the following issues.

General

1. For all muscles named in this study guide, and to the extent specified by your course or instructor, an ability to locate the muscle on a model, photograph, diagram, or dissection; to palpate the muscle or its tendons on the living body; and to state its origin, insertion, innervation, and actions

11.1 Learning Approaches (p. 264)

1. Ability to translate words commonly used in the naming of muscles and denoting such characteristics as muscle size, shape, location, number of heads, orientation, and action (table 11.1)
2. The meaning of muscle innervation; the relationship of cranial and spinal nerves to the muscles; and the common symbols for these nerves

11.2 Muscles of the Head and Neck (p. 268)

1. Two scalp muscles—the *frontalis* and *occipitalis*—and their relationship to the galea aponeurotica (table 11.2)
2. Three muscles of the ocular region—the *orbicularis oculi, levator palpebrae superioris,* and *corrugator supercilii* (table 11.2)
3. The *nasalis muscle* of the nasal region (table 11.2)
4. The location of the modiolus and how it is related to muscles of the oral and buccal regions
5. Nine muscles of the oral region—the *orbicularis oris, levator labii superioris, levator anguli oris, zygomaticus major* and *minor, risorius, depressor anguli oris,* and *depressor labii inferioris* (table 11.2)
6. Three muscles of the cheek, chin, and anterior neck—the *mentalis, buccinator,* and *platysma* (table 11.2)
7. The difference between the intrinsic and extrinsic muscles of the tongue
8. Four extrinsic tongue muscles—the *genioglossus, hyoglossus, styloglossus,* and *palatoglossus* (table 11.3)
9. Four muscles of chewing—the *temporalis, masseter, medial pterygoid,* and *lateral pterygoid* (table 11.3)
10. The difference between the suprahyoid and infrahyoid muscle groups
11. Four muscles in the suprahyoid group—the *digastric, geniohyoid, mylohyoid,* and *stylohyoid* (table 11.3)
12. Four muscles of the infrahyoid group—the *omohyoid, sternohyoid, thyrohyoid,* and *sternothyroid* (table 11.3)
13. The three *pharyngeal constrictors* and their role in swallowing (table 11.3)
14. Four flexor muscles of the neck—the *sternocleidomastoid* and three *scalene muscles*
15. Locations of the anterior and posterior triangles of the neck in relation to the sternocleidomastoid
16. Five major extensors of the neck—the *trapezius, splenius capitis, splenius cervicis, semispinalis capitis,* and *semispinalis cervicis* (table 11.4)

11.3 Muscles of the Trunk (p. 279)

1. The *diaphragm* and the three layers of intercostal muscles—*external, internal,* and *innermost intercostals* (table 11.5)
2. The contributions of other thoracic and abdominopelvic muscles to breathing
3. The four principal muscles that form the abdominal wall—the *external abdominal oblique, internal abdominal oblique, transverse abdominal,* and *rectus abdominis* (table 11.6)
4. The relationship of the abdominal muscles to the rectus sheath, inguinal ligament, linea alba, and linea semilunaris
5. The erector spinae and its three columns—the *iliocostalis, longissimus,* and *thoracis*—and the regional subdivisions of each column (lumbar, thoracic, cervical, and cephalic, with variations from one column to another) (table 11.7)
6. The *semispinalis thoracis, quadratus lumborum,* and *multifidus* (table 11.7)
7. The thoracolumbar fascia and its relationship to the erector spinae and quadratus lumborum
8. The boundaries of the perineum; its two triangles; and its three muscle compartments
9. Two muscles of the superficial perineal space—the *ischiocavernosus* and *bulbospongiosus* (table 11.8)
10. The urogenital diaphragm; two muscles in the middle compartment of the pelvic floor—the *external urethral sphincter* and *compressor urethrae* (table 11.8); how the sexes differ in this layer
11. The one muscle of the anal triangle—the *external anal sphincter*
12. The pelvic diaphragm and its two muscles—the *levator ani* and *coccygeus*

Testing Your Recall

1. Which of the following muscles is the prime mover in spitting out a mouthful of liquid?
   a. platysma
   b. buccinator
   c. risorius
   d. masseter
   e. palatoglossus

2. The word _____ in a muscle name indicates a function related to the head.
   a. cervicis
   b. carpi
   c. capitis
   d. hallucis
   e. teres

3. Which of these is *not* a suprahyoid muscle?
   a. genioglossus
   b. geniohyoid
   c. stylohyoid
   d. mylohyoid
   e. digastric

4. Which of these muscles is an extensor of the neck?
   a. external oblique
   b. sternocleidomastoid
   c. splenius capitis
   d. iliocostalis
   e. latissimus dorsi

5. Which of these muscles of the pelvic floor is the deepest?
   a. superficial transverse perineal
   b. bulbospongiosus
   c. ischiocavernosus
   d. deep transverse perineal
   e. levator ani

6. The facial nerve supplies all of the following muscles *except*
   a. the frontalis.
   b. the orbicularis oculi.
   c. the orbicularis oris.
   d. the depressor labii inferioris.
   e. the mylohyoid.

7. The _____ produce(s) lateral grinding movements of the jaw.
   a. pterygoids
   b. temporalis
   c. hyoglossus
   d. zygomaticus major and minor
   e. risorius

8. All of the following muscles act on the vertebral column *except*
   a. the serratus posterior superior.
   b. the iliocostalis thoracis.
   c. the longissimus thoracis.
   d. the spinalis thoracis.
   e. the multifidus.

9. A muscle that aids in chewing without moving the mandible is
   a. the temporalis.
   b. the mentalis.
   c. the buccinator.
   d. the levator anguli oris.
   e. the splenius cervicis.

10. Which of the following muscles raises the upper lip?
    a. levator palpebrae superioris
    b. orbicularis oris
    c. masseter
    d. zygomaticus minor
    e. mentalis

11. The prime mover of spinal extension is the _____.

12. Ejaculation results from contraction of the _____ muscle.

13. The muscle that opens your eyes is the _____.

14. As its name implies, the _____ nerve controls several muscles of the tongue.

15. The _____ muscle, named for its two bellies, opens the mouth.

16. The anterior half of the perineum is a region called the _____.

17. The abdominal aponeuroses converge on a median fibrous band on the abdomen called the _____.

18. The thyrohyoid muscle inserts on the thyroid cartilage of the _____.

19. The _____ muscles diverge like a V from the middle of the upper thorax to insertions behind the ears.

20. The largest muscle of the upper back is the _____.

Answers in the Appendix

Building Your Medical Vocabulary

State a medical meaning of each of the following word elements, and give a term in which it is used.

1. delt-
2. levat-
3. oculo-
4. apo-
5. oris
6. digito-
7. labio-
8. ipsi-
9. glosso-
10. di-

Answers in the Appendix

True or False

Determine which five of the following statements are false, and briefly explain why.

1. The origin of the sternocleidomastoid is the mastoid process.
2. The largest deep muscle of the lower back is the quadratus lumborum.
3. The muscle used to stick out your tongue is the genioglossus.
4. The abdominal oblique muscles rotate the vertebral column.
5. Exhaling requires contraction of the internal intercostal muscles.
6. The digastric muscles form the floor of the mouth.
7. The scalenes are superficial to the trapezius.
8. Cutting the phrenic nerves would paralyze the prime mover of respiration.
9. The orbicularis oculi and orbicularis oris are sphincters.
10. All of the cranial nerves innervate muscles of the head and neck.

Answers in the Appendix

Testing Your Comprehension

1. Name one antagonist of each of the following muscles: (a) orbicularis oculi, (b) genioglossus, (c) masseter, (d) sternocleidomastoid, (e) rectus abdominis.
2. Name one synergist of each of the following muscles: (a) temporalis, (b) diaphragm, (c) platysma, (d) semispinalis capitis, (e) bulbospongiosus.
3. Dental procedures, vaccination, HIV infection, and some other infections occasionally injure branches of the facial nerve and weaken or paralyze the affected muscles. Predict the problems that a person would have if the orbicularis oris and buccinator muscles were paralyzed by such a nerve lesion.
4. Removal of cancerous lymph nodes from the neck sometimes requires removal of the sternocleidomastoid on that side. How would this affect a patient's range of head movement?
5. Deeper Insight 10.2 (p. 249) remarked that the Food and Drug Administration approved Botox Cosmetic only for the treatment of frown lines between the eyebrows. From the knowledge you have gained in this chapter, name the muscles into which a physician following those guidelines would inject Botox.

Answers at www.mhhe.com/saladinha3

Improve Your Grade at **www.mhhe.com/saladinha3**

Practice quizzes, labeling activities, games, and flashcards provide fun ways to master concepts. You can also download image PowerPoint files for each chapter to create a study guide or for taking notes during lecture.

The Appendicular Musculature

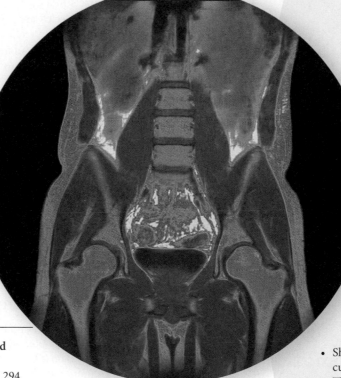

MRI scan showing muscles of the lumbar, pelvic, and upper femoral regions (frontal section)

CHAPTER OUTLINE

DEEPER INSIGHTS

BRUSHING UP

To understand this chapter, you may find it helpful to review the following concepts:
- Terminology of the limb regions (p. 14)
- Anatomy of the appendicular skeleton (chapter 8)
- Terminology of joint actions (pp. 213–218)

- Shapes of muscles (fusiform, pennate, circular, etc.) (p. 239)
- The meaning of muscle origin, insertion, and action (p. 240)
- Prime movers, synergists, antagonists, and fixators (p. 240)
- Intrinsic and extrinsic muscles (p. 241)
- Muscle innervation (p. 264)
- Greek and Latin words commonly used to name muscles (table 11.1, p. 267)

Anatomy & Physiology | **REVEALED**®
aprevealed.com

Muscular System

The appendicular musculature includes not only the muscles of the upper and lower limbs (appendages), but also several prominent muscles of the trunk that act on the limbs. In most vertebrate animals, these muscles serve for little more than locomotion, with such exceptions as digging (moles) and limited manipulation of objects (squirrels, raccoons). The upright locomotion of humans, however, has been associated with a number of evolutionary changes—large, heavily muscled lower limbs for standing, walking, and running, and upper limbs that have less muscular bulk but more mobile joints and an array of smaller muscles adapted for climbing and, more importantly, precise manipulation of objects. In this chapter, we examine the four major groups of appendicular muscles—those of the pectoral girdle, upper limb, pelvic girdle, and lower limb. We also consider several muscle injuries, which are more common in the appendicular region than in the axial region.

12.1 Muscles Acting on the Shoulder and Upper Limb

▶ **Expected Learning Outcomes**

When you have completed this section, you should be able to

- explain what muscle compartments are and describe how they are separated from each other;

- name and locate the muscles that act on the shoulder, arm, forearm, wrist, and hand;

- relate the actions of these muscles to the joint movements described in chapter 9; and

- describe the origin, insertion, and innervation of each muscle.

Muscle Groups and Compartments

The upper and lower limbs have numerous muscles that serve primarily for movement of the body and manipulation of objects. Although their number presents a learning challenge, they are arranged in logical groups that make their functional relationships easier to understand. Their names alone are often helpful indications of their functions and locations. For example, *flexor carpi ulnaris* translates "flexor of the wrist (associated with) the ulna." As a flexor, it is found on the anterior side of the forearm so it can bend the wrist anteriorly, and the name further tells us that it is on the ulnar (medial) side of the forearm. You may find it helpful to review table 11.1 (p. 265) on the naming of muscles and to pay attention to the footnotes in this chapter until such terms as *flexor* and *ulnaris* become intuitively easy to remember.

In the limbs, fibrous sheets of connective tissue, the *fasciae*, enclose spaces called **compartments.** Each compartment contains one or more functionally related muscles along with their nerve and blood supplies (fig. 12.1). We have already seen such compartmentalization in the abdominal wall and pelvic floor (chapter 11). In the ensuing tables, you will find muscles of the upper limb divided into anterior and posterior compartments, and those of the lower limb divided into anterior, posterior, medial, and lateral compartments. These major compartments are separated from each other by especially thick fasciae called **intermuscular septa,** and by the interosseous membranes of the forearm and leg (see chapter 8). In most limb regions, the muscle groups are further subdivided by thinner fasciae into superficial and deep layers. A serious problem called *compartment syndrome*, occurring when one of the muscles or blood vessels in a compartment is injured, is due partly to the fact that the muscles are so tightly bound by their fasciae (see Deeper Insight 12.1).

The upper limb is used for a broad range of both powerful and subtle actions, ranging from climbing, grasping, and throwing to writing, playing musical instruments, and manipulating small objects. Tables 12.1 through 12.5 group these into muscles that act on the scapula, those that act on the humerus and shoulder joint, those that act on the forearm and elbow joint, extrinsic (forearm) muscles that act on the wrist and hand, and intrinsic (hand) muscles that act on the fingers.

DEEPER INSIGHT   12.1

Compartment Syndrome

The fasciae of the upper and lower limb enclose the muscle compartments very snugly. If a blood vessel in a compartment is damaged by overuse or contusion (a bruising injury), blood and tissue fluid accumulate in the compartment. The fasciae prevent the compartment from expanding to relieve the pressure. Mounting pressure on the muscles, nerves, and blood vessels triggers a sequence of degenerative events called *compartment syndrome*. Blood flow to the compartment is obstructed by pressure on its arteries. If *ischemia* (poor blood flow) persists for more than 2 to 4 hours, nerves begin to die, and after 6 hours, so does muscle tissue. Nerves can regenerate after the pressure is relieved, but muscle necrosis is irreversible. The breakdown of muscle releases myoglobin into the blood. *Myoglobinuria,* the presence of myoglobin in the urine, gives the urine a dark color and is one of the key signs of compartment syndrome and some other degenerative muscle disorders. Compartment syndrome is treated by immobilizing and resting the limb and, if necessary, making an incision *(fasciotomy)* to relieve the compartment pressure.

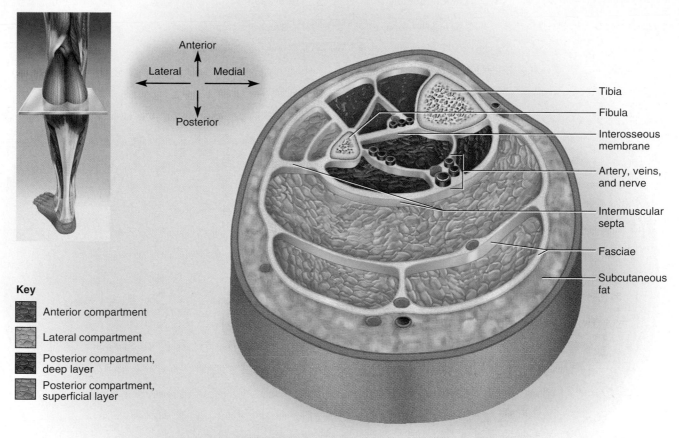

Figure 12.1 **Muscular Compartments.** A cross section of the left leg slightly above midcalf, oriented the same way as the reader's.

Key

- Anterior compartment
- Lateral compartment
- Posterior compartment, deep layer
- Posterior compartment, superficial layer

(Cross-section labels: Tibia; Fibula; Interosseous membrane; Artery, veins, and nerve; Intermuscular septa; Fasciae; Subcutaneous fat)

(Orientation: Anterior, Posterior, Lateral, Medial)

TABLE 12.1	Muscles Acting on the Shoulder

Muscles that act on the pectoral girdle originate on the axial skeleton and insert on the clavicle and scapula. The scapula is only loosely attached to the thoracic cage and is capable of considerable movement (fig. 12.2)—rotation (as in raising and lowering the apex of the shoulder), elevation and depression (as in shrugging and lowering the shoulders), and protraction and retraction (pulling the shoulders forward and back). The clavicle braces the shoulder and moderates these movements.

Anterior Group.   Muscles of the pectoral girdle fall into anterior and posterior groups (fig. 12.3 in table 12.2). The major muscles of the anterior group are the *pectoralis minor* and *serratus anterior* (see fig. 11.13b, p. 283). The pectoralis minor arises by three heads from ribs 3 to 5 and converges on the coracoid process of the scapula. The serratus anterior arises from separate heads on all or nearly all of the ribs, wraps laterally around the chest and passes across the back between the rib cage and scapula, and inserts on the medial (vertebral) border of the scapula. Thus, when it contracts, the scapula glides laterally and slightly forward around the ribs. The serratus anterior is nicknamed the "boxer's muscle" because of its role in powerful thrusting movements of the arm such as a boxer's jab.

Pectoralis Minor (PECK-toe-RAY-liss)   With serratus anterior, draws scapula laterally and forward around chest wall; with other muscles, rotates scapula and depresses apex of shoulder, as in reaching down to pick up a suitcase

  O: Ribs 3–5 and overlying fascia   **I:** Coracoid process of scapula   **N:** Medial and lateral pectoral nn.

Serratus[1] Anterior (serr-AY-tus)   With pectoralis minor, draws scapula laterally and forward around chest wall; protracts scapula, and is the prime mover in all forward-reaching and pushing actions; aids in rotating scapula to elevate apex of shoulder; fixes scapula during abduction of arm

  O: All or nearly all ribs   **I:** Medial border of scapula   **N:** Long thoracic n.

[1]*serrat* = scalloped, zigzag

TABLE 12.1 | **Muscles Acting on the Shoulder (continued)**

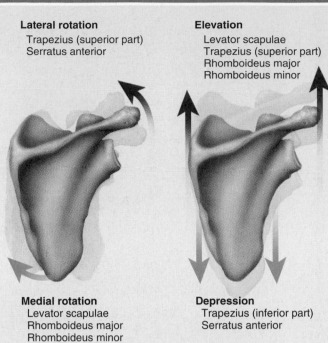

Lateral rotation
Trapezius (superior part)
Serratus anterior

Elevation
Levator scapulae
Trapezius (superior part)
Rhomboideus major
Rhomboideus minor

Medial rotation
Levator scapulae
Rhomboideus major
Rhomboideus minor

Depression
Trapezius (inferior part)
Serratus anterior

Retraction
Rhomboideus major
Rhomboideus minor
Trapezius

Protraction
Pectoralis minor
Serratus anterior

Figure 12.2 Actions of Some Thoracic Muscles on the Scapula. Note that an individual muscle can contribute to multiple actions, depending on which fibers contract and what synergists act with it.

Posterior Group. The posterior muscles that act on the scapula include the large, superficial trapezius, already discussed (table 11.4, p. 277), and three deep muscles: the *rhomboideus minor* and *rhomboideus major* (the *rhomboids*) and the *levator scapulae* (see fig. 11.14, p. 284). The action of the trapezius depends on whether its superior, middle, or inferior fibers contract and whether it acts alone or with other muscles. The levator scapulae and superior fibers of the trapezius rotate the scapula in opposite directions if either of them acts alone. If both act together, their opposite rotational effects balance each other and they elevate the scapula and shoulder, as when you lift a suitcase from the floor. Depression of the scapula occurs mainly by gravitational pull, but the trapezius and serratus anterior can depress it more rapidly and forcefully, as in swimming, hammering, and rowing.

Trapezius (tra-PEE-zee-us)   Stabilizes scapula and shoulder during arm movements; elevates apex of shoulder; acts with other muscles to rotate and retract scapula. See also roles in head and neck movements in table 11.4.

> **O:** External occipital protuberance, medial one-third of superior nuchal line, nuchal ligament, spinous processes of vertebrae C7–T3 or T4
>
> **I:** Acromion and spine of scapula, lateral one-third of clavicle
>
> **N:** Accessory n., anterior rami of C3–C4

Levator Scapulae (leh-VAY-tur SCAP-you-lee)   Elevates scapula if cervical vertebrae are fixed; flexes neck laterally if scapula is fixed; retracts scapula and braces shoulder; rotates scapula and depresses apex of shoulder

> **O:** Transverse processes of vertebrae C1–C4
>
> **I:** Superior angle to medial border of scapula
>
> **N:** C3–C4, and C5 via dorsal scapular n.

Rhomboideus Minor (rom-BOY-dee-us)   Retracts scapula and braces shoulder; fixes scapula during arm movements

> **O:** Spinous processes of vertebrae C7–T1, nuchal ligament
>
> **I:** Medial border of scapula
>
> **N:** Dorsal scapular n.

Rhomboideus Major   Same as rhomboideus minor

> **O:** Spinous processes of vertebrae T2–T5
>
> **I:** Medial border of scapula
>
> **N:** Dorsal scapular n.

TABLE 12.2	**Muscles Acting on the Arm**

Axial Muscles.   Nine muscles cross the shoulder joint and insert on the humerus. Two are considered **axial muscles** because they originate primarily on the axial skeleton—the *pectoralis major* and *latissimus dorsi* (figs. 12.3 and 12.4). The pectoralis major is the thick, fleshy muscle of the mammary region, and the latissimus dorsi is a broad muscle of the back that extends from the waist to the axilla. These muscles bear the primary responsibility for attaching the arm to the trunk and are the prime movers of the shoulder joint.

Pectoralis Major (PECK-toe-RAY-liss)   Flexes, adducts, and medially rotates humerus, as in climbing or hugging. Aids in deep inspiration.

O: Medial half of clavicle, lateral margin of sternum, costal cartilages 1–7, aponeurosis of external oblique   **I:** Lateral lip of intertubercular sulcus of humerus   **N:** Medial and lateral pectoral nn.

Latissimus Dorsi[2] (la-TISS-ih-mus DOR-sye)   Adducts and medially rotates humerus; extends the shoulder joint as in pulling on the oars of a rowboat; produces backward swing of arm in such actions as walking and bowling. With hands grasping overhead objects, pulls body forward and upward, as in climbing. Aids in deep inspiration, sudden expiration such as sneezing and coughing, and prolonged forceful expiration as in singing or blowing a sustained note on a wind instrument.

O: Vertebrae T7–L5, lower three or four ribs, iliac crest, thoracolumbar fascia   **I:** Floor of intertubercular sulcus of humerus   **N:** Thoracodorsal n.

Scapular Muscles.   The other seven muscles of the shoulder are considered **scapular muscles** because they originate on the scapula. Four of them form the rotator cuff and are treated in the next section. The most conspicuous scapular muscle is the *deltoid,* the thick, triangular muscle that caps the shoulder. This is a commonly used site for drug injections. Its anterior, lateral, and posterior fibers act like three different muscles.

Deltoid   Anterior fibers flex and medially rotate arm; lateral fibers abduct arm; posterior fibers extend and laterally rotate arm. Involved in arm swinging during such actions as walking or bowling and in adjustment of hand height for various manual tasks.

O: Acromion and spine of scapula; clavicle   **I:** Deltoid tuberosity of humerus   **N:** Axillary n.

Teres Major (TERR-eez)   Extends and medially rotates humerus; contributes to arm swinging

O: Inferior angle of scapula   **I:** Medial lip of intertubercular sulcus of humerus   **N:** Lower subscapular n.

Coracobrachialis (COR-uh-co-BRAY-kee-AL-iss)   Flexes and medially rotates arm; resists deviation of arm from frontal plane during abduction

O: Coracoid process   **I:** Medial aspect of humeral shaft   **N:** Musculocutaneous n.

Apply What You Know

Since a muscle can only pull on a bone, and not push, antagonistic muscles are needed to produce opposite actions at a joint. Reconcile this fact with the observation that the deltoid muscle both flexes and extends the shoulder.

---

[2]*latissimus* = broadest + *dorsi* = of the back

TABLE 12.2 | **Muscles Acting on the Arm (continued)**

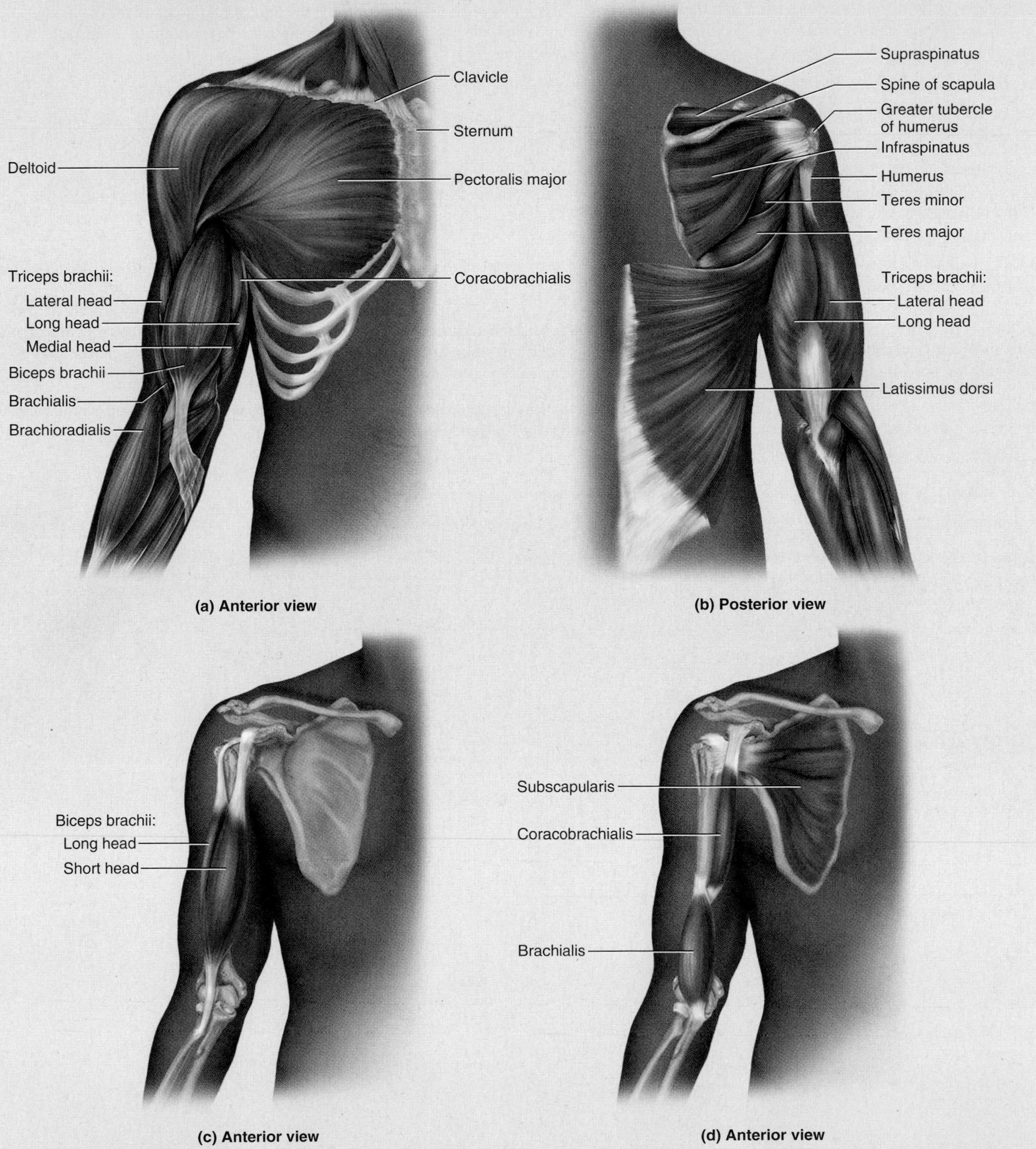

(a) Anterior view

Clavicle
Sternum
Pectoralis major
Coracobrachialis

Deltoid

Triceps brachii:
  Lateral head
  Long head
  Medial head
Biceps brachii
Brachialis
Brachioradialis

(b) Posterior view

Supraspinatus
Spine of scapula
Greater tubercle of humerus
Infraspinatus
Humerus
Teres minor
Teres major
Triceps brachii:
  Lateral head
  Long head
Latissimus dorsi

(c) Anterior view

Biceps brachii:
  Long head
  Short head

(d) Anterior view

Subscapularis
Coracobrachialis
Brachialis

Figure 12.3 **Pectoral and Brachial Muscles.** (a) Superficial muscles, anterior view. (b) Superficial muscles, posterior view (trapezius removed). (c) The biceps brachii, the superficial flexor of the elbow. (d) The brachialis, the deep flexor of the elbow, and the coracobrachialis and subscapularis, which act on the humerus.

• *What muscle serves as an antagonist of the pectoralis major?*

TABLE 12.2 | **Muscles Acting on the Arm (continued)**

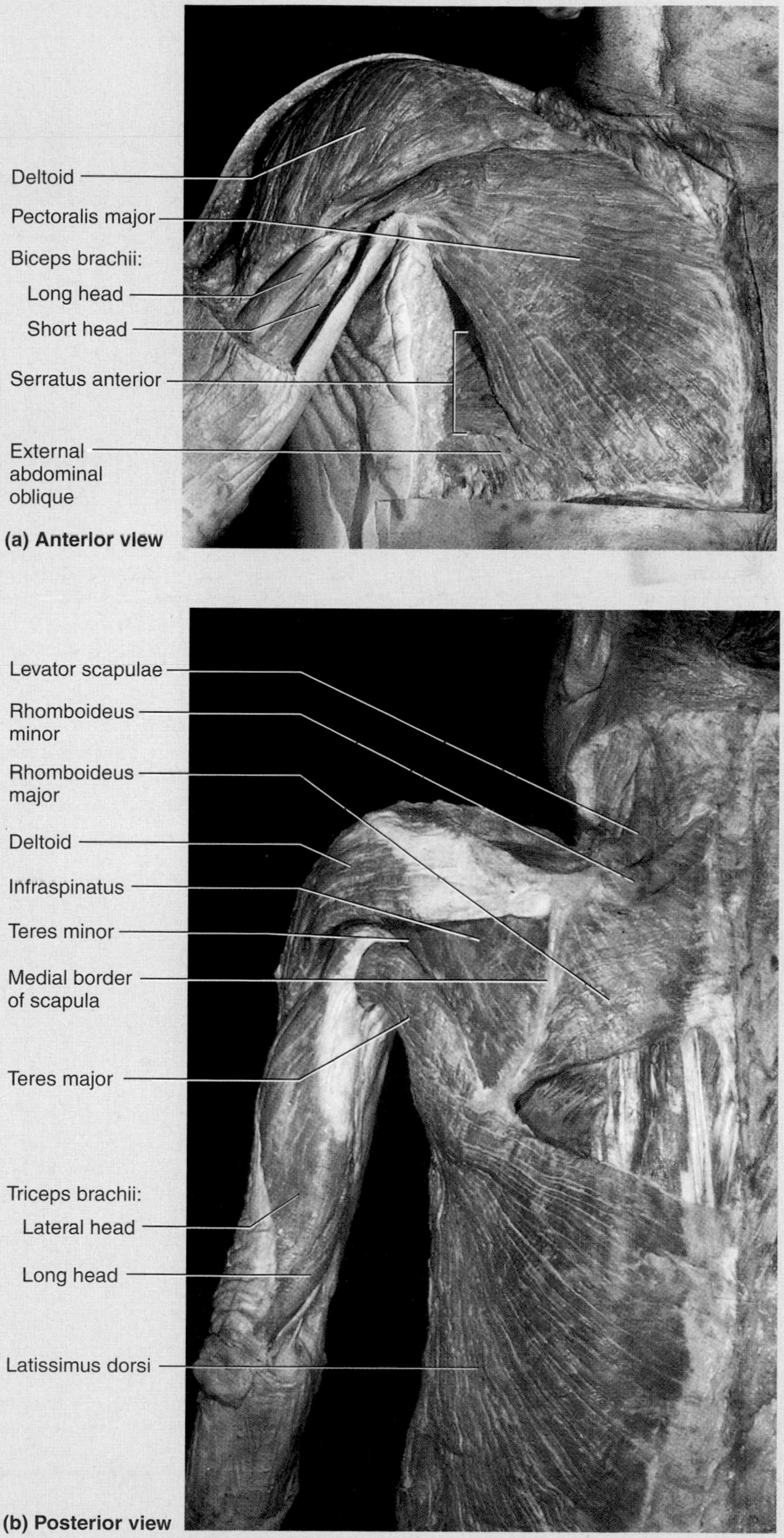

Deltoid
Pectoralis major
Biceps brachii:
Long head
Short head
Serratus anterior
External abdominal oblique

(a) Anterior view

Levator scapulae
Rhomboideus minor
Rhomboideus major
Deltoid
Infraspinatus
Teres minor
Medial border of scapula
Teres major
Triceps brachii:
Lateral head
Long head
Latissimus dorsi

(b) Posterior view

Figure 12.4 Pectoral and Brachial Muscles of the Cadaver.

TABLE 12.2	Muscles Acting on the Arm (continued)

The Rotator Cuff. Tendons of the remaining four scapular muscles form the **rotator cuff** (fig. 12.5). These muscles are nicknamed the "SITS muscles" for the first letters of their names—*supraspinatus, infraspinatus, teres minor,* and *subscapularis.* The first three muscles lie on the posterior side of the scapula (see fig. 12.3b). The supraspinatus and infraspinatus occupy the supraspinous and infraspinous fossae, above and below the scapular spine. The teres minor lies inferior to the infraspinatus. The subscapularis occupies the subscapular fossa on the anterior surface of the scapula, between the scapula and ribs (see fig. 12.3d). The tendons of these muscles merge with the joint capsule of the shoulder as they cross it en route to the humerus. They insert on the proximal end of the humerus, forming a partial sleeve around it. The rotator cuff reinforces the joint capsule and holds the head of the humerus in the glenoid cavity.

Rotator cuff injuries are common in sports and recreation. The tendon of the supraspinatus, especially, is easily damaged by strenuous circumduction (as in baseball pitching and bowling), falls (as in skiing), and hard blows from the side (as when a hockey player is slammed against the boards).

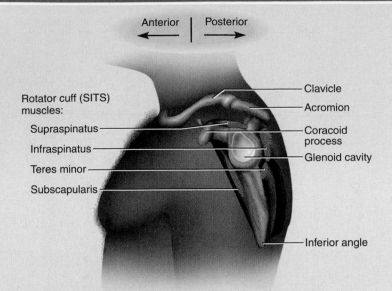

Figure 12.5 **Rotator Cuff Muscles in Relation to the Scapula.** Lateral view. For posterior and anterior views of these muscles, see figure 12.3b, d.

Supraspinatus[3] (SOO-pra-spy-NAY-tus) Aids deltoid in abduction of arm; resists downward slippage of humeral head when arm is relaxed or when carrying weight

O: Supraspinous fossa of scapula	I: Greater tubercle of humerus	N: Suprascapular n.

Infraspinatus[4] (IN-fra-spy-NAY-tus) Modulates action of deltoid, preventing humeral head from sliding upward; rotates humerus laterally

O: Infraspinous fossa of scapula	I: Greater tubercle of humerus	N: Suprascapular n.

Teres Minor (TERR-eez) Modulates action of deltoid, preventing humeral head from sliding upward as arm is abducted; rotates humerus laterally

O: Lateral border and adjacent posterior surface of scapula	I: Greater tubercle of humerus; posterior surface of joint capsule	N: Axillary n.

Subscapularis[5] (SUB-SCAP-you-LERR-iss) Modulates action of deltoid, preventing humeral head from sliding upward as arm is abducted; rotates humerus medially

O: Subscapular fossa of scapula	I: Lesser tubercle of humerus, anterior surface of joint capsule	N: Upper and lower subscapular nn.

DEEPER INSIGHT 12.2

Evolution of Limb Muscles

We can get deeper insights into human muscle functions by comparing our own muscles with those of other animals. The deltoid is a strong and prominent muscle in all hominoids (apes and humans). In its role as an abductor, it raises the arm above the shoulder—an action that a four-legged animal (quadruped) such as a horse or dog cannot perform. We take advantage of this whenever we reach for something on a high shelf or reach up for a handhold while climbing a ladder or rock wall. It is vitally important to monkeys and apes for climbing trees and swinging from overhead limbs.

The latissimus dorsi is another important muscle in climbing. We use this to pull the body up, as in rock climbing or performing pull-ups, and we condition it with "lat pulldown" exercises in fitness centers. In quadrupedal mammals, the latissimus dorsi binds the humerus to the trunk and produces a powerful backward thrust of the forelimb during walking and running.

[3]*supra* = above + *spin* = spine of scapula
[4]*infra* = below, under + *spin* = spine of scapula
[5]*sub* = below, under

TABLE 12.3	**Muscles Acting on the Forearm**

The elbow and forearm are capable of four motions—flexion, extension, pronation, and supination—carried out by muscles in both the brachium and antebrachium (arm and forearm).

Muscles with Bellies in the Arm (Brachium). The principal elbow flexors are in the anterior compartment of the arm—the brachialis and biceps brachii (see fig. 12.3c, d). The *biceps brachii* appears as a large anterior bulge and commands considerable interest among bodybuilders, but the *brachialis* underlying it generates about 50% more power and is thus the prime mover of elbow flexion. The biceps is not only a flexor but also a powerful forearm supinator. It is named for its two heads: a *short head* whose tendon arises from the coracoid process of the scapula, and a *long head* whose tendon originates on the superior margin of the glenoid cavity, loops over the shoulder, and braces the humerus against the glenoid cavity (see p. 222). The two heads converge close to the elbow on a single distal tendon that inserts on the radius and on the fascia of the medial side of the upper forearm. Note that *biceps* is the singular term; there is no such word as *bicep*. To refer to the biceps muscles of both arms, the plural is *bicipites* (by-SIP-ih-teez).

The *triceps brachii* is a three-headed muscle in the posterior compartment and is the prime mover of elbow extension (see fig. 12.3b).

Brachialis (BRAY-kee-AL-iss) Prime mover of elbow flexion

O: Anterior surface of distal half of humerus   **I:** Coronoid process and tuberosity of ulna   **N:** Musculocutaneous n., radial n.

Biceps Brachii (BY-seps BRAY-kee-eye) Rapid or forceful supination of forearm; synergist in elbow flexion; slight shoulder flexion; tendon of long head stabilizes shoulder by holding humeral head against glenoid cavity

O: Long head: superior margin of glenoid cavity   **I:** Tuberosity of radius, fascia of forearm   **N:** Musculocutaneous n.
Short head: coracoid process

Triceps Brachii (TRI-seps BRAY-kee-eye) Extends elbow; long head extends and adducts humerus

O: Long head: inferior margin of glenoid cavity and joint capsule   **I:** Olecranon, fascia of forearm   **N:** Radial n.
Lateral head: posterior surface of proximal end of humerus
Medial head: posterior surface of entire humeral shaft

Muscles with Bellies in the Forearm (Antebrachium). Most forearm muscles act on the wrist and hand, but two of them are synergists in elbow flexion and extension, and three of them function in pronation and supination. The *brachioradialis* is the large fleshy mass of the lateral (radial) side of the forearm just distal to the elbow (see figs. 12.3a and 12.7c). Its origin is on the distal end of the humerus and its insertion on the distal end of the radius. With the insertion so far from the fulcrum of the elbow, it does not generate as much force as the brachialis and biceps; it is effective mainly when those muscles have already partially flexed the elbow. The *anconeus* is a weak synergist of elbow extension on the posterior side of the elbow (see fig. 12.8e). Pronation is achieved by the *pronator quadratus* (the prime mover) near the wrist and the *pronator teres* near the elbow. Supination is usually achieved by the *supinator* of the upper forearm, with the biceps brachii aiding when additional speed or power is required (fig. 12.6).

Brachioradialis (BRAY-kee-oh-RAY-dee-AL-iss) Flexes elbow

O: Lateral supracondylar ridge of humerus   **I:** Lateral surface of radius near styloid process   **N:** Radial n.

Anconeus[6] (an-CO-nee-us) Extends elbow; may help to control ulnar movement during pronation

O: Lateral epicondyle of humerus   **I:** Olecranon and posterior surface of ulna   **N:** Radial n.

Pronator Quadratus (PRO-nay-tur quad-RAY-tus) Prime mover of forearm pronation; also resists separation of radius and ulna when force is applied to forearm through wrist, as in doing push-ups

O: Anterior surface of distal ulna   **I:** Anterior surface of distal radius   **N:** Median n.

Pronator Teres (PRO-nay-tur TERR-eez) Assists pronator quadratus in pronation, but only in rapid or forceful action; weakly flexes elbow

O: Humeral shaft near medial epicondyle; coronoid process of ulna   **I:** Lateral surface of radial shaft   **N:** Median n.

Supinator (SOO-pih-NAY-tur) Supinates forearm

O: Lateral epicondyle of humerus; supinator crest and fossa of ulna just distal to radial notch; anular and radial collateral ligaments of elbow   **I:** Proximal one-third of radius   **N:** Posterior interosseous n.

---

[6]*anconeus* = elbow

TABLE 12.3 | **Muscles Acting on the Forearm (continued)**

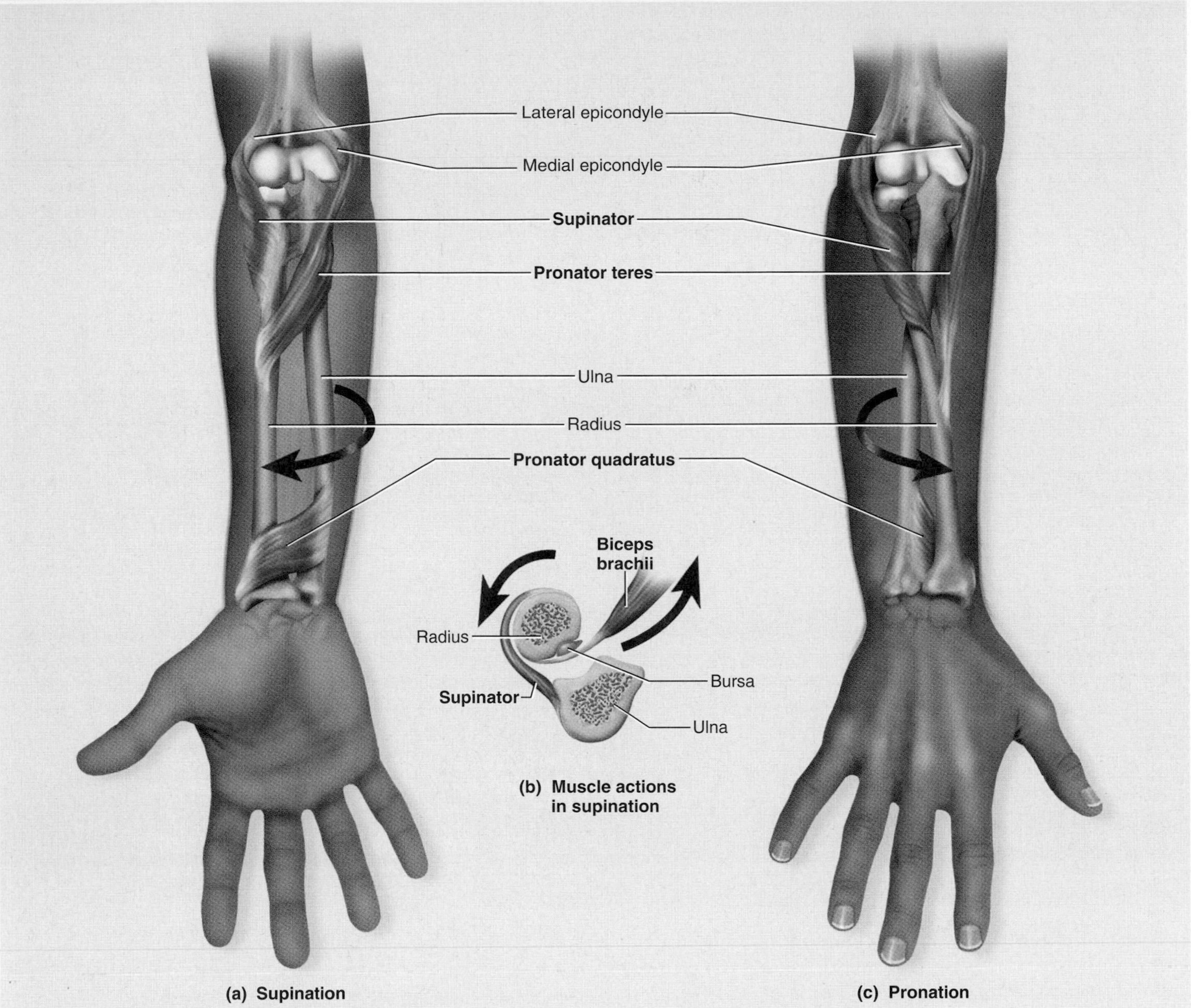

Figure 12.6 Actions of the Rotator Muscles on the Forearm. (a) Supination. (b) Cross section just distal to the elbow, showing the synergistic action of the biceps brachii and supinator. (c) Pronation.
● *What do the names of the pronator teres and pronator quadratus indicate about their shapes?*

TABLE 12.4	Muscles Acting on the Wrist and Hand

The hand is acted upon by extrinsic muscles in the forearm and intrinsic muscles in the hand itself. The bellies of the extrinsic muscles form the fleshy roundness of the upper forearm (along with the brachioradialis, table 12.3), with their tendons extending into the wrist and hand. Their actions are mainly flexion and extension of the wrist and digits, but also include radial and ulnar flexion, finger abduction and adduction, and thumb opposition. These muscles are numerous and complex, but their names often describe their location, appearance, and function.

Many of them act on the **metacarpophalangeal joints,** between the metacarpal bones of the hand and the proximal phalanges of the fingers, and the **interphalangeal joints,** between the proximal and middle or the middle and distal phalanges (or proximal–distal in the thumb, which has no middle phalanx). The metacarpophalangeal joints form the knuckles at the bases of the fingers, and the interphalangeal joints form the second and third knuckles. Some tendons cross multiple joints before inserting on a middle or distal phalanx, and can flex or extend all the joints they cross.

Most tendons of the extrinsic muscles pass under a fibrous, braceletlike sheet called the **flexor retinaculum** on the anterior side of the wrist or the **extensor retinaculum** on the posterior side. These ligaments prevent the tendons from standing up like taut bowstrings when the muscles contract. The **carpal tunnel** is a tight space between the flexor retinaculum and carpal bones. The flexor tendons passing through the tunnel are enclosed in tendon sheaths that enable them to slide back and forth quite easily, although this region is very subject to painful inflammation—*carpal tunnel syndrome*—resulting from repetitive motion (see Deeper Insight 12.3 and fig. 12.9).

Fasciae divide the forearm muscles into anterior and posterior compartments and each compartment into superficial and deep layers (fig. 12.7). The muscles will be described in these four groups.

Anterior (Flexor) Compartment, Superficial Layer. Most muscles of the anterior compartment are wrist and finger flexors that arise from a common tendon on the humerus (fig. 12.8a, b). At the distal end, the tendon of the *palmaris longus* passes over the flexor retinaculum, whereas the other tendons pass beneath it, through the carpal tunnel. The two prominent tendons you can palpate at the wrist belong to the palmaris longus on the medial side and the *flexor carpi radialis* on the lateral side (see fig. A.19a, p. 345). The latter is an important landmark for finding the radial artery, where the pulse is usually taken. The palmaris longus is absent on one or both sides (most commonly the left) in about 14% of people. To see if you have one, flex your wrist and touch the tips of your thumb and little finger together. If present, the palmaris longus tendon will stand up prominently on the wrist.

Flexor Carpi Radialis[7] **(FLEX-ur CAR-pye RAY-dee-AL-iss)**   Flexes wrist anteriorly; aids in radial flexion of wrist

O: Medial epicondyle of humerus

I: Base of metacarpals II–III

N: Median n.

Flexor Carpi Ulnaris[8] **(ul-NAY-ris)**   Flexes wrist anteriorly; aids in ulnar flexion of wrist

O: Medial epicondyle of humerus; medial margin of olecranon; posterior surface of ulna

I: Pisiform, hamate, metacarpal V

N: Ulnar n.

Flexor Digitorum Superficialis[9] **(DIDJ-ih-TOE-rum SOO-per-FISH-ee-AY-lis)**   Flexes wrist, metacarpophalangeal, and interphalangeal joints depending on action of other muscles

O: Medial epicondyle of humerus; ulnar collateral ligament; coronoid process; superior half of radius

I: Middle phalanges II–V

N: Median n.

Palmaris Longus (pal-MERR-iss)   Anchors skin and fascia of palmar region; resists shearing forces when stress is applied to skin by such actions as climbing and tool use. Weakly developed and sometimes absent.

O: Medial epicondyle of humerus

I: Flexor retinaculum, palmar aponeurosis

N: Median n.

Anterior (Flexor) Compartment, Deep Layer. The following two flexors constitute the deep layer (fig. 12.8c). The **flexor digitorum profundus** flexes fingers II–V, but the thumb (pollex) has a flexor of its own—one of several muscles serving exclusively for thumb movements.

Flexor Digitorum Profundus[10]   Flexes wrist and metacarpophalangeal and interphalangeal joints of digits II–V; sole flexor of the distal interphalangeal joints

O: Proximal three-quarters of ulna; coronoid process; interosseous membrane

I: Distal phalanges II–V

N: Median n., ulnar n.

Flexor Pollicis[11] **Longus (PAHL-ih-sis)**   Flexes phalanges of thumb

O: Radius, interosseous membrane

I: Distal phalanx I

N: Median n.

---

[7]*carpi* = of the wrist + *radialis* = of the radius
[8]*carpi* = of the wrist + *ulnaris* = of the ulna
[9]*digitorum* = of the digits + *superficialis* = shallow, near the surface
[10]*digitorum* = of the digits + *profundus* = deep
[11]*pollicis* = of the thumb (pollex)

TABLE 12.5 | Intrinsic Muscles of the Hand

The intrinsic muscles of the hand assist the flexors and extensors in the forearm and make finger movements more precise. They are divided into three groups: the **thenar group** at the base of the thumb, the **hypothenar group** at the base of the little finger, and the **midpalmar group** between these (fig. 12.10).

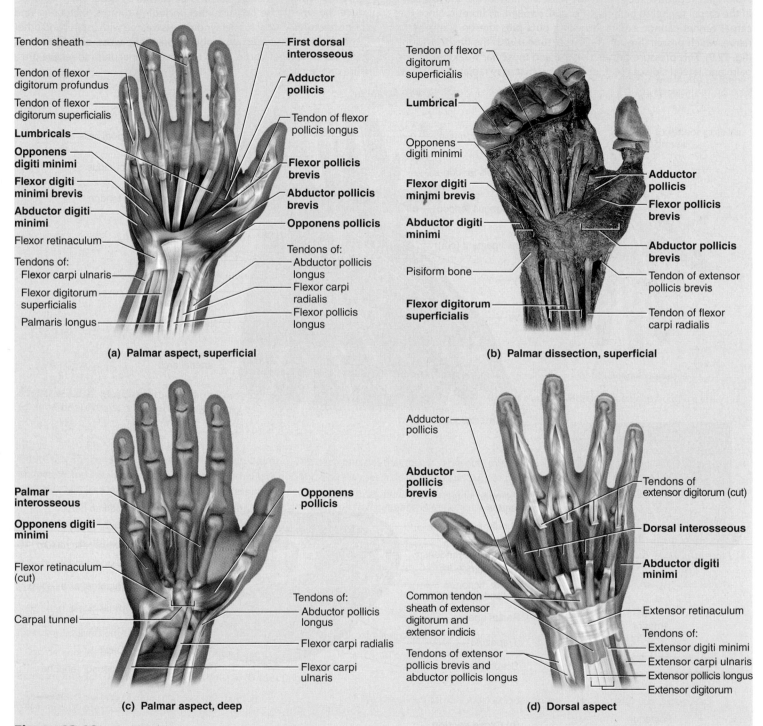

(a) Palmar aspect, superficial

(b) Palmar dissection, superficial

(c) Palmar aspect, deep

(d) Dorsal aspect

Figure 12.10 **Intrinsic Muscles of the Hand.** The boldface labels in parts (a), (c), and (d) indicate the muscles that belong to the respective layers.

TABLE 12.5	**Intrinsic Muscles of the Hand (continued)**

Thenar[14] Group.   The thenar group of muscles forms the thick, fleshy mass *(thenar eminence)* at the base of the thumb, and the *adductor pollicis* forms the web between the thumb and palm. All are concerned with thumb movements. The adductor pollicis has an *oblique head* that extends from the capitate bone of the wrist to the ulnar side of the base of the thumb, and a *transverse head* that extends from metacarpal III to the same insertion as the oblique head.

Adductor Pollicis   Draws thumb toward palm as in gripping a tool

O: Capitate; bases of metacarpals II–III; anterior ligaments of wrist; tendon sheath of flexor carpi radialis	**I:** Medial surface of base of proximal phalanx I	**N:** Ulnar n.

Abductor Pollicis Brevis   Abducts thumb in sagittal plane

O: Mainly flexor retinaculum; also scaphoid, trapezium, and abductor pollicis longus tendon	**I:** Lateral surface of proximal phalanx I	**N:** Median n.

Flexor Pollicis Brevis   Flexes metacarpophalangeal joint of thumb

O: Trapezium, trapezoid, capitate, anterior ligaments of wrist, flexor retinaculum	**I:** Proximal phalanx I	**N:** Median n., ulnar n.

Opponens Pollicis (op-PO-nenz)   Flexes metacarpal I to oppose thumb to fingertips

O: Trapezium, flexor retinaculum	**I:** Metacarpal I	**N:** Median n.

Hypothenar Group.   The hypothenar group forms the fleshy mass *(hypothenar eminence)* at the base of the little finger. All of these muscles are concerned with movement of that digit.

Abductor Digiti Minimi   Abducts little finger, as in spreading fingers apart

O: Pisiform, tendon of flexor carpi ulnaris	**I:** Medial surface of proximal phalanx V	**N:** Ulnar n.

Flexor Digiti Minimi Brevis   Flexes little finger at metacarpophalangeal joint

O: Hamulus of hamate bone, flexor retinaculum	**I:** Medial surface of proximal phalanx V	**N:** Ulnar n.

Opponens Digiti Minimi   Flexes metacarpal V at carpometacarpal joint when little finger is moved into opposition with tip of thumb; deepens palm of hand

O: Hamulus of hamate bone, flexor retinaculum	**I:** Medial surface of metacarpal V	**N:** Ulnar n.

Midpalmar Group.   The midpalmar group occupies the hollow of the palm. It has 11 small muscles divided into three groups.

Dorsal Interosseous[15] Muscles (IN-tur-OSS-ee-us) (four muscles)   Abduct fingers; strongly flex metacarpophalangeal joints but extend interphalangeal joints, depending on action of other muscles; important for grip strength

O: Each with two heads arising from facing surfaces of adjacent metacarpals	**I:** Proximal phalanges II–IV	**N:** Ulnar n.

Palmar Interosseous Muscles (Three Muscles)   Adduct fingers; other actions same as for dorsal interosseous muscles

O: Metacarpals I, II, IV, V	**I:** Proximal phalanges II, IV, V	**N:** Ulnar n.

Lumbricals[16] (LUM-brick-ulz) (Four Muscles)   Extend interphalangeal joints; contribute to ability to pinch objects between fleshy pulp of thumb and finger, instead of these digits meeting by the edges of their nails

O: Tendons of flexor digitorum profundus	**I:** Proximal phalanges II–V	**N:** Median n., ulnar n.

Before You Go On

Answer the following questions to test your understanding of the preceding section:

1. Name a muscle that inserts on the scapula and plays a significant role in each of the following actions: (a) pushing a stalled car, (b) paddling a canoe, (c) squaring the shoulders in military attention, (d) lifting the shoulder to carry a heavy box on it, and (e) lowering the shoulder to grasp a suitcase handle.

2. Describe three contrasting actions of the deltoid muscle.

3. Name the four rotator cuff muscles and describe the scapular surfaces against which they lie.

4. Name the prime movers of elbow flexion and extension.

5. Identify three functions of the biceps brachii.

6. Name three extrinsic muscles and two intrinsic muscles that flex the phalanges.

---

[14]*thenar* = of the palm
[15]*inter* = between + *osse* = bones
[16]*lumbrical* = resembling an earthworm

12.2 Muscles Acting on the Hip and Lower Limb

▶ Expected Learning Outcomes

When you have completed this section, you should be able to

- name and locate the muscles that act on the hip, knee, ankle, and toe joints;
- relate the actions of these muscles to the joint movements described in chapter 9; and
- describe the origin, insertion, and innervation of each muscle.

The largest muscles are found in the lower limb. Unlike those of the upper limb, they are adapted less for precision than for the strength needed to stand, maintain balance, walk, and run. Several of them cross and act on two or more joints, such as the hip and knee. To avoid confusion in this discussion, remember that in the anatomical sense the word *leg* refers only to that part of the limb between the knee and ankle. The term *foot* includes the tarsal region (ankle), metatarsal region, and toes. Tables 12.6 through 12.9 group the muscles of the lower limb into those that act on the femur and hip joint, those that act on the leg and knee joint, extrinsic (leg) muscles that act on the foot and ankle joint, and intrinsic (foot) muscles that act on the arches and toes.

TABLE 12.6	Muscles Acting on the Hip and Thigh

Figure 12.11 **Muscles that Act on the Hip and Femur.** Anterior view.

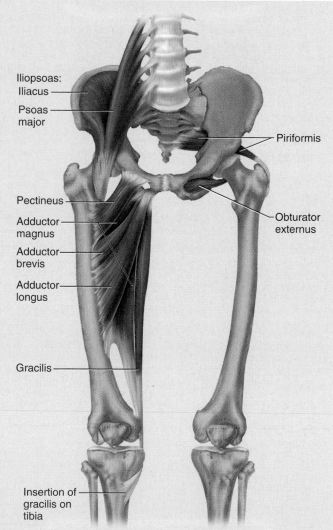

Iliopsoas:
Iliacus
Psoas major
Piriformis
Pectineus
Adductor magnus
Adductor brevis
Adductor longus
Obturator externus
Gracilis
Insertion of gracilis on tibia

Anterior Muscles of the Hip. Most muscles that act on the femur originate on the hip bone. The two principal anterior muscles are the *iliacus,* which fills most of the broad iliac fossa of the pelvis, and the *psoas major,* a thick rounded muscle that arises mainly from the lumbar vertebrae (fig. 12.11). Collectively, they are called the **iliopsoas** and share a common tendon to the femur.

Iliacus[17] **(ih-LY-uh-cus)** Flexes thigh at hip when trunk is fixed; flexes trunk at hip when thigh is fixed, as in bending forward in a chair or sitting up in bed; balances trunk during sitting

> **O:** Iliac crest and fossa, superolateral region of sacrum, anterior sacroiliac and iliolumbar ligaments **I:** Lesser trochanter and nearby shaft of femur **N:** Femoral n.

Psoas[18] **Major (SO-ass)** Same as iliacus

> **O:** Bodies and intervertebral discs of vertebrae T12–L5, transverse processes of lumbar vertebrae **I:** Lesser trochanter and nearby shaft of femur **N:** Ventral rami of lumbar spinal nn.

---

[17]*ili* = loin, flank

[18]*psoa* = loin

Lateral and Posterior Muscles of the Hip.   On the lateral and posterior sides of the hip are the *tensor fasciae latae* and three gluteal muscles. The **fascia lata** is a fibrous sheath that encircles the thigh like a subcutaneous stocking and tightly binds its muscles. On the lateral surface, it combines with the tendons of the gluteus maximus and tensor fasciae latae to form the **iliotibial band,** which extends from the iliac crest to the lateral condyle of the tibia (see figs. 12.13 and 12.14). The tensor fasciae latae tautens the iliotibial band and braces the knee, especially when the opposite foot is lifted.

    The gluteal muscles are the *gluteus maximus, gluteus medius,* and *gluteus minimus* (fig. 12.12). The gluteus maximus is the largest of these and forms most of the lean mass of the buttock. It is an extensor of the hip joint that produces the backswing of the leg in walking and provides most of the lift when you climb stairs. It generates its maximum force when the thigh is flexed at a 45° angle to the trunk. This is the advantage in starting a foot race from a crouched position. The gluteus medius is deep and lateral to the gluteus maximus. Its name refers to its size, not its position. The gluteus minimus is the smallest and deepest of the three.

Tensor Fasciae Latae[19] (TEN-sur FASH-ee-ee LAY-tee)   Extends knee, laterally rotates tibia, aids in abduction and medial rotation of femur; during standing, steadies pelvis on femoral head and steadies femoral condyles on tibia

  O: Iliac crest, anterior superior spine, deep surface of fascia lata   **I:** Lateral condyle of tibia via iliotibial band   **N:** Superior gluteal n.

Gluteus Maximus[20]   Extends thigh at hip as in stair climbing (rising to next step) or running and walking (backswing of limb); abducts thigh; elevates trunk after stooping; prevents trunk from pitching forward during walking and running; helps stabilize femur on tibia

  O: Posterior gluteal line of ilium, on posterolateral surface from iliac crest to posterior superior spine; coccyx; posterior surface of lower sacrum; aponeurosis of erector spinae   **I:** Gluteal tuberosity of femur; lateral condyle of tibia via iliotibial band   **N:** Inferior gluteal n.

Gluteus Medius and Gluteus Minimus   Abduct and medially rotate thigh; during walking, shift weight of trunk toward limb with foot on the ground as other foot is lifted

  O: Most of lateral surface of Illum between crest and acetabulum   **I:** Greater trochanter of femur   **N:** Superior gluteal n.

Lateral Rotators.   Inferior to the gluteus minimus and deep to the other two gluteal muscles are six muscles called the **lateral rotators,** named for their action on the femur (fig. 12.12). Their action is most clearly visualized when you cross your legs to rest an ankle on your knee, causing your femur to rotate and the knee to point laterally. Thus, they oppose medial rotation by the gluteus medius and minimus. Most of them also abduct or adduct the femur. The abductors are important in walking because when one lifts a foot from the ground, they shift the body weight to the other leg and prevent falling.

Gemellus[21] Superior (jeh-MEL-us)   Laterally rotates extended thigh; abducts flexed thigh. Sometimes absent.

  O: Ischial spine   **I:** Greater trochanter of femur   **N:** Nerve to obturator internus

Gemellus Inferior   Same actions as gemellus superior

  O: Ischial tuberosity   **I:** Greater trochanter of femur   **N:** Nerve to quadratus femoris

Obturator[22] Externus (OB-too-RAY-tur)   Not well understood; thought to laterally rotate thigh in climbing

  O: External surface of obturator membrane; pubic and ischial rami   **I:** Femur between head and greater trochanter   **N:** Obturator n.

Obturator Internus   Not well understood; thought to laterally rotate extended thigh and abduct flexed thigh

  O: Ramus of ischium; inferior ramus of pubis; anteromedial surface of lesser pelvis   **I:** Greater trochanter of femur   **N:** Nerve to obturator internus

Piriformis[23] (PIR-ih-FOR-mis)   Laterally rotates extended thigh; abducts flexed thigh

  O: Anterior surface of sacrum; gluteal surface of ilium; capsule of sacroiliac joint   **I:** Greater trochanter of femur   **N:** Spinal nn. L5–S2

Quadratus Femoris[24] (quad-RAY-tus FEM-oh-ris)   Laterally rotates thigh

  O: Ischial tuberosity   **I:** Intertrochanteric crest of femur   **N:** Nerve to quadratus femoris

Medial (Adductor) Compartment of the Thigh.   Fasciae divide the thigh into three compartments: the *anterior (extensor) compartment, posterior (flexor) compartment,* and *medial (adductor) compartment.* Muscles of the anterior and posterior compartments function mainly as extensors and flexors of the knee, respectively, and are treated in table 12.7. The five muscles of the medial compartment act primarily as adductors of the thigh (fig. 12.11), but some of them cross both the hip and knee joints and have additional actions as follows.

Adductor Brevis   Adducts thigh

  O: Body and inferior ramus of pubis   **I:** Linea aspera and spiral line of femur   **N:** Obturator n.

Adductor Longus   Adducts and medially rotates thigh; flexes thigh at hip

  O: Body and inferior ramus of pubis   **I:** Linea aspera of femur   **N:** Obturator n.

---

[19]*fasc* = band + *lat* = broad
[20]*glut* = buttock + *maxim* = largest
[21]*gemellus* = twin

[22]*obtur* = to close, stop up
[23]*piri* = pear + *form* = shaped
[24]*quadrat* = four-sided + *femoris* = of the thigh or femur

TABLE 12.6	Muscles Acting on the Hip and Thigh (continued)

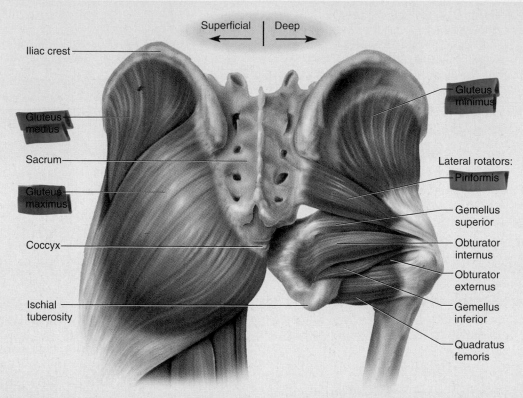

Figure 12.12 Gluteal Muscles. Posterior view. Superficial muscles are shown on the left and deep muscles on the right.
● *Describe two everyday movements of the body that employ the power of the gluteus maximus.*

Adductor Magnus   Adducts and medially rotates thigh; extends thigh at hip

O: Inferior ramus of pubis; ramus and tuberosity of ischium   **I:** Linea aspera, gluteal tuberosity, and medial supracondylar line of femur   **N:** Obturator n., tibial n.

Gracilis[25] **(GRASS-ih-lis)**   Flexes and medially rotates tibia at knee

O: Body and inferior ramus of pubis; ramus of ischium   **I:** Medial surface of tibia just below condyle   **N:** Obturator n.

Pectineus[26] **(pec-TIN-ee-us)**   Flexes and adducts thigh

O: Superior ramus of pubis   **I:** Spiral line of femur   **N:** Femoral n.

DEEPER INSIGHT   12.4

Intramuscular Injections

Muscles with thick bellies are commonly used for intramuscular (I.M.) drug injections. Since drugs injected into these muscles are absorbed into the bloodstream gradually, it is safe to administer relatively large doses (up to 5 mL) that could be dangerous or even fatal if injected directly into the bloodstream. I.M. injections also cause less tissue irritation than subcutaneous injections.

Knowledge of subsurface anatomy is necessary to avoid damaging nerves or accidentally injecting a drug into a blood vessel. Anatomical knowledge also enables a clinician to position a patient so that the muscle is relaxed, making the injection less painful.

Amounts up to 2 mL are commonly injected into the deltoid muscle about two finger widths below the acromion. A misplaced injection into the deltoid can injure the axillary nerve and cause atrophy of the muscle. Drug doses over 2 mL are commonly injected into the gluteus medius, in the superolateral quadrant of the gluteal area, at a safe distance from the sciatic nerve and major gluteal blood vessels. Injections are often given to infants and young children in the vastus lateralis of the thigh, because their deltoid and gluteal muscles are not well developed.

[25]*gracil* = slender
[26]*pectin* = comb

TABLE 12.7	Muscles Acting on the Knee and Leg

The following muscles form most of the mass of the thigh and produce their most obvious actions on the knee joint. Some of them, however, cross both the hip and knee and produce actions at both, moving the femur, tibia, and fibula.

Anterior (Extensor) Compartment of the Thigh. The anterior compartment of the thigh contains the large *quadriceps femoris muscle,* the prime mover of knee extension and the most powerful muscle of the body (figs. 12.13 and 12.14; see also fig. 12.20). As the name implies, it has four heads: the *rectus femoris, vastus lateralis, vastus medialis,* and *vastus intermedius.* All four converge on a single **quadriceps (patellar) tendon,** which extends to the patella, then continues as the **patellar ligament** and inserts on the tibial tuberosity. (Remember that a tendon usually extends from muscle to bone, and a ligament from bone to bone.) The patellar ligament is struck with a rubber reflex hammer to test the knee-jerk reflex. The quadriceps extends the knee when you stand up, take a step, or kick a ball. One head, the rectus femoris, contributes to running by acting with the iliopsoas to flex the hip in each airborne phase of the leg's cycle of motion. It also flexes the hip in such actions as high kicks, stair climbing, or simply in drawing the leg forward during a stride.

Crossing the quadriceps from the lateral side of the hip to the medial side of the knee is the narrow, straplike *sartorius,* the longest muscle of the body. It flexes the hip and knee joints and laterally rotates the thigh, as in crossing the legs. It is colloquially called the "tailor's muscle," after the cross-legged posture of a tailor supporting his work on the raised knee.

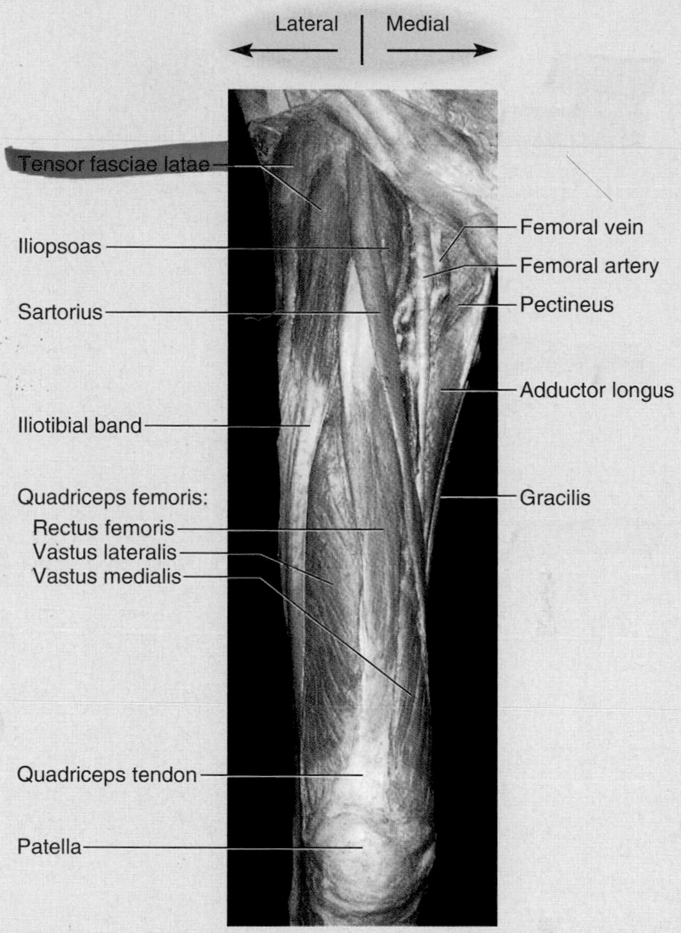

Figure 12.13 Anterior Superficial Thigh Muscles of the Cadaver. Right limb.

Quadriceps Femoris (QUAD-rih-seps FEM-oh-ris)   All heads insert on tibia through a common tendon and extend the knee, in addition to the actions of individual heads below.

Rectus Femoris   Extends knee; flexes thigh at hip; flexes trunk on hip if thigh is fixed

O: Ilium at anterior inferior spine and superior margin of acetabulum; capsule of hip joint	**I:** Patella, tibial tuberosity, lateral and medial condyles of tibia	**N:** Femoral n.

Vastus[27] Lateralis   Extends knee; retains patella in groove on femur during knee movements

O: Femur at greater trochanter and intertrochanteric line, gluteal tuberosity, and linea aspera	**I:** Same as rectus femoris	**N:** Same as rectus femoris

Vastus Medialis   Same as vastus lateralis

O: Femur at intertrochanteric line, spiral line, linea aspera, and medial supracondylar line	**I:** Same as rectus femoris	**N:** Same as rectus femoris

Vastus Intermedius   Extends knee

O: Anterior and lateral surfaces of femoral shaft	**I:** Same as rectus femoris	**N:** Same as rectus femoris

Sartorius[28]   Aids in knee and hip flexion, as in sitting or climbing; abducts and laterally rotates thigh

O: On and near anterior superior spine of ilium	**I:** Medial surface of proximal end of tibia	**N:** Femoral n.

[27]*vastus* = large, extensive
[28]*sartor* = tailor

TABLE 12.7 **Muscles Acting on the Knee and Leg (continued)**

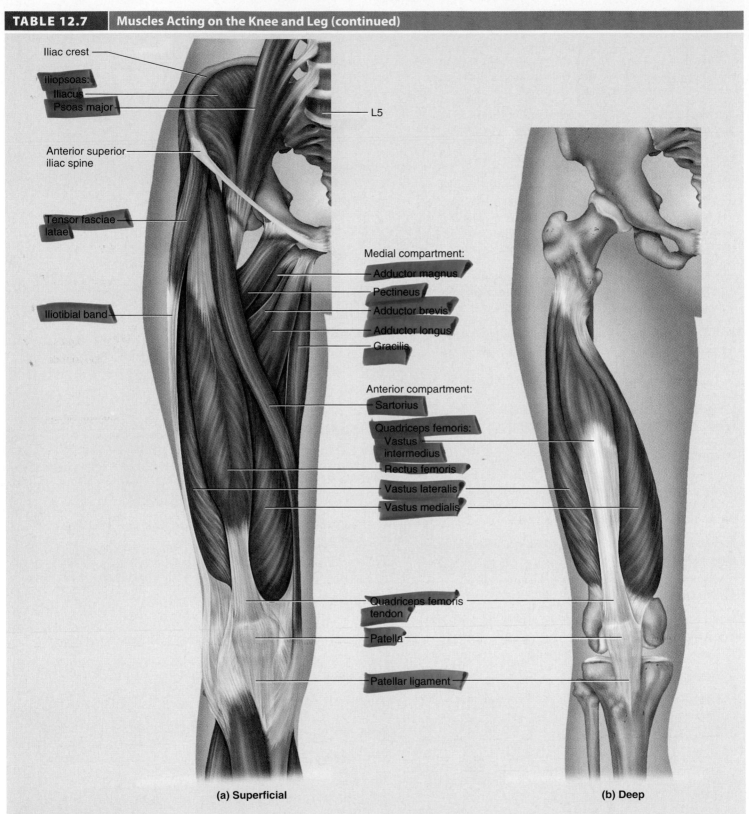

(a) Superficial

(b) Deep

Figure 12.14 Anterior Muscles of the Thigh. (a) Superficial muscles. (b) Rectus femoris and other muscles removed to expose the other three heads of the quadriceps femoris.

TABLE 12.7 | **Muscles Acting on the Knee and Leg** (continued)

Figure 12.15 Gluteal and Thigh Muscles. Posterior view. The gluteus maximus is cut to expose the hamstring muscles.
● *Note the tendinous band that crosses the semitendinous muscle. Name another muscle that is subdivided by one or more transverse tendinous bands.*

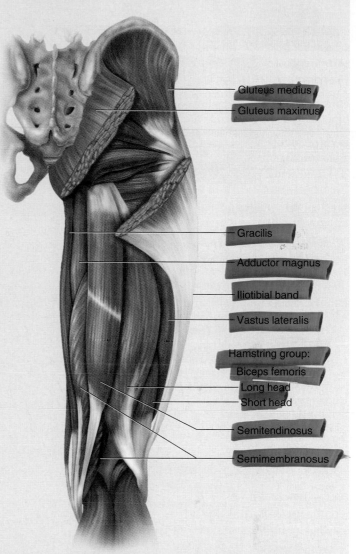

Gluteus medius
Gluteus maximus
Gracilis
Adductor magnus
Iliotibial band
Vastus lateralis
Hamstring group:
Biceps femoris
Long head
Short head
Semitendinosus
Semimembranosus

Posterior (Flexor) Compartment of the Thigh. The posterior compartment contains three muscles colloquially known as the **hamstring muscles;** from lateral to medial, they are the *biceps femortis, semitendinosus,* and *semimembranosus* (fig. 12.15). The pit at the back of the knee, known anatomically as the popliteal fossa, is colloquially called the *ham.* The tendons of these muscles can be felt as prominent cords on both sides of the pit—the biceps tendon on the lateral side and the semimembranosus and semitendinosus tendons on the medial side. When wolves attack large prey, they often attempt to sever the hamstring tendons, because this renders the prey helpless. The hamstrings flex the knee, and aided by the gluteus maximus, they extend the hip during walking and running. The semitendinosus is named for its unusually long tendon. This muscle also is usually bisected by a transverse or oblique tendinous band. The semimembranosus is named for the flat shape of its superior attachment.

Biceps Femoris Flexes knee; extends hip; elevates trunk from stooping posture; laterally rotates tibia on femur when knee is flexed; laterally rotates femur when hip is extended; counteracts forward bending at hips

O: Long head: ischial tuberosity Short head: linea aspera and lateral supracondylar line of femur	**I:** Head of fibula	**N:** Tibial n., common fibular n.

Semitendinosus[29] **(SEM-ee-TEN-din-OH-sus)** Flexes knee; medially rotates tibia on femur when knee is flexed; medially rotates femur when hip is extended; counteracts forward bending at hips

O: Ischial tuberosity	**I:** Medial surface of upper tibia	**N:** Tibial n.

Semimembranosus[30] **(SEM-ee-MEM-bran-OH-sus)** Same as semitendinosus

O: Ischial tuberosity	**I:** Medial condyle and nearby margin of tibia; intercondylar line and lateral condyle of femur; ligament of popliteal region	**N:** Tibial n.

Posterior Compartment of the Leg. Most muscles in the posterior compartment of the leg act on the ankle and foot and are reviewed in table 12.8, but the popliteus acts on the knee (see fig. 12.18b).

Popliteus[31] **(pop-LIT-ee-us)** Rotates tibia medially on femur if femur is fixed (as in sitting), or rotates femur laterally on tibia if tibia is fixed (as in standing); unlocks knee to allow flexion; may prevent forward dislocation of femur during crouching

O: Lateral condyle of femur; lateral meniscus and joint capsule	**I:** Posterior surface of upper tibia	**N:** Tibial n.

---

[29]*semi* = half + *tendinosus* = tendinous
[30]*semi* = half + *membranosus* = membranous
[31]*poplit* = ham (pit) of the knee

DEEPER INSIGHT   12.5

Hamstring Injuries

Hamstring injuries are common among sprinters, soccer players, and other athletes who depend on quick extension of the knee to kick or jump forcefully. Rapid knee extension stretches the hamstrings and often tears the proximal tendons where they originate on the ischial tu-berosity. These muscle strains are excruciatingly painful. Hamstring injuries often result from failure to warm up adequately before competition or practice.

TABLE 12.8	**Muscles Acting on the Foot**

The fleshy mass of the leg is formed by a group of **crural muscles,** which act on the foot (fig. 12.16). These muscles are tightly bound by fasciae that compress them and aid in the return of blood from the legs. The fasciae separate the crural muscles into anterior, lateral, and posterior compartments (see fig. 12.20b).

Anterior (Extensor) Compartment of the Leg.   Muscles of the anterior compartment dorsiflex the ankle and prevent the toes from scuffing the ground during walking. From lateral to medial, these muscles are the *fibularis tertius, extensor digitorum longus* (extensor of toes II–V), *extensor hallucis longus* (extensor of the great toe), and *tibialis anterior.* Their tendons are held tightly against the ankle and kept from bowing by two **extensor retinacula** similar to the one at the wrist (fig. 12.17).

Fibularis (Peroneus[32]) Tertius[33] (FIB-you-LERR-iss TUR-she-us)   Dorsiflexes and everts foot during walking, helps toes clear the ground during forward swing

O: Medial surface of lower one-third of fibula, interosseous membrane	**I:** Metatarsal V	**N:** Deep fibular (peroneal) n.

Extensor Digitorum Longus (DIDJ-ih-TOE-rum)   Extends toes II–V, dorsiflexes foot, tautens plantar aponeurosis

O: Lateral condyle of tibia, shaft of fibula, interosseous membrane	**I:** Middle and distal phalanges II–V	**N:** Deep fibular (peroneal) n.

Extensor Hallucis Longus (ha-LOO-sis)   Extends great toe, dorsiflexes foot

O: Anterior surface of middle of fibula, interosseous membrane	**I:** Distal phalanx I	**N:** Deep fibular (peroneal) n.

Tibialis[34] Anterior (TIB-ee-AY-lis)   Dorsiflexes and inverts foot; resists backward tipping of body (as when standing on a moving boat deck); helps support medial longitudinal arch of foot

O: Lateral condyle and lateral margin of proximal half of tibia; interosseous membrane	**I:** Medial cuneiform, metatarsal I	**N:** Deep fibular (peroneal) n.

Posterior (Flexor) Compartment of the Leg, Superficial Group.   The posterior compartment has superficial and deep muscle groups. The three muscles of the superficial group are plantar flexors: the *gastrocnemius, soleus,* and *plantaris* (fig. 12.18). The first two of these, collectively known as the *triceps surae,*[35] insert on the calcaneus by way of the **calcaneal (Achilles) tendon.** This is the strongest tendon of the body but is nevertheless a common site of sports injuries resulting from sudden stress. The plantaris, a weak synergist of the triceps surae, is a relatively unimportant muscle and is absent from many people; it is not tabulated here. Surgeons often use the plantaris tendon for tendon grafts needed in other parts of the body.

Gastrocnemius[36] (GAS-trock-NEE-me-us)   Plantar flexes foot, flexes knee; active in walking, running, and jumping

O: Condyles and popliteal surface of femur, lateral supracondylar line, capsule of knee joint	**I:** Calcaneus	**N:** Tibial n.

Soleus[37] (SO-lee-us)   Plantar flexes foot; steadies leg on ankle during standing

O: Posterior surface of head and proximal one-fourth of fibula; middle one-third of tibia; interosseous membrane	**I:** Calcaneus	**N:** Tibial n.

---

[32]*perone* = pinlike (fibula)
[33]*fibularis* = of the fibula + *tert* = third
[34]*tibialis* = of the tibia
[35]*sura* = calf of leg
[36]*gastro* = belly + *cnem* = leg
[37]Named for its resemblance to a flatfish (sole)

TABLE 12.8   **Muscles Acting on the Foot (continued)**

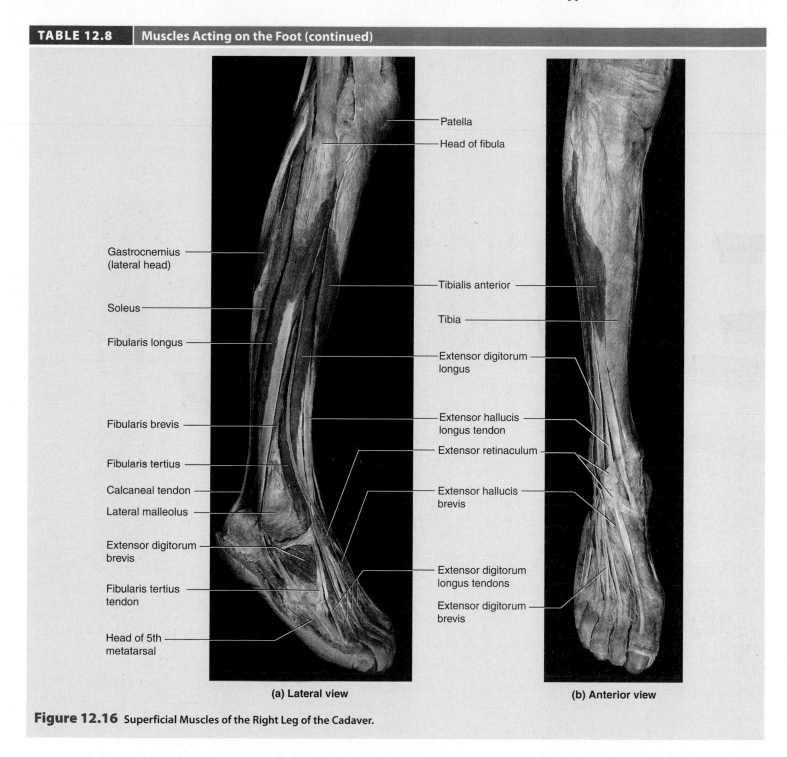

(a) Lateral view

Labels: Patella, Head of fibula, Gastrocnemius (lateral head), Soleus, Fibularis longus, Fibularis brevis, Fibularis tertius, Calcaneal tendon, Lateral malleolus, Extensor digitorum brevis, Fibularis tertius tendon, Head of 5th metatarsal, Tibialis anterior, Tibia, Extensor digitorum longus, Extensor hallucis longus tendon, Extensor retinaculum, Extensor hallucis brevis, Extensor digitorum longus tendons, Extensor digitorum brevis

(b) Anterior view

Figure 12.16 Superficial Muscles of the Right Leg of the Cadaver.

Apply What You Know

Not everyone has the same muscles. From the information provided in this chapter, identify two muscles that are lacking in some people.

TABLE 12.8	Muscles Acting on the Foot (continued)

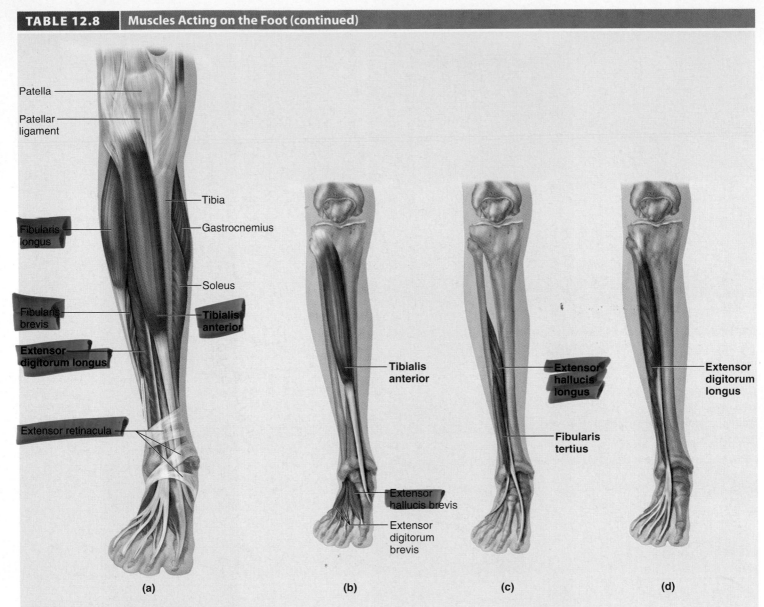

(a)   (b)   (c)   (d)

Figure 12.17 Muscles of the Leg, Anterior Compartment. Boldface labels identify muscles belonging to the anterior compartment. (a) Superficial anterior view of the leg. Some muscles of the posterior and lateral compartments are also partially visible. (b–d) Individual muscles of the anterior compartment of the leg and superior aspect of the foot.

● *Palpate the hard anterior surface of your own tibia at midshaft, then continue medially until you feel muscle. What muscle is that?*

TABLE 12.8   **Muscles Acting on the Foot (continued)**

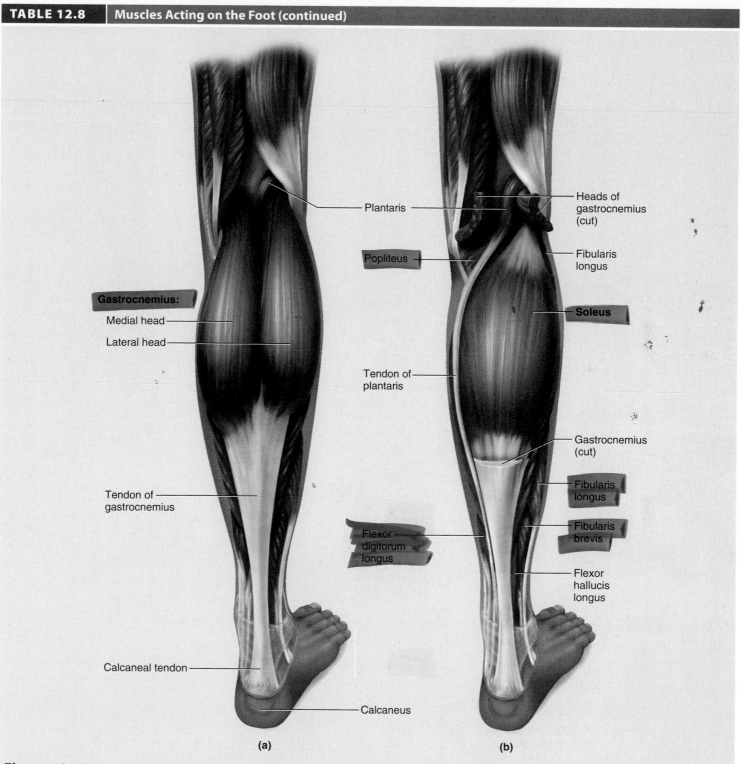

Plantaris

Heads of
gastrocnemius
(cut)

Popliteus

Fibularis
longus

Gastrocnemius:

Medial head

Lateral head

Soleus

Tendon of
plantaris

Tendon of
gastrocnemius

Gastrocnemius
(cut)

Fibularis
longus

Fibularis
brevis

Flexor
digitorum
longus

Flexor
hallucis
longus

Calcaneal tendon

Calcaneus

(a)   (b)

Figure 12.18 Superficial Muscles of the Leg, Posterior Compartment. (a) The gastrocnemius. (b) The soleus, deep to the gastrocnemius and sharing the calcaneal tendon with it. The gastrocnemius and soleus are collectively called the triceps surae.

TABLE 12.8   Muscles Acting on the Foot (continued)

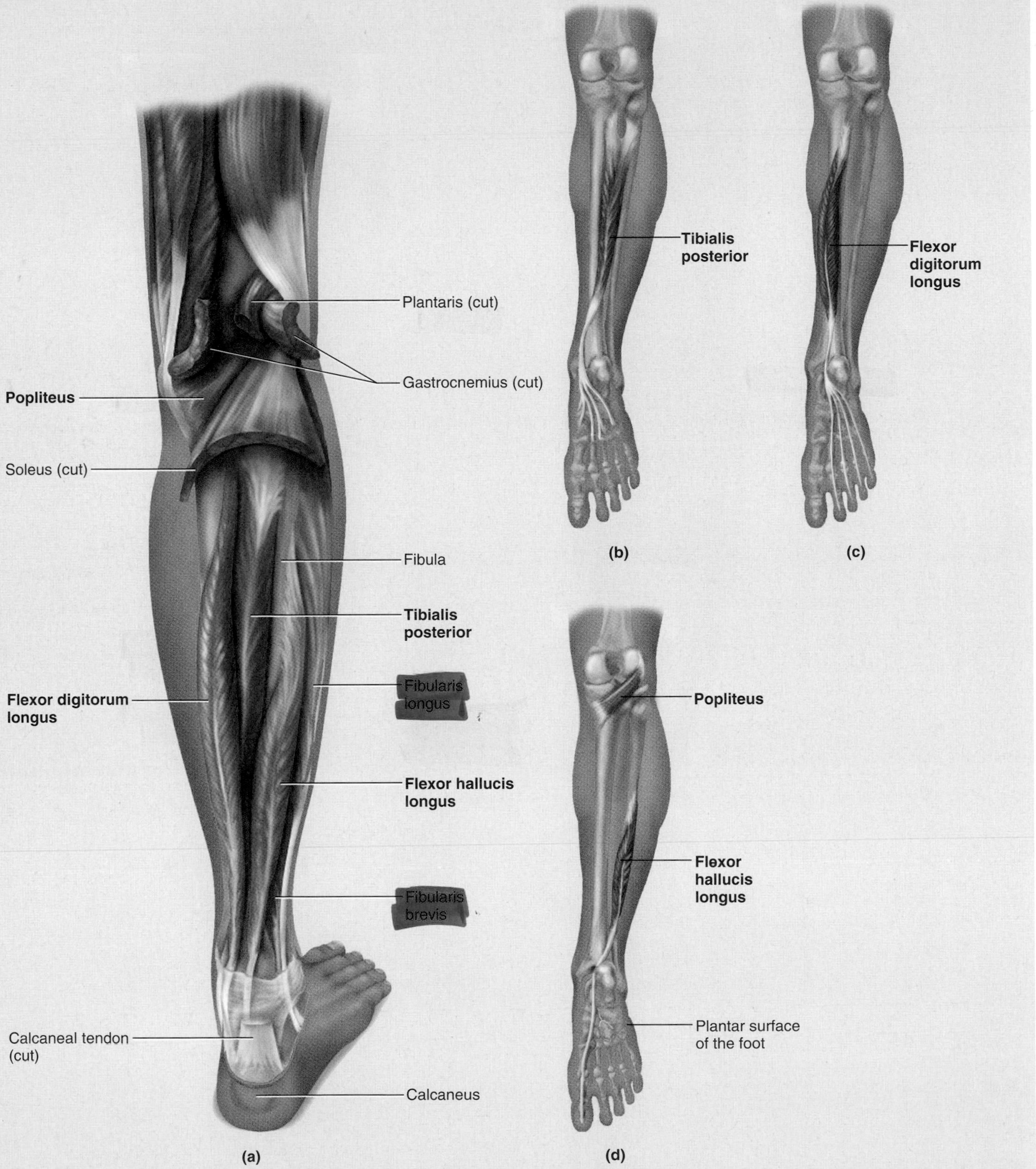

Figure 12.19 Deep Muscles of the Leg, Posterior and Lateral Compartments. (a) Muscles deep to the soleus. (b–d) Exposure of some individual deep muscles with the foot plantar flexed (sole facing viewer).

TABLE 12.8	Muscles Acting on the Foot (continued)

Posterior (Flexor) Compartment of the Leg, Deep Group.   There are four muscles in the deep group (fig. 12.19). The **flexor digitorum longus, flexor hallucis longus,** and **tibialis posterior** are plantar flexors. The fourth muscle, the **popliteus,** was described in table 12.7 because it acts on the knee rather than on the foot.

Flexor Digitorum Longus   Flexes phalanges of digits II–V as foot is raised from ground; stabilizes metatarsal heads and keeps distal pads of toes in contact with ground in toe-off and tiptoe movements

O: Posterior surface of tibial shaft   **I:** Distal phalanges II–V   **N:** Tibial n.

Flexor Hallucis Longus   Same actions as flexor digitorum longus, but for great toe (digit I)

O: Inferior two-thirds of fibula and interosseous membrane   **I:** Distal phalanx I   **N:** Tibial n.

Tibialis Posterior   Inverts foot; may assist in strong plantar flexion or control pronation of foot during walking

O: Posterior surface of proximal half of tibia, fibula, and interosseous membrane   **I:** Navicular, medial cuneiform, metatarsals II–IV   **N:** Tibial n.

Lateral (Fibular) Compartment of the Leg.   The lateral compartment includes the **fibularis brevis** and **fibularis longus** (figs. 12.16a, 12.17a, 12.20b). They plantar flex and evert the foot. Plantar flexion is important not only in standing on tiptoes but in providing lift and forward thrust each time you take a step.

Fibularis (Peroneus) Brevis   Maintains concavity of sole during toe-off and tiptoeing; may evert foot and limit inversion and help steady leg on foot

O: Lateral surface of distal two-thirds of fibula   **I:** Base of metatarsal V   **N:** Superficial fibular (peroneal) n.

Fibularis (Peroneus) Longus   Maintains concavity of sole during toe-off and tiptoeing; everts and plantar flexes foot

O: Head and lateral surface of proximal two-thirds of fibula   **I:** Medial cuneiform, metatarsal I   **N:** Superficial fibular (peroneal) n.

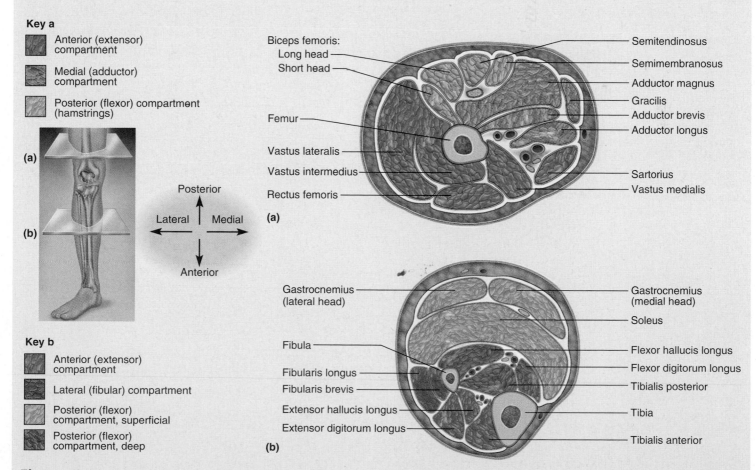

Key a

■ Anterior (extensor) compartment

■ Medial (adductor) compartment

■ Posterior (flexor) compartment (hamstrings)

(a)

(b)

Posterior

Lateral   Medial

Anterior

Key b

■ Anterior (extensor) compartment

■ Lateral (fibular) compartment

■ Posterior (flexor) compartment, superficial

■ Posterior (flexor) compartment, deep

Biceps femoris:
Long head
Short head

Femur

Vastus lateralis

Vastus intermedius

Rectus femoris

Semitendinosus
Semimembranosus
Adductor magnus
Gracilis
Adductor brevis
Adductor longus

Sartorius

Vastus medialis

(a)

Gastrocnemius (lateral head)

Fibula

Fibularis longus

Fibularis brevis

Extensor hallucis longus

Extensor digitorum longus

Gastrocnemius (medial head)

Soleus

Flexor hallucis longus

Flexor digitorum longus

Tibialis posterior

Tibia

Tibialis anterior

(b)

Figure 12.20 Serial Cross Sections Through the Lower Limb. Each section is taken at the correspondingly lettered level in the figure at the left.

TABLE 12.9	**Intrinsic Muscles of the Foot**

The intrinsic muscles of the foot help to support the arches and act on the toes in ways that aid locomotion. Several of them are similar in name and location to the intrinsic muscles of the hand.

Dorsal Aspect of Foot.   Only one of the intrinsic muscles, the *extensor digitorum brevis,* is on the dorsal (superior) side of the foot (see fig. 12.17b). The medial slip of this muscle, serving the great toe, is sometimes called the *extensor hallucis brevis.*

Extensor Digitorum Brevis   Extends proximal phalanx I and all phalanges of digits II–IV

O: Calcaneus, inferior extensor retinaculum of ankle	**I:** Proximal phalanx I, tendons of extensor digitorum longus to digits II–IV	**N:** Deep fibular (peroneal) n.

Ventral Layer 1 (Most Superficial).   All remaining intrinsic muscles are on the ventral (inferior) aspect of the foot or between the metatarsal bones. They are grouped in four layers (fig. 12.21). Dissecting into the foot from the plantar surface, one first encounters a tough fibrous sheet, the **plantar aponeurosis,** between the skin and muscles. It diverges like a fan from the calcaneus to the bases of all the toes and serves as an origin for several ventral muscles. The ventral muscles include the stout *flexor digitorum brevis* on the midline of the foot, with four tendons that supply all digits except the hallux. It is flanked by the *abductor digiti minimi* laterally and the *abductor hallucis* medially.

Flexor Digitorum Brevis   Flexes digits II–IV; supports arches of foot

O: Calcaneus, plantar aponeurosis	**I:** Middle phalanges II–V	**N:** Medial plantar n.

Abductor Digiti Minimi[38]   Abducts and flexes little toe; supports arches of foot

O: Calcaneus, plantar aponeurosis	**I:** Proximal phalanx V	**N:** Lateral plantar n.

Abductor Hallucis   Abducts great toe; supports arches of foot

O: Calcaneus, plantar aponeurosis, flexor retinaculum	**I:** Proximal phalanx I	**N:** Medial plantar n.

Ventral Layer 2.   The next deeper layer consists of the thick *quadratus plantae* in the middle of the foot and the four *lumbrical muscles* located between the metatarsals (fig. 12.21b).

Quadratus Plantae[39] **(quad-RAY-tus PLAN-tee)**   Same as flexor digitorum longus (table 12.8); flexion of digits II–V and associated locomotor functions

O: Two heads on the medial and lateral sides of calcaneus	**I:** Distal phalanges II–V via flexor digitorum longus tendons	**N:** Lateral plantar n.

Lumbricals (LUM-brick-ulz)   Flex toes II–V

O: Tendon of flexor digitorum longus	**I:** Proximal phalanges II–V	**N:** Lateral and medial plantar nn.

Ventral Layer 3.   The muscles of this layer serve only the great and little toes (fig. 12.21c). They are the *flexor digiti minimi brevis, flexor hallucis brevis,* and *adductor hallucis.* The adductor hallucis has an *oblique head* that extends diagonally from the midplantar region to the base of the great toe, and a *transverse head* that passes across the bases of digits II–IV and meets the long head at the base of the great toe.

Flexor Digiti Minimi Brevis   Flexes little toe

O: Metatarsal V, sheath of fibularis longus	**I:** Proximal phalanx V	**N:** Lateral plantar n.

Flexor Hallucis Brevis   Flexes great toe

O: Cuboid, lateral cuneiform, tibialis posterior tendon	**I:** Proximal phalanx I	**N:** Medial plantar n.

Adductor Hallucis   Adducts great toe

O: Metatarsals II–IV, fibularis longus tendon, ligaments at bases of digits III–V	**I:** Proximal phalanx I	**N:** Lateral plantar n.

Ventral Layer 4 (Deepest).   This layer consists only of the small interosseous muscles located between the metatarsal bones—four dorsal and three plantar (fig. 12.21d, e). Each *dorsal interosseous muscle* is bipennate and originates on two adjacent metatarsals. The *plantar interosseous muscles* are unipennate and originate on only one metatarsal each.

Dorsal Interosseous Muscles (Four Muscles)   Abduct toes II–IV

O: Each with two heads arising from facing surfaces of two adjacent metatarsals	**I:** Proximal phalanges II–IV	**N:** Lateral plantar n.

Plantar Interosseous Muscles (Three Muscles)   Adduct toes III–V

O: Medial aspect of metatarsals III–V	**I:** Proximal phalanges III–V	**N:** Lateral plantar n.

[38]*digiti* = of the toe + *minim* = smallest   [39]*quadrat* = four-sided + *plantae* = of the plantar region

TABLE 12.9   Intrinsic Muscles of the Foot (continued)

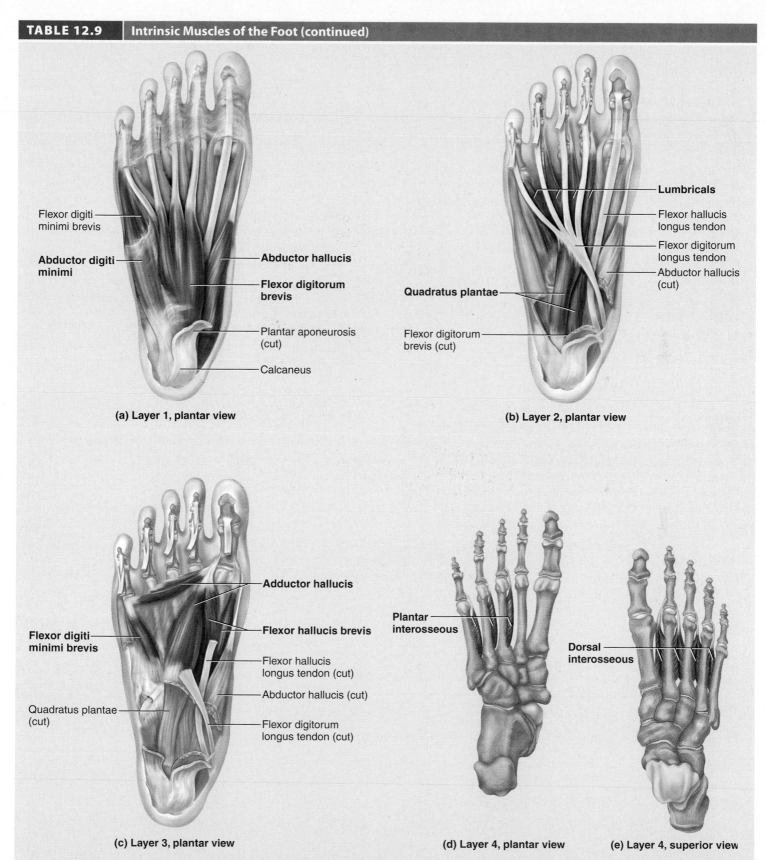

(a) Layer 1, plantar view

- Flexor digiti minimi brevis
- Abductor digiti minimi
- **Abductor hallucis**
- **Flexor digitorum brevis**
- Plantar aponeurosis (cut)
- Calcaneus

(b) Layer 2, plantar view

- **Lumbricals**
- Flexor hallucis longus tendon
- Flexor digitorum longus tendon
- Abductor hallucis (cut)
- **Quadratus plantae**
- Flexor digitorum brevis (cut)

(c) Layer 3, plantar view

- **Adductor hallucis**
- **Flexor hallucis brevis**
- Flexor digiti minimi brevis
- Flexor hallucis longus tendon (cut)
- Abductor hallucis (cut)
- Quadratus plantae (cut)
- Flexor digitorum longus tendon (cut)

(d) Layer 4, plantar view

- Plantar interosseous

(e) Layer 4, superior view

- Dorsal interosseous

Figure 12.21 Intrinsic Muscles of the Foot. (a–d) First through fourth layers, respectively, in inferior (plantar) views. (e) Fourth layer, superior view. The muscles belonging to each layer are shown in color and with boldface labels.

Before You Go On

Answer the following questions to test your understanding of the preceding section:

7. In the middle of a stride, you have one foot on the ground and you are about to swing the other leg forward. What muscles produce the movements of that leg?

8. Name the muscles that cross both the hip and knee joints and produce actions at both.

9. List the major actions of the muscles of the anterior, medial, and posterior compartments of the thigh.

10. Describe the role of plantar flexion and dorsiflexion in walking. What muscles produce these actions?

12.3 Muscle Injuries

▶ Expected Learning Outcomes

When you have completed this section, you should be able to

- explain how to reduce the risk of muscle injuries; and
- define several types of muscle injuries often incurred in sports and recreation.

Although the muscular system suffers fewer diseases than most organ systems, it is particularly vulnerable to injuries resulting from sudden and intense stress placed on muscles and tendons. Each year, thousands of athletes from high school to professional level sustain some type of muscle injury, as do increasing numbers of people who have taken up running and other forms of physical conditioning. Overzealous exertion without proper preparation and warm-up is frequently the cause. Some of the most common athletic injuries are briefly described in table 12.10. (See table 10.5, p. 259, for more general disorders of the muscular system.)

Most athletic injuries can be prevented by proper conditioning. A person who suddenly takes up vigorous exercise may not have sufficient muscle and bone mass to withstand the stresses such exercise entails. These must be developed gradually. Stretching exercises keep ligaments and joint capsules supple and therefore reduce injuries. Warm-up exercises promote more efficient and less injurious musculoskeletal function in several ways. Most of all, moderation is important, as most injuries simply result from overuse of the muscles. "No pain, no gain" is a dangerous misconception.

Muscular injuries can be treated initially with "RICE": **R**est to prevent further injury and allow repair to occur; **I**ce to reduce swelling; **C**ompression with an elastic bandage to prevent fluid accumulation and swelling; and **E**levation of an injured limb to promote drainage of blood from the affected area and limit further swelling. If these measures are not enough, anti-inflammatory drugs such as hydrocortisone and aspirin may be employed.

Before You Go On

Answer the following questions to test your understanding of the preceding section:

11. Explain why stretching exercises may reduce the incidence of muscle injuries.

12. Explain the reason for each of the four treatments in the RICE approach to muscle injuries.

TABLE 12.10	Muscle Injuries
Baseball finger	Tears in the extensor tendons of the fingers resulting from the impact of a baseball with the extended fingertip
Blocker's arm	Abnormal calcification in the lateral margin of the forearm as a result of repeated impact such as occurs in football
Pitcher's arm	Inflammation at the origin of the wrist flexors resulting from hard wrist flexion in releasing a baseball
Pulled groin	Strain in the adductor muscles of the thigh; common in gymnasts and dancers who perform splits and high kicks
Pulled hamstrings	Strained hamstring muscles or a partial tear in the tendinous origin, often with a hematoma (blood clot) in the fascia lata; frequently caused by repetitive kicking (as in football and soccer) or long, hard running
Rider's bones	Calcification in the tendons of the thigh adductors; results from prolonged abduction of the thighs when riding horses
Shinsplints	General term for several kinds of injury with pain in the crural region—tendinitis of the tibialis posterior, inflammation of the tibial periosteum, and anterior compartment syndrome. May result from unaccustomed jogging, walk-a-thons, walking on snowshoes, or any vigorous activity of the legs after a period of inactivity
Tennis elbow	Inflammation at the origin of the extensor carpi muscles on the lateral epicondyle of the humerus occurs when these muscles are repeatedly tensed during backhand strokes and then strained by sudden impact with the tennis ball. Any activity that requires rotary movements of the forearm and a firm grip of the hand (for example, using a screwdriver) can cause the symptoms of tennis elbow.
Tennis leg	Partial tear in the lateral origin of the gastrocnemius; results from repeated strains put on the muscle while supporting the body weight on the toes

Disorders Described Elsewhere

Back injuries 286	Hamstring injuries 316
Carpal tunnel syndrome 307	Hernias 289
Compartment syndrome 294	Rotator cuff injury 300

Study Guide

Assess Your Learning Outcomes

You should have a good understanding of this chapter if you can accurately address the following issues.

General

1. For all muscles named in this study guide, and to the extent specified by your course or instructor, an ability to locate the muscle on a model, photograph, diagram, or dissection; to palpate the muscle or its tendons on the living body; and to state its origin, insertion, innervation, and actions

12.1 Muscles Acting on the Shoulder and Upper Limb (p. 294)

1. Muscle compartments and the connective tissue sheets that separate them
2. Two anterior muscles that act on the pectoral girdle—the *pectoralis minor* and *serratus anterior* (table 12.1)
3. Four posterior muscles that act on the pectoral girdle—the *trapezius, levator scapulae, rhomboideus minor,* and *rhomboideus major* (table 12.1)
4. The four "SITS muscles" of the rotator cuff—the *supraspinatus, infraspinatus, teres minor,* and *subscapularis* (table 12.2)—and the vulnerability of the rotator cuff to injuries
5. Five additional muscles that, like the rotator cuff, also cross the shoulder joint and act on the humerus—the *pectoralis major, latissimus dorsi, deltoid, teres major,* and *coracobrachialis* (table 12.2)
6. Muscles of the arm that act on the elbow joint—the *brachialis, biceps brachii,* and *triceps brachii* (table 12.3)—including the individual names and courses of the two heads of the biceps and three heads of the triceps
7. Muscles of the forearm that produce forearm movements—the *brachioradialis, anconeus, pronator quadratus, pronator teres,* and *supinator* (table 12.3)
8. Superficial muscles of the anterior compartment of the forearm that act on the wrist and hand—the *flexor carpi radialis, flexor carpi ulnaris, flexor digitorum superficialis,* and *palmaris longus* (table 12.3)
9. Deep muscles of the anterior compartment of the forearm—the *flexor digitorum profundus* and *flexor pollicis longus*
10. The relationship of anterior compartment muscles to the flexor retinaculum and carpal tunnel, and the pathology of carpal tunnel syndrome
11. Superficial muscles of the posterior compartment of the forearm—the *extensor carpi radialis longus, extensor carpi radialis brevis, extensor digitorum, extensor digiti minimi,* and *extensor carpi ulnaris*
12. Deep muscles of the posterior compartment of the forearm—the *abductor pollicis longus, extensor pollicis brevis, extensor pollicis longus,* and *extensor indicis*
13. The general role of the intrinsic muscles of the hand; the three major groups into which they can be classified; and the locations of the thenar and hypothenar eminences (table 12.5)
14. The thenar group of hand muscles—the *adductor pollicis, abductor pollicis brevis, flexor pollicis brevis,* and *opponens pollicis* (table 12.5)
15. The hypothenar group of hand muscles—the *abductor digiti minimi, flexor digiti minimi brevis,* and *opponens digiti minimi* (table 12.5)
16. The midpalmar group of hand muscles—four *dorsal interosseous muscles;* three *palmar interosseous muscles;* and four *lumbrical muscles*

12.2 Muscles Acting on the Hip and Lower Limb (p. 310)

1. The *iliopsoas* of the anterior pelvic region, and the two muscles of the iliopsoas group—the *iliacus* and *psoas major* (table 12.6)
2. The three gluteal muscles—*gluteus maximus, gluteus medius,* and *gluteus minimus* (table 12.6)
3. The *tensor fasciae latae* of the hip and its relationship to the fibrous fascia lata and iliotibial band of the lateral thigh (table 12.6)
4. The deep lateral rotators of the hip—the *gemellus superior, gemellus inferior, obturator externus, obturator internus, piriformis,* and *quadratus femoris* (table 12.6)
5. Locations of the medial (adductor) compartment, anterior (extensor) compartment, and posterior (flexor) compartment of the thigh
6. Muscles of the medial compartment of the thigh—the *adductor brevis, adductor longus, adductor magnus, gracilis,* and *pectineus* (table 12.6)
7. Muscles of the anterior compartment of the thigh—the *sartorius* and the four heads of the *quadriceps femoris: rectus femoris,*

vastus lateralis, vastus medialis, and *vastus intermedius*—and the course of the quadriceps tendon and its extension, the patellar ligament (table 12.7)
8. The hamstring muscles of the posterior compartment of the thigh—the semitendinosus, semimembranosus, and *biceps femoris*—and the nearby popliteus (table 12.7)
9. Locations of the anterior, posterior, and lateral compartments of the leg (table 12.8)
10. Anterior compartment muscles of the leg—the *fibularis tertius, extensor digitorum longus, extensor hallucis longus,* and *tibialis anterior* (table 12.8)
11. Superficial posterior compartment muscles of the leg—the *gastrocnemius* and *soleus*—and their relationship to the prominent calcaneal (Achilles) tendon (table 12.8)
12. Deep posterior compartment muscles of the leg—the *flexor digitorum longus, flexor hallucis longus,* and *tibialis posterior* (table 12.8)
13. Lateral compartment muscles of the leg—the *fibularis brevis* and *fibularis longus* (table 12.8)
14. Intrinsic muscles of the foot, their general functions, and the one muscle of the superior side—the *extensor digitorum brevis* (table 12.9)
15. The inferior intrinsic muscles of the foot and their layered arrangement—the *flexor digitorum brevis, abductor digiti minimi,* and *abductor hallucis* of layer 1 (the most superficial layer); the *quadratus plantae* and four *lumbrical muscles* of layer 2; the *adductor hallucis, flexor digiti minimi brevis,* and *flexor hallucis brevis* of layer 3; and the four *dorsal interosseous muscles* and three *plantar interosseous muscles* of layer 4 (the deepest). The plantar aponeurosis and its relationship to the layer 1 muscles (table 12.9)

12.3 Muscle Injuries (p. 324)

1. The common causes, prevention, and treatment of muscle injuries
2. The anatomical sites and causes of several common muscle and tendon injuries—baseball finger, blocker's arm, pitcher's arm, pulled groin, pulled hamstrings, rider's bones, shinsplits, tennis elbow, and tennis leg (table 12.10).

Testing Your Recall

1. Which of the following muscles could you most easily do without?
   a. flexor digitorum profundus
   b. trapezius
   c. palmaris longus
   d. triceps brachii
   e. tibialis anterior

2. Which of the following has the least in common with the other four?
   a. vastus intermedius
   b. vastus lateralis
   c. vastus medialis
   d. rectus femoris
   e. biceps femoris

3. The triceps surae is a muscle group composed of
   a. the flexor hallucis longus and brevis.
   b. the gastrocnemius and soleus.
   c. lateral, medial, and long heads.
   d. the biceps brachii and triceps brachii.
   e. the vastus lateralis, medialis, and intermedius.

4. The interosseous muscles lie between
   a. the ribs.
   b. the tibia and fibula.
   c. the radius and ulna.
   d. the metacarpal bones.
   e. the phalanges.

5. Which of these muscles does *not* contribute to the rotator cuff?
   a. the supraspinatus
   b. the infraspinatus
   c. the subscapularis
   d. the teres major
   e. the teres minor

6. Which of these actions is *not* performed by the trapezius?
   a. extension of the neck
   b. depression of the scapula
   c. elevation of the scapula
   d. rotation of the scapula
   e. adduction of the humerus

7. Both the hands and feet are acted upon by a muscle or muscles called
   a. the extensor digitorum.
   b. the abductor digiti minimi.
   c. the flexor digitorum profundus.
   d. the abductor hallucis.
   e. the flexor digitorum longus.

8. Which of the following muscles does *not* extend the hip joint?
   a. rectus femoris
   b. gluteus maximus
   c. biceps femoris
   d. semitendinosus
   e. semimembranosus

9. Both the gastrocnemius and _____ muscles insert on the heel by way of the calcaneal tendon.
   a. semimembranosus
   b. tibialis posterior
   c. tibialis anterior
   d. soleus
   e. plantaris

10. Which of these is *not* in the anterior compartment of the thigh?
    a. semimembranosus
    b. rectus femoris
    c. vastus intermedius
    d. vastus lateralis
    e. sartorius

11. The major superficial muscle of the shoulder, where injections are often given, is the _____.

12. If a muscle has the word *hallucis* in its name, it must cause movement of the _____.

13. Pronation of the forearm is achieved by two muscles, the pronator _____ just distal to the elbow and the pronator _____ near the wrist.

14. The three large muscles on the posterior side of the thigh are collectively known by the colloquial name of _____ muscles.

15. Connective tissue bands called _____ prevent flexor tendons from rising like bowstrings.

16. The web between your thumb and palm consists mainly of the _____ muscle.

17. The patella is embedded in the tendon of the _____ muscle.

18. The _____ muscle, named for its origin and insertion, originates on the coracoid process of the scapula, inserts on the humerus, and adducts the arm.

19. The most medial adductor muscle of the thigh is the long, slender _____.

20. The _____ and _____ are hip flexors that originate on the pelvis and lumbar vertebrae and converge on a shared tendon that inserts on the lesser trochanter of the femur.

Answers in the Appendix

Building Your Medical Vocabulary

State a medical meaning of each of the following word elements, and give a term in which it is used.

1. serrat-
2. dorsi-
3. -ceps
4. infra-
5. teres
6. profund-
7. osse-
8. ili-
9. lat-
10. maxim-

Answers in the Appendix

True or False

Determine which five of the following statements are false, and briefly explain why.

1. All plantar flexors in the posterior compartment of the calf insert on the heel by way of the calcaneal tendon.
2. The trapezius can act as both a synergist and antagonist of the levator scapulae.
3. To push someone away from you, you would use the serratus anterior muscle more than the trapezius.
4. Both the extensor digitorum and extensor digiti minimi extend the little finger.
5. The interosseous muscles are fusiform.
6. The actions of the palmaris longus and plantaris muscles are weak and relatively dispensable.
7. The psoas major is an antagonist of the rectus femoris.
8. Rapid flexion of the knee often causes hamstring injuries.
9. Curling your toes employs the quadratus plantae muscle.
10. The tibialis posterior and tibialis anterior are synergists.

Answers in the Appendix

Testing Your Comprehension

1. Radical mastectomy, once a common treatment for breast cancer, involved removal of the pectoralis major along with the breast. What functional impairments would result from this? What synergists could a physical therapist train a patient to use to recover some lost function?
2. Table 12.6 describes a simple test for determining whether you have a palmaris longus muscle. Why do you think the other major tendon of the anterior wrist, the flexor carpi radialis tendon, does not stand out conspicuously in such a test?
3. Poorly conditioned, middle-aged people may suffer a rupture of the calcaneal tendon when the foot is suddenly dorsiflexed. Explain each of the following signs of a ruptured calcaneal tendon: (a) a prominent lump typically appears in the calf; (b) the foot can be dorsiflexed farther than usual; and (c) the patient cannot plantar flex the foot very effectively.
4. Women who habitually wear high heels may suffer painful "high heel syndrome" when they go barefoot or wear flat shoes. What muscle(s) and tendon(s) are involved? Explain.
5. A student moving out of a dormitory kneels down, in correct fashion, to lift a heavy box of books. What prime movers are involved as he straightens his legs to lift the box?

Answers at www.mhhe.com/saladinha3

Improve Your Grade at www.mhhe.com/saladinha3

Practice quizzes, labeling activities, games, and flashcards provide fun ways to master concepts. You can also download image PowerPoint files for each chapter to create a study guide or for taking notes during lecture.

Atlas of Regional and Surface Anatomy

How many muscles can you identify from their surface appearance?

ATLAS OUTLINE

Muscular System

A.1 Introduction

Regional Anatomy

On the whole, this book takes a **systems approach** to anatomy, examining the structure and function of each organ system, one at a time, regardless of which body regions it may traverse. Physicians and surgeons, however, think and act in terms of **regional anatomy.** If a patient presents with pain in the lower right quadrant (LRQ) (see fig. 1.13, p. 16), the source may be the appendix, the small intestine, an ovary or uterine tube, or an inguinal muscle, among other possibilities. The question is to think not of an entire organ system (the esophagus is probably irrelevant to the LRQ), but of what organs are present in that region and what possibilities must be considered as the cause of the pain. This atlas presents several views of the body region by region so that you can see some of the spatial relationships that exist among the organ systems considered in their separate chapters.

Surface Anatomy

In the study of human anatomy, it is easy to become so preoccupied with internal structure that we forget the importance of what we can see and feel externally. Yet external anatomy and appearance are major concerns in giving a physical examination and in many aspects of patient care. A knowledge of the body's surface landmarks is essential to one's competence in physical therapy, cardiopulmonary resuscitation, surgery, making X-rays and electrocardiograms, giving injections, drawing blood, listening to heart and respiratory sounds, measuring the pulse and blood pressure, and finding pressure points to stop arterial bleeding, among other procedures. A misguided attempt to perform some of these procedures while disregarding or misunderstanding external anatomy can be very harmful and even fatal to a patient.

Having just studied skeletal and muscular anatomy in the preceding chapters, this is an opportune time for you to study the body surface. Much of what we see there reflects the underlying structure of the superficial bones and muscles. A broad photographic overview of surface anatomy is given in chapter 1 (see fig. 1.12, p. 15), where it was necessary for providing a vocabulary for reference in the chapters that followed. This atlas shows this surface anatomy in closer detail so you can relate it to the musculoskeletal anatomy of chapters 7 through 12.

Learning Strategy

To make the most profitable use of this atlas, refer back to earlier chapters as you study these illustrations. Relate drawings of the clavicles in chapter 7 to the photograph in figure A.1, for example. Study the shape of the scapula in chapter 8 and see how much of it you can trace on the photographs in figure A.9. See if you can relate the tendons visible on the hand (see fig. A.19) to the muscles of the forearm illustrated in chapter 12, and the external landmarks of the pelvic girdle (see fig. A.15) to the bone structure in chapter 8.

For learning surface anatomy, there is a resource available to you that is far more valuable than any laboratory model or textbook illustration—your own body. For the best understanding of human structure, compare the art and photographs in this book with your body or with structures visible on a study partner. In addition to bones and muscles, you can palpate a number of superficial arteries, veins, tendons, ligaments, and cartilages, among other structures. By palpating regions such as the shoulder, elbow, or ankle, you can develop a mental image of the subsurface structures better than you can obtain by looking at two-dimensional textbook images. And the more you can study with other people, the more you will appreciate the variations in human structure and be able to apply your knowledge to your future patients or clients, who will not look quite like any textbook diagram or photograph you have ever seen. Through comparisons of art, photography, and the living body, you will get a much deeper understanding of the body than if you were to study this atlas in isolation from the earlier chapters.

At the end of this atlas, you can test your knowledge of externally visible muscle anatomy. The two photographs in figure A.25 have 30 numbered muscles and a list of 26 names, some of which are shown more than once in the photographs and some of which are not shown at all. Identify the muscles to your best ability without looking back at the previous illustrations, and then check your answers in the appendix at the back of the book.

Throughout these illustrations, the following abbreviations apply: a. = artery; m. = muscle; n. = nerve; and v. = vein. Double letters such as mm. or vv. represent the plurals.

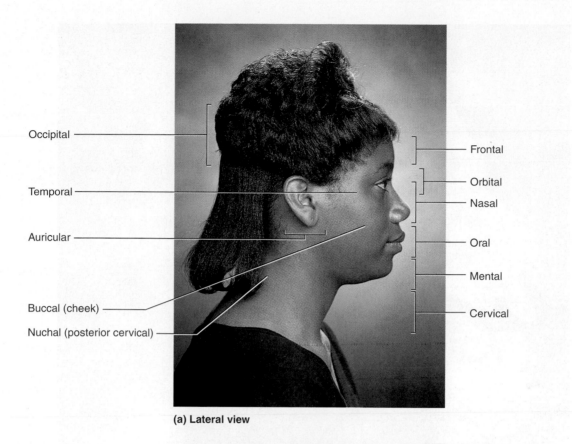

Occipital

Temporal

Auricular

Buccal (cheek)

Nuchal (posterior cervical)

Frontal

Orbital

Nasal

Oral

Mental

Cervical

(a) Lateral view

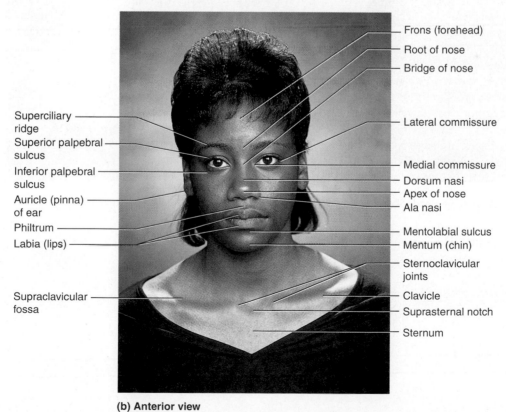

Frons (forehead)

Root of nose

Bridge of nose

Superciliary ridge

Superior palpebral sulcus

Inferior palpebral sulcus

Auricle (pinna) of ear

Philtrum

Labia (lips)

Supraclavicular fossa

Lateral commissure

Medial commissure

Dorsum nasi

Apex of nose

Ala nasi

Mentolabial sulcus

Mentum (chin)

Sternoclavicular joints

Clavicle

Suprasternal notch

Sternum

(b) Anterior view

Figure A.1 **The Head and Neck.** (a) Anatomical regions of the head. (b) Features of the facial region and upper thorax.
● *What muscle underlies the region of the philtrum? What muscle forms the slope of the shoulder?*

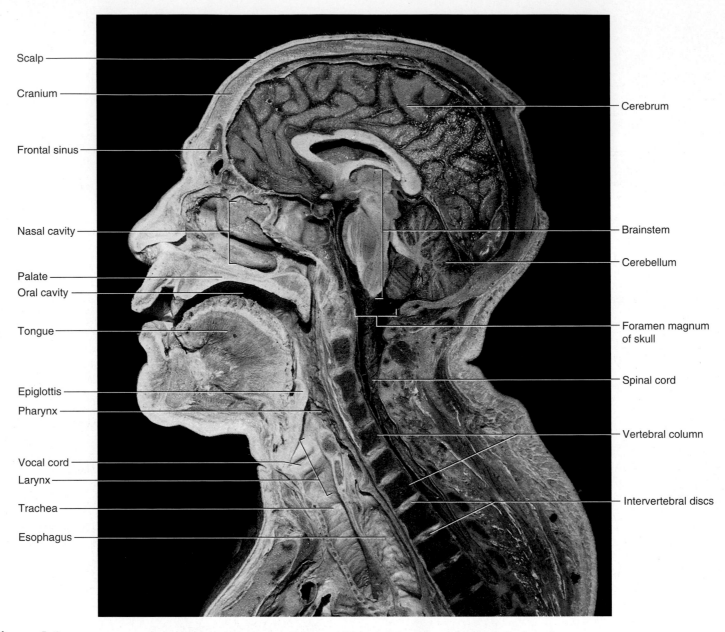

Scalp

Cranium

Frontal sinus

Nasal cavity

Palate
Oral cavity

Tongue

Epiglottis
Pharynx

Vocal cord
Larynx

Trachea

Esophagus

Cerebrum

Brainstem

Cerebellum

Foramen magnum
of skull

Spinal cord

Vertebral column

Intervertebral discs

Figure A.2 Median Section of the Head. Shows contents of the cranial, nasal, and oral cavities and vertebral canal.

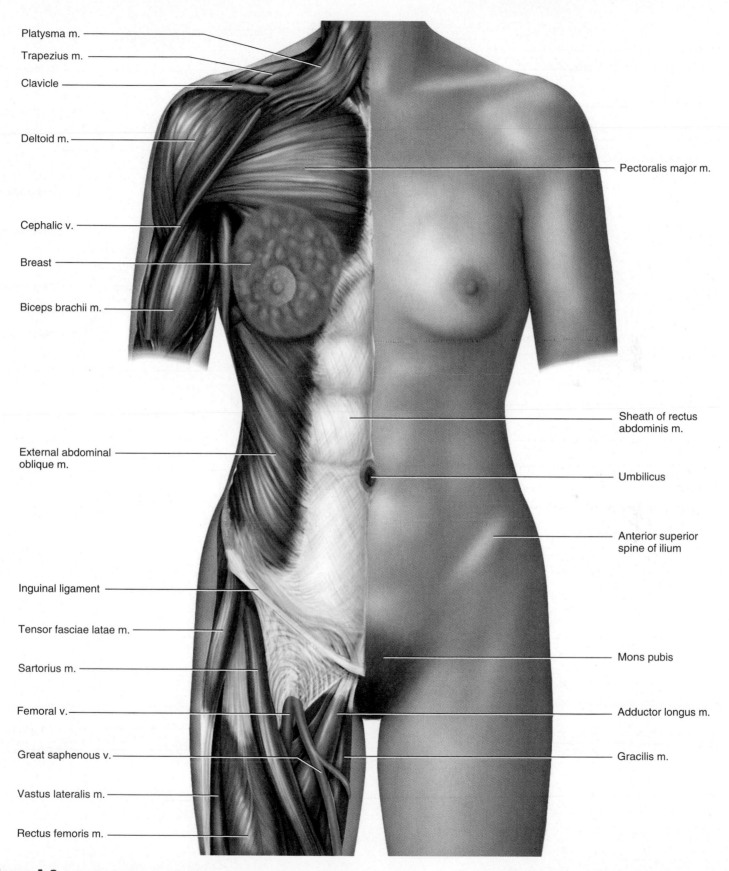

Platysma m.

Trapezius m.

Clavicle

Deltoid m.

Cephalic v.

Breast

Biceps brachii m.

External abdominal oblique m.

Inguinal ligament

Tensor fasciae latae m.

Sartorius m.

Femoral v.

Great saphenous v.

Vastus lateralis m.

Rectus femoris m.

Pectoralis major m.

Sheath of rectus abdominis m.

Umbilicus

Anterior superior spine of ilium

Mons pubis

Adductor longus m.

Gracilis m.

Figure A.3 **Superficial Anatomy of the Trunk (Female).** Surface anatomy is shown on the anatomical left, and structures immediately deep to the skin on the right.

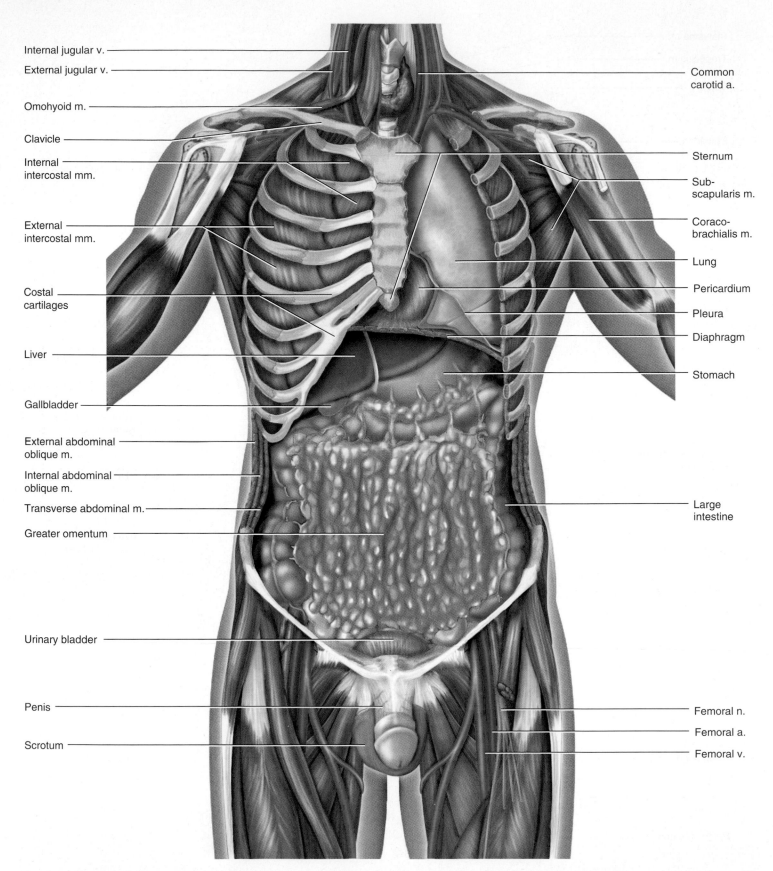

Internal jugular v.

External jugular v.

Omohyoid m.

Clavicle

Internal
intercostal mm.

External
intercostal mm.

Costal
cartilages

Liver

Gallbladder

External abdominal
oblique m.

Internal abdominal
oblique m.

Transverse abdominal m.

Greater omentum

Urinary bladder

Penis

Scrotum

Common
carotid a.

Sternum

Sub-
scapularis m.

Coraco-
brachialis m.

Lung

Pericardium

Pleura

Diaphragm

Stomach

Large
intestine

Femoral n.

Femoral a.

Femoral v.

Figure A.4 **Anatomy at the Level of the Rib Cage and Greater Omentum (Male).** The anterior body wall is removed, and the ribs, intercostal muscles, and pleura are removed from the anatomical left.

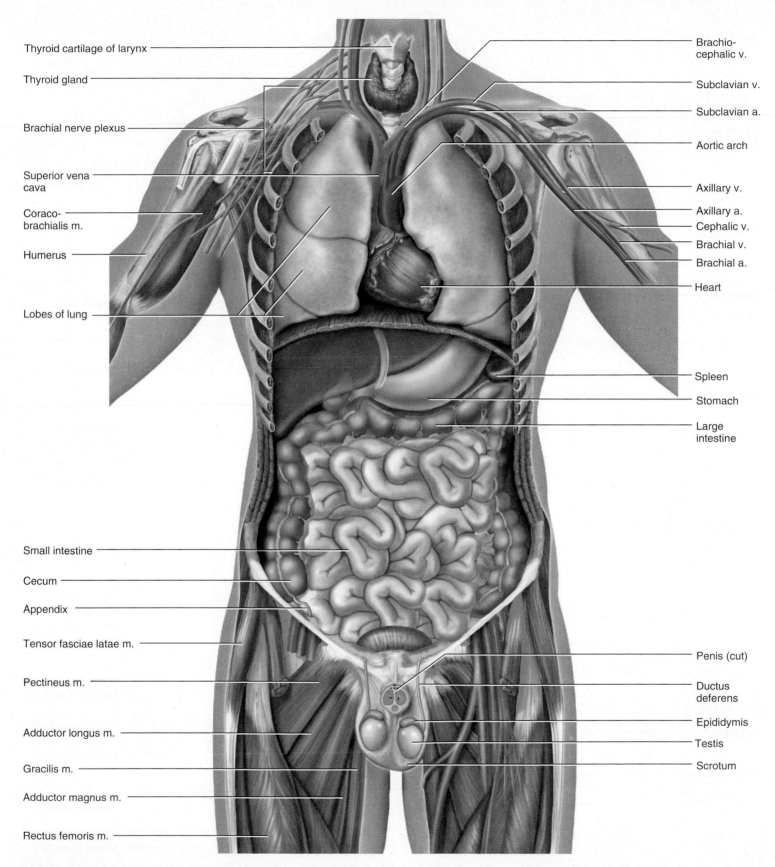

Thyroid cartilage of larynx

Thyroid gland

Brachial nerve plexus

Superior vena cava

Coraco-brachialis m.

Humerus

Lobes of lung

Small intestine

Cecum

Appendix

Tensor fasciae latae m.

Pectineus m.

Adductor longus m.

Gracilis m.

Adductor magnus m.

Rectus femoris m.

Brachio-cephalic v.

Subclavian v.

Subclavian a.

Aortic arch

Axillary v.

Axillary a.

Cephalic v.

Brachial v.

Brachial a.

Heart

Spleen

Stomach

Large intestine

Penis (cut)

Ductus deferens

Epididymis

Testis

Scrotum

Figure A.5 **Anatomy at the Level of the Lungs and Intestines (Male).** The sternum, ribs, and greater omentum are removed.
• *Name several viscera that are protected by the rib cage.*

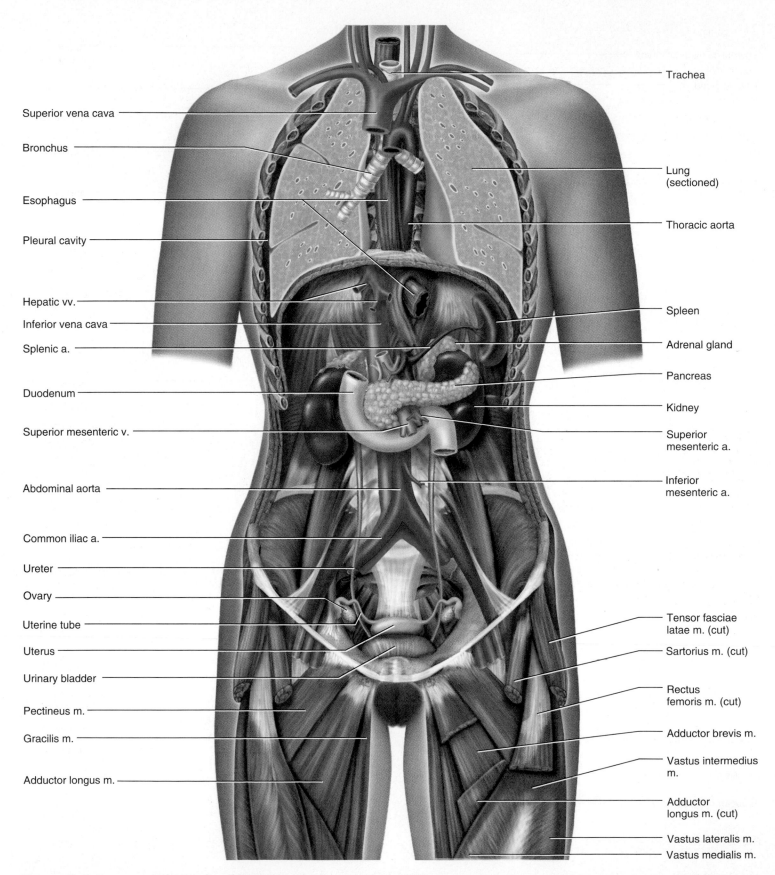

Figure A.6 **Anatomy at the Level of the Retroperitoneal Viscera (Female).** The heart is removed, the lungs are frontally sectioned, and the viscera of the peritoneal cavity and the peritoneum itself are removed.

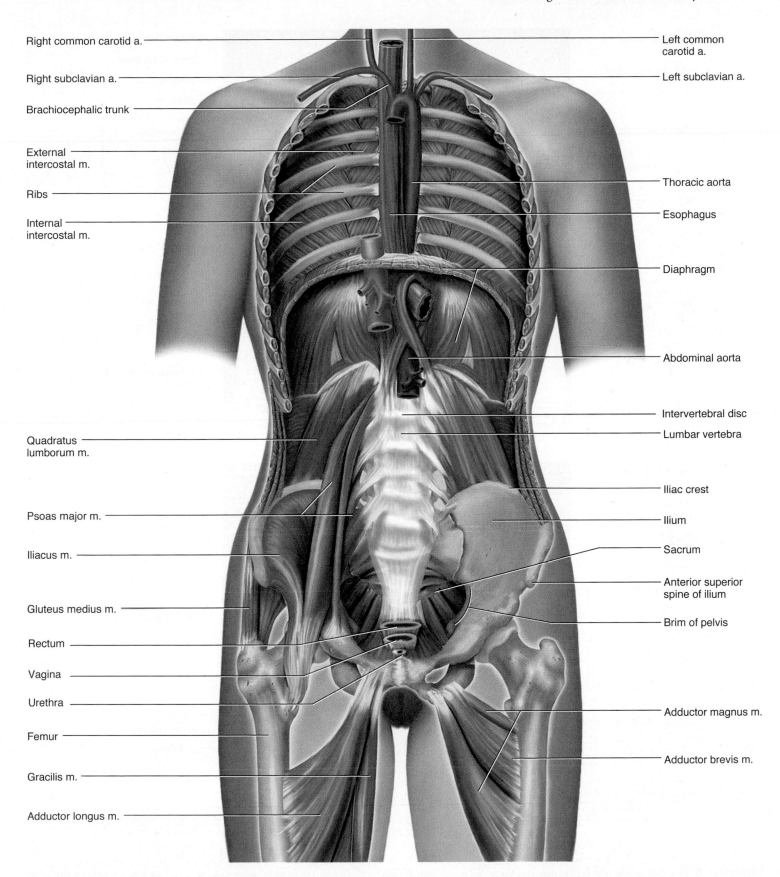

Right common carotid a.

Right subclavian a.

Brachiocephalic trunk

External intercostal m.

Ribs

Internal intercostal m.

Quadratus lumborum m.

Psoas major m.

Iliacus m.

Gluteus medius m.

Rectum

Vagina

Urethra

Femur

Gracilis m.

Adductor longus m.

Left common carotid a.

Left subclavian a.

Thoracic aorta

Esophagus

Diaphragm

Abdominal aorta

Intervertebral disc

Lumbar vertebra

Iliac crest

Ilium

Sacrum

Anterior superior spine of ilium

Brim of pelvis

Adductor magnus m.

Adductor brevis m.

Figure A.7 **Anatomy at the Level of the Posterior Body Wall (Female).** The lungs and retroperitoneal viscera are removed.

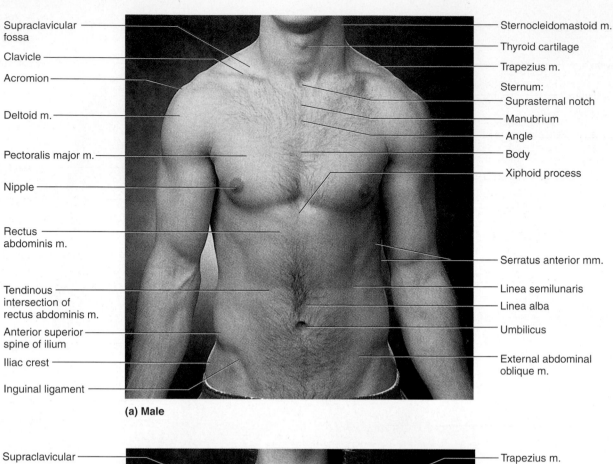

Supraclavicular fossa

Clavicle

Acromion

Deltoid m.

Pectoralis major m.

Nipple

Rectus abdominis m.

Tendinous intersection of rectus abdominis m.

Anterior superior spine of ilium

Iliac crest

Inguinal ligament

Sternocleidomastoid m.

Thyroid cartilage

Trapezius m.

Sternum:
  Suprasternal notch
  Manubrium
  Angle
  Body
  Xiphoid process

Serratus anterior mm.

Linea semilunaris

Linea alba

Umbilicus

External abdominal oblique m.

(a) Male

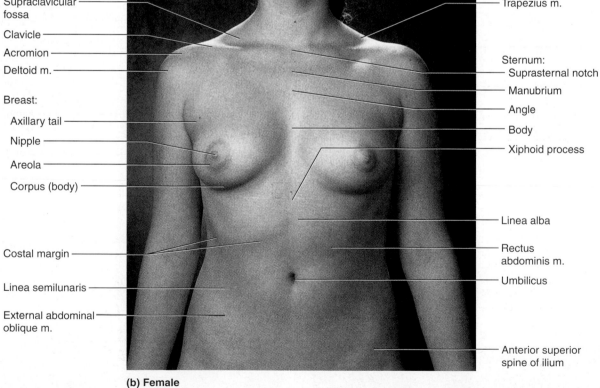

Supraclavicular fossa

Clavicle

Acromion

Deltoid m.

Breast:
  Axillary tail
  Nipple
  Areola
  Corpus (body)

Costal margin

Linea semilunaris

External abdominal oblique m.

Trapezius m.

Sternum:
  Suprasternal notch
  Manubrium
  Angle
  Body
  Xiphoid process

Linea alba

Rectus abdominis m.

Umbilicus

Anterior superior spine of ilium

(b) Female

Figure A.8 **The Thorax and Abdomen, Anterior View.** All of the features labeled are common to both sexes, though some are labeled only on the photograph that better shows them.
• *The V-shaped tendons on each side of the suprasternal notch in part (a) belong to what muscles?*

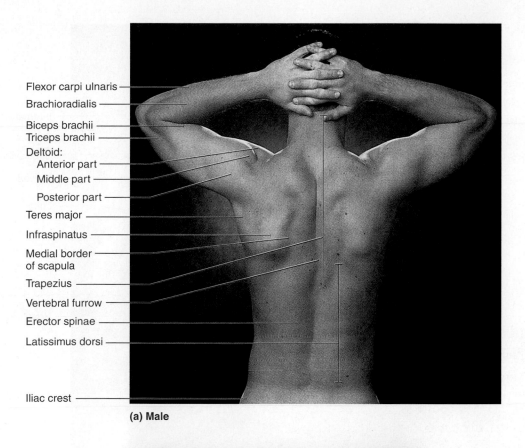

Flexor carpi ulnaris
Brachioradialis
Biceps brachii
Triceps brachii
Deltoid:
    Anterior part
    Middle part
    Posterior part
Teres major
Infraspinatus
Medial border of scapula
Trapezius
Vertebral furrow
Erector spinae
Latissimus dorsi
Iliac crest

(a) Male

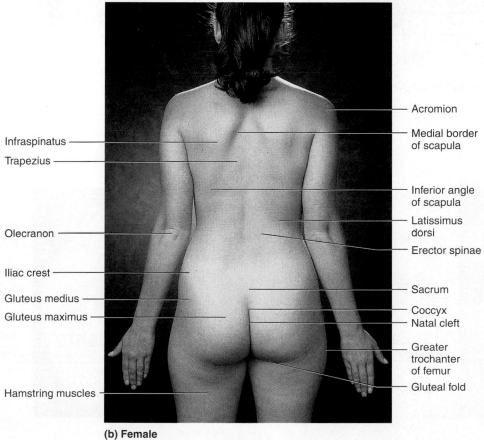

Infraspinatus
Trapezius
Olecranon
Iliac crest
Gluteus medius
Gluteus maximus
Hamstring muscles

Acromion
Medial border of scapula
Inferior angle of scapula
Latissimus dorsi
Erector spinae
Sacrum
Coccyx
Natal cleft
Greater trochanter of femur
Gluteal fold

(b) Female

Figure A.9 The Back and Gluteal Region. All of the features labeled are common to both sexes, though some are labeled only on the photograph that better shows them.

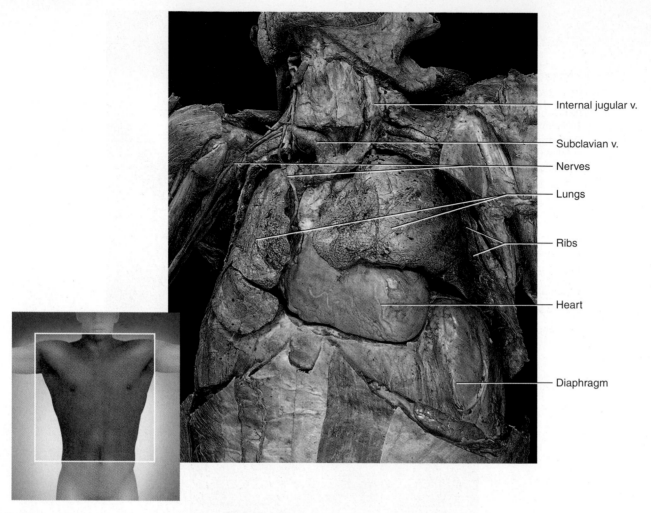

Internal jugular v.

Subclavian v.

Nerves

Lungs

Ribs

Heart

Diaphragm

Figure A.10 Frontal View of the Thoracic Cavity.

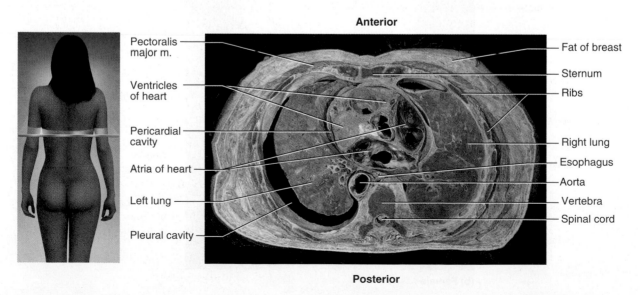

Anterior

Pectoralis major m.

Ventricles of heart

Pericardial cavity

Atria of heart

Left lung

Pleural cavity

Fat of breast

Sternum

Ribs

Right lung

Esophagus

Aorta

Vertebra

Spinal cord

Posterior

Figure A.11 Transverse Section of the Thorax. Section taken at the level shown by the inset and oriented the same as the reader's body.
• *In this section, which term best describes the position of the aorta relative to the heart: posterior, lateral, or proximal?*

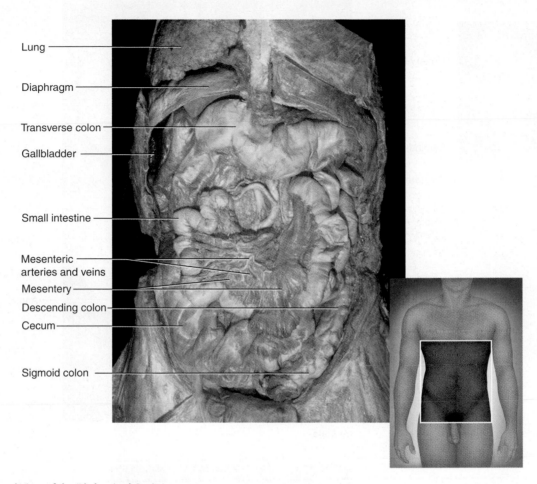

Lung

Diaphragm

Transverse colon

Gallbladder

Small intestine

Mesenteric
arteries and veins

Mesentery

Descending colon

Cecum

Sigmoid colon

Figure A.12 Frontal View of the Abdominal Cavity.

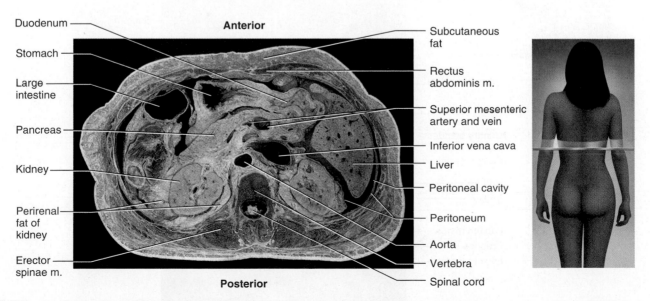

Duodenum

Stomach

Large
intestine

Pancreas

Kidney

Perirenal
fat of
kidney

Erector
spinae m.

Anterior

Posterior

Subcutaneous
fat

Rectus
abdominis m.

Superior mesenteric
artery and vein

Inferior vena cava

Liver

Peritoneal cavity

Peritoneum

Aorta

Vertebra

Spinal cord

Figure A.13 **Transverse Section of the Abdomen.** Section taken at the level shown by the inset and oriented the same as the reader's body.
● *What tissue in this photograph is immediately superficial to the rectus abdominis muscle?*

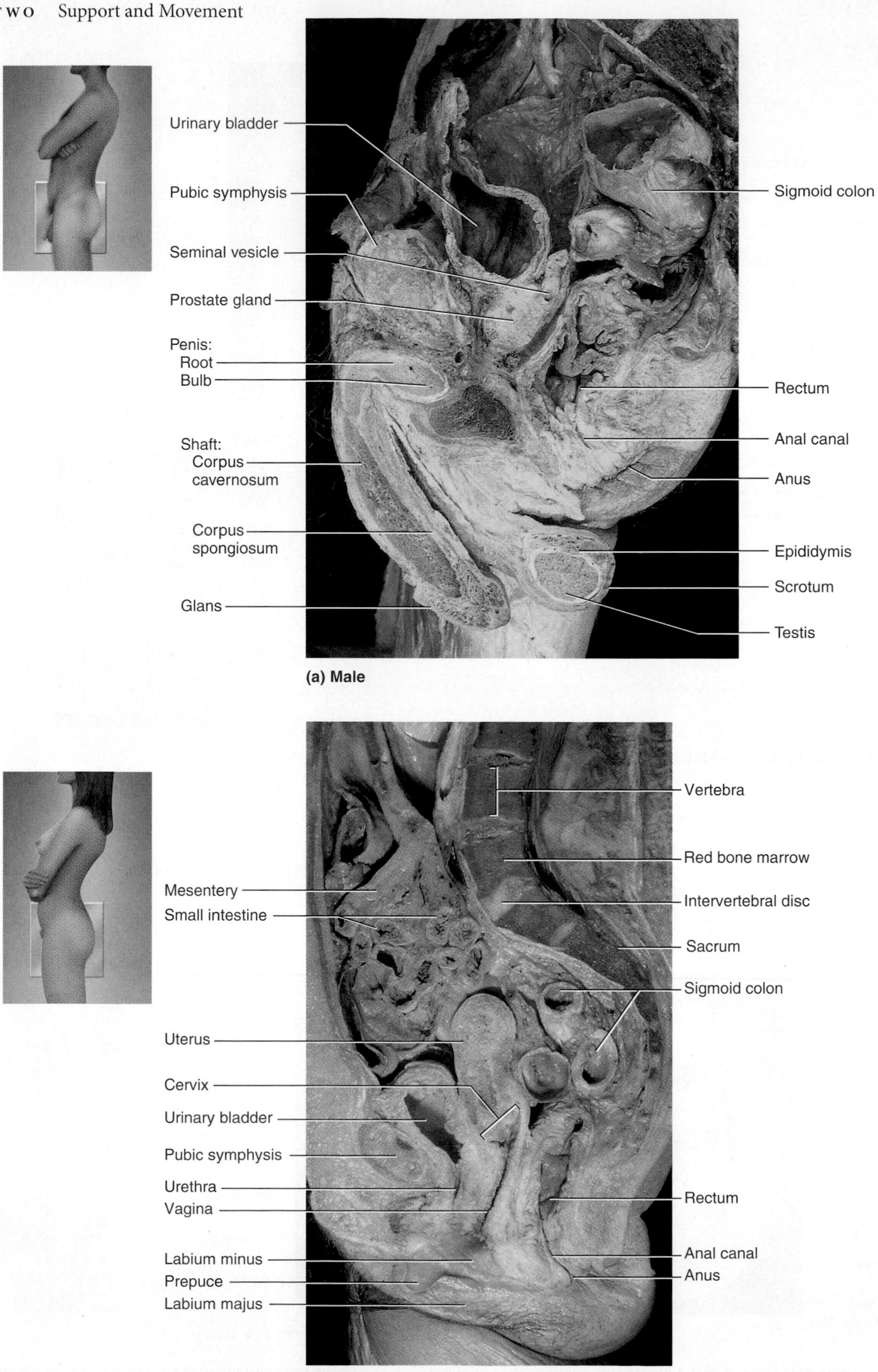

Urinary bladder

Pubic symphysis

Seminal vesicle

Prostate gland

Penis:
Root
Bulb

Shaft:
Corpus
cavernosum

Corpus
spongiosum

Glans

Sigmoid colon

Rectum

Anal canal

Anus

Epididymis

Scrotum

Testis

(a) Male

Mesentery
Small intestine

Uterus

Cervix

Urinary bladder

Pubic symphysis

Urethra

Vagina

Labium minus

Prepuce

Labium majus

Vertebra

Red bone marrow

Intervertebral disc

Sacrum

Sigmoid colon

Rectum

Anal canal
Anus

(b) Female

Figure A.14 Median Sections of the Pelvic Cavitiy. Viewed from the left.

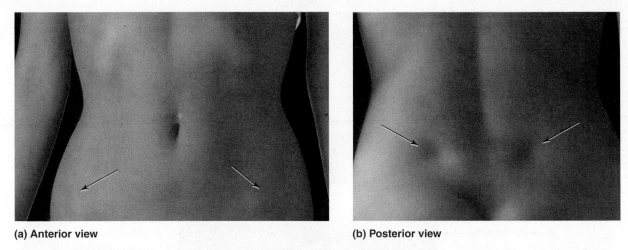

(a) Anterior view   **(b) Posterior view**

Figure A.15 Pelvic Landmarks. (a) The anterior superior spines of the ilium are marked by anterolateral protuberances (arrows) at about the location where the pockets usually open on a pair of pants. (b) The posterior superior spines are marked in some people by dimples in the sacral region (arrows).

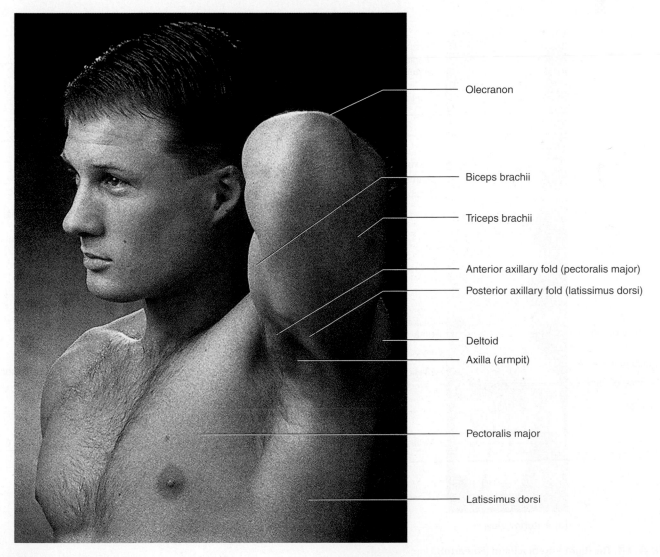

Olecranon

Biceps brachii

Triceps brachii

Anterior axillary fold (pectoralis major)

Posterior axillary fold (latissimus dorsi)

Deltoid

Axilla (armpit)

Pectoralis major

Latissimus dorsi

Figure A.16 The Axillary Region.

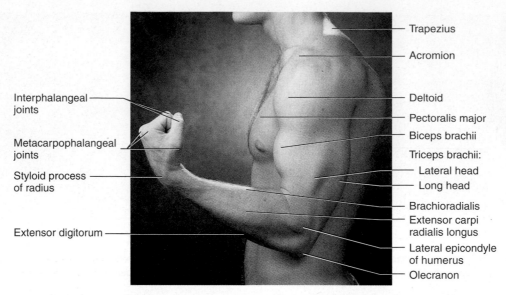

Trapezius

Acromion

Deltoid

Pectoralis major

Biceps brachii

Triceps brachii:
Lateral head
Long head

Brachioradialis
Extensor carpi
radialis longus
Lateral epicondyle
of humerus
Olecranon

Interphalangeal
joints

Metacarpophalangeal
joints

Styloid process
of radius

Extensor digitorum

Figure A.17 The Upper Limb, Lateral View.

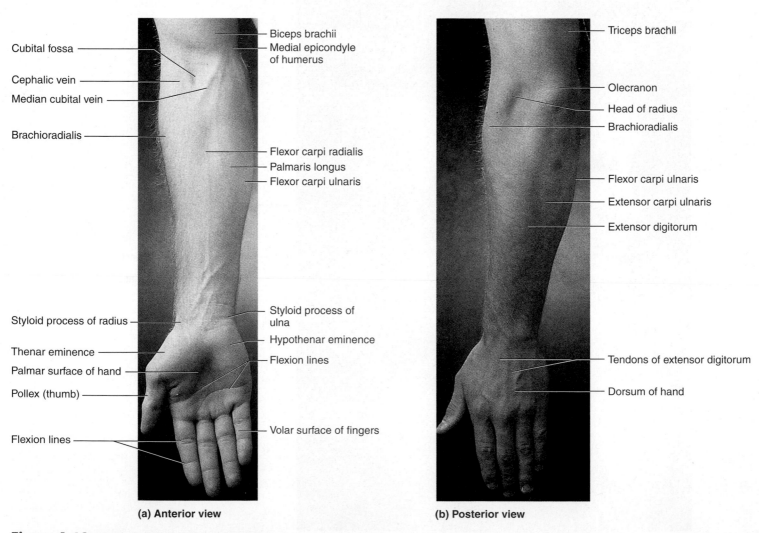

Cubital fossa

Cephalic vein

Median cubital vein

Brachioradialis

Biceps brachii
Medial epicondyle
of humerus

Flexor carpi radialis
Palmaris longus
Flexor carpi ulnaris

Styloid process of radius

Thenar eminence

Palmar surface of hand

Pollex (thumb)

Flexion lines

Styloid process of
ulna

Hypothenar eminence

Flexion lines

Volar surface of fingers

(a) Anterior view

Triceps brachii

Olecranon

Head of radius

Brachioradialis

Flexor carpi ulnaris

Extensor carpi ulnaris

Extensor digitorum

Tendons of extensor digitorum

Dorsum of hand

(b) Posterior view

Figure A.18 The Right Antebrachium (Forearm).

● *Only two tendons of the extensor digitorum are labeled, but how many tendons does this muscle have in all?*

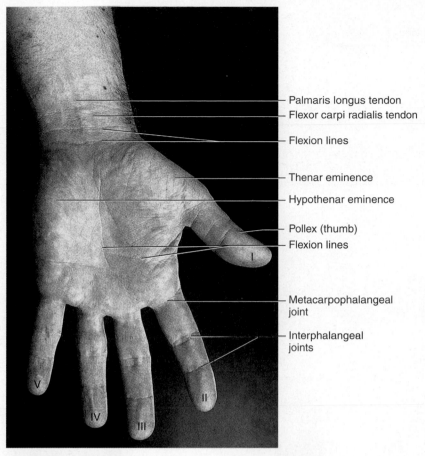

Palmaris longus tendon
Flexor carpi radialis tendon

Flexion lines

Thenar eminence

Hypothenar eminence

Pollex (thumb)
Flexion lines

Metacarpophalangeal joint

Interphalangeal joints

I

II

III

IV

V

(a) Anterior (palmar) view

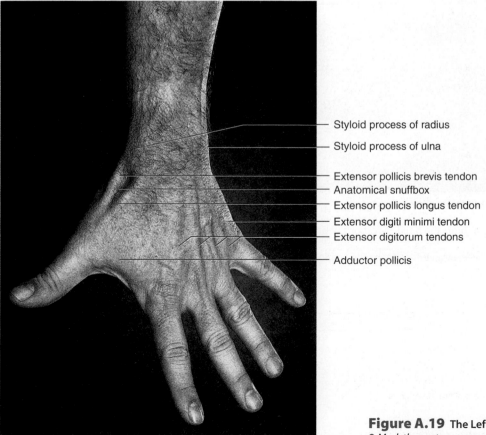

Styloid process of radius

Styloid process of ulna

Extensor pollicis brevis tendon
Anatomical snuffbox
Extensor pollicis longus tendon
Extensor digiti minimi tendon
Extensor digitorum tendons

Adductor pollicis

(b) Posterior (dorsal) view

Figure A.19 The Left Wrist and Hand.
● *Mark the spot on one or both photographs where a saddle joint can be found.*

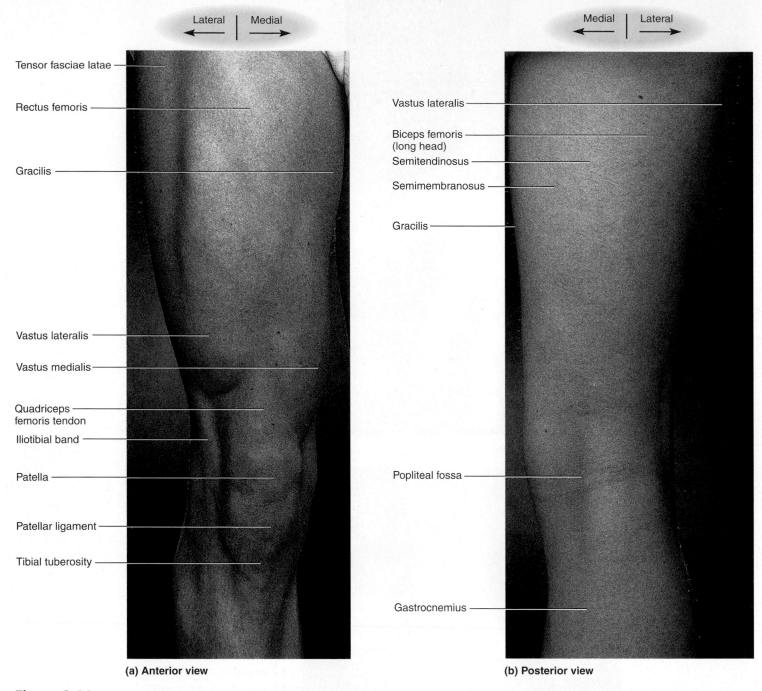

Lateral ← | → Medial

Tensor fasciae latae —

Rectus femoris —

Gracilis —

Vastus lateralis —

Vastus medialis —

Quadriceps femoris tendon —

Iliotibial band —

Patella —

Patellar ligament —

Tibial tuberosity —

(a) Anterior view

Medial ← | → Lateral

Vastus lateralis —

Biceps femoris (long head) —

Semitendinosus —

Semimembranosus —

Gracilis —

Popliteal fossa —

Gastrocnemius —

(b) Posterior view

Figure A.20 **The Right Thigh and Knee.** Locations of posterior thigh muscles are indicated, but the boundaries of the individual muscles are rarely visible on a living person.

● *Mark the spot on part (a) where the vastus intermedius would be found.*

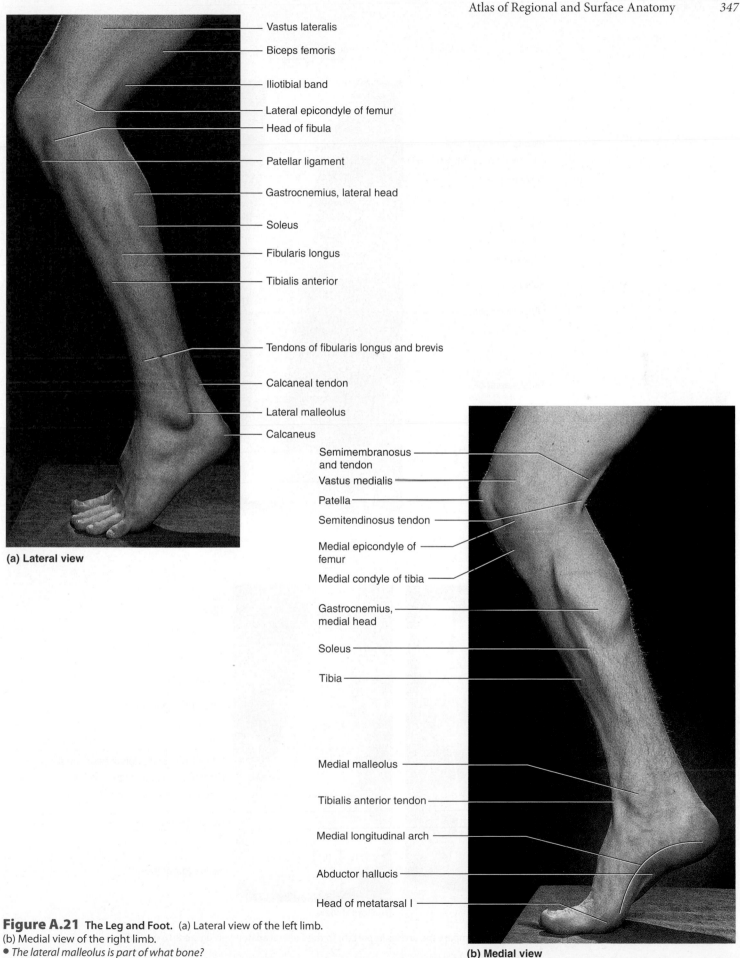

Vastus lateralis

Biceps femoris

Iliotibial band

Lateral epicondyle of femur

Head of fibula

Patellar ligament

Gastrocnemius, lateral head

Soleus

Fibularis longus

Tibialis anterior

Tendons of fibularis longus and brevis

Calcaneal tendon

Lateral malleolus

Calcaneus

(a) Lateral view

Semimembranosus
and tendon

Vastus medialis

Patella

Semitendinosus tendon

Medial epicondyle of
femur

Medial condyle of tibia

Gastrocnemius,
medial head

Soleus

Tibia

Medial malleolus

Tibialis anterior tendon

Medial longitudinal arch

Abductor hallucis

Head of metatarsal I

Figure A.21 The Leg and Foot. (a) Lateral view of the left limb.
(b) Medial view of the right limb.
• *The lateral malleolus is part of what bone?*

(b) Medial view

Medial | Lateral

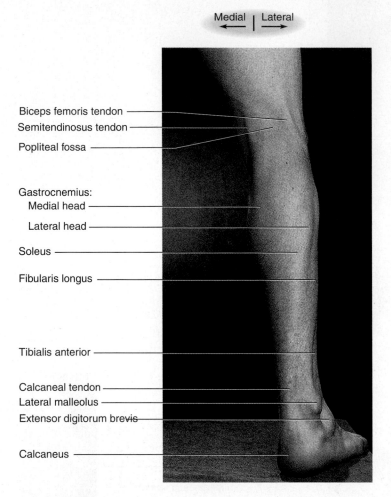

Biceps femoris tendon

Semitendinosus tendon

Popliteal fossa

Gastrocnemius:
  Medial head

  Lateral head

Soleus

Fibularis longus

Tibialis anterior

Calcaneal tendon

Lateral malleolus

Extensor digitorum brevis

Calcaneus

Figure A.22 The Right Leg and Foot, Posterior View.

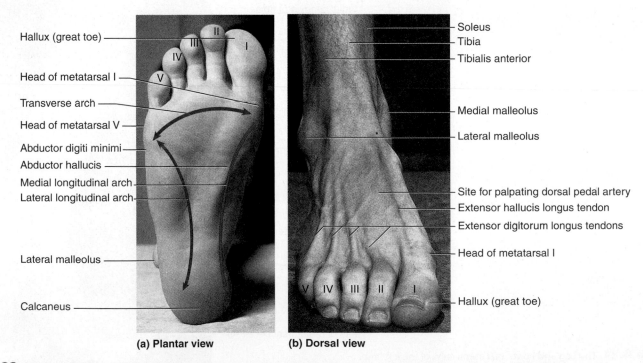

Hallux (great toe)

Head of metatarsal I

Transverse arch

Head of metatarsal V

Abductor digiti minimi

Abductor hallucis

Medial longitudinal arch

Lateral longitudinal arch

Lateral malleolus

Calcaneus

Soleus

Tibia

Tibialis anterior

Medial malleolus

Lateral malleolus

Site for palpating dorsal pedal artery

Extensor hallucis longus tendon

Extensor digitorum longus tendons

Head of metatarsal I

Hallux (great toe)

(a) Plantar view   **(b) Dorsal view**

Figure A.23 The Foot, Plantar and Dorsal Views. Compare the arches in part (b) to the skeletal anatomy in figure 8.16 (p. 198).

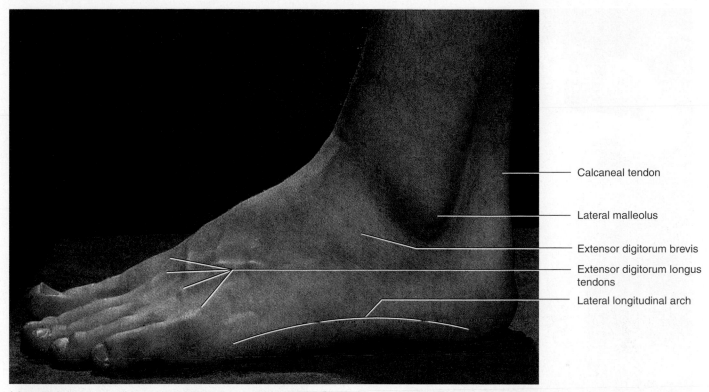

Calcaneal tendon

Lateral malleolus

Extensor digitorum brevis

Extensor digitorum longus tendons

Lateral longitudinal arch

(a) Lateral view

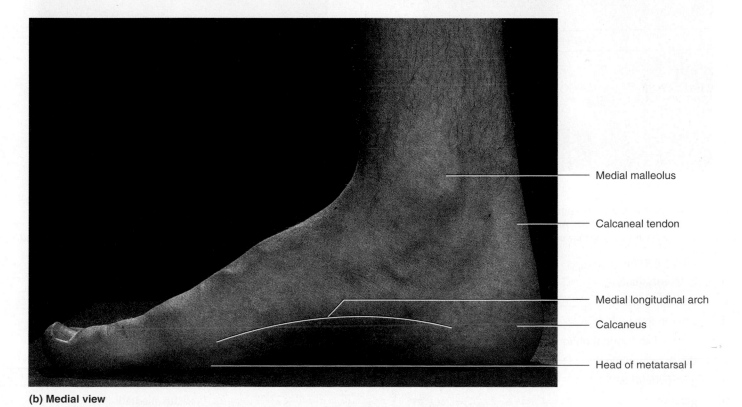

Medial malleolus

Calcaneal tendon

Medial longitudinal arch

Calcaneus

Head of metatarsal I

(b) Medial view

Figure A.24 The Foot, Lateral and Medial Views.
● *Indicate the location of middle phalanx I on each photograph.*

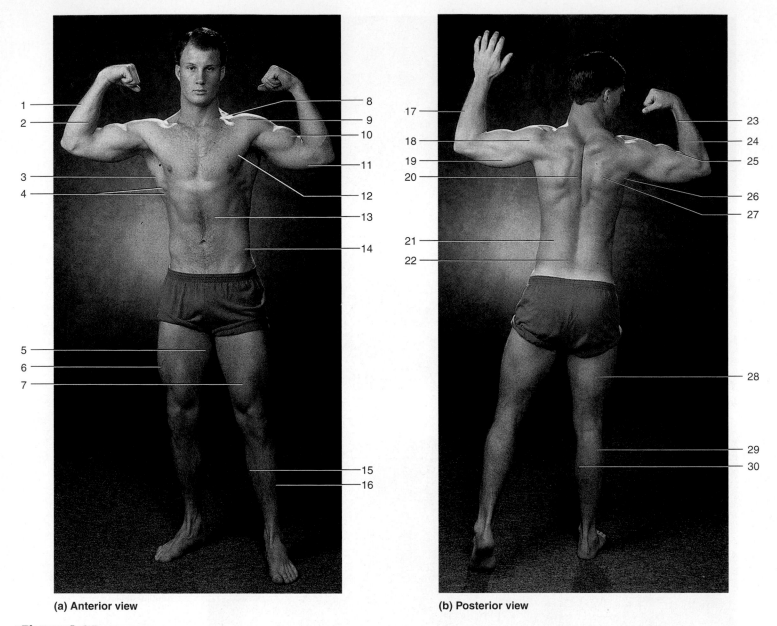

(a) Anterior view

(b) Posterior view

Figure A.25 Test of Muscle Recognition. To test your knowledge of muscle anatomy, match the 30 labeled muscles on these photographs to the following alphabetical list of muscles. Answer as many as possible without referring back to the previous illustrations. Some of these names will be used more than once since the same muscle may be shown from different perspectives, and some of these names will not be used at all. The answers are in the appendix.

a.	biceps brachii	j.	infraspinatus	s.	sternocleidomastoid
b.	brachioradialis	k.	latissimus dorsi	t.	subscapularis
c.	deltoid	l.	pectineus	u.	teres major
d.	erector spinae	m.	pectoralis major	v.	tibialis anterior
e.	external abdominal oblique	n.	rectus abdominis	w.	trapezius
f.	flexor carpi ulnaris	o.	rectus femoris	x.	triceps brachii
g.	gastrocnemius	p.	serratus anterior	y.	vastus lateralis
h.	gracilis	q.	soleus	z.	vastus medialis
i.	hamstrings	r.	splenius capitis		

Nervous Tissue

A Purkinje cell, a neuron from the cerebellum of the brain

CHAPTER OUTLINE

DEEPER INSIGHTS

BRUSHING UP

To understand this chapter, you may find it helpful to review the following concepts:
- General structure of nerve cells, especially the soma, dendrites, and axon (p. 71)
- Early embryonic development (pp. 88–91)
- Introduction to synapses and neurotransmitters (p. 247)

Anatomy & Physiology | REVEALED®
aprevealed.com

Nervous System

The next five chapters are concerned with the nervous system. This is a system of great complexity and mystery. It is the foundation of all our conscious experiences, personality, and behavior. It profoundly intrigues biologists, physicians, psychologists, and even philosophers. Understanding this fascinating system is regarded by many as the ultimate challenge facing the behavioral and life sciences.

We begin our study at the simplest organizational level—the nerve cells (*neurons*) and cells called *neuroglia* that support their function in various ways. We then progress to the organ level to examine the spinal cord (chapter 14), brain (chapter 15), autonomic nervous system (chapter 16), and sense organs (chapter 17).

13.1 Overview of the Nervous System

▶ Expected Learning Outcomes

When you have completed this section, you should be able to

- describe the function of the nervous system;
- describe the major anatomical subdivisions of the nervous system;
- explain the functional differences between these anatomical subdivisions; and
- define *nerve, ganglion, receptor,* and *effector.*

If the body is to maintain homeostasis and function effectively, its trillions of cells must work together in a coordinated fashion. If each cell behaved without regard to what others are doing, the result would be physiological chaos and death. We have two organ systems dedicated to maintaining internal coordination—the **endocrine system,** which communicates by means of chemical messengers (hormones) secreted into the blood, and the **nervous system** (fig.13.1), which employs electrical and chemical means to send messages very quickly from cell to cell. The study of the nervous system is called **neurobiology.** Its primary subdisciplines are **neuroanatomy** and **neurophysiology.**

The nervous system carries out its coordinating task in three basic steps: (1) Through sensory nerve endings, it receives information about changes in the body and external environment and it transmits coded messages to the spinal cord and brain. (2) The spinal cord and brain process this information, relate it to past experience, and determine what response, if any, is appropriate to the circumstances. (3) The spinal cord and brain issue commands primarily to muscle and gland cells to carry out such responses.

The nervous system has two major anatomical subdivisions (fig. 13.2):

- The **central nervous system (CNS)** consists of the brain and spinal cord, which are enclosed and protected by the cranium and vertebral column.

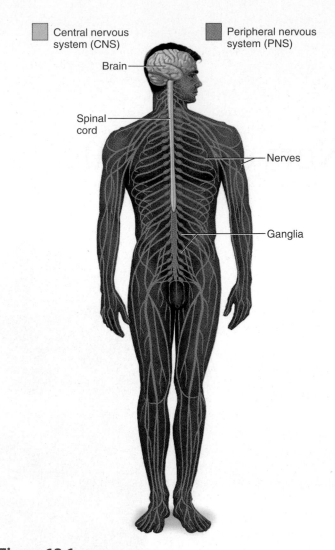

Figure 13.1 The Nervous System.
● *Which division, the CNS or PNS, is likely to suffer the most frequent injuries? Why?*

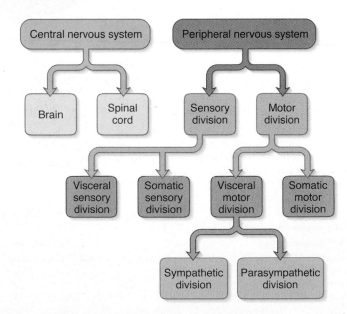

Figure 13.2 Subdivisions of the Nervous System.

- The **peripheral nervous system (PNS)** consists of all the nervous system except the brain and spinal cord. It is composed of nerves and ganglia. A **nerve** is a bundle of nerve fibers (axons) wrapped in fibrous connective tissue. Nerves emerge from the CNS through foramina of the skull and vertebral column and carry signals to and from other organs of the body. A **ganglion**[1] (plural, *ganglia*) is a knotlike swelling in a nerve where the cell bodies of neurons are concentrated.

The peripheral nervous system is functionally divided into *sensory* and *motor* divisions, and each of these is further divided into *somatic* and *visceral* divisions.

- The **sensory (afferent**[2]**) division** carries sensory signals from various **receptors** (sense organs and simple sensory nerve endings) to the CNS. This is the pathway that informs the CNS of stimuli within and around the body.
  - The **somatic**[3] **sensory division** carries signals from receptors in the skin, muscles, bones, and joints.
  - The **visceral sensory division** carries signals mainly from the viscera of the thoracic and abdominal cavities, such as the heart, lungs, stomach, and urinary bladder.
- The **motor (efferent**[4]**) division** carries signals from the CNS to gland and muscle cells that carry out the body's responses. Cells and organs that respond to commands from the nervous system are called **effectors.**
  - The **somatic motor division** carries signals to the skeletal muscles. This output produces muscular contractions that are under voluntary control, as well as involuntary muscle contractions called *somatic reflexes.*
  - The **visceral motor division (autonomic**[5] **nervous system)** carries signals to glands, cardiac muscle, and smooth muscle. We usually have no voluntary control over these effectors, and this system operates at an unconscious level. The responses of this system and its effectors are *visceral reflexes.* The autonomic nervous system has two further divisions:
    - The **sympathetic division** tends to arouse the body for action—for example, by accelerating the heartbeat and increasing respiratory airflow—but it inhibits digestion.
    - The **parasympathetic division** adapts the body for energy intake and conservation. It stimulates digestion but slows down the heartbeat and reduces respiratory airflow, for example.

The foregoing terminology may give the impression that the body has several nervous systems—central, peripheral, sensory, motor, somatic, and visceral. These are just terms of convenience, however. There is only one nervous system, and these subsystems are interconnected parts of the whole.

[1]*gangli* = knot
[2]*af* = *ad* = toward + *fer* = to carry
[3]*somat* = body + *ic* = pertaining to
[4]*ef* = *ex* = out, away + *fer* = to carry
[5]*auto* = self + *nom* = law, governance

Before You Go On

Answer the following questions to test your understanding of the preceding section:

1. Define *receptor* and *effector*. Give two examples of each.
2. Distinguish between the central and peripheral nervous systems, and between the visceral and somatic divisions of the sensory and motor systems.
3. What is another name for the visceral motor nervous system? What are the two subdivisions of this system? How do they differ in their effects on the body?

13.2 Nerve Cells (Neurons)

▶ Expected Learning Outcomes

When you have completed this section, you should be able to

- describe the properties that neurons must have to carry out their function;
- identify and define three functional classes into which all neurons fall;
- describe the structure of a representative neuron; and
- describe some variations in neuron structure.

Universal Properties of Neurons

The functional unit of the nervous system is the **nerve cell,** or **neuron;** neurons carry out the system's communicative role. These cells have three fundamental physiological properties that are necessary to this function:

1. **Excitability (irritability).** All cells possess excitability, the ability to respond to environmental changes called **stimuli.** Neurons have developed this property to the highest degree.
2. **Conductivity.** Neurons respond to stimuli by producing traveling electrical signals that quickly reach other cells at distant locations.
3. **Secretion.** When the electrical signal reaches the end of a nerve fiber, the neuron usually secretes a chemical called a *neurotransmitter* that crosses a small gap between cells and stimulates the next cell.

Apply What You Know

What basic physiological properties do a nerve cell and a muscle cell have in common? Identify one property that each cell type has that the other one lacks.

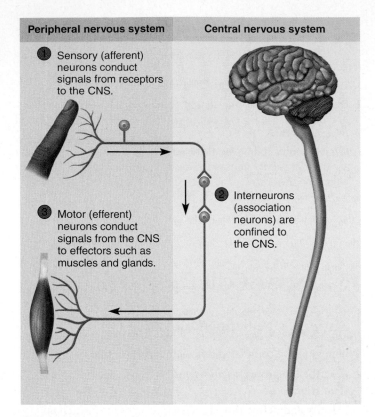

Peripheral nervous system	Central nervous system

① Sensory (afferent) neurons conduct signals from receptors to the CNS.

② Interneurons (association neurons) are confined to the CNS.

③ Motor (efferent) neurons conduct signals from the CNS to effectors such as muscles and glands.

Figure 13.3 Functional Classes of Neurons. Sensory (afferent) neurons carry signals to the central nervous system (CNS); interneurons are contained entirely within the CNS and carry signals from one neuron to another; and motor (efferent) neurons carry signals from the CNS to muscles and glands.
● *How do the origins of the words* afferent *and* efferent *relate to the functions of the respective neurons?*

Functional Classes of Neurons

There are three general classes of neurons (fig. 13.3) corresponding to the three major aspects of nervous system function listed earlier:

1. **Sensory (afferent) neurons** are specialized to detect stimuli such as light, heat, pressure, and chemicals and to transmit information about them to the CNS. These neurons can begin in almost any organ of the body but always end in the brain or spinal cord; the word *afferent* refers to signal conduction *toward* the CNS. Some sensory receptors, such as pain and smell receptors, are themselves neurons. In other cases, such as taste and hearing, the receptor is a separate cell that communicates directly with a sensory neuron.

2. **Interneurons**[6] **(association neurons)** lie entirely within the CNS. They receive signals from many other neurons and carry out the *integrative* function of the nervous system—that is, they process, store, and retrieve information and "make decisions" about how the body responds to stimuli. About 90% of human neurons are interneurons.

The word *interneuron* refers to the fact that they lie *between*, and interconnect, the incoming sensory pathways and the outgoing motor pathways of the CNS.

3. **Motor (efferent) neurons** send signals predominantly to muscle and gland cells, the effectors that carry out the body's responses to stimuli. They are called *motor* neurons because most of them lead to muscle cells, and *efferent* neurons to signify that they conduct signals *away from* the CNS.

Structure of a Neuron

There are several varieties of neurons, as we shall see, but a good starting point for discussing neuron stucture is a motor neuron of the spinal cord (fig. 13.4). The control center of the neuron is its **neurosoma,**[7] also called the **soma** or **cell body.** It has a single, centrally located nucleus with a large nucleolus. The cytoplasm contains mitochondria, lysosomes, a Golgi complex, numerous inclusions, and an extensive rough endoplasmic reticulum and cytoskeleton. The cytoskeleton consists of a dense mesh of microtubules and **neurofibrils** (bundles of actin filaments) that compartmentalize the rough ER into darkstaining regions called *Nissl*[8] *bodies,* unique to neurons (fig. 13.4d, e). Nissl bodies are a helpful clue to identifying neurons in tissue sections with mixed cell types. Mature neurons lack centrioles and apparently undergo no further mitosis after adolescence, but they are unusually long-lived cells, capable of functioning for over a hundred years. Even in old age, however, there are unspecialized stem cells in the CNS that divide and develop into new neurons.

The major cytoplasmic inclusions in a neuron are glycogen granules, lipid droplets, melanin, and a golden brown pigment called *lipofuscin*[9] (LIP-oh-FEW-sin)—produced when lysosomes digest worn-out organelles and other products. Lipofuscin collects with age and pushes the nucleus to one side of the cell. Lipofuscin granules are also called "wear-and-tear granules" because they are most abundant in old neurons.

The soma of a spinal motor neuron gives rise to a few thick processes that branch into a vast number of **dendrites**[10]—named for their striking resemblance to the bare branches of a tree in winter. The dendrites are the primary sites for receiving signals from other neurons. Some neurons have only one dendrite and some have thousands. The more dendrites a neuron has, the more information it can receive from other cells and incorporate into its decision making. As tangled as the dendrites may seem, they provide exquisitely precise pathways for the reception and processing of neural information.

On one side of the soma is a mound called the **axon hillock,** from which the **axon (nerve fiber)** originates. The axon hillock and nearby portion of the axon *(initial segment)* are collectively called the **trigger zone,** because this is usually where the neuron first generates **action potentials**—electrical changes that constitute the nerve signal. An axon is specialized for rapid conduction of nerve signals to points remote from the soma. It is cylindrical and relatively unbranched for most of its length; however, it may give rise to a few branches called

---

[6]*inter* = between

[7]*soma* = body
[8]Franz Nissl (1860–1919), German neuropathologist
[9]*lipo* = fat, lipid + *fusc* = dusky, brown
[10]*dendr* = tree, branch + *ite* = little

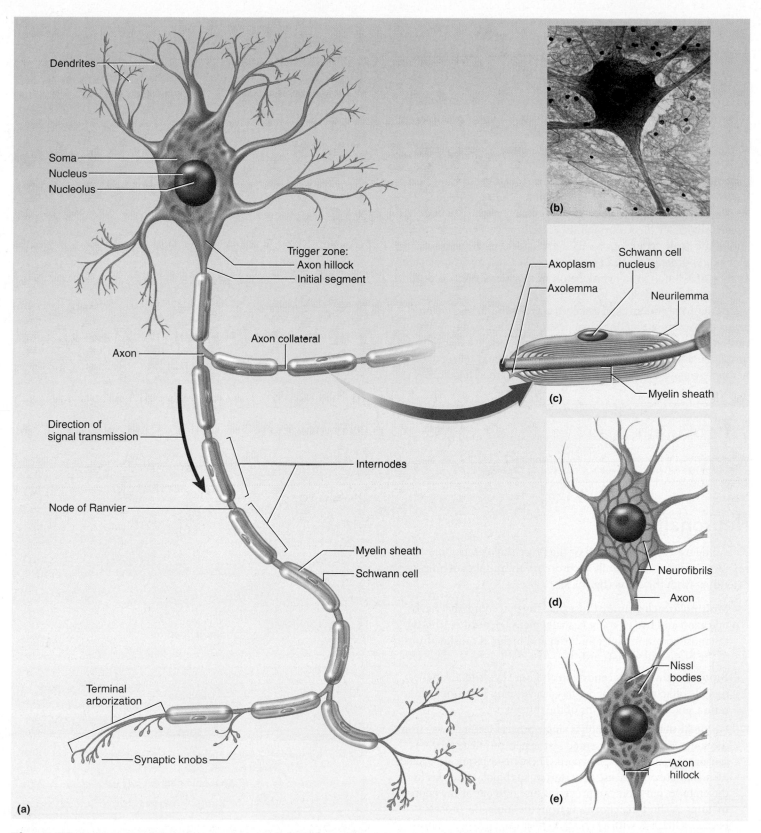

Dendrites

Soma
Nucleus
Nucleolus

Trigger zone:
Axon hillock
Initial segment

Axon collateral

Axon

Direction of
signal transmission

Internodes

Node of Ranvier

Myelin sheath

Schwann cell

Terminal
arborization

Synaptic knobs

(a)

(b)

Axoplasm
Axolemma

Schwann cell
nucleus

Neurilemma

Myelin sheath

(c)

Neurofibrils

Axon

(d)

Nissl
bodies

Axon
hillock

(e)

Figure 13.4 **A Representative Neuron.** The Schwann cells and myelin sheath are explained later in this chapter. (a) A multipolar neuron such as a spinal motor neuron. (b) Photograph of this neuron type. (c) Detail of myelin sheath. (d) Neurofibrils of the soma. (e) Nissl bodies, stained masses of rough ER separated by the bundles of neurofibrils shown in part (d).

● *What feature of this neuron is the basis for classifying it as multipolar? (Relate this figure to the discussion of fig. 13.5.)*

axon collaterals along the way, and most axons branch extensively at their distal end. An axon's cytoplasm is called the **axoplasm** and its membrane the **axolemma.**[11] A neuron never has more than one axon, and some neurons in the retina and brain have none.

Neurosomas range from 5 to 135 μm in diameter, whereas axons range from 1 to 20 μm in diameter and from a few millimeters to more than a meter long. Such dimensions are more impressive when we scale them up to the size of familiar objects. If the soma of a spinal motor neuron were the size of a tennis ball, its dendrites would form a huge bushy mass that could fill a 30-seat classroom from floor to ceiling. Its axon would be up to a mile long but a little narrower than a garden hose. This is quite a point to ponder. The neuron must assemble molecules and organelles in its "tennis ball" soma and deliver them through its "mile-long garden hose" to the end of the axon. In a process called *axonal transport,* neurons employ *motor proteins* that can carry organelles and macromolecules as they crawl along the cytoskeleton of the nerve fiber to distant destinations in the cell.

At the distal end, an axon usually has a **terminal arborization**[12]—an extensive complex of fine branches. Each branch ends in a **synaptic knob (terminal button).** The synaptic knob is a little swelling that forms a junction *(synapse*[13]*)* with a muscle cell, gland cell, or another neuron. Synapses are described in detail later in this chapter.

Apply What You Know

When proteins are needed in the synaptic knob, they must be made in the neurosoma, which is sometimes far away, and transported down the axon to the synaptic knob. Why do you think they cannot simply be made locally, in the knob itself?

Neuronal Variety

Not all neurons fit every detail of the preceding description. Neurons are classified structurally according to the number of processes extending from the soma (fig. 13.5).

- **Multipolar neurons** are those, like the preceding, that have one axon and two or more (usually many) dendrites. This is the most common type of neuron and includes most neurons of the brain and spinal cord.

- **Bipolar neurons** have one axon and one dendrite. Examples include olfactory cells of the nasal cavity, some neurons of the retina, and sensory neurons of the inner ear.

- **Unipolar neurons** have only a single process leading away from the soma. They are represented by the neurons that carry sensory signals to the spinal cord. These neurons are also called *pseudounipolar*[14] because they start out as bipolar neurons in the embryo, but their two processes fuse into one as the neuron matures. A short distance away from the soma, the process branches like a T, with a *peripheral fiber* carrying signals from

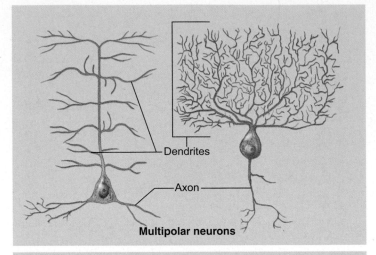

Multipolar neurons

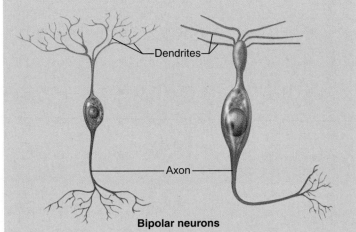

Bipolar neurons

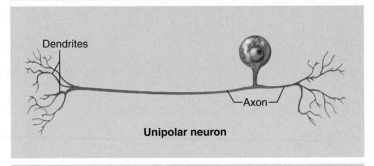

Unipolar neuron

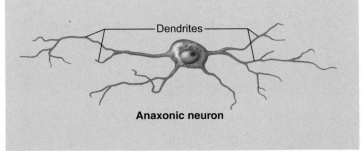

Anaxonic neuron

Figure 13.5 **Variation in Neuron Structure.** Top row, left to right: Two multipolar neurons of the brain—a pyramidal cell and a Purkinje cell. Second row, left to right: Two bipolar neurons—a bipolar cell of the retina and an olfactory neuron. Third row: A unipolar neuron of the type involved in the senses of touch and pain. Bottom row: An anaxonic neuron (amacrine cell) of the retina.

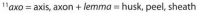

[11]*axo* = axis, axon + *lemma* = husk, peel, sheath
[12]*arbor* = treelike
[13]*syn* = together + *aps* = to touch, join
[14]*pseudo* = false

the source of sensation and a *central fiber* continuing into the spinal cord. In most other neurons, a dendrite carries signals toward the soma, and an axon carries them away. In unipolar neurons, however, there is one long fiber that bypasses the soma and carries nerve signals directly to the spinal cord. The dendrites are the branching receptive endings in the skin or other place of origin, while the rest of the fiber is considered to be the axon (defined in these neurons by the presence of myelin and the ability to generate action potentials).

- **Anaxonic neurons** have multiple dendrites but no axon. They communicate over short distances through their dendrites and do not produce action potentials. Anaxonic neurons are found in the brain, retina, and adrenal medulla. In the retina, they help in visual processes such as the perception of contrast.

Before You Go On

Answer the following questions to test your understanding of the preceding section:

4. Explain why neurons could not function without the properties of excitability, conductivity, and secretion.

5. Distinguish between sensory neurons, interneurons, and motor neurons.

6. Define each of the following and explain its importance to neuronal function: dendrites, soma, axon, and synaptic knob.

7. Sketch a multipolar, bipolar, unipolar, and anaxonic neuron and next to each sketch, state one place where such a neuron could be found.

13.3 Supportive Cells (Neuroglia)

▶ Expected Learning Outcomes

When you have completed this section, you should be able to

- name six types of cells that aid neuron function and state their respective locations and functions;

- describe the myelin sheath that is formed around certain nerve fibers;

- describe how the speed of nerve signal conduction varies with nerve fiber diameter and the presence or absence of myelin; and

- explain the relevance of neuroglia to the regeneration of damaged nerve fibers.

There are about a trillion (10^{12}) neurons in the nervous system—10 times as many neurons in one person as there are stars in our galaxy! Yet the neurons are outnumbered as much as 50 to 1 by supportive cells called **neuroglia** (noo-ROG-lee-uh), or **glial** (GLEE-ul) **cells.** Glial cells protect the neurons and aid their function. The word *glia,* which means "glue," implies one of their roles—they bind neurons

together. In the fetus, glial cells form a scaffold that guides young migrating neurons to their destinations. Wherever a mature neuron is not in synaptic contact with another cell, it is covered with glial cells. This prevents neurons from contacting each other except at points specialized for signal transmission, and thus lends precision to their conduction pathways.

Types of Neuroglia

There are six major categories of neuroglia, each with a unique function (table 13.1). Four types occur only in the central nervous system (fig. 13.6):

1. **Oligodendrocytes**[15] (OL-ih-go-DEN-dro-sites) somewhat resemble an octopus; they have a bulbous body with as many as 15 armlike processes. Each process reaches out to a nerve fiber and spirals around it like electrical tape wrapped repeatedly around a wire. This wrapping, called the *myelin sheath,* insulates the nerve fiber from the extracellular fluid and speeds up signal conduction in the nerve fiber.

TABLE 13.1	Types of Glial Cells	
Type	**Location**	**Functions**
Oligodendrocytes	CNS	Form myelin in brain and spinal cord
Ependymal cells	CNS	Line cavities of brain and spinal cord; secrete and circulate cerebrospinal fluid
Microglia	CNS	Phagocytize and destroy microorganisms, foreign matter, and dead nervous tissue
Astrocytes	CNS	Cover brain surface and nonsynaptic regions of neurons; form supportive framework for the CNS; stimulate formation of blood–brain barrier; nourish neurons and secrete growth stimulants; influence synaptic signaling between neurons; help to regulate composition of the extracellular fluid in the CNS; form scar tissue to replace damaged nervous tissue
Schwann cells	PNS	Form neurilemma around all PNS nerve fibers and myelin around most of them; aid in regeneration of damaged nerve fibers
Satellite cells	PNS	Surround somas of neurons in the ganglia; insulate them and regulate their chemical environment

[15]*oligo* = few + *dendro* = branches + *cyte* = cell

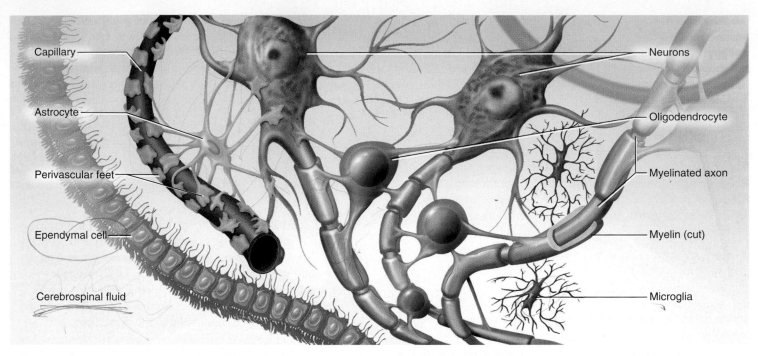

Figure 13.6 Neuroglia of the Central Nervous System.

2. **Ependymal**[16] (ep-EN-dih-mul) **cells** resemble a cuboidal epithelium lining the internal cavities of the brain and spinal cord. Unlike true epithelial cells, however, they have no basement membrane and they exhibit rootlike processes that penetrate into the underlying tissue. Ependymal cells produce a significant fraction of the *cerebrospinal fluid (CSF)*, a clear liquid that bathes the CNS and fills its internal cavities. Some ependymal cells have patches of cilia on their apical surfaces that help to circulate the CSF. Ependymal cells and CSF are considered in more detail in chapter 15 (p. 404).

3. **Microglia** are small macrophages that develop from stem cells related to the white blood cells called monocytes. They wander through the CNS, putting out fingerlike extensions to constantly probe the tissue for cellular debris or other problems. They are thought to perform a complete checkup on the brain tissue several times a day, phagocytizing dead tissue, microorganisms, and other foreign matter. They become concentrated in areas damaged by infection, trauma, or stroke. Pathologists look for clusters of microglia in histological sections of the brain as a clue to sites of injury.

4. **Astrocytes**[17] are the most abundant glial cells in the CNS and constitute over 90% of the tissue in some areas of the brain. They cover the entire brain surface and most nonsynaptic regions of the neurons in the gray matter of the CNS. They are named for their many-branched, somewhat starlike shape. They have the most diverse functions of any glia:

- They form a supportive framework for the nervous tissue.
- They have extensions called *perivascular feet*, which contact the blood capillaries and stimulate them to form a tight seal called

the *blood–brain barrier*. This barrier isolates the blood from the brain tissue and limits what substances are able to get to the brain cells, thus protecting the neurons (see p. 404).

- They convert blood glucose to lactate and supply this to the neurons for nourishment.
- They secrete proteins called *nerve growth factors* that promote neuron growth and synapse formation.
- They communicate electrically with neurons and may influence synaptic signaling between neurons.
- They regulate the chemical composition of the tissue fluid. When neurons transmit signals, they release neurotransmitters and potassium ions. Astrocytes absorb these substances and prevent them from reaching excessive levels in the tissue fluid.
- When neurons are damaged, astrocytes form hardened scar tissue and fill space formerly occupied by the neurons. This process is called *astrocytosis* or *sclerosis*.

The other two types of glial cells occur only in the peripheral nervous system:

5. **Schwann**[18] **cells** (pronounced "shwon"), or **neurilemmocytes,** envelop nerve fibers of the PNS, forming a sleeve around them called the *neurilemma* (see fig. 13.4). In most cases, a Schwann cell winds repeatedly around a nerve fiber and produces a myelin sheath between the neurilemma and nerve fiber. This is similar to the myelin sheath produced by oligodendrocytes in the CNS, but there are differences in the way myelin is produced, as described later. In addition to myelinating peripheral nerve fibers, Schwann cells assist in the regeneration of damaged fibers.

---

[16]*ependyma* = upper garment
[17]*astro* = star + *cyte* = cell

[18]Theodor Schwann (1810–82), German histologist

6. **Satellite cells** surround the neurosomas in ganglia of the PNS. They provide electrical insulation around the soma and regulate the chemical environment of the neurons.

Myelin

The **myelin** (MY-eh-lin) **sheath** is an insulating layer around a nerve fiber, somewhat like the rubber insulation on a wire. It is formed by oligodendrocytes in the central nervous system and Schwann cells

in the peripheral nervous system. Since it consists of the plasma membranes of these glial cells, its composition is like that of plasma membranes in general. It is about 20% protein and 80% lipid, the latter including phospholipids, glycolipids, and cholesterol. Myelin imparts a glistening white color to certain regions of nervous tissue, such as the *white matter* of the brain and spinal cord.

Production of the myelin sheath is called **myelination.** In the PNS, a Schwann cell spirals repeatedly around a single nerve fiber, laying down as many as a hundred compact layers of its own membrane with almost no cytoplasm between the membranes (fig. 13.7a).

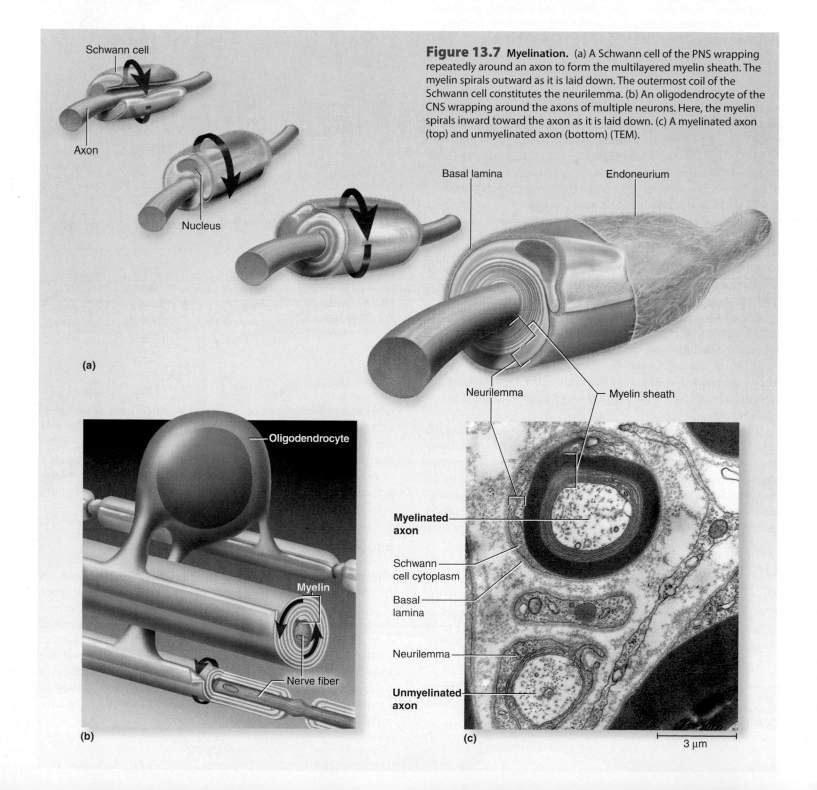

Figure 13.7 Myelination. (a) A Schwann cell of the PNS wrapping repeatedly around an axon to form the multilayered myelin sheath. The myelin spirals outward as it is laid down. The outermost coil of the Schwann cell constitutes the neurilemma. (b) An oligodendrocyte of the CNS wrapping around the axons of multiple neurons. Here, the myelin spirals inward toward the axon as it is laid down. (c) A myelinated axon (top) and unmyelinated axon (bottom) (TEM).

3 µm

DEEPER INSIGHT 13.1

Glial Cells and Brain Tumors

A tumor consists of a mass of rapidly dividing cells. Mature neurons, however, have little capacity for mitosis and seldom form tumors. Some brain tumors arise from the meninges (protective membranes of the CNS) or arise by metastasis from tumors elsewhere, such as malignant melanoma and colon cancer. Most adult brain tumors, however, are composed of glial cells, which are mitotically active throughout life. Such tumors, called *gliomas*,[19] grow rapidly and are highly malignant. Because of the blood–brain barrier, brain tumors usually do not yield to chemotherapy and must be treated with radiation or surgery.

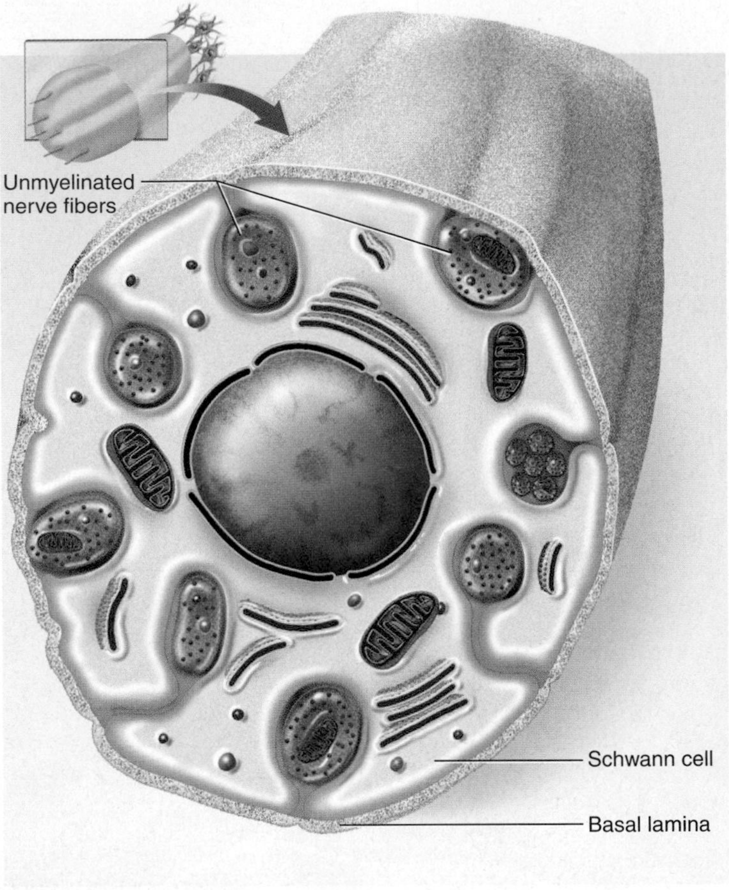

Figure 13.8 Unmyelinated Nerve Fibers. Multiple unmyelinated fibers are enclosed in channels in the surface of a single Schwann cell.
●*What is the functional disadvantage of an unmyelinated nerve fiber? What is its anatomical advantage?*

These layers constitute the myelin sheath. The Schwann cell spirals outward as it wraps the nerve fiber, finally ending with a thick outermost coil called the **neurilemma**[20] (noor-ih-LEM-ah). Here, the bulging body of the Schwann cell contains its nucleus and most of its cytoplasm. External to the neurilemma is a basal lamina and then a thin sleeve of fibrous connective tissue called the *endoneurium*.

To visualize this myelination process, imagine that you wrap an almost-empty tube of toothpaste tightly around a pencil. The pencil represents the axon, and the spiral layers of toothpaste tube (with the toothpaste squeezed out) represent the myelin. The toothpaste would be forced to one end of the tube, which would form a bulge on the external surface of the wrapping, like the body of the Schwann cell.

In the CNS, each oligodendrocyte reaches out to myelinate several nerve fibers in its immediate vicinity (fig. 13.7b). Since it is anchored to multiple nerve fibers, it cannot migrate around any one of them like a Schwann cell does. It must push newer layers of myelin under the older ones, so myelination spirals inward toward the nerve fiber. Nerve fibers of the CNS have no neurilemma or endoneurium.

In both the PNS and CNS, a nerve fiber is much longer than the reach of a single glial cell, so it requires many Schwann cells or oligodendrocytes to cover one fiber. Consequently, the myelin sheath is segmented. The gaps between the segments are called **nodes of Ranvier**[21] (RON-vee-AY), and the myelin-covered segments from one gap to the next are called **internodes.** The internodes are about 0.2 to 1.0 mm long.

Unmyelinated Nerve Fibers

Many nerve fibers in the CNS and PNS are unmyelinated. In the PNS, however, even the unmyelinated fibers are enveloped in Schwann cells. In this case, one Schwann cell harbors from 1 to 12 small nerve fibers in grooves in its surface (fig. 13.8). The Schwann cell's plasma membrane does not spiral repeatedly around the fiber as it does in a myelin sheath, but folds once around each fiber and somewhat overlaps itself along the edges. This wrapping is the neurilemma. Most nerve fibers travel through their own channels in the Schwann cell, but small fibers are sometimes bundled together in a single channel. A basal lamina surrounds the entire Schwann cell along with its nerve fibers.

Myelin and Signal Conduction

The speed at which a signal travels along a nerve fiber depends on two factors: the diameter of the fiber and the presence or absence of myelin. Signal conduction occurs along the surface of a fiber, not deep within its axoplasm. Large fibers have more surface area and conduct signals more rapidly than small fibers. Myelin further speeds signal conduction for physiological reasons beyond the scope of this book. Nerve signals travel about 0.5 to 2.0 m/sec. in small unmyelinated fibers (2–4 µm in diameter); 3 to 15 m/sec. in myelinated fibers of the same size; and as fast as 120 m/sec. in large myelinated fibers (up to 20 µm in diameter).

[19]*glia* = glial cells + *oma* = tumor
[20]*neuri* = nerve + *lemma* = husk, peel, sheath
[21]L. A. Ranvier (1835–1922), French histologist and pathologist

DEEPER INSIGHT 13.2

Diseases of the Myelin Sheath

Multiple sclerosis and Tay–Sachs disease are degenerative disorders of the myelin sheath. In *multiple sclerosis*[22] *(MS)*, the oligodendrocytes and myelin sheaths of the CNS deteriorate and are replaced by hardened scar tissue, especially between the ages of 20 and 40. Nerve conduction is disrupted with effects that depend on what part of the CNS is involved—numbness, double vision, blindness, speech defects, neurosis, or tremors. Patients experience variable cycles of milder and worse symptoms until they eventually become bedridden. The cause of MS remains uncertain; most hypotheses suggest that it is an *autoimmune disease*—a disorder in which one's immune system turns against one's own tissues—perhaps triggered by a virus in genetically susceptible individuals. There is no cure. There is conflicting evidence as to how much it shortens a patient's life expectancy, if at all. A few die within a year of diagnosis, but many people live with MS for 25 or 30 years.

Tay–Sachs[23] *disease* is a hereditary disorder seen mainly in infants of Eastern European Jewish ancestry. It results from the abnormal accumulation of a glycolipid called GM_2 (ganglioside) in the myelin sheath. GM_2 is normally decomposed by a lysosomal enzyme, but this enzyme is lacking from those who inherit the recessive Tay–Sachs gene from both parents. As GM_2 accumulates, it disrupts the conduction of nerve signals, and the victim typically suffers blindness, loss of coordination, and dementia. Signs begin to appear before the child is a year old and most victims die by the age of 3 or 4 years. Asymptomatic adult carriers can be identified by a blood test and advised by genetic counselors on the risk of their children having the disease.

One might wonder why all of our nerve fibers are not large, myelinated, and fast, but if this were so, our nervous system would be impossibly bulky or limited to far fewer fibers. Slow unmyelinated fibers are quite sufficient for processes in which quick responses are not particularly important, such as secreting stomach acid or dilating the pupil. Fast myelinated fibers are employed where speed is more important, as in motor commands to the skeletal muscles and sensory signals for vision and balance.

Schwann Cells and Nerve Regeneration

Nerve fibers in the peripheral nervous system are often damaged by cuts and other injuries, but if the cell body remains intact, an axon can often regenerate. Two things are required for regeneration of an axon: a neurilemma and an endoneurium. The Schwann cells of the neurilemma secrete *nerve growth factors* that stimulate regrowth of the axon, and the Schwann cells and endoneurium together form a regeneration tube that guides the growing axon to its destination, such as a muscle fiber. The axon grows down the middle of the tube and, if successful, it reestablishes synaptic contact with its target cell. The process is not perfect—some injured neurons fail to find the target cell, and some simply die. Nerve injuries therefore may leave a person with some loss of fine motor control. There are no Schwann cells or endoneurium in the central nervous system, and damaged CNS neurons cannot regenerate at all. However, since the CNS is enclosed in bone, it suffers less trauma than the PNS.

Before You Go On

Answer the following questions to test your understanding of the preceding section:

8. From memory, make your own table of the six kinds of glial cells and the functions of each. Which type has the most varied functions?

9. Summarize the major ways in which oligodendrocytes and Schwann cells differ in the way they produce a myelin sheath, and state where the glial cell body of each type is located relative to the myelin and nerve fiber.

10. Compare the signal conduction speed in myelinated fibers versus unmyelinated ones. Why aren't *all* nerve fibers in the body myelinated?

11. Explain why damaged nerve fibers in the PNS can regenerate but damaged fibers in the CNS cannot.

13.4 Synapses and Neural Circuits

▶ Expected Learning Outcomes

When you have completed this section, you should be able to

- describe the synaptic junctions between one neuron and another;

- describe the variety of interconnections that exist between two neurons; and

- describe four basic variations in the circuitry or "wiring patterns" of the nervous system.

No neuron functions in isolation from others; neurons work in groups of cells that are connected in patterns similar to the electrical circuits of radios and other electronic devices. In this section, we examine the connections between neurons and the functional circuits of neuron groups.

Synapses

The meeting point between a neuron and any other cell is called a **synapse.** The other cell may be an epithelial, muscular, glandular, or other cell type, but in most cases it is another neuron. Synapses

---

[22]*scler* = hard, tough + *osis* = condition
[23]Warren Tay (1843–1927), English physician; Bernard Sachs (1858–1944), American neurologist

make *neural integration* (information processing) possible; each is a "decision-making" device that determines whether a second cell will respond to signals from the first. Without synapses, signals would simply be transmitted automatically from receptors to effectors, effectors would respond to every stimulus, and the nervous system would be incapable of any decision making. But in reality, one neuron can have an enormous number of synapses and thus a great deal of information-processing capability (fig. 13.9). For example, a spinal motor neuron receives about 8,000 synaptic contacts from other neurons on its dendrites and another 2,000 on its soma. In part of the brain called the cerebellum, one neuron can have as many as 100,000 synapses. The cerebral cortex (the main information-processing tissue of the brain) is estimated to have 100 trillion (10^{14}) synapses. To get some impression of this number, imagine trying to count them. Even if you could count two synapses per second, night and day without a coffee break, and you were immortal, it would take you 1.6 million years.

A nerve signal arrives at a synapse by way of the **presynaptic neuron,** then may continue on its way via the **postsynaptic neuron** (fig. 13.10a). When a presynaptic axon ends at the dendrite of a postsynaptic neuron, the two cells are said to form an **axodendritic synapse.** When the presynaptic axon terminates on the soma of the next cell, they form an **axosomatic synapse.** When it terminates on the axon of the next cell, they form an **axoaxonic synapse** (fig. 3.10b).

Chemical Synapses and Neurotransmitters

A **chemical synapse** is a junction at which the presynaptic neuron releases a neurotransmitter to stimulate the postsynaptic cell. The neuromuscular junction (NMJ) described in chapter 10 (p. 248) is an example of this. The NMJ and many other synapses employ *acetylcholine* as a neurotransmitter. Postsynaptic neurons of the sympathetic nervous system use *norepinephrine.*

Some neurotransmitters are excitatory and tend to generate a nerve signal in the postsynaptic cell. Some widely used excitatory neurotransmitters in the central nervous system (CNS) are *glutamate* in the brain and *aspartate* in the spinal cord. Other neurotransmitters are inhibitory and suppress responses in the postsynaptic cell. The most widely used inhibitory neurotransmitters in the CNS are *gamma-aminobutyric acid (GABA)* in the brain and *glycine* in the spinal cord. Some other well-known neurotransmitters are *dopamine, serotonin, histamine,* and *beta-endorphin.* There are over 100 known neurotransmitters.

At a chemical synapse, a terminal branch of the presynaptic nerve fiber ends in a swelling, the **synaptic knob.** The knob is separated from the next cell by a 20 to 40 nm gap called the **synaptic cleft** (fig. 13.11). The knob contains membrane-bounded secretory vesicles called **synaptic vesicles,** which contain the neurotransmitter. Many of these vesicles are "docked" at release sites on the inside of the plasma membrane, ready to release their neurotransmitter on demand. A reserve pool of synaptic vesicles is located a little farther away from the membrane, clustered near the release sites and tethered to the cytoskeleton by protein microfilaments. These vesicles stand by and "step forward" to dock on the membrane and release their neurotransmitter after the previously docked vesicles have expended their contents. Synaptic vesicles are found in a few cells other than neurons, such as the sensory cells of taste, hearing, and equilibrium. They release neurotransmitter to stimulate a nearby nerve cell.

A postsynaptic neuron does not show such conspicuous specializations. At this end, a neuron has no synaptic vesicles and cannot release neurotransmitters. Its membrane does, however, contain

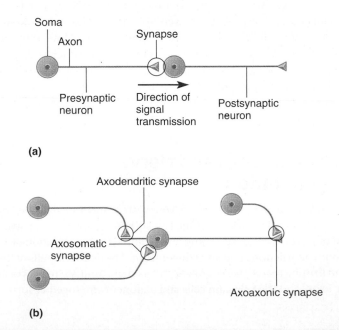

(a)

(b)

Figure 13.10 Synaptic Relationships Between Neurons. (a) Pre- and postsynaptic neurons. (b) Types of synapses defined by the site of contact on the postsynaptic neuron.

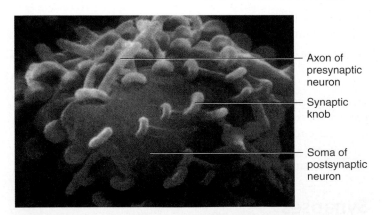

Figure 13.9 Synaptic Knobs on the Soma of a Neuron in a Marine Slug, Aplysia (SEM).
• *To which of the three classes of synapses (compare fig. 13.10) would these belong?*

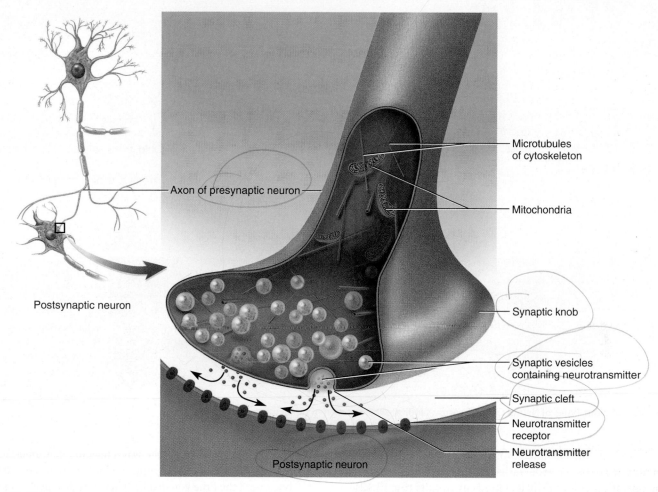

Figure 13.11 Structure of a Chemical Synapse.

proteins that function as neurotransmitter receptors, and it may be folded to increase its receptor-laden surface area and, therefore, its sensitivity to the neurotransmitter. A signal always travels in only one direction across a chemical synapse, from the presynaptic cell with synaptic vesicles to the postsynaptic cell with neurotransmitter receptors. This one-way transmission ensures the precise routing of nerve signals in the body.

Synaptic transmission begins when a nerve signal arrives at the end of the presynaptic neuron. Synaptic vesicles release neurotransmitter into the synaptic cleft, where it diffuses across to the postsynaptic cell and binds to receptors on that cell's membrane. Depending on the neurotransmitter and the type of receptor, this may either stimulate or inhibit the postsynaptic cell. The postsynaptic cell "decides" whether or not to initiate a new nerve signal based on the composite effects of excitatory and inhibitory input through the many synapses on its dendrites and soma. Therefore, a large number of presynaptic neurons may "get to vote" on whether the postsynaptic cell fires.

Apply What You Know

Of all the methods of membrane transport described in chapter 2, which one is the mechanism of neurotransmitter release?

Electrical Synapses

Another type of synapse, called an **electrical synapse,** connects some neurons, neuroglia, and cardiac and single-unit smooth muscle cells. Here, adjacent cells are joined by gap junctions that allow ions to diffuse directly from one cell into the next. These junctions have the advantage of quick transmission because there is no delay for the release and binding of neurotransmitter. Two disadvantages, however, are that they cannot integrate information and make decisions, and that signals can travel in either direction over an electrical synapse, so there is no way of directing signals from a specific origin to a specific destination.

Neural Pools and Circuits

Neurons function in ensembles called **neural pools.** One neural pool may consist of thousands to millions of interneurons concerned with a particular body function—one to control the rhythm of your breathing, one to move your limbs rhythmically as you walk, one to regulate your sense of hunger, and another to interpret smells, for example. The functioning of a neural pool hinges on the anatomical organization of its neurons, much like the functioning of a radio depends on the particular way its transistors, diodes, and

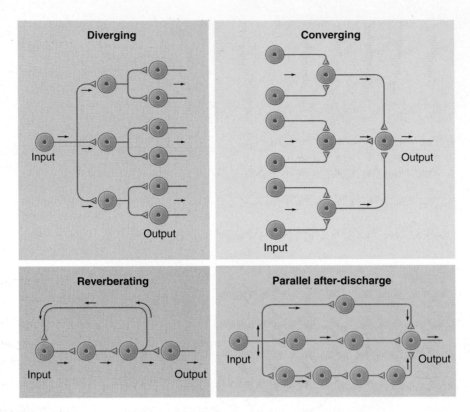

Figure 13.12 **Four Types of Neural Circuits.** Arrows indicate the direction of signal transmission.

capacitors are laid out. The interconnections between neurons are called **neural circuits.** A wide variety of neural functions result from the operation of four principal kinds of circuits (fig. 13.12):

1. In a **diverging circuit,** one nerve fiber branches and synapses with several postsynaptic cells. Each of those may synapse with several more, so input from just one neuron may produce output through dozens more. Such a circuit allows one motor neuron of the brain, for example, to ultimately stimulate thousands of muscle fibers.

2. A **converging circuit** is the opposite of a diverging circuit— input from many different sources is funneled to one neuron or neural pool. Through neural convergence, a respiratory center in the brainstem receives input from other parts of the brain, from receptors for blood chemistry in the arteries, and from stretch receptors in the lungs. The respiratory center can then produce an output that takes all of these factors into account and sets an appropriate pattern of breathing.

3. In a **reverberating circuit,** neurons stimulate each other in a linear sequence such as $A \longrightarrow B \longrightarrow C \longrightarrow D$, but neuron C sends an axon collateral back to A. As a result, every time C fires, it not only stimulates output from neuron D, but also restimulates A and starts the process over. Such a circuit produces a repetitive output that lasts until one or more neurons in the circuit fail to fire or an inhibitory signal from another source stops one of them from firing. A reverberating circuit sends repetitious signals to the diaphragm and intercostal muscles, for example, to make you inhale. When the circuit stops firing, you exhale; the next time it fires, you inhale again. Reverberating circuits may also be involved in short-term

memory (for example, in the way a telephone number "echoes" in your memory from the time you look it up in the phone book until the time you dial it). They may also play a role in the uncontrolled "storms" of neural activity that occur in epilepsy.

4. In a **parallel after-discharge circuit,** an input neuron diverges to stimulate several chains of neurons. Each chain has a different number of synapses, but eventually they all reconverge on the same output neuron. Each synapse delays a nerve signal by about 0.5 msec, so the more synapses there are in a pathway, the longer it takes a nerve signal to get through that pathway to the output neuron. The output neuron, receiving signals from multiple pathways, may go on firing for some time after the input has ceased. Unlike a reverberating circuit, this type has no feedback loop. Once all the neurons in the circuit have fired, the output ceases. Continued firing after the stimulus stops is called *after-discharge.* It explains why you can stare at a lamp, then close your eyes or turn the lamp off and continue to see an image of it for a while. Such a circuit is also important to certain reflexes—for example, when a brief pain produces a longer-lasting output to the limb muscles and causes you to draw back your hand or foot from danger. (See the discussion of *reflex arcs* in chapter 14, p. 392.)

Before You Go On

Answer the following questions to test your understanding of the preceding section:

12. At a given synapse, what features are present on the presynaptic neuron that are absent from the postsynaptic neuron?

13. In synaptic transmission, where does the neurotransmitter come from? How does it affect the postsynaptic neuron?

14. Name any four neurotransmitters and state some functional differences between them.

15. What is an electrical synapse? Where can electrical synapses be found? Identify an advantage and a disadvantage of an electrical synapse compared to a chemical synapse.

16. What is the difference between a neural pool and a neural circuit?

17. Name the four types of neural circuits and briefly describe the functional differences between them, or an advantage of each type for certain purposes.

13.5 Developmental and Clinical Perspectives

▶ Expected Learning Outcomes

When you have completed this section, you should be able to

- describe how the nervous system develops in an embryo; and
- describe a few birth defects that result from abnormalities of this developmental process.

Development of the Nervous System

Some aspects of nervous system development, or **neurulation,** were briefly discussed in chapter 4 (p. 90). Further understanding of this process will form a basis for understanding the brain and spinal cord anatomy presented in chapters 14 and 15.

The first embryonic trace of the central nervous system appears early in the third week of development. A dorsal streak called the *neuroectoderm* appears along the length of the embryo and thickens to form a **neural plate** (fig. 13.13). This is destined to give rise to most neurons and to all glial cells except microglia, which come from mesoderm. As development progresses, the neural plate sinks and the edges of it thicken, thus forming a **neural groove** with a raised **neural fold** along each side. The neural folds then fuse along the midline, somewhat like a closing zipper, beginning in the cervical (neck) region of the neural groove and progressing rostrally (toward the head) and caudally (toward the tail). By 4 weeks, this process creates a hollow channel called the **neural tube.** For a time, the neural tube is open to the amniotic fluid at the rostral and caudal ends. These openings close at 25 and 27 days, respectively. The lumen of the neural tube becomes a fluid-filled space that later constitutes the *central canal* of the spinal cord and *ventricles* of the brain.

Following closure, the neural tube separates from the overlying ectoderm, sinks a little deeper, and grows lateral processes that later give rise to motor nerve fibers. Some ectodermal cells that originally lay along the margin of the neural groove separate from the rest and form a longitudinal column on each side called the **neural**

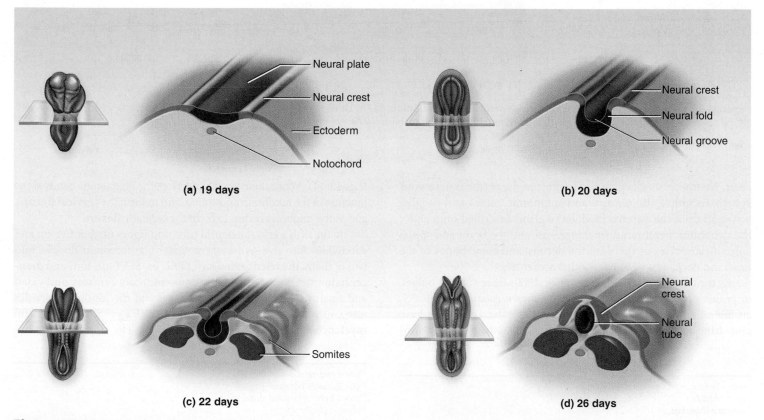

Figure 13.13 Formation of the Neural Tube. The left-hand figure in each case is a dorsal view of the embryo, and the right-hand figure is a three-dimensional representation of the tissues at the indicated level of the respective embryo. The ages are days after fertilization.

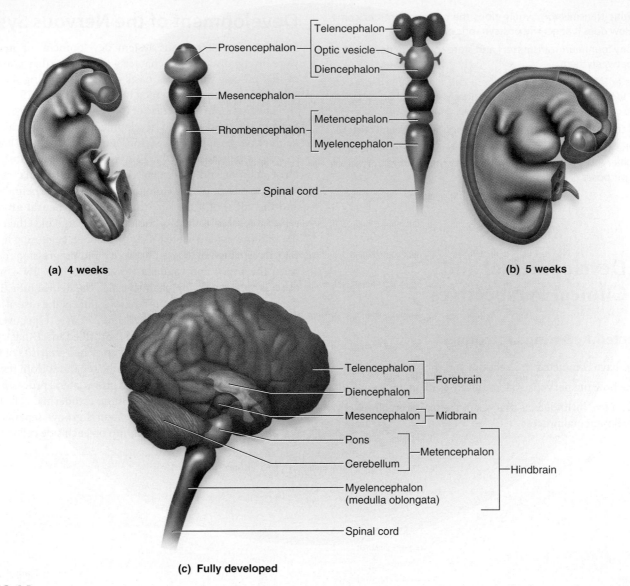

(a) 4 weeks

(b) 5 weeks

(c) Fully developed

Figure 13.14 **Primary and Secondary Vesicles of the Embryonic Brain.** (a) The three primary vesicles at 4 weeks. (b) The secondary vesicles at 5 weeks. (c) The fully developed brain, color-coded to relate its structures to the secondary embryonic vesicles.

crest. Neural crest cells give rise to most of the peripheral nervous system, including the sensory and autonomic nerves and ganglia; Schwann cells; the adrenal medulla (a gland described on p. 508); the two inner membranes *(meninges)* around the brain and spinal cord; melanocytes of the skin; the dermis; and some bones of the head and neck, as well as a few other structures.

By the fourth week, the neural tube exhibits three anterior bulges, or *primary vesicles,* called the **forebrain** *(prosencephalon[24])* (PROSS-en-SEF-uh-lon), **midbrain** *(mesencephalon[25])* (MEZ-en-SEF-uh-lon), and **hindbrain** *(rhombencephalon[26])* (ROM-ben-SEF-uh-lon)

(fig. 13.14). While these vesicles develop, the neural tube bends at the junction of the hindbrain and spinal cord to form the **cervical flexure,** and in the midbrain region to form the **cephalic flexure.**

By the fifth week, the neural tube undergoes further flexion and subdivides into five *secondary vesicles.* The forebrain divides into two of them, the **telencephalon[27]** (TEL-en-SEFF-uh-lon) and **diencephalon[28]** (DY-en-SEF-uh-lon); the midbrain remains undivided and retains the name **mesencephalon;** and the hindbrain divides into two vesicles, the **metencephalon[29]** (MET-en-SEF-uh-lon) and **myelencephalon[30]** (MY-el-en-SEF-uh-lon). The telencephalon has

[24]*pros* = before, in front + *encephal* = brain
[25]*mes* = middle
[26]*rhomb* = rhombus

[27]*tele* = end, remote
[28]*di* = through, between
[29]*met* = behind, beyond, distal to
[30]*myel* = spinal cord

a pair of lateral outgrowths that later become the *cerebral hemispheres,* and the diencephalon exhibits a pair of small cuplike *optic vesicles* that become the retinas of the eyes. Figure 13.14c shows structures of the fully developed brain that arise from each of the secondary vesicles.

In week 14, Schwann cells and oligodendrocytes begin spiraling around the nerve fibers, laying down layers of myelin and giving the fibers a white appearance. Yet very little myelin is present in the brain even at birth, and there is little visible distinction between the *gray matter* and *white matter* of the newborn brain. Myelination proceeds rapidly in infancy, and it is this, far more than the multiplication or enlargement of neurons, that accounts for most postnatal brain growth. Myelination is not completed until late adolescence. Since myelin has such a high lipid content, dietary fat is important to early nervous system development. Well-meaning parents can do their children significant harm by giving them the sort of low-fat diets (skimmed milk, etc.) that may be beneficial to an adult.

In the third month, the spinal cord extends for the full length of the embryo. As the vertebrae develop (see p. 176), *spinal nerves* arise from the cord and pass straight laterally to emerge between the vertebrae, through the intervertebral foramina. Subsequently, however, the vertebral column grows faster than the spinal cord. By birth, the cord ends in the vertebral canal of the third lumbar vertebra (L3), and by adulthood, it ends at the level of L1 to L2. As the vertebral column elongates, the spinal nerve roots elongate, so they still emerge between the same vertebrae, but the lower vertebral canal is occupied by a bundle of nerve roots instead of spinal cord. The resulting adult anatomy is described in chapter 14.

Developmental Disorders of the Nervous System

The central nervous system is subject to multiple aberrations in embryonic development. Approximately 1 out of 100 live-born infants exhibit major defects in brain development. Common among these are **neural tube defects (NTDs)** such as **spina bifida** (SPY-nuh BIF-ih-duh). Spina bifida occurs when one or more vertebrae fail to form a complete neural arch for enclosure of the spinal cord. It is especially common in the lumbosacral region. The mildest form, *spina bifida occulta,*[31] involves only one to a few vertebrae and causes no functional problems. Its only external sign is a dimple or patch of hairy pigmented skin on the lower back. *Spina bifida cystica*[32] is more serious (fig. 13.15). A sac protrudes from the spine and may contain parts of the spinal cord and nerve roots, meninges, and cerebrospinal fluid. In extreme cases, inferior spinal cord function is absent, causing paralysis of the lower limbs and urinary bladder and lack of bowel control. Bladder paralysis can lead to chronic urinary infections and renal failure. Women can significantly reduce the risk of bearing a child with spina bifida by taking supplemental folic acid (a B vitamin). However, this works only if taken habitually even before the egg is fertilized; folic acid supplements are ineffective if begun after a woman realizes she is pregnant because by then, any neural tube damage will have already occurred.

---

[31]*bifid* = divided, forked + *occult* = hidden
[32]*cyst* = sac, bladder

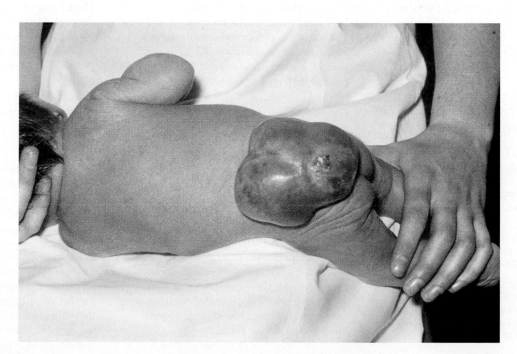

Figure 13.15 Spina Bifida Cystica.   The sac in the lumbar region is called a myelomeningocele.

Other severe neural tube defects include microcephaly and anencephaly. In **microcephaly,**[33] the face is of normal size but the brain and calvaria are abnormally small. Microcephaly is accompanied by profound mental retardation. **Anencephaly**[34] results from failure of the rostral end of the neural tube to close. This leaves the brain exposed to the amniotic fluid. The brain tissue degenerates, most of the brain is absent at birth, and the head is relatively flat or truncated above the eyes. Such infants generally die within a few hours. Neural tube defects sometimes run in families but can also be caused by teratogens and nutritional deficiencies.

---

[33]*micro* = small + *cephal* = head
[34]*an* = without + *encephal* = brain

Other disorders of the nervous system are described in chapters 14 through 17.

Before You Go On

Answer the following questions to test your understanding of the preceding section:

18. How does the neural crest originate? What cells or tissues arise from it?

19. Where does closure of the neural tube begin? What are the last regions to close?

20. What single adult structure arises from all five of the secondary vesicles of the neural tube?

21. When does myelination begin? When does it end?

Study Guide

Assess Your Learning Outcomes

You should have a good understanding of this chapter if you can accurately address the following issues.

13.1 Overview of the Nervous System (p. 352)
1. The body's two principal mechanisms of internal communication and coordination, and how they differ from each other
2. The constituents of the central nervous system (CNS) and peripheral nervous system (PNS)
3. The two divisions of the PNS and the two subdivisions of each of those, including the types of organs innervated by each subdivision
4. The autonomic nervous system, its effectors, and its two subdivisions

13.2 Nerve Cells (Neurons) (p. 353)
1. The three properties that any neuron must have to perform its function
2. The three functional classes of neurons: sensory neurons, interneurons, and motor neurons; how each is defined; and synonyms for sensory and motor neurons
3. The parts of a typical neuron
4. The four types of neurons distinguished by the number of processes that arise from the neurosoma

13.3 Supportive Cells (Neuroglia) (p. 357)
1. The meaning of *neuroglia (glial cells)* and the general purpose of the neuroglia
2. The four kinds of neuroglia found in the CNS, and their respective functions
3. The two kinds of neuroglia found in the PNS, and their respective functions
4. The structure, composition, and function of the myelin sheath; how the CNS and PNS differ with respect to the glial cells that produce their myelin; and how the two glial types do so
5. The relationship of Schwann cell to the myelin, neurilemma, basal lamina, and endoneurium
6. Why the myelin sheath consists of short internodes interrupted by nodes of Ranvier, rather than one continuous sheath
7. The roles of Schwann cells in relation to unmyelinated nerve fibers
8. How the velocity of a nerve signal varies with a nerve fiber's diameter and presence or absence of myelin
9. The necessity of the neurilemma and endoneurium for the regeneration of damaged nerve fibers; the role that they play in regeneration; and why damaged nerve fibers in the CNS cannot regenerate

13.4 Synapses and Neural Circuits (p. 361)
1. The definition of *synapse* and the function of synapses
2. How the presynaptic and postsynaptic neurons are defined at a given synapse
3. Three types of synapses defined by where the presynaptic nerve fiber ends on the postsynaptic neuron
4. The defining qualities of a chemical synapse
5. The names of several familiar neurotransmitters
6. The two types of effects that a neurotransmitter can have on a postsynaptic neuron
7. The mechanism of signal transmission across a synapse by a neurotransmitter
8. How the plasma membrane of the postsynaptic neuron differs from that of the presynaptic neuron at a synapse
9. Where electrical synapses can be found, and how they differ structurally and functionally from chemical synapses
10. The meaning of *neural pool* and some functions performed by neural pools
11. The four principal types of neural circuits—diverging, converging, reverberating, and parallel after-discharge circuits—and how they differ in function; some everyday scenarios that involve the activity of these different types of neural circuits

13.5 Developmental and Clinical Perspectives (p. 365)

1. The embryonic progression from a neural plate to a neural tube
2. The origin, location, and fate of the neural crest
3. Expansion of the neural tube into three primary vesicles, and their names; their further differentiation into five secondary vesicles, and their names; and the fate of each of the five secondary vesicles
4. The timetable of prenatal and postnatal myelination
5. The developmental explanation for why the adult spinal cord reaches only as far as the first or second lumbar vertebra
6. The meaning of *neural tube defects;* their causes; the defining attributes of spina bifida occulta, spina bifida cystica, microcephaly, and anencephaly; and what can be done to minimize the risk of neural tube defects

Testing Your Recall

1. The integrative functions of the nervous system are performed mainly by
   a. afferent neurons.
   b. efferent neurons.
   c. neuroglia.
   d. sensory neurons.
   e. interneurons.

2. Neurons arise from embryonic
   a. endoderm.
   b. epidermis.
   c. mesoderm.
   d. mesenchyme.
   e. ectoderm.

3. The soma of a mature neuron lacks
   a. a nucleus.
   b. endoplasmic reticulum.
   c. lipofuscin.
   d. centrioles.
   e. ribosomes.

4. The glial cells that destroy microorganisms in the CNS are
   a. microglia.
   b. satellite cells.
   c. ependymal cells.
   d. oligodendrocytes.
   e. astrocytes.

5. A friend takes a flash photograph of you, and you continue to see an image of the flash unit for several seconds afterward. This phenomenon is the result of a _____ circuit.
   a. diverging
   b. converging
   c. presynaptic
   d. reverberating
   e. parallel after-discharge

6. Neurotransmitters are found in
   a. the cell bodies of neurons.
   b. the dendrites.
   c. the axon hillock.
   d. the synaptic knob.
   e. the postsynaptic plasma membrane.

7. Another name for the axon of a neuron is
   a. nerve fiber.
   b. neurofibril.
   c. neurilemma.
   d. axoplasm.
   e. endoneurium.

8. Nerves that directly control the motility of the stomach or rate of the heartbeat would belong to
   a. the central nervous system.
   b. the somatic sensory division.
   c. the somatic motor division.
   d. the visceral motor division.
   e. the visceral sensory division.

9. The glial cells that guide migrating neurons in the developing fetal brain are
   a. astrocytes.
   b. oligodendrocytes.
   c. satellite cells.
   d. ependymal cells.
   e. microglia.

10. Which of the following appears earlier than all the rest in prenatal development of the nervous system?
    a. the neural groove
    b. the primary vesicles
    c. the neural plate
    d. the neural crest
    e. the neural tube

11. Neurons that convey information to the CNS are called sensory, or _____, neurons.

12. Motor effects that depend on repetitive output from a neural pool are most likely to use the _____ type of neural circuit.

13. Prenatal degeneration of the forebrain results in a birth defect called _____.

14. Neurons receive incoming signals by way of specialized processes called _____.

15. In the central nervous system, cells called _____ perform one of the same functions that Schwann cells do in the peripheral nervous system.

16. A/An _____ synapse is formed when a presynaptic neuron synapses with the cell body of a postsynaptic neuron.

17. All of the nervous system except the brain and spinal cord is called the _____.

18. The _____ and _____ are necessary for regeneration of damaged nerve fibers in the peripheral nervous system.

19. In the peripheral nervous system, the somas of the neurons are concentrated in enlarged, knotlike structures called _____ connected to the nerves.

20. At a given synapse, the _____ neuron has neurotransmitter receptors.

Answers in the Appendix

Building Your Medical Vocabulary

State a medical meaning of each of the following word elements, and give a term in which it is used.

1. -ic
2. somato-
3. neuro-
4. lipo-
5. dendro-
6. -ite
7. pseudo-
8. oligo-
9. fer-
10. sclero-

Answers in the Appendix

True or False

Determine which five of the following statements are false, and briefly explain why.

1. Adult neurons are incapable of mitosis.
2. Most neurons have more dendrites than axons.
3. Dendrites never contain synaptic vesicles.
4. Interneurons connect sense organs to the CNS.
5. Nerve signals travel faster in myelinated nerve fibers than in unmyelinated ones.
6. The myelin sheath covers the neurilemma of a nerve fiber.
7. Nodes of Ranvier are present only in myelinated fibers of the PNS.
8. Interneurons occur in the brain and spinal cord of the CNS and in ganglia of the PNS.
9. Unipolar neurons cannot produce action potentials because they have no axon.
10. There are more glial cells than neurons in the nervous system.

Answers in the Appendix

Testing Your Comprehension

1. Suppose some hypothetical disease prevented the formation of astrocytes in the fetal brain. How would you expect this to affect brain development?
2. How would nervous system function be affected if both the presynaptic and postsynaptic neurons at every synapse had both synaptic vesicles and neurotransmitter receptors?
3. What unusual characteristic of neurons can be attributed to their lack of centrioles?
4. When you cut your finger, the pain signals are conducted to the CNS by unipolar sensory neurons whose somas are near the spinal cord, amid the vertebrae. Suggest a good reason for the somas to be there, instead of in the skin closer to the origin of the pain.
5. State what division or subdivision of the peripheral nervous system would control each of the following: constriction of the pupils in bright light; the movements of your hand as you write; the sensation of a stomachache; blinking as a particle of dust is blown toward your eye; your awareness of the position of your hand as you touch your nose with your eyes closed. Briefly explain each answer.

Answers at www.mhhe.com/saladinha3

Improve Your Grade at www.mhhe.com/saladinha3

Practice quizzes, labeling activities, games, and flashcards provide fun ways to master concepts. You can also download image PowerPoint files for each chapter to create a study guide or for taking notes during lecture.

The Spinal Cord and Spinal Nerves

Cross section of a nerve showing parts of two fascicles (bundles) of nerve fibers (maroon)

BRUSHING UP

To understand this chapter, you may find it helpful to review the following concepts:
- Divisions of the nervous system (p. 352)
- Functional classes of neurons (p. 354)
- Embryonic development of the CNS (p. 365)

Nervous System

Every year in the United States, thousands of people become paralyzed by spinal cord injuries from automobile and motorcycle accidents, contact sports, and falls. Although the spinal cord is protected by the vertebral column, damage may occur as a result of vertebral fractures or viral infections. Because the spinal cord serves as a pathway for messages between the brain and the rest of the body, spinal cord injuries often have a devastating effect on one's quality of life. Their treatment is one of the most lively areas of medical research today.

As physical therapists are well aware, the consequences of spinal cord injury can vary greatly, but include paralysis of the lower limbs (*paraplegia*) or all four limbs (*quadriplegia*), respiratory paralysis, loss of sensation or motor control in more limited regions of the body, and disorders of bladder and bowel control and sexual function. Therapists who treat spinal patients must know spinal cord anatomy and function to understand their patients' functional deficits and prospects for improvement and to plan an appropriate regimen of treatment. Such anatomical knowledge is necessary, as well, for understanding *hemiplegia*, the paralysis of one or both limbs on either the left or right side of the body, even though this usually results from strokes or other brain injuries rather than spinal cord injuries. The spinal cord is the "information highway" that connects the brain with the lower body; it contains the neural routes that explain why a lesion to a specific part of the brain results in a functional loss in a specific locality in the lower body.

In this chapter, we will study not only the spinal cord but also the spinal nerves that arise from it with ladderlike regularity at intervals along its length. Thus, we will examine components of both the central and peripheral nervous systems, but these are so closely related structurally and functionally that it is appropriate to consider them together. Similarly, the brain and cranial nerves will be considered together in the following chapter. These two chapters therefore elevate our study of the nervous system from the cellular level (chapter 13) to the organ and system levels.

14.1 The Spinal Cord

▶ **Expected Learning Outcomes**

When you have completed this section, you should be able to
- enumerate the functions of the spinal cord;
- describe the surface and cross-sectional anatomy of the cord;
- explain the difference between the gray and white matter of the cord; and
- identify the major pathways (tracts) that conduct signals up and down the spinal cord, and identify the types of signals they carry.

Functions

The spinal cord serves three principal functions:

1. **Conduction.** The spinal cord contains bundles of nerve fibers that conduct information up and down the body, connecting different levels of the trunk with each other and with the brain. It enables sensory information to reach the brain, motor commands to reach the muscles and other effectors, and input received at one level of the cord to affect output from another level.

2. **Locomotion.** Walking involves repetitive, coordinated contractions of several muscle groups in the limbs. Motor neurons in the brain initiate walking and determine its speed, distance, and direction, but the simple repetitive muscle contractions that put one foot in front of another, over and over, are coordinated by groups of neurons called **central pattern generators** in the cord. These neural circuits produce the sequence of outputs to the extensor and flexor muscles that cause alternating movements of the lower limbs.

3. **Reflexes.** Reflexes are involuntary stereotyped responses to stimuli. They involve the brain, spinal cord, and peripheral nerves.

Surface Anatomy

The **spinal cord** (fig. 14.1) is a cylinder of nervous tissue that arises from the brainstem at the foramen magnum of the skull. It passes down the vertebral canal as far as the inferior margin of the first lumbar vertebra (L1) or slightly beyond. In adults, it averages about 45 cm long and 1.8 cm thick—about as thick as one's little finger. It occupies only the upper two-thirds of the vertebral canal; the lower one-third is described shortly. The cord exhibits longitudinal grooves on its anterior and posterior sides—the *anterior median fissure* and *posterior median sulcus,* respectively.

The cord gives rise to 31 pairs of spinal nerves. The first pair passes between the skull and vertebra C1, and the rest pass through the intervertebral foramina. Although the spinal cord is not visibly segmented, the part supplied by each pair of spinal nerves is called a *segment.*

The spinal cord is divided into **cervical, thoracic, lumbar,** and **sacral regions.** It may seem odd that it has a sacral region when the cord itself ends well above the sacrum. These regions, however, are named for the level of the vertebral column from which the spinal nerves emerge, not for the vertebrae that contain the cord itself.

The cord widens at two points along its course: a **cervical enlargement** in the inferior cervical region, where it gives rise to nerves of the upper limbs; and a similar **lumbar enlargement** in the lumbosacral region, where it gives rise to nerves of the pelvic region and lower limbs. Inferior to the lumbar enlargement, the cord tapers to a point called the **medullary cone (conus medullaris).** The lumbar enlargement and medullary cone give off a bundle of nerve roots that occupy the vertebral canal from L2 to S5. This bundle, named the **cauda equina**[1] (CAW-duh ee-KWY-nah) for its resemblance to a horse's tail, innervates the pelvic organs and lower limbs.

Apply What You Know

Spinal cord injuries commonly result from fractures of vertebrae C5 to C6, but never from fractures of L3 to L5. Explain both observations.

[1]*cauda* = tail + *equin* = horse

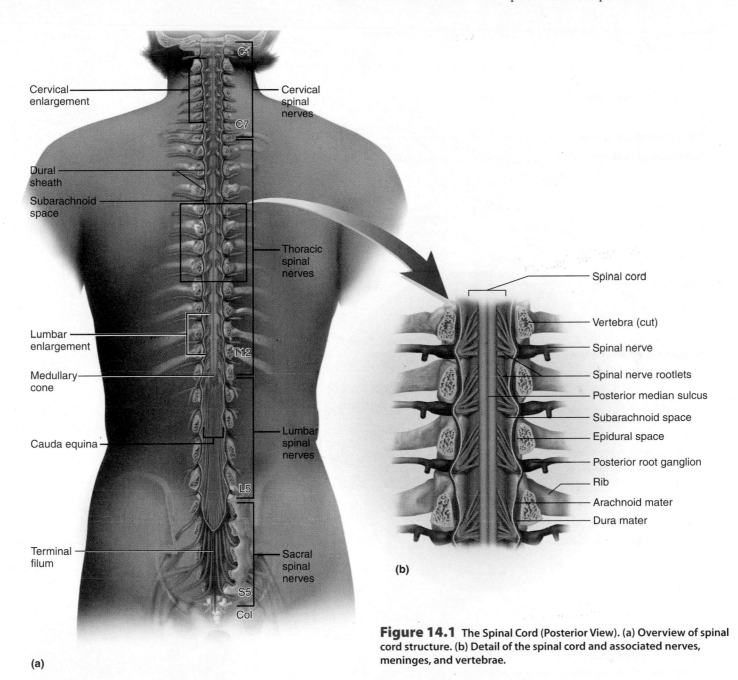

Cervical enlargement

Cervical spinal nerves

C1

C7

Dural sheath

Subarachnoid space

Thoracic spinal nerves

Lumbar enlargement

T12

Medullary cone

Lumbar spinal nerves

Cauda equina

L5

Terminal filum

Sacral spinal nerves

S5

Co1

(a)

Spinal cord

Vertebra (cut)

Spinal nerve

Spinal nerve rootlets

Posterior median sulcus

Subarachnoid space

Epidural space

Posterior root ganglion

Rib

Arachnoid mater

Dura mater

(b)

Figure 14.1 The Spinal Cord (Posterior View). (a) Overview of spinal cord structure. (b) Detail of the spinal cord and associated nerves, meninges, and vertebrae.

Meninges of the Spinal Cord

The spinal cord and brain are enclosed in three connective tissue membranes called **meninges** (meh-NIN-jeez)—singular, *meninx*[2] (MEN-inks). These membranes separate the soft tissue of the central nervous system from the bones of the vertebrae and skull. From superficial to deep, they are the dura mater, arachnoid mater, and pia mater (fig. 14.2).

The **dura mater**[3] (DOO-ruh MAH-tur) forms a loose-fitting sleeve called the **dural sheath** around the spinal cord. It is a tough collagenous membrane about as thick as a rubber kitchen glove.

The space between the sheath and vertebral bones, called the **epidural space,** is occupied by blood vessels, adipose tissue, and loose connective tissue (fig. 14.2a). Anesthetics are sometimes introduced to this space to block pain signals during childbirth or surgery; this procedure is called *epidural anesthesia.*

The **arachnoid**[4] (ah-RACK-noyd) **mater** consists of a simple squamous epithelium, the *arachnoid membrane,* adhering to the inside of the dura, and a loose mesh of collagenous and elastic fibers spanning the gap between the arachnoid membrane and the pia mater. This gap, called the **subarachnoid space,** is filled with cerebrospinal fluid (CSF), a clear liquid discussed on p. 404. Inferior to the medullary cone, the

---

[2]*menin* = membrane
[3]*dura* = tough + *mater* = mother, womb

[4]*arachn* = spider, spider web + *oid* = resembling

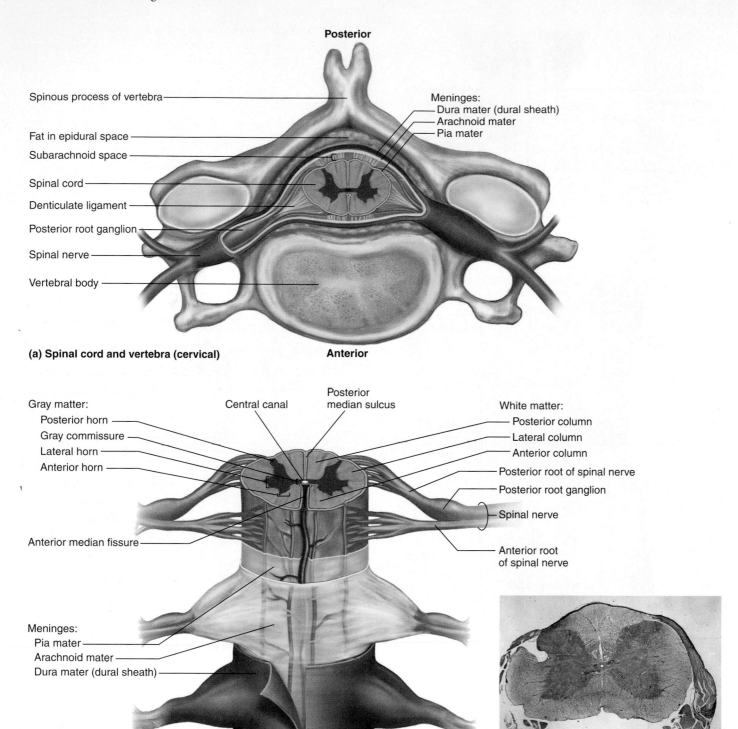

Posterior

Spinous process of vertebra

Fat in epidural space

Subarachnoid space

Spinal cord

Denticulate ligament

Posterior root ganglion

Spinal nerve

Vertebral body

Meninges:
Dura mater (dural sheath)
Arachnoid mater
Pia mater

(a) Spinal cord and vertebra (cervical)

Anterior

Gray matter:
Posterior horn
Gray commissure
Lateral horn
Anterior horn

Central canal

Posterior median sulcus

White matter:
Posterior column
Lateral column
Anterior column
Posterior root of spinal nerve
Posterior root ganglion
Spinal nerve

Anterior median fissure

Anterior root of spinal nerve

Meninges:
Pia mater
Arachnoid mater
Dura mater (dural sheath)

(b) Spinal cord and meninges (thoracic)

(c) Lumbar spinal cord

Figure 14.2 **Cross-Sectional Anatomy of the Spinal Cord.** (a) Relationship to the vertebra, meninges, and spinal nerve. (b) Detail of the spinal cord, meninges, and spinal nerves. (c) Cross section of the lumbar spinal cord with spinal nerves.

subarachnoid space is called the **lumbar cistern** and is occupied by the cauda equina and CSF (see Deeper Insight 14.1).

The **pia**[5] (PEE-uh) **mater** is a delicate, translucent membrane that closely follows the contours of the spinal cord. It continues

beyond the medullary cone as a fibrous strand, the *terminal filum,* forming part of the **coccygeal ligament** that anchors the cord to the coccyx. At regular intervals along the cord, extensions of the pia called **denticulate ligaments** extend through the arachnoid to the dura, anchoring the cord and limiting side-to-side movements.

[5]*pia* = tender, soft

DEEPER INSIGHT 14.1

Spinal Taps

Several neurological diseases are diagnosed in part by examining cerebrospinal fluid for bacteria, blood, white blood cells, or abnormalities of chemical composition. CSF is obtained by a procedure called a *spinal tap,* or *lumbar puncture.* The patient leans forward or lies on one side with the spine flexed, thus spreading the vertebral laminae and spinous processes apart. The skin over the lumbar vertebrae is anesthetized, and a needle is inserted between the spinous processes of L3 and L4 (sometimes L4 and L5). This is the safest place to obtain CSF because the spinal cord does not extend this far and is not exposed to injury by the needle. At a depth of 4 to 6 cm, the needle punctures the dura mater and enters the lumbar cistern. CSF normally drips out at a rate of about 1 drop per second. A lumbar puncture is not performed if a patient has signs of high intracranial pressure, because the sudden release of pressure (causing CSF to jet from the puncture) can cause fatal herniation of the brainstem and cerebellum into the vertebral canal.

Cross-Sectional Anatomy

Figure 14.2 shows the relationship of the spinal cord to a vertebra and the spinal nerves. The spinal cord, like the brain, consists of two kinds of nervous tissue called gray and white matter. **Gray matter** has a relatively dull color because it contains little myelin. It contains the somas, dendrites, and proximal parts of the axons of neurons. It is the site of synaptic contact between neurons and therefore the site of all synaptic integration (information processing) in the central nervous system. **White matter** contains an abundance of myelinated axons, which give it a bright, pearly white appearance. It is composed of bundles of axons, called **tracts,** that carry signals from one part of the CNS to another. In silver-stained nervous tissue sections, gray matter tends to have a brown or golden color and white matter a lighter tan to yellow color.

Gray Matter

The spinal cord has a central core of gray matter that looks somewhat butterfly- or H-shaped in cross sections. The core consists mainly of two **posterior (dorsal) horns,** which extend toward the posterolateral surfaces of the cord, and two thicker **anterior (ventral) horns,** which extend toward the anterolateral surfaces. The right and left sides are connected by a **gray commissure.** In the middle of the commissure is the **central canal,** which is collapsed in most areas of the adult spinal cord, but in some places (and in young children) remains open, lined with ependymal cells, and filled with CSF. The canal is a remnant of the lumen of the embryonic neural tube (see p. 365).

Near its attachment to the spinal cord, a spinal nerve branches into a *posterior root* and *anterior root.* The posterior root carries sensory nerve fibers, which enter the posterior horn of the cord and sometimes synapse with an interneuron there. Such interneurons are especially numerous in the cervical and lumbar enlargements and are quite evident in histological sections at these levels. The anterior horns contain the large somas of the somatic motor neurons. Axons from these neurons exit by way of the anterior root of the spinal nerve and lead to the skeletal muscles. The spinal nerve roots are described more fully later in this chapter.

In the thoracic and lumbar regions, an additional **lateral horn** is visible on each side of the gray matter. It contains neurons of the sympathetic nervous system, which send their axons out of the cord by way of the anterior root along with the somatic efferent fibers.

White Matter

The white matter of the spinal cord surrounds the gray matter. It consists of bundles of axons that course up and down the cord and provide avenues of communication between different levels of the CNS. These bundles are arranged in three pairs called **columns,** or **funiculi**[6] (few-NIC-you-lie)—a **posterior (dorsal), lateral,** and **anterior (ventral) column** on each side. Each column consists of subdivisions called **tracts,** or **fasciculi**[7] (fah-SIC-you-lye).

Spinal Tracts

Knowledge of the locations and functions of the spinal tracts is essential in diagnosing and managing spinal cord injuries. **Ascending tracts** carry sensory information up the cord and **descending tracts** conduct motor impulses down. All nerve fibers in a given tract have a similar origin, destination, and function. Many of these fibers have their origin or destination in a region called the *brainstem.* Described more fully in chapter 15 (see fig. 15.7, p. 408), this is a vertical stalk that supports the large *cerebellum* at the rear of the head and, even larger, two globes called the *cerebral hemispheres* that dominate the brain. In the following discussion, you will find references to brainstem and other regions where the spinal cord tracts begin and end. Spinal cord anatomy will grow in meaning when you study the brain.

Several of these tracts undergo **decussation**[8] (DEE-cuh-SAY-shun) as they pass up or down the brainstem and spinal cord—meaning that they cross over from the left side of the body to the right, or vice versa. As a result, the left side of the brain receives sensory information from the right side of the body and sends its motor commands to that side, whereas the right side of the brain senses and controls the left side of the body. A stroke that damages motor centers of the right side of the brain can therefore cause paralysis of the left limbs, and vice versa. When the origin and destination of a tract are on opposite sides of the body, we say they are **contralateral**[9] to each other. When a tract does not decussate, so the origin and destination of its fibers are on the same side of the body, we say they are **ipsilateral.**[10]

The major spinal cord tracts are summarized in table 14.1 and figure 14.3. Bear in mind that each tract is repeated on the right and left sides of the spinal cord.

---

[6]*funicul* = little rope, cord
[7]*fascicul* = little bundle
[8]*decuss* = to cross, form an X
[9]*contra* = opposite + *later* = side
[10]*ipsi* = the same + *later* = side

TABLE 14.1	Major Spinal Tracts		
Tract	Column	Decussation	Functions
Ascending (Sensory) Tracts			
Gracile fasciculus	Posterior	In medulla	Sensations of limb and trunk position and movement, deep discriminative touch, vibration, and visceral pain, below level T6
Cuneate fasciculus	Posterior	In medulla	Same as gracile fasciculus, but from level T6 up
Spinothalamic	Lateral and anterior	In spinal cord	Sensations of light touch, tickle, itch, temperature, pain, and pressure
Spinoreticular	Lateral and anterior	In spinal cord (some fibers)	Sensation of pain from tissue injury
Posterior spinocerebellar	Lateral	None	Feedback from muscles (proprioception)
Anterior spinocerebellar	Lateral	In spinal cord	Same as posterior spinocerebellar
Descending (Motor) Tracts			
Lateral corticospinal	Lateral	In medulla	Fine control of limbs
Anterior corticospinal	Anterior	None	Fine control of limbs
Tectospinal	Anterior	In midbrain	Reflexive head-turning in response to visual and auditory stimuli
Lateral reticulospinal	Lateral	None	Balance and posture; regulation of awareness of pain
Medial reticulospinal	Anterior	None	Same as lateral reticulospinal
Lateral vestibulospinal	Anterior	None	Balance and posture
Medial vestibulospinal	Anterior	In medulla (some fibers)	Control of head position

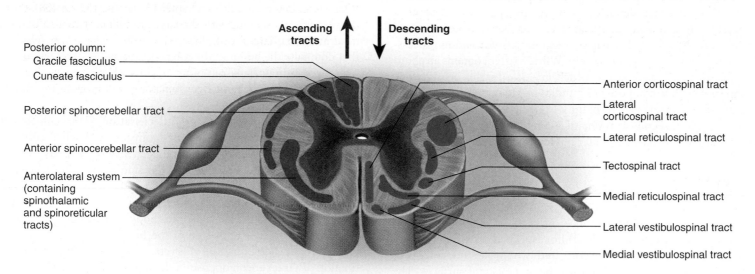

Figure 14.3 **Tracts of the Spinal Cord.** All of the illustrated tracts occur on both sides of the cord, but only the ascending sensory tracts are shown on the left (red), and only the descending motor tracts on the right (green).
● *If you were told that this is a cross section either at level T4 or T10, how could you determine which is correct?*

Ascending Tracts

Ascending tracts carry sensory signals up the spinal cord. Sensory signals typically travel across three neurons from their origin in the receptors to their destination in the sensory areas of the brain: a **first-order neuron** that detects a stimulus and conducts a signal to the spinal cord or brainstem; a **second-order neuron** that continues as far as a "gateway" called the *thalamus* at the upper end of the brainstem; and a **third-order neuron** that carries the signal the rest of the way to the sensory region of the cerebral cortex, the level of consciousness. The axons of these neurons are called the *first-* through *third-order nerve fibers.* Variations on this pattern will be noted for some of the sensory systems to follow.

The major ascending tracts are as follows. Other than the first two, their names consist of the prefix *spino-* followed by a root denoting the destination of their fibers in the brain.

- The **gracile**[11] **fasciculus** (GRAS-el fah-SIC-you-lus) (fig. 14.4a) carries signals from the midthoracic and lower parts of the body. Below vertebra T6, it composes the entire posterior column. At T6, it is joined by the cuneate fasciculus, discussed next. The gracile fasciculus consists of first-order nerve fibers that travel up the ipsilateral side of the spinal cord and terminate at the *gracile nucleus* in the medulla oblongata of the brainstem. These fibers carry signals for vibration, visceral pain, deep and discriminative touch (touch whose location one can precisely identify), and especially *proprioception*[12] from the lower limbs and lower trunk. (Proprioception is the nonvisual sense of the position and movements of the body.)

- The **cuneate**[13] (CUE-nee-ate) **fasciculus** (fig. 14.4a) joins the gracile fasciculus at the T6 level. It occupies the lateral portion of the posterior column and forces the gracile fasciculus medially. It carries the same type of sensory signals, originating from level T6 and up (from the upper limb and chest). Its fibers end in the *cuneate nucleus* on the ipsilateral side of the medulla oblongata. In the medulla, second-order fibers of the gracile and cuneate systems decussate and form the **medial lemniscus**[14] (lem-NIS-cus), a tract of nerve fibers that leads the rest of the way up the brainstem to the contralateral thalamus. Third-order fibers go from the thalamus to the cerebral cortex. Because of decussation, the signals carried by the right gracile and cuneate fasciculi ultimately go to the left cerebral hemisphere, and vice versa.

- The **spinothalamic** (SPY-no-tha-LAM-ic) **tract** (fig. 14.4b) and some smaller tracts form the *anterolateral system,* which passes up the anterior and lateral columns of the spinal cord. The spinothalamic tract carries signals for pain, temperature, pressure, tickle, itch, and light or crude touch. Light touch is the sensation produced by stroking hairless skin with a feather or cotton wisp, without indenting the skin; crude touch is touch whose location one can only vaguely identify. In this pathway, first-order neurons end in the posterior horn of the spinal cord near the point of entry. Here they synapse with second-order neurons, which decussate to the opposite side of the spinal cord and form the ascending spinothalamic tract. These fibers lead all the way to the thalamus. Third-order neurons continue from there to the cerebral cortex. Because of its decussation, the spinothalamic tract ultimately sends its signals to the contralateral cerebral hemisphere.

- The **spinoreticular tract** also travels up the anterolateral system. It carries pain signals resulting from tissue injury. The first-order sensory neurons enter the posterior horn and immediately synapse with second-order neurons. These decussate to the opposite anterolateral system, ascend the cord, and end in a loosely organized core of gray matter called the *reticular formation* in the medulla and pons. Third-order neurons continue from the pons to the thalamus, and fourth-order neurons complete the path from there to the cerebral cortex. The reticular formation is further described in chapter 15, and the role of the spinoreticular tract in pain sensation is further discussed in chapter 17.

- The **posterior** and **anterior spinocerebellar** (SPY-no-SERR-eh-BEL-ur) **tracts** travel through the lateral column and carry proprioceptive signals from the limbs and trunk to the cerebellum, a large motor control area at the rear of the brain. The first-order neurons of this system originate in the muscles and tendons and end in the posterior horn of the spinal cord. Second-order neurons send their fibers up the spinocerebellar tracts and end in the cerebellum. Fibers of the posterior tract travel up the ipsilateral side of the spinal cord. Those of the anterior tract cross over and travel up the contralateral side but then cross back in the brainstem to enter the ipsilateral side of the cerebellum. Both tracts provide the cerebellum with feedback needed to coordinate muscle action, as discussed in chapter 15.

Descending Tracts

Descending tracts carry motor signals down the brainstem and spinal cord. A descending motor pathway typically involves two neurons called the upper and lower motor neurons. The **upper motor neuron** begins with a soma in the cerebral cortex or brainstem and has an axon that terminates on a **lower motor neuron** in the brainstem or spinal cord. The axon of the lower motor neuron then leads the rest of the way to the muscle or other target organ. The names of most descending tracts consist of a word root denoting the point of origin in the brain, followed by the suffix *-spinal.* The major descending tracts are described here.

- The **corticospinal** (COR-tih-co-SPY-nul) **tracts** carry motor signals from the cerebral cortex for precise, finely coordinated limb movements. The fibers of this system form ridges called *pyramids* on the anterior surface of the medulla oblongata, so these tracts were once called *pyramidal tracts.* Most corticospinal fibers decussate in the lower medulla and form the **lateral corticospinal tract** on the contralateral side of the spinal cord. A few fibers remain uncrossed and form the **anterior corticospinal tract** on the ipsilateral side (fig. 14.5). Fibers of the anterior tract decussate lower in the spinal cord, however, so even they control contralateral muscles.

- The **tectospinal** (TEC-toe-SPY-nul) **tract** begins in a midbrain region called the *tectum* and crosses to the contralateral side of the midbrain. It descends through the brainstem to the upper spinal cord on that side, going only as far as the neck. It is involved in reflex turning of the head, especially in response to sights and sounds.

- The **lateral** and **medial reticulospinal** (reh-TIC-you-lo-SPY-nul) **tracts** originate in the reticular formation. They control muscles of the upper and lower limbs, especially to maintain posture and balance. They also contain *descending analgesic fibers* that reduce the transmission of pain signals to the brain (see p. 466).

---

[11]*gracil* = thin, slender
[12]*proprio* = one's own + *cept* = receive, sense
[13]*cune* = wedge
[14]*lemniscus* = ribbon

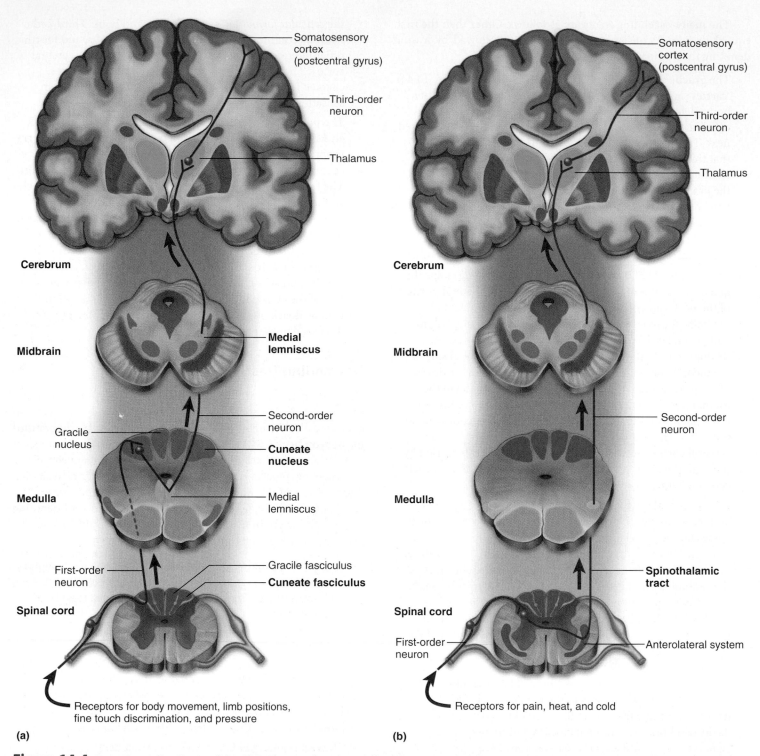

Figure 14.4 **Some Ascending Tracts of the CNS.** The spinal cord, medulla oblongata, and midbrain are shown in cross section and the cerebrum and thalamus (top) in frontal section. Nerve signals enter the spinal cord at the bottom of the figure and carry somatosensory information to the cerebral cortex at the top. (a) The cuneate fasciculus and medial lemniscus. (b) The spinothalamic tract.

● *On the basis of this figure, explain why the* right *cerebral hemisphere perceives heat and cold on the* left *side of the body. What is the name of the phenomenon that accounts for this transmission of sensory information from one side of the body to the opposite side of the brain?*

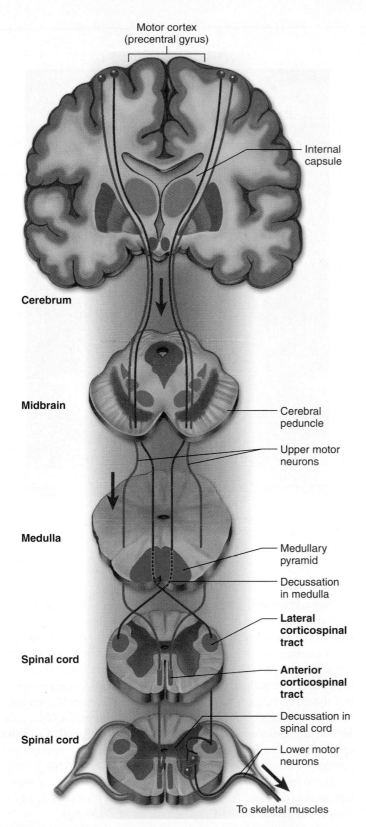

Figure 14.5 **Two Descending Tracts of the CNS.** The anterior and lateral corticospinal tracts, which carry signals for voluntary muscle contraction. Nerve signals originate in the cerebral cortex at the top of the figure and carry motor commands down the spinal cord. Pathways shown at the bottom right are duplicated on the left.

- The **lateral** and **medial vestibulospinal** (vess-TIB-you-lo-SPY-nul) **tracts** begin in the brainstem *vestibular nuclei,* which receive impulses for balance from the inner ear. The lateral vestibulospinal tract passes down the anterior column of the spinal cord and facilitates neurons that control the extensor muscles of the limbs, thus inducing the limbs to stiffen and straighten. This is an important reflex in responding to body tilt and keeping one's balance. The medial vestibulospinal tract splits into ipsilateral and contralateral fibers that descend through the anterior column on both sides of the spinal cord and terminate in the neck. It plays a role in the control of head position.

Rubrospinal tracts are prominent in other mammals, where they aid in muscle coordination. Although often pictured in illustrations of supposedly human anatomy, they are almost nonexistent in humans and have little functional importance.

Before You Go On

Answer the following questions to test your understanding of the preceding section:

1. Name the four major regions and two enlargements of the spinal cord.

2. Describe the distal (inferior) end of the spinal cord and the contents of the vertebral canal from level L2 through S5.

3. Sketch a cross section of the spinal cord showing the posterior and anterior horns. Where are the gray and white matter? Where are the columns and tracts?

4. Give an anatomical explanation as to why a stroke in the right cerebral hemisphere can paralyze the limbs on the left side of the body.

5. Identify each of the following spinal tracts with respect to whether it is ascending or descending; its origin and destination; and what sensory or motor purpose it serves: the lateral corticospinal, lateral reticulospinal, and spinothalamic tracts, and gracile fasciculus.

14.2 The Spinal Nerves

▶ Expected Learning Outcomes

When you have completed this section, you should be able to

- describe the attachment of a spinal nerve to the spinal cord;

- trace the branches of a generalized spinal nerve distal to its attachment;

- name the five plexuses of spinal nerves and describe their general anatomy;

- name some major nerves that arise from each plexus; and

- explain the relationship of dermatomes to the spinal nerves.

DEEPER INSIGHT 14.2

Poliomyelitis and Amyotrophic Lateral Sclerosis

Poliomyelitis[15] and *amyotrophic lateral sclerosis*[16] (ALS) are two diseases that result from the destruction of motor neurons. In both diseases, the skeletal muscles atrophy from lack of innervation.

Poliomyelitis is caused by the poliovirus, which destroys motor neurons in the brainstem and anterior horn of the spinal cord. Signs of polio include muscle pain, weakness, and loss of some reflexes, followed by paralysis, muscular atrophy, and sometimes respiratory arrest. The virus spreads through water contaminated by feces. Historically, polio afflicted many children who contracted the virus from contaminated public swimming pools. For a time, the polio vaccine nearly eliminated new cases, but the disease has lately begun to reemerge among children in some parts of the world.

ALS is also known as Lou Gehrig[17] disease after the baseball player who had to retire from the sport because of it. It is marked not only by the degeneration of motor neurons and atrophy of the muscles, but also sclerosis (scarring) of the lateral regions of the spinal cord—hence its name. Most cases occur when astrocytes fail to reabsorb the neurotransmitter glutamate from the tissue fluid, allowing it to accumulate to a neurotoxic level. The early signs of ALS include muscular weakness and difficulty in speaking, swallowing, and using the hands. Sensory and intellectual functions remain unaffected, as evidenced by the accomplishments of astrophysicist and best-selling author Stephen Hawking (fig. 14.6), who was stricken with ALS while he was in college. Despite near-total paralysis, he remains highly productive and communicates with the aid of a speech synthesizer and computer. Tragically, many people are quick to assume that those who have lost most of their ability to communicate their ideas and feelings have no ideas and feelings to communicate. To a victim, this may be more unbearable than the loss of motor function itself.

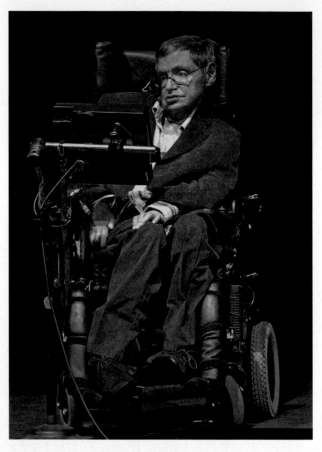

Figure 14.6 Stephen Hawking (1942–), Lucasian Professor of Mathematics at Cambridge University.

General Anatomy of Nerves and Ganglia

The spinal cord communicates with the rest of the body by way of the spinal nerves. Before we discuss those specific nerves, however, it is necessary to be familiar with the structure of nerves and ganglia in general.

A **nerve** is a cordlike organ composed of numerous nerve fibers (axons) bound together by connective tissue (fig. 14.7). If we compare a nerve fiber to a wire carrying an electrical current in one direction, a nerve would be comparable to an electrical cable composed of thousands of wires carrying currents in opposite directions. A nerve contains anywhere from a few nerve fibers to more than a million. Nerves usually have a pearly white color and resemble frayed string as they divide into smaller and smaller branches.

Nerve fibers of the peripheral nervous system are ensheathed in Schwann cells, which form a neurilemma and often a myelin sheath around the axon (see p. 360). External to the neurilemma, each fiber is surrounded by a basal lamina and then a thin sleeve of loose connective tissue called the **endoneurium.** In most nerves, the nerve fibers are gathered in bundles called **fascicles,** each wrapped in a sheath called the **perineurium.** The perineurium is composed of up to 20 layers of overlapping, squamous, epithelium-like cells. Several fascicles are then bundled together and wrapped in an outer **epineurium** to compose the nerve as a whole. The epineurium consists of dense irregular fibrous connective tissue and protects the nerve from stretching and injury. Nerves have a high metabolic rate and need a plentiful blood supply, which is furnished by blood vessels that penetrate these connective tissue coverings.

Apply What You Know

How does the structure of a nerve compare to that of a skeletal muscle? Which of the descriptive terms for nerves have similar counterparts in muscle histology?

---

[15]*polio* = gray matter + *myel* = spinal cord + *itis* = inflammation
[16]*a* = without + *myo* = muscle + *troph* = nourishment
[17]Lou Gehrig (1903–41), American baseball player

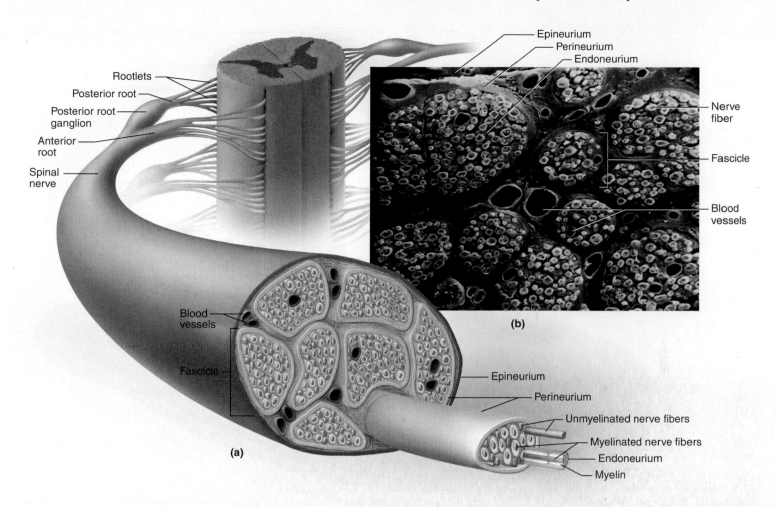

Figure 14.7 Anatomy of a Nerve. (a) A spinal nerve and its association with the spinal cord. (b) Cross section of a nerve (SEM). Myelinated nerve fibers appear in the photograph as white rings and unmyelinated fibers as solid gray.
[Part (b) From R. G. Kessel and R. H. Kardon, *Tissues and Organs: A Text-Atlas of Scanning Electron Microscopy* (W. H. Freeman, 1979)]

Peripheral nerve fibers are of two kinds: *sensory (afferent) fibers* carry signals from sensory receptors to the CNS, and *motor (efferent) fibers* carry signals from the CNS to muscles and glands. Both sensory and motor fibers can also be described as *somatic* or *visceral* and as *general* or *special* depending on the organs they innervate (table 14.2).

Purely **sensory nerves,** composed only of afferent fibers, are rare; they include the olfactory and optic nerves described in chapter 15. **Motor nerves** carry only efferent fibers. Most nerves, however, are mixed. A **mixed nerve** consists of both afferent and efferent fibers and therefore conducts signals in two directions, although any one fiber within the nerve carries signals in one direction only. Many nerves often described as motor are actually mixed because they carry sensory signals of proprioception from the muscles back to the CNS.

TABLE 14.2	The Classification of Nerve Fibers
Class	**Description**
Afferent fibers	Carry sensory signals from receptors to the CNS
Efferent fibers	Carry motor signals from the CNS to effectors
Somatic fibers	Innervate skin, skeletal muscles, bones, and joints
Visceral fibers	Innervate blood vessels, glands, and viscera
General fibers	Innervate widespread organs such as muscles, skin, glands, viscera, and blood vessels
Special fibers	Innervate more localized organs in the head, including the eyes, ears, olfactory and taste receptors, and muscles of chewing, swallowing, and facial expression

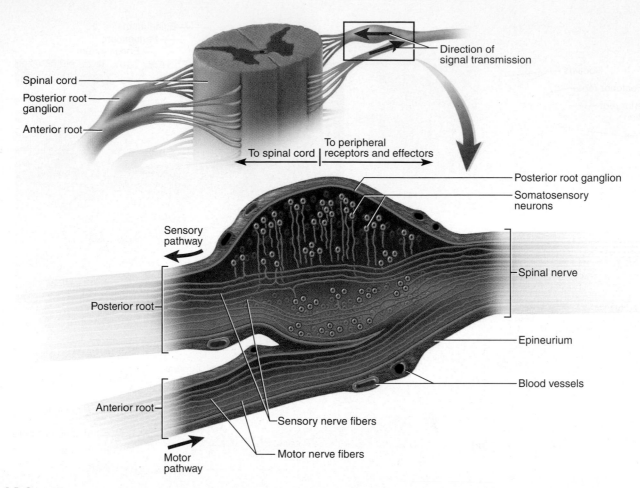

Figure 14.8 **Anatomy of a Ganglion.** Longitudinal section. The posterior root ganglion contains the somas of sensory neurons conducting signals from peripheral sense organs to the spinal cord. Below this is the anterior root of the spinal nerve, which conducts motor signals away from the spinal cord, toward muscles and other peripheral effectors. (The anterior root is not part of the ganglion.)
● *To which morphological category of neurons (see p. 356) do the somatosensory neurons in this figure belong?*

If a nerve resembles a thread, a **ganglion**[18] resembles a knot in the thread. A ganglion is a cluster of cell bodies (neurosomas) outside the CNS. It is enveloped in an epineurium continuous with that of the nerve. Among the neurosomas are bundles of nerve fibers leading into and out of the ganglion. Figure 14.8 shows a type of ganglion associated with the spinal nerves.

Spinal Nerves

There are 31 pairs of **spinal nerves:** 8 cervical (C1–C8), 12 thoracic (T1–T12), 5 lumbar (L1–L5), 5 sacral (S1–S5), and 1 coccygeal (Co) (fig. 14.9). The first cervical nerve emerges between the skull and atlas, and the others emerge through intervertebral foramina, including the anterior and posterior foramina of the sacrum and the sacral hiatus.

Proximal Branches

Each spinal nerve arises from two points of attachment to the spinal cord. In each segment of the cord, six to eight nerve **rootlets** emerge from the anterior surface and converge to form the **anterior (ventral) root** of the spinal nerve. Another six to eight rootlets emerge from the posterior surface and converge to form the **posterior (dorsal) root** (see figs. 1.1b and 14.10). A short distance away from the spinal cord, the posterior root swells into a **posterior (dorsal) root ganglion,** which contains the somas of sensory neurons (fig. 14.8). There is no corresponding ganglion on the anterior root.

Distal to the ganglion, the anterior and posterior roots merge, leave the dural sac, and form the spinal nerve proper (fig. 14.11). The nerve then exits the vertebral canal through the intervertebral foramen. The spinal nerve is a mixed nerve, carrying sensory signals to the spinal cord by way of the posterior root and ganglion, and motor signals out to more distant parts of the body.

The anterior and posterior roots are shortest in the cervical region and become longer inferiorly. The roots that arise from

---
[18]*gangli* = knot

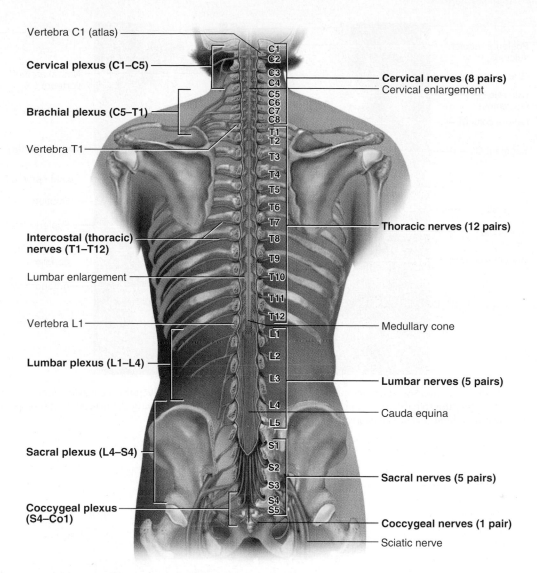

Vertebra C1 (atlas)

Cervical plexus (C1–C5)

Brachial plexus (C5–T1)

Vertebra T1

Intercostal (thoracic) nerves (T1–T12)

Lumbar enlargement

Vertebra L1

Lumbar plexus (L1–L4)

Sacral plexus (L4–S4)

Coccygeal plexus (S4–Co1)

C1
C2
C3
C4
C5
C6
C7
C8
T1
T2
T3
T4
T5
T6
T7
T8
T9
T10
T11
T12
L1
L2
L3
L4
L5
S1
S2
S3
S4
S5

Cervical nerves (8 pairs)
Cervical enlargement

Thoracic nerves (12 pairs)

Medullary cone

Lumbar nerves (5 pairs)

Cauda equina

Sacral nerves (5 pairs)

Coccygeal nerves (1 pair)

Sciatic nerve

Figure 14.9 The Spinal Nerve Roots and Plexuses (Posterior View).

segments L2 to Co of the cord form the cauda equina. Some viruses invade the central nervous system by way of these roots (see Deeper Insight 14.3).

Distal Branches

Distal to the vertebrae, the branches of a spinal nerve are more complex (fig. 14.12). Immediately after emerging from the intervertebral foramen, the nerve divides into a **posterior ramus,**[19] an **anterior ramus,** and a small **meningeal branch.** Thus, each spinal nerve branches on both ends—into posterior and anterior *roots* approaching the spinal cord, and posterior and anterior *rami* leading away from the vertebral column.

The meningeal branch (fig. 14.11) reenters the vertebral canal and innervates the meninges, vertebrae, and spinal ligaments. The posterior ramus innervates the muscles and joints in that region of the spine and the skin of the back. The anterior ramus, largest of the three, innervates the anterior and lateral skin and muscles of the trunk and gives rise to nerves of the limbs.

Apply What You Know

Do you think the meningeal branch is sensory, motor, or mixed? Explain your reasoning.

The anterior ramus differs from one region of the trunk to another. In the thoracic region, it forms an **intercostal nerve** that travels along the inferior margin of a rib and innervates the skin and intercostal muscles (thus contributing to breathing), as well as the internal oblique, external oblique, and transverse abdominal muscles. All other anterior rami form the *nerve plexuses* described next.

[19]*ramus* = branch

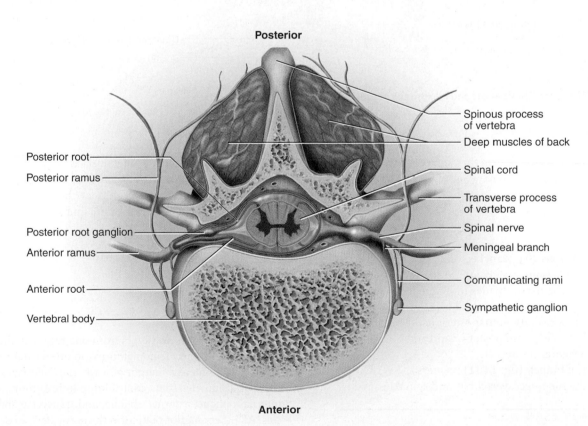

Figure 14.10 labels:

Posterior median sulcus
Gracile fasciculus
Cuneate fasciculus
Lateral column
Segment C5
Cross section
Arachnoid mater
Dura mater

Neural arch of vertebra C3 (cut)
Spinal nerve C4
Vertebral artery
Spinal nerve C5:
Rootlets
Posterior root
Posterior root ganglion
Anterior root

Figure 14.10 **The Point of Entry of Two Spinal Nerves into the Spinal Cord.** Posterior view with vertebrae cut away. Note that each posterior root divides into several rootlets that enter the spinal cord. A segment of the spinal cord is the portion receiving all the rootlets of one spinal nerve.
● *What would be the consequences of surgically cutting the spinal nerve rootlets shown in this photograph?*

Figure 14.11 labels:

Posterior
Posterior root
Posterior ramus
Posterior root ganglion
Anterior ramus
Anterior root
Vertebral body

Spinous process of vertebra
Deep muscles of back
Spinal cord
Transverse process of vertebra
Spinal nerve
Meningeal branch
Communicating rami
Sympathetic ganglion
Anterior

Figure 14.11 Branches of a Spinal Nerve in Relation to the Spinal Cord and Vertebra (Cross Section).

DEEPER INSIGHT  14.3

Shingles

Chickenpox *(varicella),* a common disease of early childhood, is caused by the *varicella-zoster* virus. It produces an itchy rash that usually clears up without complications. The virus, however, remains for life in the posterior root ganglia, kept in check by the immune system. If the immune system is compromised, however, the virus can travel down the sensory nerves by axonal transport and cause *shingles (herpes zoster).* This is particularly common after the age of 50. Shingles is characterized by a painful trail of skin discoloration and fluid-filled vesicles along the path of the nerve. These signs usually appear in the chest and waist, often on just one side of the body. There is no cure, and the vesicles generally heal spontaneously within 1 to 3 weeks. In the meantime, aspirin and steroidal ointments can help to relieve the pain and inflammation of the lesions. Antiviral drugs such as acyclovir can shorten the course of an episode of shingles, but only if taken within the first 2 to 3 days of outbreak. Even after the lesions disappear, however, some people suffer intense pain along the course of the nerve *(postherpetic neuralgia, PHN),* lasting for months or even years. PHN has proven very difficult to treat, but pain relievers and antidepressants are of some help. Childhood vaccination against varicella reduces the risk of shingles later in life. A vaccine for adults (Zostavax) has recently become available and is recommended in the United States for all healthy adults over age 60.

Nerve Plexuses

Except in the thoracic region, the anterior rami branch and anastomose (merge) repeatedly to form five weblike nerve plexuses: the small **cervical plexus** in the neck, the **brachial plexus** in the shoulder, the **lumbar plexus** of the lower back, the **sacral plexus** immediately inferior to this, and finally, the tiny **coccygeal plexus** adjacent to the lower sacrum and coccyx. A general view of these plexuses is shown in figure 14.9; they are illustrated and described in tables 14.3 through 14.6. The spinal nerve roots that give rise to each plexus are indicated in violet in each table. Some of these roots give rise to smaller branches called *trunks, anterior divisions, posterior divisions,* and *cords,* which are color-coded and explained in the individual tables.

The nerves tabulated here have somatosensory and motor functions. *Somatosensory* means that they carry sensory signals from bones, joints, muscles, and the skin, in contrast to sensory input from the viscera or from special sense organs such as the eyes and ears. (See chapter 17, p. 461, for explanation of the different modes of sensory function.) These somatosensory signals are for touch, heat, cold, stretch, pressure, pain, and other sensations. One of the most important sensory roles of these nerves is *proprioception,* in which the brain receives information about body position and movements from nerve endings in the muscles, tendons, and joints. The brain uses this information to adjust muscle actions and thereby maintain equilibrium (balance) and coordination.

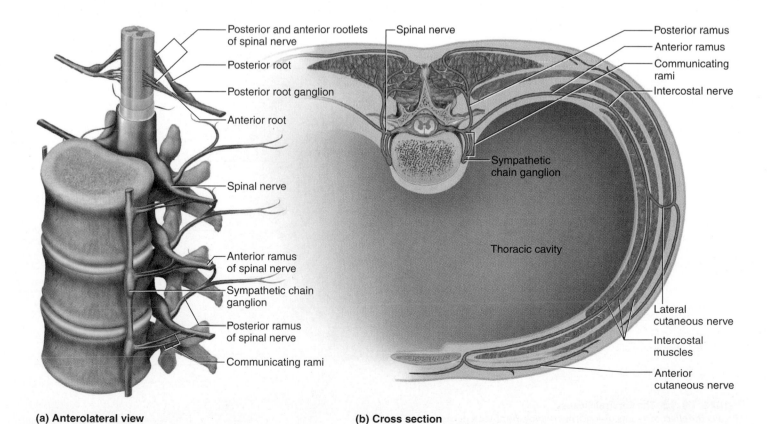

(a) Anterolateral view

(b) Cross section

Figure 14.12 Rami of the Spinal Nerves. (a) Anterolateral view of the spinal nerves and their subdivisions in relation to the spinal cord and vertebrae. (b) Cross section of the thorax showing innervation of muscles and skin of the chest and back. This section is cut through the intercostal muscles between two ribs.

The motor function of these nerves is primarily to stimulate the contraction of skeletal muscles. These nerves also carry autonomic fibers to the blood vessels of the skin, muscles, and other organs, thus adjusting blood flow to changing local needs.

The following tables identify the areas of skin innervated by the sensory fibers and the muscle groups innervated by the motor fibers of the individual nerves. The muscle tables in chapters 11 and 12 provide a more detailed breakdown of the muscles supplied by each nerve and the actions they perform. You may assume that for each muscle, these nerves also carry autonomic fibers to its blood vessels and sensory fibers from its proprioceptors. Throughout these tables, the abbreviations *n.* and *nn.* stand for *nerve* and *nerves*.

TABLE 14.3	The Cervical Plexus

The cervical plexus (fig. 14.13) receives fibers from the anterior rami of nerves C1 to C5 and gives rise to the nerves listed below, in order from superior to inferior. The most important of these are the *phrenic*[20] *nerves,* which travel down each side of the mediastinum, innervate the diaphragm, and play an essential role in breathing (see fig. 16.3, p. 446). In addition to the major nerves listed here, there are several motor branches that innervate the geniohyoid, thyrohyoid, scalene, levator scapulae, trapezius, and sternocleidomastoid muscles.

Nerve	Composition	Cutaneous and Other Sensory Innervation	Muscular Innervation (Motor and Proprioceptive)
Lesser occipital n.	Somatosensory	Upper third of medial surface of external ear, skin posterior to ear, posterolateral neck	(None)
Great auricular n.	Somatosensory	Most of the external ear, mastoid region, region from parotid salivary gland (see fig. 11.2, p. 268) to slightly inferior to angle of mandible	(None)
Transverse cervical n.	Somatosensory	Anterior and lateral neck, underside of chin	(None)
Ansa cervicalis	Mixed	(None)	Omohyoid, sternohyoid, and sternothyroid muscles
Supraclavicular nn.	Somatosensory	Lower anterior and lateral neck, shoulder, anterior chest	(None)
Phrenic n.	Mixed	Diaphragm, pleura, and pericardium	Diaphragm

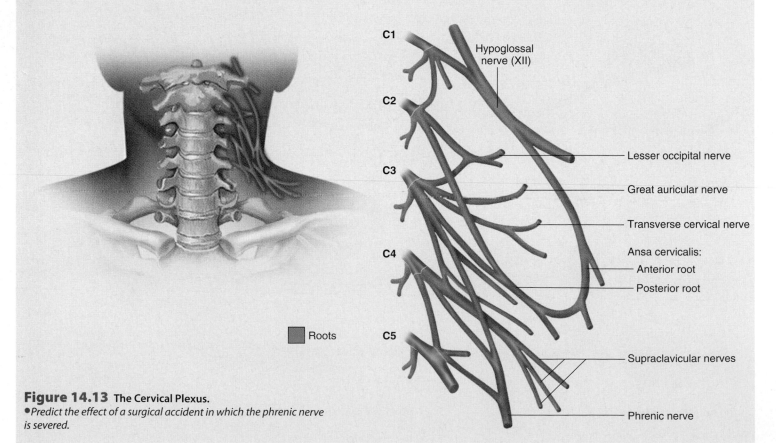

Figure 14.13 The Cervical Plexus.
• *Predict the effect of a surgical accident in which the phrenic nerve is severed.*

[20]*phren* = diaphragm

TABLE 14.4 | The Brachial Plexus

The brachial plexus (figs. 14.14 and 14.15) is formed predominantly by the anterior rami of nerves C5 to T1 (C4 and T2 make smaller contributions). It passes over the first rib into the axilla and innervates the upper limb and some muscles of the neck and shoulder. This plexus is well known for its conspicuous M or W shape when seen in cadaver dissection. The subdivisions of this plexus are called *roots, trunks, divisions,* and *cords* (color-coded in fig. 14.14). The five **roots** are the anterior rami of C5 through T1. Roots C5 and C6 converge to form the **upper trunk;** C7 continues as the **middle trunk;** and C8 and T1 converge to form the **lower trunk.** Each trunk divides into an **anterior** and **posterior division;** as the body is dissected from the anterior surface of the shoulder inward, the posterior divisions are found behind the anterior ones. Finally, the six divisions merge to form three large fiber bundles—the **lateral, posterior,** and **medial cords.** From these cords arise the following major nerves, listed in order of the illustration from superior to inferior.

Nerve	Composition	Cord of Origin	Cutaneous and Joint Innervation (Sensory)	Muscular Innervation (Motor and Proprioceptive)
Musculocutaneous n.	Mixed	Lateral	Skin of anterolateral forearm; elbow joint	Brachialis, biceps brachii, and coracobrachialis muscles
Axillary n.	Mixed	Posterior	Skin of lateral shoulder and arm; shoulder joint	Deltoid and teres minor muscles
Radial n.	Mixed	Posterior	Skin of posterior arm; posterior and lateral forearm and wrist; joints of elbow, wrist, and hand	Mainly extensor muscles of posterior arm and forearm (see tables 12.3 and 12.4)
Median n.	Mixed	Lateral and medial	Skin of lateral two-thirds of hand; tips of digits I–IV; joints of hand	Mainly forearm flexors; thenar group and lumbricals I–II of hand (see tables 12.3 to 12.5)
Ulnar n.	Mixed	Medial	Skin of palmar and medial hand and digits III–V; joints of elbow and hand	Some forearm flexors; adductor pollicis; hypothenar group; interosseous muscles; lumbricals III–IV (see tables 12.4 and 12.5)

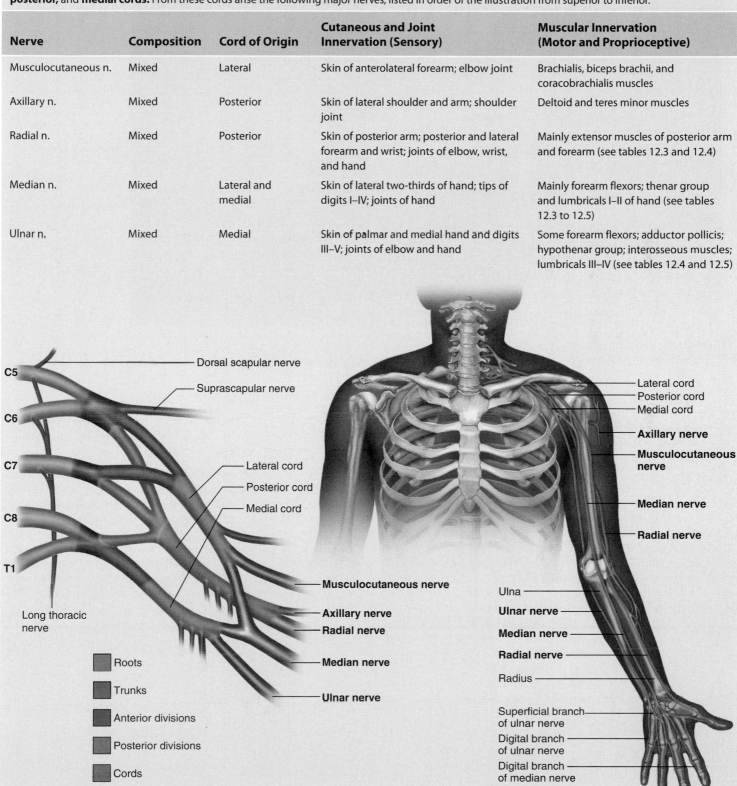

Figure 14.14 **The Brachial Plexus.** The labeled nerves innervate muscles tabulated in chapter 12, and those in boldface are further detailed in this table.

387

TABLE 14.4	The Brachial Plexus (continued)

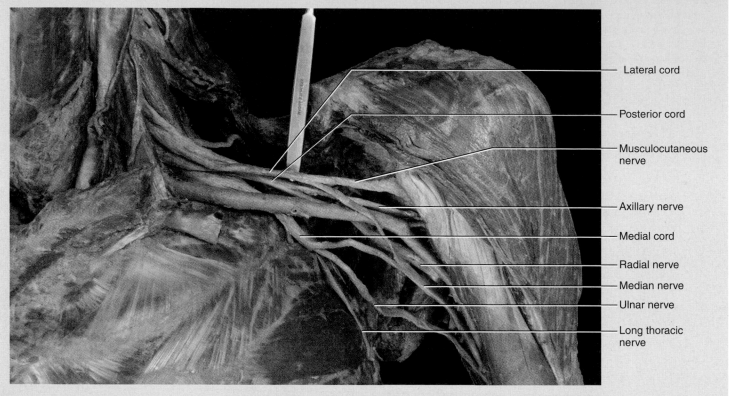

Lateral cord

Posterior cord

Musculocutaneous nerve

Axillary nerve

Medial cord

Radial nerve

Median nerve

Ulnar nerve

Long thoracic nerve

Figure 14.15 **The Brachial Plexus of a Cadaver.** Anterior view of the left shoulder.

DEEPER INSIGHT 14.4

Spinal Nerve Injuries

The radial and sciatic nerves are especially vulnerable to injury. The radial nerve, which passes through the axilla, may be compressed against the humerus by improperly adjusted crutches, causing *crutch paralysis.* A similar injury often resulted from the discredited practice of trying to correct a dislocated shoulder by putting a foot in a person's armpit and pulling on the arm. One consequence of radial nerve injury is *wrist drop*—the fingers, hand, and wrist are chronically flexed because the extensor muscles supplied by the radial nerve are paralyzed.

Because of its position and length, the sciatic nerve of the hip and thigh is the most vulnerable nerve in the body. Trauma to this nerve produces *sciatica,* a sharp pain that travels from the gluteal region along the posterior side of the thigh and leg as far as the ankle. Ninety percent of cases result from a herniated intervertebral disc or osteoarthritis of the lower spine, but sciatica can also be caused by pressure from a pregnant uterus, dislocation of the hip, injections in the wrong area of the buttock, or sitting for a long time on the edge of a hard chair. Men sometimes suffer sciatica because of the habit of sitting on a wallet carried in the hip pocket.

TABLE 14.5	The Lumbar Plexus

The lumbar plexus (fig. 14.16) is formed from the anterior rami of nerves L1 to L4 and some fibers from T12. With only five roots and two divisions, it is less complex than the brachial plexus. It gives rise to the following nerves.

Nerve	Composition	Cutaneous and Joint Innervation (Sensory)	Muscular Innervation (Motor and Proprioceptive)
Iliohypogastric n.	Mixed	Skin of lower anterior abdominal and posterolateral gluteal regions	Internal and external abdominal oblique and transverse abdominal muscles
Ilioinguinal n.	Mixed	Skin of upper medial thigh; male scrotum and root of penis; female labia majora	Internal abdominal oblique
Genitofemoral n.	Mixed	Skin of middle anterior thigh; male scrotum; female labia majora	Male cremaster muscle (see p. 707)
Lateral femoral cutaneous n.	Somatosensory	Skin of anterior and upper lateral thigh	(None)
Femoral n.	Mixed	Skin of anterior, medial, and lateral thigh and knee; skin of medial leg and foot; hip and knee joints	Iliacus, pectineus, quadriceps femoris, and sartorius muscles
Obturator n.	Mixed	Skin of medial thigh; hip and knee joints	Obturator externus; medial (adductor) thigh muscles (see table 12.6)

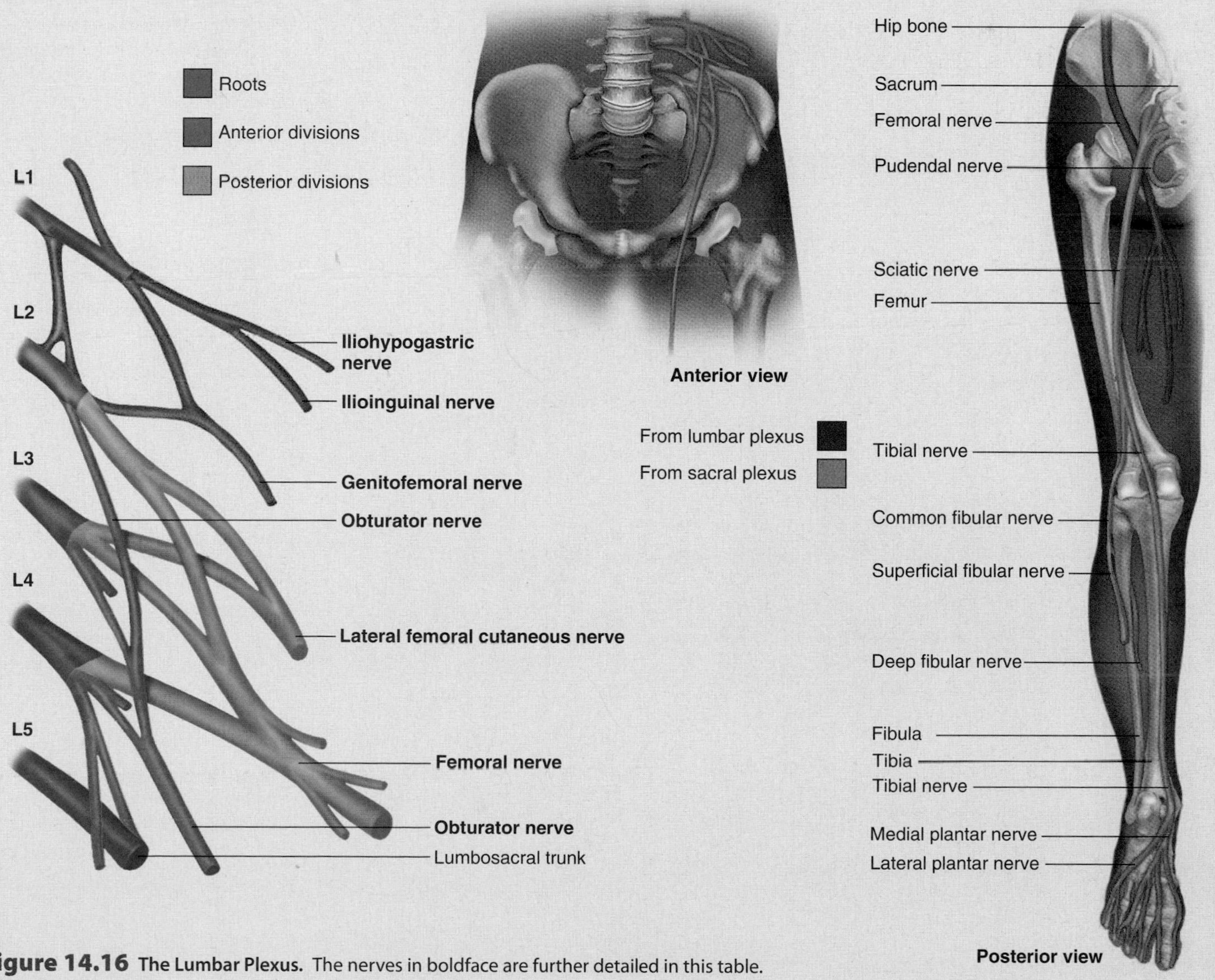

■ Roots

■ Anterior divisions

■ Posterior divisions

L1

L2

Iliohypogastric nerve

Ilioinguinal nerve

L3

Genitofemoral nerve

Obturator nerve

L4

Lateral femoral cutaneous nerve

L5

Femoral nerve

Obturator nerve

Lumbosacral trunk

Anterior view

■ From lumbar plexus

■ From sacral plexus

Hip bone

Sacrum

Femoral nerve

Pudendal nerve

Sciatic nerve

Femur

Tibial nerve

Common fibular nerve

Superficial fibular nerve

Deep fibular nerve

Fibula

Tibia

Tibial nerve

Medial plantar nerve

Lateral plantar nerve

Posterior view

Figure 14.16 **The Lumbar Plexus.** The nerves in boldface are further detailed in this table.

TABLE 14.6   The Sacral and Coccygeal Plexuses

The sacral plexus is formed from the anterior rami of nerves L4, L5, and S1 through S4. It has six roots and anterior and posterior divisions. Since it is connected to the lumbar plexus by fibers that run through the *lumbosacral trunk,* the two plexuses are sometimes referred to collectively as the *lumbosacral plexus.* The coccygeal plexus is a tiny plexus formed from the anterior rami of S4, S5, and Co (fig. 14.17).

The *tibial* and *common fibular nerves* travel together through a connective tissue sheath; they are referred to collectively as the **sciatic** (sy-AT-ic) **nerve.** The sciatic nerve passes through the greater sciatic notch of the pelvis, extends for the length of the thigh, and ends at the popliteal fossa. Here, the tibial and common fibular nerves diverge and follow their separate paths into the leg. The tibial nerve descends through the leg and then gives rise to the medial and plantar nerves in the foot. The common fibular nerve divides into deep and superficial fibular nerves. The sciatic nerve is a common focus of injury and pain (see Deeper Insight 14.4).

Nerve	Composition	Cutaneous and Joint Innervation (Sensory)	Muscular Innervation (Motor and Proprioceptive)
Superior gluteal n.	Mixed	(None)	Gluteus minimus, gluteus medius, and tensor fasciae latae muscles
Inferior gluteal n.	Mixed	(None)	Gluteus maximus muscle
Posterior cutaneous n.	Somatosensory	Skin of gluteal region, perineum, posterior and medial thigh, popliteal fossa, and upper posterior leg	(None)
Tibial n.	Mixed	Skin of posterior leg; plantar skin; knee and foot joints	Hamstring muscles; posterior muscles of leg (see tables 12.6 and 12.7); most intrinsic foot muscles (via plantar nerves) (see table 12.8)
Fibular (peroneal) n. (common, deep, and superficial)	Mixed	Skin of anterior distal third of leg, dorsum of foot, and toes I–II; knee joint	Biceps femoris muscle; anterior and lateral muscles of leg; extensor digitorum brevis muscle of foot (see tables 12.7 to 12.9)
Pudendal n.	Mixed	Skin of penis and scrotum of male; clitoris, labia majora and minora, and lower vagina of female	Muscles of perineum (see table 11.8)

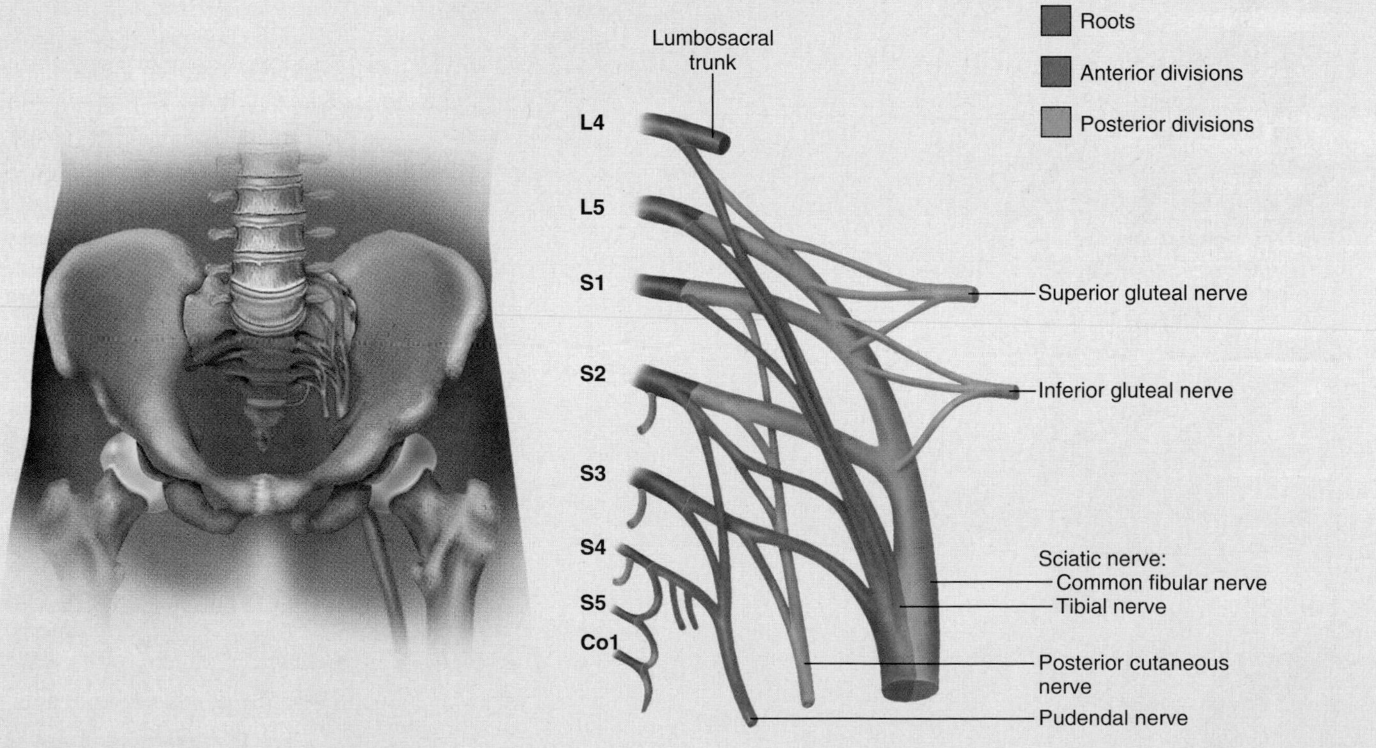

Figure 14.17 The Sacral and Coccygeal Plexuses.

Cutaneous Innervation and Dermatomes

Each spinal nerve except C1 receives sensory input from a specific area of skin called a **dermatome**,[21] derived from the embryonic dermatomes described in chapter 4. A *dermatome map* (fig. 14.18) is a diagram of the cutaneous regions innervated by each spinal nerve. Such a map is oversimplified, however, because the dermatomes overlap at their edges by as much as 50%. Therefore, sev-

---

[21]*derma* = skin + *tome* = segment, part

erance of one sensory nerve root does not entirely deaden sensation from a dermatome. It is necessary to sever or anesthetize three successive spinal nerves to produce a total loss of sensation from one dermatome. Spinal nerve damage is assessed by testing the dermatomes with pinpricks and noting areas in which the patient has no sensation.

Before You Go On

Answer the following questions to test your understanding of the preceding section:

6. What is meant by the posterior and anterior roots of a spinal nerve? Which of these is sensory, and which is motor?

7. Where are the somas of the posterior root located? Where are the somas of the anterior root?

8. List the five plexuses of spinal nerves and state where each one is located.

9. State which plexus gives rise to each of the following nerves: axillary, ilioinguinal, obturator, phrenic, pudendal, radial, and sciatic.

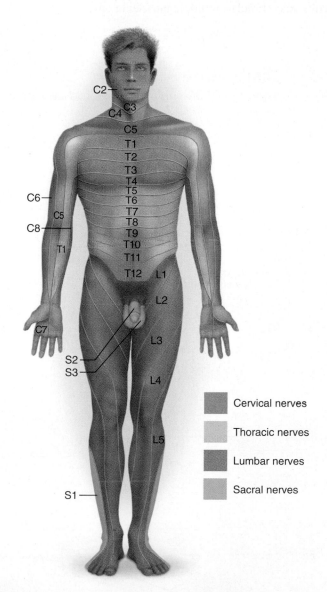

Figure 14.18 **A Dermatome Map of the Anterior Aspect of the Body.** Each zone of the skin is innervated by sensory branches of the spinal nerves indicated by the labels. Nerve C1 does not innervate the skin.

14.3 Somatic Reflexes

▶ Expected Learning Outcomes

When you have completed this section, you should be able to

- define *reflex* and explain how reflexes differ from other motor actions;
- describe the general components of a typical reflex arc; and
- describe some common variations in reflex arcs.

Reflexes are quick, involuntary, stereotyped reactions of glands or muscles to stimulation. This definition sums up four important properties of a reflex:

1. Reflexes *require stimulation*—they are not spontaneous actions but responses to sensory input.

2. Reflexes are *quick*—they generally involve few if any interneurons and minimal synaptic delay.

3. Reflexes are *involuntary*—they occur without intent, often without our awareness, and they are difficult to suppress. Given an adequate stimulus, the response is essentially automatic. You may become conscious of the stimulus that evoked a reflex, and this awareness may enable you to correct or avoid a potentially dangerous situation, but awareness is not a part of the reflex itself. It may come after the reflex action has been completed, and some reflexes occur even if the spinal cord has been severed so that no stimuli reach the brain.

4. Reflexes are *stereotyped*—they occur in essentially the same way every time; the response is very predictable.

Visceral reflexes are responses of glands, cardiac muscle, and smooth muscle. They are controlled by the autonomic nervous system and discussed in chapter 16. **Somatic reflexes** are responses of skeletal muscles, such as the quick withdrawal of your hand from a hot stove or the lifting of your foot when you step on something sharp. They are controlled by the somatic nervous system. They have traditionally been called *spinal reflexes,* although this is a misleading expression for two reasons: (1) Spinal reflexes are not exclusively somatic; the autonomic (visceral) reflexes also involve the spinal cord. (2) Some somatic reflexes are mediated more by the brain than by the spinal cord.

Somatic reflexes will be briefly discussed here from the anatomical standpoint. A somatic reflex employs a rather simple neural pathway called a **reflex arc,** from a sensory nerve ending to the spinal cord or brainstem and back to a skeletal muscle (fig. 14.19). The components of a reflex arc are as follows:

1. *Somatic receptors* in the skin, a muscle, or a tendon. These include simple nerve endings for heat and pain in the skin, specialized stretch receptors called *muscle spindles* embedded in the skeletal muscles, and other types (see chapter 17, p. 463).

2. *Afferent nerve fibers,* which carry information from these receptors into the posterior horn of the spinal cord.

3. An *integrating center,* a neural pool in the gray matter of the spinal cord or brainstem. In most reflex arcs, the integrating center has one or more interneurons. Synaptic events in the integrating center determine whether the efferent (output) neuron issues a signal to the muscle.

4. *Efferent nerve fibers,* which originate in the anterior horn of the spinal cord and carry motor impulses to the skeletal muscles.

5. *Skeletal muscles,* the somatic effectors that carry out the response.

In the simplest type of reflex arc, there is no interneuron. The afferent neuron synapses directly with an efferent neuron, so this kind of pathway is called a **monosynaptic reflex arc.** Synaptic delay is minimal, and the response is especially quick. Most reflex arcs, however, have one or more interneurons, and indeed often involve multineuron circuits with many synapses. Such reflex arcs produce more prolonged muscular responses and, by way of diverging circuits, may stimulate multiple muscles at once.

Apply What You Know

There is actually a second synapse in a "monosynaptic" reflex arc. Identify its location.

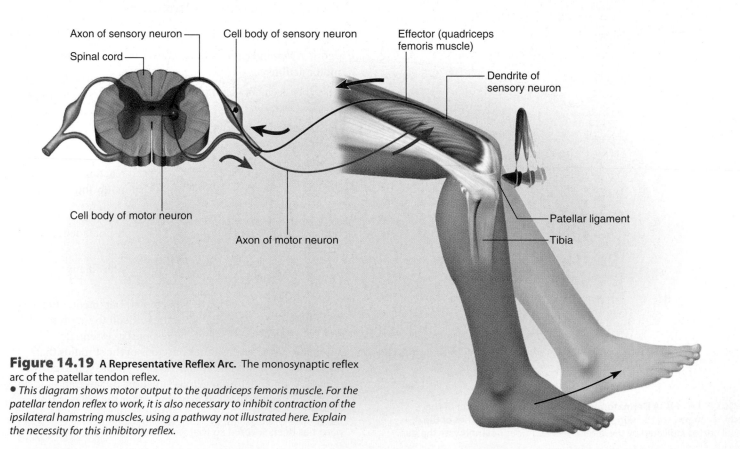

Figure 14.19 **A Representative Reflex Arc.** The monosynaptic reflex arc of the patellar tendon reflex.
• *This diagram shows motor output to the quadriceps femoris muscle. For the patellar tendon reflex to work, it is also necessary to inhibit contraction of the ipsilateral hamstring muscles, using a pathway not illustrated here. Explain the necessity for this inhibitory reflex.*

A reflex like the one diagrammed in figure 14.19 is described as an **ipsilateral reflex** because the CNS input and output are on the same side of the body. Others such as the crossed extension reflex (table 14.7) are called **contralateral reflexes** because the sensory input enters the spinal cord on one side of the body and the motor output leaves on the opposite side. In an **intersegmental reflex,** the sensory signal enters the spinal cord at one level (segment), and the motor output leaves the cord from a higher or lower level. For example, if you step on something sharp and lift your foot from the ground, some motor output leaves the spinal cord higher up and goes to trunk muscles that flex your waist. This shifts your center of gravity over the leg still on the ground, preventing you from falling over.

Table 14.7 describes several types of somatic reflexes. These reflexes are controlled primarily by the cerebrum and cerebellum of the brain, but a weak response is mediated through the spinal cord and persists even if the spinal cord is severed from the brain. The spinal component can be more pronounced if the stimulus is sudden or intense, as in the clinical testing of the knee-jerk (patellar) reflex and other stretch reflexes.

Before You Go On

Answer the following questions to test your understanding of the preceding section:

10. Define *reflex.* Distinguish between somatic and visceral reflexes.

11. List and define the five components of a typical somatic reflex arc.

12. Describe a situation in which each of the following would be functionally relevant: an ipsilateral, a contralateral, and an intersegmental reflex arc.

14.4 Clinical Perspectives

▶ Expected Learning Outcomes

When you have completed this section, you should be able to

- describe some effects of spinal cord injuries; and
- define the types of paralysis and explain the basis for their differences.

Some developmental abnormalities of the spinal cord are described in chapter 13. In children and adults, the most significant disorder of the spinal cord is trauma. Each year in the United States, 10,000 to 12,000 people become paralyzed by spinal cord trauma, usually as a result of vertebral fractures. The group at greatest risk is males from 16 to 30 years old, because of their high-risk behaviors. Fifty-five percent of their injuries are from automobile and motorcycle accidents, 18% from sports, and 15% from gunshot and stab wounds. Elderly people are also at above-average risk because of falls, and in times of war, battlefield injuries account for many cases.

Complete *transection* (severance) of the spinal cord causes immediate loss of motor control at and below the level of the injury. Victims also lose all sensation from the level of injury and below, although some patients temporarily feel burning pain within one or two dermatomes of the level of the lesion.

Apply What You Know

Respiratory paralysis typically results from spinal cord transection above level C4, but not from injuries below that level. Explain.

In the early stage, victims exhibit a syndrome (a suite of signs and symptoms) called **spinal shock.** The muscles below the level of injury exhibit flaccid paralysis (inability to contract) and an absence of reflexes because of the lack of stimulation from higher levels of the CNS. For 8 days to 8 weeks after the accident, the patient typically lacks bladder and bowel reflexes and thus retains urine and feces. Lacking sympathetic stimulation to the blood vessels, a patient may exhibit *neurogenic shock* in which the vessels dilate and blood pressure drops dangerously low. Spinal shock can last from a few days to 3 months, but typically lasts 7 to 20 days.

As spinal shock subsides, somatic reflexes begin to reappear, at first in the toes and progressing to the feet and legs. Autonomic reflexes also reappear. Contrary to the earlier urinary and fecal retention, a patient now has the opposite problem, incontinence, as the rectum and bladder empty reflexively in response to stretch. Both the somatic and autonomic nervous systems typically exhibit exaggerated reflexes, a state called *hyperreflexia* or the *mass reflex reaction.* Stimuli such as a full bladder or cutaneous touch can trigger an extreme cardiovascular reaction. The systolic blood pressure, normally about 120 mm Hg, jumps as high as 300 mm Hg, sometimes causing a stroke. Pressure receptors in the major arteries sense this rise in blood pressure and activate a reflex that slows the heart,

TABLE 14.7	Types of Somatic Reflexes
Stretch reflex	Increased muscle tension in response to stretch. Serves to maintain equilibrium and posture, stabilize joints, and make joint actions smoother and better coordinated. The knee-jerk reflex (patellar reflex, fig. 14.19) is a familiar monosynaptic spinal reflex.
Flexor reflex	Contraction of flexor muscles resulting in withdrawal of a limb from an injurious stimulus, as in withdrawal from a burn or pinprick.
Crossed extension reflex	Contraction of extensor muscles in one limb when the flexor muscles of the opposite limb contract. Stiffens one leg, for example, when the opposite leg is lifted from the ground, so that one does not fall over.
Tendon reflex	Inhibition of muscle contraction when a tendon is excessively stretched, serving to prevent tendon injuries.

sometimes to a rate as low as 30 or 40 beats/minute (*bradycardia*). Men at first lose the capacity for erection and ejaculation. They may recover these functions later and become capable of climaxing and fathering children, but still lack sexual sensation.

The most serious permanent effect of spinal cord trauma is paralysis. As reflexes reappear, they lack inhibitory control from the brain. Consequently, the initial flaccid paralysis of spinal shock changes to spastic paralysis. This typically starts with chronic flexion of the hips and knees (*flexor spasms*) and progresses to a state in which the limbs become straight and rigid (*extensor spasms*). Three forms of muscle paralysis are **paraplegia,** a paralysis of both lower limbs resulting from spinal cord lesions at levels T1 to L1; **quadriplegia,** the paralysis of all four limbs resulting from lesions above level C5; and **hemiplegia,** paralysis of one side of the body, usually resulting not from spinal cord injuries but from a stroke or other brain lesion. Spinal cord lesions from C5 to C7 can produce a state of partial quadriplegia—total paralysis of the lower limbs and partial paralysis (*paresis,* or weakness) of the upper limbs.

Treatment of spinal cord injuries is an area of intense medical research today, with hopes for recovery of spinal functions stimulated by new insights into the physiological mechanisms of spinal cord tissue death and the potential for embryonic stem cells to regenerate damaged cord tissue.

Table 14.8 describes some injuries and other disorders of the spinal cord and spinal nerves.

Before You Go On

Answer the following questions to test your understanding of the preceding section:

13. Describe the signs of spinal shock.
14. Describe the difference between flaccid paralysis and spastic paralysis.
15. Distinguish between the causes of paraplegia, quadriplegia, and hemiplegia.

TABLE 14.8	**Some Disorders of the Spinal Cord and Spinal Nerves**
Guillain-Barré syndrome	An acute demyelinating nerve disorder often triggered by viral infection, resulting in muscle weakness, elevated heart rate, unstable blood pressure, shortness of breath, and sometimes death from respiratory paralysis
Neuralgia	General term for nerve pain, often caused by pressure on spinal nerves from herniated intervertebral discs or other causes
Paresthesia	Abnormal sensations of prickling, burning, numbness, or tingling; a symptom of nerve trauma or other peripheral nerve disorders
Peripheral neuropathy	Any loss of sensory or motor function due to nerve injury; also called *nerve palsy*
Rabies (hydrophobia)	A disease usually contracted from animal bites, involving viral infection that spreads via somatic motor nerve fibers to the CNS and then out of the CNS via autonomic nerve fibers, leading to seizures, coma, and death; invariably fatal if not treated before CNS symptoms appear
Spinal meningitis	Inflammation of the spinal meninges due to viral, bacterial, or other infection

Disorders Described Elsewhere

Study Guide

Assess Your Learning Outcomes

You should have a good understanding of this chapter if you can accurately address the following issues.

14.1 The Spinal Cord (p. 372)

1. Three functions of the spinal cord and their relationship to spinal tracts, central pattern generators, and reflex arcs in the cord
2. The extent of the spinal cord in relation to the vertebrae, and the location and composition of the cauda equina
3. The four regions of the spinal cord and the basis for their names
4. What is meant by a segment of the spinal cord
5. The locations and functional significance of the cervical and lumbar enlargements of the cord
6. The three meninges associated with the spinal cord and their relationships with the epidural space, dural sheath, subarachnoid space, denticulate and coccygeal ligaments, and lumbar cistern
7. The arrangement of the gray matter and white matter of the spinal cord as seen in cross section, and the composition of each tissue
8. The locations and functional differences between the posterior horn, anterior horn, and lateral horn of the spinal gray matter
9. The columns and tracts of spinal white matter
10. The general function of the ascending tracts; the system used for naming most of them; and their individual names, locations, and functions
11. First- through third-order neurons; decussation in the ascending tracts; and the implication of this decussation for the relationship between the cerebral hemispheres and the origin of sensory signals from the lower body
12. The general function of the descending tracts; the system for naming them; and their individual names, locations, and functions
13. Upper and lower motor neurons; decussation in the descending tracts; and the implication of this decussation for the relationship between the cerebral hemispheres and motor control of the lower body

14.2 The Spinal Nerves (p. 379)

1. The structure of a nerve, including its three layers of connective tissue and how they relate to the organization of nerve fibers into fascicles
2. The differences between afferent and efferent nerve fibers; somatic and visceral fibers; and general and special fibers
3. The difference between sensory, motor, and mixed nerves
4. The structure of a ganglion; where ganglia are found; and the relationship between a ganglion and a nerve
5. The number of spinal nerves and the system for naming and numbering them
6. The structure of the proximal portion of a spinal nerve, including its posterior and anterior roots, their rootlets, and the posterior root ganglion; and how the posterior and anterior roots relate to the posterior and anterior horns of the spinal cord
7. The structure of the distal portion of a spinal nerve, particularly its division into a posterior ramus, anterior ramus, and meningeal branch, and where these three branches lead
8. The five plexuses of spinal nerves—their names, locations, and structure; the nerves that arise from them; and the structures innervated by these nerves (tables 14.3 through 14.6)

14.3 Somatic Reflexes (p. 391)

1. The general characteristics of a reflex, and how visceral reflexes differ from somatic reflexes
2. The components of a reflex arc and the path followed by the afferent and efferent nerve signals of a somatic reflex
3. The differences between a monosynaptic and a polysynaptic reflex arc
4. The differences between ipsilateral, contralateral, and intersegmental reflex arcs, and some reflexes that would employ each type
5. The nature of a stretch reflex, flexor reflex, crossed extension reflex, and tendon reflex

14.4 Clinical Perspectives (p. 393)

1. Causes and risk factors for spinal cord trauma
2. The effects of spinal cord trauma including spinal shock, hyperreflexia, and flaccid and spastic paralysis
3. The differences between paraplegia, quadriplegia, and hemiplegia with respect to their causes and parts of the body affected

Testing Your Recall

1. Below L2, the vertebral canal is occupied by a bundle of spinal nerve roots called
   a. the terminal filum.
   b. the descending tracts.
   c. the gracile fasciculus.
   d. the medullary cone.
   e. the cauda equina.

2. The brachial plexus gives rise to all of the following nerves *except*
   a. the axillary nerve.
   b. the radial nerve.
   c. the obturator nerve.
   d. the median nerve.
   e. the ulnar nerve.

3. Between the dura mater and vertebral bone, one is most likely to find
   a. arachnoid mater.
   b. denticulate ligaments.
   c. cartilage.
   d. adipose tissue.
   e. spongy bone.

4. Which of these tracts carries motor signals destined for the postural muscles?
   a. the gracile fasciculus
   b. the cuneate fasciculus
   c. spinothalamic tract
   d. vestibulospinal tract
   e. tectospinal tract

5. A patient has a gunshot wound that caused a bone fragment to nick the spinal cord. The patient now feels no pain or temperature sensations from that level of the body down. Most likely, the _____ was damaged.
   a. gracile fasciculus
   b. medial lemniscus
   c. tectospinal tract
   d. lateral corticospinal tract
   e. spinothalamic tract

6. Which of these is *not* a region of the spinal cord?
   a. cervical
   b. thoracic
   c. pelvic
   d. lumbar
   e. sacral

7. In the spinal cord, the somas of the lower motor neurons are found in
   a. the cauda equina.
   b. the posterior horns.
   c. the anterior horns.
   d. the posterior root ganglia.
   e. the fasciculi.

8. The outermost connective tissue wrapping of a nerve is called the
   a. epineurium.
   b. perineurium.
   c. endoneurium.
   d. arachnoid mater.
   e. dura mater.

9. The intercostal nerves between the ribs arise from which spinal nerve plexus?
   a. cervical
   b. brachial
   c. lumbar
   d. sacral
   e. none of them

10. All somatic reflexes share all of the following properties *except*
    a. they are quick.
    b. they are monosynaptic.
    c. they require stimulation.
    d. they are involuntary.
    e. they are stereotyped.

11. Outside the CNS, the somas of neurons are clustered in swellings called _____.

12. Distal to the intervertebral foramen, a spinal nerve branches into a posterior and anterior _____.

13. The cerebellum receives feedback from the muscles and joints by way of the _____ tracts of the spinal cord.

14. Motor innervation of the leg proper comes predominantly from the _____ plexus.

15. Neural circuits called _____ in the spinal cord produce the rhythmic muscular contractions of walking.

16. The _____ nerves arise from the cervical plexus and innervate the diaphragm.

17. The crossing of a nerve fiber or tract from the right side of the CNS to the left, or vice versa, is called _____.

18. The nonvisual awareness of the body's position and movements is called _____.

19. The _____ ganglion contains the somas of neurons that carry sensory signals to the spinal cord.

20. The sciatic nerve is a composite of two nerves, the _____ and _____.

Answers in the Appendix

Building Your Medical Vocabulary

State a medical meaning of each of the following word elements, and give a term in which it is used.

1. caudo-
2. contra-
3. later-
4. proprio-
5. gracil-
6. myelo-
7. a-
8. -itis
9. phreno-
10. tom-

Answers in the Appendix

True or False

Determine which five of the following statements are false, and briefly explain why.

1. The gracile fasciculus is a descending spinal tract.
2. At the inferior end, the adult spinal cord ends before the vertebral column does.
3. Each spinal cord segment has only one pair of spinal nerves.
4. Some spinal nerves are sensory and others are motor.
5. The dura mater adheres tightly to the bone tissue of the vertebral canal.
6. The anterior and posterior horns of the spinal cord are composed of gray matter.
7. The corticospinal tracts carry motor signals down the spinal cord.
8. The dermatomes are nonoverlapping regions of skin innervated by different spinal nerves.
9. Somatic reflexes are those that do not involve the brain.
10. The Golgi tendon reflex acts to inhibit muscle contraction.

Answers in the Appendix

Testing Your Comprehension

1. Jillian is thrown from a horse. She strikes the ground with her chin, causing severe hyperextension of the neck. Emergency medical technicians properly immobilize her neck and transport her to a hospital, but she dies 5 minutes after arrival. An autopsy shows multiple fractures of vertebrae C1, C6, and C7 and extensive damage to the spinal cord. Explain why she died rather than being left quadriplegic.

2. Wallace is the victim of a hunting accident. A bullet grazed his vertebral column, and bone fragments severed the left half of his spinal cord at segments T8 through T10. Since the accident, Wallace has had a condition called *dissociated sensory loss,* in which he feels no sensations of deep touch or limb position on the *left* side of his body below the injury and no sensations of pain or heat on the *right* side. Explain what spinal tract(s) the injury has affected and why these sensory losses are on opposite sides of the body.

3. Anthony gets into a fight between rival gangs. As an attacker comes at him with a knife, he turns to flee, but stumbles. The attacker stabs him on the medial side of the right gluteal fold, and Anthony collapses. He loses all use of his right limb, being unable to extend his hip, flex his knee, or move his foot. He never fully recovers these lost functions. Explain what nerve injury Anthony has most likely suffered.

4. Stand with your right shoulder, hip, and foot firmly against a wall. Raise your left foot from the floor without losing contact with the wall at any point. What happens? Why? What principle of this chapter does this demonstrate?

5. When a patient needs a tendon graft, surgeons sometimes use the tendon of the palmaris longus, a relatively dispensable muscle of the forearm. The median nerve lies nearby and looks very similar to this tendon. There have been cases in which a surgeon mistakenly removed a section of this nerve instead of the tendon. What effects do you think such a mistake would have on the patient?

Answers at www.mhhe.com/saladinha3

Improve Your Grade at www.mhhe.com/saladinha3

Practice quizzes, labeling activities, games, and flashcards provide fun ways to master concepts. You can also download image PowerPoint files for each chapter to create a study guide or for taking notes during lecture.

The Brain and Cranial Nerves

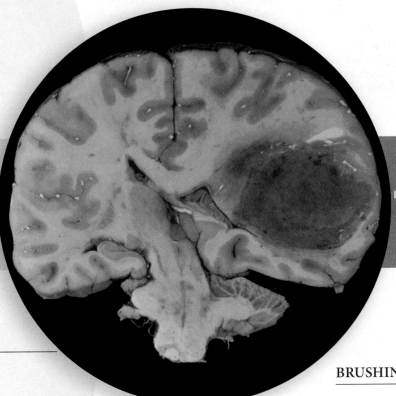

Frontal section of a brain with a large tumor (glioblastoma) in the left cerebral hemisphere

BRUSHING UP

To understand this chapter, you may find it helpful to review the following concepts:
- Anatomy of the cranium (pp. 154–162)
- Glial cells and their functions (p. 357)
- Embryonic development of the central nervous system (p. 365)
- Meninges (p. 373)
- Gray and white matter (p. 375)
- Tracts of the spinal cord (p. 375)
- Structure of nerves and ganglia (p. 380)

Anatomy & Physiology **REVEALED®**
aprevealed.com

Nervous System

The mystique of the brain continues to intrigue modern biologists and psychologists even as it did the philosophers of antiquity. Aristotle thought that the brain was merely a radiator for cooling the blood, but generations earlier, Hippocrates had expressed a more accurate view. "Men ought to know," he said, "that from the brain, and from the brain only, arise our pleasures, joy, laughter and jests, as well as our sorrows, pains, griefs and tears. Through it, in particular, we think, see, hear, and distinguish the ugly from the beautiful, the bad from the good, the pleasant from the unpleasant."

Brain function is so strongly associated with what it means to be alive and human that the cessation of brain activity is taken as a clinical criterion of death even when other organs of the body are still functioning. With its hundreds of neural pools and trillions of synapses, the brain performs sophisticated tasks beyond our present understanding. Still, all of our mental functions, no matter how complex, are ultimately based on the cellular activities described in chapter 13. The relationship of the mind or personality to the cellular function of the brain is a question that will provide fertile ground for scientific study and philosophical debate long into the future.

This chapter is a study of the brain and the cranial nerves directly connected to it. Here we will plumb some of the mysteries of motor control, sensation, emotion, thought, language, personality, memory, dreams, and plans. Your study of this chapter is one brain's attempt to understand itself.

15.1 Overview of the Brain

▶ Expected Learning Outcomes

When you have completed this section, you should be able to

- describe the major subdivisions and anatomical landmarks of the brain;

- state the locations of the gray and white matter of the brain;

- describe the meninges of the brain;

- describe the system of fluid-filled chambers within the brain;

- discuss the production, flow, and function of the cerebrospinal fluid in these chambers; and

- explain the significance of the brain barrier system.

In the evolution of the central nervous system from the simplest vertebrate animals to humans, the spinal cord has changed very little while the brain has changed a great deal. In fishes and amphibians, the brain weighs about the same as the spinal cord, but in humans, it weighs 55 times as much. It averages about 1,600 g (3.5 lb) in men and 1,450 g in women. The difference in weight is proportional to body size, not intelligence. The Neanderthal people had larger brains than modern humans do.

The human brain has a high opinion of itself: Ours is the most sophisticated brain when compared to others in terms of awareness of the environment, adaptability to environmental variation and

change, quick execution of complex decisions, fine motor control and mobility of the body, and behavioral complexity. Over the course of human evolution, the brain has shown its greatest growth in areas concerned with vision, memory, and motor control of the prehensile hand.

Major Landmarks

Before we consider the form and function of specific regions of the brain, it will help to get a general overview of its major landmarks (figs. 15.1 and 15.2). These will provide important points of reference as we progress through a more detailed study.

Two directional terms often used to describe brain anatomy are *rostral* and *caudal*. **Rostral**[1] means "toward the nose" and **caudal**[2] means "toward the tail." These are apt descriptions for an animal such as a laboratory rat, on which so much neuroscience research has been done. The terms are retained for human neuroanatomy as well, but in references to the human brain, *rostral* means toward the forehead and *caudal* means toward the spinal cord. In the spinal cord and brainstem, which are vertically oriented, *rostral* means higher and *caudal* means lower.

The brain is divided into three major portions—the *cerebrum, cerebellum,* and *brainstem.* The **cerebrum** (seh-REE-brum or SER-eh-brum) constitutes about 83% of its volume and consists of a pair of half globes called the **cerebral hemispheres**. Each hemisphere is marked by thick folds called **gyri**[3] (JY-rye; singular, *gyrus*) separated by shallow grooves called **sulci**[4] (SUL-sye; singular, *sulcus*). A very deep groove, the **longitudinal fissure,** separates the right and left hemispheres from each other. At the bottom of this fissure, the hemispheres are connected by a thick bundle of nerve fibers called the **corpus callosum**[5]—a prominent landmark for anatomical description (fig. 15.2).

The **cerebellum**[6] (SER-eh-BEL-um) lies inferior to the cerebrum and occupies the posterior cranial fossa. It is also marked by gyri, sulci, and fissures. The cerebellum is the second-largest region of the brain; it constitutes about 10% of its volume but contains over 50% of its neurons.

Authorities differ on how they define the **brainstem.** This book treats it as that which remains of the brain if the cerebrum and cerebellum are removed. Its major components, from rostral to caudal, are the *diencephalon, midbrain, pons,* and *medulla oblongata.* The most common alternative definition includes only the last three of these. In a living person, the brainstem is oriented like a vertical stalk with the cerebrum perched on top of it like a mushroom cap. Postmortem changes give it a more oblique angle in the cadaver and consequently in many medical illustrations. Caudally, the brainstem ends at the foramen magnum of the skull, and the central nervous system (CNS) continues below this as the spinal cord.

---

[1]*rostr* = nose
[2]*caud* = tail
[3]*gyr* = turn, twist
[4]*sulc* = furrow, groove
[5]*corpus* = body + *call* = thick
[6]*cereb* = brain + *ellum* = little

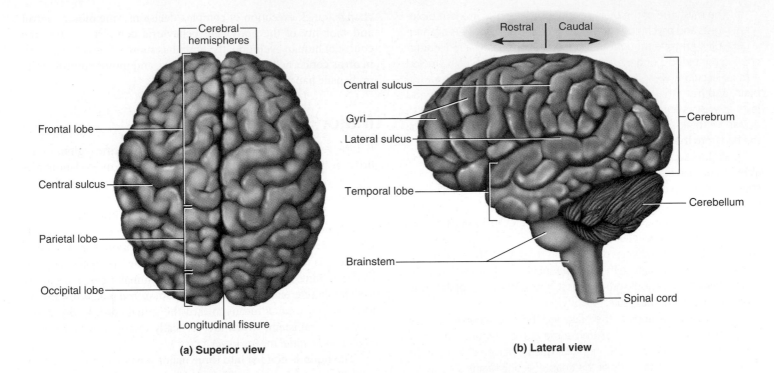

Cerebral hemispheres

Frontal lobe

Central sulcus

Parietal lobe

Occipital lobe

Longitudinal fissure

(a) Superior view

Rostral | Caudal

Central sulcus

Gyri

Lateral sulcus

Temporal lobe

Brainstem

Cerebrum

Cerebellum

Spinal cord

(b) Lateral view

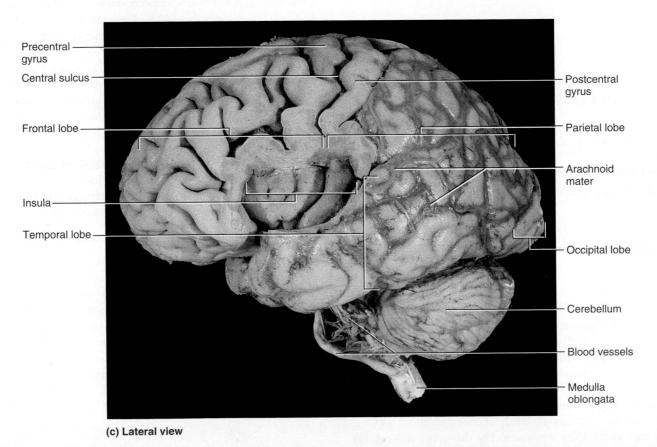

Precentral gyrus

Central sulcus

Frontal lobe

Insula

Temporal lobe

Postcentral gyrus

Parietal lobe

Arachnoid mater

Occipital lobe

Cerebellum

Blood vessels

Medulla oblongata

(c) Lateral view

Figure 15.1 Surface Anatomy of the Brain. (a) Superior view of the cerebral hemispheres. (b) Left lateral view. (c) The partially dissected brain of a cadaver. Part of the left hemisphere is cut away to expose the insula. The arachnoid mater is removed from the anterior (rostral) half of the brain to expose the gyri and sulci; the arachnoid mater with its blood vessels is seen on the posterior (caudal) half. Blood vessels of the brainstem are left in place.
• *After studying the meninges, determine which of the three had to be removed to expose any of the anatomy visible in part (c).*

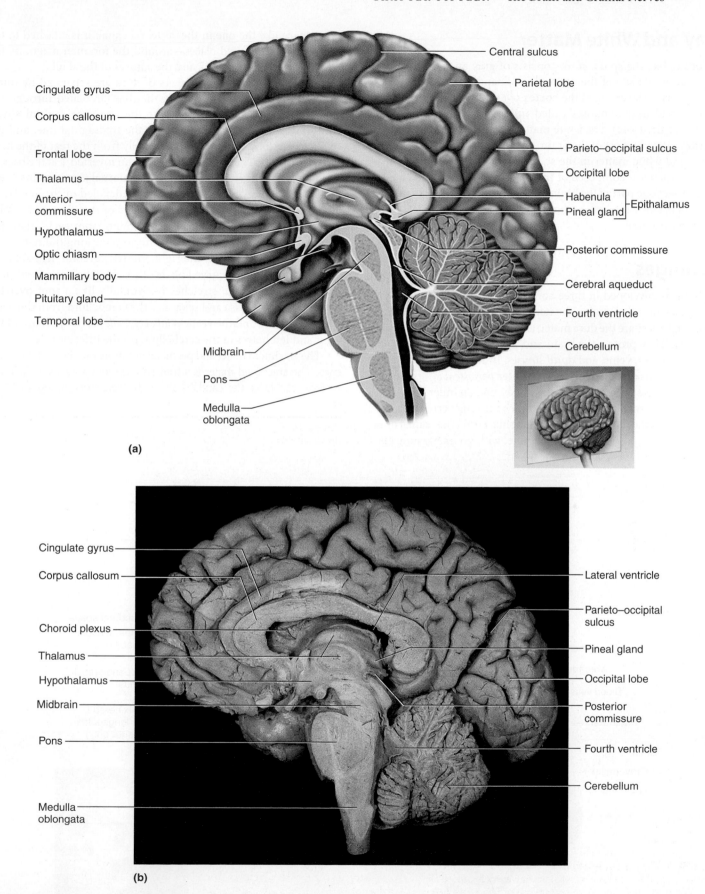

Central sulcus

Parietal lobe

Cingulate gyrus

Corpus callosum

Frontal lobe

Parieto–occipital sulcus

Occipital lobe

Thalamus

Habenula
Pineal gland ⎤ Epithalamus

Anterior commissure

Hypothalamus

Optic chiasm

Posterior commissure

Mammillary body

Cerebral aqueduct

Pituitary gland

Fourth ventricle

Temporal lobe

Cerebellum

Midbrain

Pons

Medulla oblongata

(a)

Cingulate gyrus

Corpus callosum

Lateral ventricle

Parieto–occipital sulcus

Choroid plexus

Thalamus

Pineal gland

Hypothalamus

Occipital lobe

Midbrain

Posterior commissure

Pons

Fourth ventricle

Cerebellum

Medulla oblongata

(b)

Figure 15.2 **Medial Aspect of the Brain.** (a) Median section, left lateral view. (b) Median section of the cadaver brain.

Gray and White Matter

The brain, like the spinal cord, consists of gray and white matter. Gray matter, the site of the neurosomas, dendrites, and synapses, forms a surface layer called the **cortex** over the cerebrum and cerebellum, and deeper masses called **nuclei** surrounded by white matter (see fig. 15.4c). The white matter thus lies deep to the cortical gray matter in most of the brain, opposite from the relationship of gray and white matter in the spinal cord. As in the spinal cord, the white matter is composed of **tracts,** or bundles of axons, which here connect one part of the brain to another and to the spinal cord. It gets its bright white color from myelin. The tracts are described later in more detail.

Meninges

The brain is enveloped in three connective tissue membranes, the meninges, which lie between the nervous tissue and bone. As in the spinal cord, these are the dura mater, arachnoid mater, and pia mater (fig. 15.3). They protect the brain and provide a structural framework for arteries, veins, and *dural sinuses.* In the cranial cavity, the dura mater consists of two layers—an outer *periosteal layer,* equivalent to the periosteum of the cranial bones, and an inner *meningeal layer.* Only the meningeal layer continues into the vertebral canal, where it forms the dural sac around the spinal cord. The cranial dura mater lies closely against the cranial bone, with no intervening epi-

dural space like the one in the vertebral canal. It is attached to the cranial bone in limited places—around the foramen magnum, the sella turcica, the crista galli, and the sutures of the skull.

In some places, the two layers of dura are separated by **dural sinuses,** spaces that collect blood that has circulated through the brain. Two major dural sinuses are the **superior sagittal sinus,** found just under the cranium along the midsagittal line, and the **transverse sinus,** which runs horizontally from the rear of the head toward each ear. These sinuses meet like an inverted T at the back of the brain and ultimately empty into the internal jugular veins of the neck. The anatomy of the dural sinuses is detailed in chapter 21.

In certain places, the meningeal layer of the dura mater folds inward to separate major parts of the brain from each other: the *falx*[7] *cerebri* (falks SER-eh-bry) extends into the longitudinal fissure as a vertical wall between the right and left cerebral hemispheres, and is shaped like the curved blade of a sickle; the *tentorium*[8] (ten-TOE-ree-um) *cerebelli* stretches horizontally like a roof over the posterior cranial fossa and separates the cerebellum from the overlying cerebrum; and the vertical *falx cerebelli* partially separates the right and left halves of the cerebellum on the inferior side.

The arachnoid mater and pia mater are similar to those of the spinal cord. The arachnoid mater is a transparent membrane over the brain surface, visible in the caudal half of the cerebrum in figure 15.1c.

---

[7]*falx* = sickle
[8]*tentorium* = tent

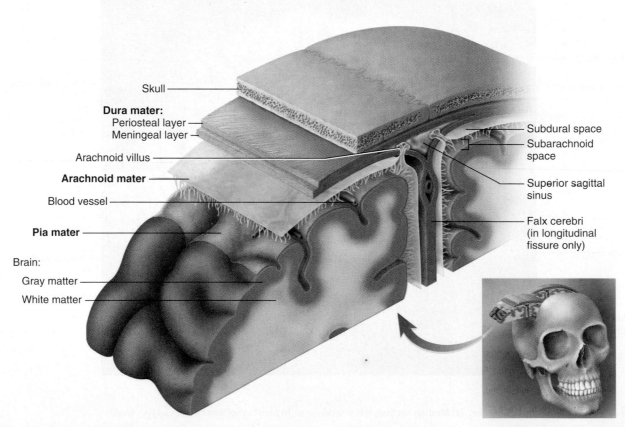

Figure 15.3 The Meninges of the Brain. Frontal section of the head.

A *subarachnoid space* separates the arachnoid from the pia, and in some places, a *subdural space* separates the dura from the arachnoid. The pia mater is a very thin, delicate membrane that closely follows all the contours of the brain surface, even dipping into the sulci. It is not usually visible without a microscope.

Ventricles and Cerebrospinal Fluid

The brain has four internal chambers called **ventricles.** The largest are the two **lateral ventricles,** which form an arc in each cerebral hemisphere (fig. 15.4). Through a pore called the **interventricular**

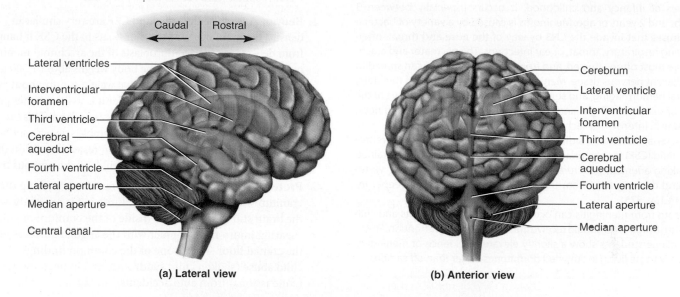

(a) Lateral view

(b) Anterior view

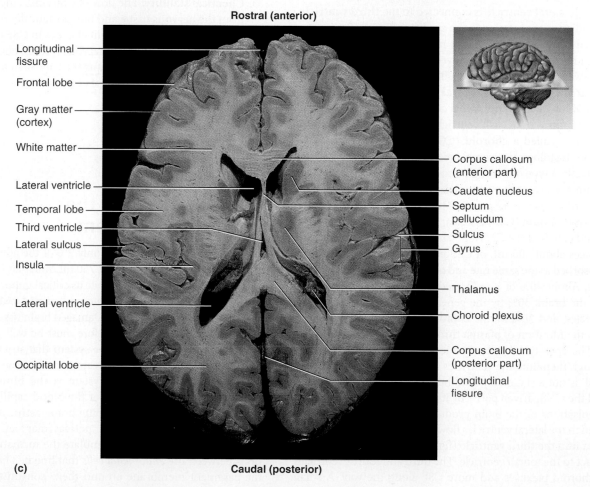

(c)

Figure 15.4 **Ventricles of the Brain.** (a) Right lateral view. (b) Anterior view. (c) Superior view of a horizontal section of the cadaver brain, showing the lateral ventricles and some other features of the cerebrum.

DEEPER INSIGHT   15.1

Meningitis

Meningitis—inflammation of the meninges—is one of the most serious diseases of infancy and childhood. It occurs especially between 3 months and 2 years of age. Meningitis is caused by a variety of bacteria and viruses that invade the CNS by way of the nose and throat, often following respiratory, throat, or ear infections. The pia mater and arachnoid are most often affected, and from here the infection can spread to the adjacent nervous tissue. Meningitis can cause swelling of the brain, cerebral hemorrhaging, and sometimes death within mere hours of the onset of symptoms. Signs and symptoms include high fever, stiff neck, drowsiness, intense headache, and vomiting.

Bacterial meningitis is diagnosed partly by examining the cerebrospinal fluid (CSF) for bacteria and white blood cells. The CSF is obtained by making a *lumbar puncture (spinal tap)* between two lumbar vertebrae and drawing fluid from the subarachnoid space (see Deeper Insight 14.1, p. 375).

Death from meningitis can occur so suddenly that infants and children with a high fever should receive immediate medical attention. Freshman college students show a slightly elevated incidence of meningitis, especially those living in crowded dormitories rather than off campus.

---

foramen, each lateral ventricle is connected to the **third ventricle,** a narrow median space inferior to the corpus callosum. From here, a canal called the **cerebral aqueduct** passes down the core of the midbrain and leads to the **fourth ventricle,** a small triangular chamber between the pons and cerebellum (see fig. 15.2). Caudally, this space narrows and forms a **central canal** that extends through the medulla oblongata into the spinal cord.

On the floor or wall of each ventricle, there is a spongy mass of blood capillaries called a **choroid** (CO-royd) **plexus** (fig. 15.4c), named for its histological resemblance to the chorion of a fetus. Ependymal cells, a type of neuroglia, cover each choroid plexus, the entire interior surface of the ventricles, and the canals of the brain and spinal cord. The choroid plexuses produce cerebrospinal fluid.

Cerebrospinal fluid (CSF) is a clear, colorless liquid that fills the ventricles and canals of the CNS and bathes its external surface. The brain produces about 500 mL of CSF per day, but the fluid is constantly reabsorbed at the same rate and only 100 to 160 mL is present at one time. About 40% of it is formed in the subarachnoid space external to the brain, 30% by the general ependymal lining of the brain ventricles, and 30% by the choroid plexuses. CSF formation begins with the filtration of plasma through the blood capillaries of the brain. The ependymal cells chemically modify the filtrate as it passes through them into the ventricles and subarachnoid space.

The CSF is not a stationary fluid but continually flows through and around the CNS, driven partly by its own pressure and partly by rhythmic pulsations of the brain produced by each heartbeat. The CSF secreted in the lateral ventricles flows through the interventricular foramina into the third ventricle (fig. 15.5), then down the cerebral aqueduct to the fourth ventricle. The third and fourth ventricles and their choroid plexuses add more CSF along the way. A small amount of CSF fills the central canal of the spinal cord, but ultimately, all of it escapes through three pores in the walls of the fourth ventricle: a *median aperture* and two *lateral apertures*. These lead into the subarachnoid space on the surface of the brain and spinal cord. CSF is absorbed from this space by **arachnoid villi,** cauliflower-like extensions of the arachnoid meninx that protrude through the dura mater into the superior sagittal sinus of the brain. CSF penetrates the walls of the arachnoid villi and mixes with the blood in the sinus.

Cerebrospinal fluid serves three purposes:

1. **Buoyancy.** Because the brain and CSF are very similar in density, the brain neither sinks nor floats in the CSF. It hangs from delicate specialized fibroblasts of the arachnoid meninx. A human brain removed from the body weighs about 1,500 g, but when suspended in CSF, its effective weight is only about 50 g. By analogy, consider how much easier it is to lift another person when you are standing in a lake than it is on land. This buoyancy effect allows the brain to attain considerable size without being impaired by its own weight. If the brain rested heavily on the floor of the cranium, the pressure would kill the nervous tissue.

2. **Protection.** CSF also protects the brain from striking the cranium when the head is jolted. If the jolt is severe, however, the brain still may strike the inside of the cranium or suffer shearing injury from contact with the angular surfaces of the cranial floor. This is one of the common findings in child abuse (shaken child syndrome) and in head injuries (concussions) from auto accidents, boxing, and the like.

3. **Chemical stability.** The flow of CSF rinses metabolic wastes from the nervous tissue and homeostatically regulates its chemical environment. Slight changes in CSF composition can cause malfunctions of the nervous system. For example, a high glycine concentration disrupts the control of temperature and blood pressure, and a high pH causes dizziness and fainting.

Apply What You Know

What effect would you expect from a small brain tumor that blocked the left interventricular foramen?

Blood Supply and the Brain Barrier System

Although the brain constitutes only 2% of the adult body weight, it receives 15% of the blood (about 750 mL/min.) and consumes 20% of the oxygen and glucose. But despite its critical importance to the brain, blood is also a source of agents such as bacterial toxins and antibodies that can harm the brain tissue. Damaged brain tissue is essentially irreplaceable, and the brain therefore must be well protected. Consequently, there is a **brain barrier system** that strictly regulates what substances get from the bloodstream into the tissue fluid of the brain.

One component of this system is the **blood–brain barrier (BBB),** which seals nearly all of the blood capillaries throughout the brain tissue. In the developing brain, astrocytes reach out and contact the capillaries with their perivascular feet. They do not fully surround the capillary, but stimulate the formation of tight junctions between the *endothelial cells* that line it. These junctions and the basement membrane around them constitute the BBB. Anything passing from the blood into the tissue fluid has to pass through the endothelial cells themselves, which are more selective than gaps between the cells can be.

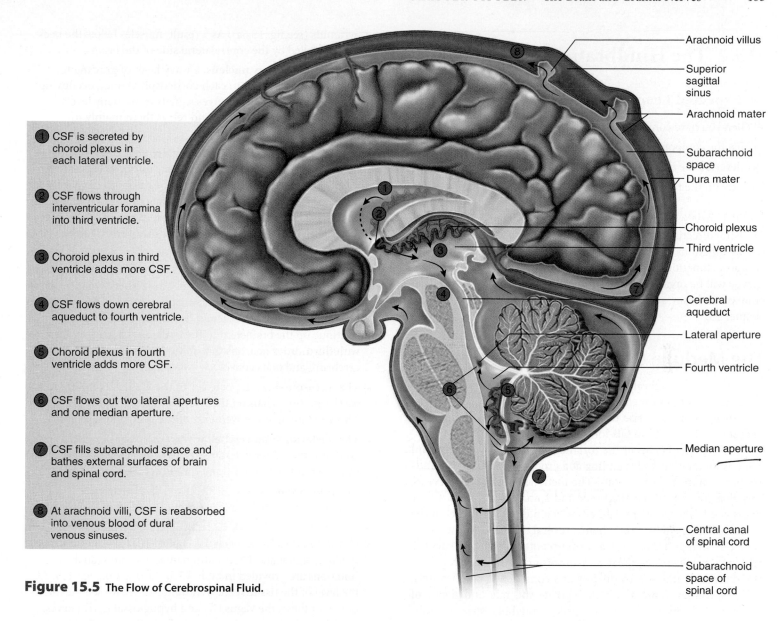

① CSF is secreted by choroid plexus in each lateral ventricle.

② CSF flows through interventricular foramina into third ventricle.

③ Choroid plexus in third ventricle adds more CSF.

④ CSF flows down cerebral aqueduct to fourth ventricle.

⑤ Choroid plexus in fourth ventricle adds more CSF.

⑥ CSF flows out two lateral apertures and one median aperture.

⑦ CSF fills subarachnoid space and bathes external surfaces of brain and spinal cord.

⑧ At arachnoid villi, CSF is reabsorbed into venous blood of dural venous sinuses.

Arachnoid villus
Superior sagittal sinus
Arachnoid mater
Subarachnoid space
Dura mater
Choroid plexus
Third ventricle
Cerebral aqueduct
Lateral aperture
Fourth ventricle
Median aperture
Central canal of spinal cord
Subarachnoid space of spinal cord

Figure 15.5 The Flow of Cerebrospinal Fluid.

At the choroid plexuses, there is a similar **blood–CSF barrier,** composed of ependymal cells joined by tight junctions. Tight junctions are absent from ependymal cells elsewhere, because it is important to allow exchanges between the brain tissue and CSF. That is, there is no brain–CSF barrier.

The brain barrier system (BBS) is highly permeable to water; glucose; lipid-soluble substances such as oxygen and carbon dioxide; and drugs such as alcohol, caffeine, nicotine, and anesthetics. While the BBS is an important protective device, it is an obstacle to the delivery of drugs such as antibiotics and cancer drugs, and thus complicates the treatment of brain diseases.

The BBB is absent from patches called **circumventricular organs (CVOs)** on the walls of the third and fourth ventricles. Here, the blood has direct access to the brain tissue, enabling the brain to monitor and respond to fluctuations in blood chemistry. Unfortunately, the CVOs also afford a route for the human immunodeficiency virus (HIV) to invade the brain.

Before You Go On

Answer the following questions to test your understanding of the preceding section:

1. List the three major parts of the brain and describe their locations.

2. Define *gyrus* and *sulcus*.

3. Name the parts of the brainstem from caudal to rostral.

4. Name the three meninges from superficial to deep.

5. Describe three functions of the cerebrospinal fluid.

6. Where does the CSF originate, and what route does it take through and around the CNS?

7. Name the two components of the brain barrier system and explain the importance of this system.

15.2 The Hindbrain and Midbrain

▶ Expected Learning Outcomes

When you have completed this section, you should be able to

- list the components of the hindbrain and midbrain;
- describe the major features of their anatomy; and
- explain the functions of each hindbrain and midbrain region.

We will survey the functional anatomy of the brain in a caudal to rostral direction, beginning with the hindbrain and its relatively simple functions, and progressing to the forebrain, the seat of such complex functions as thought, memory, and emotion. This survey will be organized around the five secondary vesicles of the embryonic brain and their mature derivatives, as described in chapter 13.

The Medulla Oblongata

The embryonic hindbrain differentiates into two subdivisions, the myelencephalon and metencephalon (see fig. 13.14, p. 366). The myelencephalon gives rise to just one structure, the **medulla oblongata** (meh-DULL-uh OB-long-GAH-ta).

The medulla begins at the foramen magnum of the skull and extends for 3 cm rostrally, ending at a groove marking the boundary between medulla and pons. The medulla contains all nerve fibers that travel between the brain and spinal cord. The last four (most caudal) pairs of cranial nerves begin or end at nuclei in the medulla. The medulla also contains several nuclei concerned with basic physiological functions: a **cardiac center**, which regulates the rate and force of the heartbeat; a **vasomotor center**, which regulates blood pressure and flow by dilating and constricting blood vessels; two **respiratory centers**, which regulate the rate and depth of breathing; and others involved in speech, coughing, sneezing, salivation, swallowing, gagging, vomiting, and sweating.

The external anatomy of the medulla oblongata is shown in figure 15.6. The anterior surface bears a pair of clublike ridges, the **pyramids.** Resembling side-by-side baseball bats, the pyramids are wider at the rostral end, taper caudally, and are separated by a longitudinal groove, the *anterior median fissure,* continuous with that of the spinal cord (fig. 15.6a). Lateral to each pyramid is a prominent bulge called the **olive.** The posterior surface of the medulla exhibits two pairs of ridges, the gracile and cuneate fasciculi, a continuation of the ones in the spinal cord (see p. 377).

Figure 15.7c shows some of the internal structure of the medulla oblongata. Cross sections at other levels would reveal different structures, but in this representative section, we see the following:

- The **corticospinal tracts,** seen anteriorly (at the bottom of the figure). These tracts occupy the pyramids and consist of nerve fibers that descend from the cerebral cortex carrying motor signals to the spinal cord. About 90% of these fibers cross over (decussate) to the opposite side of the brainstem at a point called the *pyramidal decussation,* near the caudal end of the

pyramids (see fig. 15.6a). As a result, muscles below the neck are controlled by the contralateral side of the brain.

- The **inferior olivary nucleus,** a wavy layer of gray matter immediately posterior to each corticospinal tract, occupying the olive. This nucleus receives signals from many levels of the brain and spinal cord and relays them mainly to the cerebellum.

- The **reticular formation,** a vaguely defined region of gray matter posterior to the inferior olivary nucleus. The reticular formation is an elongated body that begins in the upper spinal cord and ascends through the medulla, pons, and midbrain. It consists of numerous nuclei involved in many of the body's most basic physiological functions, some of which are discussed later in this chapter.

- The **gracile** and **cuneate nuclei,** where sensory fibers of the corresponding fasciculi terminate. These fibers synapse here with second-order nerve fibers that decussate and form the ribbonlike **medial lemniscus.**[9] In the lemniscus, they continue up the brainstem to the thalamus, synapsing there with third-order neurons that convey the signals to the cerebrum and one's conscious awareness.

- The **tectospinal tract,** posterior to the medial lemniscus, carries motor signals on their way to the cervical spinal cord. This tract mediates movements of the head and neck.

- The **posterior spinocerebellar tract** continues from the spinal cord into the posterolateral margins of the medulla. It carries sensory information destined for the cerebellum.

- The **fourth ventricle** is a CSF-filled space between the medulla and cerebellum (see fig. 15.2); the medulla forms its anterior boundary but does not completely contain it.

- Cranial nerves IX, X, part of XI, and XII begin or end in the medulla oblongata. Their Latin names and individual functions are provided in table 15.3, starting on page 429. At the level of the tissue section in figure 15.7c, we see the nuclei of two of these, the vagus (X) and hypoglossal (XII) nerves. The trigeminal nerve (V) emerges from the pons, but its nucleus extends into the medulla and is also visible in this figure. Collectively, the sensory functions of cranial nerves IX to XII include touch, pressure, temperature, taste, and pain. Their motor functions include chewing; swallowing; speech; respiration; cardiovascular control; gastrointestinal motility and secretion; and head, neck, and shoulder movements.

The Pons

The metencephalon develops into the pons and cerebellum. The **pons**[10] measures about 2.5 cm long. Most of it forms a broad anterior bulge in the brainstem just rostral to the medulla (fig. 15.6a). Posteriorly, the pons consists mainly of two pairs of thick stalks called *peduncles* that attach it to the cerebellum (fig. 15.6b); these are

---

[9]*lemn = ribbon + iscus = little*
[10]*pons = bridge*

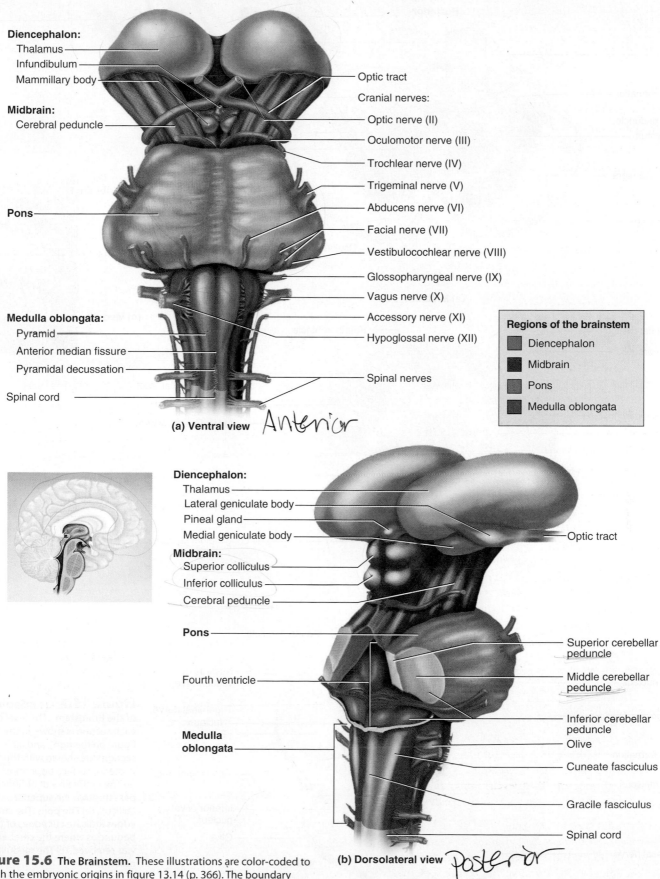

Diencephalon:
Thalamus
Infundibulum
Mammillary body

Midbrain:
Cerebral peduncle

Pons

Medulla oblongata:
Pyramid
Anterior median fissure
Pyramidal decussation

Spinal cord

Optic tract

Cranial nerves:

Optic nerve (II)

Oculomotor nerve (III)

Trochlear nerve (IV)

Trigeminal nerve (V)

Abducens nerve (VI)

Facial nerve (VII)

Vestibulocochlear nerve (VIII)

Glossopharyngeal nerve (IX)

Vagus nerve (X)

Accessory nerve (XI)

Hypoglossal nerve (XII)

Spinal nerves

Regions of the brainstem
Diencephalon
Midbrain
Pons
Medulla oblongata

(a) Ventral view   Anterior

Diencephalon:
Thalamus
Lateral geniculate body
Pineal gland
Medial geniculate body

Midbrain:
Superior colliculus
Inferior colliculus
Cerebral peduncle

Pons

Fourth ventricle

Medulla oblongata

Optic tract

Superior cerebellar peduncle

Middle cerebellar peduncle

Inferior cerebellar peduncle

Olive

Cuneate fasciculus

Gracile fasciculus

Spinal cord

(b) Dorsolateral view   Posterior

Figure 15.6 The Brainstem. These illustrations are color-coded to match the embryonic origins in figure 13.14 (p. 366). The boundary between the middle and inferior cerebellar peduncles is indistinct. Authorities vary on whether to include the diencephalon in the brainstem.

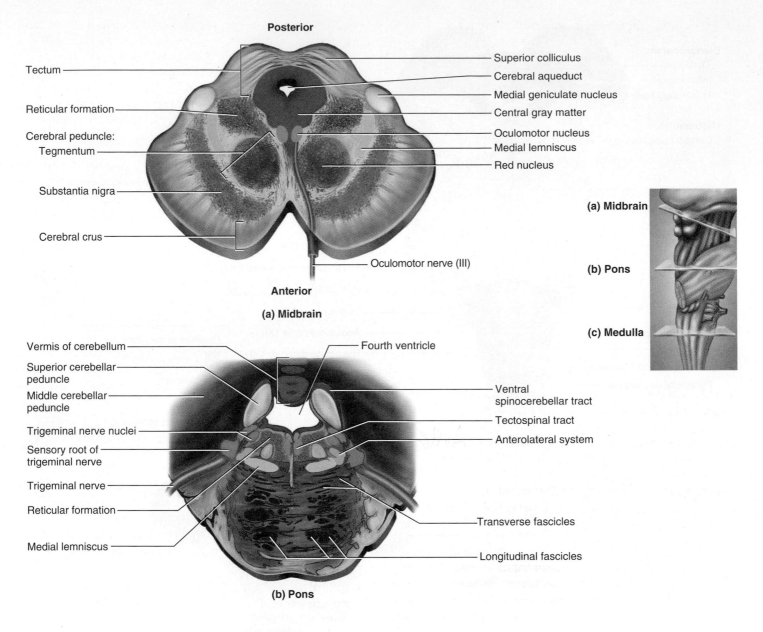

Posterior

Tectum

Reticular formation

Cerebral peduncle:
Tegmentum

Substantia nigra

Cerebral crus

Oculomotor nerve (III)

Anterior

(a) Midbrain

Superior colliculus

Cerebral aqueduct

Medial geniculate nucleus

Central gray matter

Oculomotor nucleus

Medial lemniscus

Red nucleus

(a) Midbrain

(b) Pons

(c) Medulla

Vermis of cerebellum

Superior cerebellar peduncle

Middle cerebellar peduncle

Trigeminal nerve nuclei

Sensory root of trigeminal nerve

Trigeminal nerve

Reticular formation

Medial lemniscus

Fourth ventricle

Ventral spinocerebellar tract

Tectospinal tract

Anterolateral system

Transverse fascicles

Longitudinal fascicles

(b) Pons

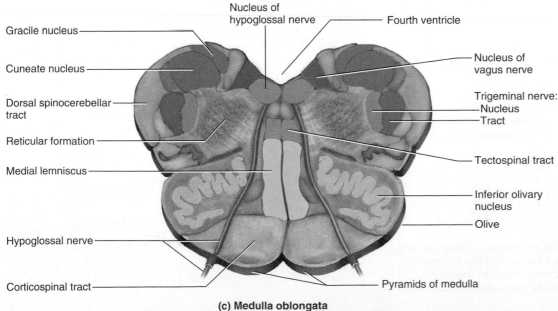

Nucleus of hypoglossal nerve

Gracile nucleus

Cuneate nucleus

Dorsal spinocerebellar tract

Reticular formation

Medial lemniscus

Hypoglossal nerve

Corticospinal tract

Fourth ventricle

Nucleus of vagus nerve

Trigeminal nerve:
Nucleus
Tract

Tectospinal tract

Inferior olivary nucleus

Olive

Pyramids of medulla

(c) Medulla oblongata

Figure 15.7 Cross Sections of the Brainstem. The level of each section is shown in the figure on the right, and all sections are shown with the posterior surface uppermost. (a) The midbrain, cut obliquely to pass through the superior colliculi. (b) The pons. The straight edges indicate cut edges of the peduncles where the cerebellum was removed. (c) The medulla oblongata.

● *Trace the route taken through all three of these sections by fibers from the gracile and cuneate fasciculi described in chapter 14.*

the cut edges of the upper half of figure 15.7b. The peduncles are further described with the cerebellum. The pons contains several nuclei involved in basic physiological functions including sleep, respiration, and bladder control.

In cross section, the pons exhibits continuations of the reticular formation, medial lemniscus, and tectospinal tract, all mentioned previously, as well as the following:

- Continuations of two spinal cord sensory tracts described in chapter 14—the **anterolateral system,** which contains the *spinothalamic tract* en route to the thalamus, and the **anterior spinocerebellar tract,** carrying signals en route to the cerebellum.

- Cranial nerves V through VIII begin or end in the pons, although we see only V (the trigeminal nerve) at the level of the cross section in figure 15.7b. The other three nerves emerge from the groove between the pons and medulla (see fig. 15.24). Their individual names and functions are provided in table 15.3. Collectively, the sensory functions of these four nerves include hearing, equilibrium, taste, and facial sensations such as touch and pain. Their motor functions include eye movements, facial expressions, chewing, swallowing, and the secretion of saliva and tears.

- Tracts of white matter in the anterior half of the pons (lower half of the figure), including transverse fascicles that cross from left to right and connect the two hemispheres of the cerebellum, and longitudinal fascicles that course up and down the pons carrying signals ascending to the thalamus and signals descending from the cerebrum to the cerebellum and medulla.

- Part of the fourth ventricle is seen bordered by the pons, the tail-like *vermis* of the cerebellum, and the superior cerebellar peduncles.

Even though the cerebellum is part of the metencephalon, it is not part of the brainstem. It will be clearer to continue this brainstem discussion with the midbrain and then return to the cerebellum.

The Midbrain

The embryonic mesencephalon produces just one mature brain structure, the **midbrain**—a short segment of the brainstem that connects the hindbrain and forebrain (see figs. 15.2 and 15.8). The midbrain, too, contains continuations of the reticular formation and medial lemniscus. The cerebral aqueduct—the channel connecting the third ventricle to the fourth—passes through the midbrain and furnishes a useful landmark in figure 15.7a. Relative to the aqueduct, this cross section of the midbrain exhibits the following structures:

- The **central (periaqueductal) gray matter,** a region of gray matter surrounding the aqueduct. This functions with the reticular formation in controlling our awareness of pain (see chapter 16).

- The **tectum,**[11] a rooflike region posterior to the aqueduct. The tectum exhibits four bulges collectively called the *corpora*

quadrigemina[12]—a rostral pair called the superior colliculi and a caudal pair called the inferior colliculi (see fig. 15.6b). The **superior colliculi**[13] (col-LIC-you-lye) function in visual attention, such as visually tracking moving objects and reflexively turning the eyes and head in response to a visual stimulus—for example, to look at something you catch sight of in your peripheral vision. The **inferior colliculi** receive and process auditory input from lower levels of the brainstem and relay it to other parts of the brain, especially the thalamus. They are sensitive to time delays between sounds heard by the two ears and thus aid in locating the source of a sound in space, and they function in auditory reflexes such as turning the head toward a sound or the startle response to a loud noise.

- The **tegmentum,**[14] the main mass of the midbrain, located anterior to the aqueduct. The tegmentum contains the **red nucleus,** named for the pink color it gets from a high density of blood vessels. Fibers from the red nucleus form the *rubrospinal tract* in most mammals, but in humans its connections go mainly to and from the cerebellum, with which it collaborates in fine motor control.

- The **substantia nigra**[15] (sub-STAN-she-uh NY-gruh), a dark gray to black nucleus pigmented with melanin, located between the tegmentum and cerebral crura (discussed next). This is a motor center that relays inhibitory signals to the thalamus and basal nuclei (both discussed later). It improves motor performance by suppressing unwanted muscle contractions. Degeneration of the substantia nigra leads to the uncontrollable muscle tremors of Parkinson disease (see p. 437).

- The **cerebral crura** (CROO-ra; singular, *crus*[16]), which anchor the cerebrum to the brainstem. The crura, tegmentum, and substantia nigra collectively form the *cerebral peduncle.* Corticospinal and other tracts from the cerebrum descend through the crura on their way to lower levels of the brainstem and spinal cord.

- Cranial nerves III and IV, both concerned with eye movements, originate in the midbrain; only III (the oculomotor nerve) is visible at the level of the section in figure 15.7a.

This figure also shows the *medial geniculate nucleus,* but this is part of the thalamus, not the midbrain. It just happens to lie in the plane of this section. The *superior cerebellar peduncles* connect to the midbrain, but are not seen at this level.

The Reticular Formation

The **reticular**[17] **formation** is a loosely organized web of gray matter that runs vertically through all levels of the brainstem and has connections with many areas of the cerebrum (fig. 15.8). It occupies

---

[11]*tectum* = roof, cover

[12]*corpora* = bodies + *quadrigemina* = quadruplets
[13]*colli* = hill + *cul* = little
[14]*tegmen* = cover
[15]*substantia* = substance + *nigra* = black
[16]*crus* = leg
[17]*ret* = network + *icul* = little

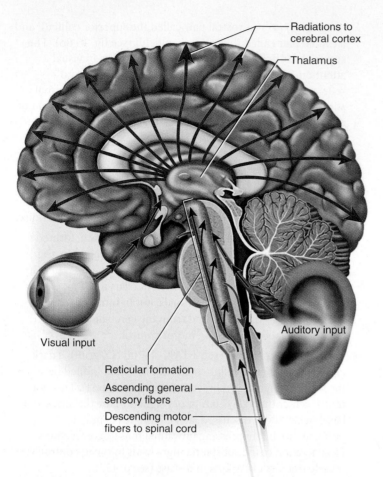

Radiations to
cerebral cortex

Thalamus

Auditory input

Visual input

Reticular formation

Ascending general
sensory fibers

Descending motor
fibers to spinal cord

Figure 15.8 **The Reticular Formation.** The formation consists of over 100 nuclei scattered through the brainstem. Red arrows indicate routes of input to the reticular formation; blue arrows indicate the radiating relay of signals from the thalamus to the cerebral cortex; and green arrows indicate output from the reticular formation to the spinal cord.
● *Locate components of the reticular formation in all three parts of figure 15.7.*

much of the space between the white fiber tracts and the more anatomically distinct brainstem nuclei. It consists of more than 100 small neural networks without well-defined boundaries. The functions of these networks include the following:

- **Somatic motor control.** Some motor neurons of the cerebral cortex and cerebellum send their axons to reticular formation nuclei, which then give rise to the *reticulospinal tracts* of the spinal cord. These tracts adjust muscle tension to maintain tone, balance, and posture, especially during body movements. The reticular formation also relays signals from the eyes and ears to the cerebellum so the cerebellum can integrate visual, auditory, and vestibular (balance and motion) stimuli into its role in motor coordination. Other reticular formation motor nuclei include *gaze centers,* which enable the eyes to track and fixate on objects, and *central pattern generators*—neural pools that produce rhythmic signals to the muscles of breathing and swallowing.

- **Cardiovascular control.** The reticular formation includes the previously mentioned cardiac and vasomotor centers of the medulla oblongata.

- **Pain modulation.** The reticular formation is one route by which pain signals from the lower body reach the cerebral cortex. It is also the origin of the *descending analgesic fibers* discussed in chapter 17 (p. 466). The nerve fibers in these pathways act in the spinal cord to block the transmission of pain signals to the brain.

- **Sleep and consciousness.** The reticular formation has projections to the cerebral cortex and thalamus that allow it some control over what sensory signals reach the cerebrum and come to our conscious attention. It plays a central role in states of consciousness such as alertness and sleep. Injury to the reticular formation can result in irreversible coma.

- **Habituation.** This is the process in which the brain learns to ignore repetitive, inconsequential stimuli while remaining sensitive to others. In a noisy city, for example, a person can sleep through traffic sounds but wake promptly to the sound of an alarm clock or a crying baby. The reticular formation screens out insignificant stimuli, preventing them from arousing cerebral centers of consciousness while it permits important or unusual sensory signals to pass. Reticular formation nuclei that modulate activity of the cerebral cortex are called the *reticular activating system* or *extrathalamic cortical modulatory system.*

The Cerebellum

The **cerebellum** is the largest part of the hindbrain and second-largest part of the brain as a whole. It consists of right and left **cerebellar hemispheres** connected by a narrow wormlike bridge, the **vermis**[18] (fig. 15.9). Each hemisphere exhibits slender, parallel folds called **folia**[19] separated by shallow sulci. The cerebellum has a surface cortex of gray matter and a deeper layer of white matter. In a sagittal section, the white matter shows a branching fernlike pattern called the **arbor vitae.**[20] Each hemisphere also has four masses of gray matter called **deep nuclei** embedded in the white matter. All cerebellar input goes to the cortex and all output comes from the deep nuclei.

The cerebellum is connected to the brainstem by three pairs of stalks called **cerebellar peduncles**[21] (peh-DUN-culs): two *inferior peduncles* connecting it to the medulla oblongata, two *middle peduncles* to the pons, and two *superior peduncles* to the midbrain (see fig. 15.6b). These consist of thick bundles of nerve fibers that carry signals into and out of the cerebellum. Connections between the cerebellum and brainstem are complex, but overlooking some exceptions, we can draw a few generalizations. Most spinal input to the cerebellum comes from the spinocerebellar tracts and travels through the inferior peduncles; input from the rest of the brain enters by way of the middle peduncles; and the superior peduncles carry most cerebellar output.

The cerebellum monitors body movements by means of information coming from the muscles and joints via the spinocerebellar tracts (fig. 15.10a). The middle peduncles carry signals from the

---

[18]*verm* = worm
[19]*foli* = leaf
[20]*arbor* = tree + *vitae* = of life
[21]*ped* = foot + *uncle* = little

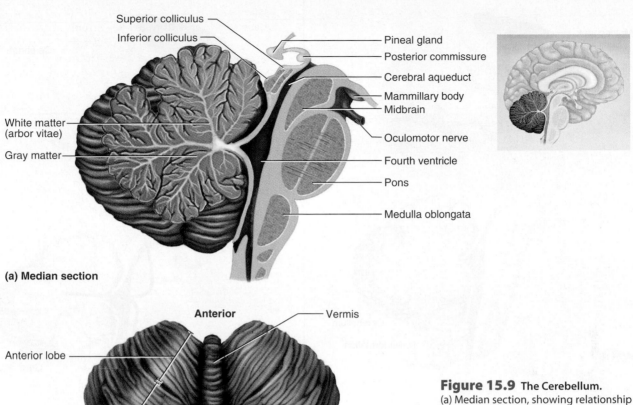

Superior colliculus
Inferior colliculus
Pineal gland
Posterior commissure
Cerebral aqueduct
Mammillary body
Midbrain
White matter (arbor vitae)
Oculomotor nerve
Gray matter
Fourth ventricle
Pons
Medulla oblongata

(a) Median section

Anterior
Vermis
Anterior lobe
Posterior lobe
Cerebellar hemisphere
Folia
Posterior

(b) Superior view

Figure 15.9 The Cerebellum.
(a) Median section, showing relationship to the brainstem. (b) Superior view.

cerebrum about what the muscles were commanded to do, enabling the cerebellum to compare the command with the performance. These peduncles also carry information from the eyes and ears for hearing and the perception of body position and movement. Output through the superior peduncles leads to various points in the midbrain and thalamus. The thalamus relays cerebellar signals to the cerebral cortex so that the cerebrum can make fine adjustments in muscle performance (fig. 15.10b).

Although it is only 10% of the mass of the brain, the cerebellum has about 60% as much surface area as the cerebral cortex and contains more than half of all brain neurons—about 100 billion of them. Its tiny, densely spaced **granule cells** are the most abundant neurons in the entire brain. Its most distinctive neurons, however, are the large globose **Purkinje**[22] (pur-KIN-jee) **cells.** These have a tremendous profusion of dendrites compressed into a single plane like a flat tree (see fig. 13.5a, p. 356, and the photograph on p. 351). Purkinje cells are arranged in single file, with these thick dendritic planes parallel to each other like books on a shelf. Their axons travel to the deep nuclei, where they synapse on output neurons that issue fibers to the brainstem.

The function of the cerebellum was unknown in the 1950s. By the 1970s, it had come to be regarded as a control center of motor coordination. Now, however, positron emission tomography (PET), functional magnetic resonance imaging (fMRI), and behavioral studies of people with cerebellar lesions have created a much more expansive view of cerebellar function. It appears that the main role of the cerebellum is the subconscious evaluation of diverse kinds of sensory input; monitoring the movements of muscles is only a part of that general role. People with cerebellar lesions exhibit serious deficits in locomotor ability, but also in several sensory, linguistic, emotional, cognitive, and other nonmotor functions. More will be said of these later in the chapter.

Table 15.1 summarizes the hindbrain and midbrain functions discussed in the last several pages.

[22]Johannes E. von Purkinje (1787–1869), Bohemian anatomist

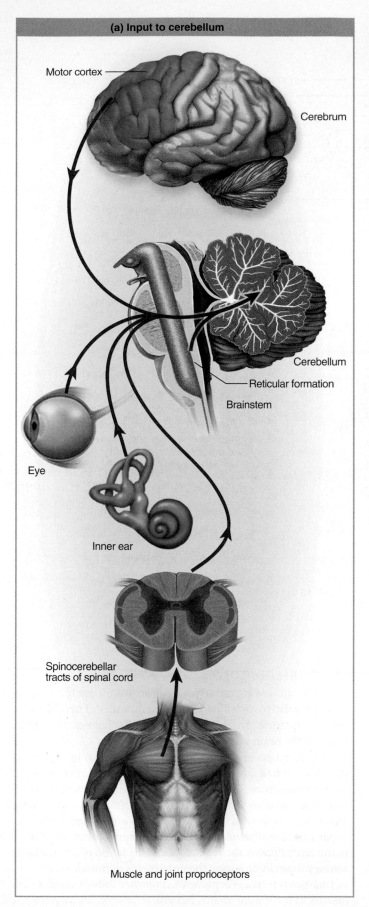
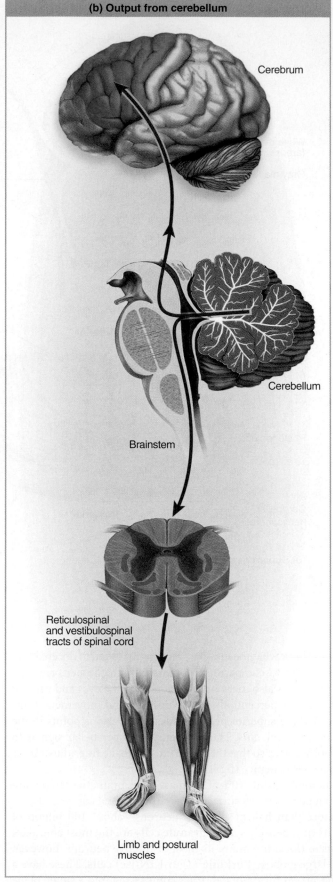

(a) Input to cerebellum

Motor cortex

Cerebrum

Cerebellum

Reticular formation

Brainstem

Eye

Inner ear

Spinocerebellar tracts of spinal cord

Muscle and joint proprioceptors

(b) Output from cerebellum

Cerebrum

Cerebellum

Brainstem

Reticulospinal and vestibulospinal tracts of spinal cord

Limb and postural muscles

Figure 15.10 Principal Pathways of Cerebellar Input and Output. (a) Afferent pathways sending input to the cerebellum. (b) Efferent pathways sending output from the cerebellum to the cerebrum and muscles.

TABLE 15.1	Hindbrain and Midbrain Functions
Medulla oblongata	Origin or termination of cranial nerves IX–XII. Sensory nuclei receive input from the taste buds, pharynx, and thoracic and abdominal viscera. Motor nuclei include the cardiac center (adjusts the rate and force of the heartbeat), vasomotor center (controls blood vessel diameter and blood pressure), two respiratory centers (control the rate and depth of breathing), and centers involved in speech, coughing, sneezing, salivation, swallowing, gagging, vomiting, sweating, gastrointestinal secretion, and movements of the tongue and head.
Pons	Sensory terminations and motor origins of cranial nerves V–VIII. Sensory nuclei receive input from the face, eye, oral and nasal cavities, sinuses, and meninges, concerned with pain, touch, temperature, taste, hearing, and equilibrium. Cranial nerve motor nuclei control chewing, swallowing, eye movements, middle- and inner-ear reflexes, facial expression, and secretion of tears and saliva. Other nuclei of pons relay signals from cerebrum to cerebellum (provide most of the input to the cerebellum) or function in sleep, respiration, bladder control, and posture.
Midbrain	Origin of cranial nerves III–IV (concerned with eye movements). Red nucleus is concerned with fine motor control. Substantia nigra relays inhibitory signals to thalamus and basal nuclei of forebrain. Central gray matter modulates awareness of pain. Superior colliculi are concerned with visual attention and tracking movements of eyes, and visual reflexes such as shifting gaze to objects seen moving in peripheral vision. Inferior colliculi relay auditory signals to thalamus and mediate auditory reflexes such as the startle response to a loud noise.
Reticular formation	A network of over 100 nuclei extending throughout brainstem, including some nuclei described earlier in this table. Involved in somatic motor control, equilibrium, visual attention, breathing, swallowing, cardiovascular control, pain modulation, sleep, and consciousness.
Cerebellum	Muscular coordination, fine motor control, muscle tone, posture, equilibrium, judging passage of time; some involvement in emotion, processing tactile input, spatial perception, and language.

Before You Go On

Answer the following questions to test your understanding of the preceding section:

8. Name several functions controlled by nuclei of the medulla.

9. Describe the anatomical and functional relationship of the pons to the cerebellum.

10. What are the functions of the corpora quadrigemina, substantia nigra, and central gray matter?

11. Describe the reticular formation and list several of its functions.

12. Describe the general functions of the cerebellum.

15.3 The Forebrain

▶ Expected Learning Outcomes

When you have completed this section, you should be able to

- name the three major components of the diencephalon and describe their locations and functions;
- identify the five lobes of the cerebrum and discuss their functions;
- describe the three types of tracts in the cerebral white matter;
- describe the distinctive cell types and histological arrangement of the cerebral cortex; and
- discuss the locations and functions of the basal nuclei and limbic system.

The forebrain consists of the diencephalon and cerebrum. As noted earlier, some authorities treat the diencephalon as the most rostral part of the brainstem, while others exclude it from the brainstem.

The Diencephalon

The embryonic diencephalon has three major derivatives: the *thalamus, hypothalamus,* and *epithalamus.* These structures surround the third ventricle of the brain.

The Thalamus

Each side of the brain has a **thalamus,**[23] an ovoid mass perched at the superior end of the brainstem beneath the cerebral hemisphere (see figs. 15.4c, 15.6, and 15.15). The thalami constitute about four-fifths of the diencephalon. They protrude medially into the third ventricle and laterally into the lateral ventricles. In about 70% of people, the two thalami are joined medially by a narrow *intermediate mass.*

The thalamus is composed of at least 23 nuclei, but we will consider only five major functional groups into which most of these are classified: the anterior, posterior, medial, lateral, and ventral groups. These regions and their functions are shown in figure 15.11a.

Generally speaking, the thalamus is the "gateway to the cerebral cortex." Nearly all sensory input and other information going to the cerebrum passes by way of synapses in the thalamic nuclei, including signals for taste; smell; hearing; equilibrium; vision; and such

---

[23]*thalamus* = chamber, inner room

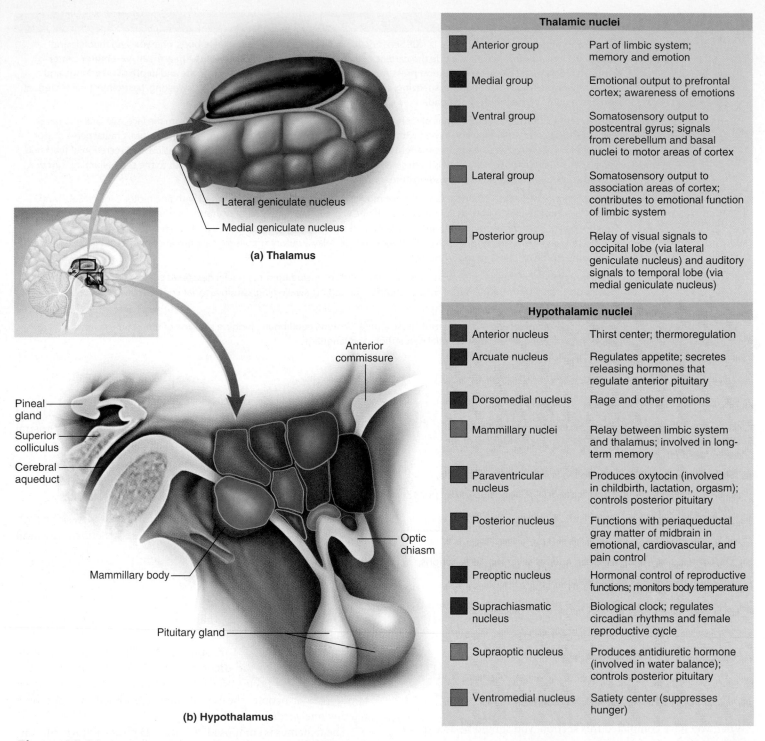

Thalamic nuclei	
Anterior group	Part of limbic system; memory and emotion
Medial group	Emotional output to prefrontal cortex; awareness of emotions
Ventral group	Somatosensory output to postcentral gyrus; signals from cerebellum and basal nuclei to motor areas of cortex
Lateral group	Somatosensory output to association areas of cortex; contributes to emotional function of limbic system
Posterior group	Relay of visual signals to occipital lobe (via lateral geniculate nucleus) and auditory signals to temporal lobe (via medial geniculate nucleus)

Hypothalamic nuclei	
Anterior nucleus	Thirst center; thermoregulation
Arcuate nucleus	Regulates appetite; secretes releasing hormones that regulate anterior pituitary
Dorsomedial nucleus	Rage and other emotions
Mammillary nuclei	Relay between limbic system and thalamus; involved in long-term memory
Paraventricular nucleus	Produces oxytocin (involved in childbirth, lactation, orgasm); controls posterior pituitary
Posterior nucleus	Functions with periaqueductal gray matter of midbrain in emotional, cardiovascular, and pain control
Preoptic nucleus	Hormonal control of reproductive functions; monitors body temperature
Suprachiasmatic nucleus	Biological clock; regulates circadian rhythms and female reproductive cycle
Supraoptic nucleus	Produces antidiuretic hormone (involved in water balance); controls posterior pituitary
Ventromedial nucleus	Satiety center (suppresses hunger)

Lateral geniculate nucleus

Medial geniculate nucleus

(a) Thalamus

Anterior commissure

Pineal gland

Superior colliculus

Cerebral aqueduct

Optic chiasm

Mammillary body

Pituitary gland

(b) Hypothalamus

Figure 15.11 The Diencephalon. Only some of the nuclei of the thalamus and hypothalamus are shown, and some of their functions listed. These lists are by no means complete. The third component of the diencephalon, the epithalamus, is not shown here; see figure 15.2a.

somatosensory functions as touch, pain, pressure, heat, and cold. The thalamic nuclei filter this information and relay only a small portion of it to the cerebral cortex.

The thalamus also plays a key role in motor control; it relays signals from the cerebellum to the cerebrum and provides feedback loops between the cerebral cortex and *basal nuclei* (deep cerebral motor centers). Finally, the thalamus is involved in the memory and emotional functions of the *limbic system*, a complex of structures that includes the cerebral cortex of the temporal and frontal lobes and some of the anterior thalamic nuclei. The role of the thalamus in motor and sensory circuits is further discussed later in this chapter and in chapter 17.

The Hypothalamus

The **hypothalamus** forms part of the walls and floor of the third ventricle. It extends anteriorly to the *optic chiasm* (ky-AZ-um), where the optic nerves meet, and posteriorly to and including a pair of humps called the *mammillary*[24] *bodies* (see fig. 15.2a). The mammillary bodies each contain three to four *mammillary nuclei*. Their primary function is to relay signals from the limbic system to the thalamus. The pituitary gland is attached to the hypothalamus by a stalk *(infundibulum)* between the optic chiasm and mammillary bodies.

The hypothalamus is the major control center of the autonomic nervous system and endocrine system and plays an essential role in the homeostatic regulation of nearly all organs of the body. Its nuclei include centers concerned with a wide variety of visceral functions (fig. 15.11b):

- **Hormone secretion.** The hypothalamus secretes hormones that control the anterior pituitary gland. Acting through the pituitary, it regulates growth, metabolism, reproduction, and stress responses. It also produces two hormones that are stored in the posterior pituitary gland—*oxytocin* concerned with labor contractions, lactation, and emotional bonding, and *antidiuretic hormone* concerned with water conservation. It sends nerve signals to the posterior pituitary to stimulate release of these hormones at appropriate times.

- **Autonomic effects.** The hypothalamus is a major integrating center for the autonomic nervous system. It sends descending fibers to nuclei lower in the brainstem that influence heart rate, blood pressure, pupillary diameter, and gastrointestinal secretion and motility, among other functions.

- **Thermoregulation.** The *hypothalamic thermostat* consists of a collection of neurons, concentrated especially in the preoptic nucleus, that monitor body temperature. When the body temperature deviates too much from normal, this center activates either a heat-loss center in the anterior hypothalamus or a heat-promoting center near the mammillary bodies in the posterior hypothalamus. Those centers activate mechanisms for raising or lowering the body temperature—shivering, sweating, or cutaneous vasoconstriction or vasodilation.

- **Food and water intake.** Neurons of the *hunger* and *satiety centers* monitor blood glucose and amino acid levels and respond to hormones from the digestive system, producing sensations of hunger and satisfaction of the appetite. Hypothalamic neurons called *osmoreceptors* monitor the osmolarity of the blood and stimulate water-seeking and drinking behavior when the body is dehydrated. Thus, our drives to eat and drink are under hypothalamic control.

- **Sleep and circadian rhythms.** The caudal part of the hypothalamus is part of the reticular formation. It contains nuclei that regulate the rhythm of sleep and waking. Superior to the optic chiasm, the hypothalamus contains a *suprachiasmatic nucleus* that controls our 24-hour *(circadian)* rhythm of activity.

- **Emotional responses.** The hypothalamus contains nuclei for a variety of emotional responses including anger, aggression, fear, pleasure, and contentment; and for sexual drive, copulation, and orgasm. The mammillary bodies provide a pathway by which emotional states can affect visceral function—for example, when anxiety accelerates the heart or upsets the stomach.

- **Memory.** In addition to their role in emotional circuits, the mammillary bodies lie in the pathway from the hippocampus to the thalamus. The hippocampus is a center for the creation of new memories—the cerebrum's "teacher"—so as intermediaries between the hippocampus and cerebral cortex, the mammillary bodies are essential for the acquisition of new memories.

The Epithalamus

The **epithalamus** consists mainly of the **pineal gland** (an endocrine gland discussed in chapter 18), the **habenula** (a relay from the limbic system to the midbrain), and a thin roof over the third ventricle (see fig. 15.2a).

The Cerebrum

The embryonic telencephalon becomes the cerebrum, the largest and most conspicuous part of the human brain. Your cerebrum enables you to turn these pages, read and comprehend the words, remember ideas, talk about them with others, and take an examination. It is the seat of sensory perception, voluntary motor actions, memory, and mental processes such as thought, judgment, and imagination, which most distinguish humans from other animals. It is the most complex and challenging frontier of neurobiology.

Gross Anatomy

The cerebrum so dwarfs and conceals the other structures that people often think of "cerebrum" and "brain" as synonymous. Its major anatomical landmarks were described on page 399 and should be reviewed if necessary (see figs. 15.1 and 15.2)—especially the two *cerebral hemispheres,* separated by the *longitudinal fissure* but connected by a prominent fiber tract, the *corpus callosum;* and the conspicuous wrinkles, or *gyri,* of each hemisphere, separated by grooves called *sulci.* The folding of the cerebral surface into gyri allows a greater amount of cortex to fit in the cranial cavity. The gyri give the cerebrum a surface area of about 2,500 cm², comparable to 4.5 pages of this book. If the cerebrum had a smooth surface, it would have only one-third as much area and proportionately less information-processing capability. This extensive folding is one of the greatest differences between the human brain and the relatively smooth-surfaced brains of most other mammals.

Some gyri have consistent and predictable anatomy; others vary from brain to brain and from the right hemisphere to the left. Certain unusually prominent sulci divide each hemisphere into five anatomically and functionally distinct lobes, listed next. The first

---

[24]*mammill* = nipple, little breast

four of these are visible superficially and are named for the cranial bones overlying them (fig. 15.12); the fifth lobe is not visible from the surface.

1. The **frontal lobe** lies immediately behind the frontal bone, superior to the eyes. From the forehead, it extends caudally to a wavy vertical groove, the **central sulcus.** It is concerned with cognition (thought) and other higher mental processes, speech, and motor control.

2. The **parietal lobe** forms the uppermost part of the brain and underlies the parietal bone. Starting at the central sulcus, it extends caudally to the **parieto–occipital sulcus,** visible on the medial surface of each hemisphere (see fig. 15.2). It is the primary site for receiving and interpreting signals of the *general senses* described later in this chapter; taste (one of the *special senses*); and some visual processing.

3. The **occipital lobe** is at the rear of the head, caudal to the parieto–occipital sulcus and underlying the occipital bone. It is the principal visual center of the brain.

4. The **temporal lobe** is a lateral, horizontal lobe deep to the temporal bone, separated from the parietal lobe above it by a deep **lateral sulcus.** Among its functions are hearing, smell, learning, memory, and some aspects of vision and emotion.

5. The **insula**[25] is a small mass of cortex deep to the lateral sulcus, made visible only by retracting or cutting away some of the overlying cerebrum (see figs. 15.1c, 15.4c, and 15.15). It is less understood than the other lobes because it is less accessible to testing in living subjects, but it apparently plays roles in taste, visceral sensation, and understanding spoken language.

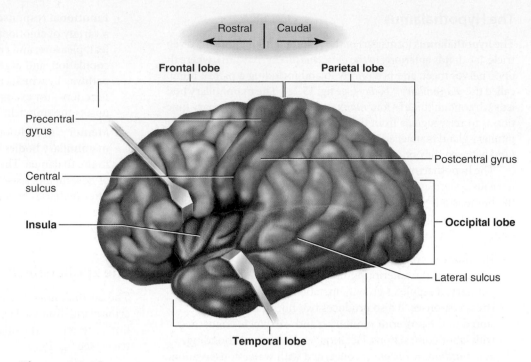

Figure 15.12 Lobes of the Cerebrum. The frontal and temporal lobes are retracted slightly to reveal the insula.

The Cerebral White Matter

Most of the volume of the cerebrum is white matter. This is composed of glia and myelinated nerve fibers that conduct signals from one region of the cerebrum to another and between the cerebrum and lower brain centers. These fibers travel in bundles called tracts. There are three types of cerebral tracts (fig. 15.13):

1. **Projection tracts** extend vertically between higher and lower brain or spinal cord centers and carry information between the cerebrum and the rest of the body. The corticospinal tracts, for example, carry motor signals from the cerebrum to the brainstem and spinal cord. Other projection tracts carry signals upward to the cerebral cortex. Superior to the

brainstem, such tracts form a dense band called the *internal capsule* between the thalamus and basal nuclei, then radiate in a diverging, fanlike array (the *corona radiata*[26]) to specific areas of the cortex.

2. **Commissural tracts** cross from one cerebral hemisphere to the other through bridges called **commissures** (COM-ih-shurs). The great majority of commissural fibers pass through the corpus callosum. A few pass through the much smaller **anterior** and **posterior commissures** (fig. 15.2a). Commissural tracts enable the two cerebral hemispheres to communicate with each other.

3. **Association tracts** connect different regions of the same hemisphere. *Long association fibers* connect different lobes to each other, whereas *short association fibers* connect different gyri within a single lobe. Among their roles, association tracts link perceptual and memory centers of the brain; for example, they enable you to smell a rose, name it, and picture what it looks like.

The Cerebral Cortex

Neural integration is carried out in the gray matter of the cerebrum, which is found in three places—the cerebral cortex, basal nuclei, and limbic system. The **cerebral cortex**[27] is a layer about 2 to 3 mm thick covering the surface of the hemispheres. It constitutes about 40% of the mass of the brain and contains 14 to 16 billion neurons.

---

[25]*insula* = island

[26]*corona* = crown + *radiata* = radiating
[27]*cortex* = bark, rind

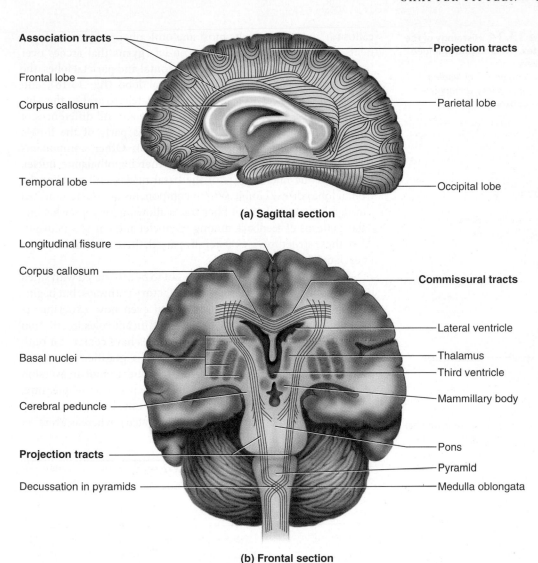

Figure 15.13 **Tracts of Cerebral White Matter.** (a) Sagittal section, showing association (red) and projection (green) tracts. (b) Frontal section, showing commissural (yellow) and projection tracts.

It possesses two principal types of neurons called stellate and pyramidal cells (fig. 15.14). **Stellate cells** are anaxonic neurons with spheroidal somas and dendrites that project for short distances in all directions. They are concerned largely with receiving sensory input and processing information on a local level. **Pyramidal cells** are tall and conical (triangular in tissue sections). Their apex points toward the brain surface and has a thick dendrite with many branches and small, knobby *dendritic spines*. The base gives rise to horizontally oriented dendrites and an axon that passes into the white matter. The axon also has collaterals that synapse with other neurons in the cortex or in deeper regions of the brain. Pyramidal cells are the output neurons of the cerebrum—they are the only cerebral neurons whose fibers leave the cortex and connect with other parts of the CNS.

About 90% of the human cerebral cortex is a six-layered tissue called **neocortex**[28] because of its relatively recent evolutionary ori-

gin. Although vertebrates have existed for about 600 million years, the neocortex did not develop significantly until about 60 million years ago, when there was a sharp increase in the diversity of mammals. It attained its highest development by far in the primates. The six layers of neocortex (fig. 15.14) vary from one part of the cerebrum to another in relative thickness, cellular composition, synaptic connections, size of the neurons, and destination of their axons. Layer IV is thickest in sensory regions and layer V in motor regions, for example. All axons that leave the cortex and enter the white matter arise from layers III, V, and VI.

Some regions of cerebral cortex have fewer than six layers. The earliest type of cortex to appear in vertebrate evolution was a one- to five-layered tissue called *paleocortex* (PALE-ee-oh-cor-tex), limited in humans to part of the insula and certain areas of the temporal lobe concerned with smell. The next to evolve was a three-layered *archicortex* (AR-kee-cor-tex), found in the human hippocampus. The neocortex was the last to evolve.

The Basal Nuclei

The **basal nuclei** (fig. 15.15) are masses of cerebral gray matter buried deep in the white matter, lateral to the thalamus. They are often called *basal ganglia*, but the word *ganglion* is best restricted to clusters of neurons outside the CNS. Neuroanatomists disagree on how many brain centers to classify as basal nuclei, but agree on at least three: the **caudate**[29] **nucleus, putamen,**[30] and **globus pallidus.**[31] The putamen and globus pallidus are also collectively called the *lentiform*[32] *nucleus,* named for its lenslike shape. The putamen and caudate nucleus are collectively called the *corpus striatum* because of their striped appearance. The basal nuclei are involved in motor control, as discussed later.

The Limbic System

The **limbic**[33] system is an important center of emotion and learning. It was originally described in the 1850s as a ring of structures on the medial side of the cerebral hemisphere, encircling the corpus

[29]*caudate* = tailed, tail-like
[30]*putam* = pod, husk
[31]*glob* = globe, ball + *pall* = pale
[32]*lenti* = lens + *form* = shape
[33]*limbus* = border

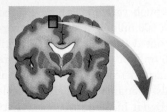

Figure 15.14 **Histology of the Neocortex.** Neurons are arranged in six layers.
● *Are the long processes leading upward from each pyramidal cell dendrites or axons?*

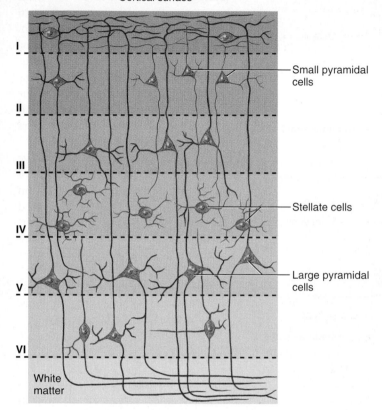

Cortical surface

I

II

III

IV

V

VI

White matter

Small pyramidal cells

Stellate cells

Large pyramidal cells

callosum and thalamus. Its most anatomically prominent components are the **cingulate**[34] (SING-you-let) **gyrus** that arches over the top of the corpus callosum in the frontal and parietal lobes, the **hippocampus**[35] in the medial temporal lobe (fig. 15.16), and the **amygdala**[36] (ah-MIG-da-luh) immediately rostral to the hippocampus, also in the temporal lobe. There are still differences of opinion on what structures to consider as parts of the limbic system, but those three are agreed upon. Other components include the mammillary bodies and other hypothalamic nuclei, some thalamic nuclei, parts of the basal nuclei, and parts of the frontal lobe cortex. Limbic system components are interconnected through a complex loop of fiber tracts allowing for somewhat circular patterns of feedback among its nuclei and cortical neurons. All of these structures are bilaterally paired; there is a limbic system in each cerebral hemisphere.

The limbic system was long thought to be associated with smell because of its close association with olfactory pathways, but beginning in the early 1900s and continuing even now, experiments have abundantly demonstrated more significant roles in emotion and memory. Most limbic system structures have centers for both gratification and aversion. Stimulation of a gratification center produces a sense of pleasure or reward; stimulation of an aversion center produces disagreeable sensations such as fear, displeasure, or sorrow. Gratification centers dominate some limbic structures, such as the *nucleus accumbens* (not illustrated), whereas aversion

---

[34]*cingul* = girdle
[35]*hippocampus* = sea horse, named for its shape
[36]*amygdala* = almond

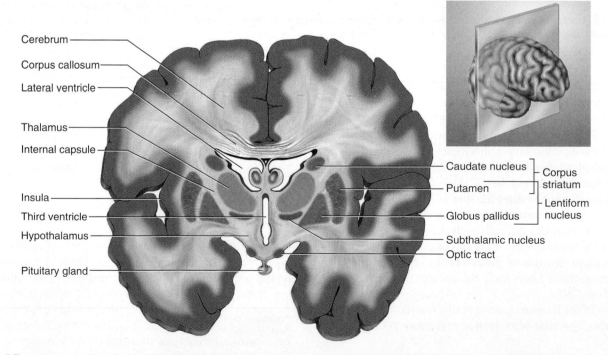

Cerebrum

Corpus callosum

Lateral ventricle

Thalamus

Internal capsule

Insula

Third ventricle

Hypothalamus

Pituitary gland

Caudate nucleus ⎱ Corpus
Putamen   ⎰ striatum

⎱ Lentiform
⎰ nucleus

Globus pallidus

Subthalamic nucleus

Optic tract

Figure 15.15 **The Basal Nuclei.** Frontal section of the brain.

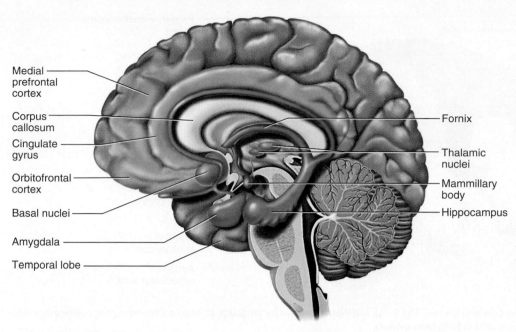

Figure 15.16 **The Limbic System.** This ring of structures includes important centers of learning and emotion. In the frontal lobe, there is no sharp rostral boundary to limbic system components.

Labels on figure:
Medial prefrontal cortex
Corpus callosum
Cingulate gyrus
Orbitofrontal cortex
Basal nuclei
Amygdala
Temporal lobe
Fornix
Thalamic nuclei
Mammillary body
Hippocampus

centers dominate others such as the amygdala. The roles of the amygdala in emotion and the hippocampus in memory are described in the coming pages.

Integrative Brain Functions

We will here examine a number of "higher" functions of the forebrain—sensory awareness, motor control, language, emotion, thought, and memory. These processes call attention especially to the cerebral cortex, but are not limited to the cerebrum; they involve also the diencephalon and cerebellum. It is impossible in many cases to assign a specific function to a specific brain region. Functions of the brain do not have such easily defined boundaries as its anatomy. Some functions overlap anatomically, some cross anatomical boundaries from one region to another, and some functions such as consciousness and memory are widely distributed through the cerebrum.

As a general principle for the functional discussion of the cerebrum, we distinguish between primary cortex and association cortex. **Primary cortex** consists of those regions that receive input directly from the sense organs or brainstem, or issue motor nerve fibers directly to the brainstem for distribution of the motor commands to cranial and spinal nerves. **Association cortex** consists of all regions other than the primary cortex, involved in integrative functions such as interpretation of sensory input, planning of motor output, cognitive (thought) processes, and the storage and retrieval of memories. About 75% of the mass of the cerebral cortex is association cortex. Typically, an area of primary cortex has an association area immediately adjacent to it, concerned with the same general function. For example, the primary visual cortex, which receives input from the eyes, is bordered by visual associa-

tion cortex, which interprets and makes cognitive sense of the visual stimuli. Some areas of association cortex are *multimodal;* instead of processing information from a single sensory source, they receive input from multiple senses and integrate this into the overall perception of one's surroundings. The association cortex of the frontal lobe is a very important center of cognitive and emotional function, the **prefrontal cortex.**

Special Senses

The *special senses* are taste, smell, hearing, equilibrium, and vision, mediated by relatively complex sense organs in the head. Signals from these organs are routed to widely separated areas of primary sensory cortex in the cerebrum. From an area of primary sensory cortex, signals are relayed to a nearby association area, where the present sensory experience is integrated with memory and made identifiable and intelligible. Here we will only briefly identify the regions of cerebral cortex concerned with each of the special senses (fig. 15.17). Chapter 17 details the pathways taken from the sense organs to these areas of the cerebrum.

- **Vision.** Visual signals are received by the **primary visual cortex** in the far posterior region of the occipital lobe. This is bordered anteriorly by the **visual association area,** which includes all the remainder of the occipital lobe, part of the parietal lobe, and much of the inferior temporal lobe. The association area integrates information so we can identify what we see.

- **Hearing.** Auditory signals are received by the **primary auditory cortex** in the superior region of the temporal lobe and in the nearby insula. The **auditory association area** occupies areas of temporal lobe inferior to the primary auditory cortex and deep within the lateral sulcus. This is where we become capable of recognizing spoken words, a familiar piece of music, or a voice on the telephone.

- **Equilibrium.** Signals from the inner ear for equilibrium project mainly to the cerebellum and several brainstem nuclei concerned with head and eye movements and visceral functions. Some fibers of this system, however, are routed through the thalamus to areas of association cortex in the roof of the lateral sulcus and near the lower end of the central sulcus. This is the seat of consciousness of our body movements and orientation in space.

- **Taste.** Gustatory signals are received by the **primary gustatory cortex** in the inferior end of the postcentral gyrus of the parietal lobe (discussed shortly) and an anterior region of the insula. The gustatory association cortex is integrated with the association area for smell.

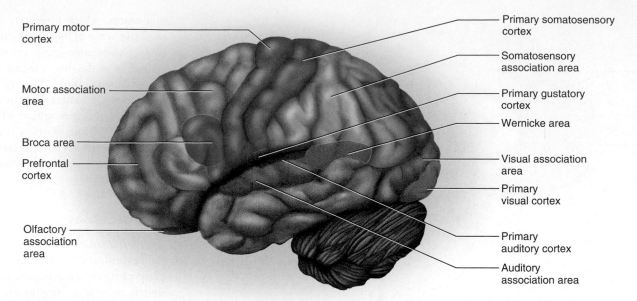

Primary motor cortex

Motor association area

Broca area

Prefrontal cortex

Olfactory association area

Primary somatosensory cortex

Somatosensory association area

Primary gustatory cortex

Wernicke area

Visual association area

Primary visual cortex

Primary auditory cortex

Auditory association area

Figure 15.17 Some Functional Regions of the Cerebral Cortex. The Broca and Wernicke areas for language abilities are found in only one hemisphere, usually the left. The other regions shown here are mirrored in both hemispheres.

- **Smell.** The primary olfactory cortex lies in the medial surface of the temporal lobe and inferior surface of the frontal lobe. The **orbitofrontal cortex** (see fig. 15.16) contains an association area that integrates gustatory, olfactory, and visual information to create a sense of the overall flavor and desirability (or rejection) of food.

General Senses

The *general (somatosensory* or *somesthetic*[37]*) senses* are widely distributed over the body and have relatively simple receptors (see chapter 17). They include such senses as touch, pressure, stretch, temperature, and pain. Somatosensory signals arising from the neck down travel up the spinal cord in the gracile and cuneate fasciculi and spinothalamic tracts. Somatosensory nerve fibers decussate in the spinal cord or medulla before reaching the thalamus (see table 14.1 and fig. 14.4). Consequently, these signals ultimately arrive in the contralateral cerebral cortex—stimuli on the right side of the body are perceived in the left cerebral hemisphere, and vice versa. The thalamus routes all somatosensory signals to one specific fold of the cerebrum, the **postcentral gyrus.** This gyrus lies immediately caudal to the central sulcus and forms the anterior border of the parietal lobe (see fig. 15.1c). It rises from the lateral sulcus up to the crown of the head and then descends into the longitudinal fissure. The cortex of this gyrus is called the **primary somatosensory cortex** (fig. 15.18).

This gyrus is like an upside-down sensory map of the contralateral side of the body, traditionally diagrammed as a *sensory homunculus*[38] (fig. 15.18b). As the diagram shows, receptors in the lower limb project to superior and medial parts of the gyrus, and recep-

tors in the face project to the inferior and lateral parts. There is a point-for-point correspondence, or **somatotopy,**[39] between an area of the body and an area of the postcentral gyrus. The reason for the bizarre, distorted appearace of the homunculus is that the amount of cerebral tissue devoted to a given body region is proportional to how richly innervated and sensitive that region is, not to its size. Thus, the hands and face are represented by a much larger region of somatosensory cortex than the trunk is.

The **somatosensory association area** is found in the parietal lobe posterior to the postcentral gyrus and in the roof of the lateral sulcus (fig. 15.17).

Motor Control

The intention to contract a skeletal muscle begins in the **motor association (premotor) area** of the frontal lobes (fig. 15.17). This is where we plan our behavior—where neurons compile a program for the degree and sequence of muscle contractions required for an action such as dancing, typing, or speaking. The program is then transmitted to neurons of the **precentral gyrus** (primary motor area), which is the most posterior gyrus of the frontal lobe, immediately anterior to the central sulcus (see fig. 15.1c). Neurons here send signals to the brainstem and spinal cord that ultimately result in muscle contractions.

The precentral gyrus, like the postcentral one, exhibits somatotopy. The neurons for toe movements, for example, are deep in the longitudinal fissure on the medial side of the gyrus. The summit of the gyrus controls the trunk, shoulder, and arm, and the inferolateral region controls the facial muscles. This map is diagrammed as a *motor homunculus* (fig. 15.19b). Like the sensory homunculus, it has a distorted look because the amount of cortex devoted to a

---

[37]*som* = body + *esthet* = sensation
[38]*hom* = man + *unculus* = little

[39]*somato* = body + *top* = place

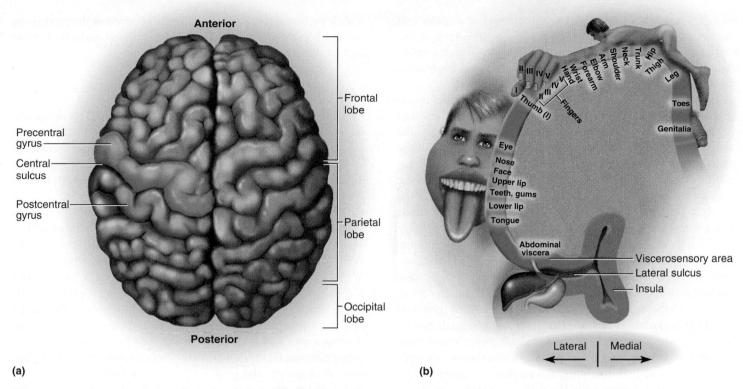

Figure 15.18 The Primary Somatosensory Area (Postcentral Gyrus). (a) Superior view of the brain showing the location of the postcentral gyrus (violet). (b) Sensory homunculus, drawn so that body parts are in proportion to the amount of cortex dedicated to their sensation.

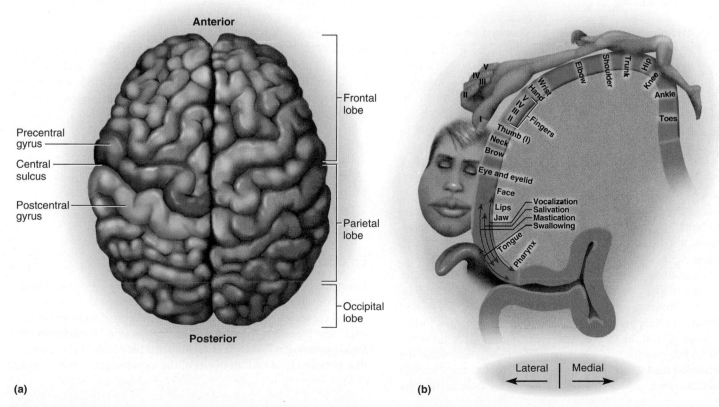

Figure 15.19 The Primary Motor Area (Precentral Gyrus). (a) Superior view of the brain showing the location of the precentral gyrus (blue). (b) Motor homunculus, drawn so that body parts are in proportion to the amount of primary motor cortex dedicated to their control.
● *Which body regions are controlled by the largest area of motor cortex—regions with a few large muscles or regions with many small muscles?*

given body region is proportional to the number of muscles and motor units in that region, not to the size of the region. Areas of fine control, such as the hands, have more muscles than such areas as the trunk and thigh, more motor units per muscle, and larger areas of motor cortex to control them.

The pyramidal cells of the precentral gyrus are called **upper motor neurons.** Their fibers project caudally, with about 19 million fibers ending in nuclei of the brainstem and 1 million forming the corticospinal tracts. These tracts decussate in the pyramids of the medulla oblongata and then continue into the spinal cord. Therefore, below the neck, each precentral gyrus controls muscles on the contralateral side of the body. In the brainstem or spinal cord, the fibers from the upper motor neurons synapse with **lower motor neurons** whose axons innervate the skeletal muscles.

Other areas of the brain important in muscle control are the basal nuclei and cerebellum. The basal nuclei receive signals from the cerebral cortex and direct their output back to the cortex. Nearly all areas of cerebral cortex, except for the primary visual and auditory areas, send signals to the basal nuclei. The basal nuclei process these and issue their output to the thalamus, which relays these signals back to the cerebral cortex—especially to the prefrontal cortex, motor association area, and precentral gyrus. The basal nuclei thus lie in a feedback circuit involved in the planning and execution of movement.

Among other functions, the basal nuclei assume control of highly practiced behaviors that one carries out with little thought—writing, typing, driving a car, using scissors, or tying one's shoes, for example. They also control the onset and cessation of planned movements and the repetitive movements at the shoulder and hip that occur during walking.

Lesions of the basal nuclei cause movement disorders called **dyskinesias.**[40] Some dyskinesias are characterized by abnormally inhibited movements—for example, difficulty rising from a chair or beginning to walk and a slow shuffling walk, as seen in Parkinson disease (see p. 437). Smooth, easy movements require the excitation of agonistic muscles and inhibition of their antagonists. In Parkinson disease, the antagonists are not inhibited. Therefore, opposing muscles at a joint fight each other, making it a struggle to move as one wishes. Other dyskinesias are characterized by exaggerated or unwanted movements, such as flailing of the limbs (*ballismus*) in Huntington disease.

The cerebellum is highly important in motor coordination. It aids in learning motor skills, maintains muscle tone and posture, smooths muscle contractions, coordinates eye and body movements, and coordinates the motions of different joints with each other (such as the shoulder and elbow in pitching a baseball). The cerebellum acts as a comparator in motor control. It receives information from the upper motor neurons of the cerebrum about what movements are intended, and information from proprioceptors in the muscles and joints about the actual performance of the movement (see fig. 15.10). The Purkinje cells of the cerebellum compare the two. If there is a discrepancy between the intent and the performance, they signal the deep cerebellar nuclei, which, in turn, relay signals to the thalamus and cerebral cortex. Motor neurons in the cortex then correct the muscle performance to match the intent. Lesions of the cerebellum can result in a clumsy, awkward gait (*ataxia*) and make some tasks such as climbing stairs virtually impossible.

Language

Language includes several abilities—reading, writing, speaking, and understanding spoken and printed words—assigned to different regions of cerebral cortex. The **Wernicke**[41] (WUR-nih-kee) **area** is responsible for the recognition of spoken and written language. It lies just posterior to the lateral sulcus, usually in the left hemisphere, at the crossroad between visual, auditory, and somatosensory areas of the cortex (see fig. 15.17). It is a sensory association area that receives input from all these neighboring regions of primary sensory cortex. The *angular gyrus,* part of the parietal lobe caudal and superior to the Wernicke area, is important in the ability to read and write.

When we intend to speak, the Wernicke area formulates phrases according to learned rules of grammar and transmits a plan of speech to the **Broca**[42] **area,** located in the inferior prefrontal cortex of the same hemisphere. PET scans show a rise in the metabolic activity of the Broca area as we prepare to speak. The Broca area generates a motor program for the muscles of the larynx, tongue, cheeks, and lips to produce speech. This program is then transmitted to the primary motor cortex, which executes it—that is, it issues commands to the lower motor neurons that supply the relevant muscles.

Lesions in the language areas of the brain tend to produce a variety of language deficits called **aphasias**[43] (ah-FAY-zee-uhs). They may include a complete inability to speak; slow speech with difficulty choosing words; invention of words that only approximate the correct ones; babbling, incomprehensible speech filled with invented words and illogical word order; inability to comprehend another person's written or spoken words; or inability to name objects that a person sees. Since cranial nerves VII and XII (the facial and hypoglossal nerves) control several of the muscles of speech, speech deficits can also result from lesions to these nerves or their brainstem nuclei.

Apply What You Know

Mr. Thompson has had a stroke that destroyed his Wernicke area. Ms. Meyers has had a stroke that destroyed her Broca area. What differences in language deficits would you expect between these two patients?

Emotion

Emotional feelings and memories are not exclusively cerebral functions, but rather result from an interaction between areas of the prefrontal cortex and the diencephalon. The hypothalamus and amygdala play especially important roles in emotion. Experiments by Swiss physiologist Walter Hess, leading to a 1949 Nobel Prize, showed that stimulation of various nuclei of the hypothalamus in cats could

---

[40]*dys* = bad, abnormal, difficult + *kines* = movement

[41]Karl Wernicke (1848–1904), German neurologist
[42]Pierre Paul Broca (1824–80), French surgeon and anthropologist
[43]*a* = without + *phas* = speech

induce rage, attack, and other emotional responses. Nuclei involved in the senses of reward and punishment have been identified in the hypothalamus of cats, rats, monkeys, and other animals.

The *amygdala,* one of the most important centers of human emotion, is a major component of the limbic system described earlier. It receives processed information from the senses of vision, hearing, taste, smell, and general somatosensory and visceral senses. Thus, it is able to mediate emotional responses to such stimuli as a disgusting odor, a foul taste, a beautiful image, pleasant music, or a stomachache. It is especially important in the sense of fear. Output from the amygdala goes in two directions of special interest: (1) Some output projects to the hypothalamus and lower brainstem and thus influences the somatic and visceral motor systems. An emotional response to a sight or sound may, through these connections, make one's heart race, make the hair stand on end (piloerection), or induce vomiting. (2) Other output projects to areas of the prefrontal cortex that mediate conscious control and expression of the emotions, such as our ability to express love or control anger.

Many important aspects of personality depend on an intact, functional amygdala and hypothalamus. When specific regions of the amygdala or hypothalamus are destroyed or artificially stimulated, humans and other animals exhibit blunted or exaggerated expressions of anger, fear, aggression, self-defense, pleasure, pain, love, sexuality, and parental affection, as well as abnormalities in learning, memory, and motivation.

Cognition

Cognition[44] is the range of mental processes by which we acquire and use knowledge—sensory perception, thought, reasoning, judg-

---

[44]*cognit* = to know, to learn

ment, memory, imagination, and intuition. Cognitive abilities of various kinds are widely distributed through the association areas of the cerebral cortex. This is the most difficult area of brain research and the most incompletely understood area of cerebral function. Much of what we know about cognitive functions of the brain has come from studies of patients with brain lesions—areas of tissue destruction resulting from cancer, stroke, and trauma. The many brain injuries incurred in World Wars I and II yielded an abundance of insights into regional brain functions. More recently, imaging methods such as positron emission tomography (PET) and functional magnetic resonance imaging (fMRI), described on page 6, have led to much more sophisticated insights into brain function. These methods allow a researcher to scan a person's brain while that person is performing various cognitive or motor tasks and to see which brain regions are most active in different mental and task states (fig. 15.20).

Attention to objects in the environment is based in the parietal lobe on the side opposite the Broca and Wernicke speech centers. Lesions here can produce *contralateral neglect syndrome,* in which a patient seems unaware of objects on one side of the body; fails even to recognize, dress, and take care of one side of his or her own body; or ignores all words on one side of a page of reading. Such patients are also unable to find their way around—say, to describe the route from home to work, or navigate within a familiar building.

The prefrontal cortex is concerned with many of our most distinctly human abilities, such as abstract thought, foresight, judgment, responsibility, a sense of purpose, and a sense of socially appropriate behavior. Lesions here tend to render a person easily distracted from a task, irresponsible, exceedingly stubborn, unable to anticipate future events, and incapable of any ambition or planning for the future (see Deeper Insight 15.2).

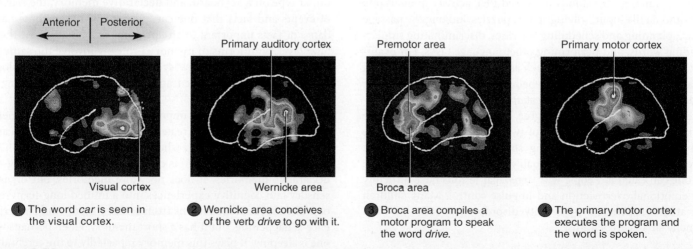

Anterior | Posterior

Primary auditory cortex

Premotor area

Primary motor cortex

Visual cortex

Wernicke area

Broca area

1 The word *car* is seen in the visual cortex.

2 Wernicke area conceives of the verb *drive* to go with it.

3 Broca area compiles a motor program to speak the word *drive.*

4 The primary motor cortex executes the program and the word is spoken.

Figure 15.20 **PET Scans of the Brain During Performance of a Language Task.** These scans represent computer-averaged images of the brains of one or more subjects who were shown a word *(car)* in print and asked to speak a word (such as *drive*) related to it. The most active brain areas are shown in red and the least active in blue. Such studies have shown that the Broca and Wernicke areas are not involved in simply repeating a word, but are active when a person must evaluate a word and choose an appropriate response. PET scans also show that different neural pools take over a task as a person practices and becomes more proficient at it.

DEEPER INSIGHT   15.2

An Accidental Lobotomy

Accidental but nonfatal destruction of parts of the brain has afforded many clues to the function of various regions. The most famous incident occurred in 1848 to Phineas Gage, a laborer on a railroad construction project in Vermont. Gage was packing blasting powder into a hole with a 3½ ft tamping iron when the powder prematurely exploded. The tamping rod was blown out of the hole and passed through Gage's maxilla, orbit, and the frontal lobe of his brain before emerging from his skull near the hairline and landing 50 ft away (fig. 15.21). Gage went into convulsions, but later sat up and conversed with his crewmates as they drove him to a physician in an oxcart. On arrival, he stepped out on his own and told the physician, "Doctor, here is business enough for you." His doctor, John Harlow, reported that he could insert his index finger all the way into Gage's wound. Yet 2 months later, Gage was walking around town carrying on his normal business.

He was not, however, the Phineas Gage people had known. Before the accident, he had been a competent, responsible, financially prudent man, well liked by his associates. In an 1868 publication on the incident, Harlow said that following the accident, Gage was "fitful, irreverent, indulging at times in the grossest profanity." He became irresponsible, lost his job, worked for a while as a circus sideshow attraction, and died a vagrant 12 years later.

A 1994 computer analysis of Gage's skull indicated that the brain injury was primarily to the inferomedial region of both frontal lobes. In Gage's time, scientists were reluctant to attribute social behavior and moral judgment to any region of the brain. These functions were strongly tied to issues of religion and ethics and were considered inaccessible to scientific analysis. Based partly on Phineas Gage and other brain injury patients like him, neuroscientists today recognize that planning, moral judgment, and emotional control are among the functions of the prefrontal cortex.

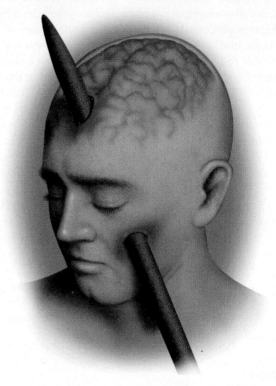

Figure 15.21   Phineas Gage's 1848 Accident. Gage suffered frontal lobe damage when a tamping bar flew upward, entered his maxilla, passed through the orbit and between the two frontal lobes, and exited the top of his head near the hairline. The accident resulted in permanent personality changes that helped to define some functions of the frontal lobes.

The cerebellum also has lately shown a surprising amount of cognitive function. It exhibits increased PET activity in analyzing visual and tactile input, solving spatial puzzles, judging the passage of time, planning and scheduling activities, discriminating similar-sounding words, and performing other language tasks. For example, if a person is given a noun such as *apple* and asked to think of a related verb such as *eat,* the cerebellum is more active than when the person is asked simply to repeat *apple.* Touching sandpaper activates the cerebellum to some degree, but not as much as when one is asked to compare the texture of two different sandpapers. The cerebellum also helps in making short-term predictions about movement, such as where a baseball will be in a second or two so that one can catch it. People with cerebellar lesions have difficulty with emotional overreaction and impulse control. Many children with attention-deficit/hyperactivity disorder (AD/HD) have abnormally small cerebellums.

Memory

Memory is one of the cognitive functions, but warrants special attention. There are two kinds of memory—**procedural memory,** the retention of motor skills such as how to tie one's shoes, play the violin, or type on a keyboard; and **declarative memory,** the retention of events and facts that one can put into words, such as names, dates, or facts important to an upcoming examination. At the cellular level, both forms of memory probably involve the same processes: the creation of new synaptic contacts and physiological changes that make synaptic transmission more efficient along certain pathways.

The limbic system has important roles in the establishment of memories. The amygdala creates emotional memories, such as the chilling fear of being stung when a wasp alights on the skin. The *hippocampus* (see fig. 15.16) is critical to the creation of long-term declarative memories. It does not store memories, but organizes sensory and cognitive experiences into a unified long-term memory. The hippocampus learns from sensory input while an experience is happening, but it has a short memory. Later, perhaps when one is sleeping, it plays this memory repeatedly to the cerebral cortex, which is a "slow learner" but forms longer-lasting memories. This process of "teaching the cerebral cortex" until a long-term memory is established is called **memory consolidation.** Lesions of the hippocampus can abolish the ability to form new declarative memories, but they do not abolish old memories or the acquisition of new procedural memories, such as the ability to learn a new motor skill.

Long-term memories are stored in various areas of the cerebral cortex. The memory of language (vocabulary and grammatical rules) resides in the Wernicke area. Our memories of faces, voices, and familiar objects are stored in the superior temporal lobe. Memories of one's social role, appropriate behavior, goals, and plans are stored in the prefrontal cortex. Procedural memories are stored in the motor association area, basal nuclei, and cerebellum.

Cerebral Lateralization

The two cerebral hemispheres look identical at a glance, but close examination reveals a number of differences. For example, in women the left temporal lobe is longer than the right. In left-handed people, the left frontal, parietal, and occipital lobes are usually wider than those on the right. The two hemispheres also differ in some of their functions (fig. 15.22). Neither hemisphere is "dominant," but each is specialized for certain tasks. This difference in function is called **cerebral lateralization**.

One hemisphere, usually the left, is called the *categorical hemisphere*. It is specialized for spoken and written language and for the sequential and analytical reasoning employed in such fields as science and mathematics. This hemisphere seems to break information into fragments and analyze it in a linear way. The other hemisphere, usually the right, is called the *representational hemisphere*. It perceives information in a more integrated, holistic way. It is the seat of imagination and insight; musical and artistic skill; perception of patterns and spatial relationships; and comparison of sights, sounds, smells, and tastes.

Figure 15.23 shows some of the forebrain structures in a coronal tissue section and a corresponding MRI of the cerebrum. Table 15.2 summarizes the forebrain functions described in the last several pages.

Before You Go On

Answer the following questions to test your understanding of the preceding section:

13. What is the role of the thalamus in sensory function?

14. List at least six functions of the hypothalamus.

15. Name the five lobes of the cerebrum and describe their locations and boundaries.

16. Where are the basal nuclei located? What is their general function?

17. Where is the limbic system located? What component of it is involved in emotion? What component is involved in memory?

18. Describe the locations and functions of the somatosensory, visual, auditory, and frontal association areas.

19. Describe the somatotopy of the primary motor cortex and primary sensory cortex.

20. What are the roles of the Wernicke area, Broca area, and precentral gyrus in language?

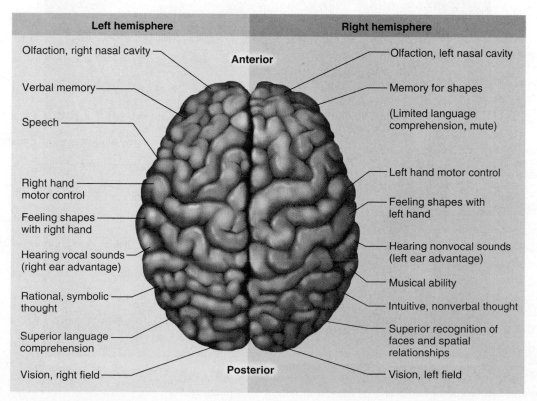

Left hemisphere — **Right hemisphere**

Olfaction, right nasal cavity — **Anterior** — Olfaction, left nasal cavity

Verbal memory — Memory for shapes

Speech — (Limited language comprehension, mute)

Right hand motor control — Left hand motor control

Feeling shapes with right hand — Feeling shapes with left hand

Hearing vocal sounds (right ear advantage) — Hearing nonvocal sounds (left ear advantage)

Rational, symbolic thought — Musical ability

— Intuitive, nonverbal thought

Superior language comprehension — Superior recognition of faces and spatial relationships

Vision, right field — **Posterior** — Vision, left field

Figure 15.22 Lateralization of Cerebral Functions. The two cerebral hemispheres are not functionally identical.
• *If a person is described as "left-brained," does this mean he or she makes little use of the right hemisphere?*

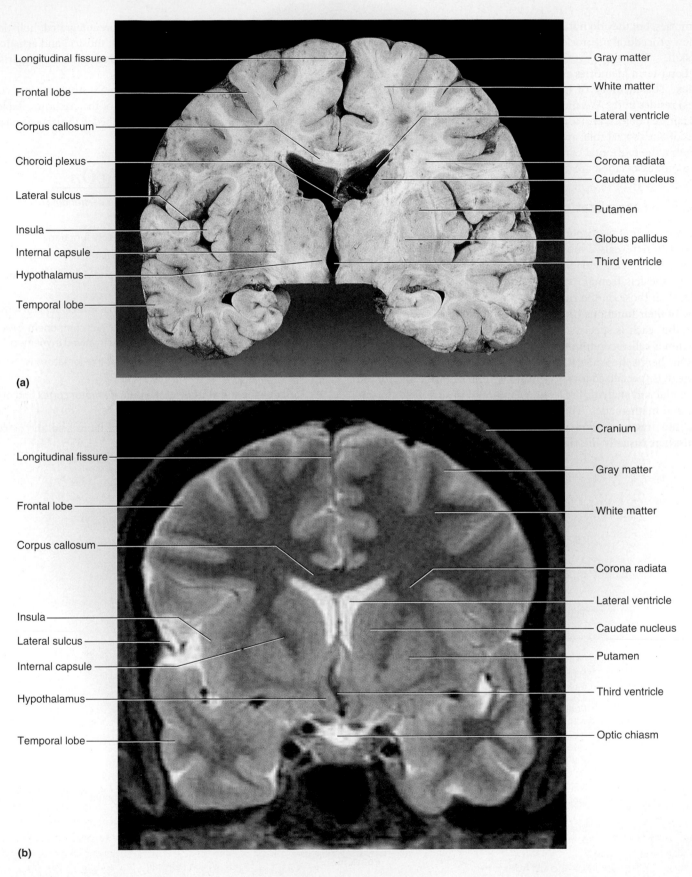

Longitudinal fissure

Frontal lobe

Corpus callosum

Choroid plexus

Lateral sulcus

Insula

Internal capsule

Hypothalamus

Temporal lobe

Gray matter

White matter

Lateral ventricle

Corona radiata

Caudate nucleus

Putamen

Globus pallidus

Third ventricle

(a)

Longitudinal fissure

Frontal lobe

Corpus callosum

Insula

Lateral sulcus

Internal capsule

Hypothalamus

Temporal lobe

Cranium

Gray matter

White matter

Corona radiata

Lateral ventricle

Caudate nucleus

Putamen

Third ventricle

Optic chiasm

(b)

Figure 15.23 Coronal Sections of the Cerebrum. (a) Tissue section. (b) A corresponding MRI.

TABLE 15.2	Forebrain Functions
Diencephalon	
Thalamus	Sensory processing; relay of sensory and other signals to cerebrum; relay of cerebral output to other parts of brain
Hypothalamus	Hormone synthesis; control of pituitary secretion; autonomic responses affecting heart rate, blood pressure, pupillary diameter, digestive secretion and motility, and other visceral functions; thermoregulation; hunger and thirst; sleep and circadian rhythms; emotional responses; sexual function; memory
Epithalamus	Hormone secretion; relay of signals between midbrain and limbic system
Cerebral lobes	
Frontal lobe	Smell; motor aspects of speech; voluntary control of skeletal muscles; procedural memory; cognitive functions such as abstract thought, judgment, responsibility, foresight, ambition, planning, and ability to stay focused on a task
Parietal lobe	Somatosensory functions, taste, awareness of body movement and orientation, language recognition, nonmotor aspects of speech
Occipital lobe	Vision
Temporal lobe	Hearing, smell, interpreting visual information, learning, memory, emotion
Insula	Hearing, taste, visceral sensation
Basal nuclei	Motor control; procedural memory
Limbic system	Learning, emotion, gratification, and aversion responses

15.4 The Cranial Nerves

▶ Expected Learning Outcomes

When you have completed this section, you should be able to

- list the 12 cranial nerves by name and number;
- identify where each cranial nerve originates and terminates; and
- state the functions of each cranial nerve.

To be functional, the brain must communicate with the rest of the body. Most of its input and output travels by way of the spinal cord, but it also communicates by way of 12 pairs of **cranial nerves** (table 15.3). These arise primarily from the base of the brain, exit the cranium through its foramina, and lead to muscles and sense organs primarily in the head and neck. They are numbered I through XII starting with the most rostral (fig. 15.24). Each nerve also has a descriptive name such as *optic nerve* and *vagus nerve*.

An Aid to Memory

Generations of biology and medical students have relied on mnemonic (memory-aiding) phrases and ditties, ranging from the sublimely silly to the unprintably ribald, to help them remember the cranial nerves and other anatomy. An old classic began, "On old Olympus' towering tops . . . ," with the first letter of each word matching the first letter of each cranial nerve (olfactory, optic, oculomotor, etc.). Some cranial nerves have changed names, however, since that passage was devised. A substitute now used by many students, taking the first letter of each nerve's name, is "Oh, once one

takes the anatomy final, very good vacation ahead." The first two letters of *ahead* represent nerves XI and XII. One of the author's former students, now a neurologist, devised a mnemonic that can remind you of the first two to four letters of most cranial nerves:

Old	**ol**factory (I)
Opie	**op**tic (II)
occasionally	**oc**ulomotor (III)
tries	**tr**ochlear (IV)
trigonometry	**trig**eminal (V)
and	**a**bducens (VI)
feels	**f**acial (VII)
very	**v**estibulocochlear (VIII)
gloomy,	**glo**ssopharyngeal (IX)
vague,	**vagu**s (X)
and	**a**ccessory (XI)
hypoactive.	**hypo**glossal (XII)

Classification

Cranial nerves are traditionally classified as sensory (I, II, and VIII), motor (III, IV, VI, XI, and XII), or mixed (V, VII, IX, and X). In reality, only cranial nerves I and II (for smell and vision) are purely sensory, whereas all the rest contain both afferent and efferent fibers and are therefore mixed nerves. Those traditionally classified as motor not only stimulate muscle contractions but also contain afferent fibers of proprioception, which provide the brain with unconscious feedback for controlling muscle contraction and which make one consciously aware of such things as the position of the tongue and orientation of the head. Cranial nerve VIII, concerned with hearing

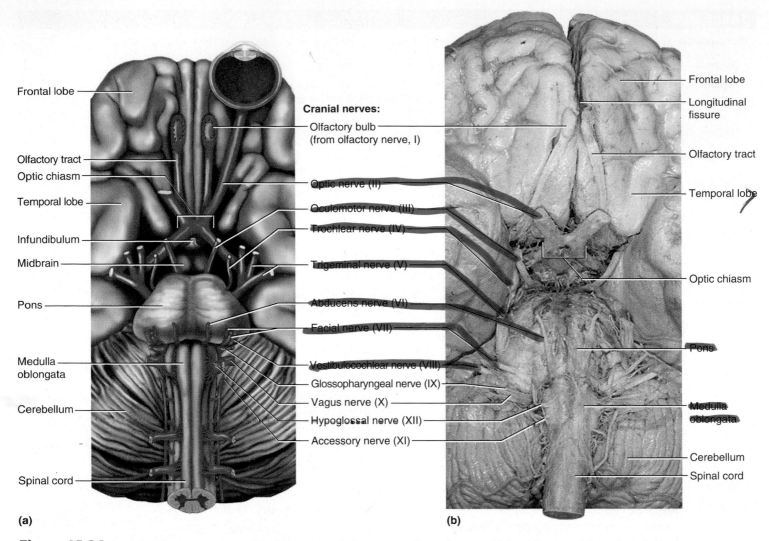

Cranial nerves:

Frontal lobe

Olfactory tract

Optic chiasm

Temporal lobe

Infundibulum

Midbrain

Pons

Medulla oblongata

Cerebellum

Spinal cord

Olfactory bulb (from olfactory nerve, I)

Optic nerve (II)

Oculomotor nerve (III)

Trochlear nerve (IV)

Trigeminal nerve (V)

Abducens nerve (VI)

Facial nerve (VII)

Vestibulocochlear nerve (VIII)

Glossopharyngeal nerve (IX)

Vagus nerve (X)

Hypoglossal nerve (XII)

Accessory nerve (XI)

Frontal lobe

Longitudinal fissure

Olfactory tract

Temporal lobe

Optic chiasm

Pons

Medulla oblongata

Cerebellum

Spinal cord

(a) (b)

Figure 15.24 **The Cranial Nerves.** (a) Base of the brain, showing the 12 pairs of cranial nerves. (b) Cranial nerves of the cadaver brain.

and equilibrium, is traditionally classified as sensory, but also has motor fibers that return signals to the inner ear and "tune" it to sharpen one's sense of hearing. The nerves traditionally classified as mixed have sensory functions quite unrelated to their motor functions—for example, the facial nerve (VII) has a sensory role in taste and a motor role in controlling facial expressions.

In order to teach the traditional classification (which is relevant for such purposes as board examinations and comparison to other books), yet remind you that all but two of these nerves are mixed, table 15.3 describes many of the nerves as *predominantly* sensory or motor.

Nerve Pathways

Most motor fibers of the cranial nerves begin in nuclei of the brainstem and lead to glands and muscles. The sensory fibers begin in receptors located mainly in the head and neck and lead mainly to the brainstem. Pathways for the special senses are described in chapter 17. Sensory fibers for proprioception begin in the muscles innervated by the motor fibers of the cranial nerves, but they often travel to the brain in a different nerve than the one that supplies the motor innervation.

Most cranial nerves carry fibers between the brainstem and ipsilateral receptors and effectors. Thus, a lesion on one side of the brainstem causes a sensory or motor deficit on the same side of the head. This contrasts with lesions to the motor and somatosensory cortex of the cerebrum, which, as we saw earlier, cause sensory and motor deficits on the contralateral side of the body. The exceptions are the optic nerve (cranial nerve II), where half the fibers decussate to the opposite side of the brain (see chapter 17), and the trochlear nerve (cranial nerve IV), in which all efferent fibers go to a muscle of the contralateral eye.

agreed by all authorities to be either mixed or purely sensory. Those classified as predominantly motor or
Motor nerves also contain proprioceptive sensory fibers, but these may come from different muscles than

he sense of smell. It consists of several separate fascicles that pass independently through the cribriform
sible on brains removed from the skull because these fascicles are severed by removal of the brain.

Termination	Cranial Passage	Effect of Damage	Clinical Test
Olfactory bulbs	Cribriform foramina of ethmoid bone	Impaired sense of smell	Determine whether subject can smell (not necessarily identify) aromatic substances such as coffee, vanilla, clove oil, or soap

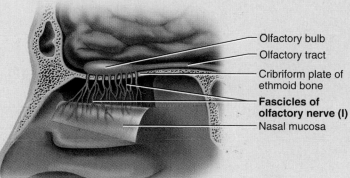

Olfactory bulb
Olfactory tract
Cribriform plate of ethmoid bone
Fascicles of olfactory nerve (I)
Nasal mucosa

Figure 15.25 The Olfactory Nerve (I).

II. Optic Nerve. This is the nerve for vision.

Composition	Function	Origin	Termination	Cranial Passage	Effect of Damage	Clinical Test
Sensory	Vision	Retina	Thalamus and midbrain	Optic foramen	Blindness in part or all of visual field	Inspect retina with ophthalmoscope; test peripheral vision and visual acuity

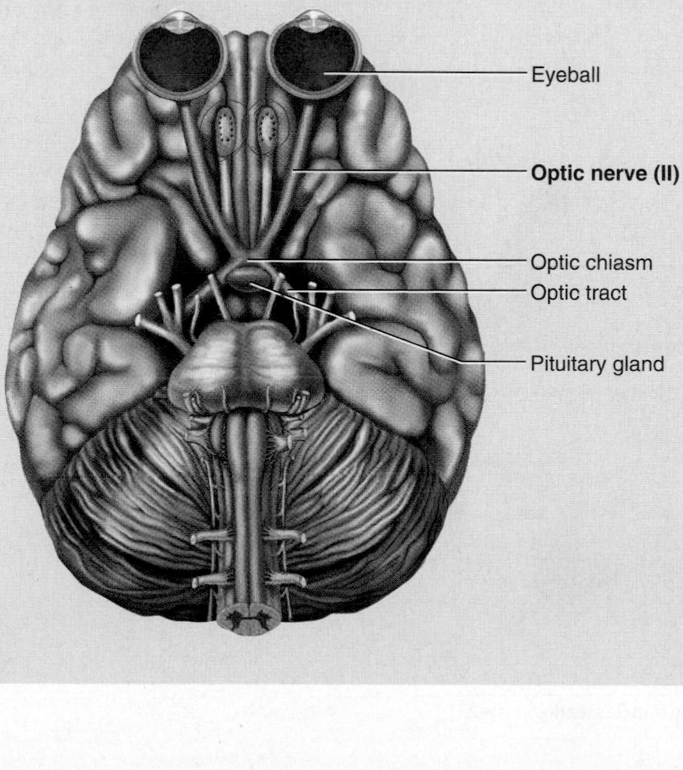

Eyeball

Optic nerve (II)

Optic chiasm
Optic tract

Pituitary gland

Figure 15.26 The Optic Nerve (II).

TABLE 15.3	**The Cranial Nerves** (continued)

III. Oculomotor[45] Nerve (OC-you-lo-MO-tur).   This nerve controls muscles that turn the eyeball up, down, and medially, as well as controlling the iris, lens, and upper eyelid.

Composition	Function	Origin	Termination	Cranial Passage	Effect of Damage	Clinical Test
Predominantly motor	Eye movements, opening of eyelid, pupillary constriction, focusing	Midbrain	Somatic fibers to levator palpebrae superioris; superior, medial, and inferior rectus muscles; and inferior oblique muscle of eye. Autonomic fibers enter eyeball and lead to constrictor of iris and ciliary muscle of lens.	Superior orbital fissure	Drooping eyelid; dilated pupil; inability to move eye in some directions; tendency of eye to rotate laterally at rest; double vision; difficulty focusing	Look for differences in size and shape of right and left pupils; test pupillary response to light; test ability to track moving objects.

Oculomotor nerve (III):
- Superior branch
- Inferior branch
- Ciliary ganglion
- Superior orbital fissure

Figure 15.27 The Oculomotor Nerve (III).

IV. Trochlear[46] Nerve (TROCK-lee-ur).   This nerve controls a muscle that rotates the eyeball medially and slightly depresses the eyeball when the head turns.

Composition	Function	Origin	Termination	Cranial Passage	Effect of Damage	Clinical Test
Predominantly motor	Eye movements	Midbrain	Superior oblique muscle of eye	Superior orbital fissure	Double vision and inability to rotate eye inferolaterally; eye points superolaterally and subject tends to tilt head toward affected side	Test ability of eye to rotate inferolaterally

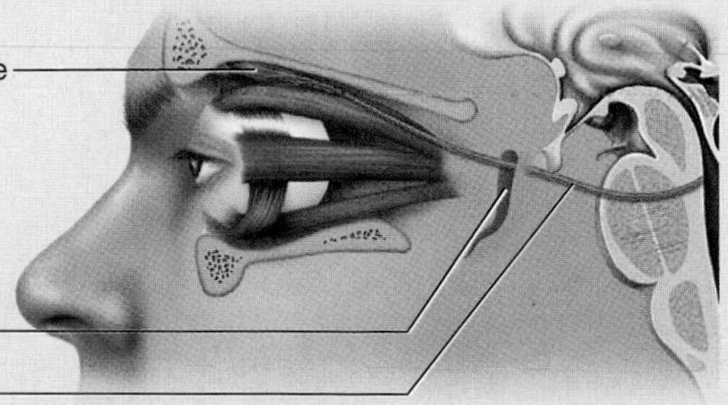

Superior oblique muscle

Superior orbital fissure

Trochlear nerve (IV)

Figure 15.28 The Trochlear Nerve (IV).

---

[45]oculo = eye + motor = mover
[46]trochlea = pulley (for a loop through which the muscle's tendon passes)

TABLE 15.3	The Cranial Nerves (continued)

V. Trigeminal[47] Nerve (tri-JEM-ih-nul). This is the most important sensory nerve of the face, and the largest cranial nerve except for the optic nerve. It forks into three divisions: *ophthalmic (V$_1$), maxillary (V$_2$),* and *mandibular (V$_3$) (see fig. 15.29 on page 432).*

Composition	Function	Origin	Termination	Cranial Passage	Effect of Damage	Clinical Test
Ophthalmic division (V$_1$)						
Sensory	Touch, temperature, and pain sensations from upper face	Superior region of face as illustrated; surface of eyeball; lacrimal (tear) gland; superior nasal mucosa; frontal and ethmoid sinuses	Pons	Superior orbital fissure	Loss of sensation from upper face	Test corneal reflex (blinking in response to light touch to eyeball)
Maxillary division (V$_2$)						
Sensory	Same as V$_1$, lower on face	Middle region of face as illustrated; nasal mucosa; maxillary sinus; palate; upper teeth and gums	Pons	Foramen rotundum and infraorbital foramen	Loss of sensation from middle face	Test sense of touch, pain, and temperature with light touch, pinpricks, and hot and cold objects
Mandibular division (V$_3$)						
Mixed	*Sensory:* Same as V$_1$ and V$_2$, lower on face *Motor:* mastication	*Sensory:* Inferior region of face as illustrated; anterior two-thirds of tongue (but not taste buds); lower teeth and gums; floor of mouth; dura mater *Motor:* Pons	*Sensory:* Pons *Motor:* Anterior belly of digastric; masseter, temporalis, mylohyoid, and pterygoid muscles; tensor tympani muscle of middle ear	Foramen ovale	Loss of sensation; impaired chewing	Assess motor functions by palpating masseter and temporalis while subject clenches teeth; test ability to move mandible from side to side and to open mouth against resistance

DEEPER INSIGHT 15.3

Some Cranial Nerve Disorders

Trigeminal neuralgia[48] (*tic douloureux*[49]) is a syndrome characterized by recurring episodes of intense stabbing pain in the trigeminal nerve. The cause is unknown; there is no visible change in the nerve. It usually occurs after the age of 50 and mostly in women. The pain lasts only a few seconds to a minute or two, but it strikes at unpredictable intervals and sometimes up to a hundred times a day. The pain usually occurs in a specific zone of the face, such as around the mouth and nose. It may be triggered by touch, drinking, tooth brushing, or washing the face. Analgesics (pain relievers) give only limited relief. Severe cases are treated by cutting the nerve, but this also deadens most other sensation in that side of the face.

Bell[50] *palsy* is a degenerative disorder of the facial nerve, probably due to a virus. It is characterized by paralysis of the facial muscles on one side with resulting distortion of the facial features, such as sagging of the mouth or lower eyelid. The paralysis may interfere with speech, prevent closure of the eye, and sometimes inhibit tear secretion. There may also be a partial loss of the sense of taste. Bell palsy may appear abruptly, sometimes overnight, and often disappears spontaneously within 3 to 5 weeks.

---

[47]*tri* = three + *gem* = born (*trigem* = triplets)
[48]*neur* = nerve + *algia* = pain
[49]*tic* = twitch, spasm + *douloureux* = painful

[50]Sir Charles Bell (1774–1842), Scottish physician

TABLE 15.3   The Cranial Nerves (continued)

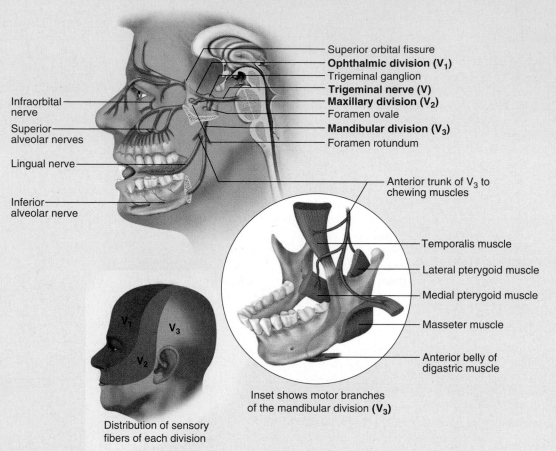

Figure 15.29 The Trigeminal Nerve (V).

VI. Abducens[51] Nerve (ab-DOO-senz).   This nerve controls a muscle that turns the eyeball laterally.

Composition	Function	Origin	Termination	Cranial Passage	Effect of Damage	Clinical Test
Predominantly motor	Lateral eye movement	Inferior pons	Lateral rectus muscle of eye	Superior orbital fissure	Inability to turn eye laterally; at rest, eye turns medially because of action of antagonistic muscles	Test lateral eye movement

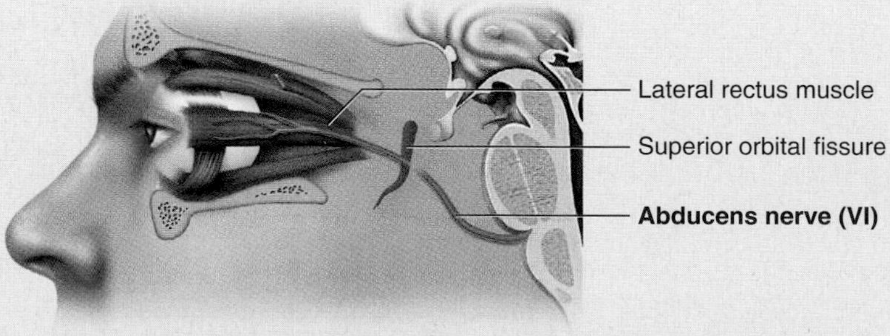

Figure 15.30 The Abducens Nerve (VI).

[51]*ab* = away + *duc* = to lead or turn

TABLE 15.3 | **The Cranial Nerves (continued)**

VII. Facial Nerve.   This is the major motor nerve of the facial muscles. It divides into five prominent branches: *temporal, zygomatic, buccal, mandibular, and cervical.*

Composition	Function	Origin	Termination	Cranial Passage	Effect of Damage	Clinical Test
Mixed	*Sensory:* Taste *Motor:* Facial expression; secretion of tears, saliva, nasal and oral mucus	*Sensory:* Taste buds of anterior two-thirds of tongue *Motor:* Pons	*Sensory:* Thalamus *Motor:* Somatic fibers to digastric muscle, stapedius muscle of middle ear, stylohyoid muscle, muscles of facial expression. Autonomic fibers to submandibular and sublingual salivary glands, tear glands, nasal and palatine glands.	Internal acoustic meatus and stylomastoid foramen	Inability to control facial muscles; sagging due to loss of muscle tone; distorted sense of taste, especially for sweets	Test anterior two-thirds of tongue with substances such as sugar, salt, vinegar, and quinine; test response of tear glands to ammonia fumes; test subject's ability to smile, frown, whistle, raise eyebrows, close eyes, etc.

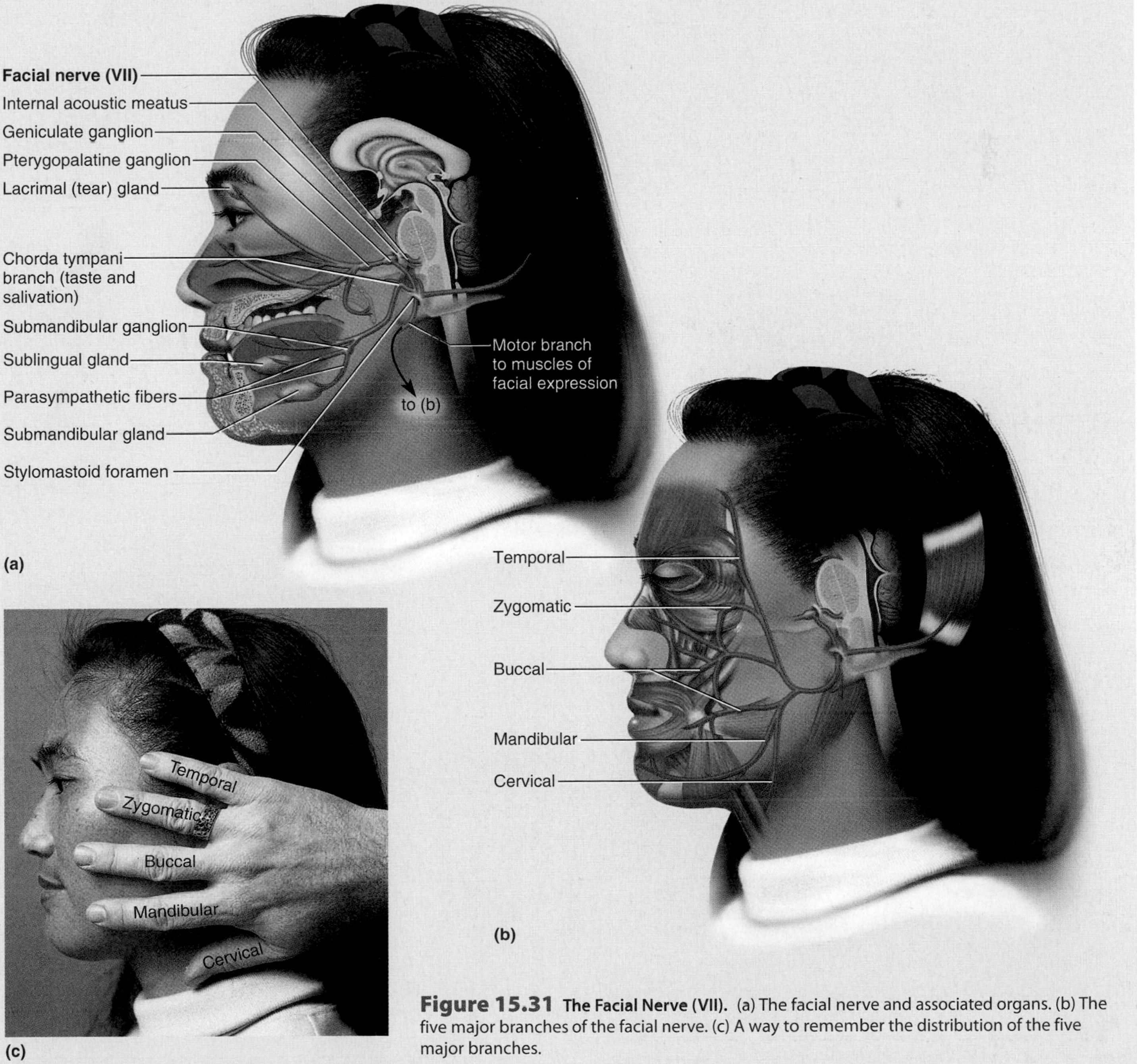

Facial nerve (VII)
Internal acoustic meatus
Geniculate ganglion
Pterygopalatine ganglion
Lacrimal (tear) gland

Chorda tympani branch (taste and salivation)
Submandibular ganglion
Sublingual gland
Parasympathetic fibers
Submandibular gland
Stylomastoid foramen

Motor branch to muscles of facial expression

to (b)

(a)

Temporal
Zygomatic
Buccal
Mandibular
Cervical

(b)

Temporal
Zygomatic
Buccal
Mandibular
Cervical

(c)

Figure 15.31 The Facial Nerve (VII). (a) The facial nerve and associated organs. (b) The five major branches of the facial nerve. (c) A way to remember the distribution of the five major branches.

TABLE 15.3 | **The Cranial Nerves (continued)**

VIII. Vestibulocochlear[52] Nerve (vess-TIB-you-lo-COC-lee-ur). This is the nerve of hearing and equilibrium, but it also has motor fibers that lead to cells of the cochlea that tune the sense of hearing (see chapter 17).

Composition	Function	Origin	Termination	Cranial Passage	Effect of Damage	Clinical Test
Predominantly sensory	Hearing and equilibrium	*Sensory:* Cochlea, vestibule, and semicircular ducts of inner ear *Motor:* Pons	*Sensory:* Fibers for hearing end in medulla; fibers for equilibrium end at junction of medulla and pons *Motor:* Outer hair cells of cochlea of inner ear	Internal acoustic meatus	Nerve deafness, dizziness, nausea, loss of balance, and nystagmus (involuntary rhythmic oscillation of eyes from side to side)	Look for nystagmus; test hearing, balance, and ability to walk a straight line

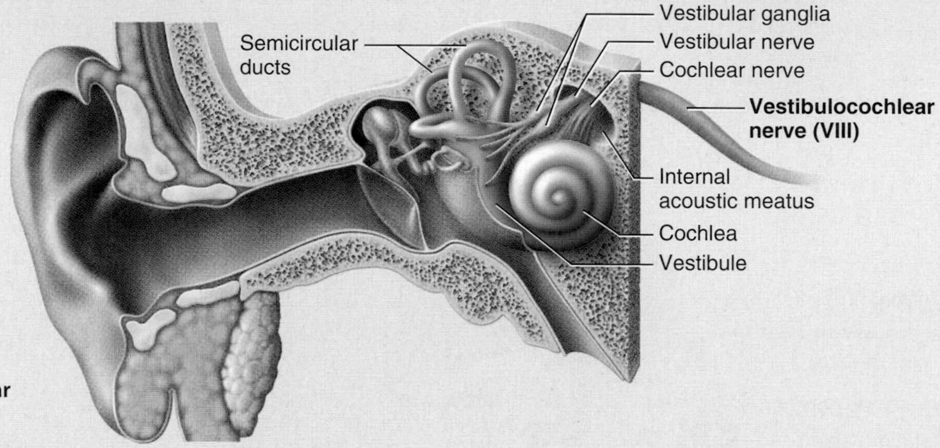

Figure 15.32 The Vestibulocochlear Nerve (VIII).

IX. Glossopharyngeal[53] Nerve (GLOSS-oh-fah-RIN-jee-ul). This is a complex, mixed nerve with numerous sensory and motor functions in the head, neck, and thoracic regions, including sensation from the tongue, throat, and outer ear; control of food ingestion; and some aspects of cardiovascular and respiratory function.

Composition	Function	Origin	Termination	Cranial Passage	Effect of Damage	Clinical Test
Mixed	*Sensory:* Taste; touch, pressure, pain and temperature sensations from tongue and outer ear; regulation of blood pressure and respiration *Motor:* Salivation, swallowing, gagging	*Sensory:* Pharynx; middle and outer ear; posterior one-third of tongue (including taste buds); internal carotid artery *Motor:* Medulla oblongata	*Sensory:* Medulla oblongata *Motor:* Parotid salivary gland; glands of posterior tongue; stylopharyngeal muscle (which dilates pharynx during swallowing)	Jugular foramen	Loss of bitter and sour taste; impaired swallowing	Test gag reflex, swallowing, and coughing; note any speech impediments; test posterior one-third of tongue with bitter and sour substances

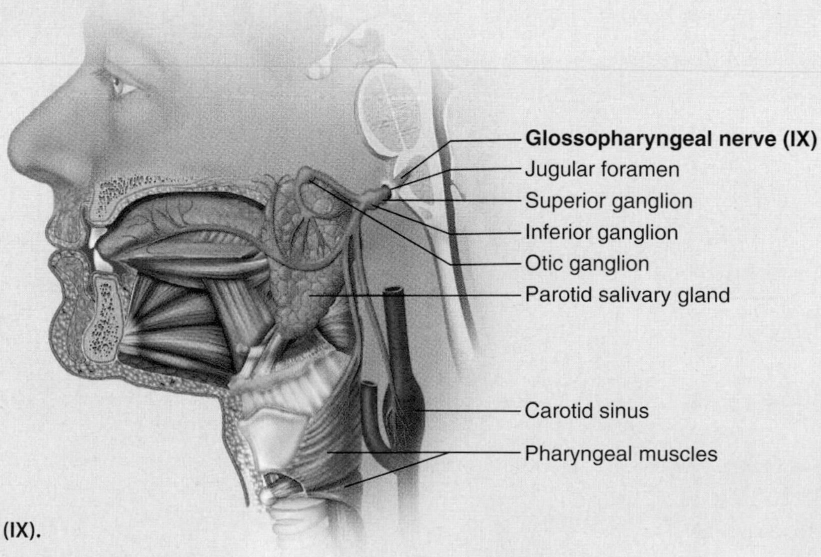

Figure 15.33 The Glossopharyngeal Nerve (IX).

[52]*vestibul* = entryway (vestibule of the inner ear) + *cochlea* = conch, snail (cochlea of the inner ear)

[53]*glosso* = tongue + *pharyng* = throat + *eal* = pertaining to

TABLE 15.3	The Cranial Nerves (continued)

X. Vagus[54] Nerve (VAY-gus).   The vagus has the most extensive distribution of any cranial nerve, supplying not only organs in the head and neck but also most viscera of the thoracic and abdominal body cavities. It plays major roles in the control of cardiac, pulmonary, digestive, and urinary functions.

Composition	Function	Origin	Termination	Cranial Passage	Effect of Damage	Clinical Test
Mixed	*Sensory:* Taste; sensations of hunger, fullness, and gastrointestinal discomfort *Motor:* Swallowing, speech, deceleration of heart, broncho-constriction, gastro-intestinal secretion and motility	*Sensory:* Thoracic and abdominal viscera, root of tongue, pharynx, larynx, epiglottis, outer ear, dura mater *Motor:* Medulla oblongata	*Sensory:* Medulla oblongata *Motor:* Tongue, palate, pharynx, larynx, lungs, heart, liver, spleen, digestive tract, kidney, ureter	Jugular foramen	Hoarseness or loss of voice; impaired swallowing and gastrointestinal motility; fatal if both vagus nerves are damaged	Examine palatal movements during speech; check for abnormalities of swallowing, absence of gag reflex, weak hoarse voice, inability to cough forcefully

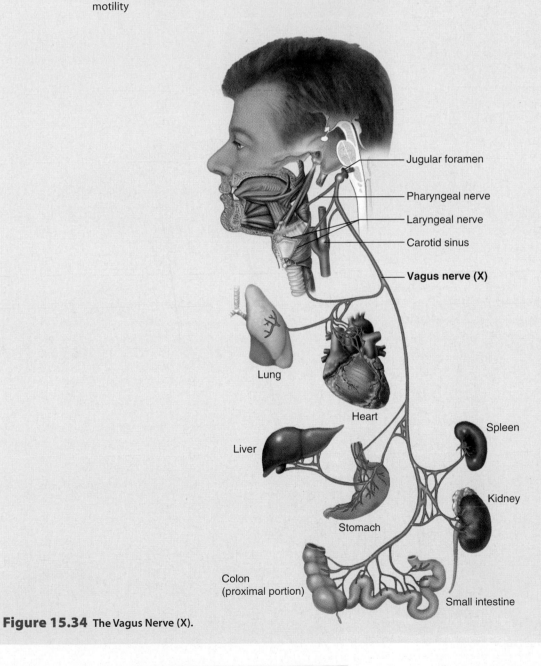

Jugular foramen

Pharyngeal nerve

Laryngeal nerve

Carotid sinus

Vagus nerve (X)

Lung

Heart

Spleen

Liver

Kidney

Stomach

Colon (proximal portion)

Small intestine

Figure 15.34 The Vagus Nerve (X).

[54]*vag* = wandering

TABLE 15.3 | The Cranial Nerves (continued)

XI. Accessory Nerve. This nerve takes an unusual path. Unlike any other spinal nerve, it does not arise entirely from the brain. A small root does arise from the medulla oblongata, but a larger root arises from the cervical spinal cord. The latter root ascends alongside the spinal cord and enters the cranial cavity through the foramen magnum. It then joins the smaller root for a short distance, and they exit the cranium together through the jugular foramen, bundled with the vagus and glossopharyngeal nerves. The accessory nerve controls mainly swallowing and neck and shoulder muscles.

Composition	Function	Origin	Termination	Cranial Passage	Effect of Damage	Clinical Test
Predominantly motor	Swallowing; head, neck, and shoulder movements	Medulla oblongata and spinal cord segments C1 to C6	Palate; pharynx; trapezius and sternocleidomastoid muscles	Jugular foramen	Impaired movement of head, neck, and shoulders; difficulty shrugging shoulder on damaged side; paralysis of sternocleidomastoid, causing head to turn toward injured side	Test ability to rotate head and shrug shoulders against resistance

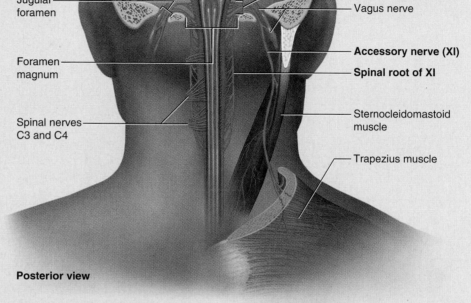

Jugular foramen
Foramen magnum
Spinal nerves C3 and C4

Cranial root of XI
Vagus nerve
Accessory nerve (XI)
Spinal root of XI
Sternocleidomastoid muscle
Trapezius muscle

Posterior view

Figure 15.35 **The Accessory Nerve (XI).** A posterior view showing the dual origin of the accessory nerve from the cervical spinal cord and medulla oblongata. The spinal root enters the cranial cavity through the foramen magnum, picks up the cranial root from the medulla, then exits the cranium through the jugular foramen and innervates neck and shoulder muscles.

XII. Hypoglossal[55] Nerve (HY-po-GLOSS-ul). This nerve controls tongue movements.

Composition	Function	Origin	Termination	Cranial Passage	Effect of Damage	Clinical Test
Predominantly motor	Tongue movements of speech, food manipulation, and swallowing	Medulla oblongata	Intrinsic and extrinsic muscles of tongue	Hypoglossal canal	Impaired speech and swallowing; inability to protrude tongue if both right and left nerves are damaged; deviation of tongue toward injured side, and atrophy on that side, if only one nerve is damaged	Note deviations of tongue as subject protrudes and retracts it; test ability to protrude tongue against resistance

Hypoglossal canal
Intrinsic muscles of the tongue
Extrinsic muscles of the tongue
Hypoglossal nerve (XII)

Figure 15.36 The Hypoglossal Nerve (XII).

[55]*hypo* = below + *gloss* = tongue + *al* = pertaining to

Before You Go On

Answer the following questions to test your understanding of the preceding section:

21. List the purely sensory cranial nerves by name and number, and state the function of each.

22. What is the only cranial nerve to extend beyond the head–neck region? In general terms, where does it lead?

23. If the oculomotor, trochlear, or abducens nerve were damaged, the effect would be similar in all three cases. What would that effect be?

24. Which cranial nerve carries sensory signals from the largest area of the face?

25. Name two cranial nerves involved in the sense of taste and describe where their sensory fibers originate.

15.5 Developmental and Clinical Perspectives

▶ Expected Learning Outcomes

When you have completed this section, you should be able to

- describe some ways in which neural function and cerebral anatomy change in old age; and

- discuss Alzheimer and Parkinson diseases at the levels of neurotransmitter function and brain anatomy.

The Aging Central Nervous System

In chapter 13, we examined development of the nervous system at the beginning of life. As so many of us are regretfully aware, the nervous system also exhibits some marked changes at the other end of the life span. It attains its peak development and efficiency around age 30. By age 75, the average brain weighs slightly less than half what it did at age 30. The cerebral gyri are narrower, the sulci are wider, the cortex is thinner, and there is more space between the brain and meninges. The remaining neurons show signs that their metabolism is slowing down, such as less rough ER and Golgi complex. Old neurons accumulate lipofuscin pigment and begin to show *neurofibrillary tangles*—dense mats of cytoskeletal elements in their cytoplasm. In the extracellular material, *senile plaques* appear, especially in people with Down syndrome and Alzheimer disease. The plaques are composed of cells and altered nerve fibers surrounding a core of *amyloid protein*.

Old neurons are also less efficient at signal conduction and transmission. The degeneration of myelin sheaths slows down conduction along the axon. The neurons have fewer synapses, and for multiple reasons, signals are not transmitted across the synapses as well as in the younger years: The neurons produce less neurotransmitter, they have fewer receptors, and the neuroglia around the synapses is more leaky and allows neurotransmitter to diffuse away. Target cells have fewer receptors for norepinephrine, and the sympathetic nervous system thus becomes less able to regulate such variables as body temperature and blood pressure.

Not all functions of the central nervous system are equally affected by aging. Language skills and long-term memory hold up better than motor coordination, intellectual function, and short-term memory. Elderly people are often better at remembering things in the distant past than remembering recent events.

Two Neurodegenerative Diseases

Like a machine with a great number of moving parts, the nervous system is highly subject to malfunctions. Neurological disorders fill many volumes of medical textbooks and can hardly be touched upon here. We have considered meningitis, one peculiar case of the effects of cerebral trauma, and two cranial nerve disorders in Deeper Insights 15.1 through 15.3; several other neurological disorders are briefly described in table 15.4. We will close with a brief look at two of the most common brain dysfunctions, Alzheimer and Parkinson diseases. Both of these relate to neurotransmitter imbalances in the brain, and are considered to be *neurodegenerative diseases*. A basic understanding of these two diseases lends added clinical relevance to some areas of the brain studied in this chapter.

Alzheimer[56] disease (AD) affects about 11% of the U.S. population over the age of 65 and 47% by age 85. It accounts for nearly half of all nursing home admissions and is a leading cause of death among the elderly. AD may begin before the age of 50 with symptoms so slight and ambiguous that early diagnosis is difficult. One of its first symptoms is memory loss, especially for recent events. As the disease progresses, patients exhibit reduced attention span and may become disoriented and lost in previously familiar places. The AD patient may become moody, confused, paranoid, combative, or hallucinatory, and may eventually lose even the ability to read, write, talk, walk, and eat. Death ensues from pneumonia or other complications of confinement and immobility.

Diagnosis of AD is confirmed on autopsy. There is atrophy of some of the gyri of the cerebral cortex and hippocampus. Neurofibrillary tangles and senile plaques are abundant (fig. 15.37). Cholinergic neurons are reduced in number and the level of acetylcholine in affected areas of the brain is consequently low. Intense research efforts are currently geared toward identifying the cause of AD and developing treatment strategies. Researchers have identified three genes on chromosomes 1, 14, and 21 for various forms of early- and late-onset AD.

Parkinson[57] disease (PD), also called *paralysis agitans* or *parkinsonism*, is a progressive loss of motor function typically beginning in a person's 50s or 60s. It is due to degeneration of dopamine-releasing neurons in the substantia nigra of the midbrain. A gene has recently been identified for a hereditary form of PD, but most cases are nonhereditary and of little-known cause. Dopamine is an inhibitory

---

[56]Alois Alzheimer (1864–1915), German neurologist
[57]James Parkinson (1755–1824), British physician

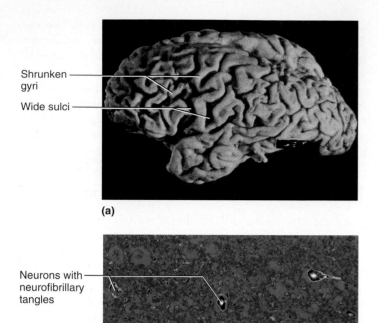

Shrunken gyri

Wide sulci

(a)

Neurons with neurofibrillary tangles

Senile plaque

(b)

Figure 15.37 **Alzheimer Disease.** (a) Brain of a person who died of AD. Note the shrunken gyri and wide sulci. (b) Cerebral tissue from a person with AD. Neurofibrillary tangles are present in the neurons, and a senile plaque is evident in the extracellular matrix.

neurotransmitter that normally prevents excessive activity in the basal nuclei. Degeneration of the dopamine-releasing neurons leads to hyperactivity of the basal nuclei and, therefore, involuntary muscle contractions. These take such forms as shaking of the hands (tremor) and compulsive "pill-rolling" motions of the thumb and fingers. In addition, the facial muscles may become rigid and produce a staring, expressionless face with a slightly open mouth. The patient's range of motion diminishes. He or she takes smaller steps and develops a slow, shuffling gait with a forward-bent posture and a tendency to fall forward. Speech becomes slurred and handwriting becomes cramped and eventually illegible. Tasks such as buttoning clothes and preparing food become increasingly laborious. Patients cannot be expected to recover from PD, but drugs, neurosurgery, and physical therapy can lessen the severity of its effects.

Before You Go On

Answer the following questions to test your understanding of the preceding section:

26. Describe two respects in which neurons function less efficiently in old age.

27. Describe some changes seen in the brain with aging.

28. Describe the neuroanatomical and behavioral changes seen in Alzheimer and Parkinson diseases.

TABLE 15.4	**Some Disorders Associated with the Brain and Cranial Nerves**
Cerebral palsy	Muscular incoordination resulting from damage to the motor areas of the brain during fetal development, birth, or infancy; causes include prenatal rubella infection, drugs, and radiation exposure; oxygen deficiency during birth; and hydrocephalus
Concussion	Damage to the brain typically resulting from a blow, often with loss of consciousness, disturbances of vision or equilibrium, and short-term amnesia
Encephalitis	Inflammation of the brain, accompanied by fever, usually caused by mosquito-borne viruses or herpes simplex virus; causes neural degeneration and necrosis; can lead to delirium, seizures, and death
Epilepsy	Disorder causing sudden, massive discharge of neurons (seizures) resulting in motor convulsions, sensory and psychic disturbances, and often impaired consciousness; may result from birth trauma, tumors, infections, drug or alcohol abuse, or congenital brain malformation
Migraine headache	Recurring headaches often accompanied by nausea, vomiting, dizziness, and aversion to light; often triggered by such factors as weather changes, stress, hunger, red wine, or noise; more common in women and sometimes hereditary
Schizophrenia	A thought disorder involving delusions, hallucinations, inappropriate emotional responses to situations, incoherent speech, and withdrawal from society, resulting from hereditary or developmental abnormalities in neural networks

Disorders Described Elsewhere

Study Guide

Assess Your Learning Outcomes

You should have a good understanding of this chapter if you can accurately address the following issues.

15.1 Overview of the Brain (p. 399)

1. The typical weight of the adult brain
2. The meanings of *rostral* and *caudal* in CNS anatomy
3. The three principal regions of the brain—cerebrum, cerebellum, and brainstem
4. Major landmarks of the brain, including the cerebral hemispheres, gyri and sulci, longitudinal fissure, and corpus callosum
5. Locations of the gray and white matter of the brain
6. The meninges of the brain; how they differ from those of the spinal cord; and the relationship of the dura mater to the dural sinuses
7. The four ventricles of the brain and their interconnections
8. The functions, sources, flow, and reabsorption of cerebrospinal fluid
9. The constituents of the brain barrier system; its significance in protecting the brain tissue; the limitations of this protection; and the significance of the barrier system as an obstacle to the treatment of some brain diseases

15.2 The Hindbrain and Midbrain (p. 406)

1. The location, anatomical features, and functions of the medulla oblongata, pons, and midbrain
2. The composition, location, and functions of the reticular formation
3. The gross anatomy and histology of the cerebellum, its routes of input and output, and its functions

15.3 The Forebrain (p. 413)

1. The two components of the forebrain and three components of the diencephalon
2. The location and general organization of the thalamus and its basic functions

3. The location and general organization of the hypothalamus and its functions
4. The location and components of the epithalamus and their functions
5. The major features of the cerebrum
6. The boundaries and functions of the five lobes of the cerebrum
7. The three types of fiber tracts in the white matter of the cerebrum, and what distinguishes the three from each other
8. The two types of neurons in the cerebral cortex, and what distinguishes the widespread neocortex from the more limited regions of paleocortex and archicortex
9. The names and locations of the basal nuclei of the cerebrum, and their general function
10. The location, major components, and general functions of the limbic system
11. The distinction between the primary cortex and the association areas of the cerebrum
12. The cerebral locations of primary and association cortex for each of the special senses: vision, hearing, equilibrium, taste, and smell
13. The somatosensory organization of the postcentral gyrus; the somatotopic correspondence between this gyrus and the contralateral side of the lower body; and the location and general function of the somatosensory association area
14. The somatic motor organization of the precentral gyrus; the somatotopic correspondence between this gyrus and the contralateral side of the lower body; and the location and general function of the motor association area
15. The locations and functional relationship of upper and lower motor neurons
16. The motor functions of the basal nuclei and cerebellum
17. The locations of the language centers of the cerebrum and how they interact in the comprehension of language, thinking of what to say or write, and speaking

18. Some causes of aphasia and some kinds of language deficit seen in aphasia
19. The roles of the hypothalamus, amygdala, and prefrontal cortex in the experience and expression of emotion
20. What is meant by cognitive function, and the roles of some brain areas in cognition
21. Forms of memory; roles of the amygdala and hippocampus in memory; and regions of cerebral cortex associated with various kinds of memory
22. The specialization of the two cerebral hemispheres for different cognitive, sensory, and motor functions, and the distinction between categorical and representational functions

15.4 The Cranial Nerves (p. 427)

1. The names and numbers of the 12 pairs of cranial nerves, and ability to identify them on the base of the brain
2. The origins, terminations, and functions of each of the cranial nerves I through XII, and of the individual subdivisions of the trigeminal nerve

15.5 Developmental and Clinical Perspectives (p. 437)

1. The common effects of aging on the central nervous system and the impact of these effects on one's functionality and quality of life
2. The signs and symptoms of Alzheimer disease (AD); the histological effects of AD that can be seen at autopsy; the relationship of AD to neurotransmitter function; and the genetic contribution to AD
3. The signs and symptoms of Parkinson disease (PD); the histological effects of PD that can be seen at autopsy; the relationship of PD to neurotransmitter function; and some treatment approaches to PD

Testing Your Recall

1. Which of these is caudal to the hypothalamus?
   a. the thalamus
   b. the optic chiasm
   c. the cerebral aqueduct
   d. the pituitary gland
   e. the corpus callosum

2. Hearing is associated mainly with
   a. the limbic system.
   b. the prefrontal cortex.
   c. the occipital lobe.
   d. the temporal lobe.
   e. the parietal lobe.

3. The blood–CSF barrier is formed by
   a. blood capillaries.
   b. endothelial cells.
   c. protoplasmic astrocytes.
   d. oligodendrocytes.
   e. ependymal cells.

4. The pyramids of the medulla oblongata contain
   a. descending corticospinal fibers.
   b. commissural fibers.
   c. ascending spinocerebellar fibers.
   d. fibers going to and from the cerebellum.
   e. ascending spinothalamic fibers.

5. Which of the following is *not* involved in vision?
   a. the temporal lobe
   b. the occipital lobe
   c. the midbrain tectum
   d. the trochlear nerve
   e. the vagus nerve

6. While studying in a noisy cafeteria, you get sleepy and doze off for a few minutes. You awaken with a start and realize that all the cafeteria sounds have just "come back." While you were dozing, this auditory input was blocked from reaching your auditory cortex by
   a. the temporal lobe.
   b. the thalamus.
   c. the reticular activating system.
   d. the medulla oblongata.
   e. the vestibulocochlear nerve.

7. Because of a brain lesion, a certain patient never feels full, but eats so excessively that she now weighs nearly 600 pounds. The lesion is most likely in her
   a. hypothalamus.
   b. amygdala.
   c. hippocampus.
   d. basal nuclei.
   e. pons.

8. The _____ is most closely associated with the cerebellum in embryonic development and remains its primary source of input fibers throughout life.
   a. telencephalon
   b. thalamus
   c. midbrain
   d. pons
   e. medulla

9. Damage to the _____ nerve could result in defects of eye movement.
   a. optic   d. facial
   b. vagus   e. abducens
   c. trigeminal

10. All of the following *except* the _____ nerve begin or end in the orbit.
    a. optic   d. abducens
    b. oculomotor   e. accessory
    c. trochlear

11. The right and left cerebral hemispheres are connected to each other by a thick C-shaped bundle of fibers called the _____.

12. The brain has four chambers called _____ filled with _____ fluid.

13. In a sagittal section, the cerebellar white matter exhibits a branching pattern called the _____.

14. Part of the limbic system involved in forming new memories is the _____.

15. Cerebrospinal fluid is secreted partly by a mass of blood capillaries called the _____ in each ventricle.

16. The primary motor area of the cerebrum is the _____ gyrus of the frontal lobe.

17. Your personality is determined mainly by which lobe of the cerebrum?

18. Areas of cerebral cortex that identify or interpret sensory information are called _____.

19. Linear, analytical, and verbal thinking occurs in the _____ hemisphere of the cerebrum, which is on the left in most people.

20. The motor pattern for speech is generated in an area of cortex called the _____ and then transmitted to the primary motor cortex to be carried out.

Answers in the Appendix

Building Your Medical Vocabulary

State a medical meaning of each of the following word elements, and give a term in which it is used.

1. gyr-
2. sulc-
3. cereb-
4. pedunc-
5. insul-
6. -ellum
7. neo-
8. quadri-
9. foli-
10. radiat-

Answers in the Appendix

True or False

Determine which five of the following statements are false, and briefly explain why.

1. The two hemispheres of the cerebellum are separated by the longitudinal fissure.
2. Degeneration of the substantia nigra causes Alzheimer disease.
3. The midbrain is caudal to the thalamus.
4. The Broca area is ipsilateral to the Wernicke area.
5. Most of the cerebrospinal fluid is produced by the choroid plexuses.
6. Hearing is a function of the occipital lobe.
7. Respiration is controlled by nuclei in both the pons and medulla oblongata.
8. The trigeminal nerve carries sensory signals from a larger area of the face than the facial nerve does.
9. Unlike other cranial nerves, the vagus nerve extends far beyond the head–neck region.
10. The optic nerve controls movements of the eye.

Answers in the Appendix

Testing Your Comprehension

1. Which cranial nerve conveys pain signals to the brain in each of the following situations: (a) sand blows into your eye; (b) you bite the rear of your tongue; and (c) your stomach hurts from eating too much?
2. How would a lesion in the cerebellum and a lesion in the basal nuclei differ in their effects on skeletal muscle function?
3. Suppose that a neuroanatomist performed two experiments on an animal species with the same basic spinal and brainstem structure as a human's: In experiment 1, he selectively transected (cut across) the pyramids on the anterior side of the medulla oblongata; and in experiment 2, he selectively transected the gracile and cuneate fasciculi on the posterior side. How would the outcomes of the two experiments differ?
4. A person can survive destruction of an entire cerebral hemisphere but cannot survive destruction of the hypothalamus, which is a much smaller mass of brain tissue. Explain this difference and describe some ways that destruction of a cerebral hemisphere would affect one's quality of life.
5. What would be the most obvious effects of lesions that destroyed each of the following: (a) the hippocampus, (b) the amygdala, (c) the Broca area, (d) the occipital lobe, and (e) the hypoglossal nerve?

Answers at www.mhhe.com/saladinha3

Improve Your Grade at **www.mhhe.com/saladinha3**

Practice quizzes, labeling activities, games, and flashcards provide fun ways to master concepts. You can also download image PowerPoint files for each chapter to create a study guide or for taking notes during lecture.

The Autonomic Nervous System and Visceral Reflexes

CHAPTER

16

Autonomic neurons in the myenteric plexus of the digestive tract

BRUSHING UP

Anatomy & Physiology **REVEALED**®
aprevealed.com
Nervous System

We are consciously aware of many of the activities of our nervous system discussed in the preceding chapters—the general and special senses, our cognitive processes and emotions, and our voluntary movements. But there is another branch of the nervous system that operates in comparative secrecy, usually without our willing it, thinking about it, or even being able to consciously modify or suppress it.

This secret agent is called the *autonomic nervous system (ANS)*. Its name means "self-governed,"[1] as it is almost fully independent of our will. Its job is to regulate such fundamental states and life processes as heart rate, blood pressure, body temperature, respiratory airflow, pupillary diameter, digestion, energy metabolism, defecation, and urination. In short, the ANS quietly manages a multitude of unconscious processes responsible for the body's homeostasis.

Walter Cannon (1871–1945), the American physiologist who coined such expressions as *homeostasis* and the *fight or flight reaction*, dedicated his career to the study of the autonomic nervous system. He found that an animal can live without a functional sympathetic nervous system (one of the two divisions of the ANS), but it must be kept warm and free of stress. It cannot regulate its body temperature, tolerate any strenuous exertion, or even survive on its own. Indeed, the ANS is more necessary for survival than are many functions of the somatic nervous system; an absence of autonomic function is fatal because the body cannot maintain homeostasis without it. Thus, for an understanding of bodily function, the mode of action of many drugs, and other aspects of health care, we must be especially aware of how the ANS works.

16.1 General Properties of the Autonomic Nervous System

▶ **Expected Learning Outcomes**

When you have completed this section, you should be able to

- explain how the autonomic and somatic nervous systems differ in form and function;

- explain what a visceral reflex is and describe some examples; and

- explain how the two divisions of the autonomic nervous system differ in general function.

General Actions

The **autonomic nervous system (ANS)** can be defined as a motor nervous system that controls glands, cardiac muscle, and smooth muscle. It is also called the **visceral motor system** to distinguish it from the somatic motor system, which controls the skeletal muscles (see fig. 13.2, p. 352). The primary target organs of the ANS are the viscera of the thoracic and abdominopelvic cavities and some structures of the body wall, including cutaneous blood vessels, sweat glands, and piloerector muscles.

The somatic and visceral motor systems are often described as voluntary and involuntary, respectively. The somatic motor system innervates voluntary (skeletal) muscle, which is usually under the control of one's will (volition). Cardiac and smooth muscle are involuntary muscles; like glands, they are not usually subject to voluntary control. This voluntary–involuntary distinction is not, however, as clear-cut as it once seemed. Some skeletal muscle responses are quite involuntary, such as the somatic reflexes, and some skeletal muscles are difficult or impossible to control, such as the middle-ear muscles. On the other hand, therapeutic uses of biofeedback (see Deeper Insight 16.1) show that some people can learn to voluntarily control such visceral functions as blood pressure.

Visceral effectors do not depend on the autonomic nervous system to function, but only to adjust (modulate) their activity to the body's changing needs. The heart, for example, goes on beating even if all autonomic nerves to it are severed, but the ANS modulates the heart rate in conditions of rest and exercise. If the somatic nerves to a skeletal muscle are severed, the muscle exhibits flaccid paralysis—it no longer contracts at all. But if the autonomic nerves to cardiac or smooth muscle are severed, the muscle often exhibits exaggerated responses *(denervation hypersensitivity)*.

Visceral Reflexes

The ANS is responsible for the body's **visceral reflexes.** These are unconscious, automatic, stereotyped responses to stimulation, much like the somatic reflexes discussed in chapter 14, but involving visceral receptors and effectors and somewhat slower responses. Some authorities regard the visceral afferent (sensory) pathways as part of the ANS, but most prefer to limit the term *ANS* to the efferent (motor) pathways. Regardless of this preference, however, autonomic activity involves a visceral reflex arc that includes receptors (nerve endings that detect stretch, tissue damage, blood chemicals, body temperature, and other internal stimuli), afferent neurons leading to the CNS, interneurons in the CNS, efferent neurons carrying motor signals away from the CNS, and finally effectors.

DEEPER INSIGHT   16.1

Biofeedback

Biofeedback is a technique in which an instrument produces auditory or visual signals in response to changes in a subject's blood pressure, heart rate, muscle tone, skin temperature, brain waves, or other physiological variables. It gives the subject awareness of changes that he or she would not ordinarily notice. Some people can be trained to control variables such as their blood pressure or heart rate in order to produce a certain tone or color of light from the apparatus. Eventually they can control them without the aid of the monitor. Biofeedback is not a quick, easy, infallible, or inexpensive cure for all ills, but it has been used successfully to treat hypertension, stress, and migraine headaches.

---

[1] *auto* = self + *nom* = rule

For example, high blood pressure activates a visceral *barore-flex*.[2] It stimulates stretch receptors called *baroreceptors* in the carotid arteries and aorta, and they transmit signals via the glossopharyngeal nerves to the medulla oblongata (fig. 16.1). The medulla integrates this input with other information and transmits efferent signals back to the heart by way of the vagus nerves. The vagus nerves slow down the heart and reduce blood pressure, thus completing a homeostatic negative feedback loop. A separate baroreflex arc accelerates the heart when blood pressure above the heart drops—for example, when we change from lying down to standing up and gravity draws blood away from the upper body.

Another example of a visceral reflex is the autonomic response to chilling of the body. Cooling of the skin stimulates nerve endings called *cold receptors.* Nerve signals travel through spinal nerve fibers

---

[2]*baro* = pressure

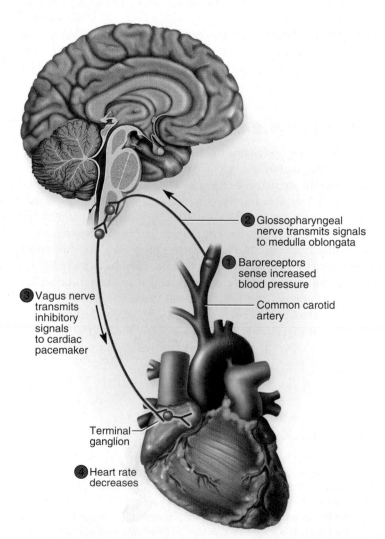

Figure 16.1 **An Autonomic Reflex Arc in the Regulation of Blood Pressure.** A rise in blood pressure is detected by baroreceptors in the carotid artery. The glossopharyngeal nerve transmits signals to the medulla oblongata, resulting in parasympathetic output from the vagus nerve that reduces the heart rate and lowers blood pressure.

to the spinal cord and then up the spinothalamic tracts to the brainstem. This input is directed to the hypothalamic thermostat, a neural pool in the preoptic nucleus of the anterior hypothalamus. Neurons here also respond directly to cooling of the blood that circulates through the hypothalamus. In response, they transmit signals to the heat-promoting center of the posterior hypothalamus. Output from the heat-promoting center travels by nerve fibers that descend through the brainstem to the spinal cord and exit the cord through sympathetic pathways described later in this chapter. Sympathetic nerve fibers ultimately reach the blood vessels of the skin, stimulating contraction of the smooth muscle of the vessel walls. The resulting vasoconstriction diverts blood away from the skin surface, thus helping to reduce heat loss. Hypothalamic output can also increase muscle tone and induce shivering to generate additional body heat, but since this involves the skeletal muscles, it is a somatic reflex rather than a visceral one. Both the visceral and somatic components of thermoregulation, however, admirably exemplify the negative feedback loops that maintain homeostasis: A deviation from the homeostatic set point is detected (chilling of the body); signals are sent to neural integrating centers (the preoptic nucleus and the heat-promoting center of the hypothalamus); and responses are activated (cutaneous vasoconstriction, increased muscle tone, shivering) that return body temperature to the set point.

Divisions of the Autonomic Nervous System

The ANS has two subsystems, the *sympathetic* and *parasympathetic divisions*. These differ in anatomy and function, but they often innervate the same target organs and may have cooperative or contrasting effects on them. The **sympathetic division** adapts the body in many ways for physical activity—it increases alertness, heart rate, blood pressure, pulmonary airflow, blood glucose concentration, and blood flow to cardiac and skeletal muscle, but at the same time, it reduces blood flow to the skin and digestive tract. Cannon referred to extreme sympathetic responses as the "fight or flight" reaction because they come into play when an animal must attack, defend itself, or flee from danger. In our own lives, this reaction occurs in many situations involving arousal, exercise, competition, stress, danger, trauma, anger, or fear. Ordinarily, however, the sympathetic division has more subtle effects that we notice barely, if at all. The **parasympathetic division,** by comparison, has a calming effect on many body functions. It is associated with reduced energy expenditure and normal bodily maintenance, including such functions as digestion and waste elimination. This can be thought of as the "resting and digesting" state.

This does not mean that the body alternates between states in which one system or the other is active. Normally, both systems are active simultaneously. They exhibit a background rate of activity called **autonomic tone,** and the balance between *sympathetic tone* and *parasympathetic tone* shifts in accordance with the body's changing needs. Parasympathetic tone, for example, maintains smooth muscle tone in the intestines and holds the resting heart rate down to about 70 to 80 beats/min. If the parasympathetic vagus nerves to the heart are cut, the heart beats at its own intrinsic rate of about 100 beats/min. Sympathetic tone keeps most blood vessels partially constricted and thus maintains blood pressure. A loss of sympathetic tone can cause such a rapid drop in blood pressure that a person goes into shock.

Neither division has universally excitatory or calming effects. The sympathetic division, for example, excites the heart but inhibits digestive and urinary functions, while the parasympathetic division has the opposite effects.

Neural Pathways

The ANS is often categorized as part of the peripheral nervous system, but it has components in both the central and peripheral nervous systems. It includes control nuclei in the hypothalamus and other regions of the brainstem, motor neurons in the spinal cord and peripheral ganglia, and nerve fibers that travel through the spinal and cranial nerves.

The autonomic motor pathway to a target organ differs significantly from somatic motor pathways. In somatic pathways, a motor neuron in the brainstem or spinal cord issues a myelinated axon that reaches all the way to a skeletal muscle. In autonomic pathways, the signal must travel across two neurons to get to the target organ, and it must cross a synapse where these neurons meet in an autonomic ganglion (fig. 16.2). The first neuron, called the **preganglionic neuron,** has a soma in the brainstem or spinal cord. Its axon terminates in a ganglion, where it synapses with a **postganglionic neuron** whose axon extends the rest of the way to the target cells.

(Some call this cell the *ganglionic neuron* since its soma is in the ganglion and only its axon is truly postganglionic.) The axons of these neurons are called the *pre-* and *postganglionic fibers.*

Differences between the somatic and autonomic nervous systems are summarized in table 16.1.

TABLE 16.1	Comparison of the Somatic and Autonomic Nervous Systems	
Feature	**Somatic**	**Autonomic**
Effectors	Skeletal muscle	Glands, smooth muscle, cardiac muscle
Control	Usually voluntary	Usually involuntary
Efferent pathways	One nerve fiber from CNS to effector; no ganglia	Two nerve fibers from CNS to effector; synapse at a ganglion
Neurotransmitters	Acetylcholine (ACh)	ACh and norepinephrine (NE)
Effect on target cells	Always excitatory	Excitatory or inhibitory
Effect of denervation	Flaccid paralysis	Denervation hypersensitivity

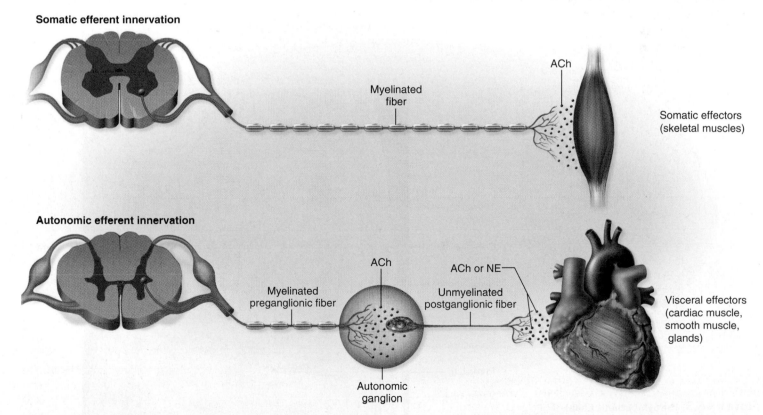

Somatic efferent innervation

ACh

Myelinated fiber

Somatic effectors (skeletal muscles)

Autonomic efferent innervation

ACh

ACh or NE

Myelinated preganglionic fiber

Unmyelinated postganglionic fiber

Visceral effectors (cardiac muscle, smooth muscle, glands)

Autonomic ganglion

Figure 16.2 **Comparison of Somatic and Autonomic Efferent Pathways.** The entire distance from CNS to effector is spanned by one neuron in the somatic system and two neurons in the autonomic system. Only acetylcholine (ACh) is employed as a neurotransmitter by the somatic neuron and the autonomic preganglionic neuron, but autonomic postganglionic neurons can employ either ACh or norepinephrine (NE).

Before You Go On

Answer the following questions to test your understanding of the preceding section:

1. How does the autonomic nervous system differ functionally and anatomically from the somatic motor system?

2. How does a visceral reflex resemble a somatic reflex? How does it differ?

3. What are the two divisions of the ANS? How do they functionally differ from each other?

4. Define *preganglionic* and *postganglionic neuron*. Why are these terms not used in describing the somatic motor system?

16.2 Anatomy of the Autonomic Nervous System

▶ Expected Learning Outcomes

When you have completed this section, you should be able to

- identify the anatomical components and nerve pathways of the sympathetic and parasympathetic divisions;

- discuss the relationship of the adrenal glands to the sympathetic nervous system; and

- describe the enteric nervous system of the digestive tract and explain its significance.

The Sympathetic Division

The sympathetic division is also called the *thoracolumbar division* because it arises from the thoracic and lumbar regions of the spinal cord. It has relatively short preganglionic and long postganglionic fibers. The preganglionic neurosomas are in the lateral horns and nearby regions of the gray matter of the spinal cord. Their fibers exit by way of spinal nerves T1 to L2 and lead to a nearby **sympathetic chain** of ganglia (**paravertebral**[3] **ganglia**). This is a longitudinal series of ganglia adjacent to each side of the vertebral column from the cervical to the coccygeal level. They are interconnected by longitudinal nerve cords (figs. 16.3 and 16.4). The number of ganglia varies from person to person, but usually there are 3 cervical (*superior, middle,* and *inferior),* 11 thoracic, 4 lumbar, 4 sacral, and 1 coccygeal ganglion in each chain.

It may seem odd that sympathetic ganglia exist in the cervical, sacral, and coccygeal regions considering that sympathetic fibers arise only from the thoracic and lumbar regions of the spinal cord (levels T1 to L2). But as shown in figure 16.4, nerve cords from the thoracic region ascend to the ganglia in the neck, and cords from the lumbar region descend to the sacral and coccygeal ganglia. Consequently, sympathetic nerve fibers are distributed to every level of the body. As a general rule, the head receives sympathetic output arising from spinal cord segment T1, the neck from T2, the thorax and upper limbs from T3 to T6, the abdomen from T7 to T11, and the lower limbs from T12 to L2. There is considerable overlap and individual variation in this pattern, however.

In the thoracolumbar region, each paravertebral ganglion is connected to a spinal nerve by two branches called *communicating*

---

[3]*para* = next to + *vertebr* = vertebral column

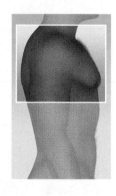

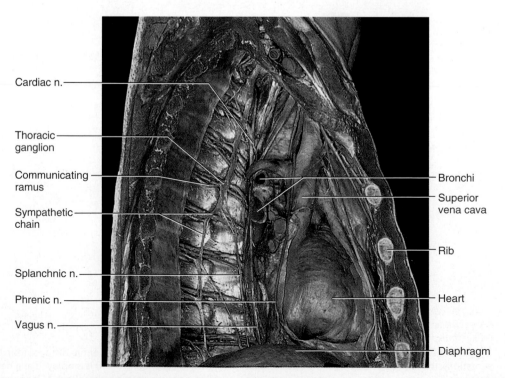

Cardiac n.

Thoracic ganglion

Communicating ramus

Sympathetic chain

Splanchnic n.

Phrenic n.

Vagus n.

Bronchi

Superior vena cava

Rib

Heart

Diaphragm

Figure 16.3 **The Sympathetic Chain Ganglia.** Right lateral view of the thoracic cavity. (n. = nerve).

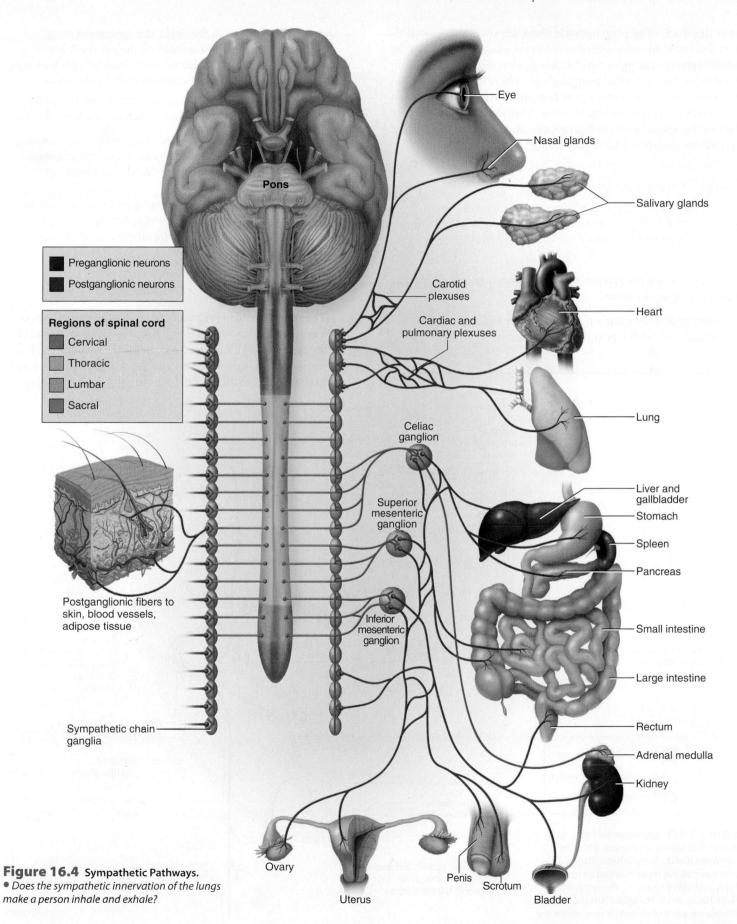

Figure 16.4 Sympathetic Pathways.
● *Does the sympathetic innervation of the lungs make a person inhale and exhale?*

Eye

Nasal glands

Salivary glands

Pons

Carotid plexuses

Cardiac and pulmonary plexuses

Heart

Lung

Preganglionic neurons

Postganglionic neurons

Regions of spinal cord

Cervical

Thoracic

Lumbar

Sacral

Celiac ganglion

Liver and gallbladder

Stomach

Superior mesenteric ganglion

Spleen

Pancreas

Postganglionic fibers to skin, blood vessels, adipose tissue

Small intestine

Inferior mesenteric ganglion

Large intestine

Rectum

Adrenal medulla

Kidney

Sympathetic chain ganglia

Ovary

Penis

Scrotum

Uterus

Bladder

rami (fig. 16.5). The preganglionic fibers are small myelinated fibers that travel from the spinal nerve to the ganglion by way of the **white communicating ramus,**[4] which gets its color and name from the myelin. Unmyelinated postganglionic fibers leave the ganglion by various routes including a **gray communicating ramus,** named for its lack of myelin and duller color. This ramus forms a bridge back to the spinal nerve. Postganglionic fibers travel by way of the gray ramus and spinal nerve to the target organs.

Apply What You Know

Would autonomic postganglionic fibers have faster or slower conduction speeds than somatic motor fibers? Why? (See hints in chapter 13. Assume no significant difference in fiber diameter.)

After entering the sympathetic chain, preganglionic fibers may follow any of three courses:

- Some end in the ganglion that they enter and synapse immediately with a postganglionic neuron.

---

[4]*ramus* = branch

- Some travel up or down the chain and synapse in ganglia at other levels. It is these fibers that link the paravertebral ganglia into a chain. They are the only route by which ganglia at the cervical, sacral, and coccygeal levels receive input.

- Some pass through the chain without synapsing and continue as *splanchnic* (SPLANK-nic) *nerves,* to be considered shortly.

Nerve fibers leave the paravertebral ganglia by three routes: spinal, sympathetic, and splanchnic nerves. These are numbered in figure 16.5 to correspond to the following descriptions:

1 **The spinal nerve route.** Some postganglionic fibers exit a ganglion by way of the gray ramus, return to the spinal nerve or its subdivisions, and travel the rest of the way to the target organ. This is the route to most sweat glands, piloerector muscles, and blood vessels of the skin and skeletal muscles.

2 **The sympathetic nerve route.** Other postganglionic fibers leave the chain by way of sympathetic nerves that extend to the heart, lungs, esophagus, and thoracic blood vessels. These nerves form a **carotid plexus** around each carotid artery and issue fibers from there to effectors in the head—including

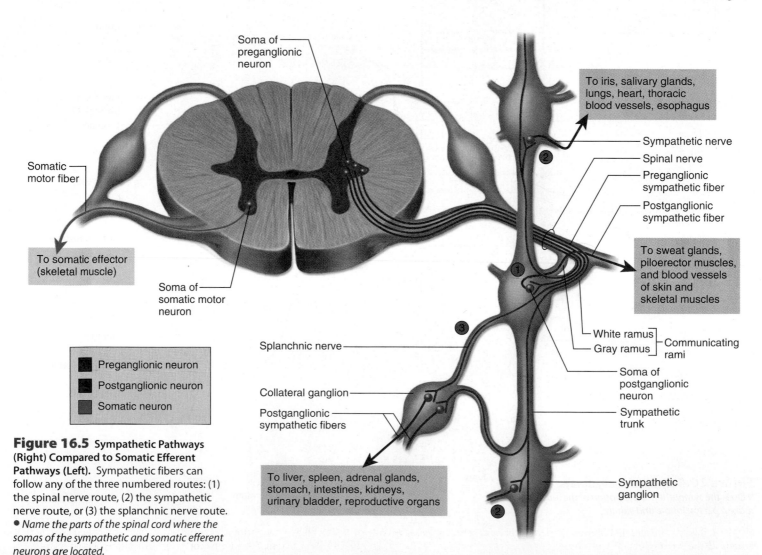

Figure 16.5 Sympathetic Pathways (Right) Compared to Somatic Efferent Pathways (Left). Sympathetic fibers can follow any of the three numbered routes: (1) the spinal nerve route, (2) the sympathetic nerve route, or (3) the splanchnic nerve route.
● *Name the parts of the spinal cord where the somas of the sympathetic and somatic efferent neurons are located.*

sweat, salivary, and nasal glands; piloerector muscles; blood vessels; and dilators of the iris. Some fibers from the superior and middle cervical ganglia form *cardiac nerves* to the heart. (The cardiac nerves also contain parasympathetic fibers.)

3. **The splanchnic[5] nerve route.** Some of the fibers that arise from spinal nerves T5 to T12 pass through the sympathetic ganglia without synapsing. Beyond the ganglia, they continue as **splanchnic nerves,** which lead to a second set of ganglia called **collateral (prevertebral) ganglia.** Here the preganglionic fibers synapse with the postganglionics.

The collateral ganglia contribute to a network, the **abdominal aortic plexus,** wrapped around the aorta (fig. 16.6). There are three major collateral ganglia in this

[5]*splanchn* = viscera

plexus—the **celiac, superior mesenteric,** and **inferior mesenteric ganglia**—located at points where arteries of the same names branch off the aorta. The postganglionic fibers accompany these arteries and their branches to the target organs. Table 16.2 summarizes the innervation to and from the three major collateral ganglia.

The term *solar plexus* is used by some authorities as a collective name for the celiac and superior mesenteric ganglia, and by others as a synonym for the celiac ganglion only. The term comes from the nerves radiating from the ganglion, like rays of the sun.

In summary, effectors in the muscles and body wall are innervated mainly by sympathetic fibers in the spinal nerves; effectors in the head and thoracic cavity by sympathetic nerves; and effectors in the abdominal cavity by splanchnic nerves.

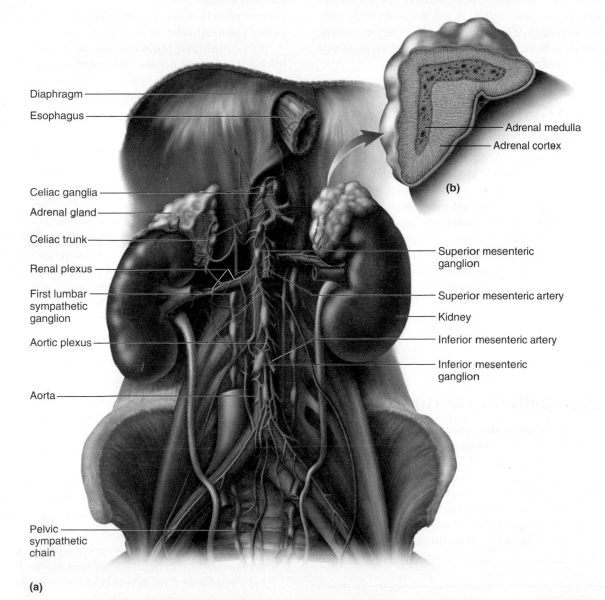

Diaphragm
Esophagus

Adrenal medulla
Adrenal cortex

(b)

Celiac ganglia
Adrenal gland
Celiac trunk
Renal plexus

Superior mesenteric ganglion
Superior mesenteric artery

First lumbar sympathetic ganglion

Kidney

Aortic plexus

Inferior mesenteric artery
Inferior mesenteric ganglion

Aorta

Pelvic sympathetic chain

(a)

Figure 16.6 **Abdominal Components of the Sympathetic Nervous System.** (a) Collateral ganglia, abdominal aortic plexus, and adrenal glands. (b) The adrenal gland, frontal section. Only the adrenal medulla plays a role in the sympathetic nervous system; the adrenal cortex has unrelated roles described in chapter 18.

Table 16.2		Innervation To and From the Collateral Ganglia		
Sympathetic Ganglia	→	**Collateral Ganglion**	→	**Postganglionic Target Organs**
Thoracic ganglia 5 to 9 or 10	→	Celiac ganglion	→	Stomach, spleen, liver, small intestine, and kidneys
Thoracic ganglia 9 and 10	→	Celiac and superior mesenteric ganglia	→	Small intestine , colon, and kidneys
Lumbar ganglia	→	Inferior mesenteric ganglion	→	Rectum, urinary bladder, and reproductive organs

There is no simple one-to-one relationship between preganglionic and postganglionic neurons in the sympathetic division. For one thing, each postganglionic cell may receive synapses from multiple preganglionic cells, thus exhibiting the principle of *neural convergence* discussed on page 364. Furthermore, each preganglionic fiber branches and synapses with multiple postganglionic fibers, thus showing *neural divergence*. Most sympathetic preganglionic neurons synapse with 10 to 20 postganglionic neurons. This means that when one preganglionic neuron fires, it can excite multiple postganglionic fibers leading to different target organs. The sympathetic division thus tends to have relatively widespread effects—as suggested by the name *sympathetic*.[6]

The Adrenal Glands

The paired **adrenal[7] (suprarenal) glands** rest like hats on the superior poles of the kidneys (fig. 16.6a). Each adrenal is actually two glands with different functions and embryonic origins. The outer rind, the **adrenal cortex,** secretes steroid hormones discussed in chapter 18. The inner core, the **adrenal medulla,** is essentially a sympathetic ganglion (fig. 16.6b). It consists of modified postganglionic neurons without dendrites or axons. Sympathetic preganglionic fibers penetrate through the cortex and terminate on these cells. The sympathetic nervous system and adrenal medulla are so closely related in development and function that they are referred to collectively as the *sympathoadrenal system.* The adrenal medulla secretes a mixture of hormones into the bloodstream—about 85% epinephrine (adrenaline), 15% norepinephrine (noradrenaline), and a trace of dopamine.

The Parasympathetic Division

The parasympathetic division is also called the *craniosacral division* because it arises from the brain and sacral region of the spinal cord; its fibers travel in certain cranial and sacral nerves. Somas of the preganglionic neurons are located in the midbrain, pons, medulla oblongata, and segments S2 to S4 of the spinal cord (fig. 16.7). They issue long preganglionic fibers that end in **terminal ganglia** in or near the target organ (see fig. 16.1). (If a terminal ganglion is embedded in the wall of a target organ, it is also called an *intramural*[8]

ganglion.) Thus, the parasympathetic division has long preganglionic fibers, reaching almost all the way to the target cells, and short postganglionic fibers that cover the rest of the distance.

There is some neural divergence in the parasympathetic division, but much less than in the sympathetic. The parasympathetic division has a ratio of fewer than five postganglionic fibers to every preganglionic fiber. Furthermore, the preganglionic fiber reaches the target organ before even this slight divergence occurs. The parasympathetic division is therefore more selective than the sympathetic in its stimulation of target organs.

Parasympathetic fibers leave the brainstem by way of the following four cranial nerves. The first three supply all parasympathetic innervation to the head, and the last one supplies viscera of the thoracic and abdominopelvic cavities.

1. **Oculomotor nerve (III).** The oculomotor nerve carries parasympathetic fibers that control the lens and pupil of the eye. The preganglionic fibers enter the orbit and terminate in the *ciliary ganglion* behind the eyeball. Postganglionic fibers enter the eyeball and innervate the *ciliary muscle,* which thickens the lens, and the *pupillary constrictor,* which narrows the pupil.

2. **Facial nerve (VII).** The facial nerve carries parasympathetic fibers that regulate the tear glands, salivary glands, and nasal glands. Soon after the facial nerve emerges from the pons, its parasympathetic fibers split away and form two smaller branches. The upper branch ends at the *pterygopalatine ganglion* near the junction of the maxilla and palatine bone. Postganglionic fibers then continue to the tear glands and glands of the nasal cavity, palate, and other areas of the oral cavity. The lower branch, called the *chorda tympani,* crosses the middle-ear cavity and ends at the *submandibular ganglion* near the angle of the mandible. Postganglionic fibers from here supply salivary glands in the floor of the mouth.

3. **Glossopharyngeal nerve (IX).** The glossopharyngeal nerve carries parasympathetic fibers concerned with salivation. The preganglionic fibers leave this nerve soon after its origin and form the *tympanic nerve.* A continuation of this nerve crosses the middle-ear cavity and ends in the *otic*[9] *ganglion* near the foramen ovale. The postganglionic fibers then follow the trigeminal nerve to the *parotid salivary gland* just in front of the earlobe.

4. **Vagus nerve (X).** The vagus nerve carries about 90% of all parasympathetic preganglionic fibers. It travels down the neck and forms three networks in the mediastinum—the

---

[6]*sym* = together + *path* = feeling
[7]*ad* = near + *ren* = kidney
[8]*intra* = within + *mur* = wall

[9]*ot* = ear + *ic* = pertaining to

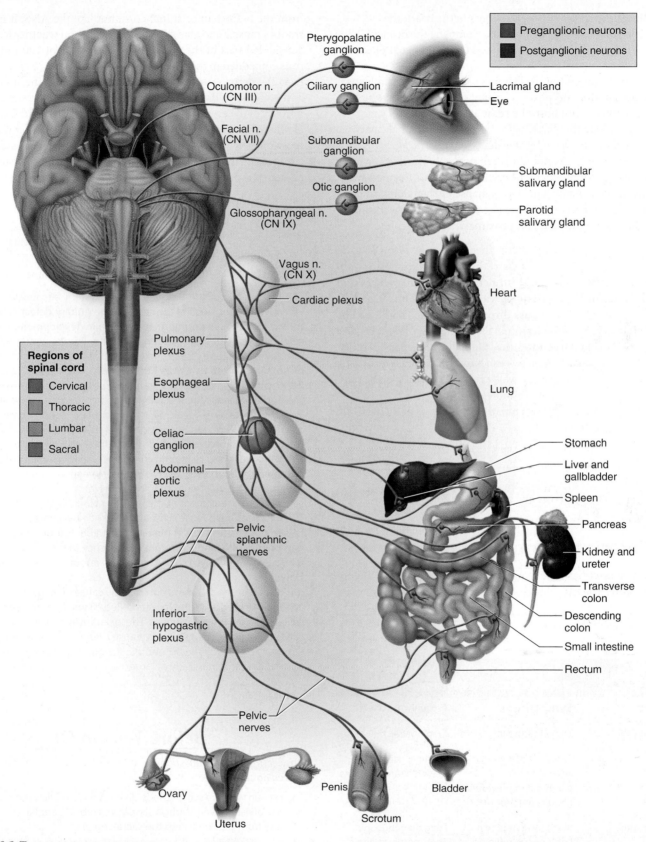

Preganglionic neurons
Postganglionic neurons

Pterygopalatine ganglion
Oculomotor n. (CN III)
Ciliary ganglion
Lacrimal gland
Eye
Facial n. (CN VII)
Submandibular ganglion
Otic ganglion
Submandibular salivary gland
Glossopharyngeal n. (CN IX)
Parotid salivary gland
Vagus n. (CN X)
Cardiac plexus
Heart
Pulmonary plexus
Esophageal plexus
Lung
Celiac ganglion
Stomach
Liver and gallbladder
Abdominal aortic plexus
Spleen
Pancreas
Pelvic splanchnic nerves
Kidney and ureter
Transverse colon
Descending colon
Inferior hypogastric plexus
Small intestine
Rectum
Pelvic nerves
Ovary
Uterus
Penis
Scrotum
Bladder

Regions of spinal cord
Cervical
Thoracic
Lumbar
Sacral

Figure 16.7 Parasympathetic Pathways.
● *Which nerve carries the most parasympathetic nerve fibers?*

cardiac plexus, which supplies fibers to the heart; the **pulmonary plexus,** whose fibers accompany the bronchi and blood vessels into the lungs; and the **esophageal plexus,** whose fibers regulate swallowing.

At the lower end of the esophagus, the esophageal plexus gives off anterior and posterior **vagal trunks,** each of which contains fibers from both the right and left vagus nerves. These penetrate the diaphragm, enter the abdominal cavity, and contribute to the extensive *abdominal aortic plexus* mentioned earlier. As noted earlier, sympathetic fibers synapse here. The parasympathetic fibers, however, pass through the plexus without synapsing. They synapse farther along, in terminal ganglia in or near the liver, pancreas, stomach, small intestine, kidney, ureter, and proximal half of the colon.

The remaining parasympathetic fibers arise from levels S2 to S4 of the spinal cord. They travel a short distance in the anterior rami of the spinal nerves and then form **pelvic splanchnic nerves** that lead to the **inferior hypogastric plexus.** Some parasympathetic fibers synapse here, but most pass through this plexus and travel by way of **pelvic nerves** to the terminal ganglia in their target organs: the distal half of the large intestine, the rectum, urinary bladder, and reproductive organs. With few exceptions, the parasympathetic system does not innervate body wall structures (sweat glands, pilo-erector muscles, or cutaneous blood vessels).

The sympathetic and parasympathetic divisions of the ANS are compared in table 16.3.

Apply What You Know

Would autonomic functions be affected if the anterior roots of the cervical spinal nerves were damaged? Why or why not?

The Enteric Nervous System

The digestive tract has a nervous system of its own called the **enteric**[10] **nervous system.** Unlike the ANS proper, it does not arise

[10]*enter* = intestines + *ic* = pertaining to

Table 16.3	**Comparison of the Sympathetic and Parasympathetic Divisions**	
Feature	**Sympathetic**	**Parasympathetic**
Origin in CNS	Thoracolumbar	Craniosacral
Location of ganglia	Paravertebral ganglia adjacent to spinal column and prevertebral ganglia anterior to it	Terminal ganglia near or within target organs
Fiber lengths	Short preganglionic Long postganglionic	Long preganglionic Short postganglionic
Neuronal divergence	Extensive	Minimal
Effects of system	Often widespread and general	More local and specific

from the brainstem or spinal cord, but like the ANS, it innervates smooth muscle and glands. Opinions differ on whether it should be considered part of the ANS. It consists of about 100 million neurons embedded in the wall of the digestive tract (see photograph on p. 442)—perhaps more neurons than there are in the spinal cord—and it has its own reflex arcs. The enteric nervous system regulates motility of the esophagus, stomach, and intestines and the secretion of digestive enzymes and acid. To function normally, however, these digestive activities also require regulation by the sympathetic and parasympathetic systems. The enteric nervous system is discussed in more detail in Deeper Insight 16.2 and chapter 24.

DEEPER INSIGHT   16.2

Megacolon

The importance of the enteric nervous system becomes vividly clear when it is absent. Such is the case in a hereditary defect called *Hirschsprung disease*.[11] During normal embryonic development, neural crest cells migrate to the large intestine and establish the enteric nervous system. In Hirschsprung disease, however, they fail to supply the distal parts of the large intestine, leaving the sigmoid colon and rectum (see fig. 24.16, p. 673) without enteric ganglia. In the absence of these ganglia, the sigmoidorectal region lacks motility, constricts permanently, and resists the passage of feces. Feces accumulate and become impacted above the constriction, resulting in *megacolon*—a massive dilation of the bowel accompanied by abdominal distension and chronic constipation. The most life-threatening complications are colonic gangrene, perforation of the bowel, and bacterial infection of the peritoneum (*peritonitis*). The treatment of choice is surgical removal of the affected segment and attachment of the healthy colon directly to the anal canal.

Hirschsprung disease is usually evident even in the newborn, which fails to have its first bowel movement. It affects four times as many infant boys as girls, and although its incidence in the general population is about 1 in 5,000 live births, it occurs in about 1 out of 10 infants with Down syndrome.

Hirschsprung disease is not the only cause of megacolon. In Central and South America, biting insects called *kissing bugs* transmit parasites called *trypanosomes* to humans. These parasites, similar to the ones that cause African sleeping sickness, cause *Chagas*[12] disease. Among other effects, they destroy the autonomic ganglia of the enteric nervous system, leading to a massively enlarged and often gangrenous colon.

Before You Go On

Answer the following questions to test your understanding of the preceding section:

5. Explain why the sympathetic division is also called the thoracolumbar division even though its paravertebral ganglia extend all the way from the cervical to the sacral region.

6. Describe or diagram the structural relationships among the following: preganglionic fiber, postganglionic fiber, anterior ramus, gray ramus, white ramus, and paravertebral ganglion.

[11]Harald Hirschsprung (1830–1916), Danish physician
[12]Carlos Chagas (1879–1934), Brazilian physician

7. Explain in anatomical terms why the parasympathetic division affects target organs more selectively than the sympathetic division does.

8. Trace the pathway of a parasympathetic fiber of the vagus nerve from the medulla oblongata to the small intestine.

- explain how the two divisions of the ANS interact when they both innervate the same organ; and

- describe how the central nervous system regulates the ANS.

16.3 Autonomic Effects

▶ **Expected Learning Outcomes**

When you have completed this section, you should be able to

- name the neurotransmitters employed by the ANS and define terms for neurons and synapses with different neurotransmitter and receptor types;

- in terms of neurotransmitters and receptors, explain why the two divisions of the ANS can have contrasting effects on the same organs;

Neurotransmitters and Receptors

As noted earlier, the two divisions of the ANS often have contrasting effects on an organ. For example, the sympathetic division accelerates the heart, and the parasympathetic division slows it down. But this does not mean the sympathetic division is stimulatory and the parasympathetic division inhibitory to every organ. Each division stimulates some organs and inhibits others. For example, the parasympathetic division stimulates the contraction of intestinal smooth muscle but inhibits cardiac muscle. The sympathetic division has the opposite effects on these two muscular tissues. Several other examples of such contrasting effects are given in table 16.4. Cases in which one division has no effect are usually because it provides little or no innervation to that tissue or organ.

Table 16.4	Effects of the Sympathetic and Parasympathetic Nervous Systems	
Target	**Effect of Sympathetic Stimulation**	**Effect of Parasympathetic Stimulation**
Pupil of eye	Dilation	Constriction
Lens of eye	Thinning for far vision	Thickening for near vision
Lacrimal (tear) glands	None	Secretion
Sweat glands	Secretion	Usually no effect but produces palmar sweating
Piloerector muscles	Hair erection	No effect
Heart rate	Increased	Decreased
Blood vessels of most viscera	Vasoconstriction	Usually no effect but dilates gastrointestinal blood vessels
Blood vessels of skeletal muscles	Vasodilation	No effect
Blood vessels of skin	Vasoconstriction	Usually no effect but dilates some facial blood vessels, causing blushing
Bronchioles	Bronchodilation	Bronchoconstriction
Kidneys	Reduced urine output	No effect
Muscle of bladder wall	No effect	Contraction, emptying bladder
Salivary glands	Thick mucous secretion	Thin serous secretion
Gastrointestinal motility	Decreased	Increased
Gastrointestinal secretion	Decreased	Increased
Liver	Glycogen breakdown	Glycogen synthesis
Pancreatic enzyme secretion	Decreased	Increased
Penis and clitoris	Loss of erection	Erection
Ejaculation (smooth muscle roles—sperm propulsion and glandular secretion)	Stimulation	No effect

How can different autonomic neurons have such contrasting effects? There are two fundamental reasons: (1) sympathetic and parasympathetic neurons secrete different neurotransmitters, and (2) cells respond in different ways even to the same neurotransmitter depending on what type of receptors they have for it. The basic categories of autonomic neurotransmitters and receptors are as follows (fig. 16.8 and table 16.5).

- **Acetylcholine (ACh).** This neurotransmitter is secreted by the preganglionic neurons in both divisions and the postganglionic neurons of the parasympathetic division. A few sympathetic postganglionics also secrete ACh—those that innervate sweat glands and some blood vessels. Any nerve fiber that secretes ACh is called a **cholinergic** (CO-li-NUR-jic) **fiber,** and any receptor that binds it is called a **cholinergic receptor.** There are two categories of cholinergic receptors:

  - **Muscarinic** (MUSS-cuh-RIN-ic) **receptors.** These are named for muscarine, a mushroom toxin used in their discovery. All cardiac muscle, smooth muscle, and gland cells that receive cholinergic innervation have muscarinic receptors. Because of different subclasses of muscarinic receptors, ACh excites some cells and inhibits others.

  - **Nicotinic** (NIC-oh-TIN-ic) **receptors.** These are named for another botanical toxin helpful to their discovery—nicotine. They occur at the synapses where autonomic preganglionic neurons stimulate the postganglionic cells; on cells of the adrenal medulla; and at the neuromuscular junctions of skeletal muscle fibers. ACh excites all cells with nicotinic receptors.

- **Norepinephrine (NE).** This neurotransmitter is secreted by nearly all sympathetic postganglionic neurons. Nerve fibers that secrete it are called **adrenergic fibers,** and the receptors for it are called **adrenergic receptors.** (NE is also called noradrenaline, the origin of the term *adrenergic*.) There are two principal categories of NE receptors:

  - **α-adrenergic receptors.** These usually have excitatory effects. For example, the binding of NE to the alpha receptors of uterine muscle promotes labor contractions.

  - **β-adrenergic receptors.** These are usually inhibitory. For example, the binding of NE to the beta receptors of the coronary arteries relaxes and dilates them.

There are exceptions to both of these adrenergic effects because of physiologically different subclasses of both α- and β-adrenergic receptors, but we will not delve into such details.

Dual Innervation

Most of the viscera receive nerve fibers from both the sympathetic and parasympathetic divisions and thus are said to have **dual innervation.** In such cases, the two divisions may have either *antagonistic* or *cooperative effects* on the same organ. Antagonistic effects oppose each other. Thus, the sympathetic division dilates the pupil, and the parasympathetic division constricts it (fig. 16.9), among other examples already discussed and contrasted in table 16.4. Cooperative effects

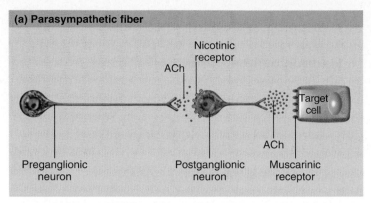

(a) Parasympathetic fiber

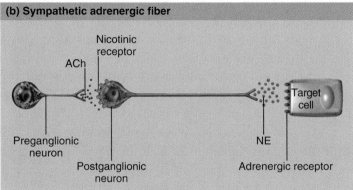

(b) Sympathetic adrenergic fiber

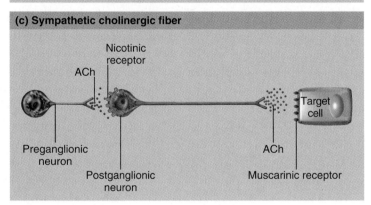

(c) Sympathetic cholinergic fiber

Figure 16.8 **Neurotransmitters and Receptors of the Autonomic Nervous System.** (a) All parasympathetic fibers are cholinergic. (b) Most sympathetic postganglionic fibers are adrenergic; they secrete norepinephrine (NE), and the target cell bears adrenergic receptors. (c) A few sympathetic postganglionic fibers are cholinergic; they secrete acetylcholine (ACh), and the target cell has cholinergic receptors of the muscarinic class.

Table 16.5	Locations of Cholinergic and Adrenergic Fibers in the ANS	
Division	**Preganglionic Fibers**	**Postganglionic Fibers**
Sympathetic	Always cholinergic	Mostly adrenergic; a few cholinergic
Parasympathetic	Always cholinergic	Always cholinergic

DEEPER INSIGHT 16.3

Drugs and the Autonomic Nervous System

The design of many drugs has been based on an understanding of autonomic neurotransmitters and receptor classes. *Sympathomimetics*[13] are drugs that enhance sympathetic action by stimulating adrenergic receptors or promoting norepinephrine release. For example, phenylephrine, found in such cold medicines as Chlor-Trimeton and Dimetapp, aids breathing by stimulating certain α-adrenergic receptors and thus dilating the bronchioles, constricting nasal blood vessels, and reducing swelling in the nasal mucosa. *Sympatholytics*[14] are drugs that suppress sympathetic action by inhibiting norepinephrine release or binding to adrenergic receptors without stimulating them. Propranolol, for example, is a *β-blocker* used to treat hypertension. It blocks the action of epinephrine and norepinephrine on β-adrenergic receptors of the heart and blood vessels.

Parasympathomimetics enhance parasympathetic effects. Pilocarpine, for example, relieves glaucoma (excessive pressure within the eye) by dilating a vessel that drains fluid from the eye. *Parasympatholytics* inhibit ACh release or block its receptors. Atropine, for example, blocks muscarinic receptors and is sometimes used to dilate the pupils for eye examinations and to dry the mucous membranes of the respiratory tract before inhalation anesthesia. It is an extract of the deadly nightshade plant, *Atropa belladonna.*[15] Women of the Middle Ages used nightshade to dilate their pupils, which was regarded as a beauty enhancement.

The branch of medicine that deals with the effects of drugs on the nervous system—especially drugs that mimic, enhance, or inhibit the action of neurotransmitters—is called *neuropharmacology.*

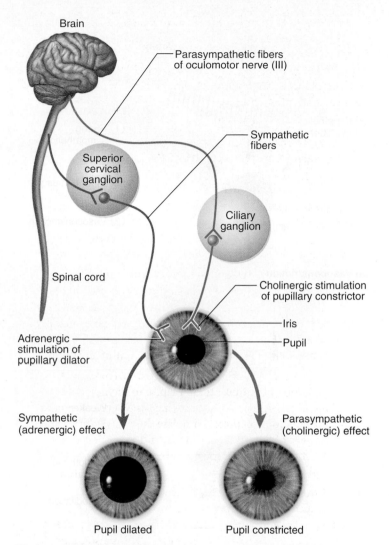

Figure 16.9 Dual Innervation of the Iris. Antagonistic effects of the sympathetic (yellow) and parasympathetic (blue) divisions on the iris.
● *If a person is in a state of fear, would you expect the pupils to be dilated or constricted? Why?*

occur when the two divisions act on different effector cells in an organ to produce a unified overall effect. For example, the parasympathetic division stimulates the secretion of salivary enzymes, and the sympathetic division stimulates the secretion of salivary mucus.

Dual innervation is not always necessary for the ANS to produce opposite effects on an organ. The adrenal medulla, piloerector muscles, sweat glands, and many blood vessels receive only sympathetic fibers. An example of control without dual innervation is the regulation of blood flow. The sympathetic fibers to a blood vessel have a baseline sympathetic tone that keeps the vessels in a state of partial constriction called *vasomotor tone* (fig. 16.10). An increase in sympathetic stimulation causes vasoconstriction by increasing smooth muscle contraction. A drop in sympathetic stimulation allows the smooth muscle to relax and the vessel to dilate.

Central Control of Autonomic Function

In spite of its name, the ANS is not an independent nervous system. All of its output originates in the CNS, and it receives input from the cerebral cortex, hypothalamus, medulla oblongata, and somatic division of the PNS.

Effects of the cerebral cortex on autonomic function are evident when anger raises the blood pressure, fear makes the heart race, thoughts of good food make the stomach rumble, and sexual thoughts or images increase blood flow to the genitals. The limbic system (p. 417) is involved in many emotional responses and has extensive connections with the hypothalamus, an important autonomic control center. Thus, the limbic system provides a pathway connecting sensory and mental experiences with the autonomic nervous system.

The hypothalamus contains many nuclei for primitive autonomic functions, including hunger, thirst, thermoregulation, and sexual response. Artificial stimulation of different regions of the hypothalamus can activate the arousal response typical of the sympathetic nervous system or have the calming effects typical of the parasympathetic. Output from the hypothalamus travels largely to nuclei in more caudal regions of the brainstem and from there to the cranial nerves and the sympathetic preganglionic neurons in the spinal cord.

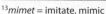

[13]*mimet* = imitate, mimic
[14]*lyt* = break down, destroy
[15]*bella* = beautiful, fine + *donna* = woman

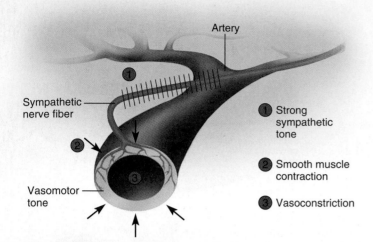

(a) Vasoconstriction

① Strong sympathetic tone

② Smooth muscle contraction

③ Vasoconstriction

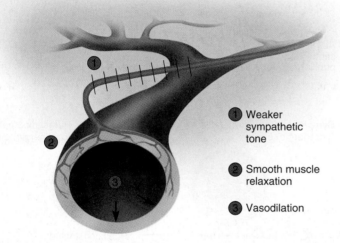

(b) Vasodilation

① Weaker sympathetic tone

② Smooth muscle relaxation

③ Vasodilation

Figure 16.10 Sympathetic and Vasomotor Tone. (a) Vasoconstriction in response to a high rate of sympathetic nerve firing. (b) Vasodilation in response to a low rate of sympathetic nerve firing. Smooth muscle relaxation allows blood pressure within the vessel to push the vessel wall outward. Black lines crossing each nerve fiber represent action potentials, with a high firing frequency in (a) and a lower frequency in (b).

The midbrain, pons, and medulla oblongata house the nuclei of cranial nerves that mediate several autonomic responses: the oculomotor nerve (pupillary constriction), facial nerve (lacrimal, nasal, palatine, and salivary gland secretion), glossopharyngeal nerve (salivation, blood pressure regulation), and vagus nerve (the chief parasympathetic supply to the thoracic and abdominal viscera).

The spinal cord also contains autonomic control nuclei. Such autonomic responses as the defecation and micturition (urination) reflexes are regulated here. Fortunately, the brain is able to inhibit these responses consciously, but when injuries sever the spinal cord from the brain, the autonomic spinal reflexes alone control the elimination of urine and feces.

Answer the following questions to test your understanding of the preceding section:

9. To what neurotransmitters do the terms *adrenergic* and *cholinergic* refer?

10. Why is a single autonomic neurotransmitter able to have opposite effects on different target cells?

11. What are the two ways in which the sympathetic and parasympathetic divisions can interact when they both innervate the same target organ? Give examples.

12. How can the sympathetic nervous system have contrasting effects in a target organ without dual innervation?

13. What system in the brain connects our conscious thoughts and feelings with the autonomic control centers of the hypothalamus?

14. List some autonomic responses that are controlled by nuclei in the hypothalamus.

15. What is the role of the midbrain, pons, and medulla in autonomic control?

16. Name some visceral reflexes controlled by the spinal cord.

16.4 Developmental and Clinical Perspectives

▶ Expected Learning Outcomes

When you have completed this section, you should be able to

- describe the embryonic origins of the autonomic neurons and ganglia;
- describe some consequences of aging of the autonomic nervous system; and
- describe a few disorders of autonomic function.

Development and Aging of the Autonomic Nervous System

Preganglionic neurons of the autonomic nervous system develop from the *neural tube* described in chapter 13; their somas remain embedded in the brainstem and spinal cord for life. Autonomic ganglia and postganglionic neurons, however, develop from the *neural crest* adjacent to the neural tube. During the fifth week of embryonic development, some neural crest cells migrate and assume positions alongside the vertebral bodies to become the sympathetic chain ganglia; others assume positions alongside the aorta to form the abdominal aortic plexus; and others migrate to the heart, lungs, digestive tract, and other viscera to form the terminal ganglia of the parasympathetic division.

The adrenal medulla arises from cells that separate from a nearby sympathetic ganglion, and thus ultimately comes from neural crest (ectodermal) cells. During its development, the medulla is surrounded by cells of mesodermal origin that produce the outer layer of the adrenal gland, the adrenal cortex (which is not part of the autonomic nervous system).

The efficiency of the ANS declines in old age, like that of the rest of the nervous system (see chapter 15). Target organs of the ANS have fewer neurotransmitter receptors in old age and are thus less responsive to autonomic stimulation. As a result, elderly people may experience dry eyes and more eye infections; slower and less effective adaptation of the eye to changing light intensities, and poorer night vision; less efficient control of blood pressure; and reduced intestinal motility and increasing constipation. Because of reduced efficiency of the baroreflex described earlier in this chapter, some elderly people experience *orthostatic hypotension,* a drop in blood pressure when they stand up, sometimes causing dizziness, loss of balance, or fainting.

Disorders of the Autonomic Nervous System

Table 16.6 describes some dysfunctions of the autonomic nervous system.

Before You Go On

Answer the following questions to test your understanding of the preceding section:

17. How do the pre- and postganglionic neurons of the ANS differ in embryonic origin?

18. Briefly state how the intestines, eyes, and blood pressure are affected in old age by the declining efficiency of the autonomic nervous system.

TABLE 16.6	Some Disorders of the Autonomic Nervous System
Achalasia[16] of the cardia	A defect in autonomic innervation of the esophagus, resulting in impaired swallowing, accompanied by failure of the lower esophageal sphincter to relax and allow food to pass into the stomach. (The region of the stomach at its junction with the esophagus is called the *cardia.*) Results in enormous dilation of the esophagus and inability to keep food down. Most common in young adults; cause remains poorly understood.
Horner[17] syndrome	Chronic unilateral pupillary constriction, sagging of the eyelid, withdrawal of the eye into the orbit, flushing of the skin, and lack of facial perspiration. Results from lesions in the cervical ganglia, upper thoracic spinal cord, or brainstem that interrupt sympathetic innervation of the head.
Raynaud[18] disease	Intermittent attacks of paleness, cyanosis, and pain in the fingers and toes, caused when cold or emotional stress triggers excessive vasoconstriction in the digits; most common in young women. In extreme cases, causes gangrene and may require amputation. Sometimes treated by severing sympathetic nerves to the affected regions.

Disorders Described Elsewhere

Autonomic effects of oculomotor and vagus nerve injuries   430, 435
Chagas disease   452
Hirschsprung disease   452
Mass reflex reaction   394
Orthostatic hypotension   457

[16]*a* = without + *chalas* = relaxation
[17]Johann F. Horner (1831–86), Swiss ophthalmologist
[18]Maurice Raynaud (1834–81), French physician

Study Guide

Assess Your Learning Outcomes

You should have a good understanding of this chapter if you can accurately address the following issues.

16.1 General Properties of the Autonomic Nervous System (p. 443)

1. The general function and effectors of the autonomic nervous system (ANS)
2. The involuntary nature of the ANS
3. The nature of visceral reflex arcs
4. The contrast between the general physiological effects of the sympathetic and parasympathetic divisions of the ANS
5. Autonomic tone and the simultaneous activity of the sympathetic and parasympathetic divisions
6. The basic anatomical components of the ANS
7. The two-neuron pathway of autonomic output

16.2 Anatomy of the Autonomic Nervous System (p. 446)

1. The spinal cord regions from which the sympathetic nervous system arises
2. The anatomy of the sympathetic chain of ganglia, including the number of ganglia at each level and why the chain extends to levels higher and lower than the spinal cord origins of sympathetic nerves

3. The varied routes and terminations of the sympathetic preganglionic nerve fibers within and sometimes beyond the sympathetic chain
4. The place of the splanchnic nerves and abdominal aortic plexus in the sympathetic nervous system, and the three ganglia of the abdominal aortic plexus
5. The degree of neural divergence in the sympathetic division and the effect of this on the stimulation of target organs by this division
6. The role of the adrenal medulla and its hormones in the sympathetic nervous system
7. Identity of the cranial and spinal nerves that carry parasympathetic output
8. The routes and destinations of the parasympathetic fibers in cranial nerves III, VII, IX, and X, including the thoracic and abdominal nerve plexuses through which the fibers of the vagus nerve pass
9. The routes and destinations of the parasympathetic fibers that arise from the sacral spinal cord, including the pelvic splanchnic nerves, inferior hypogastric plexus, and pelvic nerves
10. The location and function of the enteric nervous system

16.3 Autonomic Effects (p. 453)

1. The determination of autonomic effects on a target cell by a combination of the neurotransmitter and the receptor type involved
2. The definition of *cholinergic nerve fibers* and where they occur in the ANS
3. The two classes of cholinergic receptors and how they differ in their effects
4. The definition of *adrenergic nerve fibers* and where they occur in the ANS
5. The two classes of adrenergic receptors and how they differ in their effects
6. Cases of dual innervation of a target organ by the ANS and varied ways in which the sympathetic and parasympathetic divisions interact when dual innervation occurs
7. How the sympathetic division can exert contrasting effects on a single target organ in the absence of dual innervation
8. The multiple levels of CNS control over the autonomic nervous system

16.4 Developmental and Clinical Perspectives (p. 456)

1. Which components of the ANS arise from the embryonic neural tube and which ones arise from the neural crest
2. How ANS function changes with age, and the effects of ANS aging on functionality and quality of life

Testing Your Recall

1. The autonomic nervous system innervates all of these *except*
   a. cardiac muscle.
   b. skeletal muscle.
   c. smooth muscle.
   d. salivary glands.
   e. blood vessels.

2. Muscarinic receptors bind
   a. epinephrine.
   b. norepinephrine.
   c. acetylcholine.
   d. cholinesterase.
   e. neuropeptides.

3. All of the following cranial nerves except the _____ carry parasympathetic fibers.
   a. vagus   d. glossopharyngeal
   b. facial   e. hypoglossal
   c. oculomotor

4. Which of the following cranial nerves carries sympathetic fibers?
   a. oculomotor
   b. facial
   c. trigeminal
   d. vagus
   e. none of the cranial nerves

5. Which of these ganglia is *not* involved in the sympathetic division?
   a. intramural   d. inferior mesenteric
   b. paravertebral   e. superior cervical
   c. celiac

6. Epinephrine is secreted by
   a. sympathetic preganglionic fibers.
   b. sympathetic postganglionic fibers.
   c. parasympathetic preganglionic fibers.
   d. parasympathetic postganglionic fibers.
   e. the adrenal medulla.

7. The most important autonomic control center in the CNS is
   a. the cerebral cortex.
   b. the limbic system.
   c. the midbrain.
   d. the hypothalamus.
   e. the sympathetic chain ganglia.

8. The gray communicating ramus contains
   a. visceral sensory fibers.
   b. parasympathetic motor fibers.
   c. sympathetic preganglionic fibers.
   d. sympathetic postganglionic fibers.
   e. somatic motor fibers.

9. The neural crest gives rise to all of the following *except*
   a. sympathetic chain ganglia.
   b. the celiac ganglion.
   c. parasympathetic preganglionic neurons.
   d. parasympathetic postganglionic neurons.
   e. the adrenal medulla.

10. Which of these does *not* result from sympathetic stimulation?
    a. dilation of the pupil
    b. acceleration of the heart
    c. digestive secretion
    d. bronchodilation
    e. piloerection

11. Nerve fibers that secrete norepinephrine are called _____ fibers.

12. _____ is a state in which a target organ receives both sympathetic and parasympathetic fibers.

13. _____ is a state of continual background activity of the sympathetic and parasympathetic divisions.

14. Most parasympathetic preganglionic fibers are found in the _____ nerve.

15. The digestive tract has a semi-independent nervous system called the _____ nervous system.

16. The embryonic tissue that gives rise to all autonomic ganglia and postganglionic neurons, but not to any preganglionic neurons, is the _____.

17. The adrenal medulla consists of modified postganglionic neurons of the _____ nervous system.

18. The sympathetic nervous system has short _____ and long _____ nerve fibers.

19. Orthostatic hypotension is the result of inefficiency of the _____ reflex.

20. Sympathetic stimulation of blood vessels maintains a state of partial vasoconstriction called _____.

Answers in the Appendix

Building Your Medical Vocabulary

State a medical meaning of each of the following word elements, and give a term in which it is used.

1. nom-
2. baro-
3. splanchno-
4. reno-
5. path-
6. para-
7. lyt-
8. auto-
9. ram-
10. mur-

Answers in the Appendix

True or False

Determine which five of the following statements are false, and briefly explain why.

1. The parasympathetic nervous system shuts down when the sympathetic nervous system is active, and vice versa.

2. Some blood vessels of the skin receive parasympathetic innervation.

3. Voluntary control of the ANS is not possible.

4. The sympathetic nervous system stimulates digestion.

5. Some sympathetic postganglionic fibers are cholinergic.

6. Urination and defecation cannot occur without signals from the brain to the bladder and rectum.

7. Some parasympathetic nerve fibers are adrenergic.

8. Parasympathetic effects are more localized and specific than sympathetic effects.

9. The parasympathetic division shows less neural divergence than the sympathetic division does.

10. The two divisions of the ANS have antagonistic effects on the iris.

Answers in the Appendix

Testing Your Comprehension

1. You are dicing raw onions while preparing dinner, and the vapor makes your eyes water. Describe the afferent and efferent pathways involved in this response.

2. Suppose you are walking alone at night when you hear a dog growling close behind you. Describe the ways your sympathetic nervous system would prepare you to deal with this situation.

3. Suppose the cardiac nerves were destroyed. How would this affect the heart and the body's ability to react to a stressful situation?

4. What would be the advantage to a wolf in having its sympathetic nervous system stimulate the piloerector muscles? What happens in a human when the sympathetic system stimulates these muscles?

5. Pediatric literature has reported cases of poisoning of children with Lomotil, an antidiarrheal medicine. Lomotil works by means of the morphinelike effects of its chief ingredient, diphenoxylate, but it also contains atropine. Considering the mode of action described for atropine in Deeper Insight 16.3, why might it contribute to the antidiarrheal effect of Lomotil? In atropine poisoning, would you expect the pupils to be dilated or constricted? The skin to be moist or dry? The heart rate to be elevated or depressed? The bladder to retain urine or void uncontrollably? Explain each answer.

Answers at www.mhhe.com/saladinha3

Improve Your Grade at **www.mhhe.com/saladinha3**

Practice quizzes, labeling activities, games, and flashcards provide fun ways to master concepts. You can also download image PowerPoint files for each chapter to create a study guide or for taking notes during lecture.

Sense Organs

CHAPTER 17

A vallate papilla of the tongue, where most taste buds are located (SEM)

BRUSHING UP

To understand this chapter, you may find it helpful to review the following concepts:
- Converging circuits of neurons (p. 364)
- Spinal cord tracts (p. 375)
- Decussation (p. 375)
- Sensory areas of the cerebral cortex (p. 419)
- The cranial nerves (p. 427)

Anatomy & Physiology | REVEALED®
aprevealed.com

Nervous System

Anyone who enjoys music, art, fine food, or a good conversation appreciates the human senses. Yet their importance extends beyond deriving pleasure from the environment. In the 1950s, behavioral scientists at Princeton University studied the methods used by Soviet Communists to extract confessions from political prisoners, including solitary confinement and sensory deprivation. Student volunteers were immobilized in dark soundproof rooms or suspended in dark chambers of water. In a short time, they experienced visual, auditory, and tactile hallucinations, incoherent thought patterns, deterioration of intellectual performance, and sometimes morbid fear or panic. Similar effects are sometimes seen in burn patients who are immobilized and extensively bandaged (including the eyes) and thus suffer prolonged lack of sensory stimulation. Patients connected to life-support equipment and confined under oxygen tents sometimes become delirious. In short, sensory input is vital to the integrity of the personality and intellectual function.

Furthermore, much of the information communicated by the sense organs never comes to our conscious attention—blood pressure, body temperature, and muscle tension, for example. By monitoring such conditions, however, the sense organs initiate somatic and visceral reflexes that are indispensable to homeostasis and to our very survival in a ceaselessly changing and challenging environment.

17.1   Receptor Types and the General Senses

▶ Expected Learning Outcomes

When you have completed this section, you should be able to

- define *receptor* and *sense organ;*
- outline three ways of classifying sensory receptors;
- define *general senses,* list several types, and describe their receptors;
- explain the meaning and relevance of a sensory neuron's receptive field;
- describe the pathways that the general senses take to the cerebral cortex; and
- describe the types of pain and its projection pathways.

A sensory **receptor** is any structure specialized to detect a stimulus. Some receptors are simple nerve endings (sensory dendrites), whereas others are **sense organs**—nerve endings combined with connective, epithelial, or muscular tissues that enhance or moderate the response to a stimulus. Our eyes and ears are obvious examples of sense organs, but there are also innumerable microscopic sense organs in our skin, muscles, joints, and viscera.

Classification of Receptors

Receptors can be classified by multiple overlapping systems:

1. By **modality** (type of stimulus):
   - **Thermoreceptors** respond to heat and cold.
   - **Photoreceptors,** the eyes, respond to light.
   - **Chemoreceptors** respond to chemicals, including odors, tastes, and composition of the body fluids.
   - **Nociceptors**[1] (NO-sih-SEP-turs) are pain receptors; they respond to tissue damage resulting from trauma (blows, cuts), ischemia (poor blood flow), or excessive stimulation by agents such as heat and chemicals.
   - **Mechanoreceptors** respond to physical deformation of a tissue or cell caused by touch, pressure, stretch, tension, or vibration. They include the organs of hearing and balance and many receptors of the skin, viscera, and joints.

2. By the distribution of receptors in the body:
   - **General (somatosensory, somesthetic) senses** employ receptors that are widely distributed in the skin, muscles, tendons, joint capsules, and viscera. They detect touch, pressure, stretch, heat, cold, and pain, as well as many stimuli that we do not perceive consciously, such as blood pressure and blood chemistry. Their receptors are relatively simple in structure—sometimes nothing more than a bare dendrite.
   - **Special senses** are mediated by relatively complex sense organs of the head, innervated by the cranial nerves. They include vision, hearing, equilibrium, taste, and smell.

3. By the origins of the stimuli:
   - **Exteroceptors** sense stimuli external to the body; they include the receptors for vision, hearing, taste, smell, and the cutaneous (skin) senses such as heat, touch, and pain.
   - **Interoceptors** detect stimuli in the internal organs such as the stomach, intestines, and urinary bladder and produce such feelings as visceral pain, nausea, stretch, and pressure.
   - **Proprioceptors** sense the position and movements of the body or its parts. They occur in muscles, tendons, and joint capsules.

The General Senses

Receptors for the general senses are relatively simple in structure and physiology. They consist of one or a few nerve fibers and usually a sparse amount of connective tissue. Depending on the absence or presence of connective tissue, they are classified as unencapsulated or encapsulated nerve endings, respectively (table 17.1). Nine types of simple receptors for the general senses are described here and illustrated in figure 17.1.

---

[1]*noci* = pain

TABLE 17.1	Receptors of the General Senses	
Receptor Type	**Locations**	**Modality**
Unencapsulated endings		
Free nerve endings	Widespread, especially in epithelia and connective tissues	Pain, heat, cold
Tactile discs	Stratum basale of epidermis	Light touch, pressure
Hair receptors	Around hair follicle	Light touch, movement of hairs
Encapsulated nerve endings		
Tactile corpuscles	Dermal papillae of fingertips, palms, eyelids, lips, tongue, nipples, and genitals	Light touch, texture
End bulbs	Mucous membranes	Similar to tactile corpuscles
Bulbous corpuscles	Dermis, subcutaneous tissue, and joint capsules	Heavy continuous touch or pressure; joint movements
Lamellar corpuscles	Dermis, joint capsules, periosteum, breasts, genitals, and some viscera	Deep pressure, stretch, tickle, vibration
Muscle spindles	Skeletal muscles near tendon	Muscle stretch (proprioception)
Tendon organs	Tendons	Tension on tendons (proprioception)

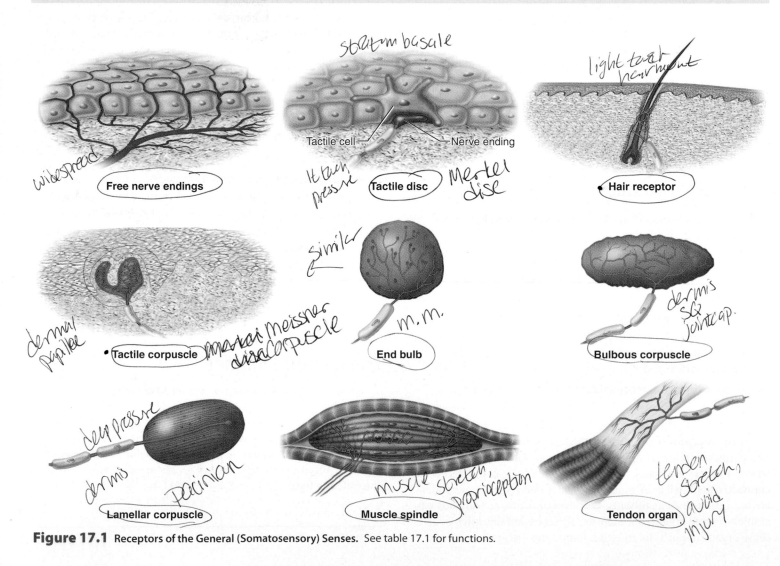

Figure 17.1 Receptors of the General (Somatosensory) Senses. See table 17.1 for functions.

Unencapsulated nerve endings are sensory dendrites that lack a connective tissue wrapping. They include the following:

- **Free nerve endings.** These include *warm receptors,* which respond to rising temperature; *cold receptors,* which respond to falling temperature; and *nociceptors,* or pain receptors. They are bare dendrites unassociated with any specific accessory cells or tissues. They are most abundant in the skin and mucous membranes. They typically show profuse, fine branches that penetrate through the connective tissue or between epithelial cells.

- **Tactile (Merkel[2]) discs.** These are receptors for light touch and pressure on the skin. A tactile disc is a flattened nerve ending associated with a specialized *tactile (Merkel) cell* at the base of the epidermis.

- **Hair receptors (peritrichial[3] endings).** These are nerve fibers entwined around a hair follicle that monitor the movements of the hair. Because they adapt quickly, we are not constantly annoyed by our clothing bending the body hairs. However, when an ant crawls across our skin, bending one hair after another, we are very aware of it.

Encapsulated nerve endings are dendrites wrapped in glial cells or connective tissue. Most of them are mechanoreceptors for touch, pressure, and stretch. The connective tissues around a sensory dendrite enhance the sensitivity of the receptor or make it more selective with respect to which type of stimulus it responds to. They include the following:

- **Tactile (Meissner[4]) corpuscles.** These are receptors for light touch and texture. They are tall, ovoid to pear-shaped, and consist of two or three nerve fibers meandering upward through a mass of flattened Schwann cells. They occur in the dermal papillae of the skin and are limited to sensitive hairless areas such as the fingertips, palms, eyelids, lips, nipples, and parts of the genitals. Drag your fingernails lightly across your forearm and then across your palm. The difference in sensation you feel is due to the high density of tactile corpuscles in your palmar skin. Tactile corpuscles enable you to tell the difference between silk and sandpaper, for example, by light strokes of your fingertips.

- **End bulbs (Krause[5] end bulbs, Krause corpuscles).** These are ovoid bodies composed of a connective tissue sheath around a sensory nerve fiber. They occur in mucous membranes of the lips and tongue, in the conjunctiva of the anterior surface of the eye, and in the epineurium of large nerves.

- **Bulbous (Ruffini[6]) corpuscles.** These are receptors for heavy touch, pressure, stretching of the skin, and joint movements. They are flattened, elongated capsules containing a few nerve fibers and are located in the dermis, subcutaneous tissue, and joint capsules.

- **Lamellar (pacinian[7]) corpuscles.** These are receptors for deep pressure, stretch, tickle, and vibration. They are large receptors, up to 1 or 2 mm long, and look like a sliced onion in cross section. A single sensory dendrite travels through the center of the corpuscle. The innermost onionlike layers around it are flattened Schwann cells, but the greater bulk of the corpuscle consists of concentric layers of fibroblasts with narrow fluid-filled spaces between them. These receptors occur in the periosteum of bone; in joint capsules; in the pancreas and some other viscera; and deep in the dermis, especially on the hands, feet, breasts, and genitals.

- **Muscle spindles.** These receptors detect stretch in a muscle and trigger various skeletal muscle (somatic) reflexes. A muscle spindle has a fibrous capsule, about 4 to 10 mm long, with a fusiform[8] shape (thick in the middle and tapered at the ends). It contains 3 to 12 modified muscle fibers called **intrafusal fibers,** which lack striations and the ability to contract except at the ends. Different types of sensory nerve fibers twine around the middle of the intrafusal fibers or have flowerlike endings that contact the ends of the muscle fibers.

- **Tendon organs.** These receptors detect stretch in a tendon and trigger a reflex that inhibits muscle contraction to avoid muscle or tendon injury. A tendon organ is about 1 mm long and consists of a tangle of knobby nerve endings squeezed into the spaces between the collagen fibers of the tendon.

The Receptive Field

The area supplied by a single sensory neuron is called its **receptive field.** Any information arriving at the CNS by way of that neuron is interpreted as coming from that field, no matter where in the field the stimulus is applied. Thus, the brain's ability to determine the precise location of a stimulus depends on the size of the field. Tactile (touch) receptors on your back, for example, have receptive fields as big as 7 cm in diameter. Any touch within that area stimulates the same neuron (fig. 17.2a), so it is difficult to tell precisely where the touch occurs. Being simultaneously touched at two points 5 cm apart within the same field would feel like a single touch. On the fingertips, by contrast, the receptive fields may be less than 1 mm in diameter—that is, there is a higher density of tactile nerve fibers. Two points of contact only 2 mm apart would thus be felt separately (fig. 17.2b). Therefore, we say that the fingertips have finer *two-point touch discrimination* than the skin on the back. This is crucial for such functions as feeling textures and manipulating small objects. The receptive field concept applies not only to touch, but also to other senses such as vision (the ability to see two points separately rather than blurred together into a single image).

Apply What You Know

Braille uses symbols composed of dots that are raised about 1 mm from the page surface and spaced about 2.5 mm apart, which a person scans with the fingertips. When a blind person reads braille, do you think he or she employs neurons with large receptive fields or small ones? Which type of sensory nerve ending do you think is most important in reading Braille? Explain.

[2]Friedrich S. Merkel (1845–1919), German anatomist and physiologist
[3]*peri* = around + *trich* = hair
[4]Georg Meissner (1829–1905), German histologist
[5]Wilhelm J. F. Krause (1833–1910), German anatomist
[6]Angelo Ruffini (1864–1929), Italian anatomist
[7]Filippo Pacini (1812–83), Italian anatomist

[8]*fusi* = spindle + *form* = shaped

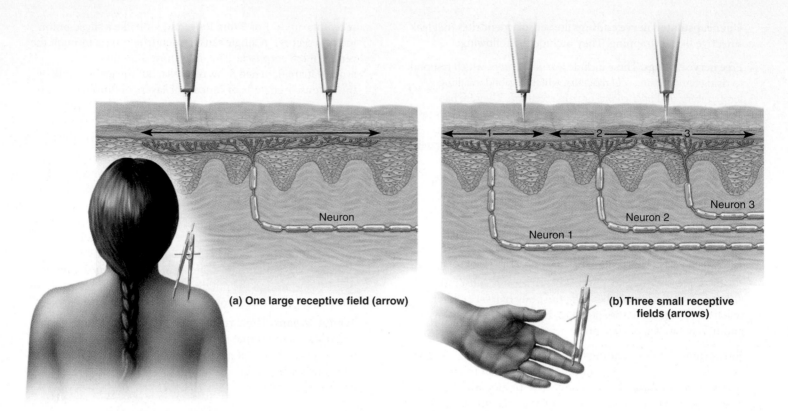

Figure 17.2 Receptive Fields of Sensory Neurons. (a) A neuron with a large receptive field, as found in the skin of the back. If the skin is touched in two nearby places within this field, the brain will sense only one point of contact. (b) Neurons with small receptive fields, as found in the fingertips. Two nearby points of contact here are likely to stimulate different neurons and be felt as separate touches.

Somatosensory Projection Pathways

The pathways followed by sensory signals to their ultimate destinations in the CNS are called **projection pathways.** From the receptor to the final destination in the brain, most somatosensory signals travel by way of three neurons called the **first-, second-,** and **third-order neurons.** Their axons are called first- through third-order nerve fibers. The first-order (afferent) fibers for touch, pressure, and proprioception are large, myelinated, and fast; those for heat and cold are small, unmyelinated or lightly myelinated, and slower.

Somatosensory signals from the head, such as facial sensations, travel by way of several cranial nerves (especially V, the trigeminal nerve) to the pons and medulla oblongata. In the brainstem, the first-order fibers of these neurons synapse with second-order neurons that decussate and end in the contralateral thalamus. Third-order neurons then complete the route to the cerebrum. Proprioceptive signals are an exception, as the second-order fibers carry these signals to the cerebellum.

Below the head, the first-order fibers enter the posterior horn of the spinal cord. Signals ascend the spinal cord in the spinothalamic and other pathways detailed in chapter 14. These pathways decussate either at or near the point of entry into the spinal cord or in the brainstem, so the primary somatosensory cortex in each cerebral hemisphere receives signals from the contralateral side of the body.

Signals for proprioception below the head travel up the spinocerebellar tracts to the cerebellum. Signals from the thoracic and abdominal viscera travel to the medulla oblongata by way of sensory fibers in the vagus nerve (cranial nerve X). Recent research has shown that visceral pain signals also ascend the spinal cord in the gracile fasciculus.

Pain

Pain is a discomfort caused by tissue injury or noxious stimulation, and typically leads to evasive action. It makes us conscious of potentially injurious situations or actual tissue injuries, allowing us to avoid injury or, failing that, to favor an injured region so that it has a better chance to heal (see Deeper Insight 17.1).

Pain is not merely an effect of overstimulation of somatosensory receptors. It has its own specialized receptors (nociceptors) and purpose. Nociceptors are especially dense in the skin and mucous membranes and occur in virtually all organs, although not in the brain. In some brain surgery, the patient must remain conscious and able to talk with the surgeon; such patients need only a local anesthetic. Nociceptors do occur in the meninges of the brain, however, and play an important role in headaches.

There are two types of nociceptors corresponding to different pain sensations. Myelinated pain fibers conduct at speeds of 12 to 30 m/sec. and produce the sensation of *fast (first) pain*—a feeling of

DEEPER INSIGHT   17.1

The Value of Pain

Although we generally regard pain as undesirable, we would be far worse off without it. Leprosy (Hansen[9] disease) provides a good example of the protective function of pain. The infection of nerves by leprosy bacteria abolishes the sense of pain from affected areas. People fail to notice minor injuries such as scrapes and splinter wounds. Their neglect of the wounds leads to serious secondary infections that damage the bone and other deeper tissues. About 25% of untreated victims suffer crippling losses of fingers or toes as a result. Diabetes mellitus is also notorious for causing nerve damage *(diabetic neuropathy)* and loss of pain, contributing to lesions that often cost people their limbs.

---

sharp, localized, stabbing pain perceived at the time of injury. Unmyelinated pain fibers conduct at speeds of 0.5 to 2.0 m/sec. and produce the *slow (second) pain* that follows—a longer-lasting, dull, diffuse feeling. Pain from the skin, muscles, and joints is called *somatic pain,* and pain from the viscera is called *visceral pain.* The latter often results from stretch, chemical irritants, or *ischemia* (poor blood flow), and it is often accompanied by nausea.

It is notoriously difficult for clinicians to locate the origin of a patient's pain because pain travels by such diverse and complex routes and the sensation can originate anywhere along the routes. Pain signals reach the brain by two main pathways, but there are multiple subroutes within each of them:

1. Pain signals from the head travel to the brainstem by way of four cranial nerves: mainly the trigeminal (V), but also the facial (VII), glossopharyngeal (IX), and vagus (X) nerves. Trigeminal fibers enter the pons and descend to synapses in the medulla. Pain fibers of the other three cranial nerves also end in the medulla. Second-order neurons arise here and ascend to the thalamus, which relays the message to the cerebral cortex. We will consider the relay from thalamus to cortex shortly.

2. Pain signals from the neck down travel by way of three of the ascending spinal cord tracts: the spinothalamic tract, spinoreticular tract, and gracile fasciculus. These pathways are described in chapter 14 and need not be repeated here (see table 14.1, p. 376, and fig. 14.4, p. 378). The spinothalamic tract is the most significant pain pathway and carries most of the somatic pain signals that ultimately reach the cerebral cortex, making us conscious of pain. The spinoreticular tract carries pain signals to the reticular formation of the brainstem, and these are ultimately relayed to the hypothalamus and limbic system. These pain signals activate visceral, emotional, and behavioral reactions to pain, such as nausea, fear, and some reflex responses. The gracile fasciculus has not been recognized as a pain pathway until recently. It carries signals to the thalamus for visceral pain, such as the pain of a stomachache or from passing a kidney stone. Figure 17.3 shows the spinothalamic and spinoreticular pain pathways.

[9]Gerhard H. A. Hansen (1841–1912), Norwegian physician

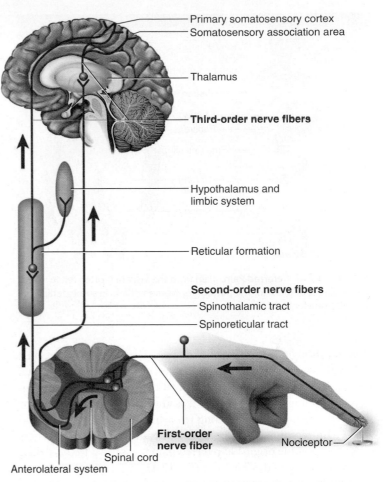

Figure 17.3 Projection Pathways for Pain. A first-order nerve fiber conducts a pain signal to the posterior horn of the spinal cord, a second-order fiber conducts it to the thalamus, and a third-order fiber conducts it to the cerebral cortex. Signals from the spinothalamic tract pass through the thalamus. Signals from the spinoreticular tract bypass the thalamus on the way to the sensory cortex.
● *What is the name of the gyrus that contains the primary somatosensory cortex? To which of the five lobes of the cerebrum does this gyrus belong?*

When the thalamus receives pain signals from the foregoing sources, it relays most of them through third-order neurons to their final destinations in the postcentral gyrus of the cerebrum. Exactly what part of this gyrus receives the signals depends on where the pain originated; recall the concept of somatotopy and the sensory homunculus in chapter 15 (p. 420). Most of this gyrus is somatosensory; that is, it receives signals for somatic pain and other senses. A region of the gyrus deep within the lateral sulcus of the brain, however, is a viscerosensory area, which receives the signals of visceral pain conveyed by the gracile fasciculus.

Pain in the viscera is often mistakenly thought to come from the skin or other superficial sites—for example, the pain of a heart attack is felt "radiating" along the left shoulder and medial side of the arm. This phenomenon, called **referred pain,** results from the convergence of neural pathways in the CNS. In the case of cardiac pain, for example, spinal cord segments T1 to T5 receive input from the heart as well as the chest and arm. Pain fibers from the heart and

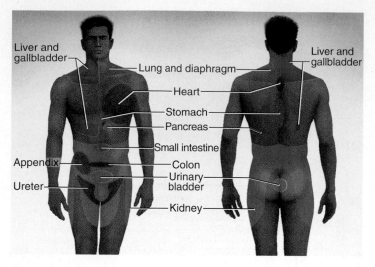

Figure 17.4 Referred Pain. Pain from the viscera is often felt in specific areas of the skin. The pattern for women differs in some details from the one for men.

skin in this region converge on the same spinal interneurons, then follow the same pathway from there to the thalamus and cerebral cortex. The brain cannot distinguish which source the arriving signals are coming from. It acts as if it assumes that signals arriving by this path are most likely coming from the skin, since the skin has more pain receptors than the heart and suffers injury more often. Knowledge of the origins of referred pain is essential to the skillful diagnosis of organ dysfunctions (fig. 17.4).

One of the most remarkable phenomena in the sense of pain is that mortally wounded people, such as soldiers in combat, often report that they feel little or no pain. This occurs because of a mechanism called the *spinal gating* of pain, in which pain signals from peripheral nerves are blocked in the spinal cord and never reach the brain.

The reticular formation gives rise to nerve fibers called **descending analgesic**[10] (pain-relieving) **fibers.** They travel down the reticulospinal tracts and synapse in the posterior horn of the spinal cord with the axons of the first-order pain neurons. Here, they secrete pain-relieving neurotransmitters called *enkephalins* and *dynorphins,* which inhibit the first-order pain neurons from releasing their own neurotransmitter. Pain signals are thus stopped at the first spinal synapse and do not reach the brain, so one feels less pain or none at all.

Before You Go On

Answer the following questions to test your understanding of the preceding section.

1. Distinguish between general and special senses.

2. Three schemes of receptor classification were presented in this section. In each scheme, how would you classify the receptors for a full bladder? How would you classify taste receptors?

---

[10]*an* = without + *alges* = pain

3. What stimulus modalities are detected by free nerve endings?

4. Name any four encapsulated nerve endings and identify the stimulus modalities for which they are specialized.

5. Where do most second-order somatosensory neurons synapse with third-order neurons?

6. How do the spinothalamic tract and reticulospinal tract differ in their roles in the perception of pain?

17.2 The Chemical Senses

▶ Expected Learning Outcomes

When you have completed this section, you should be able to

- describe the anatomy of taste and smell receptors; and
- describe the projection pathways for these two senses.

Taste and smell are the chemical senses. In both cases, environmental chemicals bind to receptor cells and trigger nerve signals in certain cranial nerves. Other chemoreceptors, not discussed in this section, are located in the brain and blood vessels and monitor the chemistry of cerebrospinal fluid and blood.

Taste

Gustation (taste) results from the action of chemicals on the **taste buds.** There are about 4,000 taste buds, mainly on the tongue but also inside the cheeks and on the soft palate, pharynx, and epiglottis. The tongue, where the sense of taste is best developed, is marked by four types of surface projections called **lingual papillae** (fig. 17.5a):

1. **Filiform**[11] **papillae** are tiny spikes without taste buds (see photograph on p. 654). They are responsible for the rough feel of a cat's tongue and are important to many mammals for grooming the fur. They are the most abundant papillae on the human tongue, but they are small and play no gustatory role. They serve, however, in one's sense of the texture of food.

2. **Foliate**[12] **papillae** are also weakly developed in humans. They form parallel ridges on the sides of the tongue about two-thirds of the way back from the tip, adjacent to the molar and premolar teeth, where most chewing occurs and most flavor chemicals are released from the food. Most of their taste buds degenerate by the age of 2 or 3 years.

3. **Fungiform**[13] (FUN-jih-form) **papillae** are shaped somewhat like mushrooms. Each one has about three taste buds, located mainly on the apex. These papillae are widely distributed but are especially concentrated at the tip and sides of the tongue.

---

[11]*fili* = thread + *form* = shaped
[12]*foli* = leaf + *ate* = like
[13]*fungi* = mushroom + *form* = shaped

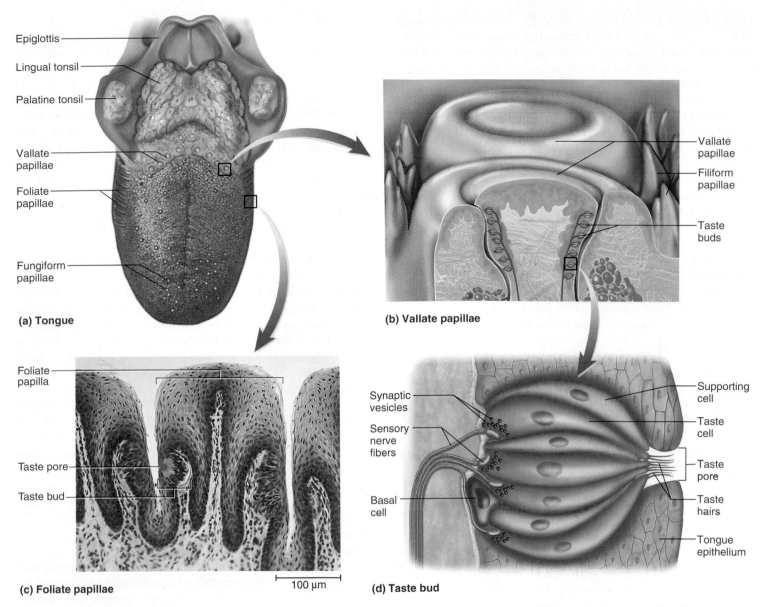

Figure 17.5 Gustatory (Taste) Receptors. (a) Superior surface of the tongue and locations of its papillae. (b) Detail of the vallate papillae. (c) Taste buds on the walls of two adjacent foliate papillae. (d) Structure of a taste bud.
● *What cell(s) in part (d) would function to replace a taste cell that dies?*

4. **Vallate**[14] **(circumvallate) papillae** are large papillae arranged in a V at the rear of the tongue. Each is surrounded by a deep circular trench. There are only 7 to 12 vallate papillae, but they contain up to half of all taste buds—around 250 each, located on the wall of the papilla facing the trench (p. 460 and fig. 17.5b).

Regardless of location and sensory specialization, all taste buds look alike (fig. 17.5c, d). They are lemon-shaped groups of 40 to 60 *taste cells, supporting cells,* and *basal cells.* **Taste (gustatory) cells** are more or less banana-shaped and have a tuft of apical microvilli called **taste hairs,** which serve as receptor surfaces for taste molecules. The hairs project into a pit called a **taste pore** on the epithelial surface of the tongue. Taste cells are epithelial cells, not neurons,

but at their bases, they synapse with sensory nerve fibers and have synaptic vesicles for the release of neurotransmitters. A taste cell lives 7 to 10 days. **Basal cells** are stem cells that multiply and replace taste cells that have died. **Supporting cells** resemble taste cells but have no taste hairs, synaptic vesicles, or sensory role.

There are five *primary taste* sensations: sweet, salty, sour, bitter, and umami. Umami, the most recently discovered primary taste, is a meaty taste stimulated by certain amino acids such as glutamate and aspartate. Pronounced "ooh-mommy," the word is Japanese and loosely means "delicious." All of the primary taste sensations can be detected throughout the tongue, but certain regions are more sensitive to one modality than to another. The tip of the tongue is most sensitive to sweets, the lateral margins to salty and sour, and the rear of the tongue (the vallate papillae) to bitter. Umami is not yet as well understood, but taste cells have been found that have umami receptors

---

[14]*vall* = wall + *ate* = like, possessing

different from the receptors for any other taste. It was once popular to show "taste maps" of the tongue indicating where these modalities were localized, but sensory physiologists long ago discarded this concept because the tongue has no regional specializations of any significance to the brain's interpretation of modality.

The many flavors we perceive are not simply a mixture of these five primary tastes but are also influenced by food texture, aroma, temperature, and appearance, as well as one's state of mind, among other things. Food technologists refer to the texture of food as *mouthfeel*. Filiform and fungiform papillae of the tongue are innervated by the lingual nerve (a branch of the trigeminal) and are sensitive to texture. Many flavors depend on smell; without its aroma, cinnamon merely has a faintly sweet taste, and coffee and peppermint are bitter. Some flavors such as pepper are due to stimulation of free endings of the trigeminal nerve.

The facial nerve (cranial nerve VII) collects sensory information from taste buds of the anterior two-thirds of the tongue; the glossopharyngeal nerve (IX) from the posterior one-third; and the vagus nerve (X) from taste buds of the palate, pharynx, and epiglottis. All taste fibers project to a site in the medulla oblongata called the *solitary nucleus*. Second-order neurons arise here and relay the signals to two destinations: (1) nuclei in the hypothalamus and amygdala that activate autonomic reflexes such as salivation, gagging, and vomiting; and (2) the thalamus, which relays signals to the insula and postcentral gyrus of the cerebrum (fig. 17.6), where we become conscious of the taste. Processed signals are further relayed to the orbitofrontal cortex (see fig. 15.16, p. 419) where they are integrated with signals from the nose and eyes and we form an overall impression of the flavor and palatability of food.

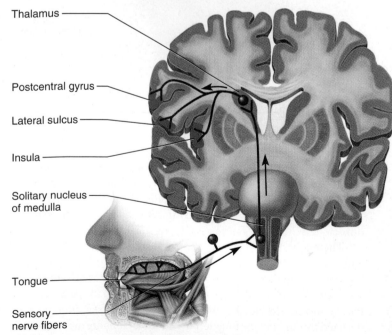

Figure 17.6 **Gustatory Projection Pathways to the Cerebral Cortex.** Other pathways not shown carry taste signals from the solitary nucleus to the hypothalamus and amygdala.

Labels on figure:
Thalamus
Postcentral gyrus
Lateral sulcus
Insula
Solitary nucleus of medulla
Tongue
Sensory nerve fibers

Smell

Olfaction (smell) resides in a patch of epithelium, the **olfactory mucosa,** on the roof of the nasal cavity (fig. 17.7). This sensory epithelium covers about 5 cm² of the superior concha and nasal septum; the rest of the nasal cavity is lined by a nonsensory *respiratory mucosa.* Their location places the olfactory cells close to the brain, but it is poorly ventilated; forcible sniffing is often needed to identify an odor or locate its source. Nevertheless, the sense of smell is highly sensitive. We can detect extremely low concentrations of odor molecules, and most people can distinguish 2,000 to 4,000 different odors; some can distinguish as many as 10,000. On average, women are more sensitive to odors than men are, and they are more sensitive to some odors near the time of ovulation than during other phases of the menstrual cycle.

The olfactory mucosa has 10 to 20 million **olfactory neurons** as well as epithelial supporting cells and basal cells. It has a yellowish tint due to lipofuscin in the supporting cells. Note that olfactory cells are neurons whereas taste cells are not. Olfactory cells are the only neurons in the body directly exposed to the external environment. Apparently this is hard on them, because they have a life span of only 60 days. Unlike most neurons, however, they are replaceable. The basal stem cells continually divide and differentiate into new olfactory cells.

An olfactory cell is shaped a little like a bowling pin. Its widest part, the soma, contains the nucleus. The neck and head of the cell are a modified dendrite with a swollen tip bearing 10 to 20 immobile cilia called **olfactory hairs.** These cilia bear the binding sites for odor molecules, and lie in a tangled mass embedded in a thin layer of mucus on the epithelial surface. The basal end of each cell tapers to become an axon. These axons collect into small fascicles that leave the nasal cavity through pores *(olfactory foramina)* in the cribriform plate of the ethmoid bone. Collectively, the fascicles are regarded as cranial nerve I (the olfactory nerve).

When olfactory fibers pass through the roof of the nose, they enter a pair of **olfactory bulbs** beneath the frontal lobes of the brain (see fig. 15.24, p. 428). Here, they synapse with the dendrites of two types of neurons higher in the bulb, called *mitral cells* and *tufted cells.* Olfactory axons reach up, and mitral and tufted cell dendrites reach down, to meet each other in spherical clusters called *glomeruli* (fig. 17.7b). All olfactory fibers leading to any one glomerulus come from cells with the same receptor type; thus, each glomerulus is dedicated to a particular odor. Higher brain centers interpret complex odors such as chocolate, perfume, or coffee by decoding signals from a combination of odor-specific glomeruli. This is similar to the way our visual system decodes all the colors of the spectrum using input from just three color-specific receptor cells of the eye.

The tufted and mitral cells carry the output from the glomeruli. Their axons form bundles called **olfactory tracts,** which course posteriorly along the underside of the frontal lobes. Most fibers of the olfactory tracts end in various neighboring regions of the inferior surface of the temporal lobe (fig. 17.8); collectively, we can regard all these regions as the **primary olfactory cortex.** It is noteworthy that olfactory signals reach the cerebral cortex directly, without first passing through the thalamus; this is not true of any of our other senses. Even in olfaction, however, some signals from the primary olfactory cortex continue to a relay in the thalamus on their way to olfactory association areas elsewhere.

From the primary olfactory cortex, signals travel to several other secondary destinations in the cerebrum and brainstem. Two important cerebral destinations are the insula and orbitofrontal cortex. The orbitofrontal cortex, which lies on the floor of the fron-

tal lobe just above the eyes, seems to be the site where we identify and discriminate among odors. It receives inputs for both taste and smell and integrates these into our overall perception of flavor. Other secondary destinations for olfactory signals include the hippocampus, amygdala, and hypothalamus. Considering the roles of these brain areas, it is not surprising that the odor of certain foods, a perfume, a hospital, or decaying flesh can evoke strong memories, emotional responses, and visceral reactions such as sneezing or coughing, the secretion of saliva and stomach acid, or vomiting.

Most areas of olfactory cortex also send fibers back to the olfactory bulbs by way of neurons called *granule cells.* Granule cells inhibit the mitral and tufted cells. An effect of this feedback is that odors can change in quality and significance under different conditions. Food may smell more appetizing when you are hungry, for example, than when you have just eaten or when you are ill.

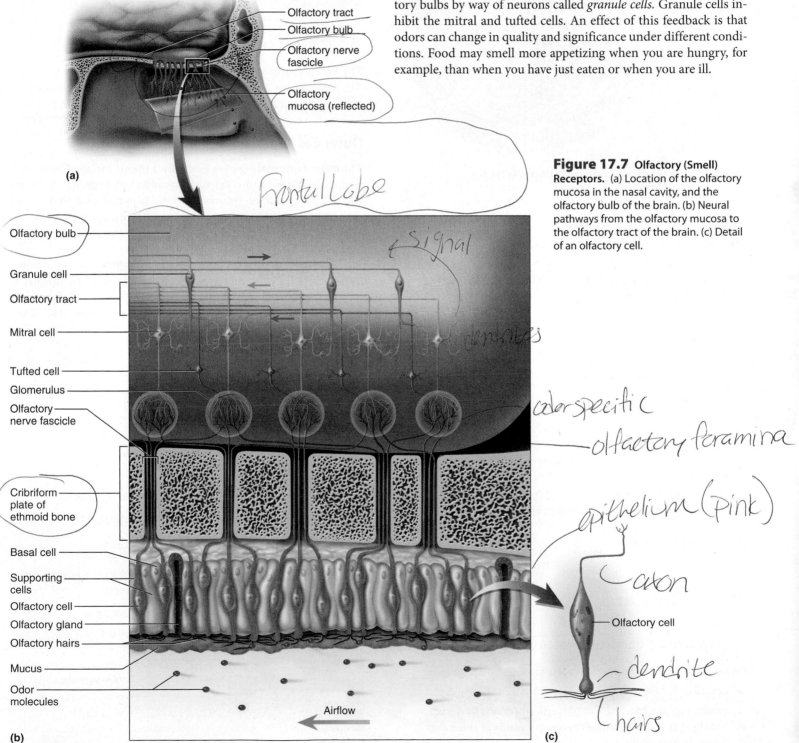

Figure 17.7 Olfactory (Smell) **Receptors.** (a) Location of the olfactory mucosa in the nasal cavity, and the olfactory bulb of the brain. (b) Neural pathways from the olfactory mucosa to the olfactory tract of the brain. (c) Detail of an olfactory cell.

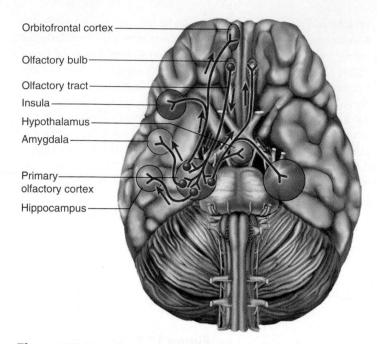

Orbitofrontal cortex

Olfactory bulb

Olfactory tract

Insula

Hypothalamus

Amygdala

Primary olfactory cortex

Hippocampus

Figure 17.8 Olfactory Projection Pathways in the Brain.

Apply What You Know

Which taste sensations could be lost after damage to (1) the facial nerve and (2) the glossopharyngeal nerve? Why? A fracture of which cranial bone would most likely eliminate the sense of smell? Why?

Before You Go On

Answer the following questions to test your understanding of the preceding section.

7. What is the difference between a lingual papilla and a taste bud? Which is visible to the naked eye?

8. Which cranial nerves carry gustatory impulses to the brain?

9. What part of an olfactory cell bears the binding sites for odor molecules?

10. What region of the brain receives subconscious input from the olfactory cells? What region receives conscious input?

17.3 The Ear

▶ Expected Learning Outcomes

When you have completed this section, you should be able to

- describe the gross and microscopic anatomy of the ear;
- briefly explain how the ear converts vibrations to nerve impulses and discriminates between sounds of different intensity and pitch;

- explain how the anatomy of the vestibular apparatus relates to our ability to interpret the body's position and movements; and

- describe the pathways taken by auditory and vestibular signals to the brain.

Hearing is a response to vibrating air molecules, and *equilibrium* is the sense of motion and balance. These senses reside in the inner ear, a maze of fluid-filled passages and sensory cells encased in the temporal bone.

Anatomy of the Ear

The ear has three sections called the *outer, middle,* and *inner ear.* The first two are concerned only with conducting sound to the inner ear, where vibration is converted to nerve signals.

Outer Ear

The **outer (external) ear** is essentially a funnel for conducting airborne vibrations to the tympanic membrane (eardrum). It begins with the fleshy **auricle (pinna)** on the side of the head, shaped and supported by elastic cartilage except for the earlobe. The auricle has a predictable arrangement of whorls and recesses that direct sound into the auditory canal (fig. 17.9).

The **auditory canal** is the passage through the temporal bone. Beginning at the external opening, the **external acoustic meatus,** it follows a slightly S-shaped course for about 3 cm to the tympanic membrane (fig. 17.10). It is lined with skin and supported by fibrocartilage at its opening and by the temporal bone for the rest of its

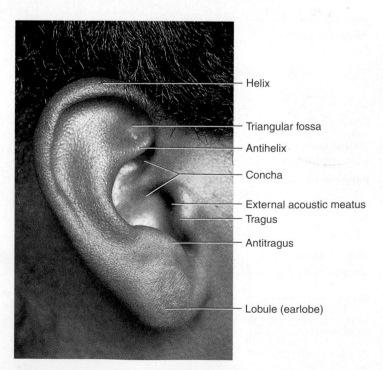

Helix

Triangular fossa

Antihelix

Concha

External acoustic meatus

Tragus

Antitragus

Lobule (earlobe)

Figure 17.9 The Auricle (Pinna) of the Ear.

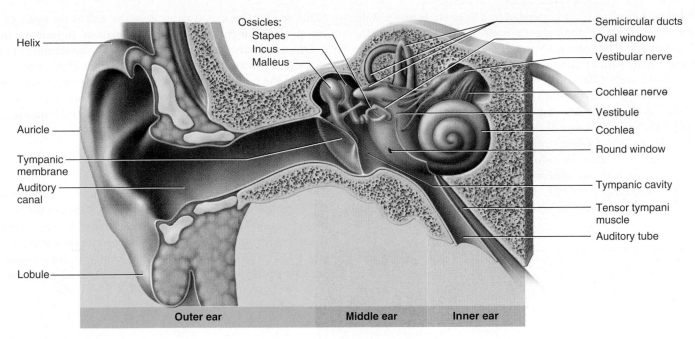

Figure 17.10 Internal Anatomy of the Ear.

length. The outer end of the canal is protected by stiff **guard hairs.** The canal has ceruminous and sebaceous glands whose secretions mix with dead skin cells and form **cerumen** (earwax). Cerumen is sticky and coats the guard hairs, making them more effective in blocking foreign particles from entering the auditory canal. Its stickiness may also be a deterrent to insects, ticks, or other pests. In addition, it contains lysozyme and has a low pH, both of which inhibit bacterial growth; it waterproofs the canal and protects its skin and the tympanic membrane from absorbing water; and it keeps the tympanic membrane pliable. Cerumen normally dries up and falls from the canal, but sometimes it becomes impacted and interferes with hearing.

Middle Ear

The **middle ear** is located in the **tympanic cavity** of the temporal bone. A space only 2 to 3 mm wide between the outer and inner ears, the tympanic cavity houses the three smallest bones and two smallest skeletal muscles of the body.

The middle ear begins with the **tympanic**[15] **membrane,** which closes the inner end of the auditory canal and separates it from the middle ear. The membrane is about 1 cm in diameter and slightly concave on its outer surface. It is suspended in a ring-shaped groove in the temporal bone and vibrates freely in response to sound. It is innervated by sensory branches of the vagus and trigeminal nerves and is highly sensitive to pain.

Posteriorly, the tympanic cavity is continuous with the mastoidal air cells in the mastoid process. The cavity is filled with air that enters by way of the **auditory (eustachian**[16]**) tube,** a passage to the nasopharynx. (Be careful not to confuse *auditory tube* with *audi-*

tory canal.) The auditory tube is normally flattened and closed, but swallowing or yawning opens it and allows air to enter or leave the tympanic cavity. This equalizes air pressure on both sides of the tympanic membrane, allowing it to vibrate freely. Excessive pressure on one side or the other muffles (dampens) the sense of hearing. Unfortunately, the auditory tube also allows throat infections to spread to the middle ear (see Deeper Insight 17.2).

The three middle-ear bones are called the **auditory ossicles**[17] (fig. 17.10). Progressing inward, the first is the **malleus,**[18] which has an elongated *handle* attached to the inner surface of the tympanic

DEEPER INSIGHT 17.2

Middle-Ear Infection

Otitis[19] *media* (middle-ear infection) is especially common in children because their auditory tubes are relatively short and horizontal. Upper respiratory infections spread easily from the throat to the tympanic cavity and mastoidal air cells. Fluid accumulates in the cavity and causes pressure, pain, and impaired hearing. If otitis media goes untreated, it may spread from the mastoidal air cells and cause meningitis, a potentially deadly infection of the meninges. Chronic otitis media can also cause fusion of the middle-ear bones, preventing them from vibrating freely and thus causing hearing loss. It is sometimes necessary to drain fluid from the tympanic cavity by lancing the tympanic membrane and inserting a tiny drainage tube—a procedure called *tympanostomy.*[20] The tube, which is eventually discharged spontaneously by the ear, relieves the pressure and permits the infection to heal.

---

[15]*tympan* = drum
[16]Bartholomeo Eustachio (1520–74), Italian anatomist

[17]*oss* = bone + *icle* = little
[18]*malleus* = hammer, mallet
[19]*ot* = ear + *itis* = inflammation
[20]*tympano* = drum + *stomy* = making an opening

membrane; a *head*, which is suspended by a ligament from the wall of the tympanic cavity; and a *short process*, which articulates with the next ossicle. The second bone is the **incus.**[21] It has a roughly triangular *body* with an articular surface where it meets the malleus; a *short limb* (not illustrated) suspended by a ligament from the wall of the tympanic cavity; and a *long limb* that articulates with the third ossicle, the stapes. The **stapes**[22] (STAY-peez) has an arch and footplate that give it an overall shape like a stirrup. The *footplate*, shaped like the sole of a steam iron, is held by a ringlike ligament in an opening called the **oval window,** where the inner ear begins.

The muscles of the middle ear are the stapedius and tensor tympani. The **stapedius** (stay-PEE-dee-us) arises from the posterior wall of the cavity and inserts on the stapes. The **tensor tympani**

(TEN-sor TIM-pan-eye) arises from the wall of the auditory tube, travels alongside it, and inserts on the malleus. In response to loud noises, these muscles contract and dampen the vibration of the ossicles, thus protecting the delicate sensory cells of the inner ear; this is called the *tympanic reflex.* The sensory cells can nevertheless be irreversibly damaged by sudden loud noises such as gunshots, and by sustained loud noise such as factory noise and loud music (see Deeper Insight 17.3).

Inner Ear

The **inner (internal) ear** is housed in a maze of passages called the **bony (osseous) labyrinth** in the petrous part of the temporal bone (see p. 160). The bony labyrinth contains a complex of fluid-filled chambers and fleshy tubes called the **membranous labyrinth** (fig. 17.11). The fluid in this labyrinth, similar to intracellular fluid, is called **endolymph.** Between the membranous labyrinth and bone

---

[21]*incus* = anvil
[22]*stapes* = stirrup

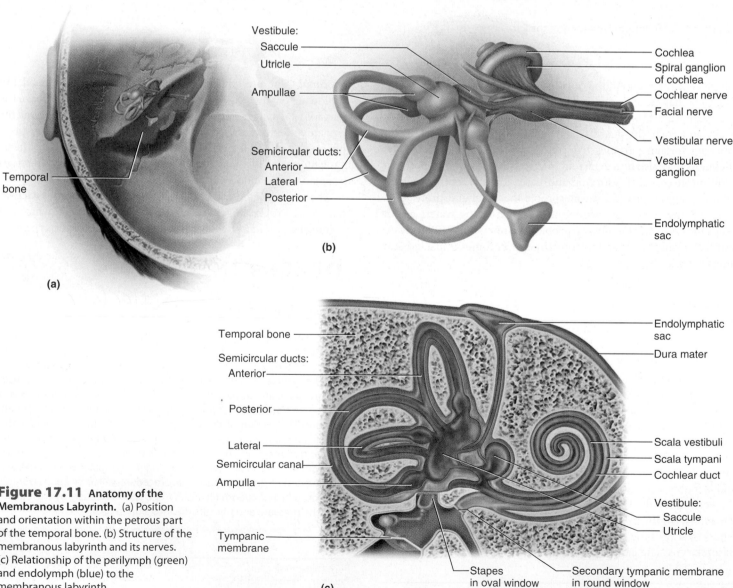

Figure 17.11 **Anatomy of the Membranous Labyrinth.** (a) Position and orientation within the petrous part of the temporal bone. (b) Structure of the membranous labyrinth and its nerves. (c) Relationship of the perilymph (green) and endolymph (blue) to the membranous labyrinth.

is another liquid, similar to cerebrospinal fluid, called **perilymph.** The membranous labyrinth within the bony labyrinth is a tube-within-a-tube structure, somewhat like a bicycle inner tube within a tire.

Proceeding inward from the oval window, the first part of the bony labyrinth is an ovoid chamber called the **vestibule,** containing two organs of balance that we will consider later—the *utricle* and *saccule.* Arising from the vestibule are three bony **semicircular canals,** containing fleshy loops called the *semicircular ducts,* which detect rotation of the head. Also arising from the vestibule is a spiral tube called the **cochlea**[23] (COC-lee-uh or COKE-lee-uh), named for its snail-like shape. This contains the organ of hearing, called the **cochlear duct,** and is therefore the focus of our immediate interest.

---

[23]*cochlea* = snail

In other vertebrates, the cochlea is straight or slightly curved. In most mammals, however, it assumes a spiral form that allows a longer cochlea to fit within a compact space of the temporal bone. In humans, it is about 9 mm wide at the base and 5 mm high. The apex points anterolaterally (fig. 17.11a). The cochlea winds for about two and a half coils around a screwlike axis of spongy bone called the **modiolus**[24] (mo-DY-oh-lus). The "threads of the screw" form a spiral platform that supports the fleshy cochlear duct.

The cochlear duct contains structures that convert sound waves to nerve signals. A vertical section cuts through the cochlea about five times (fig. 17.12a). A single cross section through it looks like figure 17.12b. It is important to realize that the structures seen in cross section actually have the form of spiral strips winding around the modiolus from base to apex. In cross section, the cochlear duct appears as a triangular space bounded by a thick **basilar membrane** below and a thin **vestibular membrane** above. Above the vestibular membrane is a space called the **scala**[25] **vestibuli** (SCAY-la vess-TIB-you-lye), and below the basilar membrane is a space called the **scala tympani** (TIM-pan-eye). These are filled with perilymph and

---

[24]*modiolus* = hub
[25]*scala* = staircase

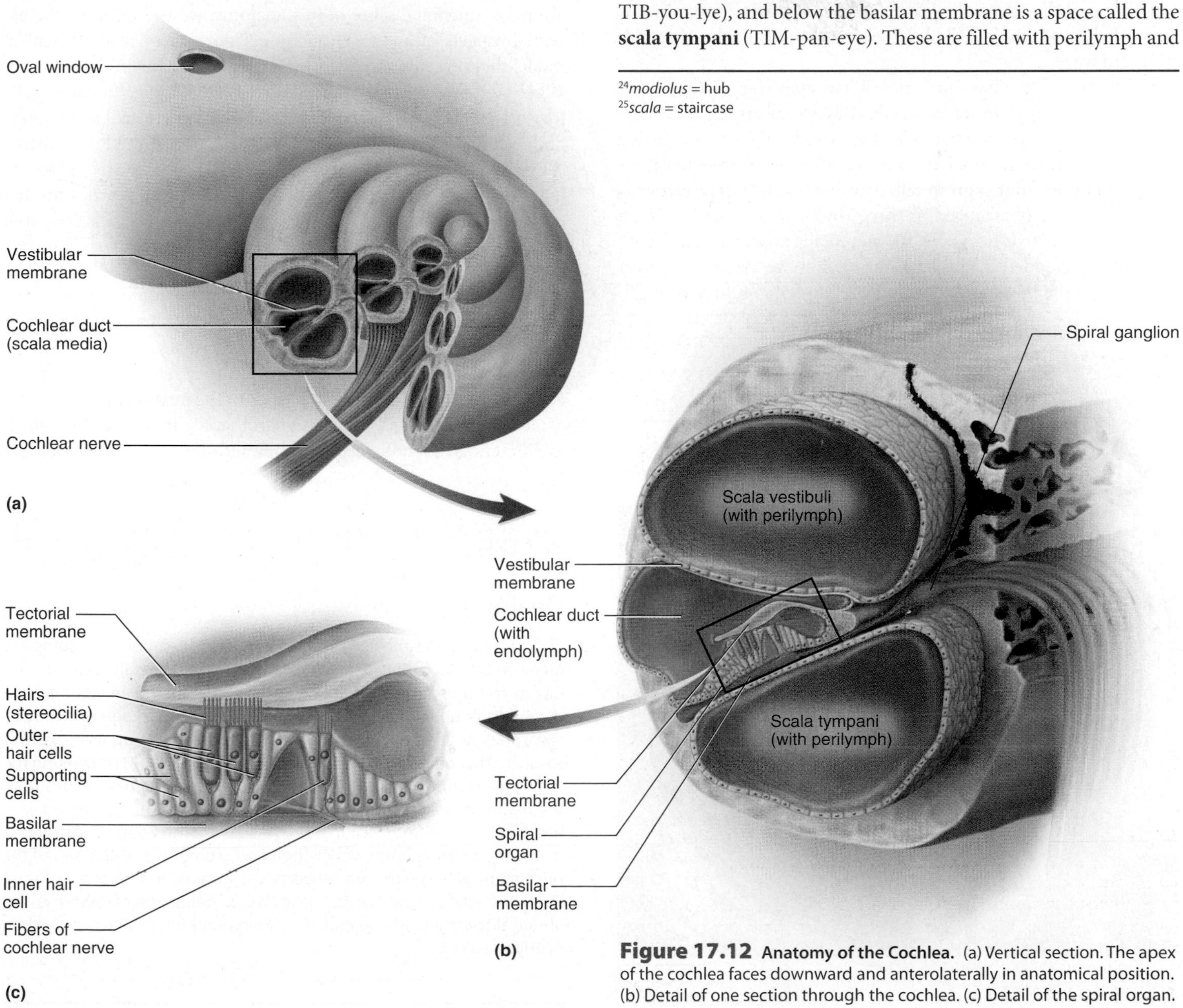

Figure 17.12 **Anatomy of the Cochlea.** (a) Vertical section. The apex of the cochlea faces downward and anterolaterally in anatomical position. (b) Detail of one section through the cochlea. (c) Detail of the spiral organ.

communicate with each other through a narrow channel at the apex of the cochlea. The scala vestibuli begins near the oval window and spirals to the apex; from there, the scala tympani spirals back down to the base and ends at the **round window** (fig. 17.11). The round window is covered by a membrane called the *secondary tympanic membrane*.

Within the cochlear duct, supported on the basilar membrane, is the **spiral organ,** also called the *acoustic organ* or *organ of Corti*[26] (COR-tee)—a thick epithelium with associated structures (the boxed area in figure 17.12b). This is the device that generates auditory nerve signals, so we must pay particular attention to its structural details.

The spiral organ has an epithelium composed of **hair cells** and supporting cells. Hair cells are named for the long, stiff microvilli called **stereocilia**[27] on their apical surfaces. (Stereocilia should not be confused with true cilia. They do not have an axoneme of microtubules as seen in cilia, and they do not move by themselves.) Resting on top of the stereocilia is a gelatinous **tectorial**[28] **membrane,** which helps to stimulate the hair cells when the cochlea responds to sound waves.

The spiral organ has four rows of hair cells (fig. 17.13). About 3,500 of these, called **inner hair cells (IHCs),** are arranged in a row by themselves on the medial side of the basilar membrane (facing the modiolus). Each IHC has a cluster of 50 to 60 stereocilia, arranged in order from short to tall. Another 20,000 **outer hair cells (OHCs)** are neatly arranged in three rows across from the inner hair cells. Each OHC has about 100 stereocilia arranged in a V, with their tips embedded in the tectorial membrane. All that we hear comes from the IHCs, which supply 90% to 95% of the sensory fi-

---

[26]Alfonso Corti (1822–88), Italian anatomist
[27]*stereo* = solid
[28]*tect* = roof

bers of the cochlear nerve. The function of the OHCs is to adjust the response of the cochlea to different frequencies and enable the IHCs to work with greater precision.

Hair cells are not neurons, but synapse with nerve fibers at their base. Inner hair cells synapse with the dendrites of sensory neurons only. Outer hair cells synapse with both sensory dendrites, which conduct signals to the brain, and axons that carry motor signals from the brainstem to the hair cells. The sensory neurons of the cochlea are bipolar neurons whose somas form a **spiral ganglion** coiled around the modiolus. The dendrites of these cells lead from the spiral organ to the ganglion; their axons pass from the ganglion down the core of the cochlea, then form the cochlear part of cranial nerve VIII, the vestibulocochlear nerve. This nerve carries auditory signals to the brainstem.

Auditory Function

To make functional sense of this anatomy, we will examine the essential aspects of auditory function. Figure 17.14 is a mechanical model that presents some of the basic mechanisms in simple form. When sound waves vibrate the tympanic membrane, the three auditory ossicles transfer these vibrations to the inner ear. The footplate of the stapes moves the fluid in the inner ear, and fluid movements cause the basilar membrane to vibrate up and down. As it does so, the hair cells on the basilar membrane are thrust up and down, while the tectorial membrane immediately above them remains relatively still. Each upward movement of the hair cells crowds the stereocilia against the tectorial membrane, forcing them to rock back and forth. This rocking motion opens membrane channels that admit bursts of potassium ions into the hair cells, exciting the cells. An excited hair cell releases a neurotransmitter from synaptic vesicles in its base, and the neurotransmitter excites the adjacent sensory nerve fiber. This creates a nerve signal, which we will trace in the next section.

The cochlea must generate signals that the brain can distinguish as differences in loudness and pitch. Loud sounds produce relatively

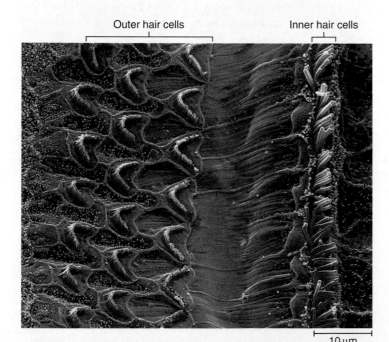

Outer hair cells   Inner hair cells

10 μm

Figure 17.13 Apical Surfaces of the Cochlear Hair Cells (SEM). All signals that we hear come from the inner hair cells on the right.

DEEPER INSIGHT   17.3

Deafness

Deafness means any hearing loss, from mild and temporary to complete and irreversible. *Conduction deafness* is a type that results from the impaired transmission of vibrations to the inner ear. It can result from a damaged tympanic membrane, otitis media, or blockage of the auditory canal. Another cause is *otosclerosis,*[29] fusion of the auditory ossicles to each other or fusion of the stapes to the oval window, preventing the bones from vibrating freely. *Sensorineural (nerve) deafness* results from the death of hair cells or any of the nervous elements concerned with hearing. It is a common occupational disease of musicians, construction workers, and others who work in noisy environments, and is part of the price many will pay for loud concerts, car stereos, and personal stereos. Deafness leads some people to develop delusions of being talked about, disparaged, or cheated. Beethoven said his deafness drove him nearly to suicide.

---

[29]*oto* = ear + *scler* = hardening + *osis* = process, condition

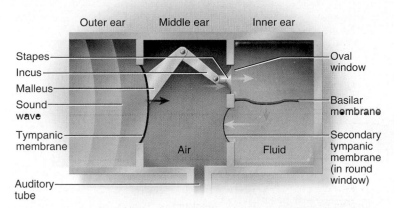

Figure 17.14 **Mechanical Model of Auditory Function.** Each inward movement of the tympanic membrane pushes inward on the auditory ossicles of the middle ear and fluid of the inner ear. The fluid pushes down on the basilar membrane, and pressure is relieved by an outward bulge of the secondary tympanic membrane. Thus the basilar membrane vibrates up and down in synchrony with the vibrations of the tympanic membrane.
● *Why would high air pressure in the middle ear reduce the movements of the basilar membrane of the inner ear?*

Labels on figure: Outer ear · Middle ear · Inner ear · Stapes · Incus · Malleus · Sound wave · Tympanic membrane · Auditory tube · Air · Fluid · Oval window · Basilar membrane · Secondary tympanic membrane (in round window)

vigorous vibrations of the basilar membrane, causing rapid firing of the cochlear nerve fibers. The nerve fibers fire more slowly in response to softer sounds. High- and low-pitched sounds vibrate different regions of the basilar membrane. High-pitched sounds especially stimulate hair cells near the base of the cochlea, and low-pitched sounds stimulate those near the tip. Thus, the brain distinguishes loudness and pitch from how fast the cochlear nerve fibers are firing and what regions of the spiral organ are generating the strongest signals.

In order to tune the cochlea and sharpen its frequency discrimination, the brainstem sends motor signals back through the cochlear nerve to the outer hair cells (OHCs). The OHCs are anchored to the basilar membrane below and anchored to the tectorial membrane through their stereocilia above. They contract in response to signals from the brain, tugging on the basilar and tectorial membranes and thus suppressing the vibration of specific regions of the basilar membrane. This enhances the ability of the brain to tell one sound frequency from another—an ability that is important for such purposes as distinguishing the words in someone else's speech.

The Auditory Projection Pathway

Sensory nerve fibers from the cochlea form the **cochlear nerve.** This joins the *vestibular nerve* (see fig. 17.11b) and the two together become the *vestibulocochlear nerve* (cranial nerve VIII). The vestibulocochlear nerve exits the internal acoustic meatus of the temporal bone, thus leaving the inner ear and entering the cranial cavity. It ends a short distance later in the *cochlear nucleus* of the medulla oblongata.

Cochlear nerve fibers synapse here with second-order neurons that lead to the nearby *superior olivary nucleus* of the pons (fig. 17.15). The superior olivary nucleus has multiple connections and functions:

- It sends signals back to the cochlea by way of cranial nerve VIII, stimulating the outer hair cells for the purpose of cochlear tuning.

- It sends signals by way of cranial nerves V_3 and VII to the tensor tympani and stapedius muscles, respectively, which are responsible for the protective tympanic reflex.

- It plays a role in *binaural*[30] *hearing*—comparing signals from the right and left ears to identify the direction from which a sound is coming.

- It issues fibers up the brainstem to the inferior colliculi of the midbrain.

The inferior colliculi aid in binaural hearing and issue fibers to the thalamus. In the thalamus, these, in turn, synapse with neurons that continue to the primary auditory cortex in the superior part of each temporal lobe. The temporal lobe is the site of conscious perception of sound; it completes the information processing essential to binaural hearing. The interpretation of sound in relation to memory—for example, the ability to recognize what a sound is—occurs in the auditory association area bordering the primary auditory cortex (see fig. 15.17, p. 420).

Extensive connections exist between the right and left nuclei of hearing throughout the brainstem, allowing for comparison of the inputs from the right and left ears and the localization of sounds in space. Thus, unlike the somatosensory cortex, the auditory cortex on each side of the brain receives signals from both ears. Because of this extensive decussation, damage to the right or left auditory cortex does not cause a unilateral loss of hearing.

The Vestibular Apparatus

The original function of the ear in vertebrate history was not hearing, but **equilibrium**—coordination, balance, and orientation in three-dimensional space. Only later did vertebrates evolve the cochlea, outer- and middle-ear structures, and auditory function of the ear. In humans, the receptors for equilibrium constitute the **vestibular apparatus,** which consists of three **semicircular ducts** and two chambers—an anterior **saccule**[31] (SAC-yule) and a posterior **utricle**[32] (YOU-trih-cul) (see fig. 17.11).

The sense of equilibrium is divided into **static equilibrium,** the perception of the orientation of the head when the body is stationary, and **dynamic equilibrium,** the perception of motion or acceleration. There are two kinds of acceleration: *linear acceleration,* a change in velocity in a straight line, as when riding in a car or elevator; and *angular acceleration,* a change in the rate of rotation, as when turning one's head. The saccule and utricle are responsible for static equilibrium and the sense of linear acceleration; the semicircular ducts detect only angular acceleration.

The Saccule and Utricle

The saccule and utricle each have a 2 by 3 mm patch of hair cells and supporting cells called a *macula.*[33] The **macula sacculi** lies nearly vertically on the wall of the saccule, and the **macula utriculi**

---

[30]*bin* = two + *aur* = ears
[31]*saccule* = little sac
[32]*utricle* = little bag
[33]*macula* = spot

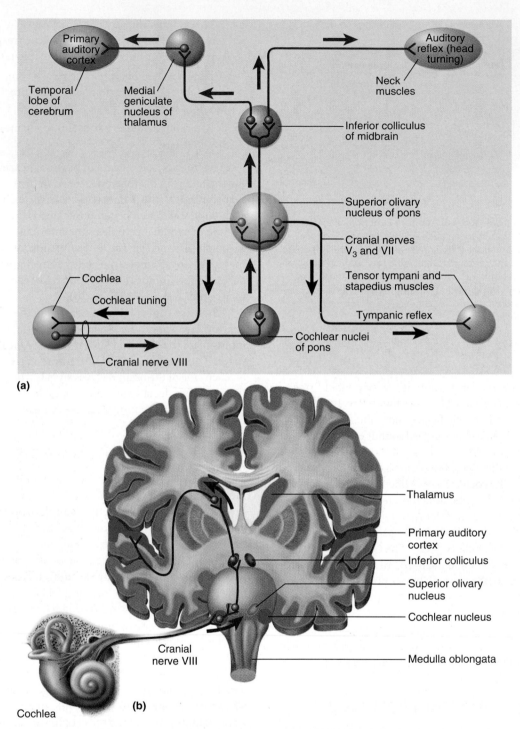

(a)

(b)

Cochlea

Figure 17.15 **Auditory Pathways in the Brain.** (a) Schematic. (b) Brainstem and frontal section of the cerebrum, showing the locations of auditory processing centers. (Cranial nerve V$_3$ = trigeminal nerve, mandibular division; CN VII = facial nerve; CN VIII = vestibulocochlear nerve)

lies nearly horizontally on the floor of the utricle (fig. 17.16a). Each hair cell of a macula has 40 to 70 stereocilia and one true cilium called a **kinocilium.**[34] The tips of the stereocilia and kinocilium are embedded in a gelatinous **otolithic membrane.** This membrane is weighted with granules called **otoliths,**[35] com-

posed of calcium carbonate and protein (fig. 17.16b). By adding to the density and inertia of the membrane, the otoliths enhance the sense of gravity and motion.

Figure 17.16c shows how the macula utriculi detects tilt of the head. With the head erect, the otolithic membrane bears directly down on the hair cells and stimulation is minimal. When the head is tilted, however, the heavy otolithic membrane sags and bends the stereocilia, stimulating the hair cells. Any orientation of the head causes a combination of stimulation to the utricles and saccules of

---

[34]*kino* = moving
[35]*oto* = ear + *lith* = stone

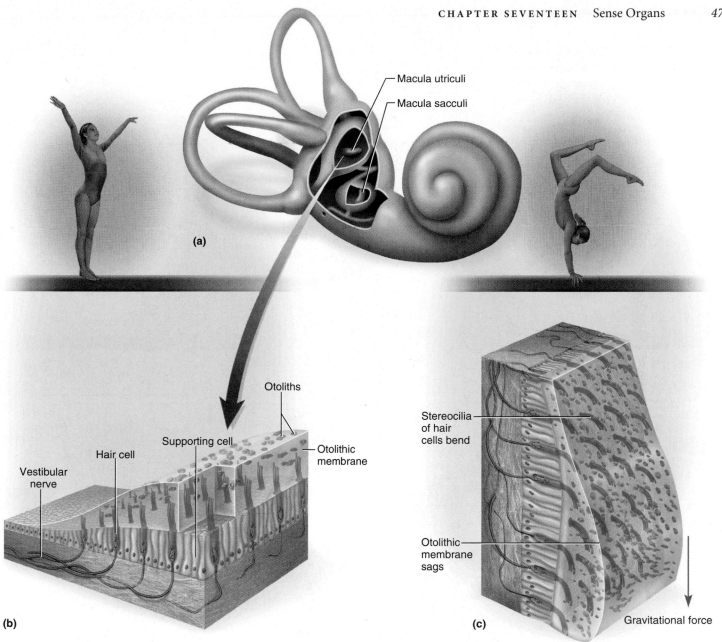

Figure 17.16 **The Saccule and Utricle.** (a) Locations of the macula sacculi and macula utriculi. (b) Structure of a macula (cutaway view). (c) Action of the otolithic membrane on the hair cells when the head is tilted.
● *Why would the macula sacculi respond more strongly than the macula utriculi to a ride in an elevator?*

the two ears. The brain interprets head orientation by comparing these inputs to each other and to other input from the eyes and stretch receptors in the neck, thereby detecting whether only the head is tilted or the entire body is tipping. The macula sacculi works similarly except that its vertical orientation makes it more responsive to up-and-down movements of the body—for example, when you stand up or jump down from a height.

The inertia of the otolithic membranes is especially important in detecting linear acceleration. Suppose you are sitting in a car at a stoplight and then begin to move. The heavy otolithic membrane of the macula utriculi briefly lags behind the rest of the tissues, bends the stereocilia backward, and stimulates the cells. When you stop at

the next intersection, the macula stops but the otolithic membrane keeps on going for a moment, bending the stereocilia forward. The hair cells convert this pattern of stimulation to nerve signals, and the brain is thus advised of changes in your linear velocity.

If you are standing in an elevator and it begins to move up, the otolithic membrane of the vertical macula sacculi lags behind briefly and pulls down on the hair cells. When the elevator stops, the otolithic membrane keeps on going for a moment and bends the hairs upward. The macula sacculi thus detects vertical acceleration. These sensations are important in such ordinary actions as sitting down, standing up, and falling, and as one's head bobs up and down during walking and running.

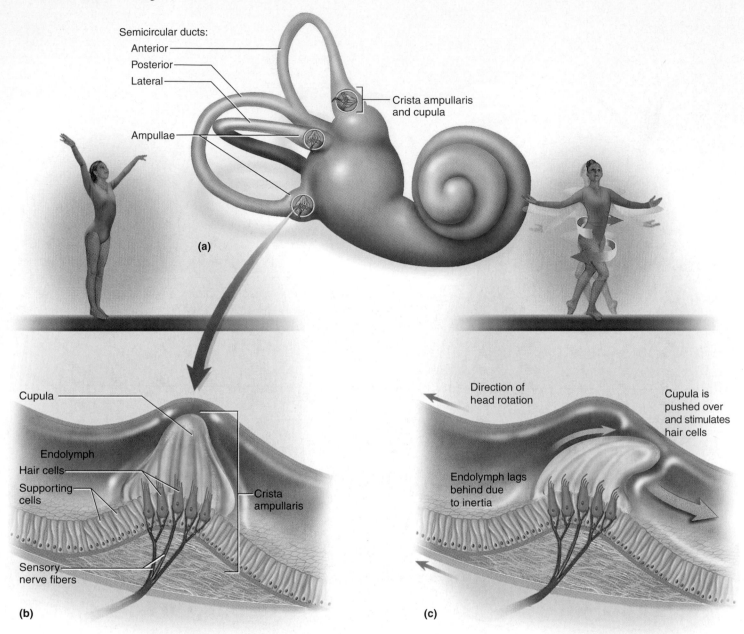

Semicircular ducts:
Anterior
Posterior
Lateral
Ampullae
Crista ampullaris and cupula

(a)

Cupula
Endolymph
Hair cells
Supporting cells
Sensory nerve fibers
Crista ampullaris

(b)

Direction of head rotation
Cupula is pushed over and stimulates hair cells
Endolymph lags behind due to inertia

(c)

Figure 17.17 **The Semicircular Ducts.** (a) Structure of the semicircular ducts, with each ampulla opened to show the crista ampullaris and cupula. (b) Detail of the crista ampullaris. (c) Action of the endolymph on the cupula and hair cells when the head is rotated.

The Semicircular Ducts

The head also experiences rotary movements, such as when you spin in a rotating chair, walk down a hall and turn a corner, bend forward to pick something up from the floor, or tilt your head laterally when you lie down in bed. Such movements are detected by the three *semicircular ducts* (fig. 17.17), each housed in an osseous *semicircular canal* of the temporal bone. The *anterior* and *posterior semicircular ducts* are positioned vertically, at right angles to each other. The *lateral semicircular duct* is about 30° from horizontal. The orientation of the ducts causes a different duct to be stimulated by rotation of the head in different planes—turning it from side to side as in gesturing "no," nodding up and down as in gesturing "yes," or tilting it from side to side as in touching your ears to your shoulders.

The semicircular ducts are filled with endolymph. Each duct opens into the utricle and has a dilated sac at one end called an **ampulla.**[36] Within the ampulla is a mound of hair cells and supporting cells called the **crista ampullaris.**[37] The hair cells have stereocilia and a kinocilium embedded in the **cupula,**[38] a gelatinous membrane that extends from the crista to the roof of the ampulla. When the head turns, the duct rotates, but the endolymph lags behind and pushes the cupula over. This bends the stereocilia and stimulates the hair cells. After 25 to 30 seconds of continual rotation, however, the endolymph catches up with the movement of the duct, and stimulation of the hair cells ceases even though motion continues.

---

[36]*ampulla* = little jar
[37]*crista* = crest, ridge + *ampullaris* = of the ampulla
[38]*cupula* = little tub

Apply What You Know

The semicircular ducts do not detect motion itself, but only acceleration—a change in the rate of motion. Explain why.

Vestibular Projection Pathways

Hair cells of the macula sacculi, macula utriculi, and semicircular ducts synapse at their bases with sensory fibers of the **vestibular nerve.** This and the cochlear nerve merge to form the vestibulocochlear nerve. Fibers of the vestibular nerve lead to a complex of four **vestibular nuclei** on each side of the pons and medulla. Nuclei on the right and left sides of the brainstem communicate extensively with each other, so each receives input from both the right and left ears. They process signals about the position and movement of the body and relay information to five targets (fig. 17.18):

1. The cerebellum, which integrates vestibular information into its control of head movements, eye movements, muscle tone, and posture.

2. Nuclei of the oculomotor, trochlear, and abducens nerves (cranial nerves III, IV, and VI). These nerves produce eye movements that compensate for movements of the head (the *vestibulo–ocular reflex*). To observe this effect, hold this book in front of you at a comfortable reading distance and fix your gaze on the middle of the page. Move the book left and right about once per second, and you will be unable to read it. Now hold the book still and shake your head from side to side at the same rate. This time you will be able to read the page because the vestibulo–ocular reflex compensates for your head movements and keeps your eyes fixed on the target. This reflex enables you to keep your vision fixed on a distant object as you walk or run toward it.

3. The reticular formation, which is thought to adjust breathing and blood circulation to changes in posture.

4. The spinal cord, where fibers descend the two vestibulospinal tracts on each side (see fig. 14.3, p. 376) and synapse on motor neurons that innervate the extensor (antigravity) muscles. This pathway allows you to make quick movements of the trunk and limbs to keep your balance.

5. The thalamus, which relays signals to two areas of the cerebral cortex. One is at the inferior end of the postcentral gyrus adjacent to sensory regions for the face. It is here that we become consciously aware of body position and movement. The other is slightly rostral to this, at the inferior end of the central sulcus in the transitional zone from primary sensory to motor cortex. This area is thought to be involved in motor control of the head and body.

Before You Go On

Answer the following questions to test your understanding of the preceding section.

11. What is the benefit of having three auditory ossicles and two muscles in the middle ear?

12. Explain how vibration of the tympanic membrane ultimately causes cochlear hair cells to release a neurotransmitter.

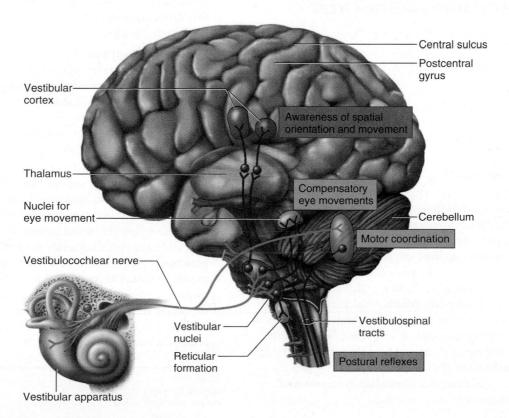

Figure 17.18 Vestibular Projection Pathways in the Brain.

13. How does the brain recognize the difference between high C and middle C of a piano? Between a loud sound and a soft one?

14. How does the function of the semicircular ducts differ from the function of the saccule and utricle?

15. How is the sensory mechanism of the semicircular ducts similar to that of the saccule and utricle?

17.4 The Eye

▶ Expected Learning Outcomes

When you have completed this section, you should be able to

- describe the anatomy of the eye and its accessory structures;
- describe the histological structure of the retina and its receptor cells;
- explain why different types of receptor cells and neural circuits are required for day and night vision; and
- trace the visual projection pathway in the brain.

Vision (sight) is the perception of objects in the environment by means of the light they emit or reflect. It employs mainly the eyes, of course, but also involves a number of accessory structures associated with the eye that protect it and aid in its function.

Accessory Structures of the Orbit

Before considering the eye itself, we will survey the accessory structures located in and around the orbit (figs. 17.19 and 17.20). These include the *eyebrows, eyelids, conjunctiva, lacrimal apparatus,* and *extrinsic eye muscles.*

- The **eyebrows** probably serve mainly to enhance facial expressions and nonverbal communication, but they may also protect the eyes from glare and help to keep perspiration from running into the eye.

- The **eyelids,** or **palpebrae** (pal-PEE-bree), close periodically to moisten the eye with tears, sweep debris from the surface, block foreign objects from the eye, and prevent visual stimuli from disturbing our sleep. The two eyelids are separated by the **palpebral fissure,** and the corners where they meet are called the **medial** and **lateral commissures (canthi).** The eyelid consists largely of the orbicularis oculi muscle covered with skin (fig. 17.20a). It also contains a supportive fibrous **tarsal plate,** which is thickened along the margin of the eyelid. Within this plate are 20 to 25 **tarsal glands** that open along the edge of the eyelid. They secrete an oil that coats the eye and reduces tear evaporation. The **eyelashes** are guard hairs that help to keep debris from the eye. Touching the eyelashes stimulates hair receptors and triggers the blink reflex.

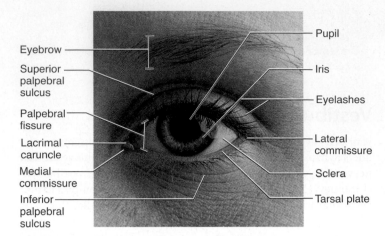

Figure 17.19 External Anatomy of the Orbital Region.

- The **conjunctiva** (CON-junk-TY-vuh) is a transparent mucous membrane that covers the inner surface of the eyelid and anterior surface of the eyeball except the clear area (cornea). The layer on the inside of the eyelid is called the *palpebral conjunctiva,* the layer on the eye surface is the *bulbar conjunctiva,* and the space between them (beneath the eyelid) is the *conjunctival sac.* The primary purpose of the conjunctiva is to secrete a thin mucous film that prevents the eyeball from drying. It is richly innervated and highly sensitive to pain. It is also very vascular (contains abundant blood vessels), which is especially evident when the vessels are dilated and the eyes are "bloodshot." Because it is vascular and the cornea is not, the conjunctiva heals more quickly than the cornea when injured.

- The **lacrimal**[39] **apparatus** (fig. 17.20b) consists of the lacrimal (tear) gland and a series of ducts that drain the tears into the nasal cavity. The **lacrimal gland,** about the size and shape of an almond, is nestled in a shallow fossa of the frontal bone in the superolateral corner of the orbit. About 12 short ducts lead from the gland to the surface of the conjunctiva. Tears function to cleanse and lubricate the eye surface and to deliver oxygen and nutrients to the conjunctiva. They also contain a bactericidal enzyme, *lysozyme,* that helps to prevent eye infections. After washing across the conjunctiva, the tears collect at the **lacrimal caruncle**[40] (CAR-un-cul), the pink fleshy mass near the medial commissure of the eye. Near the caruncle, each eyelid has a tiny pore called a lacrimal **punctum,**[41] which collects the tears and conveys them through a short **lacrimal canal** into a **lacrimal sac** in the medial wall of the orbit. From here, a **nasolacrimal duct** carries the tears to the inferior meatus of the nasal cavity—thus, an abundance of tears from crying or "watery eyes" can result in a runny nose. Normally, the tears are swallowed.

---

[39]*lacrim* = tear
[40]*car* = fleshy mass + *uncle* = little
[41]*punct* = point

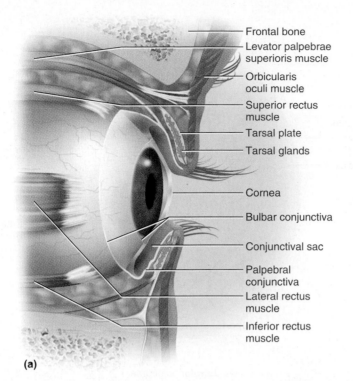

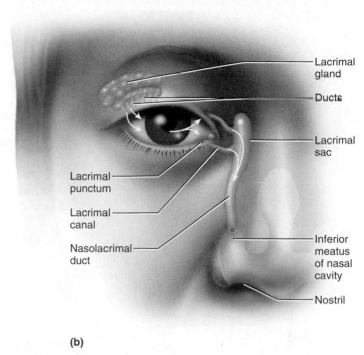

Figure 17.20 Accessory Structures of the Orbit. (a) Sagittal section of the eye and orbit. (b) The lacrimal apparatus. The arrows indicate the flow of tears from the lacrimal gland, across the front of the eye, into the lacrimal sac, and down the nasolacrimal duct.
● *What would be the effect of blockage of the lacrimal puncta?*

When you have a cold, the nasolacrimal ducts become swollen and obstructed, the tears cannot drain, and they may overflow from the brim of your eye.

- The **extrinsic eye muscles** are six muscles attached to the walls of the orbit and to the external surface of each eyeball. *Extrinsic* means arising from without, and distinguishes these from the *intrinsic* muscles inside the eyeball, discussed later. The extrinsic muscles are responsible for movements of the eye (fig. 17.21). They include four *rectus* ("straight") muscles and two *oblique* muscles. The **superior, inferior, medial,** and **lateral rectus** originate on the posterior wall of the orbit and insert on the anterior region of the eyeball, just beyond the visible "white of the eye." They move the eye up, down, medially, and laterally. The **superior oblique** travels along the medial wall of the orbit. Its tendon passes through a fibrocartilage loop, the **trochlea**[42] (TROCK-lee-uh), and inserts on the superolateral aspect of the eyeball. The **inferior oblique** extends from the medial wall of the orbit to the inferolateral aspect of the eye. To visualize the function of the oblique muscles, suppose you turn your eyes to the right. The superior oblique muscle slightly depresses your right eye, while the inferior oblique slightly elevates the left eye. The opposite occurs when you look to the left. This is the primary function of the oblique muscles, but they also rotate the eyes, turning the "twelve o'clock pole" of each eye slightly toward or away from the nose. The superior oblique muscle is

innervated by the trochlear nerve (IV), the lateral rectus muscle by the abducens nerve (VI), and the rest of these muscles by the oculomotor nerve (III).

The eye is surrounded on the sides and back by **orbital fat.** It cushions the eye, gives it freedom of motion, and protects blood vessels and nerves as they pass through the rear of the orbit.

Anatomy of the Eyeball

The eyeball itself is a sphere about 24 mm in diameter (fig. 17.22) with three principal components: (1) three layers (tunics) that form its wall; (2) optical components that admit and focus light; and (3) neural components, the retina and optic nerve. The retina is not only a neural component but also part of the inner tunic. The cornea is part of the outer tunic as well as one of the optical components.

The Tunics

There are three tunics forming the wall of the eyeball:

- The outer **fibrous layer (tunica fibrosa)** is divided into two regions, the sclera and cornea. The **sclera**[43] (white of the eye) covers most of the eye surface and consists of dense collagenous connective tissue perforated by blood vessels and nerves. The **cornea** is the anterior transparent region of modified sclera that admits light into the eye.

[42]*trochlea* = pulley

[43]*scler* = hard, tough

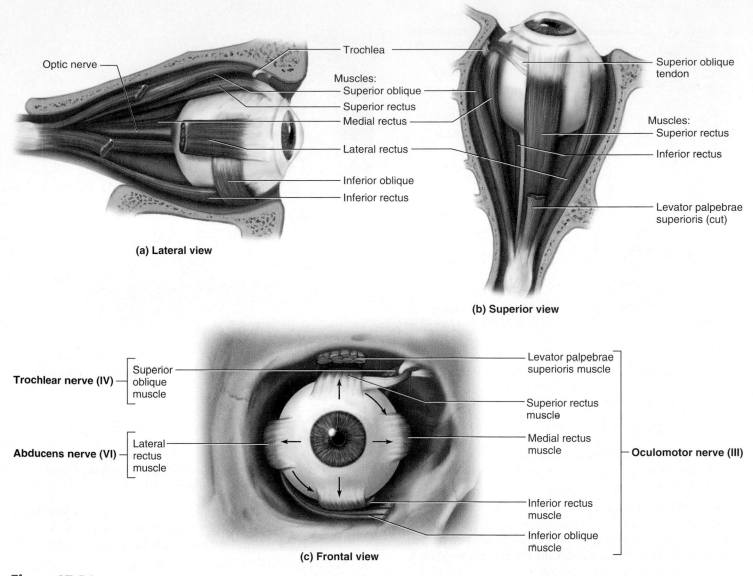

(a) Lateral view

Optic nerve

Trochlea

Muscles:
Superior oblique
Superior rectus
Medial rectus

Lateral rectus

Inferior oblique

Inferior rectus

(b) Superior view

Superior oblique tendon

Muscles:
Superior rectus

Inferior rectus

Levator palpebrae superioris (cut)

(c) Frontal view

Trochlear nerve (IV) — Superior oblique muscle

Abducens nerve (VI) — Lateral rectus muscle

Levator palpebrae superioris muscle

Superior rectus muscle

Medial rectus muscle

Inferior rectus muscle

Inferior oblique muscle

Oculomotor nerve (III)

Figure 17.21 Extrinsic Muscles of the Eye. (a) Lateral view of the right eye. The lateral rectus muscle is cut to expose a portion of the optic nerve. (b) Superior view of the right eye. (c) Innervation of the extrinsic muscles; arrows indicate the eye movement produced by each muscle.
● *Which would cause the greatest loss of visual function—trauma to cranial nerve III, IV, or VI? Why?*

• The middle **vascular layer (tunica vasculosa)** is also called the **uvea**[44] (YOU-vee-uh) because it resembles a peeled grape in fresh dissection. It consists of three regions—the choroid, ciliary body, and iris. The **choroid** (CO-royd) is a highly vascular, deeply pigmented layer of tissue behind the retina. It gets its name from a histological resemblance to the chorion of a fetus. Anteriorly, it thickens and grades into the **ciliary body,** which supports the lens and forms a muscular ring around it. The smooth muscle in the ciliary body, called the **ciliary muscle,** controls tension on the lens and is thus important in focusing it. The ciliary body also supports the iris and secretes a fluid called the aqueous humor. The **iris** is an adjustable diaphragm that controls the diameter of the **pupil,** its central opening. The iris has two pigmented layers.

One is a posterior *pigment epithelium* that blocks stray light from reaching the retina. The other is the *anterior border layer,* which contains pigmented cells called **chromatophores.**[45] High concentrations of melanin in the chromatophores give the iris a black, brown, or hazel color. If the melanin is scanty, light reflects from the posterior pigment epithelium and gives the iris a blue, green, or gray color.

The diameter of the pupil is controlled by two sets of contractile elements in the iris. The **pupillary constrictor** consists of concentric circles of smooth muscle cells around the pupil. The **pupillary dilator** consists of a spokelike arrangement of contractile processes that extend from the cells of the posterior pigment epithelium. Because of their dual epithelial and contractile nature, these cells are classified as

---

[44]*uvea* = grape

[45]*chromato* = color + *phore* = bearer

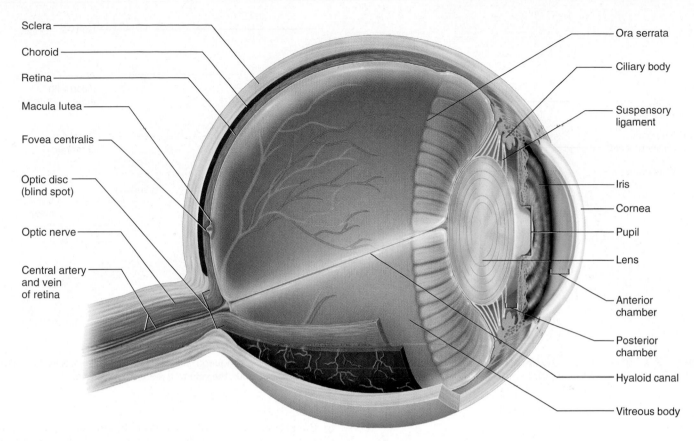

Figure 17.22 The Eye (Sagittal Section).

myoepithelial cells. When their processes contract, they dilate the pupil and admit up to five times as much light as a fully constricted pupil. The pupils constrict in response to increased light intensity and when we focus on nearby objects; they dilate in dimmer light and when we focus on more distant objects. Their response to light is called the *photopupillary reflex.*

The ciliary muscle, pupillary constrictor, and pupillary dilator are considered to be the **intrinsic eye muscles,** in contrast to the aforementioned extrinsic muscles found on the outside of the eyeball. Both sympathetic and parasympathetic nerve fibers enter the rear of the eyeball and innervate these muscles (see figs. 16.4 and 16.7, pp. 447 and 451). Sympathetic fibers from the superior cervical ganglion innervate the pupillary dilator and widen the pupil. Parasympathetic fibers reach the eye through the oculomotor nerve (cranial nerve III), innervate the pupillary constrictor, and narrow the pupil, thus antagonizing the sympathetic effect (see fig. 16.9, p. 455). The parasympathetic division also thickens the lens for near vision.

- The **inner layer (tunica interna)** consists of the *retina,* which internally lines the posterior two-thirds of the eyeball.

Optical Components

The optical components of the eye are transparent elements that admit light rays, bend (refract) them, and focus images on the retina. They include the *cornea, aqueous humor, lens,* and *vitreous body.* The cornea has been described already.

- The **aqueous humor** is a serous fluid secreted by the ciliary processes into a space between the iris and lens called the **posterior chamber** (fig. 17.23). It flows through the pupil into the **anterior chamber,** a space between the cornea and iris. From here, it is reabsorbed by a ringlike vessel called the **scleral venous sinus (canal of Schlemm[46]).** Normally, the rate of reabsorption balances the rate of secretion. (See Deeper Insight 17.4 for an important exception.)

- The **lens** is composed of flattened, tightly compressed, transparent cells called **lens fibers.** It is suspended behind the pupil by a fibrous ring called the **suspensory ligament** (figs. 17.22 and 17.24), which attaches it to the ciliary body. Tension on the ligament somewhat flattens the lens, so it is about 9.0 mm in diameter and 3.6 mm thick at the middle. When the lens is removed from the eye and not under tension, it relaxes into a more spheroidal shape and resembles a plastic bead.

- The **vitreous[47] body (vitreous humor)** is a transparent jelly that fills a large space called the **vitreous chamber** behind the lens. An oblique channel through this body, called the *hyaloid canal,* is the remnant of a *hyaloid artery* present in the embryo (see fig. 17.22). The vitreous body serves to maintain the spherical shape of the eyeball and to keep the retina smoothly pressed against the inner surface of the chamber. This is essential for the focusing of images on the retina.

---

[46]Friedrich S. Schlemm (1795–1858), German anatomist
[47]*vitre* = glassy

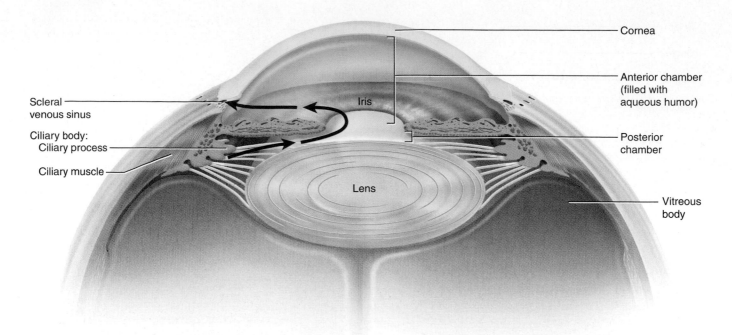

Figure 17.23 **Production and Reabsorption of Aqueous Humor.** Blue arrows indicate the flow of aqueous humor from the ciliary processes into the posterior chamber, through the pupil into the anterior chamber, and finally into the scleral venous sinus, the vein that reabsorbs the fluid.

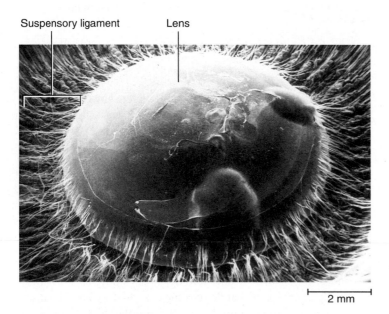

Figure 17.24 **Lens of the Eye (SEM).** Posterior view of the lens and the suspensory ligament that anchors it to the ciliary body.

Neural Components

The neural components of the eye are the retina and optic nerve. The retina is a thin transparent membrane attached at only two points—the **optic disc** where the optic nerve leaves the rear (fundus) of the eye, and its scalloped anterior margin, the **ora serrata.**[48] The rest of the retina is held smoothly against the rear of the eyeball

by the pressure of the vitreous body. It can detach (buckle away from the wall of the eyeball) because of blows to the head or insufficient pressure from the vitreous body. A *detached retina* may cause blurry areas in the field of vision. It can lead to blindness if the retina remains separated for too long from the choroid, on which it depends for oxygen, nutrition, and waste removal.

The inside rear of the eyeball, called the **fundus,** is routinely examined with an illuminating and magnifying instrument called an *ophthalmoscope.* Directly posterior to the center of the lens, on the visual axis of the eye, is a patch of cells called the **macula lutea,**[49] about 3 mm in diameter (fig. 17.25). In the center of the macula is a tiny pit, the **fovea**[50] **centralis.** It produces the most finely detailed images for a reason that will be apparent later. About 3 mm medial to the macula lutea is the optic disc. Nerve fibers converge on this point from neurons throughout the retina and leave here in a bundle that constitutes the optic nerve. Blood vessels travel through the optic nerve and enter and leave the eye at the optic disc. Eye examinations thus serve for more than evaluating the visual system; they allow for a direct, noninvasive examination of blood vessels for signs of hypertension, diabetes mellitus, atherosclerosis, and other vascular diseases.

The optic disc contains no receptor cells, so it produces a **blind spot** in the visual field of each eye. To see this effect, close your right eye and gaze straight ahead with your left; fixate on an object across the room. Now hold up a pencil about 30 cm (1 ft) from your face at eye level. Begin moving the pencil toward the left, but be sure you keep your gaze fixed on that point across the room. When the pencil is about 15° away from your line of vision, the end of it will disappear because its image falls on the blind spot of your

---

[48]*ora* = border, margin + *serrata* = notched, serrated

[49]*macula* = spot + *lutea* = yellow
[50]*fovea* = pit, depression

DEEPER INSIGHT 17.4

DEEPER INSIGHT 17.4

Cataracts and Glaucoma

The two most common causes of blindness are cataracts and glaucoma. A *cataract* is clouding of the lens. It occurs as the lens fibers darken with age, fluid-filled bubbles (vacuoles) and clefts appear between the lens fibers, and the clefts accumulate debris from degenerating fibers. Cataracts are a common complication of diabetes mellitus, but can also be induced by heavy smoking, ultraviolet radiation, radiation therapy, certain viruses and drugs, and other causes. They cause the vision to appear milky or as if one were looking from behind a waterfall.[51] Cataracts can be treated by replacing the natural lens with a plastic one. The implanted lens improves vision almost immediately, but glasses still may be needed for near vision.

Glaucoma is a state of elevated pressure within the eye that occurs when the scleral venous sinus is obstructed so aqueous humor is not reabsorbed as fast as it is secreted. Pressure in the anterior and posterior chambers drives the lens back and puts pressure on the vitreous body. The vitreous body presses the retina against the choroid and compresses the blood vessels that nourish the retina. Without a good blood supply, retinal cells die and the optic nerve may atrophy, producing blindness. Symptoms often go unnoticed until the damage is irreversible. Illusory flashes of light are an early symptom of glaucoma. Late-stage symptoms include dimness of vision,[52] a narrowed visual field, and colored halos around artificial lights. Glaucoma can be halted with drugs or surgery, but lost vision cannot be restored. This disease can be detected at an early stage in the course of regular eye examinations. The field of vision is checked, the optic nerve is examined, and the intraocular pressure is measured with an instrument called a *tonometer*.

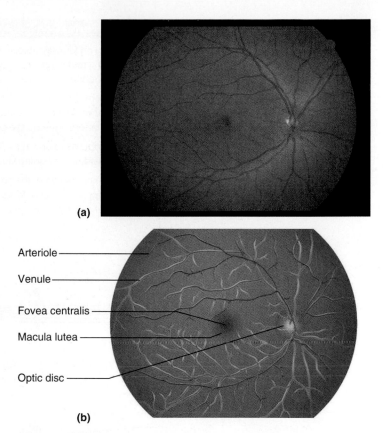

(a)

(b)

Figure 17.25 Fundus (Rear) of the Eye. (a) As seen with an ophthalmoscope. (b) Anatomical features of the fundus. Note the blood vessels diverging from the optic disc, where they enter the eye with the optic nerve. An eye examination also serves as a partial check on cardiovascular health.

[Labels: Arteriole, Venule, Fovea centralis, Macula lutea, Optic disc]

---

left eye. The reason you do not normally notice a blind patch in your visual field is that the brain uses the image surrounding the blind spot to fill in that area with similar, essentially "imaginary" information.

Formation of an Image

The visual process begins when light rays enter the eye, become focused on the retina, and produce a tiny inverted image. The cornea refracts incoming light rays toward the optical axis of the eye, and the lens makes relatively slight adjustments to fine focus the image. When you focus on something more than 6 m (20 ft) away, the lens flattens to a thickness of about 3.6 mm at the center and refracts light less. When you focus on something closer than 6 m, the lens thickens to about 4.5 mm at the center and refracts light rays more strongly. These changes in the lens are called *accommodation*. Abnormalities in lens flexibility, the shape of the cornea, or the length of the eyeball result in various deficiencies of vision explained in table 17.2 and figure 17.26.

---

[51]*cataract* = waterfall
[52]*glauco* = grayness

Structure and Function of the Retina

The conversion of light energy into action potentials occurs in the retina. Its cellular organization is shown in figure 17.27. The most posterior layer is the **pigment epithelium,** a darkly pigmented layer that serves, like the black inside of a camera, to absorb stray light so it does not degrade the visual image by reflecting back into the eye.

Apply What You Know

The vertebrate eye is often called a *camera eye* for its many resemblances to the mechanisms of a film camera. List as many comparisons as you can.

The neural apparatus of the retina consists of three principal cell layers. Progressing from the rear of the eye forward, the major retinal cells are the *photoreceptors* (mainly *rods* and *cones*), *bipolar cells,* and *ganglion cells.*

1. **Photoreceptor cells.** The photoreceptor cells include all cells that absorb light and generate a chemical or electrical signal. There are three kinds: rods, cones, and some of the ganglion cells. Only the rods and cones produce visual images; the

TABLE 17.2	Common Defects of Image Formation
Myopia[53]	Nearsightedness—difficulty focusing on faraway objects because of an abnormally elongated eyeball. Light rays come into focus before they reach the retina and begin to diverge again by the time they fall on it. Corrected with concave lenses, which cause light rays to diverge slightly before entering the eye.
Hyperopia[54]	Farsightedness—difficulty focusing on nearby objects because of an abnormally short eyeball. The retina lies in front of the focal point of the lens, and the light rays have not yet come into focus when they reach the retina. Corrected with convex lenses, which cause light rays to converge slightly before entering the eye.
Presbyopia[55]	Declining ability to focus on nearby objects as one ages. An effect of declining elasticity of the aging lens, often first noticed around age 40 to 45. Results in difficulty in reading and doing close handwork. Corrected with reading glasses or bifocal lenses.
Astigmatism[56]	Inability to simultaneously focus light rays that enter the eye on different planes. Focusing on vertical lines, such as the edge of a door, may cause horizontal lines, such as a tabletop, to go out of focus. Caused by a deviation in the shape of the cornea so that it is shaped like the back of a spoon rather than like part of a sphere. Corrected with "cylindrical" lenses, which refract light more in one plane than another.

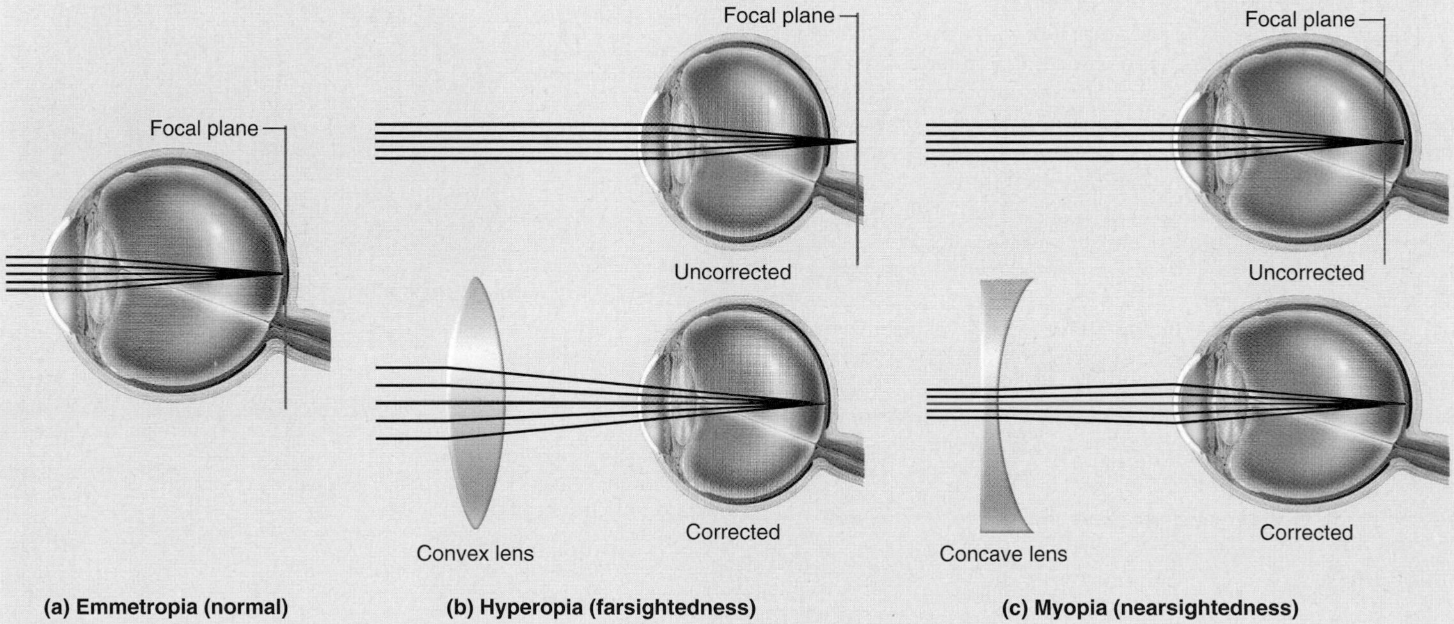

(a) Emmetropia (normal)　**(b) Hyperopia (farsightedness)**　**(c) Myopia (nearsightedness)**

Figure 17.26 **Two Common Visual Defects and the Effects of Corrective Lenses.** (a) The normal emmetropic eye, with light rays converging on the retina. (b) Hyperopia (farsightedness) and the corrective effect of a convex lens. The lens causes light rays to begin converging before they enter the eye, so they reach their focal point farther forward than usual, on the retina of the shortened eyeball. (c) Myopia (nearsightedness) and the corrective effect of a concave lens. By causing light rays to diverge before they enter the eye, this lens shifts the focal point posteriorly so that it falls on the retina of the elongated eye.

ganglion cells are discussed shortly. Each rod or cone has an **outer segment** that points toward the wall of the eye and an **inner segment** facing the interior (fig. 17.28). The two segments are separated by a narrow constriction containing nine pairs of microtubules; the outer segment is actually a highly modified cilium specialized to absorb light. The inner segment contains mitochondria and other organelles. At its base, it gives rise to a cell body, which contains the nucleus, and to processes that synapse with retinal neurons in the next layer.

In a **rod,** the outer segment is cylindrical and somewhat resembles a stack of coins in a paper roll—there is a plasma membrane around the outside and an orderly stack of about 1,000 membranous discs inside. Each disc is densely studded with globular proteins—the visual pigment *rhodopsin*. The membranes hold these pigment molecules in a position that results in the most efficient light absorption. Rod cells are responsible for **night (scotopic[57]) vision** and produce images only in shades of gray *(monochromatic vision)*. Even in ordinary indoor lighting, they are *saturated* (overstimulated) and nonfunctional.

A **cone** is similar except that the outer segment tapers to a point and the discs are not detached from the plasma membrane but are parallel infoldings of it. Cones begin to respond in light as dim as starlight and are the only receptor cells functional in daylight intensities; thus, they are entirely responsible for our **day (photopic[58]) vision.** They

[53]*my* = near + *opia* = eye condition
[54]*hyper* = beyond + *opia* = eye condition
[55]*presby* = old age + *opia* = eye condition
[56]*a* = without + *stigma* = point + *ism* = condition

[57]*scot* = dark + *op* = vision
[58]*phot* = light + *op* = vision

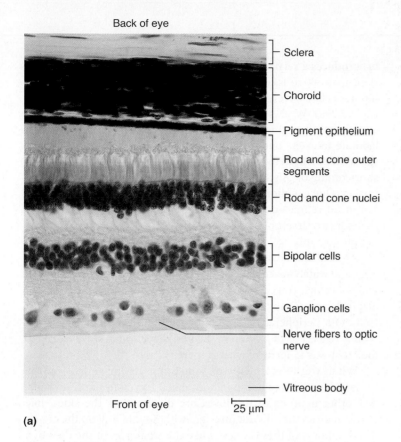

Back of eye

Sclera

Choroid

Pigment epithelium

Rod and cone outer segments

Rod and cone nuclei

Bipolar cells

Ganglion cells

Nerve fibers to optic nerve

Vitreous body

Front of eye

25 μm

(a)

Figure 17.27 **Histology of the Retina.** (a) Photomicrograph. (b) Schematic of the layers and circuitry of the retinal cells.

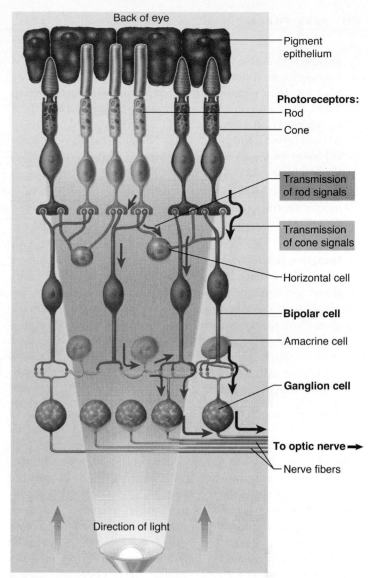

Back of eye

Pigment epithelium

Photoreceptors:
Rod

Cone

Transmission of rod signals

Transmission of cone signals

Horizontal cell

Bipolar cell

Amacrine cell

Ganglion cell

To optic nerve →

Nerve fibers

Direction of light

(b)

Figure 17.28 **Rods and Cones.** (a) Rods and cones of a salamander retina (SEM). The tall cylindrical cells are rods, and the short tapered cells in the foreground are cones. (b) Structure of human rods and cones.

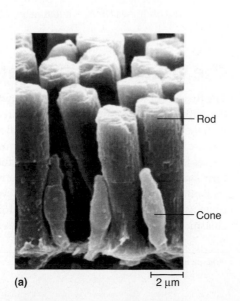

Rod

Cone

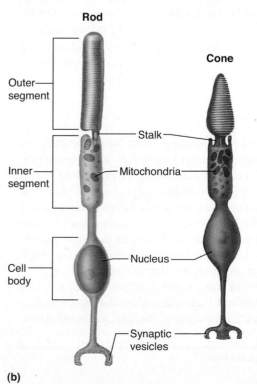

Outer segment

Inner segment

Cell body

Stalk

Mitochondria

Nucleus

Synaptic vesicles

(a)

2 μm

(b)

487

are also responsible for **color vision** *(trichromatic vision),* because unlike rods, cones do not all carry the same visual pigment. Their pigments are called *photopsins.* Some cones have a photopsin that responds best at a wavelength of 420 nanometers (nm), a deep blue light; others respond best at 531 nm (green); and still others at 558 nm (orange-yellow). All colors we see are the result of various mixtures of input to the brain from these three types of cones.

2. **Bipolar cells.** Rods and cones synapse with the dendrites of bipolar neurons, the first-order neurons of the visual pathway. These, in turn, feed directly or indirectly into the ganglion cells described next (fig. 17.27b).

3. **Ganglion cells.** These are the largest neurons of the retina, arranged in a single layer close to the vitreous body. They are the second-order neurons of the visual pathway. Most ganglion cells receive input from multiple bipolar cells. Their axons form the optic nerve. Some ganglion cells absorb light directly and conduct signals to brainstem nuclei that control pupillary diameter and the body's circadian rhythms. They do not contribute to visual images but detect only light intensity. Their sensory pigment is called *melanopsin.*

There are other retinal cells, but they do not form layers of their own. **Horizontal cells** and **amacrine**[59] (AM-ah-crin) **cells** form horizontal connections among rod, cone, and bipolar cells. Bipolar cells that carry rod signals do not synapse directly with ganglion cells, but only by way of amacrine cells. Horizontal and amacrine cells play diverse roles in enhancing the perception of contrast, the edges of objects, and changes in light intensity. In addition, much of the mass of the retina is composed of astrocytes and other types of glial cells.

There are approximately 130 million rods and 6.5 million cones in one retina, but only about 1 million nerve fibers in the optic nerve. With a ratio of 136 receptor cells to 1 optic nerve fiber, it is obvious that there must be substantial *neural convergence* and information processing in the retina itself before signals are conducted to the brain proper. Convergence begins where multiple rod or cone cells converge on one bipolar cell, and occurs again where multiple bipolar cells converge on a single ganglion cell.

The Dual Visual System

You may wonder why we have two types of photoreceptor cells, the rods and cones. Why can't we simply have one type that would produce detailed color vision, both day and night? The answer to this lies largely in the concept of neural convergence (see p. 364). The **duplicity theory** of vision holds that a single type of receptor cell cannot produce both high sensitivity and high resolution. It takes one type of cell and neural circuit, working at its maximum capacity, to provide sensitive night vision and a different type of receptor and circuit to provide high-resolution daytime vision.

The high sensitivity of rods in dim light stems partly from a cascade of chemical reactions that occurs when rhodopsin absorbs light. The cascade amplifies the effect of the light, so a small stimu-

lus produces a relatively large output from each rod. But in addition, up to 600 rods converge on each bipolar cell, and then multiple bipolar cells converge (via amacrine cells) on each ganglion cell (fig. 17.29a). Weak stimulation of many rod cells can produce an additive effect on one bipolar cell, and several bipolar cells can collaborate to excite one ganglion cell. Thus, a ganglion cell can respond in dim light that only weakly stimulates an individual rod. A shortcoming of this system is that it cannot resolve finely detailed images. One ganglion cell receives input from all the rods in about 1 mm^2 of retina—its receptive field. What the brain perceives is therefore a coarse, grainy image similar to an overenlarged newspaper photograph.

Around the edges of the retina, receptor cells are especially large and widely spaced. If you fixate on the middle of this page, you will notice that you cannot read the words near the margins. Visual acuity decreases rapidly as the image falls away from the fovea centralis. Our peripheral vision is a low-resolution system that serves mainly to alert us to motion in the periphery and to stimulate us to look that way to identify what is there.

When you look directly at something, its image falls on the fovea, which is occupied by about 4,000 tiny cones and no rods. The other neurons of the fovea are displaced to the sides, like parted hair, so they do not interfere with light falling on the cones. The smallness of these cones is like the smallness of the dots in a high-quality photograph; it is partially responsible for the high-resolution images formed at the fovea. In addition, the cones here show no neural convergence. Each cone synapses with only one bipolar cell and each bipolar cell with only one ganglion cell. This gives each foveal cone a "private line" to the brain, and each ganglion cell of the fovea reports to the brain on a receptive field of just 2 μm^2 of retinal area (fig. 17.29b). Cones distant from the fovea exhibit some neural convergence, but not nearly as much as rods do. The price of this lack of convergence at the fovea, however, is that cone cells cannot have additive effects on the ganglion cells, and the cone system therefore is less sensitive to light (requires brighter light to function).

Apply What You Know

If you look directly at a dim star in the night sky, it disappears, and if you look slightly away from it, it reappears. Why?

The Visual Projection Pathway

The first-order neurons in the visual pathway are the bipolar cells of the retina. They synapse directly or indirectly (via amacrine cells) with the second-order neurons, the retinal ganglion cells, whose axons are the fibers of the optic nerve (cranial nerve II). The optic nerves leave each orbit through the optic foramen, then converge to form an X, the **optic chiasm**[60] (KY-az-um), inferior to the hypothalamus and anterior to the pituitary. Beyond this, the fibers continue as a pair of **optic tracts** (see fig. 15.26, p. 429). Within the chiasm, half the fibers of each optic nerve cross over to the opposite

---

[59]*a* = without + *macr* = long + *in* = fiber

[60]*chiasm* = cross, X

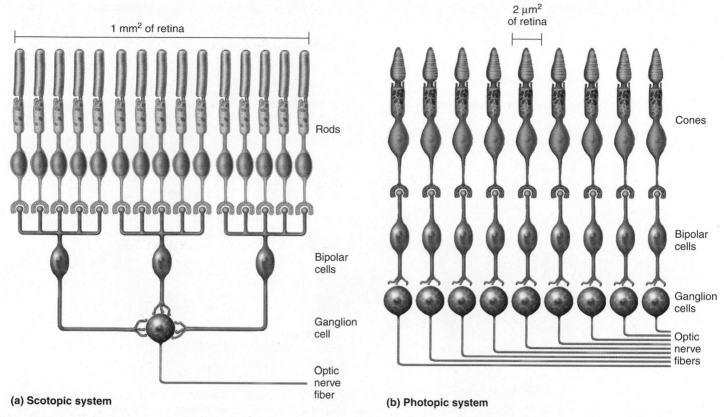

1 mm² of retina

Rods

Bipolar cells

Ganglion cell

Optic nerve fiber

(a) Scotopic system

2 μm² of retina

Cones

Bipolar cells

Ganglion cells

Optic nerve fibers

(b) Photopic system

Figure 17.29 Retinal Circuitry and Visual Sensitivity. (a) In the scotopic (night vision) system, many rod cells converge on each bipolar cell, and many bipolar cells converge on each ganglion cell. This allows the rods to combine their effects and stimulate the ganglion cell, generating a nerve signal even in dim light. However, it means that each ganglion cell (and its optic nerve fiber) represents a relatively large area of retina and produces a grainy image. (b) In the photopic (day vision) system, there is little neural convergence. In the fovea, represented here, each cone cell has a "private line" to the brain, so each optic nerve fiber represents a tiny area of retina, and vision is relatively sharp. However, the lack of convergence means that photopic vision cannot function very well in dim light because weakly stimulated cones cannot collaborate to stimulate a ganglion cell.

side of the brain (fig. 17.30). This is called **hemidecussation**,[61] since only half of the fibers decussate. As a result, the right cerebral hemisphere sees objects in the left visual field, because their images fall on the right half of each retina (the medial half of the left eye and lateral half of the right eye). You can trace the nerve fibers from each half of the retina in figure 17.30 to see that they lead to the right hemisphere. Conversely, the left hemisphere sees objects in the right visual field. Since the right brain controls motor responses on the left side of the body, and vice versa, each side of the brain sees what is on the side of the body where it exerts motor control. The visual fields of the two eyes and hemispheres overlap in the middle.

The optic tracts pass laterally around the hypothalamus, and most of their axons end in the **lateral geniculate**[62] (jeh-NIC-you-late) **nucleus** of the thalamus. Third-order neurons arise here and form the **optic radiation** of fibers in the white matter of the cerebrum. These fibers project to the primary visual cortex of the occipital lobe, where the conscious perception of an image occurs. A

stroke that destroys occipital lobe tissue can cause blindness even if the eyes are fully functional. Association tracts connect the primary visual cortex to the visual association area just anterior to it, where this sensory input is interpreted.

Optic nerve fibers from the photosensitive ganglion cells take a different route; they project to the midbrain and terminate in the superior colliculi and adjacent *pretectal nuclei*. The superior colliculi control the visual reflexes of the extrinsic muscles, and the pretectal nuclei are involved in the photopupillary and accommodation reflexes of the intrinsic muscles. Thus, these retinal ganglion cells (about 1% to 2% of the total) are not engaged in producing the images we see, but only in providing input to evoke these somatic and autonomic reflexes of the eyes.

The mechanisms of visual information processing in the brain are very complex and beyond the scope of this book. Some processing, such as contrast, brightness, motion, and stereoscopic vision, begins in the retina. The primary visual cortex in the occipital lobe is connected by association tracts to nearby visual association areas in the posterior part of the parietal lobe and inferior part of the temporal lobe. These association areas process visual data to extract information about the location, motion, color, shape, boundaries,

[61]*hemi* = half + *decuss* = to cross, form an X
[62]*geniculate* = bent like a knee

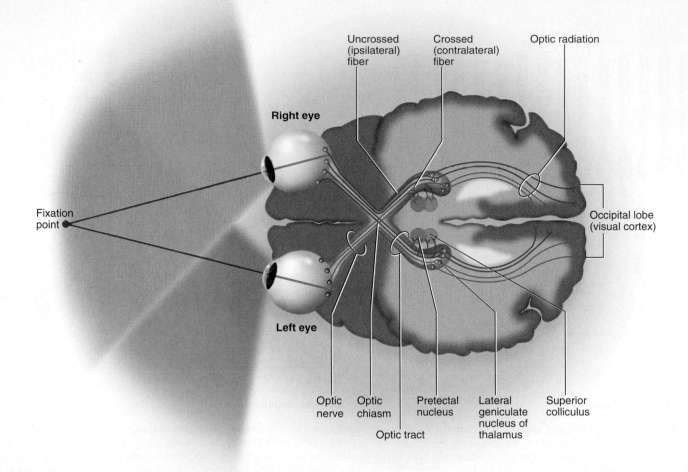

Figure 17.30 **The Visual Projection Pathway.** Note that the hemidecussation at the optic chiasm results in the occipital lobe of each cerebral hemisphere receiving input from both eyes.

and other qualities of the objects we observe. They also store visual memories and enable the brain to identify what we are seeing—for example, to recognize printed words or name the objects we see. What is yet to be learned about visual processing promises to have important implications for biology, medicine, psychology, and even philosophy.

Before You Go On

Answer the following questions to test your understanding of the preceding section.

16. List the optical components of the eye and state the role of each one in the formation of an image.

17. List as many structural and functional differences between rods and cones as you can.

18. Trace the signal pathway from the point where a retinal cell absorbs light to the point where a third-order nerve fiber ends in the occipital lobe.

19. Discuss the duplicity theory of vision, summarizing the advantage of having two types of retinal photoreceptor cells.

17.5 Developmental and Clinical Perspectives

▶ Expected Learning Outcomes

When you have completed this section, you should be able to

• describe the prenatal development of the major features of the eye and ear; and

• describe some disorders of taste, vision, hearing, equilibrium, and somatosensory function.

Development of the Ear

The ear develops from the first pharyngeal pouch (see p. 90) and adjacent tissues of the first and second pharyngeal arches. The first section to develop is the inner ear. Its earliest trace is a thickening

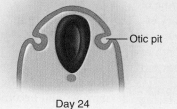
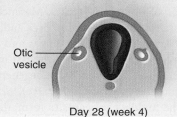
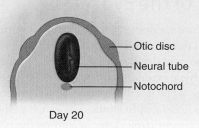

Otic disc
Neural tube
Notochord

Day 20

Otic pit

Day 24

Otic vesicle

Day 28 (week 4)

(a) Development of the otic vesicle

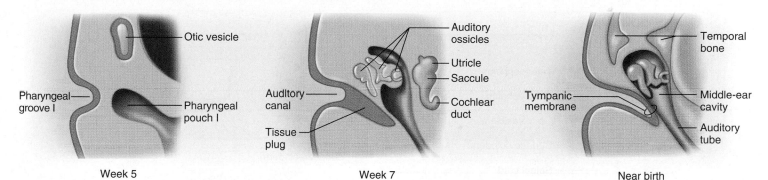

Utricle

Saccule

Week 4

Semicircular ducts

Cochlear duct

Week 5

Semicircular ducts

Cochlear duct

Week 7

(b) Development of the membranous labyrinth from the otic vesicle

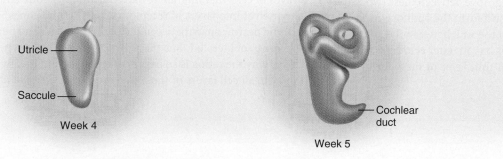

Otic vesicle

Pharyngeal groove I

Pharyngeal pouch I

Week 5

Auditory ossicles

Auditory canal

Tissue plug

Utricle

Saccule

Cochlear duct

Week 7

Temporal bone

Tympanic membrane

Middle-ear cavity

Auditory tube

Near birth

(c) Development of the auditory canal and middle ear

Figure 17.31 **Development of the Ear.** See text for explanation. In part (d), the auricular hillocks are numbered to show what parts of the fully developed auricle arise from each one.

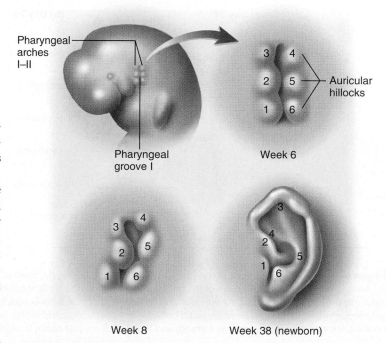

Pharyngeal arches I–II

Pharyngeal groove I

Auricular hillocks

Week 6

Week 8

Week 38 (newborn)

(d) Development of the auricle

called the **otic placode**[63] **(otic disc),** which develops in the ectoderm near the hindbrain in week 3. The placode invaginates, forming an **otic pit** and then a fully enclosed **otic vesicle,** which detaches from the overlying ectoderm by the end of week 4 (fig. 17.31a).

The otic vesicle differentiates into two chambers, the utricle and saccule. In week 5, the lower tip of the saccule elongates and begins to form the cochlear duct; soon after, three pouches grow from the utricle and begin to form the semicircular ducts (fig. 17.31b). These structures are still embedded in mesenchyme, which in week 9 chondrifies and forms a cartilaginous **otic capsule** enclosing the inner-ear structures. The capsule later ossifies to form the petrous part of the temporal bone and its bony labyrinth.

---

[63]*ot* = ear + *ic* = pertaining to + *plac* = plate + *ode* = form, shape

The middle-ear cavity and auditory tube arise by elongation of the first pharyngeal pouch, beginning in week 5. Mesenchyme of the first two pharyngeal arches gives rise to the three auditory ossicles and the two middle-ear muscles. These bones and muscles remain solidly embedded in mesenchyme until the last month of fetal development, at which time the mesenchyme degenerates and leaves the hollowed-out middle-ear cavity (fig. 17.31c).

At the same time as the middle ear begins to form, the facing edges of the first and second pharyngeal arches form three pairs of humps called *auricular hillocks.* These enlarge, fuse, and differentiate into the folds and whorls of the auricle by week 7 (fig. 17.31d). As this is happening, the first pharyngeal groove between these two arches begins to elongate, grow inward, and form the auditory canal. The tympanic membrane arises from the wall between the auditory canal and tympanic cavity, and thus has an outer ectodermal layer and inner endodermal layer, with a thin layer of mesoderm between.

Development of the Eye

An early indication of the development of the eye is the **optic vesicle,** seen by day 24 as an outgrowth on each side of the diencephalon, continuous with the neural tube (fig. 17.32a, b). As the optic vesicle reaches the overlying ectoderm, it invaginates and forms a double-walled **optic cup,** while its connection to the diencephalon narrows and becomes a hollow **optic stalk.** The outer wall of the optic cup becomes the *pigment retina,* later becoming the pigment epithelium discussed earlier. The inner wall becomes the *neural retina.* Starting around the end of week 6, the neural retina produces waves of cells that migrate toward the vitreous body and arrange themselves into layers of receptor cells (rods and cones) and neurons. The narrow space between the pigment retina and neural retina becomes obliterated, but these two walls of the cup never fuse. This is why the retina is so easily detached later in life. By the eighth month, all cell layers of the retina are present. Nerve fibers

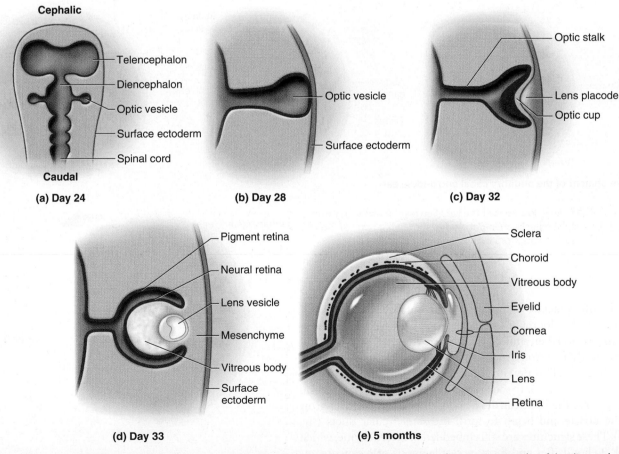

Figure 17.32 **Development of the Eye.** (a) Horizontal section of the head of an embryo. The optic vesicles show as outgrowths of the diencephalon. (b) Around day 28, the leading edge of the optic vesicle contacts the surface ectoderm. (c) The optic vesicle invaginates and becomes an optic cup, while inducing the ectoderm to form the lens placode. (d) The lens vesicle is now separated from the surface ectoderm and nestled in the optic cup, and a gelatinous vitreous body has been secreted between the lens vesicle and retina. (e) At 5 months, the lens vesicle is solidly filled with lens fibers; the sclera, choroid, iris, and cornea are partially formed; and the eyelids are just about to separate and reopen the eye.

grow from the ganglion cells into the optic stalk, occupying and eliminating its lumen as they become the optic nerve.

A seemingly peculiar aspect of the retinas of humans and other vertebrate animals is that the rods and cones face the back of the eye, *away from the incoming light.* This arrangement, called an *inverted retina,* is quite opposite the seemingly more sensible arrangement of the octopus eye, which is like ours in many respects but has its receptor cells aimed toward the light. The reason for the inverted retina of humans is that the rods and cones are homologous to the ependymal cells of the neural tube—that is, they develop from the same embryonic origin, and thus face inward toward the lumen of the optic cup just like the mature ependymal cells that face inward toward the ventricles of the brain. The homology between these retinal cells and ependymal cells is also seen in their cilia. Ependymal cells have conventional motile cilia, whereas rod and cone outer segments are modified cilia.

As the optic cup contacts the ectoderm on the surface of the embryo, it induces ectodermal cells to thicken into a **lens placode** (fig. 17.32c). The placode invaginates and forms a **lens pit,** which separates from the ectoderm by day 32 and becomes a **lens vesicle** nestled within the optic cup (fig. 17.32d). (This is quite similar to the development of the otic vesicle of the ear.) The lens vesicle is hollow at first, but becomes filled in with lens fibers by day 33. The vitreous body develops from a gelatinous secretion that accumulates in the space between the lens vesicle and optic cup.

Mesenchyme grows to completely surround and enclose the optic cup, and differentiates into the extrinsic muscles, some other accessory structures of the orbit, and some components of the eyeball including the choroid and sclera. The choroid is homologous to the pia mater and arachnoid mater, and the sclera is homologous to the dura mater—that is, they arise from the same embryonic tissues as these meninges. The mesenchyme lateral to the optic cup (near the surface) develops a split that becomes the anterior chamber of the eye. The cornea develops from both the lateral mesenchyme and the overlying ectoderm. The iris grows inward from the anterior margin of the optic cup around the end of month 3. The eyelids develop from folds of ectoderm with a mesenchymal center. The upper and lower eyelids approach each other and fuse, closing the eyes, at the end of month 3; they separate again, and the eyes open, between months 5 and 7 (fig. 17.32e).

Disorders of the Sense Organs

Several disorders of the senses have already been discussed in this chapter. These are listed in table 17.3 along with brief descriptions of five additional disorders of multiple sensory modalities.

Before You Go On

Answer the following questions to test your understanding of the preceding section.

20. Describe the contributions of the first pharyngeal pouch and the first two pharyngeal arches to the development of the ear.

21. From an embryological standpoint, explain why rod and cone outer segments face away from the incoming light.

22. How are paresthesia and tinnitus similar (table 17.3)?

23. How are anosmia and ageusia similar (table 17.3)?

TABLE 17.3	Some Sensory Disorders
Ageusia	Loss of the sense of one or more taste modalities, often due to damage to the hypoglossal nerve (loss of bitter taste) or facial nerve (loss of sweet, sour, and salty tastes).
Color blindness	Inability to distinguish certain colors from each other, such as green and orange, due to a hereditary lack of one of the three types of cones. A sex-linked recessive trait that affects more men than women.
Ménière disease	A disorder of proprioception in which one experiences episodes of vertigo (dizziness) often accompanied by nausea, tinnitus, and pressure in the ears. Usually accompanied by progressive hearing loss.
Paresthesia	Feelings of numbness, prickling, tingling, heat, or other sensations in the absence of stimulation; a symptom of nerve injuries and other neurological disorders.
Tinnitus	Perception of imaginary sounds such as whistling, buzzing, clicking, or ringing in the ear. May be temporary or permanent, intermittent or constant; typically associated with hearing loss in the high frequencies. May result from cochlear damage, aspirin or other drugs, ear infections, Ménière disease, or other causes.

Disorders Described Elsewhere

Study Guide

Assess Your Learning Outcomes

You should have a good understanding of this chapter if you can accurately address the following issues.

17.1 Receptor Types and the General Senses (p. 461)

1. The meaning of *sensory receptor* and the range of structures encompassed by the term
2. How receptors are classified by stimulus modality
3. The differences between general (somatosensory) and special senses, and which senses fall into each category
4. How receptors are classified by the origins of their stimuli
5. The types of sensory nerve endings considered to be unencapsulated, and the stimuli they are specialized to detect
6. The types of sensory nerve endings considered to be encapsulated, why they are called that, and the stimuli they are specialized to detect
7. The concept of a receptive field and what it implies about sensory effects in relation to the distance between two points of stimulation
8. The concepts of a sensory projection pathway and the first- through third-order neurons of the pathway
9. The projection pathways followed by somatosensory signals to reach the cerebral cortex and cerebellum
10. A definition of *pain*, the distinction between fast pain and slow pain, and the distinction between somatic pain and visceral pain
11. The projection pathways for pain signals from the head and for pain signals from the rest of the body
12. An explanation for referred pain, and some examples of it
13. The role of the reticular formation and descending analgesic fibers in modulating one's sensitivity to pain

17.2 The Chemical Senses (p. 466)

1. The relationship of taste buds to the lingual papillae; types of lingual papillae and where on the tongue they are located; which types of papillae have taste buds and which ones do not; and some locations of taste buds other than on the tongue
2. The structure of a taste bud and taste cells

3. The five primary taste sensations and the difference between taste and flavor
4. The projection pathway taken by taste signals to their final destinations in the cerebral cortex
5. The location and structure of the olfactory mucosa and olfactory neurons
6. The projection pathways taken by olfactory signals from the mucosa to multiple destinations in the cerebral cortex, limbic system, and hypothalamus
7. The influence of the cerebral cortex on the perception of odors, especially from the anatomical perspective

17.3 The Ear (p. 470)

1. The three principal subdivisions of the ear
2. The parts of the outer ear
3. The parts of the middle ear, including its three auditory ossicles and two muscles
4. The major components of the membranous labyrinth of the inner ear
5. The anatomy of the cochlea and the functional difference between its inner and outer hair cells
6. How vibrations in the ear lead to the generation of nerve signals
7. How cochlear function enables the brain to distinguish loudness and pitch
8. The projection pathway taken by auditory signals from the cochlea to the primary auditory cortex of the brain; the functions served by the superior olivary nucleus and inferior colliculus along the way
9. The differences between static and dynamic equilibrium and between linear and angular acceleration; the inner-ear structures specialized to detect these
10. The structure of the saccule and utricle and the macula of each
11. The action of the otolithic membrane in stimulating the macula sacculi and macula utriculi
12. How the saccules and utricles of the two ears enable one to sense the orientation of the head and distinguish between vertical and horizontal acceleration
13. The structure of the semicircular ducts and how angular movements of the head stimulate their hair cells

14. The projection pathways taken by signals of equilibrium to multiple destinations in the brainstem, cerebellum, and cerebrum

17.4 The Eye (p. 480)

1. The accessory structures of the orbit and their anatomy
2. The three layers of the wall of the eye and the constituents of each layer
3. The optical components of the eye and how they act to control light and focus images on the retina
4. The neural components of the eye and how different regions of the retina differ in function
5. The structure of rod and cone cells, where in each cell the visual pigments are located, and how the two cell types differ in function
6. The relationship of the rods and cones with bipolar cells and ganglion cells in the retina, and the origin of the optic nerve fibers
7. The photosensory but nonvisual role of some of the ganglion cells
8. The locations of horizontal cells and amacrine cells in the retina, and their roles in vision
9. The reasons why both high light sensitivity and high resolution cannot be obtained with a single type of retinal circuit (rod or cone circuit), and in relation to this, how the rod and cone circuits in the retina differ from each other
10. The projection pathways taken by retinal signals to the occipital lobe and midbrain, including the hemidecussation that occurs at the optic chiasm and how this relates to occipital lobe input from the right and left eyes
11. The functions of the midbrain and occipital lobe in vision

17.5 Developmental and Clinical Perspectives (p. 490)

1. How the principal structures of the inner ear develop from the otic placode
2. How the structures of the outer and middle ear develop from the first two pharyngeal arches and the first pharyngeal pouch
3. How the optic vesicle arises and how it leads to the development of the retina
4. How the lens, vitreous body, anterior chamber, cornea, iris, and eyelids arise in the embryo

Testing Your Recall

1. Hot and cold stimuli are detected by
   a. free nerve endings.
   b. proprioceptors.
   c. end bulbs.
   d. lamellar corpuscles.
   e. tactile corpuscles.

2. Sensory signals for all of the following except _____ must pass through the thalamus before they can reach the cerebral cortex.
   a. smell
   b. taste
   c. hearing
   d. equilibrium
   e. vision

3. The vallate papillae are more sensitive to _____ than to any of these other tastes.
   a. bitter
   b. sour
   c. sweet
   d. umami
   e. salty

4. The ear is somewhat protected from loud noises by
   a. the vestibule.
   b. the modiolus.
   c. the stapes.
   d. the stapedius.
   e. the superior rectus.

5. The sensory neurons that begin in the spiral organ of the ear end in
   a. the spiral ganglion.
   b. the cochlear nucleus.
   c. the superior olivary nucleus.
   d. the inferior colliculus.
   e. the temporal lobe.

6. The spiral organ rests on
   a. the tympanic membrane.
   b. the secondary tympanic membrane.
   c. the tectorial membrane.
   d. the vestibular membrane.
   e. the basilar membrane.

7. The acceleration you feel when an elevator begins to rise is sensed by
   a. the anterior semicircular duct.
   b. the spiral organ.
   c. the crista ampullaris.
   d. the macula sacculi.
   e. the macula utriculi.

8. The highest density of cone cells is found in
   a. the crista ampullaris.
   b. the optic disc.
   c. the fovea centralis.
   d. the chorion.
   e. the basilar membrane.

9. The dilated blood vessels seen in "bloodshot" eyes are vessels of
   a. the retina.
   b. the cornea.
   c. the conjunctiva.
   d. the sclera.
   e. the choroid.

10. A person would look cross-eyed if _____ muscles contracted at once.
    a. both medial rectus
    b. both lateral rectus
    c. the right medial rectus and left lateral rectus
    d. both superior oblique
    e. the left superior oblique and right inferior oblique

11. The most finely detailed vision occurs when an image falls on a pit in the retina called the _____.

12. Fibers of the optic nerve come from the _____ cells of the retina.

13. A sensory nerve ending specialized to detect tissue injury and produce a sensation of pain is called a/an _____.

14. The gelatinous membranes of the macula sacculi and macula utriculi are weighted by calcium carbonate–protein granules called _____.

15. Three rows of _____ in the cochlea have V-shaped arrays of stereocilia and tune the frequency sensitivity of the cochlea.

16. The _____ is a tiny bone that vibrates in the oval window and thereby transfers sound vibrations to the inner ear.

17. The _____ of the midbrain receive auditory input and trigger the head-turning auditory reflex.

18. The apical microvilli of a gustatory cell are called _____.

19. Olfactory neurons synapse with mitral cells and tufted cells in the _____, which lies inferior to the frontal lobe.

20. In the phenomenon of _____, pain from the viscera is perceived as coming from an area of the skin.

Answers in the Appendix

Building Your Medical Vocabulary

State a medical meaning of each of the following word elements, and give a term in which it is used.

1. noci-
2. an-
3. fili-
4. oto-
5. trochle-
6. fovea
7. alges-
8. -ate
9. tympano-
10. -stomy

Answers in the Appendix

True or False

Determine which five of the following statements are false, and briefly explain why.

1. Interoceptors belong to the general senses.
2. The sensory (afferent) neurons for touch end in the thalamus.
3. The right cerebral hemisphere perceives things we touch with our left hand.
4. The optic nerve is composed of the axons of the rod and cone cells.
5. Filiform papillae of the tongue have no taste buds.
6. Some chemoreceptors are exteroceptors and some are interoceptors.
7. Humans have more photoreceptor cells than taste cells.
8. Human neurons are never exposed to the external environment of the body.
9. The tympanic membrane has no nerve fibers.
10. The vitreous body occupies the posterior chamber of the eye.

Answers in the Appendix

Testing Your Comprehension

1. The principle of neural convergence was explained in chapter 13. Discuss its relevance to referred pain and scotopic vision.
2. What type of cutaneous receptor enables you to feel an insect crawling through your hair? What type enables you to palpate a patient's pulse? What type would give you a feeling that your belt is too tight?
3. Predict the consequences of a hypothetical disorder in which the eye begins to break down or reabsorb the gel of the vitreous body.
4. Suppose a virus were able to selectively invade and destroy the following nervous tissues. Predict the sensory consequences of each infection: (a) the spiral ganglion, (b) the vestibular nucleus, (c) the motor fibers of cranial nerve VIII, (d) the motor fibers of cranial nerve VII, (e) the posterior horns of spinal cord segments L3 through S5.
5. Summarize the similarities and differences between olfactory cells and taste cells.

Answers at www.mhhe.com/saladinha3

Improve Your Grade at www.mhhe.com/saladinha3

Practice quizzes, labeling activities, games, and flashcards provide fun ways to master concepts. You can also download image PowerPoint files for each chapter to create a study guide or for taking notes during lecture.

The Endocrine System

An ovarian follicle (SEM). Layered cells of the follicle wall secrete steroid hormones.

CHAPTER OUTLINE

DEEPER INSIGHTS

BRUSHING UP

To understand this chapter, you may find it helpful to review the following concepts:
- Embryonic development of the pharyngeal pouches (p. 90)
- Embryonic development of the neural crest (p. 366)
- The hypothalamus (p. 415)

Endocrine System

If the body is to function as an integrated whole, its organs must communicate with each other and coordinate their activities. Systems of internal chemical communication must have appeared very early in the history of life, as evidenced by their presence today even in very simple organisms composed of only a few cells. In humans, two such systems are especially prominent—the nervous and endocrine systems, which communicate with neurotransmitters and hormones, respectively.

Nearly everyone has heard of at least some hormones—growth hormone, thyroid hormone, estrogen, and insulin, for example. At least passingly familiar, too, are some of the glands that secrete them (the pituitary and thyroid glands, for example) and some of the disorders that result from hormone excesses, deficiencies, or dysfunctions (diabetes, goiter, dwarfism, and many others). This chapter surveys the glands and cells that produce the human hormones, with emphasis on their gross anatomy and histology. While the emphasis of this book is not on physiology or pathology, we will also include here some cursory insights into what hormones are, how they act, how they change over the life span, and a few of the many things that can go wrong with this system.

18.1 Overview of the Endocrine System

▶ Expected Learning Outcomes

When you have completed this section, you should be able to

- define *hormone* and *endocrine system*;
- describe how endocrine glands differ from exocrine glands; and
- explain the concept of target organs.

Hormones[1] may be defined as chemical messengers that are secreted into the bloodstream and stimulate physiological responses in distant organs. Their sources include a classic inventory of **endocrine glands** depicted in figure 18.1, but also include specialized secretory cells found in many tissues and organs not usually thought of as glands—including the brain, heart, small intestine, and adipose tissue. The **endocrine system** consists of all of these hormone-producing cells and glands. The study and treatment of this system are called **endocrinology.**

In chapter 3, we examined another category of glands, the *exocrine* glands. The classical distinction between exocrine and endocrine glands is the presence or absence of ducts, respectively. Most exocrine glands secrete their products onto an epithelial surface such as the skin or intestinal mucosa by way of a duct (see fig. 3.28, p. 74); sweat glands and salivary glands are good examples. By contrast, endocrine glands such as the pituitary and thyroid are ductless and release their secretions (hormones) into the bloodstream. For this reason, hormones were originally called the body's "internal secretions"; the word *endocrine*[2] still alludes to this fact. Some glands, such as the pancreas, contain both exocrine and endocrine components.

Some glands and secretory cells defy any classification as simple as exocrine versus endocrine. Liver cells, for example, behave as exocrine cells in the traditional sense when they secrete bile into ducts that lead ultimately to the small intestine. However, they also secrete hormones directly into the blood, and in this respect they act as endocrine cells. They secrete albumin and blood-clotting factors directly into the blood as well. These don't fit the traditional concept of exocrine secretions, because they are not released by way of ducts; nor do they fit the concept of endocrine secretions, because they are not hormones. Liver cells are just one of nature's myriad ways of confounding our impulse to rigidly classify things.

Endocrine glands have an unusually high density of blood capillaries, which serve to pick up and carry away their hormones. These vessels are an especially permeable type called *fenestrated capillaries,* which have patches of large pores in their walls allowing for the easy uptake of matter from the gland tissue (see fig. 21.6, p. 569).

Once a hormone enters the bloodstream, it goes wherever the blood goes; there is no way to send it selectively to a particular organ. However, the only organs or cells that respond to it are those with receptors for that hormone. We call these the **target organs** or **target cells.** Thyroid-stimulating hormone, for example, circulates everywhere the blood goes, but only the thyroid gland responds to it.

Before You Go On

Answer the following questions to test your understanding of the preceding section:

1. Define the word *hormone.* Compare hormones and neurotransmitters (see chapter 13). What do they have in common? How do they differ?

2. Compare and contrast exocrine and endocrine glands. Explain why this is an imperfect distinction.

3. What is unusual about the blood capillaries of an endocrine gland? What is the functional significance of this?

---

[1] *hormone* = to excite, set in motion

[2] *endo* = internal + *crin* = to secrete

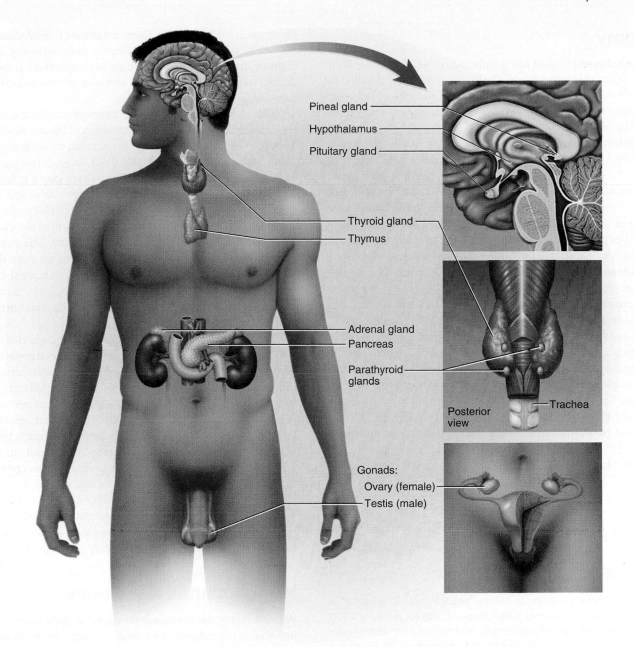

Figure 18.1 **Major Organs of the Endocrine System.** This system also includes gland cells in many organs not shown here.
● *After reading this chapter, name at least three hormone-secreting organs that are* not *shown in this illustration.*

18.2 The Hypothalamus and Pituitary Gland

▶ Expected Learning Outcomes

When you have completed this section, you should be able to

- describe the location and anatomy of the pituitary gland and its anatomical relationship with the hypothalamus;

- explain how the hypothalamus controls pituitary function; and

- list the hormones produced by the two lobes of the pituitary gland and state their functions.

There is no master control center that regulates the entire endocrine system, but the pituitary gland and a nearby region of the brain, the hypothalamus, do regulate much of it. They are an appropriate place to start our survey of endocrine glands.

Anatomy

The **hypothalamus,** shaped like a flattened funnel, forms the floor and walls of the third ventricle of the brain; it is further detailed in chapter 15 (see p. 415). It regulates primitive functions of the body ranging from water balance to sex drive and carries out many of its roles by way of the pituitary gland.

The **pituitary gland (hypophysis[3])** is attached to the hypothalamus by a stalk and is partially enclosed in the saddlelike *sella turcica* of the sphenoid bone (described on p. 161). It is an ovoid gland about 1.3 cm wide (approximately the size of a kidney bean), but grows 50% larger during pregnancy. Although it looks like a single gland on casual inspection, its histology and embryology show it to be composed of two structures—the *adenohypophysis* and *neurohypophysis*—with independent origins and separate functions.

The **adenohypophysis[4]** (AD-eh-no-hy-POFF-ih-sis) is the anterior three-quarters of the pituitary (fig. 18.2). It has two parts: a large **anterior lobe,** also called the **pars distalis** ("distal part") because it is most distal to the pituitary stalk, and the **pars tuberalis,** a small mass of cells adhering to the anterior side of the stalk. In the fetus there is also a **pars intermedia,** a strip of tissue between the anterior and posterior lobes, but during subsequent development, its cells mingle with those of the anterior lobe. There is no longer a separate pars intermedia in adults.

The adenohypophysis has three classes of cells (fig. 18.3a): **acidophils** and **basophils,** which stain darkly with acidic and basic dyes, respectively; and **chromophobes,[5]** which take up little dye and appear relatively pale. Acidophils and basophils secrete the anterior pituitary hormones. There are at least five subclasses of these cells, but they are indistinguishable with ordinary histological stains. Chromophobes exhibit little or no secretory activity and are of uncertain function. Some of them may be stem cells that give rise to acidophils and basophils.

The adenohypophysis has no nervous connection to the hypothalamus, but is linked to it by a complex of blood vessels called the **hypophyseal portal system.** In circulatory anatomy, a portal system is a route in which blood flows through two capillary beds before returning to the heart. The pituitary gland is fed by a *superior hypophyseal artery,* which breaks up into a bed of *primary capillaries* in the hypothalamus. These capillaries drain into *portal venules* (small veins) that travel down the pituitary stalk to a bed of *secondary capillaries* in the anterior pituitary. The hypothalamus controls the pituitary by secreting hormones that enter the primary capillaries, travel down the portal venules, and leave the secondary capillaries when they get to the adenohypophysis. Some of these hormones stimulate the pituitary to release its own hormones, and others inhibit it.

The **neurohypophysis** is the posterior one-quarter of the pituitary. It has three parts: the **posterior lobe (pars nervosa),** the **stalk (infundibulum)** that connects it to the hypothalamus, and the **median eminence,** an extension of the hypothalamic floor. The neurohypophysis is not a true gland but a mass of nervous tissue—neuroglia intermingled with axons that arise from certain hypothalamic neurons (fig. 18.3b). The axons travel through the stalk as a bundle called the **hypothalamo–hypophyseal tract,** and end in the posterior lobe. The neurons in the hypothalamus secrete hormones that are transported down the axons and stored in the posterior pituitary. Later, a nerve signal traveling down the same axons can trigger the release of these hormones into the blood.

Hereafter, we will limit our discussion to the anterior and posterior lobes; they secrete all of the pituitary hormones we will consider.

DEEPER INSIGHT 18.1

Empty Sella Syndrome

A radiologist normally expects to see the pituitary gland "at home" in the sella turcica, but sometimes in the course of tests for pituitary disorders, a CT or MRI scan reveals an empty sella turcica and seeming absence of a pituitary. This is called *empty sella syndrome (ESS)*. Given the vital importance of the pituitary gland, ESS might seem to be a life-threatening condition, yet it is often associated with no endocrine dysfunctions and it usually requires no treatment. More than three-quarters of ESS patients are obese women, and many are hypertensive. Many of these patients have elevated fluid pressure within the cranium, and some even seep cerebrospinal fluid from the nose. Elevated intracranial pressure can flatten the pituitary tissue against the wall of the sella turcica, creating the appearance of an empty sella even though the pituitary is present and functional. Some cases of ESS, however, result from the shrinkage of pituitary tissue following surgery, radiation therapy, or trauma. These cases are more often associated with systemic endocrine dysfunctions and may require hormone replacement therapy.

---

[3]*hypo* = below + *physis* = growth
[4]*adeno* = gland
[5]*chromo* = color, dye + *phob* = fearing, repelling

Hypothalamic Hormones

We now survey the hormones produced by these three regions—the hypothalamus, anterior lobe, and posterior lobe—including brief descriptions of the most basic functions of each hormone.

The hypothalamus produces six hormones that regulate the anterior pituitary (fig. 18.2 and table 18.1). Four of these are *releasing hormones,* which stimulate the anterior pituitary to secrete its hormones; two of them are *inhibiting hormones,* which suppress pituitary secretion. In most cases, the name of the hypothalamic hormone indicates the pituitary hormone whose secretion it stimulates or inhibits; *thyrotropin-releasing hormone,* for example, stimulates the anterior pituitary to release *thyrotropin.* Gonadotropin-releasing hormone controls both *follicle-stimulating hormone* and *luteinizing hormone,* collectively called *gonadotropins. Growth hormone–inhibiting hormone (GHIH)* is also known as *somatostatin.* The other two hypothalamic hormones—**oxytocin (OT)** and **antidiuretic hormone (ADH)**—are synthesized in the brain but stored and secreted by the posterior pituitary. They will be discussed later in connection with posterior pituitary function.

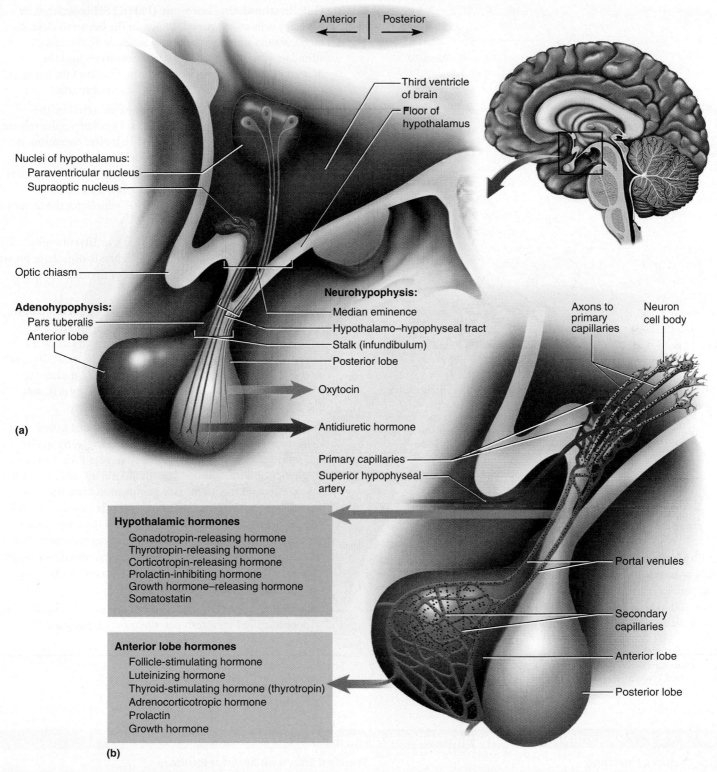

Anterior ← | → Posterior

Third ventricle of brain
Floor of hypothalamus

Nuclei of hypothalamus:
Paraventricular nucleus
Supraoptic nucleus

Optic chiasm

Adenohypophysis:
Pars tuberalis
Anterior lobe

Neurohypophysis:
Median eminence
Hypothalamo–hypophyseal tract
Stalk (infundibulum)
Posterior lobe

Oxytocin

Antidiuretic hormone

(a)

Axons to primary capillaries   Neuron cell body

Primary capillaries
Superior hypophyseal artery

Hypothalamic hormones
Gonadotropin-releasing hormone
Thyrotropin-releasing hormone
Corticotropin-releasing hormone
Prolactin-inhibiting hormone
Growth hormone–releasing hormone
Somatostatin

Anterior lobe hormones
Follicle-stimulating hormone
Luteinizing hormone
Thyroid-stimulating hormone (thyrotropin)
Adrenocorticotropic hormone
Prolactin
Growth hormone

(b)

Portal venules

Secondary capillaries

Anterior lobe

Posterior lobe

Figure 18.2 **Anatomy of the Pituitary Gland.** (a) Major structures of the pituitary and hormones of the neurohypophysis. Note that these hormones are produced by two nuclei in the hypothalamus and later released from the posterior lobe of the pituitary. (b) The hypophyseal portal system. The hormones in the violet box are secreted by the hypothalamus and travel in the portal blood vessels to the anterior pituitary. The hormones in the pink box are secreted by the anterior pituitary under the control of the hypothalamic releasers and inhibitors.
● *Which pituitary lobe is essentially composed of brain tissue?*

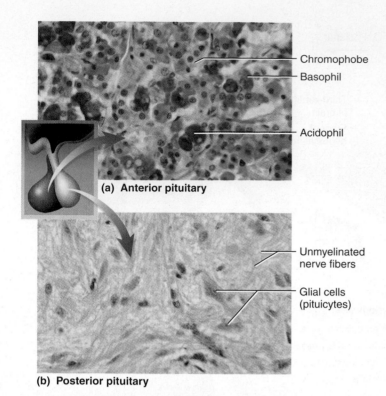

(a) **Anterior pituitary**

- Chromophobe
- Basophil
- Acidophil

(b) **Posterior pituitary**

- Unmyelinated nerve fibers
- Glial cells (pituicytes)

Figure 18.3 **Histology of the Pituitary Gland.** (a) The anterior lobe. Basophils include gonadotropes, thyrotropes, and corticotropes. Acidophils include somatotropes and lactotropes. These subtypes are not distinguishable with this histological stain The chromophobes resist staining, and their function is not yet known. (b) The posterior lobe, composed of nervous tissue.

Anterior Pituitary Hormones

The anterior lobe of the pituitary gland synthesizes and secretes six principal hormones (table 18.2). The first four of these are called **tropic**[6] or **trophic**[7] hormones because they stimulate the growth of other endocrine glands. The first two, targeted to the gonads, are called **gonadotropins.**

1. **Follicle-stimulating hormone (FSH).** FSH is secreted by pituitary cells called *gonadotropes,* in the basophil class. Its target organs are the ovaries and testes. In the ovaries, it stimulates the secretion of ovarian hormones and the development of the bubblelike *follicles* that contain the eggs. In the testes, it stimulates the production of sperm.

2. **Luteinizing hormone (LH).** LH is also secreted by the gonadotropes. In females, it stimulates *ovulation* (the release of an egg). LH is named for the fact that after ovulation, it stimulates the remainder of the follicle to develop into a yellowish body called the **corpus luteum.**[8] It also stimulates the corpus luteum to secrete progesterone, a hormone important to pregnancy. In males, LH stimulates the testes to secrete testosterone.

3. **Thyroid-stimulating hormone (TSH),** or **thyrotropin.** TSH is secreted by basophils called *thyrotropes.* It stimulates growth of the thyroid gland and the secretion of thyroid hormone, whose effects are described later.

4. **Adrenocorticotropic hormone (ACTH),** or **corticotropin.** ACTH is secreted by basophils called *corticotropes.* It is important in regulating the body's response to stress. It is named for its effect on the adrenal cortex, the outer layer of an endocrine gland near the kidney. ACTH stimulates the adrenal cortex to secrete hormones called *glucocorticoids,* discussed later.

5. **Prolactin**[9] **(PRL).** PRL is secreted by acidophils called *lactotropes (mammotropes),* which increase greatly in size and number during pregnancy. They secrete PRL during pregnancy and for as long as a woman nurses, although PRL has no effect on the mammary glands until after she gives birth. Then, it stimulates them to synthesize milk. In males, PRL has a gonadotropic effect that makes the testes more sensitive to LH. Thus, it indirectly enhances their secretion of testosterone. A tumor of the lactotropes sometimes causes milk secretion in men and in women who are not nursing; this condition is called *galactorrhea.*

6. **Growth hormone (GH),** or **somatotropin.** GH is secreted by acidophils called *somatotropes,* the most numerous cells in the

---

[6]*trop* = to turn, change
[7]*troph* = to feed, nourish

[8]*corpus* = body + *lute* = yellow
[9]*pro* = favoring + *lact* = milk

TABLE 18.1	Hypothalamic Releasing and Inhibiting Hormones That Regulate the Anterior Pituitary
Hypothalamic Hormone	**Principal Effects on Anterior Pituitary**
Thyrotropin-releasing hormone (TRH)	Promotes secretion of thyroid-stimulating hormone and prolactin
Corticotropin-releasing hormone (CRH)	Promotes secretion of adrenocorticotropic hormone
Gonadotropin-releasing hormone (GnRH)	Promotes secretion of follicle-stimulating hormone and luteinizing hormone
Growth hormone–releasing hormone (GHRH)	Promotes secretion of growth hormone
Growth hormone–inhibiting hormone (GHIH) (somatostatin)	Inhibits secretion of growth hormone and thyroid-stimulating hormone
Prolactin-inhibiting hormone (PIH)	Inhibits secretion of prolactin

TABLE 18.2	Pituitary Hormones	
Hormone	**Target Organs**	**Principal Effects**
Anterior pituitary		
Follicle-stimulating hormone (FSH)	Ovaries, testes	Female: growth of ovarian follicles and secretion of estrogen Male: sperm production
Luteinizing hormone (LH)	Ovaries, testes	Female: ovulation, production and maintenance of corpus luteum Male: testosterone secretion
Thyroid-stimulating hormone (TSH)	Thyroid gland	Growth of thyroid, secretion of thyroid hormone
Adrenocorticotropic hormone (ACTH)	Adrenal cortex	Growth of adrenal cortex, secretion of glucocorticoids
Prolactin (PRL)	Mammary glands, testes	Female: milk synthesis Male: increased LH sensitivity and testosterone secretion
Growth hormone (GH)	Liver, bone, cartilage, muscle, fat	Widespread tissue growth, especially in the stated tissues
Posterior pituitary		
Antidiuretic hormone (ADH)	Kidneys	Water retention
Oxytocin (OT)	Uterus, mammary glands, brain	Labor contractions, milk release; possibly involved in ejaculation, sperm transport in the female, sexual affection, and parent–offspring bonding

anterior pituitary. The pituitary produces at least a thousand times as much GH as any other hormone. The general effect of GH is to promote mitosis and cellular differentiation and thus to promote widespread tissue growth. Unlike the foregoing hormones, it is not targeted to one or a few organs, but has widespread effects on the body, especially on cartilage, bone, muscle, and fat. GH not only stimulates these tissues directly, but also stimulates the liver and other tissues to secrete **insulin-like growth factors (somatomedins[10])**, or **IGFs.** IGFs are hormones that stimulate the same target organs as GH (fig. 18.4). Together, they support tissue growth by mobilizing energy from fat, increasing levels of calcium and other electrolytes, and stimulating protein synthesis. Their most conspicuous effects occur during childhood and adolescence. They continue to function in energy metabolism and tissue maintenance throughout adult life, but adults suffer no ill effects from a GH deficiency.

Posterior Pituitary Hormones

The two posterior lobe hormones are ADH and OT. Both are synthesized in the hypothalamus, transported down the stalk, stored in the posterior pituitary, and released on demand (see fig. 18.2).

1. **Antidiuretic[11] hormone (ADH).** ADH is synthesized primarily in a pair of neuron clusters called the **supraoptic nuclei,** located on the left and right sides of the hypothalamus just above the optic chiasm. ADH increases water retention by the kidneys, reduces urine volume, and helps prevent

dehydration. It is also called *vasopressin* because it can cause vasoconstriction. This requires concentrations so unnaturally high for the human body, however, that this effect is of doubtful significance except in pathological states. ADH also functions as a brain neurotransmitter and is usually called vasopressin in the neuroscience literature.

2. **Oxytocin[12] (OT).** OT is synthesized primarily by a pair of hypothalamic nuclei called the **paraventricular nuclei,** located in the left and right walls of the third ventricle. It has a variety of reproductive roles. In childbirth, it stimulates labor contractions, and in lactating mothers, it stimulates the flow of milk to the nipple, making it accessible to the nursing infant. In both sexes, OT surges during sexual arousal and orgasm; this surge may play a role in the propulsion of semen through the male reproductive tract and in uterine contractions that help transport sperm up the female reproductive tract. It also evidently functions in feelings of sexual satisfaction and emotional bonding between partners and in maternal bonding with an infant. In the absence of oxytocin, female mammals tend to neglect their helpless infants.

Before You Go On

Answer the following questions to test your understanding of the preceding section:

4. What are two good reasons for considering the pituitary to be two separate glands?

5. Describe two anatomical routes by which the hypothalamus sends signals to the pituitary gland.

---

[10]Acronym for *somato*tropin *med*iating prote*in*
[11]*anti* = against + *diuret* = to pass through, urinate

[12]*oxy* = sharp, quick + *toc* = childbirth

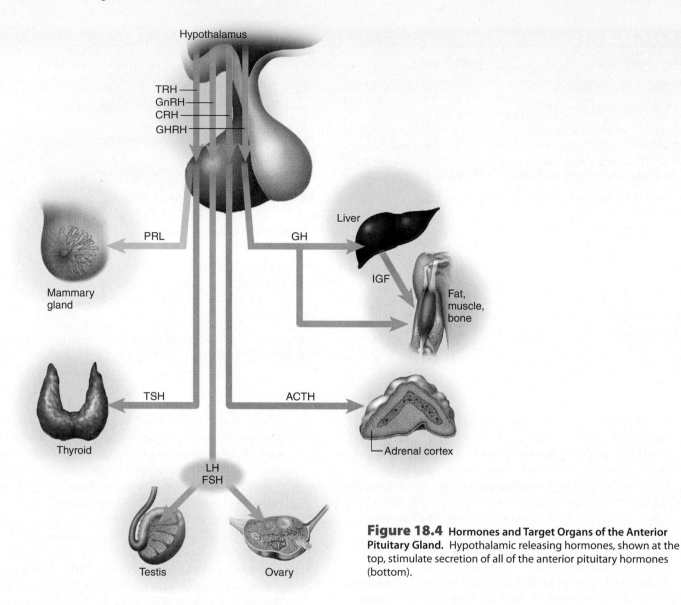

Figure 18.4 Hormones and Target Organs of the Anterior Pituitary Gland. Hypothalamic releasing hormones, shown at the top, stimulate secretion of all of the anterior pituitary hormones (bottom).

6. Construct a three-column table. In the middle column, list the six hormones of the anterior pituitary gland. On the left, list the hypothalamic hormones that control these pituitary secretions. On the right, list the target organ and effect of each anterior pituitary hormone.

7. Name the two hormones released by the posterior lobe of the pituitary, state where they are produced, and state their functions.

- name the hormones that these endocrine glands produce and state their functions; and

- discuss the roles of endocrine cells in organs other than the classic endocrine glands.

The Pineal Gland

The **pineal**[13] **gland (epiphysis cerebri)** is attached to the roof of the third ventricle beneath the posterior end of the corpus callosum (see fig. 15.2, p. 401, and fig. 18.1). Its name alludes to a shape resembling a pine cone. The philosopher René Descartes (1596–1650) thought it was the seat of the human soul. If so, children must have more soul than adults—a child's pineal gland is about 8 mm long and 5 mm wide, but after age 7, it regresses rapidly and is no more than a shrunken mass of fibrous tissue in the adult. Pineal secretion peaks between the ages of 1 and 5 years; by the end of puberty, it

18.3 Other Endocrine Glands

▶ Expected Learning Outcomes

When you have completed this section, you should be able to

- describe the structure and location of the remaining endocrine glands;

---

[13]*pineal* = pine cone

declines 75%. Such shrinkage of an organ is called **involution.** It is accompanied by the appearance of granules of calcium phosphate and calcium carbonate called *pineal sand.* These stony granules are visible on X-rays, enabling radiologists to determine the position of the gland. Displacement of the pineal from its normal location can be evidence of a brain tumor or other structural abnormality.

We no longer look for the human soul in the pineal gland, but this little organ remains an intriguing mystery with obscure functions. During the night, it synthesizes **melatonin,** which may suppress gonadotropin secretion and prevent premature sexual maturation (see p. 515). The pineal is also thought to play a role in our 24-hour *circadian rhythms* of physiological function synchronized with the cycle of daylight and darkness. It may be involved in human mood disorders such as premenstrual syndrome and seasonal affective disorder.

The Thymus

The **thymus** plays a role in three systems: endocrine, lymphatic, and immune. It is a bilobed gland in the mediastinum superior to the heart, behind the sternal manubrium. In the fetus and infant, it is enormous in comparison to adjacent organs, sometimes protruding between the lungs nearly as far as the diaphragm and extending upward into the neck (fig. 18.5a). It continues to grow until the age of 5 or 6 years, although not as fast as other thoracic organs, so its *relative* size decreases. After the age of 14, it undergoes rapid involution. It weighs about 20 g up to age 60, but it becomes increasingly fatty and less glandular with age. In the elderly, it is a small fibrous and fatty remnant barely distinguishable from the surrounding mediastinal tissues (fig. 18.5b).

The thymus is a site of maturation for certain white blood cells that are critically important for immune defense (T lymphocytes, T

for *thymus*). It secretes several hormones **(thymopoietin, thymosin,** and **thymulin)** that stimulate the development of other lymphatic organs and regulate the development and activity of the T lymphocytes. Its histology is discussed more fully in chapter 22 in relation to its immune function.

The Thyroid Gland

The **thyroid gland** is the largest endocrine gland in adults, weighing 20 to 25 g. It is composed of two lobes that wrap like a butterfly around the trachea, immediately below the larynx (fig. 18.6a). It is named for the nearby, shieldlike *thyroid*[14] *cartilage* of the larynx. Each lobe of the gland is bulbous at the inferior end and tapers superiorly. Near the inferior end, the two lobes are joined by a narrow bridge of tissue, the **isthmus,** which crosses the front of the trachea. About 50% of people have an accessory *pyramidal lobe,* usually small, growing upward from the isthmus. Some people lack an isthmus, and some have thyroid tissue embedded in the root of the tongue, the thymus, or other places in the neck.

The thyroid gland receives one of the body's highest rates of blood flow per gram of tissue. Its abundant blood vessels give the gland a dark reddish brown color. It is supplied by a pair of *superior thyroid arteries* that arise from the external carotid arteries of the neck, and a pair of *inferior thyroid arteries* that arise from the subclavian arteries near the clavicles. It is drained by two to three pairs of thyroid veins (*superior, middle,* and *inferior*), which flow into the internal jugular and brachiocephalic veins.

The main histological feature of the thyroid is that it is composed mostly of sacs called **thyroid follicles** (fig. 18.6b), which are lined by a simple cuboidal epithelium of **follicular cells** and filled

[14]*thyr* = shield + *oid* = like, resembling

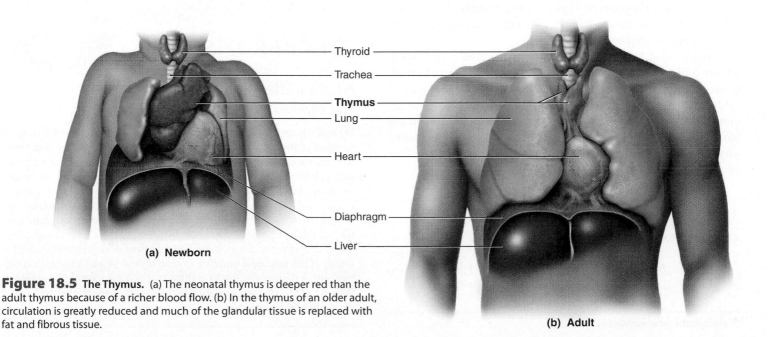

Thyroid — Trachea — Thymus — Lung — Heart — Diaphragm — Liver

(a) Newborn   **(b) Adult**

Figure 18.5 The Thymus. (a) The neonatal thymus is deeper red than the adult thymus because of a richer blood flow. (b) In the thymus of an older adult, circulation is greatly reduced and much of the glandular tissue is replaced with fat and fibrous tissue.

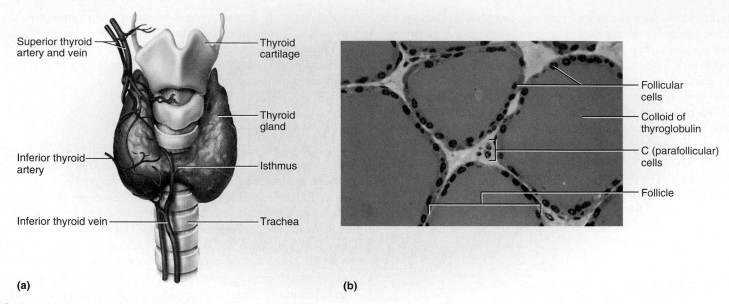

Figure 18.6 **The Thyroid Gland.** (a) Gross anatomy, anterior view. Major blood vessels are shown only on the anatomical right. (b) Histology, showing the saccular thyroid follicles (source of thyroid hormone) and a nest of C cells.
● *What is the function of the C cells?*

with a protein-rich colloid. Follicular cells secrete mainly the hormone **thyroxine,** also known as **tetraiodothyronine** (TET-ra-EYE-oh-doe-THY-ro-neen), or T_4, because it has four iodine atoms in its structure. The thyroid also produces **triiodothyronine** (try-EYE-oh-doe-THY-ro-neen), or T_3, with three iodine atoms. The expression **thyroid hormone (TH)** refers to T_4 and T_3 collectively.

Thyroid hormone stimulates prenatal and childhood brain development and bone growth, promotes pituitary secretion of growth hormone, quickens the somatic reflexes, raises the heart rate and metabolic rate, promotes intestinal absorption of carbohydrates, and lowers the plasma cholesterol level. TH secretion rises in cold weather. The resulting increase in metabolic rate typically results in a bigger appetite and greater caloric intake. The proportionate increase in heat production, called the **calorigenic**[15] **effect,** compensates for cold-weather heat loss.

Between the thyroid follicles are clusters of less numerous **C (clear) cells,** also called **parafollicular cells.** They secrete the hormone **calcitonin** when the blood calcium level rises above normal (a state called *hypercalcemia*). Calcitonin tips the balance between bone deposition and resorption in favor of deposition, and the blood calcium level falls as calcium is incorporated into the bones. This effect, however, is significant only in children; in healthy adults, the effect of calcitonin is negligible.

The Parathyroid Glands

The **parathyroid glands** are small ovoid glands in the neck measuring about 3 to 8 mm long and 2 to 5 mm wide. Usually there are four of them, but about 5% of people have more. They most often

adhere to the posterior side of the thyroid gland in the approximate positions shown in figure 18.7a, but the parathyroids are highly variable in location and not always attached to the thyroid. They can occur as far superiorly as the hyoid bone and as far inferiorly as the aortic arch. They have a thin fibrous capsule separating them from the thyroid tissue (fig. 18.7b). They are supplied with blood and drained by the same vessels as the thyroid gland.

Hypocalcemia, a calcium deficiency, stimulates the **chief cells** of the parathyroids to secrete **parathyroid hormone (PTH).** PTH raises blood calcium levels by promoting intestinal calcium absorption, inhibiting urinary calcium excretion, and indirectly stimulating osteoclasts to resorb bone.

DEEPER INSIGHT   18.2

A Lethal Effect of Postsurgical Hypoparathyroidism

Thyroid cancer and other dysfunctions sometimes require the surgical removal of thyroid tissue. Because of their variable location and small size, the parathyroid glands are sometimes accidentally removed along with it. Without hormone replacement therapy, the lack of parathyroid hormone *(hypoparathyroidism)* leads to a rapid decline in blood calcium levels. *Hypocalcemia,* a deficiency of blood-borne calcium, makes the skeletal muscles overly excitable and prone to exhibit spasmodic contractions called *hypocalcemic tetany.* One sign of this is spasmodic flexion of the hands and feet *(carpopedal spasm).* A more serious effect is tetany of the laryngeal muscles, closing off the airway and causing suffocation. A patient can die in as little as 3 or 4 days without treatment. Because of this danger, surgeons usually try to leave the posterior part of the thyroid gland intact.

---

[15]*calori* = heat + *genic* = producing

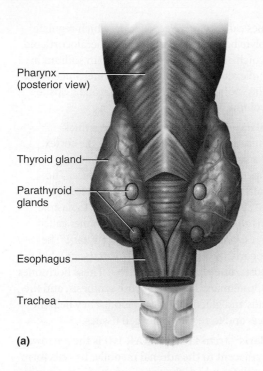

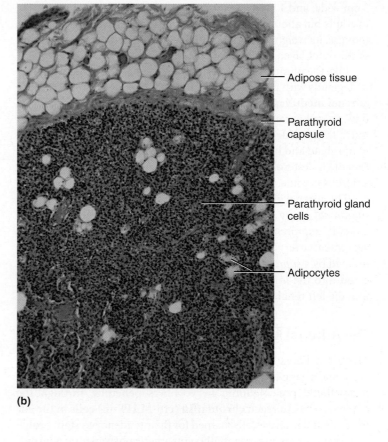

Figure 18.7 **The Parathyroid Glands.** (a) There are usually four parathyroid glands embedded in the posterior surface of the thyroid gland. (b) Histology.

Pharynx
(posterior view)

Thyroid gland

Parathyroid
glands

Esophagus

Trachea

(a)

Adipose tissue

Parathyroid
capsule

Parathyroid gland
cells

Adipocytes

(b)

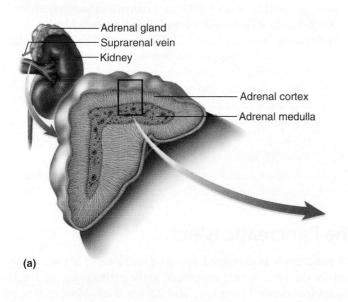

Adrenal gland
Suprarenal vein
Kidney

Adrenal cortex

Adrenal medulla

(a)

Figure 18.8 **The Adrenal Gland.** (a) Location and gross anatomy. (b) Histology.

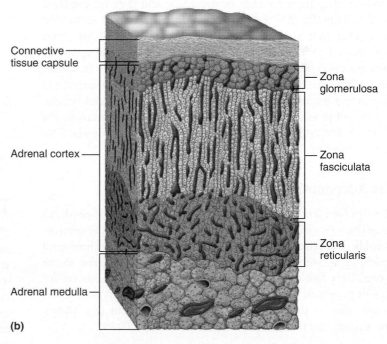

Connective
tissue capsule

Adrenal cortex

Adrenal medulla

Zona
glomerulosa

Zona
fasciculata

Zona
reticularis

(b)

The Adrenal Glands

An **adrenal (suprarenal) gland** is attached to the superior to medial aspect of each kidney (fig. 18.8). The right adrenal gland is more or less triangular and rests on the superior pole of the kidney.

The left adrenal gland is more crescent-shaped and extends from the medial indentation (hilum) of the kidney to its superior pole. Like the kidneys, the adrenal glands are retroperitoneal, lying outside the peritoneal cavity between the peritoneum and posterior body wall. Each adult adrenal gland measures about 5 cm high,

3 cm wide, and 1 cm from anterior to posterior. It weighs 7 to 10 g in adults but about twice this much in the newborn. By the age of 2 months, its weight drops about 50%, mostly because of involution of the outer layer, the adrenal cortex.

Like the pituitary, the adrenal gland forms by the merger of two fetal glands with different origins and functions. Its inner core, the **adrenal medulla,** is 10% to 20% of the gland. Depending on blood flow, its color ranges from gray to dark red. Surrounding it is a much thicker **adrenal cortex,** constituting 80% to 90% of the mass of the gland and having a yellowish color due to its high concentration of cholesterol and other lipids. The entire gland is enclosed in a fibrous capsule.

The adrenal gland receives blood from three arteries: a *superior suprarenal artery* arising from the phrenic artery of the diaphragm; a *middle suprarenal artery* arising from the aorta; and an *inferior suprarenal artery* arising from the renal artery of the kidney. It is drained by a *suprarenal vein,* which empties from the right adrenal gland into the inferior vena cava and from the left adrenal gland into the left renal vein.

The Adrenal Medulla

The adrenal medulla has a dual nature, acting as both an endocrine gland and a ganglion of the sympathetic nervous system (see p. 450). Sympathetic preganglionic nerve fibers penetrate through the adrenal cortex to reach **chromaffin** (cro-MAFF-in) **cells** in the adrenal medulla. These cells, named for their tendency to stain brown with certain dyes, are essentially sympathetic postsynaptic neurons. However, they have no axon or dendrites, and they release their products directly into the bloodstream like any other endocrine gland; thus, they are considered to be neuroendocrine cells. Upon stimulation by the nerve fibers, they secrete a mixture of **epinephrine, norepinephrine,** and **dopamine.** Secretion rises in conditions of fear, pain, and other kinds of stress, and they mimic and supplement the arousing effects of the sympathetic nervous system. Among other effects, they raise the metabolic rate, increase the heart rate and contraction strength, and mobilize glucose and fatty acids to meet the body's elevated energy requirement.

The Adrenal Cortex

The adrenal cortex surrounds the medulla on all sides. It produces more than 25 steroid hormones, known collectively as the **corticosteroids** or **corticoids.** All of them are synthesized from cholesterol which, along with other lipids, impart the yellowish color to the cortex. Only five corticosteroids are physiologically significant for present purposes. The cortex is composed of three layers of tissue, which differ in their histology and hormone output (fig. 18.8b). These layers and their principal hormones are as follows:

1. The **zona glomerulosa**[16] (glo-MER-you-LO-suh). This is a thin layer, less developed in humans than in some other mammals, located just beneath the capsule at the gland surface. The name *glomerulosa* ("full of little balls") refers to the arrangement of its cells in little round clusters. The zona glomerulosa secretes a

family of hormones called *mineralocorticoids,* which regulate the body's electrolyte balance. The principal mineralocorticoid is **aldosterone,** which acts on the kidneys to retain sodium in the body fluids and excrete potassium in the urine. Since water is retained with the sodium by osmosis, aldosterone helps to maintain blood volume and pressure.

2. The **zona fasciculata**[17] (fah-SIC-you-LAH-ta) is a thick middle layer constituting about three-quarters of the cortex. Here the cells are arranged in parallel cords, separated by blood capillaries, perpendicular to the adrenal surface. The cells of the fasciculata are called **spongiocytes** because of a foamy appearance caused by their abundance of cytoplasmic lipid droplets. This zone secretes a family of hormones called *glucocorticoids* in response to ACTH from the pituitary. The most potent glucocorticoid is **cortisol (hydrocortisone),** but the less potent **corticosterone** also has some effect. These hormones stimulate fat and protein breakdown, glucose synthesis, and the release of fatty acids and glucose into the blood. They help the body adapt to stress and to rebuild damaged tissues.

3. The **zona reticularis**[18] (reh-TIC-you-LAR-iss) is the narrow, innermost layer, adjacent to the adrenal medulla. Its cells form a branching network for which the zone is named. This layer secretes sex steroids—**androgens** and smaller amounts of **estrogens.** Androgens are present and important in both sexes but are named for their best-known effect—stimulating many aspects of male development and reproductive physiology. The androgens of the adrenal cortex are **dehydroepiandrosterone (DHEA)** (dee-HY-dro-EP-ee-an-DROSS-ter-own) and **androstenedione** (AN-dro-STEEN-di-own). These are similar to testosterone but less potent. They stimulate the prenatal development of the male reproductive tract; the growth of pubic and axillary hair and apocrine scent glands in puberty in both sexes; and the libido (sex drive) of both sexes in adolescence and adulthood.

Apply What You Know

The zona fasciculata thickens significantly in pregnant women. What do you think would be the benefit of this phenomenon?

The Pancreatic Islets

The **pancreas** is an elongated, spongy gland located below and behind the stomach, mostly superficial to the peritoneum (fig. 18.9). It is approximately 15 cm long and 2.5 cm thick. Most of it is an exocrine digestive gland, and its gross anatomy is therefore discussed in chapter 24 (see p. 678). Dispersed throughout the exocrine tissue, however, are about 1 to 2 million endocrine cell clusters called the **pancreatic islets (islets of Langerhans**[19]**).** Although they are less than 2% of the mass of the pancreas, the islets secrete hormones of vital importance to metabolism—especially insulin and glucagon. The major effect of these hormones is to regulate *glycemia,* the concentration of glucose in the blood.

---

[16]*glomerul* = little ball + *osa* = full of

[17]*fascicul* = little bundle + *ata* = possessing
[18]*reticul* = little network + *aris* = pertaining to
[19]Paul Langerhans (1847–88), German anatomist

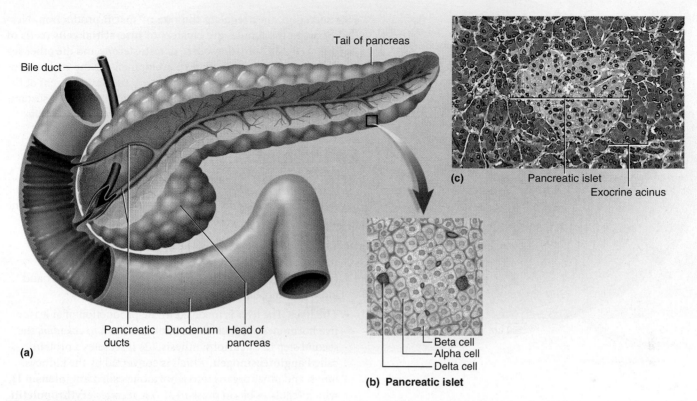

Figure 18.9 **The Pancreas.** (a) Gross anatomy and relationship to the duodenum. (b) Cells of a pancreatic islet. PP and G cells are not shown; they are few in number and cannot be distinguished with ordinary histological stains. (c) Light micrograph of a pancreatic islet amid the darker exocrine tissue.
● *What is the function of the exocrine cells in this gland?*

A typical pancreatic islet measures about 75 × 175 µm. Islets contain from a few to 3,000 cells, belonging to five classes:

1. **Alpha (α) cells, or A cells,** which secrete **glucagon.** Glucagon is secreted between meals when the blood glucose level falls. Its effects are to stimulate the release of stored glucose from the liver and fatty acids from adipose tissue, thereby providing the body with blood-borne fuel until the next meal.

2. **Beta (β) cells, or B cells,** which secrete **insulin.** Insulin is secreted during and immediately after a meal in response to rising levels of blood-borne nutrients such as glucose and amino acids. It stimulates most body tissues to absorb these nutrients and store or metabolize them.

3. **Delta (δ) cells, or D cells,** which secrete **somatostatin** (growth hormone–inhibiting hormone). Somatostatin is secreted under the same conditions as insulin. It helps to regulate the speed of digestion and nutrient absorption, and it modulates the activity of other pancreatic islet cells.

4. **PP cells, or F cells,** which secrete **pancreatic polypeptide.** This hormone inhibits gallbladder contraction and the secretion of digestive enzymes by the pancreas.

5. **G cells,** which secrete **gastrin.** This hormone stimulates acid secretion, motility, and emptying of the stomach.

The proportions of these pancreatic cells are about 20% alpha, 70% beta, 5% delta, and small numbers of PP and G cells.

The Gonads

Like the pancreas, the **gonads** (ovaries and testes) function as both endocrine and exocrine glands. Their exocrine products are eggs and sperm, and their endocrine products are the *gonadal hormones,* most of which are steroids. Their gross anatomy is described in chapter 26.

The ovary secretes chiefly **estrogen, progesterone,** and **inhibin.** Each egg develops in its own bubblelike follicle (fig. 18.10a), which is lined by a wall of **granulosa cells** and surrounded by a capsule, the **theca.** The theca and granulosa cells collaborate to produce estrogen. Midway through the monthly ovarian cycle, the follicle ovulates and begins to secrete an abundance of progesterone. Estrogen and progesterone contribute to the development of the reproductive system and feminine physique, regulate the menstrual cycle, sustain pregnancy, and prepare the mammary glands for lactation. Inhibin, which is also secreted by the follicle, is a signal from the ovary to the anterior pituitary. It inhibits the secretion of follicle-stimulating hormone (FSH). The effects of these hormones are further considered in chapter 26.

The testis secretes **testosterone,** lesser amounts of weaker androgens and estrogens, and inhibin. It consists mainly of minute tubules that produce sperm. The wall of the tubule is formed partly by **sustentacular (Sertoli[20]) cells,** the source of inhibin. By limiting

[20]Enrico Sertoli (1824–1910), Italian histologist

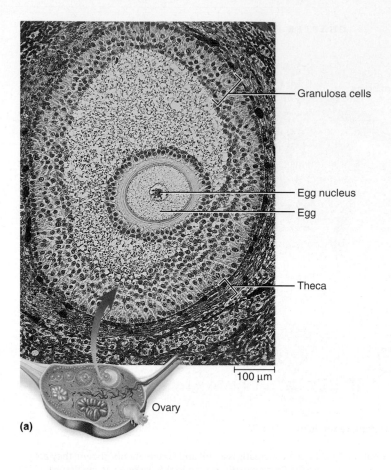

Granulosa cells

Egg nucleus

Egg

Theca

100 µm

Ovary

(a)

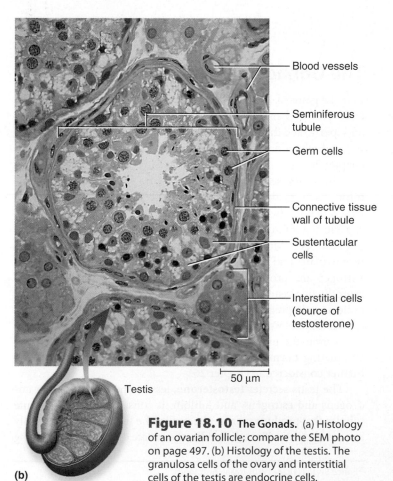

Blood vessels

Seminiferous tubule

Germ cells

Connective tissue wall of tubule

Sustentacular cells

Interstitial cells (source of testosterone)

50 µm

Testis

(b)

Figure 18.10 The Gonads. (a) Histology of an ovarian follicle; compare the SEM photo on page 497. (b) Histology of the testis. The granulosa cells of the ovary and interstitial cells of the testis are endocrine cells.

510

FSH secretion, they regulate the rate of sperm production. Nestled between the tubules are clusters of **interstitial cells (cells of Leydig**[21]) (fig. 18.10b), the source of testosterone and the other sex steroids. Testosterone stimulates development of the male reproductive system in the fetus and adolescent, the development of the masculine physique in adolescence, and the sex drive. It sustains sperm production and the sexual instinct throughout adult life.

Endocrine Cells in Other Tissues and Organs

Several other tissues and organs beyond the classical endocrine glands have hormone-secreting cells:

- **The skin.** Keratinocytes of the epidermis produce *cholecalciferol,* the first step in the synthesis of **calcitriol (vitamin D_3),** a calcium-regulating hormone. The liver and kidneys complete the process.

- **The liver.** The liver is involved in the production of at least five hormones: (1) It converts cholecalciferol to *calcidiol,* the second step in calcitriol synthesis. (2) It secretes a protein called **angiotensinogen,** which is converted by the kidneys, lungs, and other organs into a hormone called **angiotensin II,** which regulates blood pressure. (3) It secretes **erythropoietin (EPO)** (eh-RITH-ro-POY-eh-tin), a hormone that stimulates red blood cell (erythrocyte) production. (4) It secretes **hepcidin,** which regulates iron metabolism. (5) It secretes **insulin-like growth factor I (IGF-I),** which mediates the action of growth hormone.

- **The kidneys.** The kidneys secrete most (85%) of the body's erythropoietin, and carry out the third and final step in calcitriol synthesis and the second step in angiotensin II synthesis.

- **The heart.** The heart secretes **natriuretic**[22] **peptides,** hormones that enhance sodium excretion in the urine and reduce blood pressure.

- **The stomach and small intestine.** These organs have various *enteroendocrine cells,*[23] which secrete at least 10 **enteric hormones.** These regulate feeding and digestion. They include **gastrin,** which stimulates acid secretion by the stomach; **cholecystokinin** (COAL-eh-SIS-toe-KY-nin), which stimulates the secretion of bile; **ghrelin** (GREL-in), which produces a sensation of hunger; and **peptide YY,** which creates a feeling of satiety (satisfaction of the appetite) and tends to terminate eating.

- **Adipose tissue.** Fat cells secrete **leptin,** which has long-term appetite-regulating effects on the brain and plays a role in the onset of puberty.

- **Osseous tissue (bone).** Osteoblasts secrete **osteocalcin,** which enhances the secretion and action of insulin and inhibits fat deposition (weight gain).

[21]Franz von Leydig (1821–1908), German histologist
[22]*natri* = sodium + *uretic* = pertaining to urine
[23]*entero* = intestine

- **The placenta.** This organ performs many functions in pregnancy, including fetal nutrition and waste removal. But it also secretes estrogen, progesterone, and other hormones that regulate pregnancy and stimulate development of the fetus and the mother's mammary glands.

Apply What You Know

Often, two hormones have opposite *(antagonistic)* effects on the same target organs. For example, oxytocin stimulates labor contractions, and progesterone inhibits premature labor. Name some other examples of antagonistic effects among the hormones in this chapter.

You can see that the endocrine system is extensive. It includes numerous discrete glands as well as individual cells in the tissues of other organs. The endocrine organs and tissues other than the hypothalamus and pituitary are reviewed in table 18.3.

Before You Go On

Answer the following questions to test your understanding of the preceding section:

8. Name two endocrine glands that are larger in children than in adults. What are their functions?

9. What hormone increases the body's heat production in cold weather? What other functions does this hormone have?

10. Name the main hormone secreted by each layer of the adrenal cortex and one secreted by the adrenal medulla, and state the function of each.

11. What is the difference between a gonadal hormone and a gonadotropin?

12. What hormones are most important in regulating blood glucose concentration? What cells produce them? Where are these cells found?

13. Name one hormone produced by each of the following organs—the heart, liver, and placenta—and state the function of each hormone.

TABLE 18.3	Hormones from Sources Other Than the Hypothalamus and Pituitary		
Source	**Hormone**	**Target Organs and Tissues**	**Principal Effects**
Pineal gland	Melatonin	Brain	Influences mood; may regulate the timing of puberty
Thymus	Thymopoietin, thymosin, thymulin	T lymphocytes	Stimulate T lymphocyte development and activity
Thyroid gland	Thyroxine (T_4) and triiodothyronine (T_3)	Most tissues	Elevate metabolic rate and heat production; promote alertness, quicker reflexes, enhanced absorption of dietary carbohydrates, protein synthesis, fetal and childhood growth, and CNS development
	Calcitonin	Bone	Promotes net deposition of bone by inhibiting osteoclasts; reduces blood Ca^{2+} level
Parathyroid glands	Parathyroid hormone (PTH)	Bone, kidneys, small intestine	Increases blood Ca^{2+} level by stimulating bone resorption, calcitriol synthesis, and intestinal Ca^{2+} absorption, and reducing urinary Ca^{2+} excretion
Adrenal medulla	Epinephrine, norepinephrine, dopamine	Most tissues	Adaptive responses to arousal and stress
Adrenal cortex	Aldosterone	Kidney	Promotes Na^+ retention and K^+ excretion; maintains blood pressure and volume
	Cortisol and corticosterone	Most tissues	Stimulate fat and protein catabolism, gluconeogenesis, stress resistance, and tissue repair
	Androgens	Bone, muscle, integument, many other tissues	Growth of pubic and axillary hair, bone growth, sex drive, male prenatal development
Pancreatic islets	Glucagon	Primarily liver	Stimulates glucose synthesis, glycogen and fat breakdown, release of glucose and fatty acids into circulation
	Insulin	Most tissues	Stimulates glucose and amino acid uptake; lowers blood glucose level; promotes glycogen, fat, and protein synthesis
	Somatostatin	Stomach, small intestine, pancreatic islets	Inhibits digestion and nutrient absorption; inhibits glucagon and insulin secretion
	Pancreatic polypeptide	Pancreas, gallbladder	Inhibits release of bile and digestive enzymes
	Gastrin	Stomach	Stimulates acid secretion

(continued)

TABLE 18.3	Hormones from Sources Other Than the Hypothalamus and Pituitary (continued)		
Source	Hormone	Target Organs and Tissues	Principal Effects
Ovaries	Estrogen	Many tissues	Stimulates female reproductive development and adolescent growth; regulates menstrual cycle and pregnancy; prepares mammary glands for lactation
	Progesterone	Uterus, mammary glands	Regulates menstrual cycle and pregnancy; prepares mammary glands for lactation
	Inhibin	Anterior pituitary	Inhibits FSH secretion
Testes	Testosterone	Many tissues	Stimulates reproductive development, musculoskeletal growth, sperm production, and sex drive
	Inhibin	Anterior pituitary	Inhibits FSH secretion
Skin	Cholecalciferol	—	Precursor of calcitriol (see kidneys)
Liver	Calcidiol	—	Precursor of calcitriol (see kidneys)
	Angiotensinogen	—	Precursor of angiotensin II (see kidneys)
	Erythropoietin	Red bone marrow	Promotes red blood cell production
	Hepcidin	Small intestine	Promotes iron absorption
	Insulin-like growth factor I	Many tissues	Mediates action of growth hormone
Kidneys	Erythropoietin	Red bone marrow	Promotes red blood cell production
	Calcitriol	Small intestine	Increases blood calcium level mainly by promoting intestinal absorption of dietary calcium
	Angiotensinogen I	—	Precursor of angiotensin II, a vasoconstrictor
Heart	Natriuretic peptides	Kidney	Lower blood volume and pressure by promoting Na$^+$ and water loss
Stomach and small intestine	Gastrin	Stomach	Acid secretion
	Cholecystokinin	Gallbladder; brain	Bile release; appetite suppression
	Ghrelin	Brain	Sensation of hunger; initiation of feeding
	Peptide YY	Brain	Sense of satiety; termination of feeding
	Other enteric hormones	Stomach, intestines	Coordination of secretion and motility in different regions of digestive tract
Adipose tissue	Leptin	Brain	Long-term appetite regulation
Osseous tissue	Osteocalcin	Pancrease, adipose tissue	Increased insulin secretion; enhanced insulin sensitivity of target organs; reduced fat deposition
Placenta	Estrogen, progesterone, and others	Many tissues of mother and fetus	Stimulation of fetal development and maternal bodily adaptations to pregnancy; preparation for lactation

18.4 Developmental and Clinical Perspectives

▶ Expected Learning Outcomes

When you have completed this section, you should be able to

- describe the embryonic development of each major endocrine gland;
- identify which endocrine glands or hormone levels change the most in old age, and state some of the consequences of these changes; and
- describe a few common disorders of the endocrine system, especially diabetes mellitus.

Prenatal Development of the Endocrine Glands

The endocrine glands, like other glands, develop mainly from embryonic epithelia, but lose their connection to the epithelial surface as they mature—hence the absence of ducts (see fig. 3.28, p. 74). All three embryonic germ layers—ectoderm, mesoderm, and endoderm—contribute to the endocrine system.

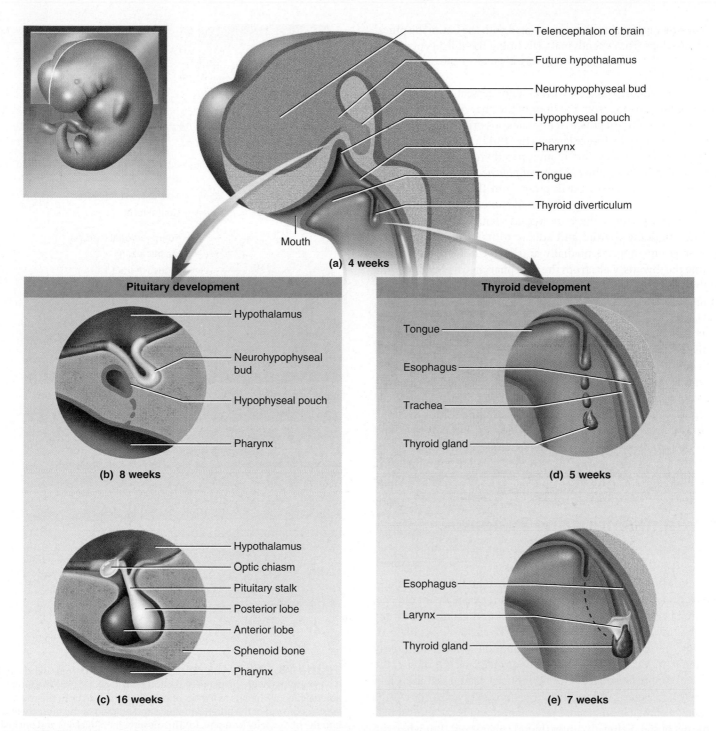

Figure 18.11 **Embryonic Development of the Pituitary and Thyroid Glands.** (a) Sagittal section of a 4-week embryo showing the early buds (future glands) of the anterior pituitary, posterior pituitary, and thyroid glands. (b) At 8 weeks, the hypophyseal pouch separates from the pharynx. (c) At 16 weeks, the structure of the pituitary is essentially complete. (d) At 5 weeks, the thyroid gland descends through the neck, and its connection to the tongue breaks down. (e) By 7 weeks, the thyroid has reached its final location on the trachea inferior to the larynx.

The pituitary has a dual origin (fig. 18.11a–c). The adenohypophysis begins with a pocket called the *hypophyseal pouch* that grows upward from the ectoderm of the pharynx. The pouch breaks away from the ectodermal surface and forms a hollow sac that continues to migrate upward. Meanwhile, growing down toward it, the neurohypophysis arises as an extension of the hypothalamus called the *neurohypophyseal bud.* The bud retains its connection to the brain throughout life as the pituitary stalk. The pouch and bud come to lie side by side and to be enclosed in the sella turcica of the sphenoid bone. They become so closely joined to each other that they look like a single gland.

The other endocrine gland associated with the brain, the pineal gland, develops from ependymal cells lining the third ventricle. A trace of the third ventricle persists as a canal in the stalk of the pineal gland.

The thyroid gland begins with an endodermal pouch *(thyroid diverticulum)* growing from the floor of the pharynx slightly posterior to the hypophyseal pouch. It migrates posteriorly to its position slightly inferior to the future larynx (fig. 18.11a, d, e).

In and near the neck, the thymus, parathyroid glands, and thyroid C cells arise from the *pharyngeal pouches* described in chapter 4 (p. 90). Cell masses break away from the third and fourth pouches and then split into posterior and anterior cell groups. The posterior groups form the parathyroid glands, which normally join the migrating thyroid and adhere to its posterior side. The anterior groups migrate medially and merge with each other to become the thymus. Cells from the fifth pharyngeal pouch become the thyroid C cells; they mingle with the rest of the thyroid tissue as the future thyroid gland migrates toward the larynx (see fig. 4.7, p. 91).

The adrenal gland, like the pituitary, has a dual origin (fig. 18.12). Recall from chapter 13 (p. 366) that ectodermal *neural crest cells* break away from the neural tube and give rise to sympathetic neurons and other cells. Some of the neural crest cells become the adrenal medulla. These neural crest masses, in turn, stimulate cell multiplication in the overlying mesothelium, the serous lining of the embryonic peritoneal cavity. The mesothelium thickens and grows around the medulla, eventually completely enclosing it and becoming the adrenal cortex. The adrenal gland is not fully developed until the age of 3 years.

The Aging Endocrine System

Endocrine glands vary widely in functionality over the life span. As we have seen in this chapter, the adrenal gland shrinks to half its size even in the first 2 months after birth. The pineal gland and thymus undergo pronounced involution after puberty. In old age, they are but shriveled vestiges of the childhood glands, with most of their secretory tissue replaced by fat or fiber.

Impaired glucose metabolism is common among the elderly and stems from causes as diverse as poor diet, lack of exercise, the presence of more fat and less muscle, reduced insulin secretion, and blunted effectiveness of insulin. Muscle ordinarily absorbs a large percentage of the blood glucose after a meal and is therefore a major means of stabilizing glycemia (blood sugar level). But when the aging individual loses muscle mass, glycemia remains elevated after a meal longer than it would in one with more muscle, and the risk of diabetes mellitus rises even if insulin secretion is normal.

Ovarian function drops precipitously at menopause, around 50 to 55 years of age, when the follicles are depleted and cease to secrete estrogen. The remaining estrogen comes mainly from the enzymatic conversion of androgens secreted by the adrenal glands, and its level is very low and noncyclic in contrast to the premenopausal state. This raises the risk of cardiovascular disease, bone loss (osteoporosis), and dementia in postmenopausal women. Androgen levels—both testosterone and adrenal DHEA—fall more gradually in men.

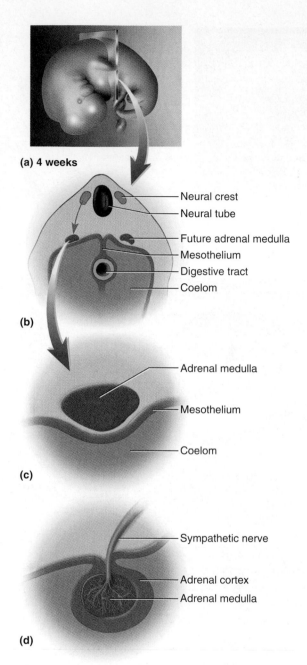

(a) 4 weeks

Neural crest
Neural tube

Future adrenal medulla
Mesothelium
Digestive tract
Coelom

(b)

Adrenal medulla

Mesothelium

Coelom

(c)

Sympathetic nerve

Adrenal cortex
Adrenal medulla

(d)

Figure 18.12 Embryonic Development of the Adrenal Gland. (a) A 4-week embryo with the plane of section seen in part (b). (b) The future adrenal medulla begins as a mass of cells that separate from the neural crest. (c) Growth of the adrenal medulla and bulging of the mesothelium into the body cavity (coelom). (d) The mesothelium thickens and encloses the adrenal medulla, giving rise to the adrenal cortex.

The circulating testosterone concentration in an 80-year-old man is about 20% of that seen in a 20-year-old. Age-related changes in reproductive function are discussed further in chapter 26.

In a number of cases, the waning of endocrine function with age results not from declining hormone secretion but from diminished effectiveness of the hormone that exists. This is the case in type 2 diabetes mellitus, which results from a deficiency or dysfunction of insulin receptors. Thyroid function declines in old age not because the gland secretes significantly less thyroxine (T_4), but because the target organs become less efficient in converting it to T_3, the more active form. The stress response also is more poorly controlled in the elderly. Normally, the anterior pituitary secretes ACTH; this stimulates the adrenal cortex to secrete cortisol; and cortisol induces the pituitary to lower the ACTH output. But in the elderly, the pituitary is less responsive to cortisol, so the ACTH level remains elevated, and the stress response lasts longer than it would in one's younger years.

Endocrine Disorders

Most disorders of the endocrine system revolve around three causes:

1. A hormone deficiency, or **hyposecretion,** exemplified by the insulin deficiency in type 1 diabetes mellitus.

2. A hormone excess, or **hypersecretion,** exemplified by pituitary gigantism, in which people grow to extraordinary heights because of excessive growth hormone secretion in childhood.

3. A **hormone insensitivity,** in which the hormone is present but a receptor defect causes the body to be unresponsive to it, as in the insulin resistance that characterizes type 2 diabetes mellitus.

These disorders are primarily physiological and therefore beyond the scope of this book, but hypo- and hypersecretion are sometimes caused by or associated with anatomical abnormalities worthy of mention here. Often, these abnormalities result from endocrine tumors or trauma.

Consider the pituitary gland, for example, snugly nestled in the sella turcica of the sphenoid bone and seemingly well protected. But surprisingly, the sphenoid is the most commonly fractured of all skull bones, with potentially grave consequences for pituitary function. A sphenoid fracture can sever the pituitary stalk, including the hypothalamo–hypophyseal tract, the portal system, or both. Such injuries cut off communication between the brain and the pituitary, thereby disrupting pituitary functions that depend on neural or hormonal signals from the hypothalamus. Anterior lobe effects can include loss of sexual functions (menstrual irregularity, sterility, and impotence due to loss of gonadotropins), and inadequate thyroid and adrenal gland function due to hyposecretion of thyroid-stimulating hormone and ACTH. Growth hormone secretion also declines markedly, although this has no clinical effect on adults. Severance of the hypothalamo–hypophyseal tract adversely affects secretion of the posterior lobe hormones oxytocin and antidiuretic hormone (see Testing Your Comprehension question 1 at the end of the chapter). Surgical accidents are another source of endocrine trauma with sometimes serious consequences (see Deeper Insight 18.2).

Tumors are another form of anatomical disorder with endocrine consequences. Pineal tumors are a case in point. They occur most often in childhood and sometimes cause premature sexual development, called *precocial puberty,* especially in boys. This lends some evidence to the theory that the pineal gland may play a role in the timing of puberty. However, precocial puberty occurs only when the tumor also affects the nearby hypothalamus, so it is difficult to say whether the effect is due specifically to the pineal tumor. Other effects of pineal tumors stem from the anatomical relationship of this gland to nearby brain structures (see Testing Your Comprehension question 2 at the end of this chapter).

A more common endocrine tumor is pheochromocytoma (FEE-oh-CRO-mo-sy-TOE-ma), a tumor of the adrenal medulla that causes hypersecretion of epinephrine and norepinephrine. The hormone excess causes nervousness, indigestion, sweating, tachycardia, hypertension, and an elevated metabolic rate. Patients often feel a sense of panic or impending doom. Most pheochromocytomas are benign (noncancerous), yet the hypertension is lifethreatening and the tumor usually calls for surgical removal.

A more conspicuous, often striking anatomical abnormality of an endocrine gland is **endemic goiter.** This is a disorder of the thyroid gland that results from a deficiency of dietary iodine. Normally, as the gland produces thyroid hormone (TH), it sends a feedback signal to the hypothalamus and anterior pituitary. In response, they modulate their secretion of thyrotropin-releasing hormone (TRH) and thyroid-stimulating hormone (TSH) and the thyroid, as a result, is not overstimulated. An iodine deficiency, however, reduces TH secretion and the feedback effect. The pituitary secretes a slightly but constantly elevated level of TSH, leading to hypertrophy of the thyroid gland. The overgrown thyroid is often visible in the neck as a goiter (fig. 18.13). Endemic goiter has become almost nonexistent in developed countries because of the addition of iodine to table salt, animal feeds, and fertilizers. It occurs most often in localities that have neither these benefits nor access to iodine-rich seafood—notably, central Africa and mountainous regions of South America, central Asia, and Indonesia. The word *endemic* refers to the occurrence of a disease in a defined geographic locality.

Some other endocrine disorders are described in table 18.4.

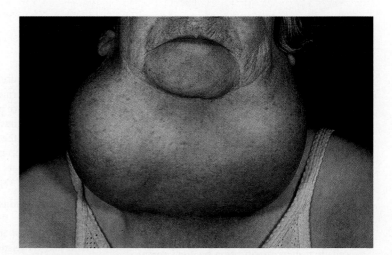

Figure 18.13 A Severe Case of Endemic Goiter.

TABLE 18.4	Disorders of the Endocrine System
Addison disease	Hyposecretion of adrenal glucocorticoids or mineralocorticoids, causing hypoglycemia, hypotension, weight loss, weakness, loss of stress resistance, darkening of the skin, and potentially fatal dehydration and electrolyte imbalances
Adrenogenital syndrome	Hypersecretion of adrenal androgens. Prenatal hypersecretion can cause girls to be born with masculinized genitalia and to be misidentified as boys. In children, it often causes enlargement of the penis or clitoris and premature puberty. In women, it causes masculinizing effects such as increased body hair, beard growth, and deepening of the voice..
Congenital hypothyroidism	Thyroid hyposecretion present from birth, resulting in stunted physical development, thickened facial features, low body temperature, lethargy, and irreversible brain damage in infancy
Cushing syndrome	Cortisol hypersecretion resulting from overactivity of the adrenal cortex. Results in disruption of carbohydrate and protein metabolism, hyperglycemia, edema, loss of bone and muscle mass, and sometimes abnormal fat deposition in the face or between the shoulders.
Myxedema	A result of severe or prolonged adult hypothyroidism, characterized by low metabolic rate, sluggishness and sleepiness, weight gain, constipation, hypertension, dry skin and hair, abnormal sensitivity to cold, and tissue swelling
Pituitary dwarfism	Abnormally short stature, with a normal proportion of limbs to trunk, resulting from growth hormone hyposecretion in childhood

Disorders Described Elsewhere

Acromegaly   146
Empty sella syndrome   500
Endemic goiter   515

Hypoparathyroidism   506
Pheochromocytoma   515

Pineal tumors   515
Pituitary trauma   515

Before You Go On

Answer the following questions to test your understanding of the preceding section:

14. With respect to embryonic development, what does the pituitary gland have in common with the thyroid gland? What does it have in common with the adrenal gland? Explain each comparison.

15. In some cases, endocrine function can decline in old age even if there is little change in the level of hormone secretion. Explain why, and support your explanation with examples.

16. Suppose an auto accident had fractured a pregnant woman's sphenoid bone and severed the pituitary stalk. Assuming that she nevertheless carried the pregnancy to full term, how might this accident affect her labor contractions? How might it affect her ability to breast-feed? Explain both effects.

17. What physical appearance might lead you to suspect that a person had a deficiency of iodine in the diet? Why would you find it more surprising to see this in a native of Tennessee than in a native of Tibet?

Study Guide

Assess Your Learning Outcomes

You should have a good understanding of this chapter if you can accurately address the following issues.

18.1 Overview of the Endocrine System (p. 498)

1. The definitions of *hormone* and *endocrine system*
2. The characteristics of an endocrine gland and how it contrasts with an exocrine gland
3. Examples of glands with dual endocrine and exocrine functions, and of secretory cells that defy classification by any simple endocrine–exocrine criteria

4. The unusual nature of the blood capillaries typical of endocrine glands, and the functional significance of this capillary type
5. How hormones travel about the body
6. The terms for the cells and organs that respond to them, and the reason that other cells and organs are unresponsive to a given hormone even if exposed to it

18.2 The Hypothalamus and Pituitary Gland (p. 499)

1. The location and anatomy of the pituitary gland and its relationship with the hypothalamus and sphenoid bone

2. The location of the adenohypophysis; its three parts in the human fetus; and which part is no longer present in adults
3. The three types of cells in the adenohypophysis and the differences between them
4. The anatomy of the hypophyseal portal system and its role in communication between the hypothalamus and adenohypophysis
5. The location of the neurohypophysis; its three parts; and the structure and function of the hypothalamo–hypophyseal tract
6. The six hypothalamic hormones that regulate the anterior pituitary, and the function of each one

7. The two hormones that are synthesized in the hypothalamus and stored in the posterior pituitary until release

8. The target organs or tissues and the functions of the anterior pituitary hormones: follicle-stimulating hormone, luteinizing hormone, thyroid-stimulating hormone, adrenocorticotropic hormone, prolactin, growth hormone, and insulin-like growth factors

9. The sites of synthesis, the target organs, and the functions of the posterior pituitary hormones: oxytocin and antidiuretic hormone

18.3 Other Endocrine Glands (p. 504)

1. The location and anatomy of the pineal gland, its changes through the life span, and the functions of melatonin

2. The location and anatomy of the thymus, its changes through the life span, its role in immunity, and the hormones that serve this role

3. The location, anatomy, and histology of the thyroid gland; its blood supply; its three hormones and which thyroid cells secrete them; and the functions of these hormones

4. The location and anatomy of the parathyroid glands; their hormone; and the function of that hormone

5. The location and anatomy of the adrenal glands, and their blood supply

6. The relationship of the adrenal medulla to the sympathetic nervous system; the hormones of the medulla; and the effects of those hormones

7. The three layers of the adrenal cortex; their histological differences; the hormones secreted by each layer; and the effects of those hormones

8. The location and structure of the pancreatic islets; the five types of islet cells; the hormone secreted by each cell type; and the functions of those hormones

9. Locations and names of the endocrine cells in the ovary and testis; their hormones; and the functions of those hormones

10. The roles of the skin, liver, kidneys, heart, stomach, small intestine, adipose tissue, bone, and placenta in synthesizing and secreting hormones, or carrying out a step in hormone synthesis; and the names and functions of these hormones

18.4 Developmental and Clinical Perspectives (p. 512)

1. Changes in insulin secretion and action with age

2. Changes in sex hormone secretion with age, and the bodily effects especially of the cessation of ovarian function at menopause

3. Some consequences of reduced hormone effectiveness in old age

4. A cause, and the effects, of trauma to the pituitary gland

5. The effects of tumors of the pineal gland and adrenal medulla

6. The cause and effects of endemic goiter

Testing Your Recall

1. Which of the following hormones is *not* synthesized by the brain?
   a. thyrotropin-releasing hormone
   b. antidiuretic hormone
   c. prolactin-releasing hormone
   d. follicle-stimulating hormone
   e. oxytocin

2. Which of the following hormones has the least in common with the others?
   a. adrenocorticotropic hormone
   b. follicle-stimulating hormone
   c. thyrotropin
   d. thyroxine
   e. prolactin

3. Which hormone would no longer be secreted if the hypothalamo–hypophyseal tract were destroyed?
   a. oxytocin
   b. follicle-stimulating hormone
   c. growth hormone
   d. adrenocorticotropic hormone
   e. corticosterone

4. Which of the following is *not* a hormone?
   a. prolactin
   b. thymosin
   c. iodine
   d. natriuretic peptide
   e. insulin-like growth factor

5. The _____ has/have no other function but hormone production, whereas the rest of these produce other secretions in addition to hormones.
   a. liver
   b. gonads
   c. salivary glands
   d. parathyroid glands
   e. pancreas

6. Which of these glands develops from the pharyngeal pouches?
   a. the anterior pituitary
   b. the posterior pituitary
   c. the thyroid gland
   d. the thymus
   e. the adrenal gland

7. Which of these glands has more exocrine than endocrine tissue?
   a. the pancreas
   b. the adenohypophysis
   c. the thyroid gland
   d. the pineal gland
   e. the adrenal gland

8. _____ leads to increased osteoclast activity and elevates the blood calcium concentration.
   a. Parathyroid hormone
   b. Calcitonin
   c. Calcitriol
   d. Aldosterone
   e. ACTH

9. Which of these endocrine glands is most directly involved in immune function?
   a. the pancreas
   b. the thymus
   c. the adenohypophysis
   d. the adrenal glands
   e. the thyroid gland

10. Both the _____ are involved in the synthesis of calcitriol and erythropoietin.
    a. anterior and posterior pituitary
    b. thyroid gland and thymus
    c. liver and kidneys
    d. parathyroids and pancreatic islets
    e. epidermis and liver

11. The _____ develops from the hypophyseal pouch of the embryo.

12. Antidiuretic hormone is produced by a group of neurons in the hypothalamus called the _____.

13. Adipocytes secrete the hormone _____, which signals the brain and helps to regulate the appetite.

14. The heart secretes two hormones called _____ that increase the urinary output of sodium and water.

15. Adrenal steroids that regulate glucose metabolism are collectively called _____.

16. The hypophyseal portal system is a means for the brain to communicate with the _____.

17. _____ cells are hormone-secreting neurons or cells derived from neurons.

18. In males, testosterone is secreted mainly by the _____ cells.

19. Cortisol is secreted by a layer of the adrenal cortex called the _____ in response to the pituitary hormone _____.

20. The hormones secreted by the stomach and small intestine are collectively called _____.

Answers in the Appendix

Building Your Medical Vocabulary

State a medical meaning of each of the following word elements, and give a term in which it is used.

1. crin-
2. oxy-
3. pro-
4. troph-
5. corpo-
6. thyro-
7. natri-
8. calori-
9. lact-
10. -phob

Answers in the Appendix

True or False

Determine which five of the following statements are false, and briefly explain why.

1. Tumors can lead to either hormone hypersecretion or hyposecretion.
2. All hormones are secreted by endocrine glands.
3. If fatty plaques of atherosclerosis blocked the arteries of the hypophyseal portal system, it could cause the ovaries and testes to malfunction.
4. The pineal gland and thymus become larger as one ages.
5. The tissue at the center of the adrenal gland is called the zona reticularis.
6. Unlike neurotransmitters, hormones cannot be selectively delivered to just one particular target organ.
7. The adenohypophysis and thyroid gland are more similar to each other in their embryonic origin than are the adenohypophysis and neurohypophysis.
8. Oxytocin and antidiuretic hormone are secreted through a duct called the pituitary stalk or infundibulum.
9. Of the endocrine glands covered in this chapter, only the adrenal glands are paired. The rest are single.
10. Enlargement of the thyroid gland would produce a swelling in the neck.

Answers in the Appendix

Testing Your Comprehension

1. A young man is involved in a motorcycle accident that fractures his sphenoid bone and severs the pituitary stalk. Shortly thereafter, he begins to excrete enormous amounts of urine, up to 30 liters per day, and suffers intense thirst. Explain how his head injury could have resulted in these effects.
2. Examine the anatomical relationship between the pineal gland and nearby brain structures, and as necessary, review the functions of those brain structures in chapter 15. In light of this information,

explain why a large pineal tumor might result in: (a) hydrocephalus; (b) paralysis of some eye movements.

3. Renal failure puts a person at risk of anemia and hypocalcemia. To prevent this, renal dialysis patients are routinely given hormone replacement therapy. Explain the hormonal connection between renal failure and each of these conditions, and identify what hormones would be administered to correct or prevent them.
4. Hormones of the adrenal cortex are steroids, whereas insulin is a protein. In

view of this, what major difference would you expect to see in the organelles of an adrenal spongiocyte and a pancreatic beta cell if you compared them with an electron microscope?

5. Selective destruction or removal of the adrenal cortex is fatal, but destruction or removal of the adrenal medulla produces no clear adverse effect. Explain why these two parts of the adrenal gland would differ so much in their necessity to life.

Answers at www.mhhe.com/saladinha3

Improve Your Grade at www.mhhe.com/saladinha3

Practice quizzes, labeling activities, games, and flashcards provide fun ways to master concepts. You can also download image PowerPoint files for each chapter to create a study guide or for taking notes during lecture.

PART FOUR

The Circulatory System I—Blood

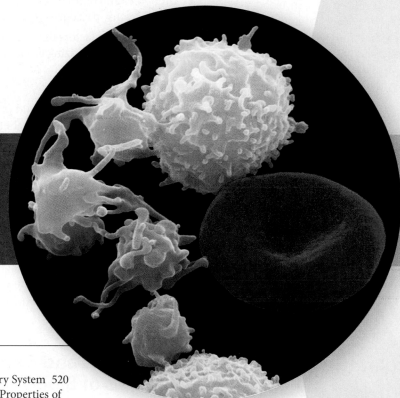

Blood cells and platelets (SEM)

CHAPTER OUTLINE

BRUSHING UP

To understand this chapter, you may find it helpful to review the following concepts:
- Osmosis (p. 34)
- Blood as a connective tissue (p. 69)
- Red bone marrow (p. 136)
- Erythropoietin (p. 510)

Anatomy & Physiology **REVEALED**®
aprevealed.com

Cardiovascular System

Blood has always had a special mystique. From time immemorial, people have seen blood flow from the body and with it, the life of the individual. People thus presumed that blood carried a mysterious "vital force," and Roman gladiators drank it to fortify themselves for battle. Even today, we become especially alarmed when we find ourselves bleeding, and the emotional impact of blood is enough to make many people faint at the sight of it. From ancient Egypt to nineteenth-century America, physicians drained "bad blood" from their patients to treat everything from gout to headaches, from menstrual cramps to mental illness. It was long thought that hereditary traits were transmitted through the blood, and people still use such unfounded expressions as "I have one-quarter Cherokee blood."

Scarcely anything meaningful was known about blood until its cells were seen with the first microscopes. Even though blood is a uniquely accessible tissue, most of what we know about it has been discovered only since the mid-twentieth century. Recent developments in this field have empowered us to save and improve the lives of countless people who would otherwise suffer or die.

19.1 Introduction

▶ Expected Learning Outcomes

When you have completed this section, you should be able to
- describe the functions and major components of the circulatory system;
- describe the components and physical properties of blood; and
- describe the composition of blood plasma.

Functions of the Circulatory System

The **circulatory system** consists of the heart, blood vessels, and blood. The term **cardiovascular system**[1] refers only to the heart and blood vessels, which are the subject of chapters 20 and 21. The study of blood, specifically, is called **hematology.**[2]

The fundamental purpose of the circulatory system is to transport substances from place to place in the blood. Blood is the liquid medium in which these materials travel; blood vessels ensure the proper routing of blood to its destinations; and the heart is the pump that keeps the blood flowing.

More specifically, the functions of the circulatory system are as follows:

Transport

- The blood carries oxygen from the lungs to all of the body's tissues, while it picks up carbon dioxide from those tissues and carries it to the lungs to be removed from the body.

- It picks up nutrients from the digestive tract and delivers them to all of the body's tissues.
- It carries metabolic wastes to the kidneys for removal.
- It carries hormones from endocrine cells to their target organs.
- It transports a variety of stem cells from the bone marrow and other origins to the tissues where they lodge and mature.

Protection

- The blood plays several roles in inflammation, a mechanism for limiting the spread of infection.
- White blood cells destroy microorganisms and cancer cells.
- Antibodies and other blood proteins neutralize toxins and help to destroy pathogens (disease agents).
- Platelets secrete factors that initiate blood clotting and other processes for minimizing blood loss.

Regulation

- By absorbing or giving off fluid under different conditions, the blood capillaries help to stabilize fluid distribution in the body.
- By buffering acids and bases, blood proteins help to stabilize the pH of the extracellular fluids.
- Shifts in blood flow help to regulate body temperature by routing blood to the skin for heat loss or retaining blood deeper in the body for heat retention.

Considering the importance of efficiently transporting nutrients, wastes, hormones, and especially oxygen from place to place, it is easy to understand why an excessive loss of blood is quickly fatal, and why the circulatory system needs mechanisms for minimizing such losses.

Components and General Properties of Blood

All of the foregoing functions depend, of course, on the characteristics of the blood. Adults generally have about 4 to 6 liters of blood. It is a liquid connective tissue composed, like other connective tissues, of cells and an extracellular matrix. The matrix is the blood **plasma,** a clear, light yellow fluid constituting a little over half of the blood volume. Suspended in the plasma are the **formed elements**—cells and cell fragments including the red blood cells, white blood cells, and platelets (fig. 19.1). The term *formed element* is used to denote that these are membrane-enclosed bodies with a definite structure visible with the microscope. Strictly speaking, they can't all be called cells because the platelets, as explained later, are merely fragments of certain bone marrow cells. The formed elements are classified as follows:

Erythrocytes[3] (red blood cells, RBCs)

Platelets

---

[1]*cardio* = heart + *vas* = vessel
[2]*hem, hemato* = blood + *logy* = study of

[3]*erythro* = red + *cyte* = cell

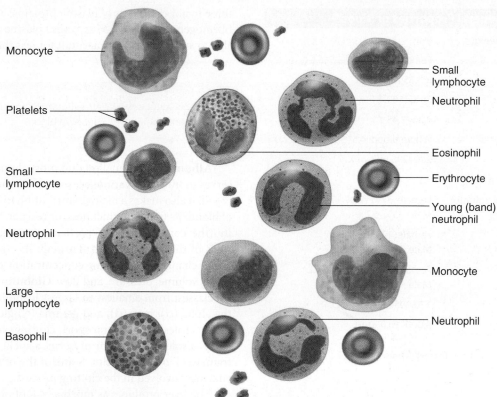

Figure 19.1 **The Formed Elements of Blood.**
• *What do erythrocytes and platelets lack that the other formed elements have?*

Leukocytes[4] (white blood cells, WBCs)

   Granulocytes

      Neutrophils

      Eosinophils

      Basophils

   Agranulocytes

      Lymphocytes

      Monocytes

Thus, there are seven kinds of formed elements: the erythrocytes, platelets, and five kinds of leukocytes. The five leukocyte types are divided into two categories, the *granulocytes* and *agranulocytes,* on grounds explained later.

The ratio of formed elements to plasma can be seen by taking a sample of blood in a tube and spinning it for a few minutes in a centrifuge (fig. 19.2). Erythrocytes, the densest elements, settle to the bottom of the tube and typically constitute about 37% to 52% of the total volume of whole blood—a value called the *hematocrit.* Leukocytes and platelets settle into a narrow cream- or buff-colored layer called the *buffy coat* just above the erythrocytes; they total 1% or less of the blood volume. At the top of the tube is the plasma, usually constituting about 47% to 63% of the volume.

Some general properties of blood are listed in table 19.1. Some of the terms in that table are defined later in this chapter.

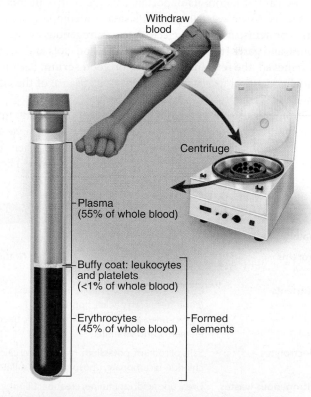

Figure 19.2 **Separating the Plasma and Formed Elements of Blood.** A small sample of blood is taken in a glass tube and spun in a centrifuge to separate the cells from the plasma. Erythrocytes are the densest components and settle to the bottom of the tube; platelets and WBCs are next; and plasma remains at the top of the tube.

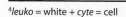

[4]*leuko* = white + *cyte* = cell

TABLE 19.1	General Properties of Blood*
Mean fraction of body weight	8%
Volume in adult body	Female: 4–5 L; male: 5–6 L
Volume/body weight	80–85 mL/kg
Mean temperature	38°C (100.4°F)
pH	7.35–7.45
Viscosity (relative to water)	Whole blood: 4.5–5.5; plasma: 2.0
Osmolarity	280–296 mOsm/L
Mean salinity (mainly NaCl)	0.9%
Hematocrit (packed cell volume)	Female: 37%–48% Male: 45%–52%
Hemoglobin	Female: 12–16 g/dL Male: 13–18 g/dL
Mean RBC count	Female: 4.2–5.4 million/µL Male: 4.6–6.2 million/µL
Platelet count	130,000–360,000/µL
Total WBC count	5,000–10,000/µL

*Values vary slightly depending on the testing methods used.

Blood Plasma

Even though blood plasma has no anatomy that we can study visually, we cannot ignore its importance as the matrix of this liquid connective tissue we call blood. Plasma is a complex mixture of water, proteins, nutrients, electrolytes, nitrogenous wastes, hormones, and gases (table 19.2). When the blood clots and the solids are removed, the remaining fluid is the blood **serum.** Serum is essentially identical to plasma except for the absence of the clotting protein fibrinogen.

Protein is the most abundant plasma solute by weight, totaling 6 to 9 grams per deciliter (g/dL). Plasma proteins play a variety of roles including clotting, defense, and transport of other solutes such as iron, copper, lipids, and hydrophobic hormones. There are

TABLE 19.2	Composition of Blood Plasma
Water	92% by weight
Proteins	Albumin, globulins, fibrinogen, other clotting factors, enzymes, and others
Nutrients	Glucose, amino acids, lactic acid, lipids (cholesterol, fatty acids, lipoproteins, triglycerides, and phospholipids), iron, trace elements, and vitamins
Electrolytes	Salts of sodium, potassium, magnesium, calcium, chloride, bicarbonate, phosphate, and sulfate
Nitrogenous wastes	Urea, uric acid, creatinine, creatine, bilirubin, and ammonia
Hormones	All hormones transported in the blood
Gases	Oxygen, carbon dioxide, and nitrogen

three major categories of plasma proteins: albumin, globulins, and fibrinogen (table 19.3). Many other plasma proteins are indispensable to survival, but they account for less than 1% of the total.

Apply What You Know

Based on your body weight, estimate the volume (in liters) and weight (in kilograms) of your own blood.

Albumin is the smallest and most abundant plasma protein. It serves to transport various plasma solutes and buffer the pH of the blood. It also makes a major contribution to two physical properties of blood: its *viscosity* (thickness, or resistance to flow) and *osmolarity* (the concentration of particles that cannot pass through the walls of the blood vessels). Through its effects on these two variables, changes in albumin concentration can significantly affect blood volume, pressure, and flow. **Globulins** are divided into three subclasses; from smallest to largest in molecular weight, they are the alpha (α), beta (β), and gamma (γ) globulins. Globulins play various roles in solute transport, clotting, and immunity. **Fibrinogen** is a soluble precursor of *fibrin,* a sticky protein that forms the framework of a blood clot. Some of the other plasma proteins are enzymes involved in the clotting process.

The liver produces as much as 4 g of plasma protein per hour, contributing all of the major proteins except gamma globulins. The gamma globulins come from *plasma cells*—connective tissue cells that are descended from white blood cells called *B lymphocytes.*

TABLE 19.3	Major Proteins of the Blood Plasma
Proteins	**Functions**
Albumin (60%)*	Major contributor to blood viscosity and osmolarity; transports lipids, hormones, calcium, and other solutes; buffers blood pH
Globulins (36%)*	
Alpha (α) globulins	
Haptoglobulin	Transports hemoglobin released by dead erythrocytes
Ceruloplasmin	Transports copper
Prothrombin	Promotes blood clotting
Others	Transport lipids, fat-soluble vitamins, and hormones
Beta (β) globulins	
Transferrin	Transports iron
Complement proteins	Aid in destruction of toxins and microorganisms
Others	Transport lipids
Gamma (γ) globulins	Antibodies; combat pathogens
Fibrinogen (4%)*	Becomes fibrin, the major component of blood clots

*Percentage of the total plasma protein by weight

Apply What You Know

How could a disease such as liver cancer or hepatitis result in impaired blood clotting?

Before You Go On

Answer the following questions to test your understanding of the preceding section:

1. List some transport, protective, and regulatory functions of the blood.

2. What are the two principal components of the blood? Outline the classification of its formed elements.

3. List the three major classes of plasma proteins. Which one is missing from blood serum?

4. What are the functions of blood albumin?

19.2 Erythrocytes

▶ Expected Learning Outcomes

When you have completed this section, you should be able to

- describe the morphology and functions of erythrocytes (RBCs);

- explain some clinical measurements of RBC and hemoglobin quantities;

- describe the structure and function of hemoglobin;

- discuss the formation, life span, death, and disposal of RBCs; and

- explain the chemical and immunologic basis and the clinical significance of blood types.

Erythrocytes, or **red blood cells (RBCs),** have two principal functions: (1) to pick up oxygen from the lungs and deliver it to tissues elsewhere, and (2) to pick up carbon dioxide from the tissues and unload it in the lungs. RBCs are the most abundant formed elements of the blood and therefore the most obvious things one sees upon its microscopic examination. They are also the most critical to survival; a severe deficiency of leukocytes or platelets can be fatal within a few days, but a severe deficiency of erythrocytes can be fatal within a few minutes. It is the lack of life-giving oxygen, carried by erythrocytes, that leads rapidly to death in cases of major trauma or hemorrhage.

Form and Function

An erythrocyte is a discoid cell with a biconcave shape—a thick rim and a thin sunken center. It is about 7.5 μm in diameter and 2.0 μm thick at the rim (fig. 19.3). While most cells, including white blood cells, have an abundance of organelles, RBCs lose their nucleus and

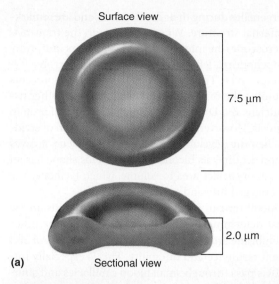

Surface view

7.5 μm

(a)   Sectional view

2.0 μm

(b)

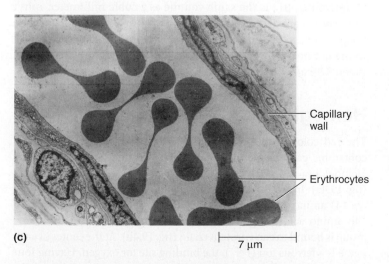

Capillary wall

Erythrocytes

(c)   7 μm

Figure 19.3 The Structure of Erythrocytes. (a) Dimensions and shape of an erythrocyte. (b) Erythrocytes on the tip of a hypodermic needle (SEM). (c) Erythrocytes in a blood capillary (TEM). Note the lack of internal structure in the cells.
● *Why do erythrocytes look caved in at the center?*

nearly all other organelles during their development and are remarkably devoid of internal structure. When viewed with the transmission electron microscope, the interior of an RBC appears uniformly gray. Lacking mitochondria, RBCs rely exclusively on anaerobic fermentation to produce ATP. The lack of aerobic respiration prevents them from consuming the oxygen they must transport to other tissues. Lacking a nucleus and DNA, RBCs also are incapable of protein synthesis and mitosis. However, the lack of a nucleus has an overriding advantage. When the developing RBC loses its nucleus, it caves in at the center and acquires its biconcave shape. This shape has an especially high ratio of surface area to volume, which facilitates the rapid diffusion of oxygen into and out of the cell.

The plasma membrane of a mature RBC has glycolipids on the outer surface that determine a person's blood type. On the inner surface of the membrane are two cytoskeletal proteins, *spectrin* and *actin,* which give it resilience and durability. This is especially important when RBCs pass through small blood capillaries and sinusoids. Many of these passages are narrower than the diameter of an RBC, forcing the RBCs to stretch, bend, and fold as they squeeze through. When they enter larger vessels, RBCs spring back to their discoid shape, like an air-filled inner tube.

With no space occupied by organelles, RBCs can carry more hemoglobin for a given cell volume. Their cytoplasm consists mainly of a 33% solution of hemoglobin (about 280 million molecules per cell). Hemoglobin is known especially for its oxygen-transport function, but it also aids in the transport of carbon dioxide and the buffering of blood pH.

Quantity of Erythrocytes

The quantity of circulating erythrocytes is critically important to health, because it determines the amount of oxygen the blood can carry. Two of the most routine measurements in hematology are measures of RBC quantity: the **RBC count** and the **hematocrit,**[5] or **packed cell volume (PCV).** The RBC count is normally 4.6 to 6.2 million RBCs/μL in men and 4.2 to 5.4 million/μL in women. (A microliter, μL, is the same volume as a cubic millimeter, mm^3; RBC counts are also expressed as $RBCs/mm^3$.) The hematocrit is the percentage of blood volume composed of RBCs (see fig. 19.2). In men, it normally ranges between 45% and 52%; in women, between 37% and 48%.

Hemoglobin

The red color of blood is due to its **hemoglobin (Hb),** an iron-containing gas-transport protein normally found only in the RBCs. Hemoglobin consists of four protein chains called **globins** (fig. 19.4a). In adult hemoglobin, two of these, the *alpha (α) chains,* are 141 amino acids long, and the other two, the *beta (β) chains,* are 146 amino acids long. A nonprotein component called the **heme** group is bound to each protein chain (fig. 19.4b). At the center of each heme is a ferrous ion (Fe^{2+}), the binding site for oxygen. Having four heme groups, each hemoglobin molecule can transport up to 4 O_2. About 5% of the CO_2 in the bloodstream is also transported by hemo-

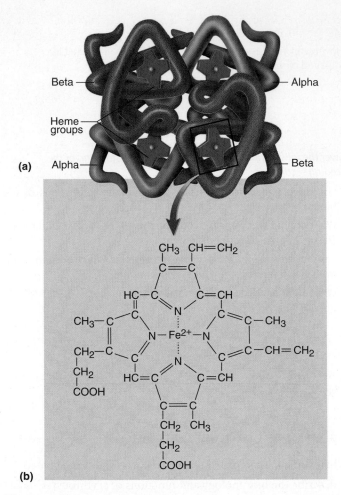

(a)

(b)

Figure 19.4 **The Structure of Hemoglobin.** (a) The hemoglobin molecule consists of two alpha proteins and two beta proteins, with a nonprotein heme group bound to each. (b) Structure of the heme group.
● *Where does oxygen bind to this molecule?*

globin, but this is bound to the globin component rather than to the heme, and a hemoglobin molecule can therefore transport both gases simultaneously. The **hemoglobin concentration** of whole blood is normally 13 to 18 g/dL in men and 12 to 16 g/dL in women.

The Erythrocyte Life Cycle

The production of red blood cells is called **erythropoiesis**[6] (eh-RITH-ro-poy-EE-sis). It is one aspect of the more general process called **hemopoiesis** (HE-mo-poy-EE-sis), the production of all formed elements of the blood. A knowledge of hemopoiesis provides a foundation for understanding leukemia, anemia, and other blood disorders. We will survey some general aspects of hemopoiesis before examining erythropoiesis specifically.

The tissues that produce blood cells are called **hemopoietic tissues.** The first hemopoietic tissues of the human embryo form in the *yolk sac,* a membrane associated with all vertebrate embryos (see fig. 4.5, p. 89). In most vertebrates (fish, amphibians, reptiles,

---

[5]*hemato* = blood + *crit* = to separate

[6]*erythro* = red + *poiesis* = formation of

and birds), this sac encloses the egg yolk, transfers yolk nutrients to the growing embryo, and produces the forerunners of the first blood cells. Even animals that don't lay eggs, however, have a yolk sac that retains its hemopoietic function. (It is also the source of cells that later produce eggs and sperm.) Cell clusters called *blood islands* form here by the third week of human development. They produce stem cells that migrate into the embryo proper and colonize the bone marrow, liver, spleen, and thymus. Here, the stem cells multiply and give rise to blood cells throughout fetal development. The liver stops producing blood cells around the time of birth. The spleen stops producing RBCs soon after, but it continues to produce lymphocytes for life.

From infancy onward, the red bone marrow produces all seven kinds of formed elements, while lymphocytes are produced not only there but also in the lymphatic tissues and organs—especially the thymus, tonsils, lymph nodes, spleen, and patches of lymphatic tissue in the mucous membranes. Blood formation in the bone marrow and lymphatic organs is called, respectively, **myeloid**[7] and **lymphoid hemopoiesis.**

All formed elements trace their origins to a common type of bone marrow stem cell, the **pluripotent stem cell (PPSC)** (or **hemopoietic stem cell;** formerly called a hemocytoblast[8]). PPSCs are so named because they have the potential to develop into multiple mature cell types. They multiply at a relatively slow rate and thus maintain a small population in the bone marrow. Some of them go on to differentiate into a variety of more specialized cells called **colony-forming units (CFUs),** each type destined to produce one or another class of formed elements. Through a series of **precursor cells,** the CFUs divide and differentiate into mature formed elements.

Erythropoiesis itself begins when a PPSC becomes an *erythrocyte colony-forming unit (ECFU)* (fig. 19.5). The hormone *erythropoietin (EPO)* stimulates the ECFU to develop into a *proerythroblast,* followed by an *erythroblast (normoblast).* Erythroblasts multiply and synthesize hemoglobin. When this task is completed, the nu-

cleus shrivels and is discharged from the cell. The cell is now called a *reticulocyte,*[9] named for a temporary network composed of clusters of ribosomes *(polyribosomes).* Reticulocytes leave the bone marrow and enter the circulating blood. When the last of the polyribosomes disintegrate and disappear, the cell is a mature erythrocyte. Normally, about 0.5% to 1.5% of the circulating RBCs are reticulocytes, but this percentage rises when the body is making RBCs especially rapidly, as when compensating for blood loss.

The entire process of transformation from PPSC to mature RBC takes 3 to 5 days and involves four major developments—a reduction in cell size, an increase in cell number, the synthesis of hemoglobin, and the loss of the nucleus and other organelles. The process normally generates about 2.5 million RBCs per second, or 20 mL of packed RBCs per day.

The average RBC lives about 120 days after its release from the bone marrow. As it ages, its membrane proteins (especially spectrin) deteriorate, and the membrane grows increasingly fragile. Without a nucleus or ribosomes, an RBC cannot synthesize new spectrin. Eventually, it ruptures as it tries to flex its way through narrow capillaries and sinusoids. The spleen has been called an "erythrocyte graveyard" because RBCs have an especially difficult time passing through its small channels. Here the old cells become trapped, broken up, and destroyed.

The 120-day life span of an RBC is often described as relatively short, owing to the fact that it lacks protein synthesis organelles and cannot repair itself. However, this is actually a relatively long life compared to other formed elements. Most leukocytes (which have nuclei) live less than a week, and platelets live about 10 days. Only monocytes and lymphocytes outlive the RBCs.

Blood Types

There are numerous genetically determined *blood groups* in the human population, each of which contains multiple **blood types.** The most familiar of these are the ABO group (with blood types A, B,

[7]*myel* = bone marrow
[8]*hemo* = blood + *cyto* = cell + *blast* = precursor

[9]*reticulo* = little network + *cyte* = cell

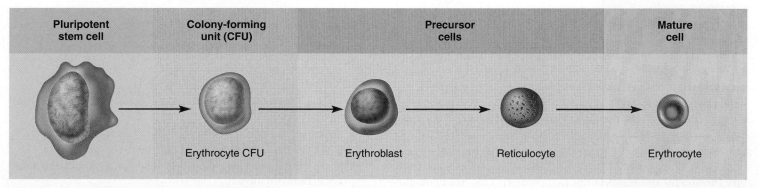

Figure 19.5 **Erythropoiesis.** Stages in the development of erythrocytes.

DEEPER INSIGHT   19.1

Bone Marrow and Cord Blood Transplants

A bone marrow transplant is one treatment option for leukemia, sickle-cell disease, some forms of anemia, and other disorders. The principle is to replace cancerous or otherwise defective marrow with donor stem cells in hopes that they will rebuild a population of normal marrow and blood cells. The patient is first given chemotherapy or radiation to destroy the defective marrow and eliminate immune cells (T cells) that would attack the donated marrow. Bone marrow is drawn from the donor's sternum or hip bone and injected into the recipient's circulatory system. Donor stem cells colonize the patient's marrow cavities and, ideally, build healthy marrow.

There are, however, several drawbacks to bone marrow transplant. For one, it is difficult to find compatible donors. Surviving T cells in the patient may attack the donor marrow, and donor T cells may attack the patient's tissues (the *graft-versus-host response*). To inhibit graft rejection, the patient must take immunosuppressant drugs for life. These drugs leave a person vulnerable to infection and have many other adverse side effects. Infections are sometimes contracted from the donated marrow itself. In short, marrow transplant is a high-risk procedure; up to one-third of patients die from complications of treatment.

An alternative with several advantages is to use blood from placentas, which are normally discarded at every childbirth. Placental blood contains more stem cells than adult bone marrow, and is less likely to carry infectious microbes. With the parents' consent, it can be harvested from the umbilical cord with a syringe; and it can be stored almost indefinitely, frozen in liquid nitrogen at cord blood banks. The immature immune cells in cord blood have less tendency to attack the recipient's tissues; thus, cord blood transplants have lower rejection rates and do not require as close a match between donor and recipient, meaning that more donors are available to a patient in need. Pioneered in the 1980s, cord blood transplants have successfully treated leukemia and a wide range of other blood diseases. Efforts are being made to further improve the procedure by stimulating placental stem cells to multiply before the transplant, and by removing placental T cells that may react against the recipient.

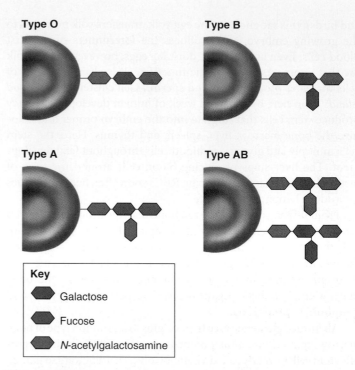

Figure 19.6 Chemical Basis of the ABO Blood Types. Blood types are determined by the chemistry of glycolipids of the RBC plasma membrane. Types A, B, AB, and O erythrocytes differ only in the terminal three to four carbohydrates of the glycolipid molecules. All of them end with galactose and fucose, but they differ in the presence or absence, and type, of additional sugar bonded to the galactose. The additional sugar is absent in type O; it is *N*-acetylgalactosamine in type A; it is another galactose in type B; and type AB cells have both the type A and B chains.

AB, and O) and Rh group (with blood types Rh-positive and Rh-negative). These types differ with respect to the chemical composition of glycolipids on the RBC surface; figure 19.6 shows how the ABO types differ in this regard. The glycolipids act as *antigens*, substances capable of evoking an immune reaction. The blood plasma contains *antibodies* that react against incompatible antigens on foreign RBCs.

RBC antigens and plasma antibodies determine the compatibility of donor and recipient blood in transfusions. For example, a person with blood type A has anti-B antibodies in the blood plasma. If this person is mistakenly given a transfusion of type B blood, those antibodies attack the donor RBCs. The RBCs agglutinate—they form clumps that obstruct the circulation in small blood vessels, with devastating consequences for such crucial organs as the brain, heart, lungs, and kidneys. The agglutinated RBCs also rupture *(hemolyze)* and release their hemoglobin. This free hemoglobin is filtered out by the kidneys, clogs the microscopic kidney tubules, and can cause death from renal failure within a week or so.

The same can happen if a type B person receives type A blood, or if a type O person receives either type A or type B. Incompatibility of Rh types between the mother and fetus sometimes causes severe anemia in the newborn infant *(hemolytic disease of the newborn).*

Apply What You Know

Why might a court of law be interested even in human blood types that have no connection to disease?

Before You Go On

Answer the following questions to test your understanding of the preceding section:

5. What are the two main functions of RBCs?

6. Define *hematocrit* and *RBC count,* and state some normal clinical values for each.

7. Describe the structure of a hemoglobin molecule. Explain where O_2 and CO_2 are carried on a hemoglobin molecule.

8. Name the stages in the production of an RBC, and state the differences between them.

9. Explain what plasma and RBC components are responsible for blood types, and why blood types are clinically important.

19.3 Leukocytes

▶ Expected Learning Outcomes

When you have completed this section, you should be able to

- describe the appearance of the five kinds of leukocytes;
- explain the function of WBCs in general and the individual roles of each WBC type;
- describe the formation and life history of WBCs; and
- describe the production, death, and disposal of leukocytes.

Form and Function

Leukocytes, or **white blood cells (WBCs),** are the least abundant formed elements, totaling only 5,000 to 10,000 WBCs/μL. Yet we cannot live long without them, because they afford protection against infectious microorganisms and other pathogens. WBCs are easily recognized in stained blood films because they have conspicuous nuclei that stain from light violet to dark purple with the most common blood stains. They are much more abundant in the body than their low number in blood films would suggest, because they spend only a few hours in the bloodstream, then migrate through the walls of the capillaries and venules and spend the rest of their lives in the connective tissues. It's as if the bloodstream were merely the subway that the WBCs take to work; in blood films, we see only the ones on their way to work, not the WBCs already in the tissues.

Leukocytes differ from erythrocytes in that they retain their organelles throughout life; thus, when viewed with the transmission electron microscope, they show a complex internal structure (fig. 19.7). Among these organelles are the usual instruments of protein synthesis—the nucleus, rough endoplasmic reticulum, ribosomes, and Golgi complex—for leukocytes must synthesize proteins in order to carry out their functions. Some of these proteins are packaged into lysosomes and other organelles, which appear as conspicuous cytoplasmic granules that distinguish one WBC type from another.

Types of Leukocytes

As outlined at the beginning of this chapter, there are five kinds of leukocytes (table 19.4). They are distinguished from each other by their relative size and abundance, the size and shape of their nuclei, the presence or absence of certain cytoplasmic granules, the coarseness and staining properties of those granules, and most importantly by their functions. All WBCs have lysosomes called **nonspecific (azurophilic[10]) granules** in the cytoplasm, so named because they absorb the blue or violet dyes of blood stains. Three of the five WBC types—neutrophils, eosinophils, and basophils—are called **granulocytes** because they also have various types of **specific granules** that stain conspicuously and distinguish each cell type from the others. Basophils are named for the fact that their

[10]*azuro* = blue + *philic* = loving

specific granules stain intensely with methylene blue, a basic dye in a common blood staining mixture called Wright's stain. Eosinophils are so named because their specific granules stain with eosin, an acidic dye in Wright's stain. The specific granules of neutrophils do not stain intensely with either dye. Specific granules contain enzymes and other chemicals employed in defense against pathogens. The two remaining WBC types—monocytes and lymphocytes—are called **agranulocytes** because they lack specific granules. Nonspecific granules are inconspicuous to the light microscope, and these cells therefore have relatively clear-looking cytoplasm.

Granulocytes

- **Neutrophils** (NEW-tro-fills) are the most abundant WBCs—generally about 4,150 cells/μL and constituting 60% to 70% of the circulating leukocytes. The nucleus is clearly visible and, in a mature neutrophil, typically consists of three to five lobes connected by slender nuclear strands. These strands are sometimes so delicate that they are scarcely visible, and the neutrophil may seem as if it has multiple nuclei. Young neutrophils have an undivided nucleus often shaped like a band; thus, they are called *band cells* or *stab[11] cells.* Neutrophils are also called *polymorphonuclear leukocytes (PMNs)* because of their varied nuclear shapes.

    The cytoplasm contains fine reddish to violet specific granules, which contain lysozyme and other antimicrobial agents. The individual granules are very small, barely visible with the light microscope, but they give the cytoplasm a pale lilac color.

    The neutrophil count rises—a condition called *neutrophilia*—in response to bacterial infections. The primary

[11]*stab* = band, bar (German)

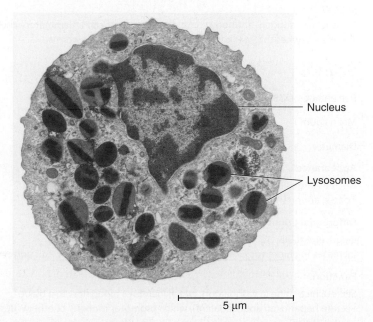

Figure 19.7 Structure of an Eosinophil. In contrast to an RBC, the WBC cytoplasm is crowded with organelles, including a nucleus.

Nucleus

Lysosomes

5 μm

TABLE 19.4	**The White Blood Cells (Leukocytes)**

Neutrophils

Percentage of WBCs	60%–70%
Mean count	4,150 cells/μL
Diameter	9–12 μm

Appearance*
Nucleus usually with 3–5 lobes in S- or C-shaped array
Fine reddish to violet granules in cytoplasm

Differential Count
Increases in bacterial infections

Functions
Phagocytize bacteria
Secrete antimicrobial chemicals

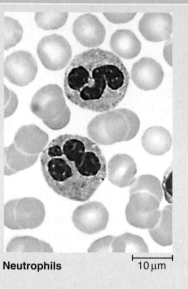

Neutrophils   10 μm

Eosinophils

Percentage of WBCs	2%–4%
Mean count	170 cells/μL
Diameter	10–14 μm

Appearance*
Nucleus usually with two large lobes connected by thin strand
Large orange-pink granules in cytoplasm

Differential Count
Fluctuates greatly from day to night, seasonally, and with phase of menstrual cycle
Increases in parasitic infections, allergies, collagen diseases, and diseases of spleen
and central nervous system

Functions
Phagocytize antigen–antibody complexes, allergens, and inflammatory chemicals
Secrete enzymes that weaken or destroy parasites such as worms

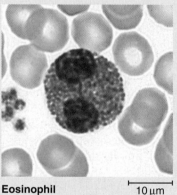

Eosinophil   10 μm

Basophils

Percentage of WBCs	<0.5%
Mean count	40 cells/μL
Diameter	8–10 μm

Appearance*
Nucleus large and irregularly shaped, but typically obscured from view
Coarse, abundant, dark violet granules in cytoplasm

Differential Count
Relatively stable
Increases in chickenpox, sinusitis, diabetes mellitus, myxedema, and polycythemia

Functions
Secrete histamine (a vasodilator), which increases blood flow to a tissue
Secrete heparin (an anticoagulant), which promotes mobility of other WBCs by preventing clotting

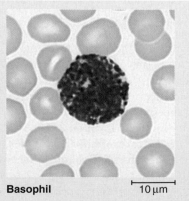

Basophil   10 μm

TABLE 19.4	The White Blood Cells (Leukocytes) (continued)

Monocytes

Percentage of WBCs	3%–8%
Mean count	460 cells/μL
Diameter	12–15 μm

Appearance*

Nucleus ovoid, kidney-shaped, or horseshoe-shaped
Abundant light violet cytoplasm with sparse, fine granules
Sometimes very large with stellate or polygonal shapes

Differential Count

Increases in viral infections and inflammation

Functions

Differentiate into macrophages (large phagocytic cells of the tissues)
Phagocytize pathogens, dead neutrophils, and debris of dead cells
Present antigens to activate other cells of immune system

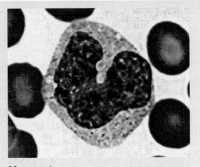
Monocyte   10 μm

Lymphocytes

Percentage of WBCs	25%–33%
Mean count	2,200 cells/μL
Diameter	
Small class	5–8 μm
Medium class	10–12 μm
Large class	14–17 μm

Appearance*

Nucleus round, ovoid, or slightly dimpled on one side, of uniform dark violet color
In small lymphocytes, nucleus fills nearly all of the cell and leaves only a scanty rim of clear, light blue cytoplasm In larger lymphocytes, cytoplasm is more abundant; large lymphocytes may be hard to differentiate from monocytes

Differential Count

Increases in diverse infections and immune responses

Functions

Several functional classes indistinguishable by light microscopy
Destroy cancer cells, cells infected with viruses, and foreign cells
Present antigens to activate other cells of immune system
Coordinate actions of other immune cells
Secrete antibodies
Serve in immune memory

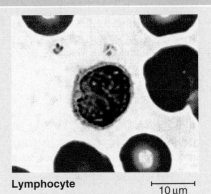
Lymphocyte   10 μm

*Appearance pertains to blood films dyed with Wright's stain.

task of the neutrophils is to destroy bacteria, which they achieve in two ways. One is to phagocytize and digest them. The other is to release a potent mix of toxic chemicals, including hypochlorite (HClO) (the active agent in household bleach) and the superoxide anion ($O_2 \bullet^-$), which reacts with hydrogen ions to produce hydrogen peroxide (H_2O_2). Just as bleach and hydrogen peroxide are often used around the home as disinfectants, they are deadly to bacteria in the tissues. These chemicals form a **killing zone** around the neutrophil, lethal to the invaders but also to the neutrophil itself. Neutrophils are thus the body's suicidal guardians against infection.

- **Eosinophils** (EE-oh-SIN-oh-fills) are harder to find in a blood film because they are only 2% to 4% of the WBC total, typically numbering about 170 cells/μL. The eosinophil count fluctuates greatly, however, from day to night, seasonally, and with the phase of the menstrual cycle. It rises (*eosinophilia*) in allergies, parasitic infections, collagen diseases, and diseases of the spleen and central nervous system. Although relatively scanty in the blood, eosinophils are abundant in the mucous membranes of the respiratory, digestive, and lower urinary tracts. The eosinophil nucleus usually has two large lobes connected by a thin strand, and the cytoplasm has an abundance of coarse rosy to orange specific granules. Eosinophils secrete chemicals that weaken or destroy relatively large parasites such as hookworms and tapeworms, too big for any one WBC to phagocytize. Eosinophils also phagocytize and dispose of inflammatory chemicals, antigen–antibody complexes (masses of antigen molecules stuck together by antibodies), and allergens (foreign antigens that trigger allergies).

- **Basophils** (BASE-oh-fills) are the rarest of the WBCs and, indeed, of all formed elements. There are about 40 basophils per microliter—usually less than 0.5% of the WBC count. They can be recognized mainly by an abundance of very coarse, dark violet specific granules. The nucleus is largely hidden from view by these granules, but is large, pale, and typically S- or U-shaped. Basophils secrete two chemicals that aid in the body's defense processes: (1) **histamine,** a vasodilator that widens the blood vessels, speeds the flow of blood to an injured tissue, and makes the blood vessels more permeable so that blood components such as neutrophils and clotting proteins can get into the connective tissues more quickly; and (2) **heparin,** an anticoagulant that inhibits blood clotting and thus promotes the mobility of other WBCs in the area. They also release chemical signals that attract eosinophils and neutrophils to a site of infection.

Agranulocytes

- **Monocytes** (MON-oh-sites) are the largest WBCs, often two or three times the diameter of an RBC. There are about 460 monocytes per microliter—about 3% to 8% of the WBC count. The nucleus is large and clearly visible, often relatively light violet, and typically ovoid, kidney-shaped, or horseshoe-shaped. The cytoplasm is abundant and contains sparse, fine granules. In prepared blood films, monocytes often assume sharply angular to spiky shapes (see fig. 19.1). The monocyte count rises in inflammation and viral infections. Monocytes go to work only after leaving the bloodstream and transforming into large tissue cells called **macrophages**[12] (MAC-ro-fay-jez). Macrophages are highly phagocytic cells that consume up to 25% of their own volume per hour. They destroy dead or dying host and foreign cells, pathogenic chemicals and microorganisms, and other foreign matter. They also chop up or "process" foreign antigens and "display" fragments of them on the cell surface to alert the immune system to the presence of a pathogen. Thus, they and a few other cells are called *antigen-presenting cells (APCs).* There are several kinds of macrophages in the body descended from monocytes or from the same hemopoietic stem cells as monocytes. Macrophages of the loose connective tissues are generally called simply *macrophages* or sometimes *histiocytes.* Macrophages in some other localities have special names:
  - **dendritic cells** in the epidermis (see p. 108) and the mucous membranes of the mouth, esophagus, and vagina;
  - **microglia,** a type of neuroglia in the central nervous system (see p. 358);

- **alveolar macrophages** in the pulmonary alveoli (see p. 643); and
- **hepatic macrophages (Kupffer cells)** in the liver sinusoids (see p. 677).

- **Lymphocytes** (LIM-fo-sites) include the smallest WBCs; at 5 to 17 μm in diameter, they range from smaller than RBCs to two and a half times as large, but those in circulating blood are generally at the small end of the range. They are second to neutrophils in abundance and are thus quickly spotted when you examine a blood film. They number about 2,200 cells/μL and are 25% to 33% of the WBC count. The lymphocyte nucleus is round, ovoid, or slightly dimpled on one side. It usually stains dark violet and fills nearly the entire cell, especially in small lymphocytes. The cytoplasm, which stains a clear light blue color, forms a narrow and often barely detectable rim around the nucleus, although it is more abundant in the larger lymphocytes. Small lymphocytes are sometimes difficult to distinguish from basophils, but most basophils are conspicuously grainy whereas the lymphocyte nucleus is uniform or merely mottled. Basophils also lack the rim of clear cytoplasm seen in most lymphocytes. Large lymphocytes can be difficult to distinguish from monocytes.

The lymphocyte count increases in diverse infections and immune responses. Some of them function in nonspecific defense of the body against viruses and cancer, but most of them are involved in *specific immunity,* a defense in which the body recognizes a certain antigen it has encountered before and mounts such a quick response that a person notices little or no illness. The various lymphocytes are not distinguishable by light microscopy, but differ in their functions. As you will see in chapter 22, there are three functional classes of lymphocytes—*NK (natural killer) cells, B cells,* and *T cells,* which attack different categories of pathogens.

The Leukocyte Life Cycle

Leukopoiesis (LOO-co-poy-EE-sis), the production of white blood cells, begins with the same pluripotent stem cells (PPSCs) as erythropoiesis. Some PPSCs differentiate into the distinct types of colony-forming units, which then go on to produce the following cell lines (fig. 19.8):

1. *Myeloblasts,* which ultimately differentiate into the three types of granulocytes (neutrophils, eosinophils, and basophils)

2. *Monoblasts,* which look identical to myeloblasts but lead ultimately to monocytes

3. *Lymphoblasts,* which give rise to three types of lymphocytes (B lymphocytes, T lymphocytes, and natural killer cells)

Granulocytes and monocytes stay in the red bone marrow until they are needed; the marrow contains 10 to 20 times as many of these cells as the circulating blood does. Lymphocytes, by contrast, begin developing in the bone marrow but then migrate elsewhere. **B lymphocytes (B cells)** mature in the bone marrow and some remain there, while others disperse and colonize the lymph nodes,

---

[12]*macro* = big + *phage* = eater

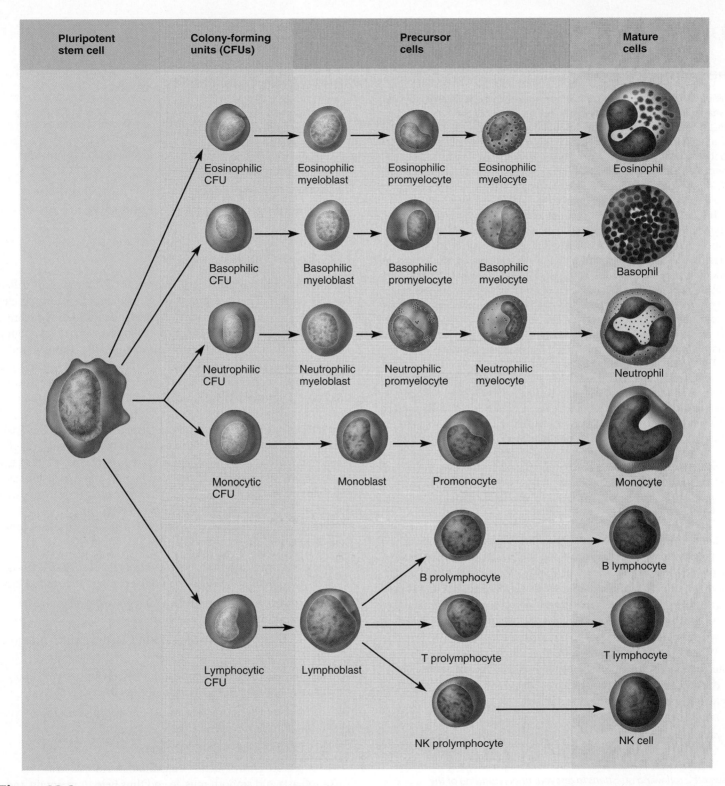

Pluripotent stem cell	Colony-forming units (CFUs)	Precursor cells			Mature cells
	Eosinophilic CFU	Eosinophilic myeloblast	Eosinophilic promyelocyte	Eosinophilic myelocyte	Eosinophil
	Basophilic CFU	Basophilic myeloblast	Basophilic promyelocyte	Basophilic myelocyte	Basophil
	Neutrophilic CFU	Neutrophilic myeloblast	Neutrophilic promyelocyte	Neutrophilic myelocyte	Neutrophil
	Monocytic CFU	Monoblast	Promonocyte		Monocyte
	Lymphocytic CFU	Lymphoblast	B prolymphocyte		B lymphocyte
			T prolymphocyte		T lymphocyte
			NK prolymphocyte		NK cell

Figure 19.8 **Leukopoiesis.** Stages in the development of leukocytes.
● *Explain the meaning and relevance of the combining form* myelo-, *seen in so many of these cell names.*

spleen, tonsils, and mucous membranes. To remember their site of maturation, it may help to think "B for bone marrow," although these cells were actually named for an organ in chickens (the *bursa of Fabricius*) where they were discovered. **T lymphocytes (T cells)** begin development in the bone marrow but then migrate to the thymus (a gland in the mediastinum just above the heart) and mature there. In this case, the *T* really does stand for *thymus*. Mature T lymphocytes disperse from the thymus and colonize the same organs as B lymphocytes. **Natural killer (NK) cells** develop in the bone marrow like B cells.

DEEPER INSIGHT   19.2

The Complete Blood Count

One of the most common clinical procedures, in both routine physical examinations and the diagnosis of disease, is a *complete blood count (CBC)*. The CBC yields a highly informative profile of data on multiple blood values: the number of RBCs, WBCs, and platelets per microliter of blood; the relative numbers (percentages) of each WBC type, called a *differential WBC count;* hematocrit; hemoglobin concentration; and various *RBC indices* such as RBC size *(mean corpuscular volume, MCV)* and hemoglobin concentration per RBC *(mean corpuscular hemoglobin, MCH).*

RBC and WBC counts used to require the microscopic examination of films of diluted blood on a calibrated slide, and a differential WBC count required examination of stained blood films. Today, most laboratories use *electronic cell counters.* These devices draw a blood sample through a very narrow tube with sensors that identify cell types and measure cell sizes and hemoglobin content. These counters give faster and more accurate results based on much larger numbers of cells than the old visual methods. However, cell counters still misidentify some cells, and a medical technologist must review the results for suspicious abnormalities and identify cells that the instrument cannot.

The wealth of information gained from a CBC is too vast to give more than a few examples here. Various forms of anemia are indicated by low RBC counts or abnormalities of RBC size, shape, and hemoglobin content. A platelet deficiency can indicate an adverse drug reaction. A high neutrophil count suggests bacterial infection, and a high eosinophil count suggests an allergy or parasitic infection. Elevated numbers of specific WBC types or WBC stem cells can indicate various forms of leukemia. If a CBC does not provide enough information or if it suggests other disorders, additional tests may be done, such as coagulation time and bone marrow biopsy.

Circulating WBCs do not stay in the blood for very long. Granulocytes circulate for 4 to 8 hours and then migrate into the tissues, where they live for another 4 or 5 days. Monocytes travel in the blood for 10 or 20 hours, then migrate into the tissues and transform into a variety of macrophages, which live as long as a few years. Lymphocytes, responsible for long-term immunity, have a life span ranging from weeks to decades.

When leukocytes die, they are generally phagocytized and digested by macrophages. Dead neutrophils, however, are responsible for the creamy color of pus, and are sometimes disposed of by the rupture of a blister onto the skin surface.

Before You Go On

Answer the following questions to test your understanding of the preceding section:

10. What is the purpose of WBCs in general?

11. Name the five kinds of WBCs and state the specific functions of each.

12. Describe the key features that enable one to microscopically identify each WBC type.

13. What are macrophages? What class of WBCs do they arise from? Name some types of macrophages.

19.4 Platelets

▶ Expected Learning Outcomes

When you have completed this section, you should be able to

- describe the structure of blood platelets;
- explain the multiple roles played by platelets in hemostasis and blood vessel maintenance;
- describe platelet production and state their longevity; and
- describe the general processes of hemostasis.

Form and Function

Platelets are not cells but small fragments of marrow cells called *megakaryocytes.* They are the second most abundant formed elements, after erythrocytes; a normal platelet count in blood from a fingerstick ranges from 130,000 to 400,000 platelets/μL (averaging about 250,000). The platelet count can vary greatly, however, under different physiological conditions and in blood samples taken from various places in the body. In spite of their numbers, platelets are so small (2 to 4 μm in diameter) that they contribute even less than the WBCs to the blood volume.

Platelets have a complex internal structure that includes lysosomes, mitochondria, microtubules and microfilaments, granules filled with platelet secretions, and a system of channels called the **open canalicular system,** which opens onto the platelet surface (fig. 19.9a). They have no nucleus. When activated, they form pseudopods and are capable of ameboid movement.

Despite their small size, platelets have a greater variety of functions than any of the true blood cells:

- They secrete *vasoconstrictors,* chemicals that cause spasmodic constriction of broken vessels and thus help reduce blood loss.
- They stick together to form temporary *platelet plugs* to seal small breaks in injured blood vessels.
- They secrete *procoagulants,* or clotting factors, which promote blood clotting.
- They initiate the formation of a clot-dissolving enzyme that dissolves blood clots that have outlasted their usefulness.
- They secrete chemicals that attract neutrophils and monocytes to sites of inflammation.
- They internalize and destroy bacteria.
- They secrete *growth factors* that stimulate mitosis in fibroblasts and smooth muscle and thus help to maintain and repair blood vessels.

Platelet Production

The production of platelets is a division of hemopoiesis called **thrombopoiesis.** (Platelets are occasionally called *thrombocytes,*[13]

---
[13]*thrombo* = clotting + *cyte* = cell

but this term is now usually reserved for nucleated true cells in other animals such as birds and reptiles.) Some pluripotent stem cells produce receptors for the hormone *thrombopoietin,* thus becoming *megakaryoblasts,* cells committed to the platelet-producing line. The megakaryoblast duplicates its DNA repeatedly without undergoing nuclear or cytoplasmic division. The result is a **megakaryocyte**[14] (meg-ah-CAR-ee-oh-site), a gigantic cell up to 150 μm in diameter, visible to the naked eye, with a huge multilobed nucleus and multiple sets of chromosomes (fig. 19.9b). Most megakaryocytes live in the red bone marrow adjacent to blood-filled spaces called sinusoids, which are lined with a thin simple squamous epithelium called the endothelium (see fig. 22.9, p. 617).

A megakaryocyte sprouts tendrils called *proplatelets* that protrude through the endothelium into the blood of the sinusoid. The blood flow shears off the proplatelets, which break up into platelets as they travel in the bloodstream. Much of this breakup is thought

---

[14]*mega* = giant + *karyo* = nucleus + *cyte* = cell

to occur in the small vessels of the lungs; proplatelets are relatively abundant in blood entering the lungs, but blood exiting the lungs shows more platelets and few proplatelets. About 25% to 40% of the platelets are stored in the spleen and released as needed. The remainder circulate freely in the blood and live for about 10 days.

Hemostasis

Hemostasis[15] is the cessation of bleeding. The details of hemostasis are beyond the scope of an anatomy textbook, but the basic roles of platelets in the process will be briefly surveyed here. Upon injury to a blood vessel, platelets release *serotonin.* This chemical stimulates *vasoconstriction,* or narrowing of the blood vessel, to reduce blood loss. Platelets also adhere to the vessel wall and to each other, forming a sticky mass called a *platelet plug.* Platelet plugs temporarily seal breaks in small blood vessels. Platelets and injured tissues around the blood vessel also release *clotting factors.* Through a series of enzymatic reactions, clotting factors convert the plasma protein *fibrinogen* into the sticky protein *fibrin.* Fibrin adheres to the wall of the blood vessel, and as blood cells and platelets arrive, many of them stick to the fibrin like insects in a spider web. The resulting mass of fibrin, platelets, and blood cells (fig. 19.10) forms a clot that ideally seals the break in the blood vessel long enough for the vessel to heal.

---

[15]*hemo* = blood + *stasis* = stability

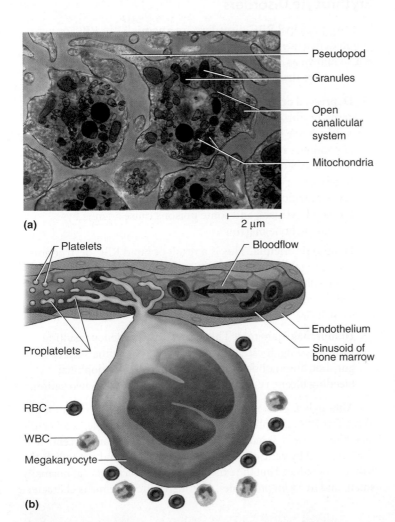

Figure 19.9 Platelets. (a) Structure of blood platelets (TEM). (b) Platelets being produced by the shearing of proplatelets from a megakaryocyte. Note the size of the megakaryocyte relative to the RBCs, WBCs, and platelets.

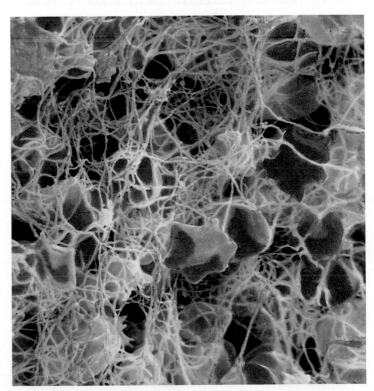

Figure 19.10 A Blood Clot (SEM). Platelets are seen trapped in a sticky protein mesh.
● *What is the name of this protein?*

Once the leak is sealed and the crisis has passed, platelets secrete *platelet-derived growth factor (PDGF)*, a substance that stimulates fibroblasts and smooth muscle to proliferate and replace the damaged tissue of the blood vessel. When tissue repair is completed and the blood clot is no longer needed, the clot must be disposed of. Platelets then secrete *factor XII*, a protein that initiates a series of reactions leading to the formation of a fibrin-digesting enzyme called *plasmin*. Plasmin dissolves the old blood clot.

Before You Go On

Answer the following questions to test your understanding of the preceding section:

14. List several functions of blood platelets.

15. How are blood platelets produced? How long do they live?

16. Briefly describe the stages in which platelets help to stop bleeding and repair a damaged blood vessel.

19.5 Clinical Perspectives

▶ Expected Learning Outcomes

When you have completed this section, you should be able to

- describe changes that occur in the blood in old age; and
- describe some common abnormalities of RBC, WBC, and platelet count and morphology, and the consequences of these abnormalities.

Hematology in Old Age

We have considered the embryonic origin and ongoing hemopoietic development of the blood in previous sections. At the other end of the life span, aging has multiple effects on the blood. Evidence suggests that the baseline rate of erythropoiesis does not change much with age; cell counts, hemoglobin concentration, and other variables are about the same among healthy people in their 70s as in the 30s. However, older people do not adapt well to stress on the hemopoietic system, perhaps because of the senescence of other organ systems. The stomach atrophies in old age, thus producing less of the *intrinsic factor* needed for the absorption of dietary vitamin B_{12}. A deficiency of vitamin B_{12} causes *pernicious anemia*. Anemia can also stem from atrophy of the kidneys in old age, since the kidneys secrete erythropoietin, the principal stimulus for RBC production. There may also be a limit to how many times the hemopoietic stem cells can divide and continue producing new blood cells.

Anemia may also result from nutritional deficiencies, inadequate exercise, disease, and other causes. The factors that cause anemia in older people are very complicated and interrelated. It is almost impossible to determine whether aging alone causes anemia in the absence of other contributing factors such as poor exercise or nutritional habits.

Thrombosis, the abnormal clotting of blood in an unbroken blood vessel, becomes increasingly problematic in old age. Plaques of atherosclerosis in the blood vessels can act as sites of blood clotting. The blood clots especially easily in the veins, where blood flow is slowest. About 25% of people over age 50 experience venous blockage by thrombosis, especially people who do not exercise regularly and people confined to a bed or wheelchair.

Disorders of the Blood

We conclude with a survey of some clinical aspects of hematology, especially disorders that affect the relative numbers of formed elements and thus the appearance of stained blood films, or that alter the appearance of the individual formed elements. Some common nonstructural blood disorders are described in table 19.5.

Erythrocyte Disorders

The two principal RBC disorders are **anemia**[16] (an RBC or hemoglobin deficiency) and **polycythemia**[17] (an RBC excess). The latter is also known as *erythrocytosis*.

There are three fundamental categories of anemia:

1. **Depressed erythropoiesis or hemoglobin synthesis.** In such cases, erythropoiesis fails to keep pace with the normal death of RBCs. We have already seen that atrophy of the stomach and kidneys in the elderly can reduce erythropoieic rates because of a deficiency of intrinsic factor and erythropoietin. At any age, a dietary deficiency of iron or certain vitamins can cause nutritional anemias (such as *iron-deficiency anemia*). Radiation, viruses, and some poisons cause anemia by destroying bone marrow.

2. **Hemolytic anemia.** This is a result of rapid RBC destruction exceeding the rate of erythropoiesis. Hemolytic anemia can result from a variety of poisons, drug reactions, sickle-cell disease, or blood-destroying parasitic infections such as malaria.

3. **Hemorrhagic anemia.** This is an RBC deficiency resulting from bleeding. It can be a consequence of trauma such as gunshot, automobile, or battlefield injuries; hemophilia; bleeding ulcers; ruptured aneurysms; or heavy menstruation.

Although anemia most obviously affects the RBC count, it can also affect RBC morphology. *Thalassemia,* for example, is a hereditary blood disease among people of Mediterranean descent. It is characterized by deficient hemoglobin synthesis, and not only is the RBC count reduced but the existing RBCs are *microcytic* (abnormally small) and *hypochromic* (pale). Iron-deficiency anemia is character-

---

[16]*an* = without + *em* = blood + *ia* = condition
[17]*poly* = many + *cyt* = cells + *hemia* = blood condition

TABLE 19.5	Some Disorders of the Blood
Disseminated intravascular coagulation (DIC)	Widespread clotting within unbroken vessels, limited to one organ or occurring throughout the body. Usually triggered by septicemia but also occurs when blood circulation slows markedly (as in cardiac arrest). Marked by widespread hemorrhaging, congestion of the vessels with clotted blood, and tissue necrosis in blood-deprived organs.
Embolism	The presence of any abnormal object (embolus) traveling in the bloodstream, such as an air bubble (air embolism), agglutinated RBCs or bacteria, or traveling blood clot (thromboembolism). Presents a danger of blocking small blood vessels and shutting off the blood flow to vital tissues, thus causing stroke, heart failure, kidney failure, or pulmonary failure.
Hemophilia	Abnormally slow blood clotting as a result of the hereditary deficiency of a clotting factor, usually factor VIII (a liver product). Prolonged bleeding results in the painful pooling of clotted blood (hematomas) in such sites as the muscles and joints, or in fatal blood loss. Treatable with injections of the missing clotting factor.
Infectious mononucleosis	Infection of B lymphocytes with Epstein–Barr virus. Usually transmitted by exchange of saliva, as in kissing; most common in adolescents and young adults. Causes fever, fatigue, sore throat, inflamed lymph nodes, and leukocytosis. Usually self-limiting and resolves within a few weeks.
Septicemia	Bacteria in the bloodstream, stemming from infection elsewhere in the body. Often causes fever, chills, and nausea, and may cause septic shock.
Thrombosis	Abnormal clotting in unbroken blood vessels, triggered by atherosclerosis and other defects, or by immobility of people confined to bed or a wheelchair. Stationary blood clots can cause stroke, heart failure, etc. (see embolism in this table), and clots can break free and cause thromboembolism.

Disorders Described Elsewhere

ized by these structural abnormalities as well as *poikilocytosis,*[18] in which RBCs assume teardrop, pencil, and other variable and abnormal shapes. Sickle-cell disease is another well-known hereditary anemia with abnormal RBC morphology (see Deeper Insight 19.3).

Whatever the cause of anemia, it may result in tiredness, lethargy, shortness of breath, or more serious consequences—organ deterioration (necrosis) stemming from oxygen deprivation (hypoxia).

Polycythemia (POL-ee-sih-THEME-ee-uh), an excess RBC count, can result from cancer of the bone marrow (*primary polycythemia*) or from a multitude of other conditions (*secondary polycythemia*) such as an abnormally high oxygen demand (as in people who engage in overzealous aerobic exercise) or low oxygen supply (as in people who live at high altitudes or suffer lung diseases such as emphysema). RBC counts can rise as high as 11 million RBCs/μL and hematocrit as high as 80%. The thick blood sludges in the vessels, tremendously increases blood pressure, and puts a dangerous strain on the cardiovascular system that can lead to heart failure or stroke.

Leukocyte Disorders

A WBC deficiency is called **leukopenia,** and can result from heavy metal poisoning, radiation exposure, and infectious diseases such

as measles, chickenpox, polio, and AIDS. Such a deficiency of disease-fighting WBCs leaves a person susceptible to *opportunistic infections,* infections that a normal person could fight off but that can be overwhelming or even life-threatening to a person with a compromised immune system. An abnormally high WBC count is called **leukocytosis.** It usually results from an infection or allergy, but can also stem from such causes as emotional stress and dehydration (WBCs become more concentrated when water is lost from the bloodstream). **Leukemia** is a cancer of the hemopoietic tissues that results in a high number of circulating WBCs. It, too, makes a person vulnerable to opportunistic infection, because even though the WBC count is high, these are immature WBCs incapable of performing their normal defensive roles. Leukemia tends to lead to anemia and thrombocytopenia (see the next section) because stem cells are diverted into the rapid production of WBCs instead of RBC and platelet production.

Platelet Disorders

Thrombocytopenia, a platelet count less than 100,000/μL, results from such causes as leukemia, radiation, or bone marrow poisoning. It results not only in impaired clotting when a vessel is injured, but also in increased bleeding because of a loss of the normal blood vessel maintenance function of platelets.

[18]*poikilo* = variable + *cyt* = cell + *osis* = condition

DEEPER INSIGHT   19.3

Sickle-Cell Disease

Sickle-cell disease is a hereditary hemoglobin defect occurring mostly among people of African descent; its symptoms occur in about 1.3% of American blacks, and about 8.3% are asymptomatic carriers with the potential to pass it to their children. The disease is caused by a defective gene that results in the substitution of valine for a glutamic acid in each beta hemoglobin chain. The abnormal hemoglobin (HbS) turns to gel at low oxygen levels, as when blood passes through the oxygen-hungry skeletal muscles. The RBCs become elongated, stiffened, and pointed (sickle-shaped; fig 19.11). These deformed, inflexible cells cannot pass freely through the tiny blood capillaries, and they tend to adhere to each other and to the capillary wall. Thus, they congregate in the small blood vessels and block the circulation. Obstruction of the circulation produces severe pain and can lead to kidney or heart failure, stroke, or paralysis, among many other effects. The spleen removes defective RBCs faster than they can be replaced, thus leading to anemia and poor physical and mental development of the individual. Without treatment, a child with sickle-cell disease has little chance of living to age 2. Advances in treatment, however, have steadily raised life expectancy to a little beyond 50 years.

Sickle-cell disease originated in areas of Africa where millions of lives are lost to malaria. Malarial parasites normally invade and reproduce in RBCs, but they cannot survive in RBCs with HbS hemoglobin. Thus the sickle-cell gene confers resistance to malaria, even in individuals who are heterozygous for it (carry only one copy of the gene) and do not have sickle-cell disease. The lives saved by HbS in Africa far outnumber the deaths from sickle-cell disease, so natural selection favors the persistence of the gene rather than its elimination. But in North America, where malaria is not prevalent, the lost lives and suffering caused by the sickle-cell gene far outweigh any of its benefits.

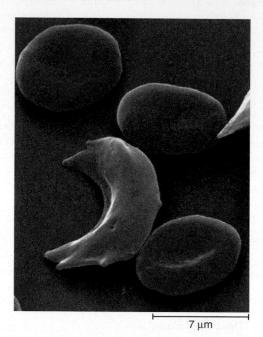

7 μm

Figure 19.11 Sickle-Cell Disease. The lower left RBC has become deformed into the pointed sickle shape diagnostic of this genetic disorder.

Before You Go On

Answer the following questions to test your understanding of the preceding section:

17. What are the terms for an excess and a deficiency of RBCs? An excess and a deficiency of WBCs?

18. What are the three basic categories of the causes of anemia?

19. In what way are leukocytosis and leukemia alike? What is the difference between them?

20. Describe some causes and effects of thrombocytopenia.

Study Guide

Assess Your Learning Outcomes

You should have a good understanding of this chapter if you can accurately address the following issues.

19.1 Introduction (p. 520)
1. The components of the circulatory system and the distinction between the circulatory system and cardiovascular system
2. Functions of the circulatory system
3. The proportion of formed elements to plasma in blood
4. The seven major types of formed elements
5. The composition of blood plasma
6. The types and sources of plasma proteins

19.2 Erythrocytes (p. 523)
1. The functions, shape, and contents of erythrocytes (RBCs)
2. The functions of hemoglobin
3. The meaning and typical values, including units of measurement, of RBC count, hematocrit, and hemoglobin concentration, and how men and women differ in these values
4. The structure of hemoglobin and where oxygen and carbon dioxide bind to the molecule
5. The meaning of *hemopoiesis,* and the organs and tissues in which it occurs both before and after birth
6. The stages of erythropoiesis and the identity of the hormone that stimulates this process

7. The life span of an RBC and the fate of expired RBCs
8. The ABO and Rh blood types; the chemical basis of differences between ABO types; and the relevance of blood types to transfusion and pregnancy

19.3 Leukocytes (p. 527)
1. The total number of leukocytes (WBCs) per microliter of blood
2. How the internal structure of a WBC differs from that of an RBC
3. The general function of WBCs
4. The distinction between granulocytes and agranulocytes
5. The three types of granulocytes, two types of agranulocytes, and how all five of these WBC types can be visually recognized in stained blood smears
6. The primary function of neutrophils and their methods of carrying it out
7. The functions of eosinophils
8. The two secretions of basophils and the function of each secretion
9. The relationship of monocytes to macrophages; the two principal functions of macrophages
10. The general functions and the three major families of lymphocytes
11. The three major cell lines of leukopoiesis

12. The roles of the red bone marrow and thymus in leukopoiesis
13. The migrations and life span of various WBC types, and the fate of expired WBCs

19.4 Platelets (p. 532)
1. The origin and structure of blood platelets, and why they are considered formed elements but not cells
2. The diverse functions of platelets
3. The process of thrombopoiesis
4. The roles that platelets play in blood clotting and the disposal of old clots

19.5 Clinical Perspectives (p. 534)
1. Ways in which the blood and hemopoietic system change in old age and some reasons for the changes
2. The meaning of *anemia;* its three basic categories; and its effects
3. The meaning of *polycythemia;* its causes; and its effects
4. The meaning of *leukopenia;* its causes; and its effects
5. The meaning and causes of *leukocytosis*
6. The meaning of *leukemia;* its effects on infection resistance; and its effects on RBC and platelet counts
7. The meaning, causes, and effects of *thrombocytopenia*

Testing Your Recall

1. Antibodies belong to a class of plasma proteins called
   a. albumins.
   b. gamma globulins.
   c. alpha globulins.
   d. procoagulants.
   e. agglutinins.

2. Serum is blood plasma minus its
   a. sodium ions.   d. albumin.
   b. calcium ions.   e. cells.
   c. fibrinogen.

3. The most abundant formed elements seen in most stained blood films are
   a. erythrocytes.   d. platelets.
   b. neutrophils.   e. monocytes.
   c. lymphocytes.

4. Heparin and histamine are secreted by
   a. plasma cells.   d. platelets.
   b. basophils.   e. neutrophils.
   c. B lymphocytes.

5. _____ have a finely granular cytoplasm and a nucleus typically divided into three to five lobes.
   a. Basophils   d. Monocytes
   b. Eosinophils   e. Neutrophils
   c. Lymphocytes

6. Platelets have all of the following functions *except*
   a. coagulation.
   b. plugging broken blood vessels.
   c. stimulating vasoconstriction.
   d. transporting oxygen.
   e. recruiting neutrophils.

7. Which of these is a granulocyte?
   a. a monocyte   d. an eosinophil
   b. a lymphocyte   e. an erythrocyte
   c. a macrophage

8. Allergies stimulate a rise in _____ count.
   a. erythrocyte   d. monocyte
   b. platelet   e. neutrophil
   c. eosinophil

9. Which of the following leads to pernicious anemia?
   a. hypoxemia
   b. iron deficiency
   c. malaria
   d. lack of intrinsic factor
   e. lack of erythropoietin

10. Oxygen binds to the _____ of a hemoglobin molecule.
    a. valine   d. spectrin
    b. Fe^{2+} e. beta chain
    c. globin

11. Production of all the formed elements of blood is called _____.

12. The percentage of blood volume composed of RBCs is called the _____.

13. Microglia and dendritic cells are two kinds of _____.

14. An excessively low WBC count is called _____.

15. _____ is the fluid that remains if all the formed elements and fibrinogen are removed from the blood.

16. The overall cessation of bleeding, involving several mechanisms, is called _____.

17. _____ results from a mutation that changes one amino acid in each beta chain of the hemoglobin molecule.

18. An excessively high RBC count is called _____.

19. Intrinsic factor enables the small intestine to absorb _____.

20. The kidney hormone _____ stimulates RBC production.

Answers in the Appendix

Building Your Medical Vocabulary

State a medical meaning of each of the following word elements, and give a term in which it is used.

1. hemato-
2. leuko-
3. -emia
4. -blast
5. erythro-
6. mega-
7. myelo-
8. thrombo-
9. macro-
10. -poiesis

Answers in the Appendix

True or False

Determine which five of the following statements are false, and briefly explain why.

1. By volume, the blood usually contains more plasma than blood cells.
2. An increase in the albumin concentration of the blood tends to affect blood pressure.
3. Anemia is caused by a low oxygen concentration in the blood.
4. The most important WBCs in combating a bacterial infection are basophils.
5. Platelets and erythrocytes lack nuclei.
6. Lymphocytes are the most abundant WBCs in the blood.
7. Platelet count is often depressed in people with leukemia.
8. All formed elements of the blood come ultimately from pluripotent stem cells.
9. Since RBCs have no nuclei, they do not live as long as the granulocytes do.
10. Leukemia is a severe deficiency of white blood cells.

Answers in the Appendix

Testing Your Comprehension

1. Considering the quantity of hemoglobin in an erythrocyte and the oxygen-binding properties of hemoglobin, calculate how many molecules of oxygen one erythrocyte could carry.

2. A patient is found to be seriously dehydrated and to have an elevated RBC count. Does the RBC count necessarily indicate a disorder of erythropoiesis? Why or why not?

3. Patients suffering from renal failure are typically placed on hemodialysis and erythropoietin (EPO) replacement therapy. Explain the reason for giving EPO, and predict what the consequences would be of not including this in the treatment regimen.

4. A leukemia patient exhibits minute hemorrhagic spots (*petechiae*) in her skin. Explain why leukemia could produce this effect.

5. Do you think platelets can synthesize proteins? Why or why not?

Answers at www.mhhe.com/saladinha3

Improve Your Grade at **www.mhhe.com/saladinha3**

Practice quizzes, labeling activities, games, and flashcards provide fun ways to master concepts. You can also download image PowerPoint files for each chapter to create a study guide or for taking notes during lecture.

The Circulatory System II—The Heart

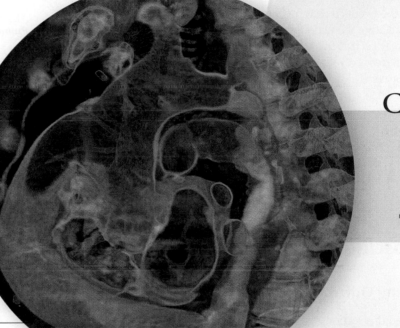

Three-dimensional CT scan of the heart. Left lateral view of a person facing left.

CHAPTER OUTLINE

DEEPER INSIGHTS

BRUSHING UP

To understand this chapter, you may find it helpful to review the following concepts:
- Thoracic cavity anatomy (p. 17)
- Desmosomes and gap junctions (p. 38)
- Endothelium (p. 76)
- Ultrastructure of striated muscle (p. 244)
- Comparisons of cardiac and skeletal muscle (p. 254)
- Anatomy of the autonomic nervous system (chapter 16)

Anatomy & Physiology | **REVEALED**
aprevealed.com

Cardiovascular System

W e are more conscious of our heart than we are of most organs, and more wary of its failure. Speculation about the heart is at least as old as written history. Some ancient Chinese, Egyptian, Greek, and Roman scholars correctly surmised that it was a pump for filling the vessels with blood. Aristotle's views, however, were a step backward. Perhaps because the heart quickens its pace when we are emotionally aroused and because grief causes "heartache," he thought it served primarily as the seat of emotion, but speculated that it might also be a source of heat to aid digestion. During the Middle Ages, Western medical schools clung dogmatically to Aristotle's ideas. In that era, perhaps the only significant advance came when thirteenth-century Arab physician Ibn an-Nafis described the role of the coronary blood vessels in nourishing the heart. The sixteenth-century dissections and anatomical charts of Vesalius, however, greatly improved knowledge of cardiovascular anatomy and set the stage for a more scientific study of the heart and treatment of its disorders.

In the early decades of the twentieth century, little could be recommended for heart disease other than bed rest. Then nitroglycerin was found to improve coronary circulation and relieve the pain resulting from physical exertion; digitalis proved effective for treating abnormal heart rhythms; and diuretics were first used to reduce hypertension. In the last several decades, such advances as coronary bypass surgery, clot-dissolving enzymes, replacement of diseased valves, heart transplants, artificial pacemakers, and artificial hearts have made cardiology one of the most dramatic and attention-getting fields of medicine.

20.1 Overview of the Cardiovascular System

▶ Expected Learning Outcomes

When you have completed this section, you should be able to

- define and distinguish between the pulmonary and systemic circuits;
- describe the general location, size, and shape of the heart; and
- describe the pericardial sac that encloses the heart.

The **cardiovascular system** consists of the heart and blood vessels. The heart functions as a muscular pump that keeps blood flowing through the vessels. The vessels deliver the blood to all the body's organs and then return it to the heart. This chapter focuses on **cardiology**[1]—a field that embraces the study of the heart, clinical evaluation of its function and disorders, and treatment of cardiac diseases. The blood vessels are discussed in chapter 21.

The Pulmonary and Systemic Circuits

The cardiovascular system has two major divisions: a **pulmonary circuit,** which carries blood to the lungs for gas exchange and returns it to the heart, and a **systemic circuit,** which supplies blood to every organ of the body (fig. 20.1), including other parts of the lungs and the wall of the heart itself.

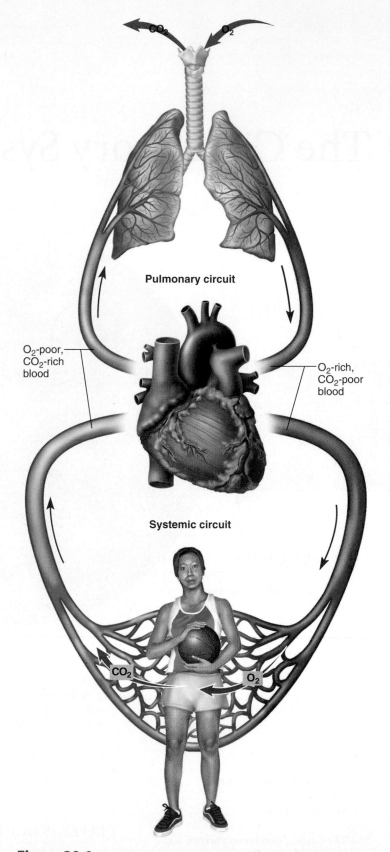

Figure 20.1 General Schematic of the Cardiovascular System.
● *Are the lungs supplied by the pulmonary circuit, the systemic circuit, or both? Explain.*

---

[1]*cardio* = heart + *logy* = study

The right half of the heart supplies the pulmonary circuit. It receives blood that has circulated through the body and pumps it into a large artery, the *pulmonary trunk*. From there, the oxygen-poor blood is distributed to the lungs, where it unloads carbon dioxide and picks up a fresh load of oxygen. It then returns to the left side of the heart by way of the *pulmonary veins* (see fig. 20.3).

The left half of the heart supplies the systemic circuit. It pumps blood into the body's largest artery, the *aorta*. The aorta gives off branches that ultimately deliver oxygen to every organ of the body and pick up their carbon dioxide and other wastes. After exchanging gases with the tissues, this blood returns to the heart by way of the body's two largest veins—the *superior vena cava*, which drains the upper body, and the *inferior vena cava*, which drains everything below the diaphragm. The pulmonary

trunk, pulmonary veins, aorta, and the two venae cavae are called the *great vessels* (*great arteries* and *veins*) because of their relatively large diameters.

Position, Size, and Shape of the Heart

The heart is located in the thoracic cavity in the mediastinum, between the lungs and deep to the sternum. From its superior to inferior midpoints, it is tilted toward the left, so about two-thirds of the heart lies to the left of the median plane (fig. 20.2; see also fig. A.10, p. 340). The broad superior portion of the heart, called the **base,** is the point of attachment for the great vessels described previously. The inferior end tapers to a blunt point, the **apex** of the heart, immediately above the diaphragm (fig. 20.3).

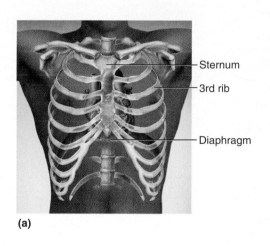

(a)

Sternum
3rd rib
Diaphragm

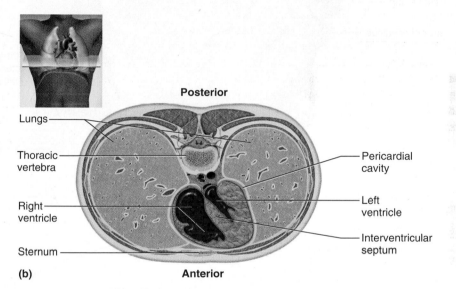

Posterior

Lungs
Thoracic vertebra
Right ventricle
Sternum

Pericardial cavity
Left ventricle
Interventricular septum

(b)

Anterior

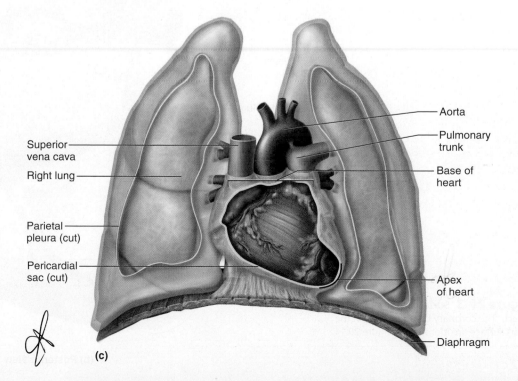

Superior vena cava
Right lung
Parietal pleura (cut)
Pericardial sac (cut)

Aorta
Pulmonary trunk
Base of heart
Apex of heart
Diaphragm

Figure 20.2 Position of the Heart in the Thoracic Cavity. (a) Relationship to the thoracic cage. (b) Cross section of the thorax at the level of the heart. (c) Frontal view of the thoracic cavity with the lungs slightly retracted and the pericardial sac opened.
• *Does most of the heart lie to the right or left of the median plane?*

(c)

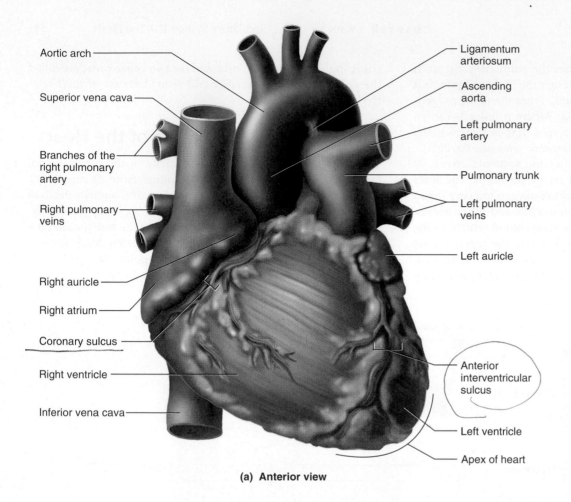

(a) Anterior view

Aortic arch

Superior vena cava

Branches of the right pulmonary artery

Right pulmonary veins

Right auricle

Right atrium

Coronary sulcus

Right ventricle

Inferior vena cava

Ligamentum arteriosum

Ascending aorta

Left pulmonary artery

Pulmonary trunk

Left pulmonary veins

Left auricle

Anterior interventricular sulcus

Left ventricle

Apex of heart

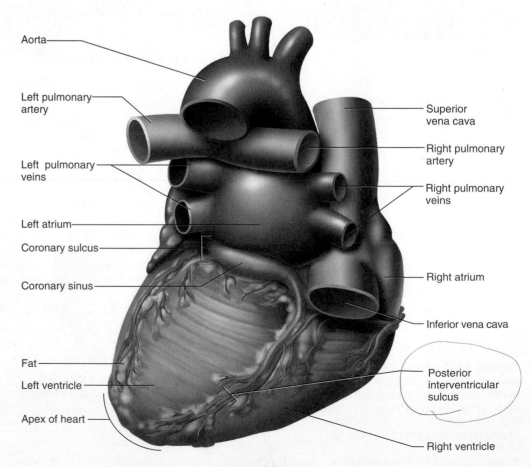

(b) Posterior view

Aorta

Left pulmonary artery

Left pulmonary veins

Left atrium

Coronary sulcus

Coronary sinus

Fat

Left ventricle

Apex of heart

Superior vena cava

Right pulmonary artery

Right pulmonary veins

Right atrium

Inferior vena cava

Posterior interventricular sulcus

Right ventricle

Figure 20.3 **Surface Anatomy of the Heart.** The coronary blood vessels on the heart surface are identified in figure 20.11.

The adult heart weighs about 300 g (10 oz) and measures about 9 cm (3.5 in.) wide at the base, 13 cm (5 in.) from base to apex, and 6 cm (2.5 in.) from anterior to posterior at its thickest point. Whatever one's body size, the heart is roughly the same size as the fist.

The Pericardium

The heart is enclosed in a double-walled sac called the **pericardium.** The outer wall, called the **pericardial sac (parietal pericardium),** has a tough, superficial *fibrous layer* of dense irregular connective tissue and a deep, thin *serous layer*. The serous layer turns inward at the base of the heart and forms the **epicardium (visceral pericardium)** of the heart surface (fig. 20.4), which is described later in the discussion of the heart wall. The pericardial sac is anchored by ligaments to the diaphragm below and the sternum anterior to it, and more loosely anchored by fibrous connective tissue to mediastinal tissue posterior to the heart.

The space between the parietal and visceral membranes is called the **pericardial cavity** (see figs. 20.2b and 20.4). It contains 5 to 30 mL of **pericardial fluid,** exuded by the serous layer of the pericardial sac. The fluid lubricates the membranes and allows the heart to beat with minimal friction. In *pericarditis*—inflammation of the pericardium—the membranes may become roughened and produce a painful *friction rub* with each heartbeat. In addition to reducing friction, the pericardium isolates the heart from other thoracic organs and allows it room to expand, yet resists excessive expansion (see *cardiac tamponade* in Deeper Insight 1.2, p. 18).

Before You Go On

Answer the following questions to test your understanding of the preceding section:

1. Distinguish between the pulmonary and systemic circuits and state which part of the heart supplies each one.

2. Make a two-color sketch of the pericardium, using one color for the pericardial sac and another for the epicardium. For the pericardial sac, label both the fibrous and serous layers. Show the relationship between the pericardium, pericardial cavity, and heart wall.

20.2 Gross Anatomy of the Heart

▶ Expected Learning Outcomes

When you have completed this section, you should be able to

- describe the three layers of the heart wall;

- identify the four chambers of the heart;

- identify the surface features of the heart and correlate them with its internal four-chambered anatomy;

- identify the four valves of the heart; and

- trace the flow of blood through the chambers and valves of the heart and the adjacent blood vessels.

The Heart Wall

The heart wall consists of three layers—a thin epicardium covering its external surface, a thick muscular myocardium in the middle, and a thin endocardium lining the interior of the chambers (fig. 20.4).

The **epicardium**[2] is a serous membrane on the heart surface. It consists mainly of a simple squamous epithelium overlying a thin layer of areolar tissue. In some places, it also includes a thick layer of adipose tissue, whereas in other areas it is fat-free and translucent, so the muscle of the myocardium shows through (fig. 20.5a). The largest branches of the coronary blood vessels travel through the epicardium. The **endocardium**[3] lines the interior of the heart chambers. Like the epicardium, it is composed of a simple squamous epithelium overlying a thin areolar tissue layer; however, it has no adipose tissue. The endocardium covers the valve surfaces and is continuous with the endothelium of the blood vessels.

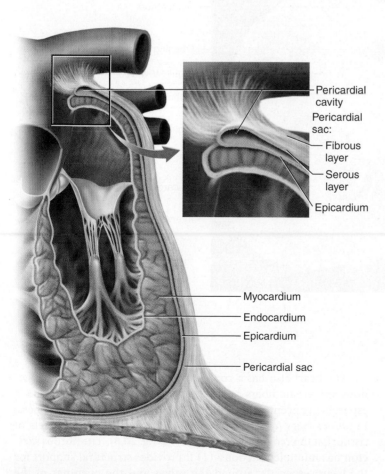

Figure 20.4 The Pericardium and Heart Wall. The inset shows layers of the wall of the left ventricle in relationship to the pericardium. The pericardium consists of the epicardium (visceral pericardium) and pericardial sac (parietal pericardium).

[2]*epi* = upon + *cardi* = heart
[3]*endo* = internal + *cardi* = heart

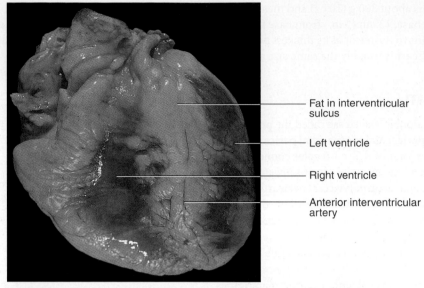

- Fat in interventricular sulcus
- Left ventricle
- Right ventricle
- Anterior interventricular artery

(a) Anterior view, external anatomy

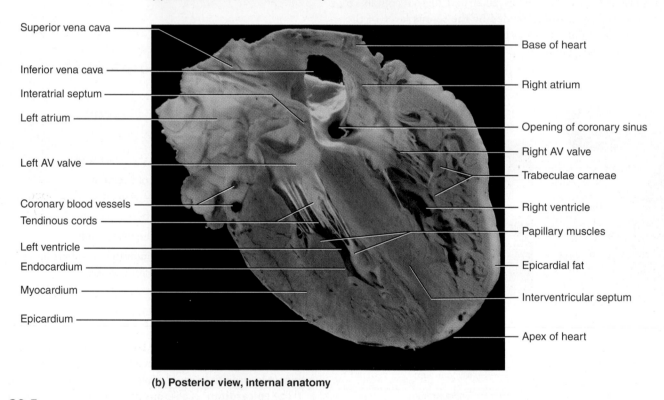

- Superior vena cava
- Inferior vena cava
- Interatrial septum
- Left atrium
- Left AV valve
- Coronary blood vessels
- Tendinous cords
- Left ventricle
- Endocardium
- Myocardium
- Epicardium

- Base of heart
- Right atrium
- Opening of coronary sinus
- Right AV valve
- Trabeculae carneae
- Right ventricle
- Papillary muscles
- Epicardial fat
- Interventricular septum
- Apex of heart

(b) Posterior view, internal anatomy

Figure 20.5 The Heart of a Human Cadaver.

The layer between these two, the **myocardium,**[4] is the thickest by far and performs the work of the heart. Its thickness is proportional to the workload on the individual heart chambers. It consists primarily of cardiac muscle arranged in a spiral called the **myocardial vortex** (fig. 20.6). This arrangement causes the heart to contract with a twisting or wringing motion that enhances the ejection of blood. The microscopic structure of the cardiac muscle cells (*cardiac myocytes* or *cardiocytes*) is detailed later.

The heart also has a connective tissue framework of collagenous and elastic fibers called the **fibrous skeleton.** This tissue is especially concentrated in the walls between the heart chambers, in *fibrous rings (anuli fibrosi)* around the valves, and in sheets of tissue that interconnect these rings (see fig. 20.8). The fibrous skeleton has multiple functions: (1) It provides structural support for the heart, especially around the valves and the openings of the great vessels; it holds these orifices open and prevents them from being excessively stretched when blood surges through them. (2) It anchors the myocytes and gives them something to pull

[4]*myo* = muscle + *cardi* = heart

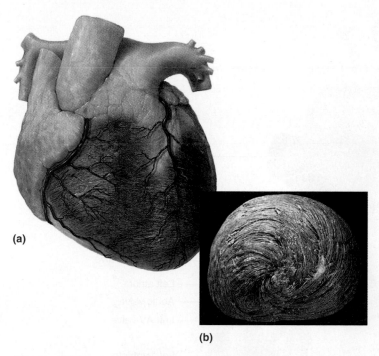

(a)

(b)

Figure 20.6 **The Myocardial Vortex.** (a) Anterior view of the heart with the epicardium rendered transparent to expose the bundles of myocardial muscle. (b) View from the apex to show the way the muscle coils around the heart. This results in a twisting motion when the ventricles contract.

against. (3) As a nonconductor of electricity, it serves as electrical insulation between the atria and the ventricles, so the atria cannot stimulate the ventricles directly. This insulation is important in the timing and coordination of electrical and contractile activity. (4) Some authorities think (though others disagree) that elastic recoil of the fibrous skeleton may aid in refilling the heart with blood after each beat, like the rubber bulb of a turkey baster that expands when you relax your grip.

Apply What You Know

Parts of the fibrous skeleton sometimes become calcified in old age. How would you expect this to affect cardiac function?

The Chambers

The heart has four chambers, best seen in frontal section (fig. 20.7). The two at the superior pole (base) of the heart are the **right** and **left atria** (AY-tree-uh; singular, *atrium*[5]). They are thin-walled receiving chambers for blood returning to the heart by way of the great veins. Most of the mass of each atrium is on the posterior side

of the heart, so only a small portion is visible from an anterior view. Here, each atrium has a small earlike extension called an **auricle**[6] that slightly increases its volume.

The two inferior heart chambers, the **right** and **left ventricles,**[7] are the pumps that eject blood into the arteries and keep it flowing around the body. The right ventricle constitutes most of the anterior portion of the heart, and the left ventricle forms the apex and inferoposterior portion.

On the surface, the boundaries of the four chambers are marked by three sulci (grooves), which are largely filled with fat and the coronary blood vessels (see fig. 20.5a). The **coronary**[8] **(atrioventricular) sulcus** encircles the heart near the base and separates the atria above from the ventricles below. It can be exposed by lifting the margins of the atria. The other two sulci extend obliquely down the heart from the coronary sulcus to the apex—one on the front of the heart called the **anterior interventricular sulcus** and one on the back called the **posterior interventricular sulcus.** These sulci overlie an internal wall, the *interventricular septum,* that divides the right ventricle from the left. The coronary sulcus and two interventricular sulci harbor the largest of the coronary blood vessels.

The atria exhibit thin flaccid walls corresponding to their light workload—all they do is pump blood into the ventricles immediately below. They are separated from each other by a wall called the **interatrial septum.** The right atrium and both auricles exhibit internal ridges of myocardium called **pectinate**[9] **muscles.** The **interventricular septum** is a much more muscular, vertical wall between the ventricles.

The right ventricle pumps blood only to the lungs and back to the left atrium, so its wall is only moderately muscular. The wall of the left ventricle is two to four times as thick because it bears the greatest workload of all four chambers, pumping blood through the entire body. Both ventricles exhibit internal ridges called **trabeculae carneae**[10] (trah-BEC-you-lee CAR-nee-ee).

The Valves

To pump blood effectively, the heart needs valves that ensure a predominantly one-way flow. There is a valve between each atrium and the corresponding ventricle and another at the exit from each ventricle into its great artery (fig. 20.7), but there are no valves where the great veins empty into the atria. Each valve consists of two or three fibrous flaps of tissue called **cusps** or **leaflets,** covered with endocardium.

The **atrioventricular (AV) valves** regulate the openings between the atria and ventricles. The **right AV (tricuspid) valve** has three cusps, and the **left AV (bicuspid) valve** has two (fig. 20.8). The left AV valve is also known as the **mitral** (MY-trul) **valve** after its resemblance to a miter, the headdress of a church bishop.

---

[5]*atrio* = entryway

[6]*auricle* = little ear
[7]*ventr* = belly, lower part + *icle* = little
[8]*coron* = crown + *ary* = pertaining to
[9]*pectin* = comb + *ate* = like
[10]*trabec* = beam + *ula* = little + *carne* = flesh, meat

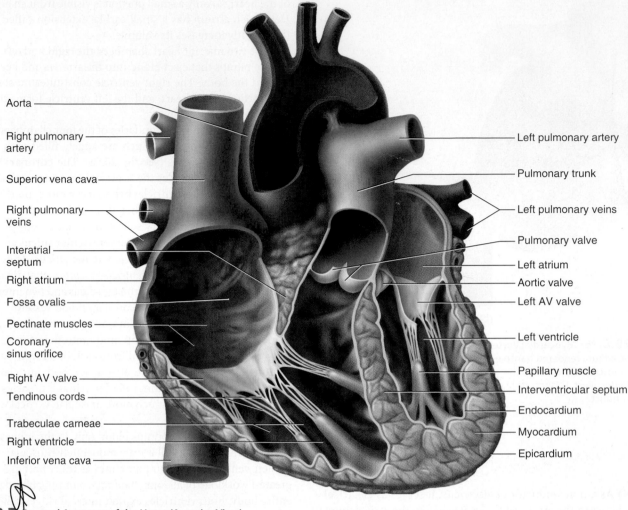

Aorta

Right pulmonary artery

Superior vena cava

Right pulmonary veins

Interatrial septum

Right atrium

Fossa ovalis

Pectinate muscles

Coronary sinus orifice

Right AV valve

Tendinous cords

Trabeculae carneae

Right ventricle

Inferior vena cava

Left pulmonary artery

Pulmonary trunk

Left pulmonary veins

Pulmonary valve

Left atrium

Aortic valve

Left AV valve

Left ventricle

Papillary muscle

Interventricular septum

Endocardium

Myocardium

Epicardium

Figure 20.7 Internal Anatomy of the Heart (Anterior View).

Stringlike **tendinous cords (chordae tendineae),** reminiscent of the shroud lines of a parachute, connect the valve cusps to conical **papillary muscles** on the floor of the ventricle.

The **semilunar**[11] **valves** (pulmonary and aortic valves) regulate the flow of blood from the ventricles into the great arteries. The **pulmonary valve** controls the opening from the right ventricle into the pulmonary trunk, and the **aortic valve** controls the opening from the left ventricle into the aorta. Each has three cusps shaped like shirt pockets. There are no tendinous cords on the semilunar valves.

The valves do not open and close by any muscular effort of their own. The cusps are simply pushed open and closed by changes in blood pressure (fig. 20.9). When the ventricles are relaxed and their pressure is low, the AV valve cusps hang down limply and both AV valves are open. Blood flows freely from the atria through the valves and into the ventricles below. When the ventricles have filled with

blood, they begin to contract and force blood upward against the underside of these valves. This pushes the valve cusps together and closes the openings, so the blood is not simply squirted back into the atria.

At the same time, the rising pressure in the ventricles forces the pulmonary and aortic valves open, and blood is ejected from the heart. As the ventricles relax again and their pressure falls below that in the arteries, arterial blood briefly flows backward and fills the pocketlike cusps of each semilunar valve. The three cusps meet in the middle of the orifice, thereby sealing it and preventing blood from reentering the heart.

Although the AV valves are attached through their tendinous cords to papillary muscles in the floor of the heart, those muscles do not help the valves open. Rather, they contract along with the rest of the ventricular myocardium and tug on the tendinous cords. This prevents the AV valves from bulging excessively into the atria or flipping inside out like windblown umbrellas. Excessive bulging due to slack tendinous cords is called *valvular prolapse* (see table 20.1).

---

[11]*semi* = half + *lun* = moon + *ar* = like

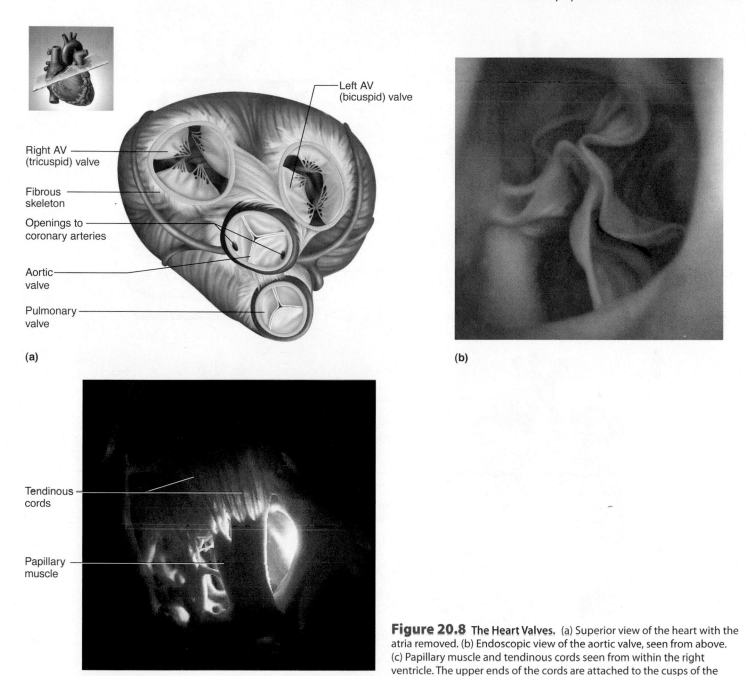

Right AV (tricuspid) valve

Fibrous skeleton

Openings to coronary arteries

Aortic valve

Pulmonary valve

Left AV (bicuspid) valve

(a)

(b)

Tendinous cords

Papillary muscle

(c)

Figure 20.8 The Heart Valves. (a) Superior view of the heart with the atria removed. (b) Endoscopic view of the aortic valve, seen from above. (c) Papillary muscle and tendinous cords seen from within the right ventricle. The upper ends of the cords are attached to the cusps of the right AV valve.

Blood Flow Through the Chambers

Until the sixteenth century, anatomists thought that blood flowed directly from the right ventricle to the left through invisible pores in the septum. This is incorrect; blood in the right and left chambers of the heart is kept entirely separate. Figure 20.10 shows the pathway of the blood as it travels from the right atrium through the body and back to the starting point.

Blood that has been through the systemic circuit returns by way of the superior and inferior venae cavae to the right atrium. It flows directly from the right atrium, through the right AV (tricuspid) valve, into the right ventricle. When the right ventricle contracts, it ejects this blood through the pulmonary valve into the pulmonary trunk, on its way to the lungs to exchange carbon dioxide for oxygen.

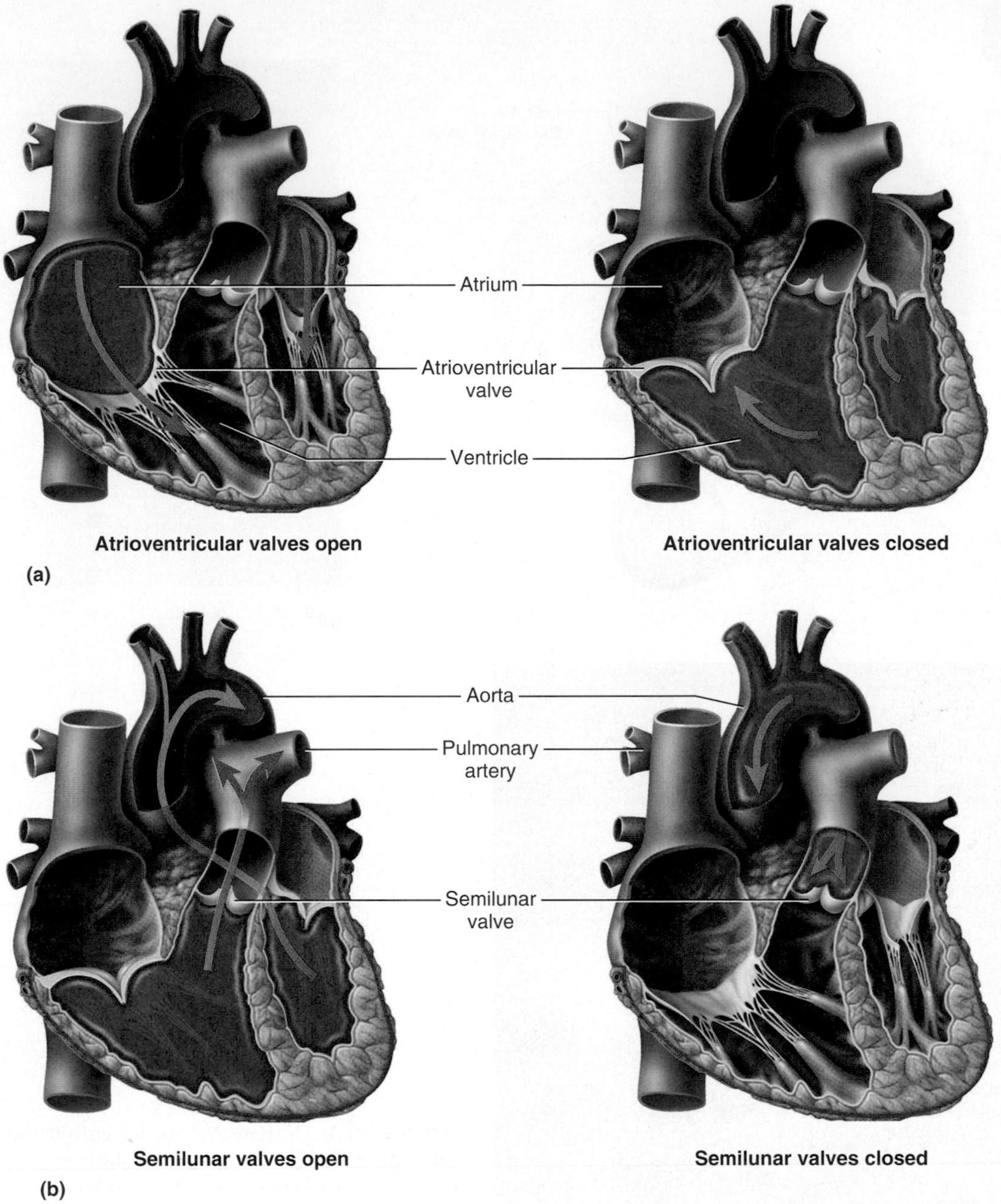

Figure 20.9 Operation of the Heart Valves. (a) The atrioventricular valves. When atrial pressure is greater than ventricular pressure, the valve opens and blood flows through. When ventricular pressure rises above atrial pressure, the blood in the ventricle pushes the valve cusps closed. (b) The semilunar valves. When the pressure in the ventricle is greater than the pressure in the artery, the valve is forced open and blood is ejected. When ventricular pressure is lower than arterial pressure, arterial blood holds the valve closed.
● *What role do the tendinous cords play?*

Blood returns from the lungs by way of two pulmonary veins on the left and two on the right; all four of these empty into the left atrium. Blood flows through the left AV (bicuspid, mitral) valve into the left ventricle. Contraction of the left ventricle ejects this blood through the aortic valve into the ascending aorta, on its way to another trip around the systemic circuit.

Before You Go On

Answer the following questions to test your understanding of the preceding section:

3. Name the three layers of the heart and describe their structural differences.

4. What are the functions of the fibrous skeleton?

5. Trace the flow of blood through the heart, naming each chamber and valve in order.

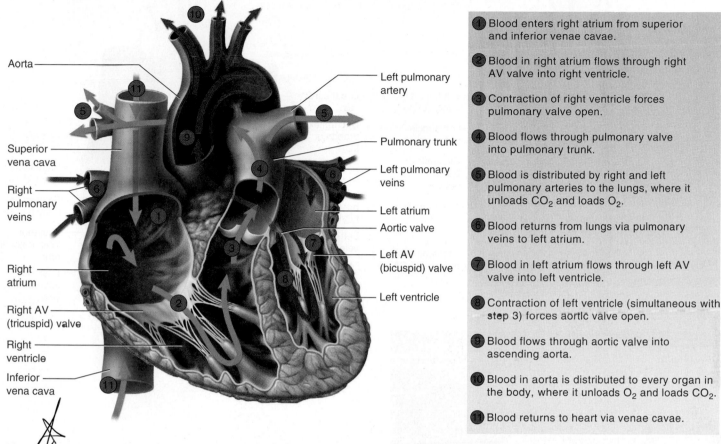

1. Blood enters right atrium from superior and inferior venae cavae.

2. Blood in right atrium flows through right AV valve into right ventricle.

3. Contraction of right ventricle forces pulmonary valve open.

4. Blood flows through pulmonary valve into pulmonary trunk.

5. Blood is distributed by right and left pulmonary arteries to the lungs, where it unloads CO_2 and loads O_2.

6. Blood returns from lungs via pulmonary veins to left atrium.

7. Blood in left atrium flows through left AV valve into left ventricle.

8. Contraction of left ventricle (simultaneous with step 3) forces aortic valve open.

9. Blood flows through aortic valve into ascending aorta.

10. Blood in aorta is distributed to every organ in the body, where it unloads O_2 and loads CO_2.

11. Blood returns to heart via venae cavae.

Figure 20.10 **The Pathway of Blood Flow Through the Heart.** The pathway from 4 to 6 is the pulmonary circuit, and the pathway from 9 to 11 is the systemic circuit. Violet arrows indicate oxygen-poor blood, and orange arrows indicate oxygen-rich blood.

20.3 Coronary Circulation

▶ Expected Learning Outcomes

When you have completed this section, you should be able to

- describe the arteries that nourish the myocardium and the veins that drain it; and

- define *myocardial infarction* and relate it to the coronary arteries.

If your heart lasts for 80 years and beats an average of 75 times a minute, it will beat more than 3 billion times and pump more than 200 million liters of blood. It is, in short, a remarkably hardworking organ, and understandably, it needs an abundant supply of oxygen and nutrients. These needs are not met to any appreciable extent by the blood in the chambers, because the diffusion of substances from there through the myocardium would be too slow. Instead, the myocardium has its own supply of arteries and capillaries that deliver blood to every myocyte. The blood vessels of the heart wall constitute the **coronary circulation.**

At rest, the coronary blood vessels supply the myocardium with about 250 mL of blood per minute. This constitutes about 5% of the circulating blood going to meet the metabolic needs of the heart alone, even though the heart is only 0.5% of the body's weight. It receives 10 times its "fair share" to sustain its strenuous workload.

Arterial Supply

The coronary circulation is the most variable aspect of cardiac anatomy. The following description covers only the pattern seen in about 70% to 85% of persons, and only the few largest vessels (compare the great density of small vessels seen in fig. 20.11c).

Immediately after the aorta leaves the left ventricle, it gives off a right and left coronary artery. The orifices of these two arteries lie deep in the pockets formed by two of the aortic valve cusps (see fig. 20.8a). The **left coronary artery (LCA)** travels through the coronary sulcus under the left auricle and divides into two branches (fig. 20.11):

1. The **anterior interventricular branch** travels down the anterior interventricular sulcus to the apex, rounds the bend, and travels a short distance up the posterior side of the heart.

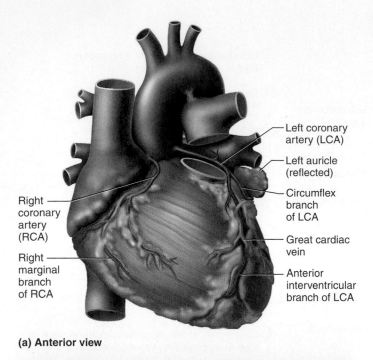

(a) Anterior view

Left coronary artery (LCA)

Left auricle (reflected)

Circumflex branch of LCA

Great cardiac vein

Anterior interventricular branch of LCA

Right coronary artery (RCA)

Right marginal branch of RCA

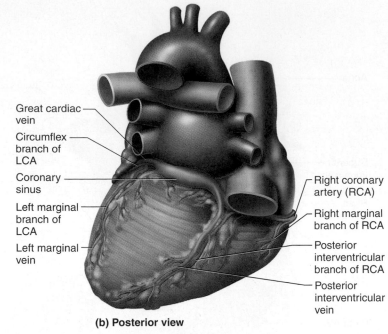

(b) Posterior view

Great cardiac vein

Circumflex branch of LCA

Coronary sinus

Left marginal branch of LCA

Left marginal vein

Right coronary artery (RCA)

Right marginal branch of RCA

Posterior interventricular branch of RCA

Posterior interventricular vein

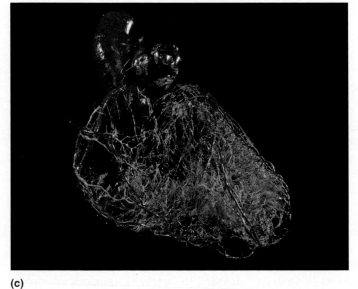

(c)

Figure 20.11 **The Principal Coronary Blood Vessels.** (a) Anterior view. (b) Posterior view. (c) A polymer cast of the coronary circulation.

There it joins the posterior interventricular branch described shortly. Clinically, it is also called the *left anterior descending (LAD) branch.* This artery supplies blood to both ventricles and the anterior two-thirds of the interventricular septum.

2. The **circumflex branch** continues around the left side of the heart in the coronary sulcus. It gives off a **left marginal branch** that passes down the left margin of the heart and furnishes blood to the left ventricle. The circumflex branch then ends on the posterior side of the heart, where it supplies blood to the left atrium and posterior wall of the left ventricle.

The **right coronary artery (RCA)** supplies the right atrium and sinoatrial node (pacemaker), then continues along the coronary sulcus under the right auricle and gives off two branches of its own:

1. The **right marginal branch** runs toward the apex of the heart and supplies the lateral aspect of the right atrium and ventricle.

2. The RCA continues around the right margin of the heart to the posterior side, sends a small branch to the atrioventricular node, then gives off a large **posterior interventricular branch.** This branch travels down the corresponding sulcus and supplies the posterior walls of both ventricles as well as the posterior portion of the interventricular septum. It ends by joining the anterior interventricular branch of the LCA.

The energy demand of the cardiac muscle is so critical that an interruption of the blood supply to any part of the myocardium can cause necrosis within minutes. A fatty deposit or blood clot in a coronary artery can produce a **myocardial infarction**[12] **(MI),** the sudden death of a patch of tissue deprived of its blood flow (see Deeper Insight 20.1). Some protection from MI is provided by

---

[12]*infarct* = to stuff

DEEPER INSIGHT 20.1

Coronary Artery Disease

Coronary artery disease (CAD) is a narrowing of the coronary arteries resulting in insufficient blood flow to maintain the myocardium. It is usually caused by *atherosclerosis,* a vascular disorder in which fatty deposits form in an arterial wall, causing arterial degeneration and obstructed blood flow. The atherosclerotic *plaque (atheroma)* is composed of lipids, smooth muscle, and scar tissue, and may progress to a calcified *complicated plaque,* causing the arterial walls to become rigid. *Myocardial infarction* (heart attack) can occur when the artery becomes so occluded that cardiac muscle begins to die from lack of oxygen. Partial obstruction of an artery can cause a temporary sense of heaviness and chest pain called *angina pectoris* when the artery constricts.

There are multiple ways in which an atheroma can lead to heart attack. The atheroma itself may block so much of the artery that blood flow is insufficient to support the cardiac muscle (fig. 20.12), especially during exercise when the metabolic need of the myocardium increases sharply. Platelets often adhere to atheromas and produce blood clots. If the vessel space (lumen) is already largely closed off by the atheroma, a blood clot may finish the job. Furthermore, a clot can break free from the atheroma and block a smaller coronary artery downstream.

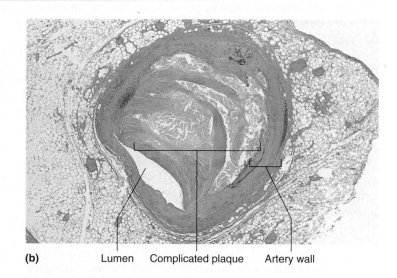

(b)   Lumen   Complicated plaque   Artery wall

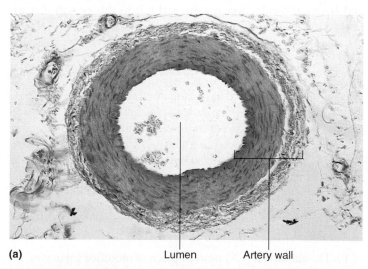

(a)   Lumen   Artery wall

Figure 20.12 **Coronary Atherosclerosis.** (a) Cross section of a healthy artery. (b) Cross section of an artery with advanced atherosclerosis. The lumen is reduced to a small space that can easily be blocked by a stationary or traveling blood clot or by vasoconstriction. Most of the original lumen is obstructed by a plaque composed of calcified scar tissue. (c) Coronary arteriogram showing 60% obstruction of the anterior interventricular artery (arrow).

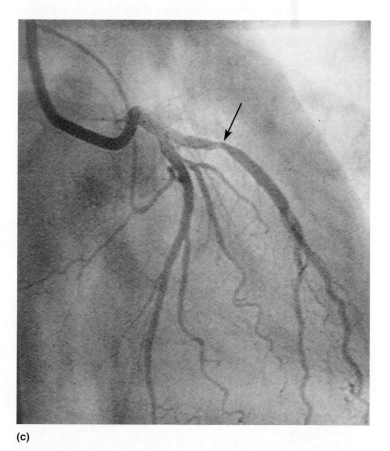

(c)

several **arterial anastomoses** (ah-NASS-tih-MO-seez), points where two arteries come together and combine their blood flow to points farther downstream. Anastomoses provide an alternative route, called **collateral circulation,** that can supply the heart tissue with blood if the primary route becomes obstructed.

In most organs, blood flow peaks when the ventricles contract and eject blood into the arteries, and diminishes when the ventricles relax and refill. The opposite is true in the coronary arteries: flow peaks when the heart relaxes. There are three reasons for this. (1) Contraction of the myocardium compresses the coronary arteries and obstructs blood flow. (2) When the ventricles contract, the aortic valve is forced open and the valve cusps cover the openings to the coronary arteries, blocking blood from flowing into them. (3) When they relax, blood in the aorta briefly surges back toward the heart. It fills the aortic valve cusps and some of it flows into the coronary arteries, like water pouring into a bucket and flowing out through a hole in the bottom. In the coronary blood vessels, therefore, blood flow increases during ventricular relaxation.

Venous Drainage

Venous drainage refers to the route by which blood leaves an organ. After flowing through capillaries of the heart wall, about 20% of the coronary blood empties directly from multiple small *thebesian*[13] *veins* into the heart chambers, especially the right ventricle. The other 80% returns to the right atrium by the following route (fig. 20.11):

- The **great cardiac vein** collects blood from the anterior aspect of the heart and travels alongside the anterior interventricular artery. It carries blood from the apex of the heart toward the coronary sulcus, then arcs around the left side of the heart and empties into the coronary sinus described below.

- The **posterior interventricular (middle cardiac) vein,** found in the posterior interventricular sulcus, collects blood from the posterior aspect of the heart. It, too, carries blood from the apex upward and drains into the same sinus.

- The **left marginal vein** travels from a point near the apex of the heart up the left margin and empties into the coronary sinus.

- The **coronary sinus,** a large transverse vein in the coronary sulcus on the posterior side of the heart, collects blood from all three of the aforementioned veins as well as some smaller ones. It empties blood into the right atrium.

Before You Go On

Answer the following questions to test your understanding of the preceding section:

6. What are the three principal branches of the left coronary artery? Where are they located on the heart surface? What are the branches of the right coronary artery, and where are they located?

7. What is the medical significance of anastomoses in the coronary arterial system?

8. Why do the coronary arteries carry a greater blood flow during ventricular relaxation than they do during ventricular contraction?

9. What are the three major veins that empty into the coronary sinus?

20.4 The Cardiac Conduction System and Cardiac Muscle

▶ **Expected Learning Outcomes**

When you have completed this section, you should be able to

- describe the heart's electrical conduction system;

- contrast the structure of cardiac and skeletal muscle; and

- describe the types and significance of intercellular junctions between cardiac muscle cells.

The most obvious physiological fact about the heart is its rhythmicity. It contracts at regular intervals, typically about 75 beats per minute (bpm) in a resting adult. Among invertebrates such as clams, crabs, and insects, each heartbeat is triggered by a pacemaker in the nervous system. The vertebrate heartbeat, however, is said to be *myogenic*[14] because the signal that triggers each heartbeat originates within the heart itself. Indeed, we can remove the heart from the body and keep it in aerated saline, and it will beat for hours. Cut the heart into little pieces, and each piece continues its own rhythmic pulsations. Thus, it obviously is not dependent on the nervous system for its rhythm. The heart has its own pacemaker and electrical conduction system, and it is to this system that we now turn our attention.

The Conduction System

Cardiac myocytes are said to be **autorhythmic**[15] because they electrically discharge, or depolarize, spontaneously at regular time intervals. Some of the myocytes lose the ability to contract and become specialized, instead, for generating and conducting these electrical signals. These cells constitute the **cardiac conduction system,** which controls the route and timing of stimulation to ensure that the four heart chambers are coordinated with each other. Electrical signals arise and travel through the conduction system in the following order (fig. 20.13):

1 The **sinoatrial (SA) node,** a patch of modified myocytes in the right atrium, just under the epicardium near the superior vena cava. This is the **pacemaker** that initiates each heartbeat and determines the heart rate.

2 Signals from the SA node spread throughout the atria, as shown by the red arrows in the figure.

3 The **atrioventricular (AV) node,** located at the lower end of the interatrial septum near the right AV valve. This node acts as an electrical gateway to the ventricles. All electrical signals traveling to the ventricles must pass through the AV

---

[13]Adam Christian Thebesius (1686–1732), German physician

[14]*myo* = muscle + *genic* = arising from
[15]*auto* = self

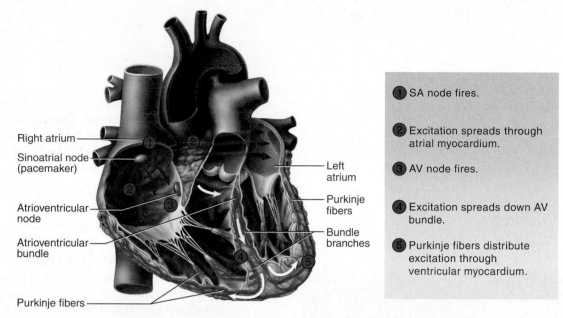

Figure 20.13 The Cardiac Conduction System. Electrical signals travel along the pathways indicated by the arrows.
● *Which atrium is the first to receive the signal that induces atrial contraction?*

The following labels appear in the figure:

Right atrium
Sinoatrial node (pacemaker)
Atrioventricular node
Atrioventricular bundle
Purkinje fibers
Left atrium
Purkinje fibers
Bundle branches

1. SA node fires.
2. Excitation spreads through atrial myocardium.
3. AV node fires.
4. Excitation spreads down AV bundle.
5. Purkinje fibers distribute excitation through ventricular myocardium.

node because the fibrous skeleton acts as an insulator that prevents currents from traveling to the ventricles by any other route.

4 The **atrioventricular (AV) bundle (bundle of His[16])**, a cord of modified myocytes by which signals leave the AV node. The bundle soon forks into **right** and **left bundle branches,** which enter the interventricular septum and descend toward the apex of the heart.

5 **Purkinje[17]** (pur-KIN-jee) **fibers,** nervelike processes that arise from the lower end of the bundle branches and turn upward to spread throughout the ventricular myocardium. Purkinje fibers distribute the electrical excitation to the myocytes of the ventricles. They form a more elaborate network in the left ventricle than in the right.

After we examine the structure of cardiac muscle, we will see how this conduction system relates to the heart's cycle of contraction and relaxation.

Structure of Cardiac Muscle

The traveling electrical signal does not end with the Purkinje fibers, and Purkinje fibers do not reach every myocyte. Rather, the myocytes pass the signal from cell to cell. This is something that skeletal muscle cannot do, so to understand how the heartbeat is coordinated, one must understand the microscopic anatomy of cardiac myocytes and how they differ from skeletal muscle fibers.

Cardiac muscle is striated like skeletal muscle but otherwise differs from it in many structural and physiological ways. Cardiac myocytes, or *cardiocytes,* are relatively short, thick cells, typically 50 to 100 μm long and 10 to 20 μm wide (fig. 20.14). The ends of the cell are slightly branched, like a log with notches in the end. Through its multiple end branches, each cardiocyte contacts several other cells, so collectively they form a network throughout a heart chamber. A cardiocyte usually has only one, centrally placed nucleus, often surrounded by a mass of the energy-storage carbohydrate, glycogen. The sarcoplasmic reticulum is less developed than in skeletal muscle; it lacks terminal cisternae, although it does have footlike sacs associated with the T tubules. The T tubules are much larger than in skeletal muscle. During excitation of the cell, they admit supplemental calcium ions from the extracellular fluid to activate muscle contraction. Cardiocytes have especially large mitochondria, which make up about 25% of the cell volume, compared to skeletal muscle mitochondria, which are much smaller and comprise only 2% of the cell volume.

DEEPER INSIGHT 20.2

Heart Block

Heart block is a condition in which electrical signals cannot travel normally through the cardiac conduction system because of disease and degeneration of the conduction system fibers. A *bundle branch block* exists when one or both of the atrioventricular bundle branches are diseased. *Total heart block* results from disease of the AV node. Heart block is one of the causes of cardiac arrhythmia, an irregularity in the heartbeat (see fig. 20.15c). In total heart block, signals from the SA node stop at the diseased AV node and cannot reach the ventricular myocardium. The ventricles then beat at their intrinsic rhythm of about 20 to 40 bpm, out of synchrony with the atria and at a rate too slow to sustain life for very long.

[16]Wilhelm His, Jr. (1863–1934), German physiologist
[17]Johannes E. Purkinje (1787–1869), Bohemian physiologist

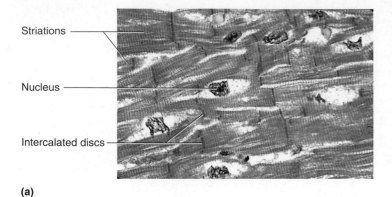

Striations

Nucleus

Intercalated discs

(a)

Striated myofibril   Glycogen   Nucleus   Mitochondria   Intercalated discs

(b)

Intercellular space

Desmosomes

Gap junctions

(c)

Figure 20.14 Cardiac Muscle. (a) Light micrograph. (b) Structure of a cardiocyte (center) and its relationship to adjacent cardiocytes. At each end, a cardiocyte is typically linked to two or more neighboring cells through the mechanical and electrical junctions of their intercalated discs. (c) Structure of an intercalated disc.

● *Which component of the intercalated disc allows a cardiocyte to electrically excite the neighboring cardiocytes?*

Apply What You Know

Why should mitochondria be larger and more abundant in cardiac muscle than in skeletal muscle?

Cardiocytes are joined end to end by thick connections called **intercalated** (in-TUR-ku-LAY-ted) **discs,** which appear as dark lines (thicker than the striations) in properly stained tissue sections. An intercalated disc is a complex steplike structure with three distinctive features not found in skeletal muscle:

1. **Interdigitating folds.** The plasma membrane at the end of the cell is folded somewhat like the bottom of an egg carton. The folds of adjoining cells interlock with each other and increase the surface area of intercellular contact.

2. **Mechanical junctions.** The cells are tightly joined by two types of mechanical junctions—the fascia adherens and desmosomes. The *fascia adherens*[18] (FASH-ee-ah ad-HEER-enz) is the most extensive. It is a broad band in which the actin of the thin myofilaments is anchored to the plasma membrane, and each cell is linked to the next by way of transmembrane proteins. Thus, the moving myofilaments of a contracting cell are able to pull indirectly on the neighboring cells. The fascia adherens is interrupted here and there by *desmosomes*. Described in more detail on page 38, desmosomes are patches of mechanical linkage between cells. They prevent the contracting cardiocytes from pulling apart.

3. **Electrical junctions.** The intercalated discs also contain *gap junctions,* which form channels that allow ions to flow from the cytoplasm of one cell directly into the next (see p. 39 for their structure). These junctions enable each myocyte to electrically stimulate its neighbors. Thus, the entire myocardium of the two atria behaves almost as if it were a single cell, as does the entire myocardium of the two ventricles. This unified action is essential for the effective pumping of a heart chamber.

Skeletal muscle contains satellite cells that can divide and replace dead muscle fibers to some extent. Cardiac muscle lacks satellite cells, however, so the repair of damaged cardiac muscle is almost entirely by fibrosis (scarring). A limited capacity for myocardial mitosis and regeneration was discovered in 2001, raising some hope that this might one day be clinically enhanced to repair hearts damaged by myocardial infarction.

Nerve Supply to the Heart

Even though the heart has its own pacemaker, it also receives both sympathetic and parasympathetic nerves. They do not activate the heartbeat, but modify its rate and contraction strength. Sympathetic nerves can raise the heart rate to as high as 230 bpm, whereas parasympathetic nerves can reduce it to as low as 20 bpm or even stop the heart for a few seconds.

The sympathetic pathway to the heart originates with neurons in the lower cervical to upper thoracic spinal cord. Efferent fibers

[18]*fascia* = band + *adherens* = adhering

from these neurons pass from the spinal cord to the sympathetic chain and travel up the chain to the three cervical ganglia. **Cardiac nerves** arise from the cervical ganglia (see fig. 16.4, p. 447) and lead mainly to the ventricular myocardium, where they increase the force of contraction. Some fibers, however, innervate the atria. Sympathetic fibers to the coronary arteries dilate them and increase coronary blood flow during exercise.

The parasympathetic pathway to the heart is through the vagus nerves. The right vagus nerve innervates mainly the SA node, and the left vagus nerve innervates mainly the AV node, although there is some cross-innervation from each nerve to both nodes. The ventricles receive little or no vagal stimulation. The vagus nerves slow the heartbeat. Without this influence, the SA node would produce

an average resting heart rate of about 100 bpm, but steady background firing of the vagus nerves *(vagal tone)* normally holds the resting rate down to about 70 to 80 bpm.

The Cardiac Cycle

The foregoing anatomy should acquire more meaning if you can relate it to the **cardiac cycle**—one complete cycle of contraction and relaxation. This will show how the structures of the heart work together to achieve blood circulation.

The electrical events of the cardiac cycle can be recorded with skin electrodes as an **electrocardiogram (ECG)** (see Deeper Insight 20.3 and fig. 20.15). Electrical excitation of a heart chamber induces contraction, or **systole** (SIS-toe-lee), which expels blood from the chamber. The relaxation of any chamber is called **diastole**

DEEPER INSIGHT   20.3

The Electrocardiogram

Next to listening to the heart sounds with a stethoscope *(auscultation)*, the most common clinical method of evaluating heart function is the *electrocardiogram (ECG* or *EKG*[19])—recording the electrical activity of the heart by means of electrodes applied to the skin. As the myocardium of the atria and ventricles electrically discharges *(depolarizes)* and recharges *(repolarizes)*, it generates electrical currents that are conducted by electrolytes in the body fluids to the skin surface. Here the activity can be recorded as small voltage changes that show as upward and downward deflections of the ECG (fig. 20.15a).

Three major events are seen in the ECG, named the *P wave,* the *QRS complex,* and the *T wave.* (The letters were arbitrarily chosen; they do not stand for any words.) Each wave and interval between waves is correlated with events of the cardiac cycle as described in the main text. The QRS complex is the largest wave because it is produced mainly by depolarization of the ventricles, which constitute the largest muscle mass of the heart and generate the greatest electrical current.

Irregularities in the ECG help to diagnose disorders such as ventricular fibrillation (fig. 20.15b), heart block (fig. 20.15c), and electrolyte imbalances. For further reflection on heart block, see Testing Your Comprehension question 4 on page 562.

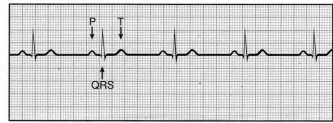

(a) Normal electrocardiogram

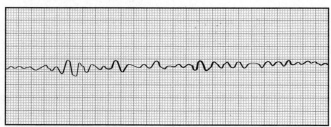

(b) Ventricular fibrillation

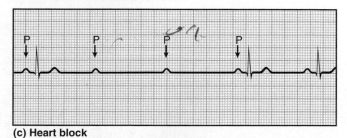

(c) Heart block

Figure 20.15 **The Electrocardiogram.** (a) A normal ECG, showing the P wave, QRS complex, and T wave. (b) The abnormal ECG of ventricular fibrillation, with irregular electrical activity and squirming, uncoordinated contractions of the ventricular myocardium. This sort of ECG is typical of a heart attack. (c) The abnormal ECG of a heart block (see Deeper Insight 20.2).

---

[19]from the German spelling, Elektrokardiogramm

(dy-ASS-toe-lee) and allows the chamber to refill. Figure 20.16 shows how the electrical and contractile activity of the heart relate to each other through the cardiac cycle.

1 Initially, all four chambers are relaxed, in diastole. The AV valves are open, and as blood flows into the heart from the venae cavae and pulmonary veins, it flows through these valves and partially fills the ventricles.

2 The sinoatrial (SA) node fires, exciting the atrial myocardium (see fig. 20.13), producing the P wave of the ECG and initiating atrial systole. The contracting atria finish filling the ventricles.

3 The atrioventricular (AV) node fires and electrical excitation spreads down the AV bundle, bundle branches, and Purkinje fibers and throughout the ventricles. Ventricular depolarization generates the QRS complex. This excitation sets off ventricular systole while the atria relax. Ventricular contraction forces the AV valves shut and the semilunar (aortic and pulmonary) valves open. The ventricles eject blood into the aorta and pulmonary trunk.

4 The ventricles repolarize (marked by the T wave) and relax; all four chambers are again in diastole. The semilunar valves reclose because of back pressure in the large arteries,

the AV valves reopen, and the ventricles begin to refill in preparation for the next cycle.

This entire cycle repeats itself at intervals normally governed by the SA node—in a resting adult heart, typically every 0.8 sec. or so, generating a heart rate of about 75 bpm. The normal heartbeat, timed by the sinoatrial node, is called a *sinus rhythm*. Figure 20.15 (b and c) shows some contrasting, abnormal records resulting from diseases of the myocardium or conduction system.

Before You Go On

Answer the following questions to test your understanding of the preceding section:

10. Why is the human heart described as myogenic? Where is its pacemaker, and what is it called?

11. List the components of the cardiac conduction system in the order traveled by signals from the pacemaker.

12. What organelle(s) is/are less developed in cardiac muscle than in skeletal muscle? What organelle(s) is/are more developed? What is the functional significance of these differences?

13. Name two types of cell junctions in the intercalated discs and explain their functional importance.

14. Identify the nerves that supply the SA node, AV node, and ventricular myocardium.

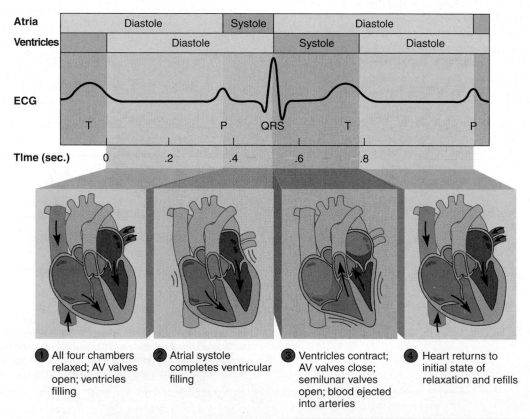

Figure 20.16 The Cardiac Cycle. Major events in one complete cycle of contraction and relaxation of the heart.

20.5 Developmental and Clinical Perspectives

▶ Expected Learning Outcomes

When you have completed this section, you should be able to

- describe the embryonic development of the human heart;
- describe how and why cardiac anatomy changes at birth;
- explain how and why the heart changes in old age; and
- define or briefly describe several of the most common heart diseases.

Prenatal Development of the Heart

The heart is one of the earliest organs to begin functioning in the embryo. The first traces of it appear in week 3; by 22 to 23 days (usually before the mother is aware that she is pregnant), the heart is already beating; by day 24, it circulates blood throughout the embryo.

In week 3, a region of mesoderm at the anterior end of the embryo condenses into a pair of longitudinal cellular cords. By day 19, these become hollow, parallel **endocardial heart tubes** (fig. 20.17a). As the embryo grows and the head region folds, these tubes are pushed closer together, the tissues dividing them break down, and they fuse into a single heart tube (fig. 20.17b). As the tubes are fusing, the surrounding mesoderm forms a primordial myocardium, responsible for the inception of the heartbeat just a few days later. The fetal heartbeat first becomes audible with a stethoscope at about 20 weeks.

With continued folding of the head region, the heart tube elongates and segments into five dilated spaces, some of them corresponding to future heart chambers. From rostral to caudal, these are the **truncus arteriosus,**[20] **bulbus cordis,**[21] **ventricle, atrium,** and **sinus venosus** (fig. 20.17c). Two of these, the ventricle and bulbus cordis, grow more rapidly than the others, causing the heart to loop into a U and then an S shape similar to a fish heart (fig. 20.17d, e). In the course of this looping, the bulbus cordis shifts caudally, the ventricle shifts to the left, and the atrium and sinus venosus shift rostrally, as indicated by the arrows in the figure. During this looping, the heart bulges into the pericardial cavity. Looping is completed by day 28 and results in the forerunners of the adult atria and ventricles assuming their final relationship to each other (the future atria are now superior, or rostral, to the future ventricles). The primordial ventricle seen at day 21 becomes the left ventricle of the adult heart, and the inferior part of the bulbus cordis becomes the right ventricle. The superior part of the bulbus cordis and the truncus arteriosus are now collectively called the **conotruncus** (fig. 20.17e). This passage soon gives rise to the aorta and pulmonary trunk.

The next phase of development is the partitioning of the heart tube into separate chambers (two atria and two ventricles) through the growth of the interatrial and interventricular septa. The interatrial septum begins to form at the end of week 4 and is well established by about 33 days, except for an opening between the atria called the *foramen ovale.* This foramen persists until after birth; its significance is discussed in the next section. The sinus venosus is initially a separate heart chamber, but it becomes extensively remodeled. Originally, it branches into a right and left horn at its inferior end. The right horn enlarges and receives systemic blood from the superior and inferior venae cavae. It eventually becomes part of the right atrium. The left horn shrinks. Parts of it become the coronary sinus, the sinoatrial node (pacemaker), and a portion of the atrioventricular node.

The interventricular septum begins to appear on the floor of the ventricle at the end of week 4 (fig. 20.17f) and increases in height as the two ventricles grow on either side of it. The septum is complete by the end of week 7. Meanwhile, the inner portion of the ventricular wall becomes honeycombed with cavities and differentiates into the trabeculae carneae, papillary muscles, and tendinous cords.

Yet another septum forms during week 5 within the bulbus cordis and truncus arteriosus, dividing this outflow passage in two along its length. The separate halves of the passage become the ascending aorta and pulmonary trunk. The passage twists about 180° as the septum forms. The evidence of this twisting shows in the way the adult pulmonary trunk twists around the aorta. This twisting must be closely coordinated with the closure of the interventricular septum so that the right ventricle will open into the pulmonary trunk and the left ventricle will open into the aorta. Developmental irregularities at this stage are responsible for many cardiac birth defects.

Changes at Birth

There is little point to pumping all of the blood through the lungs of the fetus, because the fetal lungs are not yet inflated or functional. They receive enough blood to meet their metabolic and developmental needs, but most blood bypasses the pulmonary circuit by way of two anatomical shortcuts or *shunts* (fig. 20.18). One is the **foramen ovale,** the opening through the interatrial septum. Some of the blood entering the right atrium passes through this opening directly into the left atrium and from there into the left ventricle and systemic circuit. The other shunt is a short vessel, the **ductus arteriosus,** from the base of the left pulmonary artery to the aorta. Most of the blood that the right ventricle pumps into the pulmonary trunk takes this bypass directly into the aorta instead of following the usual path to the lungs.

At birth, the lungs inflate and their resistance to blood flow drops sharply. The sudden change in pressure gradients causes a flap of tissue to seal the foramen ovale. Blood in the right atrium is no longer able to flow directly into the left and bypass the lungs. In most people, the tissues grow together and permanently seal the foramen, leaving only a depression in the right atrial wall, the *fossa ovalis,* marking its former location. The foramen remains unsealed in about 15% of adults, but the tissue flap acts as a valve that prevents blood from passing through. The ductus arteriosus normally begins to constrict around 10 to 15 hours after birth. It is effectively closed to blood flow within 2 to 4 days, and becomes a permanently closed fibrous cord (*ligamentum arteriosum*) by the age of 2 to 3 weeks (but see Deeper Insight 20.4).

---

[20]*truncus* = trunk + *arteriosus* = arterial
[21]*bulbus* = bulb + *cordis* = of the heart

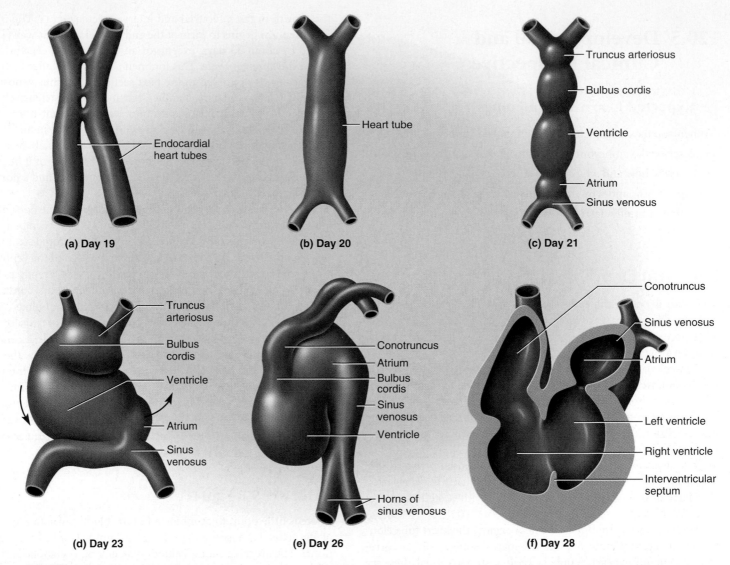

(a) Day 19

Endocardial heart tubes

(b) Day 20

Heart tube

(c) Day 21

Truncus arteriosus

Bulbus cordis

Ventricle

Atrium

Sinus venosus

(d) Day 23

Truncus arteriosus

Bulbus cordis

Ventricle

Atrium

Sinus venosus

(e) Day 26

Conotruncus

Atrium

Bulbus cordis

Sinus venosus

Ventricle

Horns of sinus venosus

(f) Day 28

Conotruncus

Sinus venosus

Atrium

Left ventricle

Right ventricle

Interventricular septum

Figure 20.17 Embryonic Development of the Heart. (a) The endocardial heart tubes beginning to fuse at day 19. (b) Complete fusion by day 20, forming the heart tube. (c) Division of the heart tube into five dilated segments by day 21. The heart begins beating about a day later. (d) The heart begins looping around day 23, with the bulbus cordis migrating caudally (left arrow) and the atrium and sinus venosus migrating rostrally (right arrow). Blood circulates throughout the embryo within a day of this stage. (e) Looping is nearly completed by day 26. (f) Frontal section of the heart at 28 days. As the interventricular septum develops, the conotruncus will divide longitudinally into the ascending aorta and pulmonary trunk, receiving blood from the left and right ventricles, respectively. The single atrium seen here divides into the right and left atria by day 33.

The Aging Heart

A noteworthy effect of aging on the cardiovascular system is a stiffening of the arteries. While that itself is not a cardiac disease, it has important repercussions on the heart. Normally, when the ventricles eject blood, the arteries expand to accommodate the surge in pressure. When arteries are stiffened by age or calcified by arteriosclerosis, they cannot do so. They resist blood flow more than younger arteries, and the heart has to work harder to overcome this resistance. Like any other muscle, when the heart works harder, it grows. The ventricles enlarge, especially the left ventricle, which has to work the hardest to overcome the most resistance. In ventricular hypertrophy, the heart wall and interventricular septum can become so thick that the space within the ventricle is severely diminished. Cardiac output sometimes declines to the point of heart failure.

Many other changes are seen in the aging heart: The valve anuli become more fibrous or even calcified, and the AV valves (especially the mitral valve) thicken and tend to prolapse. The interventricular septum sometimes deviates to the left and interferes with the ejection of blood into the aorta. The fibrous skeleton becomes less elastic, so it has less capacity to rebound in diastole and aid in the filling of the heart. There is a loss of cells from the SA node and conduction system, so impulse conduction is less efficient and more irregular. Degeneration of the conduction system increases the risk of cardiac arrhythmia or heart block. Myocytes die off

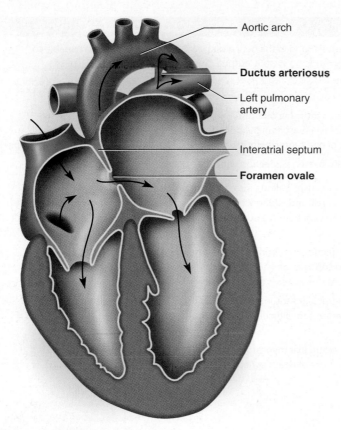

Aortic arch

Ductus arteriosus

Left pulmonary artery

Interatrial septum

Foramen ovale

Figure 20.18 The Fetal Heart. Note the two shunts (foramen ovale and ductus arteriosus) that allow most blood to bypass the nonfunctional lungs.

in the myocardium, and the heart thus becomes weaker. Exercise tolerance is further diminished by decreasing sensitivity to sympathetic stimulation in the elderly heart.

Heart Disease

Heart disease is the leading cause of death in the United States (about 30% of deaths per annum, averaged across all age groups). The most common form of heart disease is coronary atherosclerosis, often leading to myocardial infarction. However, there are a multitude of other heart diseases. The principal categories of heart disease are congenital defects in cardiac anatomy, myocardial hypertrophy or degeneration, inflammation of the pericardium and heart wall, valvular defects, and cardiac tumors. Several examples are described in the Deeper Insights in this chapter and in table 20.1.

Before You Go On

Answer the following questions to test your understanding of the preceding section:

15. When does the embryonic heart begin to beat? At what gestational age does it become audible?

16. What are the five primitive chambers that develop from the heart tube? What becomes of each as the heart continues to develop?

17. Describe the two routes (shunts) by which fetal blood bypasses the lungs. What happens to each one shortly after birth?

18. Why does the heart tend to enlarge in old age? Why does the risk of cardiac arrhythmia increase?

DEEPER INSIGHT   20.4

Patent Ductus Arteriosus

Patent[22] *ductus arteriosus (PDA)* is the failure of the ductus arteriosus to close. For a short time after birth, PDA causes no problems; but as the lungs become better inflated and more functional, pulmonary blood pressure drops below aortic blood pressure. Blood may then begin to flow from the aortic arch back into the pulmonary circuit for an immediate second trip through the lungs. Since this blood soon returns to the left ventricle, it adds markedly to the left ventricular workload. The lungs sometimes respond to the persistent high blood flow with vascular changes that increase pulmonary resistance and stress the right ventricle as well.

The signs of PDA include poor weight gain in early childhood, frequent respiratory illnesses, *dyspnea* (difficulty breathing) on exertion, and

cardiomegaly (enlargement of the heart). PDA is usually suspected at about 6 to 8 weeks of age because of a persistent "machinery-like" heart murmur; it is confirmed by X-ray and other cardiac imaging methods.

Usually, the ductus arteriosus can be stimulated to close with a prostaglandin inhibitor, but if this fails, surgery is required. Surgery is ideally performed between 1.5 and 2.5 years of age because there is a rising risk of *infective endocarditis* (see table 20.1) if it is delayed. The usual procedure is to tie off the DA with several ligatures. This is a low-risk surgery with almost no mortality. However, other methods are available for blocking blood flow through the DA that are less invasive and give an easier recovery for the young patient.

---

[22]*patent = open*

TABLE 20.1	Common Cardiac Pathologies
Cardiac tamponade	Compression of the heart by serous fluid or clotted blood in the pericardial cavity, rendering the heart unable to expand and fill completely during diastole, thus reducing systolic output. (See also Deeper Insight 1.2, p. 18.)
Cardiomyopathy	Any disease of the myocardium from causes other than valvular dysfunction or vascular diseases. Can cause atrophy or hypertrophy of the heart wall and interventricular septum, or dilation and failure of the heart.
Congestive heart failure (CHF)	Failure of either ventricle to pump as much blood as the other one, resulting in accumulation of blood and edema (congestion) in peripheral tissues. Left ventricular failure results in pulmonary congestion and right ventricular failure in systemic congestion (once called *dropsy*). Failure of one ventricle stresses the other and may lead to its subsequent failure.
Infective endocarditis	Inflammation of the endocardium, usually due to bacterial infection with streptococci or staphylococci.
Mitral valve prolapse (MVP)	A valvular defect in which one or both mitral valve cusps balloon into the atrium during ventricular contraction. Often hereditary, affecting 1 out of 40 people overall, and young women especially. Causes significant illness in only 3% of cases, including chest pain, fatigue, shortness of breath, and occasionally infective endocarditis, arrhythmia, or stroke.
Rheumatic fever	Autoimmune disease triggered by a bacterial infection. Antibodies against streptococci or other bacteria attack tissues of the heart valves, causing scarring and constriction (stenosis) of the valves, especially the mitral valve. Regurgitation of blood through the incompetent valve causes turbulence heard as a *heart murmur*.
Septal defects	Abnormal openings in the interatrial or interventricular septum, allowing blood to flow directly between right and left heart chambers. Results in pulmonary hypertension, difficulty breathing, and fatigue. Often fatal in childhood if not corrected.
Ventricular fibrillation	Squirming, uncoordinated contractions of the ventricular myocardium with no effective ejection of blood. Often caused by myocardial infarction (MI); the usual cause of death in heart attack.

Disorders Described Elsewhere

Angina pectoris   551
Arrhythmia   553
Coronary artery disease   551
Heart block   553, 555

Myocardial infarction   551
Patent ductus arteriosus   559
Pericarditis   543
Ventricular fibrillation   555

Study Guide

Assess Your Learning Outcomes

You should have a good understanding of this chapter if you can accurately address the following issues.

20.1 Overview of the Cardiovascular System (p. 540)

1. The two circuits of the cardiovascular system, the functions of each, and which side of the heart supplies each one
2. The great vessels and their associations with the heart chambers
3. The location and orientation of the heart relative to adjacent organs and the median plane of the body
4. The shape, size, and weight of the average adult heart
5. The pericardium, its parietal and visceral layers, the two tissue layers of the pericardial sac, and attachments of the pericardial sac to adjacent organs
6. The pericardial cavity and significance of the pericardial fluid

20.2 Gross Anatomy of the Heart (p. 543)

1. The names and histological composition of the three layers of the heart wall
2. The components and functions of the fibrous skeleton of the heart
3. The names, locations, and functions of the four heart chambers and the surface landmarks of the heart that correspond to the four chambers
4. Reasons for the differences in muscularity between the atria and ventricles and between the right and left ventricles
5. The great vessels and their relationships to the heart chambers

6. The internal septa that separate the heart chambers from each other
7. The names, locations, and anatomy of the two atrioventricular valves, and their relationships with the tendinous cords and papillary muscles
8. The names, locations, and anatomy of the two semilunar valves
9. The mode of operation of the heart valves
10. The path of blood flow through the heart chambers and valves

20.3 Coronary Circulation (p. 549)

1. The origins, branches, and distribution of the two coronary arteries
2. The major veins of the coronary circulation and the routes of coronary blood drainage into the right atrium

3. The cause of myocardial infarction and the protection afforded by collateral circulation in the heart

4. The reasons why coronary blood flow is greatest in diastole and diminished in systole, in contrast to arterial blood flow elsewhere in the body

20.4 The Cardiac Conduction System and Cardiac Muscle (p. 552)

1. The name and location of the pacemaker of the heart

2. The role of the atrioventricular bundle, bundle branches, Purkinje fibers, and myocardial gap junctions in conducting electrical excitation through the heart

3. The specialized structural features of cardiac myocytes, especially the components of the intercalated discs

4. The origins and terminations of sympathetic and parasympathetic nerves to the heart, and their respective effects on cardiac function

5. The four principal phases of the cardiac cycle and how they relate to cardiac anatomy

20.5 Developmental and Clinical Perspectives (p. 557)

1. The embryonic origin of the heart tube and how the tube divides and folds to produce the fully formed heart

2. The two shunts in and near the fetal heart that enable most blood to bypass the nonfunctional lungs

3. The changes in cardiac anatomy and blood flow that occur at birth when the infant begins to breathe on its own

4. The effects of aging on the heart

5. The principal categories of heart disease and essential features of some common heart diseases

Testing Your Recall

1. The cardiac conduction system includes all of the following *except*
   a. the SA node.
   b. the AV node.
   c. the bundle branches.
   d. the tendinous cords.
   e. the Purkinje fibers.

2. To get from the right atrium to the right ventricle, blood flows through the right AV, or _____, valve.
   a. pulmonary
   b. tricuspid
   c. bicuspid
   d. aortic
   e. mitral

3. There is/are _____ pulmonary vein(s) emptying into the right atrium of the heart.
   a. no
   b. one
   c. two
   d. four
   e. more than four

4. The coronary blood vessels are part of the _____ circuit of the circulatory system.
   a. cardiac
   b. pulmonary
   c. systematic
   d. systemic
   e. cardiovascular

5. The outermost layer of the heart wall is known as
   a. the pericardial sac.
   b. the epicardium.
   c. the visceral pericardium.
   d. both *a* and *c*.
   e. both *b* and *c*.

6. These are some of the points that the blood passes as it circulates through the heart chambers, listed alphabetically: (1) left atrium, (2) left ventricle, (3) mitral valve, (4) pulmonary valve, (5) right atrium, (6) right ventricle, (7) tricuspid valve. Place these in the correct order from the time that blood enters the heart from the venae cavae to the time blood leaves the heart by way of the aorta.
   a. 1–3–2–4–5–7–6
   b. 1–2–3–5–7–6–4
   c. 5–3–6–4–1–2–7
   d. 6–7–5–4–2–3–1
   e. 5–7–6–4–1–3–2

7. The ascending aorta and pulmonary trunk develop from the embryonic
   a. bulbus cordis only.
   b. truncus arteriosus only.
   c. horns of the sinus venosus.
   d. conotruncus.
   e. ventricle.

8. The _____ prevent the AV valves from bulging into the atria during ventricular systole.
   a. tendinous cords
   b. pectinate muscles
   c. trabeculae carneae
   d. AV nodes
   e. cusps

9. Blood in the anterior interventricular branch of the left coronary artery flows into myocardial blood capillaries and next drains into
   a. the superior vena cava.
   b. the great cardiac vein.
   c. the left atrium.
   d. the middle cardiac vein.
   e. the coronary sinus.

10. Which of these is *not* characteristic of the heart in old age?
    a. ventricular enlargement
    b. thickening of the atrial walls
    c. a less elastic fibrous skeleton
    d. fewer cells in the conduction system
    e. less sensitivity to norepinephrine

11. The contraction of any heart chamber is called _____, and its relaxation is called _____.

12. The circulatory route from aorta to the venae cavae is the _____ circuit.

13. The circumflex branch of the left coronary artery travels in a groove called the _____.

14. The finest passages through which electrical signals pass before reaching the ventricular myocytes are called _____.

15. Electrical signals pass quickly from one cardiac myocyte to another through the _____ of the intercalated discs.

16. The abnormal bulging of a bicuspid valve cusp into the atrium is called _____.

17. The _____ nerves innervate the heart and tend to reduce the heart rate.

18. The death of cardiac tissue from lack of blood flow is commonly known as a heart attack, but is clinically called _____.

19. Blood in the heart chambers is separated from the myocardium by a thin membrane called the _____.

20. The sinoatrial node develops from an embryonic heart tube chamber called the _____.

Answers in the Appendix

Building Your Medical Vocabulary

State a medical meaning of each of the following word elements, and give a term in which it is used.

1. cardio-
2. epi-
3. semi-
4. -ary
5. -icle
6. -genic
7. lun-
8. coron-
9. ventr-
10. fasci-

Answers in the Appendix

True or False

Determine which five of the following statements are false, and briefly explain why.

1. All blood that has circulated through the myocardium eventually flows into the coronary sinus and from there to the right atrium.
2. The aorta is the body's largest artery.
3. Normally, the only way electrical signals can get from the atria to the ventricles is to pass through the AV node and AV bundle.
4. The epicardium contains adipose tissue, but the endocardium does not.
5. If all nerves from the central nervous system to the heart were severed, the heart would stop beating.
6. The thickest myocardium is normally found in the left ventricle.
7. Many of the cardiac veins have anastomoses that ensure the myocardium will receive blood even if one of the veins becomes blocked.
8. During embryonic development, a ventricular septum grows and divides the single primordial ventricle into right and left ventricles.
9. Blood in the superior and inferior venae cavae flows through the semilunar valves as it enters the right atrium.
10. Cardiac myocytes transmit electrical signals to each other by way of their gap junctions.

Answers in the Appendix

Testing Your Comprehension

1. Mr. Jones, 78, dies of a massive myocardial infarction triggered by coronary thrombosis. Upon autopsy, necrotic myocardium is found in the lateral and posterior right ventricle and posterior interventricular septum. Based on the information in this chapter, where in the coronary circulation do you think the thrombosis occurred?
2. Becky, age 2, was born with a hole in her interventricular septum (*ventricular septal defect,* or *VSD*). Considering that the blood pressure in the left ventricle is significantly higher than blood pressure in the right ventricle, predict the effect of the VSD on Becky's pulmonary blood pressure, systemic blood pressure, and long-term changes in the ventricular walls.
3. Marcus is born with *transposition* of the great arteries, in which the aorta arises from the right ventricle and the pulmonary artery arises from the left. Assuming no other anatomical abnormalities, trace the flow of blood through the pulmonary and systemic routes in his case. Predict the consequences, if any, for the ability of Marcus's cardiovascular system to deliver oxygen to the systemic tissues. Do you think Marcus would require immediate surgical correction in early infancy; correction at the age of 2 or 3 years; or that it could be left alone and not seriously affect his life expectancy?
4. Review the condition of heart block in Deeper Insight 20.2, and examine the electrocardiogram for this condition in figure 20.15c. Explain why the second P wave in the ECG is followed by another P instead of a QRS complex.
5. In dilated cardiomyopathy of the left ventricle, the ventricle can become enormously enlarged. Explain why this might lead to regurgitation of blood through the mitral valve (blood flowing from the ventricle back into the left atrium) during ventricular systole.

Answers at www.mhhe.com/saladinha3

Improve Your Grade at www.mhhe.com/saladinha3

Practice quizzes, labeling activities, games, and flashcards provide fun ways to master concepts. You can also download image PowerPoint files for each chapter to create a study guide or for taking notes during lecture.

The Circulatory System III—Blood Vessels

Capillary beds

BRUSHING UP

To understand this chapter, you may find it helpful to review the following concepts:
- Primary germ layers of the embryo (p. 87)
- Muscle compartments of the limbs (p. 294)
- The pulmonary and systemic circuits (p. 540)
- The great vessels associated with the heart (p. 541)
- Cardiac systole and diastole (p. 555)

Anatomy & Physiology | **REVEALED**
aprevealed.com

Cardiovascular System

W here does the blood go once it leaves the heart? This question was a point of much confusion until the seventeenth century. Chinese medical history attributes a correct view of blood circulation to the mythical Yellow Emperor, Huang Ti (2697–2597 BCE), said to have noted that blood flowed in a complete circuit around the body and back to the heart. But in the second century CE, Roman physician Claudius Galen argued that it flowed back and forth in the veins, like air in the bronchial tubes. He believed that the liver received food from the small intestine and converted it to blood, the heart pumped the blood through the veins to all other organs, and those organs consumed it.

The ancient Chinese view was correct, but the first experimental demonstration of this did not come until the seventeenth century. English physician William Harvey (1578–1657) studied the filling and emptying of the heart in snakes; tied off the vessels above and below the heart to observe the effects on cardiac filling and output; measured cardiac output in a variety of living animals; and estimated the cardiac output in humans. He concluded that (1) the heart pumps more blood in half an hour than there is in the entire body; (2) not enough food is consumed to account for the continual production of so much blood; and (3) since the planets orbit the sun and he believed that the human body was modeled after the solar system, it followed that the blood must orbit the body. So for a peculiar combination of experimental and superstitious reasons, Harvey argued that the blood must return to the heart rather than being consumed by the peripheral organs. He could not explain how, since the microscope had yet to be developed to the point that allowed Marcello Malpighi (1628–1694) and Antony van Leeuwenhoek (1632–1723) to discover the blood capillaries.

Harvey published his findings in 1628 in a short but elegant book entitled *Exercitio Anatomica de Motu Cordis et Sanguinis in Animalibus* (*Anatomical Studies on the Motion of the Heart and Blood in Animals*). This landmark in the history of biology and medicine was the first experimental study of animal physiology. But so entrenched were the ideas of Aristotle and Galen in the medical community, and so strange was the idea of doing experiments on living animals, that Harvey's contemporaries rejected his ideas. Indeed, some of them regarded him as a crackpot because his conclusion flew in the face of common sense—if the blood was continually recirculated and not consumed by the tissues, they reasoned, then what purpose could it possibly serve?

Harvey lived to a ripe old age, served as physician to the kings of England, and later did important work in embryology. His case is one of the most interesting in biomedical history, for it shows how empirical science overthrows old theories and spawns better ones, and how common sense and blind allegiance to authority can interfere with acceptance of the truth. But most importantly, Harvey's contributions represent the birth of experimental physiology.

21.1 General Anatomy of the Blood Vessels

▶ Expected Learning Outcomes

When you have completed this section, you should be able to
- describe the structure of a blood vessel;

- describe the different types of arteries, capillaries, and veins;
- trace the general route usually taken by the blood from the heart and back again; and
- describe some variations on this route.

There are three principal categories of blood vessels—arteries, veins, and capillaries. **Arteries** are the efferent vessels of the cardiovascular system—that is, vessels that carry blood away from the heart. **Veins** are the afferent vessels—vessels that carry blood back to the heart. **Capillaries** are microscopic, thin-walled vessels that connect the smallest arteries to the smallest veins. Aside from their general location and direction of blood flow, these three vessel types also differ in the histological structure of their walls.

The Vessel Wall

The walls of arteries and veins are composed of three layers called *tunics* (figs. 21.1 and 21.2):

1. The **tunica interna (tunica intima)** lines the inside of the vessel and is exposed to the blood. It consists of a simple squamous epithelium called the **endothelium,** overlying a basement membrane and a sparse layer of loose connective tissue. The endothelium acts as a selectively permeable barrier to materials entering or leaving the bloodstream; it secretes chemicals that stimulate the muscle of the vessel wall to contract or relax, thus narrowing or widening the vessel; and it normally repels blood cells and platelets so that they flow freely without sticking to the vessel wall. Under special circumstances, however, platelets and blood cells do adhere to it. When the endothelium is damaged, platelets can adhere and form a blood clot; and when the tissue around a vessel is inflamed, the endothelial cells produce *cell-adhesion molecules* that induce leukocytes to adhere to the surface. This causes leukocytes to congregate in tissues where their defensive actions are needed.

2. The **tunica media,** the middle layer, is usually the thickest. It consists of smooth muscle, collagen, and in some cases, elastic tissue. The relative amounts of muscle and elastic tissue vary greatly from one vessel to another and form a basis for the classification of vessels described in the next section. The tunica media strengthens the vessels and prevents blood pressure from rupturing them, and it produces *vasomotion,* changes in vessel diameter. The widening of a vessel is called *vasodilation* and narrowing is called *vasoconstriction.*

3. The **tunica externa (tunica adventitia[1])** is the outermost layer. It consists of loose connective tissue that often merges with that of neighboring blood vessels, nerves, or other organs. It anchors the vessel and allows small nerves, lymphatic vessels, and smaller blood vessels to reach and penetrate into the tissues of a larger vessel. Small vessels called the **vasa vasorum[2]** (VAY-za vay-SO-rum) supply blood to at least the outer half of the vessel wall. Tissues of the inner half are thought to be nourished by diffusion from blood in the lumen.

---

[1]*advent* = added to
[2]*vasa* = vessels + *orum* = of

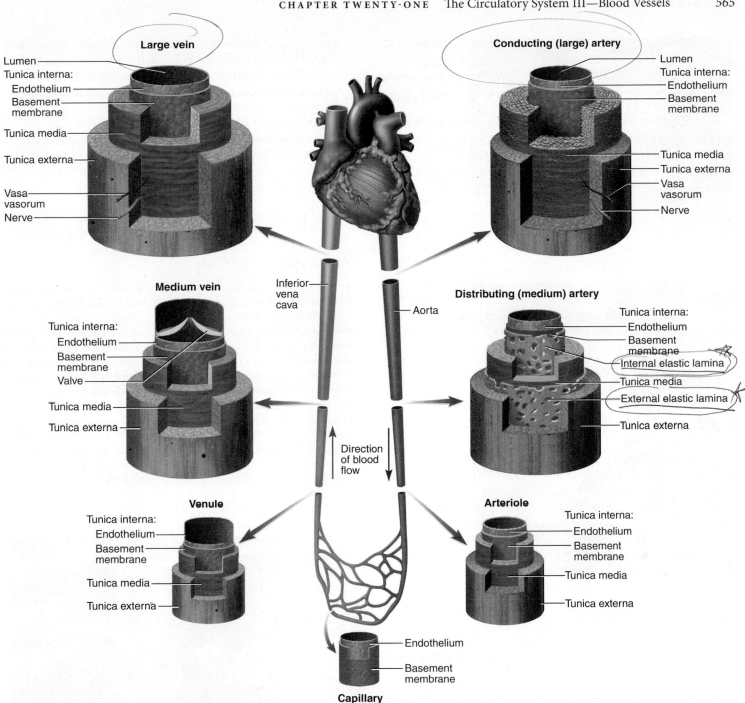

Figure 21.1 **Histological Structure of Blood Vessels.**
● *Why do the arteries have so much more elastic tissue than the veins do?*

Arteries

Arteries are considered to be the *resistance vessels* of the cardiovascular system because they have a relatively strong, resilient tissue structure that resists the high blood pressure within. Each beat of the heart creates a surge of pressure in the arteries as blood is ejected into them. Arteries are built to withstand these pressure surges. Being more muscular than veins, they retain their round shape even when empty, and they appear relatively circular in tissue sections.

Classes of Arteries

Arteries are divided into three classes by size, but of course there is a gradual transition from one class to the next.

1. **Conducting (elastic** or **large) arteries** are the biggest arteries. The aorta, common carotid and subclavian arteries, pulmonary trunk, and common iliac arteries are examples of conducting arteries. They have a layer of elastic tissue called the *internal elastic lamina* at the border between the intima and

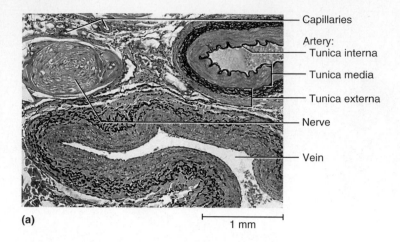

(a)

1 mm

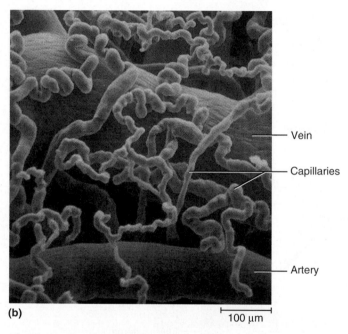

(b)

100 μm

Figure 21.2 Micrographs of Blood Vessels. (a) A neurovascular bundle, composed of a small artery, small vein, and nerve traveling together in a common sheath of connective tissue. The dark wavy line near the lumen of the artery is the internal elastic lamina. (b) A polymer cast of blood vessels of the eye (SEM).

media, but microscopically, it is incomplete and difficult to distinguish from the elastic tissue of the tunica media. The tunica media consists of 40 to 70 layers of elastic sheets, perforated like slices of Swiss cheese, alternating with thin layers of smooth muscle, collagen, and elastic fibers. In histological sections, the view is dominated by this elastic tissue. There is an *external elastic lamina* at the border between the media and externa, but it, too, is difficult to distinguish from the elastic sheets of the tunica media. The tunica externa is less than half as thick as the tunica media and relatively sparse in the largest arteries. It is well supplied with vasa vasorum.

Conducting arteries expand during ventricular systole to receive blood, and recoil during diastole. Their expansion takes some of the pressure off the blood so that smaller

arteries downstream are subjected to less systolic stress. Their recoil between heartbeats prevents the blood pressure from dropping too low while the heart is relaxing and refilling. These effects lessen the fluctuations in blood pressure that would otherwise occur. Arteries stiffened by atherosclerosis cannot expand and recoil as freely. Consequently, the downstream vessels are subjected to greater stress and are more likely to develop aneurysms (see Deeper Insight 21.1).

DEEPER INSIGHT 21.1

Aneurysm

An aneurysm is a weak point in an artery or in the heart wall. It forms a thin-walled, bulging sac that pulsates with each beat of the heart and may eventually rupture. In a *dissecting aneurysm,* blood accumulates between the tunics of an artery and separates them, usually because of degeneration of the tunica media. The most common sites of aneurysms are the abdominal aorta (fig. 21.3), renal arteries, and the arterial circle at the base of the brain. Even without hemorrhaging, aneurysms can cause pain or death by putting pressure on brain tissue, nerves, adjacent veins, pulmonary air passages, or the esophagus. Other consequences include neurological disorders, difficulty in breathing or swallowing, chronic cough, or congestion of the tissues with blood. Aneurysms sometimes result from congenital weakness of the blood vessels and sometimes from trauma or bacterial infections such as syphilis. The most common cause, however, is the combination of atherosclerosis and hypertension.

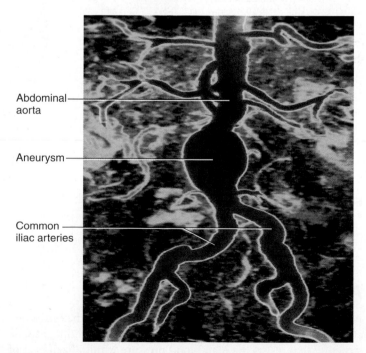

Figure 21.3 An Aortic Aneurysm. This is a magnetic resonance angiogram (MRA) of the abdominal aorta of a patient with hypertension, showing a prominent bulge (aneurysm) of the aorta immediately superior to the common iliac arteries.

2. **Distributing (muscular or medium) arteries** are smaller branches that distribute blood to specific organs. You could compare a conducting artery to an interstate highway and distributing arteries to the exit ramps and state highways that serve specific towns. Most arteries that have specific anatomical names are in these first two size classes. The brachial, femoral, renal, and splenic arteries are examples of distributing arteries. Distributing arteries typically have up to 40 layers of smooth muscle, constituting about three-quarters of the wall thickness. In histological sections, smooth muscle is more conspicuous than the elastic tissue. Both the internal and external elastic laminae, however, are thick and often conspicuous.

3. **Resistance (small) arteries** are usually too variable in number and location to be given individual names. They exhibit up to 25 layers of smooth muscle and relatively little elastic tissue. Compared to larger arteries, they have a thicker tunica media in proportion to the lumen. The smallest resistance arteries, about 40 to 200 μm in diameter and with only one to three layers of smooth muscle, are called **arterioles.** Arterioles have very little tunica externa.

Metarterioles[3] are short vessels that link arterioles and capillaries. Instead of a continuous tunica media, they have individual muscle cells spaced a short distance apart, each forming a **precapillary sphincter** that encircles the entrance to one capillary. Constriction of these sphincters reduces or shuts off blood flow through their respective capillaries and diverts blood to tissues or organs elsewhere.

Arterial Sense Organs

Certain major arteries above the heart have sensory receptors in their walls that monitor blood pressure and chemistry (fig. 21.4). These receptors transmit information to the brainstem that serves to regulate the heartbeat, vasomotion, and respiration. They are of three main kinds:

1. **Carotid sinuses.** These are *baroreceptors* (pressure sensors) that respond to changes in blood pressure. Ascending the neck on each side is a *common carotid artery* that branches near the angle of the mandible to form the *internal carotid artery* to the brain and *external carotid artery* to the face. The carotid sinuses are located in the wall of the internal carotid artery just above the branch point. The carotid sinus has a relatively thin tunica media and an abundance of glossopharyngeal nerve fibers in the tunica externa. A rise in blood pressure easily stretches the thin media and stimulates these nerve fibers. The glossopharyngeal nerve then transmits signals to the vasomotor and cardiac centers of the brainstem, and the brainstem responds by lowering the heart rate and dilating the blood vessels, thereby lowering the blood pressure. Similar baroreceptors also occur in the wall of the aortic arch.

2. **Carotid bodies.** Also located near the branch of the common carotid arteries, these are oval receptors about 3 by 5 mm, innervated by sensory fibers of the vagus and glossopharyngeal nerves. They are *chemoreceptors* that monitor changes in blood

[3]*meta* = beyond, next in a series

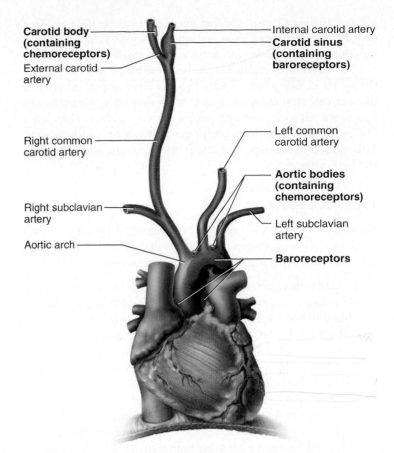

Figure 21.4 **Baroreceptors and Chemoreceptors in the Arteries Superior to the Heart.** The structures shown here are repeated in the left carotid arteries.

composition. They primarily transmit signals to the brainstem respiratory centers, which adjust breathing to stabilize the blood pH and its CO_2 and O_2 levels.

3. **Aortic bodies.** These are one to three chemoreceptors located in the aortic arch near the arteries to the head and arms. They are structurally similar to the carotid bodies and have the same function.

Capillaries

For the blood to serve any purpose, materials such as nutrients, wastes, and hormones must pass between the blood and the tissue fluids, through the walls of the vessels. There are only two places in the circulation where this occurs—the capillaries and venules. We can think of these as the "business end" of the cardiovascular system, because all the rest of the system exists to serve the exchange processes that occur here. Since capillaries greatly outnumber venules, they are the more important of the two. Capillaries are sometimes called the *exchange vessels* of the cardiovascular system.

Blood capillaries (fig. 21.5) are composed of only an endothelium and basal lamina. Their walls are as thin as 0.2 to 0.4 μm. They average about 5 μm in diameter at the proximal end (where they receive arterial blood), they widen to about 9 μm at the distal end (where they empty into a small vein), and they often branch along the way. Since erythrocytes are about 7.5 μm in diameter, they often have to stretch into elongated shapes to squeeze through the smallest capillaries.

The number of capillaries has been estimated at a billion and their total surface area at 6,300 m², but a more important point is that scarcely any cell in the body is more than 60 to 80 μm (about four to six cell widths) away from the nearest capillary. There are a few exceptions: Capillaries are scarce in tendons and ligaments, rarely found in cartilage, and absent from epithelia and the cornea and lens of the eye.

Types of Capillaries

There are three types of capillaries, distinguished by the ease with which they allow substances to pass through their walls and by structural differences that account for their greater or lesser permeability.

1. **Continuous capillaries** (fig. 21.5) occur in most tissues, such as skeletal muscle. Their endothelial cells, held together by tight junctions, form a continuous tube. A thin protein–carbohydrate layer, the **basal lamina,** surrounds the endothelium and separates it from the adjacent connective tissues. The endothelial cells are usually separated by narrow **intercellular clefts** about 4 nm wide. Small solutes such as glucose can pass through these clefts, but plasma proteins, other large molecules, and platelets and blood cells are held back. The continuous capillaries of the brain lack intercellular clefts and have more complete tight junctions that form the blood–brain barrier discussed in chapter 15 (p. 404).

   Some continuous capillaries exhibit cells called ~~pericytes~~ that lie external to the endothelium. Pericytes have elongated tendrils that wrap around the capillary. They contain the same contractile proteins as muscle, and it is thought that they contract and regulate blood flow through the capillaries. ~~They also can differentiate into endothelial and smooth muscle cells and thus contribute to vessel growth and repair.~~

2. **Fenestrated capillaries** have endothelial cells riddled with holes called **filtration pores (fenestrations[4])** (fig. 21.6). These pores are about 20 to 100 nm in diameter, and are often spanned by a thin glycoprotein membrane. They allow for the rapid passage of small molecules but still retain most proteins and larger particles in the bloodstream. Fenestrated capillaries are important in organs that engage in rapid absorption or filtration—the kidneys, endocrine glands, small intestine, and choroid plexuses of the brain, for example.

3. **Sinusoids (discontinuous capillaries)** are irregular blood-filled spaces in the liver, bone marrow, spleen, and some other organs (fig. 21.7). They are twisted, tortuous passageways, typically 30 to 40 μm wide, that conform to the shape of the surrounding tissue. The endothelial cells are separated by wide gaps with no basal lamina, and the cells also frequently have especially large fenestrations through them. Even proteins and blood cells can pass through these pores; this is how albumin, clotting factors, and other proteins synthesized by the liver enter the blood, and how newly formed blood cells enter the circulation from the bone marrow and lymphatic organs. Some sinusoids contain macrophages or other specialized cells.

Capillary Permeability

The structure of a capillary wall is closely related to its permeability, the ease with which substances pass through it from the blood to the tissue fluid, or vice versa. There are three routes that materials can travel through a capillary wall (fig. 21.8): (1) the intercellular clefts between endothelial cells; (2) the filtration pores in fenestrated capillaries; and (3) the endothelial cell plasma membrane and cytoplasm. Nonpolar molecules such as O_2, CO_2, lipids, and thyroid hormone diffuse easily through the endothelial cells. Hydrophilic substances such as glucose, electrolytes, and large molecules such as insulin pass through the intercellular clefts and filtration pores, or cross through the endothelial cells by a process called *transcytosis:* The endothelial cell internalizes molecules or fluid droplets by endocytosis on one side of the capillary wall, transports the endocytotic vesicles to the other side of the cell, and releases the substances by exocytosis on that side.

Capillary Beds

Capillaries are organized into groups called **capillary beds**—usually 10 to 100 capillaries supplied by a single metarteriole (fig. 21.9; see also photo on p. 563). Beyond the origins of the capillaries, the metarteriole continues as a **thoroughfare channel** leading directly to a venule. Capillaries empty into the distal end of the thoroughfare channel or directly into the venule.

When the precapillary sphincters are open, the capillaries are well supplied with blood and engage in exchanges with the tissue fluid. When the sphincters are closed, blood bypasses the capil-

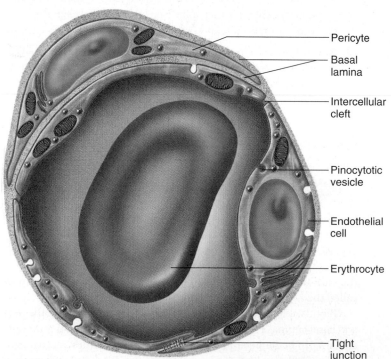

Figure 21.5 Structure of a Continuous Capillary (Cross Section).

Pericyte

Basal lamina

Intercellular cleft

Pinocytotic vesicle

Endothelial cell

Erythrocyte

Tight junction

[4]*fenestra* = window

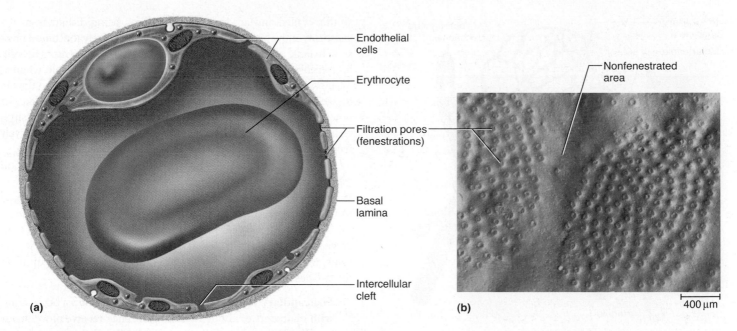

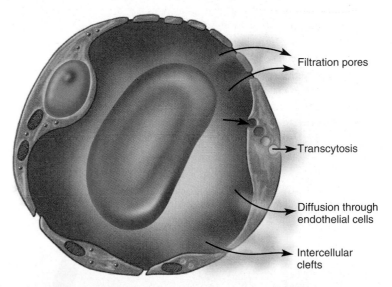

Figure 21.6 **Structure of a Fenestrated Capillary.** (a) Cross section of the capillary wall. (b) Surface view of a fenestrated endothelial cell (SEM). The cell has patches of filtration pores (fenestrations) separated by nonfenestrated areas.
● *Identify some organs that have this type of capillary rather than continuous capillaries.*

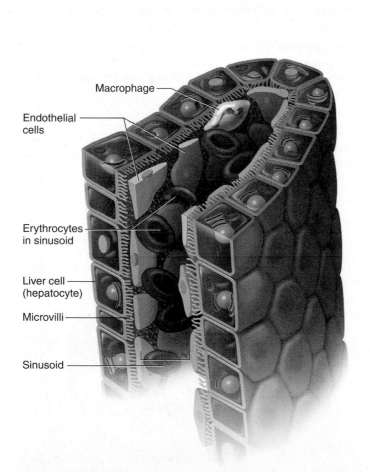

Figure 21.7 **A Sinusoid of the Liver.** Large gaps between the endothelial cells allow blood plasma to directly contact the liver cells, but retain blood cells in the lumen of the sinusoid.

Figure 21.8 **Pathways of Capillary Fluid Exchange.** Materials move through the capillary wall through filtration pores (in fenestrated capillaries only), by transcytosis, by diffusion through the endothelial cells, and through intercellular clefts. Although this figure depicts material leaving the bloodstream, material can also enter the bloodstream by the same methods.

laries, flows through the thoroughfare channel to a venule, and engages in relatively little fluid exchange. There is not enough blood in the body to fill the entire vascular system at once; consequently, about three-quarters of the body's capillaries are shut down at any given time. In the skeletal muscles, for example, about 90% of them have little to no blood flow during periods of rest. During exercise, they receive an abundant flow, while capillary beds elsewhere—for example, in the skin and intestines—shut down to compensate for it.

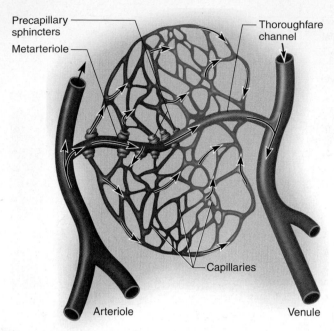

Precapillary sphincters

Metarteriole

Thoroughfare channel

Capillaries

Arteriole

Venule

(a) Sphincters open

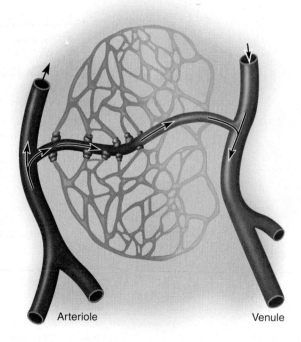

Arteriole

Venule

(b) Sphincters closed

Figure 21.9 **Regulation of Capillary Blood Flow.** (a) Precapillary sphincters dilated and capillaries abundantly supplied with blood. (b) Precapillary sphincters closed, with blood bypassing the capillaries.

Veins

Veins are regarded as the *capacitance vessels* of the cardiovascular system because they are relatively thin-walled and flaccid, and expand easily to accommodate an increased volume of blood; that is, they have a greater capacity for blood containment than arteries do. At rest, about 54% of the blood is found in the systemic veins as compared to only 11% in the systemic arteries (fig. 21.10). The reason that veins are

so thin-walled and accommodating is that, being distant from the ventricles of the heart, they are subjected to relatively low blood pressure. In large arteries, blood pressure averages 90 to 100 mm Hg (millimeters of mercury) and surges to 120 mm Hg during systole, whereas in veins, it averages about 10 mm Hg. Furthermore, the blood flow in the veins is steady, rather than pulsating with the heartbeat like the flow in the arteries. Veins therefore do not require thick, pressure-resistant walls. They collapse when empty and thus have relatively flattened, irregular shapes in histological sections (see fig. 21.2).

As we trace blood flow in the arteries, we find it splitting off repeatedly into smaller and smaller *branches* of the arterial system. In the venous system, conversely, we find small veins merging to form larger and larger ones as they approach the heart. We refer to the smaller veins as *tributaries,* by analogy to the streams that converge and act as tributaries to rivers. In examining the types of veins, we will follow the direction of blood flow, working up from the smallest to the largest vessels.

1. **Postcapillary venules** are the smallest of the veins, beginning with diameters of about 15 to 20 μm. They receive blood from capillaries directly or by way of the distal ends of the thoroughfare channels. They have a tunica interna with only a few fibroblasts around it and no muscle. Like capillaries, they are often surrounded by pericytes. Postcapillary venules are even more porous than capillaries; therefore, venules also exchange fluid with the surrounding tissues. Most leukocytes emigrate from the bloodstream through the venule walls.

2. **Muscular venules** receive blood from the postcapillary venules. They are up to 1 mm in diameter. They have a tunica media of one or two layers of smooth muscle, and a thin tunica externa.

3. **Medium veins** range up to 10 mm in diameter. Most veins with individual names are in this category, such as the radial and ulnar veins of the forearm and the small and great saphenous veins of the leg. Medium veins have a tunica interna with an endothelium, basement membrane, loose

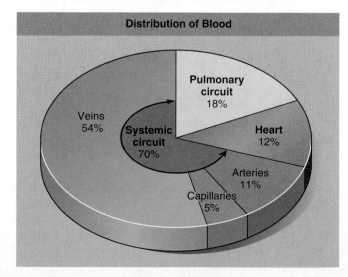

Figure 21.10 Typical Distribution of the Blood in a Resting Adult.
● *What anatomical fact allows the veins to contain so much more blood than the arteries do?*

connective tissue, and sometimes a thin internal elastic lamina. The tunica media is much thinner than it is in medium arteries; it exhibits bundles of smooth muscle, but not a continuous muscular layer as seen in arteries. The muscle is interrupted by regions of collagenous, reticular, and elastic tissue. The tunica externa is relatively thick.

Many medium veins, especially in the limbs, exhibit infoldings of the tunica interna that meet in the middle of the lumen, forming **venous valves** directed toward the heart (fig. 21.11). The pressure in the veins is not high enough to push all of the blood upward against the pull of gravity in a standing or sitting person. The upward flow of blood in these vessels depends partly on the massaging action of skeletal muscles and the ability of these valves to keep the blood from dropping down again when the muscles relax. When the muscles surrounding a vein contract, they force blood through the valves. The propulsion of blood by muscular massaging, aided by the venous valves, is a mechanism of blood flow called the *skeletal muscle pump.* Varicose veins result in part from failure of the valves (see Deeper Insight 21.2). Such valves are absent from very small and large veins, veins of the abdominal cavity, and veins of the brain.

4. **Venous sinuses** are veins with especially thin walls, large lumens, and no smooth muscle. Examples include the coronary sinus of the heart and the dural sinuses of the brain. Unlike other veins, they are not capable of vasomotion.

5. **Large veins** have diameters greater than 10 mm. They have some smooth muscle in all three tunics. The tunica media is relatively thin with only a moderate amount of smooth muscle; the tunica externa is the thickest layer and contains longitudinal bundles of muscle. Large veins include the venae cavae, pulmonary veins, internal jugular veins, and renal veins.

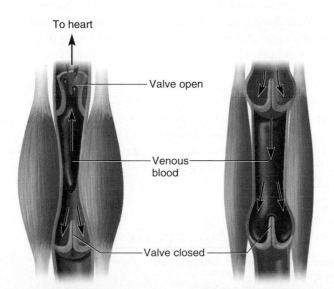

To heart

Valve open

Venous blood

Valve closed

(a) Contracted skeletal muscles   **(b) Relaxed skeletal muscles**

Figure 21.11 The Skeletal Muscle Pump. (a) Muscle contraction squeezes the deep veins and forces blood through the next valve in the direction of the heart. Valves below the point of compression prevent backflow. (b) When the muscles relax, blood flows back downward under the pull of gravity but can flow only as far as the nearest valve.

Apply What You Know

Why would venous valves be unnecessary in the jugular veins of the neck?

Circulatory Routes

The simplest and most common route of blood flow is heart $\longrightarrow$ arteries $\longrightarrow$ capillaries $\longrightarrow$ veins $\longrightarrow$ heart. Blood usually passes through only one network of capillaries from the time it leaves the heart until the time it returns (fig. 21.12a), but there are exceptions, notably portal systems and anastomoses.

A **portal system** (fig. 21.12b) is a route in which blood flows through two capillary beds—one after the other—before returning to the heart. For example, in the *hepatic portal system,* the blood picks up nutrients from a capillary bed in the small intestine, then flows through a series of veins to the liver, where there is a second capillary bed (the liver sinusoids in fig. 21.7). The blood unloads some nutrients here, picks up substances produced by the liver cells, and then flows to the heart. Other portal systems occur in the kidneys (chapter 25) and connecting the hypothalamus to the anterior pituitary gland (chapter 18).

An **anastomosis** is a point where two veins or arteries merge without intervening capillaries. In an **arteriovenous anastomosis (shunt),** blood flows from an artery directly into a vein (fig. 21.12c). Shunts occur in the fingers, palms, toes, and ears, where they reduce heat loss in cold weather by allowing warm blood to bypass these exposed surfaces. Unfortunately, this makes these blood-deprived areas more susceptible to frostbite. In an **arterial anastomosis,** two arteries merge, providing *collateral* (alternative) routes of blood supply to a tissue (fig. 21.12d). Those of the coronary circulation were mentioned in chapter 20 (page 551). They are also common around joints where movement may temporarily compress an artery and obstruct one pathway. **Venous anastomoses,** where one vein empties directly into another, are the most common. They provide several alternative routes of drainage from an organ, so blockage of a vein is rarely as life-threatening as blockage of an artery. Several arterial and venous anastomoses are described later in this chapter.

DEEPER INSIGHT   21.2

Varicose Veins

In people who stand for long periods, such as barbers and cashiers, blood tends to pool in the lower limbs and stretch the veins. This is especially true of superficial veins, which are not surrounded by supportive tissue. Stretching pulls the cusps of the venous valves farther apart until they become incapable of sealing the vessel and preventing the backflow of blood. As the veins become further distended, their walls grow weak, and they develop into *varicose veins* with irregular dilations and twisted pathways. Obesity and pregnancy also promote development of varicose veins by putting pressure on large veins of the pelvic region and obstructing drainage from the limbs. Varicose veins sometimes develop because of hereditary weakness of the valves. With less drainage of blood, tissues of the leg and foot may become edematous and painful. *Hemorrhoids* are varicose veins of the anal canal.

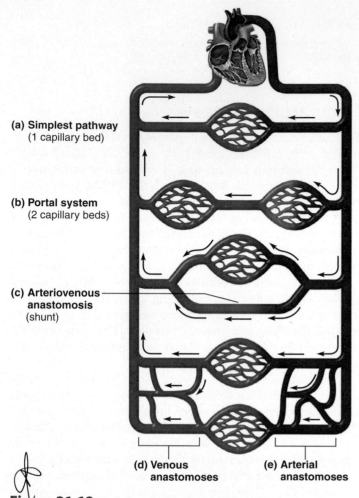

(a) Simplest pathway
(1 capillary bed)

(b) Portal system
(2 capillary beds)

(c) Arteriovenous anastomosis
(shunt)

(d) Venous anastomoses

(e) Arterial anastomoses

Figure 21.12 Variations in Circulatory Pathways.
● *After studying tables 21.1 through 21.11, identify specific sites of three arterial anastomoses and three venous anastomoses.*

Before You Go On

Answer the following questions to test your understanding of the preceding section:

1. Name the three tunics of a typical blood vessel, and explain how they differ from each other.

2. Contrast the tunica media of a conducting artery, arteriole, and venule, and explain how the histological differences are related to the functional differences between these vessels.

3. Describe the differences between a continuous capillary, a fenestrated capillary, and a sinusoid.

4. Describe the differences between a medium vein and a medium (distributing or muscular) artery. State the functional reasons for these differences.

5. Contrast an anastomosis and a portal system with the more typical pathway of blood flow.

6. Describe three routes by which blood-borne substances can pass through a capillary wall into the tissue fluid.

21.2 The Pulmonary Circuit

▶ Expected Learning Outcomes

When you have completed this section, you should be able to

- trace the route of blood through the pulmonary circuit; and

- explain the anatomical and functional differences between the pulmonary and systemic blood supplies to the lungs.

The remainder of this chapter centers on the names and pathways of the principal arteries and veins. The pulmonary circuit is described here; the section after this concerns the systemic arteries and veins of the axial region (head, neck, thoracic, and abdominopelvic regions); and the section following that concerns the systemic arteries and veins of the appendicular region (the limbs).

The pulmonary circuit (fig. 21.13) begins with the **pulmonary trunk,** a large vessel that ascends diagonally from the right ventricle and branches into the right and left **pulmonary arteries.** As it approaches the lung, the right pulmonary artery branches in two, and both branches enter the lung at a medial indentation called the *hilum* (see fig. 23.9, p. 641). The upper branch is the **superior lobar artery,** serving the superior lobe of the lung. The lower branch divides again within the lung to form the **middle lobar** and **inferior lobar arteries,** supplying the lower two lobes of that lung. The left pulmonary artery is much more variable. It gives off several superior lobar arteries to the superior lobe before entering the hilum, then enters the lung and gives off a variable number of inferior lobar arteries to the inferior lobe.

In both lungs, these arteries lead ultimately to small basketlike capillary beds that surround the pulmonary alveoli (air sacs). This is where the blood unloads CO_2 and picks up O_2. After leaving the alveolar capillaries, the pulmonary blood flows into venules and veins, ultimately leading to the main **pulmonary veins** that exit the lung at the hilum. The left atrium of the heart receives two pulmonary veins on each side (see fig. 20.3b, p. 542).

The pulmonary circuit is the only route in the body where the arteries carry oxygen-poor blood and the veins carry oxygen-rich blood; the opposite condition prevails in the systemic circuit. The purpose of the pulmonary circuit is primarily to exchange CO_2 for O_2. The lungs also receive a separate systemic blood supply by way of the *bronchial arteries* (see part I.1 in table 21.4).

Before You Go On

Answer the following questions to test your understanding of the preceding section:

7. Trace the flow of an RBC from right ventricle to left atrium, naming the vessels along the way.

8. Each lung has two separate arterial supplies. Explain their functions.

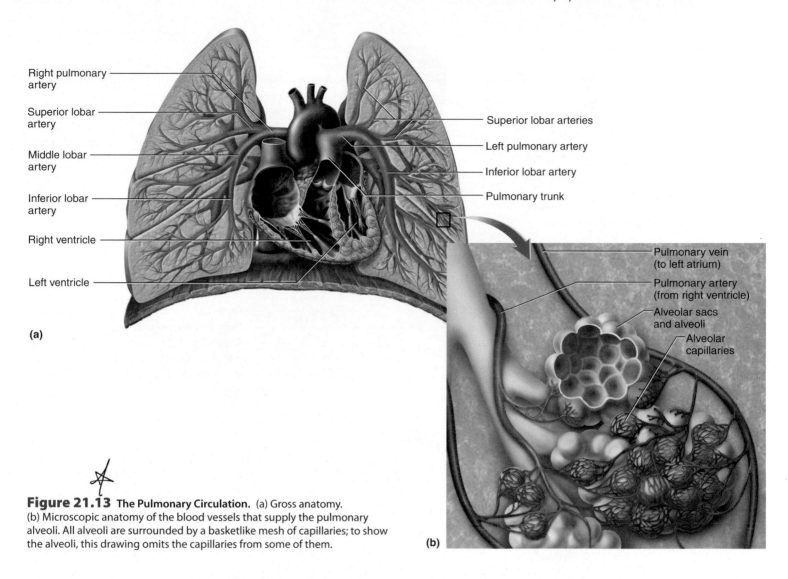

Right pulmonary artery

Superior lobar artery

Middle lobar artery

Inferior lobar artery

Right ventricle

Left ventricle

Superior lobar arteries

Left pulmonary artery

Inferior lobar artery

Pulmonary trunk

(a)

Pulmonary vein (to left atrium)

Pulmonary artery (from right ventricle)

Alveolar sacs and alveoli

Alveolar capillaries

(b)

Figure 21.13 **The Pulmonary Circulation.** (a) Gross anatomy. (b) Microscopic anatomy of the blood vessels that supply the pulmonary alveoli. All alveoli are surrounded by a basketlike mesh of capillaries; to show the alveoli, this drawing omits the capillaries from some of them.

21.3 Systemic Vessels of the Axial Region

▶ Expected Learning Outcomes

When you have completed this section, you should be able to

- identify the principal systemic arteries and veins of the axial region; and

- trace the flow of blood from the heart to any major organ of the axial region and back to the heart.

The systemic circuit (figs. 21.14 and 21.15) supplies oxygen and nutrients to all organs and removes their metabolic wastes. Part of it,

the coronary circulation, was described in chapter 20. This section surveys the remaining arteries and veins of the axial region—the head, neck, and trunk. Tables 21.1 through 21.7 trace the arterial outflow and venous return, region by region. They outline only the most common circulatory pathways; there is a great deal of anatomical variation in the circulatory system from one person to another.

The names of the blood vessels often describe their location by indicating the body region traversed (as in *axillary artery* and *brachial vein*), an adjacent bone (as in *temporal artery* and *ulnar vein*), or the organ supplied or drained by the vessel (as in *hepatic artery* and *renal vein*). In many cases, an artery and adjacent vein have similar names (*femoral artery* and *femoral vein,* for example).

As you trace blood flow in these tables, it is important to refer frequently to the illustrations. Verbal descriptions alone may often seem obscure if you do not make full use of the accompanying explanatory illustrations.

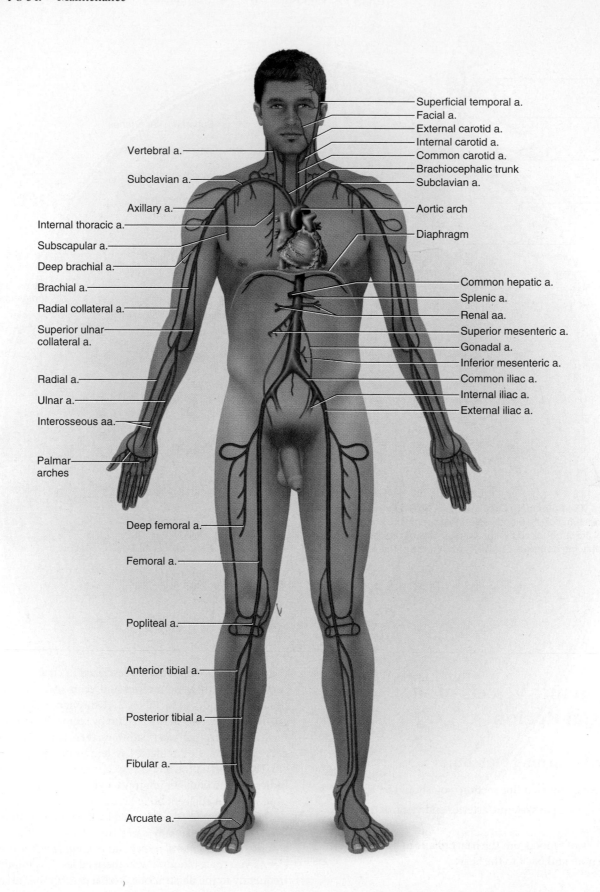

Superficial temporal a.
Facial a.
External carotid a.
Internal carotid a.
Common carotid a.
Brachiocephalic trunk
Subclavian a.
Aortic arch
Diaphragm
Common hepatic a.
Splenic a.
Renal aa.
Superior mesenteric a.
Gonadal a.
Inferior mesenteric a.
Common iliac a.
Internal iliac a.
External iliac a.

Vertebral a.
Subclavian a.
Axillary a.
Internal thoracic a.
Subscapular a.
Deep brachial a.
Brachial a.
Radial collateral a.
Superior ulnar collateral a.
Radial a.
Ulnar a.
Interosseous aa.
Palmar arches
Deep femoral a.
Femoral a.
Popliteal a.
Anterior tibial a.
Posterior tibial a.
Fibular a.
Arcuate a.

Figure 21.14 **The Major Systemic Arteries (Anterior View).** Different arteries are illustrated on the left than on the right for clarity, but nearly all of those shown occur on both sides. (a. = artery; aa. = arteries)

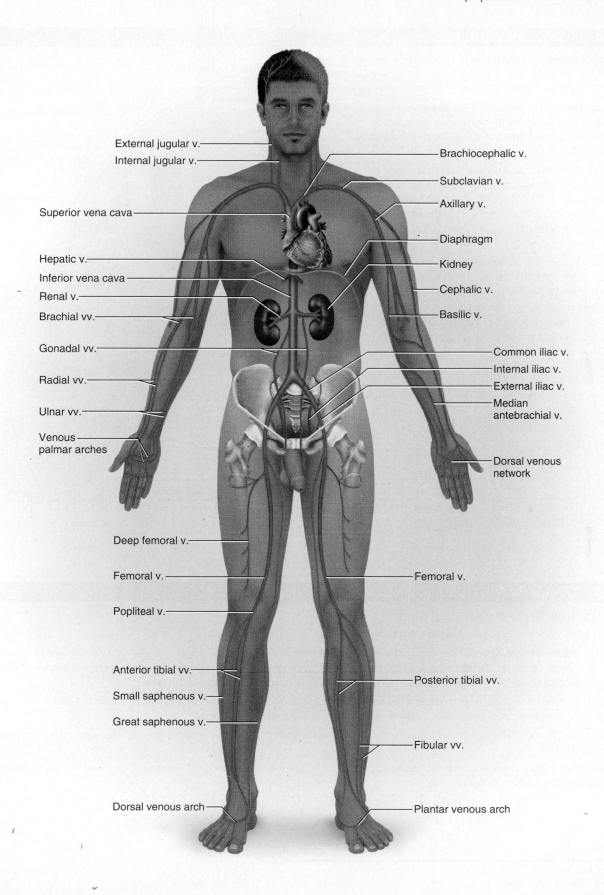

External jugular v.

Internal jugular v.

Superior vena cava

Hepatic v.

Inferior vena cava

Renal v.

Brachial vv.

Gonadal vv.

Radial vv.

Ulnar vv.

Venous palmar arches

Deep femoral v.

Femoral v.

Popliteal v.

Anterior tibial vv.

Small saphenous v.

Great saphenous v.

Dorsal venous arch

Brachiocephalic v.

Subclavian v.

Axillary v.

Diaphragm

Kidney

Cephalic v.

Basilic v.

Common iliac v.

Internal iliac v.

External iliac v.

Median antebrachial v.

Dorsal venous network

Femoral v.

Posterior tibial vv.

Fibular vv.

Plantar venous arch

Figure 21.15 **The Major Systemic Veins (Anterior View).** Different veins are illustrated on the left than on the right for clarity, but nearly all of those shown occur on both sides. (v. = vein; vv. = veins)

TABLE 21.1	The Aorta and Its Major Branches

All systemic arteries arise from the aorta, which has three principal regions (fig. 21.16):

1. The **ascending aorta** rises for about 5 cm above the left ventricle. Its only branches are the coronary arteries, which arise behind two cusps of the aortic valve. They are the origins of the coronary circulation described in chapter 20.

2. The **aortic arch** curves to the left like an inverted U superior to the heart. It gives off three major arteries in this order: the **brachiocephalic**[5] (BRAY-kee-oh-seh-FAL-ic) **trunk, left common carotid** (cah-ROT-id) **artery,** and **left subclavian**[6] (sub-CLAY-vee-un) **artery.** These are further traced in tables 21.2 and 21.8.

3. The **descending aorta** passes downward posterior to the heart, at first to the left of the vertebral column and then anterior to it, through the thoracic and abdominal cavities. It is called the *thoracic aorta* above the diaphragm and the *abdominal aorta* below it. It ends in the lower abdominal cavity by forking into the *right* and *left common iliac arteries* (see table 21.6, part IV).

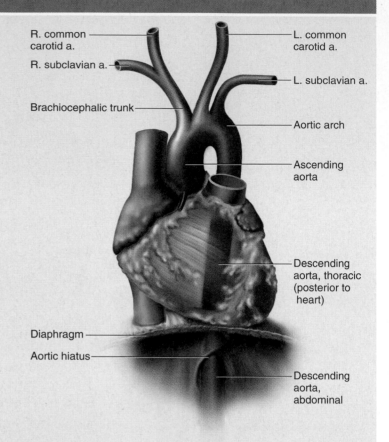

Figure 21.16 The Thoracic Aorta. (L. = left; R. = right; a. = artery)

TABLE 21.2	Arteries of the Head and Neck

I. Origins of the Head–Neck Arteries

The head and neck receive blood from four pairs of arteries (fig. 21.17):

1. The **common carotid arteries.** Shortly after leaving the aortic arch, the brachiocephalic trunk divides into the *right subclavian artery* (further traced in table 21.4) and **right common carotid artery.** A little farther along the aortic arch, the **left common carotid artery** arises independently. The common carotids pass up the anterolateral region of the neck, alongside the trachea (see part II of this table).

2. The **vertebral arteries.** These arise from the right and left subclavian arteries and travel up the neck through the transverse foramina of vertebrae C1 through C6. They enter the cranial cavity through the foramen magnum (see part III of this table).

3. The **thyrocervical**[7] **trunks.** These tiny arteries arise from the subclavian arteries lateral to the vertebral arteries; they supply the thyroid gland and some scapular muscles.

4. The **costocervical**[8] **trunks.** These arteries arise from the subclavian arteries a little farther laterally. They supply the deep neck muscles and some of the intercostal muscles of the superior rib cage.

---

[5]*brachio* = arm + *cephal* = head
[6]*sub* = below + *clavi* = clavicle, collarbone
[7]*thyro* = thyroid gland + *cerv* = neck
[8]*costo* = rib

TABLE 21.2   Arteries of the Head and Neck (continued)

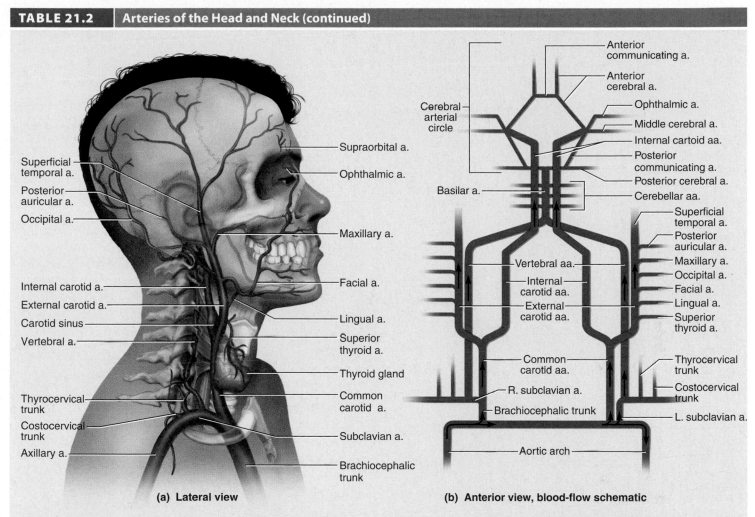

(a) Lateral view

(b) Anterior view, blood-flow schematic

Figure 21.17   Superficial (Extracranial) Arteries of the Head and Neck. The upper part of the schematic (b) depicts the cerebral circulation in figure 21.18. (a. = artery; aa. = arteries)

II. Continuation of the Common Carotid Arteries

The common carotid arteries have the most extensive distribution of all the head–neck arteries. Near the laryngeal prominence ("Adam's apple"), each common carotid branches into an *external* and *internal carotid artery*.

1. The **external carotid artery** ascends the side of the head external to the cranium and supplies most external head structures except the orbits. It gives rise to the following arteries in ascending order:

   a. the **superior thyroid artery** to the thyroid gland and larynx;

   b. the **lingual artery** to the tongue;

   c. the **facial artery** to the skin and muscles of the face;

   d. the **occipital artery** to the posterior scalp;

   e. the **maxillary artery** to the teeth, maxilla, oral cavity, and external ear; and

   f. the **superficial temporal artery** to the chewing muscles, nasal cavity, lateral aspect of the face, most of the scalp, and the dura mater.

2. The **internal carotid artery** passes medial to the angle of the mandible and enters the cranial cavity through the carotid canal of the temporal bone. It supplies the orbits and about 80% of the cerebrum. Compressing the internal carotids near the mandible can therefore cause loss of consciousness. After entering the cranial cavity, each internal carotid gives rise to the following branches:

   a. the **ophthalmic artery** to the orbit, nose, and forehead;

   b. the **anterior cerebral artery** to the medial aspect of the cerebral hemisphere (see part IV of this table); and

   c. the **middle cerebral artery,** which travels in the lateral sulcus of the cerebrum, supplies the insula, and then issues numerous branches to the lateral region of the frontal, temporal, and parietal lobes of the brain.

TABLE 21.2	**Arteries of the Head and Neck (continued)**

III. Continuation of the Vertebral Arteries

The vertebral arteries give rise to small branches that supply the spinal cord and its meninges, the cervical vertebrae, and deep muscles of the neck. They then enter the foramen magnum, supply the cranial bones and meninges, and converge to form a single **basilar artery** along the anterior aspect of the brainstem. Branches of the basilar artery supply the cerebellum, pons, and inner ear. At the pons–midbrain junction, the basilar artery divides and flows into the *cerebral arterial circle,* described next.

IV. The Cerebral Arterial Circle

Blood supply to the brain is so critical that it is furnished by several arterial anastomoses, especially an array of arteries called the **cerebral arterial circle (circle of Willis[9]),** which surrounds the pituitary gland and optic chiasm (fig. 21.18). The circle receives blood from the internal carotid and basilar arteries. Most people lack one or more of its components; only 20% have a complete arterial circle. Knowledge of the distribution of the arteries arising from the circle is crucial for understanding the effects of blood clots, aneurysms, and strokes on brain function. The anterior and posterior cerebral arteries described here and the middle cerebral artery described in part II provide the most significant blood supplies to the cerebrum. Refer to chapter 15 for reminders of the relevant brain anatomy.

1. Two **posterior cerebral arteries** arise from the basilar artery and sweep posteriorly to the rear of the brain, serving the inferior and medial regions of the temporal and occipital lobes as well as the midbrain and thalamus.

2. Two **anterior cerebral arteries** arise from the internal carotids, travel anteriorly, and then arch posteriorly over the corpus callosum as far as the posterior limit of the parietal lobe. They give off extensive branches to the frontal and parietal lobes.

3. The single **anterior communicating artery** is a short anastomosis between the right and left anterior cerebral arteries.

4. The two **posterior communicating arteries** are small anastomoses between the posterior cerebral and internal carotid arteries.

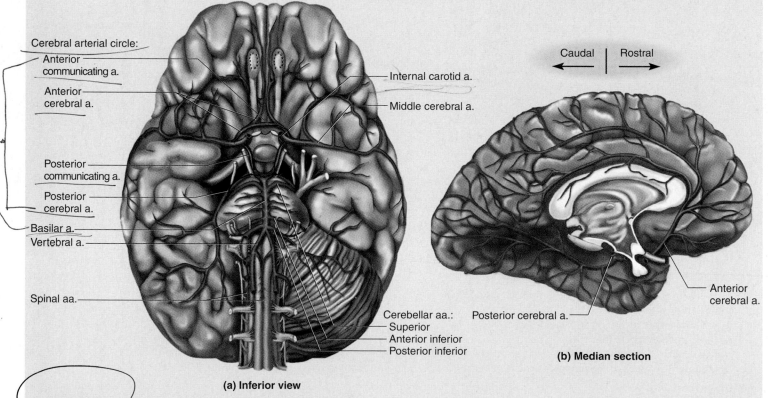

Figure 21.18 Cerebral Circulation. (a) Inferior view of the brain showing the blood supply to the brainstem, cerebellum, and cerebral arterial circle. (b) Median section of the brain showing the more distal branches of the anterior and posterior cerebral arteries. Branches of the middle cerebral artery are distributed over the lateral surface of the cerebrum (not illustrated).

[9]Thomas Willis (1621–75), English anatomist

TABLE 21.3	**Veins of the Head and Neck**

The head and neck are drained mainly by three pairs of veins—the *internal jugular, external jugular,* and *vertebral veins.* We will trace these from their origins to the *subclavian veins.*

I. Dural Venous Sinuses

After blood circulates through the brain, it collects in large thin-walled veins called **dural venous sinuses**—blood-filled spaces between the layers of the dura mater (fig. 21.19a, b). A reminder of the structure of the dura mater will be helpful in understanding these sinuses. This tough membrane between the brain and cranial bone has a periosteal layer against the bone and a meningeal layer against the brain. In a few places, a space exists between these layers to accommodate a blood-collecting sinus. Between the two cerebral hemispheres is a vertical, sickle-shaped wall of dura called the *falx cerebri,* which contains two of the sinuses. There are about 13 dural venous sinuses in all; we survey only the few most prominent ones here.

1. The **superior sagittal sinus** is contained in the superior margin of the falx cerebri and overlies the longitudinal fissure of the brain (fig. 21.19a; see also figs. 15.3 and 15.5, pp. 402 and 405). It begins anteriorly near the crista galli of the skull and extends posteriorly to the very rear of the head, ending at the level of the posterior occipital protuberance of the skull. Here it bends, usually to the right, and drains into a *transverse sinus.*

2. The **inferior sagittal sinus** is contained in the inferior margin of the falx cerebri and arches over the corpus callosum, deep in the longitudinal fissure. Posteriorly, it joins the *great cerebral vein,* and their union forms the **straight sinus,** which continues to the rear of the head (see fig. 15.5). There, the superior sagittal and straight sinuses meet in a space called the **confluence of the sinuses.**

3. Right and left **transverse sinuses** lead away from the confluence and encircle the inside of the occipital bone, leading toward the ears (fig. 21.19b); their path is marked by grooves on the inner surface of the occipital bone (see fig. 7.5b, p. 156). The right transverse sinus receives blood mainly from the superior sagittal sinus, and the left one drains mainly the straight sinus. Laterally, each transverse sinus makes an S-shaped bend, the **sigmoid sinus,** then exits the cranium through the jugular foramen. From here, the blood flows down the internal jugular vein (see part II.1 of this table).

4. The **cavernous sinuses** are honeycombs of blood-filled spaces on each side of the body of the sphenoid bone (fig. 21.19b). They receive blood from the *superior ophthalmic vein* of the orbit and the *superficial middle cerebral vein* of the brain, among other sources. They drain through several outlets including the transverse sinus, internal jugular vein, and facial vein. They are clinically important because infections can pass from the face and other superficial sites into the cranial cavity by this route. Also, inflammation of a cavernous sinus can injure important structures that pass through it, including the internal carotid artery and cranial nerves III to VI.

II. Major Veins of the Neck

Blood flows down the neck mainly through three veins on each side, all of which empty into the subclavian vein (fig. 21.19c).

1. The **internal jugular**[10] (JUG-you-lur) **vein** courses down the neck deep to the sternocleidomastoid muscle. It receives most of the blood from the brain, picks up blood from the **facial vein, superficial temporal vein,** and **superior thyroid vein** along the way, passes behind the clavicle, and joins the subclavian vein (which is further traced in table 21.5).

2. The **external jugular vein** courses down the side of the neck superficial to the sternocleidomastoid muscle and empties into the subclavian vein. It drains tributaries from the parotid gland, facial muscles, scalp, and other superficial structures. Some of this blood also follows venous anastomoses to the internal jugular vein.

3. The **vertebral vein** travels with the vertebral artery in the transverse foramina of the cervical vertebrae. Although the companion artery leads to the brain, the vertebral vein does not come from there. It drains the cervical vertebrae, spinal cord, and some of the small deep muscles of the neck, and empties into the subclavian vein.

Table 21.5 traces this blood flow the rest of the way to the heart.

---

[10]*jugul* = neck, throat

TABLE 21.3   Veins of the Head and Neck (continued)

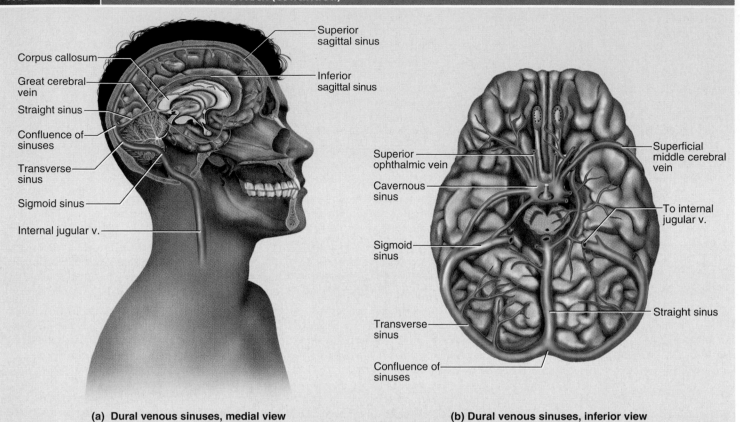

(a) **Dural venous sinuses, medial view**

Corpus callosum
Great cerebral vein
Straight sinus
Confluence of sinuses
Transverse sinus
Sigmoid sinus
Internal jugular v.
Superior sagittal sinus
Inferior sagittal sinus

(b) **Dural venous sinuses, inferior view**

Superior ophthalmic vein
Cavernous sinus
Sigmoid sinus
Transverse sinus
Confluence of sinuses
Superficial middle cerebral vein
To internal jugular v.
Straight sinus

(c) **Superficial veins of the head and neck**

Superficial temporal v.
Occipital v.
Vertebral v.
External jugular v.
Internal jugular v.
Axillary v.
Superior ophthalmic v.
Facial v.
Superior thyroid v.
Thyroid gland
Subclavian v.
Brachiocephalic v.

Figure 21.19 Veins of the Head and Neck. (a) Dural venous sinuses seen in a median section of the cerebrum. (b) Dural venous sinuses seen in an inferior view of the cerebrum. (c) Superficial (extracranial) veins of the head and neck.

TABLE 21.4	Arteries of the Thorax

The thorax is supplied by several arteries arising directly from the aorta (parts I and II of this table) and from the subclavian and axillary arteries (part III). The thoracic aorta begins distal to the aortic arch and ends at the **aortic hiatus** (hy-AY-tus), a passage through the diaphragm. Along the way, it sends off numerous small branches to the thoracic viscera and the body wall (fig. 21.20).

I. Visceral Branches of the Thoracic Aorta

These supply the viscera of the thoracic cavity:

1. **Bronchial arteries.** Although variable in number and arrangement, there are usually two of these on the left and one on the right. The right bronchial artery usually arises from one of the left bronchial arteries or from a *posterior intercostal artery* (see part II.1). The bronchial arteries supply the bronchi, bronchioles, and larger blood vessels of the lungs, the visceral pleura, the pericardium, and the esophagus.

2. **Esophageal arteries.** Four or five unpaired esophageal arteries arise from the anterior surface of the aorta and supply the esophagus.

3. **Mediastinal arteries.** Many small mediastinal arteries (not illustrated) supply structures of the posterior mediastinum.

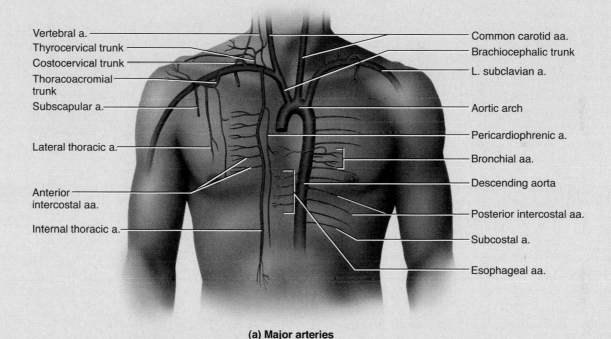

(a) Major arteries

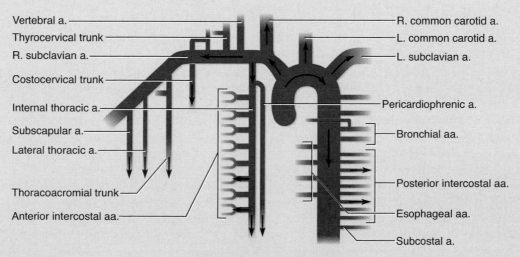

Figure 21.20 Arteries of the Thorax.   **(b) Blood-flow schematic**

TABLE 21.4   **Arteries of the Thorax (continued)**

II. Parietal Branches of the Thoracic Aorta

The following branches supply chiefly the muscles, bones, and skin of the chest wall; only the first are illustrated:

1. **Posterior intercostal arteries.** Nine pairs of these arise from the posterior surface of the aorta and course around the posterior side of the rib cage between ribs 3 through 12, then anastomose with the *anterior intercostal arteries* (see part III.1 in this table). They supply the intercostal, pectoralis, serratus anterior, and some abdominal muscles, as well as the vertebrae, spinal cord, meninges, breasts, skin, and subcutaneous tissue. They are enlarged in lactating women.

2. **Subcostal arteries.** A pair of these arise from the aorta inferior to the twelfth rib. They supply the posterior intercostal tissues, vertebrae, spinal cord, and deep muscles of the back.

3. **Superior phrenic**[11] **(FREN-ic) arteries** (not illustrated). These arteries, variable in number, arise at the aortic hiatus and supply the superior and posterior regions of the diaphragm.

III. Branches of the Subclavian and Axillary Arteries

The thoracic wall is also supplied by the following arteries, which arise in the shoulder region—the first one from the subclavian artery and the other three from its continuation, the axillary artery:

1. The **internal thoracic (mammary) artery** supplies the breast and anterior thoracic wall and issues the following branches:

   a. The **pericardiophrenic artery** supplies the pericardium and diaphragm.
   b. The **anterior intercostal arteries** arise from the thoracic artery as it descends alongside the sternum. They travel between the ribs, supply the ribs and intercostal muscles, and anastomose with the posterior intercostal arteries. Each of these sends one branch along the lower margin of the rib above and another branch along the upper margin of the rib below.

2. The **thoracoacromial**[12] **(THOR-uh-co-uh-CRO-me-ul) trunk** provides branches to the superior shoulder and pectoral regions.

3. The **lateral thoracic artery** supplies the pectoral, serratus anterior, and subscapularis muscles. It also issues branches to the breast and is larger in females than in males.

4. The **subscapular artery** is the largest branch of the axillary artery. It supplies the scapula and the latissimus dorsi, serratus anterior, teres major, deltoid, triceps brachii, and intercostal muscles.

TABLE 21.5   **Veins of the Thorax**

I. Tributaries of the Superior Vena Cava

The most prominent veins of the upper thorax are as follows. They carry blood from the shoulder region to the heart (fig. 21.21).

1. The **subclavian vein** drains the upper limb (see table 21.9). It begins at the lateral margin of the first rib and travels posterior to the clavicle. It receives the external jugular and vertebral veins, then ends where it receives the internal jugular vein.

2. The **brachiocephalic vein** is formed by union of the subclavian and internal jugular veins. The right brachiocephalic is very short, about 2.5 cm, and the left is about 6 cm long. They receive tributaries from the vertebrae, thyroid gland, and upper thoracic wall and breast, then converge to form the next vein.

3. The **superior vena cava** is formed by the union of the right and left brachiocephalic veins. It travels inferiorly for about 7 cm and empties into the right atrium of the heart. Its main tributary is the *azygos vein.* It drains all structures superior to the diaphragm except the pulmonary circuit and coronary circulation. It also receives drainage from the abdominal cavity by way of the azygos system, described next.

II. The Azygos System

The principal venous drainage of the thoracic organs is by way of the *azygos* (AZ-ih-goss) *system* (fig. 21.21). The most prominent vein of this system is the **azygos**[13] **vein,** which ascends the right side of the posterior thoracic wall and is named for the lack of a mate on the left. It receives the following tributaries, then empties into the superior vena cava at the level of vertebra T4.

1. The right **ascending lumbar vein** drains the right abdominal wall, then penetrates the diaphragm and enters the thoracic cavity. The azygos vein begins where the ascending lumbar vein meets the right **subcostal vein** beneath rib 12.

2. The right **posterior intercostal veins** drain the intercostal spaces. The first (superior) one empties into the right brachiocephalic vein; intercostals 2 and 3 join to form a *right superior intercostal vein* before emptying into the azygos; and intercostals 4 through 11 each enter the azygos vein separately.

3. The right **esophageal, mediastinal, pericardial,** and **bronchial veins** (not illustrated) drain their respective organs into the azygos.

4. The **hemiazygos**[14] **vein** ascends the posterior thoracic wall on the left. It begins where the left ascending lumbar vein, having just penetrated the diaphragm, joins the subcostal vein below rib 12. The hemiazygos then receives the lower three posterior intercostal veins, esophageal veins, and mediastinal veins. At the level of vertebra T9, it crosses to the right and empties into the azygos.

5. The **accessory hemiazygos vein** descends the posterior thoracic wall on the left. It receives drainage from posterior intercostal veins 4 through 8 and sometimes the left bronchial veins. It crosses to the right at the level of vertebra T8 and empties into the azygos vein.

The left posterior intercostal veins 1 through 3 are the only ones on this side that do not ultimately drain into the azygos vein. The first one usually drains directly into the left brachiocephalic vein. The second and third unite to form the *left superior intercostal vein,* which empties into the left brachiocephalic vein.

---

[11]*phren* = diaphragm
[12]*thoraco* = chest + *acr* = tip + *om* = shoulder

[13]unpaired; from *a* = without + *zygo* = union, mate
[14]*hemi* = half

TABLE 21.5 **Veins of the Thorax (continued)**

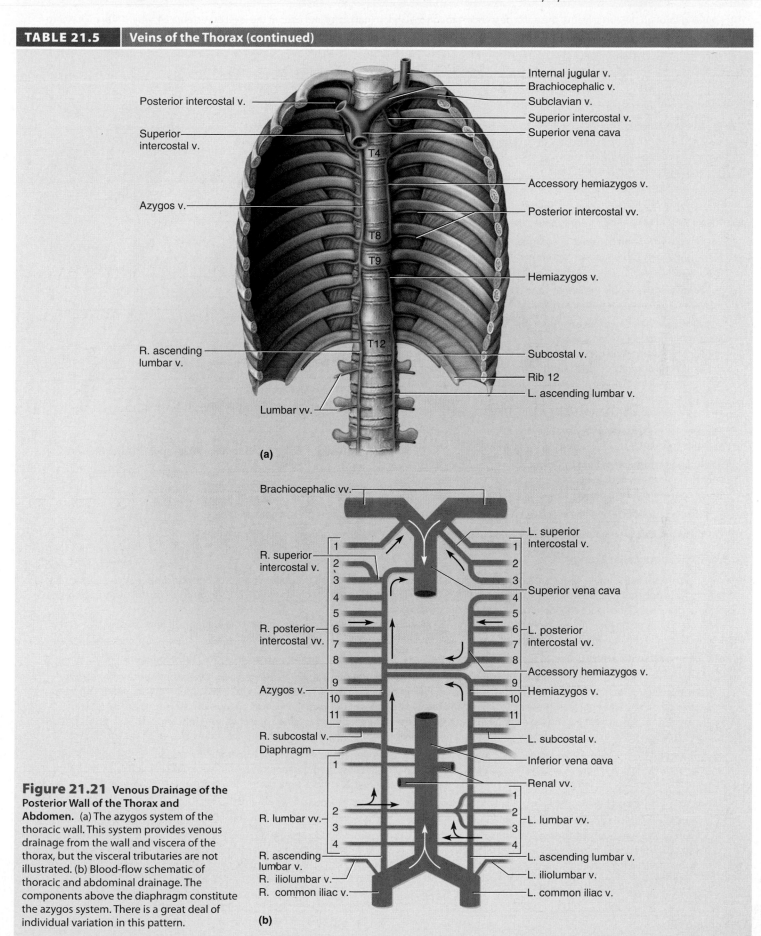

(a)

(b)

Figure 21.21 **Venous Drainage of the Posterior Wall of the Thorax and Abdomen.** (a) The azygos system of the thoracic wall. This system provides venous drainage from the wall and viscera of the thorax, but the visceral tributaries are not illustrated. (b) Blood-flow schematic of thoracic and abdominal drainage. The components above the diaphragm constitute the azygos system. There is a great deal of individual variation in this pattern.

TABLE 21.6 | **Arteries of the Abdomen and Pelvic Region**

After passing through the aortic hiatus, the aorta descends through the abdominal cavity and ends at the level of vertebra L4, where it branches into right and left *common iliac arteries*. The abdominal aorta is retroperitoneal.

I. Major Branches of the Abdominal Aorta

The abdominal aorta gives off arteries in the order listed here (fig. 21.22). Those indicated in the plural are paired right and left, and those indicated in the singular are solitary median arteries.

1. The **inferior phrenic arteries** supply the inferior surface of the diaphragm. They may arise from the aorta, celiac trunk, or renal artery. Each issues two or three small **superior suprarenal arteries** to the ipsilateral adrenal (suprarenal) gland.

2. The **celiac**[15] (SEE-lee-ac) **trunk** supplies the upper abdominal viscera (see part II of this table).

3. The **superior mesenteric artery** supplies the intestines (see part III).

4. The **middle suprarenal arteries** arise laterally from the aorta, usually at the same level as the superior mesenteric artery; they supply the adrenal glands.

5. The **renal arteries** supply the kidneys and issue a small **inferior suprarenal artery** to each adrenal gland.

6. The gonadal arteries (**ovarian arteries** in the female and **testicular arteries** in the male) are long, slender arteries that arise from the midabdominal aorta and descend along the posterior body wall to the female pelvic cavity or male scrotum. The gonads begin their embryonic development near the kidneys, and the gonadal arteries are then quite short. As the gonads descend to the pelvic cavity, these arteries grow and acquire their peculiar length and course.

7. The **inferior mesenteric artery** supplies the distal end of the large intestine (see part III).

8. The **lumbar arteries** arise from the lower aorta in four pairs. They supply the posterior abdominal wall (muscles, joints, and skin) and the spinal cord and other tissues in the vertebral canal.

9. The **median sacral artery,** a tiny medial artery at the inferior end of the aorta, supplies the sacrum and coccyx.

10. The **common iliac arteries** arise as the aorta forks at its inferior end. They are further traced in part IV of this table.

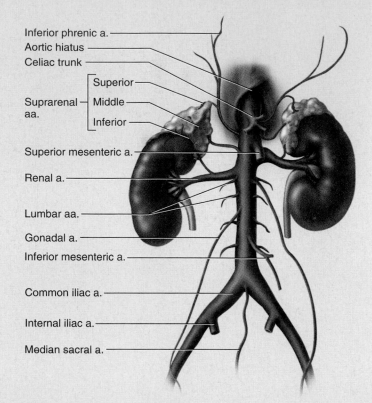

Figure 21.22 The Abdominal Aorta and Its Major Branches.

II. Branches of the Celiac Trunk

The celiac circulation to the upper abdominal viscera is perhaps the most complex route off the abdominal aorta. Because it has numerous anastomoses, the bloodstream does not follow a simple linear path but divides and rejoins itself at several points (fig. 21.23). As you study the following description, locate these branches in the figure and identify the points of anastomosis.

The short, stubby celiac trunk, barely more than 1 cm long, is a median branch of the aorta just below the diaphragm. It immediately gives rise to three branches—the *common hepatic, left gastric,* and *splenic arteries.*

1. The **common hepatic artery** passes to the right and issues two main branches—the gastroduodenal artery and the hepatic artery proper.

    a. The **gastroduodenal artery** gives off the **right gastro-omental (gastroepiploic**[16]**) artery** to the stomach. It then continues as the **superior pancreaticoduodenal** (PAN-cree-AT-ih-co-dew-ODD-eh-nul) **artery,** which splits into two branches that pass around the anterior and posterior sides of the head of the pancreas. These anastomose with the two branches of the *inferior pancreaticoduodenal artery,* discussed in part III.1.

    b. The **hepatic artery proper** ascends toward the liver. It gives off the **right gastric artery,** then branches into **right** and **left hepatic arteries.** The right hepatic artery issues a **cystic artery** to the gallbladder, then the two hepatic arteries enter the liver from below.

2. The **left gastric artery** supplies the stomach and lower esophagus, arcs around the *lesser curvature* (superomedial margin) of the stomach, and anastomoses with the right gastric artery (fig. 21.23b). Thus, the right and left gastric arteries approach from opposite directions and supply this margin of the stomach. The left gastric also has branches to the lower esophagus, and the right gastric also supplies the duodenum.

3. The **splenic artery** supplies blood to the spleen, but gives off the following branches on the way there:

    a. Several small **pancreatic arteries** supply the pancreas.

    b. The **left gastro-omental (gastroepiploic) artery** arcs around the *greater curvature* (inferolateral margin) of the stomach and anastomoses with the right gastro-omental artery. These two arteries stand off about 1 cm from the stomach itself and travel through the superior margin of the *greater omentum,* a fatty membrane suspended from the greater curvature (see figs. A.4, p. 334, and 24.3, p. 658). They furnish blood to both the stomach and omentum.

    c. The **short gastric arteries** supply the upper portion (fundus) of the stomach.

---

[15]*celi* = belly, abdomen
[16]*gastro* = stomach + *epi* = upon, above + *ploic* = pertaining to the greater omentum

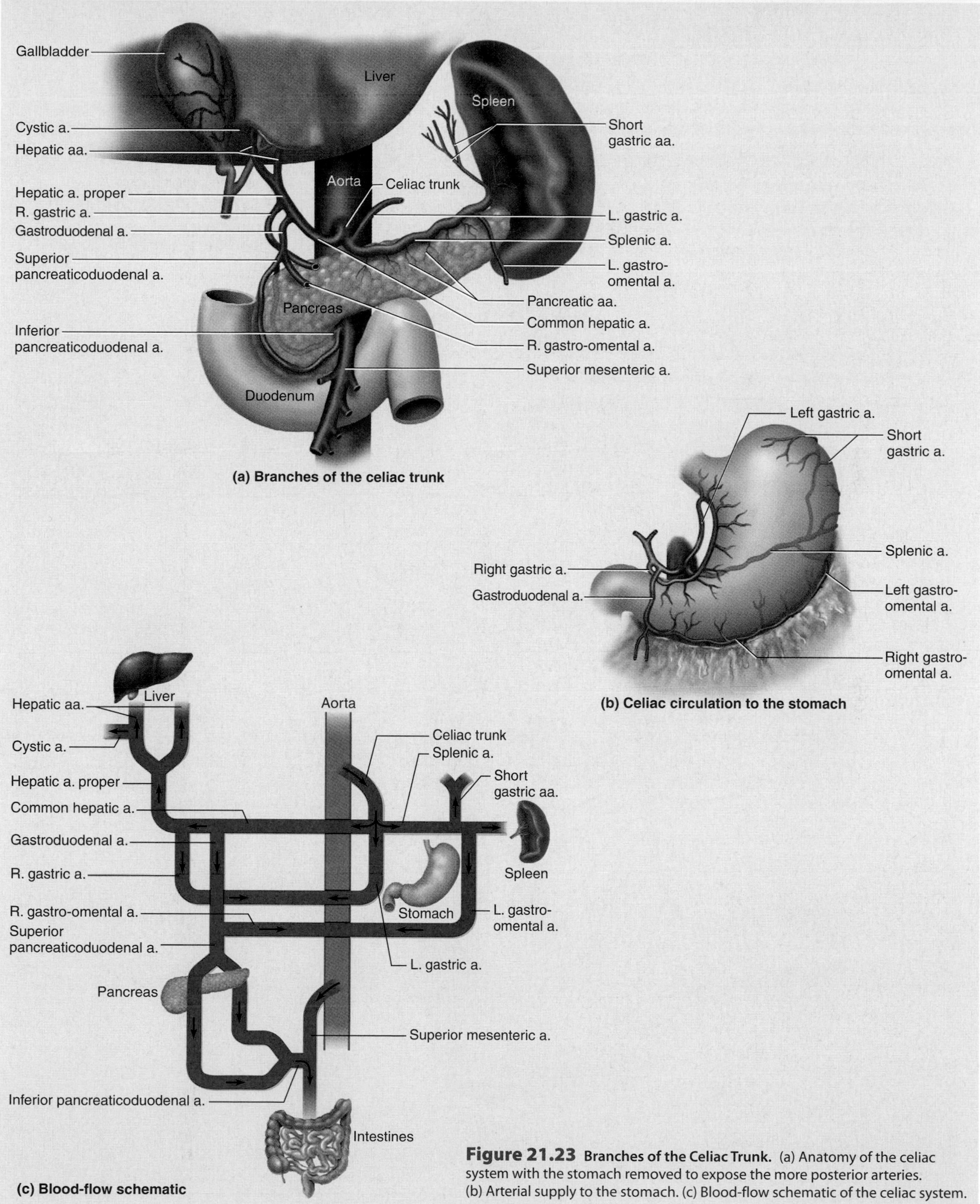

TABLE 21.6   Arteries of the Abdomen and Pelvic Region (continued)

Gallbladder

Liver

Spleen

Short gastric aa.

Cystic a.

Hepatic aa.

Aorta — Celiac trunk

Hepatic a. proper

R. gastric a.

Gastroduodenal a.

L. gastric a.

Splenic a.

Superior pancreaticoduodenal a.

L. gastro-omental a.

Pancreas

Pancreatic aa.

Common hepatic a.

Inferior pancreaticoduodenal a.

R. gastro-omental a.

Superior mesenteric a.

Duodenum

(a) Branches of the celiac trunk

Left gastric a.

Short gastric a.

Right gastric a.

Gastroduodenal a.

Splenic a.

Left gastro-omental a.

Right gastro-omental a.

(b) Celiac circulation to the stomach

Hepatic aa.   Liver

Aorta

Cystic a.

Celiac trunk

Splenic a.

Hepatic a. proper

Short gastric aa.

Common hepatic a.

Gastroduodenal a.

R. gastric a.

Spleen

R. gastro-omental a.

Superior pancreaticoduodenal a.

Stomach

L. gastro-omental a.

Pancreas

L. gastric a.

Superior mesenteric a.

Inferior pancreaticoduodenal a.

Intestines

(c) Blood-flow schematic

Figure 21.23 Branches of the Celiac Trunk. (a) Anatomy of the celiac system with the stomach removed to expose the more posterior arteries. (b) Arterial supply to the stomach. (c) Blood-flow schematic of the celiac system.

TABLE 21.6	**Arteries of the Abdomen and Pelvic Region (continued)**

III. Mesenteric Circulation

The mesentery is a translucent sheet that suspends the intestines and other abdominal viscera from the posterior body wall (see figs. 1.17, p. 19, and 24.3, p. 658). It contains numerous arteries, veins, and lymphatic vessels that supply and drain the intestines. The arterial supply arises from the *superior* and *inferior mesenteric arteries;* numerous anastomoses between these ensure adequate collateral circulation to the intestines even if one route is temporarily obstructed.

The **superior mesenteric artery** (fig. 21.24a) is the most significant intestinal blood supply, serving nearly all of the small intestine and the proximal half of the large intestine. It arises medially from the upper abdominal aorta and gives off the following branches:

1. The **inferior pancreaticoduodenal artery,** already mentioned, branches to pass around the anterior and superior sides of the pancreas and anastomose with the two branches of the superior pancreaticoduodenal artery.

2. Twelve to 15 **jejunal** and **ileal arteries** form a fanlike array that supplies nearly all of the small intestine (portions called the jejunum and ileum).

3. The **ileocolic** (ILL-ee-oh-CO-lic) **artery** supplies the ileum, appendix, and parts of the large intestine (cecum and ascending colon).

4. The **right colic artery** also supplies the ascending colon.

5. The **middle colic artery** supplies most of the transverse colon.

The **inferior mesenteric artery** arises from the lower abdominal aorta and serves the distal part of the large intestine (fig. 21.24b):

1. The **left colic artery** supplies the transverse and descending colon.

2. The **sigmoid arteries** supply the descending and sigmoid colon.

3. The **superior rectal artery** supplies the rectum.

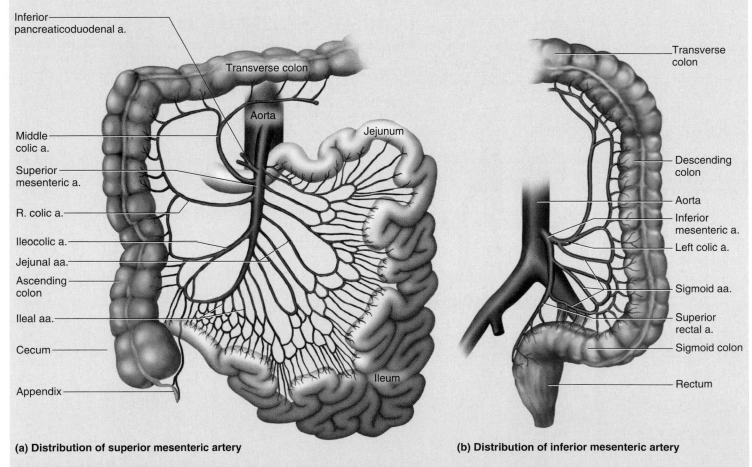

(a) Distribution of superior mesenteric artery

(b) Distribution of inferior mesenteric artery

Figure 21.24 The Mesenteric Arteries.

TABLE 21.6   Arteries of the Abdomen and Pelvic Region (continued)

IV. Arteries of the Pelvic Region

The two common iliac arteries arise by branching of the aorta, descend for another 5 cm, and then, at the level of the sacroiliac joint, each divides into an external and internal iliac artery. The external iliac supplies mainly the lower limb (see table 21.10). The **internal iliac artery** supplies mainly the pelvic wall and viscera. Its branches are shown only in schematic form in figure 21.30.

Shortly after its origin, the internal iliac divides into anterior and posterior trunks. The anterior trunk produces the following branches:

1. The **superior vesical**[17] **artery** supplies the urinary bladder and distal end of the ureter. It arises indirectly from the anterior trunk by way of a short *umbilical artery,* a remnant of the artery that supplies the fetal umbilical cord. The rest of the umbilical artery becomes a closed fibrous cord after birth.

2. In men, the **inferior vesical artery** supplies the bladder, ureter, prostate gland, and seminal vesicle. In women, the corresponding vessel is the **vaginal artery,** which supplies the vagina and part of the bladder and rectum.

3. The **middle rectal artery** supplies the rectum.

4. The **obturator artery** exits the pelvic cavity through the obturator foramen and supplies the adductor muscles of the medial thigh.

5. The **internal pudendal**[18] (pyu-DEN-dul) **artery** serves the perineum and erectile tissues of the penis and clitoris; it supplies the blood for vascular engorgement during sexual arousal.

6. In women, the **uterine artery** is the main blood supply to the uterus and supplies some blood to the vagina. It enlarges substantially in pregnancy. It passes up the uterine margin, then turns laterally at the uterine tube and anastomoses with the *ovarian artery,* thus supplying blood to the ovary as well (see part I.6 of table 21.6, and fig. 26.19, p. 724).

7. The **inferior gluteal artery** supplies the gluteal muscles and hip joint.

The posterior trunk produces the following branches:

1. The **iliolumbar artery** supplies the lumbar body wall and pelvic bones.

2. The **lateral sacral arteries** lead to tissues of the sacral canal, skin, and muscles posterior to the sacrum. There are usually two of these, superior and inferior.

3. The **superior gluteal artery** supplies the skin and muscles of the gluteal region and the muscle and bone tissues of the pelvic wall.

---

[17]*vesic* = bladder
[18]*pudend* = literally "shameful parts"; the external genitals

TABLE 21.7 | **Veins of the Abdomen and Pelvic Region**

I. Tributaries of the Inferior Vena Cava

The **inferior vena cava (IVC)** is the body's largest blood vessel, having a diameter of about 3.5 cm. It forms by the union of the right and left common iliac veins at the level of vertebra L5 and drains many of the abdominal viscera as it ascends the posterior body wall. It is retroperitoneal and lies immediately to the right of the aorta. The IVC picks up blood from numerous tributaries in the following ascending order (fig. 21.25):

1. The **internal iliac veins** drain the gluteal muscles; the medial aspect of the thigh, the urinary bladder, rectum, prostate, and ductus deferens of the male; and the uterus and vagina of the female. They unite with the *external iliac veins,* which drain the lower limb and are described in table 21.11. Their union forms the **common iliac veins,** which then converge to form the IVC.

2. Four pairs of **lumbar veins** empty into the IVC, as well as into the ascending lumbar veins described in part II.

3. The gonadal veins (**ovarian veins** in the female and **testicular veins** in the male) drain the gonads. Like the gonadal arteries, and for the same reason (table 21.6, part I.6), these are long slender vessels that end far from their origins. The left gonadal vein empties into the left renal vein, whereas the right gonadal vein empties directly into the IVC.

4. The **renal veins** drain the kidneys into the IVC. The left renal vein also receives blood from the left gonadal and left suprarenal veins. It is up to three times as long as the right renal vein, since the IVC lies to the right of the median plane of the body.

5. The **suprarenal veins** drain the adrenal (suprarenal) glands. The right suprarenal empties directly into the IVC, and the left suprarenal empties into the left renal vein.

6. The **inferior phrenic veins** drain the inferior aspect of the diaphragm.

7. The **hepatic veins** drain the liver, extending a short distance from its superior surface to the IVC.

After receiving these inputs, the IVC penetrates the diaphragm and enters the right atrium of the heart from below. It does not receive any thoracic drainage.

II. Veins of the Abdominal Wall

A pair of **ascending lumbar veins** receive blood from the common iliac veins below and the aforementioned lumbar veins of the posterior body wall (see fig. 21.21b). The ascending lumbar veins give off anastomoses with the inferior vena cava beside them as they ascend to the diaphragm. The left ascending lumbar vein passes through the diaphragm via the aortic hiatus and continues as the hemiazygos vein above. The right ascending lumbar vein passes through the diaphragm to the right of the vertebral column and continues as the azygos vein. The further paths of the azygos and hemiazygos veins are described in table 21.5.

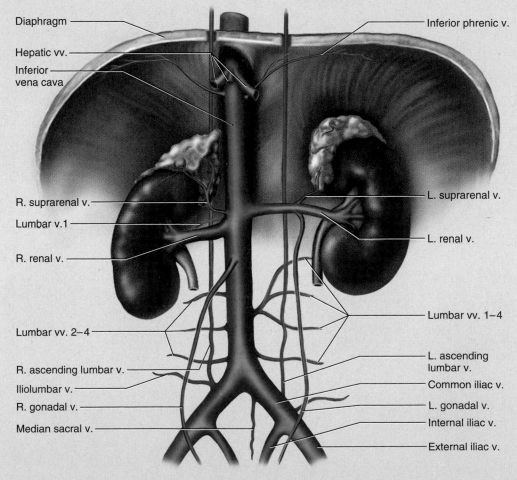

Figure 21.25 The Inferior Vena Cava and Its Tributaries. Compare the blood-flow schematic in figure 21.21b.
● *Why do the veins that drain the ovaries and testes terminate so far away from the gonads?*

TABLE 21.7   Veins of the Abdomen and Pelvic Region (continued)

III. The Hepatic Portal System

The **hepatic portal system** receives all of the blood draining from the abdominal digestive tract, as well as from the pancreas, gallbladder, and spleen (fig. 21.26). It is called a portal system because it connects capillaries of the intestines and other digestive organs to modified capillaries *(hepatic sinusoids)* of the liver; thus, the blood passes through two capillary beds in series before it returns to the heart. Intestinal blood is richly laden with nutrients for a few hours following a meal. The hepatic portal system gives the liver first claim to these nutrients before the blood is distributed to the rest of the body. It also allows the blood to be cleansed of bacteria and toxins picked up from the intestines, an important function of the liver. Its principal veins are as follows:

1. The **inferior mesenteric vein** receives blood from the rectum and distal part of the colon. It converges in a fanlike array in the mesentery and empties into the splenic vein.

2. The **superior mesenteric vein** receives blood from the entire small intestine, ascending colon, transverse colon, and stomach. It, too, exhibits a fanlike arrangement in the mesentery and then joins the splenic vein to form the hepatic portal vein.

3. The **splenic vein** drains the spleen and travels across the abdominal cavity toward the liver. Along the way, it picks up **pancreatic veins** from the pancreas, then the inferior mesenteric vein, then ends where it meets the superior mesenteric vein.

4. The **hepatic portal vein** is the continuation beyond the convergence of the splenic and superior mesenteric veins. It travels about 8 cm upward and to the right, receives the **cystic vein** from the gallbladder, then enters the inferior surface of the liver. In the liver, it ultimately leads to the innumerable microscopic hepatic sinusoids. Blood from the sinusoids empties into the hepatic veins described earlier, and they empty into the IVC. Circulation within the liver is described in more detail in chapter 24 (p. 677).

5. The left and right **gastric veins** form an arc along the lesser curvature of the stomach and empty into the hepatic portal vein.

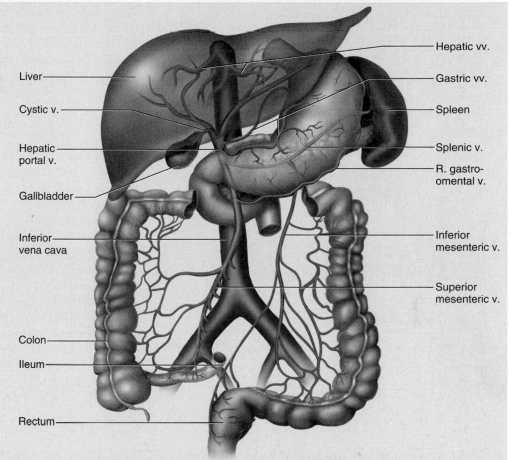

(a) Tributaries of the hepatic portal system

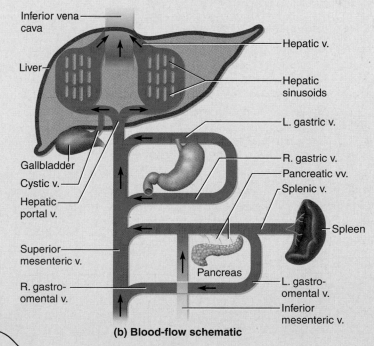

(b) Blood-flow schematic

Figure 21.26 The Hepatic Portal System.

Before You Go On

Answer the following questions to test your understanding of the preceding section:

9. Concisely contrast the destinations of the external and internal carotid arteries.

10. Briefly state the organs or parts of organs that are supplied with blood by (a) the cerebral arterial circle, (b) the celiac trunk, (c) the superior mesenteric artery, and (d) the internal iliac artery.

11. If you were dissecting a cadaver, where would you look for the internal and external jugular veins? What muscle would help you distinguish one from the other?

12. Trace a blood cell from the left lumbar body wall to the superior vena cava, naming the vessels through which it would travel.

21.4 Systemic Vessels of the Appendicular Region

▶ Expected Learning Outcomes

When you have completed this section, you should be able to

- identify the principal systemic arteries and veins of the limbs; and

- trace the flow of blood from the heart to any region of the upper or lower limb and back to the heart.

The principal vessels of the appendicular region are detailed in tables 21.8 through 21.11. While the appendicular arteries are usually deep and well protected, the veins occur in both deep and superficial groups; you may be able to see several of the superficial ones in your arms and hands. Deep veins run parallel to the arteries and often have similar names (*femoral artery* and *femoral vein,* for example). In several cases, the deep veins occur in pairs flanking the corresponding artery (such as the two *radial veins* traveling alongside the *radial artery*).

These blood vessels will be described in an order corresponding to the direction of blood flow. Thus, we will begin with the arteries in the shoulder and pelvic regions and progress to the hands and feet, whereas we will trace the veins beginning in the hands and feet and progressing toward the heart.

Venous pathways have more anastomoses than arterial pathways, so the route of flow is often not as clear. If all the anastomoses were illustrated, many of these venous pathways would look more like confusing networks than a clear route back to the heart. Therefore, most anastomoses—especially the highly variable and unnamed ones—are omitted from the figures to allow you to focus on the more general course of blood flow. The blood-flow schematics in several figures will also help to clarify these routes.

TABLE 21.8	Arteries of the Upper Limb

The upper limb is supplied by a prominent artery that changes name along its course from *subclavian* to *axillary* to *brachial,* then issues branches to the arm, forearm, and hand (fig. 21.27).

I. The Shoulder and Arm (Brachium)

1. The brachiocephalic trunk arises from the aortic arch and branches into the right common carotid artery and **left subclavian artery;** the **right subclavian artery** arises directly from the aortic arch. Each subclavian arches over the respective lung, rising as high as the base of the neck slightly superior to the clavicle. It then passes posterior to the clavicle, downward over the first rib, and ends in name only at the rib's lateral margin. In the shoulder, it gives off several small branches to the thoracic wall and viscera, described in table 21.4.

2. As the artery continues past the first rib, it is named the **axillary artery.** It continues through the axillary region, gives off small thoracic branches (see table 21.4), and ends, again in name only, at the neck of the humerus. Here, it gives off a pair of **circumflex humeral arteries,** which encircle the humerus, anastomose with each other laterally, and supply blood to the shoulder joint and deltoid muscle. Beyond this loop, the vessel is called the brachial artery.

3. The **brachial** (BRAY-kee-ul) **artery** continues down the medial and anterior sides of the humerus and ends just distal to the elbow, supplying the anterior flexor muscles of the brachium along the way. This artery is the most common site of blood pressure measurement, using an inflatable cuff that encircles the arm and compresses the artery.

4. The **deep brachial artery** arises from the proximal end of the brachial and supplies the humerus and triceps brachii muscle. About midway down the arm, it continues as the radial collateral artery.

5. The **radial collateral artery** descends in the lateral side of the arm and empties into the radial artery slightly distal to the elbow.

6. The **superior ulnar collateral artery** arises about midway along the brachial artery and descends in the medial side of the arm. It empties into the ulnar artery slightly distal to the elbow.

II. The Forearm, Wrist, and Hand

Just distal to the elbow, the brachial artery forks into the *radial* and *ulnar arteries.*

1. The **radial artery** descends the forearm laterally, alongside the radius, nourishing the lateral forearm muscles. The most common place to take a pulse is at the radial artery just proximal to the thumb.

2. The **ulnar artery** descends medially through the forearm, alongside the ulna, nourishing the medial forearm muscles.

TABLE 21.8 | **Arteries of the Upper Limb (continued)**

3. The **interosseous**[19] **arteries** of the forearm lie between the radius and ulna. They begin with a short **common interosseous artery** branching from the upper end of the ulnar artery. The common interosseous quickly divides into anterior and posterior branches. The **anterior interosseous artery** travels down the anterior side of the interosseous membrane, nourishing the radius, ulna, and deep flexor muscles. It ends distally by passing through the interosseous membrane to join the posterior interosseous artery. The **posterior interosseous artery** descends along the posterior side of the interosseous membrane and nourishes mainly the superficial extensor muscles.

4. Two U-shaped **palmar arches** arise by anastomosis of the radial and ulnar arteries at the wrist. The **deep palmar arch** is fed mainly by the radial artery and the **superficial palmar arch** mainly by the ulnar artery. The arches issue arteries to the palmar region and fingers.

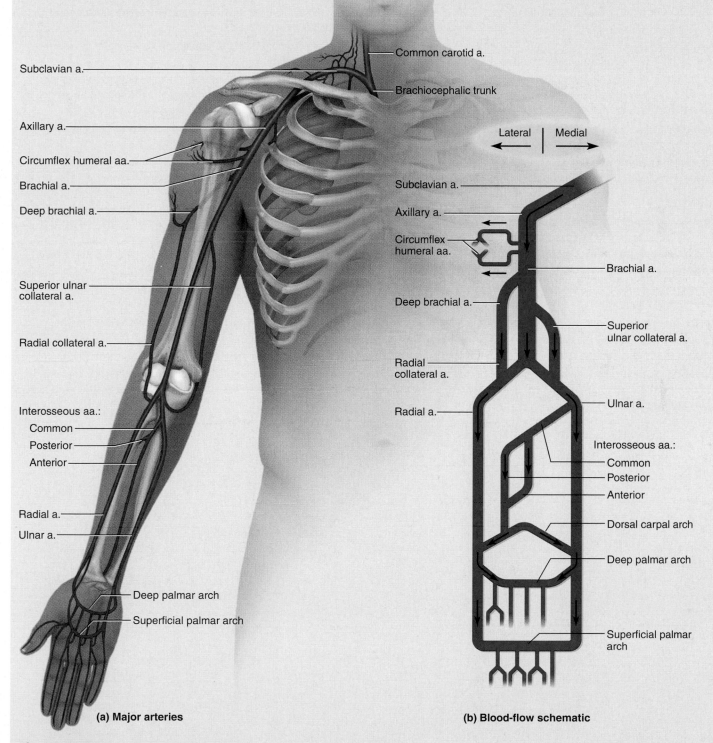

(a) **Major arteries**

(b) **Blood-flow schematic**

Figure 21.27 Arteries of the Upper Limb.
• *Why are arterial anastomoses especially common at joints such as the shoulder and elbow?*

---

[19]*inter* = between + *osse* = bones

TABLE 21.9 Veins of the Upper Limb

Both superficial and deep veins drain the upper limb, ultimately leading to axillary and subclavian veins that parallel the arteries of the same names (fig. 21.28). The superficial veins are often externally visible and are larger in diameter and carry more blood than the deep veins.

I. Superficial Veins

1. The **dorsal venous network** is a plexus of veins often visible through the skin on the back of the hand; it empties into the major superficial veins of the forearm, the cephalic and basilic.

2. The **cephalic**[20] (sef-AL-ic) **vein** arises from the lateral side of the network, travels up the lateral side of the forearm and arm to the shoulder, and joins the axillary vein there. Intravenous fluids are often administered through the distal end of this vein.

3. The **basilic**[21] (bah-SIL-ic) **vein** arises from the medial side of the network, travels up the posterior side of the forearm, and continues into the arm. It turns deeper about midway up the arm and joins the brachial vein at the axilla (see part II.4 of this table).

   As an aid to remembering which vein is cephalic and which is basilic, visualize your arm held straight away from the torso (abducted) with the thumb up, like the orientation of the embryonic limb bud. The cephalic vein runs along the upper side of the arm closer to the head (as suggested by *cephal*, "head"), and the name *basilic* is suggestive of the lower (basal) side of the arm (although not named for that reason).

4. The **median cubital vein** is a short anastomosis between the cephalic and basilic veins that obliquely crosses the cubital fossa (anterior bend of the elbow). It is often clearly visible through the skin and is the most common site for drawing blood.

5. The **median antebrachial vein** drains a network of blood vessels in the hand called the **superficial palmar venous network.** It travels up the medial forearm and terminates at the elbow, emptying variously into the basilic vein, median cubital vein, or cephalic vein.

II. Deep Veins

1. The **deep** and **superficial venous palmar arches** receive blood from the fingers and palmar region. They are anastomoses that join the radial and ulnar veins.

2. Two **radial veins** arise from the lateral side of the palmar arches and course up the forearm alongside the radius. Slightly distal to the elbow, they converge and give rise to one of the brachial veins described shortly.

3. Two **ulnar veins** arise from the medial side of the palmar arches and course up the forearm alongside the ulna. They unite near the elbow to form the other brachial vein.

4. The two **brachial veins** continue up the brachium, flanking the brachial artery, and converge into a single vein just before the axillary region.

5. The **axillary vein** forms by the union of the brachial and basilic veins. It begins at the lower margin of the teres major muscle and passes through the axillary region, picking up the cephalic vein along the way. At the lateral margin of the first rib, it changes name to the subclavian vein.

6. The **subclavian vein** continues into the shoulder posterior to the clavicle and ends where it meets the internal jugular vein of the neck. There it becomes the brachiocephalic vein. The right and left brachiocephalics converge and form the superior vena cava, which empties into the right atrium of the heart.

---

[20]*cephalic* = related to the head
[21]*basilic* = royal, prominent, important

TABLE 21.9 | **Veins of the Upper Limb (continued)**

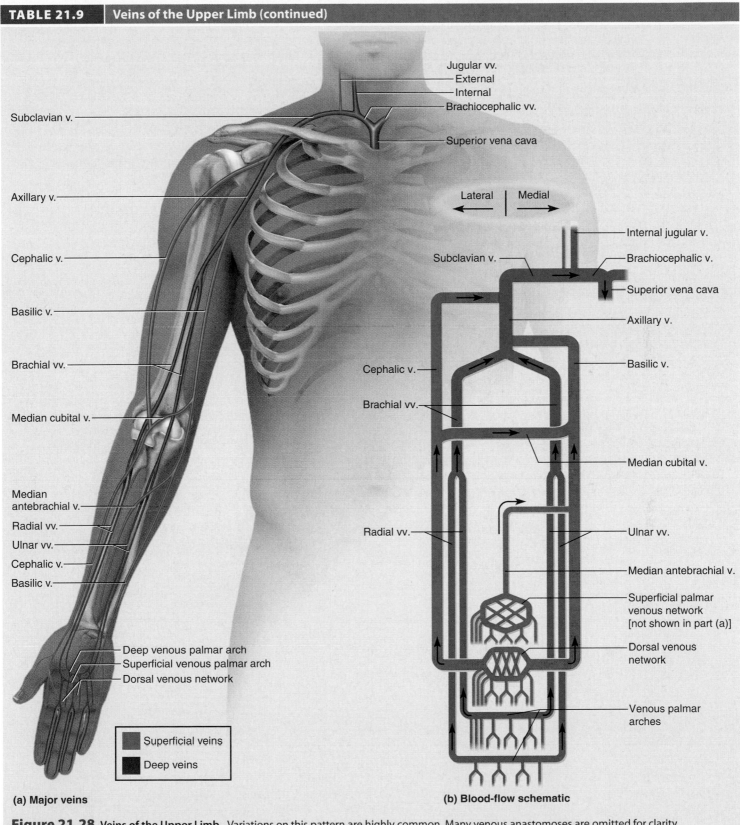

(a) Major veins

(b) Blood-flow schematic

Figure 21.28 Veins of the Upper Limb. Variations on this pattern are highly common. Many venous anastomoses are omitted for clarity.

TABLE 21.10 **Arteries of the Lower Limb**

As we have already seen, the aorta forks at its lower end into the right and left common iliac arteries, and each of these soon divides again into an internal and external iliac artery. We traced the internal iliac artery in table 21.6 (part IV), and we now trace the external iliac as it supplies the lower limb (figs. 21.29 and 21.30).

I. Arteries from the Pelvic Region to the Knee

1. The **external iliac artery** sends small branches to the skin and muscles of the abdominal wall and pelvic girdle, then passes behind the inguinal ligament and becomes the femoral artery.

2. The **femoral artery** passes through the *femoral triangle* of the upper medial thigh, where its pulse can be palpated (see Deeper Insight 21.3). In the triangle, it gives off several small arteries to the skin and then produces the following branches before descending the rest of the way to the knee.

   a. The **deep femoral artery** arises from the lateral side of the femoral, within the triangle. It is the largest branch and is the major arterial supply to the thigh muscles.

   b. Two **circumflex femoral arteries** arise from the deep femoral, encircle the head of the femur, and anastomose laterally. They supply mainly the femur, hip joint, and hamstring muscles.

3. The **popliteal artery** is a continuation of the femoral artery in the popliteal fossa at the rear of the knee. It begins where the femoral artery emerges from an opening *(adductor hiatus)* in the tendon of the adductor magnus muscle and ends where it splits into the *anterior* and *posterior tibial arteries*. As it passes through the popliteal fossa, it gives off anastomoses called **genicular**[22] **arteries** that supply the knee joint.

II. Arteries of the Leg and Foot

In the leg proper, the three most significant arteries are the anterior tibial, posterior tibial, and fibular arteries.

1. The **anterior tibial artery** arises from the popliteal artery and immediately penetrates through the interosseous membrane of the leg to the anterior compartment. There, it travels lateral to the tibia and supplies the extensor muscles. Upon reaching the ankle, it gives rise to the following dorsal arteries of the foot.

   a. The **dorsal pedal artery** traverses the ankle and upper medial surface of the foot and gives rise to the arcuate artery.

   b. The **arcuate artery** sweeps across the foot from medial to lateral and gives rise to vessels that supply the toes.

2. The **posterior tibial artery** is a continuation of the popliteal artery that passes down the leg, deep in the posterior compartment, supplying flexor muscles along the way. Inferiorly, it passes behind the medial malleolus of the ankle and into the plantar region of the foot. It gives rise to the following:

   a. The **medial** and **lateral plantar arteries** originate by bifurcation of the posterior tibial artery at the ankle. The medial plantar artery supplies mainly the great toe. The lateral plantar artery sweeps across the sole of the foot and becomes the deep plantar arch.

   b. The **deep plantar arch** gives off another set of arteries to the toes.

3. The **fibular (peroneal) artery** arises from the proximal end of the posterior tibial artery near the knee. It descends through the lateral side of the posterior compartment, supplying lateral muscles of the leg along the way, and ends in a network of arteries in the heel.

---

[22]*genic* = knee

TABLE 21.10   Arteries of the Lower Limb (continued)

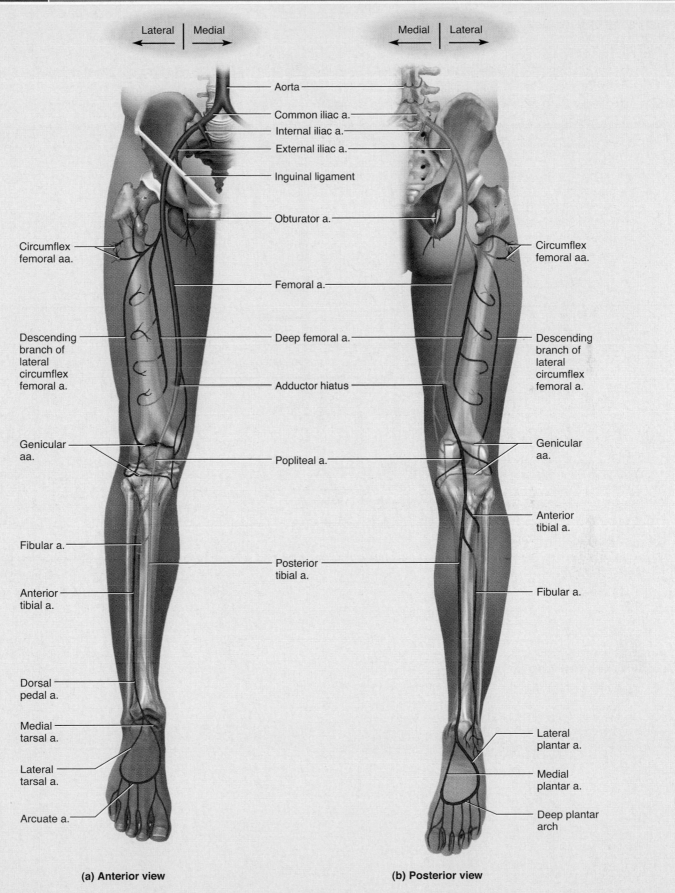

Lateral ← | → Medial   Medial ← | → Lateral

Aorta

Common iliac a.

Internal iliac a.

External iliac a.

Inguinal ligament

Obturator a.

Circumflex femoral aa.

Circumflex femoral aa.

Femoral a.

Descending branch of lateral circumflex femoral a.

Deep femoral a.

Descending branch of lateral circumflex femoral a.

Adductor hiatus

Genicular aa.

Popliteal a.

Genicular aa.

Anterior tibial a.

Fibular a.

Posterior tibial a.

Fibular a.

Anterior tibial a.

Dorsal pedal a.

Medial tarsal a.

Lateral plantar a.

Lateral tarsal a.

Medial plantar a.

Arcuate a.

Deep plantar arch

(a) Anterior view

(b) Posterior view

Figure 21.29 **Arteries of the Lower Limb.** The foot is strongly plantar flexed with the upper surface facing the viewer in part (a) and the sole facing the viewer in part (b).

TABLE 21.10 | **Arteries of the Lower Limb (continued)**

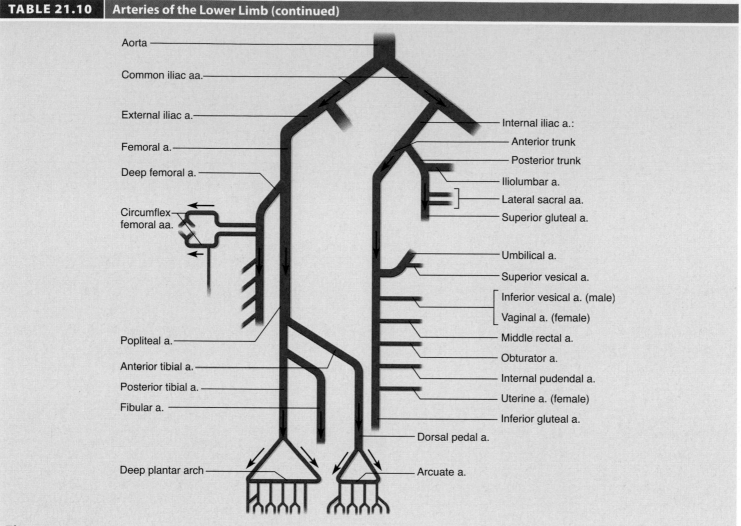

Figure 21.30 Arterial Schematic of the Pelvic Region and Lower Limb. The pelvic schematic on the right is stretched for clarity. These arteries are not located as far inferiorly as the limb arteries depicted on the left.

Apply What You Know

There are certain similarities between the arteries of the hand and foot. What arteries of the wrist and hand are most comparable in arrangement and function to the arcuate artery and deep plantar arch of the foot?

TABLE 21.11   Veins of the Lower Limb

We will follow drainage of the lower limb from the toes to the inferior vena cava (figs. 21.31 and 21.32). As in the upper limb, there are deep and superficial veins with anastomoses between them. Most of the anastomoses are omitted from the illustrations.

I. Superficial Veins

1. The **dorsal venous arch** (fig. 21.31a) is often visible through the skin on the dorsum of the foot. It collects blood from the toes and more proximal part of the foot, and has numerous anastomoses similar to the dorsal venous network of the hand. It gives rise to the following two veins.

2. The **small (short) saphenous**[23] (sah-FEE-nus) **vein** arises from the lateral side of the arch and passes up that side of the leg as far as the knee. There, it drains into the popliteal vein.

3. The **great (long) saphenous vein,** the longest vein in the body, arises from the medial side of the arch and travels all the way up the leg and thigh to the inguinal region. It empties into the femoral vein slightly inferior to the inguinal ligament. It is commonly used as a site for the long-term administration of intravenous fluids; it is a relatively accessible vein in infants and in patients in shock whose veins have collapsed. Portions of this vein are commonly used as grafts in coronary bypass surgery. The great and small saphenous veins are among the most common sites of varicose veins.

II. Deep Veins

1. The **deep plantar venous arch** (fig. 21.31b) receives blood from the toes and gives rise to **lateral** and **medial plantar veins** on the respective sides. The lateral plantar vein gives off the *fibular veins,* then crosses over to the medial side and approaches the medial plantar vein. The two plantar veins pass behind the medial malleolus of the ankle and continue as a pair of *posterior tibial veins.*

2. The two **posterior tibial veins** pass up the leg embedded deep in the calf muscles. They converge like an inverted Y into a single vein about two-thirds of the way up the tibia.

3. The two **fibular (peroneal) veins** ascend the back of the leg and similarly converge like a Y.

4. The **popliteal vein** begins near the knee by convergence of these two inverted Ys. It passes through the popliteal fossa at the back of the knee.

5. The two **anterior tibial veins** travel up the anterior compartment of the leg between the tibia and fibula (fig. 21.31a). They arise from the medial side of the dorsal venous arch, converge just distal to the knee, and then flow into the popliteal vein.

6. The **femoral vein** is a continuation of the popliteal vein into the thigh. It drains blood from the deep thigh muscles and femur.

7. The **deep femoral vein** drains the femur and muscles of the thigh supplied by the deep femoral artery. It receives four principal tributaries along the shaft of the femur and then a pair of **circumflex femoral veins** that encircle the upper femur, then finally drains into the upper femoral vein.

8. The **external iliac vein** is formed by the union of the femoral and great saphenous veins near the inguinal ligament.

9. The **internal iliac vein** follows the course of the internal iliac artery and its distribution. Its tributaries drain the gluteal muscles; the medial aspect of the thigh; the urinary bladder, rectum, prostate, and ductus deferens in the male; and the uterus and vagina in the female.

10. The **common iliac vein** is formed by the union of the external and internal iliac veins. The right and left common iliacs then unite to form the inferior vena cava.

---

[23]*saphen* = standing

TABLE 21.11 **Veins of the Lower Limb (continued)**

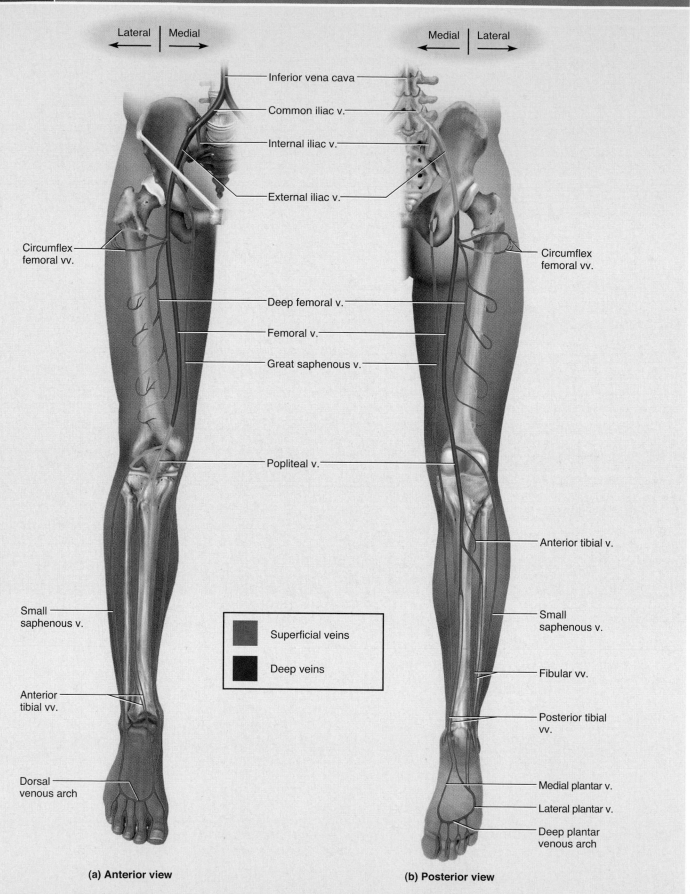

Lateral ← | → Medial

Medial ← | → Lateral

Inferior vena cava

Common iliac v.

Internal iliac v.

External iliac v.

Circumflex femoral vv.

Circumflex femoral vv.

Deep femoral v.

Femoral v.

Great saphenous v.

Popliteal v.

Anterior tibial v.

Small saphenous v.

Small saphenous v.

■ Superficial veins

■ Deep veins

Fibular vv.

Anterior tibial vv.

Posterior tibial vv.

Dorsal venous arch

Medial plantar v.

Lateral plantar v.

Deep plantar venous arch

(a) Anterior view

(b) Posterior view

Figure 21.31 Veins of the Lower Limb. The foot is strongly plantar flexed with the upper surface facing the viewer in part (a) and the sole facing the viewer in part (b).

TABLE 21.11   Veins of the Lower Limb (continued)

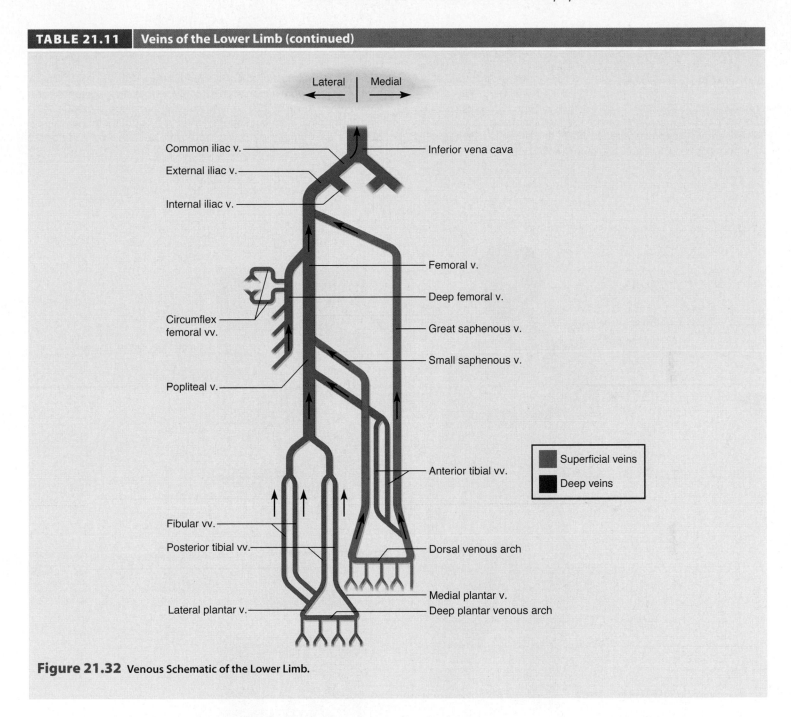

Figure 21.32 Venous Schematic of the Lower Limb.

DEEPER INSIGHT   21.3

Arterial Pressure Points

In some places, major arteries come close enough to the body surface to be palpated. These places can be used to take a pulse, and they can serve as emergency *pressure points,* where firm pressure can be applied to temporarily reduce arterial bleeding (fig. 21.33a). One of these points is the *femoral triangle* of the upper medial thigh (fig. 21.33b, c). This is an important landmark for arterial supply, venous drainage, and innervation of the lower limb. Its boundaries are the sartorius muscle laterally, the inguinal ligament superiorly, and the adductor longus muscle medially. The femoral artery, vein, and nerve run close to the surface at this point.

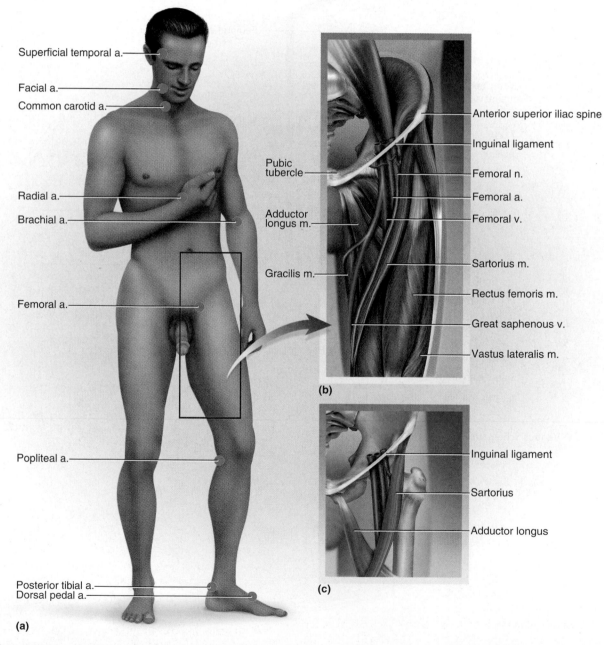

Figure 21.33 **Arterial Pressure Points.** (a) Areas where arteries lie close enough to the surface that a pulse can be palpated or pressure can be applied to reduce arterial bleeding. (b) Structures in the femoral triangle. (c) Boundaries of the femoral triangle.

Before You Go On

Before You Go On

Answer the following questions to test your understanding of the preceding section:

13. Trace one possible path of a red blood cell from the left ventricle to the toes.

14. Trace one possible path of a red blood cell from the fingers to the right atrium.

15. The subclavian, axillary, and brachial arteries are really one continuous artery. What is the reason for giving it three different names along its course?

16. State two ways in which the great saphenous vein has special clinical significance. Where is this vein located?

21.5 Developmental and Clinical Perspectives

▶ Expected Learning Outcomes

When you have completed this section, you should be able to

- describe the embryonic development of the blood vessels;

- explain how the circulatory system changes at birth; and

- describe the changes that occur in the blood vessels in old age.

Embryonic Development of the Blood Vessels

The development of blood vessels, both in the embryo and later in life, is called **angiogenesis.**[24] The first trace of embryonic angiogenesis appears at 13 to 15 days of gestation. At this time, the embryo is a three-layered disc of ectoderm, mesoderm, and endoderm, attached to the uterus by an *embryonic stalk* and associated with three membranes, the *chorion, amnion,* and *yolk sac* (see fig. 4.3, p. 88). In the yolk sac, groups of mesenchymal cells differentiate into cell masses called *blood islands.* Spaces open in the middle of a blood island, and cells in these spaces differentiate into *hemoblasts,* the forerunners of the first blood cells. Cells of the margin become *angioblasts,* which give rise to the endothelium of the future blood vessel (fig. 21.34).

As blood islands proliferate and grow, they begin to connect with each other, and their internal spaces form the lumens of the blood vessels. By the end of week 3, the yolk sac is fully vascularized. In subsequent weeks, blood islands begin to appear in the liver, spleen, and bone marrow, and those of the yolk sac and other sites external to the embryo disappear. Also during this time, the

---

[24]*angio* = vessel + *genesis* = origin

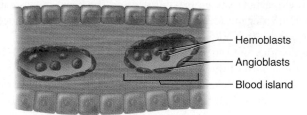

(a) Early blood islands in yolk sac

(b) Differentiation of mesenchymal cells into angioblasts and hemoblasts

(c) Merger of blood islands to form blood vessels

Figure 21.34 Development of Blood Vessels and Erythrocytes from Embryonic Blood Islands.

heart tube forms and links up with the blood vessels. Around the end of week 3, the heart begins beating, and a week later, a unidirectional blood flow is established. The blood vessels are not long uniform tubes at this time, but a network of irregularly shaped channels. Those that receive the greatest blood flow develop a tunica media and externa and become more tubular, thus becoming typical blood vessels. Those with a lesser flow either degenerate or remain composed of nothing but endothelium, thus becoming capillaries. Capillaries and the larger vessels sprout lateral branches, giving rise to the eventual anatomical circuitry of the mature cardiovascular system.

We will not examine the complex details of the embryonic development of all the major blood vessels, but primarily the major arteries and veins near the heart. Remember that the embryo forms five pairs of pharyngeal arches in weeks 4 to 5 (see chapter 4, p. 90). As these develop, a pouch called the *aortic sac* appears at the rostral end of the heart. An artery arises from each side of the sac, loops through the first pharyngeal arch, and ends in the dorsal aorta on that side. The loop is *aortic arch I*. Five more pairs of **aortic arches** (II–VI) later form between the pharyngeal pouches (fig. 21.35a). This reflects a primitive vertebrate pattern seen in fish, where the six aortic arches supply blood to the gills, but it becomes highly modified in humans and other mammals. The six arches actually never appear all at once as shown in the figure. Arches I and II degenerate before the most caudal arches appear, and arch V never develops to any great extent in mammals. Arches III, IV, and VI, however, play major roles in human development (fig. 21.35b). Arch III gives rise to the common carotid artery and the proximal portion of the internal carotid artery; the external carotid artery buds from the common carotid. The common carotids are short at first, but elongate as the embryo grows and the heart moves caudally. Arch IV degenerates on the right, but on the left it produces the aortic arch. Arch VI gives rise to the pulmonary arteries.

Initially, the embryo has two *dorsal aortae* that pass side by side for the length of the body. Caudal to the pharyngeal arches, however, these soon fuse into a single **dorsal aorta,** the forerunner of the adult descending aorta (fig. 21.36). The dorsal aorta issues about 30 pairs of *intersegmental arteries,* which supply blood to the somites and their derivatives. Most of these degenerate, but the adult intercostal arteries, lumbar arteries, and common iliac arteries are remnants of some of the embryonic intersegmental arteries.

The principal veins associated with the heart (fig. 21.36) are the *anterior cardinal vein,* which drains the head region; the *posterior cardinal vein,* which drains the body caudal to the heart; the *vitelline veins* from the yolk sac; and the *umbilical veins* from the placenta. There are initially two umbilical veins, but only the left one persists until birth. The cardinal veins provide most of the venous drainage of the embryonic body. They meet at a *common cardinal vein* just before entering the heart. The future superior vena cava develops from the right anterior cardinal and common cardinal veins. The posterior cardinal veins largely degenerate, however, leaving only the common iliac veins and part of the azygos system as remnants. The inferior vena cava develops separately, not from the posterior cardinal veins.

Changes at Birth

As we saw in chapter 20 (p. 557), the fetus has certain *shunts* that allow most blood to bypass the nonfunctional lungs: the *foramen ovale* and *ductus arteriosus*. After birth, when the lungs are functional, these shunts close, leaving a *fossa ovalis* in the interatrial

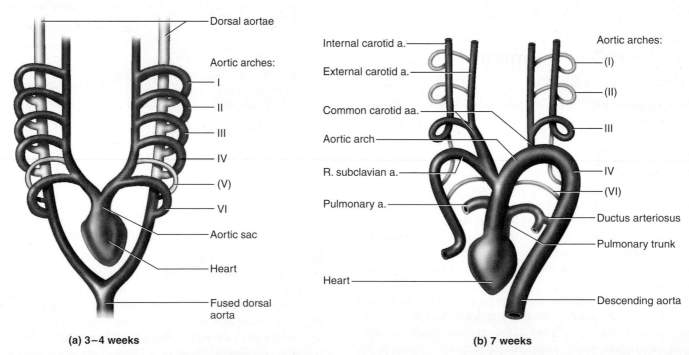

(a) 3–4 weeks　　　　　　　　　　　　　　　　**(b) 7 weeks**

Figure 21.35 **Development of Some Major Arteries from the Embryonic Aortic Arches.** (a) The six aortic arches, dorsal view. This is a composite diagram representing developments from day 22 through day 29. In reality, arch I degenerates as arches III and IV form, and arch II degenerates as arch VI forms. Aortic arch V develops very little, and sometimes not at all, in humans. (b) Remodeled arterial system at about 7 weeks. Pale colors and arch numbers in parentheses indicate the former positions of aortic arches that no longer exist at this time, for comparison to part (a). The left subclavian vein does not develop from an aortic arch and is not illustrated.

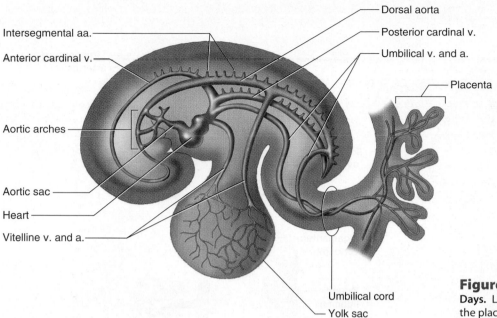

Intersegmental aa.

Anterior cardinal v.

Aortic arches

Aortic sac

Heart

Vitelline v. and a.

Dorsal aorta

Posterior cardinal v.

Umbilical v. and a.

Placenta

Umbilical cord

Yolk sac

Figure 21.36 **Major Embryonic Blood Vessels at 26 Days.** Left lateral view of the embryo, yolk sac, and part of the placenta.

septum and *ligamentum arteriosum* between the aortic arch and left pulmonary artery. Another fetal shunt called the *ductus venosus* bypasses the liver, which also is not very functional before birth. This is a vein; blood returning from the placenta enters the fetus through the umbilical vein and flows into the ductus venosus. The ductus venosus then empties into the inferior vena cava. After birth, the ductus venosus constricts, and blood is forced to flow through the liver. The ductus venosus leaves a fibrous remnant, the *ligamentum venosum,* on the inferior surface of the liver.

The two umbilical arteries become the *superior vesical arteries* to the urinary bladder. The umbilical vein becomes a fibrous cord, the *round ligament,* which attaches the liver to the anterior body wall (fig. 21.37).

The Aging Vascular System

Atherosclerosis is the principal change seen in the blood vessels with advancing age. It is such a universal phenomenon that it is difficult to isolate and identify other age-related changes independent of atherosclerosis. However, even non-atherosclerotic vessels stiffen with age, owing to increasing deposition of collagen, cross-linking of collagen molecules (the same phenomenon that stiffens the skeletal joints, lens of the eye, and tissues elsewhere), and declining resilience of the elastic fibers. This arterial stiffening is less pronounced in elderly people who engage in routine vigorous exercise. Stroke is a leading cause of death in old age, and it frequently results from atherosclerosis of the carotid arteries.

Another effect of aging is declining responsiveness of the baroreceptors, so vasomotor responses to changes in blood pres-

sure are not as quick or efficient. Among some elderly people, the blunted response causes *orthostatic hypotension:* When one goes from a lying to a sitting or standing posture, blood is drawn away from the brain by gravity. Without a prompt corrective baroreflex, the drop in cerebral blood flow can cause dizziness or even fainting and falling, which in turn presents a risk of serious bone fractures.

Vascular Diseases

Atherosclerosis, the most common vascular disease, can lead to stroke, renal failure, or heart failure—most notoriously the last of these. Therefore, it was described in chapter 20 in the context of coronary artery disease. Table 21.12 describes a few other vascular diseases. Some of the hematologic and cardiac pathologies in the preceding chapters also include aspects of vascular pathology.

Before You Go On

Answer the following questions to test your understanding of the preceding section:

17. What do angioblasts develop from, and what do they develop into?

18. Describe, in humans, what becomes of the six pairs of aortic arches typical of vertebrate embryos.

19. Name two blood vessels that close off and become fibrous cords soon after birth.

20. Describe two changes that occur in the blood vessels in old age, other than specific vascular diseases.

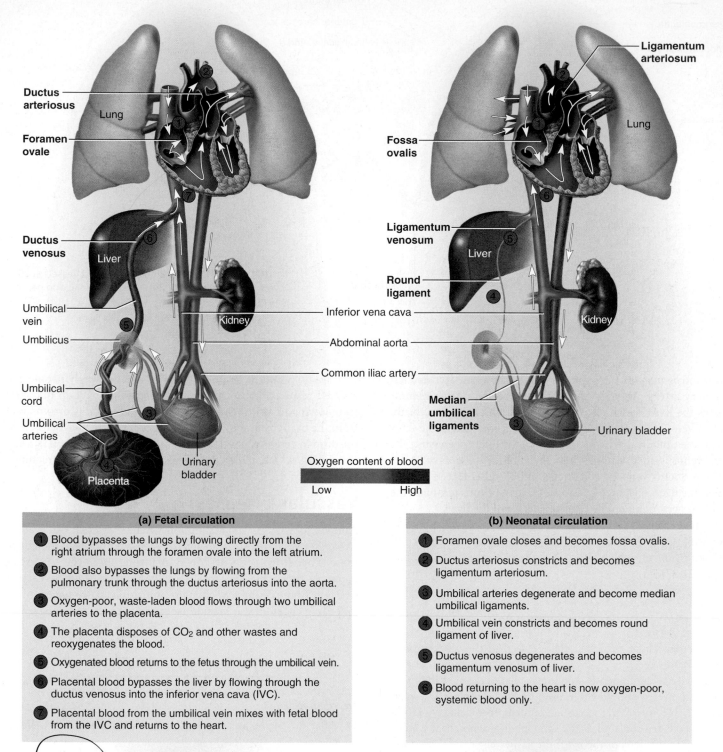

Oxygen content of blood

Low   High

(a) Fetal circulation

① Blood bypasses the lungs by flowing directly from the right atrium through the foramen ovale into the left atrium.

② Blood also bypasses the lungs by flowing from the pulmonary trunk through the ductus arteriosus into the aorta.

③ Oxygen-poor, waste-laden blood flows through two umbilical arteries to the placenta.

④ The placenta disposes of CO_2 and other wastes and reoxygenates the blood.

⑤ Oxygenated blood returns to the fetus through the umbilical vein.

⑥ Placental blood bypasses the liver by flowing through the ductus venosus into the inferior vena cava (IVC).

⑦ Placental blood from the umbilical vein mixes with fetal blood from the IVC and returns to the heart.

(b) Neonatal circulation

① Foramen ovale closes and becomes fossa ovalis.

② Ductus arteriosus constricts and becomes ligamentum arteriosum.

③ Umbilical arteries degenerate and become median umbilical ligaments.

④ Umbilical vein constricts and becomes round ligament of liver.

⑤ Ductus venosus degenerates and becomes ligamentum venosum of liver.

⑥ Blood returning to the heart is now oxygen-poor, systemic blood only.

Figure 21.37 Some Circulatory Changes Occurring at Birth. (a) Circulatory system of the full-term fetus. (b) Circulatory system of the newborn.

TABLE 21.12	Some Vascular Pathologies
Hypertension	Abnormally high blood pressure. In a young adult, a BP up to 130/85 is considered normal, BP above 140/90 is considered hypertensive, and a BP between these ranges is borderline or "high normal." About 90% of cases of hypertension (*primary hypertension*) result from a poorly understood complex of hereditary, behavioral, and other factors. Risk factors include obesity, a sedentary lifestyle, diet, smoking, sex, and race. *Secondary hypertension* (10% of cases) results from other identifiable disorders such as renal insufficiency, atherosclerosis, hyperthyroidism, and polycythemia. Treated with dietary modification, weight loss, and drugs such as beta-blockers (which reduce responsiveness of the blood vessels to sympathetic stimulation), calcium channel blockers (which relax the vascular smooth muscle), diuretics (which reduce blood volume), and ACE inhibitors (which inhibit synthesis of the vasoconstrictor angiotensin II).
Phlebitis	Inflammation of a vein, causing pain, tenderness, edema, and skin discoloration along its course. Often of unknown cause, but may follow surgery, childbirth, or infections.
Raynaud[25] disease	Occasional spasmodic contractions of the digital arteries, causing pallor, numbness, and coldness of the fingers or toes. The digits may at first appear cyanotic, but then redden, with throbbing and paresthesia (tingling, burning, or itching sensations). Repeated and severe cases can lead to brittle nails and occasionally to gangrene and a necessity for amputation. Most common in young women and often triggered by emotional stress or brief exposure to cold.
Stroke (cerebrovascular accident)	The sudden death (infarction) of brain tissue occurring when cerebral atherosclerosis, thrombosis, or hemorrhage of a cerebral aneurysm cuts off blood flow to part of the brain. Effects range from unnoticeable to fatal, depending on the extent of tissue damage and function of the affected tissue. Blindness, paralysis, loss of speech, and loss of sensation are among the sublethal effects.
Vasculitis	Inflammation of any blood vessel (see also phlebitis in this table), usually caused by an immune response or infectious pathogen, but sometimes by radiation, trauma, or toxins. Produces a wide variety of symptoms, including muscle and joint pain, fever, headache, myocardial ischemia, numbness, and blindness.

Disorders Described Elsewhere

Air embolism 535	Orthostatic hypotension 603
Aneurysm 567	Patent ductus arteriosus 559
Atherosclerosis 551	Varicose veins 571

[25]Maurice Raynaud (1834–81), French physician

Study Guide

Assess Your Learning Outcomes

You should have a good understanding of this chapter if you can accurately address the following issues.

21.1 General Anatomy of the Blood Vessels (p. 564)

1. The meaning of *arteries, capillaries,* and *veins* with respect to the route of blood flow from and back to the heart
2. The names and characteristics of the three tissue layers (tunics) of arteries and veins
3. Why arteries are called resistance vessels, and how this relates to their histology
4. The three size classes of arteries; their histological and functional differences; and examples of arteries of the two largest types

5. The location, structure, and functions of metarterioles and precapillary sphincters
6. The locations and functions of the carotid sinuses, aortic baroreceptors, carotid bodies, and aortic bodies
7. The size and general structure of blood capillaries, and the various means by which substances can pass through the capillary wall to enter or leave the bloodstream
8. The distinction between continuous capillaries, fenestrated capillaries, and sinusoids
9. The structure of a capillary bed, and its relationship to a thoroughfare channel and precapillary sphincters
10. Why veins are called capacitance vessels, and how this relates to their volume, blood pressure, and histology

11. The distinction between postcapillary venules, muscular venules, medium veins, venous sinuses, and large vein.
12. The locations, structure, and function of venous valves
13. Variations in the route of blood circulation, notably portal systems and the three kinds of anastomoses

21.2 The Pulmonary Circuit (p. 572)

1. The route of blood flow in the pulmonary circuit, including the names of its blood vessels in order from the point where blood leaves the right ventricle to the points where it enters the left atrium
2. The function of the pulmonary circuit and how this relates to the relative oxygen and carbon dioxide concentrations in its arteries and veins

3. The systemic blood supply to the lungs and how this differs functionally from the pulmonary circuit

21.3 Systemic Vessels of the Axial Region (p. 573)

1. The three principal regions of the aorta—ascending aorta, aortic arch, and descending aorta; the principal arteries that arise from the first two; the distinction between thoracic and abdominal aorta; and how the aorta ends (table 21.1)

2. The names and routes of the four arteries that ascend each side of the neck (table 21.2, part I)

3. The routes of the common carotid, external carotid, and internal carotid arteries; names of the branches given off by the external and internal carotids; and the regions supplied by those branches (table 21.2, part II)

4. The location and components of the cerebral arterial circle; the arteries that supply it; and the cerebral arteries that arise from it (table 21.2, parts III, IV)

5. The locations and names of the dural venous sinuses and their relationship with the internal jugular veins and facial veins (table 21.3, part I)

6. The locations of the internal and external jugular veins and the vertebral veins; the vessels from which these veins receive blood; and the vessels into which they ultimately drain (table 21.3, part II)

7. Branches of the thoracic aorta and the regions supplied by them (table 21.4, parts I, II)

8. The course of the subclavian and axillary arteries; the branches they give off along the way; and the regions supplied by those branches (table 21.4, part III)

9. The course of the subclavian and brachiocephalic veins; the regions drained by them; and how the superior vena cava originates and where it terminates (table 21.5, part I)

10. Locations of the azygos, hemiazygos, and accessory hemiazygos veins; the other veins that drain into this system; and where the blood of the azygos system flows next (table 21.5, part II)

11. The names, order of occurrence, and organs supplied by the arteries that branch off the abdominal aorta (table 21.6, part I)

12. The location of the celiac trunk; its three primary branches; and the organs supplied by these branches and their subdivisions (table 21.6, part II)

13. The origin of the superior mesenteric artery; its branches; and the regions of the digestive tract that it supplies (table 21.6, part III)

14. The origin of the inferior mesenteric artery; its branches; and the regions of the digestive tract that it supplies

15. The origin of the common iliac arteries; their branching into external and internal iliac arteries; the two trunks of the internal iliac; names of the branches of these two trunks; and the organs and tissues supplied by those branches (table 21.6, part IV)

16. The locations of the internal and external iliac veins; their convergence to form the common iliac vein; convergence of the two common iliac veins to form the inferior vena cava; and regions drained by the internal and external iliac veins (table 21.7, part I)

17. The tributaries that drain into the inferior vena cava, in order, as it ascends the abdominal cavity; the organs and regions drained by those tributaries; and where the inferior vena cava ends (table 21.7, part I)

18. The ascending lumbar veins and their relationships with the inferior vena cava and the azygos system (table 21.7, part II)

19. The general purpose of the hepatic portal system, and the locations of the two capillary beds that define this as a portal system (table 21.7, part III)

20. Tributaries that drain directly or indirectly into the hepatic portal vein before it enters the liver; the organs drained by these tributaries; and where the hepatic portal blood goes within the liver and after it leaves the liver (table 21.7, part III)

21.4 Systemic Vessels of the Appendicular Region (p. 590)

1. The arterial branches and alternative routes followed by the blood in getting from the subclavian artery to the palmar region, and structures supplied by their branches along the way (table 21.8)

2. The superficial veins followed by the blood in getting from the hand to the cephalic and basilic veins, and where those two veins end in the brachial region (table 21.9, part I)

3. The deep veins followed by the blood in getting from the hand to the subclavian vein, and where they pick up the blood from the cephalic and basilic veins (table 21.9, part II)

4. The arterial branches and alternative routes followed by the blood in getting from the common iliac artery to the foot, and structures supplied by their branches along the way (table 21.10)

5. The superficial veins followed by the blood in getting from the foot to the femoral vein (table 21.11, part I)

6. The deep veins followed by the blood in getting from the foot to the common iliac vein (table 21.11, part II)

21.5 Developmental and Clinical Perspectives (p. 601)

1. The origin of the first blood cells and blood vessels in the embryonic yolk sac

2. The anatomy of the embryonic aortic sac, paired dorsal aortae, and six aortic arches, and the fates of those aortic arches

3. The origin of a single dorsal aorta and of the intersegmental arteries, and some adult structures that arise from the intersegmental arteries

4. The routes of embryonic venous drainage by way of the anterior and posterior cardinal veins and the vitelline and umbilical veins, and the fate of the anterior and posterior cardinal veins

5. The shunts that allow most fetal blood to bypass the liver and lungs before birth, and what happens to these shunts shortly after birth

6. Multiple reasons why the arteries stiffen as one ages, and the health implications of this

7. The cardiovascular reason why elderly persons sometimes experience dizziness or even fainting and falling upon rising from a reclining to a sitting or standing posture

Testing Your Recall

1. Blood normally flows into a capillary bed from
   a. a distributing artery.
   b. a conducting artery.
   c. a metarteriole.
   d. a thoroughfare channel.
   e. a venule.

2. Plasma solutes enter the tissue fluid most easily from
   a. continuous capillaries.
   b. fenestrated capillaries.
   c. arteriovenous anastomoses.
   d. collateral vessels.
   e. venous anastomoses.

3. A blood vessel adapted to withstand great fluctuations in blood pressure would be expected to have
   a. an elastic tunica media.
   b. a thick tunica interna.
   c. one-way valves.
   d. a flexible endothelium.
   e. a rigid tunica media.

4. A circulatory pathway in which the blood flows through two capillary beds in series before it returns to the heart is called
   a. an arteriovenous anastomosis.
   b. an arterial anastomosis.
   c. a venous anastomosis.
   d. a venous return pathway.
   e. a portal system.

5. Intestinal blood flows into the liver by way of
   a. the superior mesenteric vein.
   b. the hepatic portal vein.
   c. the abdominal aorta.
   d. the hepatic sinusoids.
   e. the hepatic veins.

6. Blood islands first form in the embryonic
   a. spleen.
   b. yolk sac.
   c. placenta.
   d. liver.
   e. red bone marrow.

7. Most blood flowing in all of the following arteries *except* the _____ is destined to circulate through the brain before returning to the heart.
   a. vertebral arteries
   b. internal carotid arteries
   c. basilar artery
   d. superficial temporal artery
   e. anterior communicating artery

8. All of the following blood vessels *except* the _____ are located in the upper limb.
   a. cephalic vein
   b. small saphenous vein
   c. brachial artery
   d. circumflex humeral arteries
   e. metacarpal arteries

9. The adult aortic arch develops from the embryonic
   a. right aortic arch IV.
   b. left aortic arch IV.
   c. right aortic arch V.
   d. conus arteriosus.
   e. dorsal aorta.

10. To get from the posterior tibial vein to the femoral vein, blood flows through
   a. the anterior tibial vein.
   b. the popliteal vein.
   c. the internal iliac vein.
   d. the great saphenous vein.
   e. the basilic vein.

11. Filtration pores are characteristic of _____ capillaries.

12. The capillaries of skeletal muscles are of the structural type called _____.

13. The epithelium that lines the inside of a blood vessel is called _____.

14. The two _____ veins unite like an upside-down Y to form the inferior vena cava.

15. Carotid and aortic bodies are called _____ because they respond to changes in blood chemistry.

16. Movement across the capillary endothelium by the uptake and release of fluid droplets is called _____.

17. The two largest veins that empty into the right atrium are the _____ and _____.

18. The pressure sensors in the major arteries near the head are called _____.

19. Most of the blood supply to the brain comes from a ring of arterial anastomoses called the _____.

20. The major superficial veins of the arm are the _____ on the medial side and _____ on the lateral side.

Answers in the Appendix

Building Your Medical Vocabulary

State a medical meaning of each of the following word elements, and give a term in which it is used.

1. vas-
2. advent-
3. fenestr-
4. cephalo-
5. jugul-
6. angio-
7. celi-
8. genic-
9. -orum
10. vesic-

Answers in the Appendix

True or False

Determine which five of the following statements are false, and briefly explain why.

1. The lungs receive both a pulmonary and a systemic blood supply.
2. The pancreas and spleen receive their blood supply mainly from the superior mesenteric artery.
3. Veins anastomose more than arteries do.
4. From the time blood leaves the heart to the time it returns, it always passes through only one capillary bed.
5. The superior vena cava begins where the two subclavian veins meet.
6. Erythrocytes and endothelial cells arise from the same embryonic stem cells.
7. The smooth muscle of the tunica media of a large vessel is nourished mainly by the diffusion of nutrients from blood in the vessel lumen.
8. Venous blood from the intestines flows through the liver before it flows through the heart.
9. In a few unusual cases, one or more arteries of the cerebral arterial circle are lacking.
10. Arteries to the ovaries and testes originate relatively high in the abdominal cavity, near the kidneys.

Answers in the Appendix

Testing Your Comprehension

1. Suppose a posterior tibial vein was obstructed by thrombosis. Describe one or more alternative routes by which blood from the foot could get to the common iliac vein.

2. Why would a ruptured aneurysm of the basilar artery be more serious than a ruptured aneurysm of the anterior communicating artery?

3. What differences would you expect between a sample of blood taken from the superior mesenteric vein and a sample taken from a hepatic vein? Consider, especially, differences in nutrient levels and bacterial count, and look forward in the book if necessary for a preview of liver functions.

4. Why could a choke hold (a tight grip around the neck) cause a person to pass out? What arteries would be involved?

5. Why is it better to have baroreceptors in the carotid sinus rather than in some other location such as the abdominal aorta or common iliac arteries?

Answers at www.mhhe.com/saladinha3

Improve Your Grade at www.mhhe.com/saladinha3

Practice quizzes, labeling activities, games, and flashcards provide fun ways to master concepts. You can also download image PowerPoint files for each chapter to create a study guide or for taking notes during lecture.

The Lymphatic System and Immunity

Natural killer cells (yellow) attacking a human cancer cell (red)

CHAPTER OUTLINE

DEEPER INSIGHTS

BRUSHING UP

To understand this chapter, you may find it helpful to review the following concepts:
- General gland structure: capsule, septa, stroma, and parenchyma (pp. 73–74)
- Antigens and antibodies (p. 526)
- Leukocyte types, especially lymphocytes (pp. 527–530)
- Angiogenesis (p. 601)

Anatomy & Physiology | REVEALED®
aprevealed.com

Lymphatic System

The lymphatic system is a network of tissues, organs, and vessels that help to maintain the body's fluid balance, cleanse the body fluids of foreign matter, and provide immune cells for defense. Of all the body systems, it is perhaps the least familiar to most people. Yet without it, neither the circulatory system nor the immune system could function—circulation would shut down from fluid loss, and the body would be overrun by infection for lack of immunity. This chapter discusses the anatomy of the lymphatic system in relation to its roles in fluid recovery and immunity. The structure and function of the lymphatic system are so intimately tied to the immune system that this chapter will sometimes refer to them jointly as the *lymphatic–immune system.*

22.1 Lymph and Lymphatic Vessels

▶ Expected Learning Outcomes

When you have completed this section, you should be able to

- list the functions and basic components of the lymphatic system;
- explain how lymph is formed;
- describe the route that lymph takes to get into the bloodstream; and
- explain what makes lymph flow through the lymphatic vessels.

Components and Functions of the Lymphatic System

The **lymphatic**[1] **system** (fig. 22.1) consists of the following components: (1) *lymph,* the fluid the system collects from the interstitial spaces of the tissues and returns to the bloodstream; (2) *lymphatic vessels,* which transport the lymph; (3) *lymphatic tissue,* composed of aggregates of lymphocytes and macrophages that populate many organs of the body; and (4) *lymphatic organs,* in which these cells are especially concentrated and which are set off from surrounding organs by connective tissue capsules.

The functions of the lymphatic system include the following:

1. **Fluid recovery.** Fluid continually filters from the blood capillaries into the tissue spaces. The blood capillaries reabsorb about 85% of it, but the 15% they do not absorb would amount, over the course of a day, to 2 to 4 L of water and one-quarter to one-half of the plasma protein. One would die of circulatory failure within hours if this water and protein were not returned to the bloodstream. One task of the lymphatic system is to reabsorb this excess and return it to the blood. Even partial interference with lymphatic drainage can lead to severe edema and sometimes even more grotesque consequences (see Deeper Insight 22.1).

2. **Immunity.** As the lymphatic system recovers excess tissue fluid, it also picks up foreign cells and chemicals from the tissues. Some of these are **pathogens**[2]—agents with the potential to cause disease. On its way back to the bloodstream, the fluid passes through lymph nodes, where immune cells stand guard against pathogens and activate protective immune responses.

3. **Lipid absorption.** In the small intestine, special lymphatic vessels called *lacteals* absorb dietary lipids that cannot be absorbed by the intestinal blood capillaries (see p. 670).

Lymph

Lymph is usually a clear, colorless fluid, similar to blood plasma but low in protein. It originates as tissue fluid that has been taken up by the lymphatic vessels. Its composition varies substantially from place to place. After a meal, for example, lymph draining from the small intestine has a milky appearance because of its high lipid content. This intestinal lymph is called *chyle*[3] (pronounced "kile"). Lymph leaving the lymph nodes contains a large number of lymphocytes—indeed, this is the main supply of lymphocytes to the bloodstream. Lymph can also contain macrophages, hormones, bacteria, viruses, cellular debris, and even traveling cancer cells.

Lymphatic Vessels

Lymph flows through a system of **lymphatic vessels (lymphatics)** similar to blood vessels. These begin with microscopic **lymphatic capillaries (terminal lymphatics),** which penetrate nearly every tissue of the body but are absent from the central nervous system, cartilage, cornea, bone, and bone marrow. They are closely associated with blood capillaries, but unlike them, they are closed at one end (fig. 22.3). A lymphatic capillary consists of a sac of thin endothelial cells that loosely overlap each other like the shingles of a roof. The cells are tethered to surrounding tissue by protein filaments that prevent the sac from collapsing.

Unlike the endothelial cells of blood capillaries, lymphatic endothelial cells are not joined by tight junctions, nor do they have a continuous basal lamina; indeed, the gaps between them are so large that bacteria, lymphocytes, and other cells and particles can enter along with the tissue fluid. Thus, the composition of lymph arriving at a lymph node is like a report on the state of the upstream tissues. The overlapping edges of the endothelial cells act as valvular flaps that can open and close. When tissue fluid pressure is high, it pushes the flaps inward (open) and fluid flows into the capillary. When pressure is higher in the lymphatic capillary than in the tissue fluid, the flaps are pressed outward (closed).

Apply What You Know

Contrast the structure of a lymphatic capillary with that of a continuous blood capillary. Explain why their structural difference is related to their functional difference.

---

[1]*lympho* = water

[2]*patho* = disease + *gen* = producing
[3]*chyle* = juice

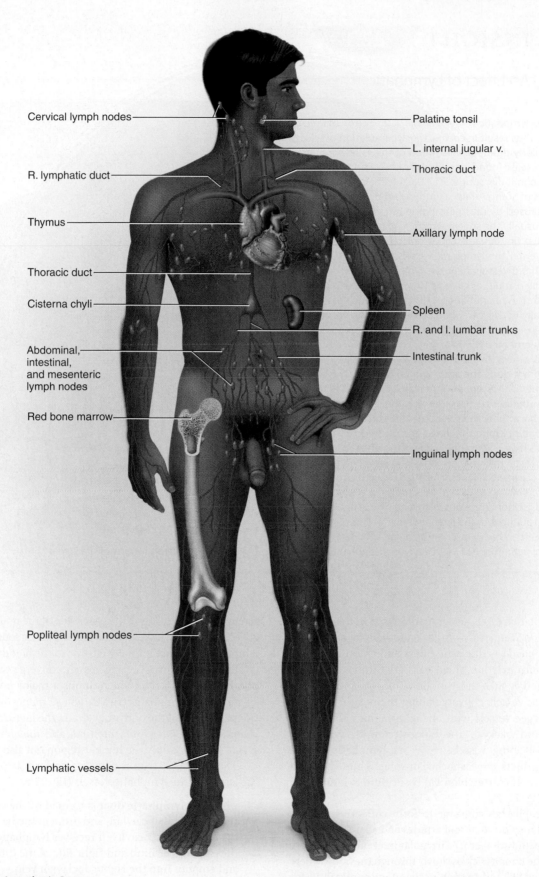

Cervical lymph nodes

Palatine tonsil

L. internal jugular v.

R. lymphatic duct

Thoracic duct

Thymus

Axillary lymph node

Thoracic duct

Cisterna chyli

Spleen

R. and l. lumbar trunks

Abdominal, intestinal, and mesenteric lymph nodes

Intestinal trunk

Red bone marrow

Inguinal lymph nodes

Popliteal lymph nodes

Lymphatic vessels

Figure 22.1 The Lymphatic System.

DEEPER INSIGHT   22.1

Elephantiasis—An Effect of Lymphatic Obstruction

Any obstruction of the lymphatic vessels can block the return of fluid to the bloodstream and thus result in *edema,* the accumulation of excess tissue fluid. A particularly dramatic illustration of this is *elephantiasis*[4] (fig. 22.2), a parasitic disease found in tropical climates worldwide, especially in Africa but also in India, southeast Asia, the Philippines, the Pacific Islands, and parts of South America (introduced by slave trading).

Elephantiasis is caused by mosquito-borne roundworms called *filariae* (fil-AIR-ee-ee), usually the species *Wuchereria bancrofti.* When an infected mosquito bites, tiny larvae escape from its proboscis, crawl into the bite wound, and enter the lymphatic vessels of the skin. They migrate to larger lymphatic vessels near the lymph nodes, where they mature into tightly coiled adults as large as 10 cm long and 0.3 cm wide. The worms cause intense inflammation of the lymphatic vessels and lymph nodes, especially in the lower half of the body. The vessels and nodes become swollen and painful, and infected males often suffer very painful edematous enlargement of the testes.

Infection with the worms, called *filariasis,* only occasionally leads to elephantiasis, but when it does, the effect can be horrible. The chronic blockage of lymph flow causes enormous enlargement and fibrosis of tissues upstream from the obstruction—notably the legs and arms, the scrotum of men *(lymph scrotum),* and sometimes the vulva and breasts of women. The skin becomes fibrotic, thickened, and cracked, so it comes to resemble an elephant's hide—hence the name of the disease.

Thousands of American military personnel who served in the Pacific theater in World War II contracted filariasis. Alarmed by pictures of extreme cases, some men feared having to carry their scrotum in a wheelbarrow. However, elephantiasis seldom develops in anyone whose first exposure to the filariae occurs in adulthood, and it requires many years of repetitive infection. Some servicemen had symptoms of filariasis for as long as 16 years after their return, but not one of them developed elephantiasis.

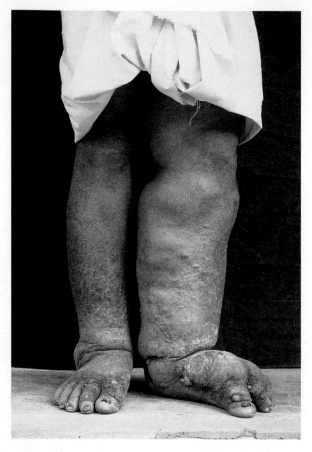

Figure 22.2 Elephantiasis of the Lower Limb.

The larger lymphatic vessels are similar to veins in their histology. They have a *tunica interna* with an endothelium and valves (fig. 22.4), a *tunica media* with elastic fibers and smooth muscle, and a thin outer *tunica externa.* Their walls are thinner and their valves are closer together than those of the veins.

As the lymphatic vessels converge along their path, they become larger and larger vessels with changing names. The route from the tissue fluid back to the bloodstream is: lymphatic capillaries ⟶ collecting vessels ⟶ six lymphatic trunks ⟶ two collecting ducts ⟶ subclavian veins. Thus, there is a continual recycling of fluid from blood to tissue fluid to lymph and back to the blood (fig. 22.5).

The lymphatic capillaries converge to form **collecting vessels.** These often travel alongside veins and arteries and share a common connective tissue sheath with them. At irregular intervals, they empty into lymph nodes. The lymph trickles slowly through the node, where bacteria are phagocytized and immune cells monitor the fluid for foreign antigens. It leaves the other side of the node through another collecting vessel, traveling on and often encountering additional lymph nodes before it finally returns to the bloodstream.

Eventually, the collecting vessels converge to form larger **lymphatic trunks,** each of which drains a major portion of the body. There are six lymphatic trunks, whose names indicate their locations and parts of the body they drain: the *jugular, subclavian, bronchomediastinal, intercostal, intestinal,* and *lumbar trunks.* The lumbar trunk drains not only the lumbar region but also the lower limbs.

The lymphatic trunks converge to form two **collecting ducts,** the largest of the lymphatic vessels (fig. 22.6):

1. The **right lymphatic duct** is formed by the convergence of the right jugular, subclavian, and bronchomediastinal trunks in the right thoracic cavity. It receives lymphatic drainage from the right upper limb and right side of the thorax and head, and empties into the right subclavian vein.

2. The **thoracic duct,** on the left, is larger and longer. It begins just below the diaphragm, anterior to the vertebral column at the level of the second lumbar vertebra. Here, the two lumbar

[4]*iasis* = medical condition

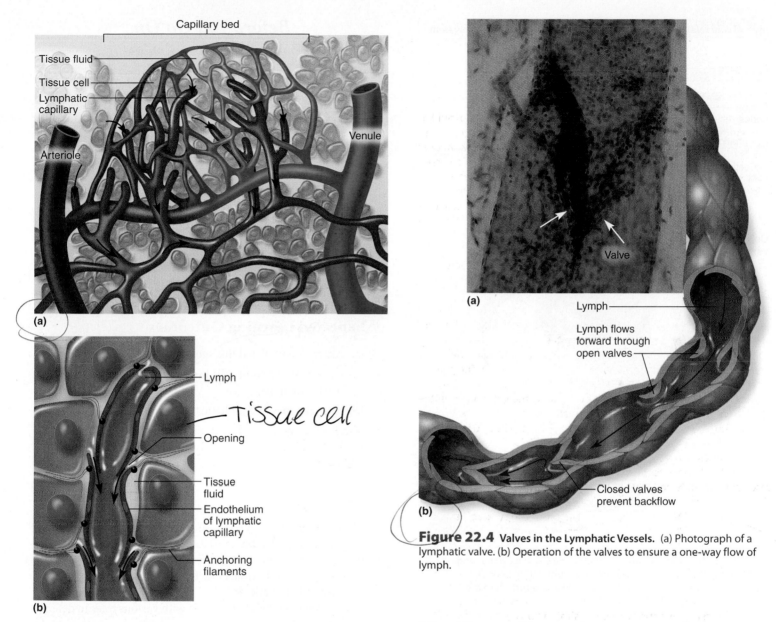

Capillary bed

Tissue fluid

Tissue cell

Lymphatic capillary

Venule

Arteriole

(a)

Valve

(a)

Lymph

Lymph flows forward through open valves

Closed valves prevent backflow

(b)

Figure 22.4 **Valves in the Lymphatic Vessels.** (a) Photograph of a lymphatic valve. (b) Operation of the valves to ensure a one-way flow of lymph.

Lymph

Tissue cell

Opening

Tissue fluid

Endothelium of lymphatic capillary

Anchoring filaments

(b)

Figure 22.3 **Lymphatic Capillaries.** (a) Relationship of the lymphatic capillaries to a bed of blood capillaries. (b) Uptake of tissue fluid by a lymphatic capillary.
● *Why can traveling (metastasizing) cancer cells get into the lymphatic system more easily than they can enter the bloodstream?*

trunks and the intestinal trunk join and form a prominent sac called the **cisterna chyli** (sis-TUR-nuh KY-lye), named for the large amount of chyle that it collects after a meal. The thoracic duct then passes through the diaphragm with the aorta and ascends the mediastinum, adjacent to the vertebral column. As it passes through the thorax, it receives additional lymph from the left bronchomediastinal, left subclavian, and left jugular trunks, then empties into the left subclavian vein. Collectively, this duct therefore drains all of the body below the diaphragm, and the left upper limb and left side of the head, neck, and thorax.

The Flow of Lymph

Lymph flows under forces similar to those that govern venous return, except that the lymphatic system has no pump like the heart, and it flows at even lower pressure and speed than venous blood. The primary mechanism of flow is rhythmic contractions of the lymphatic vessels themselves, induced when the flowing lymph stretches them. The valves of lymphatic vessels, like those of veins, prevent the fluid from flowing backward. Lymph flow is also produced by skeletal muscles squeezing the lymphatic vessels, like the skeletal muscle pump that moves venous blood. Since lymphatic vessels are often wrapped with an artery in a common connective tissue sheath, arterial pulsation may also rhythmically squeeze the lymphatic vessels and contribute to lymph flow. A thoracic (respiratory) pump promotes the flow of lymph from the abdominal to the thoracic cavity as one inhales. During inhalation, pressure in the thoracic cavity falls below the pressure in the abdominal cavity. Abdominal pressure squeezing on the abdominal lymphatic trunks and cisterna chyli causes lymph to flow upward into the thoracic

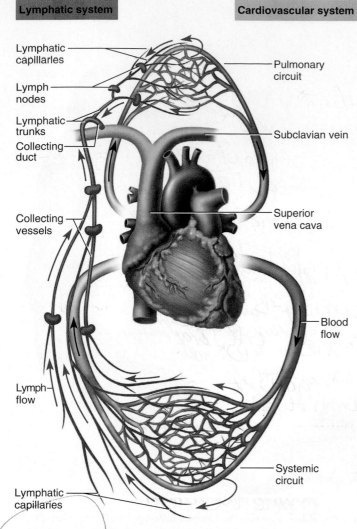

| Lymphatic system | Cardiovascular system |

Lymphatic capillaries

Lymph nodes

Lymphatic trunks

Collecting duct

Collecting vessels

Lymph flow

Lymphatic capillaries

Pulmonary circuit

Subclavian vein

Superior vena cava

Blood flow

Systemic circuit

Figure 22.5 Fluid Exchange Between the Cardiovascular and Lymphatic Systems. Blood capillaries lose fluid to the tissue spaces. The lymphatic system picks up excess tissue fluid and returns it to the bloodstream. The lymph flows from lymphatic capillaries through collecting vessels, lymphatic trunks, and collecting ducts, and is filtered through multiple lymph nodes before reentering the bloodstream at the subclavian veins.

● *Identify two benefits in having lymphatic capillaries pick up tissue fluid that is not reclaimed by the blood capillaries.*

lymphatics. Finally, at the point where the collecting ducts empty into the subclavian veins, the rapidly flowing bloodstream draws the lymph into it.

Apply What You Know

Why does it make more functional sense for the collecting ducts to connect to the subclavian veins than it would for them to connect to the subclavian arteries?

Before You Go On

Answer the following questions to test your understanding of the preceding section:

1. List the primary functions of the lymphatic system.
2. How does fluid get into the lymphatic system? What prevents it from draining back out?
3. Where does this fluid (lymph) go once it enters the lymphatic vessels? What makes it flow?

22.2 Lymphatic Cells, Tissues, and Organs

▶ **Expected Learning Outcomes**

When you have completed this section, you should be able to

- name the major types of cells in the lymphatic system and state their functions;
- describe the types of lymphatic tissue; and
- describe the anatomy and lymphatic–immune function of the red bone marrow, thymus, lymph nodes, tonsils, and spleen.

In addition to lymphatic vessels, another component of the lymphatic system is the lymphatic tissues. These range from loosely scattered cells in the mucous membranes of the digestive, respiratory, reproductive, and urinary tracts, to compact cell populations encapsulated in lymphatic organs. These tissues are composed of a variety of lymphocytes and other cells with various roles in defense and immunity.

Lymphatic Cells

Aside from cells playing purely structural roles, the lymphatic system has six principal categories of defensive cells:

1. **Natural killer (NK) cells.** These are large lymphocytes that attack and lyse bacteria, transplanted tissue cells, and *host cells* (cells of one's own body) that have either become infected with viruses or turned cancerous (see photo on p. 609). Their continual patrolling of the body "on the lookout" for abnormal cells is called *immune surveillance,* and is one of the body's most important defenses against cancer.

2. **T lymphocytes (T cells).** These are so named because they develop for a time in the thymus and later depend on thymic hormones to regulate their activity. The *T* stands for *thymus-dependent.* There are four subclasses of T cells:

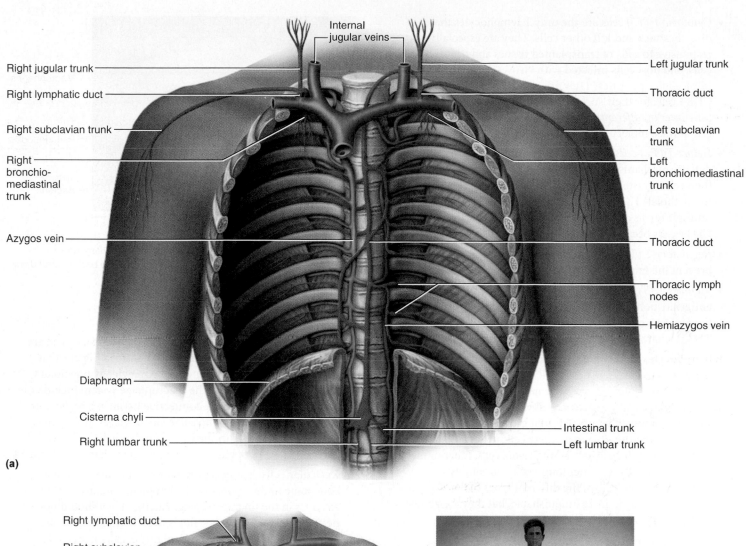

Internal
jugular veins

Right jugular trunk

Right lymphatic duct

Right subclavian trunk

Right
bronchio-
mediastinal
trunk

Azygos vein

Diaphragm

Cisterna chyli

Right lumbar trunk

Left jugular trunk

Thoracic duct

Left subclavian
trunk

Left
bronchiomediastinal
trunk

Thoracic duct

Thoracic lymph
nodes

Hemiazygos vein

Intestinal trunk

Left lumbar trunk

(a)

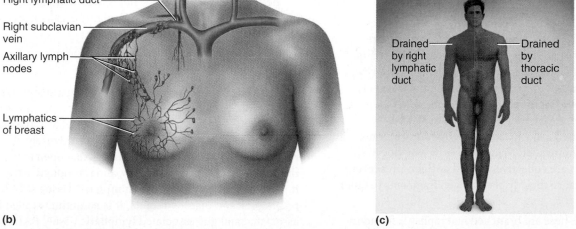

Right lymphatic duct

Right subclavian
vein

Axillary lymph
nodes

Lymphatics
of breast

Drained
by right
lymphatic
duct

Drained
by
thoracic
duct

(b)

(c)

Figure 22.6 Lymphatics of the Thoracic Region. (a) Lymphatics of the thorax and upper abdomen and their relationship to the subclavian veins, where the lymph returns to the bloodstream. (b) Lymphatic drainage of the right mammary and axillary regions. (c) Regions of the body drained by the right lymphatic duct and thoracic duct.

• *Why are the axillary lymph nodes often biopsied in cases of suspected breast cancer?*

- *Cytotoxic T (T_C) cells* are the only T lymphocytes that directly attack and kill other cells. They are especially responsive to cells of transplanted tissues and organs, cancer cells, and host cells infected with viruses, bacteria, or intracellular parasites. They are also called T8, CD8, or CD8+ cells for their surface glycoprotein, CD8. (*CD* stands for *cluster of differentiation,* a classification system for many cell-surface molecules.)

- *Helper T (T_H) cells* respond to antigens and activate various defense mechanisms, but do not carry out the attack themselves; instead, they help other immune cells respond to the threat. T_H cells play a central coordinating role in multiple forms of defense. They are also called T4, CD4, or CD4+ cells because of a surface glycoprotein called CD4.

- *Regulatory T (T_R) cells,* or *T-regs,* play an inhibitory role to prevent the immune system from running out of control.

- *Memory T (T_M) cells* provide long-lasting memory of an antigen. Upon reexposure, the immune system neutralizes the antigen so quickly that it causes no disease symptoms. This is what we mean by being immune[5] to a disease.

3. **B lymphocytes (B cells).** These lymphocytes differentiate into *plasma cells*—connective tissue cells that secrete the antibodies of the immune system. They were named for an organ in chickens (the *bursa of Fabricius*[6]) in which they were first discovered; however, you may find it more helpful to think of *B* for *bone marrow,* the site where these cells mature. Some B cells become memory B cells instead of plasma cells, functioning like memory T cells to confer long-lasting immunity.

   In stained blood films, the different types of lymphocytes are not morphologically distinguishable, but they are about 80% T cells, 15% B cells, and 5% NK and stem cells.

4. **Macrophages.** These cells develop from monocytes that have emigrated from the bloodstream. Macrophages are very large, avidly phagocytic cells. They phagocytize tissue debris, dead neutrophils, bacteria, and other foreign matter (fig. 22.7). They also process foreign matter and transport antigenically active fragments of it (*antigenic determinants*) to the cell surface, where they "display" it to T_C and T_H cells. This stimulates the T cells to launch an immune response against the foreign invader. Macrophages, B lymphocytes, and reticular cells are collectively called **antigen-presenting cells (APCs)** because they display antigen fragments to other immune cells.

5. **Dendritic cells.** These are branched macrophages found in the epidermis, mucous membranes, and lymphatic organs. (In the skin, they are often called *Langerhans*[7] *cells.*) They engulf foreign matter by receptor-mediated endocytosis rather than phagocytosis, but they otherwise function like macrophages.

   The macrophage system includes all of the body's phagocytic cells except leukocytes. Some of these phagocytes

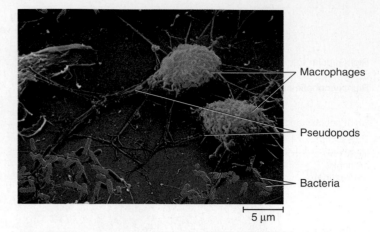

Figure 22.7 **Macrophages Attacking Bacteria.** Filamentous pseudopods of the macrophages snare the rod-shaped bacteria and draw them to the cell surface, where they are phagocytized.

are wandering cells that actively seek pathogens; others are fixed in place and phagocytize only those pathogens that come to them—although they are strategically positioned for this to occur. Cells of the macrophage system include the macrophages of the loose connective tissue, *microglia* of the central nervous system, *alveolar macrophages* in the lungs, *hepatic macrophages* in the liver, and dendritic cells. Alveolar and hepatic macrophages are described in chapters 23 and 24.

6. **Reticular cells.** These are branched stationary cells that contribute to the stroma of the lymphatic organs and act as APCs in the thymus (see fig. 22.10). (They should not be confused with reticular *fibers,* which are fine branched collagen fibers common in lymphatic organs.)

Lymphatic Tissues

Lymphatic (lymphoid) tissues are aggregations of lymphocytes in the connective tissues of mucous membranes and various organs. The simplest form is **diffuse lymphatic tissue,** in which the lymphocytes are scattered rather than densely clustered. It is particularly prevalent in body passages that open to the exterior—the respiratory, digestive, urinary, and reproductive tracts—where it is called **mucosa-associated lymphatic tissue (MALT).** (In the respiratory and digestive tracts, it is sometimes called bronchus-associated and gut-associated lymphatic tissue, BALT and GALT, respectively.)

   In some places, lymphocytes and macrophages congregate in dense masses called **lymphatic nodules (follicles)** (fig. 22.8), which come and go as pathogens invade the tissues and the immune system answers the challenge. Abundant lymphatic nodules are, however, a relatively constant feature of the lymph nodes, tonsils, and appendix. In the ileum, the distal portion of the small intestine, they form clusters called **Peyer**[8] patches.

---

[5]*immuno* = free
[6]Hieronymus Fabricius (Girolamo Fabrizzi) (1537–1619), Italian anatomist
[7]Theodor Langerhans (1839–1915), German pathologist

[8]Johann Conrad Peyer (1653–1712), Swiss anatomist

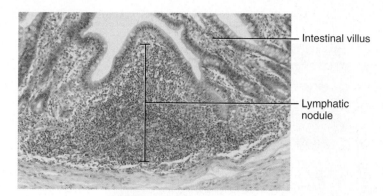

Figure 22.8 **Lymphatic Nodule in the Mucous Membrane of the Small Intestine.** This is an example of gut-associated lymphatic tissue (GALT).

Overview of Lymphatic Organs

In contrast to the diffuse lymphatic tissue, **lymphatic (lymphoid) organs** have well-defined anatomical sites and at least partial connective tissue capsules that separate the lymphatic tissue from neighboring tissues. These organs include the red bone marrow, thymus, lymph nodes, tonsils, and spleen. The red bone marrow and thymus are regarded as *primary lymphatic organs* because they are the sites where B and T lymphocytes, respectively, become *immunocompetent*—that is, able to recognize and respond to antigens. The lymph nodes, tonsils, and spleen are called *secondary lymphatic organs* because they are populated with immunocompetent lymphocytes only after the cells have matured in the primary lymphatic organs.

Red Bone Marrow

As discussed in chapter 6, there are two kinds of bone marrow—red and yellow. Red bone marrow is involved in hemopoiesis (blood formation) and immunity; yellow bone marrow can be disregarded for our present purposes. In children, red bone marrow occupies the medullary spaces of nearly the entire skeleton. In adults, it is limited to parts of the axial skeleton and the proximal heads of the humerus and femur (see fig. 6.7, p. 137). Red bone marrow is an important supplier of lymphocytes to the immune system. Its role in the life history of lymphocytes is described later.

Red bone marrow is a soft, loosely organized, highly vascular material, separated from osseous tissue by the endosteum of the bone. It produces all classes of formed elements of the blood; its red color comes from the abundance of erythrocytes. Numerous small arteries enter *nutrient foramina* on the bone surface, penetrate the bone, and empty into large *sinusoids* in the marrow (fig. 22.9). The sinusoids drain into a *central longitudinal vein* that exits the bone via the same route that the arteries entered. The sinusoids, 45 to 80 μm wide, are lined by endothelial cells, like other blood vessels, and are surrounded by reticular cells and reticular fibers. The reticular cells secrete colony-stimulating factors that induce the formation of various leukocyte types. In the long bones of the limbs, aging reticular cells accumulate fat and transform into adipose cells, eventually replacing red bone marrow with yellow bone marrow.

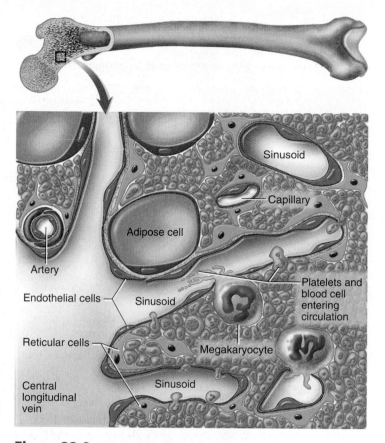

Figure 22.9 **Histology of the Red Bone Marrow.** The formed elements of blood squeeze through the endothelial cells into the sinusoids, which converge on the central longitudinal vein at the lower left.

The spaces between the sinusoids are occupied by *islands (cords)* of hemopoietic cells, composed of macrophages and blood cells in all stages of development. The macrophages destroy malformed blood cells and the nuclei discarded by developing erythrocytes. As blood cells mature, they push their way through the reticular and endothelial cells to enter the sinus and flow away in the bloodstream.

Apply What You Know

If we regard red bone marrow as a lymphatic organ and define lymphatic organs partly by the presence of a connective tissue capsule, what could we regard as the capsule of red bone marrow?

The Thymus

The **thymus** was introduced in chapter 18 because it is a member of both the endocrine and lymphatic systems. It houses developing lymphocytes and secretes hormones that regulate their later activity. It is a bilobed organ located between the sternum and aortic arch in the upper mediastinum (fig. 22.10). The two lobes are connected by a median bridge of tissue. In children, the organ is relatively firm and conical, much larger than it is in adults, and deep red due to its rich supply of blood vessels. After age 15 or so, it shrinks and contains less and less lymphatic tissue, and its color

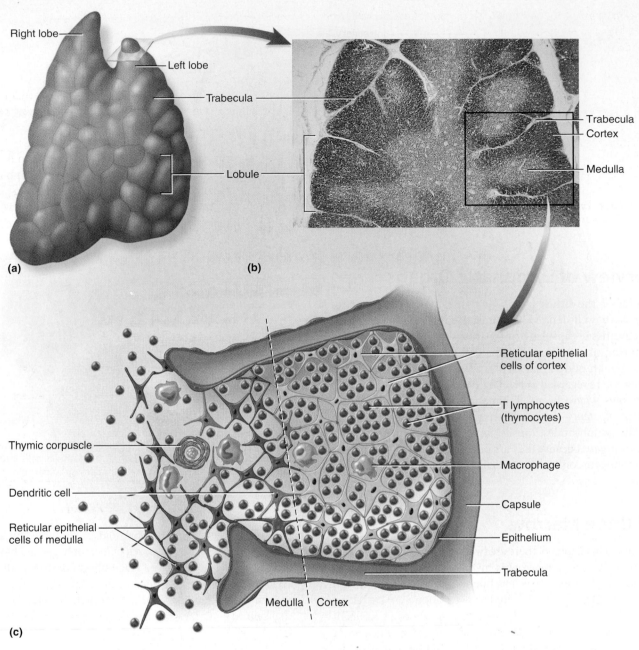

Figure 22.10 **The Thymus.** (a) Gross anatomy. (b) Histology. (c) Arrangement of the reticular epithelial cells to form the blood–thymus barrier separating the cortex from the medulla of one lobule.
● *Which of the cells in this organ secrete hormones?*

changes to gray and then yellowish as it becomes infiltrated with fat (see fig. 18.5, p. 505). In old age, it is barely distinguishable from the surrounding fat and fibrous tissue of the mediastinum.

The fibrous capsule of the thymus gives off trabeculae (septa) that penetrate into the gland and divide it into several angular lobules. Each lobule has a dense, dark-staining *cortex* and a lighter *medulla* inhabited by T lymphocytes (fig. 22.10b). **Reticular epithelial cells** seal off the cortex from the medulla and surround the blood vessels and lymphocyte clusters in the cortex. They thereby form a *blood–thymus barrier* that isolates developing lymphocytes

from blood-borne antigens. In the medulla, the reticular epithelial cells form whorls called *thymic (Hassall[9]) corpuscles,* which are useful for identifying the thymus histologically.

Besides forming the blood–thymus barrier, reticular epithelial cells secrete several signaling molecules that promote the development and action of T cells, including *thymosin, thymulin, thymopoietin, interleukins,* and *interferon.* If the thymus is removed from

---

[9]Arthur H. Hassall (1817–94), British chemist and physician

newborn mammals, they waste away and never develop immunity. Other lymphatic organs also seem to depend on thymosins or T cells and develop poorly in thymectomized animals. The relationship of T cell maturation to thymic histology is discussed later in this chapter.

Lymph Nodes

Lymph nodes are the most numerous lymphatic organs, numbering in the hundreds. They serve two functions: to cleanse the lymph and to act as a site of T and B cell activation. A lymph node is an elongated or bean-shaped structure, usually less than 3 cm long, often with an indentation called the **hilum** on one side (fig. 22.11). It is enclosed in a fibrous capsule with trabeculae that partially divide the interior of the node into compartments. Between the capsule and parenchyma is a narrow, relatively clear space called the **subcapsular sinus,** which contains reticular fibers, macrophages, and dendritic cells. Deep to this, the gland consists mainly of a stroma of reticular connective tissue (reticular fibers and reticular cells) and a parenchyma of lymphocytes and antigen-presenting cells.

The parenchyma is divided into an outer C-shaped **cortex** that encircles about four-fifths of the organ, and an inner **medulla** that extends to the surface at the hilum. The cortex consists mainly of ovoid to conical lymphatic nodules. When the lymph node is fighting a pathogen, these nodules acquire light-staining **germinal centers** where B cells multiply and differentiate into plasma cells. The

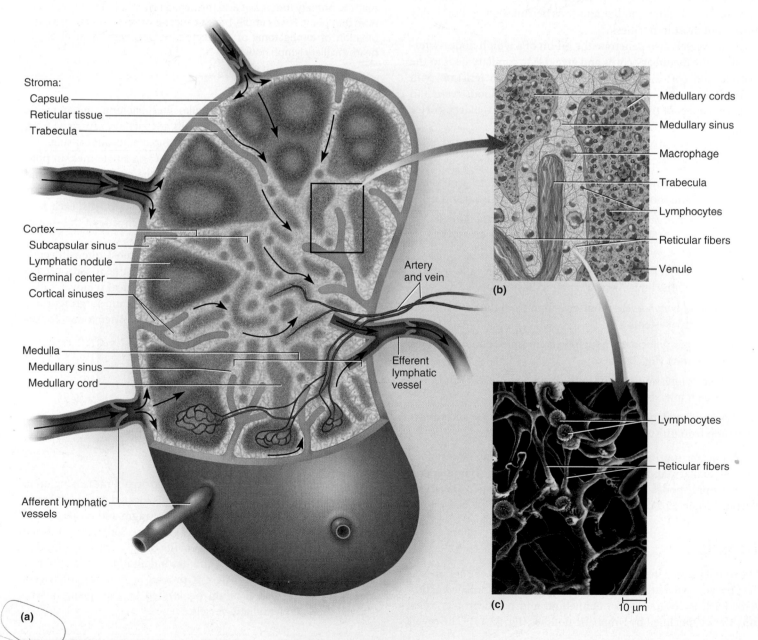

Figure 22.11 **Anatomy of a Lymph Node.** (a) Partially bisected lymph node showing pathway of lymph flow. (b) Detail of the boxed region in part (a). (c) Reticular fiber stroma and immune cells in a medullary sinus (SEM).

medulla consists largely of a branching network of *medullary cords* composed of lymphocytes, plasma cells, macrophages, reticular cells, and reticular fibers. The cortex and medulla also contain lymph-filled sinuses continuous with the subcapsular sinus.

Several **afferent lymphatic vessels** lead into the node along its convex surface. Lymph flows from these vessels into the subcapsular sinus, percolates slowly through the sinuses of the cortex and medulla, and leaves the node through one to three **efferent lymphatic vessels** that emerge from the hilum. No other lymphatic organs have afferent lymphatic vessels; lymph nodes are the only organs that filter lymph as it flows along its course. The lymph node is a bottleneck that slows down lymph flow and allows time for cleansing it of foreign matter. The macrophages and reticular cells of the sinuses remove about 99% of the impurities before the lymph leaves the node. On its way to the bloodstream, lymph flows through one lymph node after another and thus becomes quite thoroughly cleansed of most impurities.

Blood vessels also penetrate the hilum of a lymph node. Arteries follow the medullary cords and give rise to capillary beds in the medulla and cortex. In the *deep cortex* near the junction with the medulla, lymphocytes can emigrate from the bloodstream into the parenchyma of the node. Most lymphocytes in the deep cortex are T cells.

Lymph nodes are widespread but especially concentrated in the following locations:

- *Cervical lymph nodes* occur in deep and superficial groups in the neck, and monitor lymph coming from the head and neck.
- *Axillary lymph nodes* are concentrated in the armpit (axilla) and receive lymph from the upper limb and the breast (see fig. 22.6b).
- *Thoracic lymph nodes* occur in the thoracic cavity and receive lymph from the lungs, airway, and mediastinum.
- *Abdominal lymph nodes* monitor lymph from the urinary and reproductive systems.
- *Intestinal* and *mesenteric lymph nodes* monitor lymph from the digestive tract (fig. 22.12a).
- *Inguinal lymph nodes* occur in the groin (fig. 22.12b) and receive lymph from the entire lower limb.
- *Popliteal lymph nodes* occur at the back of the knee and receive lymph from the leg proper.

Physicians routinely palpate the superficial lymph nodes of the cervical, axillary, and inguinal regions for swelling (**lymphadenitis**[10]). Lymph nodes are common sites of metastatic cancer (see Deeper Insight 22.2).

Tonsils

The **tonsils** are patches of lymphatic tissue located at the entrance to the pharynx, where they guard against ingested and inhaled pathogens. Each is covered by an epithelium and has deep pits called **tonsillar crypts** lined by lymphatic nodules (fig. 22.13). The crypts

---

DEEPER INSIGHT   22.2

Lymph Nodes and Metastatic Cancer

Metastasis is a phenomenon in which cancerous cells break free of the original *primary tumor*, travel to other sites in the body, and establish new tumors. Because of the high permeability of lymphatic capillaries, metastasizing cancer cells easily enter them and travel in the lymph. They tend to lodge in the first lymph node they encounter and multiply there, eventually destroying the node. Cancerous lymph nodes are swollen but relatively firm and usually painless. Cancer of a lymph node is called *lymphoma*.[11]

Once a tumor is well established in one node, cells may emigrate from there and travel to the next. However, if the metastasis is detected early enough, cancer can sometimes be eradicated by removing not only the primary tumor, but also the nearest lymph nodes downstream from that point. For example, breast cancer is often treated with a combination of lumpectomy or mastectomy along with removal of the nearby axillary lymph nodes.

---

often contain food debris, dead leukocytes, bacteria, and antigenic chemicals. Below the crypts, the tonsils are partially separated from underlying connective tissue by an incomplete fibrous capsule.

There are three main sets of tonsils: (1) a single median **pharyngeal tonsil (adenoids)** on the wall of the pharynx just behind the nasal cavity; (2) a pair of **palatine tonsils** at the posterior margin of the oral cavity; and (3) numerous **lingual tonsils,** each with a single crypt, concentrated in a patch on each side of the root of the tongue (see fig. 24.4, p. 659).

The palatine tonsils are the largest and most often infected. *Tonsillitis* is an acute inflammation of the palatine tonsils, usually caused by a *Streptococcus* infection. Their surgical removal, called *tonsillectomy*,[12] used to be one of the most common surgical procedures performed on children, but it is done less often today. Tonsillitis is now usually treated with antibiotics.

Apply What You Know

Which tonsil(s) is or are most likely to be affected by an inhaled pathogen?

The Spleen

The **spleen** is the body's largest lymphatic organ, measuring up to 12 cm long and weighing up to 160 g. It is located in the left hypochondriac region, just inferior to the diaphragm and posterolateral to the stomach (fig. 22.14; see also fig. A.6, p. 336). It is protected by ribs 10 through 12. The spleen fits snugly between the diaphragm, stomach, and kidney and has indentations called the *gastric area* and *renal area* where it presses against these adjacent viscera. It has a medial hilum penetrated by the splenic artery, splenic vein, and lymphatic vessels.

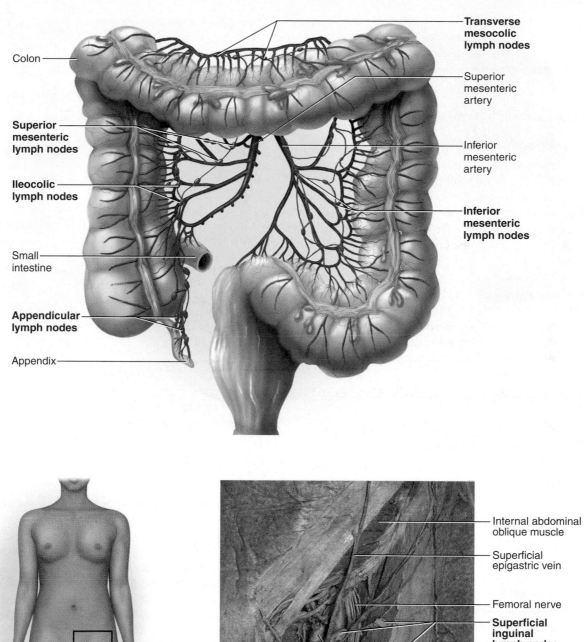

Colon

Transverse mesocolic lymph nodes

Superior mesenteric artery

Superior mesenteric lymph nodes

Ileocolic lymph nodes

Inferior mesenteric artery

Small intestine

Inferior mesenteric lymph nodes

Appendicular lymph nodes

Appendix

(b)

Internal abdominal oblique muscle

Superficial epigastric vein

Femoral nerve

Superficial inguinal lymph nodes

Femoral vein

Deep femoral artery

Femoral artery

Great saphenous vein

Sartorius muscle

Figure 22.12 **Some Areas of Lymph Node Concentration.** (a) Mesenteric lymph nodes associated with the large intestine. (b) Inguinal lymph nodes in a female cadaver.

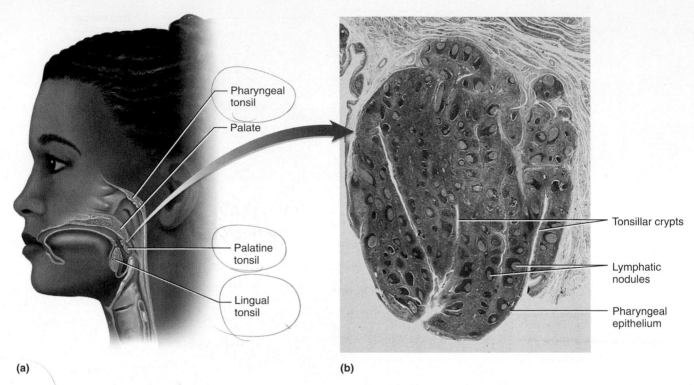

Figure 22.13 **The Tonsils.** (a) Locations of the tonsils. (b) Histology of the palatine tonsil.

The parenchyma exhibits two types of tissue named for their appearance in fresh specimens (not in stained sections): **red pulp,** which consists of sinuses gorged with concentrated erythrocytes, and **white pulp,** which consists of lymphocytes and macrophages aggregated like sleeves along small branches of the splenic artery. In tissue sections, white pulp appears as an ovoid mass of lymphocytes with an arteriole passing through it. However, it is important to bear in mind that its three-dimensional shape is not egglike but cylindrical.

These two tissue types reflect the multiple functions of the spleen. It produces blood cells in the fetus and may resume this role in adults in the event of extreme anemia. Lymphocytes and macrophages of the white pulp monitor the blood for foreign antigens, much like the lymph nodes do the lymph. The splenic blood capillaries are very permeable; they allow RBCs to leave the bloodstream, accumulate in the sinuses of the red pulp, and reenter the bloodstream later. The spleen is an "erythrocyte graveyard"—old, fragile RBCs rupture as they squeeze through the capillary walls into the sinuses. Macrophages phagocytize their remains, just as they dispose of blood-borne bacteria and other cellular debris. The spleen also helps to stabilize blood volume by transferring excess plasma from the bloodstream into the lymphatic system.

Apply What You Know

From an anatomical perspective, why are lymph nodes the only lymphatic organs that can filter the lymph?

DEEPER INSIGHT　22.3

Splenectomy

A ruptured spleen is one of the most common consequences of blows to the left thoracic or abdominal wall, as in sports injuries and automobile accidents. It is especially likely to rupture if the lower ribs are fractured, and sometimes it is nicked during abdominal surgery. The spleen is such a pulpy and vascular organ that it bleeds profusely, and its capsule is so thin and delicate that it is difficult to repair surgically. To prevent fatal hemorrhaging, it is often necessary to quickly tie off the splenic artery and remove the spleen. This procedure is called *splenectomy*.

The loss of splenic function, called *hyposplenism,* is usually not serious; its functions are adequately carried out by hepatic and bone marrow macrophages. However, it does leave a person somewhat more at risk of septicemia (bacteria in the blood) and pneumococcal infections. Therefore, if possible, surgeons try to leave some of the spleen in place; the spleen regenerates rapidly in such cases.

Some people have overactive spleens (*hypersplenism*), in which excessive phagocytosis of the formed elements of blood can lead to anemia, leukopenia, or thrombocytopenia (see chapter 19). This can be another reason for performing a splenectomy.

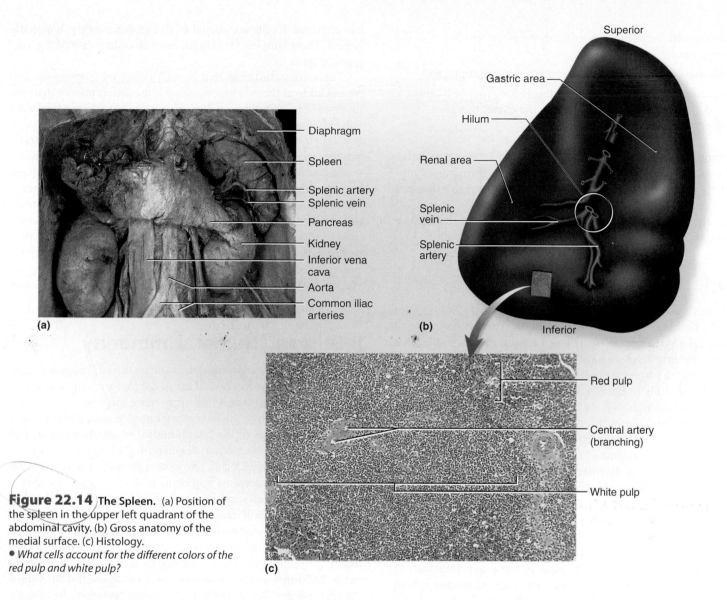

Figure 22.14 **The Spleen.** (a) Position of the spleen in the upper left quadrant of the abdominal cavity. (b) Gross anatomy of the medial surface. (c) Histology.
● *What cells account for the different colors of the red pulp and white pulp?*

Before You Go On

Answer the following questions to test your understanding of the preceding section:

4. What do T, B, and NK cells have in common? How do NK cells functionally differ from the other two? How do T and B cells functionally differ from each other?

5. What is the function of an antigen-presenting cell (APC)? Name three kinds of APCs.

6. What is a lymphatic nodule? Describe three places where lymphatic nodules can be found.

7. What are the two primary lymphatic organs? Why are they called that? Describe their collaborative relationship in producing the lymphocytes that populate other organs.

8. Describe the structural and functional differences between the cortex and medulla of a lymph node.

9. Name the three kinds of tonsils and state how they differ in number and location.

10. What are the two types of "pulp" in the spleen? What are their respective functions?

11. In what sense does the spleen serve the blood in the same way that the lymph nodes serve the lymph?

22.3 The Lymphatic System in Relation to Immunity

▶ Expected Learning Outcomes

When you have completed this section, you should be able to

- define *immune system* and explain its relationship to the lymphatic system;
- identify the body's three lines of defense against pathogens;

- distinguish between nonspecific defense and specific immunity;
- distinguish between humoral and cellular immunity; and
- describe the life histories and immune functions of B cells and T cells, and how these relate to the anatomy of the lymphatic organs.

The **immune system** is not an organ system, but rather a population of disease-fighting cells that reside in the mucous membranes, lymphatic organs, and other localities in the body. Although it does not have a specific anatomy distinct from what we have already studied in this chapter, a brief survey of immune function will enhance your understanding of the defensive role of the lymphatic system.

Modes of Defense

We have three lines of defense against pathogens: (1) a system of physical barriers to invasion, chiefly the skin and mucous membranes; (2) a system of nonspecific actions against pathogens that get past the first defense; and (3) the immune system, which not only defeats the pathogen but "remembers" it, enabling the body to defeat it so quickly in future encounters that we never notice any symptoms of disease.

The first two defenses lack the capacity to remember a particular pathogen or to react to it differently in the future. Furthermore, they defend equally against a broad range of pathogens; thus, they are called **nonspecific defenses.** In addition to the skin and mucous membranes, these nonspecific defenses include neutrophils, macrophages, natural killer (NK) cells, various antimicrobial proteins such as interferons, and such processes as inflammation, fever, and immune surveillance. The third defense confers a protection called **specific immunity**—*specific* because the body must develop a separate immunity to each pathogen. For example, immunity to one disease, such as chickenpox, does not confer immunity against another, such as measles. The ability to distinguish one pathogen from another is based on their **antigens**—complex molecules such as proteins and glycoproteins that genetically distinguish organisms, and even different members of the same species, from each other and that trigger the immune response. The agents that carry out specific immune responses are T and B lymphocytes.

Some cells play roles in both specific and nonspecific defense—notably macrophages and helper T cells. Macrophages are fairly undiscriminating in the microbes they attack, but they also act to present foreign antigens to activate the lymphocytes of specific immunity. Helper T (T_H) cells activate not only the B and T_C cells of specific immunity, but also help to mediate the nonspecific inflammatory response.

There are two forms of specific immunity called *humoral* and *cellular immunity.* **Humoral (antibody-mediated) immunity** is carried out by B lymphocytes and antibodies. It is called *humoral* because the antibodies circulate freely in the body fluids; *humor* is an archaic term for a body fluid. **Cellular (cell-mediated) immunity** is carried out by cytotoxic T cells. We will not delve into the details of these defenses, but we will take a look at how the activi-

ties of B and T cells are related to the anatomy of the lymphatic organs. These lymphocytes are the most abundant cells of the lymphatic organs.

There are some things that B and T cells have in common, and we can address these before examining the differences in their life histories. Both types begin their development as *pluripotent stem cells* (PPSCs) in the red bone marrow. PPSCs divide and give rise to *lymphocyte colony-forming units,* which ultimately produce B and T lymphocytes (see fig. 19.8, p. 531). Before they can take part in immune reactions, both types of lymphocytes must develop antigen receptors on their surfaces, giving them **immunocompetence**—the ability to recognize, bind, and respond to an antigen. In addition, the body must get rid of lymphocytes that react against its own (host) antigens so that the immune system will not attack one's own organs. The destruction or deactivation of self-reactive lymphocytes is called **negative selection.** Only about 2% of the lymphocytes survive this culling process. The developmental histories of B and T cells are contrasted in figure 22.15.

B Cells and Humoral Immunity

B cells achieve immunocompetence and go through negative selection in the red bone marrow. Many of the mature immunocompetent B cells remain there, while many more disperse and populate other sites such as the mucous membranes, spleen, and especially the cortical nodules of the lymph nodes, where they can sit and await the arrival of antigens in the incoming lymph.

When one of these B cells encounters an antigen, it internalizes and digests it, and presents fragments of the antigen to a helper T cell. The helper T cell secretes chemical *helper factors* that stimulate the B cell to divide still more. Most of its daughter cells differentiate into **plasma cells,** which are larger than B cells and have an abundance of rough endoplasmic reticulum (fig. 22.16)—as well they might, for plasma cells secrete antibodies at the astounding rate of up to 2,000 molecules per second for a life span of 4 or 5 days. Plasma cells develop mainly in the germinal centers of the nodules of the lymph nodes. About 10% of them remain there, while the rest emigrate from the lymph nodes and populate the bone marrow and other lymphatic organs and tissues. Their antibodies travel throughout the body in the blood and other fluids and react in various ways against antigens that they encounter.

Instead of becoming plasma cells, some B cells become memory cells. These live for months to years, and respond very quickly if they ever encounter the same antigen again. This provides long-lasting immunity to that pathogen.

T Cells and Cellular Immunity

T lymphocytes leave the bone marrow before reaching maturity. They migrate to the fetal thymus and colonize the cortex, where the blood–thymus barrier isolates them from premature exposure to blood-borne antigens. Reticular epithelial cells secrete thymic hormones that stimulate these T cells to develop antigen receptors, thus becoming immunocompetent. Following negative selection, the surviving T cells migrate into the medulla of the thymus, where

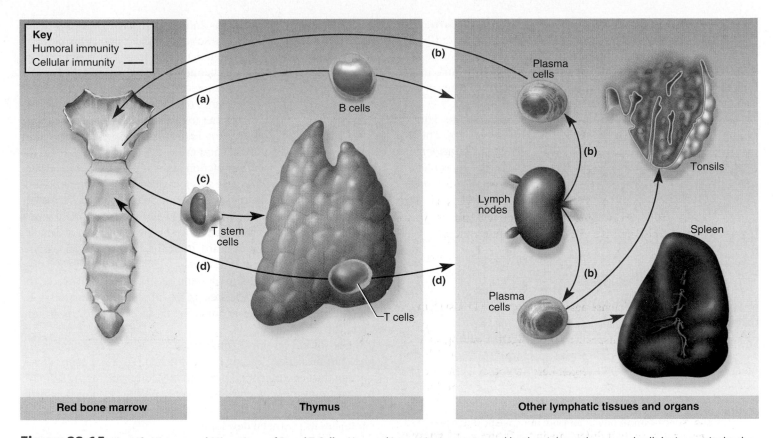

Figure 22.15 **The Life History and Migrations of B and T Cells.** Humoral immunity is represented by the violet pathways and cellular immunity by the red. (a) B cells achieve immunocompetence in the red bone marrow (left), and many emigrate to a variety of lymphatic tissues and organs, including the lymph nodes, tonsils, and spleen (right). (b) Plasma cells develop in the lymph nodes (among other sites) and emigrate to the bone marrow and other lymphatic organs, where they spend a few days secreting antibodies. (c) T stem cells emigrate from the bone marrow and attain immunocompetence in the thymus. (d) Immunocompetent T cells leave the thymus and recolonize the bone marrow or colonize various lymphatic organs (right).

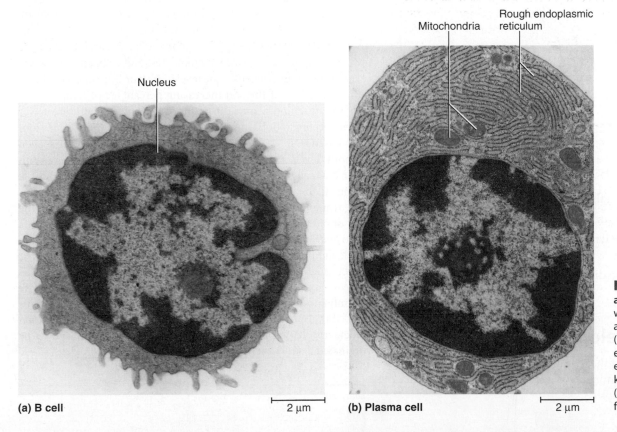

Figure 22.16 A B Cell and Plasma Cell. (a) A B cell, with a nucleus occupying almost the entire cell volume. (b) A plasma cell, showing the extreme proliferation of rough endoplasmic reticulum in keeping with its protein-(antibody-) synthesizing function.

they spend another 3 weeks. There is no blood–thymus barrier in the medulla, so T cells here can easily enter the blood and lymphatic vessels and disperse throughout the body. They colonize the same sites as B cells do, including recolonizing the red bone marrow. They become especially concentrated in the deep cortex of the lymph nodes.

When cytotoxic (T_C) cells encounter an enemy cell, they attack it directly and destroy it with a *lethal hit* of toxic chemicals. This is why immunity carried out by T cells is called *cellular (cell-mediated) immunity*. As in humoral immunity, some T cells remain as memory cells that live as long as a few decades and thus confer long-lasting protection.

Apply What You Know

Suppose a new virus emerged that selectively destroyed memory T and B cells. What would be the pathological effect of such a virus?

Before You Go On

Answer the following questions to test your understanding of the preceding section:

12. What are the three lines of defense against pathogens?

13. How does specific immunity differ from nonspecific defense?

14. What is the difference between humoral and cellular immunity?

15. Where do B cells acquire immunocompetence? Where do T cells do so?

16. What are the structural and functional differences between a B cell and a plasma cell?

22.4   Developmental and Clinical Perspectives

▶ Expected Learning Outcomes

When you have completed this section, you should be able to

- describe the embryonic origins of the lymphatic organs;
- describe changes in the lymphatic system that occur with old age; and
- describe some common disorders of the lymphatic–immune system.

Embryonic Development

Embryonic development of the thymus is described in chapter 18, page 514. Here, we will examine the development of the lymphatic vessels, lymph nodes, and spleen.

Lymphatic vessels begin as endothelium-lined channels in the mesoderm called **lymph sacs.** Some of these originate by budding from the blood vessels and then detaching from them; others originate as isolated mesodermal channels that fuse with each other and ultimately link up with the venous system. In a manner similar to blood vessel angiogenesis (see p. 601), lymph sacs proliferate, enlarge, and merge with each other to form larger and larger channels in the mesoderm (fig. 22.17a). Those with the greatest fluid flow later develop a tunica media and externa. The first to form are the *jugular lymph sacs* near the junction of the internal jugular and subclavian veins. By week 7, these sacs join the primitive veins and thus form the forerunners of the thoracic and right lymphatic ducts. The cisterna chyli arises from a *median lymph sac* that initially grows from the primitive vena cava and then breaks away from it. Smaller lymphatic vessels grow outward from the lymph sacs and follow blood vessels growing into the developing limbs.

Lymph nodes begin to develop as lymphocytes invade the lymph sacs and form cell clusters in the lumens. Blood vessels grow into these clusters, while a connective tissue capsule forms around them (fig. 22.17b).

The spleen develops from mesenchymal cells that invade the posterior mesentery leading to the stomach. Thus it remains enveloped in this mesentery and permanently connected to the stomach by a *gastrosplenic ligament*. The spleen is poorly developed at birth. Invasion of the splenic tissue by immunocompetent lymphocytes stimulates its postnatal development.

The Aging Lymphatic–Immune System

The effects of old age on the lymphatic system are seen not so much in anatomical changes as in declining immune function. There are several reasons for the reduced immune responsiveness. The quantities of red bone marrow and lymphatic tissue decline, so there are fewer hemopoietic stem cells, leukocytes, and antigen-presenting cells. As the thymus shrinks, the level of thymic hormones declines. Perhaps because of this, an increasing percentage of lymphocytes fail to mature and achieve immunocompetence. There are fewer helper T cells, so both humoral and cellular immunity suffer from their absence. T_C cells are less responsive to antigens, and even antibody levels rise more slowly in response to infection. With fewer NK cells, immune surveillance is weaker—one of multiple reasons that cancer becomes more common in old age. Paradoxically, while normal antibody responses are weaker in old age, the level of circulating *autoantibodies* rises. These are antibodies that fail to distinguish between host and foreign antigens, and therefore attack the body's own tissues, causing a variety of *autoimmune*[13] *diseases* such as rheumatoid arthritis.

With reduced immunity in old age, infectious diseases can be not only more common but also more serious. Epidemics of influenza (flu), for example, take a disproportionate toll of lives among the elderly. It becomes increasingly important in old age to be vaccinated against such acute seasonal diseases.

---

[13]*auto* = self

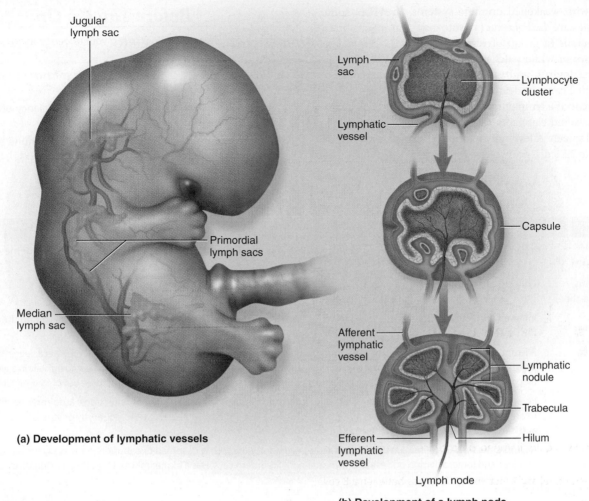

(a) Development of lymphatic vessels

Jugular lymph sac

Primordial lymph sacs

Median lymph sac

Lymph sac

Lymphocyte cluster

Lymphatic vessel

Capsule

Afferent lymphatic vessel

Lymphatic nodule

Trabecula

Efferent lymphatic vessel

Hilum

Lymph node

(b) Development of a lymph node

Figure 22.17 **Embryonic Development of the Lymphatic Vessels and Lymph Nodes.** (a) A 7-week embryo showing the right jugular lymph sac, which connects to the future subclavian vein; the primordial lymph sacs, which will merge to form the thoracic duct; and the median lymph sac, which will become the cisterna chyli. (b) Stages in the development of a lymph node. Top: Lymphocytes aggregate in a lymph sac, and blood vessels grow into the cluster. Middle: A fibrous capsule then forms around the sac as blood vessels proliferate. Bottom: Ingrowths of the capsule form trabeculae that partially subdivide the interior as the lymph node takes shape.

Lymphatic–Immune Disorders

It is a delicate balancing act for the body to discriminate between foreign and host antigens, ward off foreign pathogens, and mount immune responses that are not too weak, not too strong, and not misdirected. It comes as no surprise, therefore, that many things can go wrong. Most immune disorders can be classified into three categories: autoimmune diseases, hypersensitivity, and immunodeficiency.

Autoimmune diseases, as already mentioned, are diseases resulting from an immune attack misdirected against one's own tissues. Insulin-dependent diabetes mellitus, rheumatic fever, rheumatoid arthritis, and systemic lupus erythematosus are some examples.

Hypersensitivity is an exaggerated, harmful immune response to antigens. The most prevalent examples are *allergies*—excessive reactions to environmental antigens *(allergens)* that most people tolerate. Allergens are found in a broad range of substances such as bee and wasp venoms; toxins from poison ivy and other plants; mold; dust; pollen; animal dander; foods such as nuts, milk, eggs, and shellfish; cosmetics; latex; vaccines; and drugs such as penicil-

lin, tetracycline, and insulin. In many cases, an allergen stimulates basophils and mast cells to release histamine and other chemicals that cause a broad range of symptoms: edema, congestion, watery eyes, runny nose, hives, cramps, diarrhea, vomiting, and sometimes catastrophic circulatory failure *(anaphylactic shock).*

Immunodeficiency diseases are failures of the immune system to respond strongly enough to ward off disease. One of these is a congenital (inborn) condition—*severe combined immunodeficiency disease (SCID),* in which an infant is born without a functional immune system and must live in a sterile enclosure to avoid fatal infections. The most notorious immunodeficiency disease, of course, is AIDS *(acquired immunodeficiency syndrome).* Unlike SCID, this is not inborn but results from an infection with the human immunodeficiency virus (HIV), usually acquired by sexual intercourse or use of contaminated needles for drug injection. HIV targets especially the helper T (CD4) cells. When the T_H count drops from its normal level of 600 to 1,200 cells/µL of blood to less than 200 cells/µL, a person has "full-blown AIDS" and is highly susceptible to *opportunistic infections*—infections that become established and produce disease only in people with weakened immune systems. In AIDS, some common examples are *Toxoplasma* (a protozoan that infects brain tissue), *Pneumocystis* (a group of respiratory fungi), *Candida* (a fungus that grows in white patches on the oral mucosa), herpes simplex, cytomegalovirus, and tuberculosis. Opportunistic infection is the principal cause of death in AIDS.

More specific to the lymphatic system, this chapter has already discussed filariasis and elephantiasis, lymph node cancer, tonsillitis, and ruptured spleen. A few more lymphatic disorders are briefly described in table 22.1.

Before You Go On

Answer the following questions to test your understanding of the preceding section:

17. How do lymph sacs form in the embryo? How do lymph nodes form?

18. Describe some reasons for the declining efficiency of the immune system in old age.

19. What are the three principal categories of immune system disorders? Give an example of each.

TABLE 22.1	Some Disorders of the Lymphatic System
Lymphadenitis[14] (lim-FAD-en-EYE-tis)	Inflammation of a lymph node in response to challenge from a foreign antigen; marked by swelling and tenderness.
Lymphadenopathy[15] (lim-FAD-en-OP-a-thee)	Collective term for all diseases of the lymph nodes.
Lymphangitis[16] (LIM-fan-JY-tis)	Inflammation of a lymphatic vessel, as in filariasis (see Deeper Insight 22.1); marked by redness and pain along the course of the vessel.
Hodgkin[17] **disease**	A lymph node malignancy, with early symptoms including enlarged painful lymph nodes, especially in the neck; fever; anorexia; weight loss; night sweats; and severe itching. Diagnosis is confirmed by finding characteristic *Reed–Sternberg* cells in a lymph node biopsy. Often progresses to neighboring lymph nodes. Radiation and chemotherapy cure about three out of four patients.
Splenomegaly[18]	Enlargement of the spleen, sometimes without underlying disease but often indicating infections, autoimmune diseases, heart failure, cirrhosis, Hodgkin disease, and other cancers. The enlarged spleen may "hoard" erythrocytes, causing anemia, and it may become fragile and subject to rupture.
Non-Hodgkin lymphoma	A lymphoma similar to Hodgkin disease, but more common, with more widespread distribution in the body (including axillary, inguinal, and femoral lymph nodes), and without Reed–Sternberg cells. Has a higher mortality rate than Hodgkin disease.

Disorders Described Elsewhere

AIDS   627
Allergy   627
Autoimmune diseases   627
Cancer of lymph nodes   620
Elephantiasis   612

Filariasis   612
Opportunistic infection   627
Ruptured spleen   622
Severe combined immunodeficiency disease   627
Tonsillitis   620

---

[14]*adeno* = gland + *itis* = inflammation
[15]*adeno* = gland + *pathy* = disease
[16]*ang* = vessel + *itis* = inflammation
[17]Thomas Hodgkin (1798–1866), British physician
[18]*megaly* = enlargement

Study Guide

Assess Your Learning Outcomes

You should have a good understanding of this chapter if you can accurately address the following issues.

22.1 Lymph and Lymphatic Vessels (p. 610)

1. The anatomical components of the lymphatic system
2. Three functions of the lymphatic system
3. The appearance and composition of the lymph
4. Structure of the lymphatic capillaries and how their structure enables them to pick up large particles from the tissue fluid
5. Histology of the lymphatic collecting vessels and their anatomical and functional relationships with the lymph nodes
6. The larger lymphatic vessels including the six lymphatic trunks, the two collecting ducts, and the points where the lymph is returned to the bloodstream
7. The mechanisms for making the lymph flow through the lymphatic vessels and for preventing it from going backward

22.2 Lymphatic Cells, Tissues, and Organs (p. 614)

1. The six principal defensive cell types in the lymphatic system
2. Immune surveillance and the function of NK cells
3. The four kinds of T lymphocytes, their functions, and what the *T* in their names stands for
4. The function of B lymphocytes, their relationship to plasma cells, and what the *B* stands for

5. The functions and types of macrophages, and their relation to blood monocytes
6. The locations and functions of dendritic cells and reticular cells
7. The histology, locations, and types of diffuse lymphatic tissue
8. The feature that distinguishes lymphatic organs from lymphatic tissues
9. The distinction between primary and secondary lymphatic organs, and the members of each category
10. The structure and function of red bone marrow and the reason it can be considered an organ and not merely a tissue
11. The location, gross anatomy, histology, and functions of the thymus, and its age-related changes
12. The gross anatomy, histology, and functions of a lymph node
13. The locations of lymph nodes and the names of the body's major lymph node aggregations
14. The names, locations, and functions of the tonsils, and the structural differences between the three types of tonsils
15. The location, gross anatomy, histology, and functions of the spleen

22.3 The Lymphatic System in Relation to Immunity (p. 623)

1. The definition of *immune system* and why this is not considered an organ system
2. The body's three lines of defense against pathogens and the distinction between nonspecific defense and specific immunity

3. The nature of antigens and their role in immune function
4. The multiple mechanisms of nonspecific defense
5. A definition of *specific immunity,* the two forms of specific immunity, and the agents that carry out each of these forms
6. The origin of B and T cells and the essential processes that both of these must go through before they are capable of participating in an immune response
7. Where the B cells become immunocompetent and where they subsequently reside
8. How B cells respond when they encounter a foreign antigen, and how they provide an immune memory in humoral immunity
9. Where the T cells become immunocompetent and where they subsequently reside
10. How T cells respond when they encounter an enemy cell, and how they provide an immune memory in cellular immunity

22.4 Developmental and Clinical Perspectives (p. 626)

1. How lymphatic vessels, lymph nodes, and the spleen develop in the human embryo
2. How the lymphatic tissues and immune responses change in old age, and why elderly people are more susceptible to infectious diseases, cancer, and autoimmune diseases
3. The three principal classes of immune system disorders
4. The cause of AIDS and the essential cellular mechanism of how it produces the signs and symptoms of this disease

Testing Your Recall

1. The only lymphatic organ with both afferent and efferent lymphatic vessels is
   a. the spleen.
   b. a lymph node.
   c. a tonsil.
   d. a Peyer patch.
   e. the thymus.

2. Which of the following cells are involved in nonspecific defense but not in specific immunity?
   a. helper T cells
   b. cytotoxic T cells
   c. natural killer cells
   d. B cells
   e. plasma cells

3. The lethal hit is used by _____ to kill enemy cells.
   a. neutrophils
   b. basophils
   c. mast cells
   d. NK cells
   e. cytotoxic T cells

4. Which of these is a macrophage?
   a. microglia
   b. a plasma cell
   c. a reticular cell
   d. a helper T cell
   e. a mast cell

5. Which of these lymphatic organs has a cortex and medulla: (I) spleen; (II) lymph node; (III) thymus; (IV) red bone marrow?
   a. II only
   b. III only
   c. II and III only
   d. III and IV only
   e. I, II, and III

6. What cells form the blood–thymus barrier?
   a. astrocytes
   b. Hassall corpuscles
   c. T cells
   d. dendritic cells
   e. reticular epithelial cells

7. Where do B cells attain immunocompetence?
   a. in the red bone marrow
   b. in the germinal centers of the lymph nodes
   c. in the thymic cortex
   d. in the thymic medulla
   e. in the splenic white pulp

8. If it were not for the process of negative selection, we would expect to see more
   a. allergies.
   b. lymphatic nodules in the MALT.
   c. antigen-presenting cells.
   d. immunodeficiency diseases.
   e. autoimmune diseases.

9. Lymph nodes tend to be especially concentrated in all of these sites *except*
   a. the cervical region.
   b. the popliteal region.
   c. the carpal region.
   d. the inguinal region.
   e. the mesenteries.

10. All lymph ultimately reenters the bloodstream at what point?
    a. the right atrium
    b. the common carotid arteries
    c. the internal iliac veins
    d. the subclavian veins
    e. the inferior vena cava

11. Any organism or substance capable of causing disease is called a/an _____.

12. _____ is milky lymph, rich in fat, absorbed from the small intestine.

13. Lymphatic vessels called _____ carry lymph from one lymph node to the next.

14. The two lymphatics that empty into the subclavian veins are the _____ on the right and the _____ on the left.

15. The latter duct in question 14 begins with a sac called the _____ below the diaphragm.

16. B cells become _____ cells before they begin to secrete antibodies.

17. Any cells that process antigens and display fragments of them to activate immune reactions are called _____.

18. The _____ is a lymphatic organ composed mainly of hemopoietic islands and sinusoids.

19. The ovoid masses of lymphocytes that line the tonsillar crypts are called _____.

20. Any disease in which antibodies attack one's own tissues is called a/an _____ disease.

Answers in the Appendix

Building Your Medical Vocabulary

State a medical meaning of each of the following word elements, and give a term in which it is used.

1. -gen
2. -iasis
3. adeno-
4. -ectomy
5. -pathy
6. lympho-
7. immuno-
8. -megaly
9. -oma
10. chylo-

Answers in the Appendix

True or False

Determine which five of the following statements are false, and briefly explain why.

1. B cells play roles in both nonspecific defense and specific immunity.

2. T lymphocytes undergo negative selection in the thymus.

3. Lymphatic capillaries are more permeable than blood capillaries.

4. T lymphocytes are involved only in cellular immunity.

5. The white pulp of the spleen gets its color mainly from lymphocytes and macrophages.

6. Obstruction of a major lymphatic vessel is likely to cause edema.

7. Lymph nodes are populated by B cells but not T cells.

8. Lymphatic nodules are permanent structures enclosed in fibrous capsules.

9. Tonsillectomy is regarded as the current treatment of choice for most cases of tonsillitis.

10. Most plasma cells form in the germinal centers of the lymph nodes.

Answers in the Appendix

Testing Your Comprehension

1. About 10% of people have one or more *accessory spleens,* typically about 1 cm in diameter and located near the hilum of the main spleen or embedded in the tail of the pancreas. If a surgeon is performing a splenectomy as a treatment for hypersplenism (see Deeper Insight 22.3), why would it be important to search for and remove any accessory spleens? What might be the consequences of overlooking one of these?

2. In treating a woman for malignancy in the right breast, the surgeon removes some of her axillary lymph nodes. Following surgery, the patient experiences edema of her right arm. Explain why.

3. Explain why a detailed knowledge of the pathways of lymphatic drainage is important to the clinical management of cancer.

4. A burn research center uses mice for studies of skin grafting. To prevent graft rejection, the mice are thymectomized at birth. Even though B cells do not develop in the thymus, these mice show no humoral immune response and are very susceptible to infection. Explain why the removal of the thymus would improve the success of skin grafts but adversely affect humoral immunity.

5. Contrast the structure of a B cell with that of a plasma cell, and explain how their structural difference relates to their functional difference.

Answers at www.mhhe.com/saladinha3

Improve Your Grade at www.mhhe.com/saladinha3

Practice quizzes, labeling activities, games, and flashcards provide fun ways to master concepts. You can also download image PowerPoint files for each chapter to create a study guide or for taking notes during lecture.

The Respiratory System

CHAPTER

23

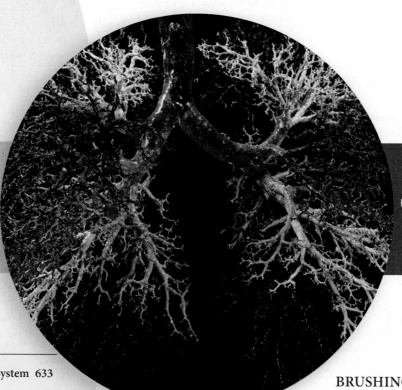

The bronchial trees, with each bronchopulmonary segment shown in a different color (polymer cast)

BRUSHING UP

To understand this chapter, you may find it helpful to review the following concepts:
- Serous and mucous membranes (p. 75)
- The ethmoid bone, maxilla, nasal bones, and vomer (pp. 161–164)
- The muscles of respiration (p. 279)
- Basic brainstem anatomy (pp. 406–410)
- Divisions of the autonomic nervous system (p. 444)
- Pulmonary blood circulation (p. 572)

Anatomy & Physiology REVEALED®
aprevealed.com

Respiratory System

Breath represents life. The first breath of a baby and the last gasp of a dying person are two of the most dramatic moments of human experience. The need to breathe is driven by cellular demands for energy provided by ATP; most ATP synthesis requires oxygen and generates carbon dioxide. The respiratory system consists essentially of tubes that deliver air to the lungs, where oxygen diffuses into the blood and carbon dioxide is removed from it.

The respiratory and cardiovascular systems have such a close functional and spatial relationship that a disorder of the lungs often has direct effects on the heart, and vice versa. The two systems are often considered jointly as the *cardiopulmonary system.*

23.1 Overview of the Respiratory System

▶ Expected Learning Outcomes

When you have completed this section, you should be able to

- state the functions of the respiratory system;
- name the principal organs of this system;
- distinguish between the conducting and respiratory divisions; and
- distinguish between the upper and lower respiratory tracts.

The **respiratory system** is an organ system specialized to provide oxygen to the blood and remove carbon dioxide from it. It has more diverse functions than are commonly supposed:

1. It provides for oxygen and carbon dioxide exchange between the blood and air.
2. It serves for speech and other vocalizations (laughing, crying).
3. It provides the sense of smell, which is important in social interactions, food selection, and avoiding danger (such as a gas leak or spoiled food).
4. By eliminating CO_2, it helps to control the pH of the body fluids. Excess CO_2 reacts with water and releases hydrogen ions:

$$CO_2 + H_2O \longrightarrow H_2CO_3 \longrightarrow HCO_3^- + H^+$$

Therefore, if the respiratory system does not keep pace with the rate of CO_2 production, H^+ accumulates and the body fluids have an abnormally low pH *(acidosis).*
5. The lungs carry out a step in the synthesis of a vasoconstrictor called *angiotensin II,* which helps to regulate blood pressure.
6. Breathing creates pressure gradients between the thorax and abdomen that promote the flow of lymph and venous blood.

7. Taking a deep breath and holding it while contracting the abdominal muscles (the *Valsalva*[1] *maneuver*) helps to expel abdominal contents during urination, defecation, and childbirth.

The principal organs of the respiratory system are the nose, pharynx, larynx, trachea, bronchi, and lungs (fig. 23.1). Within the lungs, air flows along a dead-end pathway consisting essentially of bronchi ⟶ bronchioles ⟶ alveoli (with some details to be introduced later). During **inspiration**[2] (inhaling), incoming air stops in the *alveoli* (millions of thin-walled, microscopic air sacs) and exchanges gases with the bloodstream across the alveolar wall; it flows back out during **expiration** (exhaling).

The **conducting division** of the respiratory system consists of those passages that serve only for airflow, essentially from the nostrils through the bronchioles. The **respiratory division** consists of the alveoli and other distal gas-exchange regions. The airway from the nose through the larynx is often called the **upper respiratory tract** (that is, the respiratory organs in the head and neck), and the regions from the trachea through the lungs compose

---

[1]Antonio Maria Valsalva (1666–1723), Italian anatomist
[2]*spir* = to breathe

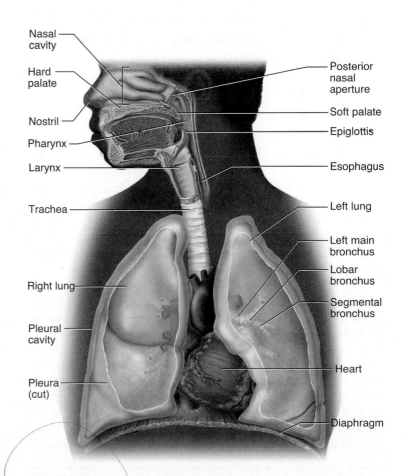

Figure 23.1 The Respiratory System.

the **lower respiratory tract** (the respiratory organs of the thorax). However, these are inexact terms and various authorities place the dividing line between the upper and lower tracts at different points.

Before You Go On

Answer the following questions to test your understanding of the preceding section:

1. What are some functions of the respiratory system other than supplying O_2 to the body and removing CO_2?

2. Which portions of the respiratory tract belong to the conducting division? What portions belong to the respiratory division? How do the two divisions differ functionally?

3. What is the distinction between the upper and lower respiratory tracts?

23.2 The Upper Respiratory Tract

▶ Expected Learning Outcomes

When you have completed this section, you should be able to

- trace the flow of air from the nose through the larynx;

- describe the anatomy of these passages;

- relate the anatomy of any portion of the upper respiratory tract to its function; and

- describe the action of the vocal cords in speech.

The Nose

The **nose** has several functions: it warms, cleanses, and humidifies inhaled air; it detects odors in the airstream; and it serves as a resonating chamber that amplifies the voice. It extends from a pair of anterior openings called the **nostrils,** or **nares** (NAIR-eze) (singular, *naris*), to a pair of posterior openings called the **posterior nasal apertures** or **choanae**[3] (co-AH-nee) (fig. 23.2b).

The facial part of the nose is shaped by bone and hyaline cartilage. Its superior half is supported by a pair of small nasal bones medially and the maxillae laterally. The inferior half is supported by the **lateral** and **alar cartilages** (fig. 23.3). By palpating your own nose, you can easily find the boundary between the bone and cartilage. The flared portion at the lower end of the nose, called the **ala nasi**[4] (AIL-ah NAZE-eye), is shaped by the alar cartilages and dense connective tissue.

The **nasal cavity** begins with a small dilated chamber called the **vestibule** just inside each nostril, bordered by the ala nasi. This space is lined with stratified squamous epithelium like the facial

skin and has stiff **guard hairs,** or **vibrissae** (vy-BRISS-ee), that block insects and large airborne particles from the nose. The nasal cavity is divided into right and left halves called **nasal fossae** (FAW-see) by a wall of bone and hyaline cartilage, the **nasal septum.** The septum has three components: the bony *vomer* forming the inferior part, the perpendicular plate of the *ethmoid bone* forming the superior part, and a hyaline *septal nasal cartilage* forming the anterior part. The roof of the nasal cavity is formed by the ethmoid and sphenoid bones, and the hard palate forms its floor. The palate separates the nasal cavity from the oral cavity and allows you to breathe while chewing food (see Deeper Insight 7.2, p. 164). The nasal cavity receives drainage from the paranasal sinuses (see p. 153) and the nasolacrimal ducts of the orbits (see p. 480).

There is not much space in the nasal cavity. Most of it is occupied by three bony scrolls covered by mucous membrane—the **superior, middle,** and **inferior nasal conchae**[5] (CON-kee), or **turbinates**—that project from the lateral walls toward the septum (see fig. 23.2). Beneath each concha is a narrow air passage called a **meatus**[6] (me-AY-tus). The narrowness of these passages and the turbulence produced by the conchae ensure that most air contacts the mucous membrane on its way through. As it does, most dust in the air sticks to the mucus, and the air picks up moisture and heat from the mucosa. The conchae thus enable the nose to cleanse, warm, and humidify the air more effectively than if the air had an unobstructed flow through a cavernous space.

Odors are detected by sensory cells in the **olfactory epithelium,** which covers a small area of the roof of the nasal fossa and adjacent parts of the septum and superior concha (see fig. 17.7, p. 469). The rest of the nasal cavity, except for the vestibule, is lined with **respiratory epithelium.** Both of these are ciliated pseudostratified columnar epithelia. However, in the olfactory epithelium, the cilia are immobile and serve to bind odor molecules. In the respiratory epithelium, they are mobile. The respiratory epithelium is similar to the one seen in figure 3.7 (p. 58). Its wineglass-shaped **goblet cells** secrete mucus, and its **ciliated cells** propel the mucus posteriorly toward the pharynx. The nasal mucosa also contains mucous glands in the lamina propria (the connective tissue layer beneath the epithelium). They supplement the mucus produced by the goblet cells. Inhaled dust, pollen, bacteria, and other foreign matter stick to the mucus and are swallowed; they are either digested or pass through the digestive tract rather than contaminating the lungs. The lamina propria is also well populated by lymphocytes that mount immune defenses against inhaled pathogens, and by plasma cells that secrete antibodies into the tissue fluid.

The lamina propria contains large blood vessels that help to warm the air. The inferior concha has an especially extensive venous plexus called the **erectile tissue (swell body).** Every 30 to 60 minutes, the erectile tissue on one side becomes engorged with blood and restricts airflow through that fossa. Most air is then directed through the other nostril and fossa, allowing the engorged side time to recover from drying. Thus, the preponderant flow of air shifts between the right and left nostrils once or twice each hour.

---

[3]*choana* = funnel
[4]*ala* = wing + *nasi* = of the nose

[5]*concha* = seashell
[6]*meatus* = passage

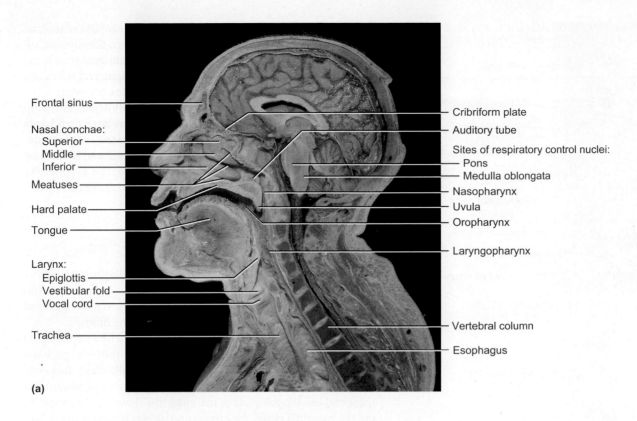

(a)

Frontal sinus

Nasal conchae:
　Superior
　Middle
　Inferior

Meatuses

Hard palate

Tongue

Larynx:
　Epiglottis
　Vestibular fold
　Vocal cord

Trachea

Cribriform plate

Auditory tube

Sites of respiratory control nuclei:
　Pons
　Medulla oblongata

Nasopharynx

Uvula

Oropharynx

Laryngopharynx

Vertebral column

Esophagus

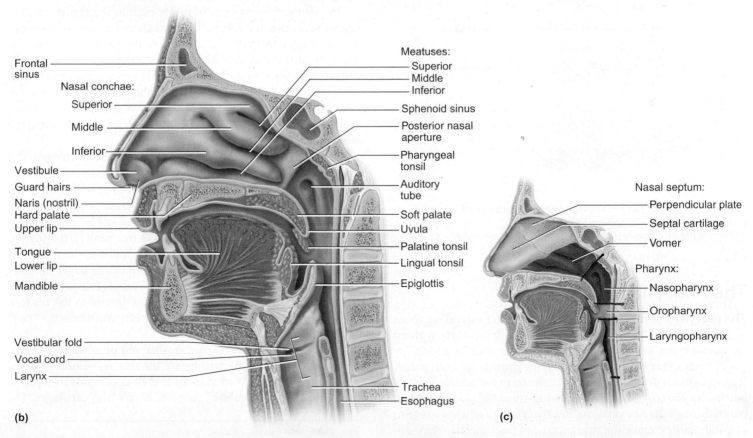

(b)

Frontal sinus

Nasal conchae:
　Superior
　Middle
　Inferior

Vestibule

Guard hairs

Naris (nostril)

Hard palate

Upper lip

Tongue

Lower lip

Mandible

Vestibular fold

Vocal cord

Larynx

Meatuses:
　Superior
　Middle
　Inferior

Sphenoid sinus

Posterior nasal aperture

Pharyngeal tonsil

Auditory tube

Soft palate

Uvula

Palatine tonsil

Lingual tonsil

Epiglottis

Trachea

Esophagus

(c)

Nasal septum:
　Perpendicular plate
　Septal cartilage
　Vomer

Pharynx:
　Nasopharynx
　Oropharynx
　Laryngopharynx

Figure 23.2 **Anatomy of the Upper Respiratory Tract.** (a) Median section of the head. (b) Internal anatomy. (c) Regions of the pharynx.
● *Draw a line across part (b) of this figure to indicate the boundary between the upper and lower respiratory tracts.*

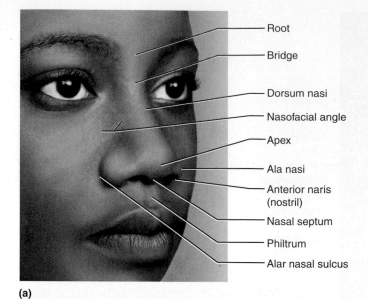

Root
Bridge
Dorsum nasi
Nasofacial angle
Apex
Ala nasi
Anterior naris (nostril)
Nasal septum
Philtrum
Alar nasal sulcus

(a)

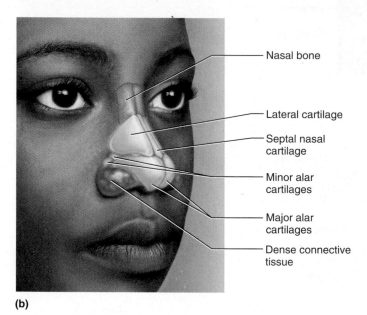

Nasal bone
Lateral cartilage
Septal nasal cartilage
Minor alar cartilages
Major alar cartilages
Dense connective tissue

(b)

Figure 23.3 **Anatomy of the Nasal Region.** (a) External anatomy. (b) Connective tissues that shape the nose.
● *Which of the cartilages in part (b) extends most deeply into the face?*

The Pharynx

The **pharynx** (FAIR-inks) is a muscular funnel extending about 13 cm (5 in.) from the choanae to the larynx. It has three regions: the *nasopharynx, oropharynx,* and *laryngopharynx* (fig. 23.2c).

The **nasopharynx** lies posterior to the choanae and soft palate. It receives the auditory (eustachian) tubes from the middle ears and houses the pharyngeal tonsil. Inhaled air turns 90° downward as it passes through the nasopharynx. Relatively large particles (>10 μm) generally cannot make the turn because of their inertia. They collide with the posterior wall of the nasopharynx and stick to the mucosa near the tonsil, which is well positioned to respond to airborne pathogens.

The **oropharynx** is a space posterior to the root of the tongue. It extends from the inferior tip of the soft palate to the superior tip of the epiglottis, and contains the palatine and lingual tonsils. Its anterior border is formed by the base of the tongue and the *fauces* (FAW-seez), the opening of the oral cavity into the pharynx.

The **laryngopharynx** (la-RIN-go-FAIR-inks) begins at the tip of the epiglottis, passes downward posterior to the larynx, and ends where the esophagus begins at the level of the *cricoid cartilage* (described shortly). The nasopharynx passes only air and is lined by pseudostratified columnar epithelium, whereas the oropharynx and laryngopharynx pass air, food, and drink and are lined by stratified squamous epithelium.

The Larynx

The **larynx** (LAIR-inks) ("voice box") is a cartilaginous chamber about 4 cm long (fig. 23.4). Its primary function is to keep food and drink out of the airway, but it has evolved the additional role of sound production (*phonation*) in many animals, including humans.

The superior opening of the larynx is guarded by a flap of tissue called the **epiglottis**[7] (figs. 23.4c and 23.5a). At rest, the epiglottis usually stands almost vertically. During swallowing, however, *extrinsic muscles* of the larynx pull the larynx upward toward the epiglottis, the tongue pushes the epiglottis downward to meet it, and the epiglottis closes the airway and directs food and drink into the esophagus behind it.

In infants, the larynx is relatively high in the throat, and the epiglottis touches the soft palate. This creates a more or less continuous airway from the nasal cavity to the larynx and allows an infant to breathe continually while swallowing. The epiglottis deflects milk away from the airstream, like rain running off a tent while it remains dry inside. By age 2, the root of the tongue becomes more muscular and forces the larynx to descend to a lower position. It then becomes impossible to breathe and swallow at the same time without choking.

The framework of the larynx consists of nine cartilages. The first three are solitary and relatively large. The most superior one, the **epiglottic cartilage,** is a spoon-shaped supportive plate in the epiglottis. The largest, the **thyroid**[8] **cartilage,** is named for its shieldlike shape. It broadly covers the anterior and lateral aspects of the larynx. The "Adam's apple" is an anterior peak of the thyroid cartilage called the *laryngeal prominence.* Testosterone stimulates the growth of this prominence, which is therefore larger in males than in females. Inferior to the thyroid cartilage is a ringlike **cricoid**[9] (CRY-coyd) **cartilage,** which connects the larynx to the trachea. The thyroid and cricoid cartilages essentially constitute the "box" of the voice box.

The remaining cartilages are smaller and occur in three pairs. Posterior to the thyroid cartilage are the two **arytenoid**[10] (AR-ih-TEE-noyd) **cartilages,** and attached to their upper ends are a pair of little horns, the **corniculate**[11] (cor-NICK-you-late) **cartilages.** The

---

[7]*epi* = above, upon + *glottis* = back of the tongue
[8]*thyr* = shield + *oid* = resembling
[9]*crico* = ring + *oid* = resembling
[10]*aryten* = ladle + *oid* = resembling
[11]*corni* = horn + *cul* = little + *ate* = possessing

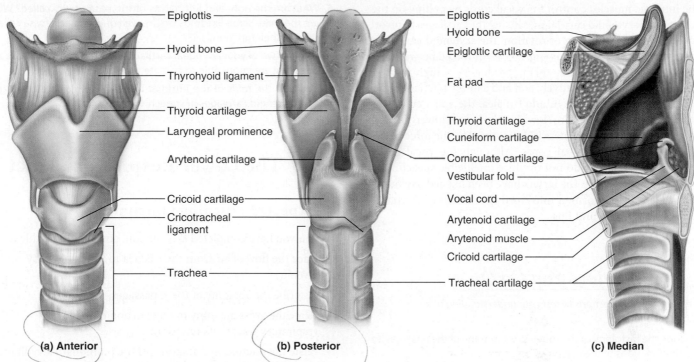

(a) Anterior

(b) Posterior

(c) Median

Figure 23.4 **Anatomy of the Larynx.** Most muscles are removed in order to show the cartilages.
● *What three cartilages in this figure are more mobile than any other?*

arytenoid and corniculate cartilages function in speech, as explained shortly. A pair of **cuneiform**[12] (cue-NEE-ih-form) **cartilages** support the soft tissues between the arytenoids and the epiglottis.

A group of fibrous ligaments bind the cartilages of the larynx together and to adjacent structures in the neck. Superiorly, a broad sheet called the **thyrohyoid ligament** joins the thyroid cartilage to the hyoid bone, and inferiorly, the **cricotracheal ligament** joins the cricoid cartilage to the trachea. These are collectively called the *extrinsic ligaments* because they link the larynx to other organs. The *intrinsic ligaments* are contained entirely within the larynx and link its nine cartilages to each other.

The walls of the larynx are also quite muscular. The deep *intrinsic muscles* operate the vocal cords, and the superficial *extrinsic muscles* connect the larynx to the hyoid bone and elevate the larynx during swallowing. The extrinsic muscles, also called the *infrahyoid group,* are named and described in table 11.3 (p. 275).

Two pairs of intrinsic ligaments, the **vestibular** and **vocal ligaments,** extend from the thyroid cartilage in front to the arytenoid cartilages in back, and support the vestibular folds and vocal cords, respectively. The vocal cords and the opening between them are collectively called the **glottis** (fig. 23.5a). The superior **vestibular folds** play no role in speech but close the glottis during swallowing. The inferior **vocal cords (vocal folds)** produce sound when air passes between them. They are covered with stratified squamous epithelium, best suited to endure vibration and contact between the cords.

---

[12]*cune* = wedge + *form* = shape

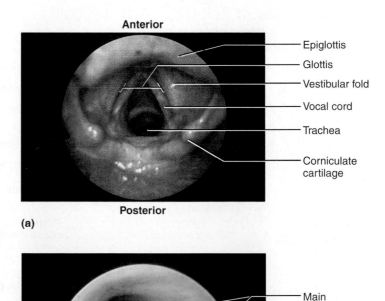

Figure 23.5 **Endoscopic Views of the Respiratory Tract.** (a) Superior view of the larynx, seen with a laryngoscope. (b) Lower end of the trachea, where it forks into the two primary bronchi, seen with a bronchoscope.

The intrinsic muscles control the vocal cords by pulling on the corniculate and arytenoid cartilages, causing the cartilages to pivot. Depending on their direction of rotation, the arytenoid cartilages abduct or adduct the vocal cords (fig. 23.6). Air forced between the adducted vocal cords vibrates them, producing a high-pitched sound when the cords are relatively taut and a lower-pitched sound when they are more relaxed. In adult males, the vocal cords are longer and thicker, vibrate more slowly, and produce lower-pitched sounds than in females. Loudness is determined by the force of the air passing between the vocal cords. Although the vocal cords produce most sound, they do not produce intelligible speech. The crude sounds coming from the larynx have been likened to a hunter's duck call. They are formed into words by actions of the pharynx, oral cavity, tongue, and lips.

Before You Go On

Answer the following questions to test your understanding of the preceding section:

4. Describe the histology of the mucous membrane of the nasal cavity and the functions of the cell types present.

5. Name the anterior and posterior openings that mark the beginning and end of the nasal cavity.

6. What are the right and left halves of the nasal cavity called? What are the three scroll-like folds on the wall of each nasal fossa called? What is their function?

7. Palpate two of your laryngeal cartilages and name them. Name the ones that are impossible to palpate on a living person.

8. Describe the roles of the intrinsic muscles, corniculate cartilages, and arytenoid cartilages in speech.

23.3 The Lower Respiratory Tract

▶ Expected Learning Outcomes

When you have completed this section, you should be able to

- trace the flow of air from the trachea to the pulmonary alveoli;

- describe the anatomy of these passages;

- relate the gross anatomy of any portion of the lower respiratory tract to its function;

- relate the microscopic anatomy of the pulmonary alveoli to their role in gas exchange; and

- describe the relationship of the pleurae to the lungs.

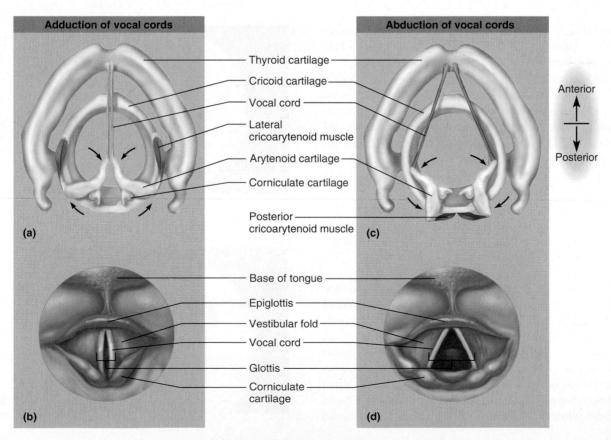

Figure 23.6 **Action of Some of the Intrinsic Laryngeal Muscles on the Vocal Cords.** (a) Adduction of the vocal cords by the lateral cricoarytenoid muscles. (b) Adducted vocal cords seen with the laryngoscope. (c) Abduction of the vocal cords by the posterior cricoarytenoid muscles. (d) Abducted vocal cords seen with the laryngoscope.

If you palpate your larynx, you will find the laryngeal prominence of the thyroid cartilage only slightly above your sternum. All the rest of the respiratory tract is in the thorax rather than the head and neck, and is thus called the lower respiratory tract. This portion extends from the trachea to the pulmonary alveoli.

The Trachea and Bronchi

The **trachea** (TRAY-kee-uh), or "windpipe," is a tube about 12 cm (4.5 in.) long and 2.5 cm (1 in.) in diameter, lying anterior to the esophagus (fig. 23.7a). It is supported by 16 to 20 **C**-shaped rings of hyaline cartilage, some of which you can palpate between your larynx and sternum. The inner lining of the trachea is a pseudostratified columnar epithelium composed mainly of mucus-secreting goblet cells, ciliated cells, and short basal stem cells (figs. 23.7b and 23.8). The mucus traps inhaled particles, and the upward beating of the cilia drives the debris-laden mucus toward the pharynx, where it is swallowed. This mechanism of debris removal is called the **mucociliary escalator.**

The connective tissue beneath the tracheal epithelium contains lymphatic nodules, mucous and serous glands, and the tracheal cartilages. Like the wire spiral in a vacuum cleaner hose, the cartilage rings reinforce the trachea and keep it from collapsing when you inhale. The word *trachea*[13] refers to the rough, corrugated surface texture imparted by these cartilage rings. The open part of the **C** faces posteriorly and allows room for the esophagus to expand as swallowed food passes by. The gap is spanned by smooth muscle tissue called the **trachealis** (fig. 23.7c). Contraction or relaxation of this muscle narrows or widens the trachea to adjust airflow for conditions of rest or exercise. The outermost layer of the trachea, called the **adventitia,** is fibrous connective tissue that blends into the adventitia of other organs of the mediastinum.

At the level of the sternal angle and the superior margin of vertebra T5, the trachea forks into right and left bronchi. The lowermost tracheal cartilage has an internal median ridge called the **carina**[14] (ca-RY-na) that directs the airflow to the right and left (see fig. 23.5b). The bronchi are further traced in the discussion of the *bronchial tree* of the lungs.

The Lungs

Each **lung** (fig. 23.9) is a somewhat conical organ with a broad, concave **base** resting on the diaphragm and a blunt peak called the **apex** projecting slightly above the clavicle. The broad **costal surface** is

[13]*trache* = rough
[14]*carina* = keel

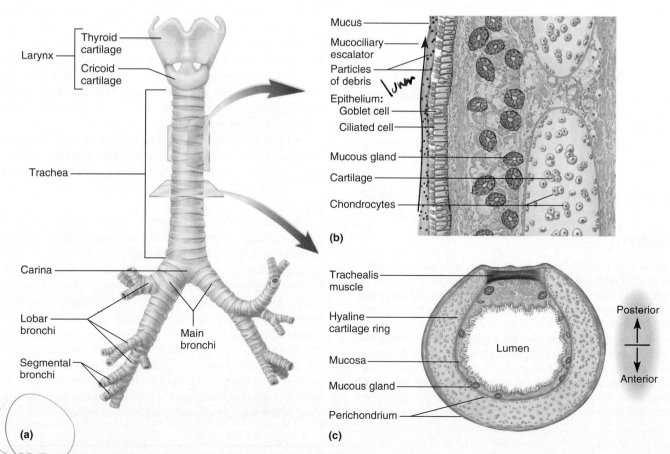

Figure 23.7 Anatomy of the Lower Respiratory Tract. (a) Anterior view. (b) Longitudinal section of the trachea showing the action of the mucociliary escalator. (c) Cross section of the trachea showing the C-shaped tracheal cartilage.

Cilia

Goblet cell

4 μm

Figure 23.8 **The Tracheal Epithelium Showing Ciliated Cells and Nonciliated Goblet Cells.** The small bumps on the goblet cells are microvilli.

● *What is the function of the goblet cells?*

pressed against the rib cage, and the smaller concave **mediastinal surface** faces medially. The mediastinal surface exhibits a slit called the **hilum** through which the lung receives the main bronchus, blood vessels, lymphatic vessels, and nerves. These structures constitute the **root** of the lung.

DEEPER INSIGHT   23.1

Tracheostomy

The functional importance of the nasal cavity becomes especially obvious when it is bypassed. If the upper airway is obstructed, it may be necessary to make a temporary opening in the trachea inferior to the larynx and insert a tube to allow airflow—a procedure called *tracheostomy.*[15] This prevents *asphyxiation,* but the inhaled air bypasses the nasal cavity and thus is not humidified. If the opening is left for long, the mucous membranes of the lower respiratory tract dry out and become encrusted, interfering with the clearance of mucus from the tract and promoting infection. When a patient is on a ventilator and air is introduced directly into the trachea, the air must be filtered and humidified by the apparatus to prevent respiratory tract damage.

The lungs are crowded by adjacent viscera and therefore neither fill the entire rib cage, nor are they symmetric. Inferior to the lungs and diaphragm, much of the space within the rib cage is occupied by the liver, spleen, and stomach (see fig. A.5, p. 335). The right lung is shorter than the left because the liver rises higher on the right. The left lung, although taller, is narrower than the right because the heart tilts toward the left and occupies more space on this side of the mediastinum. On the medial surface, the left lung has an indentation called the **cardiac impression** where the heart presses against it (fig. 23.9a); part of this is visible anteriorly as a crescent-shaped **cardiac notch** in the margin of the lung. The right lung has three lobes—**superior, middle,** and **inferior.** A deep groove called the **horizontal fissure** separates the superior lobe from the middle lobe, and a similar groove called the **oblique fissure** separates the middle and inferior lobes. The left lung has only a superior and inferior lobe and a single oblique fissure.

The Bronchial Tree

Each lung contains a branching system of air tubes called the **bronchial tree,** extending from the *main bronchus* to the *terminal bronchioles.* From the fork in the trachea, the right **main (primary) bronchus**[16] (BRON-cus) measures about 2 to 3 cm long. It is slightly wider and more vertical than the left one; consequently, inhaled (aspirated) foreign objects lodge more often in the right bronchus than in the left. Just before entering the lung, the right main bronchus gives off a **superior lobar (secondary) bronchus.** The main and lobar bronchi enter the hilum of the lung together. The superior lobar bronchus projects into the superior lobe of the lung, and the main bronchus continues a little farther and branches into **middle** and **inferior lobar bronchi** to the lower two lobes of the lung. The left main bronchus is about 5 cm long and is narrower and more horizontal than the right. It enters the hilum of the left lung before branching, then gives off superior and inferior lobar bronchi to the two lobes of that lung.

In both lungs, each lobar bronchus branches into **segmental (tertiary**[17]**) bronchi.** Each of these ventilates one functionally independent unit of lung tissue called a **bronchopulmonary segment.** There are 10 of these in the right lung and 8 in the left (see photo on p. 632).

The main bronchi are supported, like the trachea, by C-shaped hyaline cartilages, whereas the lobar and segmental bronchi are supported by overlapping crescent-shaped cartilaginous plates. All of the bronchi are lined by a ciliated pseudostratified columnar epithelium, but the cells grow shorter and the epithelium thinner as we progress distally. The lamina propria beneath the epithelium exhibits mucous glands and many aggregations of lymphocytes *(bronchus-associated lymphatic tissue, BALT),* favorably positioned to respond to inhaled pathogens. All divisions of the bronchial tree also have a substantial amount of elastic connective tissue, which contributes to the recoil that expels air from the lungs in each respiratory cycle. The mucosa also has a well-developed layer of smooth muscle, the *muscularis mucosae,* which regulates airway diameter and airflow.

[15]*stomy = making a hole*

[16]*bronch = windpipe*
[17]*tert = third*

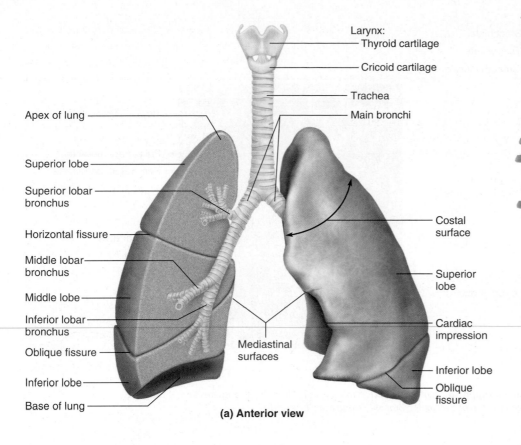

(a) Anterior view

Larynx:
Thyroid cartilage
Cricoid cartilage
Trachea
Main bronchi
Apex of lung
Superior lobe
Superior lobar bronchus
Horizontal fissure
Middle lobar bronchus
Middle lobe
Inferior lobar bronchus
Oblique fissure
Inferior lobe
Base of lung
Costal surface
Superior lobe
Cardiac impression
Mediastinal surfaces
Inferior lobe
Oblique fissure

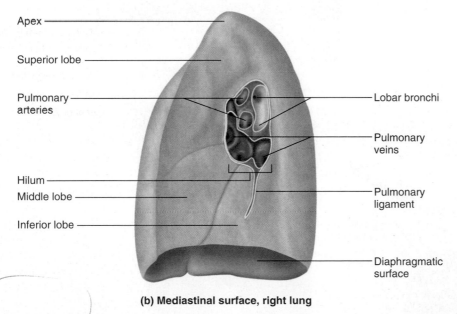

(b) Mediastinal surface, right lung

Apex
Superior lobe
Pulmonary arteries
Hilum
Middle lobe
Inferior lobe
Lobar bronchi
Pulmonary veins
Pulmonary ligament
Diaphragmatic surface

Figure 23.9 Gross Anatomy of the Lungs.

Bronchioles are continuations of the airway that lack support-ive cartilage and are 1 mm or less in diameter. The portion of the lung ventilated by one bronchiole is called a **pulmonary**[18] **lobule.** The epithelium of the bronchioles starts out as ciliated pseudostrat-

[18]*pulmon* = lung + *ary* = pertaining to

ified columnar in the larger, more proximal passages. As we progress distally, it gets thin-ner (the cells do not grow as tall) and grades into simple columnar and finally simple cuboi-dal epithelium. Bronchioles lack mucous glands and goblet cells, but they are ciliated throughout. It is an important point that the cilia continue more deeply into the airway than the mucous glands and goblet cells do. This ensures that mucus draining distally from those gland cells can still be captured by the beating cilia and cleared from the airway. Aside from the epithelium, the mucosa of the bron-chioles consists mainly of smooth muscle. Spasmodic contractions of this muscle at death cause the bronchioles to exhibit a wavy lumen in most histological sections (fig. 23.10a).

Each bronchiole divides into 50 to 80 **ter-minal bronchioles,** the final branches of the conducting division; there are about 65,000 of these in each lung. They measure 0.5 mm or less in diameter. Each terminal bronchiole gives off two or more smaller **respiratory bronchioles,** which have alveoli budding from their walls. Respiratory bronchioles are the beginning of the respiratory division. Their walls have scanty smooth muscle, and the smallest of them are nonciliated. Each respiratory bronchiole divides into 2 to 10 elongated, thin-walled passages called **alveo-lar ducts,** which also have alveoli along their walls (fig. 23.10). The alveolar ducts and smaller divisions have nonciliated simple squamous epithelia. The ducts end in **alveo-lar sacs,** which are grapelike clusters of alveoli arrayed around a central space called the **atrium** (fig. 23.10a). The distinction between an alveolar duct and atrium is their shape—an elongated passage or a space with about equal length and width. It is sometimes a sub-jective judgment whether to regard a space as an alveolar duct or atrium.

Air in the conducting division of the respi-ratory tract cannot exchange gases with the blood; the lumen of the conducting division is therefore called the *anatomical dead space.* In a state of relaxation, parasympathetic nerve fibers (from the vagus nerve) stimulate the muscularis mucosae and keep the airway par-tially constricted. This minimizes the dead space so that a greater percentage of the in-haled air goes to the alveoli, where it can oxygenate the blood. In exercise, the sympathetic nerves relax the smooth muscle and dilate the airway. Even though this increases the dead space, it enables air to flow more easily and rapidly so the alveoli can be ventilated in proportion to the demands of exercise. The increased airflow more than compensates for the increased dead space. The bronchioles

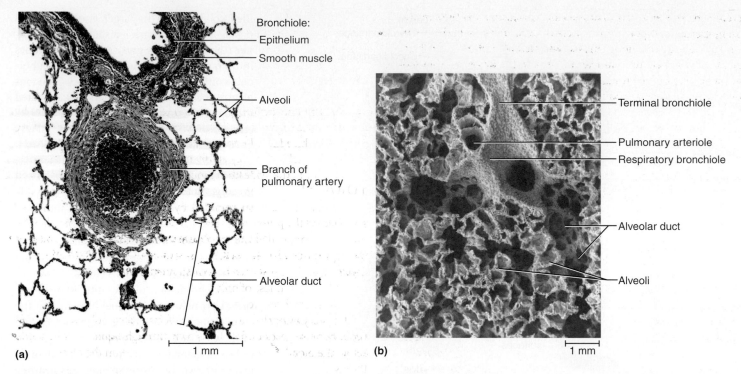

(a)

1 mm

(b)

1 mm

Bronchiole:
Epithelium
Smooth muscle

Alveoli

Branch of
pulmonary artery

Alveolar duct

Terminal bronchiole

Pulmonary arteriole
Respiratory bronchiole

Alveolar duct

Alveoli

Figure 23.10 Histology of the Lung. (a) Light micrograph. (b) Scanning electron micrograph. Note the spongy texture of the lung.
● *Histologically, how can we tell that the large passage at the top of part (a) is a bronchiole and not a bronchus?*

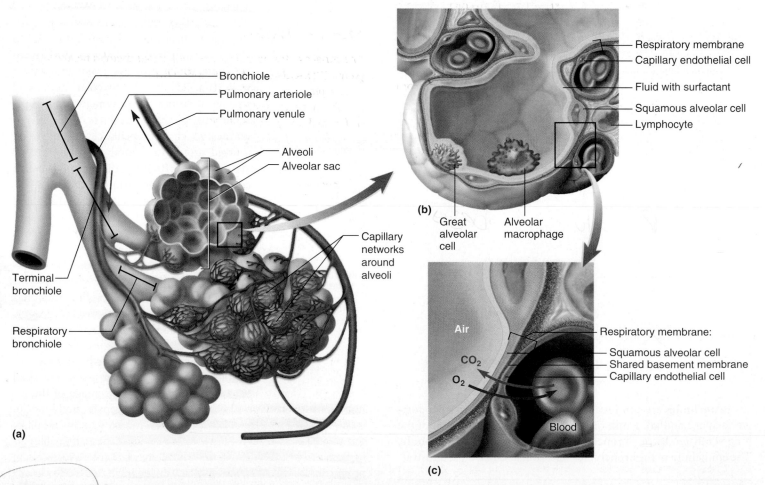

Bronchiole
Pulmonary arteriole
Pulmonary venule

Alveoli
Alveolar sac

Capillary
networks
around
alveoli

Terminal
bronchiole

Respiratory
bronchiole

(a)

Respiratory membrane
Capillary endothelial cell

Fluid with surfactant

Squamous alveolar cell

Lymphocyte

(b) Great
alveolar
cell

Alveolar
macrophage

Air

CO₂

O₂

Blood

Respiratory membrane:
Squamous alveolar cell
Shared basement membrane
Capillary endothelial cell

(c)

Figure 23.11 Pulmonary Alveoli. (a) Clusters of alveoli and their blood supply. (b) Structure of an alveolus. (c) Structure of the respiratory membrane.

exert the greatest control over airflow for two reasons: (1) they are the most numerous components of the conducting division; and (2) with their well-developed smooth muscle and lack of confining cartilage, they can change relative diameter more than the larger air passages can. Narrowing of the bronchioles is called *bronchoconstriction,* and widening is called *bronchodilation.*

As described in chapter 20, the lungs receive a blood supply from both the pulmonary arteries and bronchial arteries. Branches of the pulmonary artery closely follow the bronchial tree on their way to capillaries surrounding the alveoli (fig. 23.11), where gas exchange occurs. Branches of the bronchial arteries service the bronchi, bronchioles, and some other pulmonary tissues (see p. 581). The lungs are the only organs to receive both a pulmonary and a systemic blood supply.

Alveoli

Each human lung is a spongy mass with about 150 million little sacs, the alveoli, which provide about 70 m² of surface for gas exchange. An **alveolus**[19] (AL-vee-OH-lus) is a pouch about 0.2 to 0.5 mm in diameter (fig. 23.11). Thin, broad cells called **squamous (type I) alveolar cells** cover about 95% of the alveolar surface area. Their thinness allows for rapid gas diffusion between the alveolus and bloodstream. The other 5% is covered by round to cuboidal **great (type II) alveolar cells.** Even though they cover less surface area, great alveolar cells considerably outnumber the squamous alveolar cells. By analogy to baked goods, we could compare the shapes and surface areas of type I versus type II alveolar cells to a thinly rolled pie crust versus muffins, respectively. Great alveolar cells have two functions: (1) they repair the alveolar epithelium when the squamous alveolar cells are damaged; and (2) they secrete *pulmonary surfactant,* a mixture of phospholipids and protein that coats the alveoli and smallest bronchioles and prevents them from collapsing when one exhales. Without surfactant, the walls of a

---

[19]*alveol* = small cavity, little space

deflating alveolus would tend to cling together like sheets of wet paper, and it would be very difficult to reinflate them on the next inhalation (see Deeper Insight 23.4).

The most numerous of all cells in the lung are **alveolar macrophages (dust cells),** which wander the lumens of the alveoli and the connective tissue between them. These cells keep the alveoli free of debris by phagocytizing dust particles that escape entrapment by mucus in the higher parts of the respiratory tract. In lungs that are infected or bleeding, the macrophages also phagocytize bacteria and loose blood cells. As many as 100 million alveolar macrophages perish each day as they ride up the mucociliary escalator to be swallowed and digested, ridding the lungs of their load of debris.

Each alveolus is surrounded by a basket of blood capillaries supplied by the pulmonary artery. The barrier between the alveolar air and blood, called the **respiratory membrane,** consists only of the squamous alveolar cell, the squamous endothelial cell of the capillary, and their shared basement membrane (fig. 23.11b). These have a total thickness of only 0.5 μm, in contrast to the 7 μm diameter of the erythrocytes passing through the capillaries.

It is very important to prevent fluid from accumulating in the alveoli, because gases diffuse too slowly through liquid to sufficiently aerate the blood. Except for the film of moisture on the alveolar wall, the alveoli are kept dry by the absorption of excess liquid by the blood capillaries and abundant lymphatic capillaries of the lungs. The lungs have a more extensive lymphatic drainage than any other organ in the body. This keeps us from drowning in our own serous fluid.

The Pleurae

The surface of the lung is covered by a serous membrane, the **visceral pleura** (PLOOR-uh), which extends into the fissures. At the hilum, the visceral pleura turns back on itself and forms the **parietal pleura,** which adheres to the mediastinum, inner surface of the rib cage, and superior surface of the diaphragm (fig. 23.12). An extension of the parietal pleura, the *pulmonary ligament,* connects it to the diaphragm.

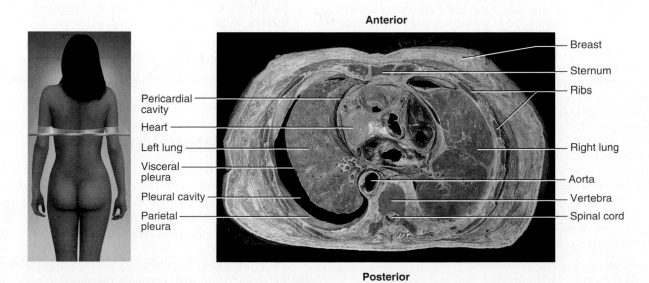

Anterior

Breast
Sternum
Ribs
Pericardial cavity
Heart
Left lung
Visceral pleura
Pleural cavity
Parietal pleura
Right lung
Aorta
Vertebra
Spinal cord

Posterior

Figure 23.12 Cross Section Through the Thoracic Cavity. This photograph is oriented the same way as the reader's body. The pleural cavity is especially evident where the left lung has shrunken away from the thoracic wall, but in a living person, the lung fully fills this space, the parietal and visceral pleurae are pressed together, and the pleural cavity is only a potential space between the membranes, as on the left side of this photograph.

The space between the parietal and visceral pleurae is called the **pleural cavity.** The two membranes are normally separated only by a film of slippery **pleural fluid;** thus, the pleural cavity is only a *potential space,* meaning there is normally no room between the membranes. Under pathological conditions, however, this space can fill with air or liquid (see *pneumothorax* in Deeper Insight 23.2).

The pleurae and pleural fluid have three functions:

1. **Reduction of friction.** Pleural fluid acts as a lubricant that enables the lungs to expand and contract with minimal friction.

2. **Creation of a pressure gradient.** During inspiration, the rib cage expands and draws the parietal pleura outward along with it. The visceral pleura clings to the parietal pleura, and since the visceral pleura is the lung surface, its outward movement expands the lung. The air pressure within the lung thus drops below the atmospheric pressure outside the body, and outside air flows down its pressure gradient into the lung.

3. **Compartmentalization.** The pleurae, mediastinum, and pericardium compartmentalize the thoracic organs and prevent infections of one organ from spreading easily to neighboring organs.

Apply What You Know

In what ways do the structure and function of the pleurae resemble the structure and function of the pericardium?

Before You Go On

Answer the following questions to test your understanding of the preceding section:

9. A dust particle is inhaled and gets into an alveolus without being trapped along the way. Describe the path it takes, naming all air passages from nostrils to alveoli. What would happen to it after arrival in an alveolus?

10. Contrast the epithelium of the bronchioles with that of the alveoli and explain how the structural difference is related to their functional differences.

11. Describe the relationship of the parietal and visceral pleurae to the lungs and thoracic wall.

DEEPER INSIGHT   23.2

Pulmonary Collapse

Pulmonary collapse (collapsed lung), or *atelectasis*[20] (AT-eh-LEC-ta-sis), is a state in which part or all of a lung is devoid of air. It is the normal state of a fetus and a newborn that has not drawn the first breath. After breathing begins, cases of pulmonary collapse fall into two categories: compression and absorption atelectasis.

Compression atelectasis is due to external pressure on the lung preventing its complete expansion. The pressure may come from blood, serous fluid, or air in the pleural cavity. Air in the pleural cavity, a state called *pneumothorax,*[21] often results from "sucking wounds" to the chest—for example, when the thoracic wall is punctured by a knife or a broken rib and inspiration sucks air through the opening. The visceral and parietal pleurae separate and the lung recoils from the thoracic wall and collapses. Pneumothorax can also occur in the absence of a chest wound if a weakened, ballooning area of the lung surface (called a *bleb*) ruptures and air flows from the lung into the pleural cavity.

Absorption atelectasis occurs when gases are absorbed into the blood and not replaced by fresh air, resulting in collapse of the alveoli. It can occur when the airway is obstructed by a mucous plug or an aspirated foreign object such as a bite of food, or when it is compressed by an adjacent tumor or pulmonary aneurysm. It also frequently occurs after surgery, especially if a patient is in pain and is reluctant to breathe deeply or change position in bed. Postoperative patients are encouraged to breathe deeply because it promotes the clearance of secretions from the lungs, the even distribution of surfactant, and the flow of air from better-ventilated alveoli into less-ventilated ones.

When one lung collapses, the positive pressure in that pleural cavity can shift the entire mediastinum (including the heart and major blood vessels) toward the other pleural cavity, compressing and partially collapsing that lung as well.

[20]*atel* = imperfect + *ectasis* = expansion
[21]*pneumo* = air, lung

23.4 Neuromuscular Aspects of Respiration

Expected Learning Outcomes

When you have completed this section, you should be able to

- identify the muscles that ventilate the lungs and describe their respective roles;

- describe the brainstem centers and peripheral nerves that control breathing, and explain their functions; and

- identify the inputs that influence the activity of those brainstem centers.

The Respiratory Muscles

The lungs do not ventilate themselves. The only muscle they contain is smooth muscle in the walls of the bronchi and bronchioles, which does not create the airflow but only affects its speed. The driving force for pulmonary ventilation comes from the skeletal muscles of the trunk, especially the diaphragm and intercostal muscles (fig. 23.13).

The prime mover of pulmonary ventilation is the **diaphragm,** the muscular dome that separates the thoracic cavity from the abdominal cavity. It alone accounts for about two-thirds of the pulmonary airflow. When relaxed, it bulges upward to its farthest ex-

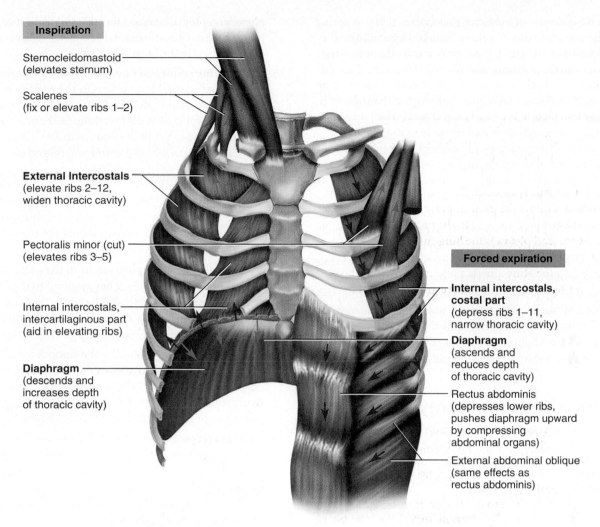

Inspiration

Sternocleidomastoid
(elevates sternum)

Scalenes
(fix or elevate ribs 1–2)

External Intercostals
(elevate ribs 2–12,
widen thoracic cavity)

Pectoralis minor (cut)
(elevates ribs 3–5)

Internal intercostals,
intercartilaginous part
(aid in elevating ribs)

Diaphragm
(descends and
increases depth
of thoracic cavity)

Forced expiration

**Internal intercostals,
costal part**
(depress ribs 1–11,
narrow thoracic cavity)

Diaphragm
(ascends and
reduces depth
of thoracic cavity)

Rectus abdominis
(depresses lower ribs,
pushes diaphragm upward
by compressing
abdominal organs)

External abdominal oblique
(same effects as
rectus abdominis)

Figure 23.13 The Respiratory Muscles. Boldface indicates the principal respiratory muscles; the others are accessory. Arrows indicate the direction of muscle pull. Muscles listed on the left are active during inspiration, and those on the right are active during forced expiration. Note that the diaphragm is active in both phases, and different parts of the internal intercostals serve for inspiration and expiration. Some other accessory muscles not shown here are discussed in the text.

tent, pressing against the base of the lungs. The lungs are then at their minimum volume. When the diaphragm contracts, it tenses and flattens somewhat, dropping about 1.5 cm in relaxed inspiration and as much as 7 cm in deep breathing. This enlarges the thoracic cavity and lungs and causes an inflow of air. When the diaphragm relaxes, it bulges upward again, compresses the lungs, and expels air.

Several other muscles aid the diaphragm as synergists. Chief among these are the **internal** and **external intercostal muscles** between the ribs. Their primary function is to stiffen the thoracic cage during respiration and prevent it from caving inward when the diaphragm descends. However, they also contribute to enlargement and contraction of the thoracic cage and add about one-third of the air that ventilates the lungs. During quiet breathing, the scalene muscles of the neck fix (hold stationary) ribs 1 and 2, while the external intercostal muscles pull the other ribs upward. Since most ribs are anchored at both ends—by their attachment to the vertebral column at the proximal (posterior) end and their attachment through the costal cartilage to the sternum at the distal (anterior)

end—they swing upward like the handle on a bucket and thrust the sternum forward. These actions increase both the transverse (left to right) and anteroposterior (AP) diameters of the chest. In deep breathing, the AP diameter increases by as much as 20%.

Other muscles of the chest and abdomen also aid in breathing, especially during forced respiration—that is, taking deeper breaths than normal. These are considered the *accessory muscles* of respiration. Deep inspiration is aided by the *erector spinae*, which arches the back and increases AP chest diameter, and by several muscles that elevate the upper ribs: the *sternocleidomastoids* and *scalenes* of the neck; the *pectoralis minor, pectoralis major,* and *serratus anterior* of the chest; and the *intercartilaginous (interchondral) part* of the internal intercostal muscles (the anterior part of the muscles between the costal cartilages). Although the scalenes merely fix the upper ribs during quiet respiration, they elevate them during forced inspiration.

Normal expiration is an energy-saving passive process. It is normally achieved by the elasticity of the lungs and thoracic cage. The bronchial tree, the attachments of the ribs to the spine and sternum, and the tendons of the diaphragm and other respiratory muscles all have a degree of elasticity that causes them to spring back when the muscles relax. As these structures recoil, the thoracic cage diminishes in size, the air pressure in the lungs rises above the atmospheric pressure, and the air flows out. The only muscular effort involved in normal expiration is a braking action—that is, the muscles relax gradually rather than abruptly, thus preventing the lungs from recoiling too suddenly. This makes the transition from inspiration to expiration smoother.

But during forced expiration—for example, when singing or shouting, coughing or sneezing, or playing a wind instrument—the *rectus abdominis* pulls down on the sternum and lower ribs, while the *interosseous (costal) part* of the internal intercostal muscles (the part between the ribs proper) pulls the other ribs downward. These actions reduce the chest diameter and help to expel air more rapidly and thoroughly than usual. Other muscles that contribute to forced expiration are the *latissimus dorsi* of the lower back, the *transverse* and *oblique abdominal muscles,* and even some muscles of the pelvic floor. They raise the pressure in the abdominal cavity and push some of the viscera, such as the stomach and liver, up against the diaphragm. This increases the pressure in the thoracic cavity and thus helps to expel air. Abdominal control of airflow is particularly important in singing, public speaking, coughing, and sneezing.

Respiratory Neuroanatomy

The heartbeat and breathing are the two most conspicuously rhythmic processes of the body; but the heart contains its own pacemaker whereas the lungs do not. As we have just seen, breathing requires the coordinated action of numerous skeletal muscles. These must be under centralized control; therefore, they depend on output from the brain. Breathing is controlled at two levels of the brain. One is cerebral and conscious, enabling us to inhale or exhale at will. Most of the time, however, we breathe without thinking about it—fortunately, for we otherwise could not go to sleep without fear of respiratory arrest (see Deeper Insight 23.3).

The automatic, unconscious cycle of breathing is controlled by three respiratory centers in the medulla oblongata and pons (fig. 23.14). Each of these is paired, left and right, with transverse connections between them.

1. The **ventral respiratory group (VRG)** is the primary generator of the respiratory rhythm. It is an elongated neural network in the medulla that issues output signals to an integrating center in the spinal cord. The left and right spinal centers issue output by way of the phrenic nerves to the diaphragm and by way of intercostal nerves to the external intercostal muscles. Periodic output from the VRG creates the basic cycle of contraction and relaxation of these muscles, causing inspiration and expiration. The following control centers alter that basic rhythm to meet ever-changing physiological needs.

2. The **dorsal respiratory group (DRG)** is an elongated mass of neurons that extends for much of the length of the medulla near the central canal, posterior to the VRG. It is apparently an integrating center that receives input from the pontine respiratory group (next); from central and peripheral chemoreceptors described shortly; and from receptors in the lungs for stretch and irritants. The DRG issues output to the VRG to modify the breathing rhythm.

3. The **pontine respiratory group (PRG)** (formerly called the *pneumotaxic center*) is a nucleus in the pons. It receives input from higher brain centers and issues output to the DRG and VRG. Its effect is to make breathing faster or slower, shallower or deeper; and to adapt to circumstances such as sleep, exercise, vocalization, and emotional responses.

These respiratory centers receive input from multiple sources:

- **Central chemoreceptors** are brainstem neurons that respond especially to changes in the pH of the cerebrospinal fluid. They are concentrated on each side of the medulla oblongata about 0.2 mm beneath its anterior surface.

- **Peripheral chemoreceptors** occur in the aortic arch and carotid bodies (fig. 23.15). They respond to the O_2 and CO_2 content and pH of the blood. The aortic bodies communicate with the medulla by way of the vagus nerves, and the carotid bodies communicate by way of the glossopharyngeal nerves.

- **Stretch receptors** are found in the smooth muscle of the bronchi and bronchioles and in the visceral pleura. They communicate with the DRG via the vagus nerves and respond to inflation of the lungs. Excessive inflation triggers a protective reflex that strongly inhibits inspiration.

- **Irritant receptors** are nerve endings in the epithelia of the airway. They respond to smoke, dust, pollen, chemical fumes, cold air, and excess mucus. They, too, transmit signals to the DRG via the vagus nerves. The DRG responds with protective reflexes such as coughing, bronchoconstriction, shallower breathing, or breath holding.

- **Higher brain centers** feed into the PRG, DRG, and spinal cord integrating centers. Thus, the limbic system, hypothalamus, and cerebral cortex influence the respiratory centers. This input allows for conscious control over breathing (as in holding one's breath) and for emotions to affect respiration—for example, in gasping, crying, and laughing,

---

DEEPER INSIGHT 23.3

Ondine's Curse

In German legend, there was a water nymph named Ondine who took a mortal lover. When her lover proved unfaithful, the king of the nymphs put a curse on him that took away his automatic physiological functions. Consequently, he had to remember to take each breath, and he could not go to sleep or he would die of suffocation—which, as exhaustion overtook him, was indeed his fate.

Some people suffer a disorder called *Ondine's curse,* in which the automatic respiratory functions are disabled—usually as a result of brainstem damage from poliomyelitis or as an accident of spinal cord surgery. Victims of Ondine's curse must remember to take each breath and cannot sleep without using a mechanical ventilator or being awakened repeatedly by episodes of *apnea* (temporary cessation of breathing).

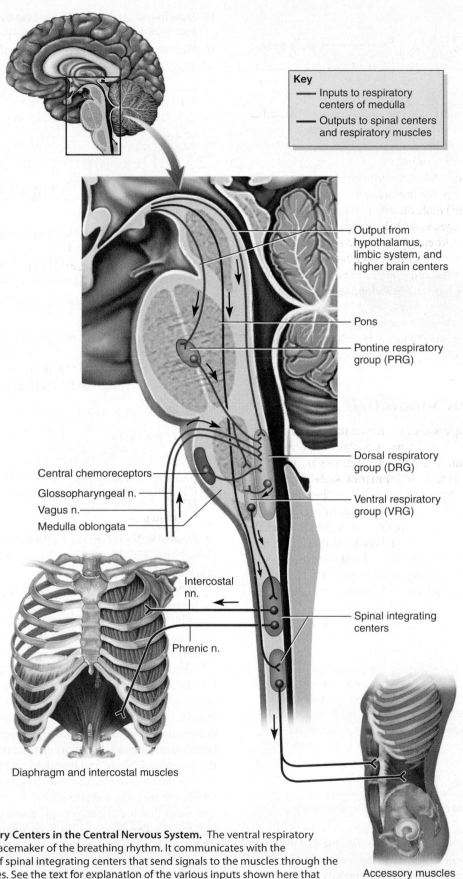

Key
—— Inputs to respiratory centers of medulla
—— Outputs to spinal centers and respiratory muscles

Output from hypothalamus, limbic system, and higher brain centers

Pons

Pontine respiratory group (PRG)

Dorsal respiratory group (DRG)

Central chemoreceptors

Glossopharyngeal n.

Vagus n.

Medulla oblongata

Ventral respiratory group (VRG)

Intercostal nn.

Spinal integrating centers

Phrenic n.

Diaphragm and intercostal muscles

Accessory muscles of respiration

Figure 23.14 Respiratory Centers in the Central Nervous System. The ventral respiratory group (VRG) is the primary pacemaker of the breathing rhythm. It communicates with the respiratory muscles by way of spinal integrating centers that send signals to the muscles through the intercostal and phrenic nerves. See the text for explanation of the various inputs shown here that modify the rhythm of the VRG. (n. = nerve, nn. = nerves)

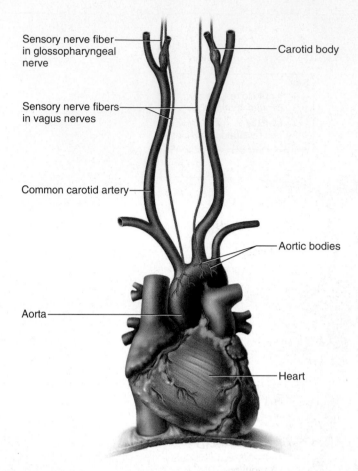

Sensory nerve fiber in glossopharyngeal nerve

Carotid body

Sensory nerve fibers in vagus nerves

Common carotid artery

Aortic bodies

Aorta

Heart

Figure 23.15 The Peripheral Chemoreceptors of Respiration.

and when anxiety provokes a bout of hyperventilation (rapid breathing in excess of physiological need). Signals for voluntary control over breathing travel down the corticospinal tracts to the respiratory neurons in the spinal cord, bypassing the brainstem centers.

Apply What You Know

Some authorities refer to the respiratory rhythm as an *autonomic* function. Discuss whether you think this is an appropriate word for it. What are the effectors of the autonomic nervous system? (See chapter 16.) What are the effectors that ventilate the lungs? What bearing might this have on the question?

Before You Go On

Answer the following questions to test your understanding of the preceding section:

12. Explain why breathing is not controlled by a pacemaker in the lungs.

13. What is the prime mover of respiration? What muscle group acts as the most important synergist? What nerves innervate each of these?

14. Name some other synergists that act during forced inspiration and some that act during forced expiration.

15. State the names and locations of the three pairs of brainstem nuclei that regulate the respiratory rhythm. What role does each one play?

16. From what sources do the respiratory nuclei receive input that influences respiration?

23.5 Developmental and Clinical Perspectives

▌ Expected Learning Outcomes

When you have completed this section, you should be able to

- describe the prenatal development of the respiratory system;
- describe the changes that occur in the respiratory system in old age; and
- describe some common respiratory disorders.

Prenatal and Neonatal Development

The first embryonic trace of the respiratory system is a small pouch in the floor of the pharynx called the **pulmonary groove,** appearing at about 3.5 weeks. The groove grows down the mediastinum as an elongated tube, the future trachea, and branches into two **lung buds** by week 4 (fig. 23.16). The lung buds branch repeatedly and grow laterally and posteriorly, occupying the space posterior to the heart. Repeated branching of the lung buds produces the bronchial tree, which is completed as far as the bronchioles by the end of month 6. For the remainder of gestation and after birth, the bronchioles bud off alveoli. The adult number of alveoli is attained around the age of 10 years.

By week 8, the lungs are isolated from the heart by the growth of the pericardium; and by week 9, the diaphragm forms and separates the lungs and pleurae from the abdominal cavity. At 28 weeks, the respiratory system is usually adequately developed to support independent life (see Deeper Insight 23.4).

By 11 weeks, the fetus begins respiratory movements called **fetal breathing,** in which it rhythmically inhales and exhales amniotic fluid for as much as 8 hours per day. Fetal breathing stimulates lung development and conditions the respiratory muscles for life outside the womb. It ceases during labor. When the newborn infant begins breathing air, the fluid in the lungs is quickly absorbed by the pulmonary blood and lymphatic capillaries.

For the newborn infant, breathing is very laborious at first. The fetal lungs are collapsed and airless, and the neonate must take very strenuous first breaths to pop the alveoli open. Once they are fully inflated, the alveoli normally never collapse again. Even the pulmonary blood vessels are collapsed in the fetus, but as the infant takes its first breaths, the drop in thoracic pressure draws blood into the pulmonary circulation and expands the vessels. As pulmonary resistance drops, the foramen ovale and ductus arteriosus close (see p. 557), and pulmonary blood flow increases to match the airflow.

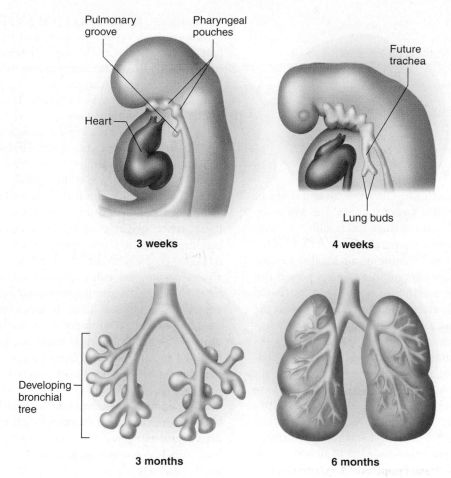

Figure 23.16 Embryonic Development of the Respiratory System.

DEEPER INSIGHT 23.4

Premature Birth and Respiratory Distress Syndrome

Premature infants often suffer from *respiratory distress syndrome (RDS)*, also called *hyaline membrane disease* (compare adult respiratory distress syndrome, table 23.1). They have not yet produced enough pulmonary surfactant to keep the alveoli open between inspirations. Consequently, the alveoli collapse during expiration, and a great effort is required to reinflate them. The infant becomes exhausted by the effort to breathe, and becomes progressively *cyanotic* (blue) because of the deficiency of oxygen in the blood (*hypoxemia*[22]). Progressive destruction of the alveolar epithelium and capillary walls leads to leakage of plasma into the alveolar air spaces and connective tissue between the alveoli. The plasma coagulates and the alveoli fill with stiff clear "membranes" of fibrin, fibrinogen, and cell debris. Eventually, the infant cannot inhale forcefully enough to inflate the alveoli again; without treatment, death ensues from hypoxemia and carbon dioxide retention (*hypercapnia*[23]).

RDS occurs in about 60% of infants born before 28 weeks of gestation, and 15% to 20% of those born between 32 and 36 weeks. It is the most common cause of neonatal death, with about 60,000 cases and 5,000 deaths per year in the United States. In addition to premature birth, some risk factors for RDS include maternal diabetes, oversedation of the mother during labor, aspiration of blood or amniotic fluid, and prenatal hypoxia caused by winding of the umbilical cord around the neck.

RDS can be treated with a ventilator that forces air into the lungs and keeps the alveoli inflated (*positive end-expiratory pressure, PEEP*) until the infant's lungs produce their own surfactant, and by giving a mist of surfactant from external sources such as calf lungs or genetically engineered bacteria. The infant may also be given oxygen therapy, but this is a limited and risky treatment because oxygen generates damaging free radicals that can cause blindness and severe bronchial problems. Oxygen toxicity can be minimized by a technique called *extracorporeal membrane oxygenation (ECMO)*, which is similar to the heart–lung bypass procedure used in surgery. Blood flows from catheters in the baby's neck to a machine that oxygenates it, warms it, and returns it to the body. ECMO is, however, a high-risk procedure used only in extreme cases.

[22]*hypo* = deficiency + *ox* = oxygen + *emia* = blood condition
[23]*hyper* = excessive + *capn* = smoke, carbon dioxide + *ia* = condition

Apply What You Know

In a certain criminal investigation, the pathologist performing an autopsy on an infant removes the lungs, places them in a pail of water, and concludes that the infant was live-born. What do you think the pathologist saw that led to this conclusion? What contrasting observation would suggest that an infant had been stillborn?

The Aging Respiratory System

Pulmonary ventilation declines steadily after the 20s and is one of several factors in a person's gradual loss of stamina. The costal cartilages and joints of the thoracic cage become less flexible, the lungs have less elastic tissue, and there are fewer alveoli in old age. There is a corresponding decline in the volume of air inhaled in each breath *(tidal volume),* the maximum amount of air a person can inhale *(vital capacity),* and the maximum speed of airflow *(forced expiratory volume).* The elderly are also less capable of clearing the lungs of irritants and pathogens, and therefore are increasingly susceptible to respiratory infections. Pneumonia causes more deaths in old age than any other communicable disease and is often contracted in hospitals and nursing homes.

Chronic obstructive pulmonary diseases (see next section) are more common in old age since they represent the cumulative effects of a lifetime of degenerative change. Declining pulmonary function also contributes to cardiovascular disease and hypoxemia, and the latter is a factor in degenerative disorders of all other organ systems. Respiratory health is therefore a major concern in aging.

Respiratory Pathology

Many of the respiratory disorders can be classified as *restrictive* or *obstructive disorders.* **Restrictive disorders** stiffen the lungs and reduce their *compliance* (ease of inflation) and vital capacity. An example is pulmonary fibrosis, in which much of the normal respiratory tissue of the lung is replaced by fibrous scar tissue. Fibrosis is an effect of such diseases as *tuberculosis* and the *black lung disease* of coal miners. **Obstructive disorders** narrow the airway and interfere with airflow, so expiration requires more effort and may be less complete than normal. Airway obstructions, bronchoconstriction, and tumors or aneurysms that compress the airways can cause obstructive disorders.

Chronic obstructive pulmonary diseases (COPDs) are disorders in which there is a long-term obstruction of airflow and a substantial reduction in pulmonary ventilation. The major COPDs are chronic bronchitis and emphysema. They are almost always caused by cigarette smoking, and are among the few leading causes of death in old age. **Chronic bronchitis** is characterized by airway congestion with *sputum,* a mixture of thick mucus and cellular debris, accompanied by chronic respiratory infection and bronchial inflammation. **Emphysema**[24] (EM-fih-SEE-muh) is characterized by a breakdown of alveolar walls and replacement of the spongy lung tissue with relatively large air-filled cavities, resulting in a reduction in alveolar surface area and reduced ability to oxygenate the blood. Any COPD can also lead to *cor pulmonale*—enlargement and potential failure of the right side of the heart due to obstruction of the pulmonary circulation.

Asthma is the most common chronic respiratory illness of children and the leading cause of school absenteeism and childhood hospitalization in the United States. About half of all cases develop before age 10 and only 15% after age 40. Asthma takes about 5,000 lives per year, and this number lately has been rising. It is usually an allergic reaction to airborne antigens (allergens) that stimulate intense bronchoconstriction and airway inflammation. The reaction commonly leads to severe coughing, wheezing, and sometimes suffocation. People who die of asthmatic suffocation typically show airways so plugged with gelatinous mucus that they could not exhale. The lungs remain hyperinflated even at autopsy. Asthma is treated with drugs that dilate the airway and relieve inflammation.

Lung cancer accounts for more deaths than any other form of cancer. The most important cause is cigarette smoking, distantly followed by air pollution. Lung cancer commonly follows or accompanies COPD. It begins with uncontrolled proliferation of cells of the surface epithelium or mucous glands of the bronchi. As the dividing epithelial cells invade the underlying tissues of the bronchial wall, the bronchus develops bleeding lesions. Dense masses of keratin and malignant cells appear in the lung parenchyma and replace functional respiratory tissue. Because of the extensive lymphatic drainage of the lungs, lung cancer quickly metastasizes to other organs—especially the pericardium, heart, bones, liver, lymph nodes, and brain. The chance of recovery is poor, with only 7% of patients surviving for 5 years after diagnosis.

Some other disorders of the respiratory system are briefly described in table 23.1.

Before You Go On

Answer the following questions to test your understanding of the preceding section:

17. When and where does the pulmonary groove appear? Concisely describe its further development.

18. What changes in the lungs occur at birth? What corresponding changes occur in the cardiovascular system?

19. Identify some reasons why a person's vital capacity declines in old age.

20. Name and compare two COPDs and describe some pathological effects they have in common.

21. In what lung tissue does lung cancer originate? How does it kill?

---

[24]*emphys* = inflamed

TABLE 23.1	Some Disorders of the Respiratory System
Acute rhinitis	The common cold. Caused by many types of viruses that infect the upper respiratory tract. Symptoms include congestion, increased nasal secretion, sneezing, and dry cough. Transmitted especially by contact of contaminated hands with mucous membranes; not transmitted orally.
Adult respiratory distress syndrome	Acute lung inflammation and alveolar injury stemming from trauma, infection, burns, aspiration of vomit, inhalation of noxious gases, drug overdoses, and other causes. Alveolar injury is accompanied by severe pulmonary edema and hemorrhage, followed by fibrosis that progressively destroys lung tissue. Fatal in about 40% of cases under age 60 and in 60% of cases over age 65.
Pneumonia	A lower respiratory infection caused by any of several viruses, fungi, or protozoans (most often the bacterium *Streptococcus pneumoniae*). Causes filling of alveoli with fluid and dead leukocytes and thickening of the respiratory membrane, which interferes with gas exchange and causes hypoxemia. Especially dangerous to infants, the elderly, and people with compromised immune systems, such as AIDS and leukemia patients.
Sleep apnea	Cessation of breathing for 10 seconds or longer during sleep; sometimes occurs hundreds of times per night, often accompanied by restlessness and snoring. Can result from altered function of CNS respiratory centers, airway obstruction, or both. Over time, may lead to daytime drowsiness, hypoxemia, polycythemia, pulmonary hypertension, congestive heart failure, and cardiac arrhythmia. Most common in obese people and in men.
Tuberculosis (TB)	Pulmonary infection with the bacterium *Mycobacterium tuberculosis,* which invades the lungs by way of air, blood, or lymph. Stimulates the lung to form fibrous nodules called tubercles around the bacteria. Progressive fibrosis compromises the elastic recoil and ventilation of the lungs. Especially common among impoverished and homeless people and becoming increasingly common among people with AIDS.

Disorders Described Elsewhere

Airway obstruction   640
Asthma   650
Atelectasis   644
Black lung disease   650

Chronic bronchitis   650
Chronic obstructive pulmonary diseases   650
Emphysema   650
Lung cancer   650

Neonatal respiratory distress syndrome   649
Ondine's curse   646
Pneumothorax   649
Pulmonary fibrosis   650

Study Guide

Assess Your Learning Outcomes

You should have a good understanding of this chapter if you can accurately address the following issues.

23.1 Overview of the Respiratory System (p. 633)

1. The multiple functions of the respiratory system
2. The purpose and components of the conducting division of the respiratory tract
3. The purpose and components of the respiratory division of the respiratory tract
4. The boundary between the upper and lower respiratory tracts, and which organs belong to each region

23.2 The Upper Respiratory Tract (p. 634)

1. Functions of the nose
2. The anterior and posterior openings into the nasal cavity; the nasal septum and fossae; and the bones and cartilages that shape the external (facial) part of the nose
3. The components of the nasal septum, and the bones that contribute to the roof and floor of the nasal cavity
4. The nasal conchae and meatuses, and their functions
5. The histology, spatial distribution, and functions of the olfactory and respiratory epithelia of the nasal cavity
6. The location and function of the erectile tissues of the nasal cavity
7. The definition and three divisions of the pharynx, and the boundaries between the divisions
8. The definition and functions of the larynx, and its superior and inferior boundaries
9. The structure and function of the glottis and its mode of operation in infants and adults
10. The nine cartilages that frame the larynx, and the two ligaments that bind the larynx to the hyoid bone above and the trachea below
11. The locations and functions of the vestibular folds and vocal cords of the larynx
12. The mode of action of the intrinsic muscles of the larynx on the vocal cords, and how this affects vocalization

23.3 The Lower Respiratory Tract (p. 638)

1. The gross anatomy and histology of the trachea, and the functional significance of its mucociliary escalator, supportive cartilages, and trachealis muscle
2. The surface anatomy of the lungs, including the apex and base; costal and mediastinal surfaces; hilum; lobes and fissures; and differences between the right and left lungs
3. All the subdivisions of the bronchial tree and the air passages, in order, from main bronchi to alveoli; the histological differences between the regions of the airway; and the functional significance of these histological differences
4. Bronchoconstriction, bronchodilation, and the reason why the bronchioles are more important than any other region of the airway in regulating pulmonary airflow

5. The pulmonary and systemic blood supplies to the lungs, and the anatomical and functional differences between these two blood supplies

6. The structure of a pulmonary alveolus; the morphology and functions of its squamous and great alveolar cells and alveolar macrophages; and the source and function of pulmonary surfactant

7. Components of the respiratory membrane that separates the alveolar air space from the capillary bloodstream, and the relevance of this membrane to gas exchange

8. The two layers of the pleura; the pleural cavity and fluid; and the functions of the pleurae

23.4 Neuromuscular Aspects of Respiration (p. 644)

1. The roles of the diaphragm and intercostal muscles in breathing; the identity and roles of other accessory muscles of respiration; and the movements of the thoracic cage that cause ventilation of the lungs

2. Differences in muscular action between quiet and forced respiration

3. The energy-saving mechanism of quiet expiration

4. The three pairs of respiratory control nuclei in the brainstem; their locations; their input and output pathways; and the function of each

5. The central and peripheral chemoreceptors, stretch receptors, irritant receptors, and higher brain centers that influence respiration

23.5 Developmental and Clinical Perspectives (p. 648)

1. The development of the embryonic pulmonary groove and its differentiation into the respiratory tract and lungs

2. The onset and functions of fetal breathing

3. The respiratory adaptations of the newborn infant

4. Reasons for the normal decline in respiratory efficiency in old age, and how this affects other organ systems

5. The difference between restrictive and obstructive disorders of the respiratory system, and examples of each

6. The chronic obstructive pulmonary diseases (COPDs); the common causes of each; their pathology; and their effect on the heart

7. How lung cancer arises; how it spreads through the lung tissue; and how it metastasizes to other organs

Testing Your Recall

1. The nasal cavity is divided by the nasal septum into right and left
   a. nares.
   b. vestibules.
   c. fossae.
   d. choanae.
   e. conchae.

2. The intrinsic laryngeal muscles regulate speech by rotating
   a. the extrinsic laryngeal muscles.
   b. the thyroid cartilage.
   c. the arytenoid cartilages.
   d. the hyoid bone.
   e. the vocal cords.

3. The largest air passages that engage in gas exchange with the blood are
   a. the respiratory bronchioles.
   b. the terminal bronchioles.
   c. the main bronchi.
   d. the alveolar ducts.
   e. the alveoli.

4. Respiratory arrest would most likely result from a tumor of the
   a. pons.
   b. midbrain.
   c. thalamus.
   d. cerebellum.
   e. medulla oblongata.

5. A deficiency of pulmonary surfactant is most likely to cause
   a. chronic obstructive pulmonary disease.
   b. atelectasis.
   c. pneumothorax.
   d. chronic bronchitis.
   e. asthma.

6. The source of pulmonary surfactant is
   a. the visceral pleura.
   b. tracheal glands.
   c. alveolar capillaries.
   d. squamous alveolar cells.
   e. great alveolar cells.

7. Which of the following are fewest in number but largest in diameter?
   a. alveoli
   b. terminal bronchioles
   c. alveolar ducts
   d. segmental bronchi
   e. respiratory bronchioles

8. The rhythm of breathing is set by neurons in
   a. the medulla oblongata.
   b. the pons.
   c. the midbrain.
   d. the hypothalamus.
   e. the cerebral cortex.

9. Which of the following muscles aid(s) in deep expiration?
   a. the scalenes
   b. the sternocleidomastoids
   c. the rectus abdominis
   d. the external intercostals
   e. the erector spinae

10. In the bronchial tree, which of the following would be found distal to all the rest?
    a. mucous glands
    b. ciliated cells
    c. alveolar ducts
    d. cartilage rings
    e. goblet cells

11. The digestive and respiratory tracts share a segment of the pharynx called the _____.

12. Within each lung, the airway forms a branching complex called the _____.

13. The flared areas of the nose lateral to the nostrils are shaped by _____ cartilages.

14. The three folds on the lateral walls of the nasal cavity are called _____.

15. _____ disorders reduce the speed of airflow through the airway.

16. Some inhaled air does not participate in gas exchange because it fills the _____ of the respiratory tract.

17. The largest cartilage of the larynx is the _____ cartilage.

18. The main pacemaker of the breathing rhythm is the _____ in the medulla oblongata.

19. The primary bronchi and pulmonary blood vessels penetrate the lung at a medial slit called the _____.

20. The last line of defense against inhaled particles are phagocytic cells called _____.

Answers in the Appendix

Building Your Medical Vocabulary

State a medical meaning of each of the following word elements, and give a term in which it is used.

1. trache-
2. alveolo-
3. -ectasis
4. capn-
5. naso-
6. spir-
7. emphys-
8. pulmo-
9. broncho-
10. pneumo-

Answers in the Appendix

True or False

Determine which five of the following statements are false, and briefly explain why.

1. The glottis is the opening from the larynx to the trachea.
2. The lungs contain more respiratory bronchioles than terminal bronchioles.
3. The lungs occupy the spaces between the parietal and visceral pleurae.
4. Expiration is normally caused by contraction of the internal intercostal muscles.
5. Atelectasis can result from causes other than pneumothorax.
6. Alveoli continue to be produced after birth.
7. Unlike bronchi, bronchioles have no cartilage.
8. Blood gases are monitored by the aortic and carotid sinuses.
9. The respiratory system begins its development by budding from the posterior side of the esophagus.
10. Extrinsic ligaments link the larynx to adjacent nonlaryngeal structures.

Answers in the Appendix

Testing Your Comprehension

1. Discuss how the different functions of the conducting division and respiratory division relate to differences in their histology.
2. From the upper to the lower end of the trachea, the ratio of goblet cells to ciliated cells gradually changes. Would you expect the highest ratio of ciliated cells to goblet cells to be in the upper trachea or the lower trachea? Give a functional rationale for your answer.
3. The bronchioles are to the airway and airflow what the arterioles are to the circulatory system and blood flow. Explain or elaborate on this comparison.
4. A patient has suffered damage to the left phrenic nerve, resulting in paralysis of the left side of the diaphragm but not the right. X-rays show that during inspiration, the right side of the diaphragm descends as normal, but the left side *rises*. Explain the unusual motion of the left diaphragm.
5. An 83-year-old woman is admitted to the hospital, where a critical-care nurse attempts to insert a nasoenteric tube ("stomach tube") for feeding. The patient begins to exhibit dyspnea, and a chest X-ray reveals air in the right pleural cavity and a collapsed right lung. The patient dies 5 days later from respiratory complications. Name the conditions revealed by the X-ray and explain how they could have resulted from the nurse's procedure.

Answers at www.mhhe.com/saladinha3

Improve Your Grade at www.mhhe.com/saladinha3

Practice quizzes, labeling activities, games, and flashcards provide fun ways to master concepts. You can also download image PowerPoint files for each chapter to create a study guide or for taking notes during lecture.

The Digestive System

Filiform papillae of the human tongue (SEM)

CHAPTER OUTLINE

DEEPER INSIGHTS

BRUSHING UP

To understand this chapter, you may find it helpful to review the following concepts:
- Brush borders and microvilli (p. 37)
- The embryonic disc and germ layers (p. 87)
- The autonomic nervous system (chapter 16)
- The celiac and mesenteric blood circulation (pp. 584–586)

Anatomy & Physiology | REVEALED®
aprevealed.com

Digestive System

Some of the most fundamental facts we know about digestion date to a grave accident in 1822. A Canadian voyageur, Alexis St. Martin, was accidentally hit by a shotgun blast while standing outside a trading post on Mackinac Island, Michigan. A frontier army doctor summoned to the scene, William Beaumont, found part of St. Martin's lung protruding through the wound, and a hole in the stomach "large enough to receive my forefinger." Surprisingly, St. Martin survived, but the wound left a permanent opening (fistula) into his stomach, covered for the rest of his life by only a loose flap of skin.

Beaumont saw this as an opportunity to learn something about digestion. Now disabled from wilderness travel, St. Martin agreed to participate in Beaumont's experiments in exchange for room and board. Working under crude frontier conditions with little idea of scientific methods, Beaumont nevertheless performed more than 200 experiments on St. Martin over a period of several years. He placed food into the stomach through the fistula and removed it hourly to observe the progress of digestion. He sent vials of gastric juice to chemists for analysis. He proved that digestion required hydrochloric acid and an unknown agent we now know to be the enzyme pepsin. St. Martin had a short temper, and during his outbursts, Beaumont observed that little digestion occurred; we now know this to be due to the inhibitory effect of the sympathetic nervous system on digestion.

In 1833, Beaumont published a book on his results that laid a foundation for modern *gastroenterology*,[1] the scientific study and medical treatment of the digestive system. Many authorities continued to believe, for a time, that the stomach acted essentially as a grinding chamber, fermentation vat, or cooking pot. Some of them even attributed digestion to a supernatural spirit in the stomach. Beaumont was proven right, however, and his work had no equal until Russian physiologist Ivan Pavlov built upon it, receiving a Nobel Prize in 1904 for his studies of digestion.

24.1 Digestive Processes and General Anatomy

▌ Expected Learning Outcomes

When you have completed this section, you should be able to

- identify the functions and major processes of the digestive system;
- list the regions and accessory organs of the digestive system;
- identify the layers of the wall of the digestive tract;
- describe the enteric nervous system; and
- name the mesenteries and describe their relationship to the digestive system.

The digestive system is essentially a disassembly line—its primary purpose is to break nutrients down into forms that can be used by the body, and to absorb them so they can be distributed to the tissues. Most of what we eat cannot be used in the form found in the food. Nutrients must be broken down into smaller components, such as amino acids and monosaccharides, that are universal to all species. Consider what happens if you eat a piece of beef, for example. The myosin of beef differs very little from that of your own muscles, but the two are not identical, and even if they were, beef myosin could not be absorbed, transported in the blood, and properly installed in your muscle cells. Like any other dietary protein, it must be broken down into amino acids before it can be used. Since beef and human proteins are made of the same 20 amino acids, those of beef proteins might indeed become part of your own myosin but could equally well wind up in your insulin, fibrinogen, collagen, or any other protein.

Digestive System Functions

The **digestive system** is the organ system that processes food, extracts nutrients from it, and eliminates the residue. It does this in five stages:

1. **ingestion,** the selective intake of food;
2. **digestion,** the mechanical and chemical breakdown of food into a form usable by the body;
3. **absorption,** the uptake of nutrients into the blood and lymph;
4. **compaction,** absorbing water and consolidating the indigestible residue into feces; and finally,
5. **defecation,** the elimination of feces.

General Anatomy

The digestive system has two anatomical subdivisions, the digestive tract and the accessory organs (fig. 24.1). The **digestive tract** is a muscular tube extending from mouth to anus, measuring about 9 m (30 ft) long in the cadaver. It is also known as the *alimentary*[2] *canal* or *gut*. It includes the mouth, pharynx, esophagus, stomach, small intestine, and large intestine. Part of it, the stomach and intestines, constitutes the *gastrointestinal (GI) tract.* The **accessory organs** are the teeth, tongue, salivary glands, liver, gallbladder, and pancreas.

The digestive tract is open to the environment at both ends. Most of the material in it has not entered any body tissues and is considered to be external to the body until it is absorbed by epithelial cells of the alimentary canal. In the strict sense, defecated food residue was never in the body.

Most of the digestive tract follows the basic structural plan shown in figure 24.2, with a wall composed of the following tissue layers, in order from the inner to the outer surface:

Mucosa

　Epithelium

　Lamina propria

　Muscularis mucosae

---

[1]*gastro* = stomach + *entero* = intestine + *logy* = study of

[2]*aliment* = food

Submucosa

Muscularis externa

    Circular layer

    Longitudinal layer

Serosa

    Areolar tissue

    Mesothelium

Slight variations on this theme are found in different regions of the tract.

The inner lining of the tract, called its **mucosa** or **mucous membrane,** consists of an epithelium, a loose connective tissue layer called the **lamina propria,** and a thin layer of smooth muscle called the **muscularis mucosae** (MUSS-cue-LAIR-is mew-CO-

see). The epithelium is simple columnar in most of the digestive tract, but the mouth, pharynx, esophagus, and anal canal differ. These upper and lower ends of the digestive tract are subject to more abrasion than the stomach and intestines, and thus have a stratified squamous epithelium. The muscularis mucosae tenses the mucosa, creating grooves and ridges that enhance its surface area and contact with food. This improves the efficiency of digestion and nutrient absorption. The mucosa often exhibits an abundance of lymphocytes and lymphatic nodules—the *mucosa-associated lymphatic tissue (MALT)* (see p. 616).

The **submucosa** is a thicker layer of loose connective tissue containing blood vessels, lymphatic vessels, a nerve plexus, and in some places, mucous glands. The MALT also extends into the submucosa in some parts of the GI tract.

The **muscularis externa** consists of usually two layers of smooth muscle near the outer surface. Cells of the inner layer encircle the tract and those of the outer layer run longitudinally. In some places, the circular layer is thickened to form valves (sphincters) that regulate the passage of material through the digestive tract. The muscularis externa is responsible for peristalsis and other forms of motility that mix food and digestive enzymes and propel material through the tract.

The **serosa** is composed of a thin layer of areolar tissue topped by a simple squamous mesothelium. It begins in the lower 3 to 4 cm of the esophagus and ends just before the rectum. The pharynx, most of the esophagus, and the rectum have no serosa but are surrounded by a fibrous connective tissue layer called the **adventitia** (AD-ven-TISH-ah), which blends into the adjacent connective tissues of other organs.

Innervation

Tongue movements, mastication, and the initial actions of swallowing employ skeletal muscles innervated by somatic motor fibers from six of the cranial nerves (V, VII, and IX–XII) and from the ansa cervicalis; these muscles and their innervation are detailed in table 11.3 (p. 273). The salivary glands are innervated by sympathetic fibers from the superior cervical ganglion and parasympathetic fibers from cranial nerves VII and IX (see figs. 15.31, p. 433; 15.33, p. 434; and 16.4, p. 447).

From the lower esophagus to the anal canal, most of the muscle is smooth muscle (the external anal sphincter is the only exception), and therefore receives only autonomic innervation. Parasympathetic innervation dominates the digestive tract and comes mainly from the vagus nerves, which supply all of the tract from esophagus to transverse colon. The descending colon and rectum receive their parasympathetic innervation from pelvic nerves arising from the inferior hypogastric plexus (see fig. 16.7, p. 451). The parasympathetic nervous system relaxes sphincter muscles and stimulates gastrointestinal motility and secretion. Thus, in general, it promotes digestion.

The sympathetic nervous system plays a lesser role, but in general it inhibits motility and secretion and keeps the GI sphincters contracted and closed. Thus, it inhibits digestion. Sympathetic efferent pathways travel through the celiac ganglion to the stomach, liver, and pancreas; through the superior mesenteric ganglion to the small intestine and most of the large intestine; and through the inferior mesenteric ganglion to the rectum (see fig. 16.4).

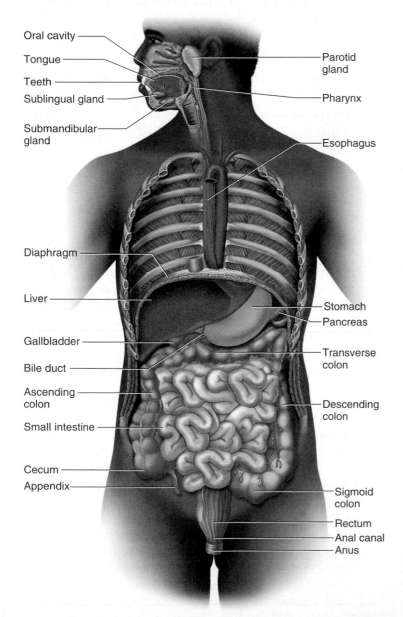

Oral cavity

Tongue

Teeth

Sublingual gland

Submandibular gland

Parotid gland

Pharynx

Esophagus

Diaphragm

Liver

Gallbladder

Bile duct

Ascending colon

Small intestine

Cecum

Appendix

Stomach

Pancreas

Transverse colon

Descending colon

Sigmoid colon

Rectum

Anal canal

Anus

Figure 24.1 The Digestive System.

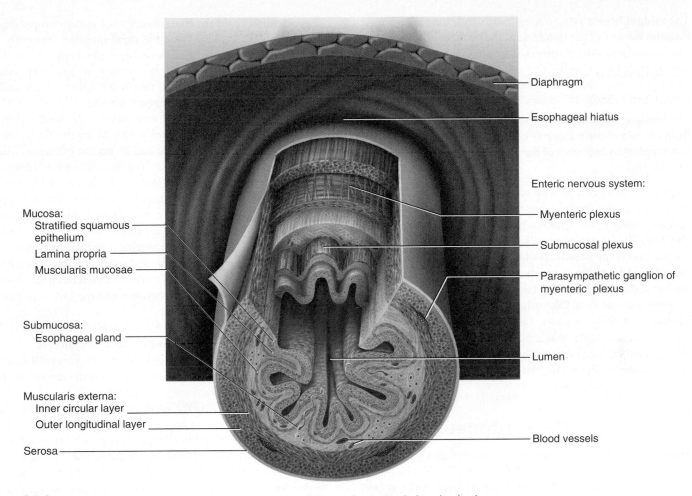

Diaphragm

Esophageal hiatus

Enteric nervous system:

Myenteric plexus

Submucosal plexus

Parasympathetic ganglion of myenteric plexus

Mucosa:
   Stratified squamous epithelium
   Lamina propria
   Muscularis mucosae

Submucosa:
   Esophageal gland

Muscularis externa:
   Inner circular layer
   Outer longitudinal layer

Serosa

Lumen

Blood vessels

Figure 24.2 **Tissue Layers of the Digestive Tract.** Cross section of the esophagus just below the diaphragm.

Even though the digestive tract receives such extensive innervation from the CNS, it can function independently even if these nerves are severed. This is because the esophagus, stomach, and intestines have their own nervous network called the **enteric**[3] **nervous system,** which is thought to have over 100 million neurons—more than the spinal cord! These include sensory neurons that monitor tension in the gut wall; interneurons; and motor neurons that activate such effectors as smooth muscle and gland cells of the gut. It also contains clusters of parasympathetic postganglionic neurons that act essentially as autonomic ganglia.

Neurons of the enteric nervous system are distributed in two networks: the **submucosal (Meissner**[4]**) plexus** in the submucosa and the **myenteric (Auerbach**[5]**) plexus** between layers of the muscularis externa (fig. 24.2). Parasympathetic preganglionic fibers terminate in the ganglia of the myenteric plexus. Postganglionic fibers arising in this plexus not only innervate the muscularis externa, but also pass through its inner circular layer and contribute to the submucosal plexus. The myenteric plexus controls peristalsis

and other contractions of the muscularis externa, while the submucosal plexus controls movements of the muscularis mucosae and glandular secretion of the mucosa.

Sensory nerve fibers monitor stretching of the GI wall and chemical conditions in the lumen. These fibers carry signals to adjacent regions of the GI tract in **short (myenteric) reflex arcs** contained in the myenteric plexus, and to the central nervous system by way of **long (vagovagal) reflex arcs,** predominantly in the vagus nerves. These visceral reflex arcs enable different regions of the GI tract to regulate each other over both short and long distances.

Circulation

We will see near the end of this chapter that the embryonic digestive tract forms in three segments: the foregut, midgut, and hindgut. These segments are defined by their arterial blood supply.

- The **foregut** includes the mouth, pharynx, esophagus, stomach, and the beginning of the duodenum (to the point where the bile duct empties into it). Above the diaphragm, the thoracic aorta gives off a series of **esophageal arteries** to the esophagus. Below the diaphragm, the foregut receives blood from branches of the **celiac trunk** (see fig. 21.23, p. 584).

[3]*enter* = intestine
[4]Georg Meissner (1829–1905), German histologist
[5]Leopold Auerbach (1828–97), German anatomist

- The **midgut** begins at the opening of the bile duct and includes the rest of the duodenum, the jejunum and ileum (the second and third portions of the small intestine), and the large intestine as far as the first two-thirds of the transverse colon. It receives blood from the **superior mesenteric artery** (fig. 21.24a, p. 586).

- The **hindgut** includes the remainder of the large intestine, from the end of the transverse colon through the anal canal. It is supplied by branches of the **inferior mesenteric artery** (fig. 21.24b).

The most noteworthy general point about the venous drainage of the GI tract is that blood from the entire tract below the diaphragm ultimately drains into the **hepatic portal vein,** which enters the liver. The system of vessels connecting the lower digestive tract to the liver is the **hepatic portal system** (table 21.7, p. 589). It routes all blood from the stomach and intestines, as well as from some other abdominal viscera, through the liver before returning it to the general circulation. Like other portal systems, this one has two capillary networks in series. Capillaries in the small intestine receive digested nutrients, and capillaries in the liver (the *hepatic sinusoids* described later) deliver these nutrients to the liver cells. This gives the liver a chance to process most nutrients and cleanse the intestinal blood of bacteria.

Relationship to the Peritoneum

In processing food, the stomach and intestines undergo such strenuous contractions that they need freedom to move in the abdominal cavity. They are not tightly bound to the abdominal wall, but over most of their length, they are loosely suspended from it by connective tissue sheets called the **mesenteries** (see figs. 1.16 and 1.17, p. 19).

The mesenteries hold the abdominal viscera in their proper relationship to each other and prevent the small intestine, especially, from becoming twisted and tangled by changes in body position and by its own contractions. Furthermore, the mesenteries provide passage for the blood vessels and nerves that supply the digestive tract, and contain many lymph nodes and lymphatic vessels.

The parietal peritoneum is a serous membrane that lines the wall of the abdominal cavity (see p. 18). Along the posterior midline of the body, it turns inward and forms the **posterior (dorsal) mesentery,** a translucent two-layered membrane extending to the digestive tract. Upon reaching an organ such as the stomach or small intestine, the two layers separate and pass around opposite sides of the organ, forming its serosa. In some places, the two layers come together again on the far side of that organ and continue as another sheet of tissue, the **anterior (ventral) mesentery.** This mesentery may hang freely in the abdominal cavity or attach to the anterior abdominal wall or other organs. The relationship between the posterior and anterior mesenteries and the serosa is shown in figure 1.16.

Two mesenteries called *omenta* are associated with the stomach (see fig. 24.11). The **lesser omentum** extends the short distance from the liver to the right superior margin *(lesser curvature)* of the stomach (fig. 24.3a). A much larger and very fatty **greater omentum** hangs like an apron from the left inferior margin *(greater curvature)* of the stomach, loosely covering the small intestine. At its inferior margin, the greater omentum turns back on itself and passes upward, behind the superficial layer, forming a deep, empty pouch between the layers like an apron turned up at the hem. At the superior margin, the upturned layer continues as a serosa that encloses the spleen and transverse colon. From the transverse colon, it continues to the posterior abdominal wall, thus anchoring the colon. This mesentery of the colon is called the **mesocolon** (fig. 24.3b).

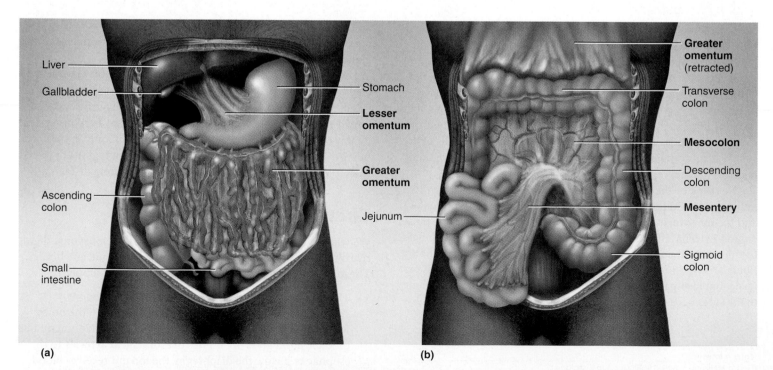

(a) (b)

Figure 24.3 **Serous Membranes Associated with the Digestive Tract.** (a) The greater and lesser omenta. (b) Greater omentum and small intestine retracted to show the mesocolon and mesentery. These membranes contain the mesenteric arteries and veins.

The omenta have a loosely organized, lacy appearance due partly to many holes or gaps in the membranes and partly to an irregular distribution of adipose tissue. They adhere to perforations or inflamed areas of the stomach or intestines, contribute immune cells to the site, and isolate infections that might otherwise give rise to peritonitis. When a person gains abdominal weight, a great deal of it is fat deposited in these mesenteries.

When an organ is enclosed by mesentery (serosa) on both sides, it is considered to be within the peritoneal cavity, or **intraperitoneal.**[6] When an organ lies against the posterior body wall and is covered by peritoneum on the anterior side only, it is said to be outside the peritoneal cavity, or **retroperitoneal.**[7] The duodenum, most of the pancreas, and parts of the large intestine are retroperitoneal. The stomach, liver, and other parts of the small and large intestines are intraperitoneal.

Before You Go On

Answer the following questions to test your understanding of the preceding section:

1. Name the principal regions of the digestive tract in order from mouth to anus.

2. What are the similarities and differences between the lamina propria and the submucosa?

3. What are the two components of the enteric nervous system? How do they differ in location and function?

4. What three major branches of the aorta supply the digestive tract? What is the relationship between these three arterial supplies and the embryonic development of the digestive tract?

5. Name the serous membranes that suspend the intestines from the posterior body wall. Name the external layer of the intestines formed by an extension of this membrane.

24.2 The Mouth Through Esophagus

▶ Expected Learning Outcomes

When you have completed this section, you should be able to

- describe the teeth, tongue, and other organs of the oral cavity;
- state the names and locations of the salivary glands;
- describe the location and function of the pharyngeal constrictor muscles; and
- describe the gross anatomy and histology of the esophagus.

[6]*intra* = within
[7]*retro* = behind

The Mouth

The **mouth** is also known as the **oral (buccal**[8]**) cavity.** Its functions include ingestion (food intake), taste and other sensory responses to food, mastication (chewing), chemical digestion, swallowing, speech, and respiration. The mouth is enclosed by the cheeks, lips, palate, and tongue (fig. 24.4). Its anterior opening between the lips is the **oral fissure,** and its posterior opening into the throat is the **fauces**[9] (FAW-seez). The mouth is lined with stratified squamous epithelium. It is keratinized in areas subject to the greatest abrasion, such as the gums and hard palate, and nonkeratinized in other areas such as the floor of the mouth, the soft palate, and the inside of the cheeks and lips.

The Cheeks and Lips

The cheeks and lips retain food and push it between the teeth for mastication, and are essential for articulate speech and for sucking and blowing actions, including suckling by infants. Their fleshiness is due mainly to subcutaneous fat, the buccinator muscles of the cheeks, and the orbicularis oris muscle of the lips. A median fold called the **labial frenulum**[10] attaches each lip to the gum, between the anterior incisors. The **vestibule** is the space between the cheeks or lips and the teeth—the space where you insert your toothbrush when brushing the outer surfaces of the teeth.

[8]*bucca* = cheek
[9]*fauces* = throat
[10]*labi* = lip + *frenulum* = little bridle

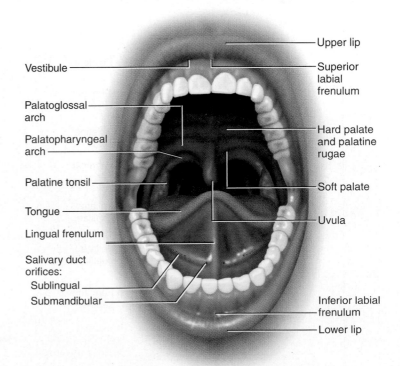

Figure 24.4 The Mouth. For a photographic medial view, see page 332.

The lips are divided into three areas: (1) The *cutaneous area* is colored like the rest of the face and has hair follicles and sebaceous glands; on the upper lip, this is where a mustache grows. (2) The *red area (vermilion),* is the hairless region where the lips meet (where some people apply lipstick). It has unusually tall dermal papillae, which allow blood capillaries and nerve endings to come closer to the epidermal surface. Thus, this area is redder and more sensitive than the cutaneous area. (3) The *labial mucosa* is the inner surface of the lip, facing the gums and teeth.

The Tongue

The tongue (fig. 24.5), although muscular and bulky, is an agile and sensitive organ with several functions: it aids in food intake; it has sensory receptors for taste, texture, and temperature that are important in the acceptance or rejection of food; it compresses and breaks up food; it maneuvers food between the teeth for mastication; it secretes mucus and enzymes; it compresses the chewed food into a soft mass, or *bolus,* that is easier to swallow; it initiates swallowing; and it is necessary for articulate speech. Its surface is covered with nonkeratinized stratified squamous epithelium and exhibits bumps and projections called **lingual papillae,** the site of the taste buds. Figure 24.5 shows the locations of the vallate, foliate, and fungiform types of papillae. Their details and relation to the taste buds are described in chapter 17 (p. 466).

The anterior two-thirds of the tongue, called the **body,** occupies the oral cavity; the posterior one-third, the **root,** occupies the oropharynx. The boundary between the body and root is marked by a V-shaped row of vallate papillae and, behind these, a groove called the **terminal sulcus.** Beneath the tongue, a median fold called the **lingual frenulum** attaches its body to the floor of the

mouth. The root of the tongue contains the lingual tonsils. Amid the tongue muscles are serous and mucous **lingual glands,** which secrete a portion of the saliva.

The muscles of the tongue, which compose most of its mass, are described in chapter 11. The **intrinsic muscles,** contained entirely within the tongue, produce the relatively subtle tongue movements of speech. The **extrinsic muscles,** with origins elsewhere and insertions in the tongue, produce most of the stronger movements of food manipulation. The extrinsic muscles include the *genioglossus, hyoglossus, styloglossus,* and *palatoglossus* (see table 11.3, p. 273).

The Palate

The palate (fig. 24.4), separating the oral cavity from the nasal cavity, makes it possible to breathe while chewing food. Its anterior portion, the **hard (bony) palate,** is supported by the palatine processes of the maxillae and by the smaller palatine bones. It has transverse ridges called *palatine rugae* that aid the tongue in holding and manipulating food. Posterior to this is the **soft palate,** which has a more spongy texture and is composed mainly of skeletal muscle and glandular tissue, but no bone. It has a conical medial projection, the **uvula,**[11] visible at the rear of the mouth. The uvula helps to retain food in the mouth until one is ready to swallow.

At the rear of the mouth, two muscular arches on each side begin at the roof near the uvula and descend to the floor. The anterior one is the **palatoglossal arch** and the posterior one is the **palatopharyngeal arch.** The latter arch marks the beginning of the pharynx. The palatine tonsils are located on the wall between the arches.

---

[11]*uvula* = little grape

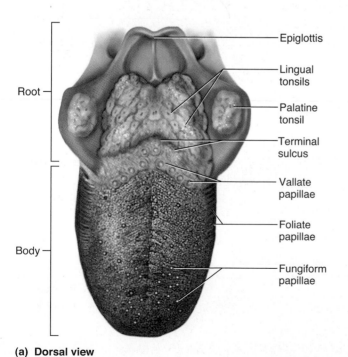

(a) **Dorsal view**

Root

Body

- Epiglottis
- Lingual tonsils
- Palatine tonsil
- Terminal sulcus
- Vallate papillae
- Foliate papillae
- Fungiform papillae

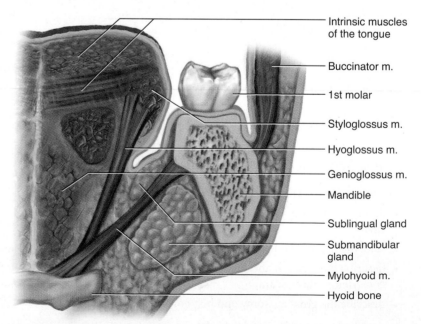

(b) **Frontal section, anterior view**

- Intrinsic muscles of the tongue
- Buccinator m.
- 1st molar
- Styloglossus m.
- Hyoglossus m.
- Genioglossus m.
- Mandible
- Sublingual gland
- Submandibular gland
- Mylohyoid m.
- Hyoid bone

Figure 24.5 **The Tongue.** For a sagittal section, see figure A.2 (p. 332).

The Teeth

The teeth are collectively called the **dentition.** They serve to masticate the food, breaking it into smaller pieces. This not only makes the food easier to swallow, but also exposes more surface area to the action of digestive enzymes and thus speeds up chemical digestion. Adults normally have 16 teeth in the mandible and 16 in the maxilla. From the midline to the rear of each jaw, there are two incisors, a canine, two premolars, and up to three molars (fig. 24.6a). The **incisors** are chisel-like cutting teeth used to bite off a piece of food. The **canines** are more pointed and act to puncture and shred it. They serve as weapons in many mammals but became reduced in the course of human evolution until they now project barely beyond the other teeth. The **premolars** and **molars** have relatively broad surfaces adapted for crushing and grinding.

Each tooth is embedded in a socket called an **alveolus,** forming a joint called a *gomphosis* between the tooth and bone (fig. 24.7). The alveolus is lined by a **periodontal** (PERR-ee-oh-DON-tul) **ligament,** a modified periosteum whose collagen fibers penetrate into the bone on one side and into the tooth on the other. This anchors the tooth firmly in the alveolus, but allows for slight movement under the stress of chewing. The gum, or **gingiva** (JIN-jih-vuh), covers the alveolar bone. Regions of a tooth are defined by their relationship to the gingiva: the **crown** is the portion above the gum, the **root** is the portion inserted into the alveolus below the gum, and the **neck** is the line where the crown, root, and gum meet. The space between the tooth and gum is the **gingival sulcus.** The hygiene of this sulcus is especially important for dental health (see Deeper Insight 24.1).

Most of a tooth consists of hard yellowish tissue called **dentin,** covered with **enamel** in the crown and **cementum** in the root. Dentin and cementum are living connective tissues with cells or cell processes embedded in a calcified matrix. Cells of the cementum *(cementocytes)* are scattered more or less randomly and occupy tiny cavities similar to the lacunae of bone. Cells of the dentin *(odontoblasts)* line the pulp

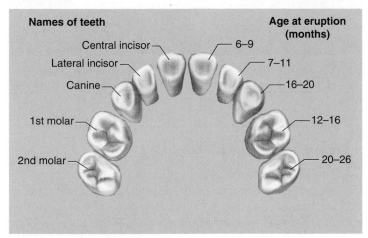

Names of teeth / **Age at eruption (months)**

- Central incisor — 6–9
- Lateral incisor — 7–11
- Canine — 16–20
- 1st molar — 12–16
- 2nd molar — 20–26

(a) Deciduous (baby) teeth

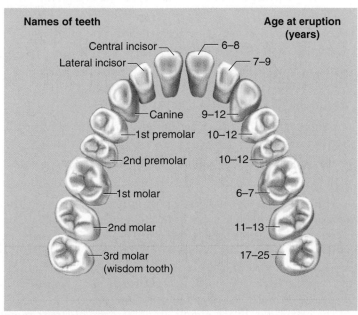

Names of teeth / **Age at eruption (years)**

- Central incisor — 6–8
- Lateral incisor — 7–9
- Canine — 9–12
- 1st premolar — 10–12
- 2nd premolar — 10–12
- 1st molar — 6–7
- 2nd molar — 11–13
- 3rd molar (wisdom tooth) — 17–25

(b) Permanent teeth

Figure 24.6 The Dentition. Each figure shows only the upper teeth. The ages at eruption are composite ages for the corresponding upper and lower teeth. Generally, the lower (mandibular) teeth erupt somewhat earlier than their upper (maxillary) counterparts.
● *Which teeth are absent from a 3-year-old child?*

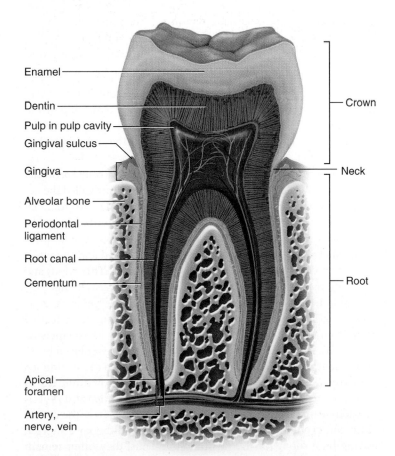

- Enamel
- Dentin
- Pulp in pulp cavity
- Gingival sulcus
- Gingiva
- Alveolar bone
- Periodontal ligament
- Root canal
- Cementum
- Apical foramen
- Artery, nerve, vein
- Crown
- Neck
- Root

Figure 24.7 Structure of a Tooth and Its Alveolus. This particular example is a molar.
● *Of all the components shown here, which is or are not living tissue(s)?*

DEEPER INSIGHT 24.1

Tooth and Gum Disease

Food leaves a sticky residue on the teeth called *plaque,* composed mainly of bacteria and sugars. If plaque is not thoroughly removed by brushing and flossing, bacteria accumulate, metabolize the sugars, and release lactic acid and other acids. These acids dissolve the minerals of enamel and dentin, and the bacteria enzymatically digest the collagen and other organic components. The eroded "cavities" of the tooth are known as *dental caries.*[12] If not repaired, caries may fully penetrate the dentin and spread to the pulp cavity. This requires either extraction of the tooth or *root canal therapy,* in which the pulp is removed and replaced with inert material.

When plaque calcifies on the tooth surface, it is called *calculus (tartar).* Calculus in the gingival sulcus wedges the tooth and gum apart and allows bacterial invasion of the sulcus. This leads to *gingivitis,* or gum inflammation. Nearly everyone has gingivitis at some time. In some cases, bacteria spread from the sulcus into the alveolar bone and begin to dissolve it, producing *periodontal disease.* About 86% of people over age 70 have periodontal disease, and many suffer tooth loss as a result. This accounts for 80% to 90% of adult tooth loss.

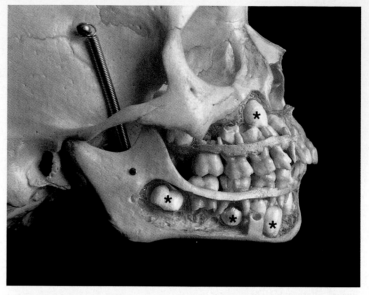

Figure 24.8 **Permanent and Deciduous Teeth in a Child's Skull.** This dissection shows erupted deciduous teeth and, deep to them and marked with asterisks, the permanent teeth waiting to erupt.

cavity and have slender processes that travel through tiny parallel tunnels in the dentin. Enamel is not a tissue but a cell-free secretion produced before the tooth emerges above the gum. Damaged dentin and cementum can regenerate, but damaged enamel cannot—it must be artificially repaired.

Internally, a tooth has a dilated **pulp cavity** in the crown and upper root, and a narrow **root canal** in the lower root. These spaces are occupied by **pulp**—a mass of loose connective tissue, blood and lymphatic vessels, and nerves. These nerves and vessels enter the tooth through a pore, the **apical foramen,** at the inferior end of each root canal.

The meeting of the teeth when the mouth closes is called *occlusion* (ah-CLUE-zhun), and the surfaces where they meet are called the **occlusal surfaces.** The occlusal surface of a premolar has two rounded bumps called **cusps;** thus, the premolars are also known as **bicuspids.** The molars have four to five cusps. Cusps of the upper and lower premolars and molars mesh when the jaws are closed and slide over each other as the jaw makes lateral chewing motions. This grinds and tears food more effectively than if the occlusal surfaces were flat.

Teeth develop beneath the gums and **erupt** (emerge) in predictable order. Twenty **deciduous teeth** ("milk teeth" or "baby teeth") erupt from the ages of 6 to 30 months, beginning with the incisors (fig. 24.6a). Between 6 and 25 years of age, these are replaced by 32 **permanent teeth.** As a permanent tooth grows deep to a deciduous tooth (fig. 24.8), the root of the deciduous tooth dissolves and leaves little more than the crown by the time it falls out. The third molars (wisdom teeth) erupt around ages 17 to 25, if at all. Over the course of human evolution, the face became flatter and the jaws shorter, leaving little room for the third molars. Thus, they often remain below the gum and become *impacted*—so crowded against neighboring teeth and bone that they cannot erupt.

The Salivary Glands

Saliva moistens the mouth; digests a small amount of starch and fat; cleanses the teeth; inhibits bacterial growth; dissolves molecules so they can stimulate the taste buds; and moistens and lubricates food and binds particles together to aid in swallowing. It is a solution of 97.0% to 99.5% water and the following solutes:

- **salivary amylase,** an enzyme that begins starch digestion in the mouth;
- **lingual lipase,** an enzyme that is activated by stomach acid and digests fat after the food is swallowed;
- **mucus,** which binds and lubricates the food mass and aids in swallowing;
- **lysozyme,** an enzyme that kills bacteria;
- **immunoglobulin A (IgA),** an antibody that inhibits bacterial growth; and
- **electrolytes,** including sodium, potassium, chloride, phosphate, and bicarbonate salts.

There are two kinds of salivary glands, intrinsic and extrinsic. The **intrinsic salivary glands** are an indefinite number of small glands dispersed amid the other oral tissues. They include *lingual glands* in the tongue, *labial glands* on the inside of the lips, and *buccal glands* on the inside of the cheeks. They secrete saliva at a fairly constant rate whether we are eating or not, but in relatively small amounts. This saliva keeps the mouth moist and inhibits bacterial growth.

The **extrinsic salivary glands** are three pairs of larger, more discrete organs located outside of the oral mucosa. They communicate with the oral cavity by way of ducts (fig. 24.9). They are:

1. The **parotid**[13] **gland,** located just beneath the skin anterior to the earlobe. Its duct passes superficially over the masseter,

---

[12]*caries* = rottenness

[13]*par* = next to + *ot* = ear

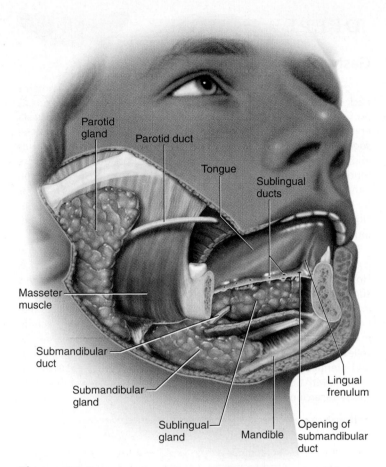

Figure 24.9 **The Extrinsic Salivary Glands.** Half of the mandible has been removed to expose the sublingual gland medial to it.

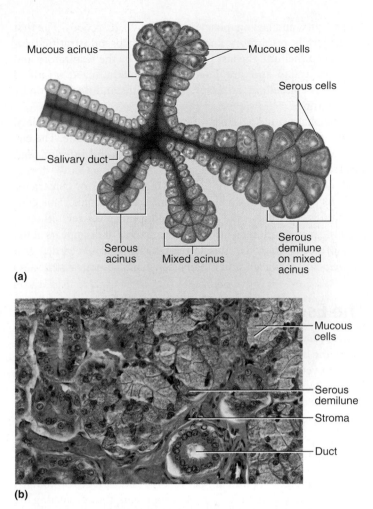

(a)

(b)

Figure 24.10 **Microscopic Anatomy of the Salivary Glands.** (a) Duct and acini of a generalized salivary gland with a mixture of mucous and serous cells. Serous cells often form crescent-shaped caps called serous demilunes over the ends of mucous acini. No one salivary gland shows all the features shown here. (b) Histology of the sublingual salivary gland.

pierces the buccinator, and opens into the mouth opposite the second upper molar tooth. *Mumps* is an inflammation and swelling of the parotid gland caused by a virus.

2. The **submandibular gland,** located halfway along the body of the mandible, medial to its margin, just deep to the mylohyoid muscle. Its duct empties into the mouth at a papilla on the side of the lingual frenulum, near the lower central incisors (see fig. 24.4).

3. The **sublingual gland,** located in the floor of the mouth. It has multiple ducts that empty into the mouth posterior to the papilla of the submandibular duct (see fig. 24.4).

These are all compound tubuloacinar glands with a treelike arrangement of branching ducts ending in acini (see p. 74). Some acini have only mucous cells, some have only serous cells, and some have a mixture of both (fig. 24.10). Typically on the last of these, called *mixed acini,* the mucous cells surround the lumen and the serous cells form a cap on one end, shaped a bit like a croissant, called the *serous demilune.* Mucous cells secrete salivary mucus, and serous cells secrete a thinner fluid rich in amylase and electrolytes.

Salivation is controlled by groups of neurons called **salivatory nuclei** in the medulla oblongata and pons. They receive signals from sensory receptors in the mouth as well as from higher brain centers that respond to the odor, sight, or thought of food. Efferent nerve pathways to the salivary glands were described earlier (p. 656). Salivation is mostly under the control of parasympathetic fibers in cranial nerves VII and IX, which stimulate the secretion of watery, enzyme-rich saliva. Sympathetic fibers from cervical ganglia stimulate the secretion of thicker, mucus-rich saliva.

Apply What You Know

Explain why the mouth may feel dry or sticky when one is nervous. (*Hint:* Draw on information about the autonomic nervous system in chapter 16.)

The Pharynx

The pharynx, as described in chapter 23, is a muscular funnel that connects the oral cavity to the esophagus and the nasal cavity to the larynx; thus, it is a point where the digestive and respiratory tracts intersect. It consists of three regions called the nasopharynx, oropharynx, and laryngopharynx (see fig. 23.2c, p. 635). The first is exclusively respiratory and is lined with pseudostratified columnar epithelium; the last two are shared by the respiratory and digestive tracts and are lined with nonkeratinized stratified squamous epithelium, an adaptation to withstanding abrasion by passing food.

The pharynx has a deep layer of longitudinally oriented skeletal muscle and a superficial layer of circular skeletal muscle. The circular muscle is divided into **superior, middle,** and **inferior pharyngeal constrictors,** which force food downward during swallowing. When one is not swallowing, the inferior constrictor remains contracted to exclude air from the esophagus. This constriction is regarded as the **upper esophageal sphincter,** although it is not an anatomical feature of the esophagus. It disappears at the time of death when the muscle relaxes. Thus, it is regarded as a *physiological sphincter* rather than a constant anatomical structure.

The Esophagus

The **esophagus** is a straight muscular tube 25 to 30 cm long, posterior to the trachea (see figs. 24.1 and 24.2). Its superior opening lies between vertebra C6 and the cricoid cartilage of the larynx. After passing downward through the mediastinum, the esophagus penetrates the diaphragm at an opening called the **esophageal hiatus,** continues another 3 to 4 cm, and meets the stomach at the level of vertebra T7. Its opening into the stomach is called the **cardiac orifice** (for its proximity to the heart). Food pauses briefly at this point before entering the stomach because of a constriction called the **lower esophageal sphincter (LES).** The LES prevents stomach contents from regurgitating into the esophagus, thus protecting the esophageal mucosa from the erosive effect of the stomach acid (see Deeper Insight 24.2).

The wall of the esophagus is organized into the tissue layers described earlier, with some regional specializations. The mucosa has a nonkeratinized stratified squamous epithelium. The submucosa contains **esophageal glands,** which secrete lubricating mucus into the lumen. When the esophagus is empty, the mucosa and submucosa are deeply folded into longitudinal ridges, giving the lumen a starlike shape in cross section.

The muscularis externa is composed of skeletal muscle in the upper one-third of the esophagus, a mixture of skeletal and smooth muscle in the middle one-third, and only smooth muscle in the lower one-third.

Most of the esophagus is in the mediastinum. Here, it is covered with a connective tissue adventitia that merges into the adventitias of the trachea and thoracic aorta. The short segment below the diaphragm is partially covered by a serosa.

Swallowing, or *deglutition* (DEE-glu-TISH-un), is a complex action involving over 22 muscles in the mouth, pharynx, and esophagus, coordinated by the **swallowing center,** a pair of nuclei in the medulla oblongata. This center communicates with muscles of the pharynx and esophagus by way of the trigeminal, facial, glossopharyngeal, and hypoglossal nerves (cranial nerves V, VII, IX, and XII), and coordinates a complex series of muscle contractions to produce swallowing without choking.

DEEPER INSIGHT   24.2

Gastroesophageal Reflux Disease

It would seem that churning of the stomach would drive its contents back up the esophagus. Such backflow, or *gastroesophageal reflux,* is normally prevented, however, by the lower esophageal sphincter. Weakening of the LES leads to repetitive or chronic backflow, called *gastroesophageal reflux disease (GERD).* Stomach acid and sometimes bile acids and pancreatic enzymes regurgitate into the esophagus and irritate the mucosa. This causes the sensation of "heartburn," named for its location although it has nothing to do with the heart. GERD affects as much as 50% of the population in the United States, especially white males. Beside sex and race, risk factors include age (middle-aged or beyond), being overweight, and going to bed too soon after eating.

The heartburn sensation can often be managed with antacids and is commonly dismissed by physicians and patients as merely a nuisance. However, in a few cases, GERD can lead to more serious complications such as scarring and narrowing of the esophagus *(stricture),* erosion and inflammation of the esophageal wall *(erosive esophagitis),* a transformation (metaplasia) of esophageal epithelium to an intestinal-type columnar epithelium *(Barrett[14] esophagus),* and a form of esophageal cancer called *adenocarcinoma.* Although most people with Barrett esophagus and adenocarcinoma have a long-term history of GERD, only 5% to 15% of those with GERD progress to Barrett esophagus and fewer than 0.1% to adenocarcinoma.

Before You Go On

Answer the following questions to test your understanding of the preceding section:

6. List as many functions of the tongue as you can.
7. Name the four types of teeth in order from the midline to the rear of the jaw. How do they differ in function?
8. What is the difference in function and location between intrinsic and extrinsic salivary glands? Name the extrinsic salivary glands and describe their locations.
9. Identify at least two histological features of the esophagus that are especially tied to its role in swallowing.

24.3 The Stomach

▶ Expected Learning Outcomes

When you have completed this section, you should be able to
- describe the gross and microscopic anatomy of the stomach;
- describe the nerve and blood supply to the stomach;

[14]Norman R. Barrett (1903–79), British surgeon

- state the function of each type of epithelial cell in the gastric mucosa; and
- explain how the stomach is protected from digesting itself.

The stomach is a muscular sac in the upper left abdominal cavity immediately inferior to the diaphragm (see fig. 24.1). It functions primarily as a food storage organ, with an internal volume of about 50 mL when empty and 1.0 to 1.5 L after a typical meal. When extremely full, it may hold up to 4 L and extend nearly as far as the pelvis. The stomach mechanically breaks up food particles, liquefies the food, and begins the chemical digestion of proteins and fat. This produces a soupy or pasty mixture of semidigested food called **chyme**[15] (pronounced "kime"). Most digestion occurs after the chyme passes on to the small intestine. The stomach also produces secretions that are important in vitamin absorption, appetite control, and coordination of different parts of the digestive tract with each other.

Gross Anatomy

The stomach is somewhat J-shaped (fig. 24.11) and vertical in tall people, whereas in short people it is more nearly horizontal. The **lesser curvature** of the stomach is the margin that extends for the short distance (about 10 cm) from esophagus to duodenum along the medial to superior aspect, facing the liver; the **greater curvature** is the longer margin (about 40 cm) from esophagus to duodenum on the lateral to inferior aspect. As described earlier, the lesser omentum extends from the lesser curvature to the liver, and the greater omentum is suspended from the greater curvature and overhangs the intestines below.

The stomach is divided into four regions: (1) The **cardiac region (cardia)** is the small area within about 3 cm of the cardiac orifice. (2) The **fundic region (fundus)** is the domelike roof superior to the esophageal attachment. (3) The **body (corpus)** makes up most of the stomach distal to the cardiac orifice. (4) The **pyloric region** is a slightly narrower pouch at the distal end; it is subdivided into a funnel-like **antrum**[16] and a narrower **pyloric canal.** The latter terminates at the **pylorus,**[17] a narrow passage into the duodenum. The pylorus is surrounded by a thick ring of smooth muscle, the **pyloric (gastroduodenal) sphincter,** which regulates the passage of chyme into the duodenum.

Microscopic Anatomy

The stomach wall has tissue layers similar to those of the esophagus, with some variations. The surface of the mucosa is a simple columnar glandular epithelium (fig. 24.12). The apical regions of its cells are filled with mucin. After it is secreted, mucin swells with water and becomes mucus. When the stomach is full, the mucosa and submucosa are flat and smooth, but as it empties, these layers fold into longitudinal wrinkles called **gastric rugae**[18] (ROO-gee). The lamina propria is almost entirely occupied by tubular glands, to be described shortly. The muscularis externa has three layers, rather than two—an outer longitudinal, middle circular, and inner oblique layer (fig. 24.11a).

Apply What You Know

Contrast the epithelium of the esophagus with that of the stomach. Why is each epithelial type best suited to the function of its respective organ?

The gastric mucosa is pocked with depressions called **gastric pits** lined with the same columnar epithelium as the mucosal surface. Two or three tubular glands open into the bottom of each gastric pit and span the rest of the lamina propria. The glands are simple wavy or coiled tubes of more or less uniform diameter, except for a constriction called the **neck** at the point where the gland opens into the pit. In the cardiac and pyloric regions, they are called **cardiac glands** and **pyloric glands,** respectively. In the rest of the stomach, they are called **gastric glands** (fig. 24.12b, c). Collectively, the glands have the following cell types:

- **Mucous cells,** which secrete mucus, predominate in the cardiac and pyloric glands. In gastric glands, they are called *mucous neck cells* and are concentrated in the neck of the gland.

- **Regenerative (stem) cells,** found in the base of the pit and neck of the gland, divide rapidly and produce a continual supply of new cells. Newly generated cells migrate upward to the gastric surface as well as downward into the glands to replace cells that die and fall off into the lumen of the stomach.

- **Parietal cells,** found mostly in the upper half of the gland, secrete hydrochloric acid (HCl), a digestive aid; *intrinsic factor,* a glycoprotein needed for the absorption of dietary vitamin B_{12}; and an appetite-regulating hormone named *ghrelin* (GREL-in). They are found mostly in the gastric glands, but a few occur in the pyloric glands.

- **Chief cells,** so named because they are the most numerous, secrete the fat-digesting enzyme *gastric lipase,* as well as *pepsinogen,* the precursor of a protein-digesting enzyme called *pepsin.* They dominate the lower half of the gastric glands but are absent from cardiac and pyloric glands.

- **Enteroendocrine cells,** concentrated especially in the lower end of a gland, secrete hormones and paracrine messengers that regulate digestion. They are found in all regions of the stomach, but are most numerous in the gastric and pyloric glands. There are at least eight kinds, each of which produces a different chemical messenger. **G cells,** for example, secrete a hormone called *gastrin,* which stimulates the exocrine cells of the gastric glands to secrete acid and enzymes.

In general, the cardiac and pyloric glands secrete mainly mucus; acid and enzyme secretion occur predominantly in the gastric glands; and hormones are secreted throughout the stomach. Table 24.1 describes the functions of the gastric gland secretions.

---

[15]*chyme* = juice
[16]*antrum* = cavity
[17]*pylorus* = gatekeeper

[18]*ruga* = fold, crease

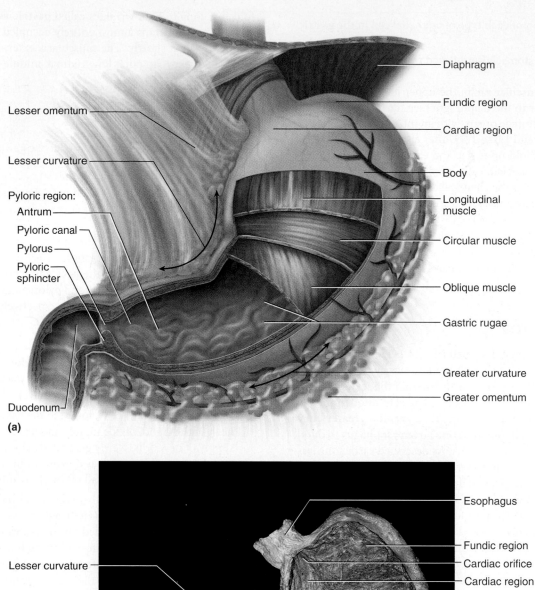

Lesser omentum

Lesser curvature

Pyloric region:
Antrum
Pyloric canal
Pylorus
Pyloric sphincter

Duodenum

(a)

Diaphragm

Fundic region

Cardiac region

Body

Longitudinal muscle

Circular muscle

Oblique muscle

Gastric rugae

Greater curvature

Greater omentum

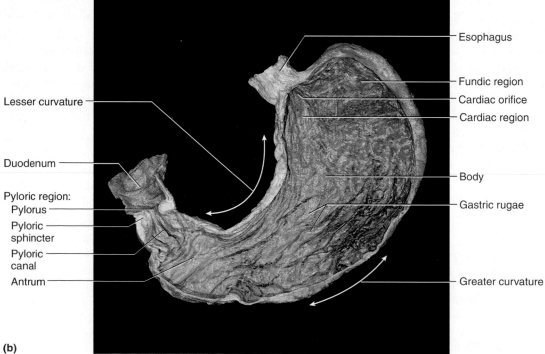

Lesser curvature

Duodenum

Pyloric region:
Pylorus
Pyloric sphincter
Pyloric canal
Antrum

(b)

Esophagus

Fundic region
Cardiac orifice
Cardiac region

Body

Gastric rugae

Greater curvature

Figure 24.11 **The Stomach.** (a) Gross anatomy. (b) Photograph of the internal surface.
• *How does the muscular wall of the stomach differ from that of the esophagus?*

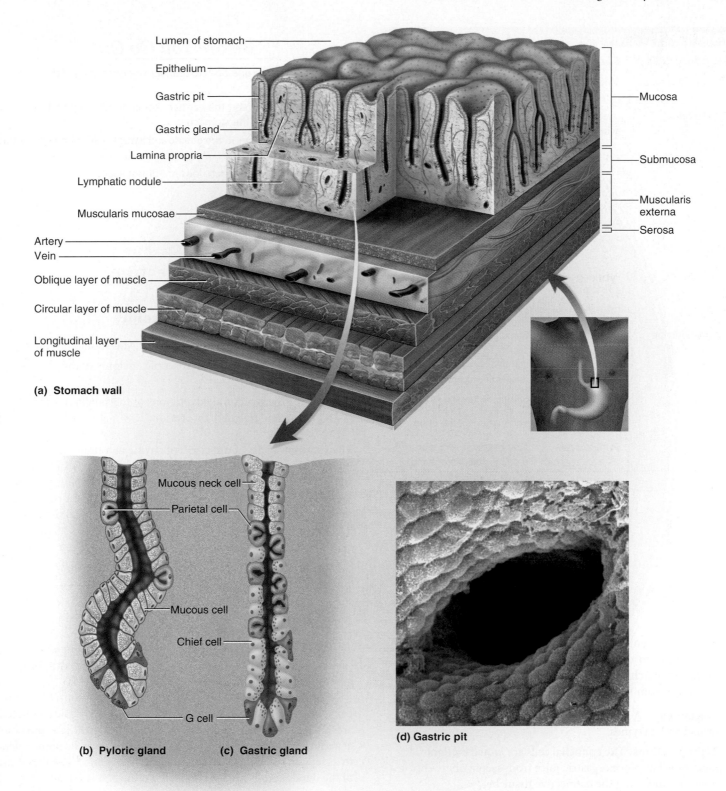

Lumen of stomach

Epithelium

Gastric pit

Gastric gland

Lamina propria

Lymphatic nodule

Muscularis mucosae

Artery

Vein

Oblique layer of muscle

Circular layer of muscle

Longitudinal layer of muscle

Mucosa

Submucosa

Muscularis externa

Serosa

(a) Stomach wall

Mucous neck cell

Parietal cell

Mucous cell

Chief cell

G cell

(b) Pyloric gland

(c) Gastric gland

(d) Gastric pit

Figure 24.12 Microscopic Anatomy of the Stomach Wall. (a) A block of tissue showing all layers from the mucosa (top) to the serosa (bottom). (b) A pyloric gland from the inferior end of the stomach. Note the absence of chief cells and relatively few parietal cells. (c) A gastric gland, the most widespread type in the stomach. (d) The opening of a gastric pit into the stomach, surrounded by the rounded apical surfaces of the columnar epithelial cells of the mucosa (SEM).

TABLE 24.1	Major Secretions of the Gastric Glands	
Secretory Cells	**Secretion**	**Function**
Mucous neck cells	Mucus	Protects mucosa from HCl and enzymes
Parietal cells	Hydrochloric acid (HCl)	Activates pepsin and lingual lipase; helps liquefy food; reduces dietary iron to usable form (Fe^{2+}); destroys ingested pathogens
	Intrinsic factor	Enables small intestine to absorb vitamin B_{12}
	Ghrelin	Secreted when stomach is empty; acts on brain to produce sensation of hunger
Chief cells	Pepsinogen	Converted to pepsin, which digests protein
	Gastric lipase	Digests fat
Enteroendocrine cells	Gastrin	Stimulates gastric glands to secrete HCl and enzymes; stimulates intestinal motility; relaxes ileocecal valve
	Serotonin	Stimulates gastric motility
	Histamine	Stimulates HCl secretion
	Somatostatin	Inhibits gastric secretion and motility; delays emptying of stomach; inhibits secretion by pancreas; inhibits gallbladder contraction and bile secretion; reduces blood circulation and nutrient absorption in small intestine

Some people enjoy haggis and tripe, made from animal stomachs, and have no difficulty digesting them. Why, then, doesn't the human stomach digest itself? The living stomach is protected in three ways from the harsh chemical environment it creates:

1. **Mucous coat.** A thick, highly alkaline mucus resists the action of acid and enzymes.

2. **Tight junctions.** The epithelial cells are joined by tight junctions that prevent gastric juice from seeping between them and digesting the connective tissue below.

3. **Epithelial cell replacement.** In spite of these other protections, the stomach's epithelial cells live only 3 to 6 days and are then sloughed off into the chyme and digested with the food. They are replaced just as rapidly, however, by the division of stem cells in the gastric pits.

The breakdown of these protective mechanisms can result in inflammation and peptic ulcer (see Deeper Insight 24.3).

Before You Go On

Answer the following questions to test your understanding of the preceding section:

10. Distinguish between the cardiac region, fundic region, body, and pyloric region of the stomach.

11. Name the cell types in the gastric and pyloric glands and state what each one secretes.

12. Explain why the stomach does not digest itself.

24.4 The Small Intestine

▶ **Expected Learning Outcomes**

When you have completed this section, you should be able to

- describe the gross and microscopic anatomy of the small intestine; and

- describe the structural adaptations of the small intestine for digestion and nutrient absorption.

The stomach "spits" about 3 mL of chyme at a time into the small intestine. Nearly all chemical digestion and nutrient absorption occur here. To perform these roles efficiently, the small intestine must have a large surface area exposed to the chyme. This surface area is imparted to it by extensive folding of the mucosa, and by the great length of the small intestine. It measures about 2.7 to 4.5 m long in a living person, but in the cadaver, where there is no muscle tone, it is 4 to 8 m long. The expression *small* intestine refers not to its length but to its diameter—about 2.5 cm (1 in.).

Gross Anatomy

The small intestine is a coiled mass filling most of the abdominal cavity inferior to the stomach and liver. It is divided into three regions: the duodenum, jejunum, and ileum (fig. 24.14).

The **duodenum** (dew-ODD-eh-num or DEW-oh-DEE-num) constitutes the first 25 cm (10 in.). It begins at the pyloric sphincter, arcs around the head of the pancreas and passes to the left, and ends at a sharp bend called the **duodenojejunal flexure.** Its name refers to its length, about equal to the width of 12 fingers.[19] The first 2 cm of the duodenum is intraperitoneal, but the rest is retroperitoneal, along with the pancreas.

Internally, the duodenum exhibits transverse to spiral ridges, up to 10 mm high, called **circular folds (plicae circulares)** (see fig. 24.19). They cause the chyme to flow on a spiral path along the mucosa, slowing its progress, causing more contact with the mucosa, and promoting thorough mixing, digestion, and nutrient absorption.

---

[19]*duoden* = 12

DEEPER INSIGHT 24.3

Peptic Ulcer

Inflammation of the stomach, called *gastritis,* can lead to a *peptic ulcer* as pepsin and hydrochloric acid erode the stomach wall (fig. 24.13). Peptic ulcers occur even more commonly in the duodenum and occasionally in the esophagus. If untreated, they can perforate the organ and cause fatal hemorrhaging or peritonitis. Most such fatalities occur in people over age 65.

There is no evidence to support the popular belief that peptic ulcers result from psychological stress. Hypersecretion of acid and pepsin is sometimes involved, but even normal secretion can cause ulceration if the mucosal defense is compromised by other causes. Many or most ulcers involve an acid-resistant bacterium, *Helicobacter pylori,* that invades the mucosa of the stomach and duodenum and opens the way to chemical damage to the tissue. Other risk factors include smoking and the use of aspirin and other nonsteroidal anti-inflammatory drugs (NSAIDs). NSAIDs suppress the synthesis of prostaglandins, which normally stimulate the secretion of protective mucus and acid-neutralizing bicarbonate. Aspirin itself is an acid that directly irritates the gastric mucosa.

Until recently, the most widely prescribed drug in the United States was Cimetidine (Tagamet), which was designed to treat peptic ulcers by reducing acid secretion. Histamine stimulates acid secretion by binding to sites on the parietal cells called H_2 receptors; Cimetidine, an H_2 blocker, prevents this binding. Lately, however, ulcers have been treated more successfully with antibiotics against *Helicobacter* combined with bismuth suspensions such as Pepto-Bismol. This is a much shorter and less expensive course of treatment and permanently cures about 90% of peptic ulcers, as compared with a cure rate of only 20% to 30% for H_2 blockers.

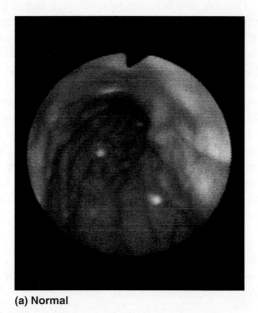

(a) Normal

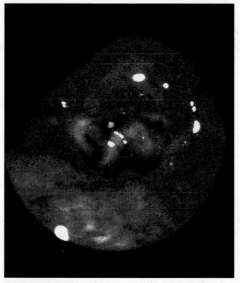

(b) Peptic ulcer

Figure 24.13 Endoscopic Views of the Gastroesophageal Junction. The esophagus can be seen opening into the cardiac stomach. (a) A view of the cardiac orifice from above, showing a healthy esophageal mucosa. The small white spots are reflections of light from the endoscope. (b) A bleeding peptic ulcer. A peptic ulcer typically has an oval shape and yellow-white color. Here the yellowish floor of the ulcer is partially obscured by black blood clots, and fresh blood is visible around the margin.

Adjacent to the head of the pancreas, the duodenal wall has a prominent wrinkle called the **major duodenal papilla** where the bile and pancreatic ducts open into the intestine. This papilla marks the boundary between the foregut and midgut. In most people, there is a smaller **minor duodenal papilla** a little proximal to this, which receives an *accessory pancreatic duct.*

The duodenum receives and mixes the stomach contents, pancreatic juice, and bile. Stomach acid is neutralized here by bicarbonate in the pancreatic juice, fats are physically broken up (emulsified) by the bile, pepsin is inactivated by the rise in pH, and pancreatic enzymes take over the job of chemical digestion.

The **jejunum** (jeh-JOO-num), by definition, is the first 40% of the small intestine beyond the duodenum—roughly 1.0 to 1.7 m in a living person. Its name refers to the fact that early anatomists typically found it to be empty.[20] The jejunum begins in the upper left quadrant of the abdomen but lies mostly within the umbilical region (see fig. 1.13, p. 16). It has large, tall, closely spaced circular folds. Its wall is relatively thick and muscular, and it has an especially rich blood supply that gives it a relatively red color. Most digestion and nutrient absorption occur here.

The **ileum**[21] forms the last 60% of the postduodenal small intestine—about 1.6 to 2.7 m. It occupies mainly the hypogastric region and part of the pelvic cavity. Compared with the jejunum, its wall is thinner, less muscular, less vascular, and has a paler pink color. Its circular folds are smaller and more sparse, and are lacking

---

[20]*jejun* = empty, dry
[21]from *eilos* = twisted

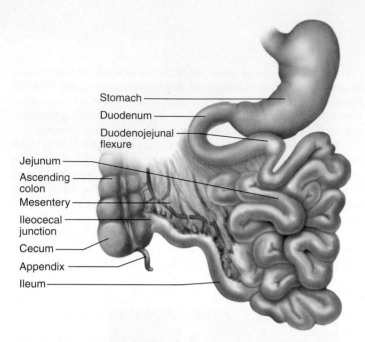

Stomach
Duodenum
Duodenojejunal flexure
Jejunum
Ascending colon
Mesentery
Ileocecal junction
Cecum
Appendix
Ileum

Figure 24.14 Gross Anatomy of the Small Intestine. The intestine is pulled aside to expose the mesentery and ileocecal junction.

from the distal end. On the side opposite from its mesenteric attachment, the ileum has prominent lymphatic nodules in clusters called **Peyer**[22] **patches,** which are readily visible to the naked eye and become progressively larger approaching the large intestine.

The end of the small intestine is the **ileocecal** (ILL-ee-oh-SEE-cul) **junction,** where the ileum joins the *cecum* of the large intestine. The muscularis of the ileum is thickened at this point to form a sphincter, the **ileocecal** (ILL-ee-oh-SEE-cul) **valve,** which protrudes into the cecum and regulates the passage of food residue into the large intestine and prevents feces from backing up into the ileum. Both the jejunum and ileum are intraperitoneal and thus covered with a serosa, which is continuous with the complex, folded mesentery that suspends the small intestine from the posterior abdominal wall.

Microscopic Anatomy

The tissue layers of the small intestine are reminiscent of those in the esophagus and stomach with modifications appropriate for nutrient digestion and absorption. The lumen is lined with simple columnar epithelium. The muscularis externa is notable for a thick inner circular layer and a thinner outer longitudinal layer.

Effective digestion and nutrient absorption require the small intestine to have a large internal surface area. This is provided by its relatively great length and by three kinds of internal folds or projections: the circular folds, villi, and microvilli. If the mucosa of the small intestine were smooth, like the inside of a hose, it would have a surface area of about 0.3 to 0.5 m², but with these surface elaborations, its actual surface area is about 200 m²—clearly a great advantage for nutrient absorption. The circular folds increase the surface area by a factor of 2 to 3; the villi by a factor of 10; and the microvilli by a factor of 20.

Circular folds, the largest of these elaborations, were described earlier. They occur from the duodenum to the middle of the ileum. They involve only the mucosa and submucosa; they are not visible on the external surface, which is smooth.

Villi (VIL-eye; singular, *villus*) are tongue- to finger-shaped projections that rise about 0.5 to 1.0 mm from the intestinal wall (fig. 24.15). They give the mucosa a fuzzy texture like a terrycloth towel. Villi are largest in the duodenum and progressively smaller in more distal regions of the small intestine. A villus is covered with two kinds of epithelial cells—columnar **enterocytes (absorptive cells)** and mucus-secreting **goblet cells.** Like epithelial cells of the stomach, those of the small intestine are joined by tight junctions that prevent digestive enzymes from seeping between them. The core of a villus is packed with areolar tissue of the lamina propria and contains a few smooth muscle cells that contract periodically. This enhances mixing of the chyme in the intestinal lumen and aids in the uptake of dietary fat.

The core also contains an arteriole, a bed of blood capillaries, a venule, and a lymphatic capillary called a **lacteal** (LAC-tee-ul) (fig. 24.15c). The blood capillaries absorb most nutrients, but the lacteal absorbs most dietary lipid. The reason lipids are little absorbed by the blood capillaries concerns their "packaging." The enterocytes package them in protein- and phospholipid-coated droplets called *chylomicrons,* then releases these into the core of the villus. Chylomicrons are too large (60 to 750 nm) to pass through the blood capillary walls into the bloodstream, but lymphatic capillaries have larger gaps between their cells (see fig. 22.3, p. 613) that allow for the uptake of these large droplets into the lymph. The lymphatic system eventually delivers them to the bloodstream. The fatty lymph in the lacteal is called *chyle.* It has a milky appearance for which the lacteal is named.[23]

Apply What You Know

Identify the exact place in the body where chylomicrons enter the blood. (*Hint:* See chapter 22.)

Microvilli are hairlike projections, about 1 μm high, that form a dense, fuzzy brush border on the surface of each enterocyte (see fig. 2.12, p. 36). Adding to the intestinal surface area is immensely important, but it is not their only function. Certain digestive enzymes called *brush border enzymes* are embedded in their plasma membranes and perform the final steps in digesting protein and carbohydrates. The chyme must come into contact with the microvilli for this to occur, so this is called *contact digestion.* This is one reason that thorough mixing of the chyme is so important. The end products of digestion are then absorbed through the membranes of the microvilli.

On the floor of the small intestine, between the bases of the villi, there are numerous pores that open into tubular glands called **intestinal crypts (crypts of Lieberkühn;**[24] LEE-ber-koohn). These crypts, similar to the gastric glands, extend as far as the muscularis mucosae. In the upper half, they consist of enterocytes and goblet

---

[22]Johann K. Peyer (1653–1712), Swiss anatomist

[23]*lact* = milk
[24]Johann N. Lieberkühn (1711–56), German anatomist

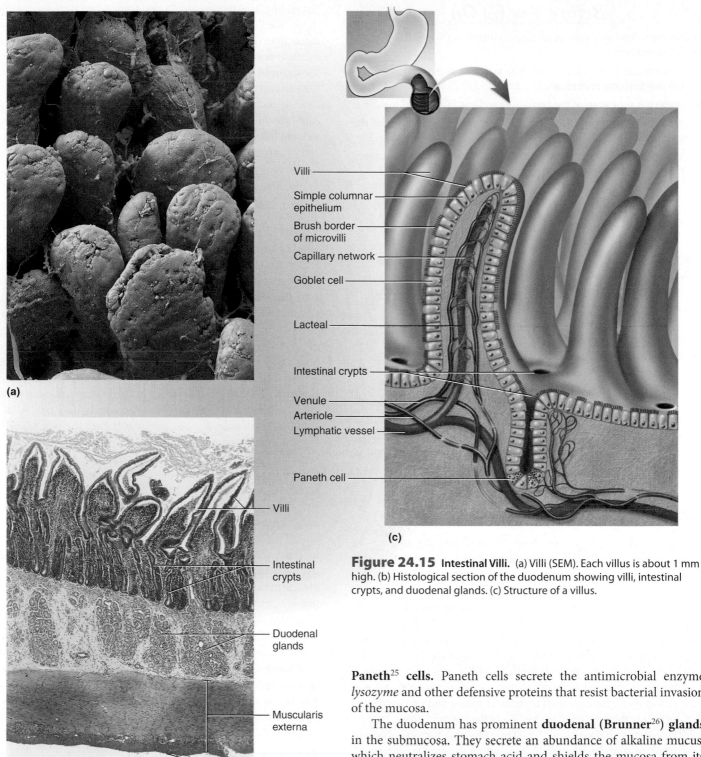

(a)

(b)

0.5 mm

Villi

Intestinal
crypts

Duodenal
glands

Muscularis
externa

Serosa

Villi

Simple columnar
epithelium

Brush border
of microvilli

Capillary network

Goblet cell

Lacteal

Intestinal crypts

Venule

Arteriole

Lymphatic vessel

Paneth cell

(c)

Figure 24.15 Intestinal Villi. (a) Villi (SEM). Each villus is about 1 mm high. (b) Histological section of the duodenum showing villi, intestinal crypts, and duodenal glands. (c) Structure of a villus.

Paneth[25] **cells.** Paneth cells secrete the antimicrobial enzyme *lysozyme* and other defensive proteins that resist bacterial invasion of the mucosa.

The duodenum has prominent **duodenal (Brunner**[26]**) glands** in the submucosa. They secrete an abundance of alkaline mucus, which neutralizes stomach acid and shields the mucosa from its erosive effects. Throughout the small intestine, the lamina propria and submucosa have a large population of lymphocytes that intercept pathogens before they can invade the bloodstream. In some places, these are aggregated into conspicuous lymphatic nodules (see fig. 22.8, p. 617), which become most conspicuous in the Peyer patches of the ileum.

cells like those of the villi. The lower half is dominated by dividing stem cells. In its life span of 3 to 6 days, an epithelial cell migrates up the crypt to the tip of the villus, where it is sloughed off and digested. Also seen deep in the crypts are enteroendocrine cells and

[25]Josef Paneth (1857–90), Austrian physician
[26]Johann C. Brunner (1653–1727), Swiss anatomist

Before You Go On

Answer the following questions to test your understanding of the preceding section:

13. Name the three regions of the small intestine and describe the distinctive features of each one.

14. Name the sphincters at the beginning and end of the small intestine.

15. What three structures increase the absorptive surface area of the small intestine?

16. Sketch a villus and label its epithelium, brush border, lamina propria, blood capillaries, and lacteal.

24.5 The Large Intestine

▶ **Expected Learning Outcomes**

When you have completed this section, you should be able to

- describe the gross and microscopic anatomy of the large intestine; and

- contrast the mucosa of the colon with that of the small intestine.

The large intestine (fig. 24.16) receives about 500 mL of indigestible food residue per day, reduces it to about 150 mL of feces by absorbing water and salts, and eliminates the feces by defecation.

Gross Anatomy

The large intestine measures about 1.5 m (5 ft) long and 6.5 cm (2.5 in.) in diameter in the cadaver. It is named for its relatively large diameter, not its length. It consists of four regions: the cecum, colon, rectum, and anal canal.

The **cecum**[27] is a pouch in the lower right abdominal quadrant inferior to the ileocecal valve. Attached to its lower end is the wormlike **appendix,** a blind tube 2 to 7 cm long. The mucosa of the appendix is densely populated with lymphocytes and is a significant source of immune cells. Herbivorous primates such as gorillas and orangutans have an enormous cecum, packed with bacteria that digest the plant fiber in their coarse diet. Humans, with their more mixed and easily digested diet, have only the appendix as a vestige of the larger cecum.

The **colon** is that portion of the large intestine between the ileocecal junction and the rectum (not including the cecum, rectum, or anal canal). It is divided into the ascending, transverse, descending, and sigmoid regions. The **ascending colon** begins at the ileocecal valve and passes up the right side of the abdominal cavity. It makes a 90° turn at the **right colic (hepatic) flexure,** near the right lobe of the liver, and becomes the **transverse colon.** This passes horizontally across the upper abdominal cavity and turns 90° downward at the **left colic (splenic) flexure** near the spleen. Here it becomes the **descending colon,** which passes down the left side of the abdominal cavity.

Ascending, transverse, and descending colons thus form a squarish, three-sided frame around the small intestine. The pelvic cavity is narrower than the abdominal cavity, so at the pelvic inlet the colon turns medially and downward, forming a roughly S-shaped portion called the **sigmoid**[28] **colon.** (Visual examination of this region is performed with an instrument called a *sigmoidoscope.*)

The **rectum,**[29] about 15 cm long, is the continuation of the large intestine into the pelvic cavity. In spite of its name, it is not perfectly straight but has three slight lateral curves as well as an anteroposterior curve. The rectal mucosa is smoother than that of the colon. It has three internal **transverse rectal folds (rectal valves)** that enable it to retain feces while passing gas. The large intestine contains about 7 to 10 L of gas, expelling about 500 mL/day anally as *flatus* and reabsorbing the rest through the colonic wall.

The final 3 cm of the large intestine is the **anal canal** (fig. 24.16b), which passes through the levator ani muscle of the pelvic floor and terminates at the anus. The mucosa of the anal canal forms longitudinal ridges called **anal columns** with depressions between them called **anal sinuses.** As feces pass through the canal, they press against the sinuses and cause them to exude extra mucus, which lubricates the canal during defecation. Large **hemorrhoidal veins** form superficial plexuses in the anal columns and around the orifice. Unlike veins in the limbs, they lack valves and are particularly subject to distension and venous pooling. *Hemorrhoids* are permanently distended veins that protrude into the anal canal or form bulges external to the anus.

The muscularis externa of the colon is unusual. Although it completely encircles the colon just as it does the small intestine, its longitudinal fibers are especially concentrated in three thickened, ribbonlike strips called the **taeniae coli** (TEE-nee-ee CO-lye). The muscle tone of the taeniae coli contracts the colon lengthwise and causes its wall to bulge, forming pouches called **haustra**[30] (HAW-stra; singular, *haustrum*). In the rectum and anal canal, however, the longitudinal muscle forms a continuous sheet and haustra are absent. The anus is normally held shut by two muscular rings—an **internal anal sphincter** composed of smooth muscle of the muscularis externa, and an **external anal sphincter** composed of skeletal muscle of the pelvic diaphragm. The internal anal sphincter is under involuntary control and relaxes automatically when the rectum is distended with feces. The external anal sphincter is under voluntary control and enables one to postpone defecation when appropriate.

The ascending and descending colon are retroperitoneal, whereas the transverse and sigmoid colon are covered with serosa and anchored to the posterior abdominal wall by the mesocolon. The serosa of these regions often has clublike fatty pouches of unknown function, called **omental (epiploic**[31]**) appendages.**

Microscopic Anatomy

The mucosa of the large intestine has a simple columnar epithelium in all regions except the lower half of the anal canal, where it has a nonkeratinized stratified squamous epithelium. The latter provides more resistance to the abrasion caused by the passage of feces. There are no circular folds or villi in the large intestine, but there

---

[27]*cec* = blind

[28]*sigm* = sigma or S + *oid* = resembling
[29]*rect* = straight
[30]*haustr* = to draw
[31]*epiploic* = pertaining to an omentum

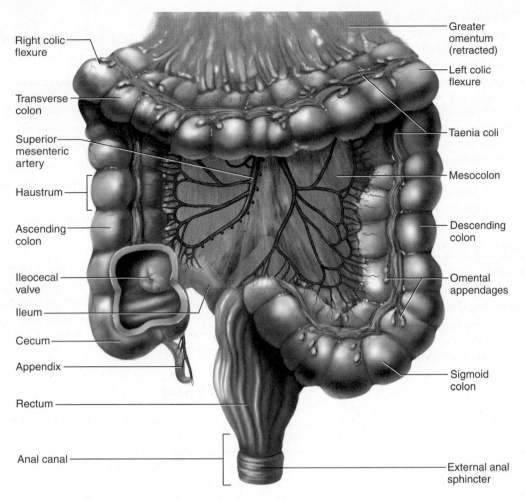

Right colic flexure

Transverse colon

Superior mesenteric artery

Haustrum

Ascending colon

Ileocecal valve

Ileum

Cecum

Appendix

Rectum

Anal canal

Greater omentum (retracted)

Left colic flexure

Taenia coli

Mesocolon

Descending colon

Omental appendages

Sigmoid colon

External anal sphincter

(a) Gross anatomy

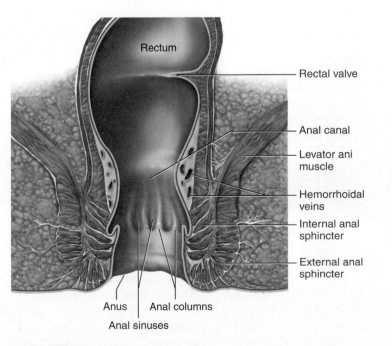

Rectum

Rectal valve

Anal canal

Levator ani muscle

Hemorrhoidal veins

Internal anal sphincter

External anal sphincter

Anus   Anal columns

Anal sinuses

(b) Anal canal

Figure 24.16 The Large Intestine.
• *Which anal sphincter is controlled by the autonomic nervous system? Which is controlled by the somatic nervous system? Explain the basis for your answers.*

are intestinal crypts. They are deeper than in the small intestine and have a greater density of goblet cells; mucus is their only significant secretion. The lamina propria and submucosa have an abundance of lymphatic tissue, providing protection from the bacteria that densely populate the large intestine.

The mucosa of the ascending and transverse colon is specialized for fluid and electrolyte absorption. During the process of digestion, a great deal of water is secreted by the salivary glands, stomach, and small intestine. Almost all of this is reabsorbed by the large intestine. When absorption is hindered, as in some bacterial infections, the result is diarrhea—an increase in the liquid eliminated with the feces. Severe diarrhea can lead to serious and sometimes fatal dehydration and electrolyte imbalances.

Before You Go On

Answer the following questions to test your understanding of the preceding section:

17. Name the regions of the large intestine in order from cecum to anus.

18. How does the mucosa of the large intestine differ from that of the small intestine? How does the muscularis externa differ?

19. How do the two anal sphincters differ in location, tissue composition, and function?

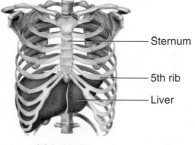

(a) Location

24.6 Accessory Glands of Digestion

▶ Expected Learning Outcomes

When you have completed this section, you should be able to

- describe the gross and microscopic anatomy of the liver, gallbladder, and bile duct system;
- describe the functions of the liver and bile;
- describe the gross and microscopic anatomy of the pancreas; and
- list the digestive secretions of the pancreas and their functions.

The small intestine receives not only chyme from the stomach but also secretions from the liver and pancreas, which empty into the digestive tract near the junction of the stomach and small intestine. These are regarded as accessory organs of digestion. Another accessory organ, the gallbladder, is associated with the liver in its anatomy and digestive function.

The Liver

The liver (fig. 24.17) is a reddish brown gland located immediately inferior to the diaphragm, filling most of the right hypochondriac and epigastric regions. It is the body's largest gland, weighing about 1.4 kg (3 lb). The liver has numerous functions (table 24.2). From their variety and importance, it is not hard to understand why liver diseases such as cirrhosis, hepatitis, and liver cancer are so serious and often fatal. Only one of these functions contributes to digestion—

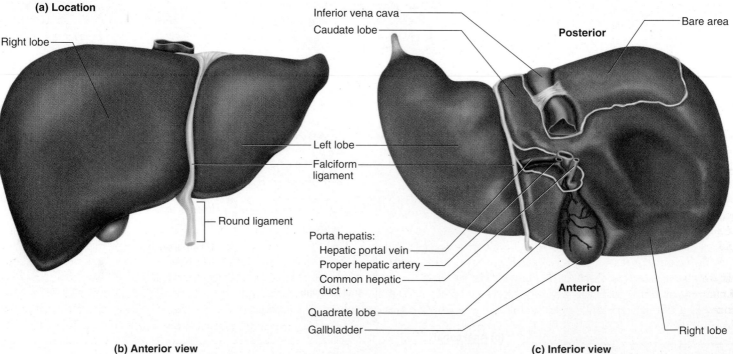

(b) Anterior view

(c) Inferior view

Figure 24.17 Gross Anatomy of the Liver.

TABLE 24.2	**Functions of the Liver**
Digestion	Synthesizes bile acids and lecithin, which emulsify fat and promote its digestion.
Carbohydrate metabolism	Converts dietary fructose and galactose to glucose. Stabilizes blood glucose concentration by storing excess glucose as glycogen *(glycogenesis),* releasing glucose from glycogen when needed *(glycogenolysis),* and synthesizing glucose from fats and amino acids *(gluconeogenesis)* when glucose demand exceeds glycogen reserves. Receives lactic acid generated by anaerobic fermentation in skeletal muscle and other tissues and converts it to pyruvic acid or glucose 6-phosphate for storage or energy-releasing metabolism.
Lipid metabolism	Degrades chylomicron remnants. Carries out most of the body's fat synthesis *(lipogenesis)* and synthesizes cholesterol and phospholipids. Produces *very-low-density lipoproteins* (VLDLs) to transport lipids to adipose tissue and other tissues for storage or use, and stores fat in its own cells. Carries out most fatty acid oxidation. Produces protein shells of *high-density lipoproteins* (HDLs), which pick up excess cholesterol from other tissues and return it to the liver; liver excretes the excess cholesterol in bile.
Protein and amino acid metabolism	Metabolizes amino acids; removes their $-NH_2$ and converts the resulting ammonia to *urea,* the major nitrogenous waste in the urine. Synthesizes some amino acids.
Vitamin and mineral metabolism	Converts vitamin D_3 to calcidiol, a step in the synthesis of the hormone calcitriol; stores a 3- to 4-month supply of vitamin D. Stores a 10-month supply of vitamin A and enough vitamin B_{12} to last one to several years. Stores iron in ferritin and releases it as needed. Excretes excess calcium by way of the bile.
Synthesis of plasma proteins	Synthesizes nearly all the proteins of blood plasma, including albumin, alpha and beta globulins, fibrinogen, prothrombin, and several other clotting factors. (Does not synthesize plasma enzymes, peptide hormones, or gamma globulins.)
Disposal of drugs, toxins, and hormones	Detoxifies alcohol, antibiotics, and many other drugs. Metabolizes bilirubin from RBC breakdown and excretes it as bile pigments. Deactivates thyroxine and steroid hormones and excretes them or converts them to a form more easily excreted by the kidneys.
Cleansing of blood	Hepatic macrophages cleanse the blood of bacteria and other foreign matter.

the secretion of bile. **Bile** is a green fluid containing minerals, cholesterol, neutral fats, phospholipids, bile pigments, and bile acids. The principal pigment is *bilirubin,* derived from the decomposition of hemoglobin. Bacteria of the large intestine metabolize bilirubin to *urobilinogen,* which is responsible for the brown color of feces. In the absence of bile secretion, the feces are grayish white and marked with streaks of undigested fat *(acholic feces). Bile acids (bile salts)* are steroids synthesized from cholesterol. Bile acids and *lecithin,* a phospholipid, emulsify fat—breaking globules of dietary fat into smaller droplets with more surface area exposed to enzyme action. Emulsification greatly enhances the efficiency of fat digestion.

Gross Anatomy

The liver is enclosed in a fibrous capsule and, external to this, most of it is covered by serosa. The serosa is absent from the *bare area* where its superior surface is attached to the diaphragm.

The liver is superficially subdivided into the right, left, quadrate, and caudate lobes. From an anterior view, we see only the large **right lobe** and smaller **left lobe.** They are separated from each other by the **falciform**[32] **ligament,** a sheet of mesentery that attaches the liver to the anterior abdominal wall. Superiorly, the falciform ligament forks into right and left **coronary**[33] **ligaments,** which suspend the liver from the diaphragm. The **round ligament (ligamentum teres),** visible anteriorly at the lower end of the falciform, is a fibrous remnant of the umbilical vein, which carries blood from the umbilical cord to the liver of a fetus.

From the inferior view, we also see a squarish anterior **quadrate lobe** next to the gallbladder and a tail-like **caudate**[34] **lobe** posterior to that. An irregular opening between these lobes, the **porta hepatis,**[35] is a point of entry for the hepatic portal vein and hepatic arteries and a point of exit for the bile passages. All of these blood vessels and bile passages travel in the lesser omentum. The gallbladder adheres to a depression on the inferior surface of the liver between the right and quadrate lobes. The posterior aspect of the liver has a deep groove (sulcus) occupied by the inferior vena cava.

Microscopic Anatomy

The interior of the liver is filled with an enormous number of tiny cylinders called **hepatic lobules,** about 2 mm long by 1 mm in diameter. A lobule consists of a **central vein** passing down its core, surrounded by radiating sheets of cuboidal cells called **hepatocytes** (fig. 24.18). Imagine spreading a book wide open until its front and back covers touch. The pages of the book would fan out around the spine somewhat like the plates of hepatocytes fan out from the central vein of a liver lobule.

Each plate of hepatocytes is an epithelium one or two cells thick. The spaces between the plates are blood-filled channels called **hepatic sinusoids.** The sinusoids are lined by a fenestrated endothelium that separates the hepatocytes from the blood cells, but allows blood plasma into the space between the hepatocytes and endothelium. The hepatocytes have a brush border of microvilli that project into this space. After a meal, as blood from the intestines circulates

---

[32]*falci* = sickle + *form* = shape
[33]*coron* = crown + *ary* = like, resembling

[34]*caud* = tail
[35]*porta* = gateway, entrance + *hepatis* = of the liver

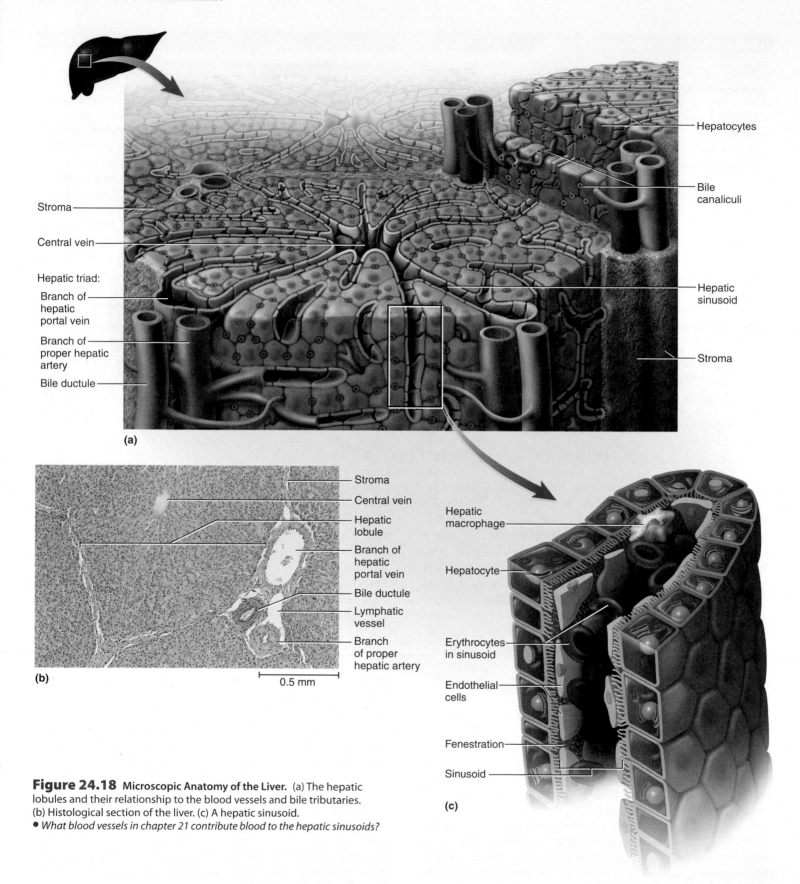

Stroma

Central vein

Hepatic triad:

Branch of
hepatic
portal vein

Branch of
proper hepatic
artery

Bile ductule

(a)

Hepatocytes

Bile
canaliculi

Hepatic
sinusoid

Stroma

Stroma

Central vein

Hepatic
lobule

Branch of
hepatic
portal vein

Bile ductule

Lymphatic
vessel

Branch
of proper
hepatic artery

(b)   0.5 mm

Hepatic
macrophage

Hepatocyte

Erythrocytes
in sinusoid

Endothelial
cells

Fenestration

Sinusoid

(c)

Figure 24.18 Microscopic Anatomy of the Liver. (a) The hepatic
lobules and their relationship to the blood vessels and bile tributaries.
(b) Histological section of the liver. (c) A hepatic sinusoid.
● *What blood vessels in chapter 21 contribute blood to the hepatic sinusoids?*

through the hepatic sinusoids, the hepatocytes rapidly remove glucose, amino acids, iron, vitamins, and other nutrients for metabolism or storage. They also remove and degrade hormones, toxins, bile pigments, and drugs. Conversely, they secrete albumin, lipoproteins, clotting factors, glucose, and other products into the blood. The sinusoids also contain phagocytic cells called **hepatic macrophages (Kupffer[36] cells),** which remove bacteria and debris from the blood.

The hepatocytes secrete bile into narrow channels, the **bile canaliculi,** between the cell plates within each lobule. Bile passes from there into the small **bile ductules** between lobules. These ductules lead ultimately to the **right** and **left hepatic ducts,** which exit the inferior surface of the liver at the porta hepatis.

The hepatic lobules are separated by a sparse connective tissue stroma. In cross sections, the stroma is especially visible in the triangular areas where three or more lobules meet. Here, there is often a **hepatic triad** of two blood vessels and a bile ductule. The blood vessels are small branches of the hepatic arteries and hepatic portal vein.

Circulation

The liver receives blood from two sources: about 70% from the **hepatic portal vein** and 30% from the **hepatic arteries.** The hepatic portal vein receives blood from veins of the stomach, intestines, pan-

---

[36]Karl W. von Kupffer (1829–1902), German anatomist

creas, and spleen, and carries it into the liver at the porta hepatis; see the hepatic portal system in table 21.7 (p. 589). All nutrients absorbed by the small intestine reach the liver by this route except for lipids (transported in the lymphatic system). Arterial blood bound for the liver exits the aorta at the celiac trunk and follows the route shown in figure 21.23 (p. 585): celiac trunk $\longrightarrow$ common hepatic artery $\longrightarrow$ hepatic artery proper $\longrightarrow$ right and left hepatic arteries, which enter the liver at the porta. These arteries deliver oxygen and other materials to the liver. Branches of the hepatic portal vein and hepatic arteries meet each other in the spaces between the liver lobules and both drain into the liver sinusoids. Hence, there is an unusual mixing of venous and arterial blood in the sinusoids. After processing by the hepatocytes, the blood collects in the central vein at the core of the lobule. Blood from the central veins ultimately converges in the right and left **hepatic veins,** which exit the superior surface of the liver and empty into the nearby inferior vena cava.

The Gallbladder and Bile Passages

Since the only digestive role of the liver is bile secretion, we will further trace the flow of bile through organs associated with the liver. The most conspicuous of these is the **gallbladder,** a pear-shaped sac on the underside of the liver that serves to store and concentrate the bile (fig. 24.19). It is about 10 cm long and internally

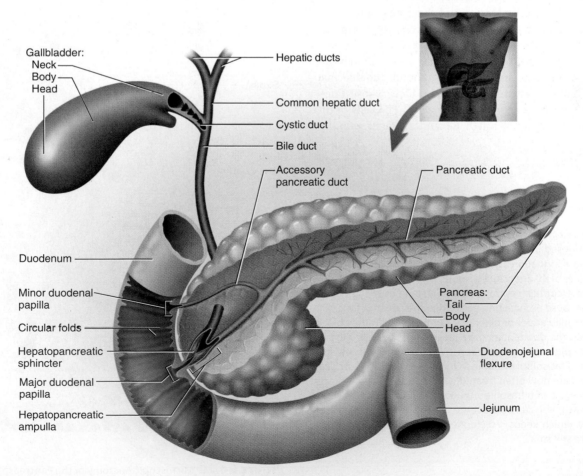

Figure 24.19 Gross Anatomy of the Gallbladder, Pancreas, and Bile Passages. The liver is omitted to show the gallbladder, which adheres to its inferior surface, and to show the hepatic ducts, which emerge from the liver tissue.

lined by a highly folded mucosa with a simple columnar epithelium. Its head *(fundus)* usually projects slightly beyond the inferior margin of the liver. Its neck *(cervix)* leads into the **cystic duct,** through which bile enters and leaves the gallbladder.

When the two hepatic ducts leave the porta hepatis, they converge almost immediately to form the **common hepatic duct.** This duct goes only a short distance before joining the cystic duct; it then becomes the **bile duct,** which descends through the lesser omentum to the duodenum. The bile duct and main pancreatic duct both approach the major duodenal papilla. Usually, just before emptying into the duodenum, the two ducts join each other and form an expanded chamber called the **hepatopancreatic ampulla.** A muscular **hepatopancreatic sphincter (sphincter of Oddi[37])** regulates the release of bile and pancreatic juice from the ampulla into the duodenum.

The Pancreas

Most digestion is carried out by pancreatic enzymes. The **pancreas** (fig. 24.19) is a spongy digestive gland posterior to the greater curvature of the stomach. It is about 15 cm long and divided into a globose *head* encircled on the right by the duodenum; a midportion called the *body;* and a blunt, tapered *tail* on the left near the spleen. It has a very thin connective tissue capsule and a nodular surface. It is retroperitoneal; its anterior surface is covered by parietal peritoneum, whereas its posterior surface contacts the aorta, left kidney, left adrenal gland, and other viscera on the posterior body wall.

The pancreas is both an endocrine and exocrine gland. Its endocrine part is the pancreatic islets, which secrete the hormones insulin and glucagon (see p. 509). Ninety-nine percent of the pancreas is exocrine tissue, which secretes enzymes and sodium bicarbonate. The exocrine pancreas is a compound tubuloacinar gland—that is, it has a system of branching ducts whose finest branches end in sacs of secretory cells, the acini. The cells of the acini exhibit a high density of rough ER and *zymogen granules,*

DEEPER INSIGHT   24.4

Gallstones

Gallstones (biliary calculi) are hard masses in the gallbladder or bile ducts, usually composed of cholesterol, calcium carbonate, and bilirubin. *Cholelithiasis,* the formation of gallstones, is most common in obese women over the age of 40 and usually results from excess cholesterol. The gallbladder may contain several gallstones, some over 1 cm in diameter. Gallstones cause excruciating pain when they obstruct the bile ducts or when the gallbladder or bile ducts contract. When they block the flow of bile into the duodenum, they cause jaundice (yellowing of the skin due to bile pigment accumulation), poor fat digestion, and impaired absorption of fat-soluble vitamins. Once treated only by surgical removal, gallstones are now often treated with stone-dissolving drugs or by *lithotripsy,* the use of ultrasonic vibration to pulverize them without surgery. Reobstruction can be prevented by inserting a stent (tube) into the bile duct, which keeps it distended and allows gallstones to pass while they are still small.

---

[37]Ruggero Oddi (1864–1913), Italian physician

which are vesicles filled with secretion (fig. 24.20). The smaller ducts converge on a main **pancreatic duct,** which runs lengthwise through the middle of the gland and joins the bile duct at the hepatopancreatic ampulla. Usually, there is a smaller **accessory pancreatic duct** that branches from the main pancreatic duct and opens independently into the duodenum at the minor duodenal papilla, proximal to the major papilla. The accessory duct bypasses the hepatopancreatic sphincter and allows pancreatic juice to be released into the duodenum even when bile is not.

The pancreas secretes 1,200 to 1,500 mL of *pancreatic juice* per day. This fluid is an alkaline mixture of water, sodium bicarbonate, other electrolytes, enzymes, and zymogens (table 24.3). Zymogens are inactive precursors of enzymes that are activated after they are secreted.

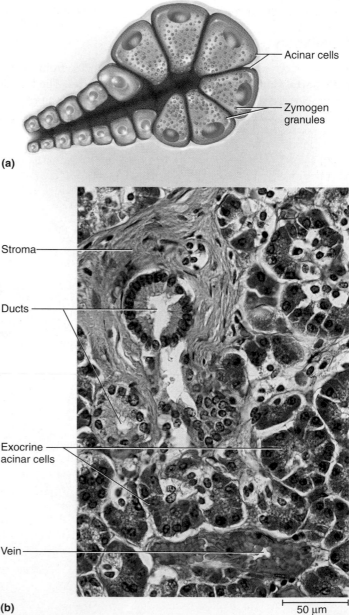

(a)

Acinar cells

Zymogen granules

Stroma

Ducts

Exocrine acinar cells

Vein

(b)

50 μm

Figure 24.20 Microscopic Anatomy of the Pancreas. (a) An acinus. (b) Histological section of the exocrine tissue and some of the connective tissue stroma.

TABLE 24.3	Exocrine Secretions of the Pancreas
Secretion	**Function**
Sodium bicarbonate	Neutralizes HCl
Zymogens	Converted to active digestive enzymes after secretion
Trypsinogen	Becomes trypsin, which digests protein
Chymotrypsinogen	Becomes chymotrypsin, which digests protein
Procarboxypeptidase	Becomes carboxypeptidase, which hydrolyzes the terminal amino acid from the carboxyl (–COOH) end of small peptides
Enzymes	
Pancreatic amylase	Digests starch
Pancreatic lipase	Digests fat
Ribonuclease	Digests RNA
Deoxyribonuclease	Digests DNA

Before You Go On

Answer the following questions to test your understanding of the preceding section:

20. What does the liver contribute to digestion? List several of its non-digestive functions.

21. Describe the structure of a hepatic lobule and the blood flow through a lobule.

22. Describe the pathway that bile takes from a hepatocyte that secretes it to the point where it enters the duodenum.

23. Describe the pathway taken by pancreatic juice from a gland acinus to the duodenum.

24. Explain why the pancreas is considered to be both an endocrine and an exocrine gland. How does the pancreas contribute to digestion?

24.6 Developmental and Clinical Perspectives

▶ Expected Learning Outcomes

When you have completed this section, you should be able to

- describe the prenatal development of the digestive tract, liver, and pancreas;

- describe the structural and functional changes in the digestive system in old age; and

- define and describe some common disorders of the digestive system.

Prenatal Development

The digestive system is one of the earliest organ systems to appear in the embryonic stage of development. Shortly after the three-layered embryonic disc is formed at 2 weeks, it elongates in the cephalocaudal (head-to-tail) direction. Endodermal pockets form at each end which become the foregut and hindgut (fig. 24.21a). Initially, there is a wide opening between the embryo and the yolk sac, but as the embryo continues to grow, the passage between them becomes constricted and a distinct tubular midgut appears. Temporarily, the midgut continues to communicate with the yolk sac through a narrow **vitelline duct.** In week 4, the anterior end of the digestive tract breaks through to form the mouth, and 3 weeks later, the posterior end breaks through to form the anus.

Growth of the embryonic body segments (*somites*) causes the lateral margins of the embryo to fold inward, changing the flat embryonic disc into a more cylindrical body and separating the embryonic body cavity from the yolk sac. By week 5, the gut is an elongated tube suspended from the body wall by the posterior (dorsal) mesentery (fig. 24.21b). Although the inner epithelial lining of the gut is endoderm, the tube is covered by a layer of mesoderm that gives rise to all other tissue layers of the digestive tract: the lamina propria, submucosa, muscle layers, and serosa.

At 4 weeks, the foregut exhibits a dilation that is the first sign of the future stomach (fig. 24.22). Further development of the digestive tract entails elongation, rotation, and differentiation of its regions into the esophagus, stomach, and small and large intestines. At 6 weeks, the body cavity is crowded by the rapidly growing liver and the intestine is so long and crowded that a loop of it herniates into the umbilical cord. This loop normally withdraws back into the enlarged body cavity in week 10, but in some tragic cases it fails to do so, resulting in severely deformed infants with part of the bowel outside the body (*omphalocele*[38]).

The liver appears in the middle of week 3 as a **liver bud,** a pocketlike outgrowth of the endodermal tube at the junction of the foregut and midgut. Its connection to the gut narrows and becomes the bile duct. A small outgrowth from the ventral side of the bile duct becomes the gallbladder and cystic duct. By week 12, the liver secretes bile into the gut, so the gut contents become dark green. The liver produces blood cells throughout most of fetal development, but this function gradually subsides in the last 2 months.

The pancreas originates as two buds, a **dorsal** and **ventral pancreatic bud,** around week 4. The ventral bud eventually rotates dorsally and merges with the dorsal bud. Pancreatic islets appear in the third month and begin secreting insulin at 5 months.

At birth, the digestive tract contains dark, sticky feces called **meconium,** which is discharged in the first few bowel movements of the neonate.

The Aging Digestive System

Like most other organ systems, the digestive system shows significant degenerative change (*senescence*) in old age. Less saliva is secreted in old age, making food less flavorful, swallowing more

---

[38]*omphalo* = navel, umbilicus + *cele* = swelling, herniation

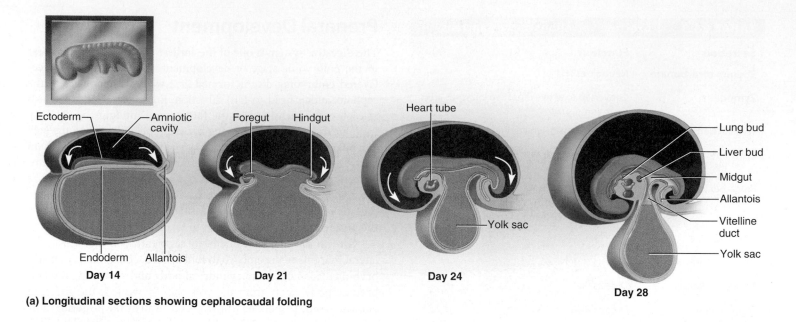

Ectoderm — Amniotic cavity

Endoderm — Allantois

Day 14

Foregut   Hindgut

Day 21

Heart tube

Yolk sac

Day 24

Lung bud
Liver bud
Midgut
Allantois
Vitelline duct
Yolk sac

Day 28

(a) Longitudinal sections showing cephalocaudal folding

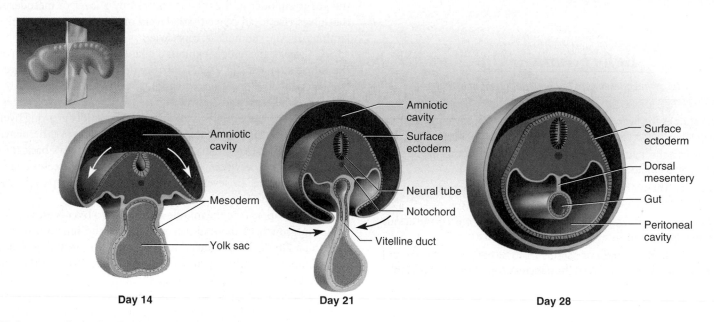

Amniotic cavity

Mesoderm

Yolk sac

Day 14

Amniotic cavity
Surface ectoderm
Neural tube
Notochord
Vitelline duct

Day 21

Surface ectoderm
Dorsal mesentery
Gut
Peritoneal cavity

Day 28

(b) Cross sections showing lateral folding

Figure 24.21 Embryonic Development of the Digestive Tract. (a) Cephalocaudal (head-to-tail) folding from days 14 to 28 produces the foregut and hindgut. The connection between the midgut and yolk sac grows progressively narrower. (b) Lateral folding over the same time period encloses the tubular gut and reduces the yolk sac connection to a narrow vitelline duct. The 28-day embryo in (b) is sectioned a little farther caudally than the 28-day embryo in (a) to show a region of hindgut already fully enclosed within the peritoneal cavity.

difficult, and the teeth more prone to caries. Many elderly people wear dentures because they have lost their teeth to caries and periodontal disease. The stratified squamous epithelium of the oral cavity and esophagus is thinner and more vulnerable to abrasion.

The gastric mucosa atrophies and secretes less acid and intrinsic factor. Acid deficiency reduces the absorption of calcium, iron, zinc, and folic acid. The declining level of intrinsic factor reduces the absorption of vitamin B_{12}. Since this vitamin is needed for hemopoiesis, the deficiency can lead to a form of anemia called *pernicious anemia*.

Heartburn becomes more common in old age as the weakening lower esophageal sphincter fails to prevent reflux into the esophagus. The most common digestive complaint of older people is constipation, which results from the reduced muscle tone and weaker peristalsis of the colon. This seems to stem from a combination of factors: atrophy of the muscularis externa, reduced sensitivity to neurotransmitters, less fiber and water in the diet, and less exercise. The liver, gallbladder, and pancreas show only slightly reduced function in old age. Any drop in liver function, however, makes it harder to detoxify drugs and can contribute to overmedication.

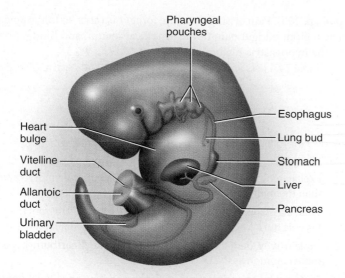

Pharyngeal pouches

Heart bulge

Vitelline duct

Allantoic duct

Urinary bladder

Esophagus

Lung bud

Stomach

Liver

Pancreas

Figure 24.22 **Lateral View of the 5-Week Embryo.** The primordial stomach is present as a foregut dilation, and the liver and pancreatic buds are present. The loop of midgut approaching the vitelline duct herniates into the umbilical cord within the next week.

Many older people eat less because of lower energy demand and appetite, because declining sensory functions make food less appealing, and because reduced mobility makes it more difficult to shop and prepare meals. However, they need fewer calories than younger people because they have lower basal metabolic rates and tend to be less physically active. Protein, vitamin, and mineral requirements remain essentially unchanged, although vitamin and mineral supplements may be needed to compensate for reduced food intake and intestinal absorption. Malnutrition is common among older people and is an important factor in anemia and reduced immunity.

Digestive Disorders

The digestive system is subject to a wide variety of disorders. Disorders of motility include difficulty swallowing (*dysphagia*), gastroesophageal reflux disease (GERD), and pyloric obstruction. Inflammatory diseases include esophagitis, gastritis, appendicitis, colitis, diverticulitis, pancreatitis, hepatitis, and cirrhosis. Cancer can occur in virtually every part of the digestive system: oral,

TABLE 24.4	Some Digestive System Diseases
Acute pancreatitis	Severe pancreatic inflammation, perhaps caused by trauma, leading to leakage of pancreatic enzymes into parenchyma, where they digest tissue and cause inflammation and hemorrhage.
Appendicitis	Inflammation of the appendix, with swelling, pain, and a threat of gangrene, perforation, and peritonitis.
Ascites	Accumulation of serous fluid in the peritoneal cavity, often causing extreme distension of the abdomen. Most often caused by cirrhosis of the liver and frequently associated with alcoholism. The diseased liver "weeps" fluid into the abdomen.
Cirrhosis[39] of the liver	An irreversible inflammatory disease of the liver often caused by alcoholism. Gives the liver a "cobbly" appearance and hard consistency due to fibrosis and nodular regeneration of damaged tissue. Obstruction of bile ducts causes jaundice, and obstruction of the circulation causes new vessels to grow and bypass the liver, leaving the liver subject to hypoxia, further necrosis, and failure.
Crohn[40] disease	Inflammation of small and large intestines, similar to ulcerative colitis in symptoms and hereditary predisposition. Produces granular lesions and fibrosis of intestine; diarrhea; and lower abdominal pain.
Diverticulitis	Presence of inflamed herniations (outpocketings, diverticula) of the colon, associated especially with low-fiber diets. Diverticula may rupture, leading to peritonitis.
Dysphagia[41]	Difficulty swallowing. Can result from esophageal obstructions (tumors, constrictions) or impaired peristalsis (due to neuromuscular disorders).
Hiatal hernia	Protrusion of part of the stomach into the thoracic cavity, where the negative thoracic pressure may cause it to balloon. Often causes gastroesophageal reflux (especially when a person is supine).
Ulcerative colitis	Chronic inflammation resulting in ulceration of the large intestine, especially the sigmoid colon and rectum. Tends to be hereditary but exact causes are not well known.

Disorders Described Elsewhere

[39]*cirrho* = orange-yellow + *osis* = condition
[40]Burrill B. Crohn (1884–1983), American gastroenterologist
[41]*dys* = bad, difficult, abnormal + *phag* = eat, swallow

esophageal, gastric, colon, rectal, hepatic, and pancreatic. Colon and pancreatic cancer are among the leading causes of cancer deaths in the United States.

Digestive disorders can be manifested in a variety of signs and symptoms: anorexia (loss of appetite), vomiting, constipation, diarrhea, abdominal pain, or gastrointestinal bleeding. Many of these are nonspecific; they do not by themselves identify a particular digestive disorder. Gastrointestinal bleeding, for example, can result from *varices* (varicose veins) in the digestive tract wall, intestinal polyps, GI inflammation, hemorrhoids, peptic ulcers, parasitic infections, or cancer. Nausea is even less specific; it may result not only from nondigestive disorders but also from such causes as tumors in the abdomen or brainstem, trauma to the urogenital organs, or inner-ear dysfunction.

It was remarked earlier that the foregut, midgut, and hindgut are defined by differences in arterial blood supply. This embryonic division also extends to the nerve supply and to the perception of pain from the digestive tract. Gastrointestinal pain is often perceived as if it were coming from the abdominal wall (see *referred pain*, p. 465). Pain arising from the foregut is referred to the epigastric region, midgut pain to the umbilical region, and hindgut pain to the hypogastric region.

Table 24.4 lists and describes some common digestive disorders.

Before You Go On

Answer the following questions to test your understanding of the preceding section:

25. Explain why the foregut and hindgut appear in the embryo earlier than the midgut.

26. What accessory digestive gland arises as a single bud from the embryonic gut? What gland arises as a pair of buds that later merge?

27. Explain why dental caries, constipation, and heartburn become more common as the digestive system ages.

28. Explain why gastrointestinal bleeding and nausea provide only inconclusive evidence for the existence and location of a digestive system disorder.

Study Guide

Assess Your Learning Outcomes

You should have a good understanding of this chapter if you can accurately address the following issues.

24.1 Digestive Processes and General Anatomy (p. 655)

1. The essential function of the digestive system and the five stages in which it performs this function

2. The distinction between the digestive tract and the accessory organs of this system, and the members of both categories

3. The tissue layers and sublayers of the wall of the digestive tract, and the histology and functional role of each layer

4. The cranial nerves that supply the digestive system; the parasympathetic and sympathetic innervation of the system; and the effects of those two divisions on digestion

5. The anatomical components of the enteric nervous system, and its role in digestion

6. Components of the foregut, midgut, and hindgut, and the blood supply to each

7. The hepatic portal system and its relationship to the digestive system

8. The relationships of the dorsal and ventral mesenteries to the digestive system, to the peritoneum, and to the serosa; anatomy of the greater and lesser omentum and the mesocolon

9. The meaning of *intraperitoneal* and *retroperitoneal* organs, and which digestive organs are in each class

24.2 The Mouth Through Esophagus (p. 659)

1. The diverse functions of the mouth (oral cavity)

2. All of the anatomical boundaries and contents of the oral cavity

3. Anatomy and functions of the cheeks and lips

4. Anatomy and functions of the tongue, and the distinction between its intrinsic and extrinsic muscles

5. The boundaries, structure, and function of the hard and soft palates; the palatoglossal and palatopharyngeal arches and their relationship with the palatine tonsils

6. Names of the adult teeth (dentition); the tissues and anatomy of an individual tooth; and the relationship of the teeth to the alveoli, periodontal ligaments, and gingivae

7. The purposes served by mastication

8. The composition and functions of saliva

9. The distinction between *intrinsic* and *extrinsic* salivary glands; the names and locations of the extrinsic salivary glands and the points at which their saliva is released into the oral cavity

10. The duct system, acini, and types of secretory cells in a salivary gland

11. The neural control of salivation

12. The anatomical regions and three constrictor muscles of the pharynx

13. The points at which the esophagus begins and ends; its upper and lower sphincters and their purposes; and the histological structure of its wall

14. The neural control of swallowing

24.3 The Stomach (p. 664)

1. The location and orientation of the stomach; its internal volume; and its functions

2. The shape and anatomical regions of the stomach; its relationship to the greater and lesser omentum; and the sphincters that regulate its entrance and exit

3. Ways in which the stomach wall differs histologically from the esophagus and other parts of the digestive tract

4. Gastric pits and their relationship with the glands of the mucosa

5. The three types of glands in the gastric mucosa; how they differ in location and function; the five kinds of cells in these glands; and the secretions of each cell type and the functions of those secretions

6. How the stomach is protected from digesting itself

24.4 The Small Intestine (p. 668)

1. The essential functions of the small intestine; and how its exceptional length is related to its functions
2. The anatomical landmarks that define the beginning and end of the duodenum; its length; its relationship to the peritoneum; the structure of its wall; its relationship to the bile and pancreatic ducts; and how its functions differ from other parts of the small intestine
3. How the jejunum is defined relative to the other parts of the small intestine; its length; how its role differs from that of the duodenum and ileum; and how its role is related to its unique appearance
4. How the ileum is defined relative to the other parts of the small intestine; its length; how its role differs from that of the duodenum and jejunum; and how its role is related to its anatomy
5. The structure of the ileocecal junction and valve, and functions of the valve
6. How the circular folds, villi, and brush border contribute to the functional efficiency of the small intestine
7. The structure of a villus; the cell types of its epithelium; the contents of its core; and how its blood capillaries and lacteal differ in their roles in nutrient absorption
8. The role of the brush border in contact digestion and nutrient absorption
9. The structure and cellular composition of the intestinal crypts of the small intestine
10. Features of the small intestine that serve to protect it from stomach acid and bacterial invasion

24.5 The Large Intestine (p. 672)

1. Functions of the large intestine
2. The four regions of the large intestine, their defining landmarks, and their total length
3. The location, structure, and functional significance of the appendix
4. The four segments of the colon and the anatomical landmarks of the transition from one to another
5. The extent, shape, and internal anatomy of the rectum and anal canal
6. The unusual musculature of the large intestine, including its taeniae coli and internal and external anal sphincters
7. The relationship of different parts of the large intestine to the peritoneum and mesocolon, and the omental appendages of its serosa
8. How the mucosa of the large intestine compares and contrasts with that of the small intestine

24.6 Accessory Glands of Digestion (p. 674)

1. The size, shape, location, and functions of the liver, and which of its functions contribute(s) to digestion
2. The composition and functional significance of bile
3. The lobes and ligaments of the liver, and the locations of the porta hepatis and gallbladder
4. The structure of the hepatic lobules and sinusoids; the relationship of the hepatocytes to the blood in the sinusoids; and the location and role of the hepatic macrophages
5. The venous and arterial blood supplies to the liver; where the venous and arterial blood mix; and the route by which blood exits the liver
6. Where bile is produced, and the route by which it is secreted from the liver
7. The location, anatomy, and function of the gallbladder; and anatomy of the bile passages that connect the liver, gallbladder, and duodenum
8. The location, gross anatomy, histology, and both endocrine and exocrine (digestive) functions of the pancreas
9. The anatomical relationships of the pancreatic ducts to the bile duct and duodenum
10. Functions of the exocrine secretions of the pancreas

24.7 Developmental and Clinical Perspectives (p. 679)

1. How the elongation and folding of the embryonic disc give rise to the foregut, hindgut, and finally the midgut
2. How the development of the digestive tract explains why its mucosal epithelium is endodermal whereas the other tissue layers of its wall are mesodermal
3. How the simple tubular embryonic gut transforms into a tract with regional differentiation into esophagus, stomach, and intestines
4. How the liver and pancreas arise from the embryonic gut
5. How aging affects salivation, the dentition, the gastrointestinal mucosa, intestinal motility, liver function, appetite, and nutrition
6. Some categories of digestive system disorders; examples in each category; and some common signs and symptoms of gastrointestinal disease

Testing Your Recall

1. All of the following are retroperitoneal *except*
   a. the liver.
   b. the pancreas.
   c. the duodenum.
   d. the ascending colon.
   e. the descending colon.

2. The falciform ligament attaches the _____ to the abdominal wall.
   a. colon   d. pancreas
   b. liver   e. stomach
   c. spleen

3. A brush border is found on the
   a. goblet cells.
   b. enterocytes.
   c. enteroendocrine cells.
   d. parietal cells.
   e. chief cells.

4. The yolk sac is connected to the embryonic _____ by way of the vitelline duct.
   a. liver bud   d. midgut
   b. stalk   e. hindgut
   c. foregut

5. Lacteals absorb dietary
   a. proteins.   d. vitamins.
   b. carbohydrates.   e. lipids.
   c. enzymes.

6. All of the following contribute to the absorptive surface area of the small intestine *except*
   a. its length.
   b. the brush border.
   c. haustra.
   d. circular folds.
   e. villi.

7. Which of the following is a periodontal tissue?
   a. gingiva   d. pulp
   b. enamel   e. dentin
   c. cementum

8. The _____ of the stomach most closely resemble the _____ of the small intestine.
   a. gastric pits, intestinal crypts
   b. pyloric glands, intestinal crypts
   c. rugae, Peyer patches
   d. parietal cells, goblet cells
   e. gastric glands, duodenal glands

9. Which of the following cells secrete digestive enzymes?
   a. chief cells
   b. mucous neck cells
   c. parietal cells
   d. goblet cells
   e. enteroendocrine cells

10. The tissue layer between the muscularis mucosae and muscularis externa of the digestive tract is
   a. the mucosa.
   b. the lamina propria.
   c. the submucosa.
   d. the serosa.
   e. the adventitia.

11. The alimentary canal has an extensive nervous network called the _____.

12. The passage of chyme from the stomach into the duodenum is controlled by a muscular ring called the _____.

13. The _____ salivary gland is named for its location near the ear.

14. The _____ is a complex of veins that carry blood from the stomach and intestines to the liver.

15. Nervous stimulation of gastrointestinal activity is mediated mainly through the parasympathetic fibers of the _____ nerves.

16. Hydrochloric acid is secreted by _____ cells of the stomach.

17. Hepatic macrophages occur in blood-filled spaces of the liver called _____.

18. The superior opening into the stomach is called the _____.

19. The root of a tooth is covered with a calcified tissue called _____.

20. Protrusions of the tongue surface, some of which bear taste buds, are called _____.

Answers in the Appendix

Building Your Medical Vocabulary

State a medical meaning of each of the following word elements, and give a term in which it is used.

1. gastro-
2. entero-
3. aliment-
4. retro-
5. bucco-
6. pyloro-
7. ruga-
8. sigmo-
9. recto-
10. hepato-

Answers in the Appendix

True or False

Determine which five of the following statements are false, and briefly explain why.

1. The liver and pancreas are retroperitoneal.
2. A tooth is composed mostly of enamel.
3. Hepatocytes secrete bile into the hepatic sinusoids.
4. The small intestine is much shorter in a living person than it is after death.
5. Peristalsis is controlled by the myenteric nerve plexus.
6. The pylorus is a gateway from the stomach to the duodenum.
7. The greater omentum suspends the stomach from the body wall.
8. Salivary acini can be composed of mucous cells, serous cells, or both.
9. In all parts of the digestive tract, the muscularis externa has two layers.
10. The external anal sphincter is under voluntary control; the internal sphincter is not.

Answers in the Appendix

Testing Your Comprehension

1. People who suffer from GERD (see Deeper Insight 24.2) when they lie down often find their heartburn is worse when they lie on the right side than when they lie on the left. Give an anatomical explanation for this effect.

2. Cystic fibrosis (CF) is characterized by unusually thick, sticky mucus that obstructs the respiratory tract and pancreatic duct. Predict the effect of CF on digestion, nutrition, and growth in childhood.

3. Explain why the small intestine would function poorly if it had the same type of mucosal epithelium as the esophagus.

4. News reports of patients (especially children) in need of organ transplants often prompt people to call and offer to donate one of their organs, such as a kidney, to save the patient's life. If you worked in an organ donor program, what would you say to a well-meaning volunteer offering to donate a liver?

5. What do you think is the functional significance of the Peyer patches in the distal ileum? Why should they be concentrated here but not found in the duodenum or jejunum?

Answers at www.mhhe.com/saladinha3

Improve Your Grade at www.mhhe.com/saladinha3

Practice quizzes, labeling activities, games, and flashcards provide fun ways to master concepts. You can also download image PowerPoint files for each chapter to create a study guide or for taking notes during lecture.

The Urinary System

The renal glomerulus, a mass of capillaries where the kidney filters the blood (SEM of a polymer cast)

CHAPTER OUTLINE

DEEPER INSIGHTS

BRUSHING UP

To understand this chapter, you may find it helpful to review the following concepts:
- Transitional epithelium (p. 60)
- General exocrine gland architecture (p. 73)
- Fenestrated capillaries (p. 568)

Anatomy & Physiology **REVEALED**
aprevealed.com

Urinary System

To live is to metabolize, and metabolism unavoidably generates a variety of waste products that are not merely unneeded by the body, but indeed toxic if they are allowed to accumulate. We rid the body of some of these wastes through the respiratory and digestive tracts and the sweat glands, but the urinary system is the principal means of waste excretion. The kidneys are glands that separate metabolic wastes from the blood. The rest of the urinary system serves only for the transport, storage, and elimination of urine.

The tasks of the kidneys go far beyond waste excretion, though. They also play indispensable roles in regulating blood volume and pressure, erythrocyte count, blood gases, blood pH, and electrolyte and acid–base balance. In performing these tasks, they have a very close physiological relationship with the endocrine, circulatory, and respiratory systems.

Anatomically, however, the urinary system is closely associated with the reproductive system. In many animals the eggs and sperm are emitted through the urinary tract, and the two systems share some aspects of their evolutionary history, embryonic development, and adult anatomy. This is reflected in humans, where the systems develop together in the embryo and, in the male, the urethra continues to serve as a passage for both urine and sperm. Thus, the urinary and reproductive systems are often collectively called the *urogenital (U-G) system,* and *urologists* treat both urinary and reproductive disorders. Because of their anatomical and developmental relationship, we consider the urinary and reproductive systems in these last two chapters.

25.1 Functions of the Urinary System

▶ **Expected Learning Outcomes**

When you have completed this section, you should be able to

- name and locate the organs of the urinary system;
- list several functions of the kidneys in addition to urine formation;
- define *excretion;* and
- identify the major nitrogenous waste excreted by the kidneys.

The **urinary system** serves primarily to cleanse the blood of metabolic wastes and eliminate them in the urine. It consists of six principal organs: two **kidneys,** two **ureters,** the **urinary bladder,** and the **urethra.** Figure 25.1 shows these organs in anterior and posterior views. The urinary tract has especially important relationships with the uterus and vagina in females and the prostate gland in males. These relationships are best appreciated from the sagittal

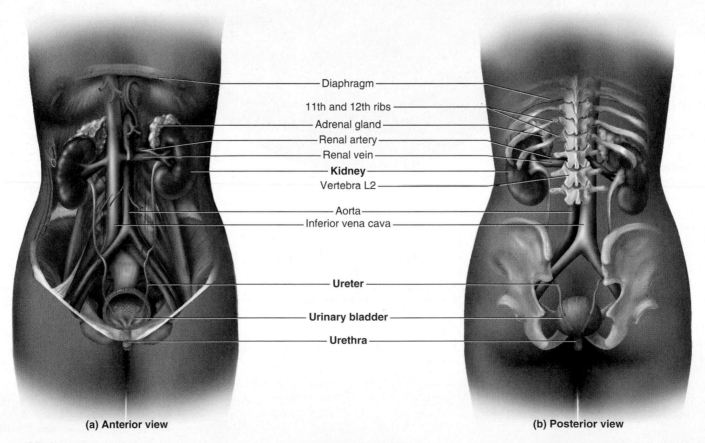

Diaphragm
11th and 12th ribs
Adrenal gland
Renal artery
Renal vein
Kidney
Vertebra L2
Aorta
Inferior vena cava
Ureter
Urinary bladder
Urethra

(a) Anterior view   (b) Posterior view

Figure 25.1 The Urinary System. Organs of the urinary system are indicated in boldface.
● *On yourself or another person, palpate the location of the kidneys. What landmark can be used to locate them?*

views in figures 26.9 (male) and 26.11 (female) (pp. 714 and 718). Most of the focus of the present chapter is on the kidneys.

Although the primary function of this system is excretion, the kidneys play more roles than are commonly realized:

- They filter blood plasma, separate wastes from the useful chemicals, and eliminate the wastes while returning the rest to the bloodstream.

- They regulate blood volume and pressure by eliminating or conserving water as necessary.

- They regulate the osmolarity of the body fluids by controlling the relative amounts of water and solutes eliminated.

- They secrete the enzyme *renin*, which activates hormonal mechanisms that control blood pressure and electrolyte balance.

- They secrete the hormone *erythropoietin*, which stimulates the production of red blood cells and thus supports the oxygen-carrying capacity of the blood.

- They carry out the final step in synthesizing the hormone *calcitriol* (vitamin D), and thereby contribute to calcium homeostasis.

- They collaborate with the lungs to regulate the CO_2 and acid–base balance of the body fluids.

- In times of extreme starvation, they convert amino acids to glucose (a process called *gluconeogenesis*) and thus help to support blood glucose level.

Excretion, the most obvious function of the urinary system, is the process of extracting wastes from the body fluids and eliminating them, thus preventing metabolic poisoning of the body. Among other things, the kidneys excrete nitrogen-containing molecules called **nitrogenous wastes.** The most abundant of these is **urea,** a product of protein metabolism. If the kidneys do not function adequately, one develops a condition called **azotemia**[1] (AZ-oh-TEE-me-uh), in which the urea concentration in the blood *(blood urea nitrogen, BUN)* is abnormally high. In severe renal failure, azotemia progresses to **uremia** (you-REE-me-uh), a syndrome of diarrhea, vomiting, dyspnea (labored breathing), and cardiac arrhythmia. Convulsions, coma, and death can follow within a few days, underscoring the importance of adequate renal function.

Before You Go On

Answer the following questions to test your understanding of the preceding section:

1. State at least four kidney functions other than forming urine.

2. What is the most abundant nitrogenous waste in the urine? What terms describe an abnormally high level of this waste in the blood, and poisoning by this waste?

25.2 Anatomy of the Kidney

▶ Expected Learning Outcomes

When you have completed this section, you should be able to

- describe the location and general appearance of the kidney, and its relationship to neighboring organs;

- identify the major external and internal features of the kidney;

- trace the flow of blood through the kidney;

- describe the nerve supply to the kidney;

- trace the flow of fluid through the renal tubules; and

- state the function of each segment of a renal tubule.

Position and Associated Structures

The kidneys lie against the posterior abdominal wall at the level of vertebrae T12 to L3. Rib 12 crosses the approximate middle of the left kidney. The right kidney is slightly lower than the left because of the space occupied by the large right lobe of the liver above it. The kidneys are retroperitoneal, along with the ureters, urinary bladder, renal artery and vein, and the adrenal[2] glands (fig. 25.2). The left adrenal gland rests on the superior pole of that kidney, and the right adrenal gland lies against the superomedial surface of its kidney. Their functions (see chapter 18, pp. 507–508) are not as directly related to the kidneys as their spatial relationship might suggest, although the kidneys and adrenals do influence each other.

Gross Anatomy

The kidney is a compound tubular gland containing about 1.2 million functional excretory units called **nephrons**[3] (NEF-rons). Each kidney weighs about 150 g and measures about 11 cm long, 6 cm wide, and 3 cm thick—about the size of a bar of bath soap. The lateral surface is convex, whereas the medial surface is concave and has a slit, the **hilum,** where it receives the renal nerves, blood vessels, lymphatics, and ureter.

The kidney is protected by three layers of connective tissue (fig. 25.2): (1) A fibrous **renal fascia,** immediately deep to the parietal peritoneum, binds the kidney and associated organs to the abdominal wall. (2) The **perirenal fat capsule,** a layer of adipose tissue, cushions the kidney and holds it in place. (3) The **fibrous capsule** encloses the kidney like a cellophane wrapper anchored at the hilum, and protects it from trauma and infection. The kidneys are suspended by collagen fibers that extend from the fibrous capsule, through the fat, to the renal fascia. The renal fascia is fused with the peritoneum anteriorly and with the fascia of the lumbar muscles posteriorly. In spite of all this, the kidneys drop about 3 cm when one goes from lying down to standing, as when getting out of bed in the morning. Under some circumstances, they become

---

[1]*azot* = nitrogen + *emia* = blood condition

[2]*ad* = to, toward, near + *ren* = kidney + *al* = pertaining to
[3]*nephro* = kidney

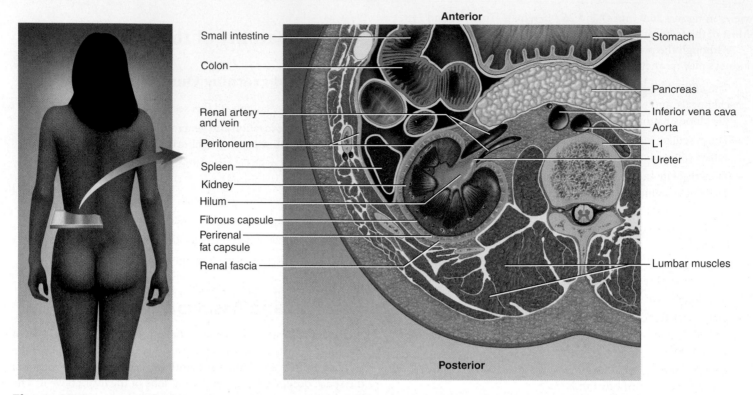

Anterior

Small intestine

Colon

Renal artery
and vein

Peritoneum

Spleen

Kidney

Hilum

Fibrous capsule

Perirenal
fat capsule

Renal fascia

Stomach

Pancreas

Inferior vena cava

Aorta

L1

Ureter

Lumbar muscles

Posterior

Figure 25.2 **Location of the Kidney.** Cross section of the abdomen at the level of vertebra L1, showing the relationship of the kidney to the body wall and peritoneum.

● *If the kidney were* not *retroperitoneal, where on this figure would you have to relocate it?*

detached and drift even lower, with pathological results (see nephroptosis, or "floating kidney," in table 25.1, p. 701).

The renal parenchyma—the glandular tissue that produces the urine—appears C-shaped in frontal section. It encircles a medial cavity, the **renal sinus,** occupied by blood and lymphatic vessels, nerves, and urine-collecting structures. Adipose tissue fills the remaining space in the sinus and holds these structures in place.

The parenchyma is divided into two zones: an outer **renal cortex** about 1 cm thick and an inner **renal medulla** facing the sinus (fig. 25.3). Extensions of the cortex called **renal columns** project toward the sinus and divide the medulla into 6 to 10 **renal pyramids.** Each pyramid is conical, with a broad base facing the cortex and a blunt point called the **renal papilla** facing the sinus. One pyramid and the overlying cortex constitute one *lobe* of the kidney.

The papilla of each renal pyramid is nestled in a cup called a **minor calyx**[4] (CAY-lix), which collects its urine. Two or three minor calyces (CAY-lih-seez) converge to form a **major calyx,** and two or three major calyces converge in the sinus to form the funnel-like **renal pelvis.**[5] The ureter is a tubular continuation of the renal pelvis that drains the urine down to the urinary bladder.

Renal Circulation

Although the kidneys account for only 0.4% of the body weight, they receive about 21% of the cardiac output (the *renal fraction*). This is a hint of how important the kidneys are in regulating blood volume and composition.

The larger divisions of the renal circulation are shown in figure 25.4. Each kidney is supplied by a **renal artery** arising from the aorta. Just before or after entering the hilum, the renal artery divides into a few **segmental arteries,** and each of these gives rise to a few **interlobar arteries.** An interlobar artery penetrates each renal column and travels between the pyramids toward the *corticomedullary junction,* the boundary between the cortex and medulla. Along the way, it branches again to form **arcuate arteries,** which make a sharp 90° bend and travel along the base of the pyramid. Each arcuate artery gives rise to several **cortical radiate (interlobular) arteries,** which pass upward into the cortex.

The finer branches of the circulation are shown in figure 25.5. As a cortical radiate artery ascends through the cortex, a series of **afferent arterioles** arise from it at nearly right angles like the limbs of a pine tree. Each afferent arteriole supplies one nephron. It leads to a spheroidal mass of capillaries called a **glomerulus**[6] (glo-MERR-you-lus), enclosed in a nephron structure called the *glomerular capsule,* to be discussed later. The glomerulus is drained by an **efferent arteriole.** The efferent arteriole usually leads to a plexus of **peritubular capillaries,** named for the fact that they form a network around the tubules of the nephron. These capillaries pick up the water and solutes reabsorbed by the tubules.

From the peritubular capillaries, blood flows to **cortical radiate (interlobular) veins, arcuate veins, interlobar veins,** and the **renal vein,** in that order. These veins travel parallel to the arteries of the same names. (There are, however, no segmental veins corresponding to the segmental arteries.) The renal vein leaves the hilum and drains into the inferior vena cava.

---

[4]*calyx* = cup
[5]*pelv* = basin

[6]*glomer* = ball + *ulus* = little

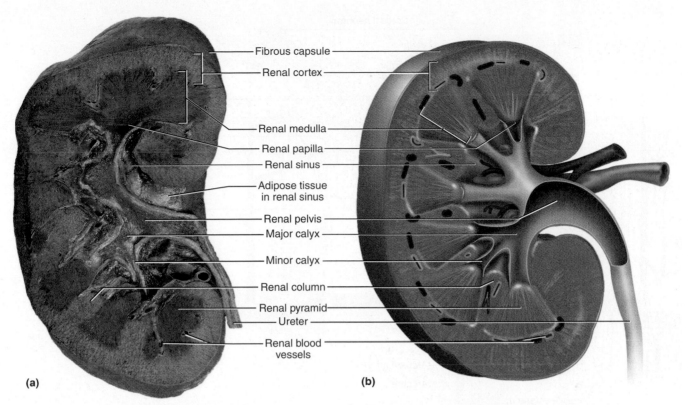

Fibrous capsule

Renal cortex

Renal medulla

Renal papilla

Renal sinus

Adipose tissue
in renal sinus

Renal pelvis

Major calyx

Minor calyx

Renal column

Renal pyramid

Ureter

Renal blood
vessels

(a)   (b)

Figure 25.3 **Gross Anatomy of the Kidney (Posterior Views).** (a) Photograph of a frontal section. (b) Major anatomical features.

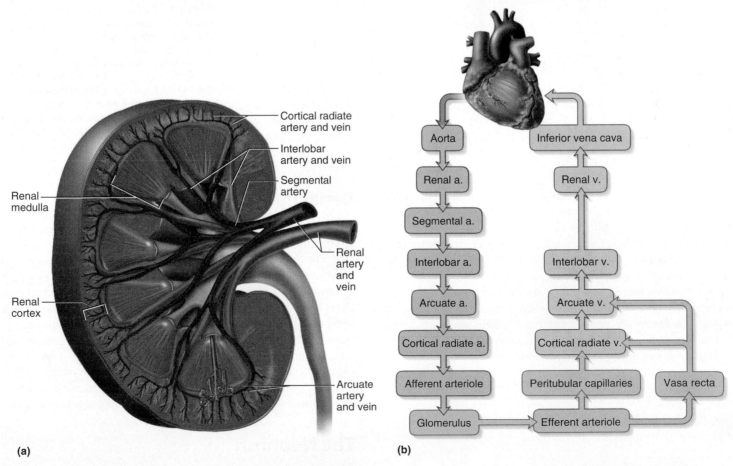

Cortical radiate
artery and vein

Interlobar
artery and vein

Segmental
artery

Renal
medulla

Renal
artery
and
vein

Renal
cortex

Arcuate
artery
and vein

(a)   (b)

Figure 25.4 **Renal Circulation.** (a) The larger blood vessels of the kidney. (b) Flow chart of renal circulation. The pathway through the vasa recta (instead of peritubular capillaries) applies only to the juxtamedullary nephrons.
• *Is the kidney in this figure shown from an anterior or posterior view? How can you tell?*

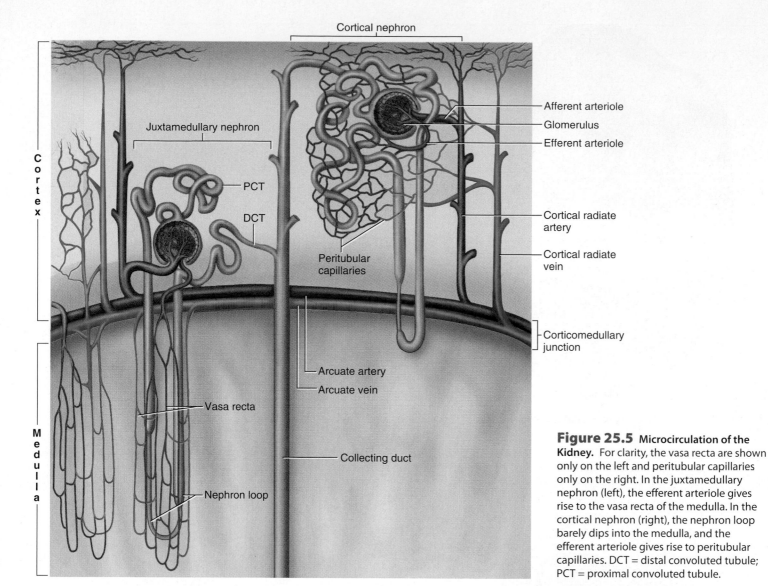

Figure 25.5 **Microcirculation of the Kidney.** For clarity, the vasa recta are shown only on the left and peritubular capillaries only on the right. In the juxtamedullary nephron (left), the efferent arteriole gives rise to the vasa recta of the medulla. In the cortical nephron (right), the nephron loop barely dips into the medulla, and the efferent arteriole gives rise to peritubular capillaries. DCT = distal convoluted tubule; PCT = proximal convoluted tubule.

The renal medulla receives only 1% to 2% of the total renal blood flow, supplied by a network of vessels called the **vasa recta.**[7] These arise from the nephrons in the deep cortex, closest to the medulla *(juxtamedullary nephrons).* Here, the efferent arterioles descend immediately into the medulla and give off the vasa recta instead of peritubular capillaries. Capillaries of the vasa recta lead into venules that ascend and empty into the arcuate and cortical radiate veins. The capillaries are wedged into the tight spaces between the medullary parts of the renal tubule, and carry away water and solutes reabsorbed by those sections of the tubule. Figure 25.4b summarizes the route of renal blood flow.

Apply What You Know

Can you identify a portal system in the renal circulation?

Renal Innervation

Wrapped around each renal artery is a **renal plexus** of nerves and ganglia (see fig. 16.6, p. 449). The plexus follows branches of the renal artery into the kidney, issuing nerve fibers to the blood vessels and convoluted tubules of the nephrons. The renal plexus includes both sympathetic and parasympathetic innervation, as well as afferent pain fibers from the kidneys en route to the spinal cord. The sympathetic nerves arise from the abdominal aortic plexus (especially its superior mesenteric and celiac ganglia). They control renal blood flow and the rate of urine production; when blood pressure falls, they stimulate the kidneys to secrete renin, an enzyme that activates hormonal mechanisms for restoring blood pressure. The parasympathetic innervation arises from the vagus nerves. Its function in the kidneys is unknown.

The Nephron

A nephron (fig. 25.6) consists of two principal parts: a *renal corpuscle,* which filters the blood plasma, and a long *renal tubule,* which converts the filtrate to urine.

---

[7]*vasa* = vessels + *recta* = straight

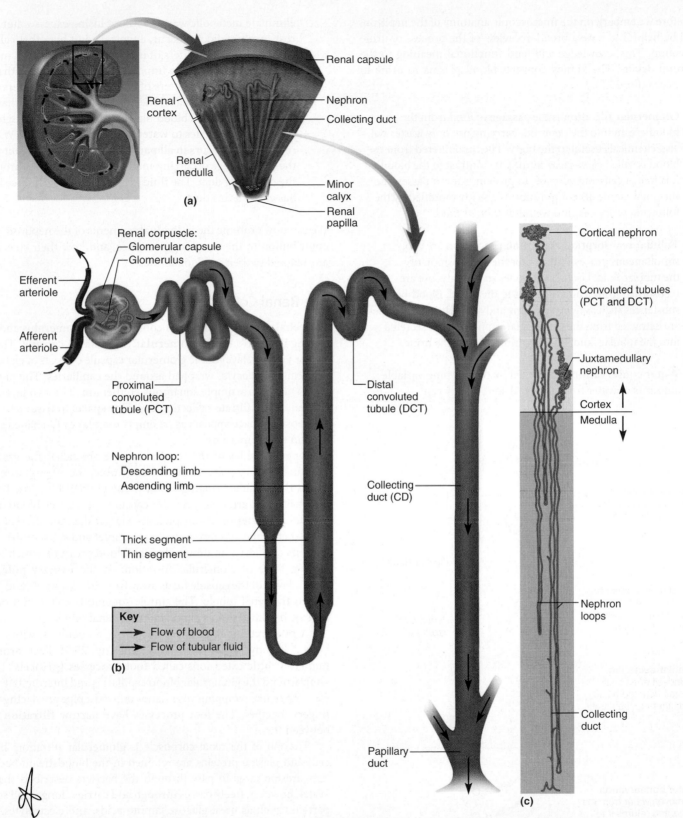

Figure 25.6 Microscopic Anatomy of the Nephron. (a) Location of the nephrons in one wedge-shaped lobe of the kidney. (b) Structure of a nephron. The nephron is stretched out to separate the convoluted tubules and is greatly shortened for the purpose of illustration. (c) The true proportions of the nephron loops relative to the convoluted tubules. Three nephrons are shown. Their proximal and distal convoluted tubules are commingled in a single mass in each nephron. Note the extreme lengths of the nephron loops.

Before we embark on the microscopic anatomy of the nephron, it will be helpful to have a broad overview of the process of urine production. This knowledge will lend functional meaning to the structural details. The kidney converts blood plasma to urine in three stages (fig. 25.7):

1. **Glomerular filtration** is the passage of fluid from the bloodstream into the nephron, carrying not only wastes but also chemicals useful to the body. The fluid filtered from the blood is called *glomerular filtrate.* In contrast to the blood, it is free of cells and very low in protein. After it passes into the renal tubule, its composition is quickly modified by the following processes, and we call it *tubular fluid.*

2. **Tubular reabsorption** and **tubular secretion** are two simultaneous processes that alter the composition of the tubular fluid. Useful substances such as glucose are reabsorbed from it and returned to the blood. Blood-borne substances such as hydrogen ions and some drugs, conversely, are extracted from the peritubular capillaries and secreted into the tubular fluid, thus becoming part of the urine.

3. **Water conservation** is achieved by reabsorbing variable amounts of water from the fluid so the body can eliminate metabolic wastes without losing excess water. If reabsorption did not occur, a typical adult hypothetically would produce 180 liters of urine per day—although in reality, this would be an impossible feat considering that we have only about 5 liters of blood and about 40 liters of total body water. Usually, the kidneys excrete urine that is **hypertonic** to the blood plasma—that is, it has a higher ratio of waste solutes to water than the plasma does. Water reabsorption occurs in all parts of the renal tubule, but is the final change occurring in the urine as it passes through the collecting duct. The fluid is regarded as urine once it has entered this duct.

We can now examine the individual segments of the nephron, their contribution to the foregoing processes, and how their structures are adapted to their individual roles.

The Renal Corpuscle

The **renal corpuscle** (fig. 25.8) consists of the glomerulus and, enclosing it, a two-layered **glomerular (Bowman[8]) capsule.** The inner, or visceral, layer of the glomerular capsule consists of elaborate cells called *podocytes* wrapped around the capillaries. The parietal (outer) layer is a simple squamous epithelium. The two layers are separated by a filtrate-collecting **capsular space.** In tissue sections, the capsular space appears as an empty circular or C-shaped space around the glomerulus.

Opposite sides of the renal corpuscle are called the vascular pole and urinary pole. At the **vascular pole,** the afferent arteriole enters the capsule, bringing blood to the glomerulus; close beside it, the efferent arteriole exits the capsule and carries blood away. The afferent arteriole is conspicuously larger than the efferent arteriole; that is, the glomerulus has a large inlet and small outlet. This gives its capillaries an unusually high blood pressure, which is the driving force of glomerular filtration. At the **urinary pole,** the parietal wall of the capsule turns away from the corpuscle and gives rise to the renal tubule. The simple squamous epithelium of the capsule becomes simple cuboidal in the renal tubule.

A **podocyte[9]** is shaped somewhat like an octopus, with a bulbous cell body and several thick arms (fig. 25.9). Each arm has numerous little extensions called **foot processes (pedicels[10])** that wrap around the glomerular blood capillaries and interdigitate with each other, like wrapping your hands around a pipe and lacing your fingers together. The foot processes have narrow **filtration slits** between them.

The job of the renal corpuscle is glomerular filtration: blood cells and plasma proteins are retained in the bloodstream because they are too large to pass through the barriers described shortly. Water, however, freely passes through and carries along small solute particles such as urea, glucose, amino acids, and electrolytes. The high blood pressure in the glomerulus drives the water and small solutes out through the capillary walls, into the capsular space. Pressure in the capsular space drives the filtrate into the renal tubule and ultimately all the way to the calyces and renal pelvis.

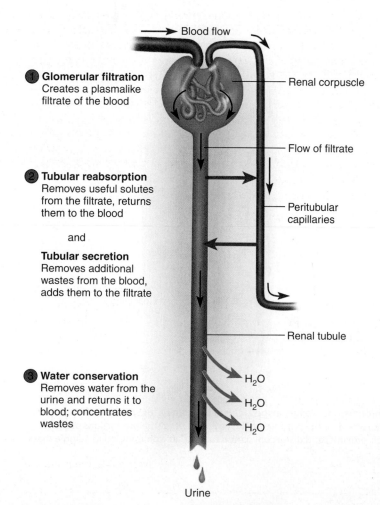

Blood flow

① **Glomerular filtration**
Creates a plasmalike filtrate of the blood

Renal corpuscle

Flow of filtrate

② **Tubular reabsorption**
Removes useful solutes from the filtrate, returns them to the blood

and

Tubular secretion
Removes additional wastes from the blood, adds them to the filtrate

Peritubular capillaries

③ **Water conservation**
Removes water from the urine and returns it to blood; concentrates wastes

Renal tubule

H_2O

H_2O

H_2O

Urine

Figure 25.7 Basic Steps in the Formation of Urine.

[8]Sir William Bowman (1816–92), British physician
[9]*podo* = foot + *cyte* = cell
[10]*pedi* = foot + *cel* = little

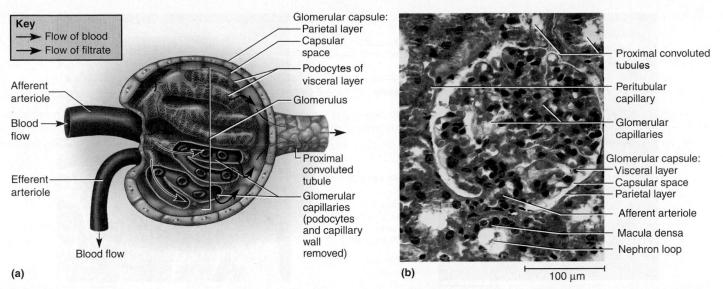

Figure 25.8 **The Renal Corpuscle.** (a) Anatomy of the corpuscle. (b) Light micrograph of the renal corpuscle and sections of the surrounding renal tubule.

Anything leaving the bloodstream must pass through a barrier called the **filtration membrane,** composed of three layers (fig. 25.9c):

1. **Capillary endothelium.** Glomerular capillaries have a fenestrated endothelium (see fig. 21.6, p. 569) honeycombed with large filtration pores about 70 to 90 nm in diameter. They are much more permeable than capillaries elsewhere, although the filtration pores are small enough to hold back blood cells.

2. **Basement membrane.** This is a layer of proteoglycan gel (a protein–carbohydrate complex) beneath the endothelial cells. For large molecules to pass through it is like trying to pass sand through a kitchen sponge. A few particles may penetrate its small spaces, but most are held back. On the basis of size alone, the basement membrane excludes any molecules larger than 8 nm. Some smaller molecules, however, are also held back by a negative electrical charge on the proteoglycans. Blood albumin is slightly less than 7 nm in diameter, but it is also negatively charged and thus repelled by the basement membrane. Therefore, the protein concentration is about 7% in the blood plasma but only 0.03% in the glomerular filtrate. The filtrate contains traces of albumin and smaller polypeptides, including some hormones.

3. **Filtration slits.** The slits between the podocyte foot processes are about 30 nm wide and are also negatively charged. This charge is a final barrier to large anions such as proteins.

Almost any molecule smaller than 3 nm passes freely through the filtration membrane. This includes water, electrolytes, glucose, fatty acids, amino acids, nitrogenous wastes, and vitamins. Such substances have about the same concentration in the filtrate as in the blood plasma. Some substances of low molecular weight are retained in the bloodstream because they are bound to plasma proteins that cannot get through the membrane. For example, most calcium, iron, and thyroid hormone in the blood are bound to plasma proteins that retard their filtration by the kidneys. The small fraction that is unbound, however, passes freely through the membrane and appears in the urine.

DEEPER INSIGHT   25.1

Blood and Protein in the Urine

Urinalysis, one of the most routine procedures performed upon patient admission and in routine medical examinations, is an analysis of the physical and chemical properties of the urine. It includes tests for blood and protein, both of which are normally lacking from urine. Damage to the filtration membrane, however, can result in blood or protein in the urine, called *hematuria*[11] and *proteinuria (albuminuria),* respectively. These can be signs of kidney infections, trauma, and other kidney diseases (see table 25.1 at the end of the chapter). They can be temporary conditions of little concern, or they can be chronic and gravely serious. Long-distance runners and swimmers often show temporary proteinuria and hematuria. Strenuous exercise reduces renal circulation as blood shifts to the muscles. With a reduced blood flow, the glomeruli deteriorate and leak protein and sometimes blood cells into the filtrate.

The Renal Tubule

The **renal (uriniferous**[12]**) tubule** is a duct that leads away from the glomerular capsule and ends at the tip of a medullary pyramid. It is about 3 cm long and divided into four major regions: the *proximal convoluted tubule, nephron loop, distal convoluted tubule,* and *collecting duct* (see fig. 25.6). Only the first three of these are parts of an individual nephron; the collecting duct receives fluid from many nephrons. Each region of the renal tubule has unique physiological properties and roles in the production of urine.

The **proximal convoluted tubule (PCT)** arises from the glomerular capsule. It is the longest and most coiled of the four regions and thus dominates histological sections of renal cortex. The PCT has a simple cuboidal epithelium with prominent microvilli (a

---

[11]*hemat* = blood + *uria* = urine condition
[12]*urin* = urine + *fer* = to carry

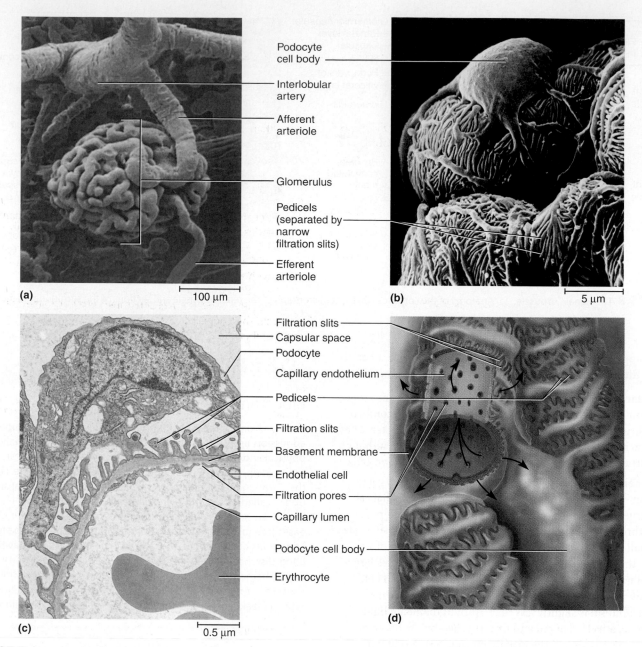

Figure 25.9 **Structure of the Glomerulus.** (a) A polymer cast of the glomerulus and nearby arteries (SEM). Note that the efferent arteriole is much narrower than the afferent arteriole, which causes blood pressure in the glomerulus to be unusually high. (b) Blood capillaries of the glomerulus closely wrapped in the spidery podocytes that form the visceral layer of the glomerular capsule (SEM). (c) A blood capillary and podocyte showing filtration pores of the capillary and filtration slits of the podocyte (TEM). (d) The production of glomerular filtrate by the passage of fluid through the endothelium and filtration slits.
[Part (a) from R. G. Kessel and R. H. Kardon, *Tissues and Organs: A Text-Atlas of Scanning Electron Microscopy* (W. H. Freeman, 1979)]
• *Which is larger, the efferent arteriole or afferent arteriole? How does this affect the function of the glomerulus?*

brush border), which attests to the great deal of absorption that occurs here. The microvilli give the epithelium a distinctively shaggy look in tissue sections.

The PCT carries out both tubular reabsorption and secretion. It reabsorbs about 65% of the glomerular filtrate, and consumes about 6% of one's daily ATP expenditure in doing so. On the surface facing the tubular fluid, the epithelial cells have a variety of membrane transport proteins that carry solutes into the cells by active transport and facilitated diffusion. These solutes and water pass through the

cell cytoplasm (the **transcellular**[13] **route**) and either diffuse out or are actively pumped out the basal and lateral cell surfaces, adjacent to peritubular blood capillaries waiting to receive them. Water and solutes also take a **paracellular**[14] **route** between the epithelial cells. Even though the cells are joined by tight junctions, these are quite leaky and allow a substantial amount of fluid to pass through.

---

[13]*trans* = across
[14]*para* = next to

Among the solutes reabsorbed by the PCT are sodium, potassium, magnesium, phosphate, chloride, bicarbonate, glucose, amino acids, lactate, protein, smaller peptides, amino acids, urea, and uric acid. Water follows by osmosis.

By tubular secretion, the PCT extracts solutes from the peritubular capillaries and secretes them into the tubular fluid, so they can be passed in the urine. Secreted solutes include hydrogen and bicarbonate ions, ammonia, urea, uric acid, creatinine, bile acids, pollutants, and some drugs (aspirin, penicillin, and morphine, for example). Notice that urea and uric acid go both ways between the blood and tubular fluid, transported by both tubular reabsorption and tubular secretion. The kidneys do not completely cleanse the blood of these wastes; indeed, they remove only about half of the urea, but this is sufficient to keep the blood urea concentration down to a safe level.

Apply What You Know

The proximal convoluted tubule exhibits some of the same structural adaptations as the small intestine, and for the same reason. Discuss what they have in common, and the reason for it.

The **nephron loop (loop of Henle**[15]) is a long U-shaped portion of the renal tubule found mostly in the medulla. It begins where the PCT straightens out and dips toward or into the medulla, forming the **descending limb** of the loop. At its deep end, the loop turns 180° and forms the **ascending limb,** which returns to the cortex, traveling parallel and close to the descending limb (see fig. 25.6c). Some parts of the loop, called **thick segments,** have a simple cuboidal epithelium, whereas the **thin segment** has a simple squamous epithelium. The thick segments engage in the active transport of salts through the tubule wall, so they have a high ATP requirement and are loaded with mitochondria to supply it. This is why the cells are relatively thick (cuboidal). The thin segment is not engaged in active transport but is relatively permeable to water. It does not need as much ATP or as many mitochondria, which is why the cells are relatively thin (squamous). The thick segments form the initial part of the descending limb and part or all of the ascending limb. The thin segment forms the lower part of the descending limb and in some nephrons, rounds the bend and continues partway up the ascending limb.

The nephron loop reabsorbs about 25% of the sodium, potassium, and chloride and 15% of the water that was in the glomerular filtrate. Its primary function, however, is to maintain a gradient of salinity in the renal medulla. It does this by pumping Na[+], K[+], and Cl[-] from the ascending limb into the medullary tissue fluid. At the corticomedullary junction, the tissue fluid is isotonic with the blood plasma (300 milliosmoles/liter), but deep in the medulla, it is four times as concentrated. The significance of this is explained later.

The nephron loops are not identical in all nephrons. Nephrons just beneath the renal capsule, close to the kidney surface, are called **cortical nephrons.** They have relatively short nephron loops that dip only slightly into the outer medulla before turning back (see fig. 25.6), or turn back even before leaving the cortex. Some cortical nephrons have no nephron loops at all. Nephrons close to the medulla are called

---

DEEPER INSIGHT   25.2

The Kidney and Life on Dry Land

Physiologists first suspected that the nephron loop plays a role in water conservation because of their studies of a variety of animal species. Animals that must conserve water have longer, more numerous nephron loops than animals with little need to conserve it. Fish and amphibians lack nephron loops and produce urine that is isotonic to their blood plasma. Aquatic mammals such as beavers have short nephron loops and only slightly hypertonic urine.

But the kangaroo rat, a desert rodent, provides an instructive contrast. It lives on seeds and other dry foods and can live without drinking any water at all. The water produced by its aerobic respiration is enough to meet its needs because its kidneys are extraordinarily efficient at conserving it. They have extremely long nephron loops and produce urine that is 10 to 14 times as concentrated as their blood plasma (compared with about 4 times, at most, in humans).

Comparative studies thus suggested a hypothesis for the function of the nephron loop that was confirmed through a long line of ensuing research. This shows how comparative anatomy provides suggestions and insights into function and why physiologists do not study human function in isolation from other species.

---

juxtamedullary[16] **nephrons.** They have very long nephron loops that extend nearly to the apex of the renal pyramid. Only 15% of the nephrons are juxtamedullary, but these are almost solely responsible for maintaining the salinity gradient of the medulla.

Immediately after reentering the renal cortex, the ascending limb of the nephron loop contacts the afferent and efferent arterioles at the vascular pole of the renal corpuscle. The two arterioles and the end of the nephron loop form the **juxtaglomerular** (JUX-tuh-glo-MER-you-lur) **apparatus**—a device for monitoring the fluid entering the distal convoluted tubule and adjusting the performance of the nephron (fig. 25.10). Three specialized cell types are found here:

1. The **macula densa**[17] is a patch of slender, closely spaced epithelial cells at the end of the loop on the side facing the afferent arteriole. These cells apparently act as sensors that monitor the flow or composition of the tubular fluid and communicate with the cells described next.

2. **Juxtaglomerular (JG) cells** are enlarged smooth muscle cells in the afferent arteriole and to some extent in the efferent arteriole, directly across from the macula densa. When stimulated by the macula, they dilate or constrict the arterioles. JG cells also secrete renin, the enzyme mentioned earlier that triggers corrective changes in blood pressure.

3. **Mesangial**[18] (mez-AN-jee-ul) **cells** occupy the cleft between the afferent and efferent arterioles and the spaces between the capillaries of the glomerulus. They are connected to the macula densa and JG cells by gap junctions and communicate

---

[15]Friedrich G. J. Henle (1809–85), German anatomist

[16]*juxta* = next to
[17]*macula* = spot, patch + *densa* = dense
[18]*mes* = in the middle + *angi* = vessel

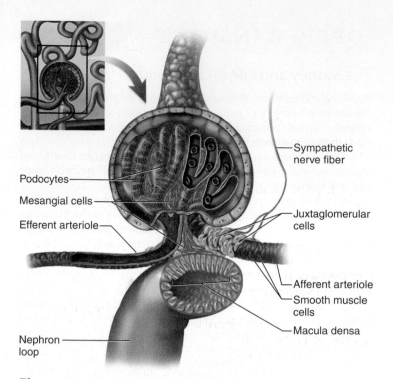

Podocytes

Mesangial cells

Efferent arteriole

Nephron loop

Sympathetic nerve fiber

Juxtaglomerular cells

Afferent arteriole

Smooth muscle cells

Macula densa

Figure 25.10 The Juxtaglomerular Apparatus.

with them by means of hormonelike secretions. They also form a supportive matrix for the glomerulus, phagocytize tissue debris, and constrict or relax the glomerular capillaries to regulate their blood flow and filtration rate.

The **distal convoluted tubule (DCT)** is a coiled part of the renal tubule located in the cortex and beginning immediately after the macula densa. It is the end of the nephron. The DCT is shorter and less convoluted than the PCT, so fewer sections of it are seen in histological sections. It has a simple cuboidal epithelium with smooth-surfaced cells nearly devoid of microvilli. It absorbs variable amounts of sodium, calcium, chloride, and water, and secretes potassium and hydrogen into the tubular fluid. Unlike the PCT, which absorbs solutes and water at a constant rate, the DCT reabsorbs these at variable rates determined by the hormone *aldosterone,* which regulates sodium and potassium excretion. *Parathyroid hormone* acts on both the PCT and DCT to regulate calcium and phosphate excretion.

The **collecting duct** is a straight tubule that passes down into the medulla. It is part of the renal tubule but not part of the nephron; the nephron and collecting duct have separate embryonic origins. The cortical part of the collecting duct receives fluid from the DCTs of several nephrons. The duct then continues into the medulla, where the greater part of it lies. Near the renal papilla, several collecting ducts converge to form a larger, short stretch called the **papillary duct.** About 30 papillary ducts drain from each papilla into a minor calyx. Once the urine drains into a minor calyx, it undergoes no further change in composition or concentration.

The collecting duct is lined with a simple cuboidal epithelium with two types of cells—intercalated ("in between") cells and principal cells. **Intercalated cells** play a role in regulating the body's acid–base balance by secreting either H^+ or bicarbonate ions

(HCO_3^-) into the urine. **Principal cells** reabsorb Na^+ and water and secrete K^+ into the urine. They represent the kidney's last chance to adjust the water content and thus the osmolarity of the urine. The principal cells also have water channels called *aquaporins* in their membranes. As the tubular fluid descends through the collecting duct, water passes by osmosis through these channels, out of the tubule and into the increasingly saline tissue fluid of the medulla. The salinity gradient created by the nephron loop makes this osmotic reabsorption of water possible. The reabsorbed water is carried away by the blood capillaries of the vasa recta.

The collecting duct is influenced by two hormones called *natriuretic peptides,* which increase sodium excretion in the urine, and another called *antidiuretic hormone,* which promotes water retention and reduces urine volume.

To summarize, the flow of fluid from the point where the glomerular filtrate is formed to the point where urine leaves the kidney is glomerular capsule $\longrightarrow$ proximal convoluted tubule $\longrightarrow$ nephron loop $\longrightarrow$ distal convoluted tubule $\longrightarrow$ collecting duct $\longrightarrow$ papillary duct $\longrightarrow$ minor calyx $\longrightarrow$ major calyx $\longrightarrow$ renal pelvis $\longrightarrow$ ureter.

Before You Go On

Answer the following questions to test your understanding of the preceding section:

3. Arrange the following in order from the most numerous to the least numerous structures in a kidney: glomeruli, major calyces, minor calyces, cortical radiate arteries, interlobar arteries.

4. Trace the path taken by one red blood cell from the renal artery to the renal vein.

5. Concisely state the functions of the glomerulus, PCT, nephron loop, DCT, and collecting duct.

6. Describe the location and appearance of podocytes and explain their function.

7. Consider one molecule of urea in the urine. Trace the route that it took from the bloodstream to the point where it left the kidney.

25.3 Anatomy of the Ureters, Urinary Bladder, and Urethra

▶ Expected Learning Outcome

When you have completed this section, you should be able to

- describe the functional anatomy of the ureters, urinary bladder, and male and female urethra.

Urine is produced continually, but fortunately it does not drain continually from the body. Urination (*micturition*) is episodic—occurring when we allow it. This is made possible by an apparatus for storing urine and by neural controls for its timely release.

DEEPER INSIGHT 25.3

Kidney Stones

A *renal calculus*[19] (kidney stone) is a hard granule of calcium, phosphate, uric acid, and protein. Renal calculi form in the renal pelvis and are usually small enough to pass unnoticed in the urine flow. Some, however, grow as large as several centimeters and block the renal pelvis or ureter, which can lead to the destruction of nephrons as pressure builds in the kidney. A large, jagged calculus passing down the ureter stimulates strong contractions that can be excruciatingly painful. It can also damage the ureter and cause hematuria. Causes of renal calculi include hypercalcemia (excess calcium in the blood), dehydration, pH imbalances, frequent urinary tract infections, or an enlarged prostate gland causing urine retention. Calculi are sometimes treated with stone-dissolving drugs, but often they require surgical removal. A nonsurgical technique called *lithotripsy*[20] uses ultrasound to pulverize the calculi into fine granules easily passed in the urine.

The Ureters

The renal pelvis funnels urine into the ureter, a retroperitoneal, muscular tube that extends to the urinary bladder. The ureter is about 25 cm long and reaches a maximum diameter of about 1.7 cm near the bladder. The ureters pass posterior to the bladder and enter it from below, penetrating obliquely through its muscular wall and opening onto its floor. A small flap of bladder mucosa acts as a valve at the opening of each ureter to keep urine from backing up into the ureter when the bladder contracts.

The ureter has three layers: an adventitia, muscularis, and mucosa. The adventitia is a connective tissue layer that binds it to the surrounding tissues. It blends with the capsule of the kidney at the superior end and with the connective tissue of the bladder wall at the inferior end. The muscularis consists of two layers of smooth muscle over most of its length, but a third layer appears in the lower ureter. The inner muscular layer consists of longitudinal muscle cells; the cells in the next layer superficial to this have a circular arrangement; and the third and outermost layer in the lower ureter is again longitudinal. Peristaltic waves of contraction in the muscularis drive urine from the renal pelvis down to the bladder. The mucosa of the ureter has a transitional epithelium that begins in the minor calyces of the kidney and extends through the urinary bladder. The lumen is very narrow and is easily obstructed by kidney stones (see Deeper Insight 25.3).

The Urinary Bladder

The urinary bladder (fig. 25.11) is a muscular sac on the floor of the pelvic cavity, inferior to the peritoneum and posterior to the pubic symphysis. It is covered by parietal peritoneum on its flattened superior surface and by a fibrous adventitia elsewhere. Its muscular layer, called the **detrusor**[21] (deh-TROO-zur) **muscle,** consists of three indistinctly separated layers of smooth muscle. The mucosa has a transitional epithelium. When the bladder is empty, this epithelium is five or six cells thick, and the mucosa has conspicuous wrinkles called **rugae**[22] (ROO-gee). When the bladder fills, the stretching smooths out the rugae, and the epithelium thins to about two or three cells thick. The bladder is a highly distensible organ, capable of holding up to 800 mL of urine.

The openings of the two ureters and the urethra mark a smooth-surfaced triangular area called the **trigone**[23] on the bladder floor. This is a common site of bladder infection (see Deeper Insight 25.4). For photographs of the relationship of the bladder and urethra to other pelvic organs in both sexes, see figure A.14 (p. 342).

The Urethra

The urethra conveys urine out of the body. In the female, it is a tube 3 to 4 cm long bound to the anterior wall of the vagina by fibrous connective tissue (fig. 25.11a). Its opening, the **external urethral orifice,** lies between the vaginal orifice and clitoris. The male urethra (fig. 25.11b) is about 18 cm long and has three regions: (1) The **prostatic urethra** begins at the urinary bladder and passes for about 2.5 cm through the prostate gland. During orgasm, it receives semen from the reproductive glands. (2) The **membranous urethra** is a short (0.5 cm), thin-walled portion where the urethra passes through the muscular floor of the pelvic cavity. (3) The **spongy (penile) urethra** is about 15 cm long and passes through the penis to the external urethral orifice. It is named for the *corpus spongiosum,* an erectile tissue that surrounds it (see p. 716). The male urethra assumes an S shape: it passes downward from the bladder, turns anteriorly as it enters the root of the penis, and then turns about 90° downward again as it enters the external, pendant part of the penis. The mucosa has a transitional epithelium near the bladder, a pseudostratified columnar epithelium for most of its length, and finally a stratified squamous epithelium near the external urethral orifice. There are mucous **urethral glands** in the wall of the penile urethra.

DEEPER INSIGHT 25.4

Urinary Tract Infections

Infection of the urinary bladder is called *cystitis*.[24] It is especially common in females because bacteria such as *Escherichia coli* can travel easily from the perineum up the short urethra. Because of this risk, young girls should be taught never to wipe the anus in a forward direction. If cystitis is untreated, bacteria can spread up the ureters and cause *pyelitis*,[25] infection of the renal pelvis. If it reaches the renal cortex and nephrons, it is called *pyelonephritis*. Kidney infections can also result from invasion by blood-borne bacteria. Urine stagnation due to renal calculi or prostate enlargement increases the risk of infection.

---

[19]*calc* = calcium, stone + *ul* = little
[20]*litho* = stone + *tripsy* = crushing
[21]*de* = down + *trus* = push

[22]*ruga* = fold, wrinkle
[23]*tri* = three + *gon* = angle
[24]*cyst* = bladder + *itis* = inflammation
[25]*pyel* = pelvis

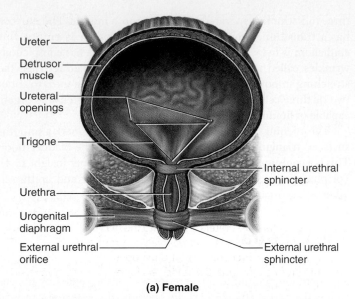

(a) Female

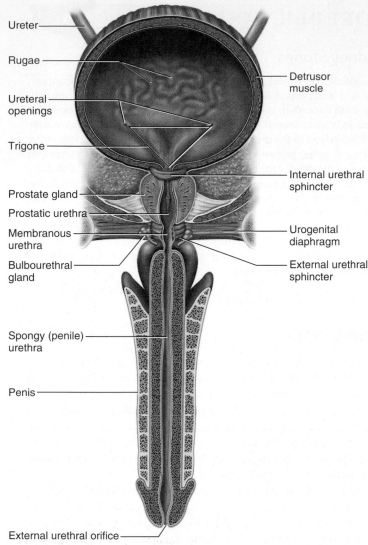

(b) Male

Figure 25.11 The Urinary Bladder and Urethra (Frontal Sections).
• *Why would an enlarged prostate gland make it more difficult to empty the bladder?*

In both sexes, the detrusor muscle is thickened near the urethra to form an **internal urethral sphincter,** which compresses the urethra and retains urine in the bladder. Since this sphincter is composed of smooth muscle, it is under involuntary control. Where the urethra passes through the pelvic floor, it is encircled by an **external urethral sphincter** of skeletal muscle, which provides voluntary control over the voiding of urine.

Before You Go On

Answer the following questions to test your understanding of the preceding section:

8. Describe the anatomical relationship of the ureter to the renal pelvis and to the bladder wall.

9. Compare and contrast the structure and function of the internal and external urethral sphincters.

10. Contrast the structure of the bladder wall when the bladder is empty with when it is full.

11. Name and define the three segments of the male urethra.

25.4 Developmental and Clinical Perspectives

▶ **Expected Learning Outcomes**

When you have completed this section, you should be able to

• describe the embryonic development of the urinary system;

• describe the degenerative changes that occur in old age;

• describe the causes and effects of renal failure; and

• briefly define or describe several urinary system diseases.

Prenatal Development

Perhaps surprisingly, the embryonic urinary system develops two pairs of primitive, temporary kidneys before "settling down" and producing the permanent pair. The system develops as if replaying the evolutionary history of the vertebrate urinary system. Early in week 4, a rudimentary kidney called the *pronephros*[26] appears in the cervical region, resembling the kidneys of many fish and amphibian embryos and larvae. The pronephros disappears by the end of that week. As it degenerates, a second kidney, the *mesonephros,*[27] appears in the thoracic to lumbar region. The mesonephros func-

---

[26]*pro* = first + *nephros* = kidney
[27]*meso* = middle + *nephros* = kidney

tions in the embryos of all vertebrates, but is of minor importance in most mammals, where wastes are eliminated via the placenta. Most of the mesonephros disappears by the end of month 2, but its collecting duct, the *mesonephric duct,* remains and contributes importantly to the male reproductive tract (see p. 729). This duct opens into an embryonic *cloaca,* a temporary rectumlike receiving chamber for the digestive, urinary, and reproductive systems. The final kidney, the *metanephros,*[28] appears in week 5 and thus overlaps the existence of the mesonephros.

The permanent urinary tract begins with a pouch called the **ureteric bud** growing from the lower end of each mesonephric duct. The closed, upper end of the bud dilates and branches to form the renal pelvis, then the major and minor calyces, and finally the collecting ducts (fig. 25.12). Each collecting duct has a cap of metanephric kidney tissue over its tip. The duct induces this cap to differentiate into an S-shaped tubule (fig. 25.13a). Blood capillaries grow into one end of the tubule and form a glomerulus as the tubule grows around it to form the double-walled glomerular capsule. The other end of the tubule breaks through to become continuous with the collecting duct. The tubule gradually lengthens and differentiates into the proximal convoluted tubule, nephron loop, and distal convoluted tubule. By the time of birth, each kidney will have formed over 1 million nephrons in this manner. No more nephrons form after birth, but the existing ones continue to grow. The kidney surface is lumpy at birth but smooths out because of nephron growth.

The kidneys originate in the pelvic region and later migrate superiorly—a movement called **ascent of the kidney.** Initially, the kidney is supplied by a pelvic branch of the aorta, but as it ascends, new arteries higher and higher on the aorta take over the job of supplying the kidney, while the lower arteries degenerate.

In weeks 4 to 7, the cloaca divides into an anterior *urogenital (U-G) sinus* and a posterior *anal canal.* The superior part of the U-G sinus forms the urinary bladder, and the inferior part forms the urethra. In infants and children, the urinary bladder is located in the abdomen. It begins to drop into the greater pelvis at about 6 years of age, but does not enter the lesser pelvis and become a true pelvic organ until after puberty. Early in its development, the bladder is connected to the allantois, an extraembryonic sac described in chapter 4 (p. 94). This connection eventually becomes a constricted passage, the *urachus* (yur-AY-kus), connecting the bladder to the umbilicus. In the adult, the urachus is reduced to a fibrous cord, the *median umbilical ligament.*

Urine production begins around week 12 of fetal development, but metabolic wastes are cleared by the placenta, not by the fetal urinary system. Fetal urine is continually recycled as the fetus voids it into the amniotic fluid, swallows it, and then excretes it again.

The Aging Urinary System

The kidneys exhibit a striking degree of atrophy in old age. From ages 25 to 85, the number of nephrons declines by 30% to 40%, and up to one-third of the remaining glomeruli become atherosclerotic, bloodless, and nonfunctional. The kidneys of a 90-year-old are 20% to 40% smaller than those of a 30-year-old and receive only half as

[28]*meta* = beyond, next in a series + *nephros* = kidney

DEEPER INSIGHT 25.5

Developmental Abnormalities of the Kidney

Several anomalies can occur during the embryonic development of the kidneys (see fig. 1.5, p. 7). *Pelvic kidney* is a condition in which the kidney fails to ascend and remains in the pelvic cavity for life. *Horseshoe kidney* is a single C-shaped kidney that arches across the lumbar region from left to right; it is formed when the ascending kidneys are crowded together and merge into one. The C typically snags on the inferior mesenteric artery, preventing the horseshoe kidney from ascending any farther. Some people have two ureters arising from a single kidney, resulting from a splitting of the ureteric bud in early embryonic development. Usually the two ureters empty into the bladder, but in rare cases, one of them empties into the uterus, vagina, urethra, or elsewhere. This requires surgical correction so that urine does not dribble continually from the urethra or vagina. Some kidneys have an *accessory renal artery* resulting from failure of one of the early, temporary renal arteries to degenerate as the permanent renal artery forms. Most such irregularities cause no functional problems and usually go unnoticed, but they may be discovered in surgery, radiography, or cadaver dissection.

much blood. They are proportionately less efficient at clearing wastes from the blood. Although baseline renal function is adequate even in old age, the kidneys have little reserve capacity; thus, other diseases can lead to surprisingly rapid renal failure. Drug doses often need to be reduced for the elderly because the kidneys cannot clear drugs from the blood as rapidly. Reduced renal function is a significant factor in overmedication of the aged.

Water balance becomes more precarious in old age because the kidneys are less responsive to antidiuretic hormone and because the sense of thirst is blunted. Even with free access to water, many elderly people do not drink enough to maintain normal blood osmolarity, and dehydration is common.

Voiding and bladder control become a problem for both men and women. About 80% of men over the age of 80 are affected by *benign prostatic hyperplasia,* a noncancerous growth of the prostate gland that compresses the urethra, reduces the force of the urine stream, and makes it harder to empty the bladder. Urine retention can cause pressure to back up in the kidneys, aggravating the failure of the nephrons. Older women become increasingly subject to *urinary incontinence* (see table 25.1), especially if their history of pregnancy and childbearing has weakened the pelvic muscles and urethral sphincters. Senescence of the sympathetic nervous system and nervous disorders such as stroke and Alzheimer disease can also cause incontinence.

Urinary System Disorders

The most serious disorder of the urinary system is renal failure. *Acute renal failure* is an abrupt decline in kidney function, often caused by trauma or by a hemorrhage or thrombosis cutting off blood flow to the nephrons. *Chronic renal failure* is a long-term, progressive, irreversible loss of functional nephrons. It can result from such causes as prolonged or repetitive kidney infections,

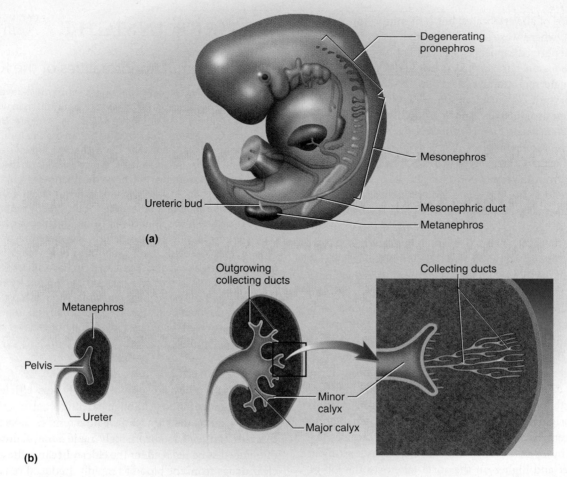

Figure 25.12   Embryonic Development of the Urinary Tract. (a) Relationship of the early ureteric bud and metanephros to the lower mesonephric duct. (b) Progression in the development of the ureter, renal pelvis, calyces, and collecting ducts, all of which arise from the ureteric bud.

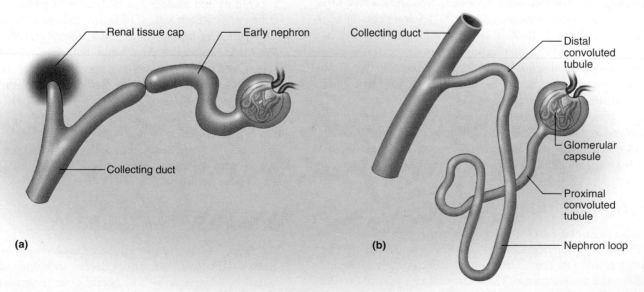

Figure 25.13   Embryonic Development of the Nephron. (a) The collecting duct induces mesoderm to differentiate into a renal (metanephric) tissue cap, as shown at the far left. This cap differentiates into an S-shaped tube that will become the nephron, as shown on the right fork of the duct (representing a later stage of development). (b) The renal tubule has begun to differentiate into proximal and distal convoluted tubules and nephron loop.

TABLE 25.1	Some Disorders of the Urinary System
Acute glomerulonephritis	An autoimmune inflammation of the glomeruli, often following a *Streptococcus* infection. Results in destruction of glomeruli leading to hematuria, proteinuria, edema, reduced glomerular filtration, and hypertension. Can progress to chronic glomerulonephritis and renal failure, but most individuals recover from acute glomerulonephritis without lasting effect.
Hydronephrosis[29]	Increase in fluid pressure in the renal pelvis and calyces owing to obstruction of the ureter by kidney stones, nephroptosis (below), or other causes. Can progress to complete cessation of glomerular filtration and atrophy of nephrons.
Nephroptosis[30] (NEFF-rop-TOE-sis)	Slippage of the kidney to an abnormally low position *(floating kidney)*. Occurs in people with too little body fat to hold the kidney in place, and in people who subject the kidneys to prolonged vibration, such as truck drivers, equestrians, and motorcyclists. Can twist or kink the ureter, which causes pain, obstructs urine flow, and potentially leads to hydronephrosis.
Nephrotic syndrome	Excretion of large amounts of protein in the urine ($\geq$ 3.5 g/day) due to glomerular injury. Can result from trauma, drugs, infections, cancer, diabetes mellitus, lupus erythematosus, and other diseases. Loss of plasma protein leads to edema, ascites (pooling of fluid in the abdominal cavity), hypotension, and susceptibility to infection (because of immunoglobulin loss).
Urinary incontinence	Inability to hold the urine; involuntary leakage from the bladder. Can result from incompetence of the urinary sphincters; bladder irritation; pressure on the bladder in pregnancy; an obstructed urinary outlet so that the bladder is constantly full and dribbles urine *(overflow incontinence)*; uncontrollable urination due to brief surges in bladder pressure, as in laughing or coughing *(stress incontinence)*; and neurological disorders such as spinal cord injuries.

Disorders Described Elsewhere

Azotemia 687	Kidney stones 697	Pyelonephritis 697
Cystitis 697	Proteinuria 693	Uremia 687
Hematuria 693	Pyelitis 697	

trauma, poisoning, atherosclerosis of the renal arteries (often in conjunction with diabetes mellitus), or an autoimmune disease called acute glomerulonephritis (see table 25.1).

Nephrons can regenerate and restore kidney function after short-term injuries, and even when some nephrons are irreversibly destroyed, others can hypertrophy and compensate for their lost function. Indeed, a person can survive on as little as one-third of one kidney. When 75% of the nephrons are lost, however, the remaining ones cannot maintain homeostasis. The result is azotemia and acidosis, and if 90% of renal function is lost, uremia is likely. Loss of nephron function also leads to anemia, because erythrocyte production depends on the hormone erythropoietin, which is secreted mainly by the kidneys.

Renal insufficiency or failure must be treated either with a kidney transplant or with *hemodialysis*. The latter is a procedure in which, usually, arterial blood is pumped through a *dialysis machine*. In the machine, the blood passes through tubes of dialysis membrane immersed in dialysis fluid. Wastes and excess water diffuse from the blood into the fluid, which is discarded, and drugs can be added to the dialysis fluid to diffuse into the blood. Another method called *continuous ambulatory peritoneal dialysis (CAPD)* frees the patient from the dialysis machine and can be carried out at home. Dialysis fluid is introduced into the peritoneal cavity through a catheter, absorbs metabolic wastes, and then is drained from the body and discarded.

Some other urinary system disorders are briefly described in Table 25.1.

Before You Go On

Answer the following questions to test your understanding of the preceding section:

12. Explain why the nephron has a different embryonic origin from the passages that extend from collecting duct to urethra.

13. How does the number of functional nephrons change in old age? What are the implications of this change for homeostasis and medication dosages?

14. What are some causes of renal failure? Describe some clinical signs of renal failure.

15. Cover the right side of table 25.1 and define or describe the disorders on the left from memory.

---

[29]*hydro* = water + *nephr* = kidney + *osis* = medical condition
[30]*nephro* = kidney + *ptosis* = sagging, falling

Study Guide

Assess Your Learning Outcomes

You should have a good understanding of this chapter if you can accurately address the following issues.

25.1 Functions of the Urinary System (p. 686)
1. The six organs of the urinary system
2. The functions of the kidneys
3. The meaning of *excretion*
4. Nitrogenous wastes, azotemia, and uremia

25.2 Anatomy of the Kidney (p. 687)
1. The location of the kidneys and their relationship to adjacent organs and the peritoneum
2. The shape of the kidney, location of the hilum, and purpose of the hilum
3. The three layers of connective tissue that surround and enclose the kidney
4. The relationships between the renal cortex, renal medulla, renal pyramids, and renal sinus
5. The relationships between a renal pyramid, minor calyx, major calyx, renal pelvis, and ureter
6. All branches of the renal artery leading to the glomerulus, peritubular capillaries, and vasa recta; and all tributaries leading from these vessels to the renal vein

7. The routes of sympathetic, parasympathetic, and sensory innervation of the kidney, and effects of the sympathetic nerves on renal function
8. The three basic steps of urine formation
9. The meaning of *nephron* and the anatomical components of this structure
10. The structural details of a renal corpuscle, especially the relationship of the podocytes to the glomerular capillaries
11. Components of the filtration membrane in the glomerulus, and the relevance of this membrane to the composition of urine
12. The path of fluid movement in glomerular filtration and flow into the renal tubule
13. Components of the renal tubule, in the order of fluid flow; which components belong to a single nephron and which are shared by multiple nephrons
14. Structural differences between the proximal convoluted tubule, nephron loop, distal convoluted tubule, and collecting duct; and the relevance of these structural differences to their functional differences
15. The contributions of each part of the renal tubule to the final composition of the urine
16. The location, components, function, and mode of action of the juxtaglomerular apparatus

25.3 Anatomy of the Ureters, Urinary Bladder, and Urethra (p. 696)
1. The anatomy, function, and mode of action of the ureters
2. The anatomy of the urinary bladder and how its structure relates to its functions
3. The anatomy of both the female and male urethra, locations of the internal and external urethral sphincters, and functional difference between the two sphincters

25.4 Developmental and Clinical Perspectives (p. 698)
1. The developmental sequence from pronephros to mesonephros to metanephros
2. The origin of the ureteric bud and the structures that arise from it
3. The pattern of development of the nephrons from mesonephric tissue
4. The cloaca, its differentiation into anal canal and urogenital sinus, and the further differentiation of the urogenital sinus into the organs of the lower urinary tract
5. Changes in the kidneys in old age, and the impact of this on one's health
6. Issues of urinary incontinence or urine retention experienced by elderly men and women, and reasons for this difference between the sexes
7. Causes, effects, and treatment of renal failure

Testing Your Recall

1. Which of these is *not* a function of the kidneys?
   a. to secrete hormones
   b. to excrete nitrogenous wastes
   c. to store urine
   d. to control blood volume
   e. to control acid–base balance

2. The compact ball of capillaries in a nephron is called
   a. the nephron loop.
   b. the peritubular plexus.
   c. the renal corpuscle.
   d. the glomerulus.
   e. the vasa recta.

3. Which of these is *not* true of the position of the kidneys in the body?
   a. They are medial to the aorta.
   b. They are retroperitoneal.
   c. The right kidney is lower than the left.
   d. They are inferior to the liver and spleen.
   e. They lie partially within the rib cage.

4. Which of these lies closest to the renal cortex?
   a. the parietal peritoneum
   b. the renal fascia
   c. the fibrous capsule
   d. the perirenal fat capsule
   e. the renal pelvis

5. The water that is reabsorbed by the collecting duct enters
   a. the nephron loop.
   b. the minor calyx.
   c. the ureter.
   d. the efferent arteriole.
   e. the vasa recta.

6. A glomerulus and glomerular capsule make up one
   a. renal capsule.
   b. renal corpuscle.
   c. kidney lobule.
   d. kidney lobe.
   e. nephron.

7. The kidney has more _____ than any of the other structures listed.
   a. arcuate arteries
   b. minor calyces
   c. medullary pyramids
   d. afferent arterioles
   e. collecting ducts

8. The _____ arises from the embryonic ureteric bud.
   a. nephron
   b. renal pelvis
   c. glomerulus
   d. urinary bladder
   e. proximal convoluted tubule

9. The _____ absorbs variable amounts of water depending on the level of antidiuretic hormone present.
   a. proximal convoluted tubule
   b. nephron loop
   c. distal convoluted tubule
   d. collecting duct
   e. urinary bladder

10. In cortical nephrons, blood of the efferent arteriole flows next into
    a. the peritubular capillaries.
    b. the arcuate artery.
    c. the arcuate vein.
    d. the vasa recta.
    e. the glomerulus.

11. The most abundant nitrogenous waste in the urine is _____.

12. The ureter, renal pelvis, calyces, and collecting duct arise from an embryonic pouch called the _____.

13. The openings of the two ureters and the urethra form the boundaries of a smooth area called the _____ on the floor of the urinary bladder.

14. The _____ is a group of epithelial cells at the distal end of the nephron loop that monitors the flow or composition of the tubular fluid.

15. To enter the capsular space, filtrate must pass between foot processes of the _____, which are cells that form the visceral layer of the glomerular capsule.

16. What part of the nephron is characterized by a brush border and especially great length?

17. Epithelial cells in the _____ of the nephron loop have few mitochondria and low metabolic activity, but they are very permeable to water.

18. The smooth muscle of the bladder wall is called the _____.

19. Each renal pyramid drains into a separate cuplike urine receptacle called a _____.

20. Blood flows through the _____ arteries just before entering the cortical radiate arteries.

Answers in the Appendix

Building Your Medical Vocabulary

State a medical meaning of each of the following word elements, and give a term in which it is used.

1. azot-
2. nephro-
3. glomerul-
4. litho-
5. podo-
6. -uria
7. macula
8. juxta-
9. pelv-
10. cysto-

Answers in the Appendix

True or False

Determine which five of the following statements are false, and briefly explain why.

1. The ureters open through pores in the roof of the urinary bladder.
2. The kidneys secrete antidiuretic hormone to promote water retention and prevent dehydration.
3. The kidney has more distal convoluted tubules than collecting ducts.
4. Tight junctions prevent material from leaking between the epithelial cells of the renal tubule.
5. Many collecting ducts empty into each minor calyx.
6. The glomerulus is a complex of blood capillaries located in the capsular space of the glomerular capsule.
7. Each cortical radiate artery serves multiple nephrons.
8. Blood-borne solutes can become incorporated into the urine by either glomerular filtration or tubular secretion.
9. The kidneys are located in the pelvic cavity.
10. The kidneys develop in the pelvic region and ascend to a higher location in the abdominal cavity during fetal development.

Answers in the Appendix

Testing Your Comprehension

1. What function could the collecting duct *not* perform if there were no nephron loops? Why?
2. Why would a simple squamous epithelium function poorly as an inner lining of the urinary bladder?
3. In some infants, the urachus fails to close; it remains as an open passage *(urachal fistula)* from the urinary bladder to the umbilicus. What would you expect to be the most obvious sign of a urachal fistula?
4. Suppose the ureters entered the bladder from above instead of from below. Which disorder in table 25.1 would you expect to result from this? Explain why.
5. In what ways do the proximal and distal convoluted tubules differ in structure? How is this related to their functional difference?

Answers at www.mhhe.com/saladinha3

Improve Your Grade at **www.mhhe.com/saladinha3**

Practice quizzes, labeling activities, games, and flashcards provide fun ways to master concepts. You can also download image PowerPoint files for each chapter to create a study guide or for taking notes during lecture.

PART FIVE

The Reproductive System

MRI scan of a fetus in the uterus at the 36th week of pregnancy

CHAPTER OUTLINE

DEEPER INSIGHTS

BRUSHING UP

To understand this chapter, you may find it helpful to review the following concepts:
- Chromosome structure (p. 47)
- Mitosis (p. 47)
- Muscles of the pelvic floor (p. 287)
- Hypothalamic releasing factors and pituitary gonadotropins (p. 500)
- Androgens (p. 508)

Anatomy & Physiology REVEALED®

aprevealed.com

Reproductive System

From all we have learned of the structure and function of the human body, it seems a wonder that it works at all! The fact is, however, that even with modern medicine we can't keep it working forever. The body inevitably suffers degenerative changes as we age, and eventually we expire. Yet our genes live on in new containers—our offspring. This final chapter concerns the means of their production—the male and female reproductive systems.

26.1 Sexual Reproduction

▶ Expected Learning Outcomes

When you have completed this section, you should be able to

- define *sexual reproduction;*
- identify the most fundamental biological distinction between male and female; and
- define *primary sex organs, secondary sex organs,* and *secondary sex characteristics.*

The Two Sexes

The essence of sexual reproduction is that it is biparental—the off-spring receive genes from two parents and therefore are not genetically identical to either one. To achieve this, the parents must produce **gametes**[1] (sex cells) that meet and combine their genes in a **zygote**[2] (fertilized egg). The gametes must have two properties for reproduction to be successful: motility so they can achieve contact, and nutrients for the developing embryo. A single cell cannot perform both of these roles optimally, because to contain ample nutrients means to be relatively large and heavy, and this is inconsistent with the need for motility. Therefore, these tasks are usually apportioned to two kinds of gametes. The small motile one—little more than DNA with a propeller—is the **sperm (spermatozoon),** and the large nutrient-laden one is the **egg (ovum).**

In any sexually reproducing species, by definition, an individual that produces eggs is female and one that produces sperm is male. This criterion is not always that simple, as we see in certain abnormalities in sexual development. Genetically, however, any human with a Y sex chromosome is classified as male and anyone lacking a Y is classified as female. Normally, a male inherits an X from the mother and Y from the father, and his sex chromosomes are thus designated XY. A female inherits an X from each parent, and therefore has an XX chromosome pair.

In mammals, the female is also the parent that provides a sheltered internal environment for the development and prenatal nutrition of the embryo. For fertilization and development to occur in the female, the male must have a copulatory organ, the penis, for introducing his gametes into the female reproductive tract and the female must have a copulatory organ, the vagina, for receiving the sperm.

Overview of the Reproductive System

The **reproductive system** in the male serves to produce sperm and introduce them into the female body. The female reproductive system produces eggs, receives the sperm, provides for the union of these gametes, harbors the fetus, gives birth, and nourishes the offspring.

In both sexes, the reproductive system consists of primary and secondary sex organs, or **genitalia.** The **primary sex organs,** or **gonads,**[3] are the organs that produce gametes—testes of the male and ovaries of the female. The **secondary sex organs** are organs other than gonads that are necessary for reproduction. In the male, they constitute a system of ducts, glands, and the penis, concerned with the storage, survival, and conveyance of sperm. In the female, they include the uterine tubes, uterus, and vagina, concerned with uniting the sperm and egg and harboring the fetus.

According to location, the reproductive organs are classified as **external** and **internal genitalia** (table 26.1). The external genitalia are located in the perineum—the diamond-shaped region marked by the pubic symphysis anteriorly, the coccyx posteriorly, and the ischial tuberosities laterally (fig. 26.1; see also fig. 11.17, p. 288). Most of them are externally visible, except for the accessory glands of the female perineum (see fig. 26.21). The internal genitalia are located mainly in the pelvic cavity, except for the male testes and some associated ducts contained in the scrotum.

Secondary sex characteristics are features that develop in adolescence, further distinguish the sexes, and play a role in mate attraction. From the call of a bullfrog to the tail of a peacock, these are well known in the animal kingdom. In humans, the identification of secondary sex characteristics rests on somewhat subjective and culturally variable judgments of what is sexually attractive. Commonly considered among the secondary sex characteristics are the pubic, axillary, and male facial hair; apocrine scent glands as-

---

[3]*gon* = seed

TABLE 26.1	The External and Internal Genitalia
External Genitalia	**Internal Genitalia**
Male	
Penis	Testes (s., testis)
Scrotum	Epididymides (s., epididymis)
	Ductus deferentes (s., ductus deferens)
	Seminal vesicles
	Prostate
	Bulbourethral glands
Female	
Mons pubis	Ovaries
Labia majora (s., labium majus)	Uterine tube
Labia minora (s., labium minus)	Uterus
Clitoris	Vagina
Vaginal orifice	
Vestibular bulbs	
Vestibular glands	
Paraurethral glands	

---

[1]*gam* = marriage, union
[2]*zygo* = yoke, union

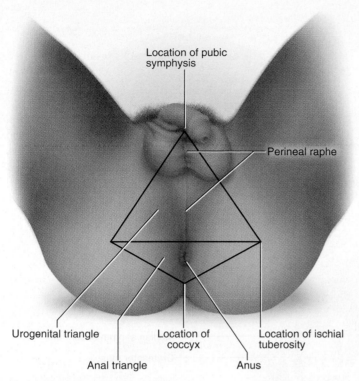

Location of pubic symphysis

Perineal raphe

Urogenital triangle

Location of coccyx

Location of ischial tuberosity

Anal triangle

Anus

Figure 26.1 The Male Perineum.

sociated with these patches of hair; differences in the texture and visibility of the hair on the limbs and trunk; the female breasts; differences in muscularity and the quantity and distribution of body fat; and differences in the pitch of the voice.

Before You Go On

Answer the following questions to test your understanding of the preceding section:

1. Define *gonad* and *gamete*. Explain the relationship between the terms.
2. Define *male, female, sperm,* and *egg.*

26.2 Male Reproductive Anatomy

▶ Expected Learning Outcomes

When you have completed this section, you should be able to

- describe the anatomy of the scrotum, testes, spermatic ducts, accessory glands, and penis;
- describe the stages in spermatogenesis, the production of sperm; and
- describe the structure of a sperm cell and the composition of semen.

We will survey the male reproductive system beginning with the scrotum and testes and following the direction of sperm transport, thus ending with the penis.

The Scrotum

The testes are contained in the **scrotum,**[4] a pouch of skin, muscle, and fibrous connective tissue (fig. 26.2). The skin of the scrotum has sebaceous glands, sparse hair, rich sensory innervation, and somewhat darker pigmentation than skin elsewhere. The scrotum is divided into right and left compartments by an internal **median septum,** which protects each testis from infections of the other one. The location of the septum is externally marked by a seam called the **perineal raphe**[5] (RAY-fee), which also extends anteriorly along the ventral side of the penis and posteriorly as far as the margin of the anus (fig. 26.1). The left testis is usually suspended lower than the right so the two are not compressed against each other between the thighs.

Posteriorly, the scrotum contains the **spermatic cord,** a bundle of fibrous connective tissue containing the *ductus deferens* (a sperm duct), blood and lymphatic vessels, and testicular nerves. It passes upward behind and superior to the testis, where it is easily palpated through the skin of the scrotum. It continues across the anterior side of the pubis and into a 4-cm-long **inguinal canal,** which leads through the muscles of the groin and emerges into the pelvic cavity. The inferior entrance into the inguinal canal is called the *external inguinal ring,* and its superior exit into the pelvic cavity is the *internal inguinal ring.*

The original reason that a scrotum evolved is a subject of debate among reproductive biologists. For whatever reason human testes reside in the scrotum, however, they have adapted to this cooler environment and cannot produce sperm at the core body temperature of 37°C; they must be held at about 35°C. The scrotum has three mechanisms for regulating the temperature of the testes:

1. The **cremaster**[6] **muscle**—strips of the internal abdominal oblique muscle that enmesh the spermatic cord. When it is cold, the cremaster contracts and draws the testes closer to the body to keep them warm. When it is warm, the cremaster relaxes and the testes are suspended farther from the body.

2. The **dartos**[7] **muscle**—a subcutaneous layer of smooth muscle. It, too, contracts when cold, making the scrotum taut and wrinkled. This reaction holds the testes snugly against the warm body and reduces the surface area of the scrotum, thus reducing heat loss.

3. The **pampiniform**[8] **plexus**—an extensive network of veins from the testis that surround the testicular artery in the spermatic cord. The plexus prevents warm arterial blood from overheating the testis, which would inhibit sperm production. It acts as a *countercurrent heat exchanger.* Blood ascending through the plexus is relatively cool (about 35°C) and absorbs

---

[4]*scrotum* = bag
[5]*raphe* = seam
[6]*cremaster* = suspender
[7]*dartos* = skinned
[8]*pampin* = tendril + *form* = shape

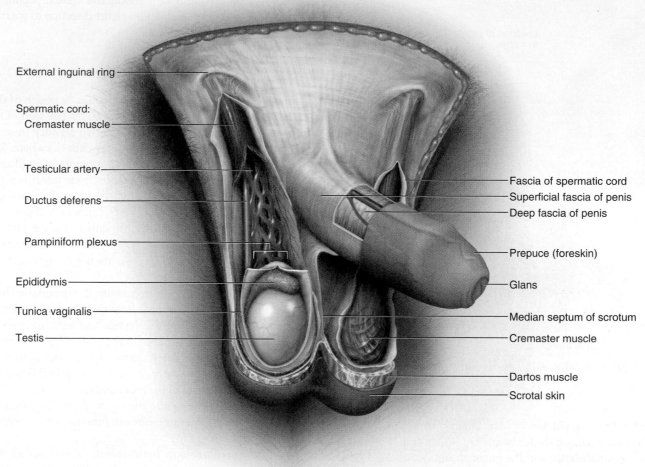

External inguinal ring

Spermatic cord:
    Cremaster muscle

Testicular artery

Ductus deferens

Pampiniform plexus

Epididymis

Tunica vaginalis

Testis

Fascia of spermatic cord
Superficial fascia of penis
Deep fascia of penis

Prepuce (foreskin)

Glans

Median septum of scrotum

Cremaster muscle

Dartos muscle
Scrotal skin

Figure 26.2 The Scrotum and Spermatic Cord.

heat from the warmer (37°C) blood descending through the testicular artery. By the time arterial blood reaches the testis, it is 1.5° to 2.5°C cooler than it was when it left the pelvic cavity.

The Testes

The **testes (testicles)** are the male gonads—combined endocrine and exocrine glands that produce sex hormones and sperm. Each testis is oval and slightly flattened, about 4 cm long, 3 cm from anterior to posterior, and 2.5 cm from left to right (fig. 26.3). Its anterior and lateral surfaces are covered by the **tunica vaginalis,**[9] a saccular extension of the peritoneum. The testis itself has a white fibrous capsule called the **tunica albuginea**[10] (TOO-nih-ca AL-byu-JIN-ee-uh). Connective tissue septa extend from the capsule into the parenchyma of the testis, dividing it into 200 to 300 wedge-shaped lobules. Each lobule contains one to three **seminiferous**[11] (SEM-ih-NIF-er-us) **tubules**—slender ducts up to 70 cm long in which the sperm are produced. Between the seminiferous tubules are clusters of **interstitial (Leydig**[12]**) cells,** the source of testosterone.

A seminiferous tubule has a narrow lumen lined by a thick **germinal epithelium** (fig. 26.4). The epithelium consists of several layers of **germ cells** in the process of becoming sperm, and a much smaller number of tall **sustentacular**[13] **(Sertoli**[14]**) cells,** which protect the germ cells and promote their development. The germ cells depend on the sustentacular cells for nutrients, waste removal, growth factors, and other needs. The sustentacular cells also secrete a hormone, *inhibin,* which regulates the rate of sperm production; and *androgen-binding protein,* which makes the testis responsive to testosterone.

A sustentacular cell is shaped a little like a tree trunk whose roots spread out over the basement membrane, forming the boundary of the tubule, and whose thick trunk reaches to the tubule lumen. Tight junctions between adjacent sustentacular cells form a **blood–testis barrier (BTB),** which prevents antibodies and other large molecules in the blood and intercellular fluid from getting to the germ cells. This is important because the germ cells, being genetically different from other cells of the body, would otherwise

---

[9]*tunica* = coat + *vagina* = sheath
[10]*alb* = white
[11]*semin* = seed, sperm + *fer* = to carry

[12]Franz von Leydig (1821–1908), German anatomist
[13]*sustentacul* = support
[14]Enrico Sertoli (1842–1910), Italian histologist

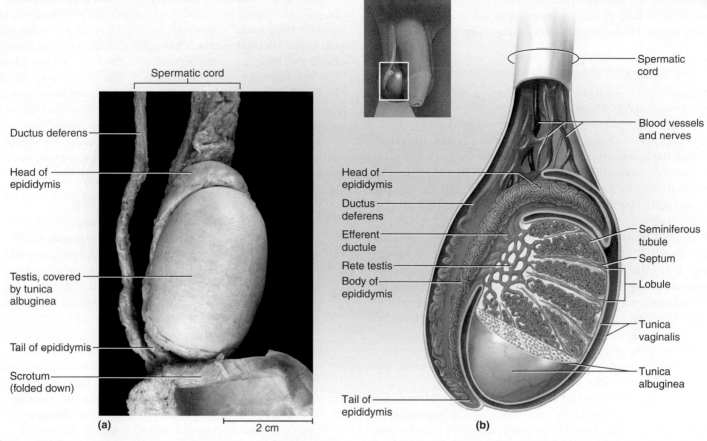

Figure 26.3 **The Testis and Associated Structures.** (a) The scrotum is opened and folded downward to reveal the testis and associated organs. (b) Anatomy of the testis, epididymis, and spermatic cord.

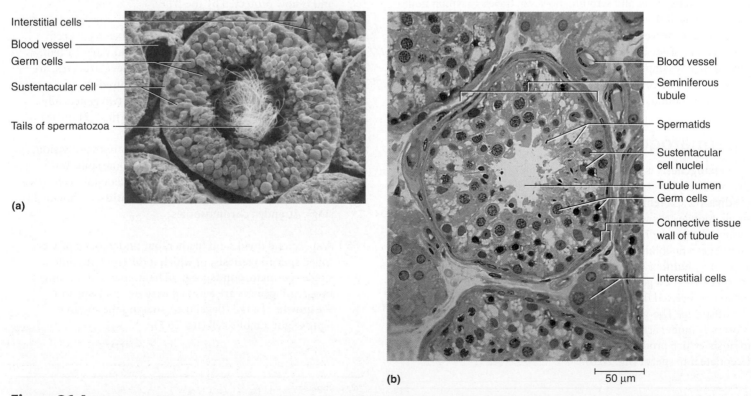

Figure 26.4 **Histology of the Testis.** (a) Scanning electron micrograph. (b) Light micrograph. Part (b) is from a region of the tubule that did not have mature sperm at the time. [Part (a) from R. G. Kessel and R. H. Kardon, *Tissues and Organs: A Text-Atlas of Scanning Electron Microscopy* (W. H. Freeman, 1979)]

be attacked by the immune system. Some cases of sterility occur when the BTB fails to form adequately in adolescence and the immune system produces antibodies against the germ cells.

Apply What You Know

Would you expect to find blood capillaries in the walls of the seminiferous tubules? Why or why not?

The seminiferous tubules lead into a network called the **rete**[15] (REE-tee) **testis,** embedded in the capsule on the posterior side. Sperm partially mature in the rete. They are moved along by the flow of fluid secreted by the sustentacular cells and by the cilia on some rete cells. Sperm do not swim while they are in the male reproductive tract.

Each testis is supplied by a **testicular artery,** which arises from the abdominal aorta just below the renal artery. This is a very long, slender artery that winds its way down the posterior abdominal wall before passing through the inguinal canal into the scrotum. Its blood pressure is very low, and this is one of the few arteries to have no pulse. Consequently, blood flow to the testes is quite meager and they receive a poor oxygen supply. The sperm appear to compensate by developing unusually large mitochondria, which may precondition them for survival in the hypoxic environment of the female reproductive tract.

Blood leaves the testis by way of the pampiniform plexus of veins. As these veins pass through the inguinal canal, they converge and form the **testicular vein.** The right testicular vein drains into the inferior vena cava and the left one drains into the left renal vein. Lymphatic vessels also drain each testis. They travel through the inguinal canal with the veins and lead to lymph nodes adjacent to the lower aorta. Lymph from the penis and scrotum, however, travels to lymph nodes adjacent to the iliac arteries and veins and in the inguinal region.

Testicular nerves lead to the gonads from spinal cord segments T10 and T11. They are mixed sensory and motor nerves containing predominantly sympathetic but also some parasympathetic fibers. The sensory fibers are concerned primarily with pain and the autonomic fibers are predominantly vasomotor, for regulation of blood flow.

Spermatogenesis and Sperm

Spermatogenesis is the process of sperm production. It occurs in the seminiferous tubules and involves three principal events: (1) division and remodeling of a relatively large germ cell into four small, mobile cells with flagella; (2) reduction of the chromosome number by one-half, so that when the sperm and egg combine, we do not get a doubling of chromosome number in every generation; and (3) a shuffling of the genes so that each chromosome of the sperm carries new gene combinations that did not exist in the chromosomes received from the parents. This ensures genetic variety in the offspring. The genetic recombination and reduction in chromosome number are achieved through a form of cell division called **meiosis,** which produces four daughter cells that subsequently differentiate into sperm (fig. 26.5).

Early in prenatal development, **primordial germ cells** form in the yolk sac, migrate by ameboid motion into the embryo itself, and colonize the *gonadal ridges* (see p. 729). Here they become stem cells called **spermatogonia.** These cells remain dormant through childhood, lying along the periphery of the seminiferous tubules near the basement membrane, outside the blood–testis barrier (BTB). These cells are **diploid**[16]—they have 46 chromosomes (23 pairs) and are genetically identical to most other cells of the body, so they require no protection from the immune system.

At puberty, testosterone secretion rises, reactivates the spermatogonia, and brings on spermatogenesis. The essential steps of spermatogenesis are as follows; see the same-numbered steps in figure 26.6.

1 Spermatogonia divide by mitosis. One daughter cell from each division remains near the tubule wall as a stem cell called the *type A spermatogonium.* Type A spermatogonia serve as a lifetime supply of stem cells, so men normally remain fertile throughout old age. The other daughter cell, called the *type B spermatogonium,* migrates slightly away from the wall on its way to becoming sperm.

2 The type B spermatogonium enlarges and becomes a **primary spermatocyte.** Since this cell is about to undergo meiosis and become genetically different from other cells of the body, it must be protected from the immune system. Ahead of the primary spermatocyte, the tight junction between two sustentacular cells is dismantled, while a new tight junction forms behind the spermatocyte. The spermatocyte moves forward toward the lumen, like passing through the double-doored airlock of a spaceship, and is now protected by the BTB.

3 The primary spermatocyte undergoes *meiosis I,* a cell division that reduces the chromosome number by half. The daughter cells, called **secondary spermatocytes,** are therefore **haploid**[17]—they have 23 unpaired chromosomes, although each chromosome consists of two genetically identical strands called chromatids (see fig. 2.21, p. 48).

4 The secondary spermatocytes undergo another division, *meiosis II,* during which each chromosome splits into separate chromatids. The result is four daughter cells (two from each spermatocyte) called **spermatids,** each with 23 single-stranded chromosomes.

5 A spermatid divides no further, but undergoes a process called **spermiogenesis,** in which it differentiates into a single spermatozoon (sperm). The fundamental changes in spermiogenesis are a loss of excess cytoplasm and the growth of a tail (flagellum), making the sperm a lightweight, mobile cell (fig. 26.7).

---

[15]*rete* = network

[16]*diplo* = double
[17]*haplo* = half

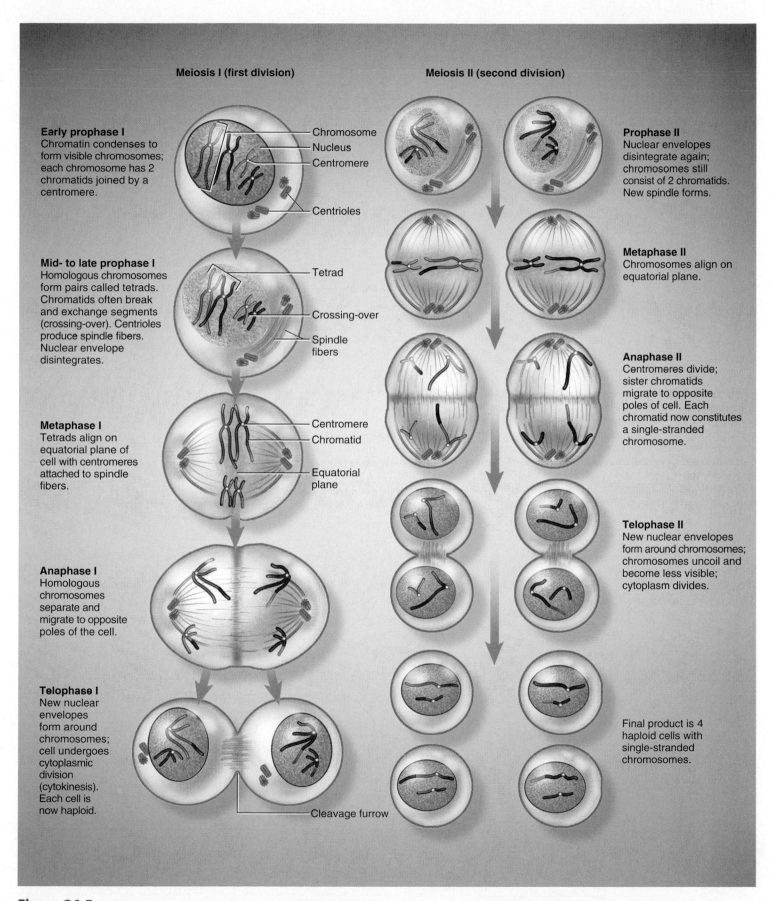

Meiosis I (first division)

Meiosis II (second division)

Early prophase I
Chromatin condenses to form visible chromosomes; each chromosome has 2 chromatids joined by a centromere.

- Chromosome
- Nucleus
- Centromere
- Centrioles

Prophase II
Nuclear envelopes disintegrate again; chromosomes still consist of 2 chromatids. New spindle forms.

Mid- to late prophase I
Homologous chromosomes form pairs called tetrads. Chromatids often break and exchange segments (crossing-over). Centrioles produce spindle fibers. Nuclear envelope disintegrates.

- Tetrad
- Crossing-over
- Spindle fibers

Metaphase II
Chromosomes align on equatorial plane.

Metaphase I
Tetrads align on equatorial plane of cell with centromeres attached to spindle fibers.

- Centromere
- Chromatid
- Equatorial plane

Anaphase II
Centromeres divide; sister chromatids migrate to opposite poles of cell. Each chromatid now constitutes a single-stranded chromosome.

Anaphase I
Homologous chromosomes separate and migrate to opposite poles of the cell.

Telophase II
New nuclear envelopes form around chromosomes; chromosomes uncoil and become less visible; cytoplasm divides.

Telophase I
New nuclear envelopes form around chromosomes; cell undergoes cytoplasmic division (cytokinesis). Each cell is now haploid.

- Cleavage furrow

Final product is 4 haploid cells with single-stranded chromosomes.

Figure 26.5 **Meiosis.** For simplicity, the cell is shown with only two pairs of homologous chromosomes. Human cells begin meiosis with 23 pairs.

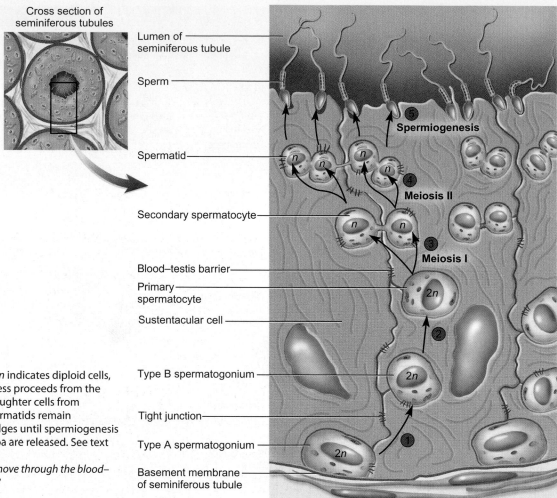

Cross section of seminiferous tubules

Lumen of seminiferous tubule

Sperm

Spermiogenesis (5)

Spermatid

Meiosis II (4)

Secondary spermatocyte

Meiosis I (3)

Blood–testis barrier

Primary spermatocyte

Sustentacular cell (2)

Type B spermatogonium

Tight junction

Type A spermatogonium (1)

Basement membrane of seminiferous tubule

Figure 26.6 Spermatogenesis. *2n* indicates diploid cells, and *n* indicates haploid cells. The process proceeds from the bottom of the figure to the top. The daughter cells from secondary spermatocytes through spermatids remain connected by slender cytoplasmic bridges until spermiogenesis is complete and Indlvldual spermatozoa are released. See text for explanation of steps 1 through 5.
● *Why must the primary spermatocyte move through the blood–testis barrier before undergoing meiosis?*

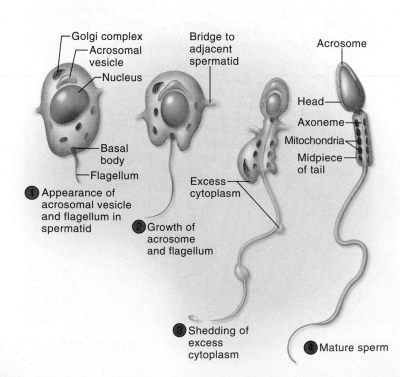

Golgi complex
Acrosomal vesicle
Nucleus

Bridge to adjacent spermatid

Acrosome

Head
Axoneme
Mitochondria
Midpiece of tail

Basal body
Flagellum

Excess cytoplasm

(1) Appearance of acrosomal vesicle and flagellum in spermatid

(2) Growth of acrosome and flagellum

(3) Shedding of excess cytoplasm

(4) Mature sperm

Figure 26.7 Spermiogenesis. In this process, the spermatids discard excess cytoplasm, grow tails, and become spermatozoa.

All stages from primary spermatocyte to spermatozoon are enfolded in tendrils of the sustentacular cells and bound to them by tight junctions and gap junctions. In addition, the daughter cells do not completely separate from each other in meiosis I and II, but remain joined to each other by narrow cytoplasmic bridges. They do not completely separate until the very end of spermiogenesis.

At the conclusion of spermiogenesis, the spermatozoa are released and washed down the tubule by fluid from the sustentacular cells. It takes about 74 days for a spermatogonium to become mature sperm. A young adult male produces about 300,000 sperm per minute, or 400 million per day.

The Spermatozoon

The spermatozoon has two parts: a pear-shaped head and a long tail (fig. 26.8). The **head,** about 4 to 5 μm long and 3 μm wide at its broadest part, contains three structures: a nucleus, acrosome, and flagellar basal body. The most important of these is the nucleus, which fills most of the head and contains a haploid set of condensed, genetically inactive chromosomes. The **acrosome**[18] is a lysosome in the form of a thin cap covering the apical half of the nucleus. It contains enzymes that are later used to penetrate the egg if the sperm is successful. The basal body of the tail flagellum is nestled in an indentation at the posterior end of the nucleus.

---

[18]*acro* = tip, peak + *some* = body

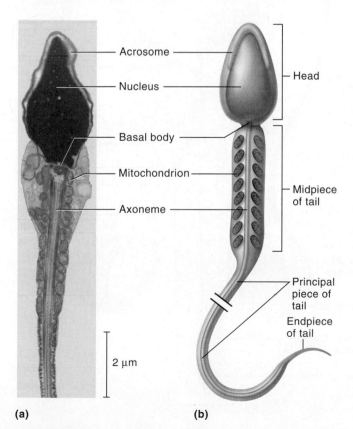

(a)   **(b)**

Figure 26.8 The Mature Spermatozoon. (a) Head and part of the tail of a spermatozoon (TEM). (b) Sperm structure.

The **tail** is divided into three regions called the midpiece, principal piece, and endpiece. The **midpiece,** a cylinder about 5 to 9 μm long and half as wide as the head, is the thickest part. It contains numerous large mitochondria that coil tightly around the axoneme of the flagellum. They produce the ATP needed for the beating of the tail when the sperm migrates up the female reproductive tract. The **principal piece,** 40 to 45 μm long, constitutes most of the tail and consists of the axoneme surrounded by a sheath of supportive fibers. The **endpiece,** 4 to 5 μm long, consists of the axoneme only and is the narrowest part of the sperm.

The Spermatic Ducts

After leaving the testis, the sperm travel through a series of *spermatic ducts* to reach the urethra. These include the following (fig. 26.9):

- **Efferent ductules.** About 12 small efferent ductules arise from the posterior side of each testis and carry sperm to the epididymis (see fig. 26.3b). They have clusters of ciliated cells that help drive the sperm along.

- **Duct of the epididymis.** The **epididymis**[19] (EP-ih-DID-ih-miss; plural, *epididymides*) is a site of sperm maturation and storage. It adheres to the posterior side of the testis, measures about 7.5 cm long, and consists of a clublike *head* at the superior end, a long middle *body,* and a slender *tail* at its inferior end. It contains a single coiled duct, about 6 m (18 ft) long, embedded in connective tissue. This duct reabsorbs about 90% of the fluid secreted by the testis. Sperm are physiologically immature (incapable of fertilizing an egg) when they leave the testis, but mature as they travel through the head and body of the epididymis. In 20 days or so, they reach the tail. They are stored here and in the adjacent portion of the ductus deferens. Stored sperm remain fertile for 40 to 60 days, but if they become too old without being ejaculated, they disintegrate and the epididymis reabsorbs them.

- **Ductus (vas) deferens.** The duct of the epididymis straightens out at the tail, turns 180°, and becomes the ductus deferens. This is a muscular tube about 45 cm long and 2.5 mm in diameter (fig. 26.9). It passes upward through the spermatic cord and inguinal canal, and enters the pelvic cavity. There, it turns medially and approaches the urinary bladder. After passing between the bladder and ureter, the duct turns downward behind the bladder and widens into a terminal **ampulla.** The ductus deferens ends by uniting with the duct of the seminal vesicle, a gland considered later. It has a very narrow lumen and a thick wall of smooth muscle well innervated by sympathetic nerve fibers.

- **Ejaculatory duct.** Where the ductus deferens and duct of the seminal vesicle meet, they form a short (2 cm) ejaculatory duct, which passes through the prostate gland and empties into the urethra. The ejaculatory duct is the last of the spermatic ducts.

---

[19]*epi* = upon + *didym* = twins, testes

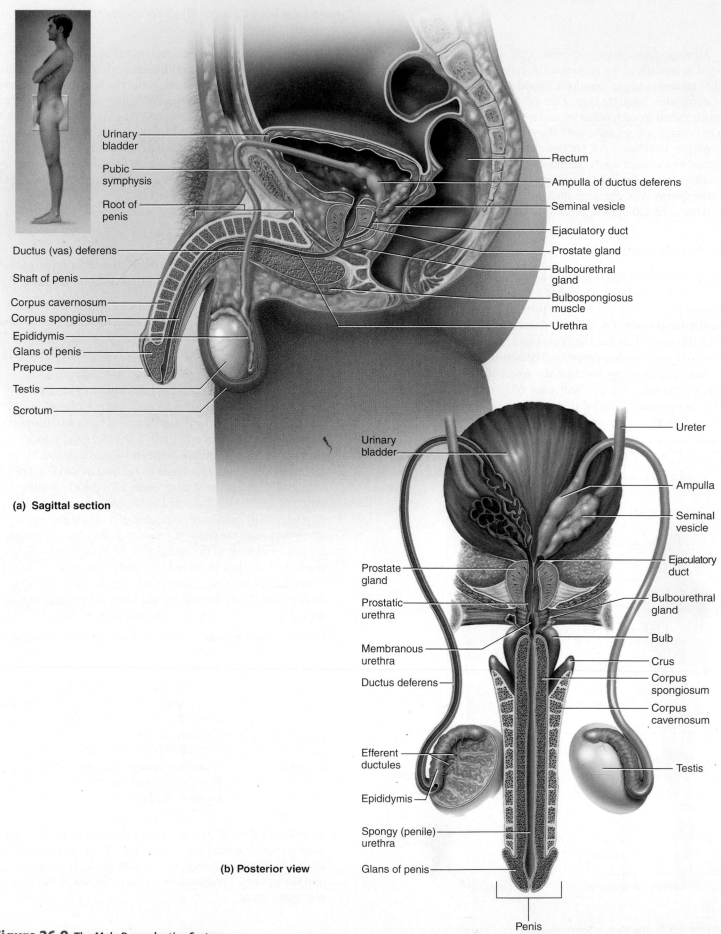

(a) Sagittal section

Urinary bladder

Pubic symphysis

Root of penis

Ductus (vas) deferens

Shaft of penis

Corpus cavernosum

Corpus spongiosum

Epididymis

Glans of penis

Prepuce

Testis

Scrotum

Rectum

Ampulla of ductus deferens

Seminal vesicle

Ejaculatory duct

Prostate gland

Bulbourethral gland

Bulbospongiosus muscle

Urethra

(b) Posterior view

Urinary bladder

Prostate gland

Prostatic urethra

Membranous urethra

Ductus deferens

Efferent ductules

Epididymis

Spongy (penile) urethra

Glans of penis

Ureter

Ampulla

Seminal vesicle

Ejaculatory duct

Bulbourethral gland

Bulb

Crus

Corpus spongiosum

Corpus cavernosum

Testis

Penis

Figure 26.9 The Male Reproductive System.

714

The male urethra is shared by the reproductive and urinary systems. It is about 20 cm long and consists of three regions: the *prostatic, membranous,* and *spongy (penile) urethra.* Although it serves both urinary and reproductive roles, it cannot pass urine and semen simultaneously. During ejaculation, the internal urethral sphincter contracts to prevent the voiding of urine and keep semen out of the urinary bladder.

The Accessory Glands

There are three sets of *accessory glands* in the male reproductive system (fig. 26.9):

- **Seminal vesicles.** These are a pair of glands posterior to the urinary bladder; one is associated with each ductus deferens. A seminal vesicle is about 5 cm long, or approximately the dimensions of one's little finger. It has a connective tissue capsule and underlying layer of smooth muscle. The secretory portion is a very convoluted duct with numerous branches that form a complex labyrinth. The duct empties into the ejaculatory duct.
- **Prostate**[20] (PROSS-tate) **gland.** This is a median structure that surrounds the urethra and ejaculatory ducts immediately inferior to the urinary bladder. It measures about 3 cm in diameter. It is actually a composite of 30 to 50 compound tubuloacinar glands enclosed in a single fibrous capsule. These glands empty into the urethra through about 20 pores in the urethral wall. The stroma of the prostate consists of connective tissue and smooth muscle, like that of the seminal vesicles. The prostate gland is the source of the two most common urogenital dysfunctions in older men (see Deeper Insight 26.1).
- **Bulbourethral (Cowper**[21]**) glands.** These are named for their position near a dilated bulb at the inner end of the penis and their association with the penile urethra. They are brownish, spherical glands about 1 cm in diameter, with a 2.5 cm duct to the urethra. During sexual arousal, they produce a clear slippery fluid that lubricates the head of the penis in preparation for intercourse. Perhaps more importantly, though, the fluid protects the sperm by neutralizing the acidity of residual urine in the urethra.

Semen

A typical ejaculation discharges 2 to 5 mL of **semen**[22] **(seminal fluid),** a complex mixture of sperm and glandular secretions. About 10% of it consists of sperm and fluids from the spermatic ducts; 30% is a thin, milky fluid from the prostate; and 60% is a viscous, yellowish fluid from the seminal vesicles. The bulbourethral glands contribute a trace of fluid. Normal semen contains about 50 to 120 million sperm per milliliter—the sperm count. A count lower than 20 to 25 million/mL is usually associated with *infertility (sterility),* the inability to fertilize an egg (see table 26.2).

[20]*pro* = before + *stat* = to stand; commonly misspelled and mispronounced "prostrate"
[21]William Cowper (1666–1709), British anatomist
[22]*semen* = seed

DEEPER INSIGHT 26.1

Prostate Diseases

The prostate gland weighs about 20 g by age 20, remains at that weight until age 45 or so, and then begins to grow slowly again. By age 70, over 90% of men show some degree of *benign prostatic hyperplasia (BPH)*—noncancerous enlargement of the gland. The major complication of BPH is that it compresses the urethra, slows the flow of urine, and sometimes promotes bladder and kidney infections.

Prostate cancer is the second most common cancer in men (after lung cancer); it affects about 9% of men over the age of 50. Prostate tumors tend to form near the periphery of the gland, where they do not obstruct urine flow; therefore, they often go unnoticed until they cause pain. Prostate cancer often metastasizes to nearby lymph nodes and then to the lungs and other organs. It is more common among American blacks than whites and is rare among the Japanese.

The position of the prostate immediately anterior to the rectum allows it to be palpated through the rectal wall to check for tumors. This procedure is called *digital rectal examination (DRE).* Prostate cancer can also be diagnosed from elevated levels of *serine protease* (also known as *prostate specific antigen, PSA*) and *acid phosphatase* (another prostatic enzyme) in the blood. Up to 80% of men with prostate cancer survive when it is detected and treated early, but only 10% to 50% survive if it spreads beyond the prostatic capsule.

Fresh semen is very sticky. This results from the action of a clotting enzyme in the prostatic fluid on a protein, *proseminogelin,* in the seminal vesicle fluid. The enzyme converts this to *seminogelin,* an adhesive protein very similar to the fibrin of a blood clot. The functional advantage of this seems to be that it ensures semen will adhere to the cervix and vagina rather than draining back out. Twenty to 30 minutes later, another prostatic enzyme, *serine protease,* breaks down seminogelin and liquifies the semen, liberating the sperm for their migration up the female reproductive tract. Sperm motility requires energy, which they get from fructose and other sugars provided by the seminal vesicles. The seminal vesicles also contribute lipids called *prostaglandins*—named for their discovery in the prostatic fluid of bulls, but later found to be even more abundant in the seminal vesicle fluid. Prostaglandins may contribute to the passage of sperm from the vagina into the uterus by thinning the mucus in the *cervical canal* (see p. 723) and perhaps inducing uterine peristaltic contractions that suck semen into the uterus.

The Penis

The **penis**[23] serves to deposit semen in the vagina. Half of it is an internal **root** and half is the externally visible **shaft** and **glans**[24] (see figs. 26.9 and 26.10). The glans is the expanded head at the distal end of the penis with the external urethral orifice at its tip. The external portion is about 8 to 10 cm (3–4 in.) long and 3 cm in diameter when flaccid (nonerect); the typical dimensions of an erect penis are 13 to 18 cm (5–7 in.) long and 4 cm in diameter.

[23]*penis* = tail
[24]*glans* = acorn

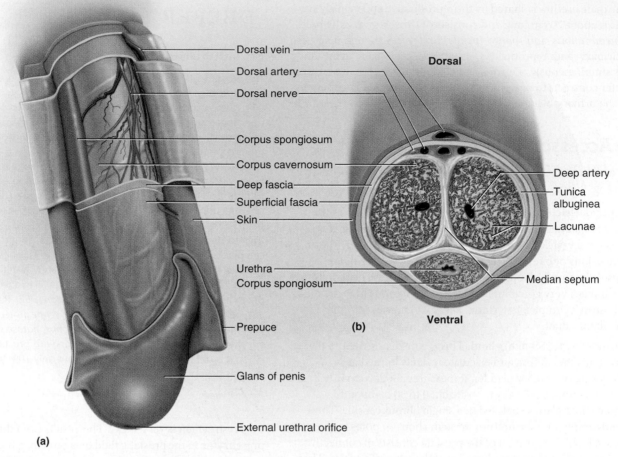

Figure 26.10 **Anatomy of the Penis.** (a) Superficial dissection of shaft, lateral view. (b) Cross section at midshaft.
● *What is the functional benefit of the corpus spongiosum not having a tunica albuginea?*

The skin is loosely attached to the shaft, allowing for expansion during erection. It continues over the glans as the **prepuce,** or foreskin, which is often removed by circumcision. A ventral fold of tissue called the *frenulum* attaches the skin to the proximal margin of the glans. The skin of the glans itself is thinner and firmly attached to the underlying erectile tissue. The glans and facing surface of the prepuce have sebaceous glands that produce a waxy secretion called **smegma.**[25]

Directional terminology may be a little confusing in the penis, because the *dorsal* side is the one that faces anteriorly, at least when the penis is flaccid, whereas the *ventral* side of the penis faces posteriorly. This is because in most mammals, the penis is horizontal, held against the abdomen by skin, and points anteriorly. The urethra passes through its lower, more obviously ventral, half. Directional terminology in the human penis follows the same convention as for other mammals, even though our bipedal posture and more pendulous penis change these anatomical relationships.

The penis consists mainly of three cylindrical bodies called **erectile tissues,** which fill with blood during sexual arousal and account for its enlargement and erection. A single erectile body, the

corpus spongiosum, passes along the ventral side of the penis and encloses the penile urethra. It expands at the distal end to fill the entire glans. Proximal to the glans, the dorsal side of the penis has a **corpus cavernosum** (plural, *corpora cavernosa*) on each side. Each is ensheathed in a tight fibrous sleeve called the **tunica albuginea,** and they are separated from each other by a **median septum.** (Note that the testes also have a tunica albuginea and the scrotum also has a median septum.)

All three cylinders of erectile tissue are spongy in appearance and contain numerous tiny blood sinuses called **lacunae.** The partitions between lacunae, called **trabeculae,** are composed of connective tissue and smooth **trabecular muscle.** In the flaccid penis, trabecular muscle tone collapses the lacunae, which appear as slits in the tissue.

At the body surface, the penis turns 90° posteriorly and continues inward as the root. The corpus spongiosum terminates at the internal end of the root as a dilated **bulb,** which is ensheathed in the bulbospongiosus muscle and attached to the lower surface of the perineal membrane in the urogenital triangle (see p. 287). The corpora cavernosa diverge like the arms of a Y. Each arm, called a **crus** (pronounced "cruss"; plural, *crura*), attaches the penis to the pubic arch of the pelvis and to the perineal membrane on its respective side. Each crus is enveloped by an ischiocavernosus muscle.

[25]*smegma* = unguent, ointment, soap

The penis receives blood from a pair of **internal pudendal (penile) arteries,** which branch from the internal iliac arteries. As each artery enters the root of the penis, it divides in two. One branch, the **dorsal artery,** travels dorsally along the penis not far beneath the skin, supplying blood to the skin, fascia, and corpus spongiosum. The other branch, the **deep artery,** travels through the core of the corpus cavernosum and gives off smaller **helicine**[26] **arteries,** which penetrate the trabeculae and empty into the lacunae. There are numerous anastomoses between the dorsal and deep arteries, so neither of them is the exclusive source of blood to any one erectile tissue.

When the deep arteries dilate, the lacunae fill with blood and the erectile tissues swell. The tunica albuginea around the corpora cavernosa cannot expand much, so pressure builds especially in these two erectile tissues and the penis becomes elongated and erect. The corpus spongiosum becomes less engorged, but swells and becomes more visible as a cordlike ridge along the ventral surface of the penis. When the penis is flaccid, most of its blood supply comes from the dorsal arteries. A median **deep dorsal vein** drains blood from the penis. It runs between the two dorsal arteries beneath the deep fascia and empties into a plexus of prostatic veins.

The penis is richly innervated by sensory and motor nerve fibers. The glans has an abundance of tactile, pressure, and temperature receptors, especially on its proximal margin and frenulum. They lead by way of a pair of prominent **dorsal nerves** to the **internal pudendal nerves,** then via the sacral plexus to spinal cord segments S2 to S4. Sensory fibers of the shaft, scrotum, perineum, and elsewhere are also highly important to sexual stimulation.

Both autonomic and somatic motor fibers carry signals from integrating centers in the spinal cord to the penis and other pelvic organs. Sympathetic nerve fibers arise from levels T12 to L2, pass through the hypogastric and pelvic nerve plexuses, and innervate the penile arteries, trabecular muscle, spermatic ducts, and accessory glands. They dilate the penile arteries and can induce erection even when the sacral region of the spinal cord is damaged. They also initiate erection in response to input from the special senses and to sexual thoughts.

Parasympathetic fibers extend from segments S2 to S4 of the spinal cord through the pudendal nerves to the arteries of the penis. They are involved in an autonomic reflex arc that causes erection in response to direct stimulation of the penis and other perineal organs.

Before You Go On

Answer the following questions to test your understanding of the preceding section:

3. Name the stages of spermatogenesis from spermatogonium to spermatozoon. How do they differ in the number of chromosomes per cell and chromatids per chromosome?

4. Name two types of cells in the testis other than the germ cells, and describe their locations and functions.

5. Describe the three major parts of a spermatozoon and state what organelles or cytoskeletal components are contained in each.

---

[26]*helic* = coil, helix

DEEPER INSIGHT 26.2

Reproductive Effects of Pollution

In recent decades, wildlife biologists have noticed increasing numbers of male birds, fish, and alligators with a variety of abnormalities in reproductive development. These deformities have been attributed to chemical pollutants called *endocrine disruptors* or *estrogen mimics.* Evidence is mounting that humans, too, are showing declining fertility and increasing anatomical abnormalities due to pollutants in water, meat, vegetables, and even breast milk and the uterine environment.

Over the last several decades, there has been an alarming increase in the incidence of *cryptorchidism* (undescended testes) and *hypospadias* (a condition in which the urethra opens on the ventral side of the penis instead of at the tip). The rate of testicular cancer has more than tripled in that time. Data on 15,000 men from several countries also show a sharp drop in average sperm count—from 113 million/mL in 1940 to only 66 million/mL in 1990. Total sperm production decreased even more, because the average volume of semen per ejaculate dropped 19% over this period.

The pollutants implicated in this trend include a wide array of common herbicides, insecticides, industrial chemicals, and breakdown products of materials ranging from plastics to dishwashing detergents. Some authorities think these chemicals act by mimicking estrogens or blocking testosterone receptors. Other scientists question the data and feel the issue may be overstated, but there is nevertheless a strong interest at this time in screening industrial chemicals for endocrine effects.

---

6. Name all the ducts the sperm follow, in order, from the time they form in the testis to the time of ejaculation.

7. Describe the locations and functions of the seminal vesicles, prostate, and bulbourethral glands.

8. Name the erectile tissues of the penis, and describe their locations relative to each other.

26.3 Female Reproductive Anatomy

▶ Expected Learning Outcomes

When you have completed this section, you should be able to

- describe the structure of the ovary;

- describe the stages of oogenesis and how these relate to changes in histology of the ovarian follicles;

- trace the female reproductive tract and describe the gross anatomy and histology of each organ;

- describe changes in the uterine lining through the menstrual cycle;

- identify the ligaments that support the female reproductive organs;

- describe the blood supply to the female reproductive tract;

- identify the external genitalia of the female; and
- describe the structure of the nonlactating breast and lactating mammary gland.

Figure 26.11 shows the female reproductive tract. The principal reproductive organs of the pelvic cavity are the *ovaries, uterine tubes, uterus,* and *vagina,* which will be described in that order.

The Ovaries

The female gonads are the **ovaries,**[27] which produce egg cells (ova) and sex hormones. The ovary is an almond-shaped organ nestled in a depression of the posterior pelvic wall called the *ovarian fossa.* It measures about 3 cm long, 1.5 cm wide, and 1 cm thick. Its cap-

---

[27]*ov* = egg + *ary* = place for

sule, like that of the testis, is called the **tunica albuginea.** The interior of the ovary is indistinctly divided into a central **medulla** and an outer **cortex** (fig. 26.12). The medulla is a core of fibrous connective tissue occupied by the principal arteries and veins of the ovary. The cortex is the site of the ovarian **follicles,** each of which consists of one developing ovum surrounded by numerous small follicular cells. The ovary does not have a system of tubules like the testis; eggs are released one at a time by the bursting of the follicles *(ovulation).* In childhood, the ovaries are smooth-surfaced. During the reproductive years, they become more corrugated because growing follicles of various stages produce bulges in the surface. After menopause, the ovaries become shrunken and composed mostly of scar tissue.

The ovaries and other internal genitalia are held in place by several connective tissue ligaments (fig. 26.13). The medial pole of the ovary is attached to the uterus by the **ovarian ligament** and its lateral pole is attached to the pelvic wall by the **suspensory ligament.** The anterior margin of the ovary is anchored by a peritoneal

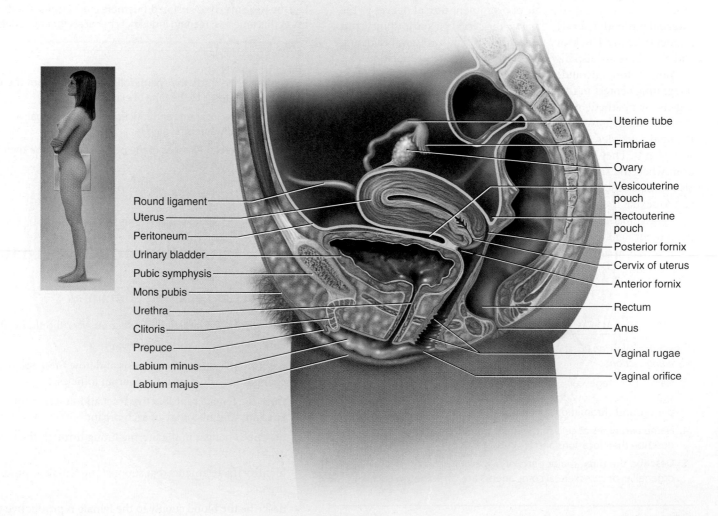

Figure 26.11 The Female Reproductive System.

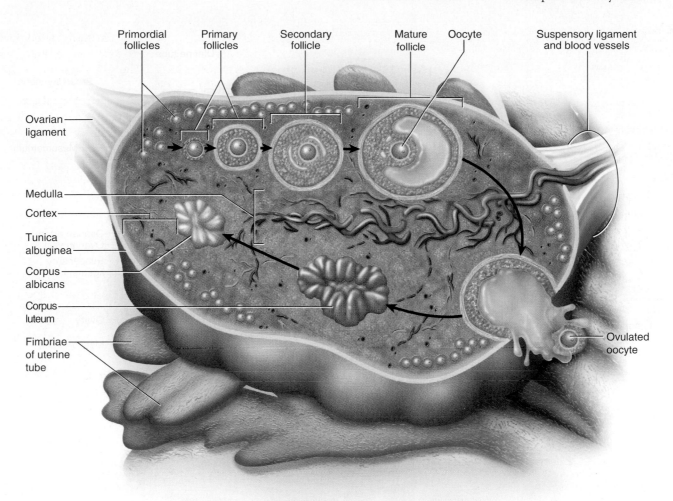

Figure 26.12 **Structure of the Ovary.** Arrows indicate the developmental sequence of the ovarian follicles; follicles do not migrate around the ovary, and not all the follicle types illustrated here are present simultaneously.

fold called the **mesovarium.**[28] This ligament extends to a sheet of peritoneum called the **broad ligament,** which flanks the uterus and encloses the uterine tube in its superior margin. If you picture these ligaments as a ⏋, the vertical bar would represent the broad ligament—flanking the uterus and enfolding the uterine tube at its upper end—and the horizontal bar would represent the mesovarium enfolding the ovary at its free end.

The ovary receives blood from two arteries: the **ovarian branch of the uterine artery,** which passes through the mesovarium and approaches the medial pole of the ovary, and the **ovarian artery,** which passes through the suspensory ligament and approaches the lateral pole (see fig. 26.19). The ovarian artery is the female equivalent of the male testicular artery, arising high on the aorta and traveling down to the gonad along the posterior body wall. The ovarian and uterine arteries anastomose along the margin of the ovary and give off multiple small arteries that enter the ovary on that side. Ovarian veins, lymphatics, and nerves also travel through the suspensory ligament. The veins and lymphatics follow a course similar to those of the testes described earlier.

Oogenesis and Folliculogenesis

The production of eggs is called **oogenesis**[29] (OH-oh-JEN-eh-sis) (fig. 26.14). Like spermatogenesis, it employs meiosis and produces haploid gametes. It differs from spermatogenesis in other respects, however: It is not a continual process, but occurs in a rhythm called the **ovarian cycle,** and for each original germ cell (oogonium), it produces only one functional gamete. The other daughter cells are tiny **polar bodies** that soon die.

The female primordial germ cells arise, like those of the male, from the yolk sac of the embryo. They colonize the gonadal ridges in the first 5 to 6 weeks and then differentiate into **oogonia** (OH-oh-GO-nee-uh). From shortly before birth until about 6 months after, many of these transform into **primary oocytes** and go as far as early meiosis I. The great majority of primary oocytes die before birth and in childhood—a "weeding out" process called *atresia*. The survivors remain dormant until puberty. Any stage from the primary oocyte to the time of fertilization can be called an egg or ovum. A girl has about 400,000 eggs at the onset of puberty, and they are considered to be her lifetime supply of germ cells.

[28]*mes* = middle + *ovari* = ovary

[29]*oo* = egg + *genesis* = production

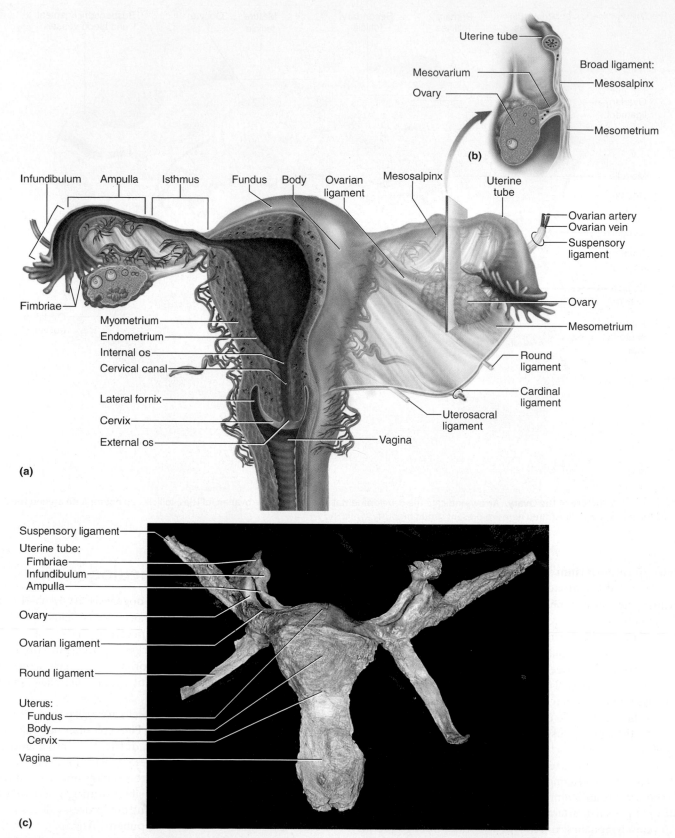

Figure 26.13 **The Female Reproductive Tract and Supportive Ligaments.** (a) Posterior view of the reproductive tract. (b) Relationship of the uterine tube and ovary to the supporting ligaments. (c) Anterior view of the major female reproductive organs from a cadaver.

Development of egg (oogenesis)

Development of follicle (folliculogenesis)

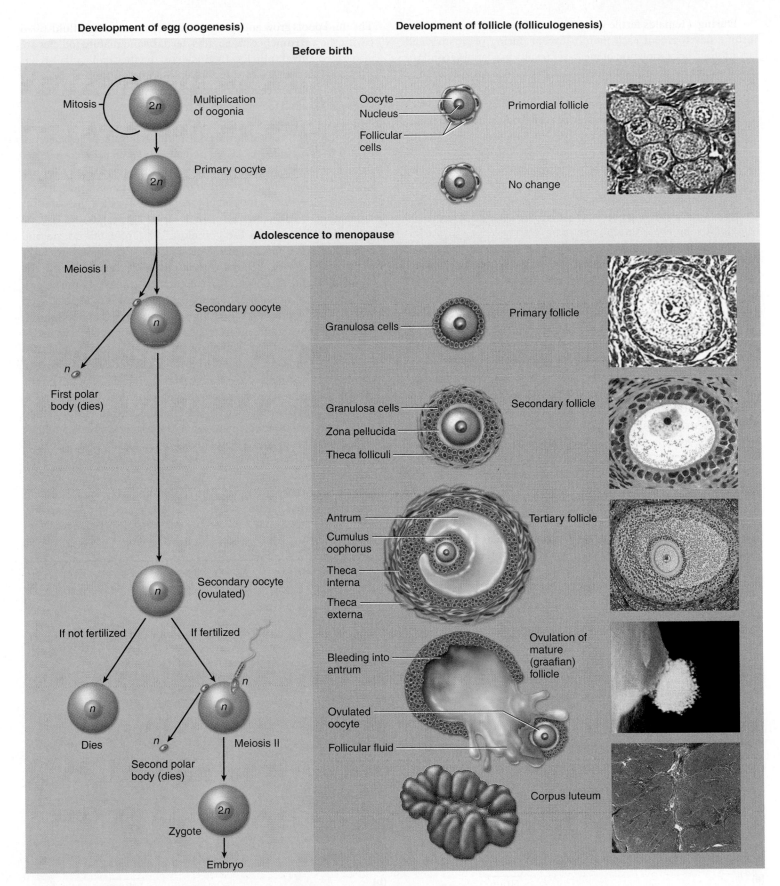

Before birth

Mitosis — 2n   Multiplication of oogonia

2n   Primary oocyte

Oocyte
Nucleus
Follicular cells
Primordial follicle

No change

Adolescence to menopause

Meiosis I

n   Secondary oocyte

n   First polar body (dies)

Granulosa cells   Primary follicle

Granulosa cells
Zona pellucida
Theca folliculi
Secondary follicle

n   Secondary oocyte (ovulated)

If not fertilized   If fertilized

n   Dies

n / n

Second polar body (dies)

Meiosis II

2n   Zygote

Embryo

Antrum
Cumulus oophorus
Theca interna
Theca externa
Tertiary follicle

Bleeding into antrum
Ovulated oocyte
Follicular fluid
Ovulation of mature (graafian) follicle

Corpus luteum

Figure 26.14 Oogenesis (Left) and Corresponding Development of the Follicle (Right).

During a female's fertile years, about 20 to 25 primary oocytes resume development each month (except during pregnancy and lactation). The completion of meiosis I results in a large **secondary oocyte** and a small *first polar body* (fig. 26.14). Meiosis I reduces the chromosome number by half, so the secondary oocyte is haploid. It retains as much of the cytoplasm as possible, so that if it is fertilized, it can divide repeatedly and produce numerous daughter cells. Splitting each oocyte into four equal but small parts would run counter to this purpose. The first polar body is simply a way of discarding the other haploid set of chromosomes; it soon dies. The secondary oocyte begins meiosis II and then goes into developmental arrest again until after ovulation. If this egg is fertilized, it completes meiosis II and produces a *second polar body.* If not fertilized, it dies and never finishes meiosis.

The stages of oogenesis are accompanied by **folliculogenesis,** a series of changes in the follicle. The primary oocyte is initially enclosed in a **primordial follicle,** composed of a single layer of squamous follicular cells applied tightly to the oocyte (fig. 26.15a). Primordial follicles are concentrated near the surface of the ovary. As the egg develops and enlarges, the follicular cells around it expand into a cuboidal shape; a follicle with a single layer of cuboidal cells around the egg is termed a **primary follicle.** While the egg enlarges still more, the follicular cells multiply and pile atop each other. Once they form two or more layers, the follicle is called a **secondary follicle.** The follicular cells are now called **granulosa cells.** They secrete a layer of glycoprotein gel, the **zona pellucida,**[30] around the egg while the connective tissue surrounding the granulosa cells condenses into a fibrous husk called the **theca**[31] **folliculi** (THEE-ca fol-IC-you-lye). The theca and granulosa cells are the female's primary source of estrogen.

Most of the primary and secondary follicles degenerate with no further development, but a few of them begin to secrete pools of estrogen-rich **follicular fluid** among the granulosa cells. Once these pockets of fluid appear, the follicle is called a **tertiary follicle.**

The fluid pools grow and merge until they form a single fluid-filled cavity, the **antrum;** because of this, tertiary and mature follicles are also called **antral follicles.** By this time, the theca folliculi also has further differentiated into two layers—an outer, fibrous *theca externa* and an inner, cellular *theca interna.* This state of development is reached around the time one's menstrual period ends, around day 5 of the menstrual cycle.

By day 10 or so, usually only one of these antral follicles survives. It enlarges to as much as 2.5 cm in diameter and bulges like a blister from the surface of the ovary. This **mature (graafian**[32]**) follicle** (fig. 26.15b) is the one destined to ovulate. Its oocyte is held against the follicular wall by a mound of granulosa cells called the **cumulus oophorus**[33] (CUE-mew-lus oh-OFF-or-us). The gelatinous zona pellucida still separates the granulosa cells from the oocyte and appears as a clear space in histological sections. The innermost layer of cumulus cells is called the **corona radiata.**[34] Microvilli from the corona cells and the oocyte span the zona pellucida. Like the blood–testis barrier of the male, the zona pellucida protects the egg from antibodies and other harmful chemicals, ensuring that nothing can get to the egg without being screened by the corona cells.

Typically around day 14, the growing follicle ruptures and releases the oocyte—an event lasting 2 to 3 minutes called **ovulation.** The oocyte and its cumulus cells emerge from a nipplelike **stigma** on the ovarian surface and are swept into the uterine tube (fig. 26.16). The rest of the follicle collapses and bleeds into the antrum. As the clotted blood is slowly absorbed, granulosa and theca interna cells multiply and fill the antrum, and a dense bed of blood capillaries grows amid them. The ovulated follicle has now become a structure called the **corpus luteum,**[35] named for a yellow lipid that accumulates in the theca interna cells (see fig. 26.12). These cells are now called **lutein cells.** The corpus luteum secretes a large amount of progesterone, which stimulates the uterus to prepare for possible pregnancy.

---

[30]*zona* = zone + *pellucid* = clear, transparent
[31]*theca* = box, case

[32]Reijnier de Graaf (1641–73), Dutch physiologist and histologist
[33]*cumulus* = little mound + *oo* = egg + *phor* = to carry
[34]*corona* = crown + *radiata* = radiating
[35]*corpus* = body + *lute* = yellow

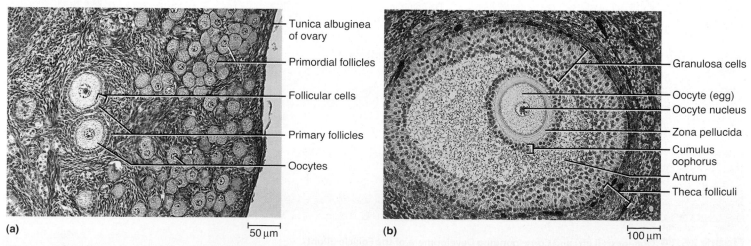

(a) — Tunica albuginea of ovary — Primordial follicles — Follicular cells — Primary follicles — Oocytes   50 µm

(b) — Granulosa cells — Oocyte (egg) — Oocyte nucleus — Zona pellucida — Cumulus oophorus — Antrum — Theca folliculi   100 µm

Figure 26.15 Ovarian Follicles. (a) Primordial and primary follicles. Note the very thin layer of squamous cells around the oocyte in a primordial follicle, and the single layer of cuboidal cells in a primary follicle. (b) A mature (graafian) follicle. Just before ovulation, this follicle will grow to as much as 2.5 cm (1 in.) in diameter.

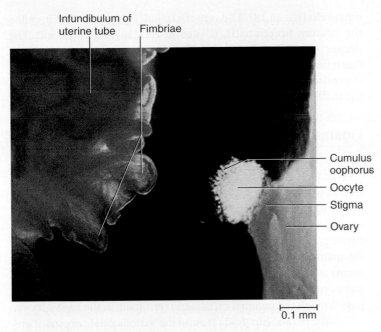

Figure 26.16 Endoscopic View of Human Ovulation. The oocyte is shrouded by a blanket of cumulus oophorus cells.

Figure 26.17 **Epithelial Lining of the Uterine Tube.** Secretory cells are shown in red and green, and cilia of the ciliated cells in yellow (SEM).

If pregnancy does not occur, the corpus luteum atrophies from days 24 to 26—a process called *involution.* By day 26 or so, involution is complete and the corpus luteum becomes an inactive scar, the **corpus albicans.**[36] If pregnancy occurs, however, the corpus luteum remains active for about 3 months, producing progesterone to sustain the early pregnancy. The placenta eventually takes over that task.

These events in the ovarian cycle are correlated with changes in uterine histology, which we will examine later.

The Uterine Tubes

An ovulated oocyte is received into the **uterine tube,** also called the **oviduct** or **fallopian**[37] **tube** (see fig. 26.11). The tube is a ciliated canal about 10 cm long leading from the ovary to the uterus. At the distal (ovarian) end, it flares into a trumpet-shaped **infundibulum**[38] with feathery projections called **fimbriae**[39] (FIM-bree-ee); the middle and longest part of the tube is the **ampulla;** and the segment near the uterus is a narrower **isthmus.** The uterine tube is enclosed in the **mesosalpinx**[40] (MEZ-oh-SAL-pinks), which is the superior margin of the broad ligament.

The wall of the uterine tube is well endowed with smooth muscle. Its mucosa is highly folded into longitudinal ridges and has an epithelium of ciliated cells and a smaller number of secretory **peg cells** (fig. 26.17). The cilia beat toward the uterus and, with the help of muscular contractions of the tube, convey the egg in that direction.

The Uterus

The **uterus**[41] is a thick muscular chamber that opens into the roof of the vagina and usually tilts forward over the urinary bladder (see figs. 26.11 and 26.13). Its function is to harbor the fetus, provide it with a source of nutrition (the placenta, composed partially of uterine tissue), and expel the fetus at the end of gestation (pregnancy). It is somewhat pear-shaped, with a broad superior curvature called the **fundus,** a midportion called the **body (corpus),** and a narrow inferior end called the **cervix.** The uterus measures about 7 cm from cervix to fundus, 4 cm wide at its broadest point on the fundus, and 2.5 cm thick, but it is somewhat larger in women who have been pregnant.

The lumen of the uterus is roughly triangular, with its two upper corners opening into the uterine tubes. In the nonpregnant uterus, the lumen is not a hollow cavity but rather a *potential space* (see p. 19); the mucous membranes of the opposite walls are pressed together with little room between them. The lumen communicates with the vagina by way of a narrow passage through the cervix called the **cervical canal.** The superior opening of this canal into the body of the uterus is the *internal os*[42] (pronounced "oz" or "ose"),

---

[36]*corpus* = body + *alb* = white
[37]Gabriele Fallopio (1528–62), Italian anatomist and physician
[38]*infundibulum* = funnel
[39]*fimbria* = fringe
[40]*meso* = mesentery + *salpin* = trumpet

[41]*uterus* = womb
[42]*os* = mouth

and its opening into the vagina is the *external os.* The canal contains **cervical glands** that secrete mucus, thought to prevent the spread of microbes from the vagina into the uterus. Near the time of ovulation, the mucus becomes thinner than usual and allows easier passage for sperm.

The Uterine Wall

The uterine wall consists of an external serosa called the *perimetrium,* a middle muscular layer called the *myometrium,* and an inner mucosa called the *endometrium.* The **perimetrium** is composed of simple squamous epithelium overlying a thin layer of areolar tissue. The **myometrium,**[43] about 1.25 cm thick in the nonpregnant uterus, constitutes most of the wall. It is composed mainly of bundles of smooth muscle that sweep downward from the fundus and spiral around the body of the uterus. The myometrium is less muscular and more fibrous near the cervix; the cervix itself is almost entirely collagenous. The muscle cells of the myometrium are about 40 μm long immediately after menstruation, twice this long at the middle of the menstrual cycle, and up to 10 times as long in pregnancy. The function of the myometrium is to produce the labor contractions that help to expel the fetus.

The inner lining of the uterus, or mucosa, is called the **endometrium.**[44] It has a simple columnar epithelium, compound tubular glands, and a stroma populated by leukocytes, macrophages, and

---

[43]*myo* = muscle + *metr* = uterus
[44]*endo* = inside + *metr* = uterus

other cells (fig. 26.18). The superficial half to two-thirds of it, called the **stratum functionalis,** is shed in each menstrual period. The deeper layer, called the **stratum basalis,** stays behind and regenerates a new functionalis in the next cycle. When pregnancy occurs, the endometrium is the site of attachment of the embryo and forms the maternal part of the placenta.

Ligaments

The uterus is supported by the muscular floor of the pelvic outlet and folds of peritoneum that form supportive ligaments around the organ, as they do for the ovary and uterine tube (see fig. 26.13). The **broad ligament** has two parts: the *mesosalpinx* mentioned earlier and the *mesometrium* on each side of the uterus. The cervix and superior part of the vagina are supported by **cardinal (lateral cervical) ligaments** extending to the pelvic wall. A pair of **uterosacral ligaments** attach the posterior surface of the uterus to the sacrum, and a pair of **round ligaments** arise from the anterior surface of the uterus, pass through the inguinal canals, and terminate in the labia majora.

As the peritoneum folds around the various pelvic organs, it creates several dead-end recesses and pouches (extensions of the peritoneal cavity). Two major ones are the **vesicouterine**[45] **pouch,** which forms the space between the uterus and urinary bladder, and the **rectouterine pouch** between the uterus and rectum (see fig. 26.11).

Blood Supply

The uterine blood supply is particularly important to the menstrual cycle and pregnancy. A **uterine artery** arises from each internal iliac artery and travels through the broad ligament to the uterus (fig. 26.19). It gives off several branches that penetrate into the myometrium and lead to **arcuate arteries.** Each arcuate artery travels in a circle around the uterus and anastomoses with the arcu-

---

[45]*vesico* = bladder

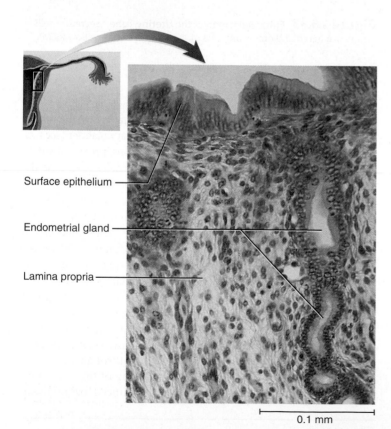

Figure 26.18 Histology of the Endometrium.

Surface epithelium

Endometrial gland

Lamina propria

0.1 mm

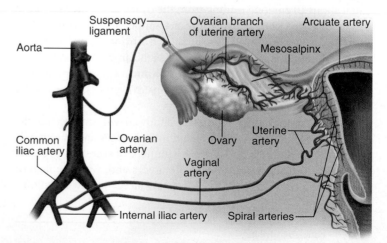

Figure 26.19 Blood Supply to the Female Reproductive Tract. The vaginal, uterine, and ovarian arteries are exaggerated in length by the perspective of the drawing, moving the aorta away from the uterus for clarity.

Suspensory ligament

Ovarian branch of uterine artery

Arcuate artery

Aorta

Mesosalpinx

Common iliac artery

Ovarian artery

Ovary

Uterine artery

Vaginal artery

Internal iliac artery

Spiral arteries

ate artery on the other side. Along its course, it gives rise to smaller arteries that penetrate the rest of the way through the myometrium, into the endometrium, and produce the **spiral arteries.** The spiral arteries coil between the endometrial glands toward the surface of the mucosa. They rhythmically constrict and dilate, making the mucosa alternately blanch and flush with blood.

Cyclic Changes in Uterine Histology

Uterine histology is not constant. In fertile women, it changes throughout the **menstrual cycle,** the monthly rhythm of endometrial buildup, breakdown, and discharge. This cycle averages 28 days long, with day 1 considered to be the first day of visible vaginal discharge of menstrual fluid. Why that discharge occurs is best understood by first considering the histological changes leading up to it.

The **proliferative phase** is a time of rebuilding of endometrial tissue lost at the last menstruation. At the end of menstruation,

around day 5, the endometrium is about 0.5 mm thick and consists only of the stratum basalis. The stratum functionalis is rebuilt by mitosis from day 6 to day 14, under the influence of estrogen from the growing ovarian follicles. By day 14, the endometrium is about 2 to 3 mm thick (fig. 26.20a).

The **secretory phase** is a period of further endometrial thickening, but results from secretion and fluid accumulation rather than mitosis. It extends from day 15 (after ovulation) to day 26 of a typical cycle, and is stimulated by progesterone from the corpus luteum. In this phase, the endometrial glands grow wider, longer, and more coiled. As a result of the coiling, a vertical section through the endometrium shows these glands with a sawtooth or zigzag appearance (fig. 26.20b). Endometrial cells and the uterine stroma accumulate glycogen during this phase. By the end of the secretory phase, the endometrium is about 5 to 6 mm thick—a soft, wet, nutritious bed available for embryonic development in the event of pregnancy.

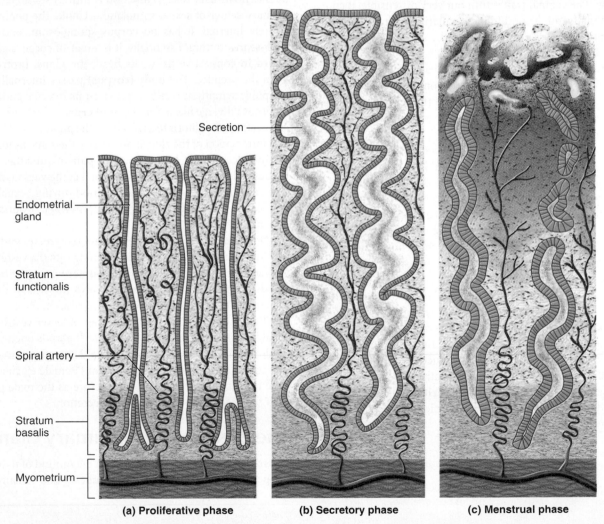

Secretion

Endometrial gland

Stratum functionalis

Spiral artery

Stratum basalis

Myometrium

(a) Proliferative phase   **(b) Secretory phase**   **(c) Menstrual phase**

Figure 26.20 **Endometrial Changes Through the Menstrual Cycle.** (a) Late proliferative phase. The endometrium is 2 to 3 mm thick and has relatively straight, narrow endometrial glands. Spiral arteries penetrate upward between the endometrial glands. (b) Secretory phase. The endometrium has thickened to 5 to 6 mm by accumulation of glycogen and mucus. The endometrial glands are much wider and more distinctly coiled, showing a zigzag or "sawtooth" appearance in histological sections. (c) Menstrual phase. Ischemic tissue has begun to die and fall away from the uterine wall, with bleeding from broken blood vessels and pooling of blood within the tissue and in the uterine cavity.

The **premenstrual phase** is a period of endometrial degeneration occuring in the last 2 days or so of the menstrual cycle. When the corpus luteum involutes, the spiral arteries exhibit spasmodic contractions that cause endometrial ischemia (interrupted blood flow). The premenstrual phase is therefore also called the **ischemic** (iss-KEE-mic) **phase.** Ischemia leads to tissue necrosis. As the endometrial glands, stroma, and blood vessels degenerate, pools of blood accumulate in the stratum functionalis. Necrotic endometrium falls away from the uterine wall, mixes with blood and serous fluid in the lumen, and forms the **menstrual fluid** (fig. 26.20c).

The **menstrual phase (menses)** begins when enough menstrual fluid has accumulated in the uterus that it begins to be discharged vaginally. The first day of external discharge marks day 1 of a new cycle.

The Vagina

The **vagina**[46] is a tube about 8 to 10 cm long that allows for the discharge of menstrual fluid, receipt of the penis and semen, and birth of a baby. The vaginal wall is thin but very distensible. It consists of an outer adventitia, a middle muscularis, and an inner mucosa. The vagina tilts posteriorly between the urethra and rectum; the urethra is bound to its anterior wall. The vagina has no glands, but it is lubricated by the *transudation* ("vaginal sweating") of serous fluid through its walls and by mucus from the cervical glands above it. The vagina extends slightly beyond the cervix and forms blind spaces called **fornices**[47] (FOR-nih-sees; singular, *fornix*) surrounding it (see fig. 26.11).

The lower end of the vagina has transverse friction ridges, or **vaginal rugae,** which stimulate the penis and help induce ejaculation. At the vaginal orifice, the mucosa folds inward and forms a membrane, the **hymen,** which stretches across the opening. The hymen has one or more openings to allow menstrual fluid to pass, but it usually must be ruptured to allow for intercourse. A little bleeding often accompanies the first act of intercourse; however, the hymen is commonly ruptured before then by tampons, medical examinations, or strenuous exercise.

The vaginal epithelium is simple cuboidal in childhood, but the estrogens of puberty transform it into a stratified squamous epithelium. This is an example of *metaplasia,* the transformation of one tissue type to another. The epithelial cells are rich in glycogen. Bacteria ferment this to lactic acid, which produces a low vaginal pH (about 3.5–4.0) that inhibits the growth of pathogens. The mucosa also has antigen-presenting cells called **dendritic cells,** which normally help in defense against infection but which also are a route by which HIV invades the female body.

Apply What You Know

Why do you think the vaginal epithelium changes type at puberty? Of all types of epithelium it might become, why stratified squamous?

The External Genitalia

The external genitalia of the female occupy most of the perineum, and are collectively also known as the **vulva**[48] or **pudendum**[49] (fig. 26.21). The perineum has the same skeletal landmarks in the female as in the male.

The **mons**[50] **pubis** (see fig. 1.12a, p. 15) consists mainly of an anterior mound of adipose tissue overlying the pubic symphysis, covered with skin and bearing pubic hair. The **labia majora**[51] (singular, *labium majus*) are a pair of thick folds of skin and adipose tissue inferior to the mons, between the thighs; the fissure between them is the *pudendal cleft.* Pubic hair grows on the lateral surfaces of the labia majora at puberty, but the medial surfaces are hairless. Medial to the labia majora are entirely hairless and much thinner **labia minora**[52] (singular, *labium minus*). The area enclosed by them, called the **vestibule,** contains the urinary and vaginal orifices. At the anterior margin of the vestibule, the labia minora meet and form a hoodlike **prepuce** over the clitoris.

The **clitoris** is structured much like the penis in many respects, but has no urinary role. Its function is entirely sensory, serving as the primary center of sexual stimulation. Unlike the penis, it is almost entirely internal, it has no corpus spongiosum, and it does not enclose the urethra. Essentially, it is a pair of corpora cavernosa enclosed in connective tissue. Its head, the **glans,** protrudes slightly from the prepuce. The **body (corpus)** passes internally, inferior to the pubic symphysis (see fig. 26.11). At its internal end, the corpora cavernosa diverge like a Y as a pair of **crura,** which, like those of the penis, attach the clitoris to each side of the pubic arch. The circulation and innervation of the clitoris are largely the same as for the penis.

Just deep to the labia majora, a pair of subcutaneous erectile tissues called the **vestibular bulbs** bracket the vagina like parentheses. They become congested with blood during sexual excitement and cause the vagina to tighten somewhat around the penis, enhancing sexual stimulation.

Next to the vagina are a pair of pea-sized **greater vestibular (Bartholin**[53]**) glands** with short ducts opening into the vestibule or lower vagina (fig. 26.21b). These are the counterpart to the bulbourethral glands of the male. They keep the vulva moist, and during sexual excitement they provide most of the lubrication for intercourse. The vestibule is also lubricated by a number of **lesser vestibular glands.** A pair of mucous **paraurethral (Skene**[54]**) glands** opens into the vestibule near the external urethral orifice. These glands may eject fluid, sometimes abundantly, during orgasm ("female ejaculation"). They arise from the same embryonic structure as the male prostate, and their fluid is similar to the prostatic secretion.

The Breasts and Mammary Glands

The mature female **breast** (fig. 26.22) is a mound of tissue overlying the pectoralis major. It enlarges at puberty and remains so for life,

---

[46]*vagina* = sheath
[47]*fornix* = arch, vault
[48]*vulva* = covering
[49]*pudend* = shameful
[50]*mons* = mountain
[51]*labi* = lip + *major* = larger, greater
[52]*minor* = smaller, lesser
[53]Caspar Bartholin (1655–1738), Danish anatomist
[54]Alexander J. C. Skene (1838–1900), American gynecologist

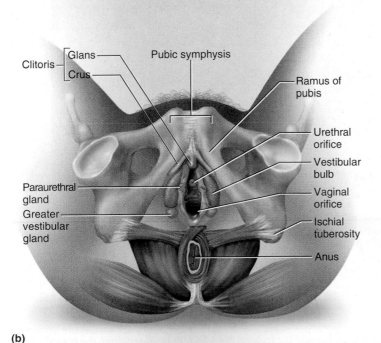

Mons pubis

Prepuce
Clitoris

Labium majus

Labium minus

Urethral orifice

Vaginal orifice

Hymen

Vestibule

Perineal raphe

Anus

(a)

Clitoris — Glans
Crus

Pubic symphysis

Ramus of pubis

Urethral orifice

Vestibular bulb

Paraurethral gland

Vaginal orifice

Greater vestibular gland

Ischial tuberosity

Anus

(b)

Figure 26.21 **The Female Perineum.** (a) Surface anatomy.
(b) Subcutaneous structures.
• *Which of the glands in part (b) is homologous to the male prostate gland?*

but most of this time it contains very little mammary gland. The **mammary gland** develops within the breast during pregnancy, remains active in the lactating breast, and atrophies when a woman ceases to nurse.

The breast has two principal regions: the conical to pendulous **body,** with the nipple at its apex, and an extension toward the armpit called the **axillary tail.** Lymphatics of the axillary tail are especially important as a route of breast cancer metastasis.

The nipple is surrounded by a circular colored zone, the **areola.** Dermal blood capillaries and nerves come closer to the surface here than in the surrounding skin and make the areola more sensitive and darker in color. In pregnancy, the areola and nipple often darken further, making them more visible to the indistinct vision of a nursing infant. Sensory nerve fibers of the areola are important in triggering a *milk ejection reflex* when an infant nurses. The areola has sparse hairs and **areolar glands,** visible as small bumps on the surface. These glands are intermediate between sweat glands and mammary glands in their degree of development. When a woman is nursing, secretions of the areolar glands and sebaceous glands protect the areola from chapping and cracking. The dermis of the areola has smooth muscle fibers that contract in response to cold, touch, and sexual arousal, wrinkling the skin and erecting the nipple.

Internally, the nonlactating breast consists mostly of adipose and collagenous tissue (fig. 26.22). Breast size is determined by the amount of adipose tissue and has no relationship to the amount of milk the mammary gland can produce. **Suspensory ligaments** attach the breast to the dermis of the overlying skin and to the fascia of the pectoralis major. The nonlactating breast contains very little glandular tissue, but it does have a system of ducts branching through its connective tissue stroma and converging on the nipple.

When the mammary gland develops during pregnancy, it exhibits 15 to 20 lobes arranged radially around the nipple, separated from each other by fibrous stroma. Each lobe is drained by a **lactiferous**[55] **duct,** which dilates to form a **lactiferous sinus** opening onto the nipple. Internally, this duct branches repeatedly with the finest branches ending in sacs called acini. The acini are organized into grapelike clusters (lobules) within each lobe of the breast. Each acinus consists of a sac of pyramidal secretory cells arranged around a central lumen (see fig. 3.29, p. 74). Like an orange in a plastic mesh bag, the acinus is surrounded by a network of contractile **myoepithelial cells** around the secretory cells (fig. 26.22d). When a woman nurses, stimulation of the nipple induces the posterior lobe of the pituitary gland to secrete oxytocin. Oxytocin stimulates the myoepithelial cells to contract, squeezing milk from the acini into the lactiferous ducts.

Before You Go On

Answer the following questions to test your understanding of the preceding section:

9. What are the stages of oogenesis? Describe all the ways in which it differs from spermatogenesis.

10. Describe the changes that occur in the ovarian follicles and uterine endometrium over the course of the ovarian and menstrual cycle.

---

[55]*lact* = milk + *fer* = to carry

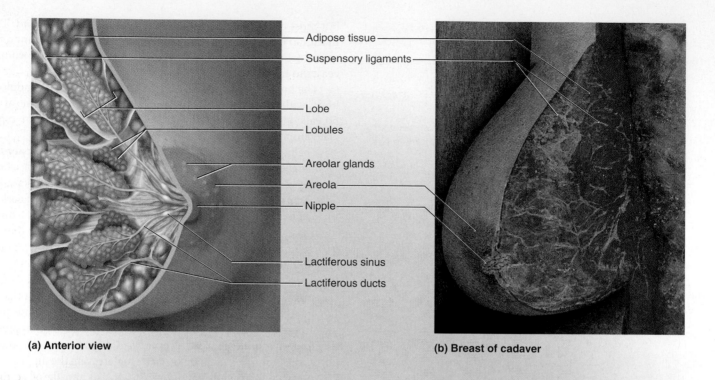

Adipose tissue
Suspensory ligaments
Lobe
Lobules
Areolar glands
Areola
Nipple
Lactiferous sinus
Lactiferous ducts

(a) Anterior view

(b) Breast of cadaver

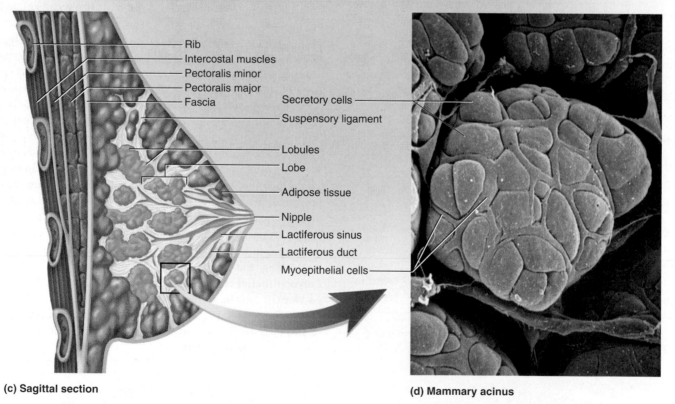

Rib
Intercostal muscles
Pectoralis minor
Pectoralis major
Fascia
Secretory cells
Suspensory ligament
Lobules
Lobe
Adipose tissue
Nipple
Lactiferous sinus
Lactiferous duct
Myoepithelial cells

(c) Sagittal section

(d) Mammary acinus

Figure 26.22 **The Breast.** Parts (a), (c), and (d) depict the breast in a lactating state. Some of the features in (a) and (c) are absent from the nonlactating breast in part (b). The cluster of lobules boxed in (c) would contain numerous microscopic acini like the one in (d). Part (d) is not to be construed as an enlargement of the entire boxed area of (c).
● *What is the function of the myoepithelial cells in (d)?*

11. How does the structure of the uterine tube mucosa relate to its function?

12. What structures are externally visible in the vulva? Describe their anatomical arrangement.

13. Name the accessory glands of the vulva and state their locations and functions.

14. How does the structure of the nonlactating breast differ from that of the lactating breast? What is the difference between the breast and the mammary gland?

26.4 Developmental and Clinical Perspectives

▶ Expected Learning Outcomes

When you have completed this section, you should be able to

- explain how sexual differentiation is determined by the sex chromosomes and prenatal hormones;

- describe the embryonic development of the reproductive system;

- describe what aspects of anatomical development the male and female have in common, and how they become differentiated from each other;

- summarize the changes in reproductive function that occur in old age; and

- describe some reproductive disorders of each sex.

Prenatal Development and Sexual Differentiation

The male and female reproductive systems begin with embryonic sex organs that are anatomically "indifferent," yet genetically destined to differentiate into the genitalia of one sex or the other. Thus, the testes and ovaries develop from an initially indistinguishable *indifferent gonad;* the glans of the penis and clitoris develop from a single embryonic *genital tubercle;* and the scrotum and labia majora develop from the embryonic *labioscrotal folds.* Organs that develop from the same embryonic precursor are said to be **homologous** to each other.

The Internal Genitalia

The gonad appears at 5 to 6 weeks as a **gonadal ridge** near the mesonephros, the primitive kidney. Adjacent to each gonadal ridge are two ducts, the **mesonephric**[56] **(wolffian**[57]**) duct** described in chapter 25 (p. 699), and the **paramesonephric**[58] **(müllerian**[59]**) duct.** In

males, the mesonephric ducts develop into parts of the reproductive tract, and the paramesonephric ducts degenerate. In females, the opposite occurs (fig. 26.23).

The differentiation of these ducts into the organs of one sex or the other is determined by an interaction between genes and hormones. If the zygote has sex chromosomes X and Y, it is normally destined to develop into a male; if it has two X chromosomes and no Y, it will develop into a female. Thus, the sex of a child is determined at conception (fertilization), depending on whether the egg (always X) is fertilized by an X- or a Y-bearing sperm.

But why? The answer lies in the Y chromosome, where a gene called **SRY** (sex-determining region of the Y) codes for a protein called **testis-determining factor (TDF).** TDF then interacts with genes on some of the other chromosomes, including a gene on the X chromosome for androgen receptors and genes that initiate the development of male anatomy. By 8 to 9 weeks, the male gonadal ridge has become a rudimentary testis whose interstitial cells begin to secrete testosterone. Testosterone stimulates the mesonephric duct to develop into the system of male reproductive ducts. Sustentacular cells of the fetal testis secrete a hormone called **müllerian-inhibiting factor (MIF),** which causes atrophy of the paramesonephric duct. Even an adult male, however, retains a tiny Y-shaped vestige of the paramesonephric ducts, like a vestigial uterus and uterine tubes, in the area of the prostatic urethra. It is named the *prostatic utricle.*[60]

In a female fetus, the absence of testosterone causes the mesonephric ducts to degenerate, and in the absence of MIF, the paramesonephric ducts develop "by default" into a female reproductive tract. Each duct differentiates into one of the uterine tubes. At their inferior end, the ducts fuse to form the single uterus and the upper one-third of the vagina. The lower two-thirds of the vagina develops as an outgrowth of the urogenital sinus described in chapter 25 (p. 699).

It may seem as if androgens should induce the formation of a male reproductive tract and estrogens induce a female reproductive tract. However, estrogen levels are always high during pregnancy, so if this mechanism were the case, it would feminize all fetuses. Thus, the development of a female results from the low level of androgens, not the presence of estrogens.

The External Genitalia

At 8 weeks, the external genitalia are represented by the following sexually undifferentiated structures (fig. 26.24):

- the **genital tubercle,** an anterior bud destined to become the glans of the penis or clitoris;

- **urogenital folds,** a pair of medial tissue folds slightly posterior to the genital tubercle; and

- **labioscrotal folds,** a larger pair of tissue folds lateral to the urogenital folds.

These organs begin to show sexual differentiation by the end of week 9, and either male or female genitalia are distinctly identifiable by the end of week 12. In the female, the three structures just listed become

---

[56]*meso* = middle + *nephr* = kidney; named for the temporary embryonic kidney
[57]Kaspar F. Wolff (1733–94), German anatomist
[58]*para* = next to
[59]Johannes P. Müller (1801–58), German physician

[60]*utricle* = little bag

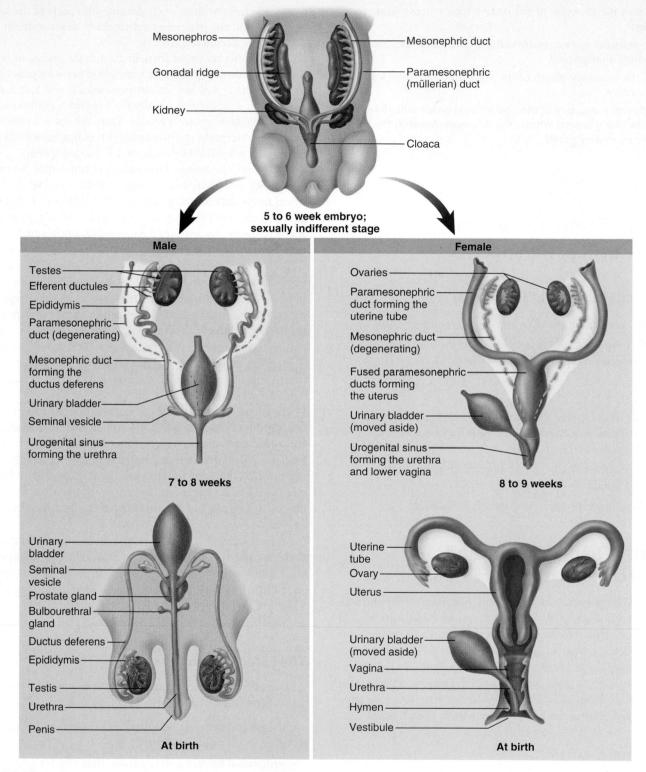

Figure 26.23 **Embryonic Development of the Male and Female Reproductive Tracts.** Note that the male tract develops from the mesonephric duct and the female tract from the paramesonephric duct; the other duct in each sex degenerates.

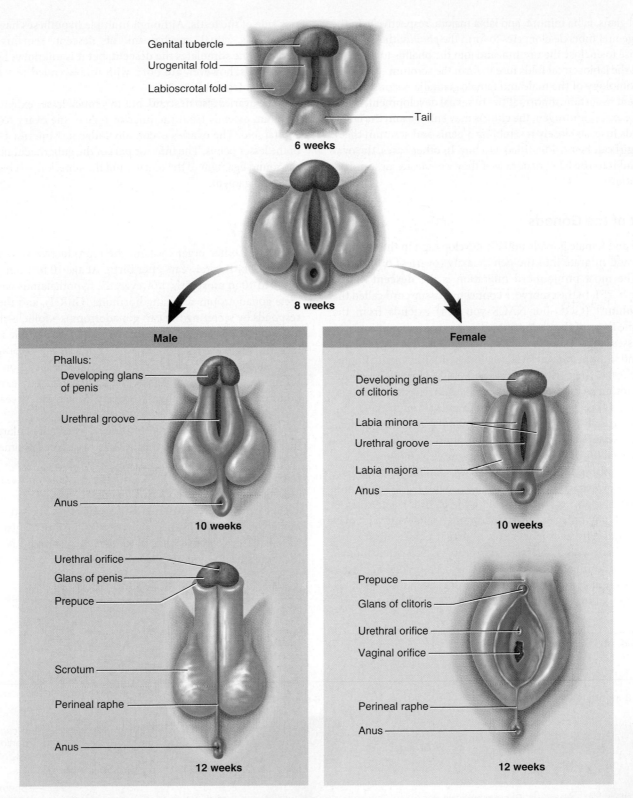

Figure 26.24 Development of the External Genitalia. By 6 weeks, the embryo has three primordial structures—the genital tubercle, urogenital folds, and labioscrotal folds—which will become the male or female genitalia. At 8 weeks, these structures have grown, but the sexes are still indistinguishable. Slight sexual differentiation is noticeable at 10 weeks. The sexes are fully distinguishable by 12 weeks. Matching colors identify homologous structures of the male and female.

the clitoral glans, labia minora, and labia majora, respectively. In the male, the genital tubercle elongates to form the *phallus;* the urogenital folds fuse to enclose the urethra, and join the phallus to form the penis; and the labioscrotal folds fuse to form the scrotum.

The homology of the male and female genitalia becomes strikingly evident in certain abnormalities of sexual development. In the presence of excess androgen, the clitoris may enlarge and the labioscrotal folds fuse, so closely resembling a penis and scrotum that a newborn girl can be misidentified as a boy. In other cases, the ovaries descend into the labia majora as if they were testes descending into a scrotum.

Descent of the Gonads

Both male and female gonads initially develop high in the abdominal cavity and migrate into the pelvic cavity (ovaries) or scrotum (testes). The most pronounced migration is the **descent of the testes** (fig. 26.25). In the embryo, a connective tissue cord called the **gubernaculum**[61] (GOO-bur-NACK-you-lum) extends from the gonad to the floor of the abdominopelvic cavity. As it continues to grow, it passes between the internal and external abdominal oblique muscles and into the scrotal swelling. Independently of any migration of the testis, the peritoneum also develops a fold that extends into the scrotum as the *vaginal process.* The gubernaculum and the vaginal process create a path of low resistance through the groin, the inguinal canal—the most common site of herniation in boys and men (*inguinal hernia;* see Deeper Insight 11.3, p. 289).

The descent of the testes begins as early as week 6. The superior part of the embryonic gonad degenerates, and its inferior part migrates downward, guided by the gubernaculum. In the seventh month, the testes abruptly pass through the inguinal canals, anterior to the pubic symphysis, into the scrotum. As they descend, they are accompanied by ever-elongating testicular arteries and veins and by lymphatic vessels, nerves, sperm ducts, and extensions of the internal abdominal oblique muscle that become the cremaster muscle. The vaginal process becomes separated from the peritoneal cavity and persists as a sac, the *tunica vaginalis,* enfolding the ante-

rior side of the testis. Although multiple hypotheses have been offered, the actual mechanism of descent remains obscure. Testosterone stimulates the descent, but it is unknown how. About 3% of boys, however, are born with undescended testes, or *cryptorchidism* (see table 26.2).

The ovaries also descend, but to a much lesser extent. A gubernaculum extends from the inferior pole of the ovary to the labioscrotal fold. The ovaries eventually lodge just inferior to the brim of the lesser pelvis. The inferior part of the gubernaculum becomes the round ligament of the uterus, and the superior part becomes the ovarian ligament.

Puberty

Unlike any other organ system, the reproductive system remains dormant for several years after birth. At age 10 to 12 in most boys and 8 to 10 in most girls, however, the hypothalamus begins to secrete gonadotropin-releasing hormone (GnRH), and the pituitary responds by secreting the two gonadotropins—follicle-stimulating hormone (FSH) and luteinizing hormone (LH). These hormones, in turn, stimulate the gonads to secrete estrogens, progesterone, and testosterone. Combined with surges in the secretion of growth hormone and other hormones, the adolescent body exhibits pronounced anatomical changes. **Puberty,**[62] the first few years of **adolescence,**[63] has begun.

In boys, the earliest sign of puberty is usually enlargement of the testes and scrotum; in girls, it is breast development, called **thelarche**[64] (thee-LAR-kee). These changes are soon followed by **pubarche** (pyu-BAR-kee), the growth of pubic and axillary hair, sebaceous glands, and apocrine glands. In girls, the third principal event is **menarche**[65] (men-AR-kee), the first menstrual period. Menarche does not immediately signify fertility. A girl's first few menstrual cycles are typically *anovulatory* (no egg is ovulated). Most girls begin ovulating regularly about a year after they begin

---

[61]*gubern* = rudder, to steer, guide

[62]*puber* = grown up
[63]*adolesc* = to grow up
[64]*thel* = breast, nipple + *arche* = beginning
[65]*men* = monthly

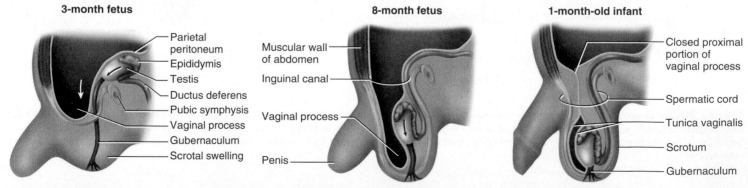

Figure 26.25 Descent of the Testis. Note that the testis and spermatic ducts are retroperitoneal. An extension of the peritoneum called the *vaginal process* follows the testis through the inguinal canal and becomes the tunica vaginalis.
● *Why is this structure of male anatomy called the tunica vaginalis?*

menstruating. The male counterpart to menarche is the first ejaculation. Puberty ends when an individual is fully capable of reproducing; adolescence extends until a person reaches full height in the late teens to early twenties.

Puberty entails many other changes too numerous for the scope of this book. The internal and external genitalia enlarge. Changes in muscularity and fat deposition bring about some of the male–female differences known as the secondary sex characteristics. The male voice deepens as the larynx enlarges. Testosterone, estrogens, and growth hormone cause rapid elongation of the long bones, and thus the adolescent growth in stature. And to the great anxiety of the parents of adolescents, the anatomical readiness for reproduction is accompanied by psychological interest in sex, the *libido,* elicited in both sexes by testosterone. (Testosterone is produced not only by the testes but also, in small amounts, by the ovaries and adrenal cortex.)

The Aging Reproductive System

Fertility and sexual function decline in and beyond middle age, owing to declining levels of testosterone and estrogen. Around ages 50 to 55, both men and women go through a period of physical and psychological change called *climacteric,* although (jokes about "male menopause" aside) only women experience *menopause,* the cessation of menses.

Male Climacteric

Male climacteric, or **andropause,** is brought on by falling levels of testosterone and inhibin. Testosterone secretion peaks at about 7 mg/day at age 20 and then declines steadily to as little as one-fifth of this level by age 80. There is a corresponding decline in the number and secretory activity of the interstitial cells (the source of testosterone) and sustentacular cells (the source of inhibin). Along with the declining testosterone level, the sperm count and libido diminish. By age 65, sperm count is typically about one-third of what it was in a man's 20s. Nevertheless, men remain capable of fathering a child throughout old age.

With reduced levels of testosterone and inhibin, the pituitary is less inhibited and it secretes elevated levels of FSH and LH. In some cases, these gonadotropins cause mood changes, hot flashes, or even illusions of suffocation—symptoms similar to those that occur in perimenopausal women. Most men, however, notice little or no effect as they pass through climacteric.

About 20% of men in their 60s and 50% of men in their 80s experience *erectile dysfunction (impotence),* the frequent inability to maintain a sufficient erection for intercourse. Erectile dysfunction can also result from hypertension, atherosclerosis, medication, diabetes mellitus, and psychological causes. Over 90% of men with erectile dysfunction nevertheless remain able to ejaculate.

Female Climacteric and Menopause

Female climacteric is brought on by declining ovarian function. It generally begins when the ovaries are down to their last 1,000 eggs or so, and the follicles and ova that remain are less responsive to gonadotropins. Consequently, the follicles secrete less estrogen and progesterone. Without these steroids, the uterus, vagina, and breasts atrophy. Intercourse may become uncomfortable, and vaginal infections more common, as the vagina becomes thinner, less distensible, and drier. The skin becomes thinner, cholesterol levels rise (increasing the risk of cardiovascular disease), and bone mass declines (increasing the risk of osteoporosis). Blood vessels constrict and dilate in response to shifting hormone balances, and the sudden dilation of cutaneous arteries may cause *hot flashes*—a spreading sense of heat from the abdomen to the thorax, neck, and face. Hot flashes may occur several times a day, sometimes accompanied by headaches resulting from the sudden vasodilation of arteries in the head. In some people, the changing hormonal profile also causes mood changes.

Female climacteric is accompanied by **menopause,** the cessation of menstruation and end of fertility. Menopause usually occurs between the ages of 45 and 55. The average age has increased steadily in the last century and is now about 52. It is difficult to precisely establish the time of menopause because the menstrual periods can stop for several months and then begin again. Menopause is generally considered to have occurred when there has been no menstruation for a year or more.

Reproductive Disorders of the Male

Prostate cancer (see Deeper Insight 26.1) is the most common cancer of the male reproductive system, but not the only one. Men are also subject to testicular, penile, and breast cancer. Testicular cancer is the most common of these three and often strikes at a relatively young age compared to prostate cancer. Testicular self-examination combined with regular physical examinations are important preventive measures. Some additional facts about these cancers and other male reproductive disorders are given in table 26.2.

Reproductive Disorders of the Female

The most important malignancies of the female reproductive system are breast and cervical cancer, although the increase in cigarette smoking has caused lung cancer to surpass both of these as a cause of female mortality in the United States. Breast cancer occurs in one out of every eight or nine American women. Breast tumors originate in cells of the mammary ducts and may metastasize to other organs by way of mammary and axillary lymphatics. Although some breast cancer is genetic, many nonhereditary risk factors are also known, including age, early menarche and late menopause, high alcohol or fat consumption, and smoking. Over 70% of cases of breast cancer, however, lack any identifiable risk factor. Early detection through regular breast X-rays (mammograms) are currently regarded as the best protection. Breast self-examination (BSE) may also be helpful, but recent research has cast doubt on whether tumors are detected early enough by BSE to significantly reduce female mortality.

Uterine cancer is of two kinds, endometrial and cervical. Cervical cancer is a slow-growing neoplasia of the lower cervical canal and can be detected by microscopic examination of squamous cells from the cervix (a Pap smear; see Deeper Insight 26.3).

TABLE 26.2	Some Male Reproductive Disorders
Breast cancer	Accounts for 0.2% of male cancers in the United States, usually seen after age 60 but sometimes in children and adolescents. (For every male who gets breast cancer, about 175 females do so.) Usually felt as a lump near the nipple, often with crusting and discharge from the nipple. Often quite advanced by the time of diagnosis, with poor prospects for recovery, because of denial and delay in seeking treatment.
Cryptorchidism[66] (crip-TOR-ki-dizm)	Failure of one or both testes to descend completely into the scrotum. Leads to infertility if not corrected, because undescended testes are too warm for spermatogenesis. In most cases, the testes descend spontaneously in the first year of infancy; otherwise, the condition can be corrected with hormone injections or surgery.
Hypospadias[67] (HY-po-SPAY-dee-us)	A congenital defect in which the urethra opens on the ventral side or base of the penis rather than at the tip; usually corrected surgically at about 1 year of age.
Infertility	Inability to fertilize an egg because of a low sperm count (< 20–25 million/mL), poor sperm motility, or a high percentage of deformed sperm (two heads, defective tails, etc.). May result from malnutrition, gonorrhea and other infections, toxins, or testosterone deficiency.
Penile cancer	Accounts for 1% of male cancers in the United States; most common in black males aged 50 to 70 and of low income. Most often seen in men with nonretractable foreskins *(phimosis)* combined with poor penile hygiene; least common in men circumcised at birth.
Testicular cancer	The most common solid tumor in men 15 to 34 years old, especially white males of middle to upper economic classes. Typically begins as a painless lump or enlargement of the testis. Highly curable if detected early. Men should routinely palpate the testes for normal size and smooth texture.
Varicocele (VAIR-ih-co-seal)	Abnormal dilation of veins of the spermatic cord, so they resemble a "bag of worms." Occurs in 10% of males in the United States. Caused by absence or incompetence of venous valves. Reduces testicular blood flow and often causes infertility.

Disorders Described Elsewhere

Benign prostatic hyperplasia 715	Erectile dsyfunction 733	Prostate cancer 715

There are many other disorders of female reproductive function, too numerous to discuss here. Pregnancy adds its own risks to women's health. Table 26.3 describes some of the more common complications of pregnancy.

More common than any of the foregoing male or female reproductive disorders are the **sexually transmitted diseases (STDs),** caused by infectious microorganisms transmitted by intercourse and other sexual activity, and to infants during or before birth. Most of these are caused by viruses and bacteria. Currently the most serious viral STDs are **AIDS** (caused by the human immunodeficiency virus, HIV) and **hepatitis C.** HIV infects helper T lymphocytes and dendritic cells, among others, thus exerting a devastating effect on the immune system and leaving a person susceptible to certain forms of cancer and opportunistic infection. The hepatitis C virus (HCV) is a common cause of liver failure and is the leading reason for liver transplants in the United States. Other viral STDs include genital herpes (usually caused by

herpes simplex virus type 2, HSV-2) and genital warts (caused by 60 or more human papillomaviruses, HPVs). Some forms of HPV are associated with cervical, vaginal, penile, and anal cancer.

Bacterial STDs include gonorrhea and syphilis, caused by the bacteria *Neisseria gonorrhoeae* and *Treponema pallidum,* respectively. Cases of these two diseases are outnumbered, however, by chlamydia (caused by *Chlamydia trachomatis*), which affects 3 to 5 million people per year in the United States.

Before You Go On

Answer the following questions to test your understanding of the preceding section:

15. What are mesonephric and paramesonephric ducts? What factors determine which one develops and which one regresses in the fetus?

16. What male structures develop from the genital tubercle and labioscrotal folds?

17. Define the *gubernaculum* and describe its function.

18. Which of the following occur in both men and women—thelarche, pubarche, climacteric, menopause, breast cancer, or cryptorchidism? Explain.

---

[66]*crypt* = hidden + *orchid* = testes
[67]*hypo* = below + *spad* = to draw off (the urine)

DEEPER INSIGHT   26.3

Cervical Cancer and Pap Smears

Cervical cancer is common among women between the ages of 30 and 50, especially those who smoke, who began sexual activity at an early age, and who have histories of frequent sexually transmitted diseases or cervical inflammation. It is often caused by the human papillomavirus (HPV), a sexually transmitted pathogen. Cervical cancer usually begins in the epithelial cells of the lower cervix, develops slowly, and remains a local, easily removed lesion for several years. If the cancerous cells spread to the subepithelial connective tissue, however, the cancer is said to be *invasive* and is much more dangerous and potentially fatal.

The best protection against cervical cancer is early detection by means of a *Pap*[68] *smear*—a procedure in which loose cells are removed from the cervix and vagina with a small flat stick and cervical brush, then microscopically examined. The pathologist looks for cells with signs of *dysplasia* (abnormal development) or carcinoma (fig. 26.26). One system of grading Pap smears classifies abnormal results in three grades of *cervical intraepithelial neoplasia (CIN)*. Findings are rated on the following scale, and further vigilance or treatment planned accordingly:

ASCUS—atypical squamous cells of undetermined significance

CIN I—mild dysplasia with cellular changes typically associated with HPV

CIN II—moderate dysplasia with precancerous lesions

CIN III—severe dysplasia, *carcinoma in situ* (preinvasive carcinoma of surface cells)

A rating of ASCUS or CIN I calls for a repeat Pap smear and visual examination of the cervix *(colposcopy*[69]*)* in 3 to 6 months. CIN II calls for a biopsy, often done with an "electric scalpel" in a procedure called LEEP (loop electrosurgical excision procedure). A cone of tissue is removed to evaluate the depth of invasion by the malignant or premalignant cells. This in itself may be curative if all margins of the specimen are normal, indicating all normal cells were removed. CIN III may be cause for *hysterectomy*[70] or radiation therapy.

An average woman is typically advised to have annual Pap smears for 3 years and may then have them less often at the discretion of her physician. Women with any of the risk factors listed may be advised to have more frequent examinations.

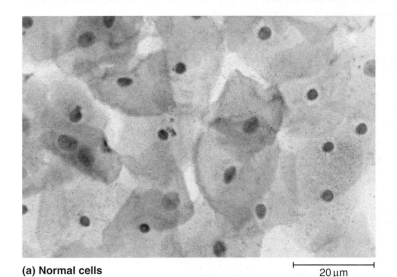

(a) Normal cells |—— 20 μm ——|

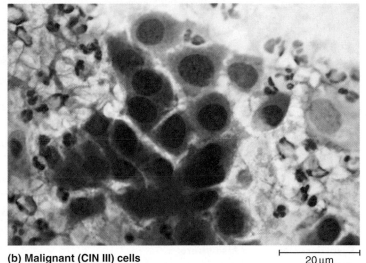

(b) Malignant (CIN III) cells |—— 20 μm ——|

Figure 26.26 **Pap Smears.** These are smears of squamous epithelial cells scraped from the cervix. In the malignant (cancerous) cells, note the loss of cell volume and the greatly enlarged nuclei.

---

[68]George N. Papanicolaou (1883–1962), Greek–American physician and cytologist

[69]*colpo* = vagina + *scopy* = viewing
[70]*hyster* = uterus + *ectomy* = cutting out

TABLE 26.3	Some Female Reproductive Disorders

Disorders of Pregnancy

Abruptio placentae[71]
Premature separation of the placenta from the uterine wall, often associated with preeclampsia or cocaine use. May require birth by cesarean section.

Ectopic[72] pregnancy
Implantation of the conceptus anywhere other than the normal location in the uterus, such as the uterine tube (over 90% of cases), cervix, or abdominal cavity; usually must be surgically terminated to prevent serious and potentially fatal hemorrhage.

Gestational diabetes
A form of diabetes mellitus that develops in about 1% to 3% of pregnant women, characterized by insulin insensitivity, hyperglycemia, glycosuria, and a risk of excessive fetal size and birth trauma. Glucose metabolism usually returns to normal after delivery of the infant, but 40% to 60% of women with gestational diabetes develop diabetes mellitus within 15 years after the pregnancy.

Hyperemesis gravidarum[73]
Prolonged vomiting, dehydration, alkalosis, and weight loss in early pregnancy, often requiring hospitalization to stabilize fluid, electrolyte, and acid–base balance; sometimes associated with liver damage.

Placenta previa[74]
Blockage of the cervical canal by the placenta, preventing birth of the infant before the placenta separates from the uterus. Requires birth by cesarean section.

Preeclampsia[75] (toxemia of pregnancy)
Rapid onset of hypertension and edema, with swelling especially of the face and hands; proteinuria and reduced glomerular filtration rate; increased blood clotting; sometimes with headaches, visual disturbances, and small cerebral infarctions. Seen in about 4% of pregnancies, especially in the third trimester of women pregnant for the first time. Can progress to *eclampsia*, with seizures and widespread vascular spasms that are sometimes fatal to the mother, fetus, or both. Eclampsia usually occurs shortly before or after childbirth.

Spontaneous abortion
Occurs in 10% to 15% of pregnancies, usually because of fetal deformities or chromosomal abnormalities incompatible with survival, but may also result from maternal abnormalities, infectious disease, and drug abuse.

Other Reproductive Disorders

Amenorrhea[76]
Absence of menstruation. Normal in pregnancy, lactation, early adolescence, and perimenopausal years, but can also result from gonadotropin hyposecretion, genetic disorders, CNS disorders, or excessively low body fat.

Dysmenorrhea[77]
Painful menstruation in the absence of pelvic disease, caused by excessive endometrial prostaglandin secretion. Prostaglandins stimulate painful contractions of myometrium and uterine blood vessels. Usually begins around age 15 or 16 and affects up to 75% of women from 15 to 25 years old.

Endometriosis
Growth of endometrial tissue in any site other than the uterus, including the uterine tubes, ovaries, urinary bladder, vagina, pelvic cavity, small or large intestine, or even the lungs or pleural cavity. May cause dysmenorrhea, abnormal vaginal bleeding, and infertility.

Leiomyomas[78] (uterine fibroids)
Benign tumors of uterine smooth muscle. Usually small and asymptomatic, but can become very large and may cause abnormal uterine bleeding and pain or heavy menstruation.

Pelvic inflammatory disease (PID)
Acute, painful inflammation due to infection of the uterus, uterine tubes, or ovaries, usually with the organisms of sexually transmitted diseases. Causes abdominopelvic pain, pain on urination, and irregular bleeding.

Disorders Described Elsewhere

---

[71]*ab* = away + *rupt* = to tear + *placentae* = of the placenta
[72]*ec* = out of + *top* = place
[73]*hyper* = excessive + *emesis* = vomiting + *gravida* = pregnant woman
[74]*pre* = before + *via* = the way (obstructing the way)
[75]*ec* = forth + *lampsia* = shining
[76]*a* = without + *meno* = monthly + *rrhea* = flow
[77]*dys* = painful, abnormal + *meno* = monthly + *rrhea* = flow
[78]*leio* = smooth + *myo* = muscle + *oma* = tumor

Study Guide

Assess Your Learning Outcomes

You should have a good understanding of this chapter if you can accurately address the following issues.

26.1 Sexual Reproduction (p. 706)

1. The definition of *sexual reproduction,* and how this relates to the necessity of gametes and a zygote
2. The division of roles between two types of gametes; how this relates to the biological definitons of *male* and *female;* and the chromosomal distinction between the sexes
3. How the sexes are distinguished by their chromosomes
4. The distinctions between primary sex organs, secondary sex organs, secondary sex characteristics, and internal and external genitalia, and examples of all of these

26.2 Male Reproductive Anatomy (p. 707)

1. The surface anatomy and contents of the scrotum
2. The spermatic cord, its contents, and its relationship to the inguinal canal and the internal and external inguinal rings
3. The relevance of temperature to sperm development and the three mechanisms for maintaining the proper temperature of the testes
4. The surface anatomy of the testis; its internal division into lobules and seminiferous tubules; the rete testis; and the interstitial cells, their function, and their relationship with the seminiferous tubules
5. The epithelium of a seminiferous tubule; its cell types; and the functions of those cells
6. The structure and purpose of the blood–testis barrier
7. The route of blood flow from aorta to testicular artery, testis, pampiniform plexus, testicular vein, and ultimately the inferior vena cava (including the right–left difference in routes)
8. The anatomical course and functions of testicular lymphatics and nerves
9. The basic events of spermatogenesis; the necessity of meiosis for sexual reproduction; and the cellular stages of spermatogenesis and their relationship to meiosis I, meiosis II, and spermiogenesis
10. The structure of a spermatozoon and the function of each of its parts

11. The series of ducts that convey sperm from the testis to the urethra; the anatomy of each; and the three regions of the male urethra
12. The three sets of male accessory glands associated with the spermatic ducts; their anatomy; and the contribution of each to semen production and intercourse
13. The volume of semen and quantity of sperm in a typical ejaculate; the chemical components of the semen; and the functions and glandular sources of each
14. The anatomical regions and dimensions of the penis, and its relationships with adjacent organs
15. The three erectile tissues and other internal structures of the penis; the structure of an individual erectile tissue and how it relates to copulatory function
16. The arterial blood supply and venous drainage of the penis, and differences in blood flow between the flaccid and erect states
17. The nerve supply to the penis and how its sympathetic, parasympathetic, and sensory nerves relate to sexual function

26.3 Female Reproductive Anatomy (p. 717)

1. The structure of the ovary
2. The supportive ligaments of the ovary; their relationship to the arterial blood supply to the ovary; and the venous and lymphatic drainage of the ovary
3. The principal events of oogenesis and how it differs from spermatogenesis
4. The cellular stages of oogenesis; how they relate to the stages of meiosis; and how they relate to ovulation and fertilization
5. The types of ovarian follicles; how they relate to oogenesis and ovulation; and the fate of a follicle after it ovulates
6. Details of the structure of a mature ovarian follicle
7. The gross anatomy and histology of the uterine tubes
8. The gross anatomy of the uterus; the names and histological characteristics of the three layers of the uterine wall
9. The supportive ligaments of the uterus; their relationship to the peritoneal pouches of the pelvic cavity; and their relationship to the blood vessels that supply the uterus
10. The histological changes that occur in the endometrium over the course of the menstrual cycle

11. The gross anatomy and histology of the vagina
12. The anatomy of the vulva, including the mons pubis; labia majora and minora; vestibule; clitoris and prepuce; vestibular bulbs; greater and lesser vestibular glands; and paraurethral glands
13. The gross anatomy of the breast; the distinction between breasts and mammary glands; and the structure of the mammary glands, particularly the acini; secretory and myoepithelial cells; and duct system

26.4 Developmental and Clinical Perspectives (p. 729)

1. Early features of genital development that are identical in males and females; the meaning of *homologous* structures, and examples; and the age at which the sexes become distinguishable
2. The mesonephric and paramesonephric ducts and the fate of each
3. How the sex chromosomes, *SRY* gene, and hormones interact to determine whether an individual develops male or female anatomy
4. Which male and female structures develop from the genital tubercle, urogenital folds, and labioscrotal folds
5. The descent of the gonads; the role of the gubernaculum; and how the descent differs in males and females
6. How puberty is hormonally induced; the distinction between puberty and adolescence; similarities and differences between boys and girls in the events of puberty; and the meaning of *thelarche, pubarche,* and *menarche* in girls
7. The hormonal causes of male and female climacteric, and the signs and symptoms of climacteric in each sex
8. The distinction between menopause and female climacteric, and why the exact time of menopause cannot be determined
9. Types of cancer that occur in the male reproductive organs, and other common male reproductive disorders
10. Types of cancer that occur in the female reproductive organs; common disorders of pregnancy; and other common female reproductive disorders
11. The most common viral and bacterial sexually transmitted diseases

Testing Your Recall

1. The ductus deferens develops from the
   _____ of the embryo.
   a. mesonephric duct
   b. paramesonephric duct
   c. phallus
   d. labioscrotal folds
   e. urogenital folds

2. Descent of the testes includes their passage
   through
   a. the inguinal canal.
   b. the spermatic cord.
   c. the ductus deferens.
   d. the seminiferous tubule.
   e. the ampulla.

3. Four spermatozoa arise from each
   a. primordial germ cell.
   b. type A spermatogonium.
   c. type B spermatogonium.
   d. secondary spermatocyte.
   e. spermatid.

4. Prior to ejaculation, sperm are stored
   primarily in
   a. the seminiferous tubules.
   b. the epididymis.
   c. the seminal vesicles.
   d. the bulb of the penis.
   e. the ejaculatory ducts.

5. Testosterone is produced primarily by
   a. the seminiferous tubules.
   b. the sustentacular cells.
   c. the interstitial cells.
   d. the seminal vesicles.
   e. the prostate gland.

6. The fluid-filled central cavity of a mature
   ovarian follicle is
   a. the antrum.
   b. the zona pellucida.
   c. the theca folliculi.
   d. the granulosa.
   e. the stigma.

7. The tissue lost in menstruation is
   a. perimetrium.
   b. myometrium.
   c. stratum basalis.
   d. stratum functionalis.
   e. stratum corneum.

8. The male scrotum is homologous to the
   female
   a. ovaries.
   b. vagina.
   c. labia majora.
   d. vestibular bulbs.
   e. clitoris.

9. The narrowest part of the uterus is
   a. the fundus.
   b. the infundibulum.
   c. the body.
   d. the ampulla.
   e. the cervix.

10. The vesicouterine pouch is a space in the
    peritoneal cavity between the uterus and
    a. the fornices.
    b. the uterine tube.
    c. the sacrum.
    d. the urinary bladder.
    e. the rectum.

11. Under the influence of androgens, the
    embryonic _____ duct develops into the
    male reproductive tract.

12. Spermatozoa obtain energy for locomotion
    from _____ in the semen.

13. The _____, a network of veins in the
    spermatic cord, helps keep the testes cooler
    than the core body temperature.

14. Each egg cell develops in its own fluid-
    filled space called a/an _____.

15. The mucosa of the uterus is called the
    _____.

16. Over half of the semen consists of
    secretions from a pair of glands called the
    _____.

17. The blood–testis barrier is formed by tight
    junctions between the _____ cells.

18. The female paraurethral glands arise from
    the same embryonic structure as the male
    _____.

19. A yellowish structure called the _____
    secretes progesterone during the secretory
    phase of the menstrual cycle.

20. The funnel-like distal end of the uterine
    tube is called the _____ and has feathery
    processes called _____.

Answers in the Appendix

Building Your Medical Vocabulary

*State a medical meaning of each of the following
word elements, and give a term in which it is
used.*

1. gameto-
2. gono-
3. hystero-
4. vagin-
5. alb-
6. ov-
7. oo-
8. -phor
9. lute-
10. metri-

Answers in the Appendix

True or False

*Determine which five of the following statements
are false, and briefly explain why.*

1. After ovulation, a follicle begins to move
   down the uterine tube to the uterus.
2. The uterine tubes develop from the
   embryonic mesonephric ducts.
3. Thelarche normally precedes menarche in
   female puberty.
4. The follicle that ovulates is called the
   primary follicle.

5. Sperm cannot develop at the core body
   temperature.
6. A high androgen level makes a fetus
   develop a male reproductive system, and
   a high estrogen level makes it develop a
   female reproductive system.
7. Most ovarian follicles degenerate before a
   girl reaches puberty.

8. The pampiniform plexus serves to keep
   the testes warm.
9. Prior to ejaculation, sperm are stored
   mainly in the epididymis.
10. The thickest layer of the uterine wall is the
    myometrium.

Answers in the Appendix

Testing Your Comprehension

1. The most common method of male sterilization is vasectomy, in which the ductus (vas) deferens is tied, cut, or both. What is the equivalent method of female sterilization? Why are the difficulty and risks of that procedure greater than for a vasectomy?

2. *Uterus bicornis* (*bicorn* = two horns) is a rare condition in which a woman has two separate uteri, each opening by its own cervix into the vagina. What abnormal event of embryonic development do you think could account for this?

3. Suppose the corpus spongiosum became as engorged with blood as the corpora cavernosa do during erection of the penis. What problem would this create for sexual function? In light of this, why is it beneficial that the corpus spongiosum is not enclosed in a tunica albuginea?

4. What male structure(s) do you think is or are homologous to the vestibular bulbs of the female? Explain your reasoning.

5. An oocyte lives for only 24 hours after ovulation if it is not fertilized. The trip down the uterine tube, from infundibulum to uterus, takes about 72 hours. In light of this, where do you think fertilization normally occurs?

Answers at www.mhhe.com/saladinha3

Improve Your Grade at www.mhhe.com/saladinha3

Practice quizzes, labeling activities, games, and flashcards provide fun ways to master concepts. You can also download image PowerPoint files for each chapter to create a study guide or for taking notes during lecture.

Answers to Study Guide Questions

Answers are provided here for the end-of-chapter Testing Your Recall, Building Your Medical Vocabulary, and True or False questions, and questions in the figure legends. Answers to Apply What You Know and Testing Your Comprehension questions are avail-

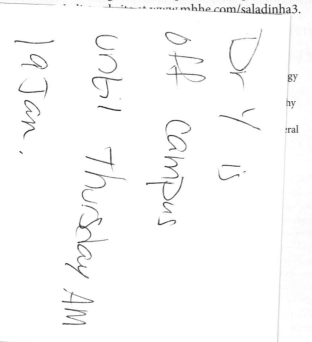

Chapter 2

Testing Your Recall

1.	e	8.	a	15.	tight junctions
2.	d	9.	d	16.	squamous
3.	b	10.	b	17.	mitochondrion, nucleus
4.	b	11.	micrometers (μm)	18.	peroxisomes, smooth ER
5.	e	12.	receptor	19.	cell-adhesion molecules
6.	a	13.	gates	20.	phagocytosis
7.	d	14.	mitochondria		

Building Your Medical Vocabulary

(Answers may vary; these are acceptable examples.)

1. cell—cytology
2. flat, scale—squamous
3. shape—fusiform
4. many—polygonal
5. loving, attracted to—hydrophilic
6. eat—phagocytosis
7. into, within—endoplasmic
8. sugar—glycocalyx
9. color—chromosome
10. beyond, next in a series—metaphase

True or False

(These items are false for the reasons given; all others are true.)

2. Proteins are only 1% to 10% of the molecules of a plasma membrane.
6. Movement down a gradient does not employ ATP.
7. Osmosis is a case of simple diffusion.
9. Desmosomes provide no channels from one cell to another.
10. The nucleolus is not an organelle.

Figure Legend Questions

2.2	The cells are thinner in the center than around the rim, so more light shines through.
2.8	The region that projects into the cytoplasm.
2.12	To stiffen and support the microvillus, anchor it to the cell, and produce contractions of the microvillus that milk its contents into the cell.
2.13	Microvilli are much smaller than cilia, they lack axonemes, and they have a supportive core of actin.
2.14	Gap junctions
2.15	Microfilaments have a supportive role in microvilli, and microtubules form the axonemes of cilia and flagella.
2.17	The nucleus

True or False

(These items are false for the reasons given; all others are true.)

3. Auscultation is listening to sounds made by the body.
4. Radiology is concerned with all methods of medical imaging.
6. Nearly every cell contains many organelles.
7. The diaphragm is inferior to the lungs.
9. There is nothing but a thin film of fluid between the parietal and visceral pleurae.

Figure Legend Questions

1.3	MRI is better than X-ray at showing soft tissues such as the eye and its muscles and the brain tissue. X-ray is better than PET at showing hard, dense structures such as the bones and teeth.
1.4	Radiography and CT use X-rays, which have the potential to cause mutations and birth defects; sonography uses no harmful radiation.
1.10	Midsagittal or median
1.17	No, the urinary bladder lies outside the peritoneal cavity; the peritoneum passes over its superior surface.

Chapter 3

Testing Your Recall

1. a
2. b
3. c
4. e
5. d
6. a
7. b
8. e
9. b
10. b
11. apoptosis (programmed cell death)
12. mesothelium
13. lacunae
14. fibers
15. collagen
16. gangrene
17. basement membrane
18. matrix
19. holocrine
20. simple

Building Your Medical Vocabulary

(Answers may vary; these are acceptable examples.)

1. tissue—histology
2. between—intercellular
3. plate—lamella
4. form, give rise to—fibroblast
5. network—reticulocyte
6. cartilage—chondrocyte
7. around—periosteum
8. nipple, female—epithelium
9. out of—exocytosis
10. dead, death—necrosis

True or False

(These items are false for the reasons given; all others are true.)

2. The noncellular components include ground substance and fibers.
5. The tongue epithelium is nonkeratinized.
6. Macrophages develop from monocytes.
8. Brown fat produces no ATP.
9. In metaplasia, one mature tissue type transforms into another.

Figure Legend Questions

3.1   The microtubule could be drawn roughly as two parallel bars, like the upper right diagram of figure 3.2, but if the globular tubulin molecules were taken into account (see fig. 2.16), each bar could be shown as a linear chain of spherical tubulin molecules.

3.13   Dense regular connective tissue

3.29   Exocytosis

3.31   The esophagus, stomach, and intestines, for example, are lined with mucous membranes; the outer surfaces of the liver, spleen, stomach, and intestines are covered with serous membranes.

Chapter 4

Testing Your Recall

1. b
2. b
3. d
4. c
5. a
6. e
7. c
8. a
9. e
10. d
11. teratogens
12. nondisjunction
13. neural groove or neural tube
14. implantation
15. chorionic villi
16. acrosome
17. uterine tube
18. somites
19. polyspermy
20. embryo

Building Your Medical Vocabulary

(Answers may vary; these are acceptable examples.)

1. half—haploid
2. sex cell—gametogenesis
3. union—zygote
4. nourish—trophoblast
5. head—hydrocephalus
6. female—gynecomastia
7. production of—embryogenesis
8. together, united—syncytiotrophoblast
9. middle—mesoderm
10. monster—teratogen

True or False

(These items are false for the reasons given; all others are true.)

1. Sperm must undergo capacitation first.
2. Fertilization occurs in the uterine tube.
3. Several sperm must digest a path for the one that fertilizes the egg.
8. Oogenesis produces one large oocyte and small discarded polar bodies.
10. The energy comes from the midpiece mitochondria.

Figure Legend Questions

4.2   An egg will have died by the time it reaches the uterus if it was not already fertilized.

4.10   At about 8 weeks

4.11   There are two arteries and one vein. Blood in the vein is more oxygenated than blood in the arteries.

4.13   If an XX egg was fertilized by a Y-bearing sperm, the result would be Klinefelter syndrome (XXY). If the egg with no sex chromosome was fertilized by a Y-bearing sperm, the resulting YO zygote would soon die for lack of critical genes found only on X chromosomes; no birth would result.

4.14   It was a female, as we can tell from the two X chromosomes at the lower right.

Chapter 5

Testing Your Recall

1. d
2. c
3. d
4. b
5. a
6. e
7. c
8. a
9. a
10. d
11. dermato-, cutane-
12. piloerector (arrector pili)
13. keratin, collagen
14. cyanosis
15. dermal papillae
16. earwax
17. sebaceous glands
18. cuticle
19. dermal papilla
20. second-degree

Building Your Medical Vocabulary

(Answers may vary; these are acceptable examples.)

1. skin—dermatology
2. above, upon—epidermal
3. below—subcutaneous
4. nipple—papillary
5. black—melanoma
6. blue—cyanosis
7. clear—stratum lucidum
8. little—papilla
9. hair—piloerector
10. cancer—carcinoma

True or False

(These items are false for the reasons given; all others are true.)

3. The dermis is mainly collagen.
4. Vitamin D synthesis begins in the keratinocytes and is completed in the liver and kidneys.
7. The hypodermis is not part of the skin.
8. People of all colors have similar densities of melanocytes.
9. Malignant melanoma is the most deadly but least common form of skin cancer.

Figure Legend Questions

5.5   Keratinocytes

5.7   The cuticle

5.10   Terminal hair; they are connected to the follicles of the coarse hairs of the pubic, axillary, and beard regions.

5.13   Asymmetry (A), border irregularity (B), and color (C). The photograph does not provide enough information to judge diameter (D).

Chapter 6

Testing Your Recall

1. e	8. e	15. parathyroid
2. a	9. b	16. articular cartilage
3. d	10. d	17. osteoblasts
4. c	11. hydroxyapatite	18. osteoporosis
5. d	12. canaliculi	19. metaphysis
6. c	13. appositional	20. intramembranous ossification
7. d	14. osteons	

Building Your Medical Vocabulary

(Answers may vary; these are acceptable examples.)
1. bone—periosteum
2. double—diploe
3. space, cavity—lacuna
4. breakdown, destroyer—osteoclast
5. medical condition—osteoporosis
6. across—diaphysis
7. study of—osteology
8. joint—articular
9. little—canaliculus
10. like, resembling—osteoid

True or False

(These items are false for the reasons given; all others are true.)
3. The most common bone disorder is osteoporosis.
4. The growth zone is the epiphyseal plate.
5. Osteoclasts develop from stem cells related to monocytes.
7. The protein of the bone matrix is collagen.
9. Only the red bone marrow is hemopoietic.

Figure Legend Questions

6.2   The wide epiphyses provide expanded surface area for bone articulation and for tendon and ligament attachments. Joints would be very unstable if they were as narrow as the diaphysis.
6.5   Spongy bone
6.7   The crest of the hip and the sternum
6.8   The parietal and frontal bones (answers may vary)
6.10 The humerus, radius, ulna, femur, tibia, and fibula (any two)
6.12 The zones of cell proliferation and hypertrophy (2 and 3)

Chapter 7

Testing Your Recall

1. b	8. a	15. anulus fibrosus
2. e	9. c	16. dens
3. a	10. c	17. auricular
4. d	11. fontanels	18. false, floating
5. a	12. temporal	19. costal cartilages
6. b	13. sutures	20. xiphoid process
7. a	14. sphenoid	

Building Your Medical Vocabulary

(Answers may vary; these are acceptable examples.)
1. skull—cranial
2. time—temporal bone
3. breast—mastoid process
4. stone—petrous
5. layer—lamina
6. wing—pterygoid
7. crest, ridge—crista galli
8. tears—lacrimal bone
9. rib—costal cartilage
10. foot—pedicle

True or False

(These items are false for the reasons given; all others are true.)
1. The vertebral bodies are derived from the sclerotomes.
2. Adults have fewer bones than children do.
4. The zygomatic processes of the temporal bone and maxilla also contribute to the arch.
5. The dura mater lies loosely against most of the cranium.
10. Lumbar vertebrae have transverse processes, but not transverse costal facets.

Figure Legend Questions

7.7   They produce turbulence in the airflow and support the mucous membranes that warm, cleanse, and humidify inhaled air.
7.8   The anterior portion of the skull would be much heavier and the skull would tend to tip forward. More effort would be required to oppose this.
7.13 The vomer and the perpendicular plate of the ethmoid.
7.16 The hyoid is a delicate, easily broken bone, and its location subjects it to fracture by a rope, hands, or other means of strangulation applied around the neck.
7.24 The dens could shift forward and severely damage the spinal cord.
7.30 Most joints of an infant are still cartilaginous and therefore not very resistant to stress.
7.36 The lumbar vertebrae and discs must bear more weight from the body and from heavy objects that one lifts, and are therefore the most likely ones to herniate. The cervical vertebrae and discs bear very little weight.

Chapter 8

Testing Your Recall

1. a	8. b	15. hamate
2. e	9. e	16. interpubic disc
3. c	10. b	17. crural
4. b	11. pollex, hallux	18. styloid
5. a	12. scapula	19. trochanters
6. d	13. 56	20. medial longitudinal
7. c	14. epicondyles	

Building Your Medical Vocabulary

(Answers may vary; these are acceptable examples.)
1. chest—pectoralis
2. peak, apex, extremity—acromion
3. little—ossicle
4. above—supraspinous
5. wrist—carpal
6. head—capitulum
7. little—acetabulum
8. beyond, next in a series—metacarpal
9. ear—auricular
10. ankle—tarsal

True or False

(These items are false for the reasons given; all others are true.)
2. Each hand and foot has 14 phalanges.
3. The upper limb is attached at the glenohumeral joint.
5. The arm contains only the humerus, but the leg contains the tibia and fibula.
7. The most frequently broken bone is the clavicle.
10. That opening is the pelvic inlet.

Figure Legend Questions

8.1   See Deeper Insight 8.1 for the reasons.
8.2   It is attached at the glenoid cavity to the humerus and attached at the acromion to the clavicle.
8.6   During birth, the fetal head must pass through the narrow pelvic inlet and outlet. It could not do so if the cranial bones were immovably fused; the infant must be born before the bones fuse at the sutures and fontanels.
8.10 Four: the head, the medial condyle, the lateral condyle, and the patellar surface
8.14 The sesamoid bone

Chapter 9

Testing Your Recall

1. c	8. d	15. gomphosis
2. b	9. b	16. serrate
3. a	10. b	17. extension
4. e	11. synovial fluid	18. range of motion
5. c	12. bursa	19. rheumatologist
6. c	13. pivot	20. menisci
7. c	14. kinesiology	

Building Your Medical Vocabulary

(Answers may vary; these are acceptable examples.)
1. joint—arthritis
2. back—reposition
3. together—symphysis
4. both—amphiarthrosis
5. growth—diaphysis
6. around—circumduction
7. away—abduction
8. toward—adduction
9. to lead—abduction
10. motion—kinesiology

True or False

(These items are false for the reasons given; all others are true.)
1. Osteoarthritis is much more common than rheumatoid arthritis.
2. A doctor who treats arthritis is a rheumatologist.
3. Synovial joints are diarthroses.
5. This action hyperextends the shoulder; the elbow cannot be hyperextended.
9. Synovial fluid fills the bursae but is secreted by the synovial membrane of the joint capsule.

Figure Legend Questions

9.1 A gomphosis is a joint between a bone and a tooth; teeth are not bones.
9.3 The interpubic disc is only the fibrocartilage pad; the pubic symphysis is the disc plus the adjacent areas of the pubic bones.
9.4 Interphalangeal joints are not subject to routine compression.
9.14 The atlas
9.17 Variable answers; for example, changing direction when walking or running, or walking on a rocky trail
9.19 The glenoid labrum

Chapter 10

Testing Your Recall

1. e	8. c	14. myosin
2. c	9. d	15. fasciae
3. c	10. b	16. myoglobin
4. a	11. synaptic vesicles	17. Z discs
5. c	12. neuromuscular junction	18. sphincter
6. e	or motor end plate	19. myoblasts
7. e	13. terminal cisternae	20. peristalsis

Building Your Medical Vocabulary

(Answers may vary; these are acceptable examples.)
1. muscle—myofibril
2. band—fascicle
3. spindle—fusiform
4. flesh, muscle—sarcolemma
5. bad, abnormal, difficult—dystrophy
6. muscle—endomysium
7. against—antagonist
8. work, action—synergist
9. segment—sarcomere
10. feather—bipennate

True or False

(These items are false for the reasons given; all others are true.)
4. The expression *slow oxidative* refers to skeletal muscle fibers.
5. One motor neuron can supply from a few to a thousand muscle fibers.
6. Calcium binds to troponin, not to myosin.
9. The blood vessels become wavy when the muscle shortens.
10. Muscle growth involves an increase in muscle fiber thickness, not in their number.

Figure Legend Questions

10.1 Unlike the other muscle types, the skeletal muscle here exhibits long parallel fibers with no branching or tapering, and many nuclei in each fiber. Its striations also distinguish it from smooth muscle.
10.3 Strength depends on both fascicle arrangement and muscle diameter. A large parallel muscle could be stronger than a small fusiform muscle.
10.4 The biceps brachii and distal end of the triceps brachii (both heads) exhibit indirect attachments. The brachialis and lateral head of the triceps brachii exhibit direct attachments.
10.5 It is a first-class lever because the fulcrum lies between the effort and resistance.
10.8 The function of the T tubule is to stimulate the opening of calcium gates in the terminal cisternae.
10.11 The neuromuscular junction is a region containing multiple neuromuscular synapses.

Chapter 11

Testing Your Recall

1. b	8. a	14. hypoglossal
2. c	9. c	15. digastric
3. a	10. d	16. urogenital triangle
4. c	11. erector spinae	17. linea alba
5. e	12. bulbospongiosus	18. larynx
6. e	13. levator palpebrae	19. sternocleidomastoid
7. a	superioris	20. trapezius

Building Your Medical Vocabulary

(Answers may vary; these are acceptable examples.)
1. triangular—deltoid
2. raise—levator labii superioris
3. eye—orbicularis oculi
4. away, apart—aponeurosis
5. mouth—orbicularis oris
6. finger, toe—flexor digitorum
7. lip—depressor labii inferioris
8. same—ipsilateral
9. tongue—hypoglossal
10. two—digastric

True or False

(These items are false for the reasons given; all others are true.)
1. The mastoid process is the insertion, not the origin.
5. Normal exhalation is passive and does not employ the internal intercostals.
6. The floor of the mouth is formed by the mylohyoid muscle.
7. The scalenes are deep to the trapezius.
10. Only cranial nerves III, V, VII, XI, and XII innervate head and neck muscles.

Figure Legend Questions

11.4 Zygomaticus major, levator palpebrae superioris, orbicularis oris (answers may vary)
11.6 If the mandible were already at its farthest right lateral excursion, the right medial pterygoid would help draw it back to the zero position or midline (medial excursion), or it could contract still more to cause left lateral excursion.
11.8 Compare figure 11.7b. A1 contains the sternohyoid and superior belly of the omohyoid; A4 contains the mylohyoid and anterior belly of the digastric; and P1 contains the scalenes, levator scapulae, and splenius capitis (answers may vary).
11.13 The pectoralis minor, subclavius, and the upper intercostal and serratus anterior muscles

Chapter 12

Testing Your Recall

1. c	8. a	15. retinacula
2. e	9. d	16. adductor pollicis
3. b	10. a	17. quadriceps femoris
4. d	11. deltoid	18. coracobrachialis
5. d	12. great toe	19. gracilis
6. e	13. teres, quadratus	20. iliacus, psoas major
7. b	14. hamstring	

Building Your Medical Vocabulary

(Answers may vary; these are acceptable examples.)
1. scalloped—serratus anterior
2. the back—latissimus dorsi
3. head—biceps
4. below—infraspinous
5. round—pronator teres
6. deep—flexor digitorum profundus
7. bone—interosseous
8. hip—iliacus
9. carry—afferent
10. largest—gluteus maximus

True or False

(These items are false for the reasons given; all others are true.)
1. The plantaris muscle inserts on the foot by a tendon of its own.
5. The interosseous muscles are pennate.
7. The psoas major and rectus femoris are synergists in flexing the hip.
8. Hamstring injuries usually result from rapid extension of the knee, not flexion.
10. These muscles are on opposite sides of the tibia and act as antagonists.

Figure Legend Questions

12.3   The deltoid
12.6   *Teres* indicates that muscle's rounded, cordlike shape; *quadratus* indicates that muscle's angular, four-sided shape.
12.7   Those two muscles are found in the distal half of the forearm, whereas this section represents the more proximal muscles.
12.12 Lifting the body to the next higher step when climbing stairs; the back-swing of the lower limb when walking or running (answers may vary)
12.17 The soleus

Atlas

Muscle Test (fig. A.25)

1. f	11. x	21. k
2. b	12. m	22. d
3. k	13. n	23. f
4. p	14. e	24. b
5. h	15. g	25. a
6. y	16. v	26. u
7. z	17. f	27. j
8. w	18. c	28. i
9. c	19. x	29. g
10. a	20. w	30. q

Figure Legend Questions

A.1   The orbicularis oris. The trapezius.
A.5   Lungs, heart, thymus, liver, stomach, spleen, kidneys
A.8   The sternocleidomastoids
A.11 Posterior
A.13 Subcutaneous fat (adipose tissue)
A.18 Four
A.19 In part (a), at the base of the first metacarpal bone, between the two leaders from the "Flexion lines" label
A.20 Deep to the rectus femoris
A.21 The fibula
A.24 There is no such bone; the great toe has only two phalanges, proximal and distal.

Chapter 13

Testing Your Recall

1. e	8. d	15. oligodendrocytes
2. e	9. a	16. axosomatic
3. d	10. c	17. peripheral nervous system
4. a	11. afferent	18. neurilemma, endoneurium
5. e	12. reverberating	19. ganglia
6. d	13. anencephaly	20. postsynaptic
7. a	14. dendrites	

Building Your Medical Vocabulary

(Answers may vary; these are acceptable examples.)
1. pertaining to—anaxonic
2. body—axosomatic
3. nerve—neurosoma
4. lipid—lipofuscin
5. tree, branch—axodendritic
6. little—dendrite
7. false—pseudounipolar
8. few—oligodendrocyte
9. carry—afferent
10. hard—multiple sclerosis

True or False

(These items are false for the reasons given; all others are true.)
4. Sensory (afferent) neurons connect sense organs to the CNS.
6. The myelin sheath is deep to the neurilemma.
7. Nodes of Ranvier also exist in the CNS.
8. Interneurons are contained entirely within the CNS.
9. Unipolar neurons have an axon and produce action potentials.

Figure Legend Questions

13.1   The PNS, because it is more exposed to trauma than the CNS. The CNS is protected by the cranial and vertebral bones.
13.3   *Afferent* is derived from *af (ad)*, meaning "toward," and *fer*, meaning "to carry." Afferent neurons carry signals toward the CNS. *Efferent* is derived from *ef (ex)*, meaning "out." Efferent neurons carry signals out of the CNS.
13.4   The presence of multiple dendrites
13.8   Unmyelinated fibers conduct signals relatively slowly, but they occupy less space.
13.9   Axosomatic

Chapter 14

Testing Your Recall

1. e	8. a	15. central pattern generators
2. c	9. e	16. phrenic
3. d	10. b	17. decussation
4. d	11. ganglia	18. proprioception
5. e	12. rami	19. posterior root
6. c	13. spinocerebellar	20. tibial, common fibular
7. c	14. sacral	

Building Your Medical Vocabulary

(Answers may vary; these are acceptable examples.)
1. tail—caudal
2. opposite, against—contralateral
3. side—anterolateral
4. of one's own—proprioceptor
5. slender—gracile fasciculus
6. spinal cord—myelin

7. without—amyotrophic
8. inflammation—poliomyelitis
9. diaphragm—phrenic
10. cut, section—dermatome

True or False

(These items are false for the reasons given; all others are true.)
1. The gracile fasciculus is a sensory (ascending) tract.
4. All spinal nerves are mixed nerves.
5. The dura mater is set off from the vertebral bone by the epidural space.
8. Dermatomes overlap each other by as much as 50%.
9. Many somatic reflexes involve the brain.

Figure Legend Questions

14.3 It would have to be T4 because the cuneate fasciculus does not exist below level T6.

14.4 Heat and cold stimuli are carried to the opposite side of the spinal cord before ascending the spinothalamic tract to the brain. The phenomenon is called decussation.

14.8 Unipolar (or pseudounipolar)

14.10 It would result in a loss of sensation within the corresponding region on that side of the body.

14.13 Respiratory arrest would occur if both the right and left phrenic nerves were severed; severing only one of them paralyzes half of the diaphragm and severely diminishes ventilation of the ipsilateral lung.

14.19 If the hamstrings contracted, they would promote flexion of the knee and thus oppose the knee-extending patellar tendon reflex.

Chapter 15

Testing Your Recall

1. c	8. d	15. choroid plexus
2. d	9. e	16. precentral
3. e	10. e	17. frontal
4. a	11. corpus callosum	18. association cortex
5. e	12. ventricles, cerebrospinal	19. categorical
6. c	13. arbor vitae	20. Broca area
7. a	14. hippocampus	

Building Your Medical Vocabulary

(Answers may vary; these are acceptable examples.)
1. turn, twist—gyrus
2. groove—sulcus
3. brain—cerebrum
4. stalk—peduncle
5. island—insula
6. little—cerebellum
7. new—neocortex
8. four—corpora quadrigemina
9. leaf—folia
10. radiating—corona radiata

True or False

(These items are false for the reasons given; all others are true.)
1. The longitudinal fissure separates the cerebral hemispheres, not cerebellar hemispheres.
2. Degeneration of the substantia nigra causes Parkinson disease.
5. The choroid plexuses produce only 30% of the CSF.
6. Hearing is a temporal lobe function.
10. The optic nerve carries visual signals, not motor signals.

Figure Legend Questions

15.1 The dura mater
15.7 (To be answered by pointing out or marking structures in the illustration)
15.8 (To be answered by pointing out or marking structures in the illustration)
15.14 Dendrites
15.19 Those with many small muscles
15.22 No; everyone makes extensive use of both hemispheres.

Chapter 16

Testing Your Recall

1. b	8. d	15. enteric
2. c	9. c	16. neural crest
3. e	10. c	17. sympathetic
4. e	11. adrenergic	18. preganglionic, postganglionic
5. a	12. dual innervation	19. baroreceptor
6. e	13. autonomic tone	20. vasomotor tone
7. d	14. vagus	

Building Your Medical Vocabulary

(Answers may vary; these are acceptable examples.)
1. rule—autonomic
2. pressure—baroreceptor
3. viscera—splanchnic nerves
4. kidney—renal
5. feeling—sympathetic
6. next to—paravertebral
7. break down, destroy—sympatholytic
8. self—autonomic
9. branch—ramus
10. wall—intramural

True or False

(These items are false for the reasons given; all others are true.)
1. Normally both divisions are active simultaneously.
3. With biofeedback and other methods, some degree of voluntary control is possible.
4. It inhibits digestion.
6. These reflexes can occur, but are less controllable, without involvement of the brain.
7. All parasympathetic fibers are cholinergic.

Figure Legend Questions

16.4 No; as explained in chapter 11, inspiration and expiration are achieved by skeletal muscles, which are not under autonomic control.

16.5 Sympathetic neurons arise from the lateral horn and somatic neurons from the anterior horn of the spinal cord.

16.7 The vagus nerve

16.9 Dilated, because fear activates the sympathetic division, which stimulates the pupillary dilator

Chapter 17

Testing Your Recall

1. a	8. c	15. outer hair cells
2. a	9. c	16. stapes
3. a	10. a	17. inferior colliculi
4. d	11. fovea centralis	18. taste hairs
5. b	12. ganglion	19. olfactory bulb
6. e	13. nociceptor	20. referred pain
7. d	14. otoliths	

Building Your Medical Vocabulary

(Answers may vary; these are acceptable examples.)
1. pain—nociceptor
2. without—anosmia
3. thread—filiform
4. ear—otitis media
5. pulley—trochlea
6. pit, depression—fovea centralis
7. pain—analgesic
8. like—foliate
9. drum—tympanic
10. making a new opening—tympanostomy

True or False

(These items are false for the reasons given; all others are true.)
2. Afferent touch fibers end in the spinal cord and medulla oblongata.
4. Pain signals are blocked after entering the spinal cord, before ascending to the brain.
8. Olfactory neurons are directly exposed to the external environment.
9. The tympanic membrane has sensory fibers of the vagus and trigeminal nerves.
10. The posterior chamber, between the iris and lens, is filled with aqueous humor.

Figure Legend Questions

17.3 Postcentral gyrus; parietal lobe
17.5 The basal cell can divide, and one of its daughter cells can become a new taste cell.
17.14 High middle-ear pressure would interfere with inward movements of the tympanic membrane and therefore reduce the transfer of vibrations to the inner ear.
17.16 The macula sacculi is vertically oriented, so up or down movements in an elevator would cause the otolithic membrane to shift up or down across the hair cells and bend their stereocilia.
17.20 Watery eyes; tears would be unable to drain from the eye surface and would spill over the eyelid.
17.21 CN III, the oculomotor nerve. This nerve controls four muscles, whereas the others each control only one. CN III is indispensable to the ability to look up, down, and sideways.

Chapter 18

Testing Your Recall

1. d	8. a	15. glucocorticoids
2. d	9. b	16. anterior pituitary
3. a	10. c	17. neuroendocrine
4. c	11. anterior pituitary	18. interstitial
5. d	12. supraoptic nucleus	19. zona fasciculata, ACTH
6. d	13. leptin	20. enteric hormones
7. a	14. natriuretic peptides	

Building Your Medical Vocabulary

(Answers may vary; these are acceptable examples.)
1. secrete, separate—endocrine
2. quick, sharp—oxytocin
3. for, favoring—prolactin
4. nourish—gonadotroph
5. body—corpus luteum
6. shield—thyroid
7. sodium—natriuretic
8. heat—calorigenic
9. milk—lactotrope
10. fear, repulsion—hydrophobic

True or False

(These items are false for the reasons given; all others are true.)
2. The heart, brain, stomach, and kidneys secrete hormones but are not usually thought of as endocrine glands.
4. The pineal gland and thymus shrink with age.
5. The center of the adrenal gland is the adrenal medulla.
8. The pituitary stalk is not a duct.
9. There are two pairs of parathyroids and one pair of gonads.

Figure Legend Questions

18.1 Heart, liver, stomach, placenta
18.2 The posterior lobe, or neurohypophysis
18.6 To secrete calcitonin
18.9 To secrete digestive enzymes

Chapter 19

Testing Your Recall

1. b	8. c	15. serum
2. c	9. d	16. hemostasis
3. a	10. b	17. sickle-cell disease
4. b	11. hemopoiesis	18. polycythemia
5. e	12. hematocrit	19. vitamin B_{12}
6. d	13. macrophages	20. erythropoietin
7. d	14. leukopenia	

Building Your Medical Vocabulary

(Answers may vary; these are acceptable examples.)
1. blood—hematology
2. white—leukocyte
3. blood condition—anemia
4. precursor—erythroblast
5. red—erythrocyte
6. large—megakaryocyte
7. bone marrow—myeloid leukemia
8. blood clot—thrombosis
9. large—macrophage
10. formation, production—hemopoiesis

True or False

(These items are false for the reasons given; all others are true.)
3. Anemia is the cause, not the result, of low blood oxygen content.
4. Neutrophils are the most actively antibacterial WBCs.
6. Neutrophils are the most abundant WBCs.
9. RBCs live longer than most WBCs.
10. WBC count is elevated in leukemia.

Figure Legend Questions

19.1 Nuclei
19.3 The developing erythrocyte sinks in at this point when its nucleus shrivels and is ejected from the cell.
19.4 To the Fe^{2+} at the center of the heme group
19.8 It means "marrow" and refers to the fact that this process occurs in the red bone marrow.
19.10 Fibrin

Chapter 20

Testing Your Recall

1. d	8. a	15. gap junctions
2. b	9. b	16. valvular prolapse
3. a	10. b	17. vagus
4. d	11. systole, diastole	18. myocardial infarction
5. e	12. systemic	19. endocardium
6. e	13. coronary sulcus	20. sinus venosus
7. d	14. Purkinje fibers	

Building Your Medical Vocabulary

(Answers may vary; these are acceptable examples.)
1. heart—cardiology
2. upon, above—epicardium
3. half—semilunar
4. pertaining to—coronary
5. small—ventricle
6. arising from—myogenic
7. moon—semilunar
8. crown—coronary
9. belly—ventricle
10. entryway—atrioventricular

True or False

(These items are false for the reasons given; all others are true.)

1. About 20% of the blood returns to the right atrium by way of thebesian veins.
5. The heart does not require nervous stimulation to beat.
7. It is anastomoses of the arteries, not veins, that serve this purpose.
8. The primordial ventricle develops into the left ventricle only.
9. There are no valves at the entrances to the right atrium.

Figure Legend Questions

20.1 Both. The pulmonary circuit delivers blood to the lungs for gas exchange. The systemic circuit delivers blood for nourishment of the pulmonary tissues.
20.2 To the left
20.9 They prevent the AV valves from prolapsing into the atria when the ventricles contract.
20.13 The right atrium
20.14 The gap junctions

Chapter 21

Testing Your Recall

1. c	8. b	15. chemoreceptors
2. b	9. b	16. transcytosis
3. a	10. b	17. superior vena cava, inferior vena cava
4. e	11. fenestrated	
5. b	12. continuous capillaries	18. carotid sinuses
6. b	13. endothelium	19. cerebral arterial circle
7. d	14. common iliac	20. basilic, cephalic

Building Your Medical Vocabulary

(Answers may vary; these are acceptable examples.)

1. vessel—vasodilation
2. added to—adventitia
3. window—fenestrated
4. head—brachiocephalic
5. neck, throat—jugular
6. vessel—angiogenesis
7. belly, abdomen—celiac trunk
8. knee—genicular artery
9. of, belonging to—vasa vasorum
10. bladder—vesical artery

True or False

(These items are false for the reasons given; all others are true.)

2. They receive blood from the celiac trunk.
4. Blood sometimes passes through portal systems (two capillary beds) or anastomoses (bypassing capillaries).
5. It is formed by the union of the two brachiocephalic veins.
7. The tunica media is nourished mainly by capillaries of the vasa vasorum.
9. One or more arteries of the circle are missing in 80% of people.

Figure Legend Questions

21.1 Arteries are subjected to greater blood pressure and must expand and recoil in phase with the heartbeat.
21.6 Endocrine glands, kidneys, small intestine (answers may vary)
21.10 Veins have thinner walls and less elastic tissue, and therefore expand more easily to accommodate a larger volume of blood.
21.12 Arterial anastomoses: the cerebral arterial circle, circumflex humeral arteries, and deep palmar arch. Venous anastomoses: union of the radial veins near the elbow, union of the left and right gastric veins, the dorsal venous arch of the foot joining the great and small saphenous veins. (Many other examples of both can be cited; answers may vary.)
21.25 Like the gonadal arteries, these veins are much shorter when the gonads begin their development high in the abdominal cavity, near the kidneys. The veins grow as the fetal gonads migrate downward.
21.27 Anastomoses allow for continued blood flow through alternate routes when joint movements temporarily compress an artery and shut off its flow.

Chapter 22

Testing Your Recall

1. b	8. e	15. cisterna chyli
2. c	9. c	16. plasma
3. e	10. d	17. antigen-presenting cells
4. a	11. pathogen	18. red bone marrow
5. c	12. chyle	19. lymphatic nodules
6. e	13. collecting vessels	20. autoimmune
7. a	14. right lymphatic duct, thoracic duct	

Building Your Medical Vocabulary

(Answers may vary; these are acceptable examples.)

1. producing—pathogen
2. medical condition—elephantiasis
3. gland—lymphadenitis
4. cutting out—tonsillectomy
5. disease—lymphadenopathy
6. water—lymphatic
7. free—immunodeficiency
8. enlargement—splenomegaly
9. mass, tumor—lymphoma
10. juice—cisterna chyli

True or False

(These items are false for the reasons given; all others are true.)

1. B cells are involved in specific immunity only.
4. Helper T cells also play a role in humoral immunity.
7. Both B and T cells populate the lymph nodes.
8. Lymphatic nodules are temporary and have no capsules.
9. Tonsillectomy is now a much less common treatment than it used to be.

Figure Legend Questions

22.3 Gaps between the endothelial cells are much larger in lymphatic capillaries than they are in blood capillaries.
22.5 (1) It prevents the accumulation of excess tissue fluid (edema). (2) It enables immune cells in the lymph nodes to continually monitor the tissue fluids for foreign matter.
22.6 Cancer cells breaking free of a breast tumor enter the lymphatics and often lodge and seed the growth of secondary (metastatic) tumors in these nearby lymph nodes.
22.10 The reticular epithelial cells
22.14 Erythrocytes account for the color of red pulp; lymphocytes and macrophages account for the color of white pulp.

Chapter 23

Testing Your Recall

1. c	8. a	15. obstructive
2. c	9. c	16. dead space
3. a	10. c	17. thyroid
4. e	11. laryngopharynx	18. ventral respiratory group
5. b	12. bronchial tree	19. hilum
6. e	13. alar	20. alveolar macrophages
7. d	14. conchae	

Building Your Medical Vocabulary

(Answers may vary; these are acceptable examples.)

1. rough—trachea
2. small cavity—alveolus
3. expansion—atelectasis
4. smoke, carbon dioxide—hypercapnia
5. nose—nasal septum
6. to breathe—inspiration
7. inflammation—emphysema
8. lung—pulmonary

9. windpipe—bronchus
10. air, lung—pneumothorax

True or False

(These items are false for the reasons given; all others are true.)

1. The glottis is at the superior end of the larynx, not the inferior (tracheal) end.
3. The space between the parietal and visceral pleura contains only a thin film of pleural fluid.
4. Normal expiration is not produced by muscular contraction.
8. The aortic and carotid sinuses monitor blood pressure.
9. The respiratory system begins as a bud arising from the floor of the pharynx.

Figure Legend Questions

23.2   The line should be drawn between the larynx and trachea labels.
23.3   The septal cartilage
23.4   The epiglottic, corniculate, and arytenoid cartilages
23.8   To secrete mucus
23.10   By their lack of cartilage

Chapter 24

Testing Your Recall

1. a	8. a	14. hepatic portal system
2. b	9. a	15. vagus
3. b	10. c	16. parietal
4. d	11. enteric nervous	17. hepatic sinusoids
5. e	system	18. cardiac orifice
6. c	12. pyloric sphincter	19. cementum
7. a	13. parotid	20. lingual papillae

Building Your Medical Vocabulary

(Answers may vary; these are acceptable examples.)

1. stomach—gastric
2. intestine—myenteric
3. food—alimentary
4. behind—retroperitoneal
5. cheek—buccal
6. gatekeeper—pylorus
7. fold, crease—ruga
8. letter S—sigmoid colon
9. straight—rectum
10. liver—hepatocyte

True or False

(These items are false for the reasons given; all others are true.)

1. The pancreas is retroperitoneal, but the liver is not.
2. A tooth is composed mainly of dentin.
3. Bile is secreted into bile canaliculi, not into the sinusoids.
7. The greater omentum is not attached to the body wall.
9. The muscularis externa of the stomach has three layers.

Figure Legend Questions

24.6   The third molars
24.7   The enamel
24.11   The stomach wall has three layers of muscle; the esophagus has two.
24.16   The autonomic nervous system controls the internal anal sphincter, and the somatic nervous system controls the external anal sphincter. This can be deduced from the fact that the internal sphincter is smooth (involuntary) muscle and the external sphincter is skeletal (voluntary) muscle.
24.18   The hepatic portal vein and the right and left hepatic arteries

Chapter 25

Testing Your Recall

1. c	8. b	15. podocytes
2. d	9. d	16. proximal convoluted
3. a	10. a	tubule
4. c	11. urea	17. thin segment
5. e	12. ureteric bud	18. detrusor
6. b	13. trigone	19. minor calyx
7. d	14. macula densa	20. arcuate

Building Your Medical Vocabulary

(Answers may vary; these are acceptable examples.)

1. nitrogen—azotemia
2. kidney—nephron
3. little ball—glomerulus
4. stone—lithotripsy
5. foot—podocyte
6. urine—anuria
7. patch—macula densa
8. next to—juxtaglomerular
9. basin—pelvis
10. bladder—cystitis

True or False

(These items are false for the reasons given; all others are true.)

1. The ureters open into the floor of the bladder.
2. ADH is secreted by the posterior lobe of the pituitary gland.
4. A substantial amount of fluid passes through the tight junctions.
6. The glomerulus is not located within the capsular space.
9. The kidneys are normally in the upper abdominal cavity.

Figure Legend Questions

25.1   The kidney itself cannot be palpated, but its position can be inferred by palpating ribs 11 and 12 and relating that to this figure.
25.2   It would have to be moved into the dark peritoneal cavity toward the upper left, where the spleen and colon are shown, anterior to the parietal peritoneum.
25.4   This is an anterior view, because the renal artery and vein lie anterior to the ureter. Compare the posterior view in figure 25.3b.
25.9   The afferent arteriole is larger than the efferent. This gives the glomerulus a large inlet, a small outlet, and consequently a high blood pressure, which is important for the filtration process.
25.11   An enlarged prostate gland compresses the prostatic urethra, the bladder's exit.

Chapter 26

Testing Your Recall

1. a	8. c	15. endometrium
2. a	9. e	16. seminal vesicles
3. c	10. d	17. sustentacular
4. b	11. mesonephric	18. prostate gland
5. c	12. fructose	19. corpus luteum
6. a	13. pampiniform plexus	20. infundibulum, fimbriae
7. d	14. follicle	

Building Your Medical Vocabulary

(Answers may vary; these are acceptable examples.)

1. marriage, union—gamete
2. seed—gonorrhea

3. uterus—hysterectomy
4. sheath—vagina
5. white—tunica albuginea
6. egg—ovary
7. egg—oogenesis
8. to carry—cumulus oophorus
9. yellow—corpus luteum
10. uterus—endometrium

True or False

(These items are false for the reasons given; all others are true.)
1. The follicle does not leave the ovary.
2. The uterine tubes develop from paramesonephric ducts.
4. The follicle that ovulates is the mature (graafian) follicle.

6. The female reproductive system develops as a result of a low androgen level.
8. The pampiniform plexus helps to keep the testes cool.

Figure Legend Questions

26.6 After meiosis, the resulting daughter cells will be genetically different from the rest of the body and would be subject to attack by blood-borne antibodies or immune cells if not protected from them.

26.10 If it were confined within a tunica albuginea, engorgement of the corpus spongiosum would probably compress the urethra and block ejaculation.

26.21 The paraurethral glands

26.22 They compress the acinus and expel milk into the lactiferous ducts.

26.25 The word root *vagin* means "sheath," which aptly describes the tunica vaginalis as a sheath or pouch of peritoneum.

This glossary defines terms likely to be most useful to the reader of this particular book, especially terms that are reintroduced most often and cannot be defined again at every introduction. Terms are defined only in the sense that they are used in this book. Some have broader meanings, even within biology and medicine, that are beyond its scope. Pronunciation guides are provided for words whose pronunciations may not be obvious. These guides should be quite intuitive, but a key at the end of the glossary indicates how to pronounce letter sequences within the guides if help is needed. Figures are cited where they will help convey the meaning of a term. Figure references such as A.3 refer to figures in the Atlas of Regional and Surface Anatomy (p. 329).

A

abdominal cavity The body cavity between the diaphragm and pelvic brim. (fig. 1.14)

abdominopelvic cavity Collective name for the abdominal and pelvic cavities, which constitute a continuous space between the diaphragm and pelvic floor. (fig. 1.14)

abduction Movement of a body part away from the median plane, as in raising the right arm away from the body to point to the right. (fig. 9.8)

accessory organ A smaller organ associated with or embedded in another and performing a related function; for example, the hair, nails, and sweat glands are accessory organs of the skin.

acetylcholine (ACh) (ASS-eh-till-CO-leen) A neurotransmitter released by somatic motor nerve fibers, parasympathetic nerve fibers, and some other neurons, composed of choline and an acetyl group.

acetylcholinesterase (AChE) (ASS-eh-till-CO-lin-ESS-ter-ase) An enzyme found in synaptic clefts and on postsynaptic cells that breaks down acetylcholine and stops synaptic signal transmission.

acidophil A cell that stains with acidic dyes, such as a pituitary acidophil. (fig. 18.3)

acinar gland (ah-SEE-nur) A gland in which the secretory cells form a dilated sac or acinus. (fig. 3.30)

acinus (ASS-in-nus) A sac of secretory cells at the inner end of a gland duct. (fig. 3.29)

acromial region The apex of the shoulder.

actin A filamentous intracellular protein that provides cytoskeletal support and interacts with other proteins, especially myosin, to cause cellular movement; important in muscle contraction and membrane actions such as phagocytosis, ameboid movement, and cytokinesis. *See also* microfilament.

action The movement produced by the contraction of a muscle.

action potential A rapid voltage change in which a plasma membrane briefly reverses electrical polarity; has a self-propagating effect that produces a traveling wave of excitation in nerve and muscle cells.

acute Pertaining to a disorder with a sudden onset, severe effects, and brief duration. *Compare* chronic.

adaptation **1.** An evolutionary process leading to the establishment of species characteristics that favor survival and reproduction. **2.** Any characteristic of anatomy, physiology, or behavior that promotes survival and reproduction. **3.** A sensory process in which a receptor adjusts its sensitivity or response to the prevailing level of stimulation, as in *dark adaptation* of the eye.

adduction (ah-DUC-shun) Movement of a body part toward the median plane, such as bringing the feet together from a spread-legged position. (fig. 9.8)

adenohypophysis The anterior two-thirds of the pituitary gland, consisting of the anterior lobe and pars tuberalis; synthesizes and secretes gonadotropins, thyrotropin, adrenocorticotropin, growth hormone, and prolactin. (fig. 18.2)

adenosine triphosphate (ATP) A molecule composed of adenine, ribose, and three phosphate groups that functions as a universal energy-transfer molecule; briefly captures energy in its phosphate bonds and transfers it to other chemical reactions, yielding adenosine diphosphate and a free phosphate group upon hydrolysis.

adipocyte A fat cell.

adipose tissue A loose connective tissue composed predominantly of adipocytes; fat. (fig. 3.18)

adrenal gland (ah-DREE-nul) An endocrine gland on the superior pole of each kidney; consists of an outer adrenal cortex and inner adrenal medulla, with separate functions and embryonic origins. (fig. 18.8)

adult stem cell Any of several kinds of undifferentiated cells that populate the body's organs, where they multiply and differentiate to replace cells that are lost to damage or normal cellular turnover. Adult stem cells have more limited developmental potential than embryonic stem cells. *See also* embryonic stem cell.

adventitia (AD-ven-TISH-uh) Loose connective tissue forming the outermost sheath around organs such as a blood vessel or the esophagus.

afferent Carrying toward, as in *afferent neurons,* which carry signals toward the central nervous system, and *afferent arterioles,* which carry blood toward a tissue.

afferent neuron *See* sensory neuron.

aging Any changes in the body that occur with the passage of time, including growth, development, and senescence.

agonist *See* prime mover.

agranulocyte Either of the two leukocyte types (lymphocytes and monocytes) that lack specific cytoplasmic granules. (fig. 19.1)

alveolus (AL-vee-OH-lus) **1.** A microscopic air sac of the lung. **2.** A gland acinus. **3.** A pit or socket in a bone, such as a tooth socket. **4.** Any small anatomical space.

Alzheimer disease (AD) (ALTS-hy-mur) A degenerative disease of the senescent brain, typically beginning with memory lapses and

progressing to severe losses of mental and motor functions and ultimately death.

ameboid movement Ameba-like crawling of cells such as leukocytes by means of pseudopods.

amnion A transparent membrane that surrounds the developing fetus and contains the amniotic fluid; the "bag of waters" that breaks during labor. (fig. 4.12)

ampulla (AM-pyu-luh) A wide or saclike portion of a tubular organ such as a semicircular duct or uterine tube.

anastomosis (ah-NASS-tih-MO-sis) An anatomical convergence, the opposite of a branch; a point where two blood vessels merge and combine their bloodstreams or where two nerves or ducts converge. (fig. 21.12)

anatomical position A reference posture on which certain standardized anatomical terminology is based. A subject in anatomical position is standing with the feet flat on the floor, arms down to the sides, and the palms and eyes directed forward. (fig. 1.8)

anatomy **1.** Structure of the body. **2.** The study of structure. *See also* morphology.

anemia A deficiency of erythrocytes or hemoglobin.

aneurysm (AN-you-riz-um) A weak, bulging point in the wall of a heart chamber or blood vessel that presents a threat of hemorrhage. (fig. 21.3)

angiogenesis The growth of new blood vessels, both prenatally and postnatally.

angiography The process of visualizing blood vessels by injecting them with a radiopaque substance and photographing them with X-rays. (fig. 1.3)

antagonist **1.** A muscle that opposes the prime mover at a joint. **2.** Any agent, such as a hormone or drug, that opposes another.

antebrachium (AN-teh-BRAY-kee-um) The region from elbow to wrist; the forearm.

anterior **1.** Pertaining to the front (facial–abdominal aspect) of the human body; also called *ventral,* especially in embryos. **2.** Pertaining to the head end of nonhuman animals.

anterior root A branch of a spinal nerve that joins the spinal cord on its anterior side, composed of motor nerve fibers; also called *ventral root.* (fig. 14.10)

antibody A protein of the gamma globulin class that reacts with an antigen; found in the blood plasma, in other body fluids, and on the surfaces of certain leukocytes and their derivatives.

antigen (AN-tih-jen) Any large molecule capable of binding to an antibody and triggering an immune response.

antigen-presenting cell (APC) A cell that phagocytizes an antigen and displays fragments of it on its surface for recognition by other cells of the immune system; chiefly macrophages and B lymphocytes.

antrum A saccular or pouchlike space, such as at the inferior end of the stomach or in an ovarian follicle.

aorta A large artery that extends from the left ventricle to the lower abdominal cavity and gives rise to all other arteries of the systemic circulation. (fig. 21.16)

aortic arch **1.** In the embryo, any of six pairs of blood vessels that arise rostral to the heart and loop mainly through the pharyngeal arches; some of these later give rise to carotid and pulmonary arteries and the permanent aortic arch. (fig. 21.35) **2.** A segment of the adult aorta that arches over the heart like an inverted U; gives rise to the brachiocephalic trunk, left common carotid artery, and left subclavian artery; then continues posterior to the heart as the descending aorta. (fig. 21.16)

apex The summit or a pointed part of an organ or body region such as the heart, lung, or shoulder.

apical surface The uppermost surface of an epithelial cell, opposite from the basement membrane, usually exposed to the lumen of an organ. (fig. 2.5)

apocrine Pertaining to certain sweat glands with large lumens and relatively thick, aromatic secretions, and to similar glands such as the mammary gland; formerly thought to form secretions by pinching off bits of apical cytoplasm. (fig. 5.10)

aponeurosis A broad, flat tendon that attaches a muscle to a bone or to other soft tissues in such locations as the abdominal wall and deep to the scalp.

apoptosis The normal death of cells that have completed their function, usually in a process involving self-destruction of the cell's DNA, shrinkage of the cell, and its phagocytosis by a macrophage; also called *programmed cell death.* *Compare* necrosis.

appendicular (AP-en-DIC-you-lur) Pertaining to the limbs and their supporting skeletal girdles. (fig. 7.1)

arcuate (AR-cue-et) Making a sharp L- or U-shaped bend (arc), or forming an arch, as in *arcuate arteries* of the kidneys and uterus.

areolar tissue (AIR-ee-OH-lur) A fibrous connective tissue with loosely organized, widely spaced fibers and cells and an abundance of fluid-filled space; found under nearly every epithelium, among other places. (fig. 3.14)

arrector pili *See* piloerector.

arteriole A small artery that empties into a metarteriole or capillary.

artery Any blood vessel that conducts blood away from the heart, or in the case of coronary arteries, away from the aorta and into the heart wall.

articular cartilage A thin layer of hyaline cartilage covering the articular surface of a bone at a synovial joint, serving to reduce friction and ease joint movement. (fig. 9.4)

articulation A skeletal joint; any point at which two bones meet; may or may not be movable.

aspect A particular view of the body or one of its structures, or a surface that faces in a particular direction, such as the anterior aspect.

association area A region of the cerebral cortex that does not directly receive sensory input or control skeletal muscles, but serves to interpret sensory information, to plan motor responses, and for memory and cognition.

atherosclerosis A degenerative disease of the blood vessels characterized by the presence of plaques on the vessel wall composed of lipid, smooth muscle, and macrophages; can lead to arterial occlusion, loss of arterial elasticity, hypertension, heart attack, kidney failure, and stroke.

ATP *See* adenosine triphosphate.

atrioventricular (AV) node (AY-tree-oh-ven-TRIC-you-lur) A group of autorhythmic cells in the interatrial septum of the heart that relays excitation from the atria to the ventricles.

atrioventricular (AV) valves The bicuspid (right) and tricuspid (left) valves between the atria and ventricles of the heart.

atrium **1.** Either of the two superior chambers of the heart, which receive systemic and pulmonary blood. **2.** The central space of an alveolar sac into which individual pulmonary alveoli open.

atrophy Shrinkage of a tissue due to age, disuse, or disease.

auditory ossicles Three small middle-ear bones that transfer vibrations from the tympanic membrane to the inner ear; the malleus, incus, and stapes.

Auerbach plexus *See* myenteric plexus.

auricle **1.** The portion of the ear external to the cranium; the pinna. **2.** An ear-shaped structure, such as the auricles of the heart.

autoantibody An antibody that fails to distinguish the body's own molecules from foreign molecules and thus attacks host tissues, causing autoimmune diseases.

autoimmune disease Any disease in which antibodies fail to distinguish between foreign and self-antigens and attack the body's own tissues; for example, systemic lupus erythematosus and rheumatic fever.

autolysis (aw-TAHL-ih-sis) Digestion of cells by their own internal enzymes.

autonomic nervous system (ANS) A motor division of the nervous system that innervates glands, smooth muscle, and cardiac muscle; consists of sympathetic and parasympathetic divisions and functions largely without voluntary control. *Compare* somatic nervous system.

autosome Any chromosome except the sex chromosomes. Genes on the autosomes are inherited without regard to the sex of the individual.

axial Pertaining to the head, neck, and trunk; the part of the body excluding the appendicular portion. (fig. 7.1)

axillary (ACK-sih-LERR-ee) Pertaining to the armpit.

axon A process of a neuron that conducts action potentials away from the soma; also called a *nerve fiber.* There is only one axon to a neuron, and it is usually much longer and much less branched than the dendrites. (fig. 13.4)

axoneme The core of a cilium or flagellum, usually composed of a "9 + 2" array of microtubules that provide support and motility. (fig. 2.13)

B

baroreceptor (BARE-oh-re-SEP-tur) Pressure sensor located in the heart, aortic arch, and carotid sinuses that triggers autonomic reflexes in response to fluctuations in blood pressure.

basal lamina A thin layer of collagen, proteoglycan, and glycoprotein that binds epithelial and other cells to adjacent connective tissue; forms part of the basement membrane of an epithelium, and surrounds some nonepithelial cells such as muscle fibers and Schwann cells. (fig. 13.8)

basal nuclei Masses of deep cerebral gray matter that play a role in the coordination of posture and movement and the performance of learned motor skills; also called *basal ganglia.* (fig. 15.15)

base The broadest part of a tapered organ such as the uterus, or the inferior aspect of an organ such as the brain.

basement membrane A thin layer of matter that underlies the deepest cells of an epithelium and binds them to the underlying connective tissue; consists of the basal lamina of the epithelial cells

and fine reticular fibers of the connective tissue. (fig. 3.31)

basophil (BASE-oh-fill) **1.** A cell that stains with basic dyes, such as a pituitary basophil. (fig. 18.3) **2.** A leukocyte with coarse cytoplasmic granules that produces heparin, histamine, and other chemicals involved in inflammation. (fig. 19.1)

belly The thick part of a skeletal muscle between its origin and insertion. (fig. 10.4)

bipedalism The habit of walking on two legs; a defining characteristic of the family Hominidae that underlies many skeletal and other characteristics of humans.

blastocyst A hollow spheroidal stage of the conceptus that implants in the uterine wall; consists of an inner cell mass, or embryoblast, enclosed in a saclike outer cell mass, or trophoblast. (fig. 4.3)

blood A liquid connective tissue composed of plasma, erythrocytes, platelets, and five kinds of leukocytes.

blood–brain barrier A barrier between the bloodstream and nervous tissue of the brain that is impermeable to many blood solutes and thus prevents them from affecting the brain tissue; formed by the tight junctions between capillary endothelial cells, the basement membrane of the endothelium, and the perivascular feet of astrocytes.

B lymphocyte A lymphocyte that functions as an antigen-presenting cell and, in humoral immunity, differentiates into an antibody-producing plasma cell; also called a *B cell.*

body **1.** The entire organism. **2.** Part of a cell, such as a neuron, containing the nucleus and most other organelles. **3.** The largest or principal part of an organ such as the stomach or uterus; also called the *corpus.*

bolus A mass of matter, especially food or feces traveling through the digestive tract.

bone **1.** A calcified connective tissue; also called *osseous tissue.* **2.** An organ of the skeleton composed of osseous tissue, fibrous connective tissue, marrow, cartilage, and other tissues.

Bowman capsule *See* glomerular capsule.

brachial (BRAY-kee-ul) Pertaining to the brachium.

brachium The region between the shoulder and elbow; the arm proper.

brainstem The stalklike lower portion of the brain, composed of all of the brain except the cerebrum and cerebellum. (Many authorities exclude the diencephalon and regard only the medulla oblongata, pons, and midbrain as the brainstem.) (fig. 15.6)

bronchiole (BRON-kee-ole) A pulmonary air passage that is usually 1 mm or less in diameter and lacks cartilage, but has relatively abundant smooth muscle, elastic tissue, and a simple cuboidal, usually ciliated epithelium.

bronchus (BRON-kus) A relatively large pulmonary air passage with supportive cartilage in the wall; any passage beginning with the primary bronchus at the fork in the trachea and ending with tertiary bronchi, from which air continues into the bronchioles.

brush border A fringe of microvilli on the apical surface of an epithelial cell, serving to enhance surface area and promote absorption. (fig. 3.6)

buccal Pertaining to the cheek.

bulb A dilated terminal part of an organ such as the penis or hair, or the olfactory bulb at the beginning of the olfactory tract.

bursa A sac filled with synovial fluid at a synovial joint, serving to facilitate muscle or joint action. (fig. 9.5)

C

calcaneal tendon (cal-CAY-nee-ul) A thick tendon at the heel that attaches the triceps surae muscles to the calcaneus; also called the *Achilles tendon*. (fig. 12.18)

calcification The hardening of a tissue due to the deposition of calcium salts; also called *mineralization*.

calculus A calcified mass, especially a renal calculus (kidney stone) or biliary calculus (gallstone).

calvaria (cal-VERR-ee-uh) The rounded bony dome that forms the roof of the cranium; the general portion of the skull superior to the eyes and ears; skullcap.

calyx (CAY-lix) (plural, *calices*) A cuplike structure, as in the kidneys. (fig. 25.3)

canal A tubular passage or tunnel such as the auditory, semicircular, or condylar canal.

canaliculus (CAN-uh-LIC-you-lus) A microscopic canal, as in osseous tissue. (fig. 6.5)

cancellous bone *See* spongy bone.

capillary (CAP-ih-LERR-ee) The narrowest type of vessel in the cardiovascular and lymphatic systems; engages in fluid exchanges with surrounding tissues.

capillary bed A network of blood capillaries that arise from a single metarteriole and converge on a thoroughfare channel or venule. (fig. 21.9)

capsule The fibrous covering of a structure such as the spleen or a synovial joint.

carbohydrate A hydrophilic organic compound composed of carbon and a 2:1 ratio of hydrogen to oxygen; includes sugars, starches, glycogen, and cellulose.

cardiac center A nucleus in the medulla oblongata that regulates autonomic reflexes for controlling the rate and strength of the heartbeat.

cardiac muscle Striated involuntary muscle of the heart. (fig. 20.14)

cardiocyte A cardiac muscle cell.

cardiopulmonary system Collective name for the heart and lungs, emphasizing their close spatial and physiological relationship.

cardiovascular system An organ system consisting of the heart and blood vessels, serving for the transport of blood. *Compare* circulatory system.

carotid body (ca-ROT-id) A small cellular mass near the branch in the common carotid artery, containing sensory cells that detect changes in the pH and the carbon dioxide and oxygen content of the blood. (fig. 21.4)

carotid sinus A dilation at the base of the internal carotid artery; contains baroreceptors, which monitor changes in blood pressure. (fig. 21.4)

carpal Pertaining to the wrist (carpus).

carrier 1. A protein that transports solutes through a cell membrane; also called a *transport protein*. 2. A person who does not exhibit a particular hereditary disorder, but who has the gene for it and may pass it to the next generation.

cartilage A connective tissue with a rubbery matrix, cells (chondrocytes) contained in lacunae, and no blood vessels; covers the joint surfaces of many bones and supports organs such as the ear and larynx.

caudal (CAW-dul) 1. Pertaining to a tail or narrow tail-like part of an organ. 2. Pertaining to the inferior part of the trunk of the body, where the tail of other animals arises. *Compare* cranial. 3. Relatively distant from the forehead, especially in reference to structures of the brain and spinal cord; for example, the medulla oblongata is caudal to the pons. *Compare* rostral.

celiac Pertaining to the abdomen.

celiac trunk An arterial trunk that arises from the abdominal aorta near the diaphragm, and quickly branches to give off arteries that supply the stomach, spleen, pancreas, liver, and other viscera of the upper abdominal cavity. (fig. 21.23)

cell The smallest subdivision of a tissue considered to be alive; consists of a plasma membrane enclosing cytoplasm and, in most cases, a nucleus.

cell body The main part of a cell, especially a neuron, where the nucleus is located; also called the *soma*.

cell junction A complex of proteins that joins adjacent cells to each other or binds a cell to extracellular material; also called an *intercellular junction* when it links two cells. Includes desmosomes, hemidesmosomes, tight junctions, gap junctions, and other types. (fig. 2.14)

central Located relatively close to the median axis of the body, as in *central nervous system;* opposite of *peripheral*.

central canal 1. A canal that passes through the core of an osteon in bone, and contains blood vessels and nerves; also called a *haversian canal* or *osteonic canal*. 2. A canal that passes through the center of the spinal cord, containing cerebrospinal fluid.

central nervous system (CNS) The brain and spinal cord. *Compare* peripheral nervous system.

central pattern generator A nucleus of neurons in the CNS that generates a repetitive motor output, producing rhythmic muscle contractions for such purposes as walking and breathing.

centriole (SEN-tree-ole) An organelle composed of a short cylinder of nine triplets of microtubules, usually paired with another centriole perpendicular to it; origin of the mitotic spindle; identical to the basal body of a cilium or flagellum. (fig. 2.17)

cephalic (seh-FAL-ic) Pertaining to the head.

cerebellum (SERR-eh-BEL-um) A large portion of the brain posterior to the brainstem and inferior to the cerebrum, responsible for equilibrium, motor coordination, some timekeeping functions, and learning of motor skills. (fig. 15.9)

cerebrospinal fluid (CSF) (SERR-eh-bro-SPY-nul, seh-REE-bro-SPY-nul) A liquid that fills the ventricles of the brain, the central canal of the spinal cord, and the space between the CNS and dura mater.

cerebrum (SERR-eh-brum, seh-REE-brum) The largest and most superior part of the brain, divided into two convoluted cerebral hemispheres separated by a deep longitudinal fissure.

cervical (SUR-vih-cul) Pertaining to the neck or any cervix.

cervix (SUR-vix) 1. The neck. 2. A narrow or necklike part of an organ such as the uterus and gallbladder. (fig. 26.11)

chemoreceptor An organ or cell specialized to detect chemicals, as in the carotid bodies and taste buds.

chief cell The majority type of cell in an organ or tissue such as the parathyroid glands or gastric glands.

choana (co-AN-ah) *See* posterior nasal aperture.

chondrocyte (CON-dro-site) A cartilage cell; a former chondroblast that has become enclosed in a lacuna in the cartilage matrix. (fig. 6.12)

chorion (CO-ree-on) A fetal membrane external to the amnion; forms part of the placenta and has diverse functions including fetal nutrition, waste removal, and hormone secretion. (fig. 4.9)

chromatid (CRO-muh-tid) One of two genetically identical rodlike bodies of a metaphase chromosome, joined to its sister chromatid at the centromere. (fig. 2.21)

chromatin (CRO-muh-tin) Filamentous material in the interphase nucleus, composed of DNA and protein; all of the chromosomes collectively.

chromosome A strand of DNA and protein carrying the genetic material of a cell's nucleus, having a fine filamentous structure during interphase and a condensed rodlike structure during mitosis and meiosis. Normally there are 46 chromosomes in the nucleus of each cell except germ cells. (fig. 2.21)

chronic Pertaining to a disorder with a gradual onset, slow progression, and long duration. *Compare* acute.

chyme (kime) A slurry of partially digested food in the stomach and small intestine.

cilium (SIL-ee-um) A hairlike process, with an axoneme, projecting from the surface of many or most cells; usually immobile, solitary, and serving a sensory or unknown role; found in large numbers on the apical surfaces of some epithelial cells (as in the respiratory tract and uterine tube), where they are motile and serve to propel matter across the surface of the epithelium. (fig. 2.13)

circulatory system An organ system consisting of the heart, blood vessels, and blood. *Compare* cardiovascular system.

circumduction A joint movement in which one end of an appendage remains relatively stationary and the other end moves in a circle. (fig. 9.11)

cisterna (sis-TUR-nuh) A fluid-filled space or sac, such as the cisterna chyli of the lymphatic system and the cisternae of the endoplasmic reticulum and Golgi complex. (fig. 2.17)

coelom A body cavity bounded on all sides by mesoderm and lined with peritoneum. The embryonic coelom becomes the thoracic and abdominopelvic cavities.

collagen (COLL-uh-jen) The most abundant protein in the body, forming the fibers of many connective tissues in places such as the dermis, tendons, and bones.

colony-forming unit (CFU) A bone marrow cell that differentiates from a pluripotent stem cell and gives rise to precursor cells, which, in turn, produce a specific class of formed elements. (fig. 19.8)

commissure (COM-ih-shur) 1. A bundle of nerve fibers that crosses from one side of the brain or spinal cord to the other. 2. A corner or angle at which the eyelids, lips, or genital labia meet; in the eye, also called the *canthus*. (fig. 17.19)

compact bone A form of osseous tissue found on bone surfaces and composed predominantly of osteons, with the tissue completely filled with mineralized matrix (other than lacunae and central canals) and leaving no room for bone marrow; also called *dense bone*. (fig. 6.5) *Compare* spongy bone.

computed tomography (CT) A method of medical imaging that uses X-rays and a computer to create an image of a thin section of the body; the image is called a *CT scan*. (fig. 1.3)

conception The fertilization of an egg, producing a zygote; the beginning of pregnancy.

conceptus All products of conception, ranging from a fertilized egg to the full-term fetus with its fetal membranes, placenta, and umbilical cord. *Compare* embryo; fetus; preembryo.

condyle (CON-dile) An articular surface on a bone, usually in the form of a knob (as on the mandible), but relatively flat on the proximal end of the tibia. (fig. 7.15)

congenital Present at birth; for example, an anatomical defect, a syphilis infection, or a hereditary disease.

congenital anomaly The abnormal structure or position of an organ at birth, resulting from a defect in prenatal development; a birth defect.

connective tissue A tissue usually composed of more extracellular than cellular volume and usually with a substantial amount of extracellular fiber; forms supportive frameworks and capsules for organs, binds structures together, holds them in place, stores energy (as in adipose tissue), or transports materials (as in blood).

contralateral On opposite sides of the body, as in reflex arcs where the stimulus comes from one side of the body and a response is given by muscles on the other side. *Compare* ipsilateral.

convergent Coming together, as in a *convergent muscle* and a converging neuronal circuit.

cornified Having a heavy surface deposit of keratin, as in the stratum corneum of the epidermis.

corona A halo- or crownlike structure, such as the corona radiata or coronal suture of the skull.

coronal plane *See* frontal plane.

corona radiata 1. An array of nerve tracts in the brain that arise mainly from the thalamus and fan out to different regions of the cerebral cortex. 2. The first layer of cuboidal cells immediately external to the zona pellucida around an egg cell.

coronary 1. Crownlike; encircling. 2. Pertaining to the heart.

coronary artery Either of two branching arteries that arise from the aorta near the heart and supply blood to the heart wall.

coronary circulation A system of blood vessels that serve the wall of the heart. (fig. 20.11)

corpus 1. A body of tissue, such as the corpus cavernosum of the penis. 2. The principal part (body) of an organ such as the uterus or stomach, as opposed to smaller regions of an organ such as its head, tail, fundus, or cervix.

corpus callosum (COR-pus ca-LO-sum) A prominent C-shaped band of nerve tracts that connect the right and left cerebral hemispheres to each other, seen superior to the third ventricle in a median section of the brain. (fig. 15.2)

corpus luteum A yellowish cellular mass that forms in the ovary from a follicle that has ovulated; secretes progesterone, hormonally regulates the second half of the menstrual cycle, and is essential to sustaining the first 7 weeks of pregnancy.

cortex (plural, *cortices*) The outer layer of some organs such as the adrenal gland, cerebrum, lymph node, and ovary; usually covers or encloses tissue called the medulla.

corticospinal tract A bundle of nerve fibers that descend through the brainstem and spinal cord and carry motor signals from the cerebral cortex to the neurons that innervate the skeletal muscles of the limbs; important in the fine control of limb movements. (fig. 14.5)

costal (COSS-tul) Pertaining to the ribs.

costal cartilage A bladelike plate of hyaline cartilage that attaches the distal end of a rib to the sternum; collectively the costal cartilages constitute much of the anterior part of the thoracic cage.

coxal Pertaining to the hip.

cranial 1. Pertaining to the cranium. 2. In a position relatively close to the head or a direction toward the head. *Compare* caudal.

cranial nerve Any of 12 pairs of nerves connected to the base of the brain and passing through foramina of the cranium.

cranium That portion of the skull that encloses the cranial cavity and protects the brain; also called the *braincase*. Comprises the frontal, parietal, temporal, occipital, sphenoid, and ethmoid bones.

crest A narrow ridge, such as the neural crest or the crest of the ilium.

cricoid cartilage The most inferior cartilage of the larynx, connecting the larynx to the trachea.

crista A crestlike structure, such as the crista galli of the ethmoid bone, crista ampullaris of the inner ear, or the crista of a mitochondrion.

cross section (c.s.) A cut perpendicular to the long axis of the body or of an organ. (fig. 3.2)

crural (CROO-rul) Pertaining to the leg proper or to the crus of a organ. *See also* crus.

crus (cruss) (plural, *crura*) 1. The region from the knee to the ankle; the leg proper. 2. A leglike extension of an organ such as the penis and clitoris. (fig. 26.9) *See also* crural.

CT scan An image of the body made by computed tomography. (fig. 1.3)

cubital region The anterior region at the bend of the elbow.

cuboidal (cue-BOY-dul) A shape that is roughly like a cube or in which the height and width are about equal.

cuneiform (cue-NEE-ih-form) Wedge-shaped, as in *cuneiform cartilages* and *cuneiform bone*.

cusp 1. One of the flaps of a valve of the heart, veins, and lymphatic vessels. 2. A conical projection on the occlusal surface of a premolar or molar tooth.

cutaneous Pertaining to the skin; integumentary.

cuticle 1. The outermost layer of a hair, consisting of a single layer of overlapping squamous cells. 2. A layer of dead epidermal cells that cover the proximal end of a nail; also called the *eponychium*.

cytokinesis (SY-toe-kih-NEE-sis) Division of the cytoplasm of a cell into two cells following nuclear division.

cytology The study of cell structure and function.

cytoplasm The contents of a cell between the plasma membrane and the nuclear envelope, consisting of cytosol, organelles, inclusions, and the cytoskeleton.

cytoskeleton A system of protein microfilaments, intermediate filaments, and microtubules in a cell, serving for physical support, cellular movement, and the routing of molecules and organelles to their destinations within the cell. (fig. 2.15)

cytosol A clear, featureless, gelatinous colloid in which the organelles and other internal structures of a cell are embedded.

cytotoxic T cell A T lymphocyte that directly attacks and destroys infected body cells, cancerous cells, and the cells of transplanted tissues.

D

daughter cells Cells that arise from a parent cell by mitosis or meiosis.

decussation (DEE-cuh-SAY-shun) The crossing of nerve fibers from the right side of the central nervous system to the left, or vice versa, especially in the spinal cord, medulla oblongata, and optic chiasm. *Compare* hemidecussation.

deep Relatively far from the body surface; opposite of *superficial*. For example, the bones are deep to the skeletal muscles.

dendrite Process of a neuron that receives information from other cells or from environmental stimuli and conducts signals to the soma. Dendrites are usually shorter, more branched, and more numerous than the axon and are incapable of producing action potentials. (fig. 13.4)

dendritic cell An antigen-presenting cell of the epidermis, vaginal mucosa, and some other epithelia.

denervation atrophy The shrinkage of skeletal muscle that occurs when the motor neuron dies or is severed from the muscle.

dense bone *See* compact bone.

dense connective tissue A connective tissue with a high density of fiber, relatively little ground substance, and scanty cells; seen in tendons and the dermis, for example. Classified as *regular* if the extracellular fibers are more or less parallel and *irregular* if the fibers travel in highly varied directions. (figs. 3.16–3.17)

depression 1. A sunken place on the surface of a bone. 2. A joint movement that lowers a body part, as in dropping the shoulders or opening the mouth. (fig. 9.9)

dermal papilla 1. A bump or ridge of dermis that extends upward to interdigitate with the epidermis, creating a wavy boundary that resists stress and slippage of the epidermis. (fig. 5.3) 2. A projection of the dermis into the bulb of a hair, supplying blood to the hair. (fig. 5.6)

dermatome 1. In the embryo, a group of mesodermal cells that arises from a somite and gives rise to the dermis on one side of one segment of the body. *Compare* myotome; sclerotome. 2. In the adult, a region of skin on the neck, trunk, or limbs that is innervated by one spinal nerve. (fig. 14.18)

dermis The deeper of the two layers of the skin, underlying the epidermis and composed of fibrous connective tissue.

desmosome (DEZ-mo-some) A patchlike intercellular junction that mechanically links two cells together. (fig. 2.14)

desquamation *See* exfoliation.

diaphragm A muscular partition that separates the thoracic cavity from the abdominal cavity and plays a major role in respiration.

diaphysis (dy-AFF-ih-sis) The shaft of a long bone. (fig. 6.2)

diarthrosis *See* synovial joint.

diencephalon (DY-en-SEFF-uh-lon) A portion of the brain between the midbrain and corpus callosum; composed of the thalamus, epithalamus, and hypothalamus. (fig. 15.11)

differentiation Development of a relatively unspecialized cell or tissue into one with a more specific structure and function.

digestive system The organ system specialized for the intake and chemical breakdown of food, absorption of nutrients, and discharge of the indigestible residue.

digit A finger or toe.

digital rays The first ridgelike traces of fingers or toes to appear in the embryonic hand plate and foot plate.

dilation (dy-LAY-shun) Widening of an organ or passageway such as a blood vessel or the pupil of the eye.

diploid *(2n)* Pertaining to a cell or organism with chromosomes in homologous pairs. All nucleated cells of the human body are diploid except for germ cells beyond the meiosis I stage of cell division.

distal Relatively distant from a point of origin or attachment; for example, the wrist is distal to the elbow. *Compare* proximal.

dizygotic (DZ) twins Two individuals who developed simultaneously in one uterus but originated from separate fertilized eggs and therefore are not genetically identical.

dorsal *See* posterior.

dorsal root *See* posterior root.

dorsal root ganglion *See* posterior root ganglion.

dorsiflexion (DOR-sih-FLEC-shun) A movement of the ankle that reduces the joint angle and raises the toes. (fig. 9.17)

dorsum The dorsal surface of a body region, especially the hand or foot surface bearing the nails.

Down syndrome *See* trisomy-21.

duct An epithelium-lined, tubular passageway, such as a semicircular duct or a gland duct.

duodenum (DEW-oh-DEE-num, dew-ODD-eh-num) The first portion of the small intestine extending for about 25 cm from the pyloric valve of the stomach to a sharp bend called the duodenojejunal flexure; receives chyme from the stomach and secretions from the liver and pancreas.

dural sheath An extension of the dura mater into the vertebral canal, loosely enclosing the spinal cord.

dura mater The thickest and most superficial of the three meninges around the brain and spinal cord.

dynein (DINE-een) A motor protein involved in the beating of cilia and flagella and in the movement of molecules and organelles within cells.

dyspnea Labored breathing.

E

ectoderm The outermost of the three primary germ layers of an embryo; gives rise to the nervous system and epidermis.

ectopic (ec-TOP-ic) In an abnormal location; for example, ectopic pregnancy and ectopic pacemakers of the heart.

edema Accumulation of excess tissue fluid, resulting in swelling of a tissue.

effector A molecule, cell, or organ that carries out a response to a stimulus.

efferent (EFF-ur-ent) Carrying away or out, such as a blood vessel that carries blood away from a tissue or a nerve fiber that conducts signals away from the central nervous system.

efferent neuron *See* motor neuron.

elastic cartilage A form of cartilage with an abundance of elastic fibers in its matrix, lending flexibility and resilience to the cartilage; found in the epiglottis and ear pinna. (fig. 3.20)

elastic fiber A connective tissue fiber, composed of the protein elastin, that stretches under tension and returns to its original length when released; responsible for the resilience of organs such as the skin and lungs.

elasticity The tendency of a stretched structure to return to its original dimensions when tension is released.

elastin A fibrous protein with the ability to stretch and recoil; found in the skin, pulmonary airway, arteries, and elastic cartilage, among other locations.

elevation A joint movement that raises a body part, as in hunching the shoulders or closing the mouth. (fig. 9.9)

embryo In humans, a developing individual from the time the ectoderm, mesoderm, and endoderm have all formed at about 16 days, through the end of 8 weeks when all organ systems are represented; preceded by the *preembryo* and followed by the *fetus*. In other animals, any unborn stage of development beginning with the two-celled stage. *Compare* conceptus; fetus; preembryo.

embryogenesis A process of prenatal development that occurs during implantation of the blastocyst and gives rise to the three primary germ layers; embryogenesis ends with the existence of an embryo.

embryology The scientific study of prenatal development, from fertilization to birth.

embryonic disc A flat plate of cells in early embryonic development, composed of initially two and then three cell layers.

embryonic stage The stage of prenatal development from day 16 through the end of week 8. *See also* embryo; fetal stage.

embryonic stem cell An undifferentiated cell from a preembryo of up to 150 cells, capable of developing into any type of embryonic or adult cell.

encapsulated nerve ending Any sensory nerve ending that is surrounded by or associated with specialized connective tissues, which enhance its sensitivity or its mode of responding to stimulation.

endocardium A tissue layer that lines the inside of the heart, composed of a simple squamous epithelium overlying a thin layer of areolar tissue.

endochondral ossification A process of bone development in which the bone is preceded by a model of hyaline cartilage in roughly the shape of the bone to come, and the cartilage is then replaced by osseous tissue. *Compare* intramembranous ossification. (fig. 6.10)

endocrine gland (EN-doe-crin) A ductless gland that secretes hormones into the bloodstream; for example, the thyroid and adrenal glands. *Compare* exocrine gland.

endocrine system A system of internal chemical communication composed of all endocrine glands and the hormone-secreting cells found in other tissues and organs.

endocytosis Any process of vesicular transport of materials from the extracellular material into a cell; includes pinocytosis, receptor-mediated endocytosis, and phagocytosis. (fig. 2.11)

endoderm The innermost of the three primary germ layers of an embryo; gives rise to the mucosae of the digestive and respiratory tracts and to their associated glands.

endogenous (en-DODJ-eh-nus) Originating internally, such as the endogenous cholesterol synthesized in the body in contrast to the exogenous cholesterol coming from the diet. *Compare* exogenous.

endometrium (EN-doe-MEE-tree-um) The mucosa of the uterus; the site of implantation and source of menstrual discharge.

endoplasmic reticulum (ER) (EN-doe-PLAZ-mic reh-TIC-you-lum) An extensive system of interconnected cytoplasmic tubules or channels; classified as *rough ER* or *smooth ER* depending on the presence or absence of ribosomes on its membrane. (fig. 2.17)

endothelium (EN-doe-THEEL-ee-um) A simple squamous epithelium that lines the lumens of the blood vessels, heart, and lymphatic vessels.

enteric (en-TERR-ic) Pertaining to the small intestine, as in *enteric hormones*.

eosinophil (EE-oh-SIN-oh-fill) A leukocyte with a large, often bilobed nucleus and coarse cytoplasmic granules that stain with eosin; phagocytizes antigen–antibody complexes, allergens, and inflammatory chemicals and secretes enzymes that combat parasitic infections. (fig. 19.1)

epiblast The layer of cells in the early embryonic disc facing the amniotic cavity. These cells migrate during gastrulation to replace the hypoblast with endoderm, then to form the mesoderm, after which the remaining surface epiblast cells are called the ectoderm.

epicardium The outermost layer of the heart wall, composed of a simple squamous epithelium overlying a thin layer of areolar tissue, and in many areas, a much thicker layer of adipose tissue; also called the *visceral pericardium*.

epicondyle A bony projection or ridge superior to a condyle, for example at the distal ends of the humerus and femur. (fig. 8.10)

epidermis A stratified squamous epithelium that constitutes the superficial layer of the skin, overlying the dermis. (fig. 5.1)

epigastric Pertaining to a medial region of the abdomen superior to the umbilical region, bordered inferiorly by the subcostal line and laterally by the midclavicular lines. (fig. 1.13)

epiglottis A flap of tissue in the pharynx that covers the glottis during swallowing, deflecting swallowed matter away from the airway and into the esophagus.

epiphyseal plate (EP-ih-FIZZ-ee-ul) A plate of hyaline cartilage between the epiphysis and diaphysis of a long bone in a child or adolescent, serving as a growth zone for bone elongation. (fig. 6.10)

epiphysis (eh-PIFF-ih-sis) **1.** The head of a long bone. (fig. 6.2) **2.** The pineal gland (epiphysis cerebri).

epithelium A type of tissue consisting of one or more layers of closely adhering cells with little intercellular material and no blood vessels; forms the coverings and linings of many organs and the parenchyma of the glands.

equilibrium **1.** The sense of balance. **2.** A state in which opposing processes occur at comparable rates and balance each other so that there is little or no net change in the system, such as a chemical equilibrium.

erectile tissue A tissue that functions by swelling with blood, as in the penis, clitoris, and inferior concha of the nasal cavity.

erythrocyte (eh-RITH-ro-site) A red blood cell.

erythropoiesis (eh-RITH-ro-poy-EE-sis) The production of erythrocytes.

eversion A movement of the foot that turns the sole laterally. (fig. 9.17)

evolution A change in the genetic composition of a population over a period of time; the mechanism that produces adaptations in human form and function. *See also* adaptation.

excitability The ability of a cell to respond to a stimulus, especially the ability of nerve and muscle cells to produce membrane voltage changes in response to stimuli; irritability.

excretion The process of eliminating metabolic waste products from a cell or from the body. *Compare* secretion.

excursion A side-to-side movement of the mandible, as in chewing. (fig. 9.15)

exfoliation The shedding of squamous cells from the surface of a stratified squamous epithelium. Also called *desquamation*. Sampling and examination of these cells, such as a Pap smear, is called *exfoliate cytology*. (fig. 3.12)

exocrine gland (EC-so-crin) A gland that secretes its products into another organ or onto the body surface, usually by way of a duct; for example, salivary and gastric glands. *Compare* endocrine gland.

exocytosis A mode of vesicular transport in which a secretory vesicle of a cell fuses with the plasma membrane and releases its contents from the cell; a mode of glandular secretion and discharge of cellular wastes. (fig. 2.11)

exogenous (ec-SODJ-eh-nus) Originating externally, such as exogenous (dietary) cholesterol; extrinsic. *Compare* endogenous.

expiration **1.** Exhaling. **2.** Dying.

extension Movement of a joint that increases the angle between articulating bones (straightens the joint). *Compare* flexion. (fig. 9.7)

external acoustic meatus   A canal in the temporal bone that conveys sound waves to the eardrum; also called the *external auditory meatus*.

exteroceptor   A sensory receptor that responds to stimuli originating outside the body, such as the eye or ear. *Compare* interoceptor.

extracellular fluid (ECF)   Any body fluid that is not contained in the cells; for example, blood, lymph, and tissue fluid.

extrinsic   (ec-STRIN-sic) **1.** Originating externally, such as extrinsic blood-clotting factors; exogenous. **2.** Not fully contained within an organ but acting on it, such as the *extrinsic muscles* of the hand and eye. *Compare* intrinsic.

F

facet   A smooth articular surface on a bone; may be flat, slightly concave, or slightly convex; for example, the articular facets of the vertebrae.

facilitated diffusion   A process of solute transport through a cellular membrane, down its concentration gradient, with the aid of a carrier protein; the carrier does not consume ATP.

fallopian tube   *See* uterine tube.

fascia   (FASH-ee-uh) A layer of connective tissue between the muscles or between the muscles and the skin. (fig. 10.2)

fascicle   (FASS-ih-cul) A bundle of muscle or nerve fibers ensheathed in connective tissue; multiple fascicles bound together constitute a muscle or nerve as a whole. Also called a *fasciculus*. (fig. 10.2)

fat   **1.** A triglyceride molecule. **2.** Adipose tissue.

female   In humans, any individual with no Y chromosome; normally, one possessing two X chromosomes in each somatic cell, and having reproductive organs that serve to produce eggs, receive sperm, provide sites of fertilization and prenatal development, expel the full-term fetus, and nourish the infant.

femoral   Pertaining to the femur or the thigh.

femoral region   The region between the hip and knee; the thigh.

femoral triangle   A triangular region of the groin bounded by the sartorius muscle, adductor longus muscle, and inguinal ligament, and through which the femoral artery, femoral vein, and femoral nerve pass close to the body surface. (fig. 21.33)

fenestrated   (FEN-eh-stray-ted) Perforated with holes or slits, as in certain blood capillaries and the elastic sheets of large arteries. (fig. 21.6)

fetal stage   The period of prenatal development from the beginning of week 9 until birth; a period in which all organ systems are represented at the outset, and grow and differentiate until capable of supporting life outside the uterus. *See also* embryonic stage; fetus.

fetus   In human development, an individual from the beginning of the ninth week when all of the organ systems are present, through the time of birth. *See also* conceptus; embryo.

fiber   **1.** In muscular histology, a skeletal muscle cell (*muscle fiber*). **2.** In neurohistology, the axon of a neuron (*nerve fiber*). **3.** Any long threadlike structure, such as a Purkinje fiber of the heart, a collagen or elastic fiber of the connective tissues, or cellulose and other digestion-resistant dietary fiber.

fibrinogen   A protein synthesized by the liver and present in blood plasma, semen, and other body fluids; precursor of the sticky protein fibrin, which forms the matrix of a clot.

fibroblast   A connective tissue cell that produces collagen fibers and ground substance; the only type of cell in tendons and ligaments.

fibrocartilage   A form of cartilage with coarse bundles of collagen fibers in the matrix, found in the intervertebral discs, joint menisci, pubic symphysis, and some tendon–bone junctions. (fig. 3.21)

fibrosis   Replacement of damaged tissue with fibrous scar tissue rather than by the original tissue type; scarring. *Compare* regeneration.

fibrous connective tissue   Any connective tissue with a preponderance of fiber, such as areolar, reticular, dense regular, and dense irregular connective tissues.

filament   A fine threadlike structure such as the myofilaments of muscle and the microfilaments and intermediate filaments of the cytoskeleton.

filtration   A process in which a fluid is physically forced through a membrane that allows water and some solutes to pass, and holds back larger particles; especially important in the emission of fluid from blood capillaries.

finger   Any of the five digits of the hand, including the thumb.

first-order neuron   An afferent (sensory) neuron that carries signals from a receptor to a second-order neuron in the spinal cord or brain. (fig. 14.4) *See also* second-order neuron; third-order neuron.

fissure   **1.** A slit through a bone, such as the orbital fissure. **2.** A deep groove, such as the longitudinal fissure between the cerebral hemispheres.

fix   **1.** To hold a structure in place, for example, by fixator muscles that prevent unwanted joint movements. **2.** To preserve a tissue by means of a fixative.

fixative   A chemical that prevents tissue decay, such as formalin or ethanol.

fixator   A muscle that minimizes or prevents bone movement in certain joint actions, such as the rhomboideus major holding the scapula stationary while the biceps brachii flexes the elbow.

flagellum   (fla-JEL-um) A long, motile, usually single hairlike extension of a cell; the tail of a sperm cell is the only functional flagellum in humans.

flat bone   A bone with a platelike shape, such as the parietal bone or sternum.

flexion   A joint movement that, in most cases, decreases the angle between two bones. (fig. 9.7) *Compare* extension.

flexor   A muscle that flexes a joint.

fMRI   *See* functional magnetic resonance imaging.

follicle   (FOLL-ih-cul) A small space, such as a hair follicle, thyroid follicle, or ovarian follicle. (fig. 26.15) *See also* lymphatic nodule.

foramen   (fo-RAY-men) A hole through a bone or other organ, in most cases providing passage for blood vessels and nerves.

foramen magnum   The largest opening into the cranial cavity, at the point where the occipital bone articulates with the vertebral column; allows passage of the spinal cord and vertebral arteries into the cranial cavity.

foramen ovale   **1.** An ovoid foramen in the sphenoid bone that allows for passage of the mandibular division of the trigeminal nerve. **2.** An opening in the fetal interatrial septum that allows blood to flow directly from the right atrium into the left atrium and bypass the pulmonary circulation.

forebrain   The most rostral part of the brain, consisting of the cerebrum and diencephalon. (fig. 13.14)

foregut   **1.** The most rostral part of the embryonic digestive tract; all of the tract rostral to the initial attachment of the yolk sac. (fig. 4.5) **2.** In adults, all of the digestive tract from the oral cavity to the major duodenal papilla, with a blood supply and innervation separate from those of the midgut and hindgut.

formed element   An erythrocyte, leukocyte, or platelet; any normal component of blood or lymph that is a cell or cell fragment, as opposed to the extracellular fluid component.

fossa   (FOSS-uh) A depression in an organ or tissue, such as the fossa ovalis of the heart or a cranial fossa of the skull.

fovea   (FOE-vee-uh) A small pit, such as the fovea capitis of the femur or fovea centralis of the retina.

free nerve ending   A bare sensory nerve ending, lacking associated connective tissue or specialized cells; includes receptors for heat, cold, and pain; also called an *unencapsulated nerve ending*.

frontal plane   An anatomical plane that passes through the body or an organ from right to left and superior to inferior; also called a *coronal plane*. (fig. 1.10)

functional magnetic resonance imaging (fMRI)   A variation on MRI that enables the visualization of moment-to-moment changes in the metabolic activity of a tissue, rather than static images; used to study quickly changing patterns of brain activity, among other diagnostic and research purposes.

fundus   The base or broadest part of certain organs such as the stomach and uterus.

funiculus   (few-NICK-you-lus) Any of the three major divisions of the white matter of the spinal cord, composed of multiple fascicles, or tracts; also called a *column*. The three funiculi on each side of the cord are the dorsal, lateral, and ventral columns.

fusiform   (FEW-zih-form) Spindle-shaped; elongated, thick in the middle, and tapered at both ends, such as the shape of a smooth muscle cell or a muscle spindle. (fig. 2.3)

G

gamete   (GAM-eet) An egg or sperm cell.

gametogenesis   (GAM-eh-toe-JEN-eh-sis) The production of eggs or sperm.

ganglion   (GANG-glee-un) A cluster of nerve cell bodies in the peripheral nervous system, often resembling a knot in a string.

gangrene   Tissue necrosis resulting from ischemia.

gap junction   A junction between two cells consisting of a pore surrounded by a ring of proteins in the plasma membrane of each cell, allowing solutes to diffuse from the cytoplasm of one cell to the next; functions include cell-to-cell nutrient transfer in the developing embryo and electrical communication between cells of cardiac and smooth muscle. (fig. 2.14)

gastric   Pertaining to the stomach.

gastrointestinal (GI) system   The part of the digestive tract composed of the stomach and intestines.

gate   A protein channel in a cellular membrane that can open or close in response to chemical, electrical, or mechanical stimuli, thus controlling when substances are allowed to pass through the membrane.

general senses   Senses such as touch, heat, cold, pain, vibration, and pressure, mediated by relatively simple sense organs that are distributed throughout the body. *See also* somesthetic; special senses.

genitalia   The pelvic reproductive organs including the *internal genitalia* in the pelvic cavity and *external genitalia* in the perineum; most of the external genitalia are externally visible, but some are subcutaneous, between the skin and the muscles of the pelvic floor.

genitourinary (G-U) system   *See* urogenital (U-G) system.

germ cell   A gamete or any precursor cell destined to become a gamete.

germ layer   Any of three tissue layers of an embryo: the ectoderm, mesoderm, or endoderm.

gestation   (jess-TAY-shun) Pregnancy.

gland   Any organ specialized for secretion or excretion; in some cases a single cell, such as a goblet cell.

glial cell   (GLEE-ul, GLY-ul) Any of the six types of supporting cells of the nervous system (oligodendrocytes, astrocytes, microglia, and ependyma in the CNS; Schwann cells and satellite cells in the PNS); constitute most of the bulk of the nervous system and perform various protective and supportive roles for the neurons. Also called *neuroglia.*

glomerular capsule   (glo-MERR-you-lur) A double-walled capsule around each glomerulus of the kidney; receives glomerular filtrate and empties into the proximal convoluted tubule. Also called the *Bowman capsule.* (fig. 25.8)

glomerulus   A spheroid mass of blood capillaries in the kidney that filters plasma and produces glomerular filtrate, which is further processed to form the urine. (fig. 25.9)

glucose   A monosaccharide ($C_6H_{12}O_6$) also known as blood sugar; glycogen, starch, cellulose, and maltose are made entirely of glucose, and glucose constitutes half of a sucrose or lactose molecule. The isomer involved in human physiology is also called *dextrose.*

gluteal   Pertaining to the buttocks.

glycocalyx   (GLY-co-CAY-licks) A layer of carbohydrate covalently bonded to the phospholipid and protein molecules of a plasma membrane; forms a surface coat on all human cells. (fig. 2.12)

glycogen   A glucose polymer synthesized by liver, muscle, uterine, and vaginal cells that serves as an energy-storage polysaccharide.

glycolipid   A phospholipid molecule with carbohydrate covalently bonded to it, found in the plasma membranes of cells.

glycoprotein   A protein–carbohydrate complex in which the protein is dominant; found in mucus and the glycocalyx of cells, for example.

glycosaminoglycan (GAG)   (GLY-cose-am-ih-no-GLY-can) A polysaccharide composed of modified sugars with amino groups; the major component of a proteoglycan. GAGs are largely responsible for the viscous consistency of tissue gel and the stiffness of cartilage.

goblet cell   A mucus-secreting gland cell, shaped somewhat like a wineglass, found in the epithelia of many mucous membranes. (fig. 3.7)

Golgi complex   (GOAL-jee) An organelle composed of several parallel cisternae, somewhat like a stack of saucers, that modifies and packages newly synthesized proteins and synthesizes carbohydrates. (fig. 2.17)

Golgi vesicle   A membrane-bounded vesicle pinched from the Golgi complex, containing its chemical product; may be retained in the cell as a lysosome or become a secretory vesicle that releases the product by exocytosis.

gonad   The ovary or testis.

gonadal ridge   The earliest trace of the embryonic gonad, a streak of tissue adjacent to the kidney, populated by the first germ cells arriving from the yolk sac at 5 to 6 weeks of gestation; also called a *genital ridge.*

granulocyte   (GRAN-you-lo-site) Any of three types of leukocytes (neutrophils, eosinophils, or basophils) with prominent cytoplasmic granules. (fig. 19.1)

granulosa cells   Cells that form a stratified cuboidal epithelium lining an ovarian follicle; source of steroid sex hormones. (fig. 26.15)

gray matter   A zone or layer of tissue in the central nervous system where the neuron cell bodies, dendrites, and synapses are found; forms the core of the spinal cord, nuclei of the brainstem, basal nuclei of the cerebrum, cerebral cortex, and cerebellar cortex. (fig. 15.4)

great toe   The large medial toe; also called the *hallux.*

great vessels   The largest of the blood vessels attached directly to the heart; the superior and inferior venae cavae, pulmonary trunk, and aorta.

gross anatomy   Bodily structure that can be observed without magnification.

ground substance   The clear, featureless material in which the fibers and cells of a connective tissue are embedded; includes the liquid plasma of the blood, tissue gel of areolar tissue, and calcified tissue of bone.

guard hairs   Coarse, stiff hairs that prevent insects, debris, or other foreign matter from entering the ear, nose, or eye; also called *vibrissae.*

gustatory   Pertaining to the sense of taste.

gyrus   (JY-rus) A wrinkle or fold in the cortex of the cerebrum or cerebellum. (fig. 15.1)

H

hair cell   A sensory cell of the cochlea, semicircular ducts, utricle, and saccule, with a fringe of surface microvilli that respond to the relative motion of a gelatinous membrane at their tips; responsible for the senses of hearing and equilibrium. (fig. 17.13)

hair follicle   An oblique epithelial pit in the skin that contains a hair and extends into the dermis or hypodermis.

hair receptor   Free sensory nerve endings entwined around a hair follicle, responsive to movement of the hair.

hallux   The great toe; the medial digit of the foot.

haploid *(n)*   Having a single set of unpaired chromosomes. In humans, the only haploid cells are germ cells past the meiosis I stage of cell division, including the mature egg and sperm.

haversian canal   *See* central canal.

head   1. The uppermost part of the human body, above the neck. 2. The expanded end of an organ such as a bone, the pancreas, or the epididymis.

helper T cell   A type of lymphocyte that performs a central coordinating role in humoral and cellular immunity; target of the human immunodeficiency virus (HIV).

hematocrit   (he-MAT-oh-crit) The percentage of blood volume that is composed of erythrocytes.

hematoma   (HE-muh-TOE-muh) A mass of clotted blood in the tissues; forms a bruise when visible through the skin.

hemidecussation   Crossing over of one half of the nerve fibers in a nerve or tract to the opposite side of the central nervous system, especially at the optic chiasm. *Compare* decussation.

hemoglobin   The red pigment of erythrocytes; binds and transports about 98.5% of the oxygen and 5% of the carbon dioxide carried in the blood.

hemopoiesis   (HE-mo-poy-EE-sis) Production of any of the formed elements of blood.

hemopoietic tissue   Any tissue in which hemopoiesis occurs, especially red bone marrow and lymphatic tissue.

hemostasis   The cessation of bleeding by the mechanisms of vascular spasm, a platelet plug, and blood clotting.

hepatic   Pertaining to the liver.

hepatic macrophage   A macrophage found in the sinusoids of the liver; also called a *Kupffer cell.*

hepatic portal system   A network of blood vessels that connect capillaries of the intestines to capillaries (sinusoids) of the liver, thus delivering newly absorbed nutrients directly to the liver.

hepatocyte   Any of the cuboidal gland cells that constitute the parenchyma of the liver.

hiatus   (hy-AY-tus) An opening or gap, such as the esophageal hiatus through the diaphragm.

hilum   (HY-lum) A point on the surface of an organ where blood vessels, lymphatic vessels, or nerves enter and leave, usually marked by a depression and slit; the midpoint of the concave surface of any organ that is roughly bean-shaped, such as the lymph nodes, kidneys, and lungs. Also called the *hilus.* (fig. 23.9)

hindbrain   The most caudal part of the brain, composed of the medulla oblongata, pons, and cerebellum. (fig. 13.14)

hindgut   1. The most caudal part of the embryonic digestive tract; all of the tract caudal to the initial attachment of the yolk sac. (fig. 4.5) 2. In adults, all of the digestive tract from the end of the transverse colon through the anal canal, with a blood supply and innervation separate from those of the foregut and midgut.

histological section   A thin slice of tissue, usually mounted on a slide and artificially stained to make its microscopic structure more visible.

histology   1. The microscopic structure of tissues and organs. 2. The study of such structure.

holocrine gland   An exocrine gland whose secretion is formed by the breakdown of entire gland cells; for example, a sebaceous gland.

homeostasis   (HO-me-oh-STAY-sis) The tendency of a living body to maintain relatively stable internal conditions in spite of greater changes in its external environment.

homologous   (ho-MOLL-oh-gus) 1. Having the same embryonic or evolutionary origin but not necessarily the same function, such as the scrotum and labia majora. 2. Pertaining to two chromosomes with identical structures and gene loci but not necessarily identical alleles; each member of the pair is inherited from a different parent.

hormone   A chemical messenger that is secreted into the blood by an endocrine gland or isolated gland cell and triggers a physiological response in distant cells with receptors for it.

hyaline cartilage   (HY-uh-lin) A form of cartilage with a relatively clear matrix and fine collagen fibers but no conspicuous elastic fibers or collagen bundles as in other types of cartilage. (fig. 3.19)

hyaluronic acid   (HY-uh-loo-RON-ic) A glycosaminoglycan that is particularly abundant in connective tissues, where it becomes hydrated and forms the tissue gel.

hydrolysis   (hy-DRAHL-ih-sis) A chemical reaction in which water is broken down into hydrogen and hydroxide ions and these are used to split a covalent bond in an organic molecule, for example in digesting starch to glucose or protein to amino acids, or breaking ATP down into ADP and phosphate.

hyperextension   A joint movement that increases the angle between two bones beyond 180°. (fig. 9.7)

hyperplasia   (HY-pur-PLAY-zhuh) The growth of a tissue through cellular multiplication, not cellular enlargement. *Compare* hypertrophy.

hypertrophy (hy-PUR-tro-fee) The growth of a tissue through cellular enlargement, not cellular multiplication; for example, the growth of muscle under the influence of exercise. *Compare* hyperplasia.

hypoblast The layer of cells in the early embryonic disc facing away from the amniotic cavity; forms the yolk sac and is then replaced during gastrulation by migrating epiblast cells.

hypochondriac Pertaining to an area on each side of the abdomen superior to the subcostal line and lateral to the midclavicular line. (fig. 1.13)

hypodermis (HY-po-DUR-miss) A layer of connective tissue deep to the skin; also called *superficial fascia, subcutaneous tissue,* or when it is predominantly adipose, *subcutaneous fat.*

hypogastric Pertaining to a medial area of the lower abdomen inferior to the intertubercular line and medial to (between) the midclavicular lines; also called the *pubic region.* (fig. 1.13)

hypophyseal portal system A circulatory pathway that connects a capillary plexus in the hypothalamus to a capillary plexus in the anterior pituitary; carries hypothalamic releasing and inhibiting hormones to the anterior pituitary. (fig. 18.2)

hypophysis The pituitary gland.

hypothalamic thermostat A nucleus of neurons in the hypothalamus responsible for the homeostatic regulation of body temperature.

hypothalamo–hypophyseal tract A bundle of nerve fibers that begin in nuclei in the hypothalamus, travel through the pituitary stalk, and terminate in the posterior lobe of the pituitary gland. They deliver the hormones oxytocin and antidiuretic hormone to the posterior pituitary for storage, and signal the pituitary when to release them into the blood. (fig. 18.2)

hypothalamus (HY-po-THAL-uh-mus) The inferior portion of the diencephalon of the brain, forming the walls and floor of the third ventricle and giving rise to the posterior pituitary gland; controls many fundamental physiological functions such as appetite, thirst, and body temperature. (fig. 15.2)

hypothesis An informed conjecture that is capable of being tested and potentially falsified by experimentation or data collection.

hypoxemia A deficiency of oxygen in the blood.

hypoxia A deficiency of oxygen in any tissue; may lead to tissue necrosis.

I

immune system A population of cells, including leukocytes and macrophages, that occur in most organs of the body and protect against foreign organisms, some foreign chemicals, and cancerous or other aberrant host cells.

immunity The ability to ward off a specific infection or disease, usually as a result of prior exposure and the body's production of antibodies or lymphocytes against a pathogen.

implantation The process in which a conceptus attaches to the uterine endometrium and then becomes embedded in it.

inclusion Any visible object in the cytoplasm of a cell other than an organelle or cytoskeletal element; usually a foreign body or a stored cell product, such as a virus, dust particle, lipid droplet, glycogen granule, or pigment.

infarction 1. The sudden death of tissue resulting from a loss of blood flow, often resulting from the occlusion of an artery; for example, cerebral infarction and myocardial infarction. 2. A region of tissue that has died from lack of blood; also called an *infarct.*

inferior Lower than another structure or point of reference from the perspective of anatomical position; for example, the stomach is inferior to the diaphragm.

infundibulum (IN-fun-DIB-you-lum) Any funnel-shaped passage or structure, such as the distal portion of the uterine tube and the stalk that attaches the pituitary gland to the hypothalamus.

inguinal (IN-gwih-nul) Pertaining to the groin. (fig. 1.13)

innervation (IN-ur-VAY-shun) The nerve supply to an organ.

insertion The point at which a muscle attaches to another tissue (usually a bone) and produces movement, opposite from its stationary origin. (fig. 10.4) *Compare* origin.

inspiration Inhaling.

integument The skin.

integumentary system An organ system consisting of the skin, cutaneous glands, hair, and nails.

interatrial septum The wall between the atria of the heart.

intercalated disc (in-TUR-kuh-LAY-ted) A complex of fascia adherens, gap junctions, and desmosomes that join two cardiac muscle cells end to end, microscopically visible as a dark line, which helps to histologically distinguish this muscle type; functions as a mechanical and electrical link between cells. (fig. 20.14)

intercellular Between cells.

intercellular junction *See* cell junction.

intercostal (IN-tur-COSS-tul) Between the ribs, as in the *intercostal* muscles, arteries, veins, and nerves.

interdigitate To fit together like the fingers of the folded hands; for example, at the dermal–epidermal boundary, podocytes of the kidney, and intercalated discs of the heart.

interneuron (IN-tur-NEW-ron) A neuron that is contained entirely in the central nervous system and, in the path of signal conduction, lies anywhere between an afferent pathway and an efferent pathway.

interoceptor A sensory receptor that responds to stimuli originating within the body. *Compare* exteroceptor.

interosseous membrane (IN-tur-OSS-ee-us) A fibrous membrane that connects the radius to the ulna and the tibia to the fibula along most of the shaft of each bone. (fig. 8.4)

interstitial (IN-tur-STISH-ul) 1. Pertaining to the extracellular spaces in a tissue. 2. Located between other structures, as in the *interstitial cells* of the testis.

interstitial fluid Fluid in the interstitial spaces of a tissue, also called *tissue fluid.*

intervertebral disc A cartilaginous pad between the bodies of two adjacent vertebrae.

intracellular Contained within a cell.

intracellular fluid (ICF) The fluid contained in the cells; one of the major fluid compartments.

intramembranous ossification A process of bone development in which there is no cartilage precursor; rather, the bone develops directly from a sheet of condensed mesenchyme. (fig. 6.8) *Compare* endochondral ossification.

intraperitoneal Within the peritoneal cavity. *Compare* retroperitoneal.

intrinsic 1. Arising from within, such as intrinsic blood-clotting factors; endogenous. 2. Fully contained within an organ, such as the intrinsic muscles of the hand and eye. *Compare* extrinsic.

inversion Movement of the foot that turns the sole medially. (fig. 9.17)

involuntary Not under conscious control, as in the case of the autonomic nervous system and cardiac and smooth muscle contraction.

involution Shrinkage of a tissue or organ by autolysis, such as shrinkage of the thymus after childhood and of the uterus after pregnancy.

ipsilateral (IP-sih-LAT-ur-ul) On the same side of the body, as in reflex arcs in which a muscular response occurs on the same side of the body as the stimulus. *Compare* contralateral.

ischemia A state in which the blood flow to a tissue is inadequate to meet its metabolic needs; may lead to tissue necrosis from hypoxia or waste accumulation.

isthmus A narrow zone of tissue connecting two larger masses; for example, at the front of the thyroid gland and connecting the uterine tube ampulla to the uterus.

J

joint *See* articulation.

K

keratin A tough protein formed by keratinocytes that constitutes the hair, nails, and stratum corneum of the epidermis.

keratinized Covered with keratin, such as the epidermis.

keratinocyte A cell of the epidermis that synthesizes keratin, then dies; most cells of the epidermis are keratinocytes, with dead ones constituting the stratum corneum.

Kupffer cell *See* hepatic macrophage.

kyphosis An exaggerated thoracic spinal curvature, often resulting from osteoporosis; also called *widow's hump* or *dowager's hump.*

L

labium (LAY-bee-um) A lip, such as those of the mouth and the labia majora and minora of the vulva.

lacrimal Pertaining to the tears or tear glands.

lacteal A lymphatic capillary located in the core of an intestinal villus, serving to absorb dietary lipids.

lacuna (la-CUE-nuh) A small cavity or depression in a tissue such as bone, cartilage, and the erectile tissues.

lamella A little plate or sheet of tissue, such as a lamella of bone. (fig. 6.5)

lamellar corpuscle A bulbous sensory receptor with one or a few dendrites enclosed in onionlike layers of Schwann cells; found in the dermis, mesenteries, pancreas, and some other viscera, and responsive to deep pressure, stretch, and high-frequency vibration. Also called a *pacinian corpuscle.* (fig. 17.1)

lamina (LAM-ih-nuh) A thin layer, such as the lamina of a vertebra or the lamina propria of a mucous membrane. (fig. 7.25)

lamina propria (PRO-pree-uh) A thin layer of areolar tissue immediately deep to the epithelium of a mucous membrane. (fig. 3.31)

laryngopharynx (la-RIN-go-FAIR-inks) The portion of the pharynx formed by the union of the oropharynx and nasopharynx, beginning at the level of the hyoid bone and extending inferiorly to the opening of the esophagus. (fig. 23.2)

larynx (LAIR-inks) A cartilaginous chamber in the neck containing the vocal cords; the voicebox. (fig. 23.4)

lateral Away from the midline of an organ or median plane of the body; toward the side. *Compare* medial.

leg 1. That part of the body between the knee and ankle; the crural region. 2. A leglike extension of an organ. *See also* crus.

lesion A circumscribed zone of tissue injury, such as a skin abrasion or myocardial infarction.

leukocyte (LOO-co-site) Any nucleated blood cell; a neutrophil, eosinophil, basophil, lymphocyte, or monocyte. Also called a *white blood cell*. (fig. 19.1)

leukopoiesis The process of leukocyte development from hemopoietic stem cells.

libido The sex drive; a psychological desire for sex.

ligament A cord or band of tough collagenous tissue binding one organ to another, especially one bone to another, and serving to hold organs in place; for example, the cruciate ligaments of the knee and falciform ligament of the liver.

light microscope (LM) A microscope that produces images with visible light.

limb 1. An appendage of the body arising from the shoulder or hip. *See also* lower limb; upper limb. 2. An appendage or extension of another structure, such as the descending limb of the nephron loop.

limb bud An outgrowth of the embryo that develops into an upper or lower limb.

limbic system A ring of brain structures that encircle the corpus callosum and thalamus, including the cingulate gyrus, hippocampus, amygdala, and other structures; functions include learning and emotion. (fig. 15.16)

line 1. Any long narrow mark. *See also* linea. 2. An elongated, slightly raised ridge on a bone, such as the nuchal lines of the skull. (fig. 7.5)

linea (LIN-ee-uh) An anatomical line, such as the linea alba of the abdomen.

lingual (LING-wul) Pertaining to the tongue.

lipid A hydrophobic organic compound with a high ratio of hydrogen to oxygen; includes steroids, fatty acids, triglycerides (fats), phospholipids, and prostaglandins.

LM 1. Light microscope. 2. Light micrograph, a photograph made through the light microscope.

load 1. To pick up oxygen or carbon dioxide for transport in the blood. 2. The resistance acted upon by a muscle.

lobe 1. A structural subdivision of an organ such as a gland, a lung, or the brain, bounded by a visible landmark such as a fissure or septum. 2. The inferior, noncartilaginous, often pendant part of the ear pinna; the earlobe.

lobule (LOB-yool) A small subdivision of an organ or of a lobe of an organ, especially of a gland.

long bone A bone such as the femur or humerus that is markedly longer than wide and that generally serves as a lever.

longitudinal section (l.s.) A cut along the longest dimension of the body or of an organ. (fig. 3.2)

loose connective tissue Areolar or reticular tissue; a connective tissue that has an abundance of ground substance and widely spaced fibers and cells.

lower limb The appendage that arises from the hip, consisting of the thigh from hip to knee; the crural region from knee to ankle; the ankle; and

the foot. Loosely called the leg, although that term properly refers only to the crural region.

lumbar Pertaining to the lower back and sides, between the thoracic cage and pelvis.

lumen The internal space of a hollow organ, such as a blood vessel or the esophagus, or a space surrounded by cells, as in a gland acinus.

lymph The fluid contained in lymphatic vessels and lymph nodes, produced by the absorption of tissue fluid.

lymphatic nodule A temporary, dense aggregation of lymphocytes in such places as mucous membranes and lymphatic organs; also called a *lymphatic follicle*. (fig. 22.8)

lymphatic system An organ system consisting of lymphatic vessels, lymph nodes, the tonsils, spleen, and thymus; functions include tissue fluid recovery and immunity.

lymph node A small organ found along the course of a lymphatic vessel; filters the lymph and contains lymphocytes and macrophages, which respond to antigens in the lymph. (fig. 22.11)

lymphocyte (LIM-foe-site) A relatively small leukocyte with numerous types and roles in nonspecific defense, humoral immunity, and cellular immunity. (fig. 19.1)

lysosome A membrane-bounded organelle containing a mixture of enzymes with a variety of intracellular and extracellular roles in digesting foreign matter, pathogens, and expired organelles. (fig. 2.17)

lysozyme An enzyme found in tears, milk, saliva, mucus, and other body fluids that destroys bacteria by digesting their cell walls; also called *muramidase*.

M

macrophage (MAC-ro-faje) Any cell of the body, other than a leukocyte, that is specialized for phagocytosis; usually derived from a blood monocyte and often functioning as an antigen-presenting cell.

macula (MAC-you-luh) A patch or spot, such as the macula lutea of the retina and macula sacculi of the inner ear.

magnetic resonance imaging (MRI) A method of producing a computerized image of the interior of the body using a strong magnetic field and radio waves. (fig. 1.3)

male In humans, any individual with a Y chromosome; normally, one possessing one X and one Y chromosome in each somatic cell, and having reproductive organs that serve to produce and deliver sperm.

mammary gland The milk-secreting gland that develops within the breast in pregnancy and lactation; only minimally developed in the breast of a nonpregnant or nonlactating woman.

mast cell A connective tissue cell, similar to a basophil, that secretes histamine, heparin, and other chemicals involved in inflammation; often concentrated along the course of a blood capillary.

matrix 1. The extracellular material of a tissue. 2. The substance or framework within which other structures are embedded, such as the fibrous matrix of a blood clot. 3. A mass of epidermal cells from which a hair root or nail root develops. 4. The fluid within a mitochondrion containing enzymes of the citric acid cycle.

meatus (me-AY-tus) An opening into a canal, such as an acoustic meatus.

mechanoreceptor A sensory nerve ending or organ specialized to detect mechanical stimuli such as touch, pressure, stretch, or vibration.

medial Toward the midline of an organ or median plane of the body. *Compare* lateral.

median plane The sagittal plane that divides the body or an organ into equal right and left halves; also called the *midsagittal plane*. (fig. 1.10) *Compare* sagittal plane.

mediastinum (ME-dee-ah-STY-num) The thick median partition of the thoracic cavity that separates one pleural cavity from the other and contains the heart, great blood vessels, and thymus. (fig. 1.14)

medical imaging Any of several noninvasive or minimally invasive methods for producing images of the interior of the body, including X-rays, MRI, PET, CT, and sonography.

medulla (meh-DUE-luh, meh-DULL-uh) Tissue deep to the cortex of certain organs such as the adrenal glands, lymph nodes, hairs, and kidneys.

medulla oblongata (OB-long-GAH-ta) The most caudal part of the brainstem, immediately superior to the foramen magnum of the skull, connecting the spinal cord to the rest of the brain. (fig. 15.2)

meiosis (my-OH-sis) A form of cell division in which a diploid cell divides twice and produces four haploid daughter cells; occurs only in gametogenesis.

Meissner plexus *See* submucosal plexus.

melanin A brown or black pigment synthesized by melanocytes and some other cells; provides color to the skin, hair, eyes, and some other organs and tissues.

melanocyte A cell of the stratum basale of the epidermis that synthesizes melanin and transfers it to the keratinocytes.

meninges (meh-NIN-jeez) (singular, *meninx*) Three fibrous membranes between the central nervous system and surrounding bone: the dura mater, arachnoid mater, and pia mater. (fig. 15.3)

merocrine (MERR-oh-crin) Pertaining to gland cells that release their product by exocytosis; also called *eccrine*. (fig. 5.10)

mesenchyme (MEZ-en-kime) A gelatinous embryonic connective tissue derived from the mesoderm; differentiates into all permanent connective tissues and most muscle.

mesentery (MEZ-en-tare-ee) A serous membrane that binds the intestines together and suspends them from the abdominal wall; the visceral continuation of the peritoneum. (fig. 24.3)

mesocolon A dorsal mesentery that anchors parts of the colon to the abdominal wall. (fig. 24.3)

mesoderm (MEZ-oh-durm) The middle layer of the three primary germ layers of an embryo; gives rise to muscle and connective tissue.

mesonephric ducts A pair of embryonic ducts that form in association with the temporary mesonephric kidney; they degenerate in the female, but in the male, they develop into parts of the reproductive tract. (fig. 26.23)

mesothelium (MEZ-oh-THEEL-ee-um) A simple squamous epithelium that covers the serous membranes.

metaphysis A growth zone at the junction between the diaphysis and epiphysis of a long bone, where cartilage is replaced by osseous tissue and the bone grows in length. (fig. 6.10)

metaplasia Transformation of one mature tissue type into another; for example, a change from pseudostratified columnar to stratified squamous epithelium in an overventilated nasal cavity.

metarteriole A short blood vessel that links an arteriole to a bed of blood capillaries, with no tunica media except for a smooth muscle precapillary sphincter at the opening to each capillary. (fig. 21.9)

metastasis (meh-TASS-tuh-sis) The spread of cancer cells from the original tumor to a new location, where they seed the development of a new tumor.

microfilament A thin filament of actin in the cytoskeleton of a cell, involved especially in the supportive core of a microvillus, the membrane skeleton just deep to the plasma membrane, and in muscle contraction. *See also* actin.

micrograph A photograph made with a microscope.

micrometer (μm) One thousandth of a millimeter, or 10^{-6} meter; a convenient unit of length for expressing the sizes of cells.

microtubule An intracellular cylinder composed of the protein tubulin, forming centrioles, the axonemes of cilia and flagella, and part of the cytoskeleton.

microvillus An outgrowth of the plasma membrane that increases the surface area of a cell and functions in absorption and some sensory processes; distinguished from cilia and flagella by its smaller size and lack of an axoneme.

midbrain A short section of the brainstem between the pons and diencephalon. (fig. 15.2)

midgut **1.** The middle part of the embryonic digestive tract, located at the attachment of the yolk sac. (fig. 4.5) **2.** In adults, all of the digestive tract from the major duodenal papilla through the end of the transverse colon, with a blood supply and innervation separate from those of the foregut and hindgut.

midsagittal plane *See* median plane.

mineralization *See* calcification.

mitochondrion (MY-toe-CON-dree-un) An organelle specialized to synthesize ATP, enclosed in a double unit membrane with infoldings of the inner membrane called cristae.

mitosis A form of cell division in which a cell divides once and produces two genetically identical daughter cells; sometimes used to refer only to the division of the genetic material or nucleus and not to include cytokinesis, the subsequent division of the cytoplasm.

mixed nerve A nerve containing both afferent (sensory) and efferent (motor) nerve fibers.

monocyte A leukocyte specialized to migrate into the tissues and transform into a macrophage. (fig. 19.1)

monozygotic (MZ) twins Two individuals that develop from the same zygote and are therefore genetically identical.

morphology Anatomy, especially as interpreted from a functional perspective.

morula A preembryonic stage of development consisting of 16 or more identical-looking cells having a bumpy surface appearance reminiscent of a mulberry. The morula develops into a blastocyst and then implants on the uterine wall.

motor neuron A neuron that transmits signals from the central nervous system to any effector (muscle or gland cell); also called an *efferent neuron.* The axon of a motor neuron is an *efferent nerve fiber.*

motor protein Any protein that produces movements of a cell or its components owing to its ability to undergo quick repetitive changes in conformation and to bind reversibly to other molecules; for example, myosin, dynein, and kinesin.

motor unit One motor neuron and all the skeletal muscle fibers innervated by it.

mouth **1.** A narrow opening into any cavity or hollow organ. **2.** The oral (buccal) cavity, bordered by the lips, cheeks, and fauces.

MRI *See* magnetic resonance imaging.

mucosa (mew-CO-suh) A tissue layer that forms the inner lining of an anatomical tract that is open to the exterior (the respiratory, digestive, urinary, and reproductive tracts). Composed of epithelium, connective tissue (lamina propria), and often smooth muscle (muscularis mucosae). Also called a *mucous membrane.* (fig. 3.31)

mucosa-associated lymphatic tissue (MALT) Aggregations of lymphocytes, including lymphatic nodules, in the mucous membranes.

mucous gland A gland that secretes mucus, such as the glands of the large intestine and nasal cavity. *Compare* serous gland.

mucous membrane *See* mucosa.

mucus A viscous, slimy or sticky secretion produced by mucous cells and mucous membranes and consisting of a hydrated glycoprotein, mucin; serves to bind particles together, such as bits of masticated food, and to protect the mucous membranes from infection and abrasion.

multipotent Pertaining to a stem cell that is capable of differentiating into multiple, but not unlimited, adult cell types; for example, bone marrow colony-forming units that can produce multiple types of leukocytes.

muscle fiber One skeletal muscle cell. *Compare* myocyte.

muscularis externa The external muscular wall of certain viscera such as the esophagus and small intestine. (fig. 24.2)

muscularis mucosae (MUSS-cue-LERR-iss mew-CO-see) A layer of smooth muscle immediately deep to the lamina propria of a mucosa. (fig. 3.31)

muscular system An organ system composed of the skeletal muscles, specialized mainly for maintaining postural support and producing movements of the bones. Cardiac and smooth muscle are not regarded as part of the muscular system.

muscular tissue A tissue composed of elongated, electrically excitable cells specialized for contraction; the three types are skeletal, cardiac, and smooth muscle.

mutagen (MEW-tuh-jen) Any agent that causes a mutation, including viruses, chemicals, and ionizing radiation.

mutation Any change in the structure of a chromosome or a DNA molecule, often resulting in a change of organismal structure or function.

myelin (MY-eh-lin) A lipid sheath around a nerve fiber, formed from closely spaced spiral layers of the plasma membrane of an oligodendrocyte or Schwann cell. (fig. 13.7)

myelination The process in which an oligodendrocyte or Schwann cell deposits myelin around a nerve fiber.

myeloid tissue Bone marrow.

myenteric plexus A plexus of parasympathetic neurons located between the layers of the muscularis externa of the digestive tract; controls peristalsis. Also called the *Auerbach plexus.*

myocardium The middle, muscular layer of the heart.

myocyte A muscle cell, especially a cell of cardiac or smooth muscle. *Compare* muscle fiber.

myoepithelial cell An epithelial cell that has become specialized to contract like a muscle cell; important in dilation of the pupil and ejection of secretions from gland acini.

myofibril (MY-oh-FY-bril) A bundle of myofilaments forming an internal subdivision of a cardiac or skeletal muscle cell. (fig. 10.8)

myofilament A protein microfilament responsible for the contraction of a muscle cell, composed mainly of myosin or actin. (fig. 10.9)

myosin A motor protein that constitutes the thick myofilaments of muscle and has globular, mobile heads of ATPase that bind to actin molecules.

myotome A group of mesodermal cells that arise from a somite in the fourth week of development and give rise to body wall muscles in that region of the trunk of the body. *Compare* dermatome; sclerotome.

N

nasal concha One of three curved or scroll-like plates of bone and mucous membrane that extends from the lateral wall toward the septum in each nasal fossa; serves to warm, cleanse, and humidify inhaled air. (fig. 23.2)

nasal septum A wall of bone and cartilage that separates the right and left nasal fossae.

nasopharynx That region of the pharynx that lies caudal to the nasal choanae and posterior or superior to the soft palate. (fig. 23.2)

natural killer (NK) cell A lymphocyte that attacks and destroys cancerous or infected cells of the body without requiring prior exposure or a specific immune response; one of the body's nonspecific defenses.

necrosis (neh-CRO-sis) Pathological tissue death due to such causes as infection, trauma, or hypoxia. *Compare* apoptosis.

neonate An infant up to 6 weeks old.

neoplasia (NEE-oh-PLAY-zee-uh) Abnormal growth of new tissue, such as a tumor, with no useful function.

nephron One of approximately 1.2 million blood-filtering, urine-producing units in each kidney; consists of a glomerulus, glomerular capsule, proximal convoluted tubule, nephron loop, and distal convoluted tubule. (fig. 25.6) *Compare* renal tubule.

nerve A cordlike organ of the peripheral nervous system composed of multiple nerve fibers ensheathed in connective tissue.

nerve fiber The axon of a single neuron.

nerve impulse A wave of self-propagating action potentials spreading along a nerve fiber; the nerve signal.

nervous system An organ system composed of the brain, spinal cord, nerves, and ganglia, specialized for rapid communication of information.

nervous tissue A tissue composed of neurons and neuroglia.

neural circuit A group of interconnected neurons that conduct signals along defined pathways to produce a sustained, repetitive, convergent, or divergent output. (fig. 13.12)

neural crest A mass of ectoderm that begins at the edges of the neural groove, then separates from the neural tube and gives rise primarily to nerves, ganglia, and the adrenal medulla. (fig. 13.13)

neural groove A longitudinal depression in the ectoderm of the embryo that closes up to form the neural tube, forerunner of the central nervous system. (fig. 13.13)

neural pool A group of interconnected neurons of the central nervous system that perform a single collective function; for example, the vasomotor center of the brainstem and speech centers of the cerebral cortex.

neural tube A dorsal hollow ectodermal tube in the embryo that develops into the central nervous system. (fig. 13.13)

neuroglia (noo-ROG-lee-uh) All cells of nervous tissue except neurons; cells that perform various supportive and protective roles for the neurons.

neurohypophysis The posterior one-third of the pituitary gland, consisting of the posterior lobe, a stalk that attaches the pituitary to the hypothalamus, and the median eminence of the hypothalamic floor; stores and secretes antidiuretic hormone and oxytocin. (fig. 18.2)

neuromuscular junction (NMJ) A synapse between a nerve fiber and a muscle cell. (fig. 10.11)

neuron (NOOR-on) A nerve cell; an electrically excitable cell specialized for producing and transmitting action potentials and secreting chemicals that stimulate adjacent cells. (fig. 13.4)

neurosoma The cell body of a neuron, containing the nucleus and usually giving rise to the axon and dendrites; also called the *soma* or *perikaryon*. (fig. 13.4)

neurotransmitter A chemical released at the distal end of an axon that stimulates an adjacent cell; for example, acetylcholine, norepinephrine, and serotonin.

neutrophil (NOO-tro-fill) A leukocyte, usually with a multilobed nucleus, that serves especially to destroy bacteria by means of phagocytosis, intracellular digestion, and secretion of bactericidal chemicals. (fig. 19.1)

nitrogenous waste Any nitrogen-containing substance produced as a metabolic waste and excreted in the urine; chiefly ammonia, urea, uric acid, and creatinine.

nociceptor (NO-sih-SEP-tur) A nerve ending specialized to detect tissue damage and produce a sensation of pain; pain receptor.

node of Ranvier A gap between adjacent segments of myelin in a myelinated nerve fiber; the point where action potentials are generated in a myelinated fiber.

nonkeratinized Pertaining to a stratified squamous epithelium that lacks a surface layer of dead compacted keratinocytes; found in the oral cavity, pharynx, esophagus, anal canal, and vagina.

notochord A middorsal supportive rod that develops in all chordate embryos, including humans; represented in the adult only by the nuclei of the intervertebral discs.

nuchal Pertaining to the back of the neck.

nuclear envelope (NEW-clee-ur) A pair of unit membranes enclosing the nucleus of a cell, with prominent pores allowing traffic of molecules between the nucleoplasm and cytoplasm. (fig. 2.17)

nuclear medicine Any use of radioisotopes to treat disease or form diagnostic images of the body.

nucleus (NEW-clee-us) 1. A cell organelle containing DNA and surrounded by a double unit membrane. 2. A mass of neurons (gray matter) surrounded by white matter of the brain, including the basal nuclei and brainstem nuclei. 3. A central structure, such as the nucleus pulposus of an intervertebral disc or nucleus of an atom.

nucleus pulposus The gelatinous center of an intervertebral disc.

O

oblique section A cut through an elongated organ on a slant, between a longitudinal and a cross section. (fig. 3.2)

occlusion 1. Meeting of the surfaces of the teeth when one bites. 2. Obstruction of an anatomical passageway, such as blockage of an artery by a thrombus or atherosclerotic plaque.

olfactory Pertaining to the sense of smell.

omentum A ventral mesentery that extends from the stomach to the liver (*lesser omentum*) or is suspended from the greater curvature of the stomach and overhangs the intestines (*greater omentum*). (fig. 24.3)

oocyte (OH-oh-site) In the development of an egg cell, a haploid stage between meiosis I and fertilization.

oogenesis (OH-oh-JEN-eh-sis) The production of a fertilizable egg cell through a series of mitotic and meiotic cell divisions; female gametogenesis.

ophthalmic (off-THAL-mic) Pertaining to the eye or vision; optic.

opposition A movement of the thumb in which it approaches or touches any fingertip of the same hand. (fig. 9.16)

optic Pertaining to the eye or vision.

optic chiasm An X-shaped point at the base of the brain, immediately rostral to the hypothalamus, where the two optic nerves meet and continue as optic tracts.

oral cavity The space enclosed by the lips anteriorly, the cheeks laterally, and the fauces posteriorly; also called the *buccal cavity*.

orbit The eye socket of the skull.

organ Any anatomical structure that is composed of at least two different tissue types, has recognizable structural boundaries, and has a discrete function different from the structures around it. Many organs are microscopic, and many organs contain smaller organs, such as the skin containing numerous microscopic sense organs.

organelle Any structure within a cell that carries out one of its metabolic roles, such as mitochondria, centrioles, endoplasmic reticulum, and the nucleus; an intracellular structure other than the cytoskeleton and inclusions.

organism Any living individual; the entire body of any living thing such as a bacterium, plant, or human.

organogenesis The prenatal developmental process in which embryonic germ layers differentiate into specific organs and organ systems; the process that converts an embryo to a fetus, occurring between day 16 and the end of week 8 of gestation.

organ system Any of 11 systems of interconnected or physiologically interrelated organs that perform one of the body's basic functions; for example, the digestive, urinary, and respiratory systems.

origin The relatively stationary attachment of a skeletal muscle. (fig. 10.4) *Compare* insertion.

oropharynx That part of the pharynx that is caudal to the fauces at the rear of the oral cavity, and anterior or inferior to the soft palate. (fig. 23.2)

osmoreceptor (OZ-mo-re-SEP-tur) A neuron of the hypothalamus that responds to changes in the osmolarity of the extracellular fluid.

osmosis The diffusion of water through a selectively permeable membrane from the side with less concentrated solutes to the side with more concentrated solutes.

osseous (OSS-ee-us) Pertaining to bone.

ossification (OSS-ih-fih-CAY-shun) Bone formation; also called *osteogenesis*. *See also* endochondral ossification; intramembranous ossification.

osteoarthritis (OA) A chronic degenerative joint disease characterized by loss of articular cartilage, growth of bone spurs, and impaired movement; occurs to various degrees in almost all people with age.

osteoblast A bone-forming cell that arises from an osteogenic cell, deposits bone matrix, and eventually becomes an osteocyte.

osteoclast A macrophage of the bone surface that dissolves the matrix and returns minerals to the extracellular fluid.

osteocyte A mature bone cell formed when an osteoblast becomes surrounded by its own matrix and entrapped in a lacuna.

osteogenesis *See* ossification.

osteon A structural unit of compact bone consisting of a central canal surrounded by concentric cylindrical lamellae of matrix. (fig. 6.5)

osteoporosis (OSS-tee-oh-pore-OH-sis) A degenerative bone disease characterized by a loss of bone mass, increasing susceptibility to spontaneous fractures, and sometimes deformity of the vertebral column; causes include aging, estrogen hyposecretion, and insufficient resistance exercise.

ovary The female gonad; produces eggs, estrogen, and progesterone.

oviduct *See* uterine tube.

ovulation The release of a mature oocyte by the bursting of an ovarian follicle.

ovum Any stage of the female gamete from the conclusion of meiosis I until fertilization; a primary oocyte; an egg.

P

pacinian corpuscle *See* lamellated corpuscle.

palate A horizontal partition between the oral and nasal cavities.

palatine Pertaining to the palate, such as palatine bones and tonsils.

palmar region The anterior surface (palm) of the hand.

pancreas A gland of the upper abdominal cavity, near the stomach, that secretes digestive enzymes and sodium bicarbonate into the duodenum and secretes hormones into the blood.

pancreatic islets (PAN-cree-AT-ic EYE-lets) Small clusters of endocrine cells in the pancreas that secrete insulin, glucagon, somatostatin, and other intercellular messengers; also called *islets of Langerhans*. (fig. 18.9)

papilla (pa-PILL-uh) A conical or nipplelike structure, such as a lingual papilla of the tongue or the papilla of a hair bulb.

papillary (PAP-ih-lerr-ee) 1. Pertaining to or shaped like a nipple, such as the papillary muscles of the heart. 2. Having papillae, such as the papillary layer of the dermis.

paramesonephric ducts A pair of embryonic ducts that form beside the mesonephric ducts; they degenerate in the male; in the female, they form the uterine tubes, uterus, and part of the vagina. (fig. 26.26)

parasympathetic nervous system (PERR-uh-SIM-pa-THET-ic) A division of the autonomic nervous system that issues efferent fibers through the cranial and sacral nerves and exerts cholinergic effects on its target organs. (fig. 16.7)

parathyroid glands (PERR-uh-THY-royd) Small endocrine glands, usually four in number, adhering to the posterior side of the thyroid gland. (fig. 18.7)

parenchyma (pa-REN-kih-muh) The tissue that performs the main physiological functions of an organ, especially a gland, as opposed to the tissues (stroma) that mainly provide structural support.

parietal (pa-RY-eh-tul) 1. Pertaining to a wall, as in the *parietal cells* of the gastric glands and *parietal bone* of the skull. 2. The outer or more superficial layer of a two-layered membrane such as the pleura, pericardium, or glomerular capsule. (fig. 1.15) *Compare* visceral.

pathogen Any disease-causing chemical or organism.

pectoral Pertaining to the chest.

pectoral girdle The circle of bones that connect the upper limb to the axial skeleton; composed of the two scapulae and the two clavicles.

pedal region The foot.

pedicel *See* pedicle.

pedicle (PED-ih-cul) A small footlike process, as in the vertebrae and the renal podocytes; also called a *pedicel.* (fig. 7.22)

pelvic cavity The space enclosed by the true (lesser) pelvis, containing the urinary bladder, rectum, and internal reproductive organs. (fig. 1.14)

pelvic girdle A ring of three bones—the two hip (coxal) bones and the sacrum—that attaches the lower limbs to the axial skeleton. *See also* pelvis. (fig. 8.6)

pelvis 1. A basinlike cradle composed of the pelvic girdle and its associated ligaments and muscles, forming the walls and floor of the lower abdominal and pelvic cavities. 2. A basinlike structure such as the renal pelvis of the kidney. (fig. 25.3)

perfusion Loosely, blood flow to any tissue or organ. More specifically, the volume of blood received by a given mass of tissue in a given unit of time, such as milliliters per gram per minute.

pericardial cavity A narrow space between the parietal and visceral layers of the pericardium, containing pericardial fluid.

pericardium A two-layered serous membrane that folds around the heart. Its visceral layer forms the heart surface *(epicardium)*, and its parietal layer forms a fibrous *pericardial sac* around the heart. (fig. 20.4)

perichondrium (PERR-ih-CON-dree-um) A layer of fibrous connective tissue covering the surface of hyaline or elastic cartilage. (fig. 3.19)

perineum (PERR-ih-NEE-um) The region between the thighs bordered by the coccyx, pubic symphysis, and ischial tuberosities; contains the orifices of the urinary, reproductive, and digestive systems. (figs. 26.1, 26.21)

periosteum (PERR-ee-OSS-tee-um) A layer of fibrous connective tissue covering the surface of a bone. (fig. 6.5)

peripheral Away from the center of the body or of an organ, as in *peripheral vision* and *peripheral blood vessels;* opposite of central.

peripheral nervous system (PNS) A subdivision of the nervous system composed of all nerves and ganglia; all of the nervous system except the central nervous system. *Compare* central nervous system.

peristalsis (PERR-ih-STAL-sis) A wave of constriction traveling along a tubular organ such as the esophagus or ureter, serving to propel its contents.

peritoneum (PERR-ih-toe-NEE-um) A serous membrane that lines the peritoneal cavity of the abdomen and covers the mesenteries and viscera.

perivascular (PERR-ih-VASS-cue-lur) Pertaining to the region surrounding a blood vessel.

peroxisome An organelle composed of a unit membrane enclosing a mixture of enzymes; serves to detoxify free radicals, alcohol, and other drugs, and to break down fatty acids; named for the hydrogen peroxide that it generates in the course of these activities.

PET *See* positron emission tomography.

phagocytosis (FAG-oh-sy-TOE-sis) A form of endocytosis in which a cell surrounds a foreign particle with pseudopods and engulfs it, enclosing it in a cytoplasmic vesicle called a phagosome. (fig. 2.11)

pharyngeal arch One of five pairs of bulbous swellings in the pharyngeal region of an embryo. (fig. 4.7)

pharyngeal pouch One of six pairs of outpocketings of the pharynx between and adjacent to the pharyngeal arches of an embryo; these form gill slits in fishes and amphibians but in humans give rise to such structures as the middle-ear cavities, palatine tonsils, thymus, parathyroid glands, and C cells of the thyroid gland. (fig. 4.7)

pharynx (FAIR-inks) A muscular passage in the throat at which the respiratory and digestive tracts cross. (fig. 23.2)

phospholipid A lipid composed of a hydrophilic head with a phosphate group and a nitrogenous group such as choline, and two hydrophobic fatty acid tails; especially important as the most numerous molecules of the plasma membrane and other unit membranes of a cell, but also involved in emulsification of dietary fat and as a component of pulmonary surfactant. (fig. 2.7)

photoreceptor Any cell or organ specialized to absorb light and generate a nerve signal; the eye and its rods, cones, and some of its ganglion cells.

phrenic (FREN-ic) 1. Pertaining to the diaphragm, as in *phrenic nerve.* 2. Pertaining to the mind, as in *schizophrenic.*

physiology 1. The functional processes of the body. 2. The study of such function.

piloerector A bundle of smooth muscle cells associated with a hair follicle, responsible for erection of the hair; also called *arrector pili.* (fig. 5.6)

pineal gland (PIN-ee-ul) A small conical endocrine gland arising from the roof of the third ventricle of the brain; produces melatonin and serotonin and may be involved in mood and timing the onset of puberty. (fig. 15.2)

pinocytosis A form of endocytosis in which the plasma membrane sinks in and internalizes a droplet of extracellular fluid in a pinocytotic vesicle. (fig. 2.11)

pituitary gland (pih-TOO-ih-terr-ee) An endocrine gland suspended from the hypothalamus and housed in the sella turcica of the sphenoid bone; secretes numerous hormones, most of which regulate the activities of other glands. (fig. 18.2)

placenta (pla-SEN-tuh) A thick discoid organ on the wall of the pregnant uterus, composed of a combination of maternal and fetal tissues, serving multiple functions in pregnancy including gas, nutrient, and waste exchange between mother and fetus. (fig. 4.11)

plantar (PLAN-tur) Pertaining to the sole of the foot.

plantar flexion A movement of the ankle that points the toes downward, as in pressing on the gas pedal of a car or standing on tiptoes. (fig. 9.17)

plaque A small scale or plate of matter, such as dental plaque, the fatty plaques of atherosclerosis, and the amyloid plaques of Alzheimer disease.

plasma The noncellular portion of the blood.

plasma cell A connective tissue cell that differentiates from a B lymphocyte and secretes antibodies. (fig. 22.16)

plasma membrane The unit membrane that encloses a cell and controls the traffic of molecules into and out of it. (fig. 2.6)

platelet A formed element of the blood derived from the peripheral cytoplasm of a megakaryocyte, known especially for its roles in stopping bleeding but also serves in dissolving blood clots, stimulating inflammation, and promoting tissue growth. (fig. 19.9)

pleura (PLOOR-uh) A two-layered serous membrane that folds around the lung. Its visceral layer forms the lung surface, and its parietal layer lines the inside of the rib cage. (fig. 23.12)

pleural cavity A narrow space between the parietal and visceral layers of the pleura, containing pleural fluid. (fig. 23.12)

plexus A network of blood vessels, lymphatic vessels, or nerves, such as a choroid plexus of the brain or the brachial plexus of nerves. (fig. 14.14)

pluripotent Pertaining to an embryonic stem cell from the morula that is capable of producing any type of embryonic or adult cell. More loosely, describing certain adult stem cells with an especially broad developmental potential, able to produce a wide variety of differentiated cell types. *See also* pluripotent stem cell.

pluripotent stem cell (PPSC) A stem cell of the bone marrow that can produce any of the formed elements of blood.

pollex The thumb.

pons The section of the brainstem between the midbrain and medulla oblongata.

popliteal (po-LIT-ee-ul) Pertaining to the posterior aspect of the knee.

portal system A circulatory pathway in which blood passes through two capillary beds in series in a single trip from the heart and back. (fig. 21.12)

positron emission tomography (PET) A method of producing a computerized image of the physiological state of a tissue using injected radioisotopes that emit positrons. (fig. 1.3)

posterior 1. Pertaining to the back aspect of the human body; also called *dorsal,* especially in embryos. 2. Pertaining to the tail end of nonhuman animals.

posterior nasal aperture The opening at the posterior limit of each nasal fossa, where air from the nasal cavity enters the pharynx; also called a *choana* or *posterior naris.* (fig. 23.2)

posterior root A branch of a spinal nerve that joins the spinal cord on its posterior side, composed of sensory nerve fibers; also called the *dorsal root.* (fig. 14.10)

posterior root ganglion A swelling in the posterior root of a spinal nerve, near the spinal cord, containing the neurosomas of the afferent neurons of the nerve; also called the *dorsal root ganglion.* (fig. 14.10)

postganglionic Pertaining to a neuron that transmits signals from a ganglion to a more distal target organ. (fig. 16.2)

postsynaptic Pertaining to a neuron or other cell that receives signals from the presynaptic neuron at a synapse. (fig. 13.10)

potential space An anatomical space that is usually obliterated by contact between two membranes but opens up if air, fluid, or other matter comes between them. Examples include the pleural cavity and the lumen of the uterus.

preembryo A developing human that has not yet formed ectoderm, mesoderm, and endoderm; when those germ layers have formed, the individual is regarded as an embryo. *Compare* conceptus; embryo.

preembryonic stage Any stage of prenatal development from fertilization through 16 days, when the primary germ layers exist and the embryonic stage begins.

preganglionic Pertaining to a neuron that transmits signals from the central nervous system to a ganglion. (fig. 16.2)

prepuce A fold of tissue over the glans of the penis or clitoris; the penile foreskin or clitoral hood.

presynaptic Pertaining to a neuron that transmits signals to a synapse. (fig. 13.10)

prime mover The muscle primarily responsible for a given joint action; agonist.

process An outgrowth of bone or other tissue, such as the mastoid process of the skull.

programmed cell death (PCD) *See* apoptosis.

projection pathway   The route taken by nerve signals from their point of origin (such as a sense organ) to their point of termination (such as the primary sensory cortex). (fig. 17.15)

pronation   A rotational movement of the forearm that turns the palm downward or posteriorly. (fig. 9.13)

prone   A position in which the body is lying face down.

proprioception   (PRO-pree-oh-SEP-shun) The nonvisual perception, usually subconscious, of the position and movements of the body, resulting from input from proprioceptors and the vestibular apparatus of the inner ear.

proprioceptor   (PRO-pree-oh-SEP-tur) A sensory receptor of the muscles, tendons, and joint capsules that detects muscle contractions and joint movements.

prostate gland   (PROSS-tate) A male reproductive gland that encircles the urethra immediately inferior to the bladder and contributes to the semen. (fig. 26.9)

protein   A polypeptide of 50 amino acids or more.

proteoglycan   A protein–carbohydrate complex in which the carbohydrate is dominant; forms a gel that binds cells and tissues together, fills the umbilical cord and eye, lubricates the joints, and forms the rubbery texture of cartilage. Formerly called *mucopolysaccharide.*

protraction   Forward movement of a body part in the horizontal plane, such as moving the mandible forward in preparation to take a bite from an apple. (fig. 9.15)

protuberance   A bony outgrowth or protruding part, such as the mental protuberance of the mandible.

proximal   Relatively near a point of origin or attachment; for example, the shoulder is proximal to the elbow. *Compare* distal.

pseudopod   (SOO-doe-pod) A temporary cytoplasmic extension of a cell used for locomotion (ameboid movement) and phagocytosis.

pseudostratified columnar epithelium   An epithelium in which every cell contacts the basement membrane, but not all of them reach the free surface, thus giving an appearance of stratification. (fig. 3.7)

pubic   Concerning the region of the genitalia. *See also* hypogastric.

pudendum   See vulva.

pulmonary   Pertaining to the lungs.

pulmonary circuit   A route of blood flow that supplies blood to the pulmonary alveoli for gas exchange and then returns it to the heart; all blood vessels between the right ventricle and the left atrium of the heart. (fig. 21.13)

R

radiography   The use of X-rays to form an image of the interior of the body. (fig. 1.3)

radiology   The branch of medicine concerned with producing images of the interior of the body, using such methods as X-rays, sonography, MRI, CT, and PET.

ramus   (RAY-mus) An anatomical branch, as in a nerve or in the pubis.

receptive field   An area of the environment or of an epithelial surface from which a given neuron receives sensory information. (fig. 17.2)

receptor   1. A cell or organ specialized to detect a stimulus, such as a taste cell or the eye. 2. A protein molecule that binds and responds to a chemical such as a hormone, neurotransmitter, or odor molecule.

receptor-mediated endocytosis   A mode of vesicular transport in which cell surface receptors bind a specific molecule in the extracellular fluid, then cluster together to be internalized by the cell. (fig. 2.11)

rectus   Straight; used in muscle names such as *rectus femoris* and *rectus abdominis.*

reflected   Folded back or away from something, often to expose another structure in anatomical demonstrations. (fig. 9.24)

reflex   A stereotyped, automatic, involuntary response to a stimulus; includes somatic reflexes, in which the effectors are skeletal muscles, and visceral (autonomic) reflexes, in which the effectors are usually visceral muscle, cardiac muscle, or glands.

reflex arc   A simple neural pathway that mediates a reflex; involves a receptor, an afferent nerve fiber, sometimes one or more interneurons, an efferent nerve fiber, and an effector. (fig. 14.19)

regeneration   Replacement of damaged tissue with new tissue of the original type. *Compare* fibrosis.

renal   (REE-nul) Pertaining to the kidney.

renal tubule   A urine-forming duct that converts glomerular filtrate to urine by processes of reabsorption and secretion of water and solutes. Consists of the proximal convoluted tubule, nephron loop, and distal convoluted tubule of an individual nephron, plus a collecting duct and papillary duct shared by multiple nephrons. (fig. 25.6) *Compare* nephron.

renin   An enzyme produced by the kidney that converts angiotensinogen to angiotensin I, the first step in producing the vasoconstrictor angiotensin II.

reproductive system   An organ system specialized for the production of offspring.

resistance   1. Opposition to the flow of fluid, such as blood in a vessel or air in a bronchiole. 2. Opposition to the movement of a joint; the load against which a muscle works. 3. A nonspecific ability to ward off an infection or disease, as opposed to the pathogen-specific defense provided by immunity.

respiratory system   An organ system specialized for the intake of air and exchange of gases with the blood, consisting of the lungs and the air passages from the nose to the bronchi.

reticular cell   (reh-TIC-you-lur) A delicate, branching cell in the reticular connective tissue of the lymphatic organs.

reticular fiber   A fine, branching collagen fiber coated with glycoprotein, found in the stroma of lymphatic organs and some other tissues and organs.

reticular tissue   A connective tissue composed of reticular cells and reticular fibers, found in bone marrow, lymphatic organs, and in lesser amounts elsewhere. (fig. 3.15)

retraction   Movement of a body part posteriorly on the horizontal plane; for example, retracting the mandible to grind food between the molars. (fig. 9.15)

retroperitoneal   Located between the peritoneum and body wall, rather than in the peritoneal cavity; descriptive of certain abdominal viscera such as the kidneys, ureters, and pancreas. (fig. 1.16) *Compare* intraperitoneal.

ribosome   A granule found free in the cytoplasm or attached to the rough endoplasmic reticulum and nuclear envelope, composed of ribosomal RNA and enzymes; specialized to read the nucleotide sequence of messenger RNA and assemble a corresponding sequence of amino acids to make a protein.

risk factor   Any environmental factor or characteristic of an individual that increases one's chance of developing a particular disease; includes such intrinsic factors as age, sex, and race and such extrinsic factors as diet, smoking, and occupation.

root   1. Part of an organ that is embedded in other tissue and therefore not externally visible, such as the root of a tooth, a hair, or the penis. *Compare* shaft. 2. The proximal end of a spinal nerve, adjacent to the spinal cord.

rostral   Relatively close to the forehead, especially in reference to structures of the brain and spinal cord; for example, the frontal lobe is rostral to the parietal lobe. *Compare* caudal.

rotation   Movement of a body part such as the humerus or forearm around its longitudinal axis. (fig. 9.14)

rough endoplasmic reticulum   Regions of endoplasmic reticulum characterized by flattened, parallel cisternae externally studded with ribosomes; involved in making proteins for export from the cell, among other functions. *See also* endoplasmic reticulum; smooth endoplasmic reticulum. (fig. 2.17)

ruga   (ROO-ga) 1. An internal fold or wrinkle in the mucosa of a hollow organ such as the stomach and urinary bladder; typically present when the organ is empty and relaxed but not when the organ is full and stretched. 2. A tissue ridge in such locations as the hard palate and vagina. (fig. 24.11)

S

sagittal plane   (SADJ-ih-tul) Any plane that extends from anterior to posterior and cephalic to caudal, and divides the body into right and left portions. (fig. 1.10) *Compare* median plane.

sarcomere   (SAR-co-meer) In skeletal and cardiac muscle, the portion of a myofibril from one Z disc to the next, constituting one contractile unit. (fig. 10.10)

sarcoplasmic reticulum (SR)   The smooth endoplasmic reticulum of a muscle cell, serving as a calcium reservoir. (fig. 10.8)

satellite cell   1. A type of glial cell found surrounding the somas of neurons in ganglia of the peripheral nervous system. 2. Stem cells of skeletal muscle that can multiply in response to muscle injury and contribute to some extent to regeneration of muscle fibers.

scanning electron microscope (SEM)   A microscope that uses an electron beam in place of light to form high-resolution, three-dimensional images of the surfaces of objects; capable of much higher magnifications than a light microscope. *Compare* transmission electron microscope.

Schwann cell   A glial cell that forms the neurilemma around all peripheral nerve fibers and the myelin sheath around many of them; also encloses neuromuscular junctions; also called a *neurilemmocyte.* (fig. 13.4)

sclerosis   (scleh-RO-sis) Hardening or stiffening of a tissue, usually with scar tissue, as in *multiple sclerosis* of the central nervous system and *atherosclerosis* of the blood vessels.

sclerotome   A group of mesodermal cells that arise from a somite in the fourth week of development and give rise to a segment of the vertebral column. *Compare* dermatome; myotome.

sebaceous gland   A holocrine gland that is usually associated with a hair follicle and produces an oily secretion, sebum. (fig. 5.10)

sebum   (SEE-bum) An oily secretion of the sebaceous glands that keeps the skin and hair pliable.

secondary sex characteristic Any feature that develops at puberty, further distinguishes the sexes from each other, and is not required for reproduction but promotes attraction between the sexes; examples include the distribution of subcutaneous fat, pitch of the voice, female breasts, male facial hair, and apocrine scent glands.

secondary sex organ An organ other than the ovaries and testes that is essential to reproduction, such as the external genitalia, internal genital ducts, and accessory reproductive glands.

second-order neuron An interneuron that receives sensory signals from a first-order neuron and relays them to a more rostral destination in the central nervous system (usually the thalamus). (fig. 14.4) *See also* first-order neuron; third-order neuron.

secretion 1. A chemical released by a cell to serve a physiological function, such as a hormone or digestive enzyme, as opposed to a waste product. 2. The process of releasing such a chemical, usually by exocytosis. *Compare* excretion.

secretory vesicle An organelle that arises from the Golgi complex and carries a secretion to the cell surface to be released by exocytosis.

section *See* histological section.

selection pressure A force of nature that favors the reproduction of some individuals over others and thus drives the evolutionary process; includes climate, predators, diseases, competition, and food supply. Human anatomy and physiology reflect adaptations to selection pressures encountered in the evolutionary history of the species.

SEM 1. Scanning electron microscope. 2. Scanning electron micrograph, a photograph taken with the scanning electron microscope. *Compare* TEM.

semen The fluid ejaculated by a male, including spermatozoa and the secretions of the prostate gland and seminal vesicles.

semicircular duct A ring-shaped, fluid-filled tube of the inner ear that detects angular acceleration of the head; enclosed in a bony passage called the semicircular canal. There are three semicircular ducts in each ear. (fig. 17.11)

semilunar valve A valve that consists of crescent-shaped cusps, including the aortic and pulmonary valves of the heart and valves of the veins and lymphatic vessels. (fig. 20.7)

sense organ Any organ that is specialized to respond to stimuli and generate a meaningful pattern of nerve signals; may be microscopic and simple, such as a tactile corpuscle, or macroscopic and complex, such as the eye or ear; may respond to stimuli arising within the body or from external sources.

sensory neuron A nerve cell that responds to a stimulus and conducts signals to the central nervous system; also called an *afferent neuron.* The axon of a sensory neuron is an *afferent nerve fiber.*

septum An anatomical wall between two structures or spaces, such as the nasal septum or the interventricular septum of the heart.

serosa *See* serous membrane.

serous fluid (SEER-us) A watery fluid similar to blood serum, formed as a filtrate of the blood or tissue fluid or as a secretion of serous gland cells; moistens the serous membranes.

serous gland A gland that secretes a relatively nonviscous product, such as the pancreas or a tear gland. *Compare* mucous gland.

serous membrane A membrane such as the peritoneum, pleura, or pericardium that lines a body cavity or covers the external surfaces of the viscera; composed of a simple squamous mesothelium and a thin layer of areolar connective tissue. (fig. 3.31)

Sertoli cell *See* sustentacular cell.

serum 1. The fluid that remains after blood has clotted and the solids have been removed; essentially the same as blood plasma except for a lack of fibrinogen. Used as a vehicle for vaccines. 2. Serous fluid.

sex chromosomes The X and Y chromosomes, which determine the sex of an individual.

shaft 1. The midpart, or diaphysis, of a long bone. 2. The external, cylindrical part of an organ such as a hair or the penis. *Compare* root.

short bone A bone that is not markedly longer than it is wide, such as the bones of the wrist and ankle.

sign An objective indication of disease that can be verified by any observor, such as cyanosis or a skin lesion. *Compare* symptom.

simple columnar epithelium An epithelium composed of a single layer of cells that are noticeably taller than they are wide. (fig. 3.6)

simple cuboidal epithelium An epithelium composed of a single layer of cells that are about equal in height and width; often, but not always, the cells appear squarish in tissue sections. (fig. 3.5)

simple diffusion Net movement of particles from a place of high concentration to a place of low concentration (down their concentration gradient), resulting from their own spontaneous motion; may or may not involve passage through a cell membrane or other membranes such as dialysis tubing.

simple squamous epithelium An epithelium composed of a single layer of thin, flat cells. (fig. 3.4)

sinoatrial (SA) node A mass of autorhythmic cells near the surface of the right atrium of the heart that serves as the pacemaker of the cardiac rhythm.

sinus 1. An air-filled space in the cranium. (fig. 7.8) 2. A modified, relatively dilated vein that lacks smooth muscle and is incapable of vasomotion, such as the dural sinuses of the cerebral circulation and coronary sinus of the heart. 3. A small fluid-filled space in an organ such as the lymph nodes. 4. Pertaining to the sinoatrial node of the heart, as in *sinus rhythm.*

sinusoid An irregularly shaped, blood-filled space in a tissue, with wide gaps between the endothelial cells; found in the liver, bone marrow, spleen, and some other organs. (fig. 22.9)

skeletal muscle Striated voluntary muscle, almost all of which is attached to the bones. (fig. 10.1)

skeletal system An organ system consisting of the bones, ligaments, bone marrow, periosteum, articular cartilages, and other tissues associated with the bones.

smear A tissue prepared for microscopic study by wiping it across a slide, rather than by sectioning; for example, blood, bone marrow, spinal cord, and Pap smears.

smooth endoplasmic reticulum Regions of endoplasmic reticulum characterized by tubular, branching cisternae lacking ribosomes; involved in detoxification, steroid synthesis, and in muscle, storage of calcium ions. *See also* endoplasmic reticulum; rough endoplasmic reticulum. (fig. 2.17)

smooth muscle Nonstriated involuntary muscle found in the walls of the blood vessels, many of the viscera, and other places. (fig. 3.27)

sodium–potassium pump An active transport protein, which, in each cycle of activity, pumps three sodium ions out of a cell and two potassium ions into the cell, with the expenditure of one ATP.

soma *See* cell body.

somatic 1. Pertaining to the body as a whole. 2. Pertaining to the skin, bones, and skeletal muscles as opposed to the viscera. 3. Pertaining to all cells other than germ cells.

somatic motor fiber A nerve fiber that innervates skeletal muscle and stimulates its contraction, as opposed to autonomic fibers.

somatic nervous system A division of the nervous system that includes afferent fibers mainly from the skin, muscles, and skeleton and efferent fibers to the skeletal muscles. *Compare* autonomic nervous system.

somatosensory 1. Pertaining to widely distributed general senses in the skin, muscles, tendons, joint capsules, and viscera, as opposed to the *special senses* found in the head only; also called *somesthetic.* 2. Pertaining to the cerebral cortex of the postcentral gyrus, which receives input from such receptors. *See also* general senses; special senses.

somatotopy A point-for-point correspondence between the locations where stimuli arise and the locations in the brain or spinal cord to which the sensory signals project, thus producing in the CNS a sensory "map" of part of the body. (fig. 15.18)

somesthetic *See* somatosensory.

somite One of the segmental blocks of embryonic mesoderm that begin to appear around day 20 and eventually number up to 44 pairs; a somite subdivides into three tissue masses—dermatome, myotome, and sclerotome, which give rise to certain aspects of the skin, muscles, and vertebrae. *See also* dermatome; myotome; sclerotome. (fig. 4.8)

sonography Production of an image of the interior of the body by means of ultrasound. (fig. 1.4)

special senses The senses of taste, smell, hearing, equilibrium, and vision, mediated by sense organs that are confined to the head and in most cases are relatively complex in structure. *See also* general senses.

sperm 1. A spermatozoon. 2. The fluid ejaculated by the male; semen. Contains spermatozoa and glandular secretions.

spermatogenesis (SPUR-ma-toe-JEN-eh-sis) The production of sperm cells through a series of mitotic and meiotic cell divisions; male gametogenesis.

spermatozoon (SPUR-ma-toe-ZOE-on) A sperm cell; the male gamete. (fig. 26.8)

sphincter (SFINK-tur) A ring of muscle that opens or closes an opening or passageway; found, for example, in the eyelids, around the urinary orifice, and at the junction of the stomach and duodenum. (fig. 24.11)

spinal column *See* vertebral column.

spinal cord The nerve cord that passes through the vertebral column and constitutes all of the central nervous system except the brain.

spinal nerve Any of the 31 pairs of nerves that arise from the spinal cord and pass through the intervertebral foramina. (fig. 14.10)

spindle 1. An elongated structure that is thick in the middle and tapered at the ends (fusiform). 2. A football-shaped complex of microtubules that guide the movement of chromosomes in mitosis and meiosis. (fig. 2.20) 3. A stretch receptor in the skeletal muscles. (fig. 17.1)

spine 1. The vertebral column. 2. A pointed process or sharp ridge on a bone, such as the styloid process of the cranium and spine of the scapula. (fig. 8.2)

spinothalamic tract A bundle of nerve fibers that ascend the spinal cord and brainstem and carry signals to the thalamus for light touch, tickle, itch, heat, cold, pain, and pressure. (fig. 14.4)

splanchnic (SPLANK-nic) Pertaining to the digestive tract.

spongy bone A form of osseous tissue found in the interiors of flat, irregular, and short bones and the epiphyses of long bones, with a matrix that forms a porous network of plates and bars, enclosing connected channels filled with bone marrow; also called *cancellous bone*. *Compare* compact bone. (fig. 6.5)

squamous Flat, scalelike, as in the surface epithelial cells of the epidermis and serous membranes. (figs. 2.3, 3.12)

stain A pigment applied to tissues to color and enhance the contrast between their nuclei, cytoplasm, extracellular material, and other tissue components.

stem cell Any undifferentiated cell that can divide and differentiate into more functionally specific cell types such as blood cells and germ cells.

stenosis The pathological constriction or narrowing of a tubular passageway or orifice of the body, such as the esophagus, uterine tube, or a valve orifice of the heart.

stereocilium An unusually long, sometimes branched microvillus lacking the axoneme and motility of a true cilium; serves such roles as absorption in the epididymis and sensory transduction in the inner ear.

sternal Pertaining to the breastbone or sternum, or the overlying region of the chest.

stimulus A chemical or physical agent in a cell's surroundings that is capable of creating a physiological response in the cell, especially agents detected by sensory cells, such as chemicals, light, and pressure.

strain The extent to which a bone or other structure is deformed when subjected to stress. *Compare* stress.

stratified 1. Layered. 2. A class of epithelia in which there are two or more cell layers, with some cells resting atop others rather than contacting the basement membrane.

stratified cuboidal epithelium An epithelium composed of two or more layers of cells in which the cells at the surface are about equal in height and width. (fig. 3.10)

stratified squamous epithelium An epithelium composed of two or more layers of cells in which the cells at the surface are flat and thin. (fig. 3.9)

stratum Any layer of tissue, such as the stratum corneum of the skin or stratum basalis of the uterus.

stratum corneum The surface layer of dead keratinocytes of the skin. (fig. 5.1)

stress 1. A mechanical force applied to any part of the body; important in stimulating bone growth, for example. *Compare* strain. 2. A condition in which any environmental influence disturbs the homeostatic equilibrium of the body and stimulates a physiological response, especially involving the increased secretion of hormones of the pituitary–adrenal axis.

striated muscle Muscular tissue in which the cells exhibit striations; skeletal and cardiac muscle. *See also* striations.

striations Alternating light and dark bands in skeletal and cardiac muscle produced by the pattern of overlapping myofilaments. (fig. 10.1)

stroma The connective tissue framework of a gland, lymphatic organ, or certain other viscera, as opposed to the tissue (parenchyma) that performs the physiological functions of the organ.

subcutaneous Beneath the skin.

submucosa A layer of loose connective tissue deep to the mucosa of an organ. (fig. 24.2)

submucosal plexus A plexus of parasympathetic neurons in the submucosa of the digestive tract, responsible for controlling glandular secretion by the mucosa and movements of the muscularis mucosae; also called the *Meissner plexus*.

sulcus A groove in the surface of an organ, as in the cerebrum, the heart, or a bone. (fig. 15.1)

superficial Relatively close to the surface; opposite of *deep*. For example, the ribs are superficial to the lungs.

superior Higher than another structure or point of reference from the perspective of anatomical position; for example, the lungs are superior to the diaphragm.

supination (SOO-pih-NAY-shun) A rotational movement of the forearm that turns the palm so that it faces upward or forward. (fig. 9.13)

supine A position in which the body is lying face up.

suprarenal Pertaining to the adrenal (suprarenal) glands, as in *suprarenal artery*.

surfactant A chemical that interferes with the formation of hydrogen bonds between water molecules, and thus reduces the cohesion of water; in the lung, a mixture of phospholipid and protein that prevents the alveoli from collapsing during expiration.

sustentacular cell 1. A cell in the wall of a seminiferous tubule of the testis that supports and protects the germ cells and secretes the hormone inhibin; also called a *Sertoli cell*. (fig. 26.6) 2. In many epithelia, such as taste buds and olfactory mucosa, any cell that supports and spaces the primary functional cells of the tissue; also called a *supporting cell*. (fig. 17.5)

suture A line along which any two bones of the skull are immovably joined, such as the coronal suture between the frontal and parietal bones. (fig. 7.6)

sympathetic nervous system A division of the autonomic nervous system that issues efferent fibers through the thoracic and lumbar nerves and usually exerts adrenergic effects on its target organs; includes a chain of paravertebral ganglia adjacent to the vertebral column, and the adrenal medulla. (fig. 16.4)

symphysis (SIM-fih-sis) A joint in which two bones are held together by fibrocartilage; for example, between bodies of the vertebrae and between the right and left pubic bones. (fig. 8.6)

symptom A subjective indication of disease that can be felt by the person who is ill but not objectively observed by another person, such as nausea or headache. *Compare* sign.

synapse (SIN-aps) 1. A junction at the end of an axon where it stimulates another cell. (fig. 13.11) 2. A gap junction between two cardiac or smooth muscle cells at which one cell electrically stimulates the other; called an *electrical synapse*. (fig. 10.17)

synaptic cleft A narrow space between the synaptic knob of an axon and the adjacent cell, across which a neurotransmitter diffuses. (fig. 13.11)

synaptic knob The swollen tip at the distal end of an axon; the site of synaptic vesicles and neurotransmitter release. (fig. 13.11)

synaptic vesicle A spheroidal organelle in a synaptic knob; contains neurotransmitter. (fig. 13.11)

syndrome A group of signs and symptoms that occur together and characterize a particular disease.

synergist (SIN-ur-jist) A muscle that works with the agonist to contribute to the same overall action at a joint.

synovial fluid (sih-NO-vee-ul) A lubricating fluid similar to egg white in consistency, found in the synovial joint cavities and bursae.

synovial joint A point where two bones are separated by a narrow, encapsulated space filled with lubricating synovial fluid; most such joints are relatively mobile. Also called a *diarthrosis*. (fig. 9.4)

systemic Widespread or pertaining to the body as a whole, as in *systemic circulation*.

systemic circuit All blood vessels that convey blood from the left ventricle to all organs of the body and back to the right atrium of the heart; all of the cardiovascular system except the heart and pulmonary circuit. (fig. 20.1)

T

TA *See* Terminologia Anatomica.

tactile Pertaining to the sense of touch.

tail 1. A slender process at one end of an organ, such as the tail of the pancreas or epididymis. (fig. 24.19) 2. In vertebrate animals, an appendage that extends beyond the anus and contains part of the vertebral column; in humans, limited to the embryo. (fig. 4.8)

target cell A cell acted upon by a nerve fiber or by a chemical messenger such as a hormone.

tarsal 1. Pertaining to the ankle. 2. Pertaining to the margin of the eyelid.

TEM 1. Transmission electron microscope. 2. Transmission electron micrograph, a photograph taken with the transmission electron microscope. *Compare* SEM.

temporal 1. Pertaining to time, as in *temporal summation* in neurons. 2. Pertaining to the side of the head, as in *temporal bone*.

tendinous cords Fibers that extend from the papillary muscles to the atrioventricular valve cusps in each ventricle of the heart, and serve to keep the valves from prolapsing during ventricular systole; also called *chordae tendineae*. (fig. 20.7)

tendon A collagenous band or cord associated with a muscle, usually attaching it to a bone and transferring muscular tension to it. *See also* aponeurosis.

teratogen Any agent capable of causing birth defects, including chemicals, infectious microorganisms, and radiation.

teres (TERR-eez) Round, cylindrical; used in the names of muscles and ligaments such as *teres major* and *ligamentum teres*.

Terminologia Anatomica A code of standard anatomical terms developed by an international committee of anatomists, the Federative Committee on Anatomical Terminology, and published in 1998; provides a worldwide standard for naming human structures.

testis The male gonad; produces spermatozoa and testosterone.

thalamus (THAL-uh-muss) The largest part of the diencephalon, located immediately inferior to the corpus callosum and bulging into each lateral ventricle; a point of synaptic relay of nearly all signals passing from lower levels of the CNS to the cerebrum. (fig. 15.11)

theory An explanatory statement, or set of statements, that concisely summarizes the state of knowledge on a phenomenon and provides direction for further study; for example, the fluid mosaic theory of the plasma membrane and the sliding filament theory of muscle contraction.

thermoreceptor A neuron specialized to respond to heat or cold, found in the skin and mucous membranes, for example.

third-order neuron An interneuron of the brain that receives sensory signals from a second-order neuron (often at the thalamus) and usually relays them to their final destination in the primary sensory cortex of the brain; in a few cases, a fourth-order neuron completes the pathway. (fig. 14.4) *See also* first-order neuron; second-order neuron.

thoracic Pertaining to the chest.

thorax A region of the trunk between the neck and the diaphragm; the chest.

thymus A lymphatic organ in the mediastinum superior to the heart; the site where T lymphocytes differentiate and become immunocompetent. (fig. 18.5)

thyroid cartilage A large shieldlike cartilage that encloses the larynx anteriorly and laterally and provides an anterior anchorage for the vocal cords and insertion for the infrahyoid muscles. (fig. 23.4)

thyroid gland An endocrine gland in the neck, partially encircling the trachea immediately inferior to the larynx. (fig. 18.6)

tight junction A zipperlike junction between epithelial cells that limits the passage of substances between them. (fig. 2.14)

tissue An aggregation of cells and extracellular materials, usually forming part of an organ and performing some discrete function for it; the four primary classes are epithelial, connective, muscular, and nervous tissue.

tissue gel The viscous colloid that forms the ground substance of many tissues; gets its consistency from hyaluronic acid or other glycosaminoglycans.

T lymphocyte A type of lymphocyte involved in nonspecific defense, humoral immunity, and cellular immunity; occurs in several forms including helper, cytotoxic, and suppressor T cells; also called a *T cell.*

trabecula (tra-BEC-you-la) A thin plate or sheet of tissue, such as the calcified trabeculae of spongy bone or the fibrous trabeculae that subdivide a gland. (fig. 6.5)

trachea (TRAY-kee-uh) A cartilage-supported tube from the inferior end of the larynx to the origin of the primary bronchi; conveys air to and from the lungs; the "windpipe."

tract 1. In the central nervous system, a bundle of nerve fibers with a similar origin, destination, and function, such as the corticospinal tracts of the spinal cord and commissural tracts of the cerebrum. 2. A continuous anatomical pathway such as the digestive tract.

transitional epithelium A stratified epithelium of the urinary tract that is capable of changing thickness and number of cell layers from relaxed to stretched states. (fig. 3.11)

transmembrane protein A protein of the plasma membrane that penetrates all the way through the membrane and contacts the intracellular and extracellular fluids. (fig. 2.8)

transmission electron microscope (TEM) A microscope that uses an electron beam in place of light to form high-resolution, two-dimensional images of ultrathin slices of cells or tissues; capable of extremely high magnification. *Compare* scanning electron microscope.

transport protein *See* carrier.

transverse plane A plane that cuts perpendicular to the long axis of an organ or passes horizontally through a human body in anatomical position. (fig. 1.10)

transverse section *See* cross section.

transverse (T) tubule A tubular extension of the plasma membrane of a muscle cell that conducts

action potentials into the sarcoplasm and excites the sarcoplasmic reticulum. (fig. 10.8)

trauma Physical injury caused by external forces such as falls, gunshot wounds, motor vehicle accidents, or burns.

trisomy-21 The presence of three copies of chromosome 21 instead of the usual two; causes variable degrees of mental retardation, a shortened life expectancy, and structural anomalies of the face and hands. Also called *Down syndrome.*

trochanter Either of two massive processes serving for muscle attachment at the proximal end of the femur.

trunk 1. That part of the body excluding the head, neck, and limbs. 2. A major blood vessel, lymphatic vessel, or nerve that gives rise to smaller branches; for example, the pulmonary trunk and spinal nerve trunks. (fig. 14.14)

T tubule *See* transverse tubule.

tubercle A rounded process on a bone, such as the greater tubercle of the humerus.

tuberosity A rough area on a bone, such as the tibial or ischial tuberosity.

tubuloacinar gland A gland in which secretory cells are found in both the tubular and acinar portions. (fig. 3.30)

tunic A layer that encircles or encloses an organ, such as the tunics of a blood vessel or eyeball; also called a *tunica.* (fig. 21.1)

tympanic membrane The eardrum.

U

ultrastructure Structure at or near the molecular level, made visible by the transmission electron microscope.

umbilical (um-BIL-ih-cul) 1. Pertaining to the cord that connects a fetus to the placenta. 2. Pertaining to the navel (umbilicus).

undifferentiated Pertaining to a cell or tissue that has not yet attained a mature functional form; capable of differentiating into one or more specialized functional cells or tissues; for example, stem cells and embryonic tissues.

unencapsulated nerve ending *See* free nerve ending.

unipotent Pertaining to a stem cell that is capable of differentiating into only one type of mature cell, such as a spermatogonium able to produce only sperm, or an epidermal basal cell able to produce only keratinocytes.

unit membrane Any cellular membrane composed of a bilayer of phospholipids and embedded proteins. A single unit membrane forms the plasma membrane and encloses many organelles of a cell, whereas double unit membranes enclose the nucleus and mitochondria. (fig. 2.6)

unmyelinated Lacking a myelin sheath. (fig. 13.7)

upper limb The appendage that arises from the shoulder, consisting of the brachium from shoulder to elbow, the antebrachium from elbow to wrist, the wrist, and the hand; loosely called the *arm,* but that term properly refers only to the brachium.

urea The most abundant nitrogenous waste in urine, formed in the liver by a reaction between ammonia and carbon dioxide.

urethra The passage that conveys urine from the urinary bladder to the outside of the body; in males, it also conveys semen and acts as part of both the urinary and reproductive tracts.

urinary system An organ system specialized to filter the blood plasma, excrete waste products from it, and regulate the body's water, acid–base, and electrolyte balance.

urogenital (U-G) system Collective term for the reproductive and urinary tracts; also called the *genitourinary (G-U) system.*

uterine tube A duct that extends from the ovary to the uterus and conveys an egg or conceptus to the uterus; also called the *fallopian tube* or *oviduct.*

V

varicose vein A vein that has become permanently distended and convoluted due to a loss of competence of the venous valves; especially common in the lower limb, esophagus, and anal canal (where they are called *hemorrhoids*).

vas (vass) (plural, *vasa*) A vessel or duct.

vascular Possessing or pertaining to blood vessels.

vasoconstriction (VAY-zo-con-STRIC-shun) The narrowing of a blood vessel due to muscular constriction of its tunica media.

vasodilation (VAY-zo-dy-LAY-shun) The widening of a blood vessel due to relaxation of the muscle of its tunica media and the outward pressure of the blood exerted against the wall.

vasomotion Any constriction or dilation of a blood vessel.

vein Any blood vessel that carries blood toward either atrium of the heart.

ventral *See* anterior.

ventral root *See* anterior root.

ventricle A fluid-filled chamber of the brain or heart. (figs. 15.4, 20.7)

venule (VEN-yool) The smallest type of vein, receiving drainage from capillaries.

vertebra (VUR-teh-bra) One of the bones of the vertebral column.

vertebral column (VUR-teh-brul) A posterior series of usually 33 vertebrae; encloses the spinal cord, supports the skull and thoracic cage, and provides attachment for the limbs and postural muscles. Also called the *spine* or *spinal column.*

vesicle 1. A fluid-filled tissue sac such as the seminal vesicle. 2. A fluid-filled spheroidal organelle such as a synaptic or secretory vesicle.

vestibular apparatus Structures of the inner ear concerned with equilibrium and the perception of the movements and orientation of the head, including the semicircular ducts, utricle, and saccule.

vestibule An anatomical receiving chamber; for example, the vestibule between the teeth and cheek, the space immediately inside the nostril, the space enclosed by the female labia majora, and the chamber of the inner ear to which the cochlea and semicircular ducts are attached.

viscera (VISS-er-uh) (singular, *viscus*) The organs contained in the body cavities, such as the brain, heart, lungs, stomach, intestines, and kidneys.

visceral 1. Pertaining to the viscera. 2. The inner or deeper layer of a two-layered membrane such as the pleura, pericardium, or glomerular capsule. (fig. 1.15) *Compare* parietal.

visceral muscle Single-unit smooth muscle found in the walls of blood vessels and the digestive, respiratory, urinary, and reproductive tracts.

volar Pertaining to the anterior surfaces of the fingers (the surfaces continuous with the palmar skin).

voluntary Under conscious control, as in skeletal muscle.

vulva The female external genitalia; the mons pubis, labia majora, and all superficial structures between the labia majora; also called the *pudendum.* (fig. 26.21)

W

white matter   White myelinated nervous tissue deep to the cortex of the cerebrum and cerebellum and superficial to the gray matter of the spinal cord. (fig. 15.4)

X

X chromosome   The larger of the two sex chromosomes; males have one X chromosome and females have two in each somatic cell.

xiphoid process   (ZIFF-oyd, ZYE-foyd) A small pointed cartilaginous or bony process at the inferior end of the sternum. (fig. 7.27)

X-ray   **1.** A high-energy, penetrating electromagnetic ray with wavelengths in the range of 0.1 to 10 nm; used in diagnosis and therapy. **2.** A photograph made with X-rays; radiograph.

Y

Y chromosome   The smaller of the two sex chromosomes, found only in males and having little if any genetic function except development of the testis.

yolk sac   An embryonic membrane that encloses the yolk in vertebrates that lay eggs and serves in humans as the origin of the first blood and germ cells. (fig. 4.5)

Z

zygomatic arch   An arch of bone anterior to the ear, formed by the zygomatic processes of the temporal, frontal, and zygomatic bones; origin of the masseter muscle. (fig. 7.5)

zygote   A single-celled, fertilized egg.

Key to Pronunciation Guides

Pronounce letter sequences in the pronunciation guides as follows:

ah	as in father
al	as in pal
ay	as in day
bry	as in bribe
byu	as in bureau
c	as in calculus
cue	as in ridiculous
cuh	as in cousin
cul	as in bicycle
cus	as in custard
dew	as in dual
eez	as in ease
eh	as in feather
err	as in merry
fal	as in fallacy
few	as in fuse
ih	as in fit
iss	as in sister
lerr	as in lair
lur	as in learn
ma	as in man
mah	as in mama
me	as in meat
merr	as in merry
mew	as in music
muh	as in mother
na	as in corona
nerr	as in nary
new	as in news
nuh	as in nothing
odj	as in dodger
oe	as in go
oh	as in home
ol	as in alcohol
oll	as in doll
ose	as in gross
oss	as in floss
perr	as in pair
pew	as in pewter
ruh	as in rugby
serr	as in serration
sterr	as in stereo
sy	as in siren
terr	as in terrain
thee	as in theme
tirr	as in tyranny
uh	as in mother
ul	as in bicycle
verr	as in very
y	as in why
zh	as in measure
zy	as in enzyme

Credits

Photographs

Chapter 1
Opener: © SPL/Photo Researchers, Inc.; **1.1a:** Neurosurg Focus © 2004 American Association of Neurological Surgeons; **1.1b:** © SPL/Photo Researchers, Inc.; **1.2:** National Library of Medicine/Peter Arnold, Inc.; **1.3a:** © U.H.B. Trust/Tony Stone Images/Getty Images; **1.3b:** Custom Medical Stock Photo, Inc.; **1.3c:** © CNRI/Phototake; **1.3d:** © Monte S. Buchsbaum, Mt. Sinai School of Medicine, New York, NY; **1.3e:** © Tony Stone Images/Getty Images; **1.4(fetus):** © Alexander Tsiaras/Photo Researchers, Inc.; **1.8-1.12d:** © The McGraw-Hill Companies, Inc./Joe DeGrandis, photographer.

Chapter 2
Opener: © Dr. Donald Fawcett & Dr. Porter/Visuals Unlimited; **2.1a-b:** Doug Bray, Biological Science, Univ. of Lethbridge; **2.2a:** © Ed Reschke; **2.2b & c:** David Phillips/Photo Researchers, Inc.; **2.6a:** David Phillips/Photo Researchers, Inc.; **2.12a:** © Ed Reschke; **2.12b:** Biophoto Associates/Photo Researchers, Inc.; **2.13a:** Custom Medical Stock Photo Inc.; **2.13c:** © Biophoto Associates/Photo Researchers, Inc.; **2.15b:** © K.G. Murti/Visuals Unlimited; **2.21(all):** © Ed Reschke; **2.22b:** © Biophoto Associates/Science Source/Photo Researchers, Inc.

Chapter 3
Opener: © Dr. Andrejs Liepins/Photo Researchers, Inc.; **3.4a & 3.5a:** © The McGraw-Hill Companies, Inc./Dennis Strete, photographer; **3.6a:** © Lester V. Bergman/Corbis; **3.7a:** © The McGraw-Hill Companies, Inc./Dennis Strete, photographer; **3.8a: 3.9a:** © The McGraw-Hill Companies, Inc./Joe DeGrandis, photographer; **3.10a:** © Ed Reschke; **3.11a:** Johnny R. Howze; **3.12:** From R.G. Kessel and R.H. Kardon, *Tissues and Organs: A Text-Atlas of Scanning Electron Microscopy* (W.H. Freeman, 1979); **3.13:** © The McGraw-Hill Companies, Inc./Rebecca Gray, photographer/Don Kincaid, dissections; **3.14a:** © The McGraw-Hill Companies, Inc./Dennis Strete, photographer; **3.15a:** © The McGraw-Hill Companies, Inc./Al Telser, photographer; **3.16a, 3.17a, 13.18a:** © The McGraw-Hill Companies, Inc./Dennis Strete, photographer; **3.19a, 3.20a:** © Ed Reschke; **3.21a:** Dr. Alvin Telser; **3.22a:** © The McGraw-Hill Companies, Inc./Dennis Strete, photographer; **3.23a, 3.24a, 3.25a, 3.26a:** © Ed Reschke.

Chapter 4
Opener: Photo Researchers, Inc.; **4.8a:** © Landrum B. Shettles, MD; **4.8b,c:** Anatomical Travelogue/Photo Researchers, Inc.; **4.11a-b:** Dr. Kurt Bentrschke; **4.12a:** © Martin Rotker/Phototake, Inc.; **4.12b:** Photo Researchers, Inc.; **4.12c:** © Landrum B. Shettles, MD; **4.14a:** Photo Researchers, Inc.; **4.14b:** Courtesy Mihalay Bartalos, from Bartalos 1967: *Medical Cytogenetics*, fig 10.2, pg. 154, Waverly, a division of Williams & Wilkins; **4.15:** Thalidomide UK.

Chapter 5
Opener: © Petit Format/Nestle/Photo Researchers, Inc.; **5.2a:** © DLILLC/Corbis; **5.2b:** © The McGraw-Hill Companies, Inc./Joe DeGrandis, photographer; **5.4a:** © The McGraw-Hill Companies, Inc./Dennis Strete, photographer; **5.4b,c:** from R.G. Kessel and R.H. Kardon, *Tissues and Organs: A Text-Atlas of Scanning Electron Microscopy* (W.H. Freeman, 1979); **5.5a(top):** © Tom & Dee Ann McCarthy/Corbis; **5.5a(bottom):** © The McGraw-Hill Companies, Inc./Dennis Strete, photographer; **5.5b(top):** © Creatas/PunchStock; **5.5b(bottom):** © The McGraw-Hill Companies, Inc./Dennis Strete, photographer; **5.6b:** © CBS/Phototake; **5.7(all):** © The McGraw-Hill Companies, Inc./Joe DeGrandis, photographer; **5.10a-c:** © The McGraw-Hill Companies, Inc./Joe DeGrandis, photographer; **5.13a:** © NMSB/Custom Medical Stock Photo, Inc.; **5.13b:** © Biophoto Associates/Photo Researchers, Inc.; **5.13c:** © James Stevenson/SPL/Photo Researchers, Inc.; **5.14a:** © SPL/Custom Medical Stock Photo, Inc.; **5.14b,c:** © John Radcliffe/Photo Researchers, Inc.

Chapter 6
Opener: © The McGraw-Hill Companies, Inc./Joe DeGrandis, photographer; **6.5a,c:** © D.W. Fawcett/Visuals Unlimited; **6.5d:** Visuals Unlimited; **6.6:** © Robert Calentine/Visuals Unlimited; **6.9:** © Ken Saladin; **6.11:** Courtesy of Utah Valley Regional Medical Center, Department of Radiology; **6.12:** Victor Eroschenko; **6.13:** © The McGraw-Hill Companies, Inc./ Joe DeGrandis, photographer; **6.14a:** Custom Medical Stock Photo, Inc.; **6.14b:** Howard Kingsnorth/Getty; **6.14c:** © Lester V. Bergman/Corbis; **6.14d:** Custom Medical Stock Photo, Inc.; **6.15:** © SIU/Visuals Unlimited; **6.17a:** © Michael Klein/Peter Arnold, Inc.; **6.17b:** © Dr. P. Marzzi/Photo Researchers, Inc.; **6.17c:** © Yoav Levy/Phototake.

Chapter 7
Opener: Mehau Kulyk/Science Photo Library/Photo Researchers, Inc.; **7.30:** © Biophoto Associates/Photo Researchers, Inc.; **7.34:** © The McGraw-Hill Companies, Inc./Bob Coyle, photographer.

Chapter 8
Opener: © NHS Trust/Stone Images/Getty Images; **8.9(top, both):** © David Hunt/specimens from the National Museum of Natural History, Smithsonian Institution; **8.9(bottom, both):** © L. Bassett/Visuals Unlimited.

Chapter 9
Opener: © NHS Trust/Tony Stone Images/Getty Images; **9.7a–9.22a:** © The McGraw-Hill Companies, Inc./Timothy L. Vacula, photographer; **9.24:** © The McGraw-Hill Companies, Inc./Rebecca Gray, photographer/Don Kincaid , dissections; **9.26b:** © L. Bassett/Visuals Unlimited; **9.27:** Dr. Ken Greer/Getty Images; **9.28a:** © SIU/Visuals Unlimited; **9.28b:** © Ron Mensching/Phototake; **9.28c:** © SIU/Peter Arnold, Inc.; **9.28d:** © Mehau Kulyk/SPL/Photo Researchers, Inc.

Chapter 10
Opener: © Don W. Fawcett/Photo Researchers, Inc.; **10.1:** © Ed Reschke; **10.2c:** © Victor Eroschenko; **10.10a:** © Visuals Unlimited; **10.11a:** © Victor B. Eichler; **10.13:** From R.G. Kessel and R.H. Kardon, *Tissues and Organs: A Text-Atlas of Scanning Electron Microscopy,* W.H. Freeman & Co., 1979; **10.16a:** © University of Kansas Medical Center, Department of Anatomy and Cell Biology; **10.16b:** © The McGraw-Hill Companies, Inc./Dennis Strete, photographer; **10.20:** Shady Awwad, MD.

Chapter 11
Opener: © Simon Fraser/Photo Researchers, Inc.; **11.2:** © The McGraw-Hill Companies, Inc./Rebecca Gray, photographer/Don Kincaid, dissections; **11.4(all)-11.8:** © The McGraw-Hill Companies, Inc./Joe DeGrandis, photographer; **11.9:** © The McGraw-Hill Companies, Inc./Rebecca Gray, photographer/Don Kincaid, dissections; **11.12:** Ralph Hutchings/Visuals Unlimited; **11.16:** Ralph Hutchings/Visuals Unlimited.

Chapter 12
Opener: Custom Medical Stock Photo, Inc.; **12.4a-b & 12.13:** © The McGraw-Hill Companies, Inc./Rebecca Gray, photographer/Don Kincaid, dissections; **12.16a-b:** © The McGraw-Hill Companies, Inc./Photo and Dissection by Christine Eckel.

Atlas

Opener: © Samantha Scott/Alamy; **A.1a-b:** © The McGraw-Hill Companies, Inc./Joe DeGrandis, photographer; **A.2:** © The McGraw-Hill Companies, Inc./Rebecca Gray, photographer/Don Kincaid, dissections; **A.8a-A.10:** © The McGraw-Hill Companies, Inc./Joe DeGrandis, photographer; **A.11-A.13:** © The McGraw-Hill Companies, Inc./Rebecca Gray, photographer/Don Kincaid, dissections; **A14.a:** © The McGraw-Hill Companies, Inc./ Dennis Strete, photographer; **A.14b:** © The McGraw-Hill Companies, Inc./ Rebecca Gray, photographer/Don Kincaid, dissections; **A.15a-A25b:** © The McGraw-Hill Companies, Inc./Joe DeGrandis, photographer.

Chapter 13

Opener: © SPL/Photo Researchers, Inc.; **13.4b:** © Ed Reschke; **13.7c:** © The McGraw-Hill Companies, Inc./Dr. Dennis Emrey, Dept. of Zoology and Genetics, Iowa State University, photographer; **13.9:** © Omikron/Science Source/Photo Researchers, Inc.; **13.15:** Biophoto Associates/Photo Researchers, Inc.

Chapter 14

Opener: © Ed Reschke; **14.2c:** Sarah Werning; **14.6:** Getty Images; **14.7b:** From Richard E. Kessel and Randy H. Kardon, *Tissues and Organs: A Text-Atlas of Scanning Electron Microscopy*, 1979, W.H. Freeman and Company; **14.10:** © From *A Stereoscopic Atlas of Anatomy* by David L. Bassett. Courtesy of Dr. Robert A. Chase, MD; **14.15:** © The McGraw-Hill Companies, Inc./Photo and Dissection by Christine Eckel.

Chapter 15

Opener: © CNRI/Photo Researchers, Inc.; **15.1c:** © The McGraw-Hill Companies, Inc./Rebecca Gray, photographer/Don Kincaid, dissections; **15.2b:** © The McGraw-Hill Companies, Inc./Dennis Strete, photographer; **15.4c:** © The McGraw-Hill Companies, Inc./Rebecca Gray, photographer/Don Kincaid, dissections; **15.20:** © Marcus E. Raichle, MD, Washington University School of Medicine, St. Louis, Missouri; **15.23a:** Photo Researchers, Inc.; **15.23b:**

Custom Medical Stock Photo; **15.24b:** © The McGraw-Hill Companies, Inc./Rebecca Gray, photographer/ Don Kincaid, dissections; **15.31c:** © The McGraw-Hill Companies, Inc./Joe DeGrandis, photographer; **15.37a:** © Custom Medical Stock Photo, Inc.; **15.37b:** © Simon Fraser/ Photo Researchers, Inc.

Chapter 16

Opener & 16.3: © From: *A Stereoscopic Atlas of Anatomy* by David L. Bassett. Courtesy of Dr. Robert A. Chase, MD.

Chapter 17

Opener: Photo Researchers, Inc.; **17.5c:** © Ed Reschke; **17.9:** © The McGraw-Hill Companies, Inc./Joe DeGrandis, photographer; **17.13:** Quest/Science Photo Library/Photo Researchers, Inc.; **17.19:** Tim Vacula, 2008; **17.24:** © Ralph C. Eagle/MD/ Photo Researchers, Inc.; **17.25a:** © Lisa Klancher; **17.27a:** © The McGraw-Hill Companies, Inc./Joe DeGrandis, photographer; **17.28a:** Courtesy of Beckman Vision Center at UCSF School of Medicine/D. Copenhagen, S. Mittman, and M. Maglio.

Chapter 18

Opener: © P. Bagavandoss/Photo Researchers, Inc.; **18.3a:** © Dr. John D. Cunningham/Visuals Unlimited; **18.3b:** © Science VU/Visuals Unlimited; **18.6b:** © Robert Calentine/ Visuals Unlimited; **18.7b:** © John Cunningham/Visuals Unlimited; **18.9c:** © Ed Reschke; **18.10a:** © Manfred Kage/Peter Arnold, Inc.; **18.10b:** © Ed Reschke; **18.13:** Courtesy of Beckman Vision Center at UCSF School of Medicine/D. Copenhagen, S. Mittman, and M. Maglio.

Chapter 19

Opener: © Juergen Berger, Max-Planck Institute/Photo Researchers, Inc.; **19.3b:** © Richard J. Poole/Polaroid International Photomicrography Competition; **19.3c:** © Don W. Fawcett/Visuals Unlimited; **Table 19.4 (Neutrophills, Eosinophil, Basophil):** © Ed Reschke; **(Monocyte, Lymphocyte):** Michael Ross/Photo Researchers, Inc.

Chapter 20

Opener: Gondelon/Photo Researchers, Inc.; **20.5a-b:** © The

McGraw-Hill Companies, Inc.; **20.6a-b:** Photo and illustration by Roy Schneider, University of Toledo. Plastinated heart model for illustration courtesy of Dr. Carlos Baptista, University of Toledo; **20.8b:** © Manfred Kage/Peter Arnold, Inc.; **20.8c:** © The McGraw-Hill Companies, Inc.; **20.11c:** Visuals Unlimited; **20.12a-b:** © Ed Reschke; **20.12c:** Custom Medical Stock Photo, Inc.; **20.14a:** © Ed Reschke.

Chapter 21

Opener: © Biophoto Associates/ Photo Researchers, Inc; **21.2a:** © The McGraw-Hill Companies, Inc./ Dennis Strete, photographer; **21.2b:** © Wolf H. Fahrenbach/Visuals Unlimited; **21.3:** © Photo Researchers, Inc.; **21.6b:** Courtesy of S. McNutt.

Chapter 22

Opener: © Eye of Science/Photo Researchers, Inc.; **22.2:** © SPL/ Custom Medical Stock Photo, Inc.; **22.4a:** © The McGraw-Hill Companies, Inc./Dennis Strete, photographer; **22.7:** Peter Arnold, Inc.; **22.8:** Custom Medical Stock Photo; **22.10b:** © The McGraw-Hill Companies, Inc./Dennis Strete, photographer; **22.11c:** Francis Leroy, Biocosmos/ Photo Researchers, Inc.; **22.12b:** © The McGraw-Hill Companies, Inc./ Rebecca Gray, photographer/Don Kincaid, dissections; **22.13b:** © Biophoto Associates/Photo Researchers, Inc.; **22.14a:** © The McGraw-Hill Companies, Inc./Dennis Strete, photographer; **22.14c:** © The McGraw-Hill Companies, Inc./Al Telser, photographer; **22.16a-b:** © Dr. Don W. Fawcett/Visuals Unlimited.

Chapter 23

Opener: Peter Arnold, Inc.; **23.2a, 23.3a-b:** © The McGraw-Hill Companies, Inc./Joe DeGrandis, photographer; **23.5a:** © Phototake; **23.5b:** © J. Siebert/Custom Medical Stock Photo; **23.8:** Custom Medical Stock Photo; **23.10a:** © Dr. Gladden Willis/Visuals Unlimited; **23.10b:** Visuals Unlimited; **23.12:** Ralph Hutchings/Visuals Unlimited.

Chapter 24

Opener: © SPL/Photo Researchers, Inc.; **24.8:** © The McGraw-Hill Companies, Inc./Rebecca Gray, photographer/Don Kincaid,

dissections; **24.10b:** © The McGraw-Hill Companies, Inc./Dennis Strete, photographer; **24.11b:** © The McGraw-Hill Companies, Inc./ Rebecca Gray, photographer/Don Kincaid, dissections; **24.12d:** Visuals Unlimited; **24.13a-b:** CNRL/SPL/ Photo Researchers, Inc.; **24.15a:** © Meckes/Ottawa/Photo Researchers, Inc.; **24.15b, 24.18b, 24.20a:** © The McGraw-Hill Companies, Inc./Dennis Strete, photographer.

Chapter 25

Opener: © The McGraw-Hill Companies, Inc./Dennis Strete, photographer; **25.3a:** Ralph Hutchings/ Visuals Unlimited; **25.8b:** © The McGraw-Hill Companies, Inc./Al Telser, photographer; **25.9a:** From R.G. Kessel and R.H. Kardon, *Tissues and Organs: A Text-Atlas of Scanning Electron Microscopy*, W.H. Freeman, 1979; **25.9b:** © Don Fawcett/Photo Researchers, Inc.; **25.9c:** © Barry F. King/Biological Photo Service.

Chapter 26

Opener: © Simon Fraser/Photo Researchers, Inc.; **26.3a:** © The McGraw-Hill Companies, Inc./ Dennis Strete, photographer; **26.4a:** From R.G. Kessel and R.H. Kardon, *Tissues and Organs: A Text-Atlas of Scanning Electron Microscopy*, W.H. Freeman, 1979; **26.4b:** © Ed Reschke; **26.8a:** Visuals Unlimited; **26.13c:** © The McGraw-Hill Companies, Inc./Rebecca Gray, photographer/Don Kincaid, dissections; **26.14(primordial, primary):** © Ed Reschke; **(secondary):** © The McGraw-Hill Companies, Inc./Al Telser, photographer; **(tertiary):** Manfred Kage/Peter Arnold, Inc.; **(graafian):** © Landrum B. Shettles, MD; **(corpus):** © The McGraw-Hill Companies, Inc./Al Telser, photographer; **26.15a:** © Ed Reschke; **26.15b:** Manfred Kage/Peter Arnold, Inc.; **26.16:** © Landrum B. Shettles, MD; **26.17:** Photo Researchers, Inc.; **26.18:** © Ed Reschke; **26.22b:** From *Anatomy & Physiology Revealed*, © The McGraw-Hill Companies, Inc./The University of Toledo, photography and dissection; **26.22d:** Courtesy Tochikazu Nagato, from *Cell and Tissue Research*, 209: 1-10, 3, pg. 6, Springer-Verlag, 1980; **26.26a-b:** © SPL/Photo Researchers, Inc.

Lexicon of Biomedical Word Elements

a- no, not, without (atom, agranulocyte)
ab- away (abducens, abduction)
acetabulo- small cup (acetabulum)
acro- tip, extremity, peak (acromion, acromegaly, acrosome)
ad- to, toward, near (adsorption, adrenal)
adeno- gland (lymphadenitis, adenohypophysis, adenoids)
aero- air, oxygen (aerobic, anaerobe, aerophagy)
af- toward (afferent)
ag- together (agglutination)
-al pertaining to (parietal, pharyngeal, temporal)
ala- wing (ala nasi)
albi- white (albicans, linea alba, albino)
algi- pain (analgesic, myalgia)
aliment- nourishment (alimentary, hyperalimentation)
allo- other, different (allele, allosteric)
amphi- both, either (amphiphilic, amphiarthrosis)
an- without (anaerobic, anemic)
ana- 1. up, build up (anabolic, anaphylaxis). 2. apart (anaphase, anatomy). 3. back (anastomosis)
andro- male (androgen)
angi- vessel (angiogram, angioplasty, hemangioma)
ante- before, in front (antebrachium)
antero- forward (anterior, anterograde)
anti- against (antidiuretic, antibody, antagonist)
apo- from, off, away, above (apocrine, aponeurosis)
arbor- tree (arboreal, arborization)
artic- 1. joint (articulation). 2. speech (articulate)
-ary pertaining to (axillary, coronary)
-ase enzyme (polymerase, kinase, amylase)
ast-, astro- star (aster, astrocyte)
-ata, -ate 1. possessing (Chordata, corniculate). **2.** plural of -*a* (stomata, carcinomata)
athero- fat (atheroma, atherosclerosis)
atrio- entryway (atrium, atrioventricular)
auri- ear (auricle, auricular)
auto- self (autolysis, autoimmune)
axi- axis, straight line (axial, axoneme, axon)
baro- pressure (baroreceptor, hyperbaric)
bene- good, well (benign, beneficial)
bi- two (bipedal, biceps, bifid)
bili- bile (biliary, bilirubin)
bio- life, living (biology, biopsy, microbial)
blasto- precursor, bud, producer (fibroblast, osteoblast, blastomere)
brachi- arm (brachium, brachialis, antebrachium)
brady- slow (bradycardia, bradypnea)
bucco- cheek (buccal, buccinator)
burso- purse (bursa, bursitis)
calc- calcium, stone (calcified, calcaneus, hypocalcemia)
callo- thick (callus, callosum)
calori- heat (calorie, calorimetry, calorigenic)
calv-, calvari- bald, skull (calvaria)
calyx cup, vessel, chalice (glycocalyx, renal calyx)
capito- head (capitis, capitate, capitulum)
capni- smoke, carbon dioxide (hypocapnia)
carcino- cancer (carcinogen, carcinoma)
cardi- heart (cardiac, cardiology, pericardium)
carot- 1. carrot (carotene). 2. stupor (carotid)
carpo- wrist, seize (carpus, metacarpal)
case- cheese (caseosa, casein)
cata- down, break down (catabolism)
cauda- tail (cauda equina, caudate nucleus)
-cel little (pedicel)
celi- belly, abdomen (celiac)
centri- center, middle (centromere, centriole)
cephalo- head (cephalic, encephalitis)
cervi- neck, narrow part (cervix, cervical)
chiasm- cross, X (optic chiasm)
choano- funnel (choana)
chole- bile (cholecystokinin, cholelithotripsy)

chondro- 1. grain (mitochondria). 2. cartilage, gristle (chondrocyte, perichondrium)
chromo- color (dichromat, chromatin, cytochrome)
chrono- time (chronotropic, chronic)
cili- eyelash (cilium, supraciliary)
circ- about, around (circadian, circumduction)
cis- cut (incision, incisor)
cisterna- reservoir (cisterna chyli)
clast- break down, destroy (osteoclast)
clavi- hammer, club (clavicle, supraclavicular)
-cle little (tubercle, corpuscle)
cleido- clavicle (sternocleidomastoid)
cnemo- lower leg (gastrocnemius)
co- together (coenzyme, cotransport)
collo- 1. hill (colliculus). 2. glue (colloid, collagen)
contra- opposite (contralateral)
corni- horn (cornified, corniculate, cornu)
corono- crown (coronary, corona, coronal)
corpo- body (corpus luteum, corpora quadrigemina)
corti- bark, rind (cortex, cortical)
costa- rib (intercostal, subcostal)
coxa- hip (os coxae, coxal)
crani- helmet (cranium, epicranius)
cribri- sieve, strainer (cribriform, area cribrosa)
crino- separate, secrete (holocrine, endocrinology)
crista- crest (crista ampullaris, mitochondrial crista)
crito- to separate (hematocrit)
cruci- cross (cruciate ligament)
-cule, -culus small (canaliculus, trabecula, auricular)
cune- wedge (cuneiform, cuneatus)
cutane-, cuti- skin (subcutaneous, cuticle)
cysto- bladder (cystitis, cholecystectomy)
cyto- cell (cytology, cytokinesis, monocyte)
de- down (defecate, deglutition, dehydration)
demi- half (demifacet, demilune)
den-, denti- tooth (dentition, dens, dental)
dendro- tree, branch (dendrite, oligodendrocyte)
derma-, dermato- skin (ectoderm, dermatology, hypodermal)
desmo- band, bond, ligament (desmosome, syndesmosis)
dia- 1. across, through, separate (diaphragm, dialysis). 2. day (circadian)
dis- 1. apart (dissect, dissociate). 2. opposite, absence (disinfect, disability)
diure- pass through, urinate (diuretic, diuresis)
dorsi- back (dorsal, dorsum, latissimus dorsi)
duc- to carry (duct, adduction, abducens)
dys- bad, abnormal, painful (dyspnea, dystrophy)
e- out (ejaculate, eversion)
-eal pertaining to (hypophyseal, arboreal)
ec-, ecto- outside, out of, external (ectopic, ectoderm, splenectomy)
ef- out of (efferent, effusion)
-el, -elle small (fontanel, organelle, micelle)
electro- electricity (electrocardiogram, electrolyte)
em- in, within (embolism, embedded)
emesi-, emeti- vomiting (emetic, hyperemesis)
-emia blood condition (anemia, hypoxemia, hypovolemia)
en- in, into (enzyme, parenchyma)
encephalo- brain (encephalitis, telencephalon)
enchymo- poured in (mesenchyme, parenchyma)
endo- within, into, internal (endocrine, endocytosis)
entero- gut, intestine (mesentery, myenteric)
epi- upon, above (epidermis, epiphysis, epididymis)
ergo- work, energy, action (allergy, adrenergic)
eryth-, erythro- red (erythema, erythrocyte)
esthesio- sensation, feeling (anesthesia, somesthetic)
eu- good, true, normal, easy (eukaryote, eupnea, aneuploidy)
exo- out (exopeptidase, exocytosis, exocrine)
facili- easy (facilitated)
fasci- band, bundle (fascia, fascicle)

fenestr- window (fenestrated, fenestra vestibuli)
fer- to carry (efferent, uriniferous)
ferri- iron (ferritin, transferrin)
fibro- fiber (fibroblast, fibrosis)
fili- thread (myofilament, filiform)
flagello- whip (flagellum)
foli- leaf (folic acid, folia)
-form shape (cuneiform, fusiform)
fove- pit, depression (fovea)
funiculo- little rope, cord (funiculus)
fusi- 1. spindle (fusiform). 2. pour out (perfusion)
gamo- marriage, union (monogamy, gamete)
gastro- belly, stomach (digastric, gastrointestinal)
-gen, -genic, -genesis producing, giving rise to (pathogen, carcinogenic, glycogenesis)
genio- chin (geniohyoid, genioglossus)
germi- 1. sprout, bud (germinal, germinativum). 2. microbe (germicide)
gero- old age (progeria, geriatrics, gerontology)
gesto- 1. to bear, carry (ingest). 2. pregnancy (gestation, progesterone)
glia- glue (neuroglia, microglia)
globu- ball, sphere (globulin, hemoglobin)
glom- ball (glomerulus)
glosso- tongue (hypoglossal, glossopharyngeal)
glyco- sugar (glycogen, glycolysis, hypoglycemia)
gono- 1. angle, corner (trigone). 2. seed, sex cell, generation (gonad, oogonium, gonorrhea)
gradi- walk, step (retrograde, gradient)
-gram recording of (sonogram, electrocardiogram)
-graph recording instrument (sonograph, electrocardiograph)
-graphy recording process (sonography, radiography)
gravi- severe, heavy (gravid, myasthenia gravis)
gyro- turn, twist (gyrus)
hallu- great toe (hallux, hallucis)
hemi- half (hemidesmosome, hemisphere, hemiazygos)
-hemia blood condition (polycythemia)
hemo- blood (hemophilia, hemoglobin, hematology)
hetero- different, other, various (heterotrophic, heterozygous)
histo- tissue, web (histology, histone)
holo- whole, entire (holistic, holocrine)
homeo- constant, unchanging, uniform (homeostasis, homeothermic)
homo- same, alike (homologous, homozygous)
hyalo- clear, glassy (hyaline, hyaluronic acid)
hydro- water (dehydration, hydrolysis, hydrophobic)
hyper- above, above normal, excessive (hyperkalemia, hypertonic)
hypo- below, below normal, deficient (hypogastric, hyponatremia, hypophysis)
-ia condition (anemia, hypocalcemia, osteomalacia)
-ic pertaining to (isotonic, hemolytic, antigenic)
-icle, -icul small (ossicle, canaliculus, reticular)
ilia- flank, loin (ilium, iliac)
-illa, -illus little (bacillus)
-in protein (trypsin, fibrin, globulin)
infra- below (infraspinous, infrared)
ino- fiber (inotropic, inositol)
insulo- island (insula, insulin)
inter- between (intercellular, intercalated, intervertebral)
intra- within (intracellular, intraocular)
iono- ion (ionotropic, cationic)
ischi- to hold back (ischium, ischemia)
-ism 1. process, state, condition (metabolism, rheumatism). **2.** doctrine, belief, theory (holism, reductionism, naturalism)
iso- same, equal (isometric, isotonic, isomer)
-issimus most, greatest (latissimus, longissimus)
-ite little (dendrite, somite)
-itis inflammation (dermatitis, gingivitis)